MW01055188

Molecular and Genome Evolution

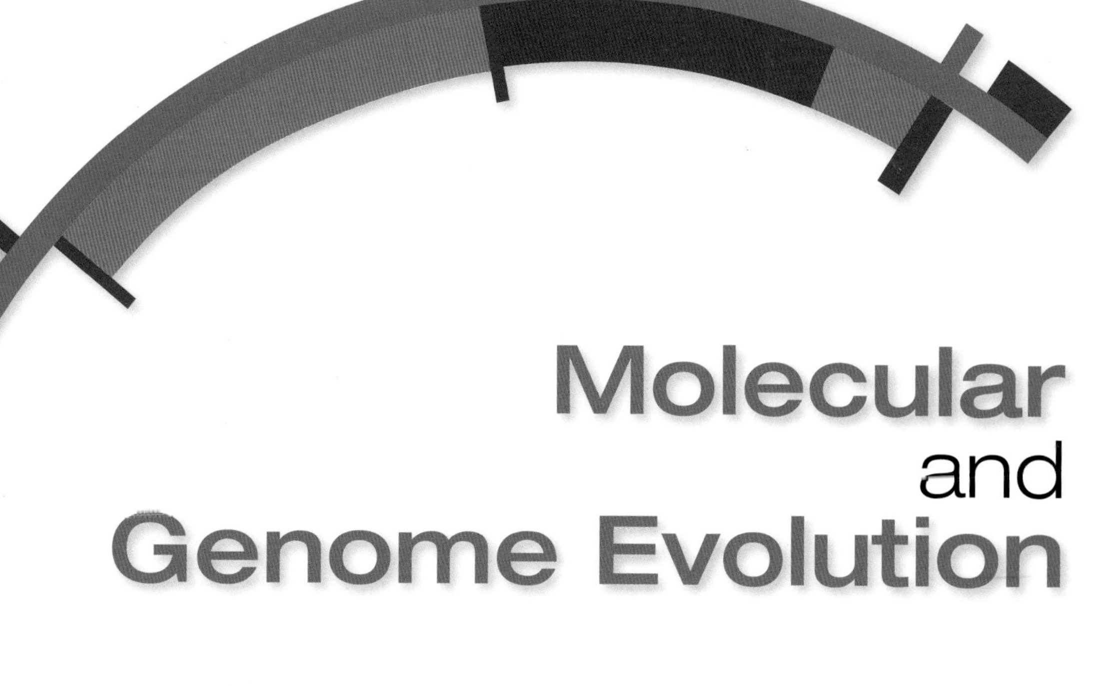

Molecular and Genome Evolution

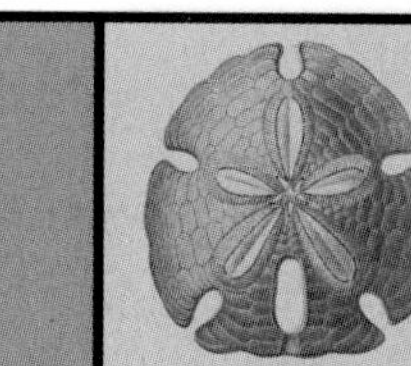
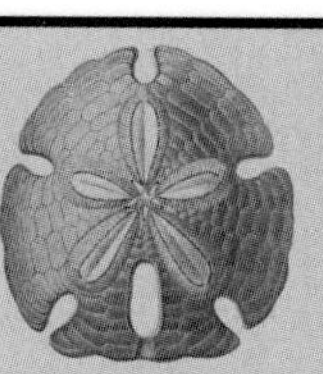
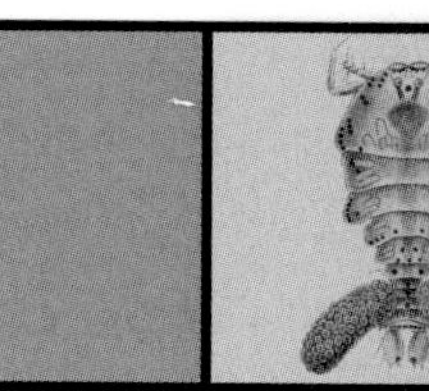

Dan Graur

University of Houston

Chapter 12, **"The Evolution of Gene Regulation"**
by Amy K. Sater, University of Houston

Chapter 13, **"Experimental Molecular Evolution"**
by Tim F. Cooper, University of Houston

Sinauer Associates, Inc. Publishers
Sunderland, Massachusetts U.S.A.

The Cover

"A Haeckelian Rhapsody in Blue with 65% of a Mitochondrial Genome"
Concept by Dan Graur and design by Joanne Delphia

Molecular and Genome Evolution

For information address:
Sinauer Associates, Inc.
P.O. Box 407
Sunderland, MA 01375 USA
FAX: 413-549-1118
publish@sinauer.com
www.sinauer.com

Lyrics on p. 1 from "Darwin's Acid" in *The Rap Guide to Evolution* © 2009 by Baba Brinkman, www.bababrinkman.com. Quoted with the kind permission of the artist.

Library of Congress Cataloging-in-Publication Data
Names: Graur, Dan, 1953- author.
Title: Molecular and genome evolution / Dan Graur, University of Houston.
Description: Sunderland, Massachusetts : Sinauer Associates, Inc., [2016]
Identifiers: LCCN 2015040070 | ISBN 9781605354699 (casebound)
Subjects: LCSH: Molecular evolution. | Genomics. | Evolution (Biology) | Phylogeny.
Classification: LCC QH325 .G697 2016 | DDC 572.8/38--dc23
LC record available at http://lccn.loc.gov/2015040070

Printed in China

5 4 3 2 1

To Mina, 1972, 32°10′27.3936″N 34°53′48.4584″E,
and other lucky numbers

The Author

Dan Graur is John and Rebecca Moores Professor in the Department of Biology and Biochemistry at the University of Houston and Professor Emeritus of Zoology at Tel Aviv University, Israel. Having earned a B.Sc. (Biology) and an M.Sc. (Zoology) at Tel Aviv University, he completed a Ph.D. (Genetics) at University of Texas Health Science Center at Houston. Coauthor of two editions of *Fundamentals of Molecular Evolution* (Sinauer Associates, 1991 and 2000), Dr. Graur has published close to 200 articles, book chapters, encyclopedia entries, commentaries, technical notes, and book reviews. He has served as Associate Editor of *Molecular Biology and Evolution* (1995–2011) and *Genome Biology and Evolution* (since 2009). Dr. Graur received an Alexander von Humboldt Research Award in 2011 and in 2015 was elected a Fellow of the American Association for the Advancement of Science. In addition to his research in molecular and genome evolution, Dr. Graur is interested in the epistemology of evolutionary biology, the societal and political implications of genetics and molecular biology, and art. He blogs at judgestarling.tumblr.com. His handle on Twitter is @dangraur. (Photograph by Chris Watts, University of Houston.)

Contents in Brief

Acknowledgments

Molecular and Genome Evolution took more than a dozen years to put together. Like all books, it could have benefited from twelve more years of contemplation, writing, and editing. However, as writer and critic Roger Angell astutely observed, a published book is one that the author lets go of in the end.

I am indebted to Bill Martin for supporting this project. His willingness to volunteer his time, his encyclopedic knowledge of all things biological, and his uncompromising love of the scientific method are greatly appreciated.

My colleagues Amy Sater and Tim Cooper wrote two chapters (Chapters 12 and 13, respectively) for this book. I wish to express my appreciation for their hard work and willingness to undertake such a thankless and time-consuming task.

My former student Yichen Zheng thoroughly reviewed and fact-checked the manuscript. I especially thank him for providing me with his unique points of view. Obviously, all remaining errors in the book are mine and mine only.

I am indebted to many mentors, colleagues, students, friends, and family members for helping me put this book together. Their comments, suggestions, corrections, discussions, and support are gratefully acknowledged. For favors great and small, I thank Joshua Akey, John Allen, Katarina Andreasen, Lara Appleby, Ricardo Azevedo, Marc Bailly-Bechet, Suganthi Balasubramanian, Michael Baudis, Dror Berel, David Bogomil, Luciano Brocchieri, Joseph Brown, Lilia Cañas, Reed Cartwright, Suneetha Chandanam, Benny Chor, Zofia Chrzanowska-Lightowlers, Laura Claghorn, Wyatt Travis Clark, Blaine Cole, Keith Cooper, Kerri Crawford, Aaron Darling, Brigitte Dauwalder, Nora Dethloff, Lou Doucette, Stuart Dryer, Peng Du, Margaret Dunn, Claire Duval, Xander Duval, Chris Elsik, J. J. Emerson, Toshinori Endo, Kiyoshi Ezawa, Cédric Feschotte, Yuriy Fofanov, George Fox, Tony Frankino, Noga Ganany, Mark Gerestein, Takashi Gojobori, Uri Gophna, Valer Gotea, Brenton Graveley, Preethi Gunaratne, Inna Gur-El, Penny Grant, Inar Graur, Lilla Graur, Mina Graur, Or Graur, Sara Graur, Matthew Hahn, Dov Herskowits, Varda Herskowits, Simon Ho, Hoang Hoang, Corey Hryc, Dorothée Huchon, Toby Hunt, Rhonda James, Spencer Johnston, Krešimir Josić, Jerzy Jurka, Aurélie Kapusta, Patrick Keeling, Erin Kelleher, Selina Khan, Marek Kimmel, Eugene Koonin, Dušan Kordiš, Shigefumi Kuwahara, Amanda Larracuente, Jennifer Leonard, William Lewis, Wen-Hsiung Li, Wenfu Li, Miao-Suey Lin, Jean Lobry, Hongan Long, Mannie Magid, Edward Max, Itay Mayrose, Michael McDonald, James McInerney, David McIntyre, Rich Meisel, Katsuhiko Mineta, Claudine Montgelard, Larry Moran, Etsuko Moriyama, Joram Mwacharo, Michael Nachman, Luay Nakhleh, Masatoshi Nei, Hoa Nguyen, Yoshihito Niimura, Norihiro Okada, Elizabeth Ostrowski, Sally Otto, Amanda Paul, Itala Paz, Steven Pennings, Nicole Perna, David Pollock, Ovidiu Popa, Mark Ptashne, Yonia Pulido, Longyi Qi, Eduardo Rocha, Matthew Rockman, Mayo Roettger, Gregg Roman, Jeffrey Ross-Ibarra, Naruya Saitou, Betsy Salazar, Rafael Sanjuán, Amy Sater, Stanley Sawyer, Manoj Samanta, Sahotra Sarkar, Yoko Satta, Marie Scavotto, Alan Schulman, Dean Scudder, Andy Sinauer, Chris Small, Kevin Spring, Arlin Stoltzfus, Dave Tang, Mallory Travis, Tamir Tuller, Yun-Huei Tzeng, Esther van der Knaap, Moshe Vardi, Lillian Warren, Hidemi Watanabe, John Welch, Dan Wells, Paul Wheeler, Thomas Wicker, William Widger, Diane Wiernasz, Carol Wigg, Ben Wilson, Melissa Wilson Sayres, Yuri Wolf, David Wool, Itai Yanai, Hsin-Yi Yeh, Hye-Jeong Yeo, Soojin Yi, Fei Yuan, Zhengdong Zhang, Lei Zhao, and Rebecca Zufall.

In the Babylonian Talmud, wisdom is likened to a tree: "Just as a small tree may set on fire a bigger tree, so too it is with scholars, the younger sharpen the minds of the older." As was the case with second century Talmudic sage Haninah, I too "have learnt much from my teachers, and from my colleagues more than from all my teachers, and from my students more than from them all." I therefore thank my former Ph.D. students, Tal Dagan, Eran Elhaik, Einat Hazkani-Covo, Giddy Landan, Ron Ophir, Nicholas Price, Tal Pupko, Niv Sabath, Shaul Shaul, and Yichen Zheng, for teaching me almost everything that I know.

At the University of Houston, I have benefited from support from a John and Rebecca Moores Professorship. I owe special gratitude to the Alexander von Humboldt Foundation, which has consistently supported my most significant scientific accomplishments since 1985.

Finally, I owe a great intellectual debt to my academic mentor and friend, David Wool, who has encouraged me and advised me for more than 37 years.

DAN GRAUR
University of Houston

Table of Contents

CHAPTER 8 / Evolution by Molecular Tinkering 339

CHAPTER 9 / Mobile Elements in Evolution 391

CHAPTER 11 / Eukaryotic Genome Evolution 491

Introduction

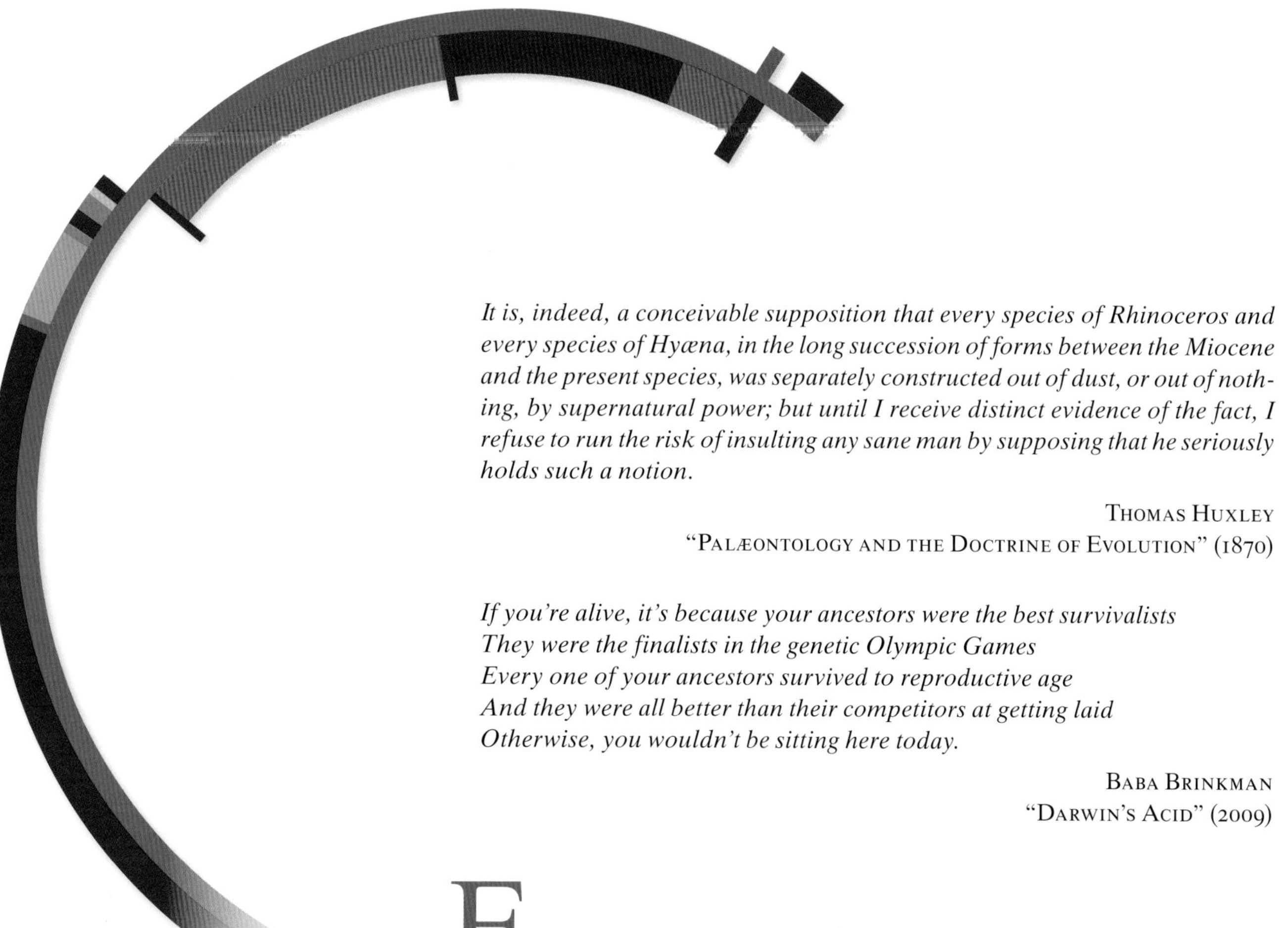

It is, indeed, a conceivable supposition that every species of Rhinoceros and every species of Hyæna, in the long succession of forms between the Miocene and the present species, was separately constructed out of dust, or out of nothing, by supernatural power; but until I receive distinct evidence of the fact, I refuse to run the risk of insulting any sane man by supposing that he seriously holds such a notion.

THOMAS HUXLEY
"PALÆONTOLOGY AND THE DOCTRINE OF EVOLUTION" (1870)

If you're alive, it's because your ancestors were the best survivalists
They were the finalists in the genetic Olympic Games
Every one of your ancestors survived to reproductive age
And they were all better than their competitors at getting laid
Otherwise, you wouldn't be sitting here today.

BABA BRINKMAN
"DARWIN'S ACID" (2009)

Evolutionary biology is a mature science. It is a coherent discipline with a handful of logical principles, each of which has repeatedly withstood rigorous empirical and observational testing. Evolution is not difficult to define. If one ignores the obfuscations of the creationists, the casuistry of the philosophers, and the ruminations of the sophisticates, evolution is merely a change in allele frequencies. There is no other mandatory attribute to the evolutionary process.

The evolutionary process may or may not be gradual; it may be imperceptibly slow or spectacularly fast. The evolutionary process may or may not be progressive. It may give rise to astonishing biological complexity or it may reduce the organism to a shadow of its former self. The only thing that can be said with certainty about evolution is that change is inevitable.

Evolution does not give rise to perfection or to anything resembling intelligent design; spinning wheels on fixed axles have never evolved in nature. Evolution does not drive the living world toward a better, fairer, more efficient, more honest, or cleverer state. Evolution merely produces changes that are not excessively deleterious. The most we can say about the products of evolution is that they work—some efficiently, some only barely, and some in a manner that is clumsy, plodding, and bureaucratic. If genes, genomes, and organisms were created by a benevolent intelligent designer, the problem of theodicy would never have arisen, and departments of pediatric oncology would never be needed.

The origin of all novelty in nature is mutation. Within the category of mutations, we include all heritable changes in the genetic information that have the potential to be transmitted from generation to generation, be they the substitution of one nucleotide by another, the integration of a piece of foreign DNA into the genome, the fusion of two chromosomes into one, or the reciprocal recombination between two homologous sequences. The greatest and most amazing novelty in evolution starts as nothing more than a single mutation arising in a single individual at a single point in time.

Mutations may or may not affect fitness. If they don't, their fate will be determined by random genetic drift; if they do, their fate will be determined by the combination of selection and random genetic drift. Only selection and random genetic drift can affect allele frequencies; hence, only selection and random genetic drift are important processes in evolution.

Evolution occurs at the population level; individuals cannot evolve. An individual can only make an evolutionary contribution by producing offspring or dying childless. A person's ability to discover the laws of motion, to solve Rubik's cube in 5.25 seconds, to teach population genetics, to prove Fermat's last theorem, or to paint *A Sunday Afternoon on the Island of La Grande Jatte* have no importance in evolution.

The efficacy of selection depends on effective population size. Effective population size, which is a historical construct, should not be confused with census population size, which is a snapshot of the present. Some species, such as humans, have huge census population sizes but very small effective population sizes. Other species, such as chimpanzees, have miniscule census population sizes (and as a result are endangered in their natural habitat), yet their effective population size is much larger than that of humans. Selection is not very potent in populations with small effective population sizes. Hence, selection has only rarely improved the human species. Because of demographic, environmental, and genetic stochasticity, however, "the best of all possible worlds" mostly isn't. All biological populations are finite; therefore, even in the absence of selection, changes in allele frequencies will occur. Thus, by definition, evolution is inevitable.

Evolution is a historical process and as such is governed by the rules of contingency, i.e., the assertion that every outcome depends on a number of prior conditions and that each of these prior conditions in turn depends on still other conditions, and so on. In other words, every evolutionary outcome ultimately depends on the details of a long chain of antecedent states, each of which exerts enormous long-term repercussions.

Evolution cannot create something out of nothing. *Creatio ex nihilo* does not exist in nature. Moreover, there is no true novelty in evolution. Digestive enzymes may evolve into venom, forelimbs may evolve into wings, hindlimbs may evolve into anchors for the muscles that maneuver the penis, and wings may evolve into halteres that function as gyroscopes; however, each new development must have a precedent. Evolution is a process that is constrained developmentally and can only act on what is already present. If a mermaid or a flying pig were shown to exist, we would have to discard the theory of evolution.

During evolution, it is easier to create equivalence than improvement. It is also easier to create functionlessness than functionality. The tens of thousands of genetic diseases that are due to gene death and loss of functionality attest to this fact. Gene death can sometimes be manageable. Thus, all large genomes are strewn with pseudogenes that have no functioning counterparts. Because of gene death, humans have to get their vitamin C from oranges and cats are not tempted by sweets. Interestingly, functionlessness may even be advantageous. Akin to an empty ecological niche, a gene's death may create opportunities for subsequent mutations.

Evolutionary biologists, like most people, have a fascination with *Homo sapiens*. Humans, however, do not occupy a privileged position in that grand scheme of things called "evolution." The human genome is smaller than that of an onion and contains fewer protein-coding genes than cassava. Human biology is infinitely more boring than that of ciliates, human population numbers are dwarfed by those of tardigrades, and human genetic variability is negligible in comparison to that of the nematode *Caenorhabditis brenneri*. Ever since Copernicus, our anthropocentric worldview has repeatedly been shown to be erroneous, yet we continue to spend billions of dollars in search of human genomic perfection, producing along the way a great many headlines and precious little science.

Finally, the same rules of evolutionary biology apply to all levels of its study. The rules at the DNA level are the same as the rules at the protein level, which in turn are the same as the rules at the morphological level. The only difference is one of resolution. Biochemists, computer jocks, medical doctors, and others who have exempted themselves from ever studying evolutionary biology have been known to invoke "paradigm shifts" in our understanding of evolution as frequently as men with prostatic hyperplasia urinate. Such bombastic proclamations may make good headlines, but they are almost never true.

Understanding a handful of principles is all there is to evolutionary biology. The rest is details. *Molecular and Genome Evolution* is about the details.

CHAPTER 1

The Molecular Basis of Biology and Evolution

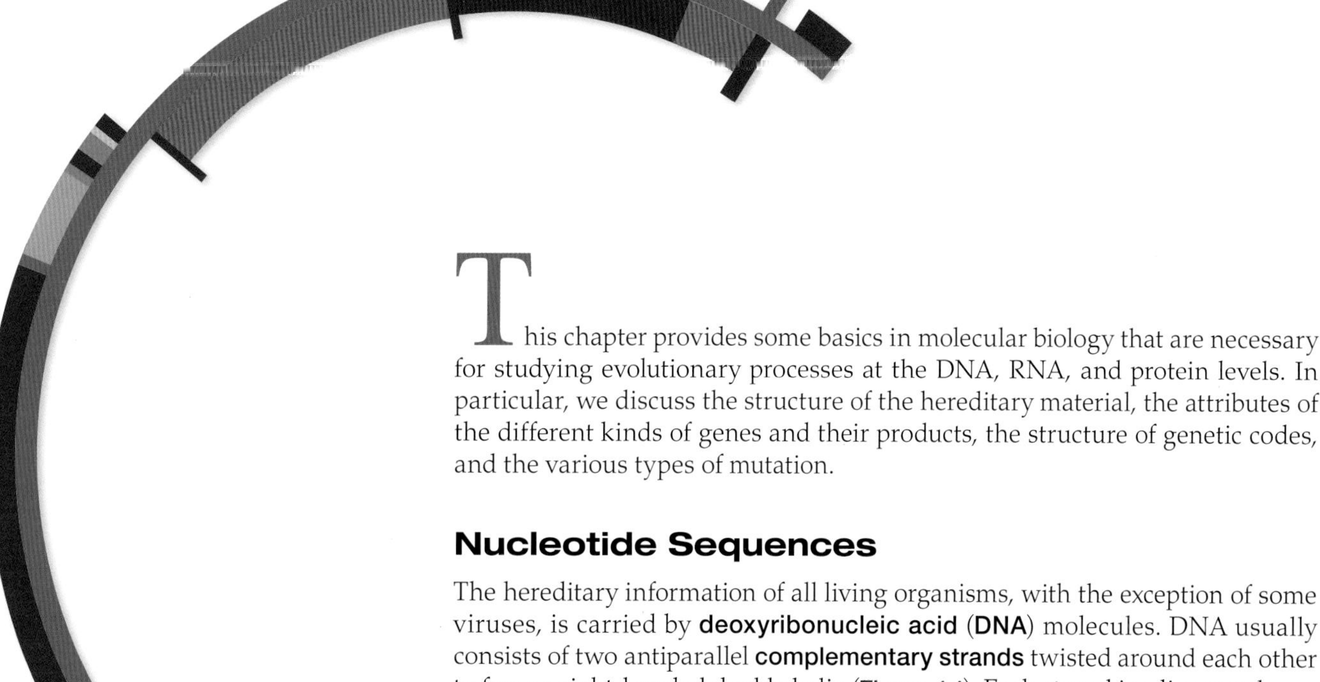

This chapter provides some basics in molecular biology that are necessary for studying evolutionary processes at the DNA, RNA, and protein levels. In particular, we discuss the structure of the hereditary material, the attributes of the different kinds of genes and their products, the structure of genetic codes, and the various types of mutation.

Nucleotide Sequences

The hereditary information of all living organisms, with the exception of some viruses, is carried by **deoxyribonucleic acid (DNA)** molecules. DNA usually consists of two antiparallel **complementary strands** twisted around each other to form a right-handed double helix (**Figure 1.1**). Each strand is a linear polynucleotide consisting of four kinds of **nucleotides**. A nucleotide is a molecule that consists of a pentose (five-carbon) sugar, a phosphate group, and a purine or pyrimidine **base**. A nucleotide without the phosphate group is called a **nucleoside**. In DNA, the nucleotides are **deoxyribonucleotides**, i.e., they contain the sugar **deoxyribose**. There are two **purines**, **adenine (A)** and **guanine (G)**, and two **pyrimidines**, **thymine (T)** and **cytosine (C)**. All one-letter abbreviations are listed in **Table 1.1**.

The **backbone** of the DNA molecule consists of deoxyribose-phosphate moieties covalently linked in tandem by asymmetrical 5′—3′ phosphodiester bonds. Consequently, the DNA molecule is **polar** (or directional), one end having a phosphoryl radical (—P) on the 5′ carbon of the terminal nucleotide, the other with a free hydroxyl (—OH) on the 3′ carbon of the terminal nucleotide. The direction of the phosphodiester bonds determines the molecule's character; thus, the sequence 5′—G—C—A—A—T—3′ is different from the sequence 3′—G—C—A—A—T—5′. As a rule, the phosphodiester bonds are omitted and the nucleotides in the DNA sequence are written from the 5′ end to the 3′ end. Hence, the sequence 5′—G—C—A—A—T—3′ is written as GCAAT. Relative to a given location of a nucleotide in a DNA sequence, the 5′ and the 3′ directions are referred to as **upstream** and **downstream**, respectively.

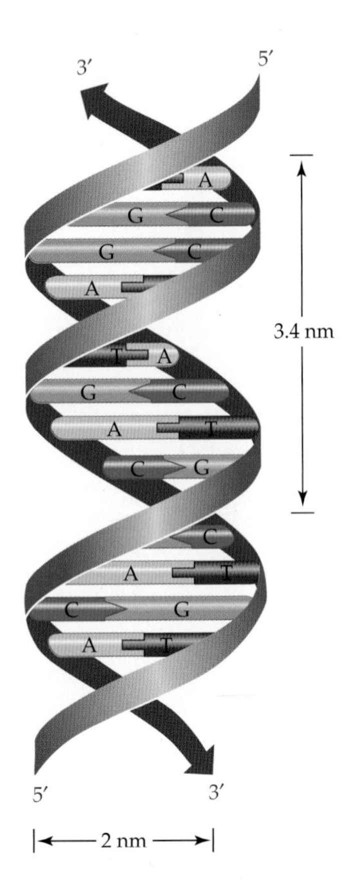

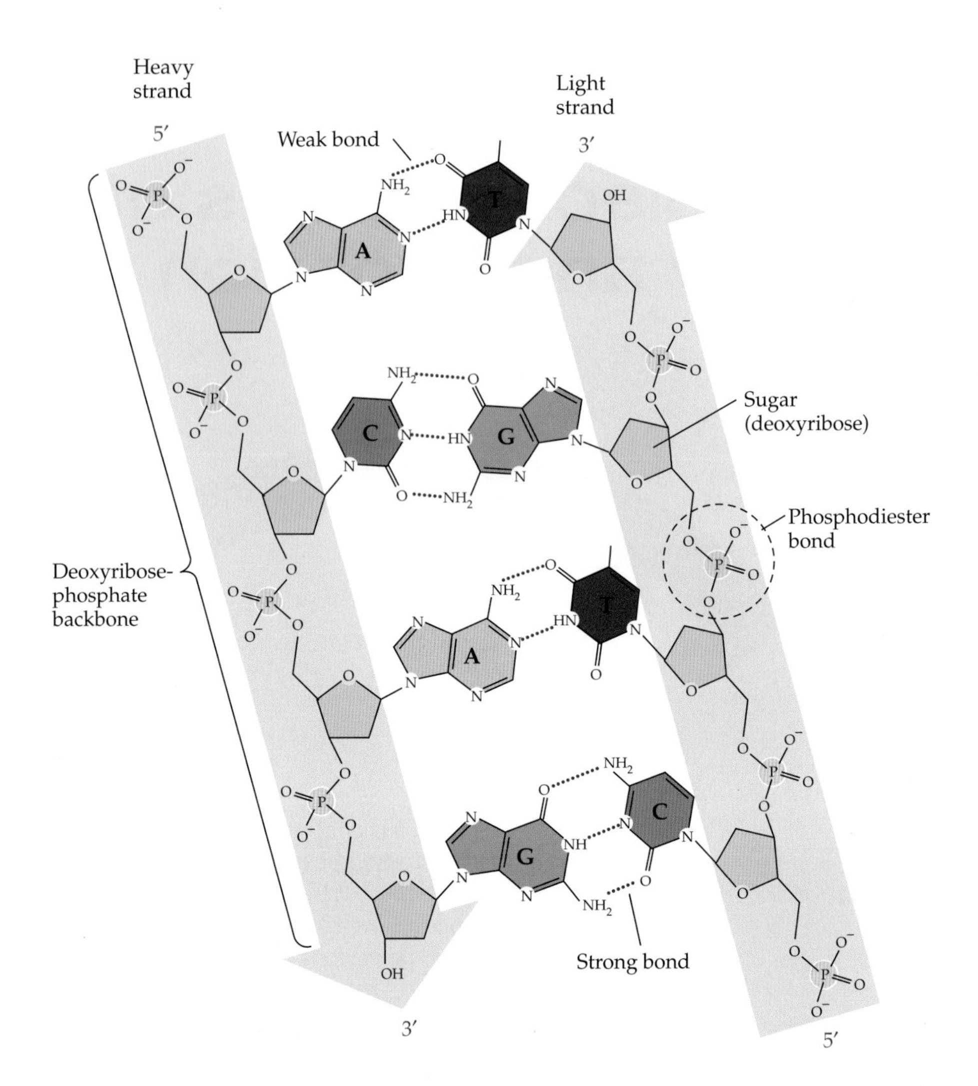

Figure 1.1 The right-handed double helix (*left*) and a schematic representation of the antiparallel structure of double-stranded DNA (*right*). The antiparallelism is emphasized by the blue arrows. Note complementary base pairing by means of hydrogen bonds (dotted lines) between thymine and adenine (weak bond) and cytosine and guanine (strong bond).

The two strands of DNA are joined throughout their lengths by hydrogen bonds between pairs of nucleotides. A purine always pairs with a pyrimidine (Figure 1.1). Specifically, adenine pairs with thymine by means of two hydrogen bonds (the **weak bond**) and guanine pairs with cytosine by means of three hydrogen bonds (the **strong bond**). These complementary pairs, commonly written as **G:C** and **A:T**, are also called the **canonical** or **Watson-Crick base pairs**. Because of complementarity, usually only one strand needs to be specified. In the double helix, the **heavy strand** is the one containing more than 50% of the heavier nucleotides, the purines A and G. The **light strand** is the one that contains more than 50% of the pyrimidines C and T.

Ribonucleic acid (**RNA**) is found as either a single- or double-stranded molecule. RNA differs from DNA by having **ribose** instead of deoxyribose as its backbone sugar moiety, and by using the nucleotide **uracil** (**U**) in place of thymine. Adenine, cytosine, guanine, thymine, and uracil are referred to as the **standard nucleotides**. Some functional RNA molecules, most notably transfer RNAs, contain **nonstandard nucleotides**, i.e., nucleotides other than A, C, T, U, and G, that have been derived by chemical modification of standard nucleotides after the transcription of the RNA. In addition to the canonical complementary base pairs, which in RNA are G:C and A:U, the G:U pair is also stable in RNA. (In DNA, stable pairing does not occur between G and T.)

The length of a single-stranded nucleic acid is measured in number of nucleotides. That of a double-stranded sequence is measured in **base pairs** (**bp**), thousands of base pairs (**kilobases**, **Kb**), millions of base pairs (**megabases**, **Mb**), or billions of base pairs (**gigabases**, **Gb**).

Genomes

The entire complement of genetic material carried by a monoploid organism, such as a prokaryote, is called the **genome**. The term "genome" (Huxley 1926) is an anglicized form of the German *genom*, which was coined by Winkler (1920) as a portmanteau of "gene" and "chromosome." In monoploids, **genome size** refers to the total amount of DNA in the genome. In diploid organisms, such as humans, the terms "genome" and "genome size" usually refer to the genetic material in a haploid gamete. For example, a human sperm (or egg) contains ~3.4 Gbp of DNA while, with few exceptions, human somatic cells contain two genomes, i.e., ~6.8 Gbp of DNA. In organisms with ploidy of three or higher (e.g., triploids, tetraploids, hexaploids), the term "genome" is more difficult to define precisely (Chapter 11). In eukaryotes, a distinction is usually made between the nuclear genome and the genomes of organelles, such as those of mitochondria and chloroplasts.

The size of a genome can be defined either as its mass in picograms (pg) of DNA (1 pg = 10^{-12} g) or as its length by using base pairs (bp), kilobase pairs (1 Kb = 10^3 bp), megabase pairs (1 Mb = 10^6 bp), or gigabase pairs (1 Gb = 10^9 bp) as the units of measurement. Conversion between mass and length is a relatively straightforward process, with 1 pg = 978 Mb and 1 Gb = 1.02 pg. In recent years the total number of DNA sequences deposited in databases has grown enormously. The total size of sequences in a database can reach values of terabase pairs (10^{12} bp) and even petabase pairs (10^{15} bp).

TABLE 1.1

One-letter abbreviations for the DNA and RNA alphabets

Symbol	Description
DNA or RNA	
A	Adenine
C	Cytosine
T	Thymine
G	Guanine
W	Weak bonds (A, T)
S	Strong bonds (C, G)
R	Purines (A, G)
Y	Pyrimidines (C, T)
K	Keto (T, G)
M	Amino (A, C)
B	C, G, or T
D	A, G, or T
H	A, C, or T
V	A, C, or G
N	A, C, T, or G (or unknown)
X	Stop codon
—	No nucleotide (gap symbol)
RNA only	
U	Uracil
I	Inosine

Genome constituents

Some parts of the genome are transcribed into RNA and some are nontranscribed (**Figure 1.2**). The entire set of transcribed sequences produced by the genome is called the **transcriptome**. Some parts of the transcriptome are translated into proteins and some are not. The entire set of proteins encoded by the genome is called the **proteome**. (In the literature, the terms "transcriptome" and "proteome" are sometimes used for a particular cell type or for a particular set of conditions.) Each of the three genomic constituents contains parts that are functional and parts that are not (Chapter 11). Thus, part of the nontranscribed genome is functional and part of it is not. The same applies to the transcribed but nontranslated part and, to a lesser extent, to the transcribed and translated part of the genome.

Somatic genome processing

As opposed to many oversimplified textbook depictions, eukaryotes do not have the same genomic content in every nucleus at every stage of life. In many eukaryotic lineages, somatic genomes may be processed in specific cells, specific nuclei, or specific phases of the life cycle. Broadly, **somatic genome processing**, also referred to as **developmentally regulated genome rearrangements**, entails modifications to a somatic genome so that it differs from the germline genome (Zufall et al. 2005). Several categories of somatic genome processing are listed below.

Genome loss entails the removal of all the genetic material from the cell. The most famous example of genome loss occurs in mammalian red blood cells. As these cells near maturity, a ring of actin filaments contracts and pinches off a segment of the cell that contains the nucleus—a type of "cell division." The expelled nucleus is then swallowed by macrophages.

Genome diminution (or **chromatin diminution**) is a process of selective elimination of portions of the genome during development. So far this process has been found in three disparate evolutionary lineages: nematodes, copepods, and

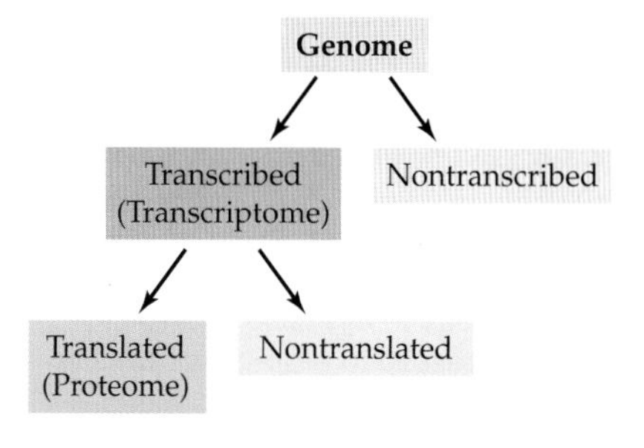

Figure 1.2 A division of the genome into three categories: nontranscribed, transcribed but not translated, and translated.

hagfish, in which nearly all heterochromatin is eliminated during chromatin diminution, such that somatic cells contain almost exclusively euchromatin. For example, in the horse parasite *Parascaris univalens*, 85% of total DNA is eliminated.

Genome-wide rearrangements are known from two groups of microbial eukaryotes: foraminiferans (or shelled amoebae) and ciliates. The rearrangements in these groups are similar to the process of chromatin diminution except that the rearrangements occur in differentiated nuclei within a single cell. In these organisms, there are two distinct nuclei: the micronucleus, which is analogous to a germline nucleus, and the transcriptionally active macronucleus, which is analogous to a somatic nucleus. During development of the macronucleus, the genome experiences a multitude of processes including deletion of specific chromosomal regions and amplification of the remaining chromosomes, such that the somatic genome may contain up to 25 million gene-size chromosomes. An example of genome-wide rearrangements is shown in **Figure 1.3a**.

A particular form of rearrangement, called **gene unscrambling**, has been found in a few ciliate taxa. In these organisms, the coding domains of the micronuclear copies of genes are interrupted by noncoding sequences and are out of order. The unscrambling entails the removal of the noncoding sequences and the rearrangement of the coding regions, thus restoring a proper reading frame.

Targeted or **local rearrangements** affect only specific loci within a genome. The particulars of the rearrangement and the loci affected vary among lineages. In **Figure 1.3b**, we illustrate the process by using the case of V(D)J processing in the vertebrate immune system.

DNA Replication

All organisms replicate their DNA before every cell division. **DNA replication** occurs at elongation rates of about 500 nucleotides per second in eubacteria and about 50 nucleotides per second in vertebrates. During DNA replication, each of the two DNA strands serves as a template for the formation of a new strand. Because each of the two daughters of a dividing cell inherits a double helix containing one "old" and one "new" strand, DNA is said to be replicated **semiconservatively**.

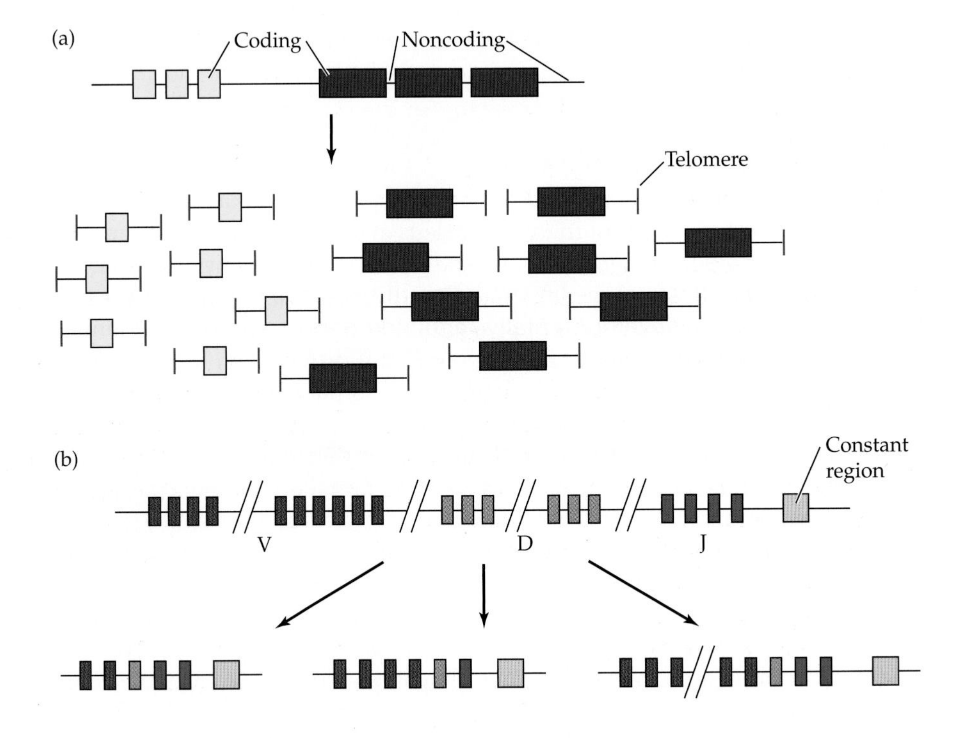

Figure 1.3 Examples of somatic genome processing. (a) Genome-wide rearrangements in some ciliates involve chromosome breakage, elimination of DNA, and amplification of processed chromosomes. Colored rectangles represent coding regions. Red vertical bars at the ends of chromosomes represent telomeres. (b) Targeted rearrangements are exemplified here by V(D)J processing in the vertebrate immune system. In this case, a single locus is processed to produce a diversity of antibody genes by joining various V, D, and J regions. Splices indicate that the distances are not drawn to scale, as well as the fact that there are more V, D, and J regions than shown. (Modified from Zufall et al. 2005.)

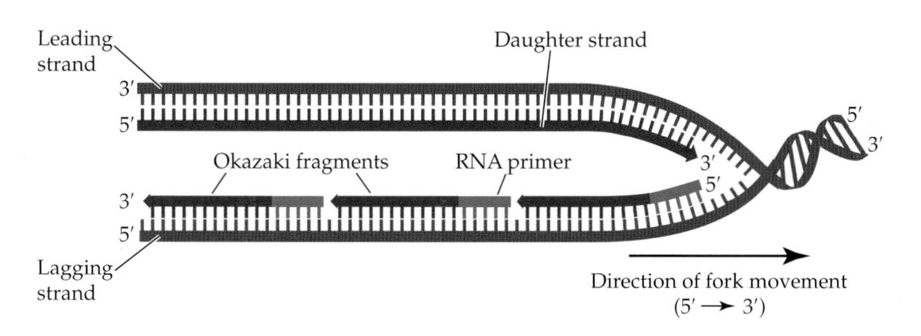

Figure 1.4 Double-stranded DNA replication. The synthesis of both daughter strands occurs only in the 5′-to-3′ direction. The leading strand serves as a template for a continuously synthesized daughter strand (long arrow). The lagging strand is replicated into discontinuous Okazaki fragments (short arrows) that are each initiated by a short RNA primer. After the removal of the primers, the fragments are enzymatically ligated to one another.

Replication starts at a structure called a **replication bubble** or **origin of replication**, a local region of the DNA molecule where the two strands of the parental DNA helix have been separated from each other. Replication proceeds in both directions, forming two **replication forks**. Because DNA replication occurs only in the 5′-to-3′ direction, one of the strands, known as the **leading strand**, serves as a template for a continuously synthesized **daughter strand** (**Figure 1.4**). The other strand, called the **lagging strand**, is replicated in discontinuous small pieces known as **Okazaki fragments** (named after Tsuneko and Reiji Okazaki, the husband-and-wife team who discovered them). Each Okazaki fragment consists of a short DNA fragment with an RNA primer at the 5′ terminus. Ligation of Okazaki fragments is preceded by the replacement of the RNA primers by DNA from the elongating neighboring fragments. Okazaki fragments are longer in prokaryotes (1,000–2,000 nucleotides) than in eukaryotes (100–200 nucleotides).

Most prokaryotic genomes have a single origin of replication, and many can replicate in less than 40 minutes. In eukaryotic cells, many replication origins exist. They are usually spaced at intervals of 30,000–300,000 bp from one another. Replication in eukaryotic cells may take several hours.

Transcription and Posttranscriptional Modifications of RNA

Transcription is the synthesis of an RNA molecule that is complementary in sequence to a DNA template. Transcription is carried out by DNA-dependent RNA polymerases. The DNA strand that provides the template for ordering the sequence of nucleotides in the transcribed RNA is called the **antisense strand**, because its sequence is complementary to the RNA. The DNA strand complementary to the antisense is called the **sense strand**, because its sequence is identical to that of the newly created RNA (except for uracil being used instead of thymine). The template DNA strand is read 3′ → 5′ by the RNA polymerase, and the new RNA strand is synthesized in the 5′ → 3′ direction. The freshly transcribed RNA is referred to as a **primary transcript**. Since the primary transcripts produced by a genome differ greatly from one another in length and molecular weight, they are collectively denoted as **heterogeneous nuclear RNA** (**hnRNA**). RNA transcripts may or may not be functional. In the case that they are not, they are referred to as **transcriptional noise**.

Primary transcripts of protein-coding genes and RNA-specifying genes may contain exons and introns. **Intervening sequences**, better known as **introns**, are those transcribed sequences that are spliced out during the processing of the primary transcript into mature RNA. All the other sequences remaining in the mature RNA following splicing are referred to as **exons**. The collection of all the exons in a genome is called the **exome**. In the literature, the term "exome" is almost invariably used to denote the collection of all exons in protein-coding genes. This is not strictly accurate, as many RNA-specifying genes also have an exon-intron structure.

According to the specific mechanism by which the intron is cleaved out, we classify introns into two main categories: self-splicing and non-self-splicing. **Self-splicing introns** (including group I, group II, group III, group IV, twintrons, and mirtrons) are

(a)

```
ATGAG-TCAATTCCATTTGAT-AAAAAACTATTAGATTATTAGGTATGAAAATGGATAC-AGAGATTTTTATAC-TTCGCAGATCGTTGGTTATTCTCAAC-AATCATAAATATAT-GGT
AUGAGCUCAAUUCCAUUUGAUCAAAAAACUAUUAGAUUAUUAGGUAUGAAAAUGGAUACCAGAGAUUUUUAUACCUUCGCAGAUCGUUGGUUAUUCUCAACCAAUCAUAAAUAUAUCGGU

ACTTTATATTTATTATTTAGTAT-GGTGCTGGTTTAGCTGGA-TAGCCTTATCTGTTAT-ATGCGTATGGAGTTAGC-AGTTGTGGTGATTTAATTTTAAATT-TAATTATCAACTTTAT
ACUUUAUAUUUAUUAUUUAGUAUCGGUGCUGGUUUAGCUGGACUAGCCUUAUCUGUUAUCAUGCGUAUGGAGUUAGCCAGUUGUGGUGAUUUAAUUUUAAAUUCUAAUUAUCAACUUUAU

AATAC-ATTGTAACTGCTCATGCC-TTATTATGATTTTTTTCTTTGTGATGC-TGCTTTAATTGGTGGGTTTGGAAATTGGTTTGTAC-TTTACTTATAGGTGCTCCTGATATGGCATTT
AAUACCAUUGUAACUGCUCAUGCCCUUAUUAUGAUUUUUUUCUUUGUGAUGCCUGCUUUAAUUGGUGGGUUUGGAAAUUGGUUUGUACCUUUACUUAUAGGUGCUCCUGAUAUGGCAUUU

CCTCGTTTAAATAATAT-AGTTTATGGTTATTACCTTCCTCTTTAATTTTACTTCTTTTATCTTCTTATGTTGAAAT-GGTGCTGGAACTGGTTGGACTGTTTATCCTCCACTATCTTCT
CCUCGUUUAAAUAAUAUCAGUUUAUGGUUAUUACCUUCCUCUUUAAUUUUACUUCUUUUAUCUUCUUAUGUUGAAAUCGGUGCUGGAACUGGUUGGACUGUUUAUCCUCCACUAUCUUCU

ATTACTGGTCATAGCGGACCATCTGTTGATTTAGCTATTTTTAGT-TGCATCTAGCTGGTGTTTCTTCTATGTTAGGTGCTAT-AATTTCATTTGTAC-ATTAAAAATATGCGT-TTAAA
AUUACUGGUCAUAGCGGACCAUCUGUUGAUUUAGCUAUUUUUAGUCUGCAUCUAGCUGGUGUUUCUUCUAUGUUAGGUGCUAUCAAUUUCAUUUGUACCAUUAAAAAUAUGCGUCUUAAA

GGATTAACAGGAGAACGTTTATCTTTATT-GTTTGGGCTGTATTAGTAACTGTGATTTTATTATTACTTTCACTGCCTGT-TTAGCAGGTGCTAT-ACTATGTTATTAACTGAT-GTAAT
GGAUUAACAGGAGAACGUUUAUCUUUAUUUGUUUGGGCUGUAUUAGUAACUGUGAUUUUAUUAUUACUUUCACUGCCUGUCUUAGCAGGUGCUAUCACUAUGUUAUUAACUGAUCGUAAU

TTTAATACATCTTCTCCTGATGCAAC-GGTGGTGGAGATCCTATTTTATATCAACATTTGCTTTGGTTTTTTGGCCATCCAGAAGTTTACATTTTAATTTTACCTGGTTTTGGTAT-GTT
UUUAAUACAUCUUUUUUUGAUGCAACCGGUGGUGGAGAUCCUAUUUUAUAUCAACAUUUGUUUUGGUUUUUUGGCCAUCCAGAAGUUUACAUUUUAAUUUUACCUGGUUUUGGUAUCGUU
                * **                                      *

TCTATTATTATTCAAGC-TATGCTAATAAAG-TATTTTTGGTTATTTAGGTATGGTGTATGCTATGTTGTCTATTGGTAT-TTGGGTTTTATAGTGTGGGCTCAT-ATATGTATACTGTA
UCUAUUAUUAUUCAAGCCUAUGCUAAUAAAGCUAUUUUUGGUUAUUUAGGUAUGGUGUAUGCUAUGUUGUCUAUUGGUAUCUUGGGUUUUAUAGUGUGGGCUCAUCAUAUGUAUACUGUA

GGATTGGATGTGGATAC--GCGCTTATTTCAC-GCTGCTACTATGAT-ATTGCTGTGCCAAC-GGTATTAAGATTTTTAG-TGGTTAGCAACCTTATTTGGTTCTTATCTTAAATTACGT
GGAUUGGAUGUGGAUACUCGCGCUUAUUUCACCGCUGCUACUAUGAUCAUUGCUGUGCCAACCGGUAUUAAGAUUUUUAGCUGGUUAGCAACCUUAUUUGGUUCUUAUCUUAAAUUACGU

ACTCCTTTACTATTCATCTTAGGTTTTCTTATTTTGTTTACATTAGGTGGTCTTAGTGGAGTTGT--TAGCTAATTCTGGTCTTGATAT-GCTTTCCATGATAC-TATTATGTTGTAGCT
ACUCCUUUACUAUUCAUCUUAGGUUUUCUUAUUUUGUUUACAUUAGGUGGUCUUAGUGGAGUUGUGUUAGCUAAUUCUGGUCUUGAUAUCGCUUUCCAUGAUACCUAUUAUGUUGUAGCU

CATTTTCATTATGT-TTATCTATGGGAGCAGT-TTTGCTGCTTTTGCTGCTTTTTAT-ATTGGTTTTCTATTATTAAATC-TTTACTACTTATGAACCACA-GAATGGGAATTTGTTTAT
CAUUUUCAUUAUGUCUUAUCUAUGGGAGCAGUCUUUGCUGCUUUUGCUGCUUUUUAUCAUUGGUUUUCUAUUAUUAAAUCCUUUACUACUUAUGAACCACACGAAUGGGAAUUUGUUUAU

AATCCTTT-GAATTACGAAAAAAAATTCCTTAT-GTCGTACTGATTTTAGTTAT-TTGAAACAGCTGGTAT-TTACATTTCATTACTAC-TTCATTGGTGTTAATCTTAC-TTCTTCCCT
AAUCCUUUCGAAUUACGAAAAAAAAUUCCUUAUCGUCGUACUGAUUUUAGUUAUCUUGAAACAGCUGGUAUCUUACAUUUCAUUACUACCUUCAUUGGUGUUAAUCUUACCUUCUUCCCU

ATGCAT-TATTAGGTCTTGCAGGAATGC-ACGCCGTATTCCAGATTATCCTGATGC-TATTTACATTTTAATTTAAT-AGTAGTTATGGTTCTTTTGTTAC-TTAGTTTCCACTATCATA
AUGCAUCUAUUAGGUCUUGCAGGAAUGCCACGCCGUAUUCCAGAUUAUCCUGAUGCCUAUUUACAUUUUAAUUUAAUCAGUAGUUAUGGUUCUUUUGUUACCUUAGUUUCCACUAUCAUA

TTCTTTATATTTGGTGT-GTTTTTAATTATAATTTATTTAT-AATGGTTTTCCTTCTATATTTTATAT-ATTAAACCTTTAATCAAAATTACTTTCTTC-GTATTAAACAAAT-ATTCAA
UUCUUUAUAUUUGGUGUCGUUUUUAAUUAUAAUUUAUUUAUCAAUGGUUUUCCUUCUAUAUUUUAUAUCAUUAAACCUUUAAUCAAAAUUACUUUCUUCCGUAUUAAACAAAUCAUUCAA

GGAAATAATGAATGTCTTAATAGTAT-TCTAAAGAATTAAC-ATCTTAATTCTTCATGTTAAACGCTTGTCTAAATTAAGCGTAAATAACTGGAGAATTGCTTAA 1721
GGAAAUAAUGAAUGUCUUAAUAGUAUCUCUAAAGAAUUAACCAUCUUAAUUCUUCAUGUUAAACGCUUGUCUAAAUUAAGCGUAAAUAACUGGAGAAUUGCUUAA 1785
```

(b)

Edited mRNA translates into:

```
M S S I P F D Q K T I R L L G M K M D T R D F Y T F A D R W L F S T N H K Y I G T L Y
F S I G A G L A G L A L S V I M R M E L A S C G D L I L N S N Y Q L Y N T I V T A H A
M I F F F V M P A L I G G F G N W F V P L L I G A P D M A F P R L N N I S L W L L P S
I L L L L S S Y V E I G A G T G W T V Y P P L S S I T G H S G P S V D L A I F S L H L
V S S M L G A I N F I C T I K N M R L K G L T G E R L S L F V W A V L V T V I L L L L
P V L A G A I T M L L T D R N F N T S F F D A T G G G D P I L Y Q H L F W F F G H P E
I L I L P G F G I V S I I I Q A Y A N K A I F G Y L G M V Y A M L S I G I L G F I V W
H M Y T V G L D V D T R A Y F T A A T M I I A V P T G I K I F S W L A T L F G S Y L K
T P L L F I L G F L I L F T L G G L S G V V L A N S G L D I A F H D T Y Y V V A H F H
L S M G A V F A A F A A F Y H W F S I I K S F T T Y E P H E W E F V Y N P F E L R K K
Y R R T D F S Y L E T A G I L H F I T T F I G V N L T F F P M H L L G L A G M P R R I
Y P D A Y L H F N L I S S Y G S F V T L V S T I I F F I F G V V F N Y N L F I N G F P
F Y I I K P L I K I T F F R I K Q I I Q G N N E C L N S I S K E L T I L I L H V K R L
L S V N N W R I A STOP
```

Unedited mRNA would translate into:

```
M S Q F H L I K N Y STOP
```

◀ **Figure 1.5** Extensive RNA editing in a mitochondrial gene of the many-headed slime mold *Physarum polycephalum.* (a) Comparison of the sense strand of the mitochondrial cytochrome *c* oxidase subunit 1 (*co1*) gene (black) with the mRNA after RNA editing (green). Insertions are indicated by upward arrows; nucleotide substitutions by asterisks. There are 60 insertions of monoribonucleotides (59 Cs and one U), 2 of dinucleotides (CU and GU), and four C → U changes. (b) Comparison of the 594-amino-acid-long protein produced by the edited mRNA and the 10-amino-acid-long protein that would have been produced by the unedited transcript. The reading frame of the unedited *co1* gene contains 36 stop codons that are "corrected" by the RNA editing. (Data from Gott et al. 1993.)

cleaved out without the help of exogenous gene products. **Non-self-splicing introns** are cleaved out by enzymes encoded elsewhere in the genome. Two major categories of non-self-splicing introns are known. Introns in nuclear protein-coding genes are called **spliceosomal introns** because they are cleaved out by means of an RNA-protein complex called the spliceosome. The second category consists of introns in nuclear tRNA-specifying genes of eukaryotes as well as archaebacterial tRNA- and rRNA-specifying genes. These introns are cleaved out by exogenous endoribonucleases that are evolutionarily unrelated to the proteins in the spliceosomal complex.

The **splicing sites** (or **junctions**) of spliceosomal introns are determined to a large extent by nucleotides at the 5′ and 3′ ends of each intron, known as **donor sites** and **acceptor sites**, respectively. The vast majority of spliceosomal intron sequences start with GU and end in AG (**GU–AG introns**). A significant minority of spliceosomal intron sequences start with AU and end in AC (**AU–AC introns**). The donor and acceptor sites have been shown to be essential for the correct splicing of introns. Exonal nucleotides adjacent to the introns are also known to contribute to the determination of the splicing sites. Following splicing, the exons abutting the intron being spliced out are ligated to each other.

In addition to intron splicing, many primary RNA transcripts, whether derived from protein-coding genes or RNA-specifying genes, undergo further modifications following transcription. Such posttranscriptional modifications may include (1) enzymatic capping of the 5′ end (i.e., the addition of a 7-methylguanosine cap), (2) degradation of nucleotides at the 3′ end, (3) polyadenylation of the 3′–end (i.e., the addition of 100–200 adenine residues), (4) insertion or deletion of standard nucleotides, (5) modification of a standard nucleotide into another (e.g., C to U), and (6) modification of a standard nucleotide into a nonstandard one, such as adenine to inosine (I). These and other processes, which alter the sequence of the primary transcript, are collectively known as **RNA editing**. In some taxa, such as the kinetoplast pathogens *Leishmania* and *Trypanosoma*, RNA editing can be quite extensive, so the resulting RNA may bear little resemblance to the DNA sequence from which it was transcribed (**Figure 1.5**). In such cases, the template gene is called a **cryptogene** (literally, "hidden gene").

Genes

When Wilhelm Johannsen coined the word "gene" in 1909, the term meant "a unit of heredity." Its material basis was unimportant for its usefulness as a concept. After the elucidation of the material basis of heredity, the term "gene" evolved in meaning until it became synonymous with a segment of DNA that is ultimately translated into a protein. Recent molecular studies, however, have altered our perception of genes, and here we adopt a more general definition: a **gene** is a sequence of genomic material (DNA or RNA) that is functional. Performance of the function may not require the gene to be translated, or even transcribed. Alternative definitions ranging from the practical to the metaphysical may be found in the literature, and there are even claims that the term "gene" has outlived its usefulness (Portin 1993; Snyder and Gerstein 2003; Griffiths and Stotz 2006; Pearson 2006; Rolston 2006; El-Hani 2007; Fox Keller and Harel 2007; Gerstein et al. 2007; Knight 2007).

Here, we classify genes into three categories: (1) **protein-coding genes**, which are transcribed into RNA and subsequently translated into proteins; (2) **RNA-specifying genes**, which are transcribed but not translated; and (3) **nontranscribed genes**. Protein-coding genes and RNA-specifying genes are also referred to as **structural** or **productive genes**. (Note that some authors restrict the definition of structural genes to protein-coding genes only.)

Protein-coding genes

A eukaryotic protein-coding gene consists of transcribed and nontranscribed regions. The nontranscribed regions are designated according to their location relative to the transcribed parts as 5′ and 3′ **flanking regions**. The 5′ flanking region contains several specific sequences that determine the initiation, tempo, timing, and tissue specificity of the transcription process. Because these 5′ flanking sequences promote the transcription process, they are also referred to as **promoters**, and the region in which they reside is called the **promoter region** (**Figure 1.6**). For example, the CAAT box and the GC box (a GC-rich sequence that may function in either orientation) control the initial binding of the RNA polymerase, whereas the TATA box controls the choice of the start point of transcription. The 3′ flanking region contains signals for the termination of the transcription process. Because of our incomplete knowledge of the regulatory elements, which may be found at considerable distances upstream or downstream from the transcribed regions of the gene, it is impossible at present to delineate with precision the points at which a gene's flanking regions begin and end. For example, in the human genome, a very important regulatory element of the lactase gene resides in an intron of an upstream neighboring gene (Chapter 2). Moreover, from the viewpoint of the genome, the 5′ flanking region of one gene is equal to the 3′ flanking region of the upstream neighboring gene, while the 3′ flanking region of a gene is one and the same as the 5′ flanking region of a downstream neighboring gene. Because of the difficulties in delineating the beginning and end of a gene, for practical purposes **gene length** is usually defined as the length of the primary transcript.

The primary transcript of a protein-coding gene is referred to as **precursor messenger RNA** (**pre-mRNA**). The transcription of protein-coding genes in eukaryotes starts at the **transcription initiation site** (the **cap site** in the RNA transcript) and ends at the **transcription termination site**, which may or may not be identical with the **polyadenylation** or **poly(A)-addition site** in the mature **messenger RNA** (**mRNA**) molecule. In other words, termination of transcription may occur further downstream from the polyadenylation site (Proudfoot 2011).

In eukaryotes, pre-mRNA may contain exons and introns. Exons or parts of exons that are translated are referred to as **protein-coding exons** or **coding regions**. The first and last coding regions of a gene are flanked by 5′ and 3′ **untranslated regions** (**UTRs**), respectively. Both the 5′ and 3′ untranslated regions are transcribed. The 3′ untranslated region is usually much longer than the 5′ untranslated region. For instance, in the human urokinase-type plasminogen activator gene (*PLAU*), the 3′ untranslated region (929 bp) is approximately eight times longer than the 5′ untranslated

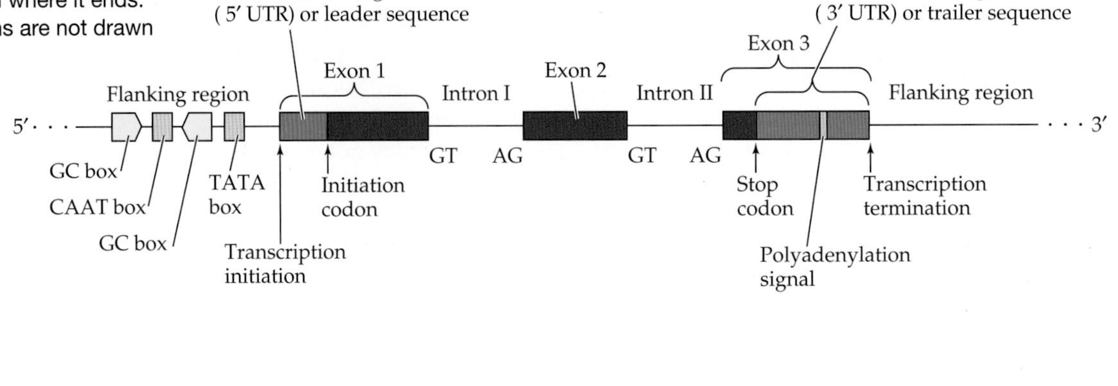

Figure 1.6 Schematic representation of a eukaryotic protein-coding gene with a single polyadenylation signal and its constituent parts. By convention, the 5′ end is at the left. Note, however, that the mRNA is transcribed from the complementary strand, not the one shown. Protein-coding regions are in red, noncoding parts of exons are in blue, and regulatory elements are in yellow. The GC boxes function in either orientation. The dotted lines indicate the impossibility of delineating with precision the flanking regions that can be said to "belong" to the gene; i.e., at the DNA level it is not possible to indicate where the gene starts and where it ends. The different regions are not drawn to scale.

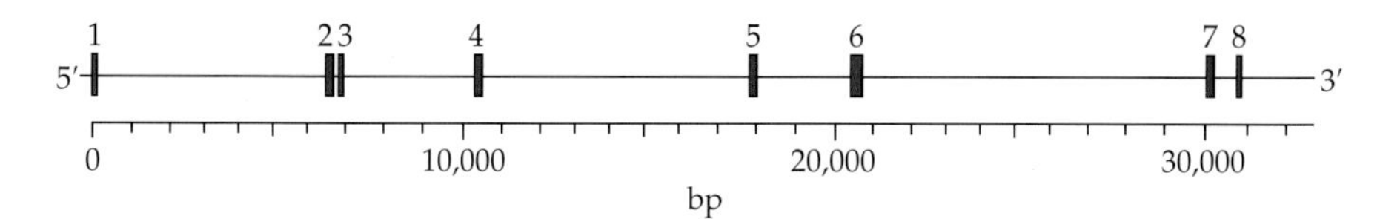

Figure 1.7 The localization of the eight exons (red bars) in the human factor IX gene. Only the transcribed region is shown. The exons and introns are drawn to scale. The total length of the exons is 1,386 nucleotides as opposed to 29,954 nucleotides for the total length of introns. The 5′ untranslated region (30 nucleotides) is much shorter than the 3′ untranslated region (1,389 nucleotides). Only about 4% of the pre-mRNA sequence actually encodes the protein. Note the closeness of several exons (e.g., 2 and 3, and 7 and 8) to each other, as opposed to the remoteness of other neighboring exons (e.g., 6 and 7). (From Graur and Li 2000.)

region (119 bp). UTRs may contain regulatory sequences involved in processing the pre-mRNA or regulating the translation process. One such sequence in eubacteria is the **polyadenylation signal**, also known as the **AATAAA** or **Proudfoot-Brownlee box**, which is located about 20 bp upstream of the transcription termination site and serves as a polyadenylation signal. Some untranslated parts may also be involved in determining the decay rate of the mRNA molecule in the cell.

The number of introns varies greatly from gene to gene. A few genes (e.g., the nebulin- and titin-coding genes) possess more than 100 introns; others (e.g., most histone and olfactory-receptor genes) are devoid of introns altogether. The distribution of intron sizes in vertebrate genes is very broad, and the longest introns extend out to hundreds of kilobases. The mean intron size in human protein-coding genes is about 2,400 bp with a standard deviation of about 5,600 bp. The distribution of exon sizes is much narrower; in humans the mean exon size is about 160 bp and the standard deviation is approximately 120 bp. Exons are not distributed evenly over the length of the gene. Some exons are clustered; others are located at great distances from neighboring exons (**Figure 1.7**).

The variation in protein-coding gene length is enormous. For example, one of the smallest protein-coding genes in the human genome is *STH*, which encodes saitohin, a protein of unknown function that may be associated with Alzheimer's disease. Evolutionarily, the protein is most probably of recent origin, as it is only found in apes (i.e., humans, chimpanzees, gorillas, orangutans, and gibbons) but not in any other primate or nonprimate mammal. It is only 445 nucleotides in length and has no introns. Its coding region is 387 nucleotides long, which translates into a 128-amino-acid-long protein. At the other extreme, one of the largest human genes is *CNTNAP2*, which encodes contactin-associated protein-like 2, molecular defects in which have been linked to several forms of epilepsy, mental retardation, and autism. The gene is over 2.3 million nucleotides in length, thus taking over 16 hours to produce a single transcript. After splicing out the 24 introns, a 9,890-nucleotide-long mRNA is produced. The coding region is 3,996 nucleotides long, producing a 1,331-amino-acid-long protein. The exon-to-intron length ratio in *CNTNAP2* is about 0.4%. In other eukaryotes (e.g., *Drosophila melanogaster*, *Arabidopsis thaliana*, and *Caenorhabditis elegans*), introns are fewer and smaller. Not all introns interrupt coding regions; some occur in untranslated regions, mainly in the region between the transcription initiation site and the translation initiation codon.

By using genomic sequences, it is impossible to tell whether or not what looks like a protein-coding gene is indeed a protein-coding gene. For this reason, the term **open reading frame** (**ORF**) is used. Strictly speaking, an open reading frame is a DNA sequence of considerable length that contains no stop codons in one of the six possible reading frames. Only experimental data can distinguish between a real gene and a gene look-alike.

Protein-coding genes in prokaryotes are different from those in eukaryotes in several respects. Most importantly, they do not contain introns, i.e., they are colinear with the protein product. A set of structural genes in prokaryotes (and more rarely in eukaryotes) may be arranged consecutively to form a unit of genetic expression that is transcribed into one polycistronic mRNA molecule and subsequently translated into different proteins. Such a unit usually contains genetic elements that control the coordinated expression of the genes belonging to the unit. This arrangement of genes is called an **operon** (Jacob et al. 1960).

TABLE 1.2

Examples of noncoding RNA classes

Class	Abbreviation	Function
Guide RNA	gRNA	Template for posttranscriptional RNA editing
Micro RNAs	miRNA	Posttranscriptional and translational regulation
Ribosomal RNA	rRNA	A component of the ribosome
Small interfering RNA	siRNA	RNA interference
Piwi-interacting RNA	piRNA	Forming RNA-protein complexes with Piwi proteins; transcriptional gene silencing of retrotransposons and other elements, particularly in male germline cells
Small nuclear RNA	snRNA	Splicing of spliceosomal introns
Small nucleolar RNA	snoRNA	Guide methylation or pseudouridylation of RNA, removal of introns from pre-mRNA, regulation of transcription factors and RNA polymerase II, maintaining telomeres
Small temporal RNA	stRNA	Regulate gene expression by preventing the mRNAs they bind to from being translated
Transfer RNA	tRNA	Adaptor molecule for amino acids in protein synthesis
Large intervening noncoding RNA	lincRNA	A diverse and heterogeneous class of noncoding RNA. Many are functionless, while others are claimed to have widely disparate roles, including involvement in X-chromosome inactivation, *trans*-activation of the androgen receptor gene, heart wall development, nuclear import, and inflammatory signaling.
Ribonucleic acid enzyme or catalytic RNA	Ribozyme	Catalysis of chemical reactions
Transfer-messenger RNA	tmRNA	Rescuing stalled ribosomes, tagging for degradation incomplete polypeptide chains, promoting degradation of aberrant mRNA
RNA in ribonucleoproteins	—	Component of RNA-protein functional complexes
Circular RNA	circRNA	Almost certainly a functionless secondary metabolite that accumulates with age

RNA-specifying genes

Both prokaryotic and eukaryotic genomes contain **RNA-specifying genes** (or simply **RNA genes**), i.e., DNA regions that are transcribed into RNAs and carry out their biological roles as RNA without being translated into proteins. These molecules are called **noncoding RNAs (ncRNAs)** or **functional RNA (fRNA)**; most are transcribed into primary RNA transcripts that are then processed into shorter mature RNA sequences.

Many classes of ncRNAs have been identified and characterized (**Table 1.2**), and many more are likely to be discovered (e.g., Weinberg et al. 2009; Salmena et al. 2011). **Ribosomal RNA (rRNA)** and **transfer RNA (tRNA)** have been known for many years. In the past two decades, a new class, **small RNAs**, has been discovered. Small RNAs are 50–200 nucleotides in length and are often involved in regulating the translation of target RNAs through RNA-RNA interactions. For example, **small nucleolar RNAs (snoRNAs)** are small RNAs whose function is to guide several posttranscriptional processes, such as methylation and alternative splicing, as well as developmental processes such as apoptosis (programmed cell death) in eukaryotes. Many small RNAs, such as U1, U2, U4, and U5, are involved in the RNA splicing processes. **Small interfering RNA (siRNA)**, sometimes known as **short interfering RNA** or **silencing RNA**, is a group of double-stranded RNA molecules, 20–25 nucleotides in length, that play a variety of roles. Most notably, some siRNAs are involved in **RNA interference (RNAi)** pathways, where they interfere with the expression of genes. As a rule, each siRNA interferes with the expression of a single gene. In addition, siRNAs can act as antiviral agents, transposon silencers, and shapers of chromatin structure. For example, **Piwi-interacting RNA** (**piRNA**), which may be 26–31 nucleotides long, forms RNA-protein complexes with Piwi proteins and is mainly involved in transcriptional and posttranscriptional gene silencing of retrotransposons and other transposable elements.

MicroRNAs (**miRNAs**) are a class of small RNAs that have been the subject of intense research in the past decade (Krol et al. 2010). In animals, miRNAs silence genes through translational inhibition; in plants, the vast majority of miRNAs mediate mRNA decay. Here, we describe the synthesis of miRNAs as it occurs in all animals except placozoans and poriferans (sponges). The synthesis of miRNA in plants differs from that in animals in many respects (Shabalina and Koonin 2008).

Like protein-coding genes, microRNA-specifying genes are transcribed by RNA polymerase II into **primary miRNA** (**pri-miRNA**) transcripts (**Figure 1.8**). If an miRNA gene resides within a protein-coding gene (e.g., in an intron) in the sense strand, it can be cotranscribed with the "host" gene. If it resides in the complementary strand or in an intergenic region, its transcription requires an independent promoter. Pri-miRNAs form **hairpins** or **stem-loop structures** that are recognized and cleaved by a microprocessor complex containing the enzymes Drosha and Pasha into **precursor miRNAs** (**pre-miRNAs**) 70–110 nucleotides in length. These pre-miRNAs are exported into the cytoplasm, where they are further processed by another RNase III, Dicer, that cleaves off the loop region of the hairpin, resulting in a 19–24 nucleotide double-stranded **miRNA:miRNA* duplex**. The miRNA strand is loaded into a ribonucleoprotein complex, the **miRNA-induced silencing complex** (**miRISC**), while the complementary miRNA* sequence is usually degraded. Rarely, miRNA* is incorporated into miRISC, where it functions as a regular micro RNA. The mature miRNA within miRISC serves as a guide for recognizing target mRNAs by partial base-pairing to a target sequence in the 3′ UTR of mRNA.

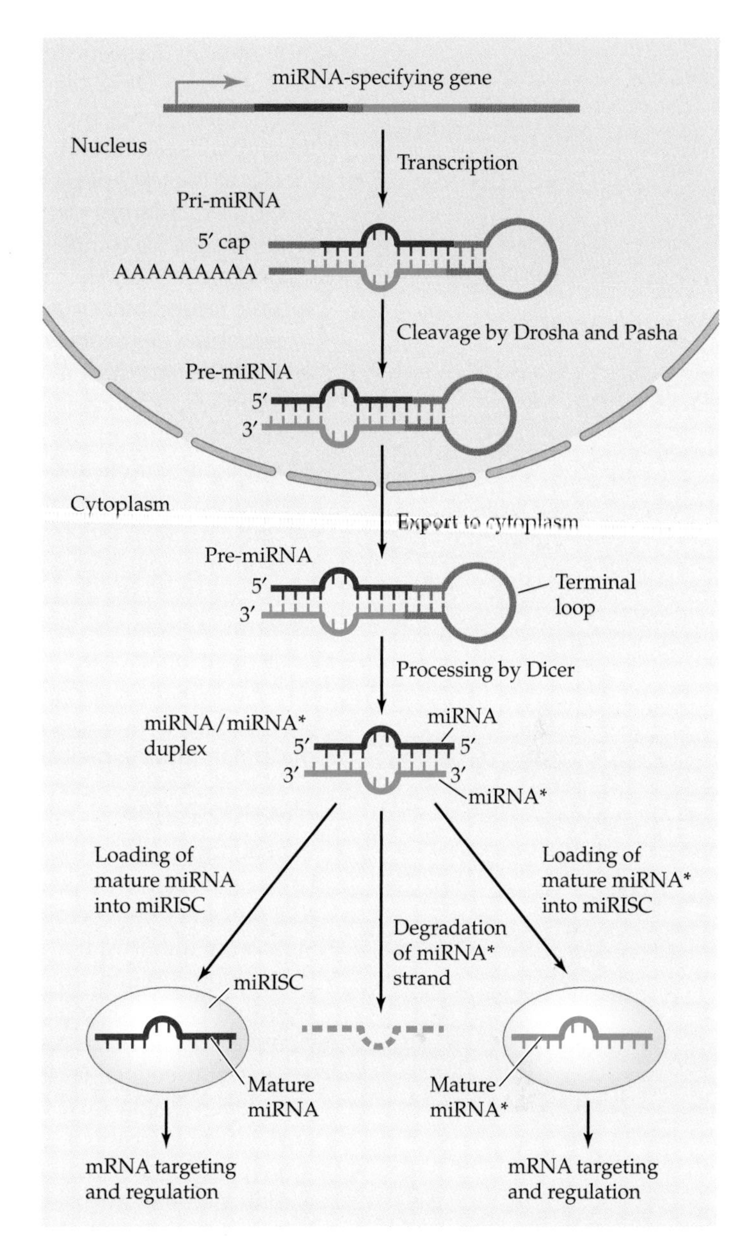

Figure 1.8 Transcription and posttranscriptional processing of microRNA from an miRNA-specifying gene. Transcription yields a primary miRNA (pri-miRNA) that is capped, polyadenylated, and subsequently cleaved into precursor miRNA (pre-miRNA). The pre-miRNA is exported to the cytoplasm, where the endoribonuclease Dicer cleaves off the loop region of the hairpin, resulting in an miRNA-miRNA* duplex. The miRNA strand is loaded into a ribonucleoprotein complex, forming a miRNA-induced silencing complex (miRISC). The complementary miRNA* sequence is either degraded or incorporated within miRISC. (Data from Berezikov 2011.)

Long noncoding RNAs (**lncRNAs**) are longer than 200 nucleotides, a somewhat arbitrary limit that distinguishes lncRNAs from the small regulatory RNAs. Although some lncRNAs may be functional, the vast majority most likely represent transcriptional noise (Wang et al. 2004; Huttenhofer et al. 2005). For only a few lncRNAs has a biologically relevant function been demonstrated.

Nontranscribed genes

Nontranscribed genes, i.e., functional parts in the genome that produce neither RNA nor proteins, can be roughly divided into **DNA regulatory elements** and **DNA structural elements**. (We note, however, that the assignment of nontranscribed genes into one of the two categories is not unequivocal.)

Many types of regulatory elements have been identified. In particular, great progress has been made in identifying **DNA binding sites** for proteins that regulate the transcription of DNA into RNA. Transcriptional regulatory elements, such as promoters, enhancers, repressors, coregulators, modifiers, attenuators, silencers, and insulators, are DNA sequences or binding sites that regulate the expression of proximal or distal genes by binding RNA, proteins, protein complexes, or secondary metabolites. These DNA binding sites can enhance, repress, or modify transcription, mostly at the level of transcription initiation.

Other regulatory elements whose function does not affect transcription include binding sites for complexes involved in DNA replication and recombination, as well as DNA sequences that are recognized by such proteins as restriction enzymes, site-specific recombinases, and methyltransferases.

Structural elements are sequences involved in the determination of structural features of the chromosome, such as supercoils and fragile sites. This category includes DNA attachment regions that bind histones and nonhistone proteins, thus determining the structure of the different forms of chromatin, as well as matrix or scaffold attachment regions and nucleosome positioning elements. **Telomeres** and **centromeres** are two structural elements that are particularly important in maintaining the integrity of linear eukaryotic chromosomes and their proper segregation to daughter cells during meiosis and mitosis. Telomeres are short repetitive sequences at the ends of many eukaryotic chromosomes; they provide protection against such eventualities as terminal exonucleolytic degradation and deter the degradation of genes near the ends of chromosomes by allowing for the shortening of chromosomes that occurs during chromosome replication. Centromeres, which also consist of short repetitive sequences, provide specific sites for attachment of the chromosomes to the spindle machinery during meiotic and mitotic segregation.

Pseudogenes

A **pseudogene** is a DNA segment that exhibits a high degree of similarity to a functional gene but which contains defects, such as nonsense and frameshift mutations, that prevent it from being expressed properly. Most pseudogenes are not transcribed; a significant minority, however, undergo transcription but not translation, and a handful of pseudogenes may even be translated (e.g., Williams et al. 2009). The main defining features of a pseudogene are (1) its similarity to a functional gene and (2) its nonfunctionality. In the literature, there are claims to the effect that some functional genes are in fact pseudogenes (Nelson 2004), or that some pseudogenes, mainly transcribed pseudogenes, may have a function (Salmena et al. 2011). Pseudogenes are ubiquitous at all genomic locations and in all organisms, although some organisms tend to harbor more pseudogenes than others. There are many types of pseudogenes. They are created by different molecular evolutionary processes and will be dealt with in Chapters 7, 8, and 9.

Amino Acids

An amino acid is a molecule containing a central carbon, called the α **carbon**, attached to (1) an **amine group** (**—NH_2**), (2) a **carboxyl group (—COOH)**, (3) a hydrogen atom, and (4) the **side chain** or **—R group** (**Figure 1.9**). The side chains vary in size, shape, charge, hydrogen-bonding capacity, composition, and chemical reactivity. It is the side chains that distinguish one amino acid from another. Indeed, the classification of amino acids is made on the basis of their —R groups.

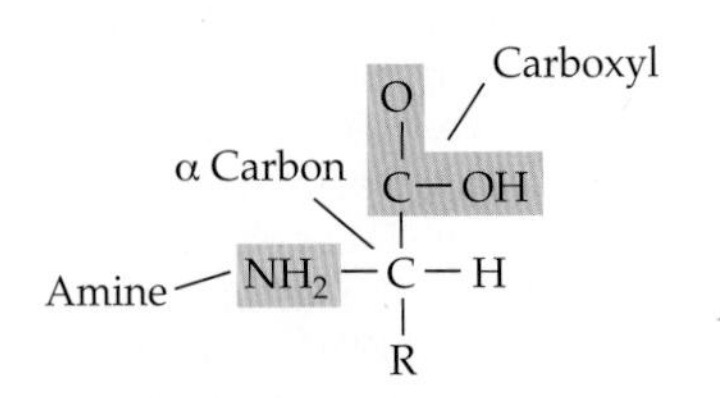

Figure 1.9 Schematic structure of a nonionized amino acid. It contains a central α carbon, an amine group, a carboxyl group, a hydrogen, and a side chain (denoted by R). Ionized structures are shown in Table 1.3.

Amino acids are divided into **proteinogenic** (literally, protein-building) and **nonproteinogenic** amino acids. Proteinogenic amino acids are those amino acids that are charged onto specific tRNAs and are added to the protein by the process of translation (described later in this chapter). There are 22 proteinogenic amino acids, of which 20 are **primary amino acids**, i.e., they are found in all proteins in all organisms (**Table 1.3**). Nonproteinogenic amino acids either are not found in proteins (e.g., the neurotransmitter 5-hydroxytryptamine, or serotonin, is an amino acid) or are the result of posttranslational modification (e.g., hydroxyproline, which is produced by posttranslational modification of proline and is a major component of the protein collagen).

TABLE 1.3

The 22 proteinogenic amino acids

Amino acid[a]			Molecular weight	Ionized structure[b]
Alanine	Ala	A	89.1	COO^- $^+H_3N-C-H$ CH_3
Arginine	Arg	R	174.2	COO^- $^+H_3N-C-H$ CH_2 CH_2 CH_2 $N-H$ $C=NH_2^+$ NH_2
Asparagine	Asn	N	132.1	COO^- $^+H_3N-C-H$ CH_2 C O NH_2
Aspartic acid	Asp	D	133.1	COO^- $^+H_3N-C-H$ CH_2 C O O^-
Cysteine	Cys	C	121.2	COO^- $^+H_3N-C-H$ CH_2 SH
Glutamic acid	Glu	E	147.1	COO^- $^+H_3N-C-H$ CH_2 CH_2 C O O^-
Glutamine	Gln	Q	146.2	COO^- $^+H_3N-C-H$ CH_2 CH_2 C O NH_2
Glycine	Gly	G	75.1	COO^- $^+H_3N-C-H$ H
Histidine	His	H	155.2	COO^- $^+H_3N-C-H$ CH_2 C=CH ^+HN NH C H
Isoleucine	Ile	I	131.2	COO^- $^+H_3N-C-H$ $H-C-CH_3$ CH_2 CH_3
Leucine	Leu	L	131.2	COO^- $^+H_3N-C-H$ CH_2 CH H_3C CH_3
Lysine	Lys	K	146.2	COO^- $^+H_3N-C-H$ CH_2 CH_2 CH_2 CH_2 NH_3^+
Methionine	Met	M	149.2	COO^- $^+H_3N-C-H$ CH_2 CH_2 S CH_3
Phenylalanine	Phe	F	165.2	COO^- $^+H_3N-C-H$ CH_2 (benzene ring)
Proline	Pro	P	115.1	COO^- $^+H_2N-C-H$ H_2C CH_2 CH_2
Serine	Ser	S	105.1	COO^- $^+H_3N-C-H$ $H-C-OH$ H
Threonine	Thr	T	119.1	COO^- $^+H_3N-C-H$ $H-C-OH$ CH_3
Tryptophan	Trp	W	204.2	COO^- $^+H_3N-C-H$ CH_2 C CH N H
Tyrosine	Tyr	Y	181.2	COO^- $^+H_3N-C-H$ CH_2 (benzene ring) OH
Valine	Val	V	117.2	COO^- $^+H_3N-C-H$ CH H_3C CH_3
Non-Primary Amino acids[c]				
Pyrrolysine	Pyl	O	255.3	COO^- $^+H_3N-C-H$ CH_2 CH_2 CH_2 CH_2 NH_2^+ $O=C$ H_3C N
Selenocysteine	Sec	U	168.1	COO^- $^+H_3N-C-H$ CH_2 SeH

[a] Amino acid names are given along with their three- and one-letter abbreviations and molecular weights (red numbers). Additional abbreviations exist for ambiguous cases. X (or Xaa) represents an unknown amino acid. B and Asx can represent either asparagine or aspartic acid, and Z and Glx can mean either glutamine or glutamic acid; these abbreviations, found mostly in early sequences derived through obsolete methodology, indicate how difficult it was to distinguish these pairs in the early day of protein sequencing. J and Xle can indicate either isoleucine or leucine, arising from the difficulty in distinguish between these two in mass spectrometry studies.

[b] Ionized structures shown are for pH 6.0–7.0

[c] Unlike the 20 primary amino acids, which are ubiquitous across all life, pyrrolysine and selenosysteine occur only in a few taxa and in a small number of proteins.

With the exception of glycine, each of the primary amino acids can form two non-superimposable, mirror-image structures around the α carbon—much like our left and right hands. These **enantiomers** can rotate plane-polarized light either clockwise (**dextrorotatory,** or D) or counterclockwise (**levorotatory,** or L). Only L-amino acids are used in the process of translation of mRNAs into proteins. D-amino acids and L-amino acids other than the primary ones are sometimes found in proteins, but these are products of posttranslational modifications.

The simplest and smallest amino acid is **glycine**, which has only a hydrogen as its side chain (molecular weight = 75). Next in size is **alanine**, having a methyl as its side chain. **Valine** has a three-carbon-long side chain. Four-carbon-long side chains are found in both **leucine** and **isoleucine**. These two large, aliphatic (i.e., containing no rings) side chains are hydrophobic (literally, "water-fearing")—that is, they have an aversion to water and tend to cluster in the internal part of a protein molecule, away from the aqueous environment of the cell. **Proline** (an imino acid rather than an amino acid) also has an aliphatic side chain, but its side chain loops back to create a second bond to the nitrogen in its amine group. This ring forces a contorted bend on the polypeptide chain.

Three amino acids have aromatic (benzene-like) side chains. **Phenylalanine** contains a phenyl ring attached to a methylene group. The aromatic ring of **tyrosine** contains a hydroxyl group, which makes this amino acid less hydrophobic than the other amino acids mentioned so far. **Tryptophan**, the largest amino acid (molecular weight = 204), has an indole ring joined to a methylene group. Phenylalanine and tryptophan are highly hydrophobic.

A sulfur atom is present in the side chains of **cysteine** and **methionine**. Both of these sulfur-containing amino acids are hydrophobic; however, the sulfhydryl in cysteine is reactive and, through oxidation, may form a disulfide bridge (cystine) with another cysteine.

Serine and **threonine** contain aliphatic hydroxyl groups and are, therefore, much more hydrophilic ("water-loving") and reactive than alanine and valine (which are the dehydroxylated versions of serine and threonine, respectively).

Five amino acids have very polar side chains and are, therefore, highly hydrophilic. **Lysine** and **arginine** are positively charged at neutral pH, and their side chains are among the largest of all amino acids. The imidazole ring of **histidine** can be either uncharged or positively charged, depending on the local environment. At pH 6.0, over 50% of histidine molecules are positively charged; at pH 7.0, less than 10% have a positive charge.

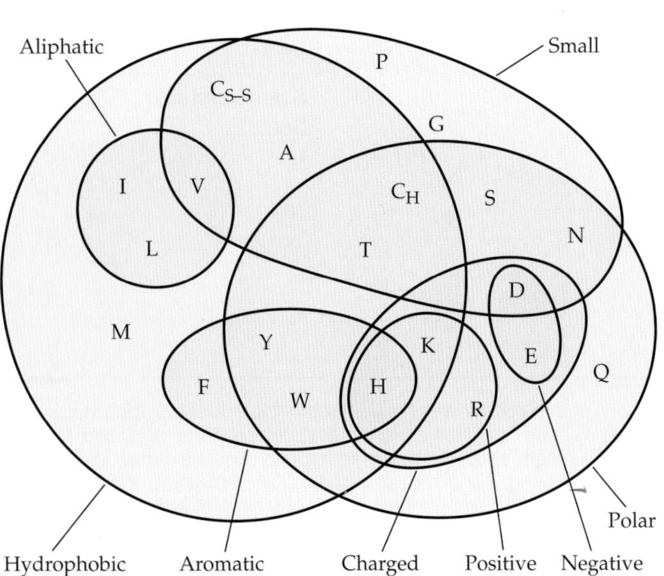

Figure 1.10 Venn diagram showing the division of the 20 primary amino acids into overlapping categories according to size, structure of the side chain, polarity, charge, and hydrophobicity. Note the placement of cysteine in two places, as reduced cysteine (C_H) and as cystine (C_{S-S}). See Table 1.3 for the one-letter abbreviations of the amino acids. (Modified from Taylor 1986.)

Two amino acids, **aspartic acid** and **glutamic acid**, have acidic side chains. They are nearly always negatively charged at physiological pH values. Uncharged derivatives of glutamic acid and aspartic acid are **glutamine** and **asparagine**, respectively, which contain a terminal amide group in place of a carboxylate.

Two proteinogenic amino acids, **pyrrolysine** and **selenocysteine**, are not considered primary amino acids because they occur in only a few taxa and in a small number of proteins. Pyrrolysine is a genetically coded amino acid used by some methanogenic archaebacteria and at least one eubacterium. It is similar to lysine but with an added pyrroline ring linked to the end of the lysine side chain. Pyrrolysine is encoded by the UAG codon, which in the standard genetic code is a stop codon. The amino acid pyrrolysine is charged onto a pyrrolysine-specific tRNA ($tRNA^{Pyl}$) by pyrrolysyl-tRNA synthetase.

Selenocysteine has a structure similar to cysteine, but with a selenium atom taking the place of sulfur. Proteins that contain one or more selenocysteine residues are called selenoproteins. Selenocysteine is encoded by the UGA codon, which in the standard genetic code is a stop codon. The UGA codon encodes selenocysteine only in the presence of a selenocysteine insertion sequence, which in eubacteria is located immediately following the UGA codon, and in archaebacteria and eukaryotes is located in the 3′ untranslated region of the mRNA. The selenocysteine-specific tRNA ($tRNA^{Sec}$) is initially charged with the amino acid serine. The enzymatic conversion of seryl-$tRNA^{Sec}$ to selenocystyl-$tRNA^{Sec}$ is accomplished prior to the incorporation of selenocysteine into the growing protein.

Amino acids may be classified into several overlapping categories by using different criteria. **Figure 1.10** is a Venn diagram of the 20 primary samino acids classified according to size, polarity, charge, and hydrophobicity.

Proteins

The term **protein**, literally the "primary substance," was suggested by Jöns Berzelius in a letter to Gerardus Mulder, published by the latter in 1938 (Vickery 1950). A protein is a macromolecule that consists of one or more polypeptide chains, each of which has a unique, precisely defined amino acid sequence. In a polypeptide chain, the α-carboxyl group of one amino acid is joined to the α-amino group of another amino acid by a **peptide bond**. For example, **Figure 1.11** shows the formation of a dipeptide, glycylalanine, from two amino acids (glycine and alanine), accompanied by the loss of a water molecule. Note that in this dipeptide, glycine acts as the carboxylic acid (its amino end is free), while alanine acts as the amine (its carboxylic end is free). If the roles were reversed, then a different dipeptide, alanylglycine, would have been formed. In other words, polypeptides are polar molecules. The standard presentation of a protein sequence is from the amino end to the carboxyl end. For instance, glycylalanine may be written GA or Gly-Ala, and alanylglycine as AG or Ala-Gly. Each amino acid in a polypeptide is called a **residue**. The "left" end of a polypeptide, containing the free amino group, is termed the **amino** or **N terminus**. The "right" end, with the free carboxyl group, is termed the **carboxyl** or **C terminus**. Polypeptides vary greatly in size, from a few amino acids to such Brobdingnagian proteins as the 26,926-amino-acid-long titin (also known as connectin), which is found in muscle cells.

Four levels of structural organization are customarily used to describe proteins. The **primary structure** is simply the linear arrangement (sequence) of amino acid residues along the polypeptide sequence. **Secondary structure** refers to the spatial arrangement, or **folding**, of amino acid residues that are near to one another in the primary structure. Some arrangements are of a regular kind, giving rise to periodical structures. A well-known periodical secondary structure is the **α helix**, a rod-like entity whose main chain coils tightly as a right-handed screw, with all the side chains sticking outward in a helical array. The very tight structure of the α helix is stabilized by same-strand hydrogen bonds between —NH groups and —CO groups spaced at four-amino-acid-residue intervals. Another periodical structural motif is the **β-pleated sheet**, in which parallel or antiparallel loosely coiled β strands are stabilized by hydrogen bonds between —NH and —CO groups from adjacent strands. Other helices and sheets are known, e.g., the π helix and the polyproline helix. Additional elements include tight turns, flexible loops, and irregular elements, in the past all referred to as **random coil**. Different proteins have varying proportions of secondary structures, and some proteins are made primarily of one type of secondary

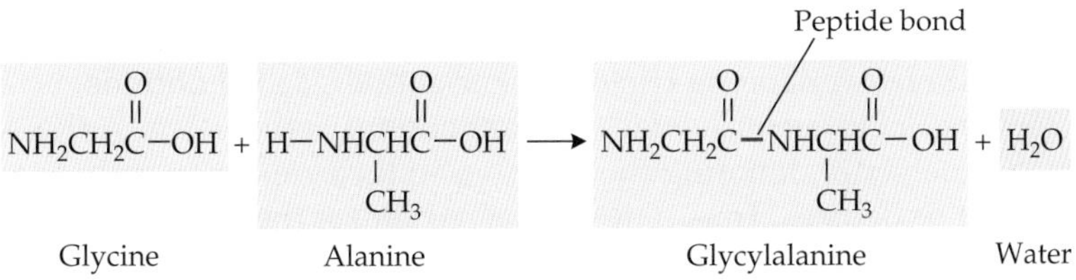

$$\underset{\text{Glycine}}{NH_2CH_2\overset{\overset{O}{\|}}{C}{-}OH} + \underset{\text{Alanine}}{H{-}NH\underset{\underset{CH_3}{|}}{C}H\overset{\overset{O}{\|}}{C}{-}OH} \longrightarrow \underset{\text{Glycylalanine}}{NH_2CH_2\overset{\overset{O}{\|}}{C}{-}NH\underset{\underset{CH_3}{|}}{C}H\overset{\overset{O}{\|}}{C}{-}OH} + \underset{\text{Water}}{H_2O}$$

Peptide bond

Figure 1.11 Formation of a dipeptide (glycylalanine) from two amino acids.

structure. For instance, the major protein in hair, keratin, is made almost entirely of α helices. In contrast, the major component of the silk produced by the moth *Bombyx mori* is a protein made entirely of β-pleated sheets.

The three-dimensional structure of a protein is termed **tertiary structure** and refers to the spatial arrangement of amino acid residues that are not close to one another in the linear primary sequence. Tertiary structure is formed by the packing of local features (such as α helices and β sheets) by covalent and noncovalent forces such as hydrogen bonds, hydrophobic interactions, and salt bridges between positively and negatively charged residues, as well as disulfide bonds between pairs of cysteines. In the process of three-dimensional folding, amino acid residues with hydrophobic side chains tend to be buried inside the protein, while those with hydrophilic side chains tend to be on the outside (p. 18). This tendency is particularly strong in globular proteins such as hemoglobin, which usually function within an aqueous environment.

Intrinsically unstructured proteins, also referred to as **naturally unfolded** or **intrinsically disordered proteins**, are characterized by lack of stable tertiary structure (Wright and Dyson 1999; Tompa 2002). The discovery of intrinsically unstructured proteins challenged the traditional paradigm, according to which the structure of a protein determines its function. Many of these proteins are functional despite the lack of a well-defined structure. A subset of such proteins can adopt a fixed three-dimensional structure after binding to other macromolecules. By 2011, roughly 600 partially or totally unstructured proteins had been identified (Dunker and Kriwacki 2011).

A protein may consist of more than one polypeptide. Such proteins are assembled after each individual polypeptide has assumed its tertiary structure. Each polypeptide chain in such a protein is called a subunit. For example, hemoglobin, the oxygen carrier in blood, consists of four subunits: two α-globins and two β-globins. Pyruvate dehydrogenase, a mitochondrial protein involved in energy metabolism, consists of 72 subunits. Proteins containing more than one polypeptide chain exhibit an additional level of structural organization. **Quaternary structure** refers to the spatial arrangement of the subunits and the nature of their contacts.

Once a polypeptide chain is formed, the macromolecule spontaneously folds itself in response to hydrogen bonds, salt bridges, and hydrophobic and hydrophilic interactions. Thus, while the secondary and tertiary structures of a protein may be said to be uniquely determined by its primary structure, the process also depends on solvent, salt concentration, temperature, and the presence of **chaperones**, i.e., proteins that assist in the folding/unfolding and assembly/disassembly of macromolecular structures.

Some proteins require nonprotein components to carry out their function. Such components are called **prosthetic groups**. The complex of a protein and its prosthetic group is called the **holoprotein**; the protein without the prosthetic group is called the **apoprotein**. Some cellular structures, such as ribosomes and telomerases are nucleoproteins, i.e., structures in which proteins are associated with nucleic acids.

Translation and Genetic Codes

The synthesis of a protein involves a process of decoding, whereby the genetic information carried by an mRNA molecule is translated into primary amino acids through the use of **transfer RNA** (**tRNA**) mediators. A transfer RNA is usually 70–90 nucleotides in length. Each of the proteinogenic amino acids has at least one type of tRNA assigned to it; most have two or more tRNAs. Each tRNA is designed to carry only one amino acid. Specific enzymes, called **aminoacyl-tRNA synthetases**, couple each amino acid to the 3′ end of its appropriate tRNA. **Translation** involves the sequential recognition of adjacent nonoverlapping triplets of nucleotides, called **codons**, by a complementary sequence of three nucleotides (the **anticodon**) in the tRNA. A tRNA is demarcated by a superscript denoting the amino acid it carries and a subscript identifying the codon it recognizes. Thus, a tRNA that recognizes the codon UUA and carries the amino acid leucine is designated $\mathrm{tRNA}^{\mathrm{Leu}}_{\mathrm{UUA}}$.

Translation starts at an **initiation codon** and proceeds until a **stop codon**, or **termination codon**, is encountered. Because codons consist of three nucleotides, each sequence can in principle be read in three phases. The phase in which an RNA sequence is translated is determined by the initiation codon and is referred to as the **reading frame**. In the translational machinery at the interface between the ribosome, the charged tRNA, and the mRNA, each codon is translated into a specific amino acid, which is subsequently added to the elongating polypeptide. Stop codons are recognized by proteins called **release factors**, which terminate the translation process.

Following translation, the creation of a functional protein may involve posttranslational processes, such as folding into a three-dimensional structure, modifications of primary amino acids, removal of terminal sequences at both ends, intra- or intercellular transport, the addition of subunits and prosthetic groups, and protein splicing.

The correspondence between the codons and the amino acids is determined by a set of rules called the **genetic code**. The genetic code is composed of nucleotide triplets, the codons, each of which specify either one amino acid or a translational stop. The code is **nonoverlapping**, i.e., successive codons are read in order and each nucleotide is part of only one codon in a reading frame. The code is also **commaless**, i.e., adjacent codons are not separated by noncoding bases or groups of bases.

With few exceptions, the genetic code for nuclear protein-coding genes is universal, i.e., the translation of almost all eukaryotic and prokaryotic nuclear genes is determined by the same set of rules (**Table 1.4**). Because no genetic code is truly universal, however, the appellation **standard genetic code** is preferred.

Since a codon consists of three nucleotides, and since there are four different types of nucleotides, there are $4^3 = 64$ possible codons. In the standard genetic code, 61 of these code for specific amino acids and are called **sense codons**; the remaining 3 are stop codons. In the standard genetic code, the three stop codons are UAA, UAG, and UGA.

TABLE 1.4

The standard genetic code

Codon	Amino acid	Codon	Amino acid	Codon	Amino acid	Codon	Amino acid
UUU	Phe	UCU	Ser	UAU	Tyr	UGU	Cys
UUC	Phe	UCC	Ser	UAC	Tyr	UGC	Cys
UUA	Leu	UCA	Ser	UAA	Stop	UGA	Stop
UUG	Leu	UCG	Ser	UAG	Stop	UGG	Trp
CUU	Leu	CCU	Pro	CAU	His	CGU	Arg
CUC	Leu	CCC	Pro	CAC	His	CGC	Arg
CUA	Leu	CCA	Pro	CAA	Gln	CGA	Arg
CUG	Leu	CCG	Pro	CAG	Gln	CGG	Arg
AUU	Ile	ACU	Thr	AAU	Asn	AGU	Ser
AUC	Ile	ACC	Thr	AAC	Asn	AGC	Ser
AUA	Ile	ACA	Thr	AAA	Lys	AGA	Arg
AUG	Met	ACG	Thr	AAG	Lys	AGG	Arg
GUU	Val	GCU	Ala	GAU	Asp	GGU	Gly
GUC	Val	GCC	Ala	GAC	Asp	GGC	Gly
GUA	Val	GCA	Ala	GAA	Glu	GGA	Gly
GUG	Val	GCG	Ala	GAG	Glu	GGG	Gly

TABLE 1.5

Watson-Crick and wobble base pairing rules

First-position anticodon	Third-position codon[a]
A	A, **U**, I, k²C, Um, cmnm⁵U, mam⁵s²U
C	A, U, **G**, I
U	**A**, U, G, I
G	A, **C**, U, Cm, I, Um, cmnm⁵U, mam⁵s²U

Source: Data from Crick (1966), Murphy and Ramakrishnan (2004), Agris et al. (2007), and Alkatib et al. (2012). Modified nucleotide abbreviations are from www.ddbj.nig.ac.jp/sub/ref4-e.html.

[a]*Abbreviations*: A, adenine; C, cytosine; U, uracil; G, guanine; I, inosine; k²C, lysidine; Cm, 2′-O-methylcytidine; Um, 2′-O-methyluracil; cmnm⁵U, 5-carboxymethylaminomethyl uracil; mam⁵s²U, 5-methylaminomethyl-2-thiouracil. Numerous modified nucleotides are known in addition to the ones shown here. Codon-anticodon pairing rules are often tRNA- or taxon-specific and may be affected by nucleotide modifications at the tRNA position immediately downstream of the anticodon. First-anticodon positions that pair with third-codon positions according to the Watson-Crick base pair rules are shown in bold blue.

The proper translation of a codon depends on the tRNA recognizing correctly an appropriate codon, as well as on the recognition by the proper aminoacyl-tRNA synthetase of its cognate tRNA. For example, the codon CUU (Leu) must be recognized by the proper tRNA, say one with an AAG anticodon. This tRNA, in turn, must be recognized by leucyl-tRNA synthetase. For translation, each of the 61 sense codons in the standard genetic code requires a tRNA molecule with a complementary anticodon. If pairing between the codon and the anticodon were to follow only canonical Watson-Crick base pairing, 61 species of tRNA would be required. Since most organisms have at most 41 species of tRNA, some tRNA species must pair with more than one codon. Crick (1966) proposed the **wobble hypothesis** to account for the pairing of one anticodon to more than one codon. He postulated that the first base of the anticodon, which binds to the third base of the codon, was not as spatially confined as the other two bases, and thus could engage in nonstandard base pairing, or **wobbling**. Crick's (1966) wobble rules have since been extended, mostly to account for the effects of posttranscriptional modifications affecting the first anticodon position in tRNAs (**Table 1.5**).

With very rare exceptions, the genetic code is **unambiguous**, i.e., one codon can code for only one amino acid. One such exception involves the CUG codon in the yeast *Candida maltosa* and its relatives, which is translated into either serine or leucine (Suzuki et al. 1997). Another case involves the UGA codon in the ciliate *Euplotes crassus*, which specifies either cysteine or selenocysteine, even within the same gene (Turanov et al. 2009). Such codons are termed **polysemous** (literally, "possessing a multiplicity of meanings"). Other exceptions involve termination codons, which are sometimes translated by charged tRNAs, resulting in translation continuing beyond the original termination codon until a downstream stop codon is encountered. This process is called **translational readthrough**, and the tRNAs involved in the readthrough are called **termination suppressors**. In several organisms, natural termination suppressors are known. For example, in *Escherichia coli*, the UGA codon can direct the incorporation of the unusual amino acid selenocysteine (Table 1.3) instead of signaling the termination of the translation process.

Since there are 61 sense codons and only 20 primary amino acids, most amino acids (18 out of 20) are encoded by more than one codon. Such a code is referred to as a **degenerate code**. Three amino acids are encoded by six codons, five by four codons, one by three codons, nine by two codons, and two by one codon. The different codons specifying the same amino acid are called **synonymous codons**. The synonymous codons that differ from each other at the third position only comprise a **codon family**. For example, the four codons for valine (GUU, GUC, GUA, and GUG) form a four-codon family. In contrast, the six codons for leucine are divided into a four-codon family (CUU, CUC, CUA, and CUG) and a two-codon family (UUA and UUG).

The first amino acid in most eukaryotic and archaebacterial proteins is a methionine encoded by the initiation codon AUG. This amino acid is usually removed in the mature protein. Most eubacterial genes also use the AUG codon for initiation, but the amino acid initiating the translation process is a methionine derivative called **formylmethionine**. Alternative initiation codons are known in both prokaryotes and eukaryotes. For example, in *E. coli* 83% of all initiation codons are AUG, 14% are GUG, and 3% are UUG.

The standard genetic code is also used in the independent process of translation employed by the genomes of plastids, such as the chloroplasts of vascular plants. In contrast, the mitochondrial genomes of vertebrates use codes that are different from the standard genetic code; this vertebrate mitochondrial code is shown in **Table 1.6**. Note that two of the codons that specify serine in the standard genetic code are used as termination codons, and that tryptophan and methionine are each encoded by two codons rather than one.

A few prokaryotic genomes have been shown to use alternative genetic codes. For example, eubacterial species belonging to the genus *Mycoplasma* use UGA to code tryptophan. Deviations from the standard genetic code have also been observed in the nuclear genomes of a few eukaryotes. For example, ciliates belonging to the genera *Oxytricha, Paramecium, Stylonychia,* and *Tetrahymena* use UAA and UAG to code for glutamine, and in five species of the yeast genus *Candida,* CUG codes for serine. As a rule, there are only minor differences between the standard genetic code and the seventeen alternative genetic codes described so far (Chapter 10).

In some organisms, some codons may never appear in protein-coding genes. These are called **absent codons**. Some codons are not recognized by either an appropriate tRNA or a release factor. These are called **unassigned** or **hungry codons**. For example, in the genome of the Gram-positive eubacterium *Micrococcus luteus,* the codons AGA and AUA are unassigned. A similar situation was found in *Mycoplasma capricolum,* in which codon CGG is unassigned. Unassigned codons differ from stop codons by not being recognized by release factors. When an in-frame unassigned codon is encountered, translation may stall, and the nascent polypeptide, no longer

TABLE 1.6

The vertebrate mitochondrial genetic code[a]

Codon	Amino acid	Codon	Amino acid	Codon	Amino acid	Codon	Amino acid
UUU	Phe	UCU	Ser	UAU	Tyr	UGU	Cys
UUC	Phe	UCC	Ser	UAC	Tyr	UGC	Cys
UUA	Leu	UCA	Ser	UAA	Stop	UGA	Trp
UUG	Leu	UCG	Ser	UAG	Stop	UGG	Trp
CUU	Leu	CCU	Pro	CAU	His	CGU	Arg
CUC	Leu	CCC	Pro	CAC	His	CGC	Arg
CUA	Leu	CCA	Pro	CAA	Gln	CGA	Arg
CUG	Leu	CCG	Pro	CAG	Gln	CGG	Arg
AUU	Ile	ACU	Thr	AAU	Asn	AGU	Ser
AUC	Ile	ACC	Thr	AAC	Asn	AGC	Ser
AUA	Met	ACA	Thr	AAA	Lys	AGA	Unassigned
AUG	Met	ACG	Thr	AAG	Lys	AGG	Unassigned
GUU	Val	GCU	Ala	GAU	Asp	GGU	Gly
GUC	Val	GCC	Ala	GAC	Asp	GGC	Gly
GUA	Val	GCA	Ala	GAA	Glu	GGA	Gly
GUG	Val	GCG	Ala	GAG	Glu	GGG	Gly

[a]Differences from the standard genetic code are shown in red type.

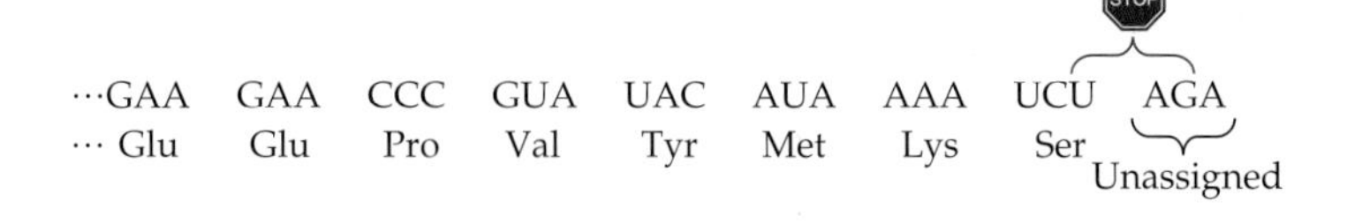

Figure 1.12 The reading frame of human mitochondrial cytochrome c oxidase I ends in an inframe unassigned codon (AGA) preceded by codon UCU. The mitochondrial ribosome has been shown to employ –1 frameshifting upon encountering this unassigned codon, yielding a standard stop codon, UAG.

being extended, may fall off the ribosome and be degraded. Alternatively, a –1 or a +1 frameshift may occur. One such example involves the AGA and AGG codons in human mitochondria (Lightowlers and Chrzanowska-Lightowlers 2010; Temperley et al. 2010). Initially, these codons were thought to be stop codons; however, only one mitochondrial release factor is known to exist (mtRF1a), and this release factor recognizes only the UAA and UAG stop codons, which are used in 11 of the 13 protein-coding genes. Interestingly, the AGA and AGG triplets that terminate the coding region of two mitochondrial genes are preceded by a codon that ends in U. It has been shown that the mitochondrial ribosome employs –1 frameshifting upon encountering the unassigned codons AGA and AGG, thus turning them into the regular stop codon UAG (**Figure 1.12**).

Information Flow among DNA, RNA, and Proteins

In principle, there can be nine possible routs of information flow among DNA, RNA, and proteins (**Figure 1.13a**). Only five of these are realized in nature (**Figure 1.13b**). These are DNA replication, RNA replication, transcription, reverse transcription, and translation. One pathway of information flow, usually worded as "DNA makes RNA makes protein," is frequently and incorrectly referred to as the **central dogma** (e.g., Lesk 2010). The discovery of **reverse transcription**, i.e., the fact that information can flow from RNA to DNA, is understood to have invalidated the dogma. In reality, the central dogma simply states that "once information has got into a protein it can't get out again," which is as true now as it was when it was first stated (Crick 1956, 1970).

Mutation

DNA is a highly stable molecule in vivo, and it is usually replicated with extraordinary accuracy. Rarely, however, errors may occur either during DNA replication or at other times, thereby giving rise to new sequences. These errors are called **mutations**. Mutations can occur in either somatic or germline cells. Since **somatic mutations** are not inherited by the progeny, from an evolutionary point of view they can be ignored, and throughout this book the term "mutation" will be used exclusively to denote mutations occurring in germline cells. Some organisms (e.g., vascular plants) do not have a sequestered germline, so the distinction between somatic and germline mutations is not absolute. In an evolutionary context, a mutation is defined as heritable change in the genetic information that is passed onto the next generation.

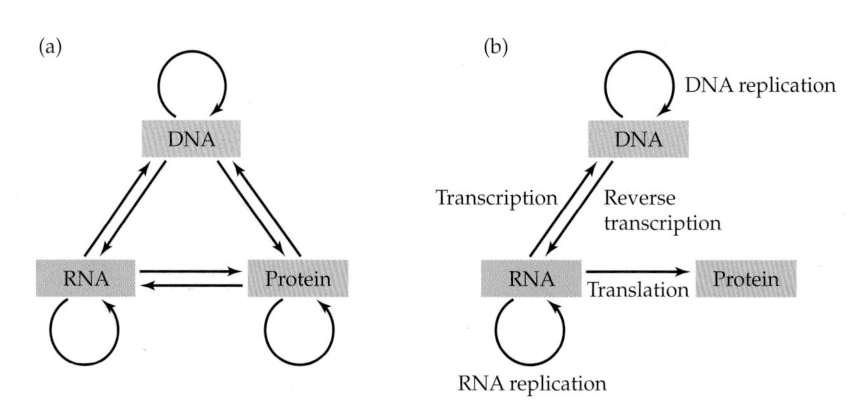

Figure 1.13 (a) Nine possible routes for information flow among DNA, RNA, and proteins. (b) Five routes of information flow found in nature.

Mutations are the ultimate source of genetic variation and novelty in evolution. Unfortunately, many mutations are deleterious. Thus, a mutational "balancing act" needs to be maintained in nature. Too many mutations may spell death for the organism; too few mutations may bring evolution to a halt.

Four aspects of the mutational process will be dealt with in this section: (1) the classification of mutations into different types, (2) the effects that a mutation may have on the hereditary material and its products, (3) the spatial and temporal attributes of mutations, and (4) the randomness of mutation.

Classification of mutations

Mutations that occur during DNA replication are called **replication-dependent mutations**; mutations occurring at other times in the cell cycle are called **replication-independent mutations**. These two types of mutations exhibit strikingly different evolutionary rates and patterns (Chapter 4). Mutations may also be divided into **induced mutations** and **spontaneous mutations**, according to whether or not the mutation was caused (induced) by an external mutagen such as caffeine, nitrous acid, ethidium bromide, or ultraviolet radiation.

Mutations may or may not affect function. Mutations that do affect function may be classified by their effect on function (Muller 1932). An **amorphic mutation** is one that results in the complete loss of gene function. Such mutations are also called **loss-of-function mutations**, **null mutations**, or **knock-out mutations**. A **hypomorphic mutation** results in the partial loss of gene function. For example, a hypomorphic mutation may decrease gene transcription or may quantitatively reduce protein production. A **hypermorphic mutation** causes a quantitative increase in normal gene function.

A **neomorphic mutation** results in a gene function that is different from that of the unmutated gene. Such mutations are also called **gain-of-function mutations**. Gain-of-function mutations are increasingly found to affect fitness. For example, a point mutation from A to G occurring in the intergenic region between the embryonic hemoglobin-chain gene ζ and its upstream neighbor, μ, changes the sequence TAATAA to TGATAA, thus creating a new binding site for a transcription factor. This binding site causes significant downregulation of the downstream μ, α_2, and α_1 genes, thereby causing α-thalassemia, a fitness-reducing disease (De Gobbi et al. 2006).

Antimorphic mutations constitute a special case of neomorphism, whereby the mutant has a contrary or antagonistic effect to the unmutated gene. In diploid organisms, antimorphic mutations are also referred to as **dominant negative mutations**. Antimorphic mutations may occur in genes encoding subunits of homopolymers (quaternary structures made out of identical subunits) that only function if all subunits are functional. If both copies of the gene in a diploid organism encode functional proteins, then all homopolymers will be functional. If, on the other hand, one of the copies encodes a defective protein, then the vast majority of the homopolymers will be dysfunctional. For example, if the functional structure is composed of two identical subunits (a homodimer), then only 25% of the homodimers will be functional. If the functional structure is a homotetramer (four identical subunits), then only 6% of the homotetramers will be functional.

Mutations that do not affect function may be referred to as **homomorphic**. Some mutations, called **conditional mutations**, produce changes in phenotype in one set of environmental conditions (called **restrictive conditions**) but not in others (called **permissive conditions**).

Genes and gene products do not usually act in isolation, but rather interact with one another within a network (Chapter 4), sometimes referred to as an **interactome**. Thus, a mutation only rarely affects the function of a gene or a gene product by itself; usually its interactions with other genes or gene products are affected (**Figure 1.14**). A mutation in a gene that obliterates all its interactions is called **node removal**. Short of node removal, we refer to all other mutations that affect interactions as **edgetic mutations** or **edgetic perturbations** (Rual et al. 2005; Sahni et al. 2013). Edgetic mutations may entail either the loss of some interactions or the gain of new interactions.

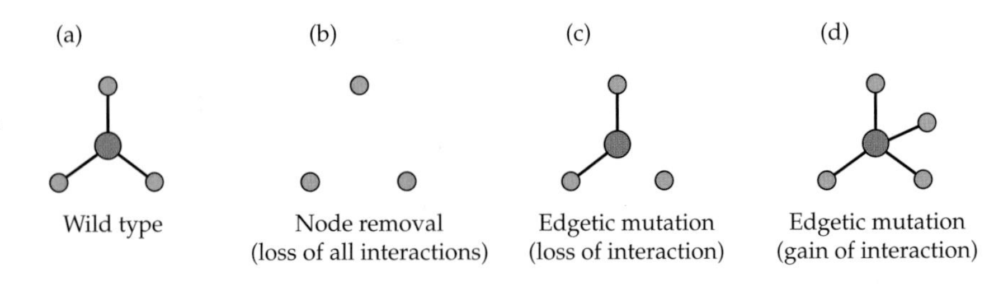

Figure 1.14 (a) A simple interaction graph in which a gene or gene product in the wild type (solid circle) interacts with three other biological entities. Node removal (b) entails the loss of all the interactions. Edgetic mutations (c, d) are classified into losses of interactions (c) or gains of interactions (d).

Mutations may also be classified by the length of the DNA sequence affected by the mutational event. For instance, mutations may affect a single nucleotide through the substitution of one nucleotide by another (**point mutations**), or they may affect the order of nucleotides (**segmental mutations**).

Point Mutations

Point mutations are divided into transitions and transversions. **Transitions** are changes between A and G (purines), or between C and T (pyrimidines). **Transversions** are changes between a purine and a pyrimidine. There are four types of transitions, A → G, G → A, C → T, and T → C. There are eight types of transversions, A → C, A → T, C → A, C → G, T → A, T → C, G → C, and G → T.

Point mutations are thought to arise mainly from the mispairing of bases during DNA replication, such as A:C, which is a purine-pyrimidine mispair, and A:A, which is a purine-purine mispair (**Figure 1.15**). Mispairing is made possible by one of the bases assuming an unfavored tautomeric form, i.e., enol instead of keto in the case of guanine and thymine, or imino instead of amino in the case of adenine and cytosine (**Figure 1.16**). There are many other mechanisms that generate point mutations. For example, guanine oxidation into 8-oxo-7,8-dihydroguanine (8-oxo-G) may result in a G → T transversion. Another frequent source of point mutations is the breaking of the bond between the sugar and the base in the nucleotide (**depurination** or **depyrimidination**). The resulting abasic (apurinic or apyrimidinic) nucleotide is problematic

Figure 1.15 A GC → AT mutation caused by guanine assuming an unfavored tautomeric form, enol, instead of its common keto form. This enol form pairs with thymine instead of cytosine. During the subsequent replication, the newly incorporated thymine pairs normally with adenine. It is also possible for guanine to convert to its rare enol form during its incorporation opposite a template strand, rather than at the time of replication (not shown). In this case, guanine will be located opposite a thymine and replication will result in an AT → GC mutation. (Modified from Johnston 2006.)

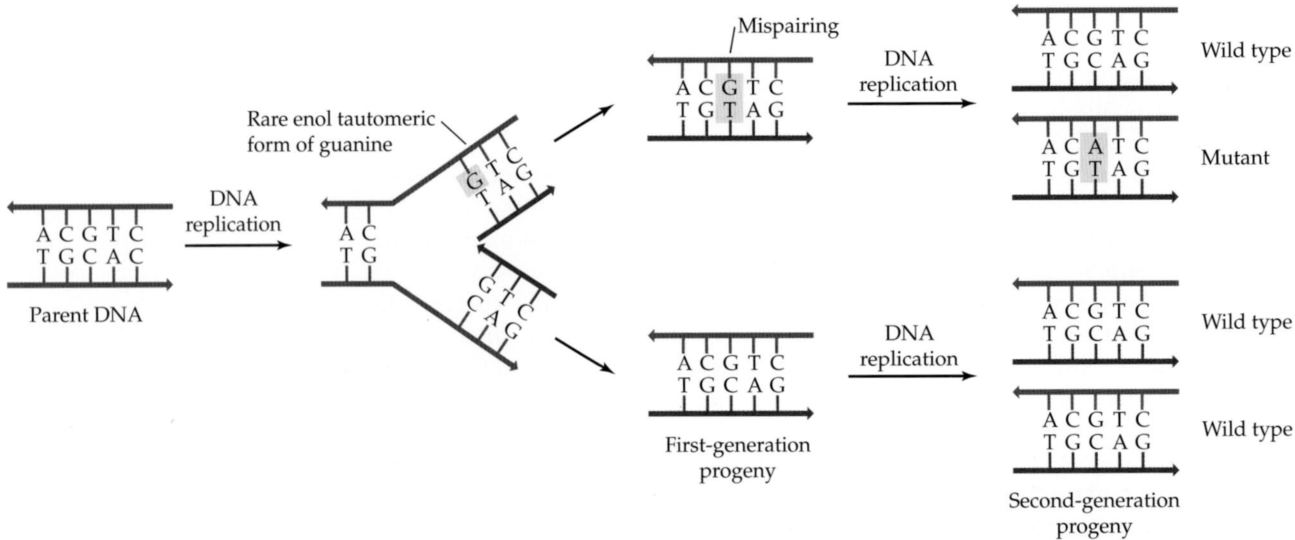

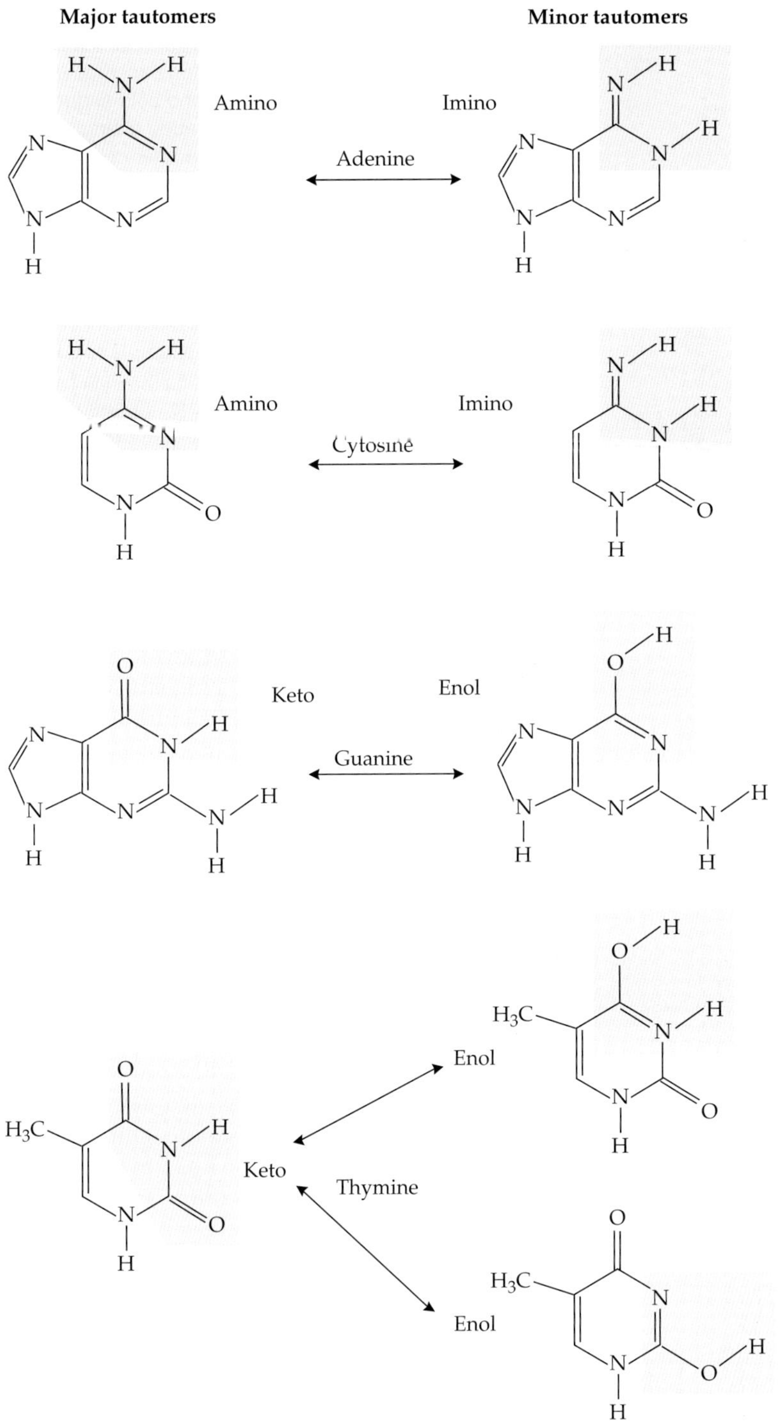

Major tautomers

Minor tautomers

Amino Imino Adenine

Amino Imino Cytosine

Keto Enol Guanine

Keto Enol Enol Thymine

Figure 1.16 Amino ↔ imino and keto ↔ enol tautomerisms. Adenine and cytosine are usually found in the amino form but rarely will assume the imino configuration. Guanine and thymine are usually found in the keto form but rarely form the enol configuration. Thymine has two enol tautomers. Each of the minor tautomers can assume two rotational forms (not shown). (Modified from Mathews and van Holde 1990.)

in the process of DNA replication as the polymerase cannot determine the identity of the missing base it has to copy. Adenine is frequently placed opposite the abasic nucleotide. Thus, when the missing base is a thymine, the placement of the adenine will restore the original sequence.

Point mutations occurring in protein-coding regions may be classified according to their effect on the product of translation—the protein (**Figure 1.17a**). A **synonymous mutation** results in a codon that specifies the same amino acid as the unmutated codon (**Figure 1.17b**); the term is also applied to the change of a stop codon into another stop codon (**Figure 1.17c**). The terms "synonymous mutation" and "silent mutation" are often used interchangeably because in the great majority of cases, synonymous changes do not alter the amino acid sequence of the protein. However,

(a)

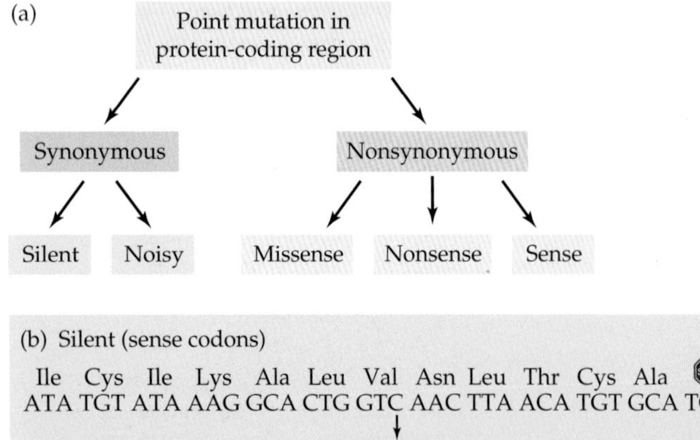

(b) Silent (sense codons)

Ile Cys Ile Lys Ala Leu Val Asn Leu Thr Cys Ala STOP
ATA TGT ATA AAG GCA CTG GTC AAC TTA ACA TGT GCA TGA AGGGTATTTTAA

ATA TGT ATA AAG GCA CTG GTA AAC TTA ACA TGT GCA TGA AGGGTATTTTAA
Ile Cys Ile Lys Ala Leu Val Asn Leu Thr Cys Ala STOP

(c) Silent (stop codons)

Ile Cys Ile Lys Ala Leu Val Asn Leu Thr Cys Ala STOP
ATA TGT ATA AAG GCA CTG GTC AAC TTA ACA TGT GCA TGA AGGGTATTTTAA

ATA TGT ATA AAG GCA CTG GTA AAC TTA ACA TGT GCA TAA AGGGTATTTTAA
Ile Cys Ile Lys Ala Leu Val Asn Leu Thr Cys Ala STOP

(d) Missense

Ile Cys Ile Lys Ala Leu Val Asn Leu Thr Cys Ala STOP
ATA TGT ATA AAG GCA CTG GTC AAC TTA ACA TGT GCA TGA AGGGTATTTTAA

ATA TGT ATA AAG GCA CTG TTA AAC TTA ACA TGT GCA TGA AGGGTATTTTAA
Ile Cys Ile Lys Ala Leu Phe Asn Leu Thr Cys Ala STOP

(e) Nonsense

Ile Cys Ile Lys Ala Leu Val Asn Leu Thr Cys Ala STOP
ATA TGT ATA AAG GCA CTG GTC AAC TTA ACA TGT GCA TGA AGGGTATTTTAA

ATA TGT ATA TAG GCACTGGTAAACTTAACATGTGCATGAAGGGTATTTTAA
Ile Cys Ile STOP

(f) Sense

Ile Cys Ile Lys Ala Leu Val Asn Leu Thr Cys Ala STOP
ATA TGT ATA AAG GCA CTG GTC AAC TTA ACA TGT GCA TGA AGGGTATTTTAA

ATA TGT ATA AAG GCA CTG GTA AAC TTA ACA TGT GCA TGT AGG GTA TTT TAA
Ile Cys Ile Lys Ala Leu Val Asn Leu Thr Cys Ala Cys Arg Val Phe STOP

Figure 1.17 Types of point mutations in coding regions. (a) A schematic division of point mutations according to their effects on the amino acid sequence. (b) A synonymous mutation changing a codon for valine into another codon for valine. (c) A synonymous mutation from one stop codon to another. (d) A missense mutation from a codon for valine to one for phenylalanine. (e) A nonsense mutation from a codon for lysine. (f) A sense nonsynonymous mutation causing protein elongation. An example of a noisy synonymous mutation is shown in Figure 1.17. In panels (b)–(f), the upper and lower DNA and amino acid sequences represent the situation before and after mutation, respectively.

a synonymous mutation may not always be silent. It may, for instance, create a new splicing site, obliterate a splicing site, or disturb an exonic-splice enhancer, thereby turning an exonic sequence into an intron, or vice versa, and causing a different polypeptide to be produced (for a review, see Cartegni et al. 2002). It is therefore useful to distinguish between **silent mutations**, which are synonymous mutations whose effects are not detectable at the amino acid level, and **noisy mutations**, which are synonymous mutations that result in a change at the protein level. For example, a synonymous change from the glycine codon GGT to the glycine codon GGA in codon 25 of the first exon of β-globin has been shown to create a new splice junction, resulting in the production of a frameshifted protein of abnormal length (**Figure 1.18**). This synonymous change results in the blood disorder β^{+}-thalassemia. Synonymous mutations may also influence the way the mRNA folds, which can in turn perturb the translation process; they can interfere with the binding of microRNAs, possibly disrupting normal regulation; and they can affect how fast or how accurately the mRNA is translated. Interestingly, through effects on the timing of cotranslational folding, synonymous changes may even change the way the protein folds without changing its primary sequence. Such noisy synonymous mutations are obviously not silent. Of course, the converse is always true: all silent mutations in a protein-coding gene are synonymous.

A **nonsynonymous** or **amino acid-altering mutation** occurs when the mutated codon specifies a different amino acid from the one specified by the original codon. A change in an amino acid resulting from a nonsynonymous mutation is called a **replacement**. Nonsynonymous mutations are further classified into missense, nonsense, and sense mutations.

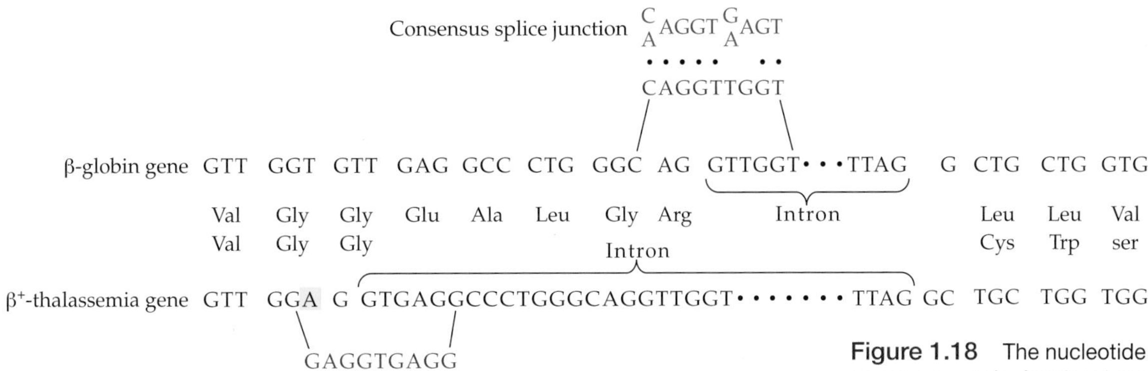

Figure 1.18 The nucleotide sequences at the borders between exon 1 and intron 1, and intron 1 and exon 2 in the β-globin gene from a normal individual and a patient with β+-thalassemia. The mutated nucleotide is boxed in gold. Splicing junctions are shown in color. The splice junction of each gene is compared with the sequence of the consensus splice junction, and dots denote where nucleotides of the splice junction are identical to those of the consensus sequence. Note that the mutation in the β+-thalassemia gene is synonymous because both GGT and GGA code for the amino acid glycine. It is a noisy mutation, however, because it activates a new splicing site in the β+-thalassemia gene that results in the production of a frameshifted protein. (Modified from Goldsmith et al. 1983.)

A **missense mutation** changes the affected codon into a codon that specifies a different amino acid from the one previously encoded (**Figure 1.17d**). Missense mutations in initiation codons constitute a special case because the effects are usually more profound than those of missense mutations elsewhere, and they can vary from drastically reducing or completely obliterating translation to activating downstream or upstream initiation codons that may or may not cause frameshifting.

A **nonsense mutation** changes a sense codon into a termination codon, thus prematurely ending the translation process and ultimately resulting in the production of a truncated protein (**Figure 1.17e**). Codons that can mutate to a termination codon by a single nucleotide change, e.g., UGC (Tyr), are called **pretermination codons**. In the standard genetic code there are 18 pretermination codons (UUA, UUG, UCA, UCG, UGG, UAU, UAC, CAA, CAG, AAA, AAG, GAA, GAG, UGU, UGC, CGA, AGA, and GGA). Five pretermination codons (UUA, UCA, UGG, UAU, and UAC) can change into termination codons through either of two point mutations. For example, UUA can change to two termination codons (UAA and UGA). Some nonsense mutations may affect splicing by activating cryptic donor or acceptor sites. One such case in humans results in Hutchinson-Gilford progeria syndrome, a disorder characterized by premature aging (Kudlow et al. 2007).

A nonsynonymous **sense mutation** is a mutation that changes a stop codon into a sense codon (**Figure 1.17f**). Such a mutation converts the 3′ UTR into a coding region and causes the translation to continue, sometimes up to the polyadenylation site. Sense mutations belong to a category of **protein-elongating mutations**, together with mutations that obliterate the initiation codon while activating an upstream inframe initiation codon.

Each of the sense codons can mutate to nine other codons by means of a single nucleotide change. For example, CCU (Pro) can experience six nonsynonymous mutations, to UCU (Ser), ACU (Thr), GCU (Ala), CUU (Leu), CAU (His), or CGU (Arg), and three synonymous mutations, to CCC, CCA, or CCG. Since the standard genetic code consists of 61 sense codons, there are $61 \times 9 = 549$ possible point mutations. If we assume they occur with equal frequency, and that all codons are equally frequent in coding regions, we can compute the expected proportion of the different types of point mutations from the genetic code. These are shown in **Table 1.7**.

Because of the structure of the genetic code, synonymous changes occur mainly at the third position of codons. Indeed, almost 70% of all the possible nucleotide changes at the third position are synonymous. In contrast, all the nucleotide changes at the second position of codons are nonsynonymous, and so are most nucleotide changes at the first position (96%).

In the vast majority of cases, the exchange of a codon by a synonym, i.e., one that codes for the same amino acid, requires only one or at most two synonymous nucleotide changes. The only exception to this rule is the exchange of a serine codon belong-

TABLE 1.7

Relative frequencies of different types of point mutations in a random protein-coding sequence

Mutations	Number	Percent
Total in all codons	**549**	**100**
Synonymous	134	25
Nonsynonymous	415	75
Missense	392	71
Nonsense	23	4
Total in first codons	**183**	**100**
Synonymous	8	4
Nonsynonymous	175	96
Missense	166	91
Nonsense	9	5
Total in second codons	**183**	**100**
Synonymous	0	0
Nonsynonymous	183	100
Missense	176	96
Nonsense	7	4
Total in third codons	**183**	**100**
Synonymous	126	69
Nonsynonymous	57	31
Missense	50	27
Nonsense	7	4

ing to the four-codon family (UCU, UCC, UCA, and UCG) by one belonging to the two-codon family (AGU and AGC). Such an event requires two nonsynonymous nucleotide changes.

Mutations in regulatory sequences, particularly *cis*-acting regulatory sequences, can have far-reaching effects on gene expression, such as overexpression, underexpression, expression in a different tissue, expression in a different developmental stage, alterations in coexpressions with other genes and gene products, gene inactivation, and even de novo gene activation.

Segmental Mutations

Segmental mutations affect the order of nucleotides, and the vast majority of them, with the exception of single nucleotide deletions or insertions, affect strings of several adjacent nucleotides. In the following sections, we will briefly describe (1) **recombination**, the exchange of one sequence with another; (2) **deletion**, the removal of one or more nucleotides from the DNA; (3) **insertion**, the addition of one or more nucleotides to the sequence; and (4) **inversion**, the rotation by 180° of a double-stranded DNA segment comprising two or more base pairs (**Figure 1.19**). Other segmental mutation types are known. Tandem duplication, i.e., the creation of an identical copy of a DNA sequence in the immediate proximity of the original sequence, will be dealt with in Chapter 7. Transposition, i.e., the transfer of a DNA segment to a new position on either the same chromosome or a different chromosome, will be dealt with in Chapter 9. Microscopic mutations, i.e., mutations that can be detected through optical microscopy, will be dealt with in Chapter 11.

Recombination

Genetic recombination is a process by which a DNA sequence is broken and then joined to a different one. If the recombination event involves two homologous DNA sequences, it is referred to as **homologous** or **generalized recombination**; if it involves two nonhomologous sequences, it is referred to as **site-specific recombination**. There are two types of homologous recombination (**Figure 1.20**): **crossing over** or **cross-over** (**reciprocal recombination**) and **gene conversion** (**nonreciprocal recombination**). Crossing over involves the even exchange of homologous sequences between homologous chromosomes, thereby producing new combinations of adjacent sequences while at the same time retaining both variants involved in the recombination event. Gene conversion, on the other hand, involves the uneven replacement of one sequence by another, a process resulting in the loss of one of the variant sequences involved in the recombination event.

Site-specific recombination involves the exchange of a sequence (usually a very short one, comprising no more than a

(a) Original sequence

AGGCAAACCTACTGGTCTTAT

(b) Recombination

AGGCAAACCTACTGCAAACAT

GTCTT

(c) Deletion

ACCTA

AGGCAACTGGTCTTAT

(d) Insertion

AGGCAAACCTACTAAAGCGGTCTTAT

(e) Inversion

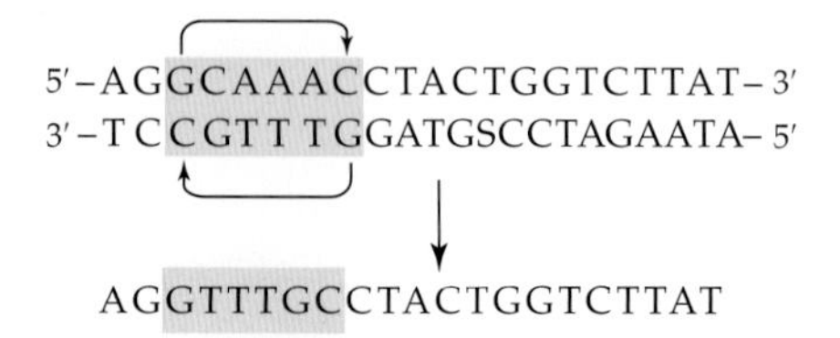

Figure 1.19 Types of segmental mutations. (a) Original sequence. (b) Recombination, the exchange of the sequence GTCTT by CAAAC. (c) Deletion of the sequence ACCTA. (d) Insertion of the sequence AAAGC. (e) To visualize inversion, it is necessary to draw the two strands of the DNA. Thus, the inversion of 5′—GCAAAC—3′ yields a 5′—GTTTGC—3′ sequence. (Modified from Graur and Li 2000.)

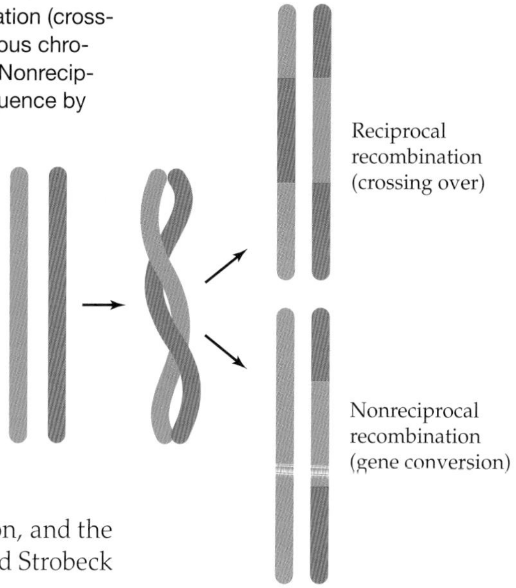

Figure 1.20 The two types of homologous recombination. Reciprocal recombination (crossing over) involves the even exchange of homologous sequences between homologous chromosomes, in which both variants involved in the recombination event are retained. Nonreciprocal recombination (gene conversion) involves the uneven replacement of one sequence by another, a process resulting in the loss of variation.

Reciprocal recombination (crossing over)

Nonreciprocal recombination (gene conversion)

few nucleotides) by another (usually a long one that bears no similarity to the original sequence). Site-specific recombination is responsible, for instance, for the integration of phage genomes into eubacterial chromosomes. From a mutational point of view, site-specific recombination may be approximately dealt with as a type of insertion.

Reciprocal recombination in conjunction with point mutations is a powerful generator of genetic variability. For example, a recombination between sequences 5′—AACT—3′ and 5′—CACG—3′ may result in two new sequences: 5′—AACG—3′ and 5′—CACT—3′. The more alleles there are, the more alleles will come into being through recombination, and the rate of generating new genetic variants may become quite high. Golding and Strobeck (1983) referred to this phenomenon as "variation begets variation."

Deletions and insertions

Deletions and insertions can occur by several mechanisms. One mechanism is **unequal crossing over** (**Figure 1.21**). Unequal crossing over between two chromosomes results in the deletion of a DNA segment in one chromosome and a reciprocal insertion in the other. Because of the higher probability of misalignment, the chance of unequal crossing over is greatly increased if a DNA segment is duplicated in tandem.

Deletion can also occur through **intrastrand deletion** (**Figure 1.22**), a type of site-specific recombination that arises when a repeated sequence pairs with another in the same orientation on the same chromatid and an intrachromosomal crossing-over event occurs as a consequence. The precise excision of transposable elements frequently involves recombination between short direct repeats flanking the element (Chapter 9). Similarly, intrastrand deletion is responsible for the reduction in the number of repeats in tandem arrays, such as simple repetitive DNA and satellite DNA (Chapter 11).

Another mechanism is **replication slippage**, or **slipped-strand mispairing.** This type of event occurs in DNA regions that contain contiguous short repeats (**Figure 1.23**). Some triplet and quadruplet repeats, such as $(CAG)_n$ and $(CCTG)_n$, where the subscript denotes the number of times the sequence within parentheses is repeated, are hotspots of repeat expansion and contraction. Slipped strand mispairing can also cause insertions and deletions in nonreplicating DNA (Huttley et al. 2000).

A fourth mechanism responsible for the insertion or deletion of DNA sequences is DNA transposition (Chapter 7).

Deletions and insertions are collectively referred to as **indels** (short for **in**sertion-or-**del**etion) because when two sequences are compared, it is impossible to tell whether a deletion occurred in one sequence or an insertion occurred in the other (Chapter 3). The number of nucleotides in an indel ranges from one or a few nucleotides to contiguous stretches involving thousands of nucleotides. Short indels (up to 30 nucleotides) are most often caused by errors in the process of DNA replication, such

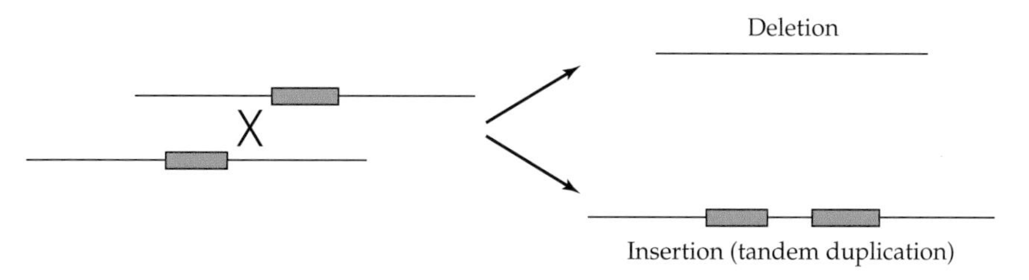

Figure 1.21 Unequal crossing over resulting in the deletion of a DNA sequence in one of the daughter strands and the duplication of the same sequence in the other. A box denotes a particular stretch of DNA.

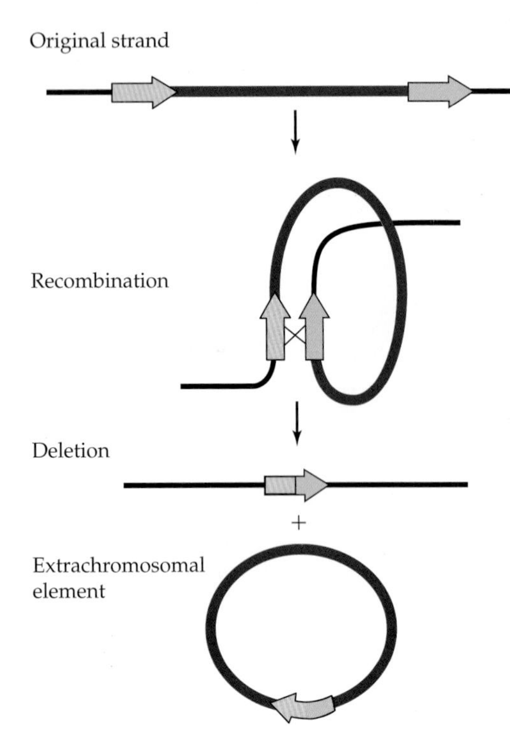

Figure 1.22 Generation of a deletion by the intrastrand deletion process. The repeated sequences (arrows) that are oriented in the same direction recombine (×) to produce a genomic deletion and an extrachromosomal element that will eventually be lost. (Modified from Graur and Li 2000.)

as slipped-strand mispairing; long insertions or deletions are due to unequal crossing over, site-specific recombination, DNA transposition, or horizontal gene transfer (Chapter 9).

In a coding region, an indel that is not a multiple of three nucleotides causes a shift in the reading frame, so the coding sequence downstream of the gap will be read in the wrong phase. Such a mutation is known as a **frameshift mutation**. Consequently, indels may not only introduce numerous amino acid changes, but also obliterate the termination codon or bring into phase a new stop codon, resulting in a protein of abnormal length (**Figure 1.24**).

An indel that is a multiple of three nucleotides is called a **nonframeshifting indel** (**Figure 1.25**). A nonframeshifting indel can occur between two codons (**phase 0 indel**), after the first nucleotide of a codon (**phase 1 indel**), or after the second nucleotide of a codon (**phase 2 indel**). Nonframeshifting indels can be conservative or nonconservative. A **conservative nonframeshifting indel** does not affect the amino acids encoded by the adjacent upstream and downstream codons. A **nonconservative nonframeshifting indel** affects the amino acids encoded by the adjacent codons.

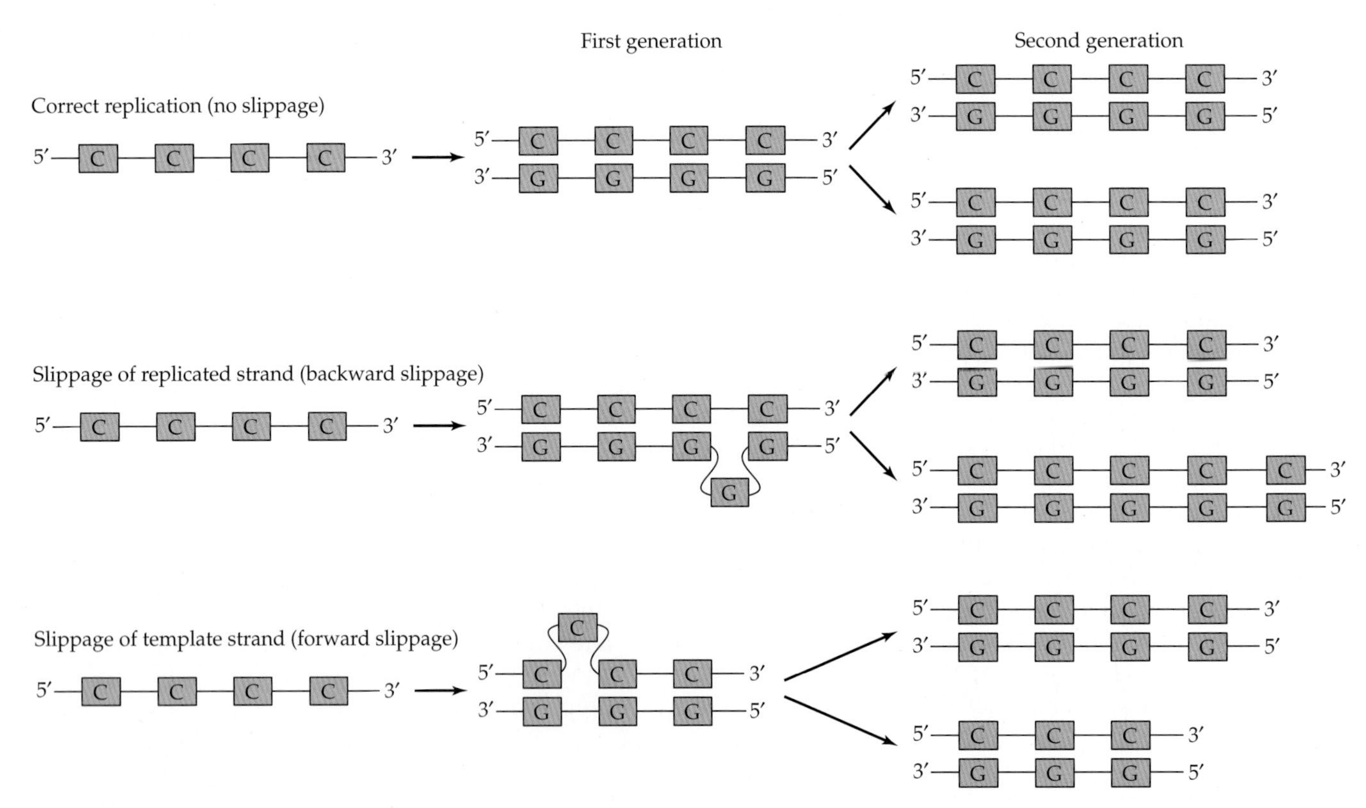

Figure 1.23 Slipped-strand mispairing during DNA replication can give rise to deletions and insertions. Illegitimate base pairing in regions of repetitive DNA during replication, coupled with inadequate DNA mismatch repair, can produce deletions or insertions of one or more repeated units. The figure shows a strand of DNA being carried through two rounds of replication. Slippage involves a region of nonpairing (shown as a bulge) containing one or more repeats of the newly replicated strand (backward slippage) or of the parental template strand (forward slippage), causing, respectively, an insertion or a deletion on the newly synthesized strand. In the two generations, the lower and upper strands represent the parental DNA strand and the newly synthesized complementary strand, respectively. (Modified from Thomson et al. 2003.)

Figure 1.24 Examples of frameshifts in reading frames. (a) Deletion of a G causes premature termination. (b) Deletion of an A obliterates a stop codon. (c) Insertion of a G obliterates a stop codon. (d) Insertion of the dinucleotide GA causes premature termination of translation. Stop codons are shown in bold type. (From Graur and Li 2000.)

(a)

Lys Ala Leu Val Leu Leu Thr Ile Cys Ile STOP
AAG GCA CTG GTC CTG TTA ACA ATA TGT ATA **TAA** TACCATCGCAATATGAAAATC
↓
G

AAG GCA CTG TCC TGT **TAA** CAATATGTATATAATACCATCGCAATATGAAAATC
Lys Ala Leu Phe Cys STOP

(b)

Lys Ala Leu Val Leu Leu Thr Ile Cys Ile STOP
AAG GCA CTG GTC CTG TTA ACA ATA TGT ATA **TAA** TACCATCGCAATATGAAAATC
↓
A

AAG GCA CTG GTC CTG TTA ACA ATA TGT ATT AAT ACC ATC GCA ATA **TGA** AAA
Lys Ala Leu Val Leu Leu Thr Ile Cys Ile Asn Thr Ile Ala Ile STOP

(c)

Lys Ala Axn Val Leu Leu Thr Ile Cys Ile STOP
AAG GCA AAC GTC CTG TTA ACA ATA TGT ATA **TAA** TACCATCGCAATAGGG
↑
G

AAG GCA AAC GGT CCT GTT AAC AAT ATG TAT ATA ATA CCA TCG CAA **TAG** GG
Lys Ala Asn Gly Pro Val Asn Asn Met Tyr Ile Ile Pro Ser Gln STOP

(d)

Lys Ala Asn Val Leu Leu Thr Ile Cys Ile STOP
AAG GCA AAC GTC CTG TTA ACA ATA TGT ATA **TAA** TACCATCGCAATAGGG
↑
GA

AAG GCA AAC GAG TCC TGT **TAA** CAATATGTATATAATACCATCGCAATAGGG
Lys Ala Asn Glu Ser Cys STOP

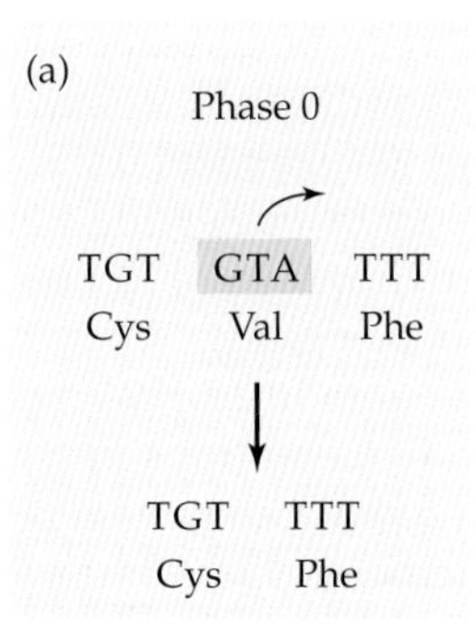

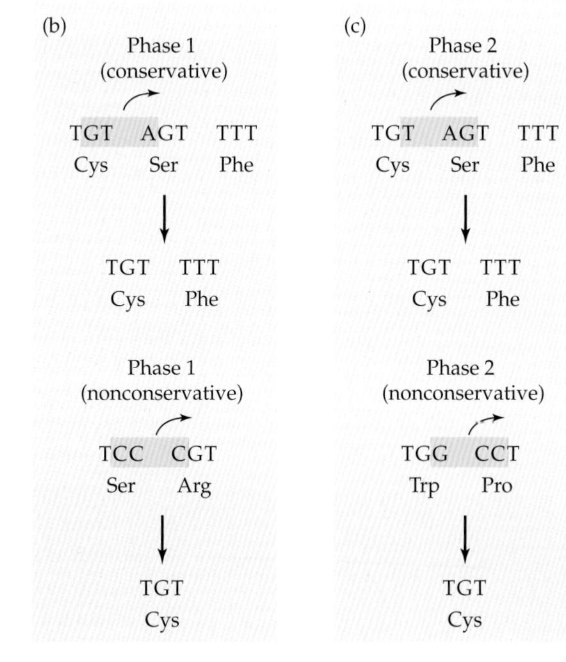

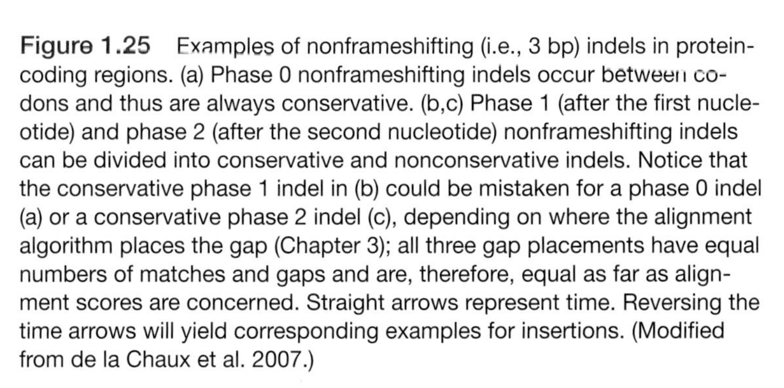

Figure 1.25 Examples of nonframeshifting (i.e., 3 bp) indels in protein-coding regions. (a) Phase 0 nonframeshifting indels occur between codons and thus are always conservative. (b,c) Phase 1 (after the first nucleotide) and phase 2 (after the second nucleotide) nonframeshifting indels can be divided into conservative and nonconservative indels. Notice that the conservative phase 1 indel in (b) could be mistaken for a phase 0 indel (a) or a conservative phase 2 indel (c), depending on where the alignment algorithm places the gap (Chapter 3); all three gap placements have equal numbers of matches and gaps and are, therefore, equal as far as alignment scores are concerned. Straight arrows represent time. Reversing the time arrows will yield corresponding examples for insertions. (Modified from de la Chaux et al. 2007.)

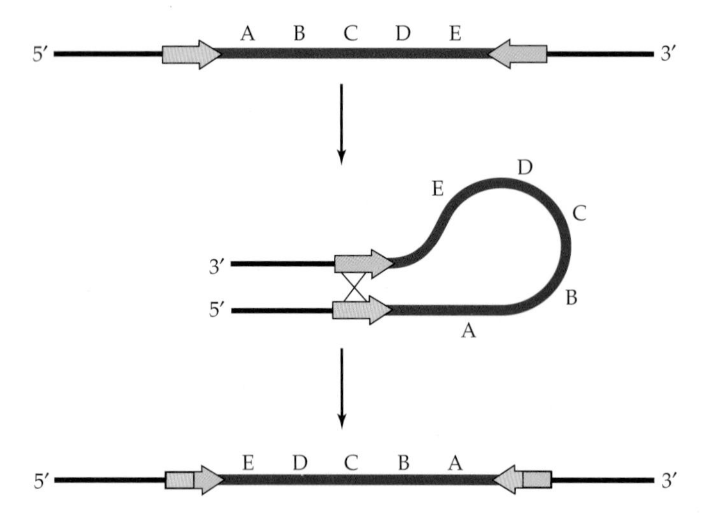

Figure 1.26 Crossing over (×) between homologous segments (arrows) oriented in opposite directions on the same chromosome results in an inversion involving the DNA sequence between the homologous inverted repeats.

Inversions

Inversion is a type of DNA rearrangement that can occur following several molecular processes, chief among them intrachromosomal crossing over between two homologous segments that are oriented in opposite directions (**Figure 1.26**). The vast majority of known inversions involve very long stretches of DNA, usually hundreds or thousands of nucleotides in length.

Spatial distribution of mutations

Mutations do not occur randomly throughout the genome. Some regions are more prone to mutate than others, and they are called **hotspots of mutation**. Some regions are less likely to mutate than others and are called **coldspots of mutation**. The dinucleotide 5′—CG—3′ (often denoted CpG) is a hotspot of mutation in animal genomes because the cytosine in CpG is frequently methylated and changed to 5′—TG—3′ (TpG) or 5′—CA—3′ (CpA). The dinucleotide 5′—TT—3′ is a hotspot of mutation in prokaryotes but usually not in eukaryotes. In eubacteria, regions within the DNA containing short **molecular palindromes** (i.e., sequences that read the same on the complementary strand, such as 5′—GCCGGC—3′, 5′—GGCGCC—3′, and 5′—GGGCCC—3′) were found to be more prone to mutate than other regions. In eukaryotic genomes, short tandem repeats are often hotspots for deletions and insertions, probably as a result of slipped-strand mispairing. Sequences containing runs of alternating purine-pyrimidine dimers, such as GC, AT, and GT, are capable of adopting a left-handed DNA conformation called Z-DNA, which constitutes a hotspot for deletions involving even numbers of base pairs.

Are Mutations Random?

Mutations are commonly said to occur "randomly." However, empirical studies show unambiguously that mutations do not occur at random with respect to either genomic location or mutation type, nor are all types of mutations equally frequent. For example, transitions mostly occur more frequently than transversions, and nonadvantageous mutations occur more frequently than advantageous ones. Finally, mutation rates and patterns, as well as mutational fitness effects, are known to be species-specific, gender-specific, developmental stage-specific, strand-specific, and genome-specific and to be affected by nucleotide composition, neighboring nucleotides, genetic background, and environmental factors, such as exposure to mutagens. So, which aspect of mutation is random?

Mutations are random in one respect: a mutation is expected to occur with the same frequency under conditions in which it confers an advantage on the organism carrying it, confers no advantage, or is deleterious (Luria and Delbrück 1943; Lederberg and Lederberg 1952). "It may seem a deplorable imperfection of nature," declared evolutionary biologist Theodosius Dobzhansky (1970), "that mutability is not restricted to changes that enhance the adeptness of their carriers." Notwithstanding this "imperfect" state of affairs, it would be difficult to imagine how a gene could know how and when to mutate so that the result would be beneficial for the organism carrying the gene.

The theoretical construct **adaptive mutation** is defined as a mutation arising as an immediate and direct response to selective pressure. The existence of adaptive mutations is sporadically debated in the literature, frequently with ferocious intensity. So far, no unambiguous example of adaptive mutation has ever been found, and the hypothesis according to which mutations occur randomly regardless of the utility of the mutation to the organism remains valid.

CHAPTER 2

Allele Dynamics in Populations

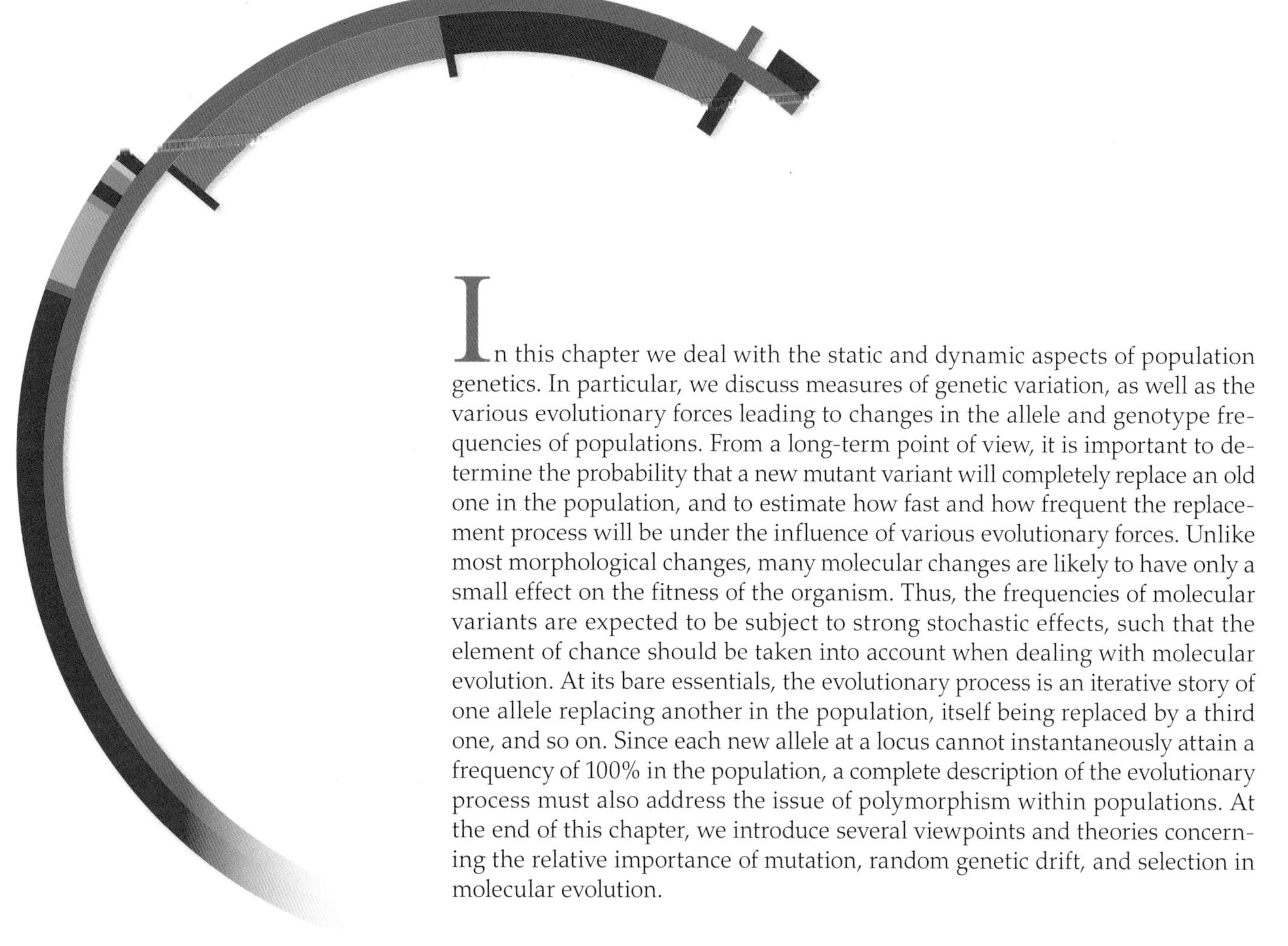

In this chapter we deal with the static and dynamic aspects of population genetics. In particular, we discuss measures of genetic variation, as well as the various evolutionary forces leading to changes in the allele and genotype frequencies of populations. From a long-term point of view, it is important to determine the probability that a new mutant variant will completely replace an old one in the population, and to estimate how fast and how frequent the replacement process will be under the influence of various evolutionary forces. Unlike most morphological changes, many molecular changes are likely to have only a small effect on the fitness of the organism. Thus, the frequencies of molecular variants are expected to be subject to strong stochastic effects, such that the element of chance should be taken into account when dealing with molecular evolution. At its bare essentials, the evolutionary process is an iterative story of one allele replacing another in the population, itself being replaced by a third one, and so on. Since each new allele at a locus cannot instantaneously attain a frequency of 100% in the population, a complete description of the evolutionary process must also address the issue of polymorphism within populations. At the end of this chapter, we introduce several viewpoints and theories concerning the relative importance of mutation, random genetic drift, and selection in molecular evolution.

Standing Genetic Variation

In biology, a **population** is defined as a group of contemporaneous individuals that occupy a defined space, and among which genetic information may be exchanged. (We note that this definition only applies to sexual organisms; in asexual taxa, populations and species are defined by phenotypic and genetic similarity.)

The chromosomal or genomic location of a gene or any other genetic element is called a **locus** (plural: **loci**), and alternative DNA sequences at a locus are called **alleles**. In a population, more than one allele may be present at a locus. The relative proportion of an allele in a population is referred to as the **allele frequency** or **gene frequency**. For example, let us assume that in a haploid population of size N individuals, two alleles, A_1 and A_2, are present at a certain locus and that the number of copies of alleles A_1 and A_2 in the population are n_1 and n_2, respectively. The allele frequencies in this population are n_1/N and n_2/N for alleles A_1 and A_2, respectively. Note that $n_1 + n_2 = N$, and $(n_1/N) + (n_2/N) = 1$. The set of all alleles existing in a population at all loci is called the **gene pool**.

The genetic constitution of an individual is called the **genotype**. The term is usually used with reference to a specific locus under consideration. The relative proportion of a genotype in a population is referred to as the **genotype frequency**. A set of alleles at adjacent or **linked loci** is referred to as the **haplotype**. (The term "haplotype" is a portmanteau of *haplo*id and geno*type*.) In the absence of recombination, the haplotype constituents are transmitted together from generation to generation. The term "haplotype" may refer to the allelic makeup at two loci, three or more loci, or all the loci in a chromosome. In genomic studies, "haplotype" frequently refers to a number of **single-nucleotide polymorphisms** (**SNPs**) that are associated with one another. In clinical studies, haplotypes are frequently correlated with nonmolecular traits (e.g., disease or morphological traits) and are used in attempts to identify the genetic basis of such traits. In genome evolutionary studies, haplotypes are sometimes used to identify the age of an allelic variant and the evolutionary forces leading to its spread in the population.

In haploid organisms, the allele and the genotype are one and the same, so a change in genotype frequencies is equivalent to a change in allele frequencies. In diploids and polyploids (organisms with a ploidy number higher than 2), a change in genotype frequencies is possible without a change in allele frequencies (**Figure 2.1**). In evolutionary studies (as opposed to population genetics), we are mainly interested in changes in allele frequencies, and not so much in changes in genotype frequencies that are not accompanied by changes in allele frequencies. Thus, biological factors that mainly affect genotype frequencies, such as inbreeding, assortative mating, migration, and population subdivision (or substructuring), will not be discussed in this book.

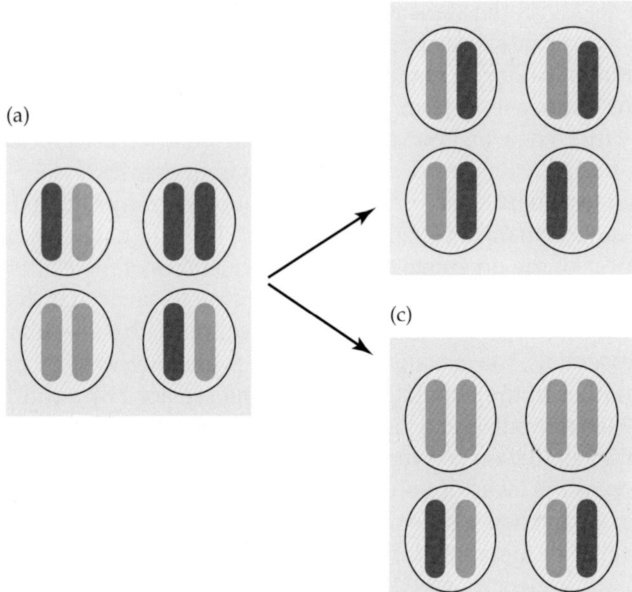

Figure 2.1 Schematic illustration of possible changes in the genetic makeup of a population. (a) In the initial population of four individuals, two are homozygotes (one for the orange and one for the brown allele) and two are heterozygotes. The frequency of the two alleles is thus 0.50. (b) All four individuals are heterozygotes, so the genotype frequencies have changed while the allele frequencies remain the same as in (a). That is, a change in genotype frequency has occurred without any change in allele frequency. (c) Here, two individuals are heterozygotes and two are homozygotes, as in (a); however, both homozygotes in (c) have the same genotype. Consequently, the genotype frequencies are different from those in (a), and the frequencies of the orange and brown alleles are 0.75 and 0.25, respectively.

A population is **monomorphic** at a locus if there exists only one allele at this locus in the population. A locus is said to be **polymorphic** if two or more alleles coexist in the population. In the literature, a distinction is usually made between common variants, rare variants, and private variants. **Common variants** are usually defined in the literature as alleles that appear in the sample at frequencies above 5%. **Rare variants** are alleles whose frequencies are below 5%. **Private variants** are alleles that are found only once in the sample. We note, however, that if one of the alleles at a locus has a very high frequency in the population, say 99% or more, then none of the other alleles are likely to be observed in a sample unless the sample size is very large. Thus, for practical purposes, a locus is commonly defined as polymorphic only if the frequency of its most common allele is less than 99%. The above definitions are obviously arbitrary, and in the literature one may find different thresholds.

Gene diversity

One of the simplest ways to measure the extent of polymorphism in a population is to compute the average proportion of polymorphic loci (P) by dividing the number of polymorphic loci by the total number of loci sampled. For example, if 4 of the 20 loci are polymorphic, then $P = 4/20 = 0.20$. This measure, however, is dependent on the number of individuals studied, because the smaller the sample size, the more difficult it is to identify polymorphic loci as such.

A more appropriate measure of genetic variability is the **mean expected heterozygosity**, or **gene diversity**. This measure (1) does not depend on an arbitrary delineation of polymorphism, (2) can be computed directly from knowledge of the allele frequencies, and (3) is less affected by sampling effects. Gene diversity at a locus, or **single-locus expected heterozygosity**, is defined as

$$h = 1 - \sum_{i=1}^{m} x_i^2 \tag{2.1}$$

where x_i is the frequency of allele i, and m is the total number of alleles at the locus. For any given locus, h is the probability that two alleles chosen at random from the population are different from each other. The average of the h values over all the loci studied, H, can be used as an estimate of the extent of genetic variability within the population. That is,

$$H = \frac{1}{n} \sum_{i=1}^{n} h_i \tag{2.2}$$

where h_i is the gene diversity at locus i, and n is the number of loci. Gene diversity is a useful measure of genetic diversity because it has a clear biological meaning and is applicable to all organisms, regardless of ploidy.

Nucleotide diversity

The gene-diversity measures h and H are useful when the number of distinct alleles is small compared with sample size. Thus, these measures have been used extensively for genetic variation revealed by protein electrophoresis, restriction fragment length polymorphism (RFLP), variable number of tandem repeats (VNTR), and loci consisting of small numbers of nucleotides. However, gene-diversity measures are mostly unsuitable for DNA sequence data, since the extent of genetic variation at the DNA level in nature is quite extensive. In particular, when long sequences are considered, each sequence in the sample is likely to be different by one or more nucleotides from the other sequences, and in most cases both h and H will be close to 1. Gene diversity considers whether or not two randomly chosen sequences are different from one another but ignores the magnitude of the difference. Thus, when the number of alleles is large and h is close to 1, the usefulness of gene-diversity measures as informative measures of polymorphism is greatly reduced. Consider, for instance, the two groups of sequences in **Figure 2.2**. Intuitively, we would think that the sequences in Figure 2.2a should be less polymorphic than the ones in Figure 2.2b. However, h will be the same for both groups of sequences.

For DNA sequence data, a more appropriate measure of polymorphism in a population is the average number of nucleotide differences per site between any two randomly chosen sequences. This measure is called **nucleotide diversity** (Nei and Li 1979) and is denoted by Π:

$$\Pi = \sum_{ij} x_i x_j \pi_{ji} \tag{2.3}$$

where x_i and x_j are the frequencies of the ith and jth type of DNA sequences, respectively, and π_{ij} is the proportion of different nucleotides between the ith and jth types. The Π values for the sequences in Figure 2.2 are 0.031 and 0.109, i.e., the nucleotide diversity measures are in agreement with our intuitive perception that group (a) is less variable than group (b). (Note that for the case of $\pi_{ij} = 1$, the value of Π will be the same

(a)	(b)
GAGGTGCAACAG	GAGGTGCAACAG
GCGGTGCAACAG	GAGGACCAACAG
GTGGTGCAACAG	GTGGTGCATCAA
GGGGTGCAACAG	GGGGTGGAACAG

Figure 2.2 Two groups of four DNA sequences. In (a), each sequence differs from any other sequence at a single nucleotide site (shaded). In (b), each sequence differs from any other sequence at two or more nucleotide sites. However, in both cases, each sequence is represented in its group only once, so the values of the single-locus diversity measure will be the same for both groups.

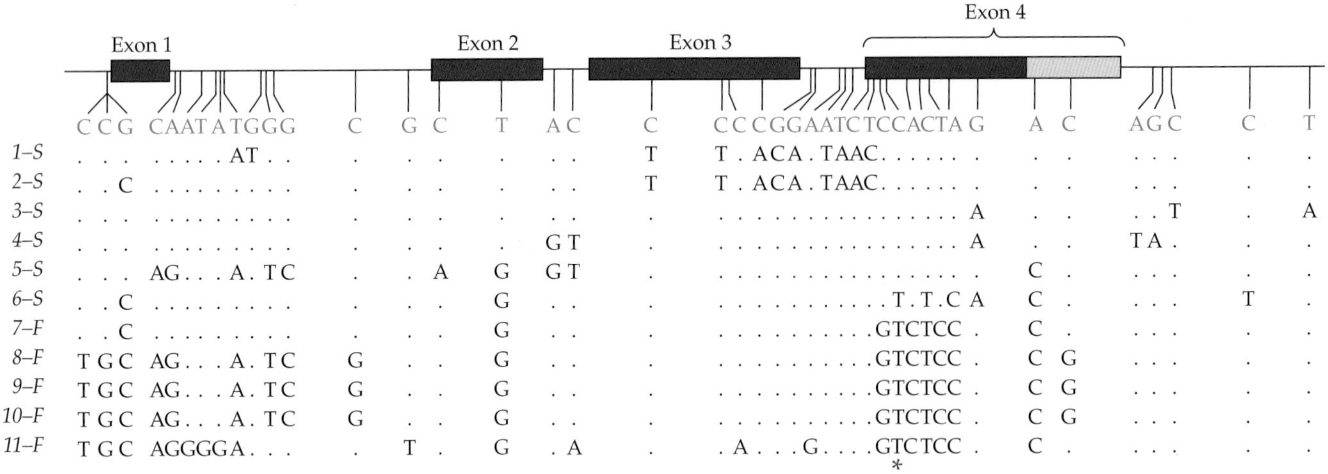

Figure 2.3 Polymorphic nucleotide sites among 11 sequences of the alcohol dehydrogenase gene in *Drosophila melanogaster*. Exons are shown as boxes; translated and untranslated regions are in red and blue, respectively. The consensus sequence is shown in blue type. Differences from the consensus sequence are shown; the dots indicate identity with the consensus sequence. The red asterisk in the exon 4 sequence indicates the site of the lysine → threonine replacement that is responsible for the mobility difference between the fast (*F*) and slow (*S*) electrophoretic alleles. (Modified from Hartl and Clark 1997.)

as that of *h* in Equation 2.1.) When dealing with small samples and/or very short sequences, calculating Π requires correction for stochastic error (Tajima 1983; Nei 1987).

One of the first studies on nucleotide diversity at the DNA sequence level concerned the alcohol dehydrogenase (*Adh*) locus in *Drosophila melanogaster.* Eleven sequences spanning the *Adh* region were sequenced by Kreitman (1983). The aligned sequences were 2,379 nucleotides long. Disregarding deletions and insertions, there were nine different alleles, one of which was represented in the sample by three sequences (*8-F, 9-F,* and *10-F*); the rest were represented by one sequence each (**Figure 2.3**). Thus, the frequencies $x_1 - x_8$ were each 1/11, and the frequency x_9 was 3/11.

Forty-three nucleotide sites were polymorphic. We first calculate the proportion of different nucleotides for each pair of alleles. For example, alleles *1-S* and *2-S* differ from each other by three nucleotides out of 2,379, or $\pi_{12} = 0.13\%$. The π_{ij} values for all the pairs in the sample are listed in **Table 2.1**. By using Equation 2.3, the nucleotide diversity is estimated to be $\Pi = 0.007$. Six of the alleles studied were slow-migrating electrophoretic variants (*S*), and five were fast-migrating (*F*). The products of *S* and *F* alleles are distinguished from each other by one amino acid replacement that confers a different electrophoretic mobility to the proteins. The nucleotide diversity for each of these electrophoretic classes was calculated separately. We obtain $\Pi = 0.006$ for the *S* class, and $\Pi = 0.003$ for *F,* i.e., the *S* alleles are twice as variable as the *F* alleles.

Structural variation

So far we have only discussed genetic variation due to point mutations. Genetic variation, however, can be due to other types of mutation. **Structural variation** is the variation resulting from segmental mutations, such as deletions, duplications, insertions, inversions, and translocations. In the literature a distinction is made between submicroscopic and microscopic variation. **Submicroscopic variation** involves genomic regions of less than 3 Mb in length and is detectable by various sequencing methods. **Microscopic variation** involves regions of lengths above 3 Mb and may be detected though microscopic examination of karyotypes. **Copy number variants** (**CNVs**) are submicroscopic structural variations that involve duplications or losses of genomic segments. CNVs result from deletion, duplication, and replicative transposition. If the variation in number of repeated sequences occurs in tandem, it is referred to as **variable number of tandem repeats** (**VNTRs**).

What Is Evolution?

One of the commonest misconceptions among the uninitiated is that individual organisms can evolve during their lifespan. This is, of course, not true; populations, not individuals, evolve. Changes in an individual over the course of its lifetime may be

TABLE 2.1

Pairwise percent nucleotide differences among 11 alleles of the alcohol dehydrogenase locus in *Drosophila melanogaster*[a]

Allele	*1-S*	*2-S*	*3-S*	*4-S*	*5-S*	*6-S*	*7-F*	*8-F*	*9-F*	*10-F*
1-S										
2-S	0.13									
3-S	0.59	0.55								
4-S	0.67	0.63	0.25							
5-S	0.80	0.84	0.55	0.46						
6-S	0.80	0.67	0.38	0.46	0.59					
7-F	0.84	0.71	0.50	0.59	0.63	0.21				
8-F	1.13	1.10	0.88	0.97	0.59	0.59	0.38			
9-F	1.13	1.10	0.88	0.97	0.59	0.59	0.38	0.00		
10-F	1.13	1.10	0.88	0.97	0.59	0.59	0.38	0.00	0.00	
11-F	1.22	1.18	0.97	1.05	0.84	0.67	0.46	0.42	0.42	0.42

Source: From Nei (1987).

[a] Total number of compared sites is 2,379. *S* and *F* denote the slow- and fast-migrating electrophoretic alleles, respectively.

(1) developmental, (2) caused by the environment, or (3) caused by the environment interacting with the genetic endowment of the individual. However, these acquired changes are not passed on to one's offspring. While it would be convenient for the environment to cause adaptive and heritable changes to our genome, evolution just doesn't work in a Lamarckian fashion.

So what is evolution? The lexical literature abounds with confusing and often contradictory definitions. For example, the *Oxford English Dictionary* defines evolution as the "transformation of animals, plants, and other living organisms into different forms by the accumulation of changes over successive generations," leading eventually to the "origination or transformation" of organisms, organs, physiological processes, and biological molecules. According to the *Merriam-Webster Dictionary*, evolution is "a process of continuous change from a lower, simpler, or worse to a higher, more complex, or better state." Such definitions are, of course, too fuzzy and inexact to be useful in the practice of science, which requires simple, practical, and unambiguous definitions.

In this book, we will use a very simple and unequivocal definition. **Evolution**, according to this definition, is the process of change in the genetic makeup of populations, that is, a change in genotype and allele frequencies. By providing an exact and measurable definition of evolution, we can apply strict scientific methodology to what is essentially a historical process.

Changes in Allele Frequencies

We will start our schematic description of the dynamic part of the evolutionary process with a new mutation (i.e., a new allele at a locus) arising in one individual in a population. For this new mutation to become significant from an evolutionary point of view, it must increase in frequency and ultimately become fixed in the population (i.e., all the individuals in a subsequent generation will share the same mutant allele). If it does not increase in frequency, a mutation will have little effect on the evolutionary history of the species. There are two major factors affecting the frequency of an allele in the population: selection and random genetic drift. In classical evolutionary studies involving morphological traits, selection has been considered as the major driving force in evolution. In contrast, random genetic drift is thought to have played an important role in evolution at the molecular level.

There are two mathematical approaches to studying genetic changes in populations: deterministic and stochastic. The **deterministic model** is simpler. It assumes that changes in the frequencies of alleles in a population from generation to generation occur in a unique manner and can be unambiguously predicted from knowledge of initial conditions. Strictly speaking, this approach applies only when two conditions are met: (1) the population is infinite in size, and (2) the direction and magnitude of selection either remain constant with time or change according to deterministic rules. These conditions are obviously never met in nature, and therefore a purely deterministic approach may not be sufficient to describe the temporal changes in allele frequencies. Random or unpredictable fluctuations in allele frequencies must also be taken into account, and dealing with random fluctuations requires a different mathematical approach.

Stochastic models assume that changes in allele frequencies occur in a probabilistic manner. That is, from knowledge of the conditions in one generation, one cannot predict unambiguously the allele frequencies in the next generation; one can only determine the probabilities with which certain allele frequencies are likely to be attained.

Obviously, stochastic models are preferable to deterministic ones, since they are based on more realistic assumptions. However, deterministic models are much easier to treat mathematically, and under certain circumstances, they yield sufficiently accurate approximations. The following discussion deals with selection in a deterministic fashion.

Selection

Selection is defined as the differential reproduction of genetically distinct individuals or genotypes within a population. Differential reproduction is caused by differences among individuals in such traits as mortality, fertility, fecundity, mating success, and the viability of the offspring. Selection is predicated on the availability of genetic variation among individuals in characters related to reproductive success. When a population consists of individuals that do not differ from one another in such traits, it is not subject to selection. Selection may lead to changes in allele frequencies over time. However, a mere change in allele frequencies from generation to generation does not necessarily indicate that selection is at work. Other processes, such as random genetic drift (discussed later in this chapter), can bring about temporal changes in allele frequencies as well. Interestingly, the opposite is also true: a lack of change in allele frequencies does not necessarily indicate that selection is absent.

In the literature, the terms **natural selection** and **artificial selection** are used to distinguish between selection driven by the natural biotic and abiotic environment, on the one hand, and the process through which humans favor or disfavor specific traits and use breeding schemes to achieve changes in the genetic pool of the selected population, on the other. As far as the dynamics of genes and alleles in the population is concerned, there is no difference between natural and artificial selection, and here, we will mostly use the term "selection" without qualifiers.

The **fitness** of a genotype, commonly denoted by w, is a measure of the individual's ability to survive and reproduce. However, since the size of a population is usually constrained by the carrying capacity of the environment in which the population resides, the evolutionary success of an individual is determined not by its **absolute fitness**, but by its **relative fitness** in comparison to the other genotypes in the population. In nature, the fitness of a genotype is not expected to remain constant for all generations and under all environmental circumstances. However, by assigning a constant value of fitness to each genotype, we are able to formulate simple theories or models, which are useful for understanding the dynamics of change in the genetic structure of a population brought about by selection. In the simplest class of models, we assume that the fitness of the organism is determined solely by its genetic makeup.

We also assume that all loci contribute independently to the fitness of the individual (i.e., that the different loci do not interact with one another in any manner that affects the fitness of the organism) so that each locus can be dealt with separately.

Mutations may or may not affect the organism's phenotype. In the case that they do, they may or may not affect the fitness of the organism that carries the mutation. Mutations span the entire range of fitness effects, from lethal to mildly deleterious to neutral to beneficial. Mutations that reduce the fitness of their carriers are called **deleterious**; if the mutation reduces the fitness to 0, it is called **lethal**. Such mutations will be selected against. This type of selection is called **negative** or **purifying selection**. (Note that mutations resulting in "lethal" genetic disorders, such as Huntington's disease (formerly called Huntington's chorea), which affects individuals only after the reproductive stage of life, are not strictly speaking lethal, since the mutation does not appreciably affect components of fitness.)

If a new mutation is as fit as the best allele in the population, then such a mutation is selectively **neutral**, and its fate is not determined by selection. Most mutations arising in a population are either deleterious or neutral. In exceedingly rare cases, a mutation may arise that increases the fitness of its carriers. Such a mutation is called **advantageous**, and it will be subjected to **advantageous** or **positive selection**. For historical emphasis, positive selection is sometimes referred to as **positive Darwinian selection**.

"Positive selection" is sometimes synonymized with "adaptation." **Adaptation** is the evolutionary process whereby an organism becomes better able to survive and reproduce in its habitat. The state of being adapted, that is, the degree to which an organism is able to survive and reproduce in a given habitat, is called **adaptedness**. An **adaptive trait** is an aspect of the genotype or the phenotype of an organism that enables or enhances its probability of surviving and reproducing. Although the term "adaptation" is widely used in the scientific and popular literature, it is impossible to quantify. Thus, "adaptation" and its derivatives should be used sparsely, mainly in the context of hypotheses that yield testable and refutable quantitative predictions.

Let us now consider the case of one locus with two alleles, A_1 and A_2. Each allele can be assigned an intrinsic fitness value; it can be advantageous, deleterious, or neutral. However, this assignment is only applicable to haploid organisms. In diploid organisms, fitness is ultimately determined by the interaction between the two alleles at a locus. With two alleles, there are three possible diploid genotypes, A_1A_1, A_1A_2, and A_2A_2, and their fitnesses can be denoted by w_{11}, w_{12}, and w_{22}, respectively. Given that the frequency of allele A_1 in a population is p and the frequency of the complementary allele, A_2, is $q = 1 - p$, we can show that, under random mating, the frequencies of the A_1A_1, A_1A_2, and A_2A_2 genotypes are p^2, $2pq$, and q^2, respectively. A population in which such genotypic ratios are maintained is said to be at **Hardy-Weinberg equilibrium**. The three possible genotypes at a locus with two alleles can be assigned the following fitness values and initial frequencies in a population in Hardy-Weinberg equilibrium:

Genotype	A_1A_1	A_1A_2	A_2A_{22}
Fitness	w_{11}	w_{12}	w_{22}
Frequency	p_2	$2pq$	q_2

Let us now consider the dynamics of allele frequency changes following selection. The relative contribution of each genotype to the subsequent generation is the product of its initial frequency and its fitness. For the frequencies of the three genotypes and their fitnesses given above, the relative contributions of the three genotypes to the next generation will be p^2w_{11}, $2pqw_{12}$, and q^2w_{22} for A_1A_1, A_1A_2, and A_2A_2, respectively. Since half of the alleles carried by A_1A_2 individuals and all of the alleles carried by A_2A_2 individuals are A_2, the frequency of allele A_2 in the next generation (q_{t+1}) will become

$$q_{t+1} = \frac{pqw_{12} + q^2w_{22}}{p^2w_{11} + 2pqw_{12} + q^2w_{22}} \tag{2.4}$$

The extent of change in the frequency of allele A_2 per generation is denoted as Δq. We can show that

$$\Delta q = q_{t+1} - q_t = \frac{pq\,[p\,(w_{12} - w_{11}) + q\,(w_{22} - w_{12})]}{p^2 w_{11} + 2pq w_{12} + q^2 w_{22}} \tag{2.5}$$

Let us assume that A_1 is the original or "old" allele in the population. We will also assume that the population is diploid, and therefore the initial population consists of only one genotype, A_1A_1. We then consider the dynamics of change in allele frequencies following the appearance of a new mutant allele, A_2, and the consequent creation of two new genotypes, A_1A_2 and A_2A_2. For mathematical convenience, we will assign a relative fitness value of 1 to the A_1A_1 genotype. The fitness of the newly created genotypes, A_1A_2 and A_2A_2, will depend on the mode of interaction between A_1 and A_2.

For example, if A_2 is completely dominant over A_1, then w_{11}, w_{12}, and w_{22} will be written as 1, 1 + *s*, and 1 + *s*, respectively, where *s* is the **selection coefficient**, or the difference between the fitness of a genotype under study and the fitness of a reference genotype. A positive value of *s* denotes an increase in fitness (**selective advantage**) in comparison with A_1A_1, while a negative value denotes a decrease in fitness (**selective disadvantage**). When $s = 0$, the fitness of an A_2-carrying genotype will be the same as that of A_1A_1 (**selective neutrality**).

The term "dominance" has two meaning. The first refers to Mendelian phenotypic dominance, whereby A_1 is said to be dominant if the phenotype of A_1A_1 is the same as that of A_1A_2. Here, we are interested in the second meaning, which refers to fitness and has nothing to do with phenotype. That is, if the fitness of A_1A_1 is the same as that of A_1A_2, then in the evolutionary context, A_1 is said to be the dominant allele. Five common modes of allelic interaction will be considered: (1) codominance, (2) complete recessiveness, (3) complete dominance, (4) overdominance, and (5) underdominance.

Codominance

In **codominant selection**, the two homozygotes have different fitness values, whereas the fitness of the heterozygote is the mean of the fitnesses of the two homozygous genotypes. The relative fitness values for the three genotypes can be written as

Genotype	A_1A_1	A_1A_2	A_2A_{22}
Fitness	1	1 + *s*	1 + 2*s*

From Equation 2.5, we obtain the following change in the frequency of allele A_2 per generation under codominance:

$$\Delta q = \frac{spq}{1 + 2spq + 2sq^2} = \frac{spq}{1 + 2sq\,(p + q)} = \frac{spq}{1 + 2sq} \tag{2.6}$$

By iteration, Equation 2.6 can be used to compute the frequency of A_2 in any generation. However, the following approximation leads to a much more convenient solution. Note that if *s* is small, as is usually the case, the denominator in Equation 2.6 is approximately 1, and the equation is reduced to $\Delta q = spq$, which can be approximated by the differential equation

$$\frac{dq}{dt} = spq = sq\,(1 - q) \tag{2.7}$$

The solution of Equation 2.7 is given by

$$q_t = \frac{1}{1 + \left(\frac{1 - q_0}{q_0}\right) e^{-st}} \tag{2.8}$$

where q_0 and q_t are the frequencies of A_2 in generations 0 and *t*, respectively.

When q_0 is very small, as in the case of a new mutation arising in a population, Equation 2.8 reduces to

$$q_t = \frac{q_0 e^{st}}{1 + q_0 e^{st}} \quad (2.9)$$

In Equation 2.8, the frequency q_t is expressed as a function of time t. Alternatively, t can be expressed as a function of q as

$$t = \frac{1}{s} \ln \frac{q_t(1-q_0)}{q_0(1-q_t)} \quad (2.10)$$

where ln denotes the natural logarithmic function. From this equation, one can calculate the number of generations required for the frequency of A_2 to change from one value (q_0) to another (q_t).

Figure 2.4a (p. 44) illustrates the increase in the frequency of allele A_2 for $s = 0.01$. We see that codominant selection always increases the frequency of one allele at the expense of the other, regardless of the relative allele frequencies in the population. Therefore, codominance is a type of **directional selection**. Note, however, that at low frequencies, selection for a codominant allele is not very efficient (i.e., the change in allele frequencies is slow). The reason is that at low frequencies, the proportion of A_2 alleles residing in heterozygotes is large. For example, when the frequency of A_2 is 0.5, 50% of the A_2 alleles are carried by heterozygotes, whereas when the frequency of A_2 is 0.01, 99% of all A_2 alleles reside in heterozygotes. Because heterozygotes are subject to weaker selective pressures than are A_2A_2 homozygotes (s versus $2s$), the overall change in allele frequencies at low values of q will be small.

Dominance and recessiveness

In dominant selection, the two homozygotes have different fitness values, whereas the heterozygote has the same fitness as of one of the two homozygous genotypes. Here, we distinguish between two cases. In the first case, the new allele, A_2, is **dominant** over the old allele, A_1, and the fitness of the heterozygote, A_1A_2, is identical to the fitness of the A_2A_2 homozygote. The relative fitness values of the three genotypes are

Genotype	A_1A_1	A_1A_2	A_2A_{22}
Fitness	1	$1 + 2s$	$1 + 2s$

From Equation 2.5, we obtain the change in frequency of A_2 per generation:

$$\Delta q = \frac{2sp^2q}{1 + 2s - 2sp^2} \quad (2.11)$$

In the second case, A_1 is dominant over A_2, and the fitness of the heterozygote, A_1A_2, is identical to that of the A_1A_1 homozygote. That is, the new allele, A_2, is **recessive**. The relative fitness values are now

Genotype	A_1A_1	A_1A_2	A_2A_{22}
Fitness	1	1	$1 + 2s$

From Equation 2.5, we obtain the change in frequency of A_2 per generation:

$$\Delta q = \frac{2spq^2}{1 + 2sq^2} \quad (2.12)$$

Figures 2.4b and c illustrate the increase in the frequency of allele A_2 for $s = 0.01$ for dominant and recessive A_2 alleles, respectively. The two types of selection are also directional because the frequency of A_2 increases at the expense of A_1, regardless of the allele frequencies in the population. In **Figure 2.4b**, the new allele is advantageous and dominant over the old one, so selection is very efficient and the frequency of A_2 increases rapidly. In contrast, it takes many generations for the recessive allele A_2 to reach substantial frequencies in the population (**Figure 2.4c**).

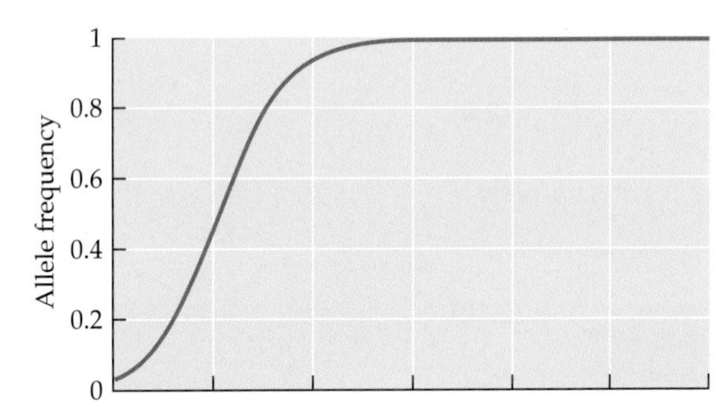

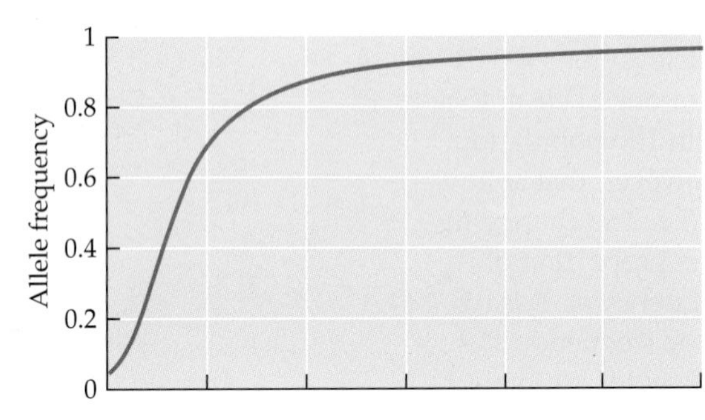

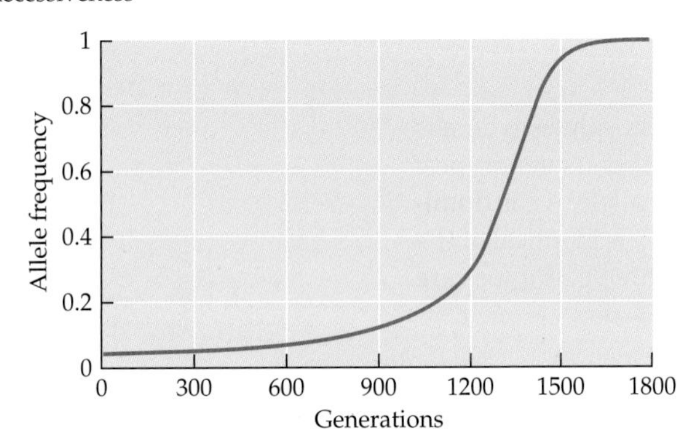

Figure 2.4 Changes in the frequency of an advantageous allele A_2 under (a) codominant selection; (b) dominant selection, where the new allele A_2 is dominant over A_1; and (c) dominant selection, where the new allele A_2 is recessive to A_1. In all cases, the frequency of A_2 at time 0 is $q = 0.04$, and the selective advantage is $s = 0.01$.

What about dominant and recessive alleles that are deleterious or even lethal? Dominant deleterious alleles are expected to be rapidly eliminated from the population, which is why dominant deleterious alleles are only seen in natural populations when they occur de novo. Human diseases that we think of as caused by dominant alleles, such as type 1 amyotrophic lateral sclerosis (sometimes referred to as "Lou Gehrig's disease" after the famous American baseball player who was diagnosed with the disorder) and Huntington's disease, are only dominant phenotypically. In terms of fitness, they are neither dominant nor deleterious: because these diseases have late ages of onset, the debilitating symptoms occur postreproductively, and the fitness of those affected is not significantly lower than that of the noncarriers.

The case of deleterious recessive alleles is very different. We note that the less frequent allele A_2 is, the larger the proportion of A_2 alleles in a heterozygous state will be. Thus, as far as selection against recessive homozygotes is concerned, at low frequencies only very few alleles in each generation will be subject to selection, and their frequency in the population will not change much. In other words, selection will not be very efficient in increasing the frequencies of beneficial alleles or, conversely, in reducing the frequencies of deleterious ones, as long as the frequency of the deleterious allele in a population is low. For this reason, it is almost impossible to rid a population of recessive deleterious alleles, even when such alleles are lethal in a homozygous state (i.e., $s = -1$). The persistence of such genetic diseases as Tay-Sachs syndrome and cystic fibrosis in human populations attests to the inefficiency of directional selection at low allele frequencies. In fact, each human genotype carries on average approximately 1,000 deleterious alleles (Sunyaev et al. 2001). Moreover, as we will see below, some "recessive diseases" are only recessive phenotypically, while in terms of fitness they are not recessive at all.

Overdominance and underdominance

In **overdominant selection**, the heterozygote has the highest fitness. Thus,

Genotype	A_1A_1	A_1A_2	A_2A_{22}
Fitness	1	$1 + s$	$1 + t$

Here t and s are used to denote different selection coefficient values. In overdominance, $s > 0$ and $s > t$. Depending on whether the fitness of A_2A_2 is higher than, equal to, or lower than that of A_1A_1, t can be positive, zero, or negative, respectively. The change in allele frequencies is expressed as

$$\Delta q = \frac{pq(s - 2sq + tq)}{1 + 2spq + tq^2} \tag{2.13}$$

Figure 2.5a illustrates the changes in the frequency of allele A_2.

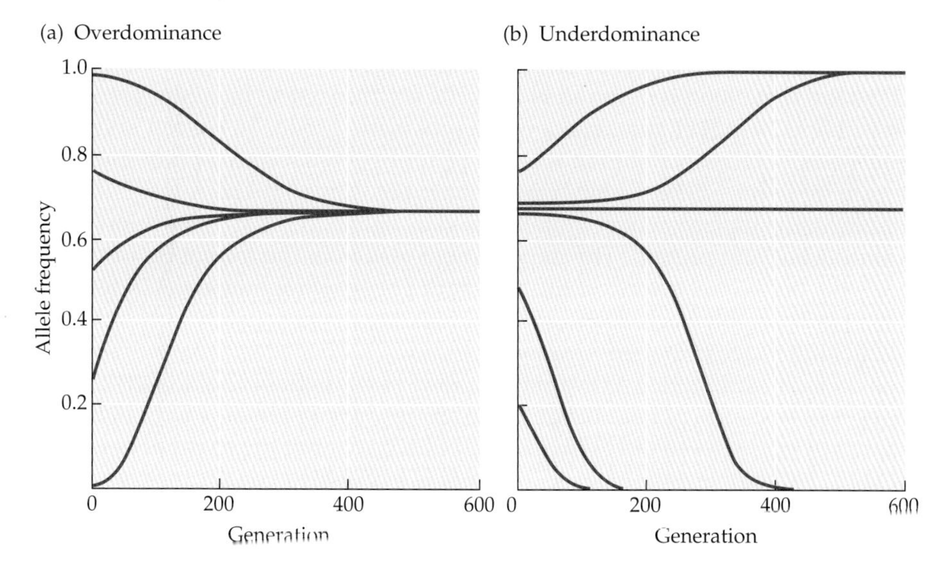

Figure 2.5 (a) Changes in the frequency of an allele subject to overdominant selection. Initial frequencies from top to bottom curves: 0.99, 0.75, 0.50, 0.25, and 0.01; $s = 0.04$ and $t = 0.02$. Since the s and t values are exceptionally large, the change in allele frequency is rapid. Note that there is a stable equilibrium at $q = 0.667$. (b) Changes in the frequency of an allele subject to underdominant selection. Initial frequencies from top to bottom curves: 0.75, 0.668, 0.667, 0.666, 0.50, and 0.20; $s = -0.02$ and $t = -0.01$. Again, because of the large values of s and t, the change in allele frequency is rapid. Note that there is an equilibrium at $q = 0.667$. This equilibrium, however, is unstable, since even the slightest deviation from it will cause one of the alleles to be eliminated from the population.

In contrast to the codominant or dominant selection regimes, in which allele A_1 is eventually eliminated from the population, under overdominant selection the population sooner or later reaches a **stable equilibrium** in which alleles A_1 and A_2 coexist. The equilibrium is stable because in case of a deviation from equilibrium, selection will quickly restore the equilibrium frequencies. After equilibrium is reached, no further change in allele frequencies will be observed (i.e., $\Delta q = 0$). Thus, overdominant selection belongs to a class of selection regimes called **balancing** or **stabilizing selection**. Balancing selection, in which two or more alleles are maintained in a population by selection, is predicted to lead to high levels of genetic diversity and to alleles that coexist in the population for very long periods of time. Sometimes, the alleles may be older than the species, in which case several species may share the same polymorphic alleles. Such sharing of alleles among species is referred to as **transspecies** or **transspecific polymorphism** (Figure 2.6).

The frequency of allele A_2 at equilibrium, $\hat{q}$, is obtained by solving Equation 2.13 for $\Delta q = 0$:

$$\hat{q} = \frac{s}{2s - t} \quad (2.14)$$

When $t = 0$ (i.e., both homozygotes have identical fitness values), the equilibrium frequencies of both alleles will be 50%.

In **underdominant selection**, the heterozygote has the lowest fitness, i.e., $s < 0$ and $s < t$. As in the case of overdominant selection, here too the change in the frequency of A_2 is described by Equation 2.13 and the frequency of allele A_2 at equilibrium by Equation 2.14. However, in this case the population reaches an **unstable equilibrium**, i.e., any deviation from the equilibrium frequencies in Equation 2.14 will drive the allele below the equilibrium frequency to extinction, while the other allele will be fixed (Figure 2.5b). Underdominant selection exemplifies a situation whereby a new allele, A_2, may be eliminated from the population even if the fitness of its homozygote (A_2A_2) is much higher than that of the prevailing homozygote (A_1A_1). Since a new allele arising in the population by mutation always has a very low frequency (i.e., much below the equilibrium frequency), new underdominant mutations will always be eliminated from the population. Underdominant selection is thus a powerful illustration of the fallacy of intuitive evolutionary thinking, according to which an allele that increases the fitness of its carriers will always replace one that does not.

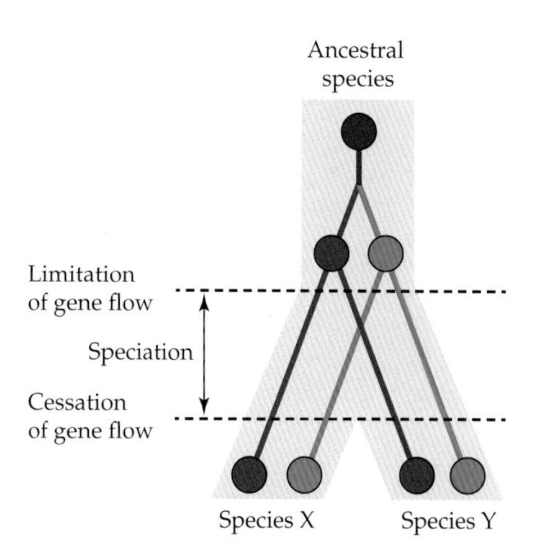

Figure 2.6 The concept of transspecies polymorphism. Two alleles (blue and red) at a single locus are passed on from the ancestral species through the speciation phase to the descendant species X and Y. (Modified from Klein et al. 2007.)

PATHOLOGICAL RECESSIVENESS, PHENOTYPIC CODOMINANCE, AND FITNESS OVERDOMINANCE: THE CASE OF SICKLE CELL ANEMIA Hemoglobin is an oxygen-transport metalloprotein in red blood cells (erythrocytes). In adults, it consists of two α chains, two β chains, and a heme prosthetic group. Sickle cell anemia is a phenotypically recessive human disease resulting from a single point mutation in the reading frame of the β-globin gene on chromosome 11. The A to T substitution results in the replacement of glutamic acid by valine in the sixth position of the 146-amino-acid-long protein. This replacement results in the formation of "sticky patches" that bind to receptors on the α-globin. The resulting hemoglobin S (HbS) has the capacity to polymerize by forming fibrous precipitates that distort the erythrocyte, making it sickle shaped. Homozygotes for the *HbS* allele have a drastically reduced lifespan and, hence, presumably a lower fitness than the non-sickle-cell homozygotes. In contrast, heterozygotes are largely asymptomatic, although both normal red blood cells and sickle-shaped cells are present in the blood of heterozygotes. Thus, pathologically, sickle cell anemia is a recessive genetic disease, while phenotypically it is a codominant trait.

We know, however, that genetic diseases that lower the fitness and are recessive cannot attain sizable frequencies in the population. Yet, there are populations in equatorial Africa in which the frequency of the HbS-producing allele may exceed 20%. Moreover, the *HbS* allele is known to have independently arisen and to have spread in different human populations at least five times. These findings indicate that in terms of fitness, the sickle cell allele is advantageous; more specifically the *HbS* gene seems to be subject to overdominant selection.

The selective agent seems to be the frequently deadly disease malaria, in particular the form caused by *Plasmodium falciparum*. *HbS* heterozygotes do not succumb to malaria, most probably because of increased host tolerance due to interference with the parasite's ability to remodel the actin cytoskeleton within the cytoplasm (Ferreira et al. 2011; Cyrklaff et al. 2011). From a population genetics point of view, the net result is that we have a balancing-selection regime in which the heterozygotes have a higher fitness than the non-sickle-cell homozygote, because they are protected from malaria, and a higher fitness than the sickle cell homozygote, because they do not suffer the consequences of sickle cell anemia.

Allison (1954, 2002) estimated the fitness of the *HbS* homozygote (*SS*) to be one-quarter of the fitness of the *Hb* homozygote (*AA*) and reported the highest observed frequency of the *S* allele in Africa to be more than 20%. Thus, $\hat{q} = 0.2$ and $t = -0.75$, and by using Equation 2.14, we can estimate s (the selective advantage of *HbS*-carrying heterozygotes) to be +0.25. The relative fitness values of the three genotypes are

Genotype	*AA*	*AS* (sickle cell trait)	*SS* (sickle cell anemia)
Fitness	1	1.25	0.25

Hemoglobin C (HbC) is a variant of β-globin in which the glutamic acid residue at the sixth amino acid position was replaced by a lysine. Hemoglobin C is particularly common among the Yoruba and the Dogon people of West Africa and their descendants in the Americas. *HbC* homozygotes only suffer from a very mild form of hemolytic anemia and, hence, are not at a great disadvantage in comparison with the wild-type homozygotes. The heterozygotes, moreover, are completely asymptomatic. HbC was found to be associated with a 29% reduction in risk of clinical malaria in *HbC* heterozygotes and a 93% reduction in *HbC* homozygotes. Thus, the protective effect against malaria associated with HbC may be greater than that of HbS in some populations (Agarwal et al. 2000). Interestingly, the mechanism by which HbC protects against malaria is identical to that of HbS (Cyrklaff et al. 2011).

These findings, together with the limited pathology of HbC compared with that of HbS, support the hypothesis that, in the long term and in the absence of malarial control, *HbC* would replace *HbS* in central West Africa.

SEX-SPECIFIC SELECTION: THE MOTHER'S CURSE

In previous sections, we assumed that selection operates equally in males and females. Here we describe a case where this assumption does not hold.

The mitochondrial genome in animals is transmitted maternally. That is, it is passed only from the mother to the offspring of both sexes. Thus, viewed from mitochondrial perspective, males are an evolutionary dead end (Parsch 2011). Regardless of the effect of the mitochondria on male survival, reproductive success, and any other fitness component, selection cannot operate in males. This implies that selection will be powerless in removing mitochondrial mutations that have a negative impact on the fitness of males only as long as the female fitness is not impacted. This phenomenon is referred to as the **male mutational load** or the mother's curse (Gemmell et al. 2004).

Innocenti et al. (2011) provided experimental validation for the male mutational load phenomenon by showing that deleterious mutations in mitochondria affected the expression of 1,171 nuclear genes in males only, but only six genes in females (and one gene in both males and females). Within the nuclear genes whose expression was affected in males only, there was a significant overrepresentation of male-biased genes, i.e., genes that are expressed at much higher levels in males than in females, particularly genes expressed in the testes.

Random Genetic Drift

As noted above, selection is not the only factor that can cause changes in allele frequencies. Allele frequency changes can also occur by chance, in which case the changes are not directional but random. An important factor in producing random fluctuations in allele frequencies is the **random sampling of gametes** in the process of reproduction (**Figure 2.7**). Sampling occurs because, in the vast majority of cases in nature, the number of gametes available in any generation is much larger than the number of adult individuals produced in the next generation. In other words, only a minute fraction of gametes succeeds in developing into adults. (In a diploid population subject to Mendelian segregation, sampling can still occur even if there is no excess of gametes, i.e., even if each individual contributes exactly two gametes to the next generation. The reason is that heterozygotes can produce two types of gametes, but the two gametes passing on to the next generation may by chance be of the same type.)

To see the effect of sampling, let us consider an idealized situation in which all individuals in the population have the same fitness, and selection does not operate. We further simplify the problem by considering a population with nonoverlapping generations (e.g.,

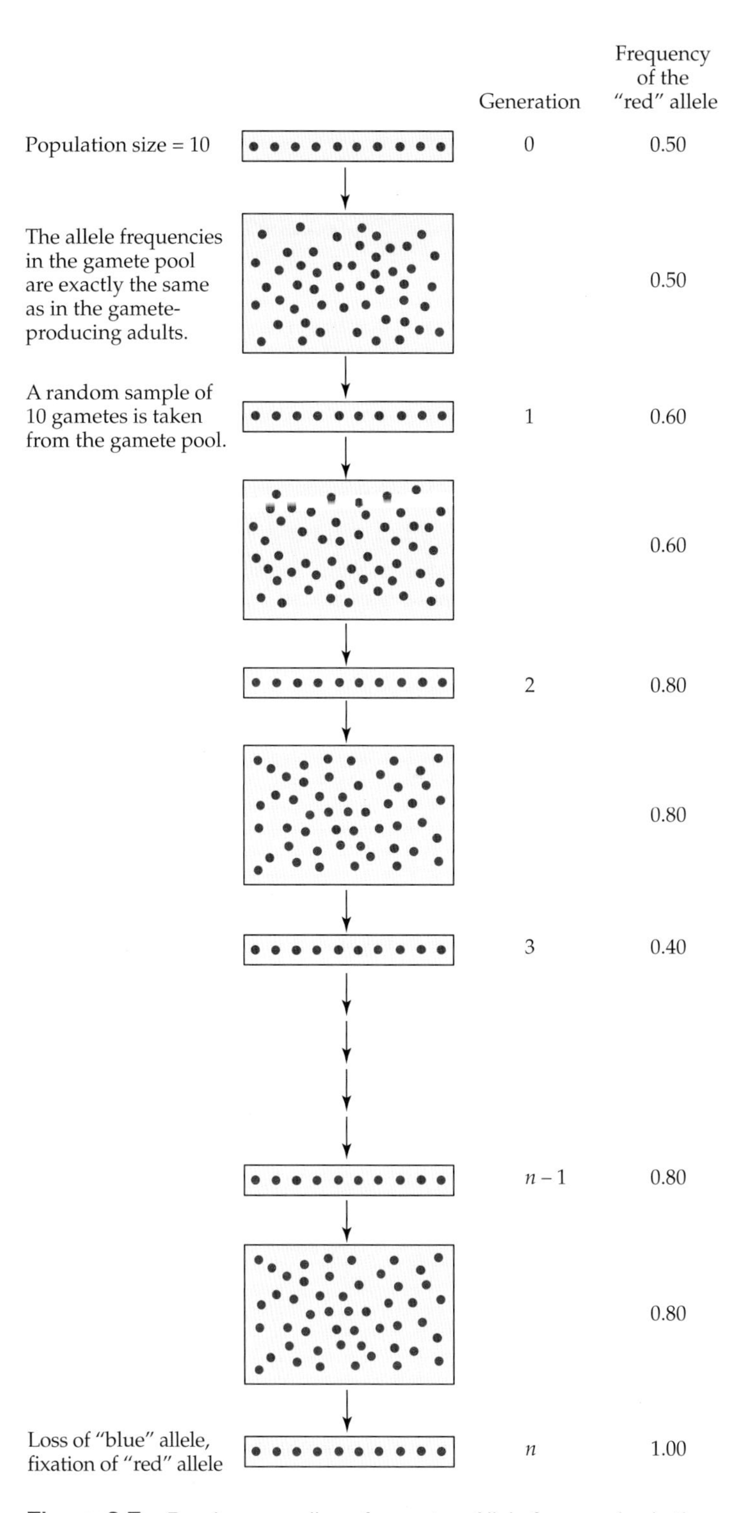

Figure 2.7 Random sampling of gametes. Allele frequencies in the gamete pools (deep boxes) in each generation are assumed to reflect exactly the allele frequencies in the adults of the parental generation (smaller boxes). Since the population size is finite (10 individuals), allele frequencies fluctuate up and down. If the random sampling process continues for long periods of time, by chance one allele may reach fixation, and the population will not change in subsequent generations unless a new allele enters the population by either mutation or migration.

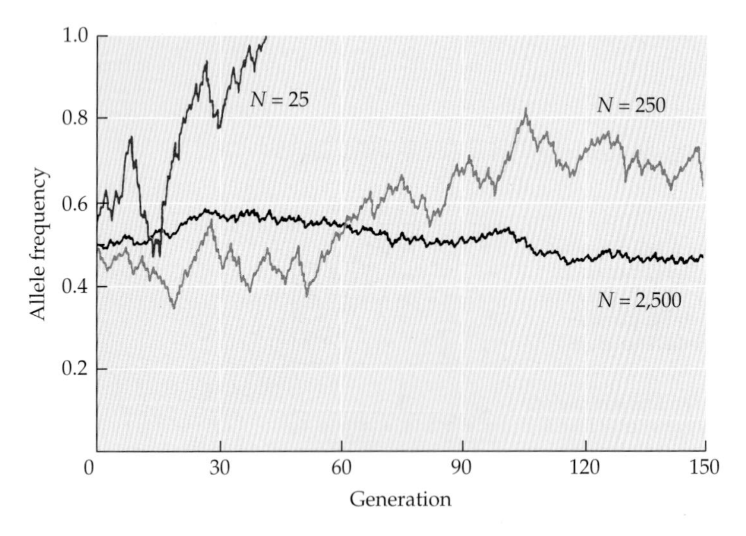

Figure 2.8 Changes in frequencies of alleles subject to random genetic drift in populations of different sizes. The smallest population ($N = 25$) reached fixation after 42 generations. The other two populations were still polymorphic after 150 generations but will ultimately reach fixation if the experiment is continued long enough. (Modified from Bodmer and Cavalli-Sforza 1976.)

Figure 2.9 Two possible outcomes of random genetic drift in populations of size 25 and $p_0 = 0.5$. In each generation, 25 alleles were sampled with replacement from the previous generation. In the population represented by the blue line, the allele becomes fixed in generation 27; in the other population, the allele is lost in generation 49. (Modified from Li 1997.)

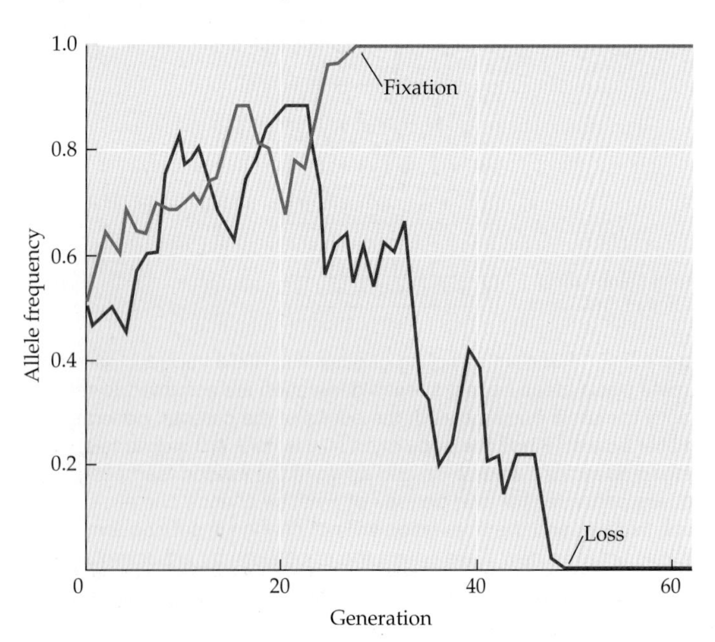

a group of individuals that reproduce simultaneously and die immediately afterward) such that any given generation can be unambiguously distinguished from both previous and subsequent generations. Finally, we assume that the population size does not change from generation to generation.

The population under consideration is diploid and consists of N individuals, so at any given locus, the population contains $2N$ genes. Let us again consider the simple case of one locus with two alleles, A_1 and A_2, with frequencies p and $q = 1 - p$, respectively. When $2N$ gametes are sampled from the infinite gamete pool, the probability, P_i, that the sample contains exactly i alleles of type A_1 is given by the binomial probability function

$$P_i = \frac{(2N)!}{i!(2N-i)!} p^i q^{2N-i} \tag{2.15}$$

where ! denotes the factorial function, i.e., $(2N)! = 1 \times 2 \times 3 \times \ldots \times (2N)$. Since P_i is always greater than 0 for populations in which two alleles coexist (i.e., $0 < p < 1$), the allele frequencies may change from generation to generation without the aid of selection. The process of change in allele frequency due solely to chance effects is called **random genetic drift**.

In **Figure 2.8** we illustrate the effects of random sampling on the frequencies of alleles in populations of different sizes. The allele frequencies change from generation to generation, but the direction of the change is random at any point in time. An obvious feature of random genetic drift is that fluctuations in allele frequencies are much more pronounced in small populations than in larger ones. In **Figure 2.9** we show two possible outcomes of random genetic drift in a population of size $N = 25$. In one replicate, allele A_2 is fixed in generation 27; in the other replicate, allele A_2 is lost in generation 49.

Let us follow the dynamics of change in allele frequencies due to the process of random genetic drift in succeeding generations. The frequencies of allele A_1 are written as $p_0, p_1, p_2, \ldots, p_t$, where the subscripts denote the generation number. The initial frequency of allele A_1 is p_0. In the absence of selection, we expect p_1 to be equal to p_0, and so on for all subsequent generations. The fact that the population is finite, however, means that p_1 will be equal to p_0 only on the average (i.e., when repeating the sampling process an infinite number of times). The evolutionary history of a population is a unique sequence of events. Thus, in real life, sampling occurs only once in each generation, and p_1 is usually different from p_0. In the second generation, the frequency p_2 will no longer depend on p_0 but only on p_1. Similarly, in the third generation, the frequency p_3 will depend on neither p_0 nor p_1 but only on p_2. Thus, the most important property of random genetic drift is its cumulative behavior; from generation to generation, the frequency of an allele will, on average, deviate more and more from its initial frequency.

In mathematical terms, the mean and variance of the frequency of allele A_1 at generation t, denoted by $\bar{p}_t$ and $V(p_t)$, respectively, are given by

$$\bar{p}_t = p_0 \tag{2.16}$$

and

$$V(p_t) = p_0(1-p_0)\left[1-\left(1-\frac{1}{2N}\right)^t\right] \approx p_0(1-p_0)\left[1-e^{-t/(2N)}\right] \tag{2.17}$$

where p_0 denotes the initial frequency of A_1. Note that although the mean frequency does not change with time, the variance increases with time; that is, with each passing generation the allele frequencies will tend to deviate further and further from their initial values. However, the change in allele frequencies will not be systematic in its direction. We also note that as N increases, the variance will decrease.

To see the cumulative effect of random genetic drift, let us consider the following numerical example. A certain population is composed of five diploid individuals in which the frequencies of the two alleles at a locus, A_1 and A_2, are each 50%. What is the probability of obtaining the same allele frequencies in the next generation? By using Equation 2.15, we obtain a probability of 25%. In other words, in 75% of the cases the allele frequencies in the second generation will be different from the initial allele frequencies. Moreover, the probability of retaining the initial allele frequencies in subsequent generations will no longer be 0.25, but will become progressively smaller. For example, the probability of having an equal number of alleles A_1 and A_2 in the population in the third generation (i.e., $p_3 = q_3 = 0.5$) is about 18%. The probability drops to only about 5% in the tenth generation (**Figure 2.10**). Concomitantly, the probability of either A_1 or A_2 being lost increases with time because in every generation there is a finite probability that all the chosen gametes happen to carry the same allele. In the above example, the probability of one of the alleles being lost is already about 0.1% in the first generation, and this probability increases dramatically in subsequent generations.

Once the frequency of an allele reaches either 0 or 1, its frequency will not change in subsequent generations. The first case is referred to as **loss** or **extinction**, and the second as **fixation**. If the process of sampling continues for long periods of time, the probability of such an eventuality reaches certainty. Thus, the ultimate result of random genetic drift is the fixation of one allele and the loss of all the others. This will happen unless there is an input of alleles into the population by such processes as mutation or migration, or unless polymorphism is actively maintained by a balancing type of selection.

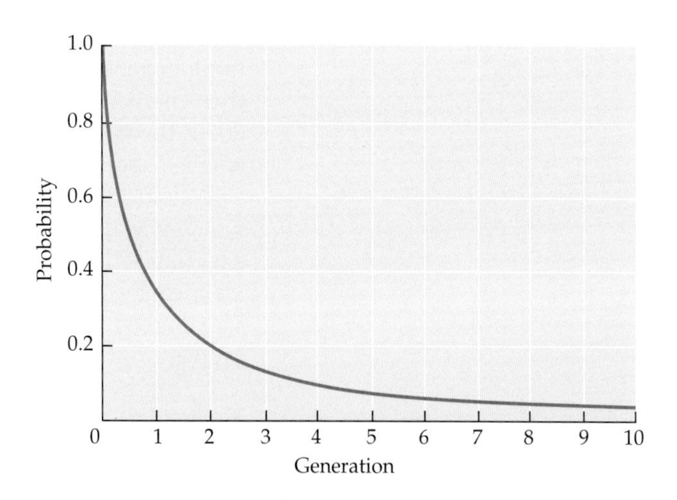

Figure 2.10 Probability of maintaining the same initial allele frequencies over time for two selectively neutral alleles. $N = 5$ and $p = 0.5$. (From Graur and Li 2000.)

Census Population Size and Effective Population Size

Population size is one of the most fundamental parameters in ecology, conservation biology, population genetics, and evolutionary biology. It is a primary determinant of population dynamics, and it is used to assess the relative importance of deterministic and stochastic evolutionary forces. In biological management, it is frequently the basis for conservation policy. In population genetics and evolutionary biology, it is used to determine the relative importance of random genetic drift in a population, and rate of loss of genetic variation, and the probability that advantageous mutations will arise in a population.

There are several ways to define population size. In the fields of ecology and conservation biology, we are often interested in the number of individuals in a population occupying a spatially defined area or belonging to a particular evolutionary lineage at a specific point in time or time range. This basic parameter is called the **census population size** (N). N is important for understanding demographic factors such as extinction risks.

Evolutionary biologists and population geneticists, on the other hand, are typically more interested in the evolutionary consequences of population size. Thus, from the point of view of population genetics and molecular evolution—in particular, for modeling the effects of random genetic drift—the relevant number of individuals to be considered consists of only those individuals that actively participate in reproduction, i.e., pass genes from one generation to the next. This part is called the **effective population size** and is denoted by N_e. Unfortunately, there is no such thing as "*the* effective size" of a population; the effective population size of choice depends on the scientific question one is trying to address, and the estimates differ from one another not only because of the measures used by the particular methods, but also because of the temporal and spatial frameworks (e.g., Fraser et al. 2007).

Many different concepts of N_e have been proposed in the literature; here, we will only be concerned with the three most commonly used: inbreeding effective population size, variance effective population size, and coalescence effective population size.

The various concepts of effective population size are usually defined in reference to so-called ideal population models, such as the **Wright-Fisher population** (Fisher 1922; Wright 1931). A Wright-Fisher population (sometimes referred to as a Fisher-Wright population) consists of a fixed number of randomly mating diploid hermaphroditic individuals (N). The population is closed, i.e., there is no migration to or from this population, and each individual has an equal chance of mating with any individual in the population, including itself, as well as an equal probability of producing offspring. Generations are discrete, i.e., all members of the population reach maturity in one generation and die after the opportunity to reproduce once. Thus, new individuals are formed each generation by the random sampling with replacement of gametes produced by the parental generation. In consequence, all parents have the same probability of contributing a gamete to an individual that survives to breed in the next generation. In other words, the alleles of generation $t + 1$ are a random sample of alleles from generation t. Wright-Fisher populations do not exist in nature, although populations of hermaphroditic marine organisms, which shed large numbers of eggs and sperm that fuse randomly to make new zygotes, may come somewhat close to such an idealized situation (Charlesworth 2009). A Wright-Fisher population can be described by a single variable—its size.

An effective population size is the size of the Wright-Fisher population that would experience the effects of genetic drift to the same degree as the population under study. For example, if an actual population of 50 organisms experiences the effects of random genetic drift at the same rate as a Wright-Fisher population of 20 organisms, then the population of 50 organisms is said to have an effective population size of 20. In the vast majority of cases, but not in all cases, the effective population size is smaller, sometimes much smaller, than the census population size.

Wright (1931) introduced the concept of **inbreeding effective population size**, which was defined as the size of a Wright-Fisher idealized population that would experience the same increase in the mean probability that two alleles at a locus are identical by descent (i.e., the **inbreeding coefficient**) as would the actual population. The inbreeding effective population size, thus, predicts the rate of increase in homozygosity, or the rate of decrease in heterozygosity, in a finite population.

N_e can be estimated at different temporal scales. These estimates can be roughly divided into short-term effective population size and long-term effective population size.

Short-term effective population size

The "short-term" in **short-term effective population size** refers to the fact that this measure is derived from considerations related to one or a few generations of genetic sampling. Here, we deal with estimating the short-term effective population size by using the approach of Nei and Tajima (1981) as applied to the variance effective population size. The **variance effective population size** (Crow 1954) is defined as the size of a Wright-Fisher idealized population that would experience the same increase in the allele-frequency variance as would the actual population.

The basic protocol for estimating short-term N_e involves two samples separated by a defined number of generations (usually a single generation). Consider, for instance, a population with a known census size N, and assume that the frequency of allele A_1 at generation t is p. If the number of individuals taking part in reproduction is N, then the variance of the frequency of allele A_1 in the next generation, p_{t+1}, may be obtained from Equation 2.17 by setting $t = 1$:

$$V(p_{t+1}) = \frac{p_t(1-p_t)}{2N} \quad (2.18)$$

In practice, since not all individuals in the population have an equal probability of taking part in the reproductive process, the observed variance will be larger than that obtained from Equation 2.18. The effective population size (N_e) is the value that is substituted for N in order to satisfy Equation 2.18, i.e.,

$$V(p_{t+1}) = \frac{p_t(1-p_t)}{2N_e} \quad (2.19)$$

Because estimating short-term effective population size requires sampling the population at least twice, short-term N_e is mostly applicable to laboratory populations or populations maintained in highly artificial settings. Under such settings, the short-term effective population size tends to be lower that the census population size, but not by many orders of magnitude as in the case of long-term N_e (see below). For example, in a laboratory population of the Mediterranean fruit fly (*Ceratitis capitata*), the short-term effective population size hovered around 40% of the census population size (Debouzie 1980). In humans, the short-term N_e is only slightly larger than $N/3$, whether the calculation is done for the entire human population (Nei and Imaizumi 1966) or for an isolate (Mourali-Chebil and Heyer 2006).

Coalescence and long-term effective population size

The genetic variation within a population is shaped not only by what happens in the time span of one or two generations; population structure is also shaped by the long-term interactions among the evolutionary forces of selection, migration, drift, and mutation. In other words, we need to consider the **long-term effective population size**, i.e., follow the evolution of all the alleles in the population back to the time of their origin from a common allele.

Coalescent theory deals with retrospective models of population genetics that attempt to trace all alleles of a gene in a population to a single ancestral copy, known as the **most recent common ancestor** (**MRCA**). **Coalescence** refers to the process in which the evolutionary lineages of two alleles in a population trace back to a common ancestral allele in a past generation. The relationships among all the alleles in a sample are represented by a graphic device called **gene genealogy** (or **coalescent),** which is similar in form to a phylogenetic tree. Understanding the statistical properties of the coalescent under different assumptions forms the basis of coalescent theory. In the simplest case, coalescent theory assumes no recombination, no selection, no gene flow into the population, and no internal subdivision of the population.

Let us consider an ideal Wright-Fisher population of constant size, N, in which each generation t is formed by sampling from generation $t - 1$ (**Figure 2.11**). Because this is a Wright-Fisher population, the census size of the population is equal to the effective population size, i.e., $N = N_e$. From the present population ($t = 0$), we sample $i \ll N$ alleles and track backward the ancestry of each of them. Our sample can be viewed as a collection of lineages, in which each allele randomly selects its parental lineage from the previous generation. Thus, each allele in each generation is con-

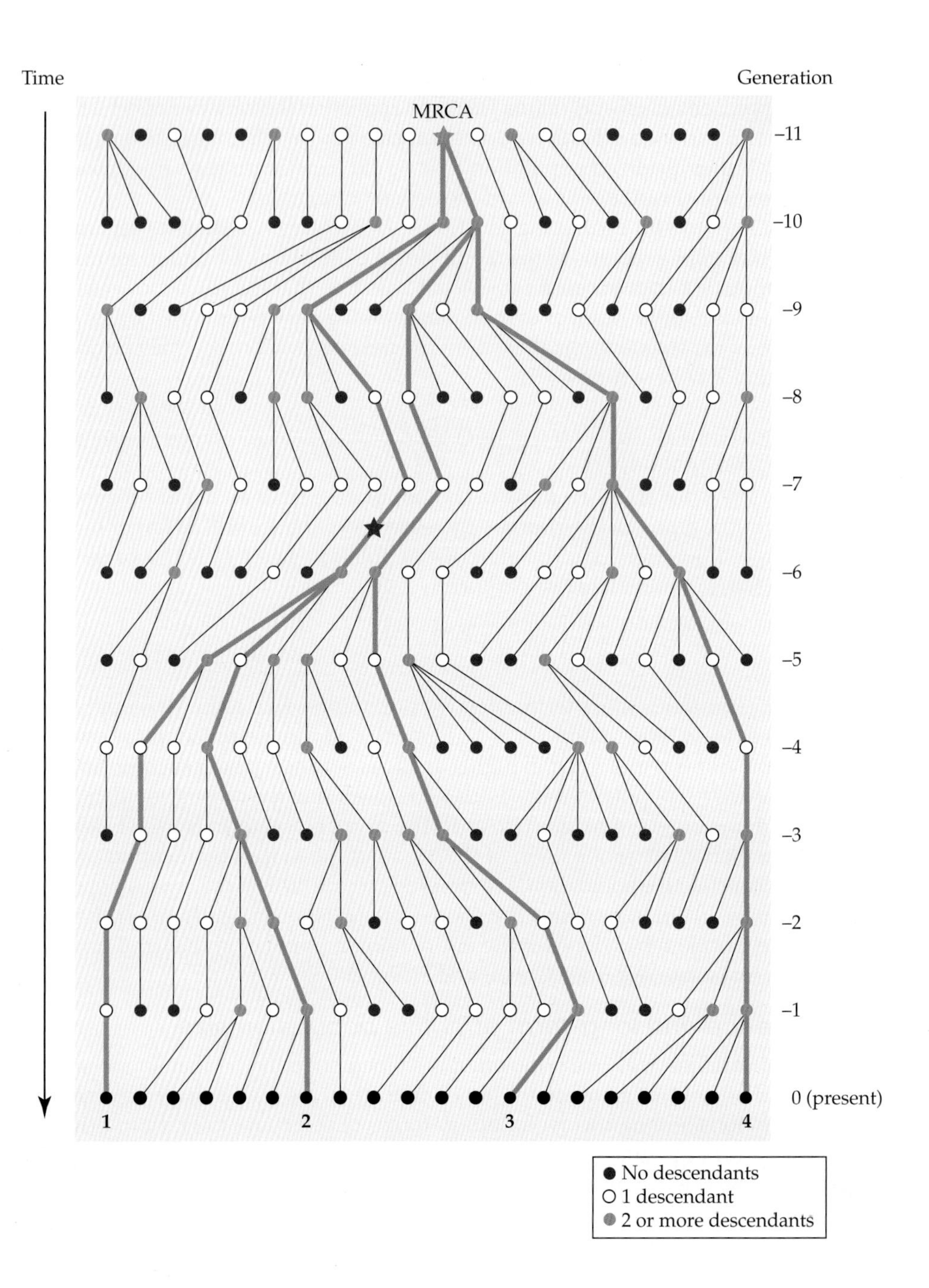

Figure 2.11 An example of the coalescence process in an ideal Wright-Fisher population of size 20. Individual alleles in the population are marked by circles, and the ancestor-descendant relationships among them are shown as lines. An allele that has no descendants is marked in red; if it has a single descendant, it is white; ancestor alleles with two or more descendants are in blue. The alleles at the present population (generation 0) are black. Consider a sample of four alleles (i = 4) at a genetic locus that have been obtained from the present population (bottom of chart). If we trace the ancestry of these four alleles back in time 11 generations, they will form a coalescent (green lines). Two of the alleles in the sample (1 and 2) coalesce at generation –6. The other coalescence events occur at generations –10 (alleles 3 and 4) and –11 (the ancestor of alleles 1 and 2, and the ancestor of alleles 3 and 4). One point mutation (red star) is indicated; mutations are supposed to be rare enough so that no more than one is observed at a given locus. All alleles in the present population trace their ancestry to a single ancestral copy, known as the most recent common ancestor (MRCA; blue star).

nected with exactly one allele in any previous generation. On the other hand, an allele in one generation may be connected to no allele, one allele, or two or more alleles in any subsequent generation. Whenever two alleles in a generation connect to the same parent allele in a previous generation, their lineages are said to coalesce. The number of coalescent events in a coalescent is equal to the number of alleles in the study.

Let us now consider a sample of i = 4 alleles: the first coalescence merges the contemporary alleles into i – 1 = 3 lineages at generation –6; the second coalescence merges these into i – 2 = 2 lineages at generation –10; and the third coalescence merges these into i – 3 = 1 lineage at generation –11.

It turns out that the mean time required for two allelic lineages to coalesce depends on the effective population size (N_e) and the number of allelic lineages (i) in the sample. The time to coalescence follows an exponential distribution, with a standard deviation equal to the mean. With i alleles present in a diploid population, the expected time back to the next coalescence event is given by

$$t_{i \to i-1} = \frac{4N_e}{i(i-1)} \tag{2.20}$$

For example, starting with four alleles, it will take on average $N_e/3$ generations to coalesce into three lineages. The time it will take for three alleles to coalesce into two lineages will be $2N_e/3$ generations, and the time it will take for two alleles to coalesce into one allele will be $2N_e$ generations.

For a sample of *i* alleles, the time to the coalescence of all the alleles in the population, i.e., the last time the sample of alleles shared a common ancestor, is

$$t_{i \to i-1} = \frac{4N}{i(i-1)} \tag{2.21}$$

(Wakeley 2008). From Equation 2.21, we can see that the expected time for all of the alleles in an ideal Wright-Fisher population to coalesce into a single common ancestor, i.e., when $i = 2N$, is approximately $4N$ generations.

We note that the coalescent cannot be observed; however, the results of this Gedankenexperiment are relevant to data on population samples, because variation in a sample of allelic sequences reflects mutations that have arisen in different branches of the tree since the most recent common ancestor. To model a sample, we need to allow mutations to occur.

Let us envisage the simplest possible situation, that of a Wright-Fisher population in which the genetic variation at a selectively unconstrained locus is determined by the balance between the mutational input of new alleles and the loss of existing alleles by random genetic drift. Let us further assume the **infinite sites model**, which assumes that the mutation rate probability per generation per site, μ, is very low, so at most one mutation arises per site. Under this model each new mutation is unique. These assumptions allow us to derive formulas for predicting the expected values of some commonly used measures of variability, such as nucleotide diversity, which is the frequency with which a pair of randomly selected orthologous sequences differ at a nucleotide site (p. 37).

Consider a pair of alleles taken randomly from the sample. There is a time, *t*, connecting each of them to their common ancestor. They will be identical at a site if no mutation has arisen over the time separating them from each other, which is $2t$. The probability that a mutation has arisen at that site and caused them to differ from each other is $2t\mu$. Since *t* for the entire coalescent has an expected value of $t_{N \to 1} = 2N$, then the net probability of a nucleotide difference at a given site is $4N\mu$. Averaging over all pairs of alleles in the sample, and over a large number of sites, gives the expected value of the nucleotide site diversity for a sample (Π) as

$$\Pi = 4N\mu \tag{2.22}$$

where Π is the expected value of nucleotide diversity in the population, *N* is the population size (which in the case of a Wright-Fisher population is equal to the effective population size), and μ is the neutral mutation rate per generation.

When information about *N* and μ is available, the expected value of Π can be calculated. In non-Wright-Fisher populations, i.e., in real populations, in which N_e is usually much smaller than *N*, the observed Π turns out to be much smaller than the expected one. N_e is the value that should be substituted for *N* in Equation 2.22 so that the equality holds:

$$\Pi = 4N_e\mu \tag{2.23}$$

and

$$N_e = \frac{\Pi}{4\mu} \tag{2.24}$$

The meaning of these results is simple: populations with large effective population sizes can maintain a large amount of genetic variability, or put another way, the effective population size determines the amount of genetic variation that can be maintained in the population.

When μ is unknown, we can still obtain estimates of relative values for N_e from different populations when the same markers and methods are used to estimate their Π values (Wang 2005).

A large body of software exists for simulating data sets under the coalescent process, as well as for inferring population parameters from genetic data (e.g., Librado and Rozas 2009; Excoffier and Lischer 2010).

Factors conspiring to reduce the effective population size relative to the census population size

It is a general observation that long-term N_e is smaller, sometimes considerably smaller, than N (Frankham 1995). For example, the effective population size of the mosquito *Anopheles gambiae* in Kenya has been estimated to be about 2,000—about six orders of magnitude smaller than the census population size (Lehmann et al. 1998). In **Table 2.2**, we list several long-term effective population size estimates. An interesting observation is that N_e values do not bear a good relationship to current population sizes. For example, the census size of the world's human populations exceeds 7 billion individuals, while tentative estimates of the total size of bonobo populations range from ~15,000 to 100,000, but the two species have very similar long-term effective population sizes.

Various factors can contribute to the difference between census population size and effective population size. For example, because of the division into two sexes, the situation may arise in which a small number of individuals of one sex can greatly reduce the effective population size below the total number of breeding individuals. This type of reduction in N_e in comparison to the census size is frequently observed in polygamous species, such as social mammals and territorial birds (which practice polygyny) or honeybees (which practice polyandry). If a population consists of N_m males and N_f females ($N = N_m + N_f$), N_e is given by

$$N_e = \frac{4N_m N_f}{N_m + N_f} \tag{2.25}$$

Note that unless the number of females taking part in the reproductive process equals that of males, N_e will always be smaller than N. As an extreme example, let us assume that in a population of size N in which the sexes are equal in number, all females ($N/2$) but only one male take part in the reproductive process. From Equation 2.25, we get $N_e = 2N / (1 + N/2)$. If N is considerably larger than 1, such that $N/2 + 1 \approx N/2$, then N_e becomes ~4, regardless of the census population size. Thus, although the population of a healthy hive in midsummer can average between 40,000 and 80,000 honeybees (*Apis mellifera*), the existence of a nonreproductive caste and the practice of polyandry may yield an N_e value not much larger than 4 (i.e., lower by four orders of magnitude).

The effective population size can also be much reduced by long-term variations in the population size, which in turn are caused by such factors as environmental catastrophes, cyclical modes of reproduction, and local extinction and recolonization events. Episodes of low population size have a disproportionate effect on the overall value of N_e. The long-term effective population size in a species for a period of t generations is given by

$$N_e = \frac{t}{\sum_{i=0}^{t} \frac{1}{N_i}} = \frac{t}{\frac{1}{N_1} + \frac{1}{N_2} + \ldots + \frac{1}{N_t}} \tag{2.26}$$

Table 2.2

Long-term effective population size estimates from DNA sequence diversities

Species	N_e
Human (*Homo sapiens*)	10,400[a]
Bonobo (*Pan paniscus*)	12,300
Chimpanzee (*Pan troglodytes*)	21,300
Gorilla (*Gorilla gorilla*)	25,200
Drosophila melanogaster	1,150,000
Caenorhabditis elegans	80,000
Plasmodium falciparum	210,000–300,000
Escherichia coli	25,000,000

Source: Modified from Charlesworth (2009).

[a] The human long-term effective population size has been a subject of much debate (e.g., Hawks 2008).

where N_i is the population size of the ith generation. In other words, N_e equals the harmonic mean of the N_i values, and consequently it is closer to the smallest value of N_i than to the largest one. Similarly, if a population goes through a bottleneck, the long-term effective population size is greatly reduced even if the population has long regained its prebottleneck census size.

The mode of inheritance also affects N_e. For example, the values of N_e for autosomal alleles, which are carried with equal probabilities by both sexes, are larger than the values for X-linked loci, which are carried with unequal probabilities by the two sexes. The lowest values of N_e are expected in Y-linked or organelle loci, which are passed on from generation to generation uniparentally.

Other factors that reduce N_e in comparison with N include inbreeding, i.e., correlation between the maternal and paternal alleles of an individual; a large variation in offspring number among individuals; age structure in the population; limited migration between partly or wholly isolated populations; and directional selection. The long-term maintenance of two or more alleles by balancing selection results in an elevation in N_e relative to N at sites that are closely linked to the target of selection (the **Hill-Robertson effect**).

Gene Substitution

Gene or **allele substitution** is defined as the process whereby a mutant allele completely replaces the old or **wild-type** allele in a population. In this process, a mutant allele arises in a population as a single copy and becomes fixed after a certain number of generations. Not all mutations, however, reach fixation. In fact, the majority of them are lost after a few generations. Thus, we need to address the issue of **fixation probability** and discuss the factors affecting the chance that a new mutant allele will reach fixation in a population.

The time it takes for a new allele to become fixed is called the **fixation time**. In the following sections we will identify the factors that affect the speed with which a new mutant allele replaces an old one in a population.

New mutations arise continuously within populations. Consequently, gene substitutions occur in succession, with one allele replacing another and in time being itself replaced by a newer allele. Thus, we can speak of the **rate of gene substitution**, i.e., the number of fixations of new alleles per unit time.

Fixation probability

The probability that a particular allele will become fixed in a population depends on (1) its present frequency, (2) its selective advantage or disadvantage, s, and (3) the effective population size, N_e. In this section, we consider the case of codominance and assume that the relative fitness of the three genotypes A_1A_1, A_1A_2, and A_2A_2 are 1, $1 + s$, and $1 + 2s$, respectively.

Kimura (1962) showed that the probability of fixation of A_2 is

$$P = \frac{1 - e^{-4N_e sq}}{1 - e^{-4N_e s}} \tag{2.27}$$

where q is the initial frequency of allele A_2. Since $e^{-x} \approx 1 - x$ for small values of x, Equation 2.27 reduces to $P \approx q$ as s approaches 0. Thus, for a neutral allele, the fixation probability equals its frequency in the population. For example, a neutral allele with a frequency of 40% will become fixed in 40% of the cases and will be lost in 60% of the cases. This is intuitively understandable because in the case of neutral alleles, fixation occurs by random genetic drift, which favors neither allele.

We note that a new mutation arising as a single copy in a diploid population of size N has an initial frequency of $1/(2N)$. The probability of fixation of an individual mutant allele, P, is thus obtained by replacing q with $1/(2N)$ in Equation 2.27. When $s \neq 0$,

$$P = \frac{1 - e^{-(2N_e s/N)}}{1 - e^{-4N_e s}} \tag{2.28}$$

For a neutral mutation, i.e., $s = 0$, Equation 2.28 becomes

$$P = \frac{1}{2N} \tag{2.29}$$

If the population size is equal to the effective population size, Equation 2.28 reduces to

$$P = \frac{1 - e^{-2s}}{1 - e^{-4Ns}} \tag{2.30}$$

If the absolute value of s is small, we obtain

$$P = \frac{2s}{1 - e^{-4Ns}} \tag{2.31}$$

For positive values of s and large values of N, Equation 2.31 reduces to

$$P \approx 2s \tag{2.32}$$

Thus, if an advantageous mutation arises in a large population and its selective advantage over the rest of the alleles is small, say up to 5%, the probability of its fixation is approximately twice its selective advantage. For example, if a new codominant mutation with $s = 0.01$ arises in a population, the probability of its eventual fixation is 2%.

Let us now consider a numerical example. A new mutation arises in a population of 1,000 individuals. What is the probability that this allele will become fixed in the population if (1) it is neutral, (2) it confers a selective advantage of 0.01, or (3) it has a selective disadvantage of 0.001? For simplicity, we assume that $N = N_e$. For the neutral case, the probability of fixation as calculated by using Equation 2.29 is 0.05%. From Equations 2.30 and 2.32, we obtain probabilities of 2% and 0.004% for the advantageous and deleterious mutations, respectively. These results are noteworthy, since they essentially mean that an advantageous mutation does not always become fixed in the population. In fact, 98% of all the mutations with a selective advantage of 0.01 will be lost by chance. This theoretical finding is of great importance, since it shows that the perception of adaptive evolution as a process in which advantageous mutations arise in populations and invariably take over the population in subsequent generations is a naïve concept. Moreover, even deleterious mutations have a finite probability of becoming fixed in a population, albeit a small one. However, the mere fact that a deleterious allele may become fixed in a population at the expense of "better" alleles illustrates in a powerful way the importance of chance events in determining the fate of mutations during evolution. Selection is sometimes erroneously perceived as omnipotent; however, selection can neither ensure the fixation of every beneficial mutation, nor sieve out every deleterious one.

As the population size becomes larger, the chance effects become smaller. For instance, in the above example, if the effective population size is 10,000 rather than 1,000, then the fixation probabilities become 0.005%, 2%, and $\sim 10^{-20}$ for neutral, advantageous, and deleterious mutations, respectively. Thus, while the fixation probability for the advantageous mutation remains approximately the same, the fixation probability for the neutral mutation becomes smaller, and that for the deleterious allele becomes indistinguishable from zero.

Fixation time

The time required for the fixation or loss of an allele depends on (1) the frequency of the allele, (2) its selective advantage or disadvantage, and (3) the size of the population. The mean time to fixation or loss becomes shorter as the frequency of the allele approaches 1 or 0, respectively.

When dealing with new mutations, it is more convenient to treat fixation and loss separately. Here, we deal with the mean fixation time of those mutations that will eventually become fixed in the population. This variable is called the **conditional fixation time.** In the case of a new mutation whose initial frequency in a diploid population is by definition $q = 1/(2N)$, the mean conditional fixation time, $\bar{t}_t$, was calculated by Kimura and Ohta (1969). For a codominant mutation with a strong selective advantage, i.e., $s \gg 0$, $\bar{t}_t$ is approximated by

$$\bar{t} = \frac{2\ln(2N_e - 1)}{s} \tag{2.33}$$

For a neutral mutation or for the case of very weak selection, i.e., $s < 1/(2N_e)$, $\bar{t}_t$ is approximated by

$$\bar{t} = 4N_e \text{ generations} \tag{2.34}$$

To illustrate the difference between different types of mutation, let us assume that a species has an effective population size of about 10^6 and a mean generation time of 2 years. Under these conditions, it will take a neutral mutation, on average, 8 million years to become fixed in the population. In comparison, a mutation with a selective advantage of 1% will become fixed in the same population in only about 5,800 years. Interestingly, the conditional fixation time for a deleterious allele with a selective disadvantage $-s$ is the same as that for an advantageous allele with a selective advantage $+s$ (Maruyama and Kimura 1974). This is intuitively understandable given the high probability of loss for a deleterious allele. That is, for a deleterious allele to become fixed in a population, fixation must occur very quickly.

In **Figure 2.12**, we present in a schematic manner the dynamics of gene substitution for advantageous, neutral, deleterious, and overdominant mutations. We note that advantageous mutations are either rapidly lost or rapidly fixed in the population. In contrast, the frequency changes for neutral alleles are slow, and the fixation time is much longer than that for advantageous mutations. Genetic polymorphism due to overdominant alleles can last for very long periods of evolutionary time as the alleles are seldom fixed or lost.

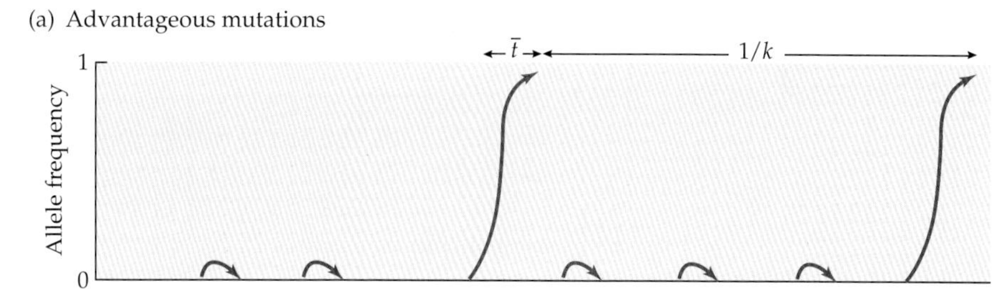

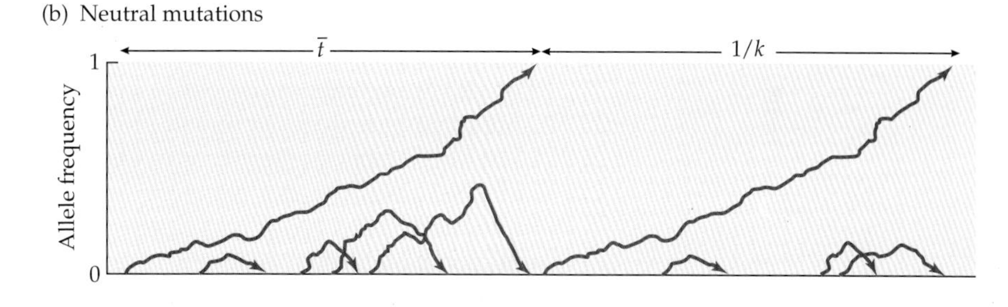

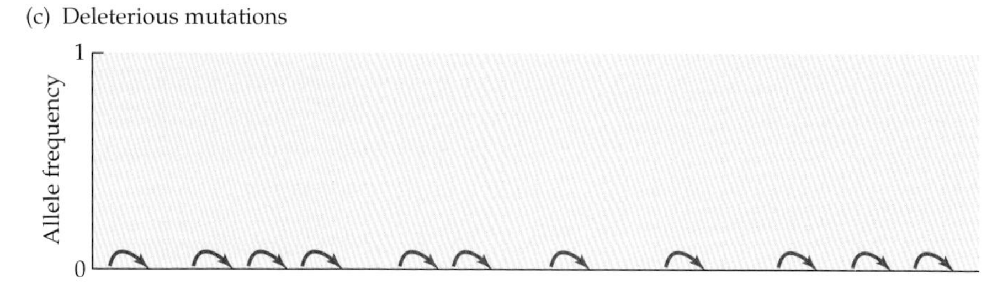

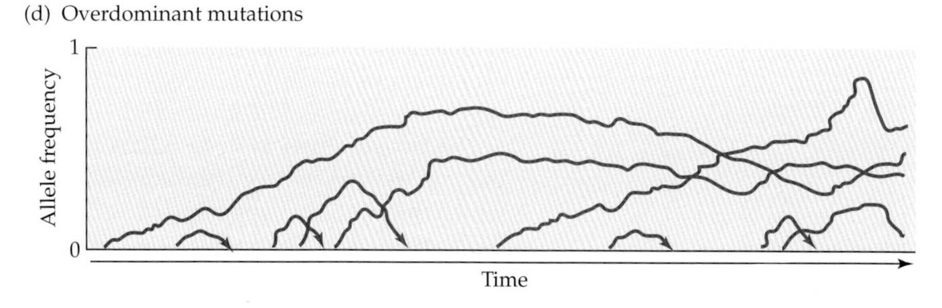

Figure 2.12 Schematic representation of the dynamics of gene substitution for advantageous, neutral, deleterious, and overdominant mutations. (a) Advantageous mutations are either quickly lost from the population or quickly fixed, so their contribution to genetic polymorphism is small. (b) The frequency of neutral alleles changes very slowly by comparison, so a large amount of transient polymorphism is generated. (c) Deleterious alleles are much less likely to undergo fixation, but when they do, the conditional fixation time is as short as for advantageous mutations. (d) Overdominant mutations very seldom become fixed; however, they maintain stable polymorphisms for very long periods of evolutionary time. The conditional fixation time is $\bar{t}$, and $1/k$ is the mean time between two consecutive fixation events.

Rate of gene substitution

Let us now consider the **rate of gene substitution**, defined as the number of mutations reaching fixation per unit time. We will first consider neutral mutations. If neutral mutations occur at a rate of u per gene per generation, then the number of mutations arising at a locus in a diploid population of size N is $2Nu$ per generation. Since the probability of fixation for each of these mutations is $1/(2N)$, we obtain the rate of substitution, K, of neutral alleles by multiplying the total number of mutations by the probability of their fixation:

$$K = 2Nu\left(\frac{1}{2N}\right) = u \quad (2.35)$$

Thus, for neutral mutations, the rate of substitution is equal to the rate of mutation—a remarkably simple and important result (Kimura 1968b). This result can be intuitively understood by noting that, in a large population, the number of mutations arising every generation is high but the fixation probability of each mutation is low. In comparison, in a small population, the number of mutations arising every generation is low, but the fixation probability of each mutation is high. As a consequence, the rate of substitution for neutral mutations is independent of population size.

For advantageous mutations, the rate of substitution can also be obtained by multiplying the rate of mutation by the probability of fixation for advantageous alleles as given in Equation 2.32. For codominance with $s > 0$, we obtain

$$K = 4Nsu \quad (2.36)$$

In other words, the rate of substitution for the case of codominance depends on population size (N), selective advantage (s), and the rate of mutation (u).

The inverse of K (i.e., $1/K$) is the **mean time between two consecutive fixation events** (Figure 2.12).

Mutational meltdown: The double jeopardy of small populations

We have already seen that it is possible for deleterious alleles to become fixed via genetic drift. We also know that deleterious mutations occur much more frequently than advantageous mutations. Finally, we have shown that in small populations random genetic drift is a much more important evolutionary force than selection. These three facts suggest that in small populations deleterious mutations may accumulate, resulting in a decrease in mean fitness with evolutionary time. In small populations, the risk of decreasing fitness may be significant and may result in a **mutational meltdown**, whereby extinction results from the accumulation of deleterious alleles (Lynch and Gabriel 1990; Lynch et al. 1995). Furthermore, because mutations appear only rarely in small populations and because advantageous mutations constitute only a tiny fraction of the total number of mutations, the population may not have enough advantageous alleles to adapt to a changing environment. Thus, a "double jeopardy" situation arises due to the fact that a population with a small effective population size fixes more deleterious and fewer beneficial alleles over time than a larger population.

Nearly neutral mutations

In previous sections we discussed mutations as if they belong to two discrete categories: (1) mutations that have a nonzero selection coefficient ($s \neq 0$), i.e., advantageous or deleterious mutations whose fate is mainly determined by natural selection, and (2) mutations that do not affect the fitness of their carriers either way and are hence selectively neutral. This discrete division is merely an approximation that can only be applied to extremely large populations. In practice, the fate of a mutation depends not only on its intrinsic contribution to fitness but also on the size of the population in which this mutation appeared.

From Equation 2.27, which describes the probability of fixation of any codominant allele in a finite diploid population, we note that the probability of fixation does not depend only on the coefficient of selection (s), but is a function of the product $N_e s$ where N_e is the effective population size and s the selective advantage or disadvantage.

Because the variable $N_e s$ is continuous, all attempts to classify the possible values of $N_e s$ into discrete categories will be arbitrary. Such categorization can, however, serve a useful purpose as a concise summary. Following Tachida (1991), we can distinguish between three categories: (1) **selected mutations** ($|N_e s| > 0.75$), (2) **nearly neutral mutations** ($0.05 < |N_e s| < 0.75$), and (3) **effectively neutral mutations** ($|N_e s| < 0.05$). Nearly neutral mutations can be divided into **slightly deleterious mutations** ($-0.75 < N_e s < -0.05$) and **slightly advantageous mutations** ($-0.05 < N_e s < -0.75$).

Let us now consider a slightly deleterious mutation whose fate will be determined by the interplay between purifying selection and random genetic drift. Purifying selection acts by decreasing the frequency of the deleterious allele at a rate that depends upon its selective disadvantage, s, where $s < 0$. Allele frequencies can also change because of random genetic drift, although this change is nondirectional. The rate of change in allele frequencies due to genetic drift is proportional to $1/N_e$, where N_e is the effective population size. Thus, whether a deleterious mutation is quickly lost, or whether it gets the same chance of becoming fixed that a neutral mutation has, is determined by its selective disadvantage relative to the effective size of the population into which it is introduced.

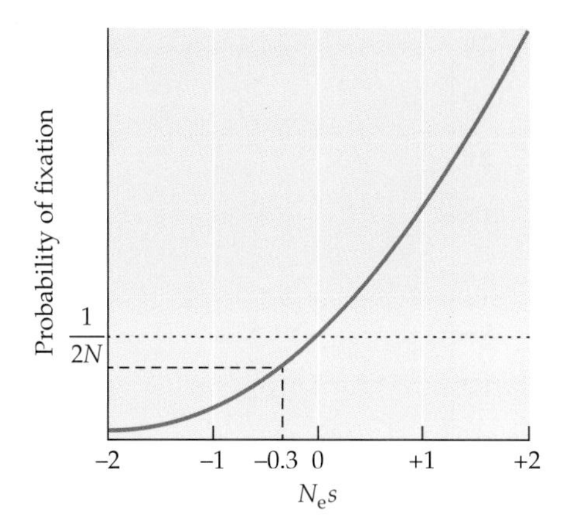

Figure 2.13 The fixation probability of a mutant as a function of the product of the effective population size and the selection coefficient relative to the completely neutral case. The probability of fixation of a strictly neutral mutation in a diploid population is equal to its initial frequency, $1/(2N)$ (dotted line). A mutation with $N_e s = -0.3$ (dashed line) has a fixation probability that is ~86% of that of a strictly neutral allele. (Modified from Yi 2006.)

In **Figure 2.13** we see that the fixation probability is a continuous monotone function of $N_e s$. Because of the dependency on population size, a mutation with the same s can have very different fixation probabilities in different populations. A deleterious mutation may be selected against in a very large population, whereas in a population with a small effective size, it may behave in a manner that is indistinguishable from a strictly neutral mutation. The same applies to advantageous mutations.

To quantify the joint effect of natural selection and genetic drift on a new deleterious mutation, it is convenient to consider its fixation probability relative to that of a neutral allele, which can be computed as $4N_e s / \left(1 - e^{-4N_e s}\right)$. For example, if a mutation with a slight selective disadvantage ($s = -10^{-5}$) arises in a population with $N_e = 1{,}000{,}000$, its probability of becoming fixed relative to a neutral mutation is approximately 10^{-16}, or practically no chance at all. In other words, it is almost impossible for a deleterious mutation to spread if $N_e s \gg 1$. However, if a mutation with the same selective disadvantage occurs in a population with $N_e = 30{,}000$ (so that $N_e s = -0.3$), the fixation probability is as much as 86% of that of a neutral allele (Figure 2.13). In other words, despite its selective disadvantage, such a mutation will behave like a neutral mutation.

The key point here is that a mutation can be deleterious and strongly selected against in one population, but in another, smaller population it will behave like a neutral mutation that may reach fixation via random genetic drift. If $|4N_e s| \ll 1$, drift will dominate the dynamics of allele frequency changes, and the probability of fixations of a deleterious allele can be approximated by its initial frequency, i.e., $1/(2N_e)$. If, on the other hand, $|4N_e s| \gg 1$, selection will dominate, and the probability of fixation of a deleterious allele will asymptotically approach zero.

In **Figure 2.14** we show the relationship between the rate of substitution and the level of nucleotide diversity, on the one hand, and $N_e s$, on the other. One interesting observation is that slightly deleterious alleles can make substantial contributions to genetic variation despite the fact that their probability of fixation is quite low. The reason is that when their frequencies are low, slightly deleterious alleles mostly reside in heterozygotes and selection against them is even weaker, so random drift predominates in their population dynamics. Selection against deleterious alleles becomes more effective if the slightly deleterious alleles attain intermediate frequencies. Thus, while their contribution to genetic variation is considerable, the probability of their fixation is very low.

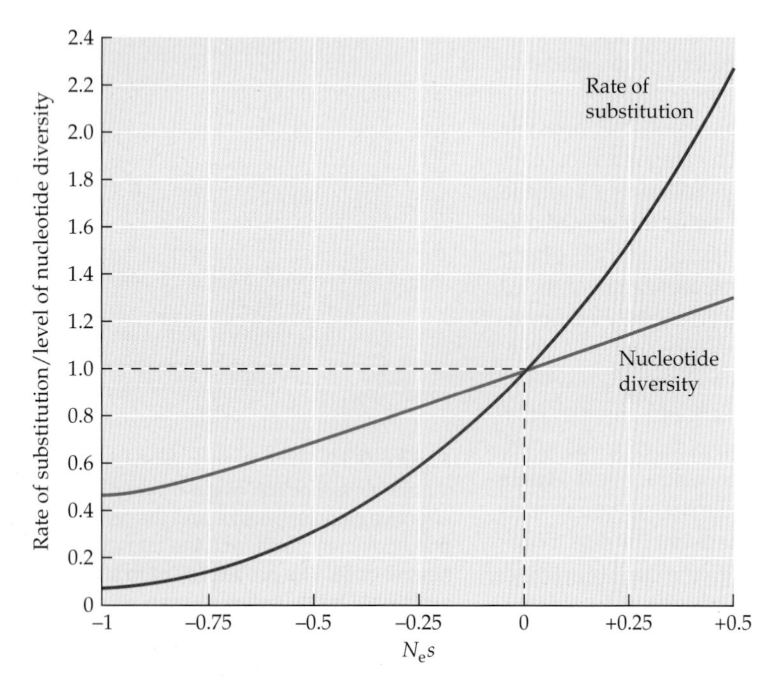

Figure 2.14 The effect of weak selection ($-1 < N_es < 0.5$) on the rate of substitution and on levels of nucleotide diversity relative to the strictly neutral case. The *y*-axis represents the values of the two variables relative to the strictly neutral case, so when $N_es = 0$, both curves pass through 1.0 on the *y*-axis. Note that for positive selection, the effect on the rate of substitution is much greater than the impact of selection on nucleotide diversity. When mutations are deleterious, the rate of substitution falls rapidly to near zero while the population can still retain appreciable levels of genetic variability. (Modified from Hartl and Clark 2007.)

Second-Order Selection

Can selection discriminate among genotypes that do not differ in fitness from one another? Can selection operate on an allele that does not alter the fitness of the individual that carries it but has the potential to affect the fitness of the future descendants of this individual? In other words, can selection be anticipatory? Raising such questions may sound abstruse, yet under certain circumstances such questions are not as nonsensical as they seem. Selection for or against an allele that does not result in an increase in the fitness of its carriers but has the potential to increase the fitness of the carriers' descendants is called **second-order selection**. Second-order selection is usually invoked to explain the evolution of components of the evolutionary process, such as mutation rates, recombination rates, mutational robustness, **evolvability** (the propensity to evolve), and **adaptability** (the propensity to adapt). In **Figure 2.15** we illustrate the concept of second-order selection through the effects of mutation rate and mutational robustness on the evolution of populations.

On theoretical grounds, the strength of second-order selection (s) has been shown to be at most of the same magnitude as the deleterious mutation rate per generation (μ_{del}), that is,

$$s = \mu_{del} \quad (2.37)$$

(Kimura 1967; Proulx and Phillips 2005). We note that the selection coefficient in second-order selection lumps all deleterious mutations together and is independent of the magnitude of the deleterious effect on fitness.

Determining the deleterious mutation rate is not as easy as determining the neutral mutation rate (Chapter 4). There are, however, several methods for estimating this rate (Baer et al. 2007), of which the most straightforward involves mutation accumulation experiments (Chapter 13). In mutation accumulation experiments, only one or a few individuals are used to generate a new generation of individuals, thereby ensuring that the effective population size remains close to the absolute possible minimum. In such populations, selection is weak and ineffective, and with the exception of lethal mutations, the fate of all other mutations is controlled mainly by chance so that they have an equal probability of being transmitted to the next generation. By using a setup similar to the one described above, Kibota and Lynch (1996) estimated the rate of deleterious mutation in *Escherichia coli* to be 0.002 mutations per genome per generation.

The evolution of mutation rates

It has been known for more than a century that the vast majority of nonneutral mutations are deleterious (Morgan 1903). This fact led Alfred Sturtevant to raise the question, "Why does the mutation rate not become reduced to zero?" (Sturtevant 1937). His somewhat unsatisfying answer was that such a reduction was not possible because of the "nature" of the genes. Sturtevant, however, noticed that considerable variation in mutation rates exists within and among species, and this observation led him to conclude that mutation rates are evolvable.

We should note here that a zero mutation rate would lead to extinction. That is, without genetic variation, there can be no evolution, so a nonzero mutation rate is essential for populations to adapt to the ever-changing environmental circumstances

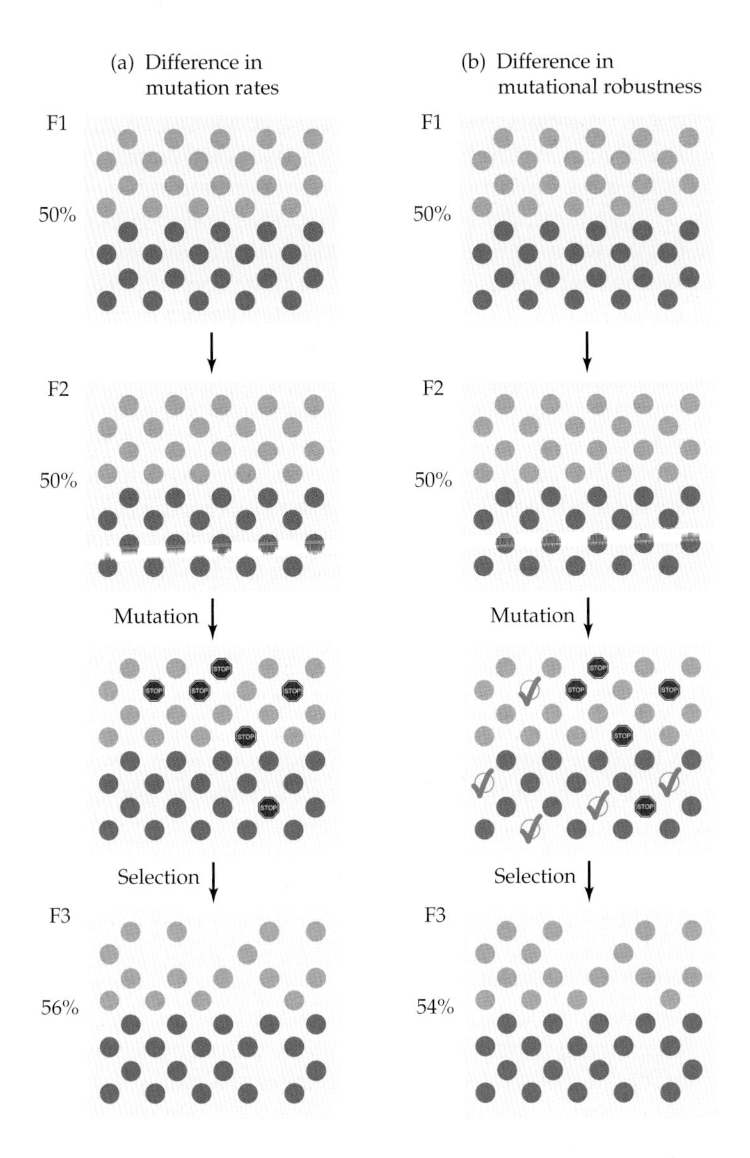

Figure 2.15 Second-order selection. (a) The evolution of a population in which 50% of the individuals in the first generation (F1) have a high mutation rate (orange circles) and 50% have a low mutation rate (blue circles). The two groups of individuals do not differ from each other in fitness, and hence, in the absence of random genetic drift, the two genotypes are expected to maintain their initial frequencies in the next generation (F2). The number of mutants, however, is expected to differ between the two groups, and since most mutations are deleterious (red stop signs), the population with the higher mutation rate is expected to give rise to more low-fitness individuals than the population with the low mutation rate. As a result, the frequency of low-mutation-rate individuals in the general population is expected to increase at the expense of high-mutation-rate individuals. In this illustration, we assumed for simplicity that all deleterious mutations are lethal. Percentages denote the frequencies of individuals with low mutation rates. (b) The evolution of a population in which 50% of the individuals in the first generation (F1) have a low degree of mutational robustness (orange circles) and 50% have a high degree of robustness (blue circles). As in the population in (a), the two genotypes do not differ from each other in their fitness, and hence selection will not affect the allele frequencies in the next generation (F2). The number of mutants is expected to be equal in both groups; however, the proportion of individuals carrying deleterious mutations (stop signs) will be higher among those with low mutational robustness than among individuals with high mutational robustness. Conversely, neutral mutants (check marks) will be more common in the high mutational robustness group. As a consequence, the frequency of high mutational robustness individuals in F3 is expected to increase at the expense of individuals with low mutational robustness. Percentages denote the frequencies of high mutational robustness individuals.

to which every species is exposed throughout its evolutionary history. Nonetheless, it is reasonable to inquire whether or not any observed mutation rate is the lowest mutation rate possible under given circumstances. This question is especially relevant given that in some microbial systems, the mutation rate could be lowered experimentally (e.g., Quiñones and Piechocki 1985; Loh et al. 2010), so the existing mutation rates were by no means the lowest possible.

Kimura (1967) was of the opinion that the mutation rate cannot be substantially reduced, because of a **cost of fidelity**. In other words, he envisioned the mutation rate in a lineage to be a compromise between the benefits of complete fidelity in the replication of the genetic material and the cost of achieving such fidelity. Drake (1991) suggested that the rate of mutation "is likely to be determined by deep general forces, perhaps by a balance between the usually deleterious effects of mutation and the physiological costs of further reducing mutation rates." The implicit assumption was that organisms under strong selection for rapid replication cannot maximize the fidelity of DNA replication without limiting the rate of DNA synthesis. Furió et al. (2005) studied the cost of fidelity in the vesicular stomatitis RNA virus and discovered that there is indeed a cost of replication fidelity. That is, only by slowing down the process of reverse transcription—the means by which the virus reproduces—can a virus achieve increased fidelity in replication. That trade-off between replication fidelity and replication speed was also observed in other taxa (Martincorena 2012).

An alternative hypothesis has been put forward by Lynch (2008, 2010, 2011). The impetus for Lynch's hypothesis was the observation that prokaryotes have lower mutation rates per generation than unicellular eukaryotic organisms, which in turn have lower rates than invertebrates, which in turn have lower rates than vertebrates. Lynch (2010) raised the possibility that the lower bound on the mutation rate is not set by physiological or biochemical limitations, but by the inability of selection to operate on slightly advantageous mutations when the effective population size is low.

For simplicity, let us assume that there are two types of mutations that can affect the mutation rate: **mutators** increase the mutation rate; **antimutators** decrease it. Here, we only concern ourselves with antimutators. Since selection on mutation rate is a second-order selection, then the magnitude of selection is at most of the same magnitude as the deleterious mutation rate, μ_{del}. When an antimutator arises, the deleterious mutation rate per generation, U_{del}, decreases in value. This in turn means that the coefficient of selection decreases too. In other words, any lowering of the mutation rate will constitute a slightly advantageous mutation or even an effectively neutral mutation. Under this model, in sexually reproducing populations, selection will be effective only if $2N_eU_{del} \gg 1$, where N_e is the effective population size. Positive selection on antimutators will be influenced by positive selection only in species with huge effective population sizes and by random genetic drift in species with smaller effective population sizes. Moreover, positive selection is expected to decrease in intensity as additional antimutators will lower mutation rates even further.

Lynch (2006) estimated the average effective population sizes to be approximately 10^8, 10^7, 10^6, and 10^4 in free-living prokaryotes, free-living unicellular eukaryotes, invertebrates, and vertebrates, respectively. In plants, the average effective population size was approximately 10^6 for annual species and 10^4 in trees. Selection on slightly advantageous mutations would, for instance, explain why the observed pattern of mutation rates per generation is lowest in free-living bacteria, which have very large populations sizes, rather than in vertebrates, which have very small effective population sizes.

The evolution of mutational robustness

Mutational robustness is the extent to which the function, the phenotype, and most importantly the fitness of a genotype remain unchanged in the face of mutations. In a genotype with a high mutational robustness, a smaller fraction of all mutations occurring in it will be deleterious than in a genotype with a low mutational robustness (Figure 2.15). Because a smaller fraction of all mutations will be deleterious, a larger fraction will be neutral.

Mutational robustness can be empirically measured by introducing mutations and determining the proportion of mutants that retain the same phenotype, function, or fitness. Proteins thus investigated have shown a relatively constant degree of mutational robustness. That is, roughly two-thirds of mutations in protein-coding regions are neutral. Mutational robustness at the level of the genome varies widely among taxa. For example, more than 95% of point mutations in *C. elegans* have no detectable fitness effects, while the proportion of nonlethal (deleterious and neutral) mutations in the single-stranded DNA virus ΦX174 turned out to be merely 20%.

Can selection increase mutational robustness? We note that increases in mutational robustness will confer no fitness advantage or disadvantage on the organism in which these changes occur.

As mentioned previously, the strength of selection for mutational robustness has been shown to be at most of the same magnitude as the deleterious mutation rate, μ_{del}. For a DNA, RNA, or protein molecule, the deleterious mutation rate is given by

$$\mu_{del} = \mu L P_{del} \tag{2.38}$$

where μ is the mutation rate per site per generation, L is the length of the sequence, and P_{del} is the probability that a mutation is deleterious. (Note that under the assumption that advantageous mutations occur only very rarely, $1 - P_{del}$ is a measure of neutrality or mutational robustness.) Let us consider, for example, the human gene encoding the enzyme 3-methyladenine DNA glycosylase, which has a length of 894 bp and a P_{del} value of 26% (Price et al. 2011). Assuming that $\mu = 2.5 \times 10^8$ mutations per nucleotide per generation, μ_{del} is estimated to be approximately 5.8×10^{-6}. Thus, selection for mutational robustness in this fairly typical gene is expected to be very weak.

Given that the rate of mutation is very low and the deleterious mutation rate is even lower, the main factor determining the extent to which mutational robustness will respond to direct selection is the effective population size, N_e. For example, a diploid population is expected to respond to selection for mutational robustness provided that it obeys the condition $2N_e\mu_{del} \gg 1$ (Wright 1931; Kimura 1968b; Li 1978). Therefore, according to theory, mutational robustness should only evolve under direct selection in taxa with high values of $N_e\mu_{del}$, such as those found in certain RNA viruses, a few prokaryotes, and extremely large artificial populations generated by computer simulations. In agreement with this prediction, experimental evidence for evolution of mutational robustness under direct selection has so far only been observed in a single RNA virus (Sanjuán et al. 2004).

Can mutational robustness evolve in eukaryotes? It can, but not as a consequence of direct selection. It can evolve as a by-product of selection for another trait with which mutational robustness is positively correlated, say, the thermodynamic stability of an RNA molecule or the binding affinity of an enzyme for its ligand.

Violations of Mendel's Laws of Inheritance

In previous sections, we assumed that all alleles in the population are passed on from one generation to the next in a manner consistent with Gregor Mendel's two laws of inheritance: the law of segregation and the law of independent assortment (Bateson 1909). The **law of segregation** states that every diploid individual contains two alleles at each locus and that the probability of a haploid gamete inheriting one allele is equal to the probability of it inheriting the other. Mendel's **law of independent assortment** states that separate alleles at separate loci are passed from parents to offspring independently of one another. That is, the four possible combinations of alleles at these two loci are passed onto the offspring with equal probabilities. More precisely, the law states that during gamete formation, the alleles at one locus assort independently of the alleles at another locus. Below we describe violations of the law of segregation through transmission ratio distortion and violations of the law of independent assortment through genetic linkage.

Transmission Ratio Distortion

According to Mendelian rules, the two parental alleles at a locus should have the same probability of being segregated into the haploid gametes. Indeed, most alleles follow a "fair" pattern of Mendelian segregation, resulting in each of the homologous alleles being found in approximately 50% of the gametes. However, there exist many mechanisms that can subvert the "fairness" of the Mendelian process and lead to an unequal transmission of parental alleles from heterozygous parents to their offspring (Crow 1979). In the absence of knowledge of the exact mechanisms involved, the most appropriate description of such phenomena is **transmission ratio distortion** (Dunn and Bennett 1968).

Genetic elements responsible for distorting transmission ratios are remarkably diverse—some exist in multiple locations in the genome, others at unique sites; some are nuclear genes, while others reside in organelle or symbiont genomes. Hurst and Werren (2001) subdivided such elements into five types according to the mechanism by which each increases its frequency in the population: (1) segregation distorters, (2) postsegregation distorters, (3) converting elements, (4) sex allocation distorters, and (5) autonomous replicating elements. In addition to these categories, one should add a sixth category for instances of transmission ratio distortion whose causes are not known yet.

Segregation distortion

The first category in Hurst and Werren's (2001) scheme is segregation distortion. Rhoades (1942), who worked on maize, was the first to describe one such pattern of unequal transmission of alleles, which he called **preferential segregation**. Seventeen years later, Sandler et al. (1959) described a similar occurrence in *Drosophila* and referred to the phenomenon as **segregation distortion**—a term that has since largely replaced Rhoades's nomenclature.

Segregation distortion can occur through several mechanisms. One mechanism, which has so far been found only in the genus *Drosophila,* involves an interaction between two linked loci, a **driver** called *Segregation distorter* (*Sd*) and a **target** called *Responder* (*Rsp*). At the *Segregation distorter* locus, there are two alleles. The Sd^+ allele consists of a single gene that encodes an intact protein called RanGAP (596 amino acids). This protein is involved in the transport of other proteins from the cytosol to the nucleus. The *Sd* allele consists of two protein-coding genes: one that produces RanGAP and one that originated as a partial duplication of this RanGAP-coding gene and produces a truncated protein called Sd-RanGAP (363 amino acids). Sd-RanGAP retains the same enzyme activity as RanGAP but its activity is mislocalized to the nucleus (Larracuente and Presgraves 2012).

The *Responder* locus contains a large block of tandemly repeated sequences. Each *Rsp* repeat unit is organized as a dimer with "left" and "right" 120 bp AT-rich repeats. An XbaI restriction site (TCTAGA) separates the dimeric units from each other. The sensitivity to segregation distortion correlates with the number of repeats in the *Rsp* locus: supersensitive Rsp^{ss} alleles have ~2,500 copies, sensitive Rsp^{s} alleles have ~700 copies, semi-sensitive Rsp^{semi} alleles have ~300 copies, insensitive Rsp^{i} alleles associated with Sd^+ have 100–200 copies, and Rsp^{i} alleles associated with *Sd* have less than 20 copies of the dimeric repeat.

The truncated Sd-RanGAP protein interferes to varying degrees with the normal processing of sperm bearing sensitive *Rsp* alleles. The *Sd* and Rsp^{i} alleles are often found coupled together. If one chromosome carries *Sd* and Rsp^{i} and the homologous chromosome carries a sensitive Rsp^{s} allele, then the vast majority of the sperm cells that bear Rsp^{s} will degenerate, and the chromosomes that bear *Sd* and Rsp^{i} will be inherited by up to 99% of all the progeny. Thus, the *Sd* allele will cause a deviation from the normal segregation pattern toward a pattern favoring itself.

Since *Sd* frequencies of 3–4% are sufficient to cause the fixation of Rsp^{i} alleles (Charlesworth and Hartl 1978; Crow 1979), an obvious question to ask is why do sensitive Rsp^{s} alleles persist in natural populations. In a competition experiment involving a mixed population consisting of flies with 700 copies of the *Rsp* repeat and flies with only 20 copies, it was observed that the change in the frequency of the 20-repeat alleles depended on the presence or absence of the *Sd* allele (Wu et al. 1989). In the presence of the *Sd* allele, the 20-repeat alleles increased in frequency. In the absence of the *Sd* allele, flies that carried 700 copies had a higher fitness than flies with only 20 copies. Thus, the *Sd* allele favors an allele that is otherwise inferior in fitness to the one it disfavors.

Segregation distortion complexes are often found in inverted regions of chromosomes. This is probably because these systems frequently involve two loci. Inversions restrict recombination and hence protect against the decoupling of the *Segregation distorter* and *Responder* alleles by recombination, which may lead to *Sd* and Rsp^{s} being located on the same "suicidal" chromosome.

Segregation distorters are difficult to observe unless detectable genetic markers are present and unless the driving element exhibits polymorphism within the species. Segregation distorters are most easily detected when they occur on the sex chromosomes, because a sex-ratio bias is observed as a result of an excess of gametes with either the X chromosome (X drive against Y) or the Y chromosome (Y drive against X). Segregation distortion systems involving sex chromosomes are referred to as **sex-ratio distorters**. Sex-ratio distorters have been observed in various organisms, including flies, mammals, and plants, and are likely to be widespread in nature, although the incidence of sex-ratio distorters in non-model organisms is not known.

Supernumerary **B chromosomes** represent another class of segregation distorters. These chromosomes, which are widespread in eukaryotes, are not essential for viability and can be either absent or present in one or more copies per individual. They spread through populations and persist by virtue of their capacity to be inherited by greater than 50% of progeny as a result of preferential segregation at meiosis, as well as through accumulation of mitotic events before gamete formation (Beukeboom 1994).

Other mechanisms of segregation distortion use the asymmetry of meiosis in females, in which the diploid oogonium—the diploid germ cell precursor of female gametes—gives rise to a single functional oocyte, which may subsequently take part in reproduction, as well as three polar bodies that represent developmental and evolutionary dead ends. Segregation distortion occurs if one allele ends up preferentially (i.e., with a probability of more than 0.5) in the oocyte whereas its homolog ends up preferentially in the polar bodies. Because this type of segregation distortion occurs during meiosis, Sandler and Novitski (1957) referred to this process as **meiotic drive**. In time, however, this term became almost synonymous with "segregation distortion." As a consequence, in contemporary genetic literature, what used to be called meiotic drive is sometimes referred to as **true meiotic drive** (e.g., Brandvain and Coop 2012).

Segregation distortion occurs in many taxa (Jaenike 1996; Kozielska et al. 2010). Segregation distortion is advantageous at the gene level, as distorter alleles have a transmission advantage and their frequency in the population will increase. Many distorters in nature show almost complete distortion when unsuppressed (a distorter allele may be present in more than 90% of functional gametes). However, considerable variation exists among populations and among different distorters, with effective distortion rates ranging from just above 0.5 to almost 1.

We note, however, that the presence of a segregation distortion allele is usually not neutral with respect to individual fitness. In some well-studied cases, homozygosity for a segregation distorter allele was found to cause sterility in males or even lethality in both sexes (e.g., Burt and Trivers 2006). In particular, sex segregation distorters may lead to extremely lopsided sex ratios in the population to such an extent that the population may be put at risk of becoming extinct (Jaenike 2001).

Postsegregation distortion

Postsegregation distorters are genomic elements that reduce the frequency of noncarrier individuals after fertilization and the commencement of development. Several of these elements act by killing individuals that have not received the selfish element. In the flour beetle, *Tribolium castaneum*, the *Medea* locus (*m*aternal-*e*ffect *d*ominant *e*mbryonic *a*rrest, a clever acronym referencing a character in Greek mythology who slew her own children) involves a maternal effect allele that kills progeny that do not inherit the allele (Beeman et al. 1992). Cytoplasmic incompatibility, induced by the cytoplasmically inherited bacterium *Wolbachia* (which is widespread in insects, arachnids, crustaceans, and nematodes), also involves the killing of uninfected zygotes, but in this case the effect has a paternal rather than a maternal component (e.g., Stouthamer et al. 1999).

Converting elements

Converting elements induce biased gene conversion (Chapter 8) and can also cause transmission ratio distortions. One example of these elements is a class of genes encoding homing endonucleases (Gimble and Thorner 1992). Found in both organelle and nuclear genomes, these elements encode an enzyme that introduces a double-stranded break at recognition motifs. The break is not repaired by direct re-ligation but by using the sequence that contains the homing endonuclease gene as a template. The end result is a conversion of the target sequence to one that contains the converting element. The repair also splits the recognition motif, thus preventing future self-cleavage. Homing endonuclease sequences, therefore, are overrepresented among the gametes of heterozygous individuals and may increase in frequency in the population, often to fixation.

Sex allocation distortion

Nuclear sequences are transmitted equally through male and female gametes, and selection in panmictic populations favors equality of allocation to sons and daughters. By contrast, cytoplasmic sequences, that is, genes encoded by organelles or endosymbionts, are often inherited uniparentally, most often through female gametes only. Selection on cytoplasmic sequences favors the variants that increase allocation to the sex that can transmit them (females) over the sex that cannot (Cosmides and Tooby 1981). Such cytoplasmic sequences are referred to as **sex allocation distorters**. Examples of sex allocation distorters include mitochondria that induce cytoplasmic male sterility in plants, cytoplasmic microorganisms that feminize hosts, male-killing microorganisms, and parthenogenesis-inducing microorganisms (Hurst 1993).

Autonomous replicating elements

Autonomous replicating elements, which include most transposable and mobile elements, have the ability to multiply within the genome and to occupy new positions in the genome (Chapter 9). Early studies assumed that transposable elements had some beneficial function; however, Hickey (1982) discovered that transposable elements can spread through the population even if the excision and insertion events required for their transposition cause harmful mutations. Although transposable elements might rarely induce beneficial mutations, their spread is most parsimoniously explained by their ability to replicate in the genome as "genomic parasites." Genetic elements that multiply within genomes and increase their frequencies within a population are also referred to as selfish genetic elements or selfish DNA (Chapter 11).

Linkage Equilibrium and Disequilibrium

Under Mendel's law of independent assortment the genetic transmission of an allele at one locus was assumed to be independent of the transmission of another allele at a different locus. Under this assumption, we could treat each locus separately. In practice, however, the transmission of an allele at a locus may be dependent on the transmission of alleles at other loci. The most common cause for this lack of independence is **genetic linkage**, i.e., the close physical proximity of two loci on the same chromosome and the finite rate of meiotic recombination in the sequence separating the two loci from each other. In consequence, a realistic treatment of the transmission of alleles from generation to generation must involve multilocus multiallele models.

We restrict the following discussion to the very simple case of a diploid organism with two autosomal loci, A and B, each with two alleles, A_1 and A_2 at locus A, and B_1 and B_2 at locus B. The frequencies of alleles A_1, A_2, B_1, and B_2 are p_1, q_1, p_2, and q_2, respectively, where $p_1 + q_1 = p_2 + q_2 = 1$. **Linkage equilibrium** occurs if the association between the alleles at the two loci is random. In our simple model, the frequencies of the four possible genotypic types—A_1B_1, A_1B_2, A_2B_1, and A_2B_2—at linkage equilibrium will be p_1p_2, p_1q_2, q_1p_2, and q_1q_2, respectively, where $p_1p_2 + p_1q_2 + q_1p_2 + q_1q_2 = 1$.

Let us assume that in a certain population, the observed frequencies of the four possible genotypic types, A_1B_1, A_1B_2, A_2B_1, and A_2B_2, are x_{11}, x_{12}, x_{21}, and x_{22}, respectively, where $x_{11} + x_{12} + x_{21} + x_{22} = 1$. From these observed values, one can easily compute the frequencies of the four alleles, A_1, A_2, B_1, and B_2, as $p_1 = x_{11} + x_{12}$, $q_1 = x_{21} + x_{22}$, $p_2 = x_{11} + x_{21}$, and $q_2 = x_{12} + x_{22}$. The expectation under linkage equilibrium is that $x_{11} = p_1p_2$, $x_{12} = p_1q_2$, $x_{21} = q_1p_2$, and $x_{22} = q_1q_2$. Deviations from these gametic expectations are indicative of **linkage disequilibrium**. In other words, linkage disequilibrium describes a situation in which some combinations of alleles or genetic markers occur significantly more or significantly less frequently in a population than would be expected from the random formation of haplotypes from the allele frequencies at the different loci. The difference between expected and observed frequencies (e.g., $D = x_{11} - p_1q_1$) is called the **linkage disequilibrium parameter** (Lewontin and Kojima 1960). The most common expression of D is

$$D = x_{11}x_{22} - x_{12}x_{21} \tag{2.39}$$

where x_{11} and x_{22} are referred to as the **coupling genotypes**, and x_{12} and x_{21} are referred to as the **repulsion genotypes**.

At its minimum, when $D = 0$, there is no disequilibrium. Unfortunately, the maximum value of D is not universal; it depends on the particular allele frequencies in each population. D is influenced by many factors, including genetic linkage, the rate of recombination, mutation rate, genetic drift, nonrandom mating, and population substructure. In some organisms, such as bacteria, linkage disequilibrium may be very pronounced throughout the entire genome because they reproduce asexually and there is little recombination to break down the linkage disequilibrium.

Let us now examine the dynamics of change in genotype frequencies with time. We start with a population, in which the frequencies of the four genotypic types, A_1B_1, A_1B_2, A_2B_1, and A_2B_2, in the first generation are $x_{11(1)}$, $x_{12(1)}$, $x_{21(1)}$, and $x_{22(1)}$, respectively. The frequencies of the four gametes in the next generation will depend on the **recombination frequency**, r, between loci A and B, which can assume a value ranging from 0 (no recombination) to 0.5 (independent assortment, i.e., half of the gametes will be the products of recombination). If the initial frequency of the genotype A_1B_1 is $x_{11(1)}$, then the frequency of this gamete in the next generation, $x_{11(2)}$, will be

$$x_{11(2)} = (1-r)x_{11(1)} + rp_1p_2 \tag{2.40}$$

Equation 2.40 indicates that in the second generation an A_1B_1 gamete can be either a nonrecombinant descendant of A_1B_1 or the result of a recombination event between A_1B_2 and A_2B_1. The probability of nonrecombination is $1 - r$. Thus, the probability that A_1B_1 is not a recombinant is $(1 - r)x_{11(1)}$. If, on the other hand, A_1B_1 is the product of recombination, then the probability is rp_1p_2. In a similar manner, we can calculate the frequencies of the other three genotypes. Finally, we can look at change in the linkage disequilibrium parameter with time. From Equation 2.40, one can easily deduce that linkage disequilibrium will decay with time due to recombination occurring between the two loci. Of course, linkage disequilibrium does not disappear at once. The temporal dynamics of linkage equilibrium decay will depend on the degree of recombination, i.e., the value of r between the two loci. This change in D with time can be approximated as

$$D_t = e^{-rt}D_0 \tag{2.41}$$

where $D_t = e^{-rt}D_0$ and D_t are the linkage disequilibrium parameters at times 0 and t, respectively, and r is the recombination frequency.

The initial D value in **Figure 2.16** is the maximum possible for the case of two loci, each with two alleles whose frequencies are 0.5 each. With no linkage or under independent assortment ($r = 0.5$), most of the disequilibrium is lost in 5 to 6 generations. In contrast, with tight linkage, a substantial proportion of the initial disequilibrium will remain even after many generations. For example, when $r = 0.001$, only approximately 2.5% of the linkage disequilibrium will be lost after 25 generations.

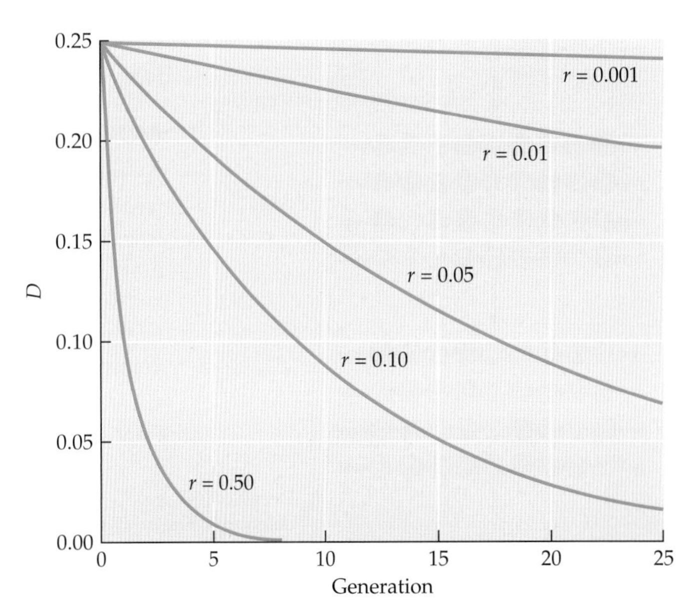

Figure 2.16 Decay of the linkage disequilibrium parameter, D, as a function of time (measured in generations) and the frequency of recombination, r. (Modified from Hedrick 2005.)

Hitchhiking and Selective Sweep

Let us consider a population in which two neutral haplotypes, A_2B_1 and A_2B_2, coexist with frequencies of p_2 and q_2, respectively. Let us further assume that an advantageous mutation, A_1, arises on the haplotype carrying the B_1 allele. Of course, the advantageous mutation could have arisen on the haplotype carrying the B_2 allele. The fact that it occurred where it occurred is entirely accidental. Without the advantageous allele arising at locus A,

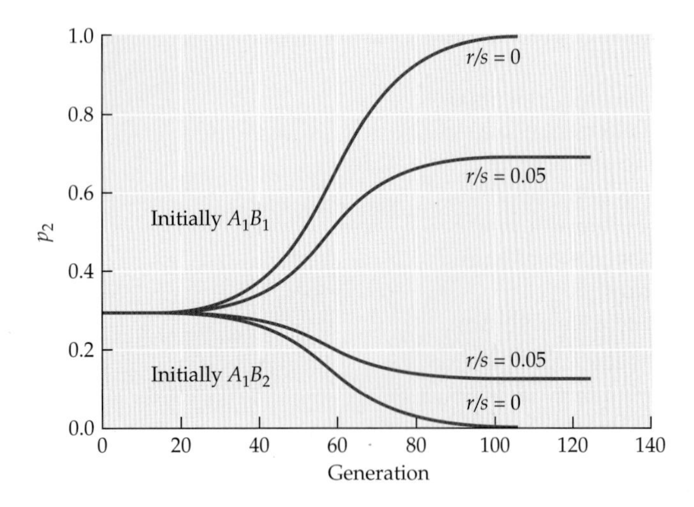

Figure 2.17 The frequency of the B_2 allele under different hitchhiking scenarios. For the upper two curves, the advantageous A_1 mutation was initially linked to the B_1 allele; in the bottom two, it was linked to the B_2 allele. The selective advantage was $s = 0.2$ for all scenarios. (Modified from Gillespie 2004.)

the probability of fixation for alleles B_1 and B_2 would have been p_2 and q_2, respectively. The linkage to the advantageous allele A_1, however, alters these expectations, because on its way to fixation, the advantageous mutation A_1 will carry along the linked B_1 allele and will ultimately render the population monomorphic at locus B. The process through which an advantageous mutation reduces or eliminates genetic variation at genetically linked sites is called **selective sweep**. The process by which a neutral or even deleterious allele that is sufficiently tightly linked to the positively selected allele increases its frequency or is swept to fixation is called **genetic hitchhiking**. With hitchhiking, only the initial conditions are stochastic; once the process of fixation begins, selection takes over. Because of the stochastic component of hitchhiking, the process is sometimes referred to as **genetic draft** (Gillespie 2000).

Figure 2.17 gives some examples of trajectories for the B_1 allele as a function of the ratio between the recombination frequency (r) between loci A and B and the selective advantage (s) of allele A_1. If the selection coefficient is large and the recombination rate is small, then the neutral allele B_1 will be quickly fixed on the coattails of the advantageous allele A_1 before the linkage is disrupted. If, on the other hand, the selection advantage is not very large and recombination is frequent, the linkage between B_1 and A_1 will be disrupted before the fixation of B_1. Because linkage disequilibrium decays with time, the existence of sequence disequilibrium may only be used to detect recent selective sweeps.

Through succeeding generations, the descendants of the positively selected allele and nearby hitchhiking alleles on the same haplotype will become more and more common in the population. However, the entire haplotype is not passed down from generation to generation as a unit. Rather, because of recombination, the segment of the haplotype that includes the selected allele is slowly reduced in size (**Figure 2.18**).

Molecular signatures of selective sweeps

As mentioned previously, a positively selected allele will not increase its frequency in the population, but through hitchhiking and selective sweep, it will increase the frequencies of linked alleles. Selective sweeps leave several characteristic molecular signatures in the population. First, a selective sweep will eliminate nucleotide variation in the region of the genome close to the beneficial allele, resulting in a diminished degree of nucleotide diversity relative to other regions in the genome that have not experienced a

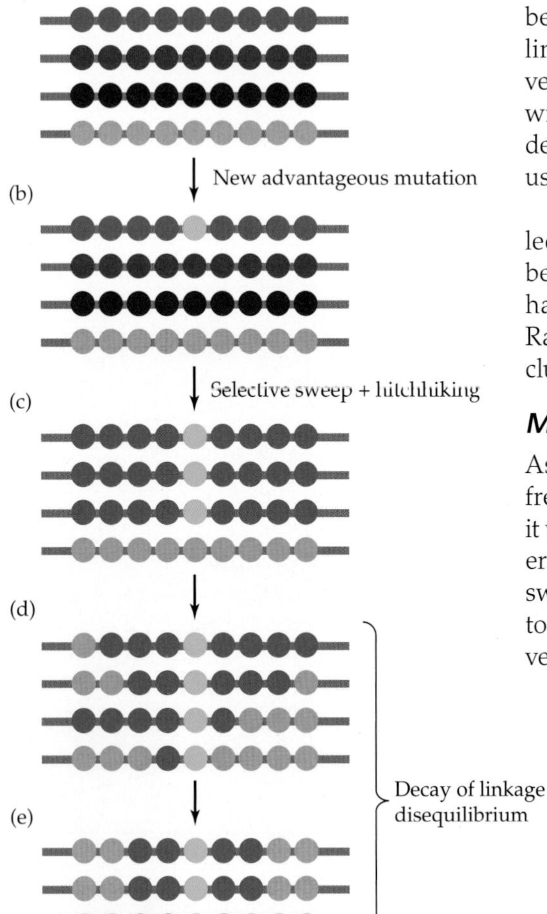

Figure 2.18 Selective sweep, hitchhiking, and decay of linkage disequilibrium. (a) The initial population consists of several different haplotypes (different color circles), each of which is comprised of linked loci. (b) An advantageous mutation (green) arises in the blue haplotype. (c) As a consequence of its selective advantage, the frequency of the advantageous allele increases in the population. Through hitchhiking, neutral alleles that are genetically linked to the advantageous allele will also increase in frequency. (d–e) Because of recombination, the segment of the original (blue) haplotype that includes the selected allele is slowly reduced in size.

selective sweep. Second, a selective sweep will result in an excess of high-frequency derived alleles. Third, an allele under positive selection is expected to increase in frequency so rapidly that long-range associations with neighboring loci polymorphisms (i.e., the **long-range haplotype**) do not have sufficient time to be disrupted by recombination. Thus, a selective sweep will lead to creation of linkage disequilibrium over large swaths of DNA around the positively selected variant. Finally, the difference in allele frequencies among isolated populations will be larger if one or more populations experience positive selection than between populations evolving under selective neutrality. We note, however, that the molecular telltale signs of positive selection will rapidly decay in time, mainly through recombination, but also through mutation. For example, the length of the sequence that exhibits linkage disequilibrium will become shorter and shorter, and as a consequence, linkage disequilibrium can only provide evidence for recent episodes of positive selection (Ohta and Kimura 1975). A selective sweep takes approximately $2\ln(2N_e/s)$ generations, where N_e is the effective population size and s is the selection coefficient (Stephan et al. 1992). In addition, the signature of positive selection will be identifiable for an additional amount of time, depending on the rates of mutation and recombination in the relevant region.

Using the estimated human effective population size of 10,000 (Sabeti et al. 2002) and assuming the generation time is 25 years, sweeps with strong selection coefficients (e.g., $s = 5\%$) will be complete in ~10,000 years. Alleles with a lower selection coefficient will take longer to reach fixation in the population. For example, an allele with $s = 1\%$ will take ~50,000 years to fix in the population. The time to completion of a selective sweep is largely dependent on the strength of selection. Therefore, sweeps that are presently incomplete are either recent sweeps with a strong selection coefficient or old sweeps with a weaker selection coefficient.

A particularly convenient method to detect recent bouts of positive selection is to look for high-frequency haplotypes in the population that exhibit high linkage disequilibrium over long distances between chromosomal locations. The observation of unusually frequent, long haplotypes within a population is indicative of selective sweeps around positively selected alleles (**Figure 2.19**). One such example concerns the phenomenon of industrial melanism in the peppered moth, *Biston betularia.* Industrial melanism is widely recognized to have occurred very recently and very rapidly. That is, the black form (*carbonaria*) increased in frequency at the expense of the light form (*typica*) from less than 1% to close to 100% in about 50 years (Haldane 1956). Indeed, at the genomic level, *carbonaria* has been shown to have a single mutational origin, and the genomic sequence around the *carbonaria* locus carries the signature of very strong and very recent selection (van't Hof et al. 2011).

There are many approaches to detecting recent bouts of positive selection based on different combinations of haplotype length and variation within and between populations (e.g., Sabeti et al. 2002, 2006; Wang et al. 2004; Zhang et al. 2006; Tang et al. 2007; see reviews by Nielsen 2005; Biswas and Akey 2006). Unfortunately, many methods for identifying positive selection by using linkage disequilibrium tend, under certain circumstances, to overestimate the number of sites subject to positive selection (e.g., Jensen et al. 2007). That said, in silico genome-wide searches for segments harboring signals of positive selection (e.g., Grossman et al. 2013) are useful in the sense that they yield candidate loci for further in vitro and in vivo studies.

		Haplotype frequency	
		Low	High
Haplotype length	Short	Old neutral or deleterious allele	Old neutral or advantageous allele
	Long	New neutral, deleterious, or advantageous allele	New advantageous allele

Figure 2.19 The use of haplotype length and haplotype frequency to detect positive selection. A new mutation, regardless of its effect on fitness, is always associated with the long haplotype on which it occurred. If the mutation is neutral, by the time it increases its frequency or reaches fixation, the linked haplotype will become shorter through recombination. Because of the rapidity with which they increase their frequency in the population, recent advantageous mutations will be associated with high-frequency long haplotypes (lower right quadrant).

The evolution of lactase persistence in Africa and Europe

The digestion of the disaccharide lactose, the primary sugar present in milk, into its monosaccharide constituents, glucose and galactose, is catalyzed by a small-intestine enzyme called lactase-phlorizin hydrolase (LPH), or lactase for short. In most mammals (including humans), levels of lactase decline rapidly after weaning, and adults are not able to digest lactose. In humans, individuals who are unable to digest lactose as adults are commonly referred to as lactose intolerant, and the trait is referred to as lactase nonpersistence. Digestion of fresh milk in individuals who are lactose intolerant can result in severe abdominal distress, including diarrhea, which for most of human history resulted in severe health consequences. In populations in which the only source of milk is the mother, lactase nonpersistence is a selectively advantageous trait, since breastfeeding is a potent, albeit imperfect, contraceptive that promotes amenorrhea (absence of menstruation) and delays resumption of ovulation. However, in some populations, a derived genetic trait has appeared, in which the ability to digest lactose is maintained in adults. Such individuals are lactose tolerant due to lactase persistence. This trait is particularly common in populations that have traditionally practiced dairying, i.e., in populations in which milk can be obtained extramaternally.

The highest frequencies of lactase persistence are found in northern European populations (>90% in Swedes and Danes), with decreasing frequencies across southern Europe and the Middle East (~50% in Spanish, French, and Arab populations). Lactase persistence is also common among pastoralist populations from Africa who have a history of drinking fresh milk (~90% in the Tutsi). The lowest frequencies of lactase persistence are found in Native Americans and Pacific Islanders, as well as in most sub-Saharan African and Southeast Asian populations (~1% in Chinese populations). Based on the correlation between the prevalence of the lactase persistence and the cultural practice of dairying, lactose tolerance is considered an example of gene-culture coadaptation in populations that consume milk. The reasons that adult milk consumption is adaptive may include nutritional benefits from milk such as protein, vitamin D, and calcium. Calcium and vitamin D absorption could be important in northern latitude populations, where individuals have less sunlight exposure and may be susceptible to rickets and osteomalacia. In lactose-tolerant African populations, the selective advantage of adult milk consumption could be as a source of water in arid zones.

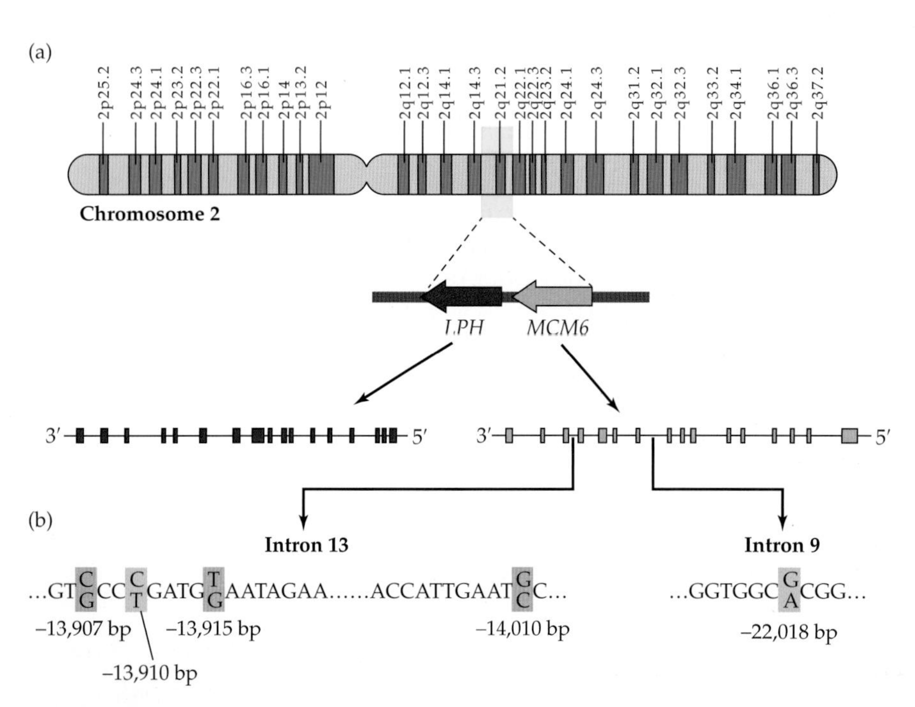

Figure 2.20 Single-nucleotide polymorphisms associated with lactase persistence in European and African populations. (a) The 17-exon 49.3 Kb *LPH* gene is on chromosome 2, located ~3.1 Kb downstream of the 17-exon, 36.2 Kb *MCM6* gene. (b) Five SNPs located in the *MCM6* gene are associated with lactase persistence: −13910 and −22018 in European populations (marked in green), and −13907, −13915, and −14010 in African populations (marked in tan). Haplotypic evidence suggests that the T_{-13910} and C_{-14010} variants have experienced recent strong selection in European and African populations, respectively. (Modified from Ranciaro and Tishkoff 2010.)

Pedigree studies have indicated that lactase persistence is inherited in an autosomal dominant manner. The 17-exon *LPH* gene was mapped to the long arm of chromosome 2 (**Figure 2.20**). Two single-nucleotide polymorphisms (SNPs) upstream of the *LPH* gene were found to be almost perfectly associated with lactase persistence in Finnish and other northern European populations. Interestingly, each of these SNPs occurred in an intron of the upstream *MCM6* gene, which encodes a highly conserved minichromosome maintenance protein that is essential for the initiation of genome replication. In intron 9 of *MCM6*, at a distance of –22018 nucleotides upstream of the *LPH* gene, a G to A substitution was associated with lactase persistence, and a similar association was found with a C to T substitution in intron 13, at a distance of –13910 nucleotides upstream of the *LPH* gene. Further studies have shown that the T_{-13910} variant up-regulates *LPH*-gene expression and that people who carry the T_{-13910} variant have similar haplotypes extending as far as 1 million base pairs away, indicating that this haplotype has been under strong selection within the last 5,000–10,000 years (Bersaglieri et al. 2004). Studies on lactose tolerance in African populations indicated that lactase persistence in these populations had evolved independently of the European populations. Three SNPs upstream of the *LPH* gene were found to be associated with lactase persistence in African populations. All three were located in intron 13 of *MCM6* (C to G at position –13907, T to G at position –13915, and G to C at position –14010). One of these variants, C_{-14010}, turned out to be associated with a very long haplotype, consistent with a selective sweep over the past 7,000 years or so (Tishkoff et al. 2007). Lactase persistence has, thus, in parallel and independently evolved at least twice in geographically distinct human populations.

Background Selection

In the case of strong negative selection on a locus, genetically linked neutral or even advantageous variants will also be removed, producing a decrease in the level of variation surrounding the locus under purifying selection (Charlesworth et al. 1993). This process of purging nondeleterious alleles from the population due to spatial proximity to deleterious alleles is called **background selection**. In a sense, background selection is the opposite of selective sweep. However, because the deleterious mutations driving background selection are removed from the population, and because they can occur randomly in any haplotype, background selection produces no linkage disequilibrium and is, hence, extremely difficult to detect.

Epistasis

In previous sections, we assumed that each locus contributes independently to the fitness of the individual (i.e., that the different loci do not interact with one another in any manner that affects the fitness of the organism), so each locus can be dealt with separately. This is not, however, always the case. **Epistasis** refers to interactions among alleles at different loci resulting in "nonindependent effects" of the different loci. In other words, epistasis occurs when the effects of an allele at one locus are modified by one or several alleles at other loci.

Epistasis may be defined at the fitness level or at the level of the phenotype. We distinguish between **functional epistasis**, in which alleles at different loci produce nonindependent phenotypic effects, and **fitness epistasis**, in which alleles at different loci nonindependently determine the fitness of their carrier, whether or not epistasis is detectable at the level of the phenotype. Here, we are only interested in epistasis in its evolutionary context, i.e., in the relationship between genotype and fitness, without regard to the relationship between genotype and phenotype. The **genetic background effect**, according to which a mutation may have different effects on fitness depending on the genome in which it occurs, may be regarded as a generalized kind of fitness epistasis.

There have been many different uses of the term "epistasis" since its original definition (Bateson 1909), and what is meant by "independent effects," i.e., the fitness expectation in the absence of epistasis, is not defined consistently in the literature. Here, we define independence as **additivity**, i.e., in the absence of epistasis, the combined effect on fitness of two alleles at two loci is equal to the sum of the individual effects. There are many other definitions of epistatic independence in the literature (e.g., Gao et al. 2010). In particular, independence is often defined as **multiplicativity**, i.e., in the absence of epistasis, the combined effect on fitness of two alleles at two loci is expected to be equal to the product of their individual effects (Cordell 2002).

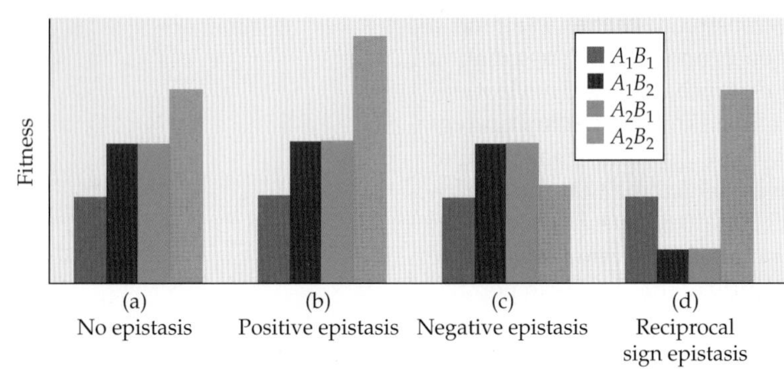

Figure 2.21 Types of interactions between alleles A_1 and A_2 at the A locus and B_1 and B_2 at the B locus. (a) Additive interaction (no epistasis), (b) positive epistasis, (c) negative epistasis, and (d) reciprocal sign epistasis. (Courtesy of Yichen Zheng.)

Let us now deal with the simple case of a haploid organism with two loci, A and B, each segregating two alleles, A_1 and A_2, and B_1 and B_2, respectively. There can be four genotypes: A_1B_1, A_1B_2, A_2B_1, and A_2B_2. Let us assume that A_1B_1 is the reference genotype and its fitness is $w_{A1B1} = 1$. Let us now assume that replacement of allele A_1 by allele A_2 increases the fitness of the carrier by 0.1, and that the replacement of allele B_1 by allele B_2 also increases fitness by 0.1. If we define epistasis as a nonadditive effect on the fitness, then if the fitness of A_1B_2 is $w_{A1B2} = 1.1$, that of A_2B_1 is $w_{A2B1} = 1.1$, and that of A_2B_2 is $w_{A2B2} = 1.2$, then there is no epistasis, because the combined effects of alleles A_2 and B_2 equal the sum of their individual effects (**Figure 2.21a**).

If, on the other hand, the fitness of A_1B_2 is $w_{A1B2} = 1.1$, that of A_2B_1 is $w_{A2B1} = 1.1$, and that of A_2B_2 is $w_{A2B2} = 1.3$, then the two-locus interactions represent a case of **positive epistasis** (**Figure 2.21b**), i.e., the combination of alleles A_2 and B_2 increases fitness more than they would individually. That is,

$$w_{A1B1} + w_{A2B2} > w_{A1B2} + w_{A2B1} \tag{2.42a}$$

and

$$w_{A2B2} > w_{A1B2} + w_{A2B1} - w_{A1B1} \tag{2.42b}$$

Alternatively, if the fitness of A_1B_2 is $w_{A1B2} = 1.1$, that of A_2B_1 is $w_{A2B1} = 1.1$, and that of A_2B_2 is $w_{A2B2} = 1$, then the two-locus interactions represent a case of **negative epistasis** (**Figure 2.21c**), i.e., the combination of alleles A_2 and B_2 increases fitness less than they would individually. In other words,

$$w_{A1B1} + w_{A2B2} < w_{A1B2} + w_{A2B1} \tag{2.43a}$$

and

$$w_{A2B2} < w_{A1B2} + w_{A2B1} - w_{A1B1} \tag{2.43b}$$

Positive and negative epistases are also referred to as **magnitude epistasis** because the combined effect of the two alleles only affects the magnitude of the fitness effect. Another category of epistasis is **sign epistasis**, or **reciprocal sign epistasis** (**Figure 2.21d**). In sign epistasis, the sign of the combined fitness effect is the opposite of that of the individual effects. For example, let us start again with A_1B_1 whose fitness is $w_{A1B1} = 1$. Let us now assume that both A_2 and B_2 have negative effects on fitness, such that the fitness of A_1B_2 is $w_{A1B2} = 0.8$ and that of A_2B_1 is $w_{A2B1} = 0.8$. If the fitness of A_2B_2 is $w_{A2B2} = 1.2$, then the combined effect of A_2 and B_2 on fitness is positive.

The Driving Forces in Evolution

Evolutionary explanations can be broadly classified according to the relative importance assigned to random genetic drift versus various forms of selection. **Mutationist hypotheses** emphasize mutational input and random genetic drift. **Neutralist hypotheses** explain evolutionary phenomena by stressing the effects of mutation, random genetic drift, and purifying selection. **Selectionist hypotheses** give emphasis to the effects of the advantageous and balancing modes of selection as the main driving forces in the evolutionary process.

The neo-Darwinian theory and the neutral mutation hypothesis

Darwin proposed his theory of evolution by natural selection without knowledge of the sources of variation in populations. After Mendel's laws were rediscovered and genetic variation was shown to be generated by mutation, Darwinism and Mendelism were combined into a theory variously referred to as the **synthetic theory of evolution** or **neo-Darwinism**. In this theory, positive Darwinian selection is given the dominant role in shaping the genetic makeup of populations and in the process of gene substitution.

Up to the 1970s, neo-Darwinism dominated evolutionary biology. Factors such as mutation and random genetic drift were thought of as minor contributors at best. This particular brand of neo-Darwinism was called **panselectionism** (Dietrich 1994) or the **Panglossian paradigm** (Gould and Lewontin 1979), an eponymous term derived from Pangloss, the fictional professor of metaphysico-theologo-cosmonigology in Voltaire's *Candide*, who teaches his pupils that they live in the "best of all possible worlds" and that "all is for the best."

According to the selectionist perception of the evolutionary process, gene substitution is the end result of a positive adaptive process whereby a new allele takes over future generations of the population if and only if it improves the fitness of its carriers. Polymorphism, on the other hand, is maintained by some form of balancing selection. Thus, substitution and polymorphism were thought of as two separate phenomena driven by different evolutionary forces. Neo-Darwinian theories maintain that most genetic polymorphism in nature is stable, i.e., the same alleles are maintained at constant frequencies for long periods of evolutionary time.

The late 1960s witnessed a revolution in population genetics. The availability of protein sequence data removed the species boundaries in population genetics studies and for the first time provided adequate empirical data for examining theories pertaining to the process of gene substitution. In 1968, Kimura postulated that the majority of molecular changes in evolution are due to the random fixation of neutral or nearly neutral mutations (Kimura 1968a; see also King and Jukes 1969). This hypothesis, now known as the **neutral theory of molecular evolution**, contends that at the molecular level the majority of evolutionary changes and much of the variability within species are caused neither by positive selection of advantageous alleles nor by balancing selection, but by random genetic drift of mutant alleles that are selectively neutral (or nearly so). Neutrality, in the sense of the theory, does not imply strict equality in fitness for all alleles. It only means that the fate of alleles is determined largely by random genetic drift. In other words, selection may operate, but its intensity is too weak to offset chance effects. For this to be true, the absolute value of the selective advantage or disadvantage of an allele, $|s|$, must be smaller than $1/(2N_e)$, where N_e is the effective population size.

According to the neutral theory, the frequency of alleles is determined largely by stochastic rules, and the picture that we obtain at any given time is merely a transient state representing a temporary frame from an ongoing dynamic process. Consequently, polymorphic loci consist of alleles that either are on their way to fixation or are about to become extinct. Viewed from this perspective, all molecular manifestations that are relevant to the evolutionary process should be regarded as the result of a continuous process of mutational input and a concomitant random extinction or fixation of alleles. Thus, the neutral theory regards substitution and polymorphism as two facets of the same phenomenon. Substitution is a long and gradual process whereby the frequencies of mutant alleles increase or decrease randomly until the alleles are ultimately fixed or lost by chance. At any given time, some loci will possess alleles at frequencies that are neither 0% nor 100%. These are the polymorphic loci. Moreover, most genetic polymorphism in populations is unstable and transient, i.e., allele frequencies fluctuate with time and the alleles are replaced continuously.

Interestingly, the neutral theory, even in its strictest form, does not preclude adaptation. According to Kimura (1983), a population that is free from selection can accumulate many polymorphic neutral alleles. Then, if a change in ecological circumstances occurs, some of the neutral alleles will cease to be neutral and will become deleterious, against which purifying selection will operate. After these alleles are removed, the population will become more adapted to its new circumstances than before. Thus, theoretically at least, adaptive evolution may occur without positive selection.

An important modification to the neutral mutation theory, the **slightly deleterious mutation hypothesis,** was introduced by Ohta (1973). According to this theory, a major role is played by borderline mutations, whose evolutionary behavior depends on population size. These variants behave like deleterious mutations in large populations and like neutral mutations in small populations. Kimura (1979) suggested a role for **slightly advantageous mutations**, i.e., mutations that behave like advantageous mutations in large populations and like neutral mutations in small populations. Currently, the two modifications of the neutral theory are combined into the **nearly neutral mutation theory** (Ohta and Tachida 1990; Ohta 1992).

The heated controversy over the neutral mutation hypothesis during the 1970s and 1980s has had a significant impact on the field of molecular evolution. First, it has led to the general recognition that the effect of random genetic drift cannot be neglected when considering the evolutionary dynamics of molecular changes. Second, the synthesis of molecular biology and population genetics has been greatly strengthened by the introduction of the concept that molecular evolution and genetic polymorphism are but two aspects of the same phenomenon. The controversy has quieted down considerably in the current millennium, as it is now recognized that any adequate theory of evolution at the molecular level should take into account both selection and drift. In fact, without the neutral theory as a null hypothesis, the selectionist paradigm comes perilously close to being "a theory which explains nothing because it explains everything" (Lewontin 1974).

The distribution of fitness effects

At its core, the historical dispute between neutralists and selectionists concerns the distribution of fitness values of mutant alleles. Both theories agree that most new mutations are deleterious and that these mutations are quickly removed from the population, so they contribute neither to the rate of substitution nor to the amount of polymorphism within populations. The difference concerns the relative proportion of neutral mutations among nondeleterious mutations. While selectionists claimed that very few mutations are selectively neutral, neutralists maintained that the majority of nondeleterious mutations are effectively neutral.

Mutations are sometimes divided for convenience into three categories: harmful, neutral, and advantageous. In reality, however, mutations exhibit a continuum of selective effects, from lethal, through strongly and weakly deleterious, to neutral, to mildly or highly adaptive. Here we are interested in the relative frequencies of adaptive mutations—the **distribution of fitness effects**.

The most direct method to study the distribution of fitness effects is to collect spontaneous mutations and then assay their fitness effects. One drawback of this direct method is that it can only identify large effects on fitness, usually only mutations that increase or decrease fitness by more than 1% ($s > |0.01|$). Furthermore, such experiments are time-consuming, so they have been done in only a handful of species that have short generation times, such as viruses, certain bacteria, and yeasts. The results of one such study (Sanjuán et al. 2004) are shown in **Figure 2.22**. In this experiment, 48 mutant clones of the RNA vesicular stomatitis virus (VSV) were created by site-directed mutagenesis. Each clone contained a single point mutation relative to the ancestral strain. The mutations were distributed evenly along the genome. As expected, the vast majority of mutations were lethal (40%) or deleterious but

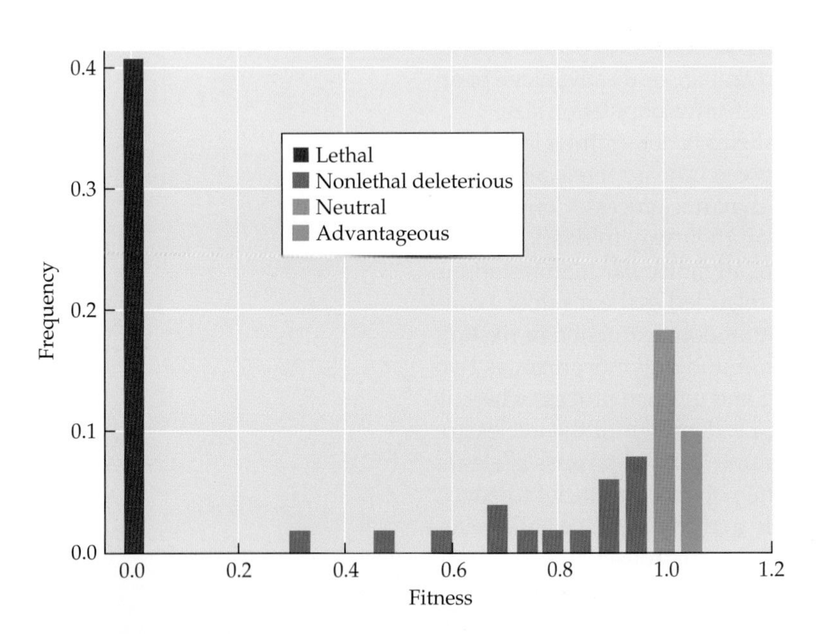

Figure 2.22 Fitness effects of randomly introduced single point mutations in 48 clones of vesicular stomatitis virus (VSV). Random mutations were introduced and the fitnesses of the mutants were compared against the wild-type parental strain. A fitness of 0 (red bar) indicates no progeny were recovered, i.e., the mutation was lethal. A fitness of less than 1 (blue bars) indicates that the mutant was less fit than the wild type, i.e, the mutation was deleterious. A fitness greater than 1 (green bar) indicates that the mutant was more fit than the wild type. (Modified from Sanjuán et al. 2004.)

not lethal (29%). The rest of the mutations were neutral (27%), with only a few mutations being advantageous (4%). Perfeito et al. (2007) suggested that most mutations in *E. coli* have extremely small effects on fitness, while only 0.7% of mutations are beneficial.

Of course, the proportion of neutral mutations depends on population size; what may be neutral in a small population size may be advantageous or deleterious in a large population. Similarly, the fitness effect of a mutation may be affected by genomic location, the specific organism under study, developmental stage, and countless other factors. That said, from published studies so far, one may conclude that the overwhelming majority of mutations are deleterious or neutral, with the proportion of advantageous mutations ranging from 0% to 15% (reviewed in Eyre-Walker and Keightley 2007).

A test of neutrality based on genetic polymorphism

Before describing tests of neutrality, we must first clarify the meaning of "neutrality" in the context of these tests. According to the neutral mutation hypothesis, the "majority of mutations" that have contributed significantly to the genetic variation in natural populations are neutral or nearly neutral. This definition is not sufficiently quantitative, and accordingly for the purposes of the tests, we will use the assumption of **strict neutrality**, i.e., that all mutations that contribute to nucleotide diversity within species and to differences among species are selectively neutral.

Many tests of neutrality have been proposed in the literature; here we will only concern ourselves with tests that use genetic polymorphism data. (In subsequent chapters, we present tests based on rates of substitution, phylogenetic trees, and linkage disequilibrium.) According to the neutral mutation hypothesis, polymorphism is a transient phase of molecular evolution, and the rate of evolution is positively correlated with the level of within-population variation (Kimura and Ohta 1971). One may, therefore, test the neutral mutation hypothesis by comparing the degree of DNA sequence variation within populations with the degree of variation between populations. Many such tests have been developed (e.g., Kreitman and Aguadé 1986; Hudson et al. 1987; Sawyer and Hartl 1992; Innan 2006; Achaz 2008). Below we present a simple method proposed by McDonald and Kreitman (1991).

Consider two samples of protein-coding sequences from species 1 and 2. A nucleotide site in the sequences is said to be polymorphic if it exhibits any variation in one or both species. A site is deemed to represent a fixed difference between the two species if it shows no intraspecific variation within either species but differs between the species. All other sites are monomorphic and are not used in the analysis.

The polymorphic and fixed-difference sites are further divided into two categories, synonymous and nonsynonymous. The McDonald-Kreitman method uses a 2 × 2 contingency table to test the independence of one classification (polymorphic versus fixed) from the other (synonymous versus nonsynonymous). The test is based on the following assumptions: (1) only nonsynonymous mutations may be adaptive, (2) synonymous mutations are always neutral, and (3) a selectively advantageous mutation will be fixed in the population much more rapidly than a neutral mutation, and hence it is less likely to be found in a polymorphic state. Under the neutral mutation hypothesis, the expectation is that the ratio of fixed nonsynonymous differences (D_n) to fixed synonymous differences (D_s) will be the same as the ratio of nonsynonymous polymorphisms (P_n) to synonymous polymorphisms (P_s).

$$\frac{D_n}{D_s} = \frac{P_n}{P_s} \tag{2.44}$$

The reason for this expectation is that in the neutral mutation theory, fixation is merely the end result of a process of allele frequency changes in the population. A significant difference between the two ratios can therefore be used to reject the neutral mutation hypothesis.

Table 2.3 shows the numbers of synonymous and nonsynonymous fixed differences and polymorphisms at the glucose-6-phosphate dehydrogenase (*G6PD*) gene

TABLE 2.3

Fixed differences and polymorphisms at the glucose-6-phosphate dehydrogenase locus of *Drosophila melanogaster* and *D. simulans*[a]

Type of change	Fixed differences	Polymorphisms
Synonymous	26	36
Nonsynonymous	21	2

Source: Data from Eanes et al. (1993).

[a] The comparisons are based on 32 sequences from *D. melanogaster* and 12 sequences from *D. simulans*, with an aligned length of 1,705 bp.

between *Drosophila melanogaster* and *D. simulans* as an example of the use of the McDonald-Kreitman method to detect deviations from neutral evolution (Eanes et al. 1993). The ratio of fixed nonsynonymous differences to fixed synonymous differences is 21/26 = 0.81, whereas the ratio of nonsynonymous polymorphisms to synonymous polymorphisms is only 2/36 = 0.06. This highly significant difference implies a tenfold excess of nonsynonymous changes over that expected if the *G6PD* gene had evolved in a strictly neutral fashion. Thus, strict neutrality was rejected for this locus in these two *Drosophila* species.

Let us denote the proportion of nonsynonymous substitutions that have been fixed by positive selection by α. Charlesworth (1994) showed that it is possible to estimate α by using the data from the McDonald-Kreitman method as

$$\alpha = 1 - \frac{D_s P_n}{D_n P_s} \tag{2.45}$$

(Eyre-Walker 2006). Several methods have been proposed in the literature to estimate the average value of α across genes (e.g., Fay et al. 2001; Bierne and Eyre-Walker 2004). By using Equation 2.45 on the data in Table 2.3, we obtain $\alpha = 0.93$. That is, 93% of all nonsynonymous substitutions in *G6PD* that have occurred since the divergence of the two *Drosophila* species have reached fixation by positive selection. In a sample of 44 orthologous protein-coding genes in *D. melanogaster* and *D. simulans*, the mean fraction of nonsynonymous substitutions that have reached fixation by positive selection was estimated to be 45% (Bierne and Eyre-Walker 2004). Interestingly, a sample of 330 protein-coding genes from human and mouse revealed no evidence for positive selection (i.e., $\alpha = 0$), as was the case for 12 genes from the plants *Arabidopsis thaliana* and *A. lyrata* (Zhang and Li 2005; Bustamante et al. 2002).

Consequences of Explosive Population Growth: Single-Nucleotide Variation in Humans

Homo sapiens has experienced an unprecedented population explosion in the last 6,000 years. In merely 200–300 generations, human population increased a thousandfold, from about 7 million people in the year 4000 BCE to 7 billion in 2013. As a consequence of this rapid growth, a huge gap has been created between the census population size and the effective population size. Given the number of humans currently populating the planet, the number of mutations arising each generation is enormous—at least 100 billion new mutations. Thus, the vast majority of genetic variants observed in the human population are expected to be very recent in origin.

Fu et al. (2013) examined 15,336 protein-coding genes from 6,515 people and inferred the ages of 1,146,401 autosomal single-nucleotide variants. Most of the genetic variants were estimated to have arisen in the last 5,000 years or so, despite the fact that anatomically modern humans have existed for at least 100,000 years. Specifically, of the more than 700,000 genetic variants found in people of European ancestry, more than 81% arose in the past 5,000 years. In people of African ancestry, 58% of the approximately 650,000 variants were found to be less than 5,000 years old.

Fu et al. (2013) also estimated that ~86% of the variants are selectively neutral and ~14% are deleterious. Only 32 putatively advantageous variants (0.003%) were found in their sample. Since the vast majority of mutations in protein-coding exons are either deleterious or neutral and since most mutations are very recent, most variants are expected to be either rare variants or private ones, with newly arisen common variants possibly representing instances of advantageous mutation. Deleterious variants in the population are expected to have lower frequencies and to be younger than neutral variants. Indeed, Fu et al. estimated that ~73% of all protein-coding single-nucleotide variants and 86% of deleterious single-nucleotide variants arose in the past 5,000–10,000 years. Furthermore, people of European descent have an excess of deleterious variants compared with people of African descent, consistent with weaker purifying selection due to a serious decrease in population size (bottleneck) during the out-of-Africa dispersal.

Another consequence of explosive population growth concerns advantageous mutations. The current size of the human population relative to its effective population size implies that positive selection should not be very effective in fixing advantageous mutations. Indeed, hundreds of thousands of new advantageous mutations may be circulating within human populations, each with the potential to endow us with new capabilities. Unfortunately, the fixation probability of even the most advantageous mutation is very small and, hence, most advantageous mutations will be lost.

CHAPTER 3

DNA and Amino Acid Sequence Evolution

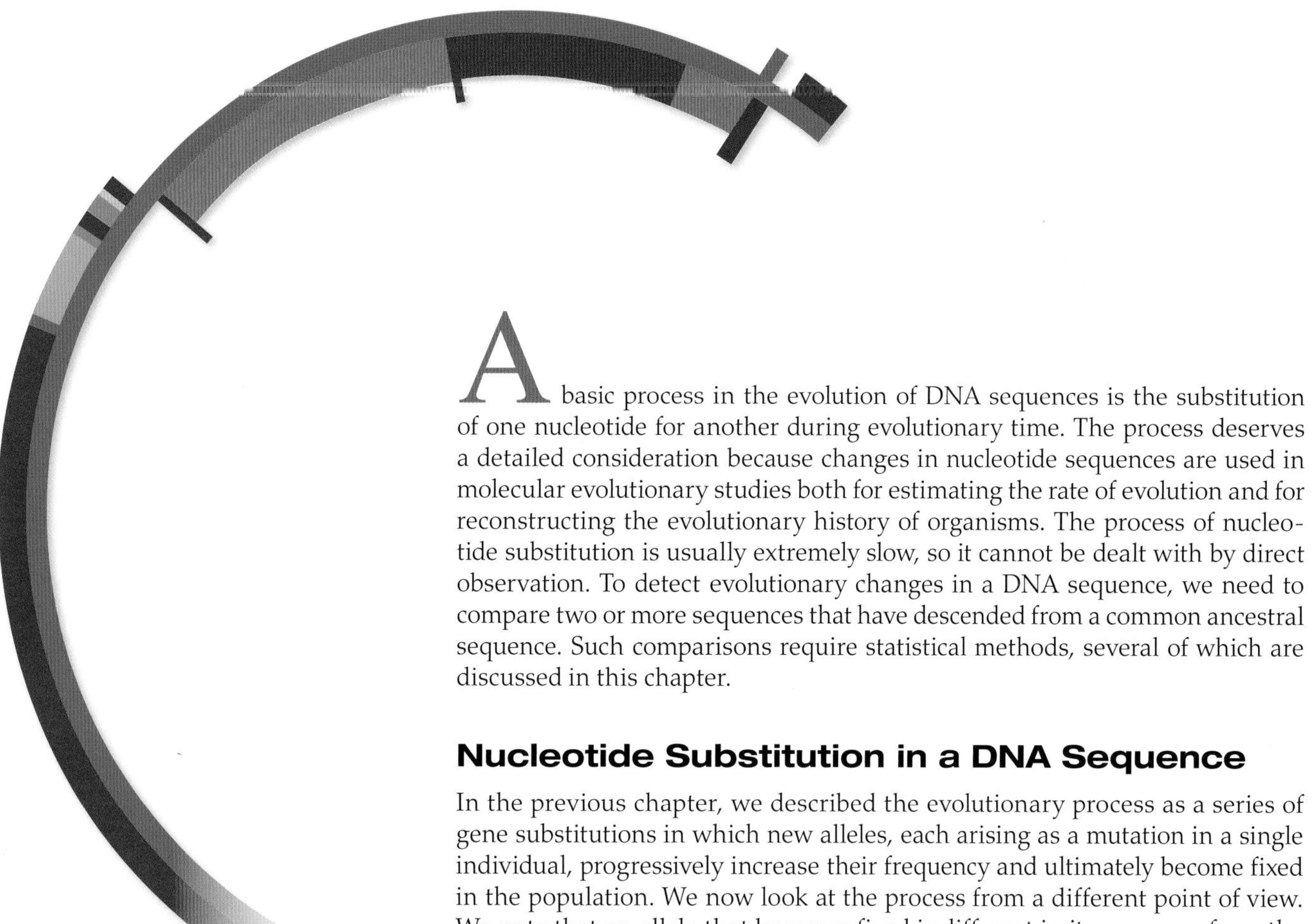

A basic process in the evolution of DNA sequences is the substitution of one nucleotide for another during evolutionary time. The process deserves a detailed consideration because changes in nucleotide sequences are used in molecular evolutionary studies both for estimating the rate of evolution and for reconstructing the evolutionary history of organisms. The process of nucleotide substitution is usually extremely slow, so it cannot be dealt with by direct observation. To detect evolutionary changes in a DNA sequence, we need to compare two or more sequences that have descended from a common ancestral sequence. Such comparisons require statistical methods, several of which are discussed in this chapter.

Nucleotide Substitution in a DNA Sequence

In the previous chapter, we described the evolutionary process as a series of gene substitutions in which new alleles, each arising as a mutation in a single individual, progressively increase their frequency and ultimately become fixed in the population. We now look at the process from a different point of view. We note that an allele that becomes fixed is different in its sequence from the allele that it replaces. That is, the substitution of a new allele for an old one is the substitution of a new sequence for a previous sequence. If we use a time scale in which one time unit is larger than the time of fixation, then the DNA sequence at any given locus will appear to change continuously. For this reason, it is interesting to study how the nucleotides within a DNA sequence change with time. As explained later, the results of these studies can be used to develop methods for estimating the number of substitutions between two sequences.

To study the dynamics of nucleotide substitution, we must make several assumptions regarding the probability of the substitution of one nucleotide by another. Numerous such mathematical schemes have been proposed (for a review, see Li 1997). We will restrict our discussion to the simplest and most frequently used: Jukes and Cantor's one-parameter model, and Kimura's two-parameter model. More complicated models of nucleotide substitution will be described only briefly.

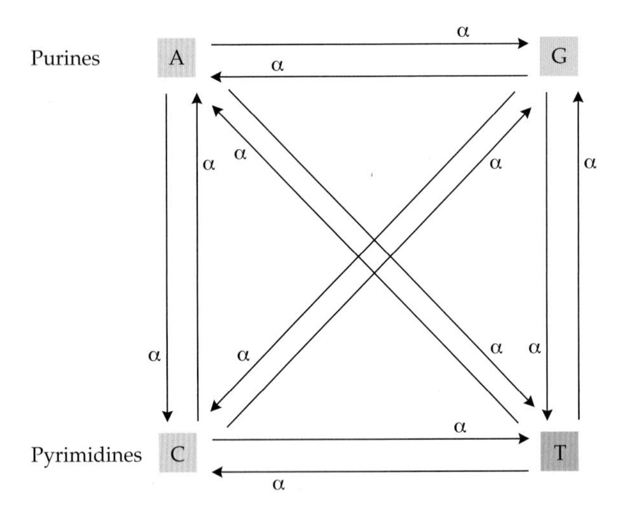

Figure 3.1 One-parameter model of nucleotide substitution. The rate of substitution in each direction is α.

Jukes and Cantor's one-parameter model

The substitution scheme of Jukes and Cantor's (1969) model is shown in **Figure 3.1**. This simple model assumes that substitutions occur with equal probability among the four nucleotides. In other words, there is no bias in the direction of change. For example, if the nucleotide under consideration is A, it will change to T, C, or G with equal probability. Conversely, the probabilities of T or C or G changing to A are also equal to one another. In this model, the rate of substitution for each nucleotide is 3α per unit time, and the rate of substitution in each of the three possible directions of change is α. Because the model involves a single parameter, α, it is called the **one-parameter model**.

Let us assume that the nucleotide residing at a certain site in a DNA sequence is A at time 0. First, we ask what is the probability that this site will be occupied by A at time t. This probability is denoted by $P_{A(t)}$. Since we start with A, the probability that this site is occupied by A at time 0 is $P_{A(0)} = 1$. At time 1, the probability of still having A at this site is given by

$$P_{A(1)} = 1 - 3\alpha \tag{3.1}$$

in which 3α is the probability that A will change to T, C, or G, and $1 - 3\alpha$ is the probability that A will remain unchanged.

The probability of having A at time 2 is

$$P_{A(2)} = (1 - 3\alpha)P_{A(1)} + \alpha\left[1 - P_{A(1)}\right] \tag{3.2}$$

To derive this equation, we considered two possible scenarios: (1) the nucleotide will remain unchanged from time 0 to time 2, and (2) the nucleotide will change to T, C, or G at time 1 but subsequently revert to A at time 2 (**Figure 3.2**). The probability that the nucleotide will be A at time 1 is $P_{A(1)}$, and the probability that it will remain A at time 2 is $1 - 3\alpha$. The product of these two independent variables gives us the probability for the first scenario, which constitutes the first term in Equation 3.2. The probability that the nucleotide will not be A at time 1 is $1 - P_{A(1)}$, and the probability that it will change back to A at time 2 is α. The product of these two variables gives us the probability for the second scenario and constitutes the second term in Equation 3.2.

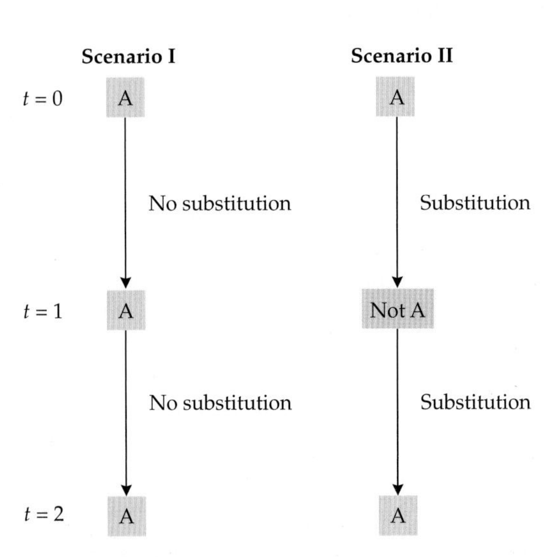

Figure 3.2 Two possible scenarios according to the one-parameter model for having adenine (A) at a site at time $t = 2$, given that the site had A at time 0.

Using the above formulation, we can show that the following recurrence equation applies to any t:

$$P_{A(t+1)} = (1 - 3\alpha)P_{A(t)} + \alpha\left[1 - P_{A(t)}\right] \tag{3.3}$$

We note that Equation 3.3 will also hold for $t = 0$, because $P_{A(0)} = 1$, and hence $P_{A(0+1)} = (1 - 3\alpha)P_{A(0)} + \alpha[1 - P_{A(0)}] = 1 - 3\alpha$, which is identical with Equation 3.1. We can rewrite Equation 3.3 in terms of the amount of change in $P_{A(t)}$ per unit time as

$$\Delta P_{A(t)} = P_{A(t+1)} - P_{A(t)} = -3\alpha P_{A(t)} + \alpha[1 - P_{A(t)}] = -4\alpha P_{A(t)} + \alpha \tag{3.4}$$

So far we have considered a discrete-time process. We can, however, approximate this process by a continuous-time model, by regarding $\Delta P_{A(t)}$ as the rate of change at time t. With this approximation, Equation 3.4 is rewritten as

$$\frac{dP_{A(t)}}{dt} = -4\alpha P_{A(t)} + \alpha \tag{3.5}$$

This is a first-order linear differential equation, and the solution is given by

$$P_{A(t)} = \frac{1}{4} + \left(P_{A(0)} - \frac{1}{4}\right)e^{-4\alpha t} \tag{3.6}$$

Since we started with A, the probability that the site has A at time 0 is 1. Thus, $P_{A(0)}$ = 1, and consequently,

$$P_{A(t)} = \frac{1}{4} + \frac{3}{4} e^{-4\alpha t} \quad (3.7)$$

Actually, Equation 3.6 holds regardless of the initial conditions. For example, if the initial nucleotide is not A, then $P_{A(0)} = 0$, and the probability of having A at this position at time t is

$$P_{A(t)} = \frac{1}{4} - \frac{1}{4} e^{-4\alpha t} \quad (3.8)$$

Equations 3.7 and 3.8 are sufficient for describing the substitution process. From Equation 3.7, we can see that if the initial nucleotide is A, then $P_{A(t)}$ decreases exponentially from 1 to 1⁄4 (**Figure 3.3**). On the other hand, from Equation 3.8 we see that if the initial nucleotide is not A, then $P_{A(t)}$ will increase monotonically from 0 to 1⁄4. This also holds true for T, C, and G.

Under the Jukes-Cantor model, the probability of each of the four nucleotides at equilibrium is 1⁄4. After equilibrium is reached, there will be no further changes in probabilities, i.e., $P_{A(t)} = P_{T(t)} = P_{C(t)} = P_{G(t)} = 1/4$ for all subsequent times.

So far, we have focused on a particular nucleotide site and treated $P_{A(t)}$ as a probability. However, $P_{A(t)}$ can also be interpreted as the frequency of A in a DNA sequence at time t. For example, if we start with a sequence made entirely of adenines, then $P_{A(0)}$ = 1, and $P_{A(t)}$ is the expected frequency of A in the sequence at time t. The expected frequency of A in the sequence at equilibrium will be 1⁄4, and so will the expected frequencies of T, C, and G. After equilibrium is reached, no further change in the nucleotide frequencies is expected to occur. However, the actual frequencies of the nucleotides will remain unchanged only in DNA sequences of infinite length. In practice, the lengths of DNA sequences are finite, so fluctuations in nucleotide frequencies are likely to occur.

Equation 3.7 can be rewritten in a more explicit form to take into account the facts that the initial nucleotide is A and the nucleotide at time t is also A.

$$P_{AA(t)} = \frac{1}{4} + \frac{3}{4} e^{-4\alpha t} \quad (3.9)$$

If the initial nucleotide is G instead of A, then from Equation 3.8 we obtain

$$P_{GA(t)} = \frac{1}{4} - \frac{1}{4} e^{-4\alpha t} \quad (3.10)$$

Since all the nucleotides are equivalent under the Jukes-Cantor model, $P_{GA(t)} = P_{CA(t)} = P_{TA(t)}$. In fact, we can consider the general probability, $P_{ij(t)}$, that a nucleotide will

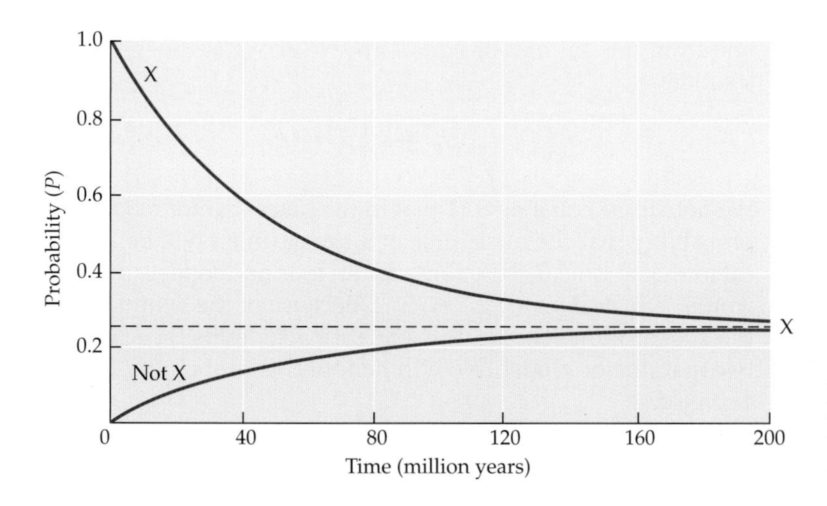

Figure 3.3 Temporal changes in the probability, P, of having nucleotide X at a position starting with either X (upper line) or a different nucleotide (lower line). The dashed line denotes the equilibrium frequency (P = 0.25); $\alpha = 5 \times 10^{-9}$ substitutions per site per year.

become j at time t, given that it was i at time 0. By using this generalized notation and Equation 3.9, we obtain

$$P_{ii(t)} = \frac{1}{4} + \frac{3}{4}e^{-4\alpha t} \quad (3.11)$$

In a similar manner, from Equation 3.10 we obtain

$$P_{ij(t)} = \frac{1}{4} - \frac{1}{4}e^{-4\alpha t} \quad (3.12)$$

where $i \neq j$.

Kimura's two-parameter model

The assumption that all nucleotide substitutions occur with equal probability, as in Jukes and Cantor's model, is unrealistic in most cases. For example, transitions (i.e., changes between A and G or between C and T) are generally more frequent than transversions. To take this fact into account, Kimura (1980) proposed a **two-parameter model** (**Figure 3.4**). In this scheme, the rate of transitional substitution at each nucleotide site is α per unit time, whereas the rate of each type of transversional substitution is β per unit time.

Let us first consider the probability that a site that has A at time 0 will have A at time t. After one time unit, the probability that A will change to G is α, and the probability that A will change to either C or T is 2β. Thus, the probability that A will remain unchanged after one time unit is

$$P_{AA(1)} = 1 - \alpha - 2\beta \quad (3.13)$$

At time 2, the probability of having A at this site is given by the sum of the probabilities of four different scenarios: (1) A remained unchanged at $t = 1$ and $t = 2$; (2) A changed into G at $t = 1$ and reverted by a transition to A at $t = 2$; (3) A changed into C at $t = 1$ and reverted by a transversion to A at $t = 2$; and (4) A changed into T at $t = 1$ and reverted by a transversion to A at $t = 2$ (**Figure 3.5**). Hence,

$$P_{AA(2)} = (1 - \alpha - 2\beta)P_{AA(1)} + \beta P_{TA(1)} + \beta P_{CA(1)} + \alpha P_{GA(1)} \quad (3.14)$$

By extension we obtain the following recurrence equation for the general case:

$$P_{AA(t+1)} = (1 - \alpha - 2\beta)P_{AA(t)} + \beta P_{TA(t)} + \beta P_{CA(t)} + \alpha P_{GA(t)} \quad (3.15)$$

After rewriting this equation as the amount of change in $P_{AA(t)}$ per unit time, and after approximating the discrete-time model by using the continuous-time model, we obtain the following differential equation:

$$\frac{dP_{AA(t)}}{dt} = -(\alpha + 2\beta)P_{AA(t)} + \beta P_{TA(t)} + \beta P_{CA(t)} + \alpha P_{GA(t)} \quad (3.16)$$

Similarly, we can obtain equations for $P_{TA(t)}$, $P_{CA(t)}$, and $P_{GA(t)}$, and from this set of four equations, we arrive at the following solution:

$$P_{AA(t)} = \frac{1}{4} + \frac{1}{4}e^{-4\beta t} + \frac{1}{2}e^{-2(\alpha+\beta)t} \quad (3.17)$$

We note from Equation 3.11 that in the Jukes-Cantor model, the probability that the nucleotide at a site at time t will be identical to that at time 0 is the same for all four nucleotides. In other words, $P_{AA(t)} = P_{GG(t)} = P_{CC(t)} = P_{TT(t)}$. Because of the symmetry of the substitution scheme, this equality also holds for Kimura's two-parameter model. We will denote this probability by $X_{(t)}$. Therefore,

$$X_{(t)} = \frac{1}{4} + \frac{1}{4}e^{-4\beta t} + \frac{1}{2}e^{-2(\alpha+\beta)t} \quad (3.18)$$

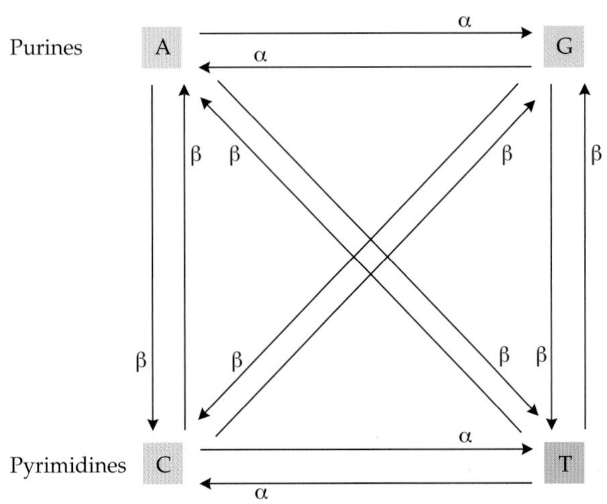

Figure 3.4 Two-parameter model of nucleotide substitution. The rate of transition (α) may or may not be equal to the rate of transversion (β).

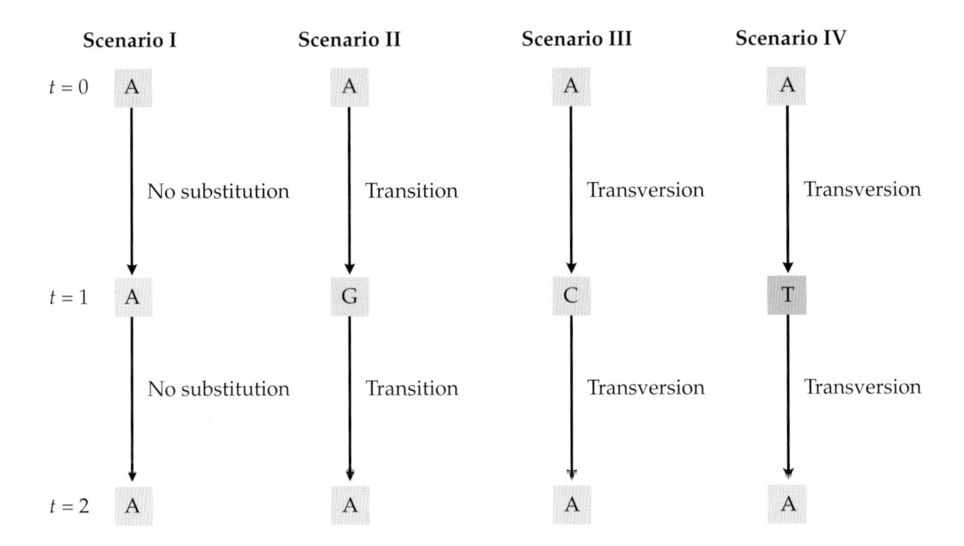

Figure 3.5 Four possible scenarios, according to Kimura's (1980) two-parameter model, for having A at a site at time t = 2, given that the site had A at time 0.

At equilibrium, i.e., at $t = \infty$, Equation 3.18 reduces to $X_{(\infty)} = 1/4$. Thus, as in the case of Jukes and Cantor's model, the equilibrium frequencies of the four nucleotides are 1/4.

Under the Jukes-Cantor model, Equation 3.12 holds regardless of whether the change from nucleotide i to nucleotide j is a transition or a transversion. In contrast, in Kimura's two-parameter model, we need to distinguish between transitional and transversional changes. We denote by $Y_{(t)}$ the probability that the initial nucleotide and the nucleotide at time t will differ from each other by a transition. Because of the symmetry of the substitution scheme, $Y_{(t)} = P_{AG(t)} = P_{GA(t)} = P_{TC(t)} = P_{CT(t)}$. It can be shown that

$$Y_{(t)} = \frac{1}{4} + \frac{1}{4}e^{-4\beta t} - \frac{1}{2}e^{-2(\alpha+\beta)t} \tag{3.19}$$

The probability, $Z_{(t)}$, that the nucleotide at time t and the initial nucleotide will differ by a specific type of transversion is given by

$$Z_{(t)} = \frac{1}{4} - \frac{1}{4}e^{-4\beta t} \tag{3.20}$$

Note that each nucleotide is subject to two types of transversion, but only one type of transition. For example, if the initial nucleotide is A, then the two possible transversional changes are A → C and A → T. Therefore, the probability that the initial nucleotide and the nucleotide at time t will differ by one of the two types of transversion is twice the probability given in Equation 3.20. Note also that $X_{(t)} + Y_{(t)} + 2Z_{(t)} = 1$.

Number of Nucleotide Substitutions between Two DNA Sequences

The substitution of alleles in a population generally takes thousands or even millions of years to complete (Chapter 2). For this reason, we cannot deal with the process of nucleotide substitution by direct observation, and nucleotide substitutions are always inferred from pairwise comparisons of DNA molecules that share a common evolutionary origin. After two nucleotide sequences diverge from each other, each of them will start accumulating nucleotide substitutions. Thus, the number of nucleotide substitutions that have occurred since two sequences diverged is the most basic and commonly used variable in molecular evolution.

If two sequences of length N differ from each other at n sites, then the proportion of differences, n/N, is referred to as the **degree of divergence** or **Hamming distance**. Degrees of divergence are usually expressed as percentages ($n/N \times 100\%$). When the

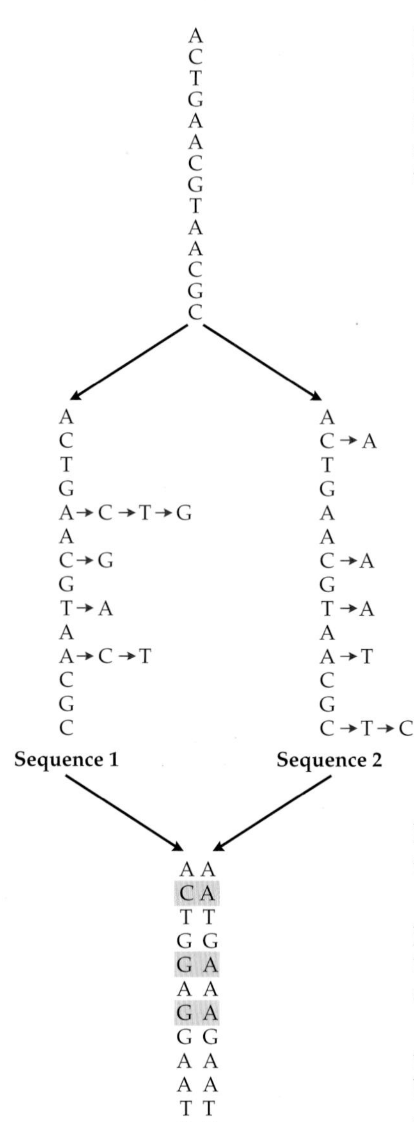

Figure 3.6 Two homologous DNA sequences that descended from an ancestral sequence have accumulated mutations since their divergence from each other. Note that although 13 mutations have occurred (red arrows), differences can be detected at only three nucleotide sites (gold boxes). Note further that "sequential substitutions," "coincidental substitutions," "parallel substitutions," "convergent substitutions," and "back substitutions" all involve multiple substitutions at the same site, though perhaps in different lineages. (From Graur and Li 2000.)

degree of divergence between the two sequences is small, the chance for more than one substitution to have occurred at any site is negligible, and the number of observed differences between the two sequences should be close to the actual number of substitutions. However, if the degree of divergence is substantial, then the observed number of differences is likely to be smaller than the actual number of substitutions because of **multiple substitutions** or **multiple hits** at the same site. For example, if the nucleotide at a certain site changed from A to C and then to T in one sequence, and from A to T in the other sequence, then the two sequences under comparison would be identical at this site, despite the fact that three substitutions occurred in their evolutionary history (**Figure 3.6**). Many methods have been proposed in the literature to correct for multiple substitutions (e.g., Jukes and Cantor 1969; Holmquist 1972; Holmquist and Pearl 1980; Kimura 1980, 1981; Kaplan and Risko 1982; Lanave et al. 1984). In the following sections we will review some of the most frequently used methods.

The number of nucleotide substitutions between two sequences is usually expressed in terms of the number of substitutions per nucleotide site rather than the total number of substitutions between the two sequences. This facilitates comparisons among sequence pairs that differ in length.

Protein-coding and noncoding sequences should be treated separately because they usually evolve at different rates. In the former case, it is advisable to distinguish between synonymous and nonsynonymous substitutions, since they are known to evolve at markedly different rates and because a distinction between the two types of substitution may provide us additional insight into the mechanisms of molecular evolution (Chapter 4).

Number of substitutions between two noncoding sequences

The results that we obtained earlier in this chapter for a single DNA sequence can be applied to studying nucleotide divergence between two sequences that share a common origin. We assume that all the sites in a sequence evolve at the same rate and follow the same substitution scheme. The number of sites compared between two sequences is denoted by L. Deletions and insertions are excluded from the analyses.

Let us start with the one-parameter model. In this model, it is sufficient to consider only $I_{(t)}$, which is the probability that the nucleotide at a given site at time t is the same in both sequences. Suppose that the nucleotide at a given site was A at time 0. At time t, the probability that a descendant sequence will have A at this site is $P_{AA(t)}$, and consequently the probability that two descendant sequences have A at this site is $P^2_{AA(t)}$. Similarly, the probabilities that both sequences will have T, C, or G at this site are $P^2_{TA(t)}$, $P^2_{AC(t)}$, and $P^2_{AG(t)}$, respectively. Therefore,

$$I_{(t)} = P^2_{AA(t)} + P^2_{AT(t)} + P^2_{AC(t)} + P^2_{AG(t)} \tag{3.21}$$

From Equations 3.11 and 3.12, we obtain

$$I_{(t)} = \frac{1}{4} + \frac{3}{4}e^{-8\alpha t} \tag{3.22}$$

Equation 3.22 also holds for T, C, or G. Therefore, regardless of the initial nucleotide at a site, $I_{(t)}$ represents the proportion of identical nucleotides between two sequences that diverged t time units ago. Note that the probability that the two sequences are different at a site at time t is $p = 1 - I_{(t)}$. Thus,

$$p = \frac{3}{4}\left(1 - e^{-8\alpha t}\right) \tag{3.23}$$

or

$$8\alpha t = -\ln\left(1 - \frac{4}{3}p\right) \tag{3.24}$$

The time of divergence between two sequences is usually not known, and thus we cannot estimate α. Instead, we compute K, which is the number of substitutions per site since the time of divergence between the two sequences. In the case of the one-parameter model, $K = 2(3\alpha t)$, where $3\alpha t$ is the number of substitutions per site in a single lineage. By using Equation 3.24 we can calculate K as

$$K = -\frac{3}{4}\ln\left(1 - \frac{4}{3}p\right) \tag{3.25}$$

where p is the observed proportion of different nucleotides between the two sequences (Jukes and Cantor 1969). For sequences of length L, the sampling variance of K, $V(K)$, is approximately given by

$$V(K) = \frac{p - p^2}{L\left(1 - \frac{4}{3}p\right)^2} \tag{3.26}$$

(Kimura and Ohta 1972). Equation 3.26 is only applicable for large values of L.

In the case of the two-parameter model (Kimura 1980), the differences between two sequences are classified into transitions and transversions. Let P and Q be the proportions of transitional and transversional differences between the two sequences, respectively. Then the number of nucleotide substitutions per site between the two sequences, K, is estimated by

$$K = \frac{1}{2}\ln\left(\frac{1}{1 - 2P - Q}\right) + \frac{1}{4}\ln\left(\frac{1}{1 - 2Q}\right) \tag{3.27}$$

Note that if we make no distinction between transitional and transversional differences, i.e., $p = P + Q$, then Equation 3.27 reduces to Equation 3.25, as in Jukes and Cantor's model. The sampling variance is approximately given by

$$V(K) = \frac{1}{L}\left[P\left(\frac{1}{1-2P-Q}\right)^2 + Q\left(\frac{1}{2-4P-2Q} + \frac{1}{2-4Q}\right)^2 - \left(\frac{P}{1-2P-Q} + \frac{Q}{2-4P-2Q} + \frac{Q}{2-4Q}\right)^2\right] \tag{3.28}$$

Let us now consider a hypothetical numerical example of two sequences of length 200 nucleotides that differ from each other by 20 transitions and 4 transversions. Thus, $L = 200$, $P = 20/200 = 0.10$, and $Q = 4/200 = 0.02$. According to the two-parameter model, we obtain $K \approx 0.13$. The total number of substitutions can be obtained by multiplying the number of substitutions per site, K, by the number of sites, L. In this case we obtain an estimate of about 26 substitutions, resulting in 24 observed differences between the two sequences. According to the one-parameter model, $p = 24/200 = 0.12$, and $K \approx 0.13$.

In the above example, the two models give essentially the same estimate because the degree of divergence is small enough that the corrected degree of divergence (i.e., the number of nucleotide substitutions) is only slightly larger than the uncorrected value (i.e., the number of nucleotide differences). In such cases, one may use Jukes and Cantor's model, which is simpler.

TABLE 3.1

General matrix of nucleotide substitution[a]

	A	T	C	G
A	$1 - \alpha_{12} - \alpha_{13} - \alpha_{14}$	α_{12}	α_{13}	α_{14}
T	α_{21}	$1 - \alpha_{21} - \alpha_{23} - \alpha_{24}$	α_{23}	α_{24}
C	α_{31}	α_{32}	$1 - \alpha_{31} - \alpha_{32} - \alpha_{34}$	α_{34}
G	α_{41}	α_{42}	α_{43}	$1 - \alpha_{41} - \alpha_{42} - \alpha_{43}$

[a]Values on the diagonal (**bold blue**) represent the probability of no change per unit time.

On the other hand, when the degree of divergence between two sequences is large, the estimates obtained by the two models may differ considerably. For example, consider two sequences with $L = 200$ that differ from each other by 50 transitions and 16 transversions. Thus, $P = 50/200 = 0.25$, and $Q = 16/200 = 0.08$. For the two-parameter model, $K \approx 0.48$. For the one-parameter model, $p = 66/200 = 0.33$, and $K \approx 0.43$, which is 10% smaller than the value obtained by using the two-parameter model. When the degree of divergence between two sequences is large, and especially in cases where there are prior reasons to believe that the rate of transition differs considerably from the rate of transversion, the two-parameter model tends to be more accurate than the one-parameter model.

Substitution schemes with more than two parameters

Since there are four types of nucleotides, and each of them can be substituted by any of the other three, there are 12 possible types of substitutions. Each of these substitution types has a certain probability of occurring. These substitution probabilities can be written in the form of a matrix, the elements of which, α_{ij}, denote the probability of substitution of nucleotide i for nucleotide j per unit time (**Table 3.1**). For example, the substitution schemes in Figure 3.1 can also be represented in matrix form as in **Table 3.2a**. The numerous mathematical models that have been used to study the dynamics of nucleotide substitution are different from each other in the particular probabilities

TABLE 3.2

The one-parameter (Jukes and Cantor 1969) and four-parameter (Blaisdell 1985) schemes of nucleotide substitution in matrix form[a]

	A	T	C	G
(a) One parameter				
A	$1 - 3\alpha$	α	α	α
T	α	$1 - 3\alpha$	α	α
C	α	α	$1 - 3\alpha$	α
G	α	α	α	$1 - 3\alpha$
(b) Four parameters				
A	$1 - \alpha - 2\gamma$	γ	γ	α
T	δ	$1 - \alpha - 2\delta$	α	δ
C	δ	β	$1 - \alpha - 2\delta$	δ
G	β	γ	γ	$1 - \beta - 2\gamma$

[a]Values on the diagonals (**bold blue**) represent the probability of no change per unit time.

assigned to the different types of substitution. In **Table 3.2b** we present Blaisdell's (1985) four-parameter substitution scheme. In this scheme, there are two different rates of transition, for A ↔ G and T ↔ C, respectively, and two different rates of transversion.

In the above formulations we have used the Markov chain process, which is a mathematical model for the evolution of a memory-less system, that is, one for which the likelihood of a given future state depends only on its present state and not on any past states. In other words, only the information at the present time is useful for predicting the outcomes in the future; the past is irrelevant. The matrices in Tables 3.1 and 3.2 belong to a class known as **stochastic matrices**, in which all elements are nonnegative and the sum of the elements in each row is equal to 1 (for detailed descriptions, see Feller 1968, 1971; Cox and Miller 1977).

A general question arises: Which model should we use to calculate the number of substitutions between two sequences? Intuitively, it would seem that a model with a large number of parameters should perform better than one with fewer parameters. In practice, this is not necessarily the case, for several reasons. One reason is that in addition to the substitution scheme, it is usually necessary to make further assumptions in the formulation of each model. For example, to estimate the number of substitutions between two sequences according to six-parameter models, we need to assume that the common ancestral sequence was at equilibrium in terms of its nucleotide frequencies (Kimura 1981). Such an assumption is not necessary for the solution of either the one-parameter or two-parameter model. Obviously, the addition of assumptions (and parameters that need to be estimated) may greatly increase the estimation errors.

The second and probably more important reason is that sampling errors, which arise because the number of nucleotides compared is finite, can render a method inapplicable. This is because all estimators of K, such as Equations 3.25 and 3.27, involve logarithmic functions, and the argument can become zero or negative. This undesirable effect becomes more problematic with the increase in the number of parameters involved. The more two sequences are divergent from each other, the less applicable the multiple-parameter models will be. For example, when $K = 2.0$, four- and six-parameter methods may not be applicable to the vast majority of sequence comparisons. By contrast, the one-parameter method has been found to be inapplicable in only very rare cases. The proportion of inapplicable cases decreases with the length of the sequence, L. However, even for large values of L (e.g., $L = 3{,}000$), four-parameter models are rendered inapplicable in about 25% of the cases when $K = 2.0$, and this proportion is even higher for six-parameter models.

Finally, we note that for closely related sequences, the estimates obtained by the different methods are quite similar to one another.

Violation of assumptions

In the distance measures discussed so far, several assumptions have been made that may not be met by the sequences under study. For example, the rate of nucleotide substitution was assumed to be the same for all sites. This assumption may not hold, as the rate may vary greatly from site to site (Chapter 4). An additional assumption was that substitutions occur in an independent manner, i.e., that the probability that a certain substitution will occur at a site is not affected by (1) the context of the surrounding nucleotides, (2) the occurrence of a substitution at a different site, or (3) the history of substitutions at the site in question. The assumption of independence may not always apply in nature. For example, hairpin structures require nucleotide changes in one part of the sequence to be compensated for by complementary changes elsewhere.

Finally, in the methods described above, the substitution matrix was assumed not to change in time, so the nucleotide frequencies are maintained at a constant equilibrium value throughout their evolution. This may not be the case, especially in protein-coding sequences exhibiting extreme codon-usage biases (Chapter 4). For each of these cases, special distance methods and corrections have been developed (Yang 2006).

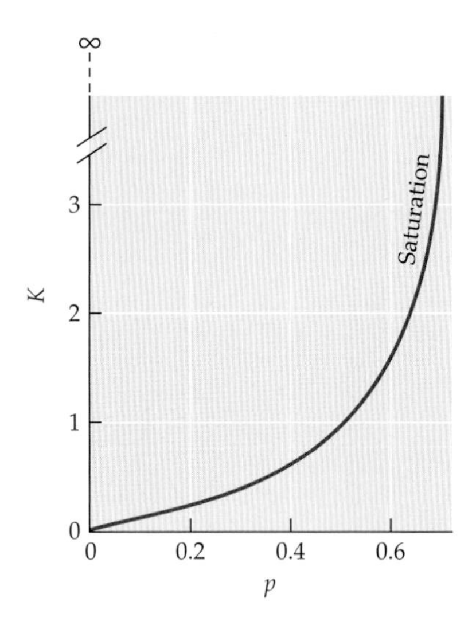

Figure 3.7 Relationship between *p* (the number of nucleotide differences between two sequences) and *K* (number of nucleotide substitutions between two sequences). As *p* increases, the curve reaches saturation, where the value of *p* is no longer related to *K*. In the literature, this relationship is usually shown with *K* in the abscissa and *p* in the ordinate, despite the fact that *p* and *K* are the independent and dependent variables, respectively.

Saturation

Because of multiple substitutions, the inferred number of substitutions between two sequences usually underestimates the true value of the number of substitutions. **Figure 3.7** illustrates the relationship between the proportion of differences per site between two sequences (*p*, the independent variable) and the number of substitutions per site between two sequences (*K*, the dependent variable) as inferred by using the Jukes and Cantor one-parameter model. At low *p* values, the underestimation of *K* is usually corrected by the estimation models. As the *p* value approaches 0.75 (the maximum possible value), *K* approaches infinity (∞), a situation referred to as **saturation**. At $p > 0.75$, $K = \infty$. The precise shape of the curve in Figure 3.7 depends, of course, on the details of the substitution model; however, saturation is an impassable obstacle even for the most sophisticated estimation models.

Number of Substitutions between Two Protein-Coding Genes

Computing the number of substitutions between two protein-coding sequences is generally more complicated than computing the number of substitutions between two noncoding sequences, because a distinction should be made between synonymous and nonsynonymous substitutions. Many methods have been proposed in the literature for estimating the numbers of synonymous and nonsynonymous substitutions (e.g., Miyata and Yasunaga 1980; Li et al. 1985b; Nei and Gojobori 1986; Li 1993; Pamilo and Bianchi 1993; Goldman and Yang 1994; Comeron 1995; Halpern and Bruno 1998; Yang and Nielsen 2000).

In studying protein-coding sequences, we usually exclude the initiation and the termination codons from analysis, because these two codons seldom change with time.

Computing the number of substitutions per site for synonymous and nonsynonymous substitutions separately requires determination of the appropriate nominators, i.e., the number of synonymous and nonsynymous substitutions, as well as the appropriate denominators, i.e., the numbers of synonymous and nonsynonymous sites. Neither the numerators nor the denominators are easy to compute. Let us start with the denominators. One difficulty is that the classification of a site into a synonymous or a nonsynonymous site changes with context. For example, the third position of CGG (Arg) is synonymous, i.e., any change in this position will result in a codon for arginine. However, if the first position changes to T, then the third position of the resulting codon, TGG (Trp), becomes nonsynonymous, i.e., any change in this position will result in a codon for an amino acid other than tryptophan. The second difficulty is that many sites are neither completely synonymous nor completely nonsynonymous. For example, a transition in the third position of codon GAT (Asp) is synonymous, while a transversion to either GAG or GAA is nonsynonymous. Consequently, many sites must be counted as part synonymous and part nonsynonymous. Note also that transitions result in synonymous substitutions more frequently than do transversions; therefore, the substitution scheme used to determine the number of synonymous and nonsynonymous substitutions must take into account this fact.

Two different approaches have been proposed for estimating the number of synonymous substitutions per synonymous site (K_S) and the number of nonsynonymous substitutions per nonsynonymous site (K_A). In the codon-based approach, a codon site rather than a nucleotide site is considered as the unit of evolution. Goldman and Yang (1994), for instance, proposed one such model with 61 states for the 61 sense codons. The reason for the exclusion of the three nonsense (stop) codons is that mutations to or from stop codons can be assumed to drastically affect the structure and function of the protein and, therefore, to rarely survive. Readers interested in codon-based models should consult Cannarozzi and Schneider (2012). Here, we will only deal with nucleotide-site-based approaches, i.e., methods in which each nucleotide position is dealt with independently. We note, however, that computer programs and software packages have been developed for both approaches.

One way to count the number of synonymous sites (N_S) and the number of nonsynonymous sites (N_A) in a coding sequence was proposed by Miyata and Yasunaga (1980) and Nei and Gojobori (1986). In this method, nucleotide sites are classified as follows. Consider a particular position in a codon. Let i be the number of possible synonymous changes at this site. Then this site is counted as $i/3$ synonymous and $(3 - i)/3$ nonsynonymous. For example, in the codon TTT (Phe), the first two positions are counted as nonsynonymous because no synonymous change can occur at these positions, and the third position is counted as one-third synonymous and two-thirds nonsynonymous because one of the three possible changes at this position is synonymous. As another example, the codon ACT (Thr) has two nonsynonymous sites (the first two positions) and one synonymous site (the third position), because all possible changes at the first two positions are nonsynonymous while all possible changes at the third position are synonymous. When comparing two sequences, we first count N_S and N_A in each sequence and then compute the averages between the two sequences.

Next, we classify nucleotide differences into synonymous and nonsynonymous differences. For two codons that differ by only one nucleotide, the difference is easily inferred. For example, the difference between the two codons GTC (Val) and GTT (Val) is synonymous, while the difference between the two codons GTC (Val) and GCC (Ala) is nonsynonymous. For two codons that differ by more than one nucleotide, the problem of estimating the numbers of synonymous and nonsynonymous substitutions becomes more complicated because we need to determine the order in which the substitutions occurred.

Let us consider the case in which two codons differ from each other by two substitutions. For example, for the two codons CCC (Pro) and CAA (Gln), there are two possible pathways:

Pathway I: CCC (Pro) ↔ CCA (Pro) ↔ CAA (Gln)

Pathway II: CCC (Pro) ↔ CAC (His) ↔ CAA (Gln)

Pathway I requires one synonymous and one nonsynonymous change, whereas pathway II requires two nonsynonymous changes.

There are basically two ways to deal with multiple substitutions at a codon. The first one assumes that all pathways are equally probable (Nei and Gojobori 1986), so we average the numbers of the different types of substitutions for all the possible scenarios. This scheme is called the **unweighted method**. For example, if we assume that the two pathways shown above are equally likely, then the number of nonsynonymous differences is $(1 + 2)/2 = 1.5$, and the number of synonymous differences is $(1 + 0)/2 = 0.5$.

The second method, the **weighted method**, employs a priori criteria to decide which pathway is more probable. For instance, it is known that synonymous substitutions occur considerably more frequently than nonsynonymous substitutions (Chapter 4), and so, in the above example, pathway I is more probable that pathway II. If we give a weight of 0.8 for pathway I and a weight of 0.2 for pathway II, then the number of nonsynonymous differences between the two codons is estimated as $(0.8 \times 1) + (0.2 \times 2) = 1.2$. The number of synonymous differences is estimated as $(0.8 \times 1) + (0.2 \times 0) = 0.8$. The weights used here are hypothetical. To determine quantitatively and realistically the relative probabilities of the different pathways, adequate information on the relative likelihood of all possible codon changes is necessary. These values have been estimated empirically by Miyata and Yasunaga (1980) from protein sequence data and by Li et al. (1985b) from DNA sequence data.

In practice, the weighted and unweighted methods usually yield quite similar estimates (Nei and Gojobori 1986), but they can be important for genes coding for highly conservative proteins. One such example is the mammalian glucagon (Lopez et al. 1984). The arginine at the seventeenth amino acid position is encoded by CGC in hamster and by AGG in cow. There are two possible pathways between these two codons:

Pathway I: CGC (Arg) ↔ CGG (Arg) ↔ AGG (Arg)
Pathway II: CGC (Arg) ↔ AGC (Ser) ↔ AGG (Arg)

The first pathway requires two synonymous substitutions; the second requires two nonsynonymous substitutions. Under the assumption that both pathways are equally probable, we would infer that one nonsynonymous substitution and one synonymous substitution have occurred at this position since the divergence between hamster and cow. Since glucagon consists of only 29 amino acids, this inference will greatly inflate the rate of nonsynonymous substitution at the expense of the synonymous substitution rate. Such errors may also occur when dealing with leucine codons belonging to different codon families, such as TTA and CTT.

Finally, the situation is even more complex when two codons differ at all three positions. For example, if one sequence has CTT at a codon site and the second sequence has AGG, we have to consider six possible pathways:

Pathway I: CTT (Leu) ↔ ATT (Ile) ↔ AGT (Ser) ↔ AGG (Arg)
Pathway II: CTT (Leu) ↔ ATT (Ile) ↔ ATG (Met) ↔ AGG (Arg)
Pathway III: CTT (Leu) ↔ CGT (Arg) ↔ AGT (Ser) ↔ AGG (Arg)
Pathway IV: CTT (Leu) ↔ CTG (Leu) ↔ ATG (Met) ↔ AGG (Arg)
Pathway V: CTT (Leu) ↔ CGT (Arg) ↔ CGG (Arg) ↔ AGG (Arg)
Pathway VI: CTT (Leu) ↔ CTG (Leu) ↔ CGG (Arg) ↔ AGG (Arg)

The numbers of synonymous and nonsynonymous differences between two protein-coding sequences are denoted by M_S and M_A. We can therefore compute the number of synonymous differences per synonymous site as $p_S = M_S/N_S$ and the number of nonsynonymous differences per nonsynonymous site as $p_A = M_A/N_A$. These formulas obviously do not take into account the effect of multiple hits at the same site. We can make such corrections by using Jukes and Cantor's formula (Equation 3.25):

$$K_S = -\frac{3}{4}\ln\left(1-\frac{4M_S}{3N_S}\right) \tag{3.29}$$

and

$$K_A = -\frac{3}{4}\ln\left(1-\frac{4M_A}{3N_A}\right) \tag{3.30}$$

An alternative method for calculating K_S and K_A was proposed by Li et al. (1985b). According to this method, we first classify the nucleotide sites into **nondegenerate** (or **zerofold degenerate**), **twofold degenerate**, and **fourfold degenerate** sites. A site is nondegenerate if all possible changes at this site are nonsynonymous, twofold degenerate if one of the three possible changes is synonymous, and fourfold degenerate if all possible changes at the site are synonymous. For example, the first two positions of codon TTT (Phe) are nondegenerate, the third position of TTT is twofold degenerate, and the third position of codon GTT (Val) is fourfold degenerate (see Chapter 1, Table 1.3). In the standard genetic code, the third positions of the three isoleucine codons are treated for simplicity as twofold degenerate sites, although in reality the degree of degeneracy at these positions is threefold. In vertebrate mitochondrial genes, there are only two codons for isoleucine, and the third position in these codons is indeed a twofold degenerate site (see Chapter 1, Table 1.4). Using the above rules, we first count the numbers of the three types of sites for each of the two sequences, and then compute the averages, denoting them by L_0 (nondegenerate), L_2 (twofold degenerate), and L_4 (fourfold degenerate).

From the above classification of nucleotide sites, we can calculate the numbers of substitutions between two coding sequences for the three types of sites separately. The nucleotide differences in each class are further classified into transitional (*Si*) and transversional (V_i) differences, where $i = 0$, 2, and 4 denote nondegeneracy, twofold degeneracy and fourfold degeneracy, respectively. Note that by definition, all the substitutions at nondegenerate sites are nonsynonymous. Similarly, all the substitutions at fourfold degenerate sites are synonymous. At twofold degenerate sites, transitional changes (C ↔ T and A ↔ G) are synonymous, whereas all the other changes, which are transversions, are nonsynonymous. There are no exceptions to this rule in the vertebrate mitochondrial genetic code. In the standard genetic code, on the other hand, there are two exceptions: (1) the first position of four arginine codons (CGA, CGG, AGA, and AGG), in which one type of transversion is synonymous while the other type is nonsynonymous; and (2) the last position in the three isoleucine codons (ATT, ATC, and ATA). In the first case, C ↔ A transversions in the first codon position are included in S_2, and C ↔ T and C ↔ G changes are included in V_2. Tzeng et al. (2004) proposed a correction for the problem with the arginine codons. Accordingly, T ↔ C, T ↔ A, and C ↔ A changes in the third codon position are included in S_2, and T ↔ G, C ↔ G, and A ↔ G changes are included in V_2. Similar methodological adjustments may be required for other genetic codes.

The proportion of transitional differences at i-fold degenerate sites between two sequences is calculated as

$$P_i = \frac{S_i}{L_i} \tag{3.31}$$

Similarly, the proportion of transversional differences at i-fold degenerate sites between two sequences is

$$Q_i = \frac{V_i}{L_i} \tag{3.32}$$

Kimura's (1980) two-parameter method is used to estimate the numbers of transitional (A_i) and transversional (B_i) substitutions per ith type site. The means are given by

$$A_i = \frac{1}{2}\ln(a_i) - \frac{1}{4}\ln(b_i) \tag{3.33}$$

and

$$B_i = \frac{1}{2}\ln(b_i) \tag{3.34}$$

The variances are given by

$$V(A_i) = \frac{a_i^2 P_i + c_i^2 Q_i - \left(a_i P_i + c_i Q_i\right)^2}{L_i} \tag{3.35}$$

and

$$V(B_i) = \frac{b_i^2 Q_i \left(1 - Q_i\right)}{L_i} \tag{3.36}$$

where L_i is the number of i-class degeneracy sites, $a_i = 1/(1 - 2P_i - Q_i)$, $b_i = 1/(1 - 2Q_i)$, and $c_i = (a_i - b_i)/2$. The total number of substitutions per ith type degenerate site, K_i, is given by

$$K_i = A_i + B_i \tag{3.37}$$

with an approximate sampling variance of

$$V(K_i) = \frac{a_i^2 P_i + d_i^2 Q_i - \left(a_i P_i + d_i Q_i\right)^2}{L_i} \tag{3.38}$$

where $d_i = b_i + c_i$. We note that A_2 and B_2 denote the numbers of synonymous and nonsynonymous substitutions per twofold degenerate site, respectively; $K_4 = A_4 + B_4$ denotes the number of synonymous substitutions per fourfold degenerate site; and

$K_0 = A_0 + B_0$ denotes the numbers of nonsynonymous substitutions per nondegenerate site. These formulas can be used to compare the rates of substitution among the three different types of sites.

We denote by K_S the number of synonymous substitutions per synonymous site, and by K_A the number of nonsynonymous substitutions per nonsynonymous site. Since one-third of a twofold degenerate site is synonymous and two-thirds are nonsynonymous, K_S and K_A may be obtained by

$$K_S = \frac{3(L_2A_2 + L_4K_4)}{L_2 + 3L_4} \tag{3.39}$$

and

$$K_A = \frac{3(L_2B_2 + L_0K_0)}{2L_2 + 3L_0} \tag{3.40}$$

Since transitional substitutions tend to occur more often than transversional substitutions, and since most transitional changes at twofold degenerate sites are synonymous, counting a twofold degenerate site as a one-third synonymous site will tend to underestimate the number of synonymous sites and to overestimate the number of nonsynonymous sites, thereby overestimating K_S and underestimating K_A. To overcome these problems, Li (1993) and Pamilo and Bianchi (1993) proposed to calculate the number of synonymous substitutions by taking $(L_2A_2 + L_4A_4) / (L_2 + L_4)$, i.e., the weighted average of A_2 and A_4, as an estimate of the transitional component of nucleotide substitution at twofold and fourfold degenerate sites. Similarly, the weighted average $(L_0B_0 + L_2B_2) / (L_0 + L_2)$ is used as an estimate of the mean transversional number of substitutions at nondegenerate and twofold degenerate sites. Tzeng et al. (2004) found that this modification does not work well in general. Instead, they proposed the following modifications:

$$K_S = \frac{L_2A_2 + L_4K_4}{\frac{(k-1)L_2}{(k-1)+2} + L_4} \text{ when } k \geq 2 \tag{3.41a}$$

$$K_S = \frac{L_2A_2 + L_4K_4}{\frac{L_2}{3} + L_4} \text{ when } k \leq 2 \tag{3.41b}$$

and

$$K_A = \frac{L_2B_2 + L_0K_0}{\frac{2L_2}{(k-1)+2} + L_0} \text{ when } k \geq 2 \tag{3.42a}$$

$$K_A = \frac{L_2B_2 + L_0K_0}{\frac{2L_2}{3} + L_0} \text{ when } k \leq 2 \tag{3.41b}$$

where $k = t_i/t_v$ = transition/transversion rate ratio. When $k = 2$, Equations 3.41a and b and 3.42a and b are the same as Equations 3.39 and 3.40, respectively.

The approximate variances of K_S and K_A are given by

$$V(K_S) = \frac{L_2^2V(A_2) + L_4^2V(A_4)}{(L_2 + L_4)^2} + V(B_4) - \frac{2b_4Q_4\left[a_4P_4 - c_4(1 - Q_4)\right]}{L_2 + L_4} \tag{3.43a}$$

$$V(K_A) = V(A_0) + \frac{L_0^2V(B_0) + L_2^2V(B_2)}{(L_0 + L_2)^2} - \frac{2b_0Q_0\left[a_0P_0 - c_0(1 - Q_0)\right]}{L_0 + L_2} \tag{3.43b}$$

where n is the number of amino acid differences between the two sequences and L is the length of the aligned sequences.

A simple model that can be used to convert p into the number of amino acid replacements between two sequences is the Poisson process. The number of amino acid replacements per site, d, is estimated as

Number of Amino Acid Replacements between Two Proteins

From the comparison of two amino acid sequences, we can calculate the observed proportion of amino acids differences between the two sequences as

$$p = n/L \quad (3.44)$$

where n is the number of amino acid differences between the two sequences and L is the length of the aligned sequences.

The number of amino acid replacements, d, and the variance, $V(d)$ can be estimated by the Poisson process as

$$d = -\ln(1 - p) \quad (3.45)$$

$$V(d) = \frac{p}{L(1-p)} \quad (3.46)$$

Alignment of Nucleotide and Amino Acid Sequences

Comparison of two or more homologous sequences involves the pairing of homologous residues (nucleotides or amino acids) and requires the identification of the locations of deletions and insertions that might have occurred in any of the lineages under study since their divergence from a common ancestor. This process is referred to as **sequence alignment**. In evolutionary terms, a sequence alignment represents a hypothesis concerning **positional homology**, i.e., a claim to the effect that the aligned residues represent descendants from a common ancestral residue. In **Figure 3.8** we show an alignment of two DNA sequences that reflects accurately the evolutionary changes that had accumulated independently in the two lineages since the divergence of the sequences from each other. Such an alignment is called the **true alignment**. Of course, in reality, we do not have any information on the ancestral sequence or the changes that occurred in the branches leading to the extant sequences. By using certain algorithms and criteria (below), we obtain an **inferred** or **realized alignment** that may or may not be identical with the true alignment. An error in an alignment

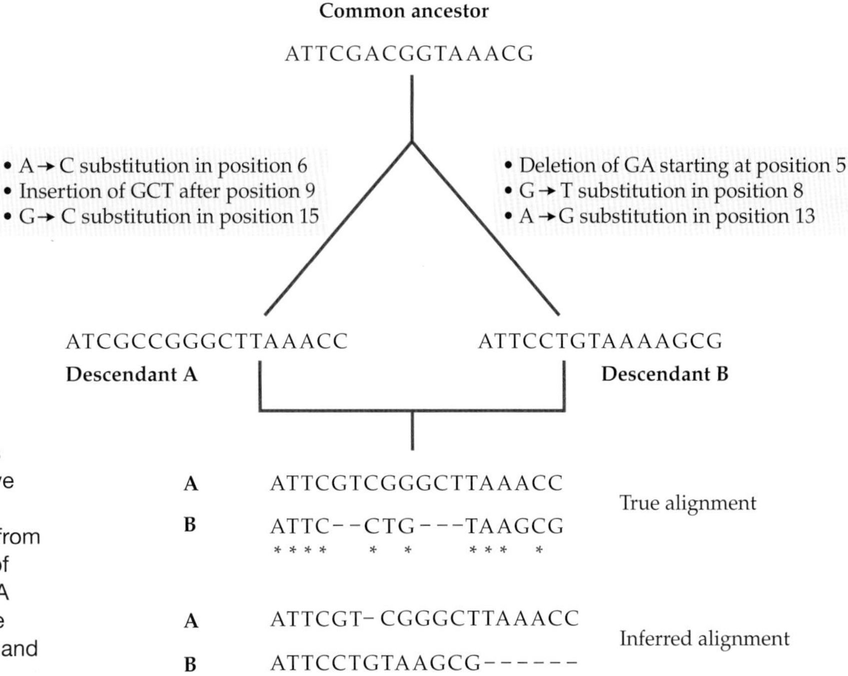

Figure 3.8 Two homologous DNA sequences that descended from an ancestral sequence have each independently accumulated substitutions, deletions, and insertions since their divergence from each other. In the true alignment, the locations of the insertions and deletions (indels) are shown. A realized alignment may be different from the true alignment. Dashes and asterisks indicate indels and matches, respectively.

means that an ancestral position and its descendants have not been identified correctly. Given that sequence alignment is the first step in many evolutionary studies, and that errors in alignment tend to amplify in later computational stages, it is extremely important to carefully assess the quality of inferred alignments before using them in subsequent analyses.

There are two categories of sequence alignment algorithms. **Local alignment** methods aim to determine whether or not subsegments of sequence A are present in sequence B. Local alignment methods have their greatest utility in database searching and retrieval of sequences (e.g., BLAST, Altschul et al. 1990). They may also be of use in detecting nonhomologous sequences with a certain degree of similarity to one another, possibly due to convergence. In evolutionary studies, we are mainly interested in sequence comparisons in which two or more sequences are aligned along their entire lengths. This type of alignment is called **global alignment**. In global alignment algorithms, each element of one sequence is compared with each element in all the other sequences.

We will illustrate the process of alignment mainly by using DNA sequences, but the same principles and procedures can be used to align amino acid sequences. As a matter of fact, one usually obtains more reliable alignments by using amino acid sequences than by using DNA sequences. There are two reasons for this: (1) amino acids change less frequently during evolution than nucleotides, and (2) there are 20 amino acids and only four nucleotides, so the probability for two sites to be identical by chance is lower at the amino acid level than at the nucleotide level.

Pairwise alignment

For reasons of computational complexity, **pairwise alignment** (the alignment of two sequences) is treated separately from **multiple-sequence alignment** (the alignment of three or more sequences). The rationale for this distinction lies in the fact that pairwise alignment problems have exact solutions that can be calculated in reasonable computational time, whereas exact solutions for multiple-sequence alignment problems take impracticably long times to compute. As a consequence, algorithms of multiple-sequence alignments use approximate approaches (heuristics).

A pairwise DNA sequence alignment consists of a series of paired bases, one base from each sequence. There are three types of aligned pairs: **matches**, **mismatches**, and **gaps**. A matched pair is one in which the same nucleotide appears in both sequences, i.e., it is assumed that the nucleotide at this site has not changed since the divergence between the two sequences. A mismatched pair is a pair in which different nucleotides are found in the two sequences, i.e., at least one substitution has occurred in one of the sequences since their divergence from each other. A gap is a pair consisting of a base from one sequence and a **null base** from the other. Null bases are denoted by –. A gap indicates that a deletion has occurred in one sequence or an insertion has occurred in the other. Since alignments do not allow us to distinguish between these two possibilities, deletions and insertions are collectively referred to as **indels** (short for **in**sertion-or-**del**etion).

Consider the case of two DNA sequences, A and B, of lengths m and n, respectively. If we denote the number of matched pairs by x, the number of mismatched pairs by y, and the total number of pairs containing a null base by z, we obtain

$$n+m=2(x+y)+z \tag{3.47}$$

A distinction is sometimes made between **terminal gaps** and **internal gaps**. For example, when aligning a partial sequence of a gene with a complete sequence of a homologous gene, it makes no sense to include the terminal gaps (i.e., the missing data from the first sequence) in the calculation. Special treatment is sometimes also needed where internal gaps are concerned, for instance, in alignments between genomic sequences that contain introns and processed mRNA sequences that do not. This can be achieved through assigning particularly lenient gap-extension penalties (pp. 98–100).

Manual alignment

When there are very few gaps and the two sequences are not too different from each other in any other respect, a reasonable alignment can be obtained by visual inspection using either specialized alignment editors or plain text editors. The advantages of this method include (1) the use of the most powerful and trainable of all tools—the brain, and (2) the ability to directly integrate additional data, such as information of domain structure (Chapter 7). The disadvantages of this method are its **subjectivity** (i.e, the inability to formally specify the algorithm), **irreproducibility** (i.e., the inability of two researchers to reach the same result), **unscalability** (i.e., the inability to apply the method to long sequences), and **incommensurability** (i.e., the inability to compare the results to those derived from other methods).

The dot matrix method

In the **dot matrix method** (Gibbs and McIntyre 1970), the two sequences to be aligned are written out as column and row headings of a two-dimensional matrix (**Figure 3.9**). A **dot** (or **hit**) is put in the **dot matrix plot** at all positions where the nucleotides in the two sequences are identical. That is, a dot plotted at point (*x*,*y*) indicates that the nucleotide at position *x* in the first sequence is the same as the nucleotide at position *y* in the second sequence.

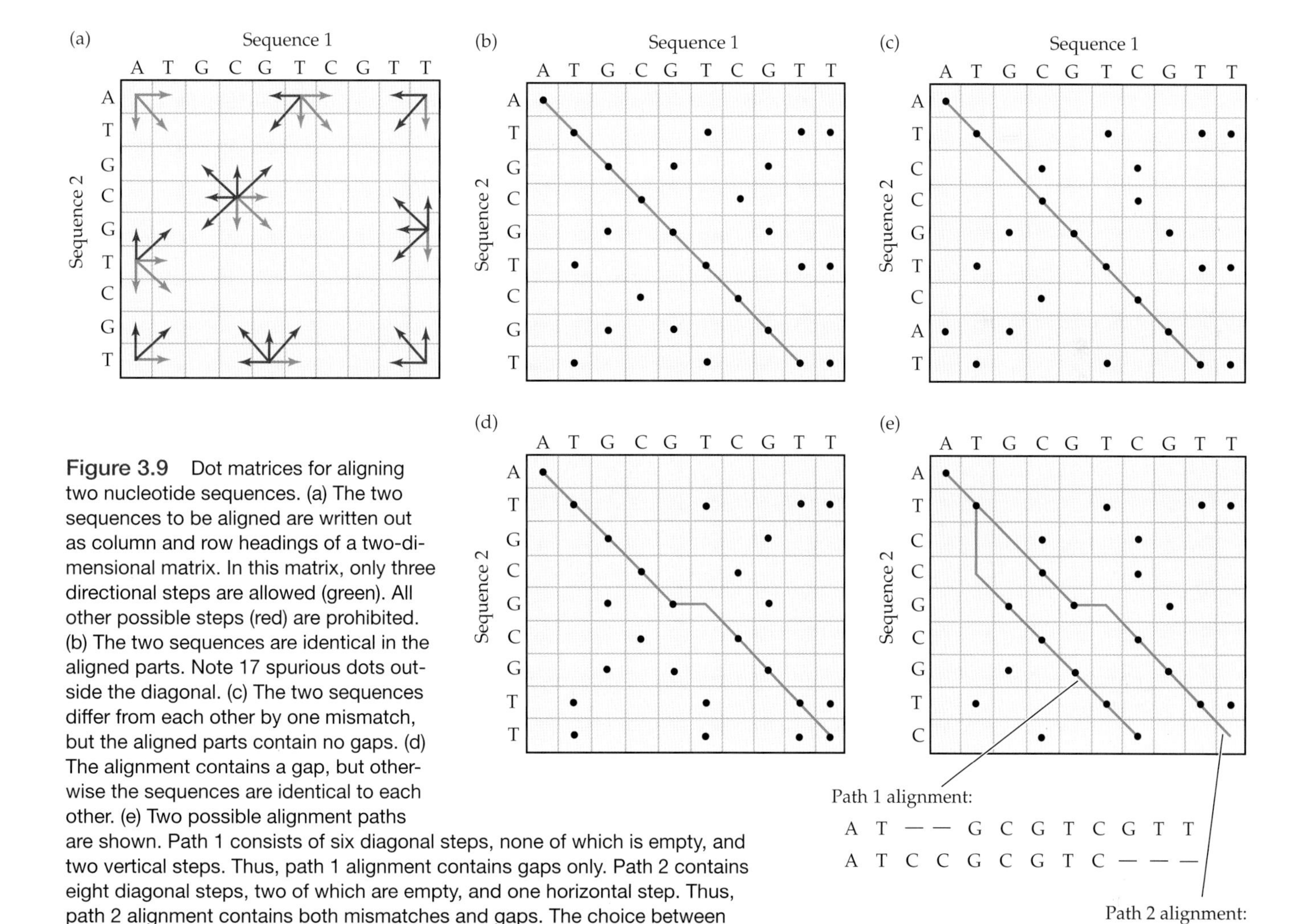

Figure 3.9 Dot matrices for aligning two nucleotide sequences. (a) The two sequences to be aligned are written out as column and row headings of a two-dimensional matrix. In this matrix, only three directional steps are allowed (green). All other possible steps (red) are prohibited. (b) The two sequences are identical in the aligned parts. Note 17 spurious dots outside the diagonal. (c) The two sequences differ from each other by one mismatch, but the aligned parts contain no gaps. (d) The alignment contains a gap, but otherwise the sequences are identical to each other. (e) Two possible alignment paths are shown. Path 1 consists of six diagonal steps, none of which is empty, and two vertical steps. Thus, path 1 alignment contains gaps only. Path 2 contains eight diagonal steps, two of which are empty, and one horizontal step. Thus, path 2 alignment contains both mismatches and gaps. The choice between the alignment in path 1 and that in path 2 depends on whether or not terminal gaps are allowed, as well as on the type of gap penalty used, i.e., which evolutionary sequence of events is more probable: a two-nucleotide gap (–) as in path 1, or a one-nucleotide gap and two substitutions (*) as in path 2.

The alignment is defined by a path through the matrix that starts with either the upper-left element or a nearby element in the topmost row or leftmost column and that ends in either the lower-right element or a nearby element in the bottommost row or the rightmost column. There are three allowed directional steps in this matrix: a diagonal step from left to right, a vertical step from top to bottom, and a horizontal step from left to right (Figure 3.9a). All other possible steps are disallowed. A diagonal step through a dot indicates a match, a diagonal step through an empty element of the matrix indicates a mismatch, a horizontal step indicates a null nucleotide in the sequence on the left of the matrix, and a vertical step indicates a null nucleotide in the sequence on the top of the matrix.

If the two sequences are completely identical (or if a sequence is compared to itself), there will be dots in all the diagonal elements of the matrix (Figure 3.9b). If the two sequences differ from each other only by nucleotide substitutions, there will be dots in most of the matrix elements on the diagonal (Figure 3.9c). If an insertion occurred in one of the two sequences but there are no substitutions, there will be an area within the matrix where the alignment diagonal will be shifted either vertically or horizontally (Figure 3.9d). In each of these three cases, the alignment is self-evident. If the two sequences differ from each other by both gaps and nucleotide substitutions, however, it may be difficult to identify the location of the gaps and to choose between several alternative alignments (Figure 3.9e). In such cases, visual inspection and the dot matrix may not be reliable, so more-rigorous methods are required.

As seen in **Figure 3.10a**, a dot matrix may become very cluttered. That is, many elements in the matrix, other than those representing the real alignment, are occupied by dots, thereby obscuring the alignment. The reason is that when two DNA sequences

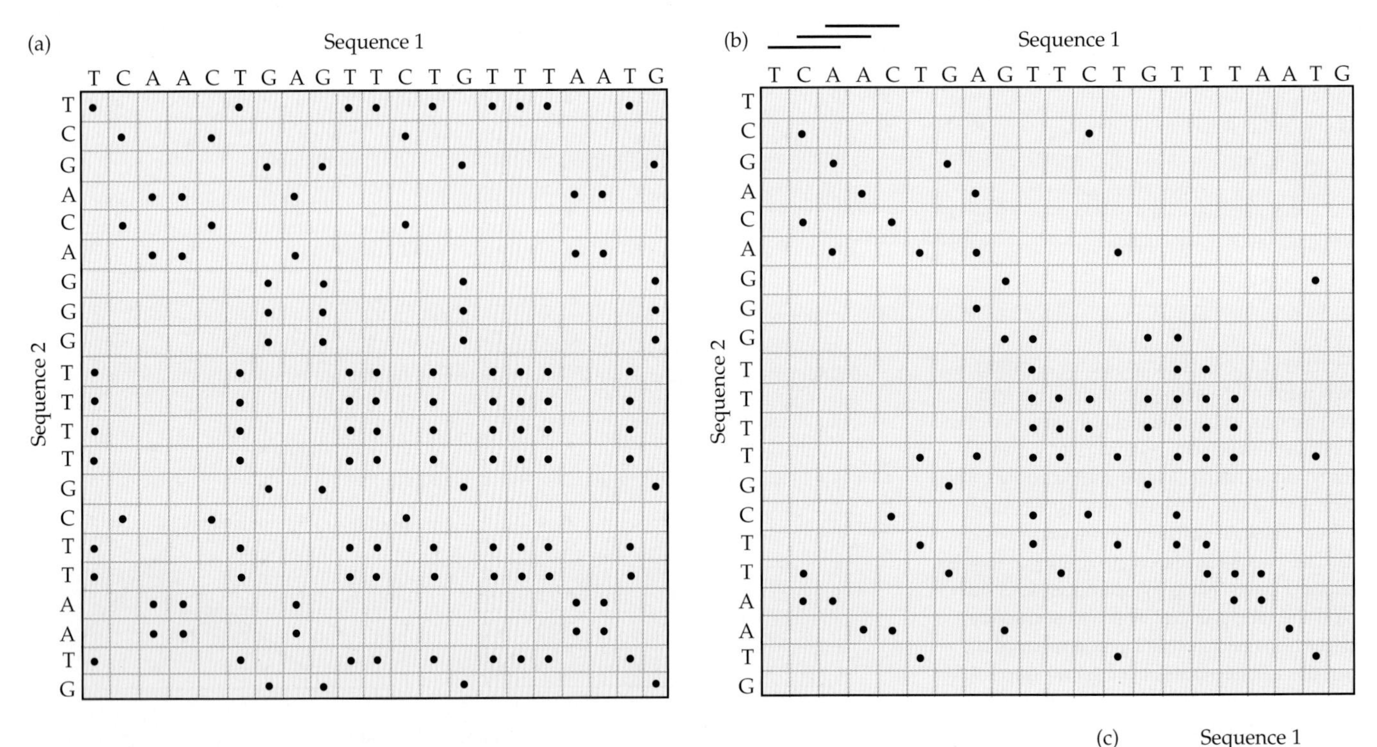

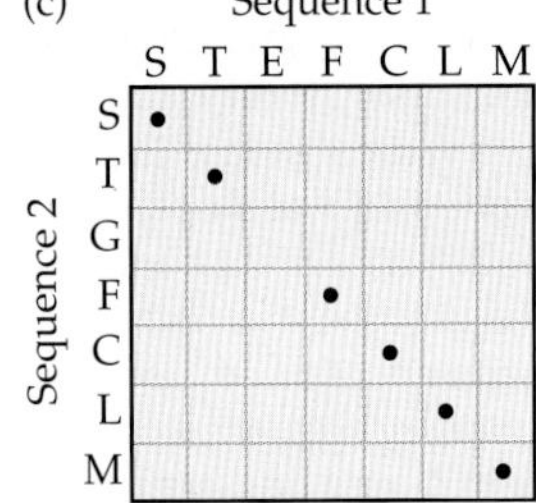

Figure 3.10 (a) Dot matrix for two nucleotide sequences. The true alignment is obscured by many spurious dots in the matrix. (b) Dot matrix for the two nucleotide sequences in (a), obtained by using a sliding window three nucleotides long and setting up a stringency threshold of two out of three matches for putting a dot in the middle of the window. The first three windows on the column headings are shown (overlapping horizontal lines). This filtering method clears many of the spurious dots, and the alignment now stands out on a less cluttered background. (c) Dot matrix for the two amino acid sequences obtained by translating the nucleotide sequences in (a). The alignment is now unambiguous. (Modified from Graur and Li 2000.)

are compared, approximately 25% of the elements in the matrix will be occupied by dots by chance alone.

One way to filter out spurious matches is to apply a **sliding-window smoothing** protocol. Thus, instead of single-nucleotide sites, overlapping fixed-length windows are made to slide along the two sequences. Three such windows are shown in **Figure 3.10b**. To qualify as a match, each comparison of these sliding windows is required to achieve a certain minimal threshold.

There are three parameters that affect the number of spurious matches and, hence, the resolution of the dot matrix plot: **alphabet size** (i.e., the number of letters in the alphabet used to write down the sequences), **window size**, and **stringency** (i.e., the minimum number of matches out of the length of the window size to qualify as a hit). The window size must be an odd number since the dot needs to be positioned in the middle element.

In Figure 3.10b, we use the DNA alphabet of four letters, a window size that is three nucleotides long, and a minimum stringency of 2/3, i.e., a dot is put in the matrix only when at least two of the three nucleotides show a match between the two sequences. Note that the alignment path is now more evident on a less cluttered background. For protein-coding regions, another way to enhance the visibility of the alignment path is to use the amino acid sequences instead of the DNA sequences. The increase in alphabet size from 4 to 20 and the decrease in the lengths of the sequences to one-third greatly decreases the number of spurious dots (**Figure 3.10c**).

We note, however, that dot matrix methods rely on the power of human cognition to recognize patterns indicative of similarity and to add gaps to the sequences to achieve an alignment. Moreover, it is very difficult to be sure that the obtained alignment is the best possible alignment. For this reason, scientists developed criteria for quantifying alignments as well as algorithms for identifying the best alignment according to these criteria.

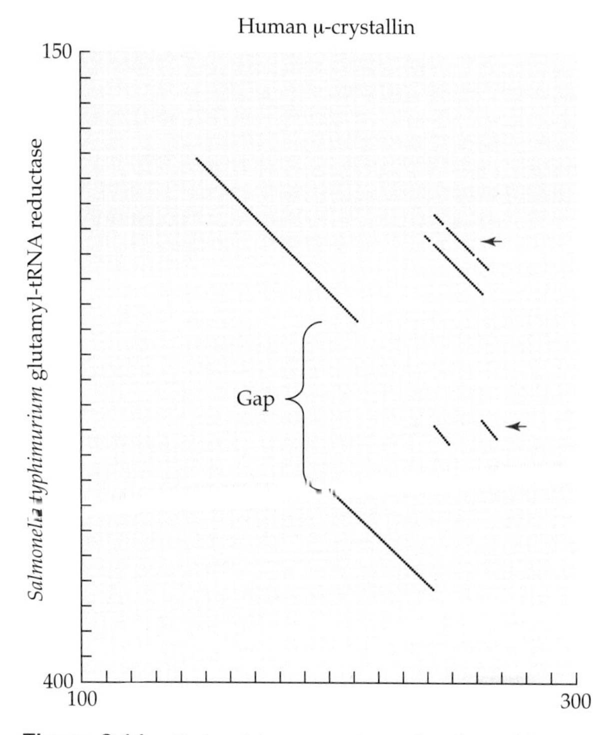

Figure 3.11 Dot matrix comparison of amino acid sequences between human μ-crystallin and glutamyl-tRNA reductase from *Salmonella typhimurium*. Only the C-terminal ends of the proteins were used. The window size was 60 amino acids and the stringency was 24 matches. Diagonals show regions of similarity. The vertical gap indicates that a coding region corresponding to approximately 75 amino acids has either been deleted from the human gene or inserted into the bacterial gene. The diagonally oriented parallel lines indicate duplicated regions within the genes (red arrows).

It should be added, however, that dot matrix plots are very useful in their own right in unraveling important information on the evolution of sequences. They make it very easy to visually identify insertions, deletions, inversions, repeated sequences, and inverted repeated sequences. For example, a dot matrix comparison between the amino acid sequences of human μ-crystallin (a lens protein) and glutamyl-tRNA reductase (an enzyme involved in the synthesis of the pigment porphyrin) from the bacterium *Salmonella typhimurium* indicates quite clearly that these two outwardly unrelated proteins share a common evolutionary origin (**Figure 3.11**). Moreover, it is very easy to see that (1) a coding region corresponding to approximately 75 amino acids in the protein has either been deleted from the human gene or inserted into the bacterial gene and (2) an internal duplication has occurred in the bacterial gene or one repeat out of two was deleted from the human gene.

Scoring matrices and gap penalties

As mentioned previously, the true alignment between two sequences is one that reflects accurately the evolutionary relationships between the sequences. Since the true alignment is unknown, in practice we search for an **optimal alignment**, i.e., the alignment in which the numbers of mismatches and gaps are minimized according to certain criteria (see below). Unfortunately, reducing the number of mismatches usually results in an increase in the number of gaps, and vice versa. For example, consider the following two sequences, A and B:

A: T C A G A C G A T T G $L_A = 11$

B: T C G G A G C T G $L_B = 9$

We can reduce the number of mismatches to zero as follows:

```
      T C A G – A C G – A T T G
(I)
      T C – G G A – G C – T – G
```

The number of gaps in this case is 6. Conversely, the number of gaps can be reduced to a single gap having the minimum possible size of $|L_A - L_B| = 2$ nucleotides, with a consequent increase in the number of mismatches:

```
      T C A G A C G A T T G
(II)      *     * * * *
      T C G G A G C T G – –
```

In this example, there is only one gap (two nucleotides in length), but the number of mismatches (indicated by asterisks) has increased to 5 (1 transition and 4 transversions).

Alternatively, we can choose an alignment that minimizes neither the number of gaps nor the number of mismatches. For example,

```
       T C A G – A C G A T T G
(III)                  *   *
       T C – G G A – G C T G –
```

In this case the number of mismatches is 2 (both transversions), and the number of gaps is 4.

So, which of the three alignments is preferable? It is obvious that comparing mismatches with gaps is like comparing apples with oranges. As a consequence, we must find a common denominator with which to compare gaps and mismatches. The common denominator is called the **gap penalty** or **gap cost**. The gap penalty is a factor (or a set of factors) by which gap values (e.g., the number of gaps, the lengths of the gaps, the distances between neighboring gaps) are multiplied to make the gaps equivalent in value to the mismatches. The gap penalties are based on our assessment of how frequently different types of insertions and deletions occur in evolution in comparison with the frequency of occurrence of point substitutions. Of course, we must also assign **mismatch penalties**, i.e., an assessment of how frequently different substitutions occur. In many cases, gap penalties are defined in terms of mismatch penalties in the form of "one gap equals X mismatches," i.e., the gap penalty is defined as a relative penalty. It should be emphasized that all gap and mismatch penalty systems are merely mathematical models that only imperfectly emulate and quantify relationships among different changes in DNA and protein sequences along different evolutionary lineages.

In principle, for each alignment, we must have a **scoring scheme**. The scoring scheme comprises a gap penalty (see below) and a **scoring matrix,** $\mathbf{M}(a,b)$ that specifies the score for each type of match ($a = b$) or mismatch ($a \neq b$). The units in a scoring matrix may be the nucleotides in DNA or RNA sequences, the codons in protein-coding regions, or the amino acids in protein sequences.

DNA scoring matrices are usually simple. In the simplest scheme all mismatches are given the same penalty (analogous to the one-parameter model). That is, $\mathbf{M}(a,b)$ is positive if $a = b$, and negative otherwise. In more complicated matrices a distinction may be made between transition and transversion mismatches (analogous to the two-parameter model). Finally, each type of mismatch may be penalized differently.

The scoring matrices for protein sequences are usually more complex than those for DNA sequences. Because replacements between amino acids with similar biochemical properties occur more frequently than replacements between amino acids that are dissimilar from each other (Chapter 4), amino acid mismatches are penalized either by the degree of conservation of chemical properties between the two amino acids in a mismatched pair, or by taking into account empirical information on the observed relative frequencies of replacements among the 20 amino acid residues.

One of the earliest scoring systems for amino acids was PAM (originally denoting *p*oint *a*ccepted *m*utations, but also referred to as *p*ercent *a*ccepted *m*utations). PAM was

	A	R	N	D	C	D	E	C	H	I	L	K	M	F	P	S	T	W	Y	V	B	Z	X	STOP
A	4	–1	–2	–2	0	–1	–1	0	–2	–1	–1	–1	–1	–2	–1	1	0	–3	–2	0	–2	–1	0	–4
R	–1	5	0	–2	–3	1	0	–2	0	–3	–2	2	–1	–3	–2	–1	–1	–3	–2	–3	–1	0	–1	–4
N	–2	0	6	1	–3	0	0	0	–1	–3	–3	0	–2	–3	–2	1	0	–4	–2	–3	3	0	–1	–4
D	–2	–2	1	6	–3	0	2	–1	–1	–3	–4	–1	–3	–3	–1	0	–1	–4	–3	–3	4	1	–1	–4
C	0	–3	–3	–3	9	–3	–4	–3	–3	–1	–1	–3	–1	–2	–3	–1	–1	–2	–2	–1	–3	–3	–2	–4
Q	–1	1	0	0	–3	5	2	–2	0	–3	–2	1	0	–3	–1	0	–1	–2	–1	–2	0	3	–1	–4
E	–1	0	0	2	–4	2	5	–2	0	–3	–3	1	–2	–3	–1	0	–1	–3	–2	–2	1	4	–1	–4
G	0	–2	0	–1	–3	–2	–2	6	–2	–4	–4	–2	–3	–3	–2	0	–2	–2	–3	–3	–1	–2	–1	–4
H	–2	0	1	–1	–3	0	0	–2	8	–3	–3	–1	–2	–1	–2	–1	–2	–2	2	–3	0	0	–1	–4
I	–1	–3	–3	–3	–1	–3	–3	–4	–3	4	2	–3	1	0	–3	–2	–1	–3	–1	3	–3	–3	–1	–4
L	–1	–2	–3	–4	–1	–2	–3	–4	–3	2	4	–2	2	0	–3	–2	–1	–2	–1	1	–4	–3	–1	–4
K	–1	2	0	–1	–3	1	1	–2	–1	–3	–2	5	–1	–3	–1	0	–1	–3	–2	–2	0	1	–1	–4
M	–1	–1	–2	–3	–1	0	–2	–3	–2	1	2	–1	5	0	–2	–1	–1	–1	–1	1	–3	–1	–1	–4
F	–2	–3	–3	–3	–2	–3	–3	–3	–1	0	0	–3	0	6	–4	–2	–2	1	3	–1	–3	–3	–1	–4
P	–1	–2	–2	–1	–3	–1	–1	–2	–2	–3	–3	–1	–2	–4	7	–1	–1	–4	–3	–2	–2	–1	–2	–4
S	1	–1	1	0	–1	0	0	0	–1	–2	–2	0	–1	–2	–1	4	1	–3	–2	–2	0	0	0	–4
T	0	–1	0	–1	–1	–1	–1	–2	–2	–1	–1	–1	–1	–2	–1	1	5	–2	–2	0	–1	–1	0	–4
W	–3	–3	–4	–4	–2	–2	–3	–2	–2	3	–2	–3	1	1	4	3	2	11	2	–3	–4	3	2	4
Y	–2	–2	–2	–3	–2	–1	–2	–3	2	–1	–1	–2	–1	3	–3	–2	–2	–2	7	–1	–3	–2	–1	–4
V	0	–3	–3	–3	–1	–2	–2	–3	–3	3	1	–2	1	–1	–2	–2	0	–3	–1	4	–3	–2	–1	–4
B	–2	–1	3	–4	–3	0	1	–1	0	–3	–4	0	–3	–3	–2	0	–1	–4	–3	–3	4	1	–1	–4
Z	–1	0	0	–1	–3	3	–4	–2	0	–3	–3	1	–1	–3	–1	0	–1	–3	–2	–2	1	4	–1	–4
X	0	–1	–1	–1	–2	–1	–1	–1	–1	–1	–1	–1	–1	–1	–2	0	0	–2	–1	–1	–1	–1	–1	–4
STOP	–4	–4	–4	–1	–4	–4	–4	–4	–4	–4	–4	–4	–4	–4	–4	–4	–4	–4	–4	–4	–4	–4	–4	1

Figure 3.12 The BLOSUM62 scoring matrix for amino acids. The one-letter abbreviation system for amino acids is used (Chapter 1, Table 1.3). B denotes aspartic acid or asparagine, Z denotes glutamic acid or glutamine, X denotes any amino acid or an unknown amino acid, and stop signs denote a termination codon.

put forward by Dayhoff et al. (1978). Here, we will present a class of empirical scoring matrices called BLOSUM (*blo*cks of *a*mino acid *su*bstitution *m*atrix) proposed by Henikoff and Henikoff (1992). BLOSUM scoring matrices are based on the frequencies of amino acids and the frequencies of each type of amino acid replacement observed in a database of gapless blocks of conserved to very conserved regions within protein families. Each match and each mismatch is assigned a score. Several BLOSUMs exist and they are denoted by a numeric suffix, such as BLOSUM45. The numeric suffix indicates the degree of similarity between the pairs of sequences on which the matrix was based. For example, BLOSUM62 is a matrix calculated from comparisons of sequences with no less than 62% divergence. BLOSUMs with high numeric suffixes are useful for comparing closely related sequences, while BLOSUMs with low numeric suffixes are designed for comparing distantly related sequences. Thus, for instance, BLOSUM80 should be used for less divergent alignments than BLOSUM45.

One such matrix, BLOSUM62, is shown in **Figure 3.12**. Note that all numbers on the diagonal are positive, i.e., matches are rewarded, but that the scores are different, implying, for instance, that during evolution alanine, isoleucine, leucine, serine, and valine change more frequently (low positive scores) than cysteine, histidine, and tryptophan (high positive scores). Note also that the matrix is symmetrical, i.e., a change from amino acid X to another amino acid Y has the same score as a change from Y to X. In most cases, mismatches are penalized; however, some mismatches are not (zero scores), and some are even rewarded (positive scores). One such case is the mismatch arginine (R) to lysine (K), which is assigned a positive score (+2).

A gap penalty is the cost of inserting or deleting k consecutive moieties (nucleotides or amino acids). The gap penalty has two components, a **gap-opening** (or **gap-open**) **penalty** (i.e., the cost of opening a one-nucleotide or one-amino-acid gap) and a **gap-extension penalty** (i.e., the cost of extending the gap by one nucleotide). There are many systems of assigning gap-extension penalties. In **Figure 3.13**, we present three systems. In the **fixed gap penalty system**, there are no gap-extension costs. In the **affine** or **linear gap penalty system**, the gap-extension cost is calculated by multiplying the gap length minus 1 (no gap-extension penalty is needed for gaps of size 1) by a constant representing the cost of increasing the gap by 1 moiety. For example, for a gap of length 1, the gap cost consists of the gap-opening penalty only; for a gap of length 3, the gap cost consists of the gap-opening penalty plus twice the gap extension penalty. Because the gap-extension cost for long gaps can become very large in the

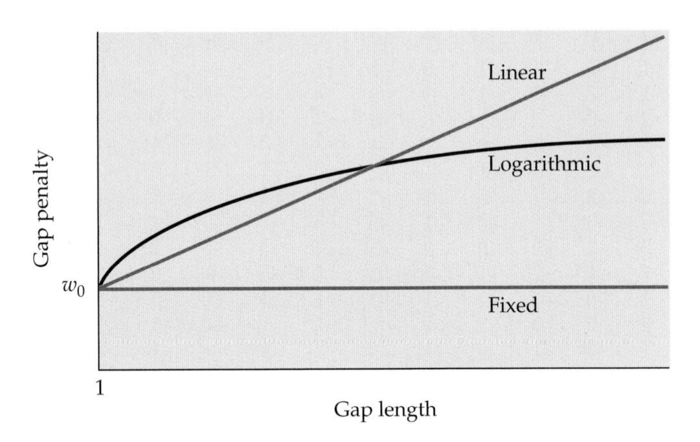

Figure 3.13 Three gap penalty systems. The gap-opening penalty in all three systems is w_0. In the linear gap penalty system, the gap-extension cost is the product of the gap length minus 1 and the gap-extension penalty for increasing the gap by 1. In the logarithmic gap penalty system, the gap-extension penalty increases with the logarithm of the gap length. In the fixed gap penalty system, there are no gap-extension costs. (Modified from Graur and Li 2000.)

linear system, some researchers have proposed the **logarithmic gap penalty** system as a way of reducing the cost of long gaps (e.g., Gu and Li 1995). In this system, the gap-extension penalty increases more slowly with gap length.

Let us now compare alignments I, II, and III (pp. 97–98) by using a scoring system in which matches and mismatches have a score of 0, a linear gap penalty system with the gap-opening penalty of –1, and a gap-extension penalty of –6. The scores for alignments I, II, and III are –6, –7, and –4, respectively. Therefore, out of these three alignments, alignment III is judged to be the best. By using a different scoring system, in which matches and mismatches have a score of 0, the gap-opening penalty is –1, and the gap-extension penalty is 0, we obtain scores of –6, –3, and –4 for alignments I, II, and III, respectively. Thus, with this system, alignment II is judged to be the best.

Alignment algorithms

The purpose of an alignment algorithm is to choose the alignment associated with the best score from among all possible alignments (the **optimal alignment**). The score associated with the optimal alignment is called the **optimal score**. Of course, the optimal alignment depends on the scoring scheme, and it may differ from optimal alignments obtained by using other scoring schemes. We note, however, that the number of possible alignments may be very large even for short sequences; for sequences usually encountered in molecular evolutionary studies, the number of possible alignments may literally be astronomical.

The number of pairwise alignments is given by

$$\binom{n+m}{\min(n,m)} = \frac{(n+m)!}{n!m!} \approx \sqrt{\frac{n+m}{2\pi nm}} \times \frac{(n+m)^{n+m}}{n^n m^m} \tag{3.48}$$

where n and m are the lengths of the two sequences, ! is the factorial, and π is the irrational mathematical constant 3.141593… . For example, when two 200-residue-long DNA sequences are compared, there are more than 10^{153} possible alignments (Torres et al. 2003). (In comparison, the number of protons in the universe is merely 10^{80}.) Fortunately, there are computer algorithms for searching for the optimal alignment between two sequences that do not require an exhaustive search of all the possibilities. Among the most frequently used methods are those of Needleman and Wunsch (1970) and Sellers (1974).

The **Needleman-Wunsch algorithm** uses **dynamic programming**, which is a general computational technique used in many fields of study. It is applicable when large computational tasks can be divided into a succession of small stages such that (1) the solution of the initial step is trivial, (2) each partial solution in a later stage can be calculated by reference to only a small number of solutions in earlier stages, and (3) the final stage contains the overall solution. Dynamic programming can be applied to alignment problems because alignment scores obey the following rule:

$$\max S_{1\to x,\to y} = \max\left[\left(\max S_{1\to x-1,\,1\to y} + S_{x,\,y|x-1,y}\right), \left(\max S_{1\to x,\,1\to y-1} + S_{x,\,y|x,y-1}\right), \left(\max S_{1\to x-1,\,1\to y-1} + S_{x,\,y|x-1,y-1}\right)\right] \tag{3.49}$$

in which $\max S_{1\to x,1\to y}$ is the best alignment score for the two sequences up to residue x in the first sequence and residue y in the second sequence, and $S_{x,y|x,y-1}$ is the score for aligning residue x in the first sequence with residue y in the second sequence conditional on the previous matrix element (the pointer) being $(x,y–1)$.

Dynamic programming for pairwise sequence alignment requires three steps: (1) **matrix initialization**, (2) **matrix fill**, and (3) **traceback**. In the matrix-initialization step, a two-dimensional matrix with $m + 1$ columns and $n + 1$ rows is created, where m and n correspond to the lengths of the two sequences to be aligned. The sequences are, then, arranged as column and row headings with the topmost row and the leftmost column filled with zeros. Next, we proceed to fill in the elements of the matrix by starting in the upperleft corner and finding the best score, $\max S_{x,y}$, for each element in the matrix. The best score for an element (x, y) is determined by adding the appropriate scores to the those from the element to the left $(x - 1, y)$, above $(x, y - 1)$, and above and to the left diagonally $(x - 1, y - 1)$. As in the dot matrix method, only three directions are allowed: from left to right and from top to bottom (denoting gaps) and diagonally from top left to bottom right (denoting matches and mismatches). For each element in the matrix the best score is calculated. Concomitantly, the position of the best alignment score in the previous matrix element is stored. This stored position is called a **pointer**. The relationship between the new $S_{1\rightarrow x,1\rightarrow y}$ value and the pointer is represented by an **arrow**. In the last stage, called the traceback, we proceed from right to left and follow the best pointers. The graph of pointers in the traceback, also referred to as the **path graph**, defines the paths through the matrix that correspond to the optimal alignment. Before the traceback stage, however, we need to decide whether we will allow terminal gaps or not. If we do not allow terminal gaps, then the traceback starts at the bottom-right corner of the matrix. If we allow terminal gaps, then the traceback starts with the highest score in either the rightmost column or the bottommost row.

Figure 3.14 shows an example of a dynamic programming alignment of two sequences, GAATTCAGT and GGATCGA. The score for all matches was set at +5, the penalty for all mismatches at –3, and the penalty for a gap at –4 regardless of length (i.e., the gap-extension penalty was set to zero). The matrix initialization and the allowable steps are shown in Figure 3.14a. The filled-in matrix and the arrows are shown in Figure 3.14e. We note that an element may have more than a single pointer, and that arrows from many elements may lead to the same pointer. Next, we need to make a decision about the allowability of terminal gaps. In this example, if we do not allow terminal gaps, the traceback starts with element (7,9). If we allow terminal gaps, then the traceback starts with element (7,7). The completed traceback for the no-terminal-gaps case is shown in Figure 3.14f; two paths are defined by the traceback, representing two optimal alignments. The completed traceback for the terminal-gaps-allowed case is in Figure 3.14g; again, two equally optimal alignments are identified.

One of the most important things to remember is that the resulting alignment depends on the choice of score as well as on the gap penalties, which in turn depend on crucial assumptions about the pattern of substitution and the frequency of gap events relative to the frequency of point substitutions. For example, consider the alignment of human pancreatic hormone precursor and chicken pancreatic hormone (**Figure 3.15**). When the two sequences are aligned with no gap penalties, the similarity between these homologous sequences is not evident (Figure 3.15a). However, when gap penalties are applied, the similarity becomes clear (Figure 3.15b). The resemblance becomes even more enhanced if we consider not only perfect matches (**identities**) between the amino acids but also imperfect matches (**similarities**) between them (Figure 3.15c).

Finally, we need to emphasize yet again that the meaning of "optimal alignment" is the alignment with the highest score according to a predetermined scoring scheme. It is not necessarily the most biologically meaningful alignment.

Multiple-sequence alignment

Multiple-sequence alignment can be viewed as an extension of pairwise sequence alignment, but the complexity of the computation grows exponentially with the number of sequences being considered and their lengths, and therefore it is not feasible to exhaustively search for the optimal alignment even for a modest number of short sequences. For example, there are approximately 10^{38} possible alignments that can be produced from five 10-nucleotide-long sequences (Slowinski 1998). Even if examining each alignment will take a second, it will take significantly longer than the age of the universe to complete an exhaustive search for the optimal alignment. Unfortunately,

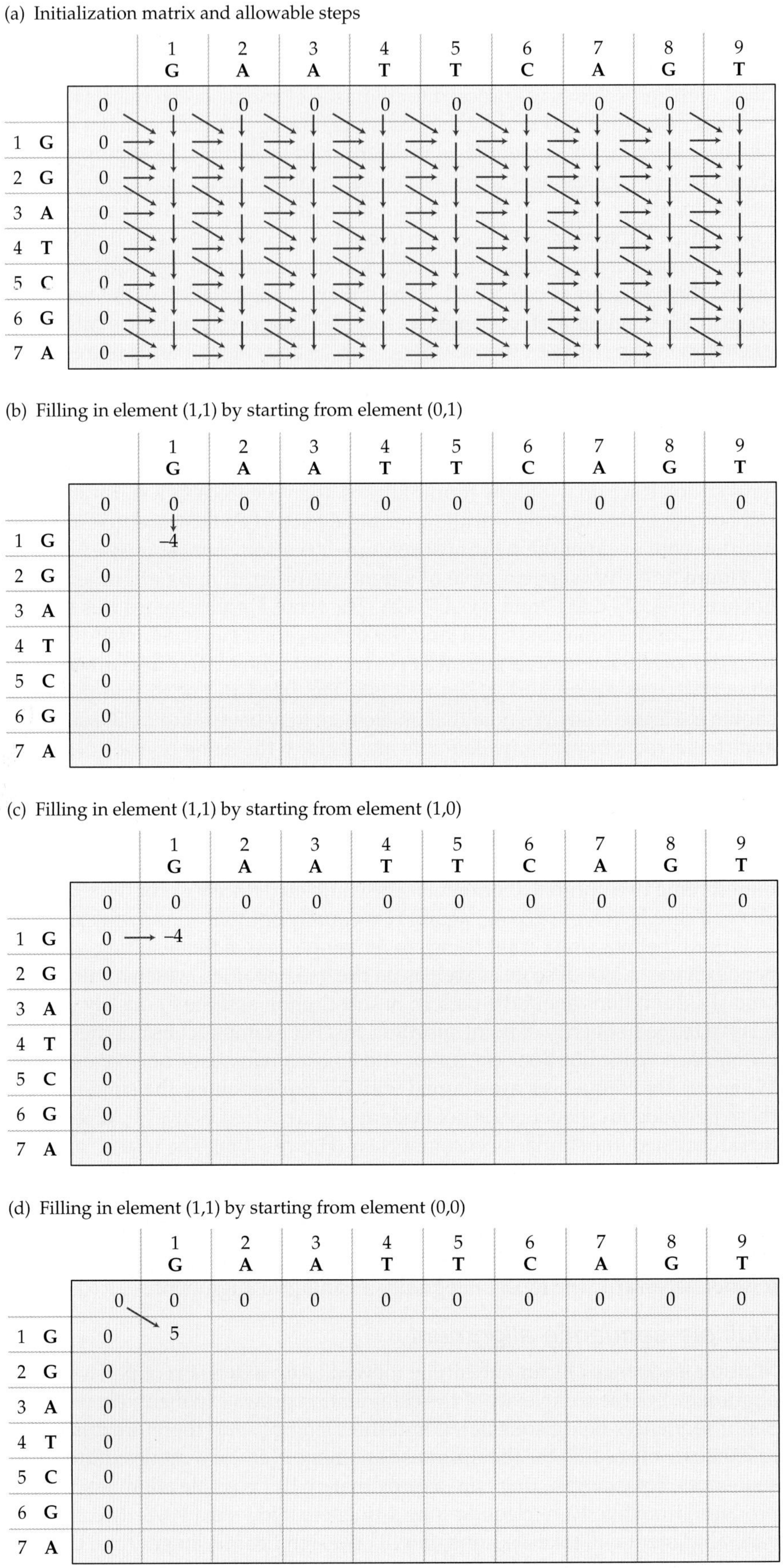

Figure 3.14 Pairwise alignment by dynamic programming of two sequences, GAATTCAGT and GGATCGA. The scores for matches and mismatches were set at +5 and –3, respectively. The gap-opening and the gap-extension penalties were set at –4 and 0, respectively. (a) The matrix initialization and the allowable steps. (b) Filling in element (1,1) by starting from element (0,1) requires opening a gap, thus the score is –4. (c) Filling in element (1,1) by starting from element (1,0) requires opening a gap, thus the score is –4. (d) Filling in element (1,1) by starting from element (0,0) results in a match, thus the score is +5. Obviously, this score is the highest of the three possible scores, so element (0,0) becomes the pointer. (e) The completed matrix and the arrows. (f) Under the assumption that no terminal gaps are allowed, we start the traceback at element (9,7). There are two path graphs, indicating that the two alignments below the matrix are equally optimal. (g) If we allow terminal gaps, then the traceback starts with element (7,7), which is the highest score from among all elements in the rightmost column and the bottommost row. Again, two path graphs are identified, indicating that the two alignments below the matrix are equally optimal.

(a) Initialization matrix and allowable steps

		1 G	2 A	3 A	4 T	5 T	6 C	7 A	8 G	9 T
	0	0	0	0	0	0	0	0	0	0
1 G	0									
2 G	0									
3 A	0									
4 T	0									
5 C	0									
6 G	0									
7 A	0									

(b) Filling in element (1,1) by starting from element (0,1)

		1 G	2 A	3 A	4 T	5 T	6 C	7 A	8 G	9 T
	0	0	0	0	0	0	0	0	0	0
1 G	0	–4								
2 G	0									
3 A	0									
4 T	0									
5 C	0									
6 G	0									
7 A	0									

(c) Filling in element (1,1) by starting from element (1,0)

		1 G	2 A	3 A	4 T	5 T	6 C	7 A	8 G	9 T
	0	0	0	0	0	0	0	0	0	0
1 G	0	–4								
2 G	0									
3 A	0									
4 T	0									
5 C	0									
6 G	0									
7 A	0									

(d) Filling in element (1,1) by starting from element (0,0)

		1 G	2 A	3 A	4 T	5 T	6 C	7 A	8 G	9 T
	0	0	0	0	0	0	0	0	0	0
1 G	0	5								
2 G	0									
3 A	0									
4 T	0									
5 C	0									
6 G	0									
7 A	0									

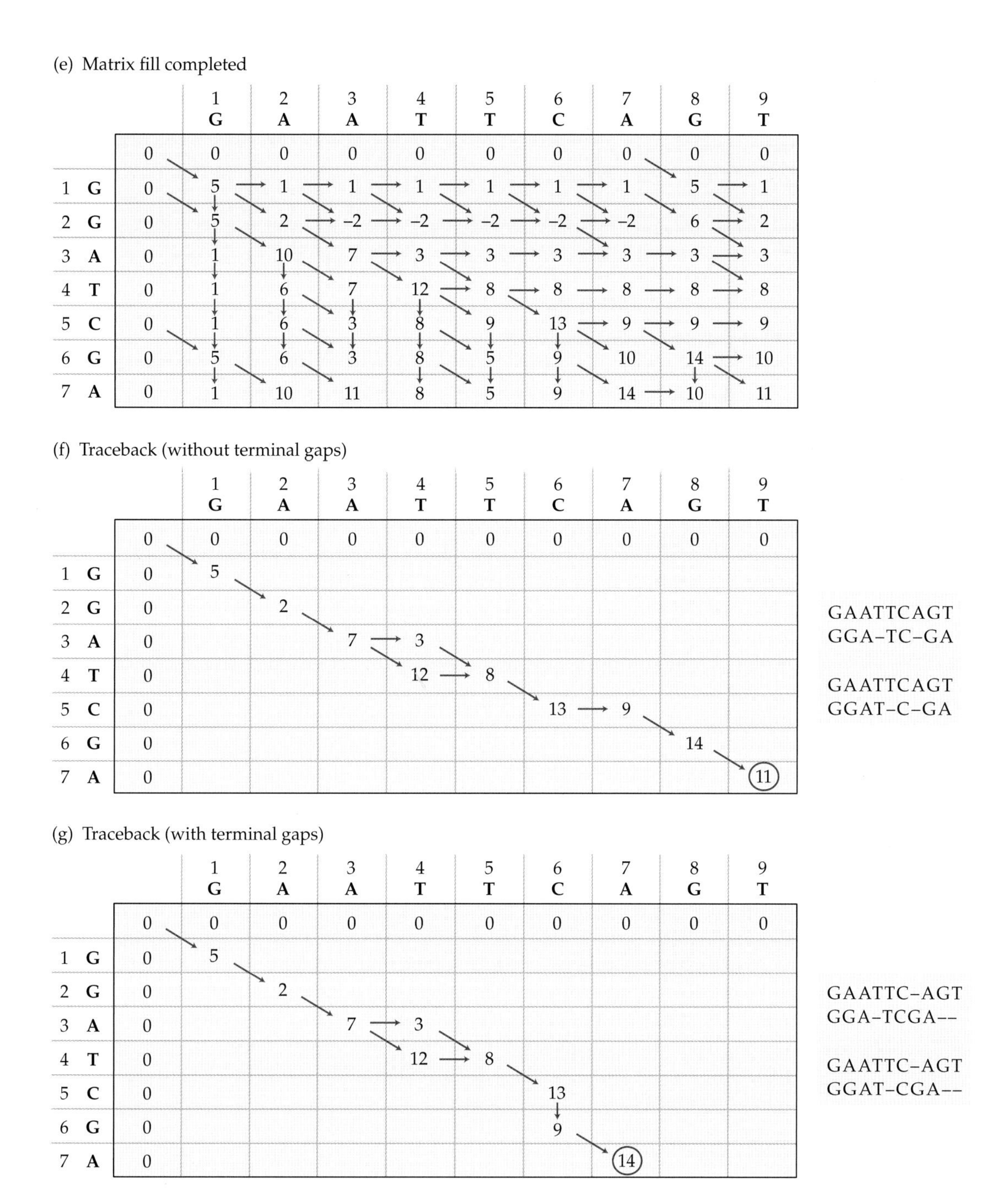

shortcuts, such as dynamic programming, do not work for alignment problems involving three sequences or more.

Many heuristic multiple-alignment methods have been proposed in the literature, and for the vast majority of them computer programs are available. The most popular such programs are CLUSTAL (Higgins and Sharp 1988; Higgins et al. 1992; Higgins 1994; Thompson et al. 1994, 1997; Higgins et al. 1996; Jeanmougin et al. 1998; Larkin et al. 2007; Sievers et al. 2011), COFFEE (Notredame et al. 1998, 2000; O'Sullivan et al. 2004; Wallace et al. 2006; Moretti et al. 2007), DIALIGN (Morgenstern et al. 1998a, 1998b; Morgenstern 1999, 2004; Subramanian et al. 2005, 2008), MAFFT (Katoh et al. 2002, 2005a, 2005b; Katoh and Toh 2008), MUSCLE (Edgar 2004), and PRANK (Löytynoja and Goldman 2005). All these programs use some sort of "quick and dirty" progressive algorithms (e.g., Feng and Doolittle 1987), in which a new sequence is added to a group of already aligned sequences in order of decreasing similarity.

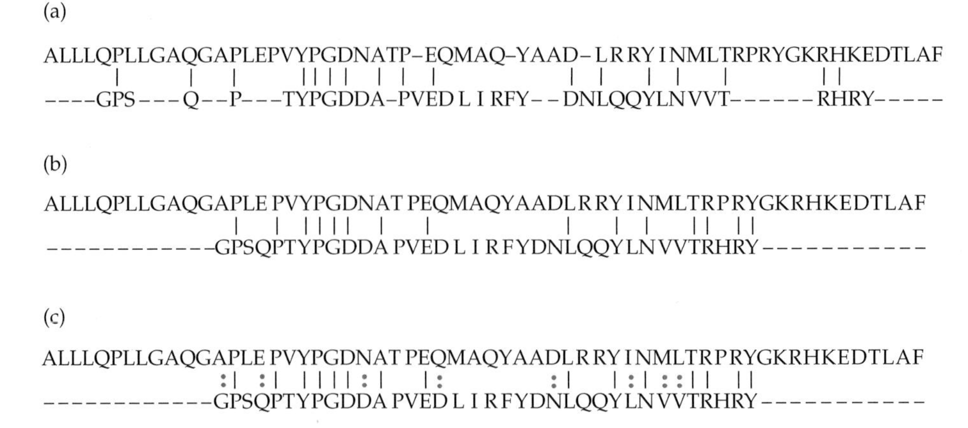

Figure 3.15 The effect of gap penalties on pairwise alignment between two proteins. The alignment of human pancreatic hormone precursor and chicken pancreatic hormone is shown. Perfect matches (identities) are indicated by vertical straight lines. (a) The penalty for gaps is 0. (b) The gap penalty for a gap of size k nucleotides was set at $wk = 1 + 0.1k$. (c) The same alignment as in (b), but the similarity between the two sequences is enhanced by showing pairs of biochemically similar amino acids (blue dots). (From Graur and Li 2000.)

Here, we describe in broad brushstrokes the multiple-alignment strategy used in the CLUSTAL programs. There are three distinct stages in the strategy (see **Figure 3.16** for a flowchart). In the first stage, the degrees of similarity (or scores) between all possible sequence pairs are computed according to a predetermined scheme. In the second stage, a **guide tree** is constructed from the similarity matrix generated in the first stage. (Different methodologies of converting similarity matrices into trees are described in Chapter 5.) Because we are interested in handling a large number of sequences, a speedy method of tree reconstruction should be chosen. In the third stage, sequences are added one by one to the growing multiple-sequence alignment. The order in which the sequences are added is determined by the guide tree. When a new sequence is added to an existing alignment (also called a **cluster** or a **profile**), the cluster is regarded as fixed and any gaps introduced into it are placed as columns of gaps in all the sequences in the cluster. A mismatch between a cluster and a newly added sequence is scored as the average of scores between the newly added sequence and each of the sequences in the cluster.

A simple example involving three very short amino acid sequences is shown in **Figure 3.17**. With three sequences, there are three possible pairwise comparisons. With a scoring scheme of +1 for a match and –1 for a gap regardless of size, sequences 1 and 2 turn out to be the most similar pair and will, therefore, cluster first in the guide tree. We note that in the alignment of sequences 1 and 2, there is a gap of two amino acids starting at position 2 in sequence 2. In the next step we align the cluster with sequence 3. This alignment results in a gap of one amino acid following position 16 in the cluster.

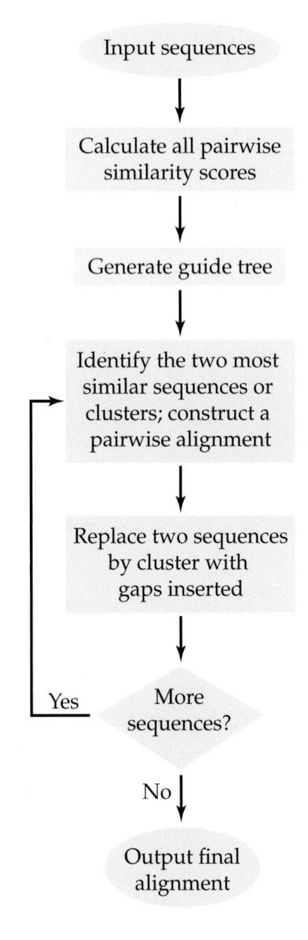

Figure 3.16 Flowchart of the multiple-alignment strategy used in CLUSTAL. (Modified from Higgins and Sharp 1988.)

Quality of alignments

Because of the heuristic nature of multiple-sequence alignment methods, and the fact that most alignment programs report only a single alignment, the accuracy of multiple alignments should not be taken for granted. Unfortunately, little thought is usually devoted to the possibility that multiple-sequence alignment methods may yield artifactual results (Morrison 2009), and few people pay attention to the caveat that multiple-sequence alignments are "a very useful starting point for manual refinement" (Thompson et al. 1994). The vast majority of multiple-sequence alignments are treated in a manner reminiscent of laboratory disposables; that is, they are produced automatically and discarded unthinkingly on the road to some other goal, such as a phylogenetic tree or a three-dimensional structure. We conjecture that more than 99% of all multiple-sequence alignments that are used to produce publishable results are never even seen by a human being. Yet, when a rare alignment is actually inspected by a researcher, it is usually found wanting. Moreover, errors in multiple-sequence alignment tend to be amplified in subsequent analyses. For example, multiple-sequence alignment errors

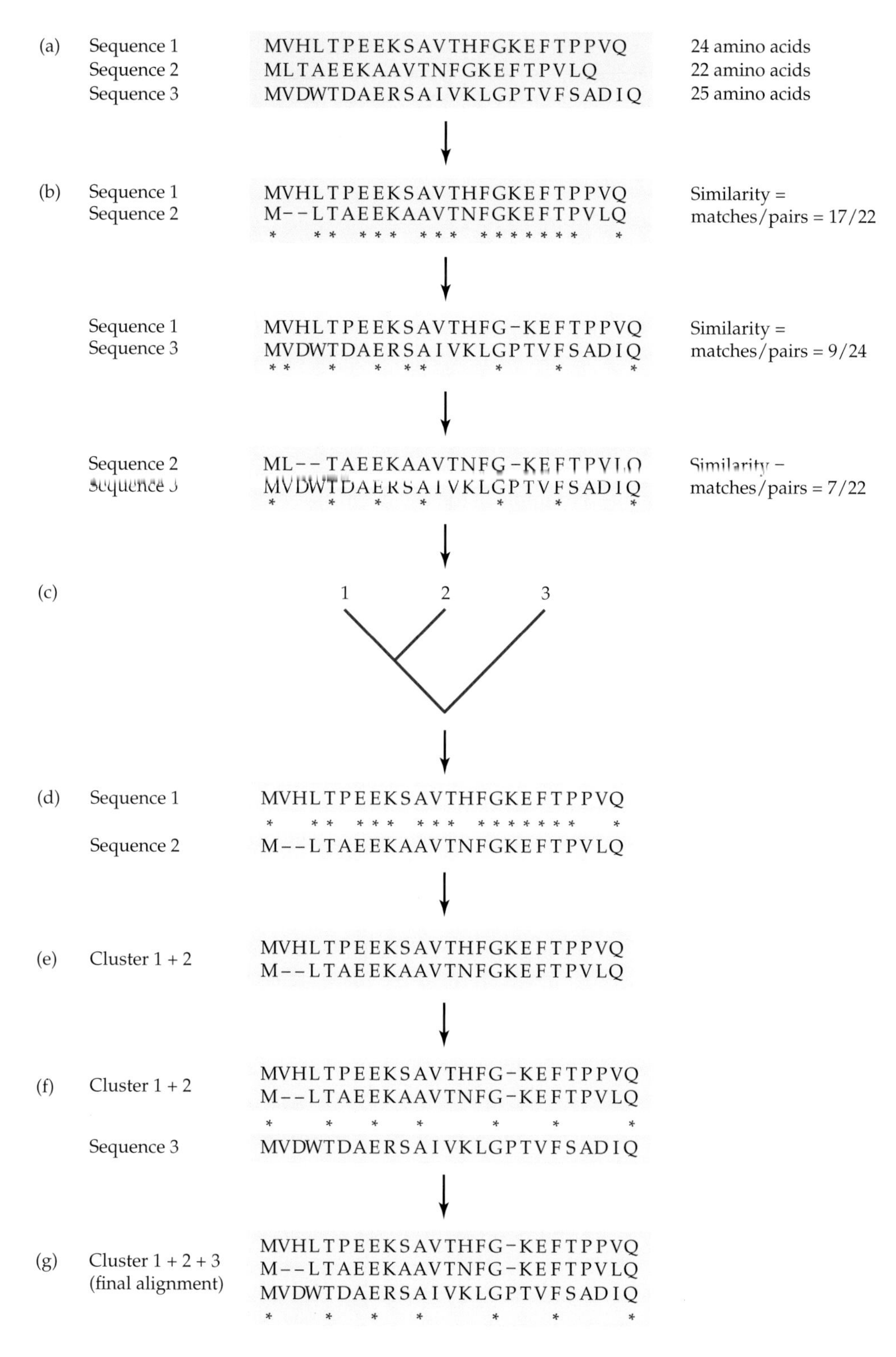

Figure 3.17 A simple example illustrating the algorithmic steps in aligning three sequences. (a) We start with three sequences of 24, 22, and 25 amino acids, respectively. (b) Three pairwise alignments are constructed and the pairwise sequence similarities are computed. (c) The pairwise similarities are used to construct the guide tree. (d) Following the guide tree, sequences 1 and 2 are the first to be aligned. (e) Aligned sequences 1 and 2 are replaced by cluster 1+2. (f) Pairwise alignment of cluster 1+2 with sequence 3. (g) The alignment of cluster 1+2 with sequence 3 is replaced by cluster 1+2+3. Since there are no more sequences to align, cluster 1+2+3 is the final alignment. Asterisks and dashes denote matches and gaps, respectively.

can have significant downstream effects on phylogenetic topological reconstructions (Ogden and Rosenberg, 2006). Multiple-sequence alignments are so notoriously inadequate that the literature is littered with phrases such as "the alignment was subsequently corrected by visual or manual inspection" (e.g., O'Callaghan et al. 1999; Kawasaki et al. 2000). Unfortunately, visual inspection is neither an objective nor a reproducible method, while science strives for both objectivity and reproducibility.

Here we describe one simple method for identifying and quantifying the uncertainties in multiple-sequence alignments. The method, called **HoT** (**H**eads **o**r **T**ails),

(a) "Head" alignment

```
AAGT---GTAGACGGAT
||||   || ||  |||
AAGTGTGGTGGAGTGAT
```

(b) Reversed head alignment

```
TAGGCA---GATGTGAA
|||  |   | ||||||
TAGTGAGGTGGTGTGAA
```

(c) Tail alignment

```
AAGTGTAG---ACGGAT
|||||| |   |  |||
AAGTGTGGTGGAGTGAT
```

(d) HoT comparison

```
AAGT---GTAGACGGAT
||||   || ||  |||
AAGTGTGGTGGAGTGAT

AAGTGTAG---ACGGAT
|||||| |   |  |||
AAGTGTGGTGGAGTGAT
```

Figure 3.18 The HoT method for assessing alignment reliability. (a) The two sequences are aligned in their original orientation, the Head alignment. (b) The sequences are then aligned in the reverse orientation. (c) The alignment in (b) is unreversed, resulting in the Tail alignment. (d) A comparison of the two alignments reveals two identical regions of alignment (shaded). These are the reliable parts of the alignment. The total length of the alignment is 17 residues, of which 10 are reliably aligned (~59%) and 7 are not.

is based upon the a priori expectation that sequence alignment results should be independent of the orientation of the input sequences (Landan and Graur 2007). Thus, reversing residue order prior to alignment should yield the exact reverse of the alignment obtained by using the unreversed sequences. Such "ideal" alignments, however, are the exception in real-life settings, and the two alignments, which we term the heads and tails alignments, are usually different to a greater or lesser degree. The degree of agreement (or disagreement) between these two alignments may be used to assess the reliability of the sequence alignment.

In **Figure 3.18** we show an example of the use of HoT to assess alignment reliability. Other methods for assessing alignment reliability have been proposed in the literature (e.g., Penn et al. 2010; Kim and Ma 2011; Wu et al. 2012).

Alignment of Genomic Sequences

As genomes evolve, they undergo large-scale evolutionary changes that present a challenge to sequence alignment not posed by short sequences. **Genome alignment** must simultaneously account for genome rearrangements, horizontal gene transfer, deletions, insertions, duplications, nucleotide substitutions, and massive shuffling of segments. Algorithms for genome alignment usually go through three main steps (Ureta-Vidal et al. 2003). In the first step, the algorithm identifies **seeds** (i.e., short, exact or nearly exact matching strings of nucleotides found in both sequences to be aligned). In the next step, the seeds are used as **points of extension** or **anchoring points** to partition the genome alignment into many regular **subalignment** problems. Finally, global alignment is performed on each of the subalignments. Particular genome alignment algorithms (e.g., Bray et al. 2003; Darling et al. 2004, 2010) differ from one another in the order in which the three main steps are taken, the extent of recursiveness, and the details of additional steps, such as the deletion of sequence blocks to allow extension of alignment colinearity. An example of a pairwise genome alignment between two archaebacterial genomes is given in **Figure 3.19**.

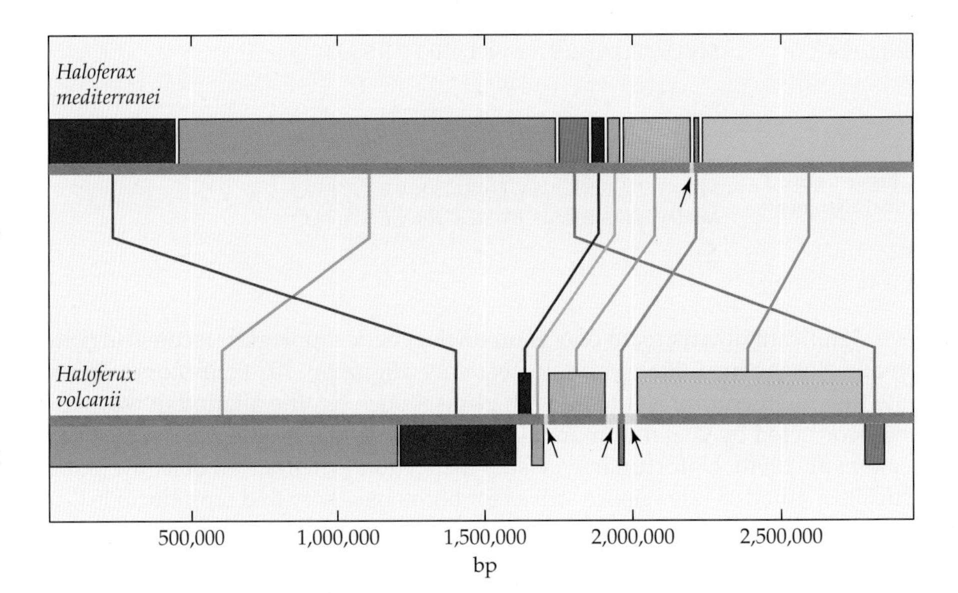

Figure 3.19 A pairwise genome alignment between the halophile (literally, salt-loving) archaebacteria *Haloferax mediterranei* (from the Mediterranean Sea) and *H. volcanii* (from the Dead Sea). The alignment algorithm used was MAUVE (Darling et al. 2004). Eight homologous blocks have been identified (different colors). A change in polarity due to inversion relative to *H. mediterranei* is denoted by the placement of the block under the chromosome (gray bar). Lines of appropriate colors connect the homologous genomic blocks, also referred to as syntenic regions. Three small regions in *H. volcanii* and one region in *H. mediterranei* (yellow, arrows) have no homologous counterparts in the other species. (Courtesy of Aaron Darling.)

CHAPTER 4

Rates and Patterns of Molecular Evolution

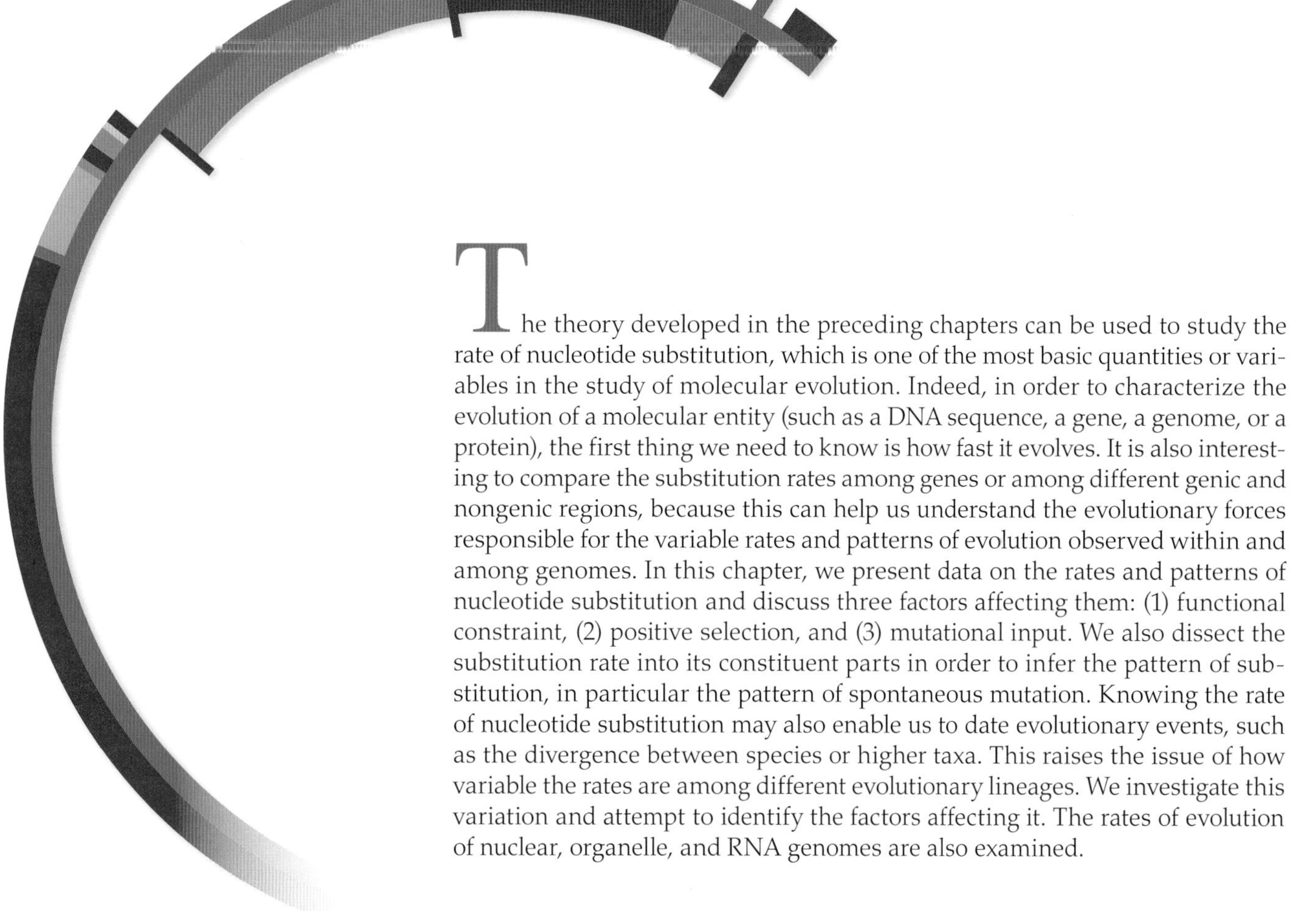

The theory developed in the preceding chapters can be used to study the rate of nucleotide substitution, which is one of the most basic quantities or variables in the study of molecular evolution. Indeed, in order to characterize the evolution of a molecular entity (such as a DNA sequence, a gene, a genome, or a protein), the first thing we need to know is how fast it evolves. It is also interesting to compare the substitution rates among genes or among different genic and nongenic regions, because this can help us understand the evolutionary forces responsible for the variable rates and patterns of evolution observed within and among genomes. In this chapter, we present data on the rates and patterns of nucleotide substitution and discuss three factors affecting them: (1) functional constraint, (2) positive selection, and (3) mutational input. We also dissect the substitution rate into its constituent parts in order to infer the pattern of substitution, in particular the pattern of spontaneous mutation. Knowing the rate of nucleotide substitution may also enable us to date evolutionary events, such as the divergence between species or higher taxa. This raises the issue of how variable the rates are among different evolutionary lineages. We investigate this variation and attempt to identify the factors affecting it. The rates of evolution of nuclear, organelle, and RNA genomes are also examined.

Rates of Point Mutation

Mutation is the ultimate source of genetic novelty. For this reason, the rate at which new mutations arise is a central issue in genetics and evolution. At the molecular level, the rate of mutation is defined as the number of new sequence variants arising in a predefined target region per unit time. Target regions are usually nucleotides (in single-stranded DNA and RNA viruses), base pairs (in organisms with double-stranded genomes), genes, gametes, chromosomes, or genomes. Time units may be taxon-specific, such as replication time or generation time, or they may be absolute, such as chronological time. Because mutations occur infrequently and because many of them are highly deleterious or lethal (and therefore unobservable), estimating rates of mutation used to be extraordinarily difficult. Prior to the development of rapid sequencing methods, only very few estimates of mutation rates existed in the literature (Nachman 2004).

There are essentially four main approaches to estimating mutation rates. The first and oldest of these approaches was developed by Danforth (1923) and Haldane (1927, 1935). This approach was based on the assumption that deleterious alleles in a population exist because of a balance between the process of mutation that creates these alleles and purifying selection that eliminates them. This assumption provided a straightforward method to estimate the mutation rate for deleterious alleles in cases in which the frequency of an allele could be measured and the strength of purifying selection could be estimated. Of course, although many mutations are harmful, not all of them are; thus, the inferred rates of mutation for deleterious alleles will be an underestimate of the total mutation rate. For autosomal dominant mutations, the equilibrium allele frequency, q, can be calculated as

$$q = \frac{u}{|s|} \tag{4.1}$$

where μ is the mutation rate and $|s|$ is the absolute value of the selection coefficient associated with the deleterious mutant. The mutation rate is, therefore,

$$\mu = q|s| \tag{4.2}$$

For X-linked recessive mutations, the mutation rate is

$$\mu = \frac{q|s|}{3} \tag{4.3}$$

Let us now consider the case of hemophilia A, which is caused by deleterious mutations in the X-linked gene that encodes coagulation factor VIII. Let us assume that most people with hemophilia A do not reproduce, i.e., that their fitness is approximately 0 ($w \approx 0$ and $s \approx -1$). Given that one-third of the X chromosomes in the population are carried by males and that males are chiefly the ones expressing the disease phenotype, then the mutation rate should roughly be three times the frequency of male carriers. By using empirically determined values of s and q, Haldane (1935) inferred that mutations causing hemophilia A arise at a mean rate of roughly 2×10^{-5} per generation.

The second method of estimating mutation rate is based on the theoretical result that the neutral mutation rate is equal to the rate of substitution for neutral alleles (Chapter 2; Kimura 1968b). Thus, homologous stretches of nonfunctional DNA, which presumably evolve solely by random genetic drift, can be compared between two species to calculate the amount of sequence divergence. If the generation time and the time since the two species diverged from each other are known, then the mutation rate per generation can be easily estimated. This approach has been used for synonymous sites and for pseudogenes in comparisons between human and chimpanzee, and these studies yielded estimates of neutral mutation rates of about $1 - 2 \times 10^{-8}$ per nucleotide site per generation (Kondrashov and Crow 1993; Drake et al. 1998; Nachman and Crowell 2000). Given that the total length of the human diploid genome is approximately 7×10^9 base pairs, we can multiply the mutation rate per site by the genome size and deduce that each newborn in the human population carries around 100 new point mutations not found in their parents.

The third method of inferring mutation rate is the most direct. In this method, large nonrecombining stretches of DNA, such as on human Y chromosomes, are sequenced in individuals related to one another by descent. One such pedigree is shown in **Figure 4.1**. In this example (Xue et al. 2009), about 10 Mb of Y-chromosome DNA was sequenced from two males separated by 13 generations. The common ancestor of the two individuals was born in 1805, i.e., about 200 years before the analysis. Four mutations were discovered. Thus, the mean rate of mutation in the human nuclear genome was estimated to be $4 / (10^7 \times 200 \times 2) = 1 \times 10^{-9}$ mutations per site per year.

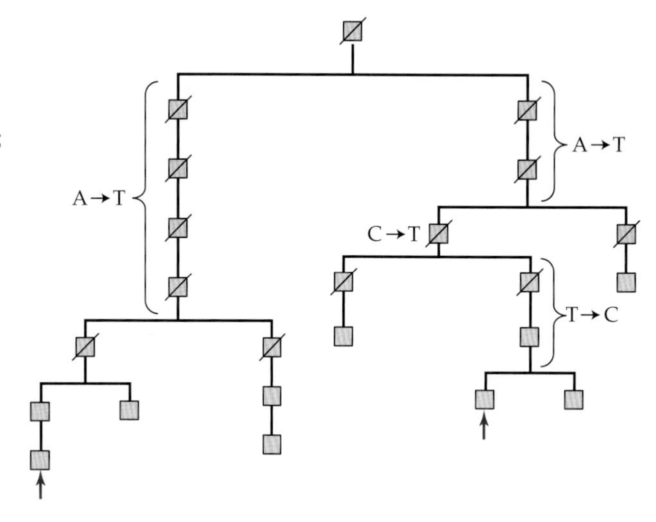

Figure 4.1 Pedigree of two male probands (marked by red arrows) separated by 13 generations (~400 years) from which chromosomes Y were sequenced. Four point mutations were inferred, two transitions (C → T and T → C) and two transversions (A → T). The occurrence of only one mutation (C → T) could be attributed to a particular individual; three mutations were inferred to have occurred on branches containing multiple individuals (brackets). Squares indicate live male individuals from which DNA was obtained. Crossed squares denote deceased individuals. (Modified from Xue et al. 2009.)

Obviously, the confidence interval of an estimate based on so few data points should be very large.

With the dramatic drop in the cost of sequencing, from about $3 billion per human genome in the 1990s to less than $5,000 per genome in the last 5 years (Hayden 2012), it has become economically feasible to sequence whole genomes of parents and children and identify all mutations. We note, however, that fast sequencing methods are error prone, so the sequencing error rate may at times exceed the rate of mutation. Be that as it may, current estimates of the average mutation rate in humans converge on a value of approximately 1×10^{-8} mutations per site per generation (Roach et al. 2010; Campbell et al. 2012; Kong et al. 2012; Michaelson et al. 2012; Sun et al. 2012), which translates into a mutation rate of about 3×10^{-10} mutations per site per year under the assumption that the generation time in humans is 30 years. These rates indicate that each newborn in the human population carries on average ~70 new mutations in his or her diploid genome (Roach et al. 2010).

In humans, mutation rates have been found to vary widely throughout the genome (Wolfe and Li 2003; Michaelson et al. 2012), with 0.02% and 0.5% of the genome being located in mutational hotspots and coldspots, respectively (Michaelson et al. 2012). In particular, transitions at CpG sites are much more frequent than at other sites (Sommer 1995; Nachman and Crowell 2000; Kondrashov 2003). Interestingly, while mutation rates varied widely, the transition-to-transversion ratio of about 2.3 remained approximately constant throughout the genome (Roach et al. 2010). We note that mutations having large deleterious effects on fitness cannot be observed through the sequencing of the genomes of adults. The rate of lethal and extremely deleterious mutations can be obtained by sequencing the haploid DNA of gametes, such as sperm cells, from an individual whose somatic genome is known (e.g., Wang et al. 2012).

The fourth approach is only applicable to experimental organisms with short generation times. In this approach one can overcome the problem of detecting deleterious mutations, by using the fact that a mutation will behave as a neutral mutation as long as its selective disadvantage, s, is smaller than $1/(2N_e)$, where N_e is the effective population size (Chapter 2). By keeping N_e at the absolute minimum, one can ensure that all but the most deleterious mutations will be observed. For instance, Denver et al. (2004) used a population of the hermaphrodite nematode, *Caenorhabditis elegans*, in which a single individual in each generation was used to produce the succeeding generation. This experiment, which lasted hundreds of generations, yielded a mean estimate of about 10^{-6} point mutations per site per year; that is, the mean mutation rate in *C. elegans* was inferred to be about three orders of magnitude larger that that in the nuclear genome of mammals. Ossowski et al. (2010) used a similar experimental setup in the plant *Arabidopsis thaliana* and inferred a spontaneous mutation rate of 7×10^{-7} nucleotide substitutions per site per generation, the majority of which were G:C → A:T transitions.

The rate of mutation in mammalian mitochondrial DNA has been estimated to be at least ten times higher than the average nuclear rate. The mean rate of mutation in the mitochondrial genome of *Drosophila* has been estimated to be approximately 6 × 10^{-8} per site per generation, which translates into a rate of about 10^{-6} per site per year (Haag-Liautard et al. 2008).

Mutation rates in viruses span a range of approximately six orders of magnitude, from the single-stranded RNA swine vesicular stomatitis virus at about 10^{-3} mutations per site per year, to the double-stranded DNA herpesvirus at about 10^{-9} mutations per site per year (Duffy et al. 2008).

Rates of Segmental Mutations

In the literature, there are as yet only few estimates of spontaneous rates of mutation for insertions, deletions, recombinations, and inversions. One of the first studies on the rates and patterns of deletions and insertions was conducted by Ophir and Graur (1997). By using 156 independently derived processed pseudogenes from humans, mice, and rats, they concluded that deletions occur approximately three times faster than insertions. Deletions were found to occur on average once every 40 nucleotide substitutions, whereas insertions were much rarer, occurring once every 100 substitutions, indicating that the mechanisms involved in deletion formation are most probably different from those responsible for the formation of insertions. As with point mutations (p. 155), deletions and insertions also occurred at a higher rate in murids (mouse and rat) than in humans. The age of the pseudogene, however, explained less than 20% of the variation in the number of deletions and insertions per site, indicating that factors other than evolutionary time may play a significant role in the evolutionary dynamics of indel accumulation. In the nuclear genomes of mammals, insertions and deletions are known to occur much less frequently than nucleotide substitutions. Kondrashov (2003) estimated that deletion and insertion mutations occur at a rate that is approximately 25 times slower than that for point mutations. In the human nuclear genome, short deletions (≤60 bp) occur more frequently than short insertions by a ratio of about 3:1 (**Figure 4.2**; Zhang and Gerstein 2003).

By using the genomes of a mother, a father, and their two children, Roach et al. (2010) identified 155 crossover events occurring during parental meioses. The events were not distributed randomly along the genome; 92 of them occurred in hotspots of recombination. Crossover events in maternal meioses occurred almost two times more frequently than in paternal meioses, in contrast to the tendency of point mutations to occur more frequently in males than in females (pp. 126–128).

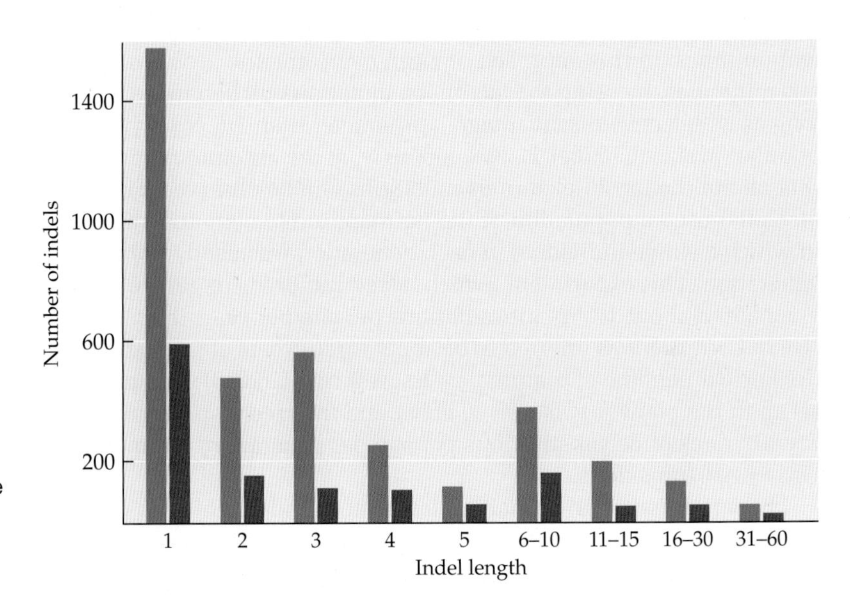

Figure 4.2 Length distribution of indels from 1,726 human processed pseudogenes comprising more than 700,000 nucleotides. Only deletions and insertions shorter than 60 bp are shown. The sample consists of 3,740 deletions and 1,291 insertions. (Data from Zhang and Gerstein 2003. Courtesy of Nicholas Price.)

Rates of Nucleotide Substitution

The **rate of nucleotide substitution**, r, is defined as the number of substitutions per site per year. The mean rate of substitution can be calculated by dividing the number of substitutions, K, between two homologous sequences by $2T$, where T is the time of divergence between the two sequences (**Figure 4.3**). That is,

$$r = \frac{K}{2T} \tag{4.4}$$

T is assumed to be the same as the time of divergence between the two species from which the two sequences were taken, and it is usually inferred from paleontological and biogeographical data. When Equation 4.4 is applied to closely related species, such as humans and chimpanzees, it is essential that we take into account the allelic divergence (polymorphism) that had existed within the ancestral population prior to the divergence event (Takahata and Satta 1997; Edwards and Beerli 2000; Nichols 2001).

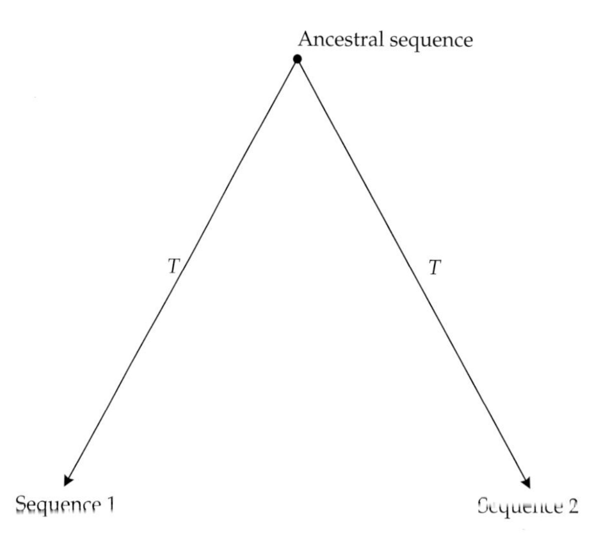

Figure 4.3 Divergence of two homologous sequences from a common ancestral sequence T years ago.

In this section we will deal with the issue of rate variation among genes and among different regions in a gene. For this purpose, it is advisable to use the same species pair for all the genes under consideration. The reason is twofold. First, there are usually considerable uncertainties about paleontological estimates of divergence times. By using the same pair of species, we can compare rates of substitution among genes without knowledge of the divergence time. Second, the rate of substitution may vary considerably among lineages. In this case, differences in rates between two genes may be due to differences between lineages rather than to differences that are attributable to the genes themselves.

Obtaining a reliable estimate of the rate of nucleotide substitution requires that the degree of sequence divergence be neither too small nor too large. If it is too small, the rate estimate will be influenced by large chance effects, whereas if it is too large, the estimate may be unreliable because of the difficulties in correcting for multiple substitutions at the same site (Chapter 3).

Rates of substitution in protein-coding sequences

For reasons of functionality and medical importance, protein-coding genes have traditionally attracted a great deal of attention from both molecular and evolutionary biologists. Because of the difficulties in determining which parts of the 5′ and 3′ flanking regions can be meaningfully associated with a gene, in the following discussion we will address only the transcribed parts, i.e., 5′ and 3′ untranslated regions, coding regions, and introns. In dealing with coding regions, it is important to discriminate between nucleotide changes that affect the primary structure of the protein, i.e., nonsynonymous substitutions, and changes that do not affect the protein, i.e., synonymous changes.

In **Table 4.1** we list the rates of synonymous and nonsynonymous substitution for 48 protein-coding genes. The genes are arranged in order of increasing rate of nonsynonymous substitution. The rates were obtained from comparisons between human and mouse homologous genes. Times of divergence of placental mammals are hotly disputed in the literature (Graur and Martin 2004). Here, we set the human-mouse divergence event at 80 million years ago, which is close to the minimum age estimate in the literature.

We note that the rate of nonsynonymous substitution is extremely variable among genes. In our sample, it ranges from effectively zero in the core histones H4 and H3 to about 3.1×10^{-9} substitutions per nonsynonymous site per year in interferon γ. Nonsynonymous nucleotide substitutions are, of course, reflected in the rates of protein evolution, which may vary widely. As is well known, certain proteins (e.g., histones, actins, ribosomal proteins, and ubiquitins) are extremely conservative. Ubiquitin, for instance, is completely conserved between human and *Drosophila*, and it differs

Table 4.1

Rates of synonymous and nonsynonymous nucleotide substitutions in various mammalian protein-coding genes

Gene	Number of codons compared	Nonsynonymous rate[a]	Synonymous rate[a]
Histone H4	102	0.00	3.94
Histone H3	376	0.00	4.52
Ubiquitin	76	0.00	7.35
Actin α	376	0.01	2.92
Ribosomal protein S14	150	0.02	2.16
Translation elongation factor 2	857	0.02	4.37
Ribosomal protein S17	134	0.06	2.69
Aldolase A	363	0.09	2.78
Myosin β heavy chain	1933	0.10	2.15
Preprosomatostatin	348	0.10	2.67
Hydroxanthine phosphoribosyltransferase	217	0.12	1.57
Creatine kinase M	380	0.15	2.72
Lipoprotein lipase	437	0.19	2.95
Lactate dehydrogenase A	331	0.19	4.06
Glyceraldehyde 3-phosphate dehydrogenase	332	0.20	2.30
Glutamine synthetase	371	0.23	2.95
Prion protein	224	0.29	3.89
Preproglucagon	540	0.33	2.81
Thymidine kinase	232	0.43	3.93
β1 Adrenergic receptor	412	0.45	3.07
α-Globin	141	0.56	4.38
Insulin-like growth factor II	179	0.57	2.01
Myoglobin	153	0.57	4.10
Fibrinogen γ	411	0.58	4.13

among animals, plants, and fungi by only 2 or 3 of 76 amino acid residues. Hemoglobins, myoglobin, and some carbonic anhydrases have evolved at intermediate rates, while apolipoproteins, immunoglobulins, interferons, and interleukins have evolved very rapidly.

The rate of synonymous substitution also varies considerably from gene to gene, though much less than the nonsynonymous rate. To demonstrate this difference, let us use the coefficient of variation, which is defined as the ratio of the standard deviation to the mean and which measures the extent of variability as a percentage of the mean of the population. The coefficient of variation for the nonsynonymous rates in Table 4.1 is 95%, whereas that for the synonymous rates is only 31%.

In the vast majority of genes, the synonymous substitution rate greatly exceeds the nonsynonymous rate. For example, the mean rate of nonsynonymous substitution for the genes in Table 4.1 is 0.75×10^{-9} substitutions per nonsynonymous site per year,

Table 4.1 *(continued)*

Gene	Number of codons compared	Nonsynonymous rate[a]	Synonymous rate[a]
Amylase	506	0.63	3.42
Preproinsulin	110	0.64	5.25
Adenine phosphoribosyltransferase	179	0.68	3.56
Atrial natriuretic factor	149	0.72	3.38
Tumor necrosis factor	231	0.76	2.91
Erythropoietin	191	0.77	3.56
β-Globin	146	0.78	2.58
Carbonic anhydrase I	260	0.84	3.22
Albumin	590	0.92	5.16
Apolipoprotein β	1,035	0.95	3.29
Luteinizing hormone	140	1.05	2.90
Apolipoprotein E	291	1.10	3.72
Immunoglobulin Ig V_H	100	1.10	4.76
Preproparathyroid hormone	345	1.27	3.30
β-Low-density lipoprotein receptor binding domain	273	1.32	3.64
Urokinase-plasminogen activator	430	1.34	3.11
Growth hormone	189	1.34	3.79
Interferon α1	166	1.47	3.24
Interleukin-1	265	1.50	3.27
Apolipoprotein A-I	235	1.64	3.97
Immunoglobulin Ig κ	106	2.03	5.56
Interferon β1	159	2.38	5.33
Preprorelaxin	549	2.43	6.23
Interferon γ	136	3.06	5.50
Average[b]		0.75 (0.71)	3.65 (1.15)

Source: Modified from Li (1997).

[a]All rates are based on comparisons between human and mouse or rat genes. The time of divergence was set at 80 million years ago. Rates are in units of substitutions per site per 10^9 years.

[b]The average is the arithmetic mean; values in parentheses are the standard deviations computed over all genes.

whereas the mean rate of synonymous substitution is 3.65×10^{-9} substitutions per synonymous site per year, i.e., about five times higher. An extreme example illustrating the difference between synonymous and nonsynonymous substitution rates is shown in **Figure 4.4**. In this comparison of ubiquitin genes from human and yeast, the inferred number of synonymous substitutions per synonymous site is 5.86 (which almost certainly is indicative of saturation), and the inferred number of nonsynonymous substitutions per nonsynonymous site is 0.03. Thus, synonymous substitutions are inferred to have accumulated about 200 times faster than nonsynonymous substitutions. The distribution of K_A-to-K_S ratios in over 13,000 orthologous protein-coding genes from human and chimpanzee are shown in **Figure 4.5**.

(a) Nucleotide sequences

Human	ATG	CAG	ATC	TTC	GTC	AAG	ACC	CTG	ACT	GGT	AAG	ACC	ATC	ACT	CTC	GAA	GTG	GAG	CCG	AGT
Yeast		A	T		C	A	T	A	A	G		T	A	C	A	G	T	A	T T	TCC

Human	GAC	ACC	ATT	GAG	AAT	GTC	AAG	GCA	AAG	ATC	CAA	GAC	AAG	GAA	GGC	ATC	CCT	CCT	GAC	CAG
Yeast		T		C	C		A	AGT	A	T		T	A		T			G	T	A

Human	CAG	AGG	TTG	ATC	TTT	GCT	GGG	AAA	CAG	CTG	GAA	GAT	GGA	CGC	ACC	CTG	TCT	GAC	TAC	AAC
Yeast		A		T			T	G	A	A			T	A A		T				

Human	ATC	CAG	AAA	GAG	TCC	ACC	CTG	CAC	CTG	GTG	CTC	CGT	CTT	AGA	GGT	GGG
Yeast		A	G	A	T	T	T		T		T G	A A	G			T

(b) Amino acid sequences

Human	MQIFVKTLTG	KTITLEVEPS	DTIENVKAKI	QDKEGIPPDQ	QRLIFAGKQL	EDGRTLLSDYN	IQKESTLHLV	LRLRGG
Yeast		S	D S					

Figure 4.4 Preponderance of synonymous substitutions over nonsynonymous substitutions as revealed by the alignment of ubiquitin genes (a) and proteins (b) from human and yeast (*Saccharomyces cerevisiae*). The human sequences (tan) are shown in their entirety; for the yeast sequences (blue), only sites that differ from the human sequences are shown. The two genes differ from each other at 58 nucleotide positions (a), but at only three amino acid positions (b).

At fourfold degenerate sites (Chapter 3) it is possible to compare the rate of transitional substitution with that of transversional substitution (**Table 4.2**), since both types of substitution are synonymous. The rate of transition tends to be higher than that of transversion, although at each fourfold degenerate site, two types of transversional change and only one type of transitional change can occur. This observation can be largely explained by the fact that in mammals, transitional mutations occur more frequently than transversional ones. We note, however, that this mutational bias is not universal (Keller et al. 2007). At twofold degenerate sites, the rate of transitional substitution is slightly lower than that at fourfold degenerate sites, but the rate of transversional substitution is only about a quarter the corresponding rate at fourfold degenerate sites. The rate of transversion at twofold degenerate sites is low because the vast majority of these transversional changes are nonsynonymous. (The exceptions to this rule involve the following synonymous pairs: UUA ↔ CUA, UUG ↔ CUG, CGA ↔ AGA, and CGG ↔ AGG). At nondegenerate sites, at which all changes are nonsynonymous, the rates of transitional and transversional substitution are on average about the same

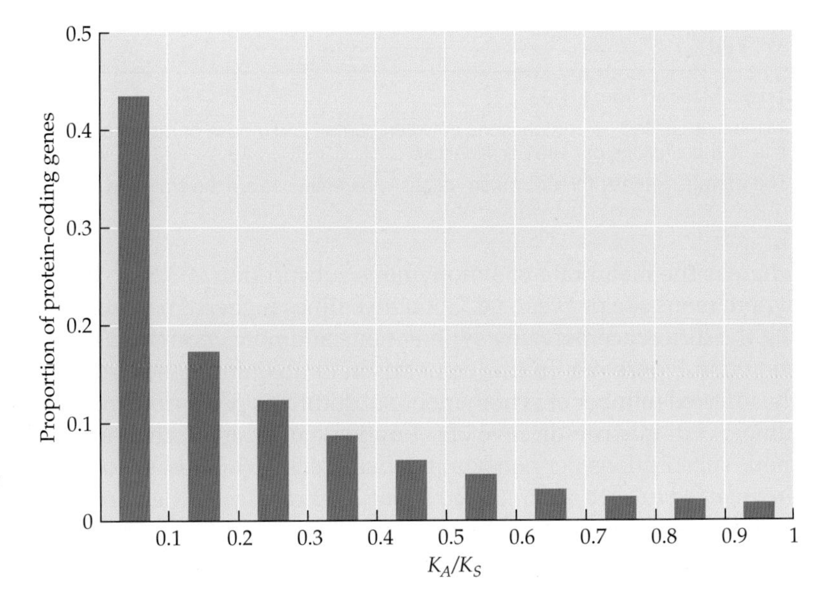

Figure 4.5 Histogram of ratios of nonsynonymous substitution (K_A) to synonymous substitution (K_S) rates estimated from 11,218 pairs of orthologous protein-coding genes from human and chimpanzee (Mikkelsen et al. 2005). The number of pairs in Mikkelsen et al. was 13,454, but 136 could not be used because their K_S was equal to 0. (Courtesy of Yichen Zheng.)

Table 4.2

Rates of transitional and transversional substitutions at nondegenerate, twofold degenerate, and fourfold degenerate codon sites[a]

Type of substitution	Nondegenerate	Twofold degenerate	Fourfold degenerate
Transitions	0.40	1.86	2.24
Transversions	0.38	0.38	1.47
All substitutions	0.78	2.24	3.71

Source: From Li (1997).

[a]All rates are based on comparisons between human and mouse or rat genes. The time of divergence was set at 80 million years ago. Rates are in units of substitutions per site per 10^9 years.

and are usually considerably lower than the corresponding values at fourfold degenerate sites. As a rule, the rates of nucleotide substitution are lowest at nondegenerate sites, intermediate at twofold degenerate sites, and highest at fourfold degenerate sites.

Rates of substitution in noncoding regions

In the following discussion, we distinguish among five classes of noncoding regions: 5′ and 3′ untranslated regions, introns, intergenic regions, and pseudogenes. Pseudogenes are DNA sequences that were derived from functional genes but have been rendered nonfunctional by various mechanisms (Chapters 1, 7, and 9). Here, we treat pseudogenes and intergenic regions separately because we are certain that the vast majority of sequences defined as "pseudogenes" in the literature are indeed devoid of function, whereas for many intergenic parts, we cannot be as certain.

In **Table 4.3**, we present numbers of substitutions per site in different coding and noncoding regions based on a genomic comparison between human and chimpanzee. We can roughly classify the rates into groups. The first group consists of pseudogenes, which have on average the highest rate of substitution. The second group consists of the fast-evolving introns and intergenic regions. The third group, which evolves at slightly slower rates than the first group, consists of 5′ and 3′ untranslated regions and fourfold degenerate sites. The fourth group, which evolves at the slowest rate, consists of coding regions, in particular nondegenerate sites, which evolve nine times slower than pseudogenes. We note, however, that considerable variation exists within each coding and noncoding category as well as among lineages. For example, first introns in primates are significantly longer and evolve much faster than introns with higher ordinal numbers (Gazave et al. 2007), while first introns in murids and ciliates are longer but evolve more slowly than the rest of the introns (Gaffney and Keightley 2006; Zheng et al. 2013).

On the basis of a comparison between 3,165 human and mouse genes, Waterston et al. (2002) discovered that the rates of substitution were not uniform within any of the gene regions. There were peaks of conservation at the transition from one region to another, for example at the border between the 5′ untranslated region and the initiation codon, or at the border between the stop

Table 4.3

Mean number of nucleotide substitutions per aligned nucleotide site ($\overline{K}$) between human and chimpanzee

Region	Number of contiguous regions	Aligned length (Kb)	$\overline{K}$
5′ Untranslated	582	68.7	0.001
Fourfold degenerate sites	1,226	61.9	0.011
Nondegenerate sites	1,226	272.8	0.002
Coding	1,226	426.7	0.005
Introns	2,691	597.7	0.012
3′ Untranslated	1,071	321.2	0.009
Intergenic	5,604	1268.1	0.013
Pseudogenes	737	752.4	0.020

Source: Data from Nachman and Crowell (2000), Hellmann et al. (2003), and Balasubramanian et al. (2009).

codon and the 3′ untranslated region in the last exon. Within the coding region, the slowest-evolving codon was the initiation codon.

Causes of Variation in Substitution Rates

To infer the causes underlying the observed variation in substitution rates among DNA regions, we note that the rate of substitution is determined by two factors: (1) the rate of mutation and (2) the probability of fixation of a mutation (Chapter 2). The latter depends on whether the mutation is advantageous, neutral, or deleterious. Since the rate of mutation is unlikely to vary much within a gene but may vary among genes, we will deal with the rate variation among different regions of a gene and the variation among genes separately.

The concept of functional constraint

The intensity of purifying selection is determined by the degree of intolerance characteristic of a site or a genomic region toward mutations. At the DNA level, this **functional** or **selective constraint** defines the range of alternative nucleotides that are acceptable at a site without negatively affecting the function or structure of the gene or the gene product (Miyata et al. 1980; Jukes and Kimura 1984). DNA regions in which a mutation is likely to affect function have a more stringent functional constraint than regions devoid of function. The stronger the functional constraint is, the slower the rate of substitution will be.

Kimura (1977, 1983) has illustrated this principle by means of a simple model. Suppose that a certain fraction, f_0, of all mutations in a certain molecule is selectively neutral or nearly neutral and the rest are deleterious. (Advantageous mutations are assumed to occur only very rarely, such that their relative frequency is effectively zero, and they mostly do not contribute significantly to the overall rate of molecular evolution.) If we denote by v_T the total mutation rate per unit time, then the rate of neutral mutation is

$$v_0 = v_T f_0 \tag{4.5}$$

According to the neutral theory of molecular evolution (Chapter 2), the rate of substitution is $K = v_0$. Hence,

$$K = v_T f_0 \tag{4.6}$$

From Equation 4.6, we see that the highest rate of substitution is expected to occur in a sequence that does not have any function, such that all mutations in it are neutral (i.e., $f_0 = 1$). Indeed, pseudogenes, which are devoid of function, and introns and intergenic regions, which are mostly devoid of function, have the highest rates of nucleotide substitution (Table 4.3). Thus, although the model is clearly oversimplified, it is helpful for explaining the rate differences among different DNA regions.

As far as protein-coding genes are concerned, numerous comparative studies of both protein and DNA sequence data have led to the general conclusion that an inverse relationship exists between stringency of functional constraint and the rate of evolution. We note that following Kimura (1983), "functional constraint" is frequently equated in the literature with "importance." This conceptual equivalence is not always as straightforward as it seems (Graur 1985).

Quantifying the degree of protein tolerance toward amino acid replacements

By using random mutagenesis, it is currently possible to create in the laboratory huge numbers of mutants that differ form the wild-type protein by a single random amino acid replacement. These variants can then be used to quantify the tolerance of a protein to random amino acid change. Guo et al. (2004) studied the probability that a random amino acid replacement will lead to a protein's functional inactivation. For the human DNA repair enzyme 3-methyladenine DNA glycosylase, they obtained a probability of 34 ± 6%.

By using this experimental approach, Besenmatter et al. (2007) showed that thermostable proteins (i.e., proteins that can function at high temperatures) are better at tolerating amino acid replacements than are mesostable proteins (i.e., proteins that function only at temperatures close to room temperature).

Synonymous versus nonsynonymous rates

Because the rates of mutation (v_T) at synonymous and nonsynonymous sites within a gene should be the same, or at least very similar, the difference in substitution rates between synonymous and nonsynonymous sites can only be attributed to differences in the intensity of purifying selection between the two types of sites. This is understandable in light of the neutral theory of molecular evolution (Chapter 2). Mutations that result in an amino acid replacement have a greater chance of causing deleterious effects on the function of the protein than do synonymous changes. Ophir et al. (1999) estimated that only 10% of all substitutions at second codon positions, which are always nonsynonymous, are neutral, the rest being deleterious. Consequently, the majority of nonsynonymous mutations will be eliminated from the population by purifying selection. The result will be a reduction in the rate of substitution at nonsynonymous sites. In contrast, synonymous changes have a better chance of being neutral, and a larger proportion of them will be fixed in the population.

Of course, nonsynonymous substitutions may have a better chance of improving the function of a protein. Therefore, if advantageous selection plays a major role in the evolution of proteins, the rate of nonsynonymous substitution should exceed that of synonymous substitution (Chapter 2).

Since v_T is the same for both synonymous and nonsynonymous sites, the almost universal observation that synonymous substitution rates are higher than nonsynonymous substitution rates is a result of the higher f_0 values for synonymous sites than for nonsynonymous ones. The contrast between synonymous and nonsynonymous rates in protein-coding genes serves as an illuminating demonstration of the inverse relationship between the intensity of the functional constraint and the rate of molecular evolution. That the average rate of synonymous substitution is somewhat lower than the rate of substitution in pseudogenes would seem to indicate that some synonymous mutations are selected against. One reason for this may be that not all synonymous substitutions are "silent" in terms of fitness, at the mRNA level, the amino acid level, or the level of gene expression (Parmley and Hurst 2007). There is now considerable evidence that not all synonymous mutations are silent. Synonymous mutations may, at times, disrupt splicing, interfere with microRNA binding, alter mRNA stability, affect protein abundance, or induce translational pausing, thereby affecting protein folding and activity (Chapter 1). Another reason may be that some codons are preferred over their synonyms (pp. 139–142).

Variation among different gene regions

In principle, fast rates of evolution may be explained by either strong positive Darwinian selection or total relaxation of selection. Pseudogenes are the fastest-evolving elements in the genome. Since pseudogenes are, by definition, devoid of function, the reason for their fast rates of evolution must be a lack of functional constraint. Indeed, pseudogenes are the paradigm of neutral evolution (Li et al. 1981). The observation that fourfold degenerate sites, 5′ and 3′ untranslated regions, introns, and intergenic regions have lower substitution rates than the rate of substitution in pseudogenes leads us to conclude that these regions are somewhat functionally constrained. Indeed, in the clinical literature many deleterious mutations in untranslated regions (Campuzano et al. 1996; Naukkarinen et al. 2005; Li 2006) and fourfold degenerate sites (Chamary et al. 2006) have been described.

Within a protein, the different structural or functional domains are likely to be subject to different functional constraints and to evolve at different rates. A classic example is provided by insulin, a dimeric hormone secreted by the β cells of the pancreatic islets of Langerhans. The precursor of insulin, preproinsulin, is a chain of 86 amino acids

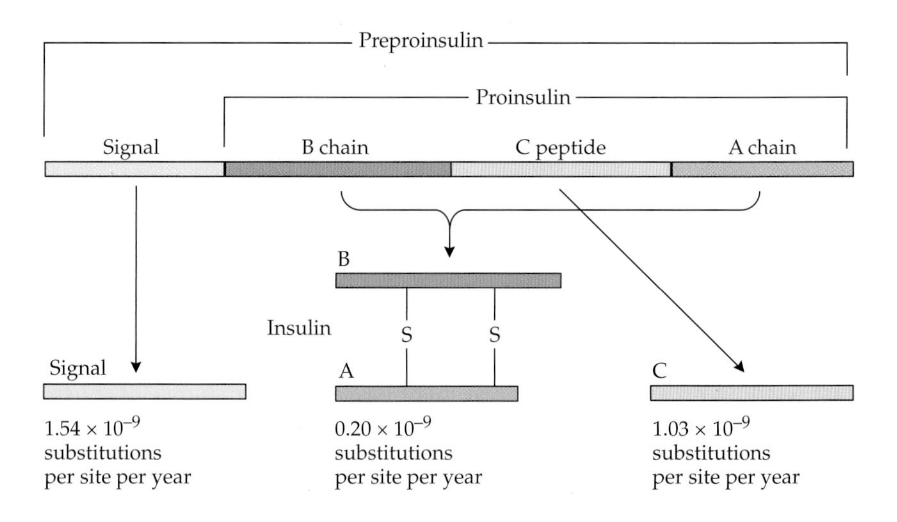

Figure 4.6 Comparison of the rates of nonsynonymous nucleotide substitution among three DNA regions coding for preproinsulin (functional insulin, the C peptide, and the signal peptide). The mature insulin consists of A and B chains, linked by two disulfide (S—S) bonds. The rates are based on comparisons between human and mouse genes. The time of divergence was set at 80 million years ago. (Courtesy of Yichen Zheng and Fei Yuan.)

consisting of four segments: A, B, C, and a signal peptide (**Figure 4.6**). After the signal peptide is removed, the remaining 62-amino-acid-long proinsulin folds into a specific three-dimensional structure stabilized by two disulfide bonds. Internal cuts then excise the 31-amino-acid-long C peptide, which resides in the middle of the proinsulin chain, to create the active insulin hormone out of the two remaining segments, A and B. Neither the signal peptide nor the C peptide takes part in the hormonal activity of insulin. The signal peptide is thought to facilitate the secretion of preproinsulin, whereas the C peptide is necessary for the creation of the proper tertiary structure of the hormone. Presumably, therefore, the signal peptide and the C peptide are less constrained than the A and B chains. Indeed, the nonsynonymous substitution rate for the region coding for the signal peptide is almost 8 times higher than the nonsynonymous rate for the mature insulin and almost 13 times higher than that for the A chain, which is the most conserved polypeptide in preproinsulin. However, considerable constraints must still operate on the C peptide and the signal peptide, because their nonsynonymous rates are rather low in comparison with other protein-coding genes.

An additional example involves hemoglobin, a tetrameric protein composed of two α and two β chains. The surface of the hemoglobin molecule performs no specific function and is constrained only by the requirement that it must be hydrophilic. On the other hand, the internal residues, especially the amino acids lining the heme pocket, play an important role in the normal function of the molecule. The functional constraints on the heme pocket are quite stringent. Enzymatic substrates must be able to enter and leave the heme pocket and yet must be held in the proper conformation for reactivity. Oxygen must be able to move freely into and out of the hemoglobin during transport and delivery and yet must remain bound to iron until release is appropriate.

Indeed, the rates of substitution on the surfaces of the α and β chains are 1.35×10^{-9} and 2.73×10^{-9} amino acid substitutions per site per year, respectively, while the rates of substitution in the interior are only 0.17×10^{-9} and 0.24×10^{-9} for α and β, respectively. Purifying selection seems to be the mechanism responsible for the fact that in both the α and the β chains, the residues on the surface of the molecule evolve 8–11 times faster than the residues in the interior (Kimura and Ohta 1973). Indeed, mutations affecting the interior of hemoglobin were shown to cause particularly harmful abnormalities, whereas replacements of amino acids on the surface of the molecule often do not exhibit any clinical effects (e.g., Perutz 1983). Interestingly, this disparity in evolutionary rates between residues buried in the interior of a protein and solvent-exposed residues on the surface of the protein was observed in many other proteins and seems to be independent of function (Lin et al. 2007).

Many genes are initially translated into long proteins (e.g., preproinsulin), which are posttranslationally cleaved to produce smaller active molecules (e.g., insulin). Signal peptides direct mature peptides to their appropriate cellular locations, after which

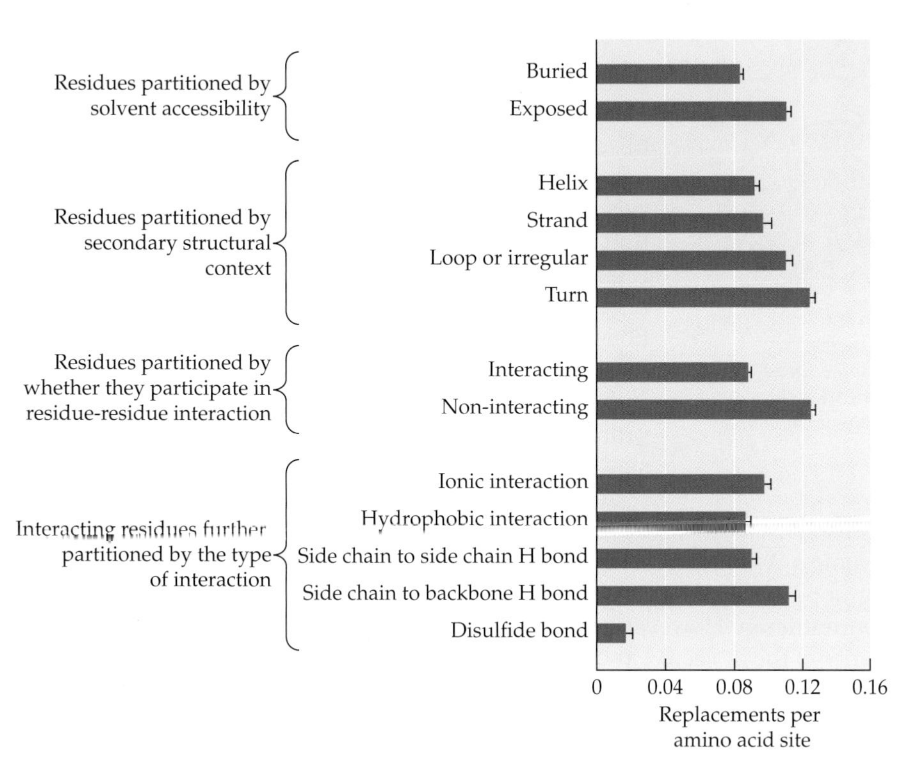

Figure 4.7 Effect of structural context on protein sequence evolution. Numbers of replacements per amino acid site were calculated from a set of 767 rat-human orthologous proteins with known three-dimensional structures in human. Note that constrained positions (e.g., buried and interacting residues) evolve more slowly than less constrained positions (e.g., exposed and noninteracting residues). (Modified from Choi et al. 2006.)

they are cleaved off. Since, as far as signal peptides are concerned, many alternative amino acid sequences are functionally equivalent, i.e., the original sequence can be exchanged for many other sequences with no noticeable effects (e.g., Shields and Blobel 1977), signal peptides may not be very stringently constrained. Indeed, Williams et al. (2000) found that signal peptides on average evolve approximately five times faster than the flanking mature protein. They do, however, evolve more slowly than pseudogenes, indicating that they are subject to some purifying selection.

Rates of evolution in proteins were also found to vary with secondary structure, relative position in the three-dimensional structure, and residue-residue interactions (**Figure 4.7**).

Variation among genes

To explain the large variation in the rates of nonsynonymous substitution among genes, we must consider two possible culprits: (1) the rate of mutation and (2) the intensity of selection. The assumption of equal rates of mutation for different genes may not hold, because different regions of the genome may have different propensities to mutate. However, the difference in mutation rates among different genomic regions is far too small to account for the more than 1,000-fold range in nonsynonymous substitution rates. Thus, the most important factor in determining the rates of nonsynonymous substitution seems to be the intensity of purifying selection, which in turn is determined by functional constraints.

To illustrate the effect of functional constraints on the evolution of different genes, let us consider lipoproteins and the core histones, which exhibit markedly different rates of nonsynonymous substitution. Lipoproteins, which are derived from apolipoprotein precursors following the cleavage of the signal peptides, are water-soluble proteins that serve as the major carriers of lipids in the blood and other tissue fluids of vertebrates. The lipid-binding domains of lipoproteins consist mostly of hydrophobic amino acid residues. Comparative analyses of lipoprotein sequences from various mammalian orders suggest that in these domains, exchanges among hydrophobic amino acids (e.g., valine for leucine) are acceptable at many sites (Luo et al. 1989). This lax structural requirement may explain the fairly high nonsynonymous rate in these genes.

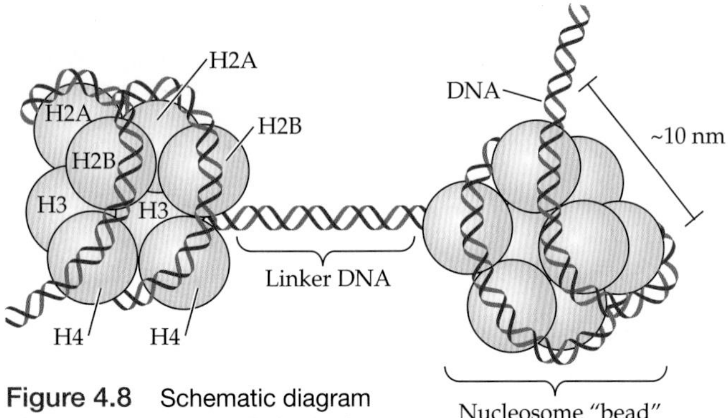

Figure 4.8 Schematic diagram of a nucleosome. The DNA double helix is wound around the core histones (two each of histones H2A, H2B, H3, and H4). Histone H1 (not shown) binds to the outside of this core particle and to the linker DNA. (From Becker et al. 2009.)

At the other extreme, we have the core histones (H2A, H2B, H3, and H4). Since most amino acids in each core histone interact directly with either the DNA or other core histones in the formation of the nucleosome (**Figure 4.8**), it is reasonable to assume that there are very few possible substitutions that can occur without hindering the function of the protein. In addition, core histones must retain their strict compactness and high alkalinity, which are necessary for their intrastrand location and their interaction with the acidic DNA molecule. As a consequence, core histones are very intolerant of most amino acid changes. Indeed, these proteins are among the slowest-evolving proteins known, evolving more than 1,000 times more slowly than the lipoproteins.

Another illuminating example of the importance of functional constraint can be derived from comparing vertebrate hemoglobins and cytochrome c. As oxygen carriers, hemoglobins require the attachment of heme prosthetic groups and have the capability to respond structurally to changes in pH and CO_2 concentration. However, most of their functional requirements are restricted to the interior of the molecule, and as we have seen in the previous section, many amino acid–altering mutations, especially on the surface of this globular protein, are acceptable. Cytochrome c also binds heme, carries oxygen, and responds structurally to changes in physiological conditions. In addition to these hemoglobin-like functions, this protein also interacts at its surface with two very large enzymes: cytochrome oxidase and cytochrome reductase. Thus, a higher proportion of surface amino acids in cytochrome c take part in specific functions, and its rate of amino acid substitution is consequently lower than that of hemoglobin.

Why the rate of synonymous substitution also varies from gene to gene is less clear. There may be two reasons for this variation. In part, the variation may represent stochastic fluctuations (Kumar and Subramanian 2002); in part, it may be due to other factors, such as variation in the rate of mutation among different regions of the genome (e.g., Wolfe et al. 1989a) and selection operating on synonymous mutations (e.g., Mouchiroud et al. 1995). If indeed the rate of mutation varies among different genomic regions, then the variation in rates of synonymous substitution may simply reflect the chromosomal position of the gene. As for selection, it is possible, for instance, that not all synonymous codons have equal effects on fitness, and some synonymous substitutions may be selected against. However, although purifying selection has been shown to affect the synonymous substitution rate, as well as the pattern of usage of synonymous codons in the genomes of bacteria, yeast, and *Drosophila*, the intensity of this type of selection in mammals is negligible (Price and Graur 2015). In fact, in some studies (Williams and Hurst 2002; Chuang and Li 2007), the null hypothesis of stochastic variation in synonymous rates in mammals could not be rejected.

It has also been noted that there is a positive correlation between synonymous and nonsynonymous substitution rates in a gene (Graur 1985; Li et al. 1985b; Mouchiroud et al. 1995). This may be explained by assuming either that the rate of mutation varies along the genome and among genes (and hence some genes will have high rates of both synonymous and nonsynonymous substitution), or that the extent of selection at synonymous sites is affected by the nucleotide composition at adjacent nonsynonymous positions (Ticher and Graur 1989).

Variables associated with protein evolutionary rates

So far, we have treated selective constraint as the range of alternative residues at a site that do not adversely affect function. We will now discuss particular variables that have been proposed to have important and independent roles in determining rates of protein evolution. One of the earliest suggestions concerned the number and complexity of interactions of a protein with other proteins as well as with other molecules

(Dickerson 1971). According to this suggestion, the number and extent of interactions should be negatively correlated with the rate of amino acid replacements. More recently, this hypothesis has been tested in the context of **protein-protein interaction networks**. In **Figure 4.9**, we present a simple example of a protein-protein interaction network consisting of five proteins (A–E), represented by the nodes, each of which interacts with at least one other protein. There are five interactions, denoted by the links. In biological networks, three variables are usually studied: **degree centrality** or **connectedness** (i.e., the number of interactions for a protein), **betweenness centrality** (i.e., the number of times that a node appears on the shortest path between all pairs of nodes), and **closeness centrality** (i.e., the mean number of links connecting a protein to all other proteins in the network). So far, only very week correlations have been found between these variables and the rate of amino acid replacement (Rocha 2006).

Another variable that has been suggested to influence the rate of evolution is **functional density** (Zuckerkandl 1976b). The functional density of a gene, F, is defined as n_s/N, where n_s is the number of sites committed to specific functions and N is the total number of sites. F, therefore, is the proportion of amino acids that are subject to stringent functional constraints. The higher the functional density, the lower the rate of substitution is expected to be. Thus, if active sites constitute only 1% of the sequence of a protein, it will be less constrained and will therefore evolve more quickly than a protein that devotes 50% of its sequence to performing specific biochemical or physiological tasks. Unfortunately, there is no consensus on how to measure functional density in proteins quantitatively (Rocha 2006), although qualitatively some rough negative associations between functional density and rates of amino acid–altering substitution have been reported.

Many other variables have been implicated as determinants of evolutionary rate in proteins. In particular, the literature abounds in studies looking into the effects of expression levels (measured by mRNA abundance or protein concentrations in cells), expression breadth (number of different tissues expressing the gene in question), essentiality (the ability of the organism to grow after gene knockout), dispensability (the decrease in growth rate after gene knockout), epistasis (the effects of the genetic background), developmental timing of expression, recombination rates, cost of biosynthesis, protein length, functional category, and the number and length of introns (e.g., Drummond et al. 2005; Rocha 2006; Singh et al. 2009; Breen et al. 2012).

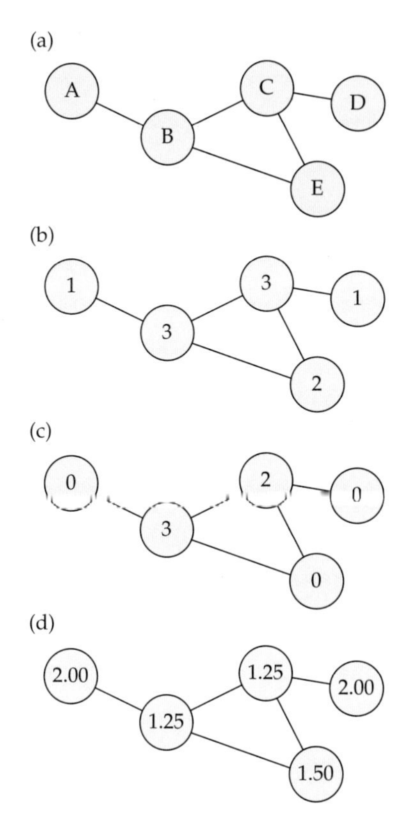

Figure 4.9 A simple example of a protein-protein interaction network. (a) The network consists of five proteins (nodes) and five interactions (links). The nodes are denoted by circles and interactions by connecting lines. (b) The connectedness of a node is defined as the number of links emanating from a node. In our example, node A has a single link, so its connectedness is 1, while node C is connected to B, D, and E, so its connectedness is 3. (c) Betweenness centrality is defined as the number of times a node appears on the shortest path between all other pairs of nodes. For example, node A is never found on the shortest path of any pair of nodes, i.e., its betweenness centrality is 0. In contrast, node B is found on the shortest paths connecting A and C, A and D, and A and E. Its betweenness is, therefore, 3. (d) Closeness centrality is the mean number of links connecting a protein to all other proteins. For example, protein A is connected to proteins B, C, D, and E by 1, 2, 3, and 2 links, respectively. Its closeness is, therefore, (1 + 2 + 3 + 2)/4 = 2.00.

Evolutionary conservation and disease

According to the neutral theory of evolution, the rate of substitution (as inferred from between-species comparisons) should positively correlate with the degree of genetic polymorphism (as inferred from comparisons among individuals within one species). An interesting corollary of this hypothesis is that we should observe very little variation at the population level at evolutionarily conserved positions. Moreover, some of the variation at conserved positions should be deleterious (i.e., associated with disease). This association between evolutionary conservation and disease has invariably been shown to be true (e.g., Ng and Henikoff 2001; Sunyaev et al. 2001; Subramanian and Kumar 2006).

We can illustrate this principle by using the example of Gaucher disease, which is an autosomal recessive lysosomal storage disorder due to deficient activity of an enzyme called acid β-glucosidase (also known a β-glucocerebrosidase). There are many subtypes of Gaucher disease, with fitness effects ranging from slight reduction in fitness to perinatal lethality, in which death occurs during the period from approximately 5 months of gestation to about 7 days after birth. We aligned the amino acid sequences of acid β-glucosidase from nine placental mammals (human, chimpanzee, Sumatran orangutan, bovine, pig, dog, horse, rat, and mouse). The length of the alignment (excluding one gap due to a codon deletion in the ancestor of mouse and rat) was 496 amino acids, of which 387 were identical in all nine species and 109 were variable. Thirty-six single amino acid replacements resulting in Gaucher disease are

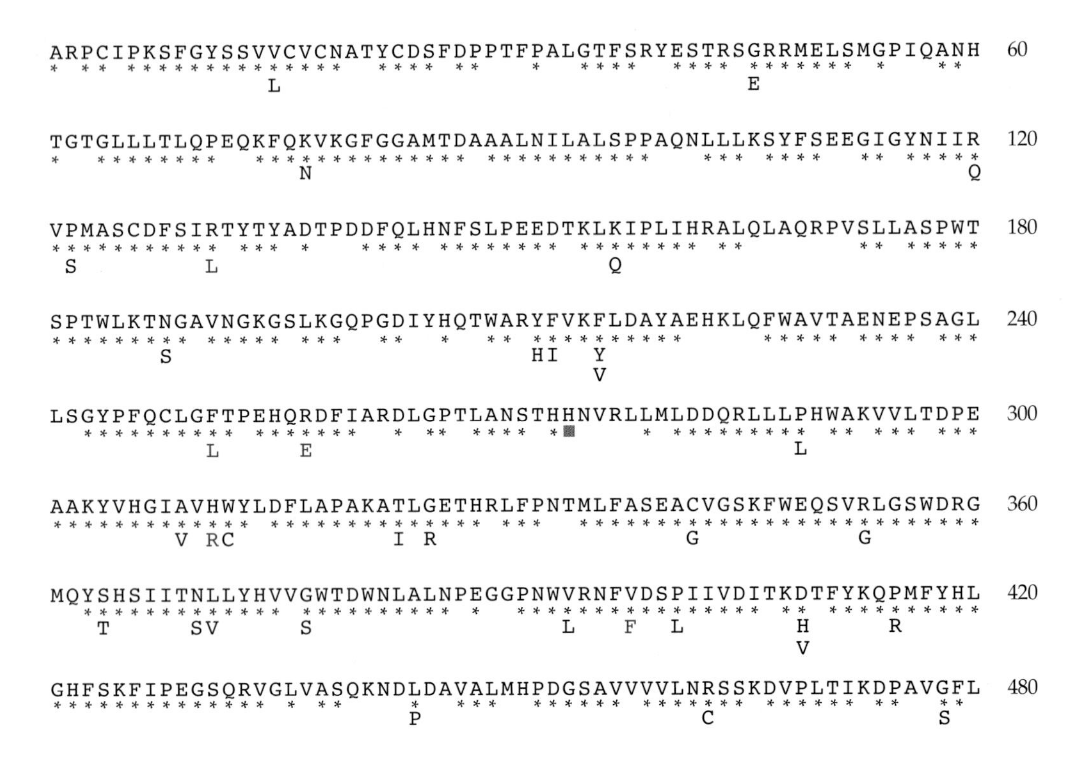

Figure 4.10 Association between evolutionary conservation and disease in the case of Gaucher disease, which is caused by mutations in the gene encoding acid β-glucosidase (GBA). The amino acid sequence of GBA from human is shown. Amino acid sites that are invariant in human, chimpanzee, Sumatran orangutan, bovine, pig, dog, horse, rat, and mouse are marked with asterisks. A gap due to an amino acid deletion in mouse and rat at position 273 is indicated by a blue box. Thirty-six single amino acid replacements (at 34 amino acid positions) resulting in Gaucher disease are shown. For example, both the amino acid replacements D → H and D → V at amino acid position 409 will result in Gaucher disease. Lethal mutations resulting in perinatal death are shown in red. All deleterious (fitness-reducing) mutations occur at completely conserved sites (letters below asterisks).

described in the literature, and all of them were found to occur at completely conserved sites (**Figure 4.10**). This statistically significant nonrandom association between disease and evolutionary conservation ($p = 0.0002$) indicates that invariable sites are conserved because they evolve under extremely stringent constraints and cannot tolerate change.

Relaxation of selection

If the rate of substitution is indeed inversely related to the stringency of the functional constraint, as claimed by the neutral theory, then we should observe an increase in the rate of nucleotide substitution in genes that have lost their function. We have already seen that pseudogenes, in which all constraints have presumably been removed, are the fastest-evolving sequences in the mammalian nuclear genome. Let us now examine what happens when selection constraints are only partially, rather than entirely, removed. Such a phenomenon is called **relaxation of selection**. One of the processes that invariably leads to relaxation of selection is domestication. Studies on dogs, maize, rice, yeast, cattle, chickens, yaks, Bactrian camels, and, unsurprisingly, humans have uncovered hundreds of cases in which the selective constraint on a certain gene or a group of genes has diminished because of changes in the environment related to domestication. Domestication, of course, also entails a permanent or temporary reduction in population size, which reduces the efficacy of selection and boosts the effects of random genetic drift (Chapter 2).

An illustration of the phenomenon of selection relaxation in evolution is provided by the evolution of the single-copy αA-crystallin gene in the subterranean blind mole rat (*Spalax ehrenbergi*). In vertebrates, the crystallins function mainly as structural components of the eye lens. In the blind mole rat, however, αA-crystallin has long lost this functional role; its vestigial lenses contain disorganized, vacuolated cells expressing α-crystallins at very low levels. The highly degenerated visual system of *Spalax* is presumably a by-product of its adaptation to a fossorial lifestyle in the subterranean environment for the past 25 million years. The loss of vision in *Spalax* can be construed as a relaxation of selection, which is expected to result in an acceleration of the evolutionary rate. Since αA-crystallin is usually a slow-evolving protein, an increase in the rate of substitution should be readily detectable.

Hendriks et al. (1987) sequenced the αA-crystallin gene in the blind mole rat and compared its rate of substitution with those in other rodents, such as mouse, rat, gerbil, hamster, and squirrel, which possess fully functional eyes. The αA-crystallin gene of the blind mole rat turned out to possess all the prerequisites for normal function and expression, including the proper signals for alternative splicing, but its nonsynonymous rate of substitution was found to be exceptionally high, almost 20 times faster than the rate in rat. Nevertheless, the nonsynonymous rate was still much lower than that in pseudogenes, indicating that functional constraints may still be operating. Moreover, Smulders et al. (2002) studied the structure and function of mole rat αA-crystallin and found that the protein has viable secondary and quaternary structures as well as normal thermostability.

Hendriks et al. (1987) suggested that αA-crystallin might not have lost all of its vision-related functions, such as photoperiod perception and adaptation to seasonal changes. However, the fact that the atrophied eye of *Spalax* does not respond to light argues against this explanation. An alternative explanation for the slower rate in the αA-crystallin gene than in pseudogenes may be that the blind mole rat lost its vision more recently than 25 million years ago. Consequently, the rate of nonsynonymous substitution after nonfunctionalization may have been underestimated. This argument, however, fails to explain why the gene is still intact as far as the essential molecular structures for its expression are concerned. The most likely explanation at this point is that the αA-crystallin gene product may also serve a function unrelated to that of the eye. This possibility is supported by the fact that αA-crystallin has been found in tissues other than the lens. In particular, however, the promoter of the αA-crystallin gene in *Spalax* was found to be very active in muscle cells (Hough et al. 2002). Such multiple roles are known for several proteins, including many crystallins (Chapter 8). Support for this hypothesis was provided by the discovery that αA-crystallin also functions as a molecular chaperone (Bova et al. 1997; Mornon et al. 1998), albeit one whose activity is considerably lower than that of αA-crystallin in rat (Smulders et al. 2002). Therefore, the functional constraints on αA-crystallin in the mole rat cannot be assumed to have been completely removed with the atrophy of the eye. It is quite certain that αA-crystallin represents a case of partial loss of function resulting in an increased rate of nucleotide substitution due to substantial relaxation of selection.

Selective intolerance toward indels

Through the alignment of more than 8,000 orthologous protein-coding genes from human, mouse, and rat, Taylor et al. (2004) identified 1,743 indel events in rodents. Deletions turned out to be more frequent than insertions, particularly for the rat lineage (a 70% excess). As expected from their functional importance, the internal structural cores of proteins were less tolerant than loop regions as far as both deletions and insertions were concerned. Transmembrane domains and signal peptides were the least tolerant toward indels, while **low-complexity regions**, i.e., protein sequences composed of one or a very small number of amino acids, were the most tolerant. Sequence slippage (Chapter 1) accounted for about half of the insertions and about a third of the deletions.

Identifying positive and purifying selection

Several methods exist for quantitatively estimating the intensity of selection on protein-coding genes (e.g., Bustamante et al. 2002). One of the more common methods relies on the ratio of nonsynonymous to synonymous rates of substitution.

The assumptions of this method are that (1) all synonymous mutations are neutral, i.e., they do not affect the fitness of the organism; (2) nonsynonymous mutations can be advantageous, neutral, or deleterious; and (3) nonsynonymous mutations that are advantageous will undergo fixation in a population much more rapidly than neutral mutations (Chapter 2). Thus, if advantageous selection plays a major role in the evolution of a protein, then the rate of nonsynonymous substitution should exceed that of synonymous substitution. Therefore, one way to detect positive Darwinian selection is to show that the number of substitutions per nonsynonymous site is significantly greater than the number of substitutions per synonymous site ($K_A/K_S > 1$). If, on the other hand, evolution proceeds according to the neutral theory of molecular evolution, i.e., if the only forces determining the fate of a new mutant are purifying selection and random genetic drift, then the ratio K_A/K_S is expected to be smaller than 1. Under strict neutrality, i.e., if selection does not operate at all, the ratio is expected to equal 1. A ratio of 1 may cast serious doubt on the actual functionality of the gene under study, for if the function of the gene cannot be destroyed by nonsynonymous mutations, i.e., if the amino acid sequence encoded by the gene is not important, then it is doubtful that the gene has a function to begin with.

The **nonsynonymous-to-synonymous ratio test** was pioneered by Li and Gojobori (1983) and Hill and Hastie (1987) and was put forward explicitly by Hughes and Nei (1988). Several versions have been proposed in the literature to test for this difference. Here we present two simple tests.

In the first test, we use the number of substitutions per nonsynonymous site (K_A) and the number of substitutions per synonymous site (K_S) along with their variances, which have been calculated by Equations 3.29 and 3.30 (Chapter 3). The test statistic is Student's t:

$$t = \frac{K_A - K_S}{\sqrt{V(K_A) + V(K_S)}} \quad (4.7)$$

where V denotes the variance. Assuming that the statistic follows the t distribution, we may perform a one-tailed test of the null hypothesis that $K_A \leq K_S$ against the alternative hypothesis that $K_A > K_S$. Rejection of the null hypothesis would mean that positive selection has played a major role in the evolution of the sequences under study.

When the numbers of synonymous and nonsynonymous substitutions are small, the statistic is unlikely to follow the t distribution and we are likely to reject the null hypothesis more often than expected by chance. To overcome this problem, Zhang et al. (1997) proposed the following test. In this test, we use the numbers of synonymous and nonsynonymous differences between the two protein-coding sequences, M_S and M_A, and the average numbers of synonymous and nonsynonymous sites, N_S and N_A (Chapter 3). Under the null hypothesis of neutral evolution, i.e., no positive selection, we expect $M_S/N_S = M_A/N_A$. We can therefore use a 2 × 2 table as follows:

	Nonsynonymous	Synonymous	Total
Changes	M_A	M_S	$M_A + M_S$
No changes	$N_A - M_A$	$N_S - M_S$	$L - (M_A + M_S)$
Total	N_A	N_S	L

where L is the length of the aligned sequence. We can test the null hypothesis by using the χ^2 test or the exact binomial distribution test (Fisher's exact test).

The fraction of genes that exhibit $K_A/K_S > 1$ is usually very small. Only six out 7,645 homologous genes from human, chimpanzee, and mouse (0.08%) could be shown to have experienced positive Darwinian selection (Clark et al. 2003). We note that the K_A/K_S test can only detect positive selection if the proportion of nonsynonymous substitutions that increase fitness is very large, say in excess of 70% (Eyre-Walker 2006).

Estimating the intensity of purifying selection

The nonsynonymous-to-synonymous ratio tests assume that synonymous substitutions are not subject to selection. This may not always be true (Wolf et al. 2009). For example, the fitness effects of synonymous and nonsynonymous mutations were experimentally shown to exhibit almost identical distributions in two protein-coding genes of *Salmonella typhimurium* (Lind et al. 2010). In other words, a synonymous mutation was as likely to be deleterious as a nonsynonymous one.

Here, we present a very simple method (Ophir et al. 1999) that does not make this assumption. Instead, it uses pseudogenes as the neutral standard. The method requires three homologous sequences: a pseudogene (ψA), a functional homologous gene from the same species (A), and a functional homolog from a different species (B). For the analysis, we must be certain that ψA diverged from A after the divergence of A from B. For each such trio (**Figure 4.11**) and by using the relative rate test (p. 151), it is possible to calculate the numbers of nucleotide substitutions along the branches leading to ψA and A, i.e., $K_{\psi A}$ and K_A. Following Kimura's (1977) model in Equation 4.6, the numbers of nucleotide substitutions along the two branches leading to ψA and A are given by

$$K_{\psi A} = v_{\psi A} f_{\psi A} \quad (4.8)$$

and

$$K_A = v_A f_A \quad (4.9)$$

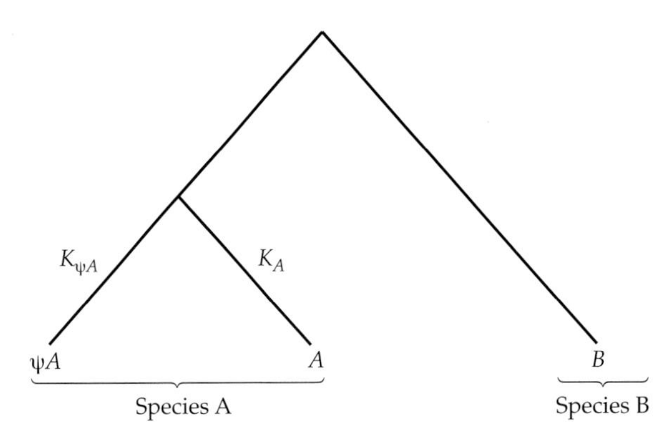

Figure 4.11 Phylogenetic tree for three homologous sequences used to quantify the intensity of purifying selection on a protein-coding gene. The sequences are: a pseudogene (ψA), a functional homologous gene (A) from the same species, and a functional homolog (B) from a different species. $K_{\psi A}$ and K_A denote the rates of substitution along the branches leading to ψA and A, respectively. (From Graur and Li 2000.)

where v is the total mutation rate per unit time, f is the fraction of mutations that are selectively neutral or nearly so, and subscripts identify the branches.

If we assume that the mutation rate is the same in the gene and the pseudogene, i.e., $v_A = v_{\psi A}$, and if we further assume that mutations occurring in a pseudogene do not affect the fitness of the organism, i.e., $f_{\psi A} = 1$, we obtain

$$f_A = \frac{K_A}{K_{\psi A}} \quad (4.10)$$

By definition, the fraction of deleterious mutations that are subject to purifying selection (or the intensity of selection) is $1 - f_A$.

As expected, Ophir et al. (1999) found that the least constrained codon position is the third, followed by the first and second codon positions. In the second codon position, in which all possible mutations are nonsynonymous, they found that in humans approximately 7% of all mutations were neutral. The corresponding value for the first codon position was 21%. Interestingly, only about 42% of all the mutations occurring in the third codon position of protein-coding genes were found to be neutral in their sample of human genes. Given that about 70% of all possible mutations in this position are synonymous (Chapter 1, Table 1.5), these results indicate that a significant fraction of synonymous mutations are selected against. Similar results were obtained by Bustamante et al. (2002) by using human, mouse, and rat pseudogenes.

Are slowly evolving regions always important?

In the absence of positive Darwinian selection, the universal observation is that important sequences tend to evolve more slowly than less important ones. The opposite, however, is not always true. That is, conserved regions in the genome may not always be important. Of course, defining "importance" is not a trivial undertaking. Hurst and Smith (1999) tested the relationship between rate of substitution and dispensability (as a proxy for importance). Approximately two-thirds of all knockouts of individual mouse genes give rise to viable fertile mice. These genes have thus been termed "nonessential," in contrast to "essential" genes, the knockouts of which result in death or infertility. Although nonessential genes are also likely to be under selection that favors sequence conservation, it is predicted that they will be subject to lesser intensities of purifying selection and should therefore evolve faster. In a comparison of 74 nonessential genes with 64 essential ones, the rate of substitution was found not to correlate with the severity of the knockout phenotype. To account for differences in function, Hurst and Smith (1999) restricted their analysis to neuron-specific genes, which have significantly lower rates of substitution than other genes. They could find no difference between the rates of substitution of 16 essential neuron-specific genes and 18 nonessential ones.

More recently, Nóbrega et al. (2004) deleted two megabase-long sequences from the mouse genome: a 1,817 kb region mapping to mouse chromosome 3 and a 983 kb region mapping to chromosome 19. (Orthologous regions of about the same size are present on human chromosomes 1 and 10, respectively.) The two regions were defined as gene deserts (Chapter 11), i.e., they lacked annotated genes and evidence of transcription. Together, the two regions contained 1,243 sequences, 100 bp or more in length, that were conserved between human and mouse. Surprisingly, viable mice homozygous for the deletions were generated and were indistinguishable from wild-type littermates with regard to morphology, reproductive fitness, growth, longevity, and general homeostasis. Further analysis of the expression of multiple genes bracketing the deletions revealed only minor expression differences between homozygous-deletion mice and wild-type mice. Thus, the deletion of so many sequences that have been conserved for such a long period of time resulted in no reduction in fitness.

Ahituv et al. (2007) went one step further and removed from the mouse genome four **ultraconserved** elements—sequences of 200 base pairs or longer that are 100% identical among human, mouse, and rat. Remarkably, all four resulting lines of mice lacking these ultraconserved elements were viable and fertile, and they exhibited no phenotypic or developmental abnormalities. These results indicate that extreme sequence conservation may not necessarily reflect extreme evolutionary constraints; other evolutionary forces (e.g., mutational coldspots, gene conversion) may promote sequence conservation.

Male-Driven Evolution: Mutational Input and Slow-X Evolution

From Equation 4.6 it is clear that the rate of substitution should be influenced not only by functional constraints, but also by the rate of mutation. An interesting example of the effect of the mutation rate on the rate of substitution is the so-called **male-driven evolution** or **slow-X evolution**, which is the hypothesis that the male germline contributes inordinately more mutations to subsequent generations than the female germline. In the 1930s and 1940s, pioneering geneticist J. B. S. Haldane noticed a peculiar inheritance pattern in families with long histories of the X-linked disease hemophilia. The faulty mutation responsible for the blood clotting disorder tended to originate on the X chromosomes that fathers passed down to their daughters rather than on the X chromosomes from the mothers. Haldane (1947) subsequently proposed that children inherit more mutations from their fathers than from their mothers: "If mutation is due to faulty copying of genes at the nuclear division, we might expect it to be commoner in males than in females." He then added, however, "It is difficult to see how this could be proved or disproved for many years to come."

Because mammalian oogenesis (egg production) differs fundamentally from the process of spermatogenesis (sperm production), the number of germ cell divisions from one generation to the next in males, n_m, is usually much larger than that in females, n_f. In humans, the number of cell divisions from zygote to fertilized oocyte in females is constant because all of the oocytes are essentially formed by the fifth month of development, and only two further cell divisions are required to produce the zygote. The estimated number of successive cell divisions in the female from zygote to mature egg is 24 (Vogel and Motulsky 1996). In males, there are ~30 cell divisions from zygote to stem spermatogonia at puberty. Five subsequent cell divisions are required for spermatogenesis, but thereafter the spermatogenesis cycle occurs approximately every 16 days, or 23 times per year. This means that in males, the number of cell divisions required to produce sperm is age-dependent. If an average age of 13 is taken for onset of puberty, then the total number of cell divisions at age 20 years is ~30 + 5 + [23 × (20 – 13)], or about 196 divisions. This number increases to 426 by age 30, and to 1,346 by age 70. Given that errors in DNA replication provide the great majority of mutations, one might expect that the male mutation rate would not only be substantially greater than the female rate, but would also increase with paternal age.

In organisms with shorter generation times than humans, the difference between males and females in the number of cell divisions is not as large. In mice, for instance, $n_m = 57$ and $n_f = 28$, and in rats, $n_m = 58$ and $n_f = 29$.

There are two approaches to investigating Haldane's hypothesis concerning the rate of mutation in males being larger than that in females. One method requires ingenuity and clear mathematical reasoning; the other requires massive amounts of DNA sequencing.

Let us start with the approach of Miyata et al. (1987) for testing Haldane's hypothesis. Let u_m and u_f be the mutation rates in males and females, respectively, and α be the ratio of male to female mutation rates. That is,

$$\alpha = \frac{u_m}{u_f} \tag{4.11}$$

Since an autosomal sequence is derived from the father or the mother with equal probabilities, the mutation rate per generation for an autosomal sequence is

$$A = \frac{u_m + u_f}{2} \tag{4.12}$$

An X-linked sequence is carried two-thirds of the time by females and one-third of the time by males. Therefore, the mutation rate per generation for a sequence located on the X chromosome is

$$X = \frac{u_m + 2u_f}{3} \tag{4.13}$$

A Y-linked sequence is only carried by males, so its rate of mutation per generation is

$$Y = u_m \tag{4.14}$$

From these three equations it is easy to see that the ratio of Y to A is

$$Y/A = \frac{2\alpha}{1+\alpha} \tag{4.15}$$

Similarly,

$$X/A = \frac{2(2+\alpha)}{3(1+\alpha)} \tag{4.16}$$

and

$$Y/X = \frac{3\alpha}{2+\alpha} \tag{4.17}$$

Equations 4.15–4.17 allow us to infer α in real-life cases. The zinc finger protein–coding genes are a good case to use for studying the ratio of male to female mutation rates because in all eutherian mammals studied to date, there are at least two homologous genes, an X-linked gene (*Zfx*) and a Y-linked one (*Zfy*). Shimmin et al. (1993) sequenced the last intron of the human, orangutan, baboon, and squirrel monkey *Zfx* and *Zfy* genes, which are highly similar. As seen previously, there are very few functional constraints on introns, and therefore we may disregard selective forces in this case. For all pairwise comparisons, Shimmin et al. (1993) found that the Y sequences were more divergent, i.e., have evolved faster, than their X-linked homologs. The mean Y/X ratio was 2.25, which by using Equation 4.16 translates into an estimate of $\alpha \approx 6$. This indicates that in simians (sometimes incorrectly referred to as "higher monkeys"), the mutation rate is considerably higher in the male germline than in the female germline, i.e., evolution is "male-driven." Interestingly, the ratio of the number of germ cell divisions from one generation to the next in males, n_m, to that in females, n_f, is also approximately $200/33 \approx 6$.

To avoid the possible effect of reduced mutation on the X chromosome (see below), Makova and Li (2002) and Li et al. (2002) compared the divergence of homologous sequences on chromosome 3 and the Y chromosome. The average α among human, bonobo, gorilla, siamang, and gibbon was 5.3. In rat, mouse, hamster, and fox, the mutation rate in males was found to be twice as large as that in females (Lanfear and Holland 1991; Chang et al. 1994), which agrees with the $n_m/n_f = 2$ ratio in these species.

McVean and Hurst (1997) raised the possibility that the higher rate of nucleotide substitution in Y-linked sequences than in X-linked ones is not due to male-driven evolution but due to a reduction in the mutation rate in the X chromosome compared with the mutation rates in the Y chromosome and the autosomes. One possible factor may be that the male germline genome is heavily methylated (Driscoll and Migeon 1990), which can result in high mutation rates. Another factor may be that selection may favor a lower mutation rate on the X chromosome than on the autosomes, owing to the exposure of deleterious recessive mutations on **hemizygous chromosomes** (i.e., chromosomes that exist in a single copy in one sex but in two copies in the other). To test this possibility against the hypothesis of male-driven evolution, Ellegren and Fridolfsson (1997) and Axelsson et al. (2004) studied rates of mutation in birds. As opposed to the situation in mammals, in which the females are homogametic (XX) and the males are heterogametic (XY), in birds males are homogametic (WW) and females are heterogametic (WZ). The rate of substitution in the Z chromosome was found to be higher than that in autosomes, which in turn was higher than that in the W chromosome, consistent with the idea that the rate differences are due to mutation rates being higher in males than in females. Under McVean and Hurst's (1997) hypothesis, this high male-to-female ratio would have to be explained by a reduction in the mutation rate on the Z chromosome, which is equivalent to the Y chromosome in mammals, in comparison with that on the W chromosome (the X equivalent). This explanation, however, would amount to special pleading, and it is much simpler to assume that male-driven evolution occurs in both mammals and birds. Male-driven evolution has also been discovered in other vertebrate species (e.g., Ellegren and Fridlofsson 2003).

In recent years, the idea that the vast majority of mutations originate in the father has been tested directly through whole-genome sequencing efforts. For example, Kong et al. (2012) sequenced the entire genomes of 78 mother-father-offspring trios. The average mutation rate was inferred to be 1.2×10^{-8} per nucleotide per generation, with fathers passing on to their offspring nearly four times as many mutations as mothers. Moreover, in humans the mutation rate was found to correlate positively with the age of the father, so a 36-year-old will pass on to his children twice as many mutations as a 20-year-old, and a 70-year-old will pass on eight times as many mutations. No such age effect was detected as far as mothers were concerned. Paternal age effect was found to explain close to half of the variation in genome-wide mutation rates. These findings indicate quite conclusively that point mutations are replication-dependent events. The difference in mutation rates between males and females seems to apply not only to point mutations, but also to insertions, deletions, and gene conversions (Webster et al. 2005).

Mutation at CpG sites exhibit very weak male mutation bias, consistent with a replication-independent mechanism generating mutations at CpG sites (Taylor et al. 2006).

Rates of Evolution under Positive Selection

The rates of nucleotide substitution in the vast majority of genes and nongenic regions of the genome can be explained by a combination of (1) mutational input, (2) random genetic drift of neutral or nearly neutral alleles, and (3) purifying selection against deleterious alleles. In a few cases, however, positive selection was found to play an important role in the molecular evolution of genes. Positive selection is sometimes referred to as adaptive selection. We believe that the term "adaptive selection" should only be used when it can be shown that selection has led to a demonstrable adaptation of the organism under study to its biotic or abiotic environment.

Using the ratio of nonsynonymous to synonymous substitutions, scientists have found only a handful of cases in which they needed to invoke positive selection. K_A/K_S values for 5,641 mouse-rat orthologous protein-coding genes are shown in **Figure 4.12**. The vast majority of genes are subject to purifying selection, a few genes evolve under strict neutrality, and only a negligible minority experience positive selection.

This test and similar ones have been used extensively in the literature to detect departures from the neutral mode of molecular evolution. Interestingly, elevated K_A/K_S ratios are frequently found in sex-related genes, for example, genes involved in mating behavior, fertilization, spermatogenesis, ejaculation, sperm motility, sex determination, and fertility (Nielsen 2005; Schully and Hellberg 2006; Gibbs et al. 2007; Clark 2008). Indeed, the highest ratio of nonsynonymous to synonymous substitution (K_A/K_S = 5.15) for a full-length protein was found in the 18-kilodalton protein in the acrosomal vesicle at the anterior of the sperm cell of several abalone species (Vacquier et al. 1997). There are several forces hypothesized to drive the adaptive evolution by positive selection of proteins involved in sexual reproduction. One thought is that sex-related genes are subject to positive selection for short periods of time during speciation as a means of erecting reproductive barriers that restrict gene flow between the speciating populations (Civetta and Singh 1998). Other factors associated with positive selection include sexual selection, sperm competition, sexual conflict, reinforcement of sexual isolation, and pathogen defense of sperm.

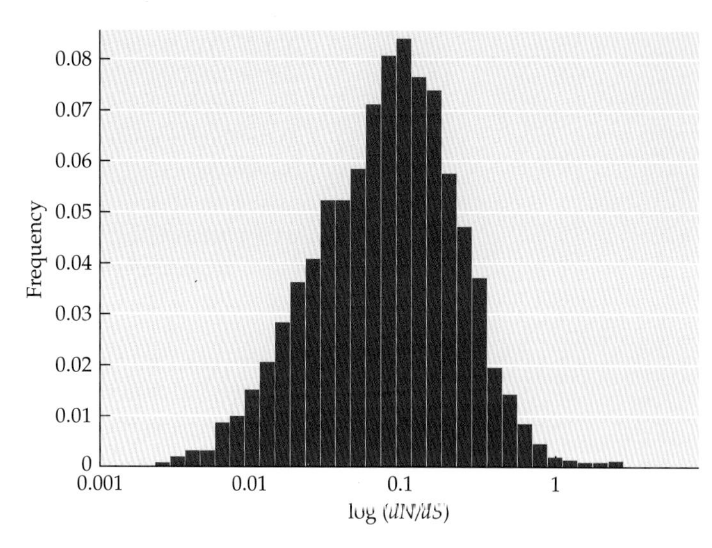

Figure 4.12 Distribution of K_A/K_S in a comparison of 5,641 orthologous protein-coding genes from mouse and rat. Genes with K_A = 0 or K_S = 0 were omitted. Note the logarithmic scale on the *x*-axis. (Modified from Gibbs et al. 2007.)

Prevalence of positive selection

A general consensus in the study of molecular evolution is that while the vast majority of protein-coding genes evolve by random genetic drift of neutral or nearly neutral alleles, it is the genes exhibiting signs of positive Darwinian selection that are "interesting" because they allow us some insight into function and changes in functional requirements during evolution. Indeed, much of the molecular evolution literature deals with detecting positive selection at different scales of resolution, from the individual nucleotide site to the entire genome. Here, we restrict our vista to studies in which positive Darwinian selection has been inferred through the K_A/K_S ratio test.

To our knowledge, Endo et al. (1996) were the first to study the prevalence of positive selection by using this test. In their study, only 17 out of 3,595 groups of homologous sequences from eukaryotes, prokaryotes, and viruses (about 0.45%) had a ratio of nonsynonymous to synonymous substitution that was significantly larger than 1. More recently, by use of this ratio, it has been shown that the proportion of positively selected genes varies widely among and within taxonomic lineages. However, the results vary widely depending on taxonomic sampling, sequence sampling, quality filtering, and estimation model. For example, the proportion of positively selected genes in humans was estimated to be 0.02% (Hughes and Nei 1988), 1.1% (Bakewell et al. 2007), 1.5% (Arbiza et al. 2006), 2.4% (Mikkelsen et al. 2005; Kosiol et al. 2008), 5.2% (Jørgensen et al. 2005), 5.3% (Jørgensen et al. 2005), and 8.7% (Clark et al. 2003)—a 435-fold range.

Are these estimates of prevalence of positive selection reliable? Many studies indicate that most estimates in the literature constitute overestimates of the true prevalence. The reasons for the inflated estimates may be (1) confounding biological effects that mimic the effects of positive Darwinian selection, such as biased nucleotide and codon usage (Albu et al. 2008) and biased gene conversion (Berglund et al. 2009; Galtier et al. 2009a; Hurst 2009); (2) deficient data, such as low sequence quality, misannotation of coding regions, and ambiguities in multiple sequence alignment (Schneider et al. 2009; Markova-Raina and Petrov 2011; Privman et al. 2012); (3) variability in the actual rates of neutral substitution (Rubinstein et al. 2011); and (4) inherent biases in methodology (e.g., Zhang and Li 2005; Nozawa et al. 2009). Thus, positive selection is most probably much less prevalent than some estimates would lead us to believe.

We must note, however, that the K_A/K_S ratio test can detect positive selection only when many sites in the gene are subject to selection. Positive selection affecting only parts of genes or individual sites may go undetected. An additional problem is that the method requires divergent protein-coding sequences. The K_A/K_S ratio test is inapplicable to closely related sequences, such as samples within a species. Finally, the method can only be applied to protein-coding genes. Detecting selection affecting the evolution of other functional elements in the genome requires different approaches.

Finally, it is therefore important to keep in mind that the mathematical definition and estimation of selection coefficients are but crude attempts at representing nature, and we should not assign too much significance to small selection coefficients (Nei et al. 2010).

Fast-X evolution

As seen previously, because of male mutational bias, the X chromosome is expected to accumulate neutral nucleotide substitutions at a slower rate than autosomes. The situation is reversed for advantageous mutations. Let us consider new autosomal mutations, which are initially low in frequency and primarily present in heterozygotes. If these mutations are either wholly or partially recessive, they will be obscured by the ancestral allele and will only rarely be exposed to selective forces. In contrast, novel recessive beneficial mutations on the X chromosome will be directly exposed to selection in the hemizygous sex. The probabilities of fixation for new beneficial mutations that are recessive or partially recessive are higher for X-linked loci than for autosomal ones (Charlesworth et al. 1987). Selection would therefore fix beneficial mutations and weed out deleterious mutations faster on sex chromosomes than on autosomes, a situation referred to as **fast-X evolution**.

Slow-X evolution has been detected in most primate lineages, whereas fast-X evolution has been detected in only a few of them (Xu et al. 2012). Similar results have been obtained in birds, where males are homogametic (ZZ) and females are heterogametic (ZW). That is, hemizygous exposure in females was found to increase the rate of fixation for beneficial nondominant mutations (Mank et al. 2007).

Rates of Evolution under Balancing Selection

Balancing selection can maintain transspecies polymorphism (Chapter 2). It can, thus, be detected by identifying shared polymorphic alleles in two species. A genome-wide scan for shared polymorphic alleles in the genomes of human and chimpanzee populations yielded tentative evidence of balancing selection for 125 loci. For 6 of these loci, there is convincing evidence for an ancestral polymorphism that predated the human-chimpanzee divergence and persisted in the present-day populations of these two species (Leffler et al. 2013). Some of the genes under balancing selection seem to be involved in host-pathogen interactions.

Patterns of Substitution and Replacement

There are 12 possible types of nucleotide substitution (e.g., A → G, A → C, T → A). To calculate the number of nucleotide substitutions between two DNA sequences, we pool all 12 types together. However, it is sometimes interesting to determine separately the frequency with which each type of substitution occurs. The **pattern of nucleotide substitution** is defined as the relative frequency with which a certain nucleotide changes into another during evolution. The pattern is usually shown in the form of a 4 × 4 matrix, in which each of the 12 elements of the matrix (excluding the four diagonal elements, which represent the case of no substitution) denote the number of changes from a certain nucleotide to another. In practice, we distinguish between a **directional pattern of substitution** (in which the direction of substitution is known, e.g., A → T) and a **nondirectional pattern of substitution** (in which only the nucleotides involved in the substitution are know, but the direction of the substitution is not, e.g., A ↔ T).

Let P_{ij} be the proportion of base changes from the *i*th type to the *j*th type of nucleotide (i, j = A, T, C, or G, and $i \neq j$). This proportion is calculated as

$$P_{ij} = \frac{n_{ij}}{n_i} \tag{4.18}$$

where n_{ij} is the number of substitutions from i to j, and n_i is the number of the i nucleotides in the ancestral sequence. To be able to compare the patterns of nucleotide substitution between sequences, we define f_{ij}, the relative substitution frequency from nucleotide i to nucleotide j, as

$$f_{ij} = \frac{P_{ij}}{\sum_i \sum_{j \neq i} P_{ij}} \times 100 \tag{4.19}$$

Thus, f_{ij} represents the expected number of base changes from the *i*th type of nucleotide to the *j*th type among every 100 substitutions in a random sequence (i.e., in a sequence in which the four bases are equally frequent). We note that although the pattern of substitution is determined by comparing single-stranded sequences, DNA is actually double-stranded, so each change from one nucleotide to another represents in fact a change from one base pair to another base pair. Thus, the shorthand f_{AT} actually represents a change from A:T to T:A.

Patterns of spontaneous mutation

Because point mutation is one of the most important factors in the evolution of DNA sequences, molecular evolutionists have long been interested in knowing the **pattern of spontaneous mutation** (e.g., Beale and Lehmann 1965; Fitch 1967; Zuckerkandl et al. 1971; Vogel and Kopun 1977). This pattern can serve as the standard for inferring how far the observed frequencies of interchange between nucleotides in any given DNA sequence have deviated from the values expected under the assumption of no selection, or strict selective neutrality.

One way to study the pattern of point mutation is to examine the pattern of substitution in regions of DNA that are subject to no selective constraint. Pseudogenes are particularly useful in this respect. Since they are devoid of function, all mutations occurring in pseudogenes are selectively neutral and become fixed in the population with equal probability. Thus, the rate of nucleotide substitution in pseudogenes is expected to equal the rate of mutation. Similarly, the pattern of nucleotide substitution in pseudogenes is expected to reflect the pattern of spontaneous point mutation.

Figure 4.13 shows a simple method for inferring the nucleotide substitutions in a pseudogene sequence (Gojobori et al. 1982). Sequence 1 is a pseudogene, sequence 2 is a functional counterpart from the same species, and sequence 3 is a functional sequence from a different species that diverged from the first species before the emergence of the pseudogene. Suppose that at a certain nucleotide site, sequences 1 and 2 have A and G, respectively. Then we can assume that the nucleotide in the pseudogene sequence has changed from G to A if sequence 3 has G, but that the nucleotide in sequence 2 has changed from A to G if sequence 3 has A. However, if sequence 3 has T or C, then we cannot decide the direction of change, and in this case the site is excluded from the comparison. In general, a substitution from nucleotide X to nucleotide Y is inferred when both the functional genes have nucleotide X at a homologous site, but the pseudogene has Y at the homologous site. In all other cases, the changes cannot be determined uniquely, and these sites are excluded from the analysis. Similarly, deletions and insertions should also be excluded. Since the rate of substitution is usually much higher in pseudogenes than in the homologous functional genes,

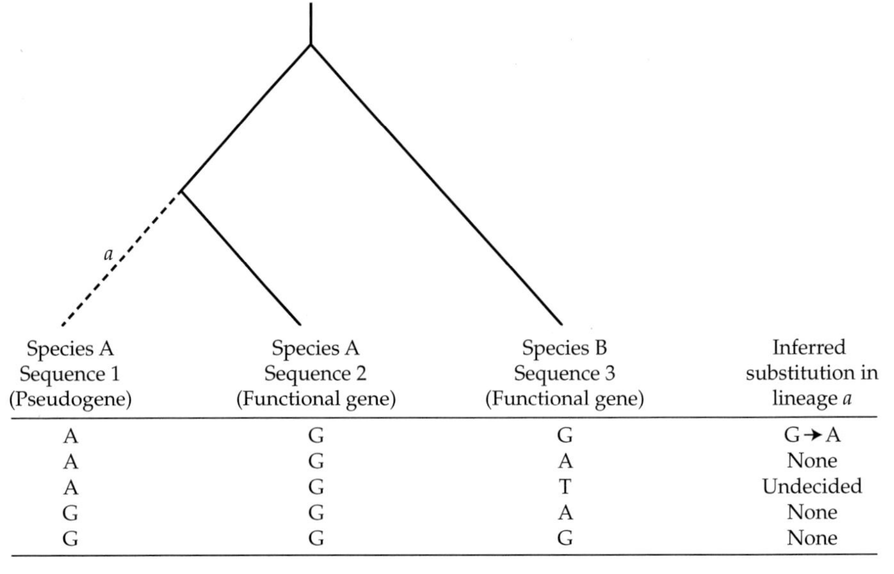

Species A Sequence 1 (Pseudogene)	Species A Sequence 2 (Functional gene)	Species B Sequence 3 (Functional gene)	Inferred substitution in lineage *a*
A	G	G	G→A
A	G	A	None
A	G	T	Undecided
G	G	A	None
G	G	G	None

Figure 4.13 A tree for inferring the pattern of nucleotide substitution in a pseudogene sequence. The dashed line *a* implies that the sequence is nonfunctional. In cases where the nucleotides occupying homologous sites in sequences 2 and 3 are identical but are different from the nucleotide in sequence 1, the type of substitution in lineage *a* can be unambiguously inferred. (Modified from Graur and Li 2000.)

differences in the nucleotide sequence between a gene and a pseudogene are explained in the majority of cases by substitutions that have occurred in the pseudogene rather than by substitutions in the functional gene. This inference is made more certain by using sites that have not changed in the functional genes since the divergence between the two species.

Empirical findings indicate that f_{ij} values usually vary greatly from one pseudogene to another. This variation has been attributed largely to chance effects. Thus, to obtain a reliable estimate of the mutation pattern from the pattern of substitution in pseudogenes, we need to use many sequences. The matrix in **Table 4.4a** represents the combined pattern of substitution inferred from 105 processed pseudogene sequences from humans. We note that the direction of mutation is nonrandom. For example, A changes more often to G than to either T or C. The four elements from the upper right corner to the lower left corner are the f_{ij} values for transitions, while the other eight elements represent transversions. All transitions, and in particular C to T and G to A, occur more often than transversions. The sum of the relative frequencies of transitions is 68% (66% if CG dinucleotides are excluded; see below). Because there are four types of transitions and eight types of transversions, the expected proportion of transitions under the assumption that all possible mutations occur with equal frequencies is 33%. The observed proportion (67%) is about twice the expected value. Thus, in vertebrate nuclear genomes, transitional mutations occur twice as frequently as transversions. In animal mitochondrial genomes, the ratio of transitions to transversions ranges from 15 to 20. However, the inequality between transitions and transversions is not a general phenomenon in nature; in the wingless grasshopper *Podisma pedestris* there is no significant difference between transition and transversion rates (Keller et al. 2007).

Some nucleotides are more mutable than others. In the last column of Table 4.4a, we list the relative frequencies of all mutations from A, T, C, and G to any other nucleotide. Were all the four nucleotides equally mutable, we would expect a value of 25% in each of the column's elements. In reality, we see that G mutates with a relative frequency of 30% (32% if CG dinucleotides are excluded), i.e., G is a highly mutable nucleotide, while A mutates with a relative frequency of 20% (i.e., it is least mutable). In the bottom row, we list the relative frequencies of all mutations that result in A, T, C, or G. Note that 56% of all mutations result in either A or T, while the expectation for the case of equal probabilities of mutation in all directions is 50%. Since there is a tendency for C and G to change to A or T, and since A and T are not as mutable as C and G, the pseudogenes in our sample are expected to become richer in A and T.

Genes are, of course, constrained and may only change to the extent that this does not interfere with biochemical function or result in a less stable pattern of codon usage. Though genes have amino acid compositions distinct from those of intergenic DNA sequences that have been "translated" in silico, they clearly reflect the underlying levels of nucleotide usage in the genomes as a whole.

It is known that, in addition to base mispairing, the transition from C to T can also arise from conversion of methylated C residues to T residues upon deamination (Coulondre et al. 1978; Razin and Riggs 1980). The effect will elevate the frequencies of C:G → T:A and G:C → A:T, i.e., f_{CT} and f_{GA}. Since about 90% of methylated C residues in vertebrate DNA occur at 5′–CG–3′ dinucleotides (Razin and Riggs 1980), this effect should be expressed mainly as changes of the CG dinucleotides to TG or CA. After a gene becomes a pseudogene, such changes will no longer be subject to any functional

Table 4.4

Pattern of substitution in human and microbial pseudogenes

(a) Human pseudogenes[a]

	To				
From	A	T	C	G	Row totals
A	—	3.4 ± 0.7	4.5 ± 0.8	**12.5 ± 1.1**	20.3
		(3.6 ± 0.7)	(4.8 ± 0.9)	**(13.3 ± 1.1)**	(21.6)
T	3.3 ± 0.6	—	**13.8 ± 1.9**	3.3 ± 0.6	20.4
	(3.5 ± 0.6)		**(14.7 ± 2.0)**	(3.5 ± 0.6)	(21.7)
C	4.2 ± 0.5	**20.7 ± 1.3**	—	4.6 ± 0.6	29.5
	(4.2 ± 0.5)	**(16.4 ± 1.3)**		(4.4 ± 0.6)	(25.1)
G	**20.4 ± 1.4**	4.4 ± 0.6	4.9 ± 0.7	—	29.7
	(21.9 ± 1.5)	(4.6 ± 0.6)	(5.2 ± 0.8)		(31.6)
Column totals	27.9	28.5	23.2	20.5	
	(29.5)	(24.6)	(23.2)	(21.3)	

(b) Microbial pseudogenes[b]

	To				
From	A	T	C	G	Row totals
A	—	2.9 ± 2.4	3.7 ± 2.4	**8.1 ± 3.4**	14.7
		(3.1 ± 2.6)	(4.0 ± 2.7)	**(8.7 ± 3.7)**	(15.8)
T	2.1 ± 1.8	—	**5.9 ± 2.9**	3.5 ± 2.5	11.5
	(2.2 ± 2.0)		**(6.3 ± 3.2)**	(3.7 ± 2.7)	(12.2)
C	7.4 ± 2.6	**26.6 ± 6.1**	—	8.5 ± 2.8	42.5
	(7.1 ± 2.9)	**(25.7 ± 6.3)**		(8.6 ± 3.2)	(41.4)
G	**15.8 ± 4.1**	8.5 ± 2.6	7.0 ± 2.3	—	31.3
	(15.6 ± 4.6)	(8.1 ± 2.8)	(6.9 ± 2.7)		(30.6)
Column totals	25.3	38.0	16.6	20.1	
	(24.9)	(36.9)	(17.2)	(21.0)	

[a] Table entries are the inferred percentages (*fij*) of nucleotide changes from *i* to *j* (± standard deviations) based on 105 processed pseudogene sequences from humans with mouse or rat as outgroup. Values in parentheses were obtained by excluding all CG dinucleotides from the comparison. Transitions are in **bold blue**. Data courtesy of Ron Ophir.

[b] Table entries are the inferred percentages (*fij*) of nucleotide changes from *i* to *j* in the leading strand based on analyses of 569 unitary pseudogene sequences from *Mycobacterium leprae*. Since the *M. leprae* genome does not contain functional homologs of these pseudogenes, we used functional homologs from *M. tuberculosis* in the comparison. Outgroup orthologs were derived from the genome of *Streptomyces coelicolor*. Values in parentheses were obtained by excluding all CG dinucleotides from the comparison. Transitions are in **bold blue**. Data from Mitchell and Graur (2005).

constraint and can contribute significantly to C → T and G → A transitions if the frequency of CG is relatively high before the silencing of the gene (i.e., its loss of function). The substitution pattern obtained by excluding all nucleotide sites where the CG dinucleotides occur is given in parentheses in Table 4.4a. This pattern is probably more suitable for predicting the pattern of mutations in a sequence that has not been subject to functional constraints for a long time (e.g., some parts of an intron), because in such a sequence few CG dinucleotides would exist to begin with. The pattern obtained after excluding the CG dinucleotides is somewhat different from that obtained otherwise. In particular, the relative frequencies of the transition C → T is lower by about 20%.

The discovery of pseudogenes in bacterial genomes (Andersson and Andersson 1999a,b; Cole et al. 2001) presented us with the opportunity to infer strand-specific patterns of mutation in bacteria (Mitchell and Graur 2005). Because the bacterial pseudogenes are almost without exception unitary pseudogenes (Chapter 7), i.e., they have no homologous functional counterparts in the genome, the methodology in Figure 4.13 needs to be modified somewhat: sequence 1 is the pseudogene as before, but sequence 2 is a functional gene from a closely related species, and sequence 3 is from an outgroup species. In **Table 4.4b**, we present a substitution matrix based on an analysis of 569 pseudogenes. Again, the direction of mutation is nonrandom; transitions constitute more than 56% of all inferred mutations. The transitional bias is mainly due to C → T and G → A. The most frequent mutation is C → T; it accounts for more than a quarter of all mutations in the leading strand, as opposed to the 8.3% expectation under equal probabilities for all mutations.

Patterns of mutation and strand asymmetry

Because the mechanisms for replicating the leading and the lagging strands of DNA are different (Okazaki et al. 1967, 1968; Ogawa and Okazaki 1980), as are their spatial organization within cells (White et al. 2008), it is, in principle, possible that the patterns of mutation differ too. In the absence of selection, this would lead to a difference in the pattern of substitution. Wu and Maeda (1987) proposed a method to test strand asymmetry in the patterns of mutation. The test can be performed regardless of whether or not we have information on the manner in which a strand in question is replicated. The only difference between the two cases is that when we have no information on the manner of replication of a strand, the test can only tell us whether or not a difference between the two strands exists, whereas if we know the manner of replication, we can infer the direction of the difference. Here, we assume that the manner of replication of each of the two strands is not known.

Let us designate the two strands as 1 and 2 (**Figure 4.14**). Nucleotide substitutions are scored on strand 1. A change from T to C actually means that a T → C change occurred on strand 1 and an A → G change occurred on the complementary strand 2. The relative substitution frequency from T to C, designated f_{TC}, is the sum of the frequencies of T → C on strand 1 and A → G on strand 2. Similarly, f_{AG} is the sum of the frequencies of A → G on strand 1 and T → C on strand 2. If the patterns and rates of replication error are identical, we expect the equality $f_{TC} = f_{AG}$ to hold for both strands 1 and 2. Similarly, we expect $f_{CT} = f_{GA}$, $f_{CA} = f_{GT}$, $f_{AC} = f_{TG}$, $f_{GC} = f_{CG}$, and $f_{AT} = f_{TA}$.

Initial studies on primate mitochondrial DNA (mtDNA) revealed an extreme transitional bias in the pattern of nucleotide substitution (Brown et al. 1982). Tamura and Nei (1993) studied the pattern of substitution in the control region of mtDNA using extensive data from humans and chimpanzees. With the exception of a conservative central portion, which was excluded from the analysis, the control region is thought to be devoid of functional constraints, and as such its pattern of substitution may reflect the pattern of spontaneous mutation in mtDNA. The estimated pattern indicated that transversions occur with extremely low frequencies; the average transition/transversion ratio was 15.7, i.e., about 95% of all mutations were transitions. Moreover the relative frequency of transition between pyrimidines (C ↔ T) was almost twice as large as that between purines (G ↔ A), violating the two equalities, $f_{CT} = f_{GA}$ and $f_{TC} = f_{AG}$ and suggesting that the patterns and rates of mutation may be different be-

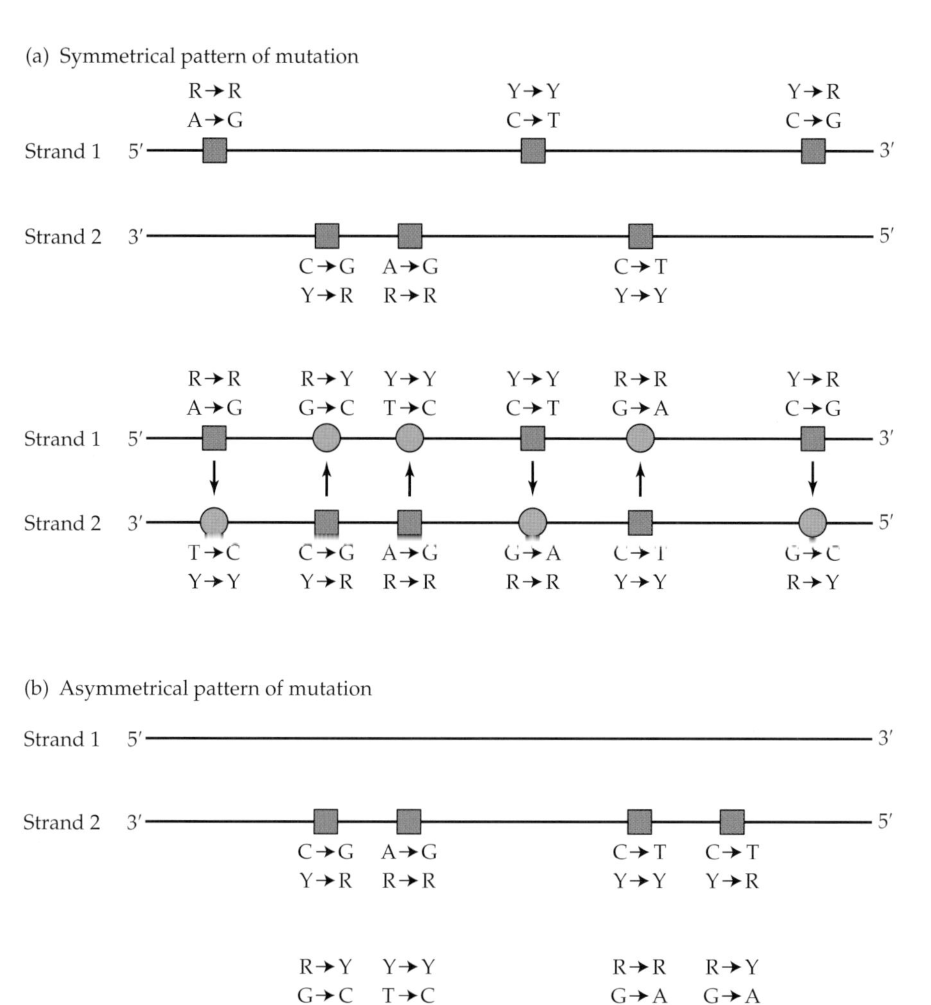

Figure 4.14 Symmetrical and asymmetrical patterns of mutation. Each mutation is accompanied by a double notation: from a nucleotide to another (e.g., C → G) and from a base type to another (e.g., Y → R, where Y stands for a pyrimidine and R stands for a purine). (a) The pattern of mutation is the same in the two strands. That is, the same replication errors (squares) that occur in strand 1 (C → G, A → G, and C → T) also occur in strand 2. The replication errors in one strand induce (arrows) complementary changes (circles) in the complementary strand. Because of this symmetry, we observe the following equalities in strand 1: $f_{TC} = f_{AG}$, $f_{CT} = f_{GA}$, $f_{GC} = f_{CG}$, $f_{RR} = f_{YY}$, and $f_{RY} = f_{YR}$. The same equalities are observed in strand 2. (b) The pattern of mutations differs between the two strands. Here we show an extreme case of asymmetry, in which all replication errors occur on strand 2. As a result, some of the equalities in (a) do not hold. For example, in strand 1, $f_{RY} = 2$, while $f_{YR} = 0$.

tween the two strands. In comparison, no evidence for strand inequality was obtained from the nuclear pattern in Table 4.4a, where the relative frequency of transition between pyrimidines (34.5%) was about the same as that between purines (32.9%).

From Table 4.4b, it is clear that in bacteria the mutation pattern is strongly asymmetrical between the two differently replicating strands. The relative frequency of transition between pyrimidines (32.5%) is 40% higher than between purines (23.9%).

Clustered multinucleotide substitutions: Positive selection or nonrandomness of mutation?

A widely held assumption in molecular analyses is that point mutations occur independently of one another, each caused by a separate mutational event. The occurrence of multiple closely spaced substitutions that appear to violate the assumption of mutational independence is often interpreted as evidence for positive selection, balancing selection, or compensatory evolution.

Here, we define **clustered multinucleotide substitutions** as two or more nucleotide substitutions occurring in close proximity to one another, for example, substitutions affecting adjacent nucleotides or substitutions occurring in the same codon. In comparative studies, it is possible to distinguish between two scenarios: (1) the two sub-

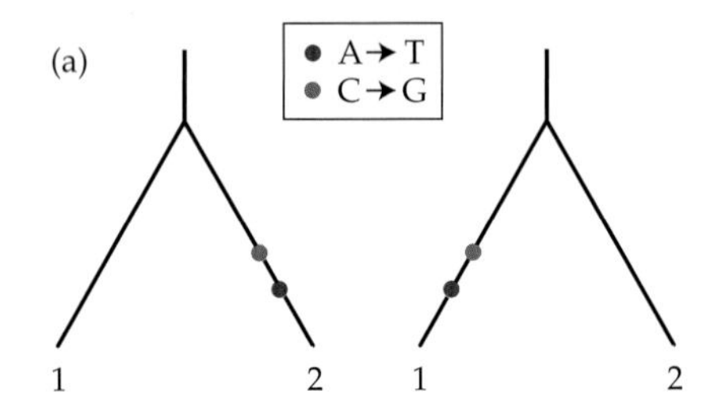

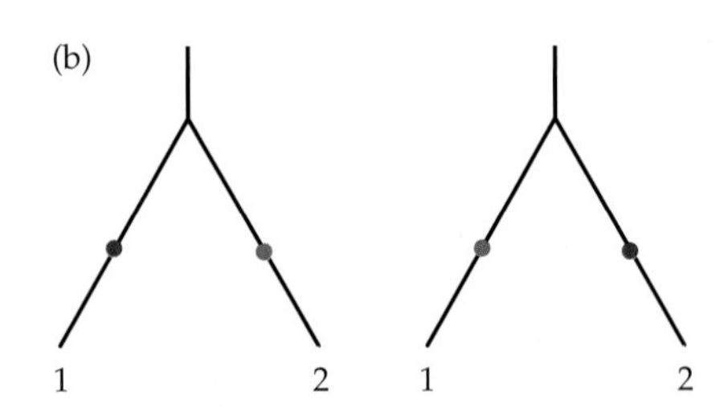

Figure 4.15 Multinucleotide substitutions. The red and blue circles represent individual substitutions (e.g., A → T and C → G, respectively). (a) Two ways that two substitutions can occur on the same haplotypic lineage. (b) Two ways that two substitutions can occur on different lineages. In practice, the two ways in (b) cannot be distinguished from each other. If all substitutions are independent, the number of pairs occurring on the same lineage (a) should equal the number of pairs occurring on different lineages (b).

stitutions occurred in the same lineage or (2) one substitution occurred in one lineage and the other occurred in the second linage (**Figure 4.15**). If all substitutions are independent from one another, the number of pairs occurring on the same lineage should equal the number of pairs occurring in different lineages. In real data, the assumption of independence seems to be frequently violated. For example, Bazykin et al. (2004) compared 2,894 pairs of orthologous codons from mouse and rat that differ from each other at two nonsynonymous positions, in which neither possible intermediate codon was a stop codon. In a phylogenetic reconstruction with human orthologous genes as outgroups, the two nonsynonymous substitutions occurred in the same lineage 1,847 times (64%) and in different lineages only 1,047 times (36%). These numbers clearly deviate from parity. There can be two explanations for such results. Bazykin et al. (2004) favor an explanation that involves positive selection acting on closely spaced sites and resulting in successive amino acid replacements. An alternative explanation was suggested by

TABLE 4.5

Physicochemical distances between pairs of amino acids[a]

Arg	Leu	Pro	Thr	Ala	Val	Gly	Ile	Phe	Tyr	Cys	His	Gln	Asn	Lys	Asp	Glu	Met	Trp	
110	145	74	58	99	124	56	142	155	144	112	89	68	46	121	65	80	135	177	Ser
	102	103	71	112	96	125	97	97	77	180	29	43	86	26	96	54	91	101	Arg
		98	92	96	32	138	**5**	22	36	198	99	113	153	107	172	138	15	61	Leu
			38	27	68	42	95	114	110	169	77	76	91	103	108	93	87	147	Pro
				58	69	59	89	103	92	149	47	42	65	78	85	63	81	128	Thr
					64	60	94	113	112	195	86	91	111	106	126	107	84	148	Ala
						109	29	50	55	192	84	96	133	97	152	121	21	88	Val
							135	153	147	159	98	87	80	127	94	98	127	184	Gly
								21	33	198	94	109	149	102	168	134	10	61	Ile
									22	205	100	116	158	102	177	140	28	40	Phe
										194	83	99	143	85	160	122	36	37	Tyr
											174	154	139	202	154	170	196	**215**	Cys
												24	68	32	81	40	87	115	His
													46	53	61	29	101	130	Gln
														94	23	42	142	174	Asn
															101	56	95	110	Lys
																45	160	181	Asp
																	126	152	Glu
																		67	Met

Source: Grantham (1974).

[a]Mean distance is 100. The largest and smallest distances are highlighted.

Schrider et al. (2011). According to this explanation, mutations do not occur independently of one another, but rather exhibit a clumped distribution within short stretches of DNA. Clustered multinucleotide mutational events have been detected in all taxa so far studied. A number of mechanisms may be invoked to explain such mutational clustering, including hypermutation due to incorrectly transcribed or translated DNA polymerases, or due to one mutation increasing the probability of an additional mutation occurring at a nearby site.

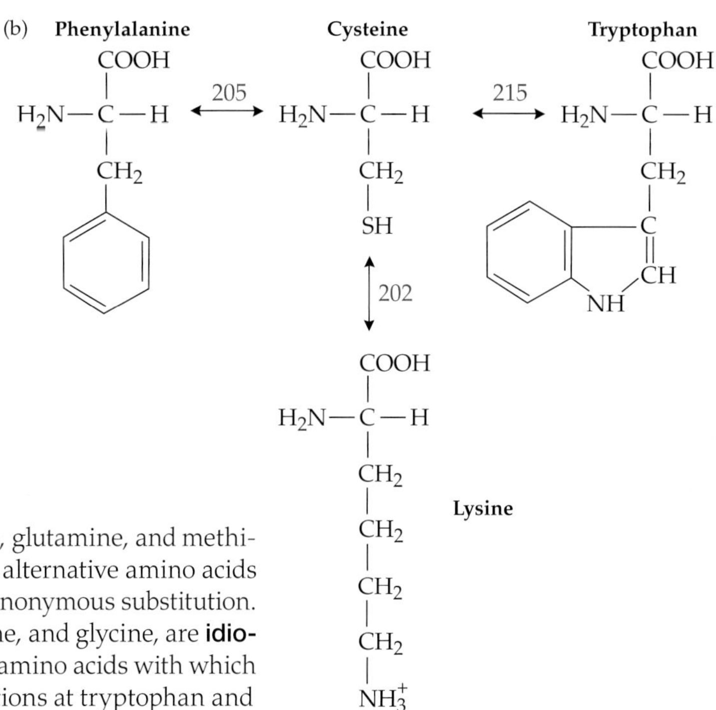

Figure 4.16 Similarity of amino acid pairs. Red numerals represent Grantham's (1974) physicochemical distances. (a) The most similar amino acid pairs are leucine and isoleucine, and leucine and methionine. (b) The most dissimilar amino acid pairs are cysteine and tryptophan, cysteine and phenylalanine, and cysteine and lysine. Note that all three of the most dissimilar pairs involve the idiosyncratic amino acid cysteine. (Modified from Graur and Li 2000.)

Patterns of amino acid replacement

There are many measures in the literature aimed at quantifying the similarity or dissimilarity between two amino acids (e.g., Sneath 1966; Grantham 1974; Miyata et al. 1979; Yampolsky and Stoltzfus 2005.). These so-called **physicochemical distances** are based on such properties of the amino acids as polarity, molecular volume, and chemical composition. Grantham's (1974) physicochemical distances are shown in **Table 4.5**. A replacement of an amino acid by a similar one (i.e., an exchange between two amino acids separated by a small distance) is called a **conservative replacement** (**Figure 4.16a**). A replacement of an amino acid by a dissimilar one (i.e., an exchange between two amino acids separated by a large distance) is called a **radical replacement** (**Figure 4.16b**). Some amino acids, such as leucine, isoleucine, glutamine, and methionine, are **typical amino acids**; there are several similar alternative amino acids with which these can be replaced through a single nonsynonymous substitution. Other amino acids, such as cysteine, tryptophan, tyrosine, and glycine, are **idiosyncratic amino acids**; there are few similar alternative amino acids with which they can be replaced. Indeed, nonsynonymous substitutions at tryptophan and cysteine codons have the highest probability of causing disease (Vitkup et al. 2003). Graur (1985) devised a **stability index**, which is the mean physicochemical distance between an amino acid and its mutational derivatives that can be produced through a single nucleotide substitution. The stability index can be used to predict the evolutionary propensity of amino acids to undergo replacement. Other indices have been suggested in the literature (e.g., Tourasse and Li 2000; Tang et al. 2004).

It has been known since the early work of Zuckerkandl and Pauling (1965) that conservative replacements occur more frequently than radical replacements in protein evolution. The conservative nature of amino acid replacement is evident when the relative frequencies of amino acid replacements are plotted against physicochemical distances (**Figure 4.17**). From the standpoint of the neutral theory, this phenomenon can be easily explained by using Equation 4.6. Conservative replacements are likely to be less disruptive than radical ones; hence, the probability that a mutational change will be selectively neutral (as opposed to deleterious) is greater if the amino acid replacement occurs between two similar amino acids than if it occurs between two dissimilar ones. However, in some cases the codons of similar amino acids differ by more than one nucleotide, and so a conservative amino acid replacement may be less probable than a more radical replacement.

Argyle (1980) devised a circular graphical representation of amino acid exchangeability. A modified version by Pieber and Tohá (1983) is shown in **Figure**

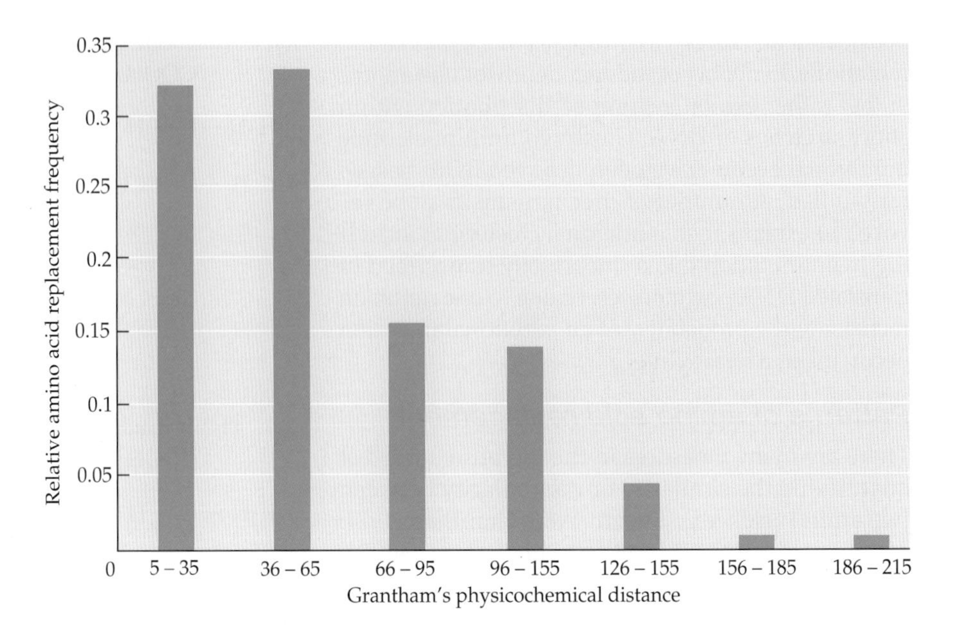

Figure 4.17 Relationship between Grantham's (1974) physicochemical distance and relative amino acid replacement frequency in 20 mammalian proteins (ryanodine receptor, dystrophin, ataxia-telangiectasia locus protein, coagulation factors VIII and IX, cystic fibrosis transmembrane conductance regulator, α-glucosidase, low-density lipoprotein receptor, pyruvate kinase, butyrylcholinesterase, hexoseaminidases A and B, glucocerebrosidase, phenylalanine hydroxylase, fumarylacetoacetate hydrolase, galactose-1-phosophate uridyltransferase, peripherin, uroporphyrinogene III synthase, CD40 ligand, and von Hippel-Lindau disease tumor suppressor). Replacement frequency is much higher among similar amino acids, as measured by Grantham's distance. (Courtesy of Inna Gur-El.)

4.18. Depending on the protein, 60–90% of observed amino acid replacements involve the nearest or second-nearest neighbors on the ring.

What protein properties are conserved in protein evolution?

The evolution of each protein-coding gene is constrained by the functional requirements of the specific protein it produces. However, it may be interesting to find out whether or not there are also some general properties that are constrained during evolution in all proteins. The answer seems to be that several properties are indeed conserved during protein evolution (Soto and Tohác 1983). The two most highly conserved properties are **bulkiness** (volume) and **refractive index** (a measure of protein density). **Hydrophobicity** and **polarity** also seem to be moderately well conserved, whereas **optical rotation** seems to be an irrelevant property in the evolution of proteins. Surprisingly, **charge**, which might be expected to be an important factor determining protein evolution, is one of the least conserved properties (Leunissen et al. 1990). Other conserved properties are **shape** and **bond-forming ability**.

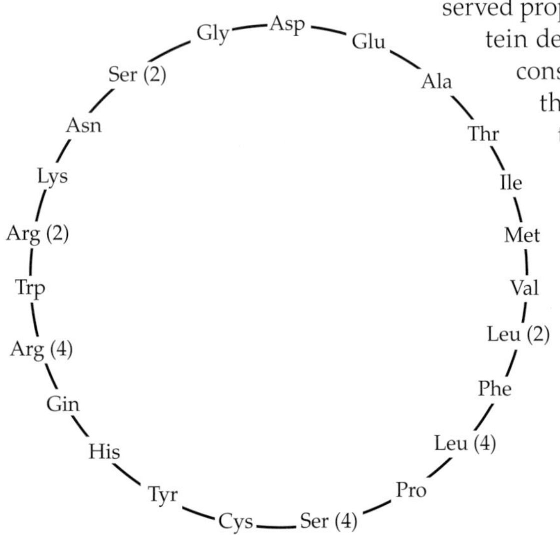

Figure 4.18 A circular graphical representation of amino acid exchangeability according to Argyle's (1980) method. Numbers in parentheses denote number of codons in a family in cases where an amino acid is encoded by two codon families. About 60–90% of the observed amino acid replacements involve the nearest or second nearest neighbors on the ring. (Data from Pieber and Tohá 1983.)

As a consequence of the conservation of bulkiness and hydrophobicity, some amino acids tend to be relatively impervious to replacement during evolution. Indeed, glycine, the smallest amino acid, tends to be extremely conserved during evolution regardless of its proximity to functionally active sites. Conceivably, replacements at sites occupied by glycine will invariably introduce into the polypeptide chain a much bulkier amino acid. Since such a structural disruption has a high probability of adversely affecting the function of the protein regardless of the location of glycine relative to the active site, such mutations are frequently selected against. Consequently, genes that encode proteins that contain large proportions of glycine residues will tend to evolve more slowly than proteins that are glycine-poor (Graur 1985). In addition to glycine, other amino acids (e.g., lysine, cysteine, and proline) are also consistently conserved (Naor et al. 1996). Lysine and cysteine are most probably conserved because of their involvement in cross-linking between polypeptide chains, while the imino acid proline is conserved because of its unique contribution to the bending of proteins.

Heterotachy

So far we have explained the variation in substitution rates among sites by differences in functional constraints. In these analyses, we usually assumed that the mutation rates and the functional constraints at a site remain constant during evolution. These assumptions do not always hold. In fact, in many cases, we observe different rates of substitution at the same site between evolutionary lineages. **Heterotachy** (Greek for "different speed") refers to shifts in site-specific evolutionary rates over time. The shifts may be due to lineage-specific changes in mutation rate or to changes in selective constraint.

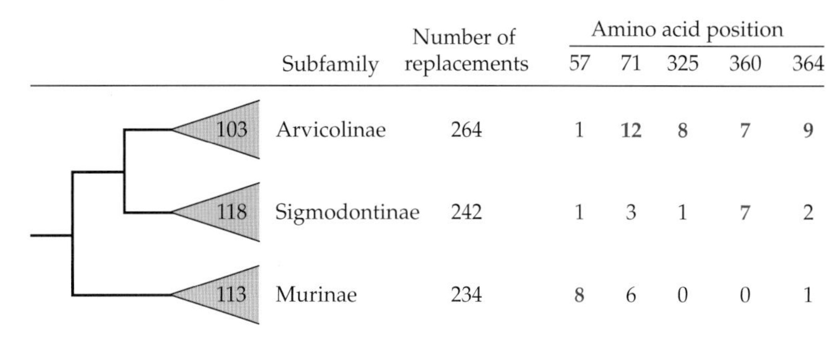

	Subfamily	Number of replacements	Amino acid position 57	71	325	360	364
103	Arvicolinae	264	1	12	8	7	9
118	Sigmodontinae	242	1	3	1	7	2
113	Murinae	234	8	6	0	0	1

Figure 4.19 Five heterotachous amino acid positions in the cytochrome b sequences of murids. Three monophyletic taxa (subfamilies) whose phylogenetic relationships are displayed on the left were used to build the data set. The number of species in each subfamily is shown on the phylogenetic tree. Each position is described by the number of replacements it underwent in each subfamily. A constant substitution rate can be rejected for the five positions displayed here. These positions were chosen from among the most variable ones, so the rejection is highly significant statistically. Within each amino acid position, the highest number of replacements is displayed in red. Note that different amino acid positions exhibit rate accelerations in the different taxa. Heterotachy is thus different from a global evolutionary acceleration, which would occur in all amino acid positions in a taxon and is unrelated to molecular clock issues. (Modified from Lopez et al. 2002.)

By using amino acid sequences of mitochondrial cytochrome b, for which the function is likely the same in all vertebrates, Lopez et al. (2002) showed that 95% of the variable positions in this protein are heterotachous, i.e., they underwent dramatic variations of substitution rate among vertebrate lineages. In **Figure 4.19**, we illustrate heterotachy in three rodent lineages at five amino acid positions within cytochrome b. At position 325, we see a high rate of amino acid replacement in members of subfamily Arvicolinae (voles, lemmings, and muskrats), while not even one change occurred in subfamily Murinae (Old World rats and mice). In comparison, at position 57, the rate of amino acid replacement in Murinae is high, whereas the rates in Arvicolinae and Sigmodontinae (New World rats and mice) are extremely low.

A large number of sequences for which the evolutionary relationships are known are required to efficiently detect heterotachy (Wang et al. 2011). Thus, at the present time, the extent of this phenomenon can only be estimated in a small number of cases.

Nonrandom Usage of Synonymous Codons

Because of the degeneracy of all genetic codes, most of the amino acids are encoded by more than one codon (Chapter 1). Since most synonymous mutations are silent and do not affect the amino acid sequence of the encoded protein, and since natural selection was thought to operate predominantly at the protein level, synonymous mutations were at first thought to be strictly neutral (Kimura 1968b; King and Jukes 1969). According to this view, and assuming a symmetrical pattern of mutation, the expectation was that all the synonymous codons for an amino acid would be used more or less with equal frequencies. As DNA sequence data accumulated, however, it soon became evident that the usage of synonymous codons is distinctly nonrandom in many prokaryotic and eukaryotic genes (Grantham et al. 1980). For example, 21 of the 23 leucine residues in the *Escherichia coli* outer membrane protein II (*ompA*) are encoded by the codon CUG, although five other codons for leucine are available. **Table 4.6** shows part of a compilation of codon usage by Sharp et al. (1988). For each group of synonymous codons, if the usage is equal, then the relative frequency of each codon should be $1/n$, where n is the number of synonymous codons for the amino acid in question. This is clearly not so in the majority of cases. Moreover, in some organisms, such as *E. coli* and yeast, codon usage bias is much stronger in highly expressed genes than in lowly expressed ones.

In the following discussion, we outline several measures for quantifying the degree of codon usage bias and discuss the evolutionary forces shaping codon usage.

Table 4.6

Codon usage in *Escherichia coli* and *Saccharomyces cerevisiae*[a]

		E. coli		*S. cerevisiae*	
Amino acid	Codon	High	Low	High	Low
Leu	UUA	0.01	0.21	0.09	0.25
	UUG	0.01	0.15	**0.89**	0.25
	CUU	0.02	0.12	0.00	0.12
	CUC	0.03	0.11	0.00	0.09
	CUA	0.01	0.05	0.03	0.16
	CUG	**0.92**	0.37	0.00	0.14
Val	GUU	**0.60**	0.27	**0.52**	0.28
	GUC	0.02	0.25	**0.48**	0.19
	GUA	0.28	0.16	0.00	0.30
	GUG	0.10	0.29	0.00	0.23
Ile	AUU	0.16	0.46	**0.42**	0.43
	AUC	**0.84**	0.33	**0.58**	0.22
	AUA	0.00	0.17	0.00	0.35
Phe	UUU	0.17	0.67	0.09	0.69
	UUC	**0.83**	0.33	**0.91**	0.31

Source: From Sharp et al. (1988).

[a]Under equal codon usage, the relative frequencies for each codon in a group should be $1/n$, where n is the number of synonymous codons for the amino acid in question. "High" and "low" denote genes with high and low expression levels, respectively. Preferred codons in highly expressed genes are highlighted.

Measures of codon usage bias

There are dozens of measures for quantifying nonrandom synonymous codon usage (Roth et al. 2012). Here we only present some simple measures. The **relative synonymous codon usage** (**RSCU**) is one such measure. The *RSCU* value for a codon is the number of occurrences of that codon in the gene divided by the number of occurrences expected under the assumption of equal codon usage.

$$RSCU_i = \frac{X_i}{\frac{1}{n}\sum_{i=1}^{n} X_i} \tag{4.20}$$

where n is the number of synonymous codons ($1 \leq n \leq 6$) for the amino acid under study, i is the codon, and X_i is the number of occurrences of codon i (Sharp et al. 1986). For each amino acid, the sum of its *RSCU* values will equal the number of its synonymous codons. If the synonymous codons for an amino acid are used with equal frequencies, then their *RSCU* values will all equal 1. **Figure 4.20** shows the *RSCU* values for all 64 codons in the γ-proteobacterium *Pseudomonas aeruginosa*. We note that with the trivial exception of the single codon families TGG and ATG, for tryptophan and methionine, respectively, most other codons have *RSCU* values that deviate considerably from 1.

If certain codons are used preferentially in an organism, it is useful to devise a measure with which to distinguish between genes that use the preferred codons and those that do not. Sharp and Li (1987b) devised a **codon adaptation index** (**CAI**). In the first step of computing *CAI* values, a reference table of *RSCU* values for highly ex-

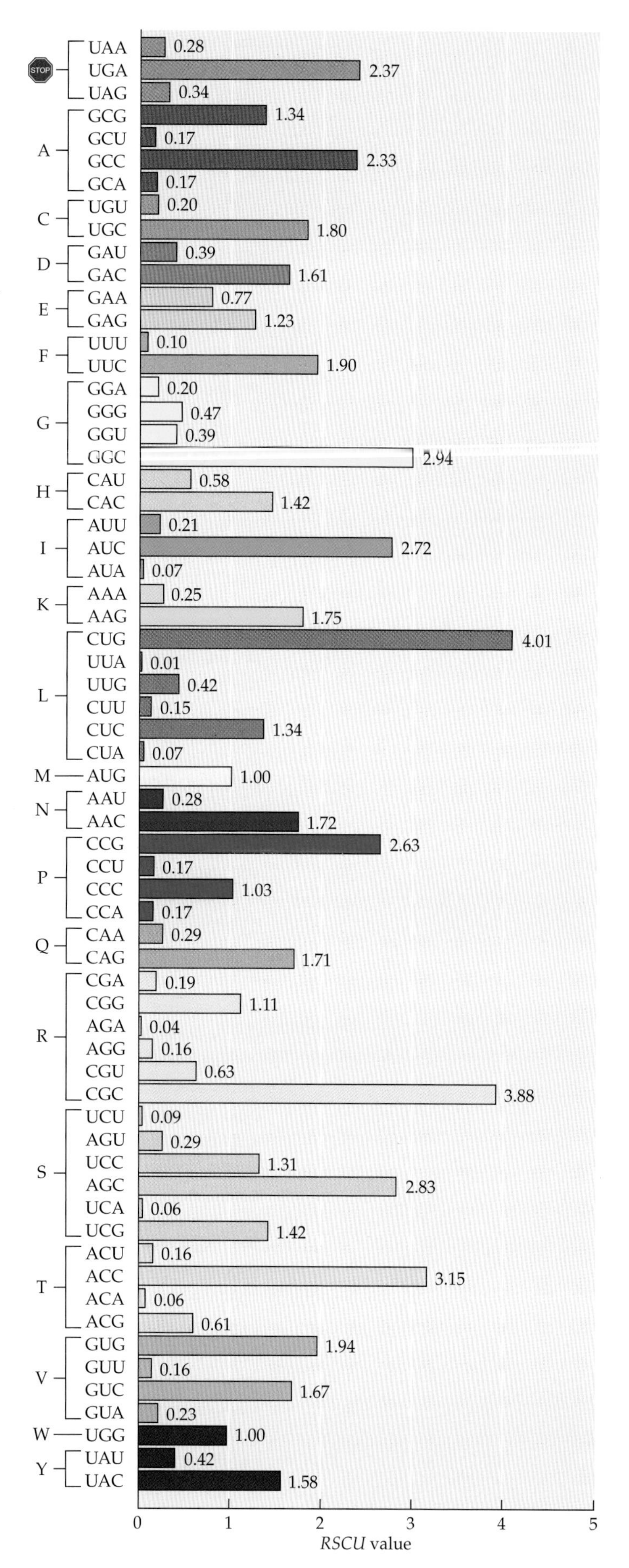

Figure 4.20 *RSCU* values of all 64 codons in the coding regions of *Pseudomonas aeruginosa* (1,865,980 codons). Amino acids are shown in different colors and identified by their one-letter abbreviations (Chapter 1; Table 1.3).

pressed genes is compiled. From this table, it is possible to identify the codons that are most frequently used for each amino acid. The **relative adaptedness** of a codon, w_i, is computed as

$$w_i = \frac{RSCU_i}{RSCU_{max}} \tag{4.21}$$

where $RSCU_{max}$ is the *RSCU* value for the most frequently used codon for an amino acid.

The *CAI* value for a gene is calculated as the geometric mean of w_i values for all the codons used in that gene:

$$CAI = \left(\prod_{i=1}^{L} w_i \right)^{\frac{1}{L}} \tag{4.22}$$

where L is the number of codons.

An alternative to CAI is the frequency of optimal codons, F_{OP}, which is the simplest measure of codon usage bias:

$$F_{OP} = \frac{x_{OP}}{x_{OP} + x_{NON}} \tag{4.23}$$

where x_{OP} and x_{NON} are the number of optimal and nonoptimal codons in a gene, respectively.

The two measures, *CAI* and F_{OP}, presuppose advance knowledge of the optimal codons for each amino acid. In contrast, the **effective number of codons (ENC)** measures bias from equal usage of codons without knowledge of the optimal codons. It is estimated by

$$ENC = 2 + \frac{9}{F_2} + \frac{1}{F_3} + \frac{5}{F_4} + \frac{3}{F_6} \tag{4.24}$$

where F_i (i = 2, 3, 4, or 6) is the average probability that two randomly chosen codons for an amino acid with i codons will be identical (Wright 1990). (Note the analogy with nucleotide diversity presented in Chapter 2.) In genes translated by the standard genetic code, *ENC* values may range from 20 (the number of amino acids) to 61 (the number of sense codons). That is, *ENC* assumes the maximum value when only one codon is used for each amino acid, and the minimum value when there is no codon usage bias.

Species-specific and universal patterns of codon usage

An observation that has been helpful for understanding the phenomenon of nonrandom codon usage is that genes in an organism generally show the same pattern of choices among synonymous codons (Grantham et al. 1980). The choices, however, differ from one organism to another. In light of these facts, Grantham et al. (1980)

proposed the **genome hypothesis**, according to which the genes in any given genome use the same strategy with respect to choices among synonymous codons; in other words, the bias in codon usage is species-specific. The genome hypothesis turned out to be strictly true only in a few organisms, and even in these organisms there is considerable heterogeneity in codon usage among genes within a genome. However, the hypothesis was useful in encouraging a torrent of studies on the subject.

Are there features of codon usage shared among all organisms? So far, all suggestions concerning universal patterns have been refuted; there seem to be no universals of codon usage shared by all organisms or even a majority of them. There are no universally preferred or universally avoided codons, although the claim has been made that there may be some universal preferences and avoidances as far as codon neighbor pairs are concerned. For example, the pair NNG GNN, where "N" stands for all four possible nucleotides, seems to be preferred, while the pair NNG CNN seems to be avoided in many eukaryotes and prokaryotes (Tats et al. 2008).

Determinants of Codon Usage

Biases in synonymous codon usage can in principle be caused by (1) mutational biases, (2) selection favoring preferred codons, (3) purifying selection against unfavorable codons, or (4) selection unrelated to particular codons that nonetheless produces a pattern of biased codon usage. In mutational explanations, codon biases arise from properties of the mutational process, for example, biases in mutation patterns. Mutational explanations are neutral because they do not require any fitness advantage or disadvantage to be associated with alternative synonymous codons (Plotkin and Kudla 2011). Codon usage varies among taxa, within a genome, and within a gene. In the following sections we will treat the interspecific variation separately of intragenomic and intragenic variation.

Interspecific variation in codon usage and amino acid usage

The most important single determinant of codon usage variation across species is GC content. In fact, differences in codon usage between bacterial species can be quite accurately predicted from the nucleotide content in their non-protein-coding regions (e.g., Hershberg and Petrov 2008). Since genomic GC content is mostly determined by mutational processes, most interspecific variation in codon usage can be attributed to mutation. In species in which mutation patterns vary across the genome, such as in mammals, the mutation hypothesis predicts a dependence of codon usage on the nucleotide composition in the immediate vicinity of the gene. Such context dependence has indeed been observed (Fedorov et al. 2002).

King and Jukes (1969) were the first to discover a positive correlation between genomic GC content and amino acid composition, indicating that within each protein, only few amino acids are under selective constraint. According to Knight et al. (2001), within each protein-coding gene, approximately 70–90% of the variation in codon usage is explainable by GC content. As far as the amino acid composition of a protein is concerned, about 70–80% of the variation is explained by the GC content of its gene.

Intragenomic variation in codon usage

The variation in codon usage among genes within a genome is often quite large. This variation cannot usually be explained by mutational biases, for if codon usage is due to biases in mutation patterns, then the expectation is that the magnitudes and the directions of the biases will be more or less the same for all codon families and for all genes, regardless of function or expression level. Let us assume that the mutation pattern in an organism tends to result in AT-rich sequences. Under such a mutational regime, it is expected that all fourfold degenerate codon families will exhibit a preference for codons ending in A or T. Thus, the preferred codons for valine should be GTA and GTT, and the preferred codons for arginine should be CGA and CGT. Some bacteria, such as *Mycoplasma capricolum*, exhibit this type of codon usage bias; in the eight codon families with four members, codons ending in A or T and codons ending in C or G are used 95.5% and 4.5% of the time, respectively. Thus, mutation bias seems to be the predominant force affecting codon usage bias in this species. Mutation bias, however, cannot explain cases in which the preferred codon for valine is GUU, while the preferred codon for leucine is CUG, as is the case in *E. coli*. In such cases, we need to invoke selection. Yang and Nielsen (2008) tested the mutation bias hypothesis in mammalian genomes and found that it can be rejected in the vast majority of cases.

The bulk of intragenic variation in codon usage must, therefore, be attributed to selection. Is the selection leading to biased codon usage positive or purifying? That is, are preferred codons selected for or are nonpreferred codons selected against? We note that positive selection is expected to accelerate the rate of evolution, while purifying selection is expected to slow it down (Chapter 2). In model organisms, such as *E. coli*, *Saccharomyces cerevisiae*, *C. elegans*, *A. thaliana*, and *Drosophila melanogaster*, as well as in many mammalian species, including humans, there is a significant negative correlation between the degree of codon bias and the rate of synonymous substitution (e.g., Bierne and Eyre-Walker 2003; Plotkin and Kudla 2011). This finding points to purifying selection, rather than positive selection, as the driving selective force determining nonrandom codon usage.

Numerous selective factors have been invoked to explain codon usage bias. In the following sections, we will only discuss factors related to (1) translational efficiency, (2) translational accuracy, and (3) selection unrelated to codons that nonetheless affect codon usage.

Translational efficiency and translation accuracy

Upon discovering that ribosomal protein-coding genes in *E. coli* preferentially use synonymous codons that are recognized by the most abundant isoaccepting tRNA, Post et al. (1979) suggested that using codons that are translated by abundant tRNA species will increase translational efficiency and accuracy, thereby conveying a selective advantage. According to Post et al. (1979), preferential codon usage may reflect the cell's need that the translation of abundant proteins be more efficient than that of less abundant proteins. Alternatively, they suggested, the preferential codon usage could be the consequence of selection to minimize translational errors, as a small number of tRNA molecules available for translating a codon may limit the rate of protein synthesis and increase translational errors, presumably due to more effective competition by incorrect tRNAs.

Let us now examine these two hypotheses: (1) selection related to translational efficiency (i.e., selection against translational inefficiency), and (2) selection related to translational fidelity or accuracy (i.e., selection against translational infidelity). According to the translational efficiency hypothesis, codon bias is more extreme in highly expressed genes to match a skew in isoaccepting tRNAs and, hence, increase the speed of translation and protein yields. Moreover, speedy and efficient translation has the added benefit of reducing protein misfolding that occurs when translation is slow (Kimchi-Sarfaty et al. 2007). From an evolutionary perspective, the translational efficiency hypothesis is problematic because if high protein levels are desirable, it would seem easier for selection to increase the rate of transcription or the rate of translation initiation by improving one or two regulatory elements rather than "fiddling" with hundreds of synonymous mutations to affect translation elongation.

According to the translational fidelity hypothesis, selection keeps the fraction of mistranslated proteins to a minimum by reducing the proportion of codons that are decoded by rare tRNA species, which are frequently mistranslated. Drummond and Wilke (2008) have shown that even at a modest mistranslation rate of of 5×10^4, a 400-residue protein can be expected to contain at least one mistranslated amino acid 18% of the time. Such a substantial proportion of defective protein molecules is likely to destabilize the network of biological pathways under protein control that are essential for life (Powers and Balch 2008).

The vast majority of experimental tests in the literature can only validate a connection between codon usage and the process of translation; they cannot distinguish between the efficiency hypothesis and the fidelity hypothesis. For example, by examining the relative abundances of tRNAs in *E. coli* and *S. cerevisiae*, Ikemura (1981, 1982) showed that, in both species, a positive correlation exists between the relative frequencies of the synonymous codons in a gene and the relative abundances of their cognate tRNA species. The correlation was very strong for highly expressed genes, but weak to nonexistent in lowly expressed genes. Here we will examine two examples of the relationship that exists between tRNA abundance and codon usage. (We will use the notation $\text{tRNA}^{\text{aa}}_{\text{NNN}}$, where "aa" denotes the amino acid and "NNN" denotes the anticodon.) In *E. coli*, the most abundant tRNA for leucine is $\text{tRNA}^{\text{Leu}}_{\text{CAG}}$, which pairs with codon CUG. In highly expressed genes, this codon is used in excess of 90% of the time, whereas the other five codons are used only rarely. In contradistinction, in lowly expressed genes, CUG is used only about 40% of the time. In *S. cerevisae*, the most abundant tRNA for leucine is $\text{tRNA}^{\text{Leu}}_{\text{CAA}}$, and indeed in highly expressed genes, UUG (the codon recognized by this tRNA) is much more frequently used than the other five codons. In contrast, $\text{tRNA}^{\text{Leu}}_{\text{GAG}}$ is very rare, and indeed the two codons recognized by this tRNA, CUC and CUU, appears only very rarely in highly expressed genes. A simple explanation for these findings is that in highly expressed genes, purifying selection for translational efficiency and accuracy is strong, so the codon usage bias is pronounced. In lowly expressed genes, on the other hand, selection is relatively weak, so the usage pattern is mainly affected by mutation pressure and random genetic drift and is consequently less skewed (Sharp and Li 1986; dos Reis et al. 2003).

The importance of translational efficiency and accuracy in determining the codon usage pattern in highly expressed genes is further supported by the following observation (Ikemura 1981): It is known that codon-anticodon pairing may involve wobbling between the third codon position and the first anticodon position, that is, pairings other than U:A, A:U, C:G, and G:C, where the letter preceding the colon denotes the nucleotide in the third position of the codon, and the letter after the colon denotes the nucleotide in the first position of the anticodon. For example, U in the first position of anticodons can pair not only with A but also with G. Similarly, G in the first position of anticodons can pair with both C and U. On the other hand, C in the first anticodon position can only pair with G at the third position of codons, and A can only pair with U. Wobbling also involves modified nucleotides at the first anticodon position of some tRNAs, and these can recognize more than one codon (Chapter 1, Table 1.5). For example, inosine (a modified adenine) can pair with any of the three bases U, C, and A. Interestingly, most tRNAs that can recognize more than one codon exhibit differential preferences for one of them. For example, 4-thiouridine (S^4U) in the wobble position of an anticodon can recognize both A and G; however, it has a marked preference for A-terminated codons over G-terminated ones. Such a preference should be reflected in highly expressed genes. The two codons for lysine in *E. coli* are recognized by a tRNA molecule that has S^4U in the wobble position of the anticodon, and indeed, in highly expressed *E. coli* genes, AAA codons are preferred over AAG.

How can we distinguish between selection against inefficiency and selection against infidelity? There are several lines of evidence that discriminate between the efficiency and accuracy hypotheses. One of the most compelling pieces of evidence in support of the efficiency hypothesis is the finding that the most highly expressed protein-coding genes in the genome, i.e., the ribosomal protein-coding genes, exhibit more pronounced codon usage biases than other genes. **Figure 4.21** shows the distribution of the effective numbers of codons (*ENC*) for 3,907 *D. melanogaster* genes, as well as for a subset of 94 ribosomal protein-coding genes (Moriyama and Powell 1997). The vast majority of genes show codon usage biases, as indicated by a mean *ENC* value of approximately 49. In the ribosomal protein-coding genes, the codon usage bias is far stronger (mean $ENC \approx 39$), indicating that selection for translational efficiency plays an important role in determining the choice of synonymous codons in genes that encode high-yield proteins.

Another piece of evidence in favor of the efficiency theory is the observed correlation between the minimum generation time of a species and the degree of codon adaptation in its highly expressed genes (e.g., Sharp et al. 2005).

The best evidence in favor of the accuracy model was provided by Akashi (1994) and Akashi and Schaeffer (1997), who found a greater tendency for tRNA-adapted codons at residues that are strongly conserved across divergent *Drosophila* species; this suggests that sites that are under strong negative selection at the amino acid level also show stronger codon adaptation, presumably to reduce mistranslation. Similar results were obtained for other organisms (e.g., Stoletzki and Eyre-Walker 2007; Zhou et al. 2009).

Another line of evidence arises from the positive correlation between codon adaptation and gene length in bacteria (Eyre-Walker 1996; Moriyama and Powell 1998), which was interpreted to reflect a greater energetic cost of missense and nonsense translation errors in long proteins, especially if the mistranslations occur near the 3′ end. However, the relationship with gene length does not hold for all organisms.

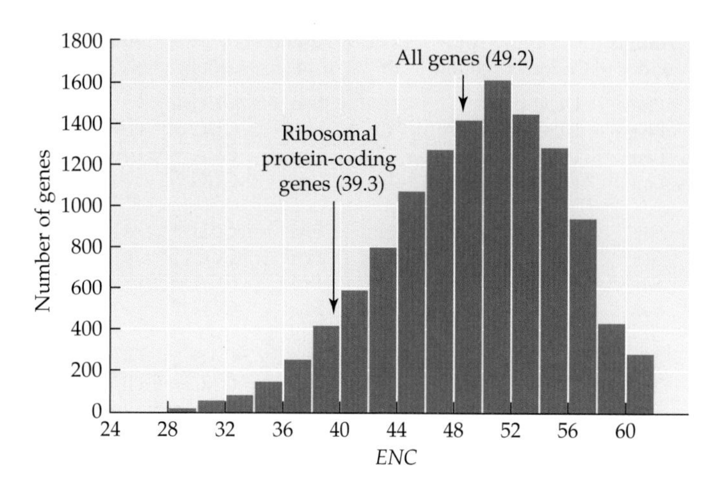

Figure 4.21 Distribution of *ENC* (effective number of codons) values for 3,907 genes from *Drosophila melanogaster*. The mean *ENC* value for all the genes is 49.2; that for a subset of 94 ribosomal protein-coding genes with more than 100 codons is 39.3. (Courtesy of Etsuko Moriyama.)

It is important to note that the accuracy and efficiency hypotheses are not mutually exclusive. Indeed, it is possible to imagine how both efficiency and accuracy can be simultaneously optimized. The two hypotheses, however, assume that codons corresponding to more abundant tRNAs are better than codons corresponding to less abundant tRNAs, and this may not always be true. For instance, Shah and Gilchrist (2010) demonstrated that codons corresponding to more abundant tRNAs do not always produce lower missense errors, as has been commonly assumed. Moreover, they found that for some amino acids, codon choice for elongation speed is different from codon choice for elongation accuracy.

Finally, large deviations from optimal codon usage are not always deleterious. There are cases in which nonoptimal codon usage is of great functional importance. The protein-coding genes of the red bread mold *Neurospora crassa* exhibit strong codon bias, with the third position of almost every codon exhibiting a preference, in the order C > G > T > A. Indeed, the genes encoding the top 100 most abundant proteins exhibit much stronger codon bias than the rest of the protein-coding genes. The FREQUENCY protein, which is a clock protein and responsible for circadian rhythms, is a lowly expressed protein that is encoded by the *frq* gene that exhibits almost no codon bias. Laboratory optimization of the codon usage of *frq* not only resulted in increased protein levels, but also abolished circadian rhythms (Zhou et al. 2013). Nonoptimal codon usage seems to be an important factor in determining circadian rhythms in many organisms (e.g., Xu et al. 2013), which shows that nonoptimal codon usage in a genomic context may in some instances be optimal in terms of function.

The tRNA adaptation index

So far, we have established an almost universal connection between the efficiency and accuracy of translation and codon usage, whereby highly expressed genes utilize codons that are recognized by the most abundant tRNA molecules in the cell. Interestingly, tRNA abundance seems to be determined primarily by tRNA copy number (e.g., Ikemura 1981). That is, tRNAs specified by many genes are more abundant than tRNAs specified by fewer genes. In the yeast genome, for instance, there are seven genes that specify $\text{tRNA}^{\text{Arg}}_{\text{ACG}}$, and they recognize codons CGT, CGC, and CGA. In contrast, there is a single gene specifying $\text{tRNA}^{\text{Arg}}_{\text{CCG}}$, and this gene recognizes CGG. Indeed, the frequencies of the codons recognized by $\text{tRNA}^{\text{Arg}}_{\text{ACG}}$ and $\text{tRNA}^{\text{Arg}}_{\text{CCG}}$ are ~87%

Amino acid	Codon	Anti-codon	*N*
Phe	UUU		
Phe	UUC	GAA	10
Leu	UUA	UAA	7
Leu	UUG	CAA	10
Ser	UCU	AGA	11
Ser	UCC		
Ser	UCA	UGA	3
Ser	UCG	CGA	1
Tyr	UAU		
Tyr	UAC	GUA	8
Cys	UGU		
Cys	UGC	GCA	4
Trp	UGG	CCA	6

Amino acid	Codon	Anti-codon	*N*
Leu	CUU		
Leu	CUC	GAG	1
Leu	CUA	UAG	3
Leu	CUG		
Pro	CCU	AGG	2
Pro	CCC		
Pro	CCA	UGG	10
Pro	CCG		
His	CAU		
His	CAC	GUG	7
Gln	CAA	UUG	9
Gln	CAG	CUG	1
Arg	CGU	ACG	7
Arg	CGC		
Arg	CGA		
Arg	CGG	CCG	1

Amino acid	Codon	Anti-codon	*N*
Ile	AUU	AAU	13
Ile	AUC		
Ile	AUA	UAU	2
Met	AUG	CAU	10
Thr	ACU	AGU	11
Thr	ACC		
Thr	ACA	UGU	4
Thr	ACG	CGU	1
Asn	AAU		
Asn	AAC	GUU	11
Lys	AAA	UUU	7
Lys	AAG	CUU	14
Ser	AGU		
Ser	AGC	GCU	2
Arg	AGA	UCU	12
Arg	AGG	CCU	1

Amino acid	Codon	Anti-codon	*N*
Val	GUU	AAC	14
Val	GUC		
Val	GUA	UAC	2
Val	GUG	CAC	2
Ala	GCU	AGC	11
Ala	GCC		
Ala	GCA	UGC	5
Ala	GCG		
Asp	GAU		
Asp	GAC	GUC	17
Glu	GAA	UUC	15
Glu	GAG	CUC	2
Gly	GGU		
Gly	GGC	GCC	16
Gly	GGA	UCC	3
Gly	GGG	CCC	2

Figure 4.22 Codon-anticodon recognition rules and number of tRNA-specifying genes (*N*) in the nuclear genome of *Saccharomyces cerevisae*. Theoretically, other codon-anticodon interactions are possible, such as between codon UUU and anticodon AAA. However, no tRNA with the AAA anticodon is specified by the *S. cerevisae* nuclear genome.

and ~14%, respectively. The relationship between codon usage and tRNA-gene number compels us to abandon the genetic code as the exclusive point of reference, and instead look at the process of translation from the viewpoint of the tRNA and the anticodon.

In **Figure 4.22**, we list the anticodons recognizing all possible codons in the yeast genome. There are some interesting features emerging from this figure. First, not all possible anticodons are specified by the genome. For example, the anticodon AAA, which can recognize codon UUU for phenylalanine, does not exist; instead the UUU codon is recognized by anticodon GAA through the use of the G:U pairing in the wobble position. (We note, however, that missing anticodons are a species-specific phenomenon; the AAA anticodon is found in the cow genome, but not in the human genome, which in this respect resembles the yeast genome.) Second, some anticodons recognize more than one codon, while others recognize a single one. The number of codons recognized by the anticodon range from one to three for unmodified nucleotides, and from one to four for modified ones. Third, some codons are recognized by a single anticodon, while others are recognized by two anticodons. Fourth, some codon families are recognized by a single anticodon, while others utilize two or more anticodons. For example, the cysteine codon family (UGT and UGC) uses the single anticodon GCA, whereas the lysine codon family (AAA and AAG) uses two anticodons, UUU, recognizing both AAA and AAG, and CUU, recognizing AAG only. (In yeast the maximum number of anticodons for a codon family is three; the maximum possible number is four.) Finally, we note that in cases in which one anticodon recognizes two or more codons, then the nonwobble codon-anticodon pairings are preferred over the wobble ones.

If codon usage is influenced by the efficiency of translation, and if the efficiency of translation is ultimately determined by tRNA abundance, then in addition to the codon adaptation index, we need a tRNA adaptation index to quantify the degree of correlation between codons and tRNA abundance. The **tRNA adaptation index** (tAI) is a measure that quantifies the optimality of the tRNA usage by the coding sequence of a gene. It is analogous to and inspired by the codon adaptation index presented above. Because of the highly positive correlation between tRNA abundance and tRNA-gene copy number, we use the latter as a proxy for the former.

We first define W_i as the absolute adaptiveness of codon i as

$$W_i = \sum_{j=1}^{n_i}(1 - s_{ij})T_{ij} \quad (4.25)$$

where n_i is the number of tRNA isoacceptors that recognize the ith codon, T_{ij} is the gene copy number of the jth tRNA that recognizes the the ith codon, and s_{ij} is a selective constraint on the efficiency of the codon-anticodon coupling. By using the data in Figure 4.22, the values of n_i and T_{ij} for yeast can be easily calculated. For example, for codon UUG, n_i is 2 ($tRNA^{Leu}_{CAA}$ and $tRNA^{Arg}_{UAA}$). T_{ij} for $tRNA^{Arg}_{CCG}$ is 10 and T_{ij} for $tRNA^{Leu}_{CAA}$ is 7. Calculating s_{ij} values is much more complicated, since in practice they are calculated iteratively so as to maximize the correlation between *tAI* values and either mRNA levels or protein abundance levels (dos Reis et al. 2003; Tuller et al. 2010). For purposes of illustration, we will use a simple scheme, whereby $s_{ij} = 0$ for nonwobble pairings and $s_{ij} = 0.5$ for wobble pairings, i.e., we assume that the selective constraint is higher for imperfect (wobble) pairings than for nonwobble pairings. By using Equation 4.25 for codon UUG in yeast, we obtain $W_{UUG} = (1 - 0) \times 10 + (1 - 0.5) \times 7 = 13.5$.

After computing the absolute adaptiveness of all the sense codons, we can compute the relative adaptiveness as

$$w_i = \frac{W_i}{W_{max}} \quad (4.26)$$

where W_{max} is the highest of all W_i values.

The tRNA adaptation index of gene g, tAI_g, is defined as the geometric mean of the relative adaptiveness values of its codons,

$$tAI_g = \left(\prod_{k=1}^{l_g} w_{i_{kg}}\right)^{1/l_g} \quad (4.27)$$

where i_{kg} is the codon defined by the kth triplet in gene g, and l_g is the length of the gene as measured in number of sense codons. So tAI_g estimates the degree of adaptation of gene g to the genomic gene pool.

Intragenic variation in codon usage

So far, we have dealt with codon usage as a property of a gene, essentially assuming that codon usage bias is uniform along the length of the coding region within an mRNA sequence. In reality, codon usage can vary dramatically intragenically. We note, however, that most deviations from uniform codon usage are either gene- or taxon-specific, or they are driven by selection on features unrelated to codon usage itself (see below). Only two universal selection regimes for producing systematic intragenic variation in codon usage have been proposed so far.

The first mechanism was described by Tuller et al. (2010), who found that codons recognized by low-abundance tRNAs are overrepresented in the 5′ region of many highly expressed genes. This pattern suggests that ribosomes translate more slowly over the initial 30–50 codons (the so-called **ramp stage**) and then translate the remainder of the mRNA at full speed. What purpose could this ramp play in translation? Slowing translation elongation immediately after initiation effectively generates more uniform spacing between ribosomes farther down the mRNA, which prevents ribosome congestion and such subsequent phenomena as translation stalling and termination. Another potential role for the ramp involves protein folding. The length of the ramp corresponds quite well to the length of the polypeptide needed to fill the exit tunnel of the ribosome, so the nascent peptide chain should emerge from the ribosome as it transitions from the slow ramp stage to the fast stage of elongation. This raises the possibility that the slowdown in the ramp might increase the fraction of correctly folded product (e.g., Siller et al. 2010).

A second mechanism was described by Cannarozzi et al. (2010), who studied patterns of synonymous codon usage along the gene. In the **autocorrelated pattern**

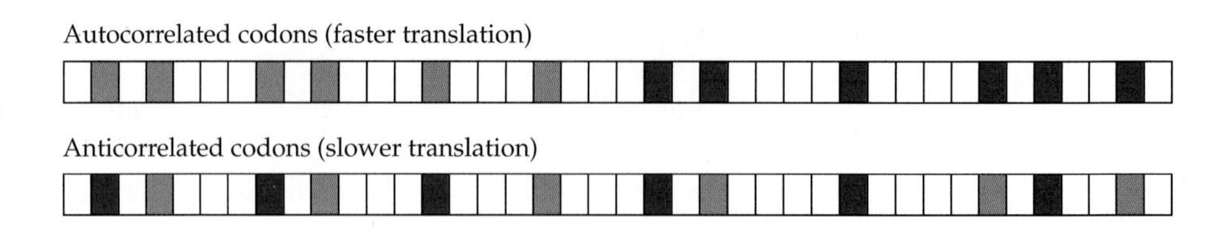

Figure 4.23 The arrangement of synonymous codons along an mRNA influences translation speed. Shown is a simple example of two mRNA sequences of 20 codons each (squares). Twelve of these codons encode a certain amino acid. This amino acid is specified by a two-codon family. In the top sequence, the two codons (red and green) are arranged along the mRNA in an autocorrelated manner, and as a consequence translation is faster than in the bottom sequence in which they are arranged in an anticorrelated manner.

(**Figure 4.23**), when an amino acid recurs in the protein, there is a strong propensity for the same codon to be used for both occurrences of the amino acid. In the **anticorrelated pattern**, when an amino acid recurs in the protein, there is a strong tendency for a different codon to be used the second time. Cannarozzi et al. (2010) found that the translation of mRNAs that have an autocorrelated pattern is faster than that for mRNAs that have an anticorrelated pattern of codon usage. In other words, selection for fast or slow translation speed may affect codon usage.

Indirect selection on codon usage

Codon usage within a gene may be driven by the fact that many synonymous mutations may experience selection for reasons unrelated to codon usage itself. First, some synonymous mutations may disrupt motifs that are recognized by transcriptional or posttranscriptional mechanisms. Second, sites that require ribosomal pausing for proper cotranslational protein folding or ubiquitin modification may experience selection against changes of codons that are poorly adapted to the tRNA pool. Third, codon choice that promotes proper nucleosome positioning is selectively advantageous, especially in 5′ regions. Fourth, synonymous mutations can create spurious splice sites or disrupt splicing control elements (Chapter 1).

Finally, codon usage may be influenced by selection related to mRNA folding, i.e., the forming of double-stranded structures in the mRNA. Strong secondary structures close to the 5′ end of the reading frame are selected against, as such structures my disrupt translation initiation by inhibiting ribosomal initiation (Bulmer 1991; Eyre-Walker and Bulmer 1993; Gu et al. 2010) and ultimately translation levels (Kudla et al. 2009).

Why do only some organisms have biased codon usages?

Some organisms, such as the bacterium *E. coli*, the yeast *S. cerevisae*, and the nematode *C. elegans*, show marked codon biases that can be attributed to selection. Others, such as the bacterium *Helicobacter pylori* and humans, present little evidence of translational selection. A possible explanation for this difference was suggested by dos Reis et al. (2003), who hypothesized that tRNA-gene redundancy and genome size are interacting forces in determining translational selection and codon usage bias. In their work, they suggested that an optimal combination of these factors exists for which the action of translational selection is maximal. Roughly, the magnitude of selection is maximal in genomes 1–30 Mb in size that contain 150–600 tRNA-specifying genes. Both *H. pylori* and humans fall outside this range. The genome of *H. pylori* contains only 36 tRNA-specifying genes with only one tRNA gene having two copies, while the haploid genome size of humans is approximately 3,500 Mb.

Sharp et al. (2005) discovered that species selected for rapid growth have more rRNA and tRNA genes and stronger codon usage biases. One such example is the bacterium *Clostridium perfringens*, whose codon usage was found to be under the most stringent selection discovered so far. It also has a generation time as short as 7 minutes.

Codon usage in unicellular and multicellular organisms

The relationship between codon usage and tRNA abundance is very clear in many unicellular organisms. In multicellular organisms, the situation is less clear because different cells produce different proteins. Similarly, large tissue-specific differences in tRNA abundance have been documented (e.g., Dittmar et al. 2006). Therefore, a simple relationship between codon usage and tRNA abundance is not expected. In some cases, there is evidence for tissue-specific adaptations of the number of isoacceptor tRNA species and codon frequencies. For example, the middle part of the silk gland in the silk moth *Bombyx mori* secretes a protein called sericin (31% serine), while the posterior part of the gland produces fibroin (46% glycine, 29% alanine, and 12% serine). Interestingly, the populations of isoaccepting $tRNA^{Ser}$ species are different for the two parts of the gland. While cells in the middle part contain mainly tRNA molecules that recognize AGU and AGC codons, the posterior part contains mainly tRNAs that recognize UCA (Hentzen et al. 1981). This example notwithstanding, in most cases the correlation between the relative quantity of isoacceptor tRNA species and codon frequencies is far from being convincing.

Are constraints on codon usage a property of the genome as a whole, or do they vary intragenomically? Over 200 homologous reading frames from human, chimpanzee, mouse, rat, and dog ranging in length from 60 to 180 codons were found to exhibit identical codon usage biases (Schattner and Diekhans 2006). These regions were located within protein-coding genes with overall synonymous substitution rates not significantly higher or lower than those of other genes. This finding suggests that the phenomenon of codon usage conservation may also be due to localized constraints rather than gene-wide or genome-wide constraints.

Codon usage and population size

The general observation is that the strength of selection against nonpreferred codons is very weak, so most mutations are nearly neutral. In fact, selection may be so weak that selection would be the dominant force in the evolution of codon bias only in species with very large effective population sizes, while random genetic drift would dominate the evolutionary dynamics of codon substitution in species with small effective population sizes. Indeed, in comparisons of codon usage biases among *Drosophila* species (Akashi 1997; Vicario et al. 2007; Petit and Barbadilla 2009), the general finding is that species with large effective population sizes have stronger codon usage biases than species with small population sizes. Moreover, even in species in which selection plays the principal role in determining codon usage bias, not all genes would be under substantial selective constraints; genes with low expression should exhibit codon biases that merely reflect mutational biases.

Molecular Clocks

In their comparative studies of hemoglobin and cytochrome c protein sequences from different species, Zuckerkandl and Pauling (1962, 1965) and Margoliash (1963) first noticed that the rates of amino acid replacement were approximately the same among various mammalian lineages (**Figure 4.24**). Zuckerkandl and Pauling (1965) therefore proposed that for any given protein, the rate of molecular evolution is approximately constant over time in all lineages—that is, there exists a **molecular clock**. The proposal immediately stimulated a great deal of interest in the use of macromolecules in evolutionary studies. Indeed, if proteins evolve at constant rates, they can be used to determine dates of species divergence and to reconstruct phylogenetic relationships among organisms. This practice would be analogous to the dating of geological times by measuring the decay of radioactive elements.

Under the molecular clock assumption, Equation 4.4 can be used to estimate the rate of substitution, r, for any given protein or DNA region by using a species pair for which the date of divergence can be established on the basis of paleontological data. This estimated rate could then be used to date the divergence time between two species for which paleontological data on their divergence time are lacking. For example, let us assume that the rate of nonsynonymous substitution for the α chain of hemoglobin is 0.56×10^{-9} substitutions per site per year, and that α-globins from rat and human differ by 0.093 substitutions per site. Then, under the molecular clock

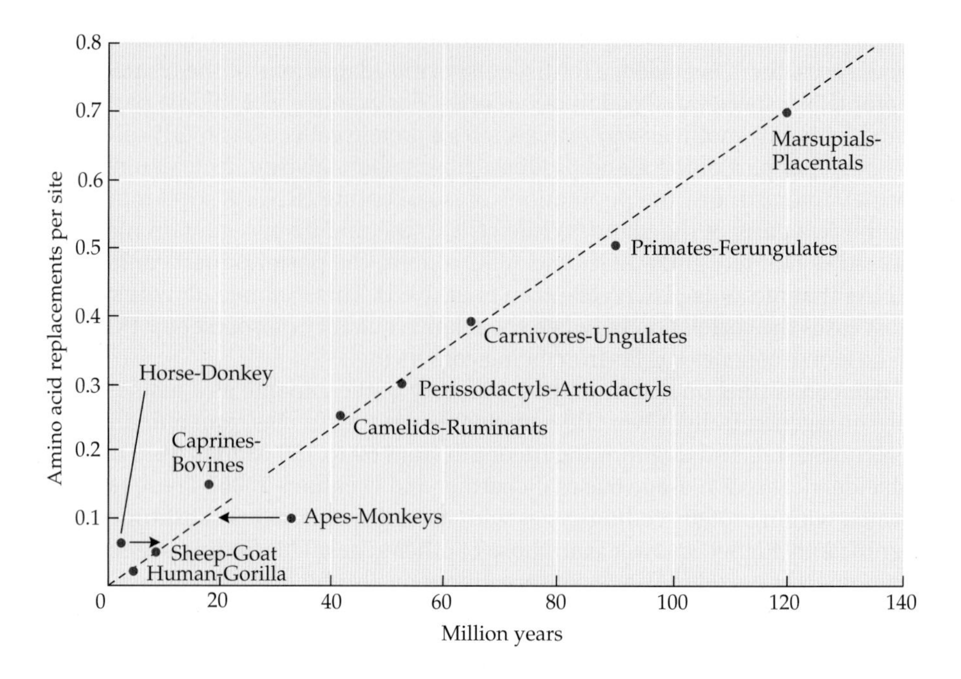

Figure 4.24 Number of amino acid replacements per amino acid site (inferred from a concatenated sequence consisting of hemoglobins α and β, cytochrome c, and fibrinopeptide A) among various mammalian groups plotted against geological estimates of divergence times. The dashed line represents the molecular clock expectation of equal rates of amino acid replacement in all evolutionary lineages. There are three large deviations (arrows) of the observed values from the molecular clock line. (Data from Langley and Fitch 1974.)

hypothesis, the divergence time between the human and rat lineages is estimated to be approximately $0.093/(2 \times 0.56 \times 10^{-9}) = 83$ million years ago.

In the 1970s, protein sequence data became available, and many statistical analyses were conducted to test the molecular clock hypothesis (Dickerson 1971; Ohta and Kimura 1971; Langley and Fitch 1974; Wilson et al. 1977). A general finding was that a rough linear relationship exists between the estimated number of amino acid replacements and divergence time. This finding led to the proposal of the following evolutionary rule by Kimura (1983): "For each protein, the rate of evolution in terms of amino acid substitutions is approximately constant per year per site for various lines, as long as the function and tertiary structure of the molecule remain essentially unaltered." The condition appended to the rule was meant to take care of instances in which, in a particular evolutionary lineage, a gene loses some or all of its functions (p. 122) or acquires a new biological role (Chapter 7). In such cases, the functional constraints operating on the gene are altered, and as a consequence the gene will no longer evolve at the same rate as the homologous genes in other organisms, which retain their original functions.

The molecular clock hypothesis stimulated a great deal of controversy. Classical evolutionists, for instance, argued against it because the suggestion of rate constancy did not sit well with the erratic tempo of evolution at the morphological and physiological levels. The hypothesis met with particularly heated opposition when the assumption of rate constancy was used to obtain an estimate of 5 million years for the divergence between humans and the African apes (Sarich and Wilson 1967), in sharp contrast to the then-prevailing view among paleontologists that humans and apes diverged at least 15 million years ago. Many molecular evolutionists have also challenged the validity of the molecular clock hypothesis. In particular, Goodman (1981a,b) and his associates (Czelusniak et al. 1982) contended that the rate of evolution often accelerates following gene duplication. For example, they claimed that extremely high rates of amino acid replacement occurred following the gene duplication that gave rise to the α- and β-hemoglobins and that the high rates were due to advantageous mutations that improved the function of these globin chains. Moreover, Goodman and his associates also objected to the molecular clock on grounds that protein sequence evolution often proceeds much more rapidly at speciation events than during periods in which no speciation occurs. For instance, Goodman (1981b) and Goodman et al. (1974, 1975) claimed that an increase in the rate of amino acid replacement in cytochrome c occurred at least three times during the evolution of this protein: in the period immediately after the teleost-tetrapod divergence (~400 million years ago), in the early phases of divergence among eutherian orders (~80 million years ago), and during the initial stages of divergence among primates (less than 55 million years ago).

It should be noted, however, that from the beginning, proponents of the molecular clock hypothesis recognized the existence of exceptional cases. Thus, while a given protein usually evolves at a characteristic rate regardless of the organismic lineage, in some particular lineage it may evolve at a markedly different rate. A well-known case was that of insulin, which has evolved much faster in the evolutionary lineage leading to the guinea pig than along other lines (King and Jukes 1969).

Despite the fact that the rate constancy assumption has always been controversial, it has been widely used in the estimation of divergence times and the reconstruction of phylogenetic trees. Thus, the question of the validity of the molecular clock is an important issue in molecular evolution.

Relative Rate Tests

The controversy over the molecular clock hypothesis often involves disagreements on dates of species divergence. For example, in Figure 4.24, there are three large deviations from expectations under a strict molecular clock. Such deviations may indicate real slowdowns or accelerations of evolutionary rates, or they may indicate that the paleontological estimates of divergence times are wrong. Unfortunately, if we are allowed to question the time estimates and move left or right any point that deviates from the expectations of the molecular clock (arrows in Figure 4.24), then we will always be able to fit any set of data to a straight line, and we will never be able to refute the assumption of rate constancy, thereby depriving it of its scientific status.

To avoid such circular conundrums, several tests that do not require knowledge of divergence times have been developed. We present two variants of one such test. Many other methods have been suggested in the literature (e.g., Felsenstein 1988; Muse and Weir 1992; Goldman 1993; Tajima 1993; Takezaki et al. 1995; Robinson et al. 1998; Kumar and Filipski 2001; Xia 2009).

The **relative rate test** of Margoliash (1963) and Sarich and Wilson (1973) is illustrated in **Figure 4.25**. Suppose that we want to compare the rates in lineages A and B. Then, we use a third species, C, as an outgroup reference. The choice of the outgroup species is very important. On the one hand, we should be certain that the outgroup species branched off earlier than the divergence of species A and B, and on the other, we should choose a closely related species. For example, to compare the rates in the human and chimpanzee lineages, we can use orangutan as reference (Chapter 5).

From Figure 4.25, it is easy to see that the number of substitutions between species A and B, K_{AB}, is equal to the sum of substitutions that have occurred from point O to point A (K_{OA}) and from point O to point B (K_{OB}). That is,

$$K_{AB} = K_{OA} + K_{OB} \quad (4.28)$$

Similarly,

$$K_{AC} = K_{OA} + K_{OC} \quad (4.29)$$

and

$$K_{BC} = K_{OB} + K_{OC} \quad (4.30)$$

Since K_{AC}, K_{BC}, and K_{AB} can be directly estimated from the nucleotide sequences (Chapter 3), we can easily solve the three equations to find the values of K_{OA}, K_{OB}, and K_{OC}:

$$K_{OA} = \frac{K_{AC} + K_{AB} - K_{BC}}{2} \quad (4.31)$$

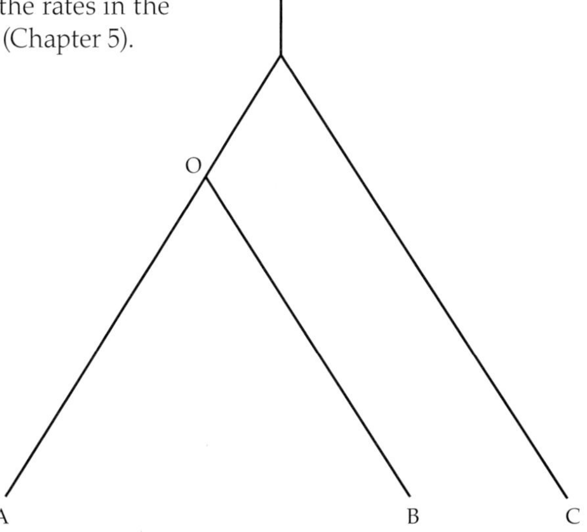

Figure 4.25 Phylogenetic tree used in the relative rate test. O denotes the common ancestor of species A and B. C is the outgroup. (From Graur and Li 2000.)

$$K_{OB} = \frac{K_{AB} + K_{BC} - K_{AC}}{2} \quad (4.32)$$

$$K_{OC} = \frac{K_{AC} + K_{BC} - K_{AB}}{2} \quad (4.33)$$

We can now decide whether the rates of substitution in lineages A and B are equal to each other by comparing the value of K_{OA} with that of K_{OB}. The time that has passed since species A and B last shared a common ancestor is by definition equal for both lineages. Thus, according to the molecular clock hypothesis, K_{OA} and K_{OB} should be equal to each other, i.e., $K_{OA} - K_{OB} = 0$. From the above equations, we note that $K_{OA} - K_{OB} = K_{AC} - K_{BC}$. Therefore, we can compare the rates of substitution in A and B directly from K_{AC} and K_{BC}. In statistical terminology, the above approach means that we use $K_{AC} - K_{BC}$ as an estimator of $K_{OA} - K_{OB}$. $K_{OA} - K_{OB}$ represents the difference in branch length between the two lineages leading from node O to species A and B. We denote this difference by d. A positive d value means that the molecule has evolved faster in lineage A than in lineage B; a negative d value means that the molecule evolved faster in lineage B.

The variance of d is given by

$$V(d) = V(K_{AC}) + V(K_{BC}) - 2V(K_{OC}) \quad (4.34)$$

Under the one-parameter model, $V(K_{AC})$ and $V(K_{BC})$ can be obtained from Equation 3.26 (Chapter 3). $V(K_{OC})$ can be obtained by putting

$$p = \frac{3}{4}\left(1 - e^{-4/3K_{OC}}\right) \quad (4.35)$$

into Equation 3.27. For a more detailed explanation of the derivation of the formulas, see Wu and Li (1985).

A simple way to test whether an observed d value is significantly different from 0 is to compare it with its standard error. For example, an absolute d value larger than two times the standard error may be considered significantly different from 0 at the 5% level. Similarly, an absolute d value larger than 2.6 times the standard error may be considered significantly different from 0 at the 1% level.

The relative rate test can be used even if the order of divergence among species A, B, and C is not known for certain. We note that Equations 4.28–4.33 also hold for the two alternative cases: (1) A and C diverged from each other after the divergence of B, and (2) B and C diverged from each other after the divergence of A. In other words, even when the order of divergence among three species is unknown, two of the species must be more closely related to each other than either is to the third. In this case we can compare the values of K_{OA}, K_{OB}, and K_{OC} and find out whether any two values are identical. Failure to find even one identity among the three comparisons would cast doubt on the molecular clock hypothesis for the three species under consideration.

Although the relative rate test of Margoliash (1963) and Sarich and Wilson (1973) does not rely on knowledge of divergence times, it does assume (1) that the substitution model (e.g., Jukes and Cantor's one-parameter method) is known, and (2) that the substitution rates among different sites vary according to some prespecified distribution, for example, the Γ distribution (Jin and Nei 1990).

Tajima (1993) suggested several simple, albeit less powerful, methods for overcoming these difficulties. Here, we will only discuss the **one degree of freedom**, or **1D method**. We start with three aligned nucleotide sequences, 1, 2, and 3. Let n_{ijk} be the observed number of sites where sequences 1, 2, and 3 have nucleotides i, j, and k, respectively, where i, j, and k can be nucleotides A, G, C, or T. If sequence 3 is the outgroup, then the expectation, E, of n_{ijk} should be equal to that of n_{jik}, i.e.,

$$E(n_{ijk}) = E(n_{jik}) \quad (4.36)$$

This equality holds regardless of the substitution model or the pattern of variation in substitution rates among sites.

We define m_1 as follows:

$$\begin{aligned} m_1 &= \sum n_{ijj} \\ &= n_{\text{AGG}} + n_{\text{ACC}} + n_{\text{ATT}} + n_{\text{GAA}} + n_{\text{GCC}} + n_{\text{GTT}} \\ &+ n_{\text{CAA}} + n_{\text{CGG}} + n_{\text{CTT}} + n_{\text{TAA}} + n_{\text{TGG}} + n_{\text{TCC}} \end{aligned} \quad (4.37)$$

Similarly, we define m_2 as

$$\begin{aligned} m_2 &= \sum n_{jij} \\ &= n_{\text{AGA}} + n_{\text{ACA}} + n_{\text{ATA}} + n_{\text{GAG}} + n_{\text{GCG}} + n_{\text{GTG}} \\ &+ n_{\text{CAC}} + n_{\text{CGC}} + n_{\text{CTC}} + n_{\text{TAT}} + n_{\text{TGT}} + n_{\text{TCT}} \end{aligned} \quad (4.38)$$

Note that only sites in which exactly two types of nucleotides exist in the three sequences are used in this analysis.

When sequence 3 is the outgroup, the expectation of m_1 under the molecular clock is equal to that of m_2:

$$E(m_1) = E(m_2) \quad (4.39)$$

The equality can be tested by using χ^2 with one degree of freedom, namely,

$$\chi^2 = \frac{(m_1 - m_2)^2}{m_1 + m_2} \quad (4.40)$$

The molecular clock hypothesis can also be tested by comparing homologous genes that originated through a gene duplication event (Chapter 7). If the gene duplication occurred prior to (but not too long before) the split of the two species (**Figure 4.26**), then we can determine whether or not the homologous genes evolve at the same rate in the two lineages by comparing the number of substitutions between the two duplicate genes in one lineage with the number of substitutions between the duplicate genes in the second lineage (Wu and Li 1985).

If we denote the number of substitutions between sequence i and sequence j by K_{ij}, we see from Figure 4.26 that the number of substitutions between A_1 and B_1 (K_{A1B1}) can be calculated as

$$K_{A_1B_1} = K_{AA_1} + K_{OA} + K_{OB} + K_{BB_1} \quad (4.41)$$

Similarly, the number of substitutions between A_2 and B_2 is

$$K_{A_2B_2} = K_{AA_2} + K_{OA} + K_{OB} + K_{BB_2} \quad (4.42)$$

From Equations 4.41 and 4.42 we obtain

$$K_{A_1B_1} - K_{A_2B_2} = K_{AA_1} + K_{BB_1} - K_{AA_2} - K_{BB_2} \quad (4.43)$$

If A_1 evolves at the same rate as A_2, and B_1 evolves at the same rate as B_2 (i.e., if the assumption of rate constancy holds for both genes), then $K_{AA_1} = K_{AA_2}$ and $K_{BB_1} = K_{BB_2}$. Therefore, under the molecular clock hypothesis, $K_{A_1B_1} = K_{A_2B_2} = 0$. Note that genes A_1 and A_2 serve as outgroups for comparing the rates between the lineages leading to B_1 and B_2, and vice versa. This test neither requires knowledge of the divergence time between the two species under study, nor assumes that the rates of substitution of A_1 and A_2 are equal to those of B_1 and B_2. The

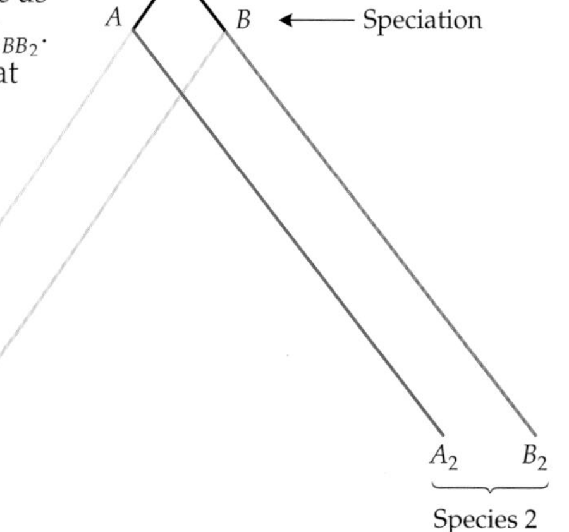

Figure 4.26 Phylogenetic tree used to test constancy of rates among lineages A and B by using duplicate genes. *O* denotes the common ancestor of all four genes. *A* is the ancestor of the duplicated genes A_1 and A_2, *B* is the ancestor of the duplicated genes B_1 and B_2, and the subscripts denote the species. (From Graur and Li 2000.)

drawback is that gene conversion may occur between the duplicated genes (Chapter 7) and distort the results. Thus, this method must be used with caution.

Local Clocks

Sometimes we are interested in whether or not a molecular clock exists for a particular group of organisms. Such a clock, if found, would be applicable only to the group of organisms under study and to no other. Such clocks are called **local clocks**. In the following discussion, we will review the question of local clocks in several taxa by using relative rate tests.

Nearly equal rates in mice and rats

Table 4.7 shows a comparison of the rates of synonymous and nonsynonymous substitution in mice and rats, using the relative rate test. In all cases, species A is mouse, species B is rat, and species C is hamster. A positive sign for the value of $K_{AC} - K_{BC}$ means that the rate in mice is higher than that in rats, whereas a negative sign indicates the opposite. The sequence data consist of 28 genes with a combined total aligned length of 11,295 nucleotides (O'hUigin and Li 1992). The values of $K_{AC} - K_{BC}$ are not significantly different from 0 for either synonymous or nonsynonymous substitutions and, therefore, the null hypothesis of equal substitution rates in mice and rats cannot be rejected. An analysis based on a much larger data set of orthologous genes from mouse and rat, with human as outgroup, seems to indicate that the rates of synonymous and nonsynonymous substitution in rat are significantly higher than those in mouse but that the difference is merely 2–5% (Friedman and Hughes 2005).

Lower rates in humans than in monkeys

On the basis of immunological distances and protein sequence data, Goodman (1961) and Goodman et al. (1971) suggested that a rate slowdown occurred in hominoids (humans and apes) after their separation from the cercopithecines (Old World monkeys). Wilson et al. (1977), however, contended that the slowdown was an artifact owing to the use of an erroneous estimate of the ape-human divergence time. They conducted relative rate tests using both immunological distance data and protein sequence data and concluded that there was no evidence for a hominoid slowdown.

In **Table 4.8**, K_{AC} is the distance between human (*Homo sapiens*) and a New World monkey, the South American squirrel monkey (*Saimiri sciureus*). K_{BC} is the distance between an Old World monkey, the yellow baboon (*Papio cynocephalus*), and *S. sciureus*. The average $K_{AC} - K_{BC}$ is significantly lower than 0 and we may, therefore, conclude that the rate of substitution in introns, which mostly reflects the rate of mutation, is lower in humans than in the African monkeys. Interestingly, the molecular clock in humans was found to run about 3–4% slower than that in the chimpanzees and about 10% slower than that in gorillas, indicating that the slowdown in the rate of evolution started after the divergence of apes from the Old World monkeys, with a further slowdown in the lineage of the common ancestor of chimpanzees and humans following

Table 4.7

Differences in the number of nucleotide substitutions per 100 sites between mice (species A) and rats (species B), with hamsters (species C) as a reference

Type of substitution	Number of sites compared	K_{AB}[a]	K_{AC}	K_{BC}	$K_{AC} - K_{BC}$
Synonymous	4,855	19.9 ± 0.7	31.1 ± 0.9	32.4 ± 1.0	–1.3 ± 7.9
Nonsynonymous	17,440	1.9 ± 0.1	2.9 ± 0.1	2.7 ± 0.1	0.3 ± 1.3

Source: Modified from O'hUigin and Li (1992).

[a]Mean ± standard error. K_{ij} = number of substitutions per 100 sites between species *i* and *j*.

Table 4.8

Differences in the number of nucleotide substitutions per 100 sites between humans (species A) and an Old World monkey lineage (species B), with a New World monkey lineage (species C) as a reference

Sequence length (bp)	K_{AB}[a]	K_{AC}	K_{BC}	$K_{AC} - K_{BC}$	Ratio[b]
15,304	6.85	12.98	13.95	-0.98 ± 0.25***	1.33

Source: Data from Yi et al. (2002).

[a] K_{ij} = number of substitutions per 100 sites in 29 introns between species *i* and *j*.

[b] The ratio of the substitution rate in the Old World monkey lineage to that in the human lineage.

*** Significant at the 0.1% level.

the split from the gorilla lineage. The final slowdown occurred in the human lineage after its split from chimpanzee (Elango et al. 2006).

Higher rates in rodents than in other mammals

Based on DNA-DNA hybridization data, Laird et al. (1969), Kohne (1970), Kohne et al. (1972), and Rice (1972) concluded that the rate of nucleotide substitution in rat and mouse is much higher than that in human and chimpanzee. Their findings were challenged by Sarich (1972), Sarich and Wilson (1973), and Wilson et al. (1977), who argued that some of the studies were based on questionable estimates of divergence time. However, some of the observed differences were too large to be explained by inaccuracies in the estimation of divergence times.

Gu and Li (1992) compared the rates of amino acid replacement in the rodent and human lineages with chicken as an outgroup. The reason amino acid sequences were used instead of DNA is that the chicken and mammalian lineages diverged about 300 million years ago, so it is difficult to obtain reliable estimates of divergence at synonymous sites. We denote by N_R the number of residues at which the human and rodent sequences are different but the human and the chicken sequences are identical; this case represents an instance in which an amino acid replacement presumably occurred in the rodent lineage. We denote by N_H the number of residues at which the human and rodent sequences are different but the rodent and the chicken sequences are identical; this case represents an instance in which an amino acid replacement presumably occurred in the human lineage. Under the null assumption of equal rates in the human and rodent lineages, N_R is expected to be equal to N_H. For the 54 proteins used in Gu and Li's (1992) study, $N_R = 600$ and $N_H = 416$ ($p < 0.001$). Therefore, there is good evidence for an overall faster substitution rate in the lineage leading to mouse and rat than in the human lineage.

Gu and Li (1992) also compared the frequencies of insertions and deletions (gaps) in the human and rodent sequences, using chicken as a reference. The total length of gaps in the rodent lineage was 385 amino acids, compared with only 108 in the human lineage ($p < 0.001$). The number of gaps was also higher in the rodent lineage (44 gaps) than in the human lineage (31 gaps), but because of the small sample size, the difference was not statistically significant ($0.05 < p < 0.1$).

The null hypothesis of equal rates in rodents and primates has also been tested by using comparisons between duplicated-gene pairs. **Table 4.9** shows the results from an analysis of the hemoglobin and aldolase duplicate genes in humans and mice (Li et al. 1987a). In the case of the globin genes, duplicate copies in the mouse are always more divergent than those in humans. This is true for both synonymous and nonsynonymous substitutions. For the three gene pairs considered, the ratios of the number of substitutions in mouse to those in human are 1.24, 1.56, and 1.71 for synonymous substitutions, and 1.13, 1.17, and 1.50 for nonsynonymous substitutions. In the case of the aldolase *A* and *B* genes, the degree of divergence at synonymous sites is 1.24 times higher in rat than in human, but the degrees of divergence at nonsynonymous sites are not significantly different between the two species.

Table 4.9

Differences in the number of nucleotide substitutions per synonymous site (K_S) and per nonsynonymous site (K_A) between duplicated genes in humans and rodents

Duplicated gene pair	K_S	K_A
β-like hemoglobin genes[a]		
Human adult–Human fetal	0.73	0.18
Mouse adult–Mouse fetal	0.90	0.21
Human adult–Human embryonic	0.62	0.16
Mouse adult–Mouse embryonic	0.97	0.18
Human fetal–Human embryonic	0.56	0.10
Mouse fetal–Mouse embryonic	0.96	0.15
Aldolase *A* and *B* genes		
Human *A*–Human *B*	1.55	0.21
Rat *A*–Rat *B*	1.92	0.21

Source: From Li et al. (1987a).

[a]The adult hemoglobin genes are β in human and β_{maj} in mouse; the fetal genes are $^{A}\gamma$ in human and β_{H1} in mouse; and the embryonic genes are ε in human and *y*2 in mouse.

Thus, the results of tests of the molecular clock hypothesis using duplicate genes also support the view that the rate of nucleotide substitution is higher in rodents than in humans. We must remember, however, that the ratios are underestimates because the rates of substitution in the two species are not entirely independent of each other. For example, from Figure 4.26, we see that genes A_1 and A_2 share the same evolutionary history between the time of gene duplication (O) and the time of species divergence (A). Similarly genes B_1 and B_2 share the same evolutionary history between the time of gene duplication (O) and the time of species divergence (B). In order to obtain reliable estimates for the ratio of substitution rates in two lineages, we must use genes that had been duplicated shortly before the time of speciation, i.e., we have to make sure that K_{OA} and K_{OB} are much smaller than K_{AA_1}, K_{AA_2}, K_{BB_1}, and K_{BB_1}.

Is the rate of evolution in rodents exceptionally rapid, or are primates exceptionally slow? Li and Makova (2008) compared the rates of nucleotide substitution among representatives of six eutherian taxa: human (Primates); mouse or rat (Rodentia); rabbit (Lagomorpha); cat, dog, or ferret (Carnivora); horse (Perissodactyla); and cow, sheep, or pig (Ruminantia). As outgroup, opossum, kangaroo, or potoroo (Metatheria) were used. **Table 4.10** shows that the nonsynonymous rate is significantly higher in the rodent lineage than in all the other lineages. Note that these values are averages over the long period of time since the divergence of the rodent lineage from the other eutherian lineages. As the rate of evolution would have been similar among lineages during the early stage of divergence, the rate differences among lineages in more recent times are likely much larger than the long-term averages. Thus, rodents have a much faster molecular clock than do any other eutherians in this sample.

Evaluation of the molecular clock hypothesis

In its most extreme form, the molecular clock hypothesis postulates that homologous stretches of DNA evolve at essentially the same rate along all evolutionary lineages for as long as they maintain their original function (e.g., Wilson et al. 1987). Analyses of DNA sequences from many orders of animals indicate that a global molecular clock exists in neither vertebrates nor invertebrates.

Table 4.10

Differences in the number of nonsynonymous nucleotide substitutions per 100 sites between the rodent lineage (species A) and a nonrodent lineage (species B), with the marsupial lineage (species C) as a reference[a]

Species B	Number of genes	Number of sites	K_{AB}[a]	K_{AC}	K_{BC}	$K_{AC} - K_{BC}$	Ratio[b]
Human	34	34,067	7.1	12.8	11.6	1.2**	1.4
Rabbit	13	13,851	9.3	15.9	14.4	1.5**	1.4
Carnivore	8	5,563	12.3	20.1	17.9	2.3**	1.5
Horse	9	5,729	9.0	17.1	15.3	1.9**	1.5
Artiodactyl	24	22,316	10.7	16.6	15.4	1.3**	1.4

Source: Modified from Li and Makova (2008).

[a]K_{ij} = number of substitutions per 100 sites between species *i* and *j*.

[b]The ratio of the substitution rate in the rodent lineage to that in the non-rodent lineage.

**Significant at the 1% level.

Thus, a truly universal molecular clock that applies to all organisms does not exist. Indeed, the substitution rate among distantly or even not so distantly related organisms has been shown to vary greatly among lineages. For example, in invertebrates, rate variation is observable not only between distantly related organisms, but also between closely related genera and species (Thomas et al. 2006). Therefore, caution must be exercised when using the molecular clock assumption to infer times of divergence, more so when distantly related species are concerned (Chapter 5).

Of course, there are many local clocks that tick fairly regularly for many groups of closely related species, such as murids. Moreover, one must note the remarkable fact that while parameters related to the life histories of animal and plant species are different from each other by many orders of magnitude, the mean synonymous substitution rates in the nuclear genomes of these organisms differ by no more than one or two orders of magnitude. Therefore, the molecular clock can still be used to estimate times of species divergence with a fair degree of confidence, provided appropriate corrections for the unequal rates of molecular evolution among lineages are made.

"Primitive" versus "advanced": A question of rates

In the literature one often encounters the adjectives "primitive" or "lower" attached to the names of organisms or taxonomic groups. For example, the flagellated protozoan *Giardia* is often said to be a "primitive eukaryote." In a similar vein, sponges are defined as "primitive metazoans," and the whisk fern, *Psilotum nudum*, is defined as a "primitive vascular plant." Other organisms are defined as "advanced" or "higher." We note, however, that the terms "primitive" and "advanced" cannot be defined unambiguously. We further note that all the organisms mentioned above are extant organisms and as such they are as "primitive" or as "advanced" as other extant organisms. The ranking of organisms started with the Aristotelian Great Chain of Being and was used by Linnaeus in his *Systema Naturae* to rank animals into Primates (humans and monkeys), Secundates (mammals), and Tertiates (all others). Only the term "Primates" is still in use.

The relative rate tests allow us to attach an objective meaning to the terms "primitive" and "advanced." In **Figure 4.27**, we show an ancestral taxon that gave rise to two descendant taxa. The lineage leading to descendant 1 evolved faster (i.e., accumulated more substitutions) than the lineage leading to descendant 2. Therefore, descendant 2 resembles the ancestral taxon more than descendant 1. Since, by temporal placement, the ancestor predates the descendants (i.e., the ancestor is the primitive taxon), we may conclude that descendant 2 is more "primitive" than descendant 1. The question of "primitive" versus "advanced" is, therefore, a question of rates of evolution. The slower a lineage evolves, the more primitive its descendants are.

We now note that the lineage leading to humans has experienced the slowest average rate of nuclear gene evolution from among all placental mammalian lineages, and most probably from among most lineages in the animal kingdom. Is *Homo sapiens* the most "primitive" animal?

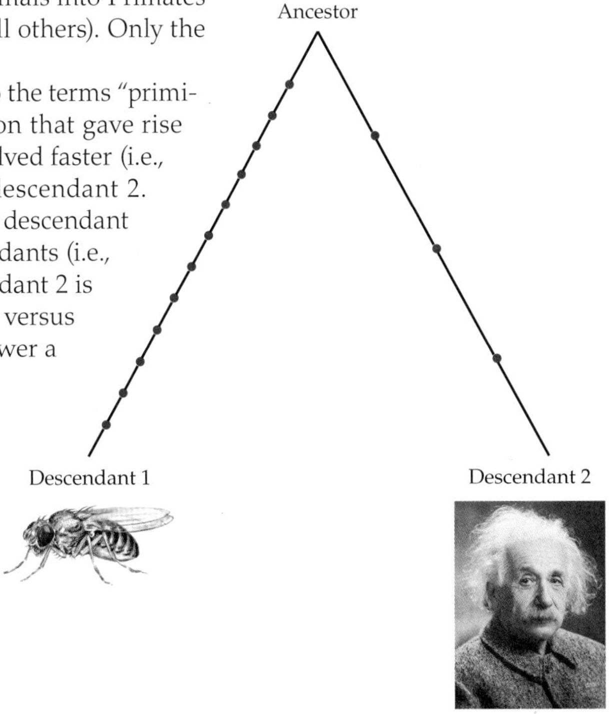

Figure 4.27 An ancestral taxon gives rise to two descendant taxa. The branch leading to descendant 1 has accumulated more substitutions (dots) than the branch leading to descendant 2. Descendant 2, therefore, resembles the ancestral taxon more than does descendant 1. Descendant 2 is, hence, more "primitive" than taxon 1.

Causes of Variation in Substitution Rates among Evolutionary Lineages

Many hypotheses have been proposed to account for the differences in the rate of nucleotide substitution among evolutionary lineages (Sniegowski et al. 2000; Baer et al. 2007). Here we review four hypotheses. These hypotheses are not mutually exclusive. Rather, all the factors detailed below may contribute to varying extents to the rate variation among lineages.

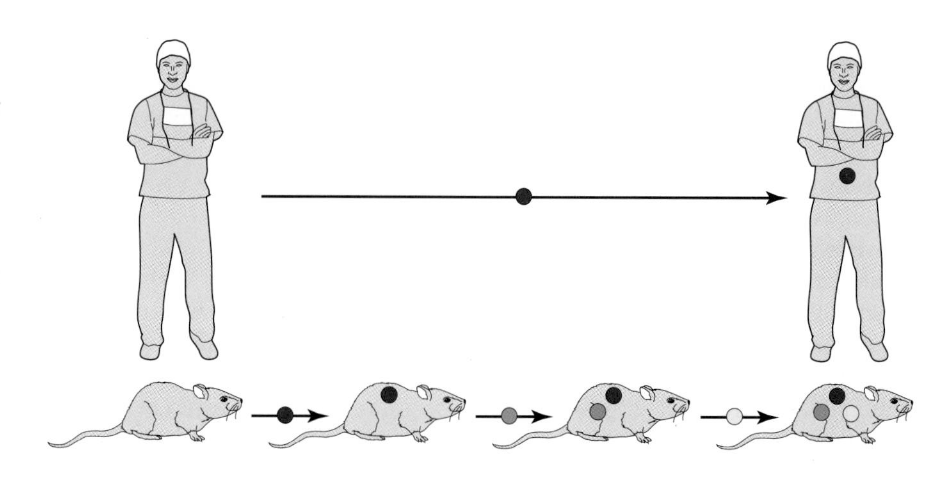

Figure 4.28 Schematic representation of the generation-time effect on the rate of mutation. Under the assumptions that mutations (circles) occur proportionally to the number of DNA replications and that the number of DNA replications from one generation to the next is the same in human and rodent, then the number of mutations that will accumulate within a time unit will be larger in rodent than in human. In this schematic example, DNA is replicated once between one generation and the next and in each replication the DNA accumulates one mutation. The length of one generation in humans (long arrow) is equal to that of three generations in rodents (short arrows). Thus, rodents will accumulate three times more mutations than humans.

The DNA repair hypothesis

The **DNA repair hypothesis** posits that divergent substitution rates can also partly be accounted for by differences in the efficiency of the DNA repair system (Britten 1986). There are limited data indicating that rodents have a less efficient DNA repair system than humans and consequently accumulate more mutations per replication cycle. One such example involves the removal of hypoxanthine residues, a product of adenine deamination that constitutes an important source of transition mutations. The removal of hypoxanthine from DNA is carried on by a group of enzymes called 3-methyladenine DNA glucosylases. Saparbaev and Laval (1994) have shown that 3-methyladenine DNA glucosylases from humans are more efficient than their counterparts in rats.

The generation-time effect hypothesis

The higher substitution rates in monkeys than in humans and the higher rates in rodents than in the other eutherian orders can be explained by the so-called **generation-time effect hypothesis** (Goodman 1961, 1962; Laird et al. 1969; Kohne 1970). The generation time in rodents is much shorter than that in humans, whereas the numbers of germline DNA replication cycles per generation are not very different (**Figure 4.28**). Therefore, the number of germline replications per unit time could be many times higher in rodents than in humans. Assuming that mutations accumulate chiefly during the process of DNA replication, the more cycles of replication there are, the more mutational errors will occur. In the absence of differences in selection intensities, rodents are expected to have higher substitution rates than humans. Similarly, monkeys have a shorter generation time than humans and so are expected to have a higher substitution rate. In mammals, the rates of substitution in both the nuclear and mitochondrial genomes were found to behave largely according to the expectations of the generation-time effect hypothesis (**Figure 4.29**). The generation-time hypothesis is also supported by findings that the rates of substitution in the nuclear and chloroplast genomes are higher in annual plants than in perennials (e.g., Gaut et al. 1992; Baldwin and Andreasen 2001).

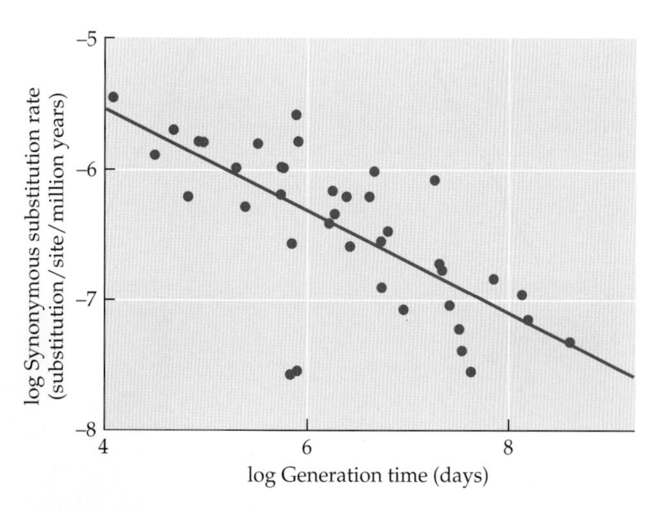

Figure 4.29 Mitochondrial synonymous substitution rates in mammals exhibit a negative correlation with generation time. The median of age at first birth was used as a proxy for generation time. The nine longest mitochondrial protein-coding genes (*ATP6*, *COI*, *COII*, *COIII*, *CYTB*, *ND1*, *ND2*, *ND4*, and *ND5*) from 160 species were used. (Data from Welch et al. 2008.)

An interesting finding pertaining to the generation-time effect hypothesis concerns a comparison between the rate of evolution of CpG and non-CpG sites in primates (Kim et al. 2006). While non-CpG substitutions show clear generation-time dependency, mirroring their origin as errors of replication, CpG substitutions occur relatively constantly with chronological time, reflecting the fact that the methylation process that gives rise to such mutations is replication independent. It is, therefore, possible that estimates of divergence times may be more accurate if they are based on substitutions at CpG sites.

The metabolic rate hypothesis

The **metabolic rate hypothesis** suggests that metabolic rates (defined as amounts of O_2 consumed per weight unit per unit time) influence the rate of nucleotide substitution (Martin and Palumbi 1993). Since metabolic rate is negatively correlated with body mass, the expectation is that large organisms will have lower rates of substitution than small organisms, and that poikilotherms (cold-blooded organisms) will have lower rates of nucleotide substitution than homeotherms (warm-blooded organisms) of comparable body mass (**Figure 4.30**). According to Martin and Palumbi's (1993) suggestion, the correlation between metabolic rate and the rate of nucleotide substitution is mediated by (1) the mutagenic effects of oxygen radicals that are abundant by-products of aerobic respiration and whose production is proportional to the metabolic rate, and (2) increased rates of DNA turnover (i.e., DNA synthesis, repair, and degradation) in organisms with high metabolic rates. Gillooly et al. (2005, 2007) presented some evidence that the variation in synonymous rates of nucleotide substitution, as well as the variation in amino acid replacement rate in proteins, may be predicted by the mass-specific metabolic rate, which in turn depends on body mass and ambient temperature.

We note that in animals a positive correlation exists between body mass and generation time; thus, distinguishing between the generation-time effect hypothesis and the metabolic rate hypothesis is anything but trivial. Moreover, several studies have indicated that the correlation between metabolic and evolutionary rates may not be a general phenomenon (e.g., Seddon et al. 1998; Thomas et al. 2006; Lanfear et al. 2007; Galtier et al. 2009b).

The varying-selection hypothesis

All the factors discussed so far involve variations in mutation rates. However, rates of substitutions may also vary among evolutionary lineages because of differences in selection intensities. For example, in large populations, a greater proportion of the mutations may be affected by purifying selection (Chapter 2). This will result in a decrease in the proportion of neutral mutations and a consequent decrease in the rates of substitution (e.g., Ohta 1973, 1995). Alternatively, one species may be subjected to

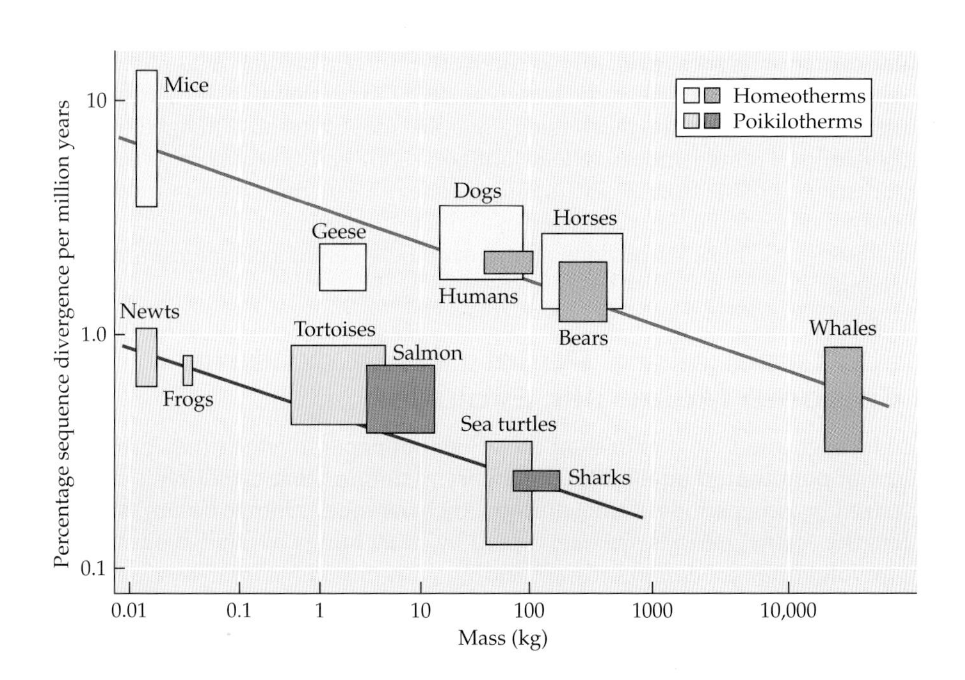

Figure 4.30 The relationship between body mass (as a proxy for metabolic rates) and rates of substitution in warm blooded (homeothermic) and cold blooded (poikilothermic) vertebrates. Rectangles represent the range of substitution rates and body sizes for a given taxon. (Modified from Martin and Palumbi 1993.)

selective constraints that are on average stronger than those operating on another species, and as a consequence a species may evolve faster or slower than another. It is very difficult to assess the relative contribution of each factor to the rate of molecular evolution. For example, large animals usually have longer longevities, longer generation times, smaller population sizes, and lower metabolic rates than small animals.

Are Living Fossils Molecular Fossils Too?

The question we attempt to address here is this: Are rates of morphological evolution correlated with rates of molecular evolution? In other words, is morphological stasis always accompanied by molecular stasis. Such studies require a definition of living fossils.

Living fossils are extant taxa that have not changed morphologically for long periods of time. Thus, a living fossil is characterized by the retention of a great number of plesiomorphies (ancestral traits) of great antiquity. How many plesiomorphies are required and what constitutes "great antiquity" should be specified in each case under study. In the literature, time periods over 100 million years are usually considered "great antiquity." Some researchers find the term "living fossil" objectionable a priori, as it tends to color subsequent inferences. Nonetheless, it is possible to define a taxon objectively as a living fossil on a case-by-case basis. Interestingly, on careful inspection, some famous "living fossils," such as the coelacanth (*Latimeria chalumnae*), turn out not to be living fossils at all (Forey 1984; Casane and Laurenti 2013).

Some ferns, such as *Marattia* and *Angiopteris*, as well as the tree ferns (order Cyatheales) have experienced a relative morphological stasis for the past 165–200 million years. Interestingly, all these clades of ferns have undergone dramatic decelerations in their rates of molecular evolution and are, hence, "molecular living fossils" in addition to being living fossils.

In some other cases, we find no relationship between the rates of molecular and morphological evolution. For example, the mitochondrial DNA of the alligator (*Alligator mississippiensis*), which is considered to be a living fossil, evolves much faster than that of birds, which presumably appeared on the evolutionary stage much more recently (Janke and Árnason 1997). Turtles, on the other hand, which have remained morphologically unchanged since Triassic times, seem to evolve at the molecular equivalent of a "turtle's pace" (Avise et al. 1992). The current holder of the title "slowest-evolving vertebrate" is a living fossil called ratfish or elephant shark, *Callorhinchus milii* (Ilves and Randall 2007; Venkatesh et al. 2014).

The most spectacular example of a lack of relationship between morphology and molecular evolution is most probably found in the horseshoe crab (class Merostomata), which despite its name is more closely related to scorpions, spiders, and mites than to crustaceans. While the morphology of horseshoe crabs has changed little in the last 500 million years—indeed, one extant horseshoe crab, *Limulus polyphemus*, is almost indistinguishable from its extinct Jurassic relatives—their rates of molecular evolution are unexceptional (e.g., Nguyen et al. 1986; Tokugana et al. 1993).

Thus, there seems to be no obvious relationship between morphological and molecular change.

Phyletic Gradualism, Punctuated Equilibria, and Episodic Molecular Evolution

In 1972 Eldredge and Gould published an influential paper in which they proposed the theory of **punctuated equilibria**. The theory maintains that evolution is mostly characterized by long periods of time during which very little change occurs (**stasis**). These lengthy "calm" periods are interrupted by short bursts of rapid evolutionary change (**revolutions**), predominantly coinciding with periods of intense speciation. The theory challenged what Eldredge and Gould (1972) called **phyletic gradualism**, which regards evolution as a gradual process, albeit with some variation in rates. In principle, a third mode of evolution may be imagined, one in which the rates of evolution vary widely but are uncorrelated with speciation events. Sudden increases or decreases in the rate of evolution that are not associated with speciation are sometimes referred to as **episodic evolution** (Liu et al. 2001) or **pulsed evolution** (Hunt 2008). We note that all punctuated events are episodic, but not all episodic accelerations or decelerations of evolutionary rates match the definition of punctuated equilibria.

As noted before, the rate of evolution at the molecular level exhibits a certain degree of variation, but much of the variation can be accommodated within the phyletic gradualist view (Gillespie 1991). In some cases, however, we observe prominent accelerations or decelerations of substitution rates along particular lineages. In the case of sudden increases in the rate of evolution, it is usually difficult to tell whether the rate variation represents a case of episodic evolution or a punctuated event. Pagel et al. (2006) devised a phylogenetic method for identifying punctuated evolution, i.e., one in which increases in rates of evolution are associated with the internal nodes of a rooted tree. The method uses path length (i.e., number of substitutions) along a path from the root to each terminal node. Under punctuated evolution, we expect a positive correlation between the number of nodes and the length of the path. No such correlation is expected if the mode of evolution is gradual or if accelerations or decelerations in rate are unrelated to speciation. Pagel et al. (2006) analyzed 89 phylogenetic trees and detected punctuated effects in 27% of them. The percentage of nucleotide substitutions that could be attributed to punctuational episodes was 16%. Interestingly, the frequencies of puctuational effects among plants (44%) and fungi (60%) were at least double the frequency in animals (18%).

Rates of Substitution in Organelle DNA

The vast majority of eukaryotes possess at least one extranuclear genome, which is replicated independently of the nuclear genome. Organelle genomes, such as those of mitochondria and chloroplasts, are invariably smaller than nuclear genomes and are almost always inherited uniparentally. Animal mitochondria are inherited maternally, although some paternal "leaking" has been detected. In plants, three modes of transmission of organelles have been described: maternal, paternal, and biparental.

Mitochondrial rates of evolution

The synonymous rate of substitution in most animal mitochondrial protein-coding genes has been estimated to be much higher than that for nuclear protein-coding genes (Brown et al. 1979, 1982; Hellberg 2006; Nabholz et al. 2008). The principal reason for the high rate of substitution in animal mitochondrial DNA (mtDNA) seems to be a high rate of mutation relative to the nuclear rate. Several causes singly or in combination may contribute to the elevated mutation rates: (1) a low fidelity of the DNA replication process in mitochondria, (2) an inefficient repair mechanism, (3) a high concentration of mutagens (e.g., superoxide radicals, O_2^-) resulting from the metabolic functions performed by mitochondria, and (4) an effective population size that is one-quarter that for nuclear genes.

Nabholz et al. (2008) used a database of mitochondrial cytochrome b genes from 1,696 mammalian species and concluded (1) that the mean mitochondrial rate of synonymous substitution is nearly 60 times higher than the nuclear one; (2) that the synonymous rate variation among lineages spans two orders of magnitude, from ~0.007 substitutions per synonymous site per year in fin whales to ~0.7 substitutions per synonymous site per year in gerbils; and (3) that the rate of mutation is negatively correlated with longevity.

As with all correlations, there are three possibilities: (1) a reduced mutation rate is responsible for increased longevity, (2) increased longevity is responsible for decreasing mutation rates, or (3) both longevity and mutation rates are correlated with a third factor such as generations time. Nabholz et al. (2008) promote the first possibility by invoking the **longevity hypothesis**, according to which natural selection tends to decrease the mitochondrial mutation rate in long-lived species. The longevity hypothesis is based on Harman's (1956) **mitochondrial theory of aging**, also referred to as the **free radical theory of aging**. According to this theory a vicious cycle exists in the mitochondria, whereby mutations in mtDNA cause respiratory chain dysfunctions that enhance the production of DNA-damaging oxygen radicals, which in turn result in the accumulation of further mtDNA mutations, resulting in a bioenergetic crisis that ultimately leads to aging and death. In support of the longevity hypothesis, Nabholz et al. (2008) put particular emphasis on a comparison between rodents and bats in which a kind of decoupling exists between metabolic rates (with body mass as proxy) and generation time (with mean longevity as proxy). The average body mass of bats is about 15% that of rodents, yet their average longevity is 2.4 times higher. In agreement with the longevity hypothesis, the rate of synonymous substitution in bats is about half that in rodents. We note, however, that selection for reduced mutation rates constitutes a type of second-order selection (Chapter 2) that requires large effective population sizes. It is doubtful that such huge effective population sizes have ever existed in mammals.

Not all mtDNAs evolve rapidly. Rates of substitution are slow in the vast majority of plants and fungi and in some animals, including sponges and anthozoans (corals, anemones, and sea pens). Although most plant mitochondrial sequences evolve slowly, some cases of highly accelerated rates have been found. Comparison of the fastest and slowest lineages shows that synonymous substitution rates vary by four orders of magnitude across seed plants. In other words, some plant mitochondrial lineages accumulate more synonymous change in 10,000 years than do others in 100 million years. Studies by Hellberg (2006) and Mower et al. (2007) concluded that slow mtDNA evolution is the primordial state in nature. Substitution rates of mtDNA switched from slow to fast abruptly many times in the evolution of eukaryotes, especially in plants. Two such switches are also documented in animals.

Interestingly, in many slow-evolving mtDNAs, there is little mutational bias favoring transitions over transversions. In contrast, in almost all fast-evolving mtDNAs, transitions predominate over transversions, sometimes by orders of magnitude.

Plastid rates of evolution

Since the plant and animal kingdoms diverged about 1 billion years ago, the pattern of evolution in plants might have become very different from that in animals. Indeed, plants differ from animals in the organization of their organelle DNA by having a much larger and structurally more variable mitochondrial genome, and by having a third independent genome, the chloroplast genome (Palmer 1985).

The chloroplast genome in vascular plants is circular and varies in size from about 50,000 bp in some nonphotosynthetic plants to about 250,000 bp, with an average size of about 150,000 bp. Angiosperm chloroplast genomes are generally highly conserved in gene order, gene content, and gene organization. Chloroplast genes in photosynthetic plants have low rates of nucleotide substitution and the ratio K_A to K_S is much smaller than 1. In contrast, the chloroplast genomes of nonphotosynthetic plants exhibit elevated rates of substitution and much higher K_A to K_S ratios (e.g., Wicket et al. 2008). These findings are consistent with the neutral expectations following relaxation of functional constraints or total or partial nonfunctionalization (p. 122). In some *Geranium* species increased rates of nucleotide substitution in the chloroplast have been observed, although their photosynthetic apparatus is intact (Guisinger et al. 2008).

Substitution and rearrangement rates

Do the rates of nucleotide substitution correlate with the rate of structural changes in the genomes of organelles? As far as mammals are concerned, the answer seems negative, as mitochondrial DNA evolves very rapidly in terms of nucleotide substitutions, but the spatial arrangement of genes and the size of the genome are fairly constant among species. On the other hand, the mitochondrial genome of plants undergoes frequent structural changes in terms of size and gene order, but the rate of nucleotide substitution is extremely low. In chloroplast DNA and in mitochondrial DNA of arthropods, the rates of nucleotide substitution and genome rearrangement are correlated. All in all, the evidence suggests that the two processes occur independently.

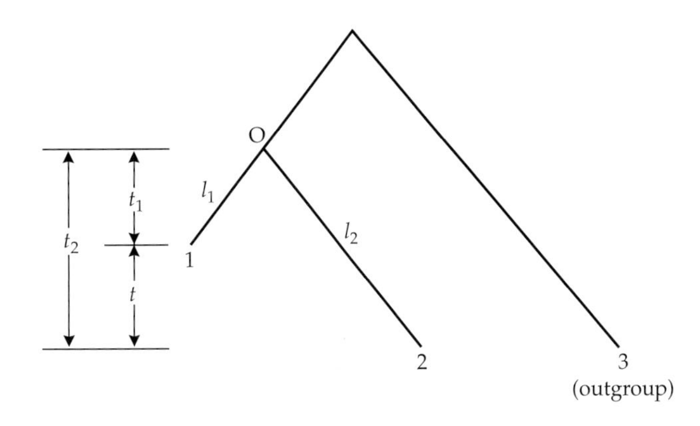

Figure 4.31 Model tree for estimating the rate of nucleotide substitution in RNA viruses. The expected number of substitutions on the branches leading to sequences 1 and 2 are denoted by l_1 and l_2, respectively. Sequence 1, which was isolated at t_1, was collected t years earlier than sequence 2, which was isolated at t_2. (Modified from Graur and Li 2000.)

Rates of Substitution in Viruses

Some viruses evolve at exceptionally high rates (Holland et al. 1982), reaching rates that are about 1 million times faster than those for animal DNA genomes. Therefore, significant numbers of nucleotide substitutions accumulate over short time periods, and differences in nucleotide sequences between viral strains isolated at relatively short time intervals are easily detectable. This property allows for a different approach to estimating evolutionary rates than that used previously (Chapter 3).

Several methods have been proposed in the literature for estimating the rate of nucleotide substitution in serially sampled viruses (Buonagurio et al. 1986; Saitou and Nei 1986; Li et al. 1988). In all models it is assumed that the rate of substitution is constant over time and that no further mutation occurs after the virus is isolated. Here we describe a method used by Li et al. (1988).

Figure 4.31 shows the model tree used in the calculations. Sequence 1 was isolated t years earlier than sequence 2, r is the rate of substitution per nucleotide site per year, and l_1 and l_2 are the expected numbers of substitutions per site from the ancestral node O to the time of isolation of sequences 1 and 2, respectively. Therefore

$$l_2 - l_1 = rt_2 - rt_1 = rt \quad (4.44)$$

By using sequence 3 as an outgroup reference, we obtain

$$l_2 - l_1 = d_{23} - d_{13} \quad (4.45)$$

where d_{ij} denotes the number of substitutions per site between sequences i and j. Combining Equations 4.44 and 4.45, we obtain

$$r = \frac{d_{23} - d_{13}}{t} \quad (4.46)$$

Methods for calculating the variance of r can be found in Wu and Li (1985) and Li and Tanimura (1987a,b). Note that one may use multiple outgroup references, but the outgroup references should be closely related to sequences 1 and 2, otherwise the variance of r may become too large. The best situation in which to apply this model is when t is relatively large but t_1 is small. If t were small, $l_2 - l_1$ and $d_{23} - d_{13}$ would also be small, and then even a small error in the estimation of these variables would lead to a large error in the estimation of r. However, t should not be too large, because it is difficult to obtain reliable estimates when the true values of d_{ij} are larger than 1. On the other hand, if t_1 is large, then l_1 will also be large, as will the variances of l_1, $l_2 - l_1$, and $d_{23} - d_{13}$. An example of the application of this method is given in the next section.

Human immunodeficiency viruses

Acquired immune deficiency syndrome (AIDS) is caused by a retrovirus called human immunodeficiency virus (HIV). Two major types, HIV-1 and HIV-2, are known. It is interesting to know the rates of nucleotide substitution in various regions of the HIV genome because, presumably, these rates determine the rate of change in viral pathogenicity and antigenicity (Rabson and Martin 1985; Gallo 1987). A number of authors (e.g., Hahn et al. 1986; Yokoyama and Gojobori 1987) have studied the rate of nucleotide substitution in HIV-1. The results below are from Li et al. (1988).

Table 4.11

Rates of synonymous (K_S) and nonsynonymous (K_A) nucleotide substitution per site per year (× 10^{-3}) in various coding regions of the HIV-1 genome

Coding region	Function	K_S (range)	K_A (range)
gag	Group-specific antigen	9.7 (6.5–13.1)	1.7 (1.1–2.3)
pol	Polymerase	11.0 (7.4–14.8)	1.6 (1.0–2.1)
sor	Infectivity	9.1 (6.1–12.3)	4.7 (3.1–6.3)
tat (exon 2)	Regulatory	7.0 (4.7–9.5)	8.3 (5.6–11.2)
art (exon 3)	Regulatory	7.4 (5.0–10.0)	6.6 (4.5–8.9)
gp120	Outer membrane protein	8.1 (5.5–10.9)	3.3 (2.2–4.4)
envhv	Hypervariable region	17.2 (11.6–23.2)	14.0 (9.4–18.8)
gp41	Transmembrane protein	9.8 (6.6–13.2)	5.1 (3.5–6.9)
env	Envelope	9.2 (6.2–12.4)	5.1 (3.5–6.9)
p27	Capsid protein	7.9 (5.3–10.7)	5.9 (4.0–8.0)
Average[a]		9.64 (2.92)	5.63 (3.60)

Source: Modified from Li et al. (1988).

[a]The average is the arithmetic mean, and values in parentheses are the standard deviations computed over all genes.

Two strains of the HIV-1 virus, denoted WMJ1 and WMJ2, were isolated from a 2-year-old child on October 3, 1984, and January 15, 1985, respectively. The child was presumed to have been infected only once (perinatally by her mother) by a single strain of HIV. In the following discussion, WMJ1 and WMJ2 are sequences 1 and 2, respectively, and $t = 3.4$ months (0.28 years). By using a number of reference outgroup strains, Li et al. (1988) calculated the rate of nucleotide substitution in the WMJ strains. The mean distance of the reference strains from WMJ1, d_{13}, was 0.0655, and from WMJ2, $d_{23} = 0.0675$. Therefore, $d_{23} - d_{13} = 0.0020$. From Equation 4.46, we obtain $r = 0.0020/0.28 = 7.1 \times 10^{-3}$ substitutions per site per year.

Table 4.11 lists the synonymous (K_S) and nonsynonymous (K_A) rates of substitution for the different genes in HIV-1. In nine of ten genes, the rates of synonymous substitution are higher than the nonsynonymous rates. However, while the mean K_S/K_A ratio in mammalian nuclear genes is about 5 and in some highly conservative genes it can reach values in excess of 100 (Table 4.1), the mean K_S/K_A ratio in HIV is less than 2, and even in the two most conservative genes, *gag* and *pol*, the ratio is less than 7. Thus, the extent of purifying selection in HIV genes is very weak.

The estimates for the average synonymous and nonsynonymous rates for the HIV genome are more than 10^6 times greater than the corresponding rates in mammalian genes and are similar to the values obtained for other RNA viruses (Holland et al. 1982; Gojobori and Yokoyama 1985; Hayashida et al. 1985; Buonagurio et al. 1986; Saitou and Nei 1986). The high substitution rates in HIV have been suggested to be mainly due to errors in the reverse transcription from RNA to DNA (Rabson and Martin 1985; Coffin 1986; Hahn et al. 1986). As pointed out by several authors, the high substitution rates may result in rapid modification of viral properties such as tissue tropism and sensitivity to antiviral drug therapy. In particular, the extremely high nonsynonymous rate in the hypervariable regions of the *env* gene may lead to extremely rapid changes in viral antigenicity.

CHAPTER 5

Molecular Phylogenetics and Phylogenetic Trees

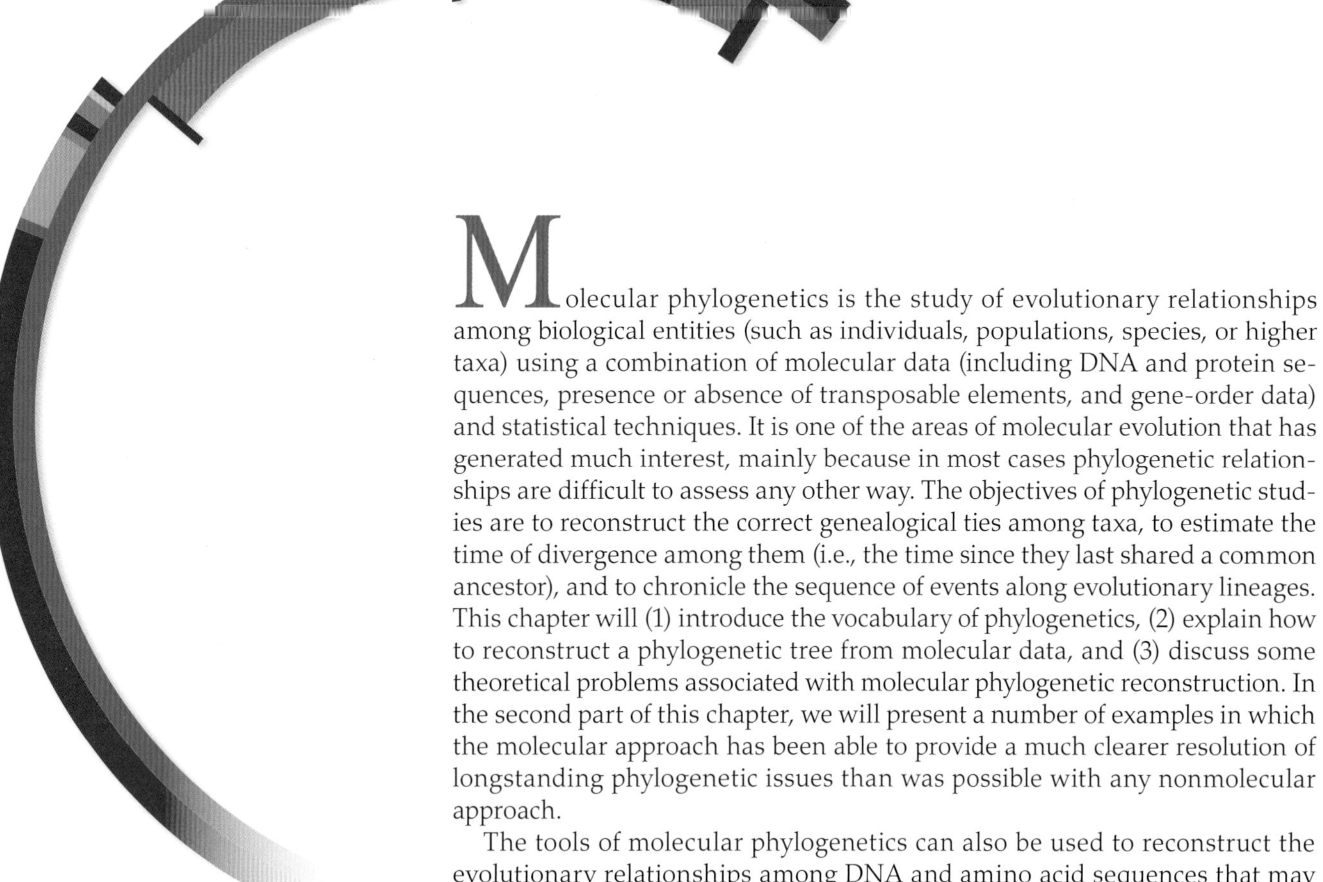

Molecular phylogenetics is the study of evolutionary relationships among biological entities (such as individuals, populations, species, or higher taxa) using a combination of molecular data (including DNA and protein sequences, presence or absence of transposable elements, and gene-order data) and statistical techniques. It is one of the areas of molecular evolution that has generated much interest, mainly because in most cases phylogenetic relationships are difficult to assess any other way. The objectives of phylogenetic studies are to reconstruct the correct genealogical ties among taxa, to estimate the time of divergence among them (i.e., the time since they last shared a common ancestor), and to chronicle the sequence of events along evolutionary lineages. This chapter will (1) introduce the vocabulary of phylogenetics, (2) explain how to reconstruct a phylogenetic tree from molecular data, and (3) discuss some theoretical problems associated with molecular phylogenetic reconstruction. In the second part of this chapter, we will present a number of examples in which the molecular approach has been able to provide a much clearer resolution of longstanding phylogenetic issues than was possible with any nonmolecular approach.

The tools of molecular phylogenetics can also be used to reconstruct the evolutionary relationships among DNA and amino acid sequences that may be related to one another by processes such as gene duplication (Chapter 7) or horizontal gene transfer (Chapter 9). In this chapter we confine ourselves to evolutionary events that can be modeled using phylogenetic trees; phylogenetic reconstructions that necessitate the use of networks will be dealt with in Chapter 6.

Impacts of Molecular Data on Phylogenetic Studies

The study of molecular phylogenetics began at the turn of the twentieth century with a series of immunochemical studies showing that serological cross-reactions were stronger for closely related organisms than for distantly related ones. The evolutionary implications of these findings were used by Nuttall (1902,

1904) to infer the phylogenetic relationships among various groups of animals, such as eutherians (placental mammals), primates, and ungulates (hoofed mammals). For example, he determined that the closest relatives of humans are the apes, followed in order of decreasing relatedness by the Old World monkeys, the New World monkeys, and the prosimians.

Since the late 1950s, various techniques have been developed in molecular biology, and this started the extensive use of molecular data in phylogenetic research. In particular, molecular phylogenetics progressed tremendously in the 1960s and 1970s as a result of the development of protein-sequencing methodologies. Less expensive and more expedient methods such as protein electrophoresis, DNA–DNA hybridization, and immunological methods, though less accurate than protein sequencing, were extensively used to study the phylogenetic relationships among populations or closely related species (e.g., Bain and Deutsch 1947; Sibley 1960; Goodman 1962; Sarich and Wilson 1967; De Ley 1974; Ayala 1976; Wilson et al. 1977). The application of these methods also stimulated the development of measures of genetic distance and tree-making methods (e.g., Fitch and Margoliash 1967; Nei 1975; Felsenstein 1988; Miyamoto and Cracraft 1991; Swofford et al. 1996).

The rapid accumulation of DNA sequence data since the late 1970s has had a great impact on molecular phylogenetics and has led to an unprecedented level of activity in the field. Indeed, molecular data have been used, on the one hand, to infer the phylogenetic relationships among closely related populations or species, such as the relationships among human populations (Cann et al. 1987; Vigilant et al. 1991; Hedges et al. 1992; Templeton 1992; Horai et al. 1993; Torroni et al. 1993; Bailliet et al. 1994) and the relationships among apes; and on the other hand, they have been used to study very ancient evolutionary occurrences, such as the origin of mitochondria and chloroplasts and the divergence of phyla and kingdoms (Takabe and Akazawa 1975; Woese and Fox 1977a; Gray 1983; Gray et al. 1984; Hasegawa et al. 1985; Cedergren et al. 1988; Giovannovi et al. 1988; Lockhart et al. 1994).

The late 1990s witnessed the beginning of **phylogenomics**, a scientific discipline at the intersection between phylogenetics and genomics (Eisen 1998). The term "phylogenomics" has had multiple meanings. Originally, it referred to the prediction of the function of genes based on their phylogenetic history. Subsequently, the term became associated with the process of reconstructing species trees by combining the information from many genes or entire genomes (Philippe et al. 2004; Delsuc et al. 2005; Jeffroy et al. 2006) as well as, in particular, with the integration of genomics and evolutionary biology for purposes of reconstructing the temporal dynamics of genomes (Eisen and Hanawalt 1999; Eisen and Fraser 2003; Whitaker et al. 2009).

Future research directions in phylogenetics are likely to be influenced by four factors: (1) the amount of genomic (and metagenomic) data being generated by emerging sequencing technologies; (2) the availability of population-level data, requiring the integration of tools from population genetics into phylogenetic analysis; (3) the development of fast and accurate algorithms to reconstruct and visualize evolutionary histories involving enormous numbers of taxa; and (4) the input from such areas as developmental biology and structural biology that are bound to provide us novel insights into the phylogenetics of organismal and molecular entities.

In an 1857 letter to Thomas Huxley, Charles Darwin wrote, "The time will come, I believe, though I shall not live to see it, when we shall have fairly true genealogical trees of each great kingdom of Nature." Molecular data have proved so powerful in the study of evolutionary history that Darwin's dream may be within our grasp (even if the trees he imagined may not always be a suitable metaphor for all evolutionary processes). Of course, we should not abandon other means of evolutionary inquiry, such as morphology, anatomy, physiology, structural biology, and paleontology. Rather, different approaches provide complementary data. Indeed, our mere ability to identify the subjects of our molecular inquiry is based almost exclusively on morphological and anatomical information. Furthermore, paleontological information is one of only a handful of methods that can provide a time frame for evolutionary events.

Advantages of Molecular Data in Phylogenetic Studies

There are several reasons why molecular data, particularly DNA and amino acid sequences, may be more suitable for evolutionary studies than morphological data (Graur 1993). First, DNA and protein sequences are strictly heritable entities. This may not be true for many morphological traits that can be influenced to varying extents by environmental factors. Second, the description of molecular characters and character states is unambiguous. Thus, the third amino acid in the preproinsulin of the rabbit (*Oryctolagus cuniculus*) can be unambiguously identified as serine, and the homologous position in the preproinsulin of the golden hamster (*Mesocricetus auratus*) as leucine. In contrast, morphological descriptions frequently contain such ambiguous modifiers as "thin," "reduced," "slightly elongated," "partially enclosed," and "somewhat flattened." Third, molecular traits generally evolve in a much more regular manner than do morphological and physiological characters and therefore can provide a clearer picture of the relationships among organisms. Fourth, molecular data are often much more amenable to quantitative treatments than are morphological data. In fact, sophisticated mathematical and statistical theories have been developed for the quantitative analysis of DNA sequence data, whereas morphological studies retain a great deal of qualitative argumentation and special pleading. Fifth, homology assessment is easier with molecular data than with morphological traits because quantitative statements concerning sequence similarity can be translated into qualitative statements on homology through the statistical rejection of alternative hypotheses. Sixth, some molecular data are robust, i.e., can be used to assess evolutionary relationships among very distantly related organisms. For example, numerous protein and ribosomal RNA sequences can be used to reconstruct evolutionary relationships among such distantly related organisms as fungi, plants, and animals. In contrast, there are few morphological characters that can be used for such a purpose. Finally, molecular data are much more abundant than morphological data. This abundance is especially useful when working with organisms such as microbial prokaryotes and eukaryotes, which possess only a limited number of morphological or physiological characters that can be used in phylogenetic studies.

Species and Speciation

How many species are there in the world? Currently there are approximately 1.4 million described species, yet this number is probably far from being close to the actual number of extant species. Many estimates of the number of extant species in nature can be found in the literature, the lowest being 2–10 million species and the highest 50–100 million species.

Biologists think of species as the most important and basic evolutionary unit, yet ever since the term was coined at the beginning of the seventeenth century, the exact definition of "species" and the criteria by which species are to be distinguished from one another have been the subjects of much debate. Similarly, very little is known about the evolutionary processes by which new biological species arise.

The species concept

There is no agreement among biologists on the meaning of the term "species," except that it is a most important and basic concept. In the literature, one can find more than a dozen different **species concepts** embracing different genetic, ecological, evolutionary, and epistemological principles (de Queiroz 2007). Here, we will briefly deal with four main categories of concepts: the typological, the phenetic, the biological, and the phylogenetic. We will also consider a pragmatic approach, which is not so much a concept as a practical realization that all species concepts have major faults and lack universal applicability (Richards 2007).

One version of the **typological species concept**, which dates back to Aristotle, defines a species as a group of organisms that can be defined by a set of necessary and sufficient unchanging properties defined by an ideal "type." In current practice, a **type** is a particular specimen of an organism to which the scientific name of that organism is formally attached. In other words, a type serves to anchor and centralize the defining features of that particular taxon. Although the typological species concept is not only non-evolutionary but also anti-evolutionary, it is the most widely used concept in taxonomic nomenclature and classification. For example, a beetle can be called *Lagocheirus delestali* only if it possesses the same set of properties as the type specimen labeled "Costa Rica, Alajuela, Estación de la Reserva Biológica Alberto Manuel Brenes, Alt. 850 m, 21 de abril de 2006, José Rafael Esteban Durán" that has been deposited in the Instituto Nacional de Investigación y Tecnología Agraria y Alimentaria, Madrid, and was described in a formal manner by Toledo and Esteban Durán (2008).

Most typological species are defined on the basis of morphology and are, hence, called **morphospecies**. Some microbial species, especially bacteria, may be defined on the basis of other traits, such as cell surface antigens (serotype), in which case they are called **serospecies**. With the advent of metagenomics, i.e., the sequencing of genetic material recovered directly from environmental samples, it has become common to define species by their genotype, even if no morphology can be associated with it. Such typological definitions are referred to as **genospecies**.

The **phenetic species concept** (Sokal and Crovello 1970) can be thought of as relaxation of the typological concept by dispensing with the type. The phenetic species approach defines a species as a set of organisms that resemble one other more than they resemble members of another set. The degree of similarity is measured by a distance statistic.

The most popular species concept in textbooks and the most difficult one to apply in practice is the **biological species concept**, in which species are defined as "groups of actually or potentially interbreeding natural populations, which are reproductively isolated from other such groups" (Mayr 1942).

The **phylogenetic species concept** is a historical, lineage-based concept that defines a species as a monophyletic group in a phylogeny. That is, a phylogenetic species is the smallest set of organisms that share an ancestor not shared by any other group.

Of course, none of these definitions are satisfactory under all (or even most) circumstances. For instance, the biological species concept is not applicable to either fossils or asexually reproducing organisms. As a consequence, we can either join the endless and mostly futile debates on the meaning of the term "species," or we can adopt Darwin's pragmatic approach to the species problem, as phrased in *On the Origin of Species*: "No one definition has satisfied all naturalists; yet every naturalist knows vaguely what he means when he speaks of a species."

Speciation

Speciation is the evolutionary process by which new species arise. Darwin referred to the origin of species as a "mystery of mysteries," and despite more than a century and a half of intense work, the study of speciation is rich in convoluted terminology and poor in facts. Evolutionary biology is nowhere near explaining the mechanisms that have produced the enormous diversity of life forms in nature.

A new species may arise by **cladogenesis**, i.e., the splitting of lineages, by **anagenesis**, i.e., a change within a lineage, or by **genome hybridization**, i.e., the merging of lineages. Cook (1906) recognized that speciation in sexual organisms occurs mainly through cladogenesis, and here the terms "speciation" and "cladogenesis" are used interchangeably. From the viewpoint of genetics, speciation equals reproductive isolation. If the two speciating populations occupy nonoverlapping ranges, for example, if they are separated by physical barriers, the process of speciation is referred to as **allopatric speciation**. **Sympatric speciation** occurs when the populations become reproductively isolated in spite of the fact that they occupy overlapping ranges. In sexual species, two types of isolating mechanisms are known: **premating** or **prezygotic barriers**, such as temporal, behavioral, or mechanical isolation, and **postmating** or **postzygotic barriers**, such as gametic incompatibility, hybrid inviability, or hybrid sterility.

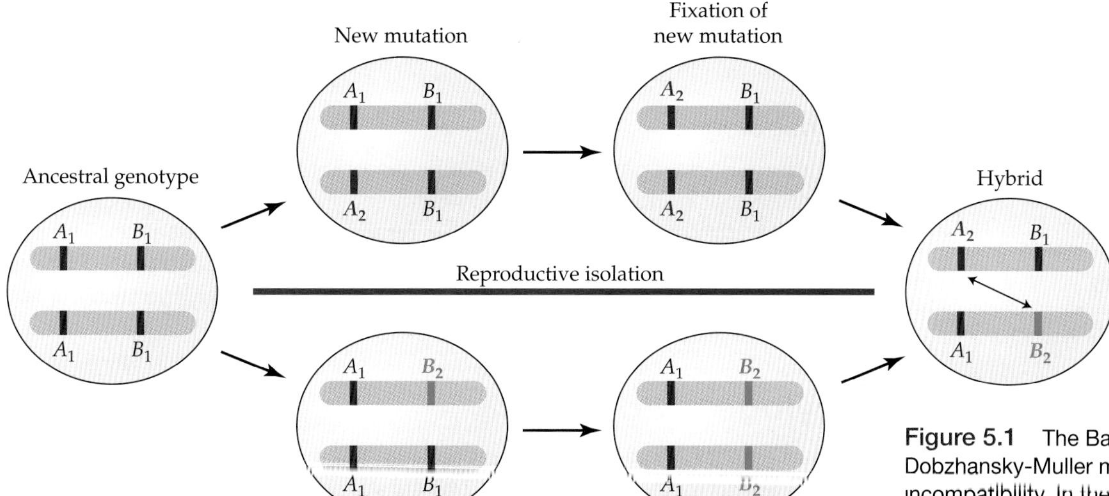

Figure 5.1 The Bateson-Dobzhansky-Muller model of hybrid incompatibility. In the ancestral population, the genotype is $A_1A_1B_1B_1$. After the population splits into two, one new mutation, A_2, arises in one population, and another, B_2, in the second population. Because of strong negative epistatic interactions, A_2 and B_2 are in effect mutually incompatible. In the hybrid, the incompatibility is indicated by the double-headed arrow. Note that the incompatibility is an incidental by-product of the divergence process. (Modified from Johnson 2008.)

Bateson (1909), Dobzhansky (1937), and Muller (1940, 1942) independently suggested an evolutionary mechanism for allopatric speciation. In this scenario, a population is split into two isolated populations which then evolve independently, i.e., different alleles may become fixed at different loci. In the simplest Bateson-Dobzhansky-Muller model (also referred to as the Dobzhansky-Muller model), two loci, A and B, are involved in the speciation process (**Figure 5.1**). In the original population, locus A is fixed for the A_1 allele and locus B is fixed for the B_1 allele. After reproductive isolation, the A locus of population 1 becomes fixed for a new A_2 allele, while the B locus of population 2 becomes fixed for a new B_2 allele. Speciation occurs if the hybrid offspring from these two populations suffers considerable reduction in viability or another fitness component due to negative epistatic interactions between alleles A_2 and B_2. The Bateson-Dobzhansky-Muller model of incompatibility is significant in the history of science because a specific and testable genetic mechanism for the evolution of reproductive isolation was proposed for first time.

Barbash et al. (2003) and Brideau et al. (2006) were the first to describe a pair of genes that behave in the manner predicted by the Bateson-Dobzhansky-Muller model. Their experiments involved two closely related sibling species, *Drosophila melanogaster* and *D. simulans*. When *D. melanogaster* females are crossed with *D. simulans* males, hybrid daughters survive whereas hybrid sons die; in the reciprocal cross, *D. simulans* females × *D. melanogaster* males, hybrid sons survive whereas hybrid daughters die. Thus, despite the morphological and molecular similarity between *D. melanogaster* and *D. simulans*, these two species are reproductively isolated. Molecular genetic analyses of inviability in *D. melanogaster* × *D. simulans* hybrids revealed that the lethality of hybrid males (*D. melanogaster* mothers × *D. simulans* fathers) is caused by two loci: *Hybrid male rescue* (*Hmr*) on chromosome X of *D. melanogaster* and *Lethal hybrid rescue* (*Lhr*) on chromosome 2 of *D. simulans*. Both *Hmr* and *Lhr* encode proteins with DNA-binding domains, and both have experienced rapid (possibly adaptive) evolution since the divergence of the two species from each other.

The simple case described above seems to be quite rare in nature; pairs of genes being solely responsible for speciation have not been found in other systems. Rather, speciation seems to usually rely on more complex incompatibilities involving numerous alleles at many loci (e.g., Kao et al. 2010).

Is speciation an event that can be assigned a particular time in evolution? When we claim that two species have diverged from one another x million years ago, does our statement mean that at time $x - 1$ hour there existed a single species and at time $x + 1$ hour there were two species? Speciation is a process that takes time, and reproductive barriers are not absolute for long periods of time. For example, it has been proposed

(Patterson et al. 2006) that the human and chimpanzee lineages first diverged (i.e., ceased to interbreed) and then later interbred again before permanent reproductive barriers were ultimately raised.

Terminology

The terminology of phylogenetics is quite discombobulated. The naming of organisms is referred to as **nomenclature**. The assignment of organisms and groups of organisms to biologically coherent groups (taxa) is referred to as **classification**. The practice of nomenclature and classification is collectively referred to as **taxonomy**. **Phylogenetics** refers to the evolutionary relationships among taxa. Finally, the study of taxonomy and phylogenetics is referred to collectively as **systematics**.

Phylogenetic Trees

In mathematics, a **graph** is a representation of a set of objects in which some pairs of objects are connected by links. The objects are represented by mathematical abstractions called **nodes** (or **vertices**), and the links that connect some pairs of vertices are called **branches** (or **edges** or **arcs**). A **path** in a graph is a sequence of branches that connect any two nodes. A graph can be connected or disconnected. A graph that contains one or more nodes that are unreachable from at least one other node in the graph is called a **disconnected graph** (**Figure 5.2a**). Disconnected graphs have no use in phylogenetics. A graph that is in one piece, i.e., one in which every node can be reached from any other node, is said to be a **connected graph**. Connected graphs come in two flavors. A **tree** is a connected graph in which any two nodes are connected by a single path (**Figure 5.2b**). The term "network" does not have an unambiguous mathematical definition, and is used here in its phylogenetic sense. That is, a **network** is a connected graph in which at least two nodes are connected by two or more pathways (**Figure 5.2c**). In the vast majority of phylogenetic studies, the evolutionary relationships among organisms in a group are illustrated by means of trees (or **dendrograms**). Networks are used in phylogeny to represent reticulations (Chapter 6).

In a tree, the nodes represent the **taxonomic units**. The taxonomic units represented by the nodes can be DNA sequences, genes, proteins, individuals, populations, species, or higher taxa. The branches define the relationships among the taxonomic units in terms of descent and ancestry. The branching pattern of a tree is called its **topology**.

We distinguish between **internal nodes** and **terminal nodes** (or **leaves**), and **external** (or **peripheral**) **branches** (branches that connect an internal node and a terminal node) and **internal branches** (branches that connect two internal nodes). For example, in **Figure 5.3**, nodes A, B, C, D, and E are terminal, whereas all others are internal. Branches AF, BF, CG, DG, and EI in Figure 5.3 are external; all others are internal. Terminal nodes represent the actual data under study and are, hence, also referred to as **operational taxonomic units** (**OTUs**). Internal nodes represent inferred ancestral units, and since we have no empirical data pertaining to these entities, they are referred to as **hypothetical taxonomic units** (**HTUs**).

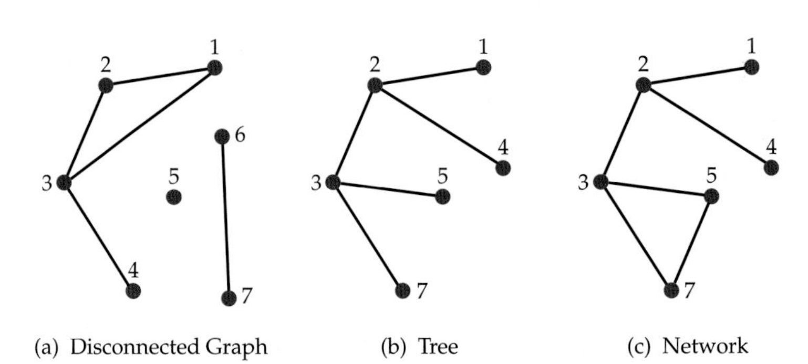

Figure 5.2 Graphs, trees, and networks. (a) A disconnected graph with seven nodes. Node 6, for example, cannot be reached from node 2. Note also that one part of the graph, node 5, contains no edges. (b) A tree is a connected graph in which there is a unique path between any two nodes. For example, the only path connecting nodes 1 and 5 passes through nodes 2 and 3. No other path between nodes 1 and 5 exists. (c) A network is a connected graph in which one or more pairs of nodes are connected by multiple paths. For example, there are two paths between nodes 2 and 5 (one through node 3, and one through nodes 3 and 7).

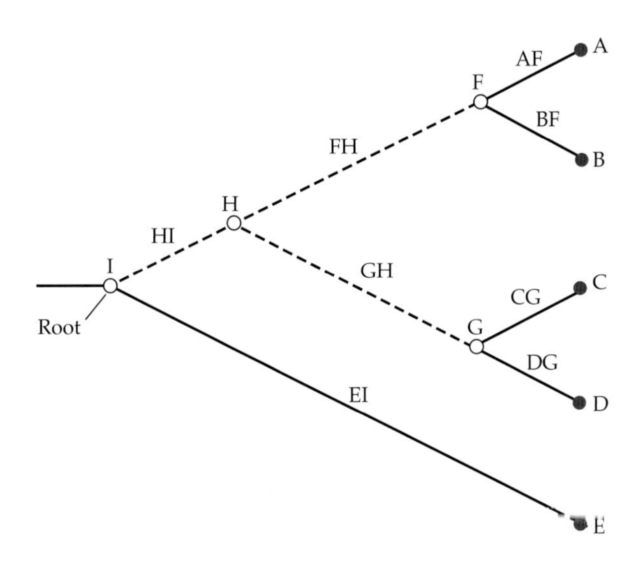

Figure 5.3 A phylogenetic tree illustrating the evolutionary relationships among five operational taxonomic units (A–E). Solid and white circles denote terminal and internal nodes, respectively. Solid and dashed lines denote terminal branches (AF, BF, CG, DG, and EI) and internal branches (FH, GH, and HI), respectively. The internal nodes (F–H) represent the hypothetical taxonomic units. I is the root.

A node is **bifurcating** (or **dichotomous**) if it has only two immediate descendant lineages, but **multifurcating** (or **polytomous**) if it has more than two immediate descendant lineages (**Figure 5.4**). In a strictly bifurcating tree, each internal node is incident to exactly three branches, two derived and one ancestral. In evolutionary studies we assume that the process of speciation is usually a binary one, i.e., that speciation results in the formation of not more than two species from a single stock at any one time. Thus, the common representation of phylogenies employs bifurcating trees, in which each ancestral taxon splits into two descendant taxa. There are two possible interpretations for a **multifurcation** (or **polytomy**) in a tree: either it represents the true sequence of events (**hard polytomy**), whereby an ancestral taxon gave rise to three or more descendant taxa simultaneously, or it represents an instance in which the exact order of two or more bifurcations cannot be determined unambiguously with the available data (**soft polytomy**). An **unresolved** or **partially resolved tree** will have at least one internal node that is incident to four or more branches. In the following, we assume that speciation is always a bifurcating process, and multifurcations will be assumed to represent soft polytomies, i.e., cases in which the exact temporal sequence of several bifurcations cannot be determined with the data at hand.

The use of trees to model evolutionary relationships can be traced back to Lamarck (1809); however, it was Haeckel (1866) who popularized the iconography of trees (Ragan 2009).

Rooted and unrooted trees

Trees can be rooted or unrooted. In a **rooted tree** there exists a particular node, called the **root**, from which a unique path leads to any other node (**Figure 5.5a**). The direction of each path corresponds to evolutionary time, and the root is the common ancestor of all the taxonomic units under study. An **unrooted** tree is a tree that only specifies the degree of kinship among the taxonomic units but does not define the evolutionary path (**Figure 5.5b**). Thus, strictly speaking, an unrooted tree may not

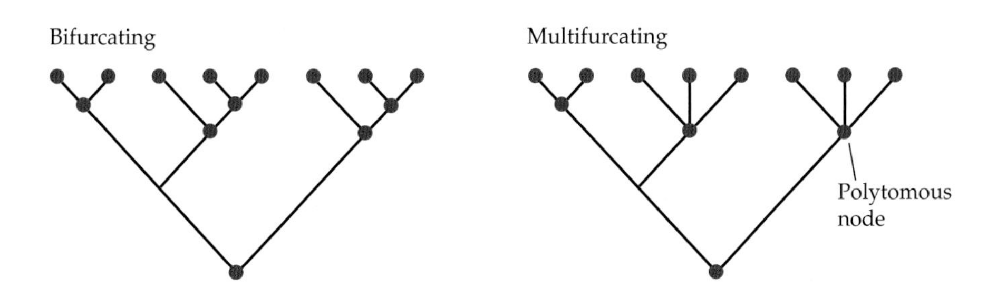

Figure 5.4 A node is bifurcating if it has only two immediate descendant lineages (a), and multifurcating if it has more than two immediate descendant lineages (b).

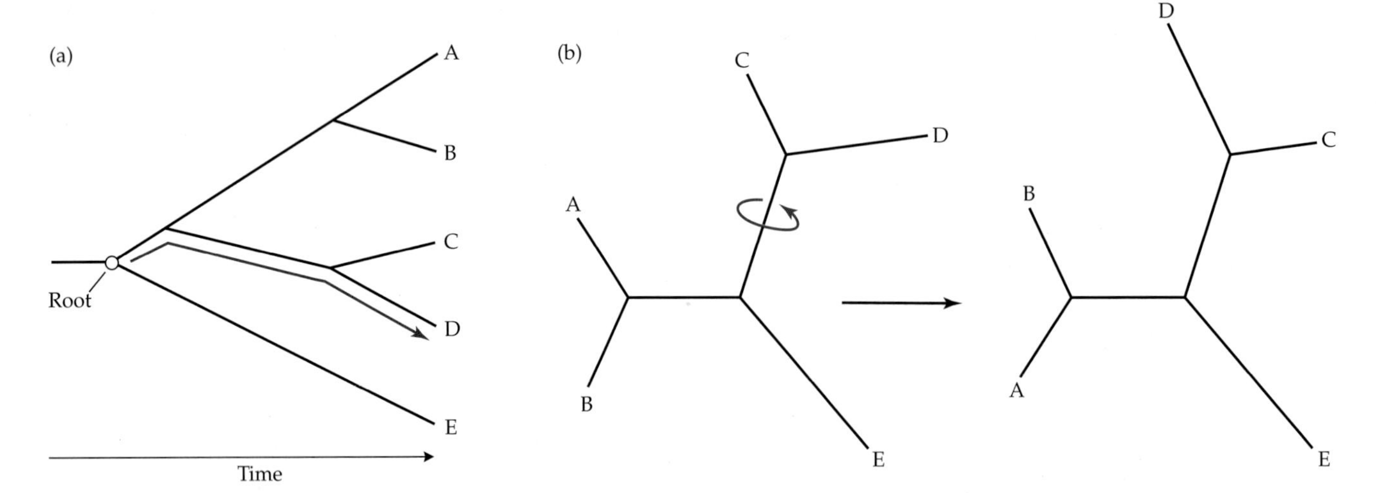

Figure 5.5 Rooted (a) and unrooted (b) trees. The red arrow indicates the unique path leading from the root to OTU D. The topology of the tree does not change by rotating branches around a node. The two trees in (b) have identical topologies. (Modified from Graur and Li 2000.)

in itself be considered a phylogenetic tree, since the time arrow is not specified. Unrooted trees neither make assumptions nor require knowledge about common ancestors. Rooted and unrooted trees are sometimes called directed and undirected trees, respectively. The vast majority of tree-making methods yield unrooted trees.

An unrooted tree of n taxa has n terminal nodes representing the OTUs (taxa) and $n - 2$ internal nodes. Such a tree has $2n - 3$ branches, of which $n - 3$ are internal and n are external. In a rooted tree, there are n terminal nodes and $n - 1$ internal ones, as well as $2n - 2$ branches, of which $n - 2$ are internal and n are external. Two OTUs separated by a single node are called **neighbors** or **neighboring taxa**. In an unrooted tree with four external nodes, the internal branch is frequently referred to as the **central branch**.

The topology of either rooted or unrooted trees is not affected by rotating branches around an internal node (Figure 5.5b).

Scaled and unscaled trees

Figure 5.6 illustrates two common ways of drawing a phylogenetic tree. In Figure 5.6a, the branches are **unscaled**; their lengths are not proportional to the number of changes, which are indicated on the branches. This type of presentation allows us to line up the extant OTUs and to place the internal nodes representing divergence events on a time scale when the times of divergence are known or have been estimated. Unscaled

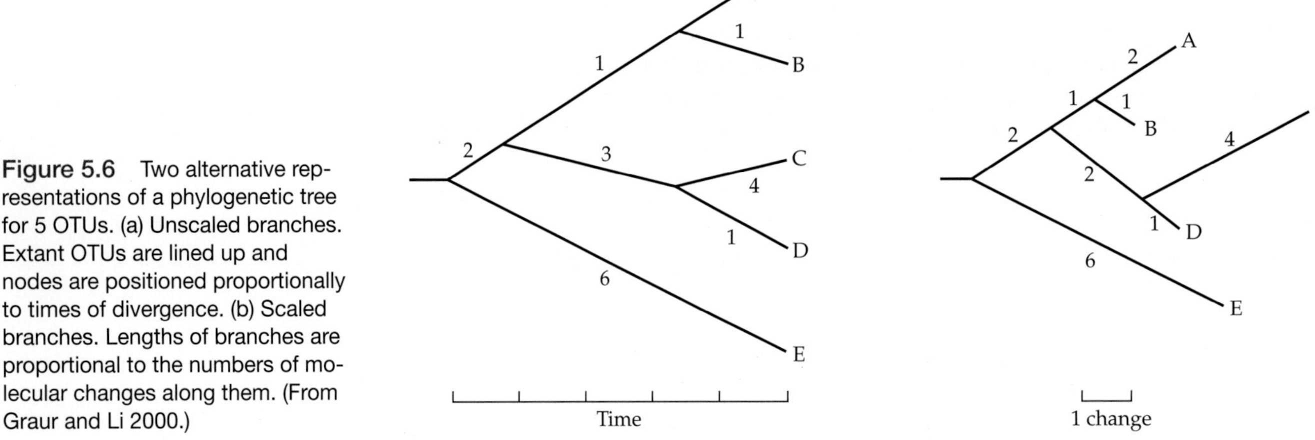

Figure 5.6 Two alternative representations of a phylogenetic tree for 5 OTUs. (a) Unscaled branches. Extant OTUs are lined up and nodes are positioned proportionally to times of divergence. (b) Scaled branches. Lengths of branches are proportional to the numbers of molecular changes along them. (From Graur and Li 2000.)

trees are also called **cladograms**. In Figure 5.6b, the branches are **scaled**; each branch is proportional to the number of changes that have occurred along that branch. Scaled trees are also called **phylograms**. Thus, the branches of phylograms are proportional to the amount of change on the branch, while cladograms "measure" time.

The topology of a tree is not affected by the absolute or relative lengths of the branches. The number of changes (e.g., nucleotide substitutions) that have occurred along a branch is called the **branch length**. The total number of changes in a particular tree is called the **tree length**.

The Newick format

In previous sections, we drew phylogenetic trees as two-dimensional graphs. In computer programs, trees are represented by a linear (one-dimensional) notation called the **Newick format**. The originator of this format in mathematics was Cayley (1857). The Newick format for phylogenetic trees was adopted on June 26, 1986, at an informal meeting convened by Joseph Felsenstein at Newick's Restaurant in Dover, New Hampshire. Since that time, it has become the most widely used representation of trees in bioinformatics, currently serving as the de facto standard for almost all phylogenetic software tools. Unfortunately, the format has never been formally described; it was first mentioned in a scientific publication by Olsen et al. (1992).

In the Newick format, trees are represented by a string of nested parentheses enclosing OTU names separated by commas. Optionally, branch lengths, HTU names, measures of statistical support for internal branches, and the position of the root can be specified. Nowadays, some formats using the Newick standard can also handle missing information. They are referred to as **flexible**. Because phylogenetic trees may contain numerous OTUs, the Newick notation may require multiple coding lines. To be able to distinguish unambiguously between trees listed one after another, each phylogenetic tree in the Newick format ends in a semicolon (;).

In the Newick format, the pattern of the parentheses indicates the topology of the tree: each pair of parentheses enclose all members of a clade, in the case of a rooted tree, or a group of neighboring taxa, in the case of an unrooted tree.

In the case of a rooted tree, the root is placed on the branch connecting the two taxa or the two groups of taxa separated by the comma within the external parentheses. For example, the rooted tree in **Figure 5.7a** is written as (((((A,B),C),D),E),F);. Within its most external parentheses, a comma separates taxon F from the rest of the taxa (A, B, C, D, and E).

A characteristic of an unrooted tree with five or more taxa is that it contains a **trifurcation** (three-way split). In the Newick format, this trifurcation is represented by a pair of parentheses that contains three taxa or three pairs of taxa. For example, the unrooted tree in **Figure 5.7b** is written in the Newick format as ((A,B),(C,D),(E,F));. The external parentheses contain three groups of taxa—(A,B), (C,D), and (E,F)—separated by two commas.

The topology of every tree does not yield a single unique rendering in the Newick format. There are two reasons for this state of affairs. The first reason is trivial. The left-to-right order of OTUs is biologically uninteresting. Thus, the tree in Figure 5.7a can also be written as (F,(E,(D,(C,(B,A))))); or (F,(E,(D,(C,(A,B))))); or ((E,(D,(C,(B,A)))),F); and so on. In rooted trees, we must make sure that the root is placed inside the external parentheses. In the case of unrooted trees, this restriction does not apply, so the number of alternative Newick formats is larger than that for rooted trees.

In the Newick format, scaled trees are written with the branch lengths placed immediately after the group descended from that branch and separated by a colon (Figure 5.7c). There are two ways to include measures of statistical support for internal branches (e.g., bootstrap values). They may be either placed in brackets after the internal branch length or placed in front of the internal branch length followed by a colon. Only the first way is shown in Figure 5.7d.

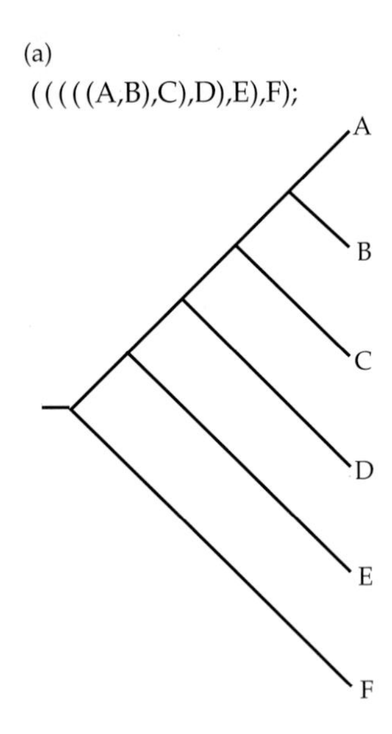

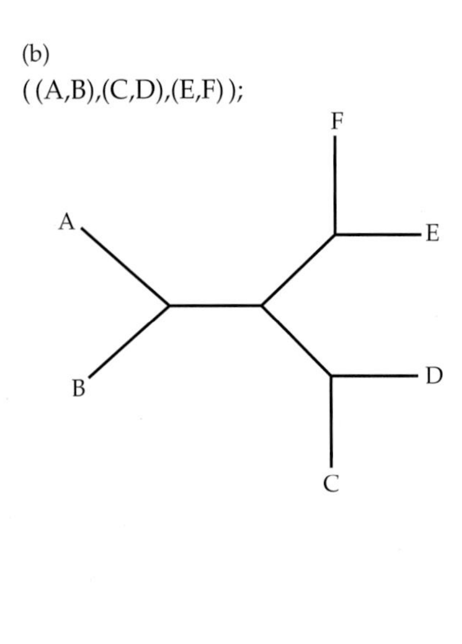

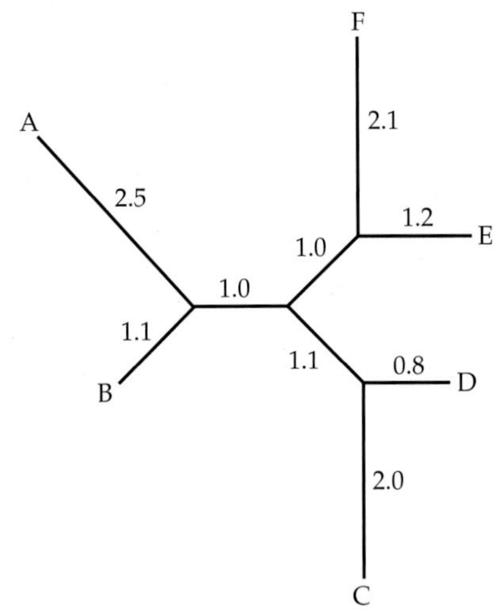

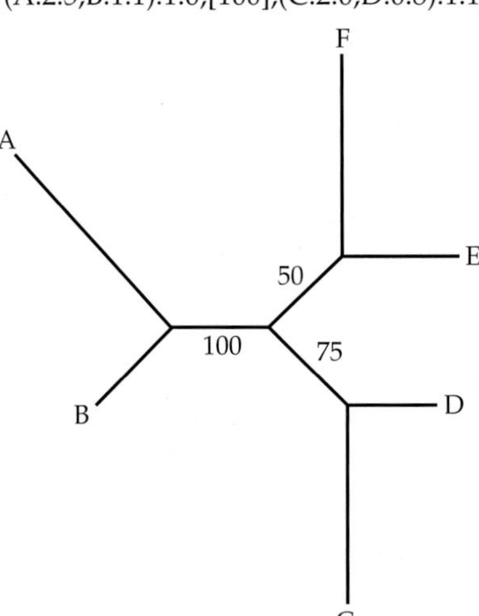

Figure 5.7 Newick format representations for rooted (a), unrooted unscaled (b), and unrooted scaled (c and d) trees. The numbers in (c) represent branch lengths. The numbers in (d) represent bootstrap values. Note that alternative renderings of the Newick format are possible for each tree. (Modified from Graur and Li 2000.)

Number of possible phylogenetic trees

For three OTUs, A, B, and C, there exists only one possible unrooted tree (**Figure 5.8a**). There are, however, 3 different rooted trees (**Figure 5.8b**). For four OTUs, there are 3 possible unrooted trees (**Figure 5.8c**) and 15 rooted ones (**Figure 5.8d**). The number of bifurcating rooted trees (N_R) for n OTUs is given by

$$N_R = \frac{(2n-3)!}{2^{n-2}(n-2)!} \tag{5.1}$$

when $n \geq 2$ (Cavalli-Sforza and Edwards 1967). The number of bifurcating unrooted trees (N_U) for $n \geq 3$ is

$$N_U = \frac{(2n-5)!}{2^{n-3}(n-3)!} \tag{5.2}$$

Note that the number of possible unrooted trees for n OTUs is equal to the number of possible rooted trees for $n - 1$ OTUs, i.e., rooting an unrooted tree is equivalent to adding one branch to one of its existing branches. The numbers of possible rooted and unrooted trees for up to 20 OTUs are given in **Table 5.1**. We see that both N_R and N_U increase very rapidly with n, and for 10 OTUs there are already more than 2 million

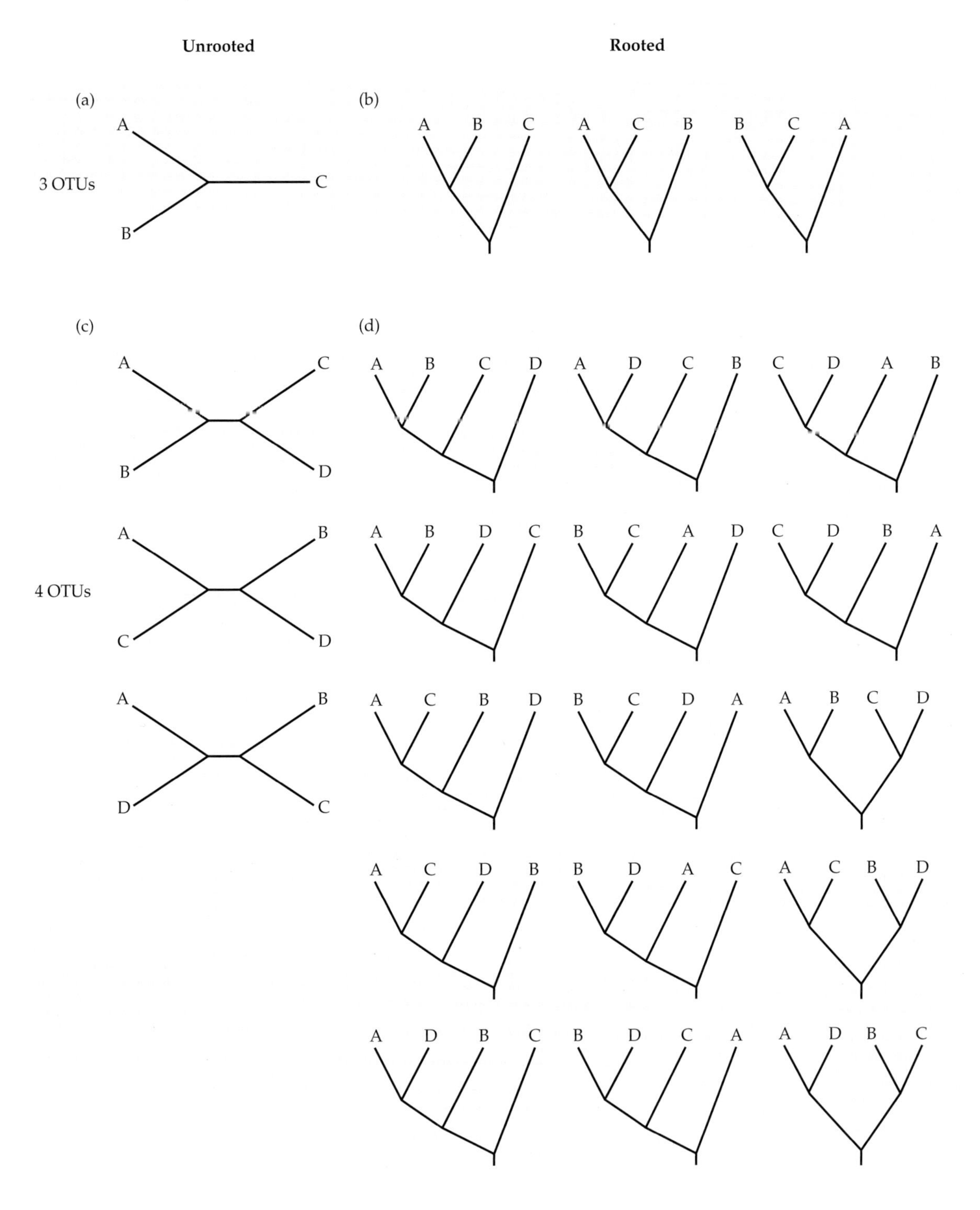

Figure 5.8 From three OTUs it is possible to construct only a single unrooted tree (a) but 3 different rooted ones (b). From four OTUs it is possible to construct 3 unrooted trees (c) and 15 rooted ones (d). (From Graur and Li 2000.)

bifurcating unrooted trees and close to 35 million rooted ones. For 20 OTUs there are close to 10^{22} rooted trees. Since only one of these trees correctly represents the true evolutionary relationships among the OTUs, it is usually very difficult to identify the true phylogenetic tree when n is large.

Tree balance

Tree balance is a measure of the degree of symmetry of a rooted phylogenetic tree. It serves as an indication of the pattern of speciation events in the group of taxa under study. The intuition behind this measure is easily seen by contrasting the case of a completely **balanced** tree and a completely **unbalanced** (or **pectinate**, i.e., comblike)

TABLE 5.1

Numbers of possible rooted and unrooted trees for up to 20 OTUs

Number of OTUs	Number of rooted trees	Number of unrooted trees
2	1	1
3	3	1
4	15	3
5	105	15
6	945	105
7	10,395	945
8	135,135	10,395
9	2,027,025	135,135
10	34,459,425	2,027,025
11	654,729,075	34,459,425
12	13,749,310,575	654,729,075
13	316,234,143,225	13,749,310,575
14	7,905,853,580,625	316,234,143,225
15	213,458,046,676,875	7,905,853,580,625
16	6,190,283,353,629,375	213,458,046,676,875
17	191,898,783,962,510,625	6,190,283,353,629,375
18	6,332,659,870,762,850,625	191,898,783,962,510,625
19	221,643,095,476,699,771,875	6,332,659,870,762,850,625
20	8,200,794.532,637,891,559,375	221,643,095,476,699,771,875

Source: Data from Felsenstein (1978b).

one (**Figure 5.9**). In a perfectly unbalanced tree, only one descendant of an interior node continues to speciate after a splitting event. In a perfectly balanced tree, by contrast, all descendants of an interior node participate equally in cladogenesis.

There are many measures of tree balance (Shao and Sokal 1990; Kirkpatrick and Slatkin 1993). Here, we present a single measure, the *C* index, which was proposed by Colless (1982, 1995). A rooted tree has $n - 1$ internal nodes (including the root), where n is the number of OTUs. Each internal (ancestral) node, i, partitions the OTUs that descend from it into two groups of sizes r_i and s_i, where $r_i \geq s_i$. The *C* index is based on the difference between r_i and s_i, summed over all interior nodes:

$$C = \frac{2}{n(n-3)+2} \sum_{i=1}^{n-1} (r_i - s_i) \quad \textbf{(5.3)}$$

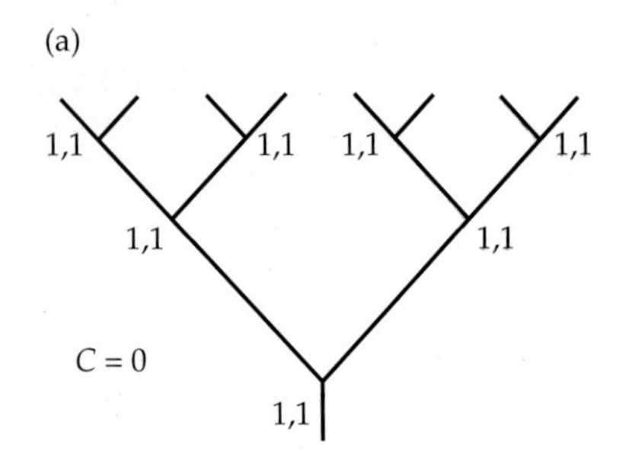

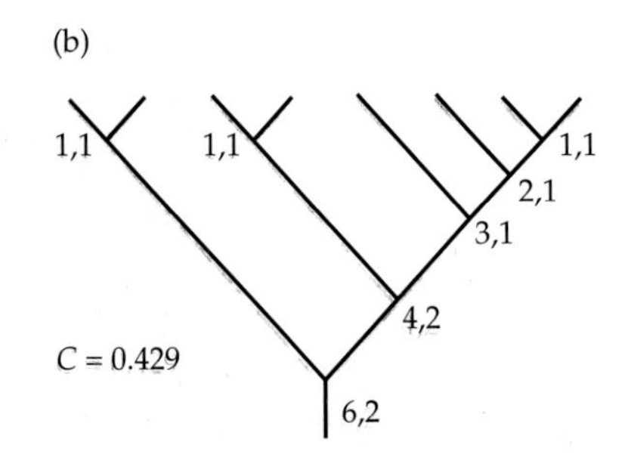

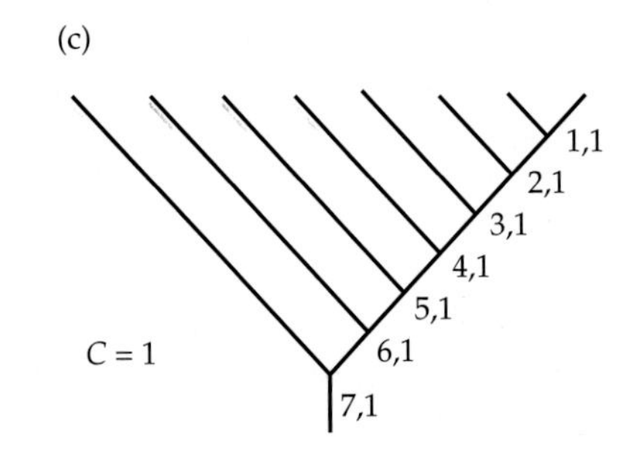

Figure 5.9 Three rooted phylogenetic trees ranging from completely balanced (a) through an intermediate degree of balance (b) to completely unbalanced (c). *C* is the *C* index of Colless (1982, 1995), calculated for the three trees by using Equation 5.3. In the three eight-OTU rooted trees, there are seven internal nodes, each partitioning the descendant nodes into two groups of sizes r_i and s_i, where $r_i \geq s_i$. Adjacent to each node, the values of r_i and s_i are to the left and the right of the comma, respectively.

The C index increases from a value of 0 for a perfectly balanced tree (Figure 5.9a) to a value of 1 for a perfectly pectinate one (Figure 5.9c). An intermediate tree with a C index of 0.429 is shown in Figure 5.9b. Tree balance measures can be used to infer patterns of speciation, provided the taxonomic representation in the sample under study is exhaustive or very close to complete. Tree balance measures are also very important as indicators of the ease of phylogenetic reconstruction. Because, by definition, unbalanced trees contain long branches, they are more difficult to reconstruct phylogenetically than balanced trees (Moreira and Philippe 2000). In fact, unbalanced and balanced tree are sometimes referred to as "good" and "bad" trees, respectively (e.g., Sackin 1972).

True and inferred trees

The sequence of speciation events that has led to the formation of any group of OTUs is historically unique. Thus, only one of all the possible trees that can be built with a given number of OTUs represents the true evolutionary history. Such a phylogenetic tree is called the **true tree**. A tree that is obtained by using a certain set of data and a certain method of tree reconstruction is called an **inferred tree**. An inferred tree may or may not be identical to the true tree.

Gene trees and species trees

Phylogeny is the representation of the branching history of the routes of inheritance of organisms. At every locus, if we trace back the history of any two alleles from any two populations, we will eventually reach a common ancestral allele from which both contemporary alleles have been derived. The routes of inheritance represent the passage of genes from parents to offspring, and the branching pattern depicts a **gene tree**. Different genes, however, may have different evolutionary histories, i.e., different routes of inheritance. We note, however, that the routes of inheritance are mostly confined by reproductive barriers—that is, gene flow occurs only within the species. A species is therefore like a bundle of genetic connections, in which many entangled parent-offspring lines form the ties that bundle individuals together into a species lineage.

A **species tree** represents the actual evolutionary relationships among species. In a species tree, a bifurcation represents the time of speciation, i.e., the time when the two species became distinct and reproductively isolated from one another. An unspoken assumption in molecular evolution is that a gene tree obtained by combining the information from many genes or entire genomes will be identical to the species tree.

A gene tree can differ from a species tree in two respects. First, the divergence of two alleles, each sampled from a different species, may have predated the divergence of the species from each other. In **Figure 5.10**, we see that the allele divergence event that gave rise to alleles *a* and *f* occurred at time D_1. By using *a* and *f* as representatives of species 1 and 2, we will overestimate the time of speciation. Moreover, an allele from one species may be more closely related to an allele from a different species than to any other conspecific allele. This possibility is exemplified by allele *d* in species 2, which is more closely related to alleles *b* and *c* from species 1 than to the conspecific alleles *e* and *f*. The reason for this "anomaly" may be the existence of genetic polymorphism in the ancestral species or **incomplete lineage sorting**.

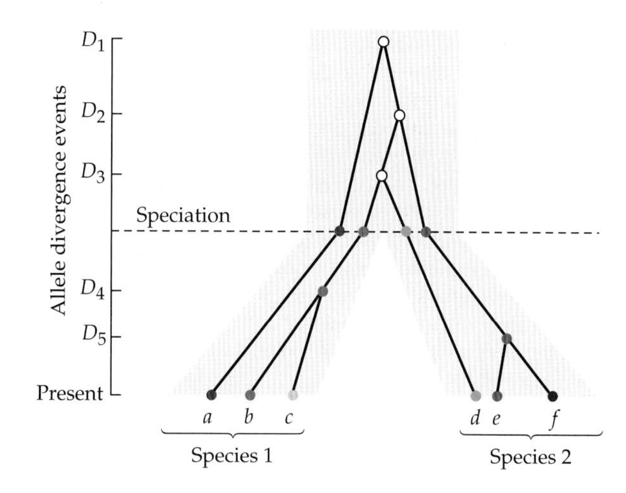

Figure 5.10 Diagram showing that in a genetically polymorphic population, allele divergence events (D_1–D_5) may occur before or after the speciation event. The evolutionary history of allele divergence resulting in the six alleles (*a*–*f*) found in species 1 and 2 is shown in solid lines. At the time of speciation (i.e., population splitting), the ancestral population consisted of four alleles (circles): *a*, the ancestor of *b* and *c*, *d*, and the ancestor of *e* and *f*. (Modified from Nei 1987.)

Incomplete lineage sorting occurs when an ancestral species undergoes several speciation events in a short period of time. If, for a given gene, the ancestral polymorphism is not fully resolved into two monophyletic lineages when the second speciation occurs, then a non-zero probability exists that the gene tree will be different from the species tree. An interesting example of incomplete lineage sorting was discovered in a comparison of the human, chimpanzee, and gorilla genomes. Although humans and

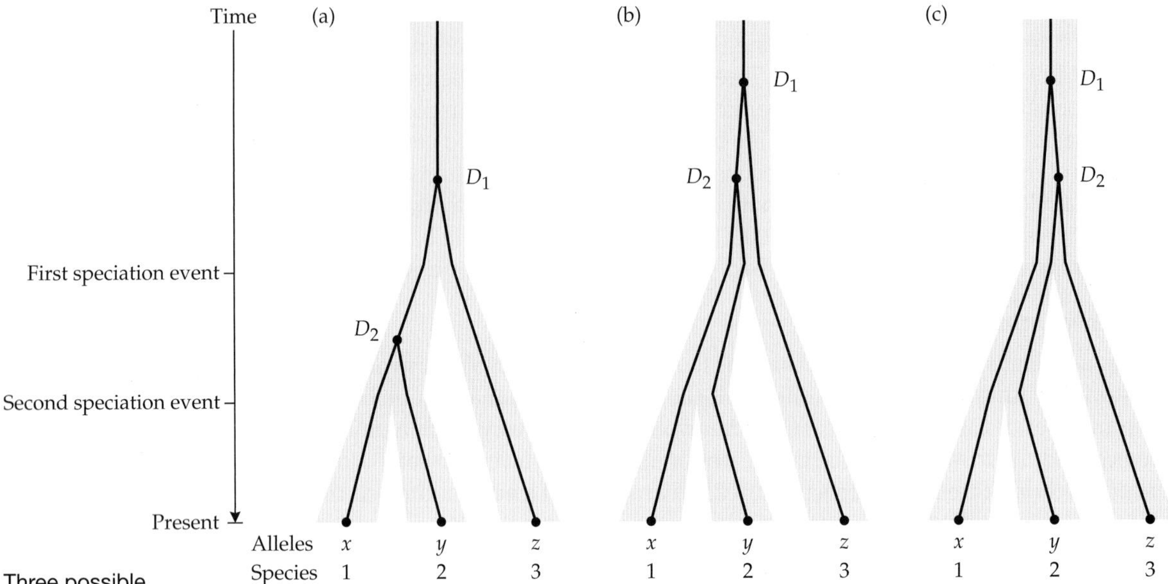

Figure 5.11 Three possible relationships between a species tree (yellow) and a gene tree (black lines). The species trees in (a), (b), and (c) are identical, as they all consist of two divergence events that occurred at the same time. The divergence times among the three alleles are denoted by D_1 and D_2 for the first and second divergence events, respectively. In (a) and (b), the topologies of the species trees are identical to those of the gene trees. Note that in (a) the times of divergence among the alleles are only slightly larger than the times of divergence among the species. In (b), on the other hand, the time of divergence between alleles *x* and *y* greatly predates the time of divergence (t_1) between their respective populations. The topology of the gene tree in (c) is different from that of the species tree because allele *y* from species 2 is more closely related to allele *z* from species 3 than either allele is to allele *x* from species 1, whereas species 1 is more closely related to species 2 than either is to species 3. (Modified from Nei 1987.)

chimpanzees are expected to have shared a more recent common ancestor with each other than either does with gorillas, Scally et al. (2012) found that in 30% of the gorilla genome, the DNA sequences are more similar to the homologous sequences from human or chimpanzee than the sequences of human and chimpanzee are to each other.

We note that ancestral genetic polymorphism will not represent a serious impediment to obtaining the correct phylogenetic tree if we are only concerned with long intervals between successive speciation events.

Figure 5.11 shows three different possible relationships between the gene tree and species trees. The topologies of gene trees (a) and (b) are identical with those of the corresponding species trees (e.g., OTUs *x* and *y* are closer phylogenetically to each other than either of them is to *z*, and species 1 and 2 are closer to each other than either of them is to 3). Gene tree (c), however, is different from the species tree, since alleles *y* and *z* are closer to each other than either is to allele *x*, while the populations from which the three alleles were derived exhibit different phylogenetic relationships. The probability of obtaining the erroneous species topology (c) is quite high when the interval between the first and second speciation event is short (Pamilo and Nei 1988), as is probably the case with the phylogenetic relationships among humans, chimpanzees, and gorillas. This probability cannot be substantially decreased by increasing the number of alleles sampled at one locus. To avoid this type of error, one needs to use many unlinked genes in the reconstruction of a phylogeny. A large amount of data is also required to avoid stochastic errors. Other reasons the gene tree and the species trees may be different from each other may be gene duplication (see Chapter 7) and reticulate evolution (Chapter 9). We should, therefore, exercise great caution in inferring species trees from gene trees.

Taxa and clades

A **taxon** is a species or a group of species (e.g., genus, family, order, class) that has been given a name, for example, *Homo sapiens* (the species name for modern humans), Lepidoptera (the order of insects comprising butterflies and moths), Ungulata (hoofed mammals), and Vermes (worms). Codes of biological nomenclature attempt to ensure that every taxon has a single and stable name, and that every name is used for only one taxon.

One of the main aims of phylogenetic studies is to establish the evolutionary relationships among different taxa. In particular, we are interested in the identification of **natural clades** (or **monophyletic groups**). Strictly speaking, a clade is defined as a

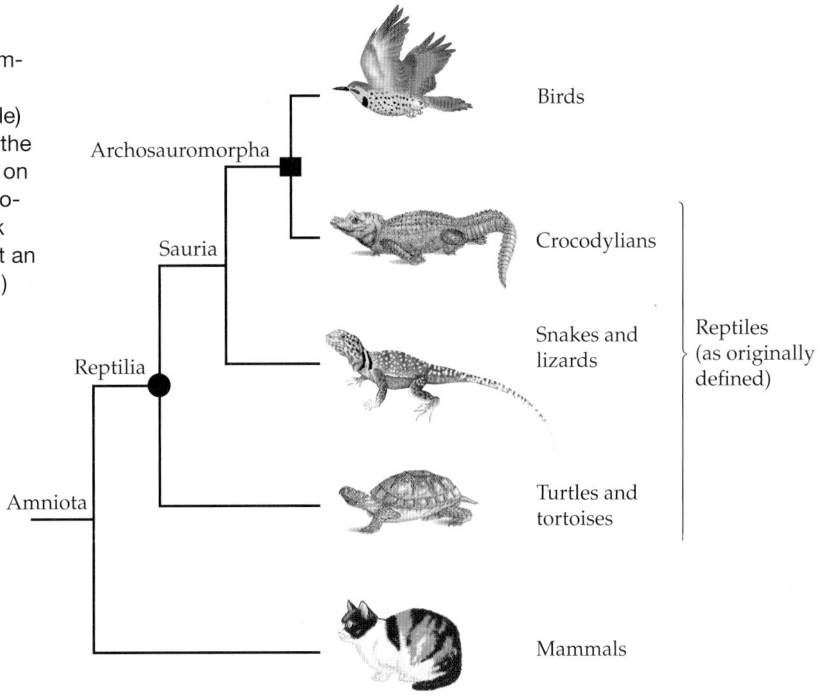

Figure 5.12 Cladogram of birds, reptiles, and mammals. The reptiles do not constitute a natural clade, since their most recent common ancestor (black circle) also gave rise to the birds, which are not included in the original definition of reptiles. Birds and crocodylians, on the other hand, constitute a natural clade (Archosauromorpha), since they share a common ancestor (black square) that is not shared by any organism that is not an archosauromorph. (Modified from Graur and Li 2000.)

Archosauromorpha
Sauria
Reptilia
Amniota
Birds
Crocodylians
Snakes and lizards
Turtles and tortoises
Mammals
Reptiles (as originally defined)

group of all the taxa that have been derived from a common ancestor, plus the common ancestor itself. In molecular phylogenetics, it is common to use the term "clade" for any group of taxa under study that share a common ancestor not shared by any species outside the group. If a clade is composed of two taxa, they are referred to as **sister taxa**. A taxonomic group whose common ancestor is shared by any other taxon is **paraphyletic**.

Figure 5.12 shows one possible evolutionary tree for the three amniote taxa: birds, reptiles, and mammals. Let us examine whether or not the taxon "reptiles" (consisting of crocodylians, snakes, lizards, turtles, and tortoises) fits the definition of clade. We note that the three groups of reptiles share a common ancestor with another group, the birds, which is not included within the definition of "reptiles." The reptiles are therefore paraphyletic. In this tree, birds and crocodylians constitute a natural clade, Archosauromorpha, because they share a common ancestor not shared by any extant organism other than birds and crocodylians. Similarly, all birds and all reptiles taken together constitute a natural clade, Reptilia.

Named taxonomic groups that do not constitute natural clades, such as Pisces (fishes), Protista (unicellular eukaryotes), Insectivora, Ungulata, and Vermes—groups that are not monophyletic—are called **convenience taxa.**

An implied assumption in all phylogenetic studies is that all life forms, both extant and extinct, share a common origin, and their ancestries can be traced back to one or a few organisms that lived approximately 4 billion years ago. Thus, all animals, plants, and bacteria are related by descent. Closely related organisms are descended from more recent common ancestors than are distantly related ones.

Types of Molecular Homology

Two or more nucleotide or amino acid sequences are said to be **homologous** if they are related by descent. Homology is often ascertained on the basis of sequence similarity. Thus, if two or more sequences exhibit high degrees of similarity, it is likely (but not always certain) that they are homologous. Sequence similarity may also arise without common ancestry: by chance or because of convergence driven by similar selective pressures. Such sequences, which are similar but not homologous, are said to be **analogous**. It is important to distinguish between homology, which is a qualitative statement, and similarity, which is a quantitative and hence quantifiable variable that can be expressed by such measures as percent similarity or percent identity. Sequence similarity is measurable; homology is a hypothesis, which at the molecular level may be supported by sequence similarity. Of course, as with any other scientific hypothesis, homology between two sequences may be tested and every so often rejected (Fitch 2000).

In **Figure 5.13** we illustrate various types of molecular homology. Homologous sequences are **orthologous** if they are separated by one or more speciation events (Fitch 1970). In other words, when an ancestral species gives rise to two descendant species,

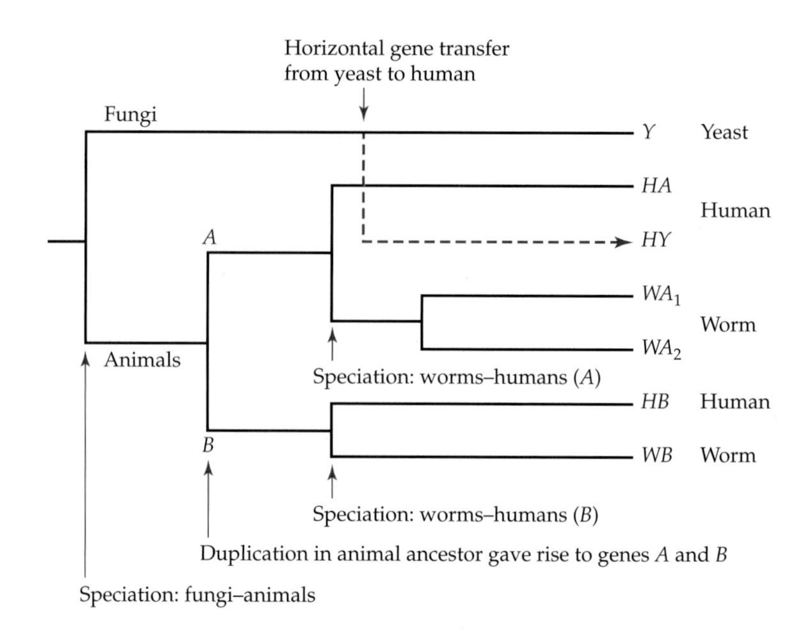

Figure 5.13 Types of homology. Consider a gene that existed in the ancestor of yeast and animals. The gene was duplicated early in the animal lineage, before the human-worm split, into genes *A* and *B*. After the human-worm split, the *A* form was duplicated in the worm. In addition, a copy of the yeast gene *Y* was horizontally transferred into the human lineage (dashed arrow). In this scenario, the yeast gene is orthologous to all worm and human genes, with the exception of *HY*. *HY* and *Y* are paralogs because they are derived from a type of gene duplication. *HA* and *HY* are xenologs because they are related via a horizontal gene transfer event. When the human and worm genes are compared, *HA* is orthologous with both WA_1 and WA_2 genes, which are inparalogs of each other. By contrast, genes *HB* and *HA* are outparalogs in the comparison of human with worm. However, *HB* and *HA* are inparalogs in reference to the human-yeast divergence event, because the animal-yeast split predated the duplication that gave rise to *HA* and *HB*. (Modified from Sonnhammer and Koonin 2002.)

the copies of a single gene in the resulting species are said to be orthologous. Orthologs are similar to one another because they originated by vertical descent from a single gene in the last common ancestor. In phylogenetic studies concerned with the evolutionary reconstruction of relationships among taxa, one must use only orthologous sequences.

Homologous sequences are **paralogous** if they were separated by a gene duplication event (Chapter 7), that is, if a gene in an organism is duplicated, then the two copies are paralogs (Fitch 1970). Paralogous sequences provide useful insight into the way genes evolve, and phylogenetic methodology may be used to reconstruct the sequence of evolutionary duplications giving rise to a group of paralogous genes. If two genes are related to each other by both a speciation event and a gene duplication event, they are still paralogous. In such cases a distinction has been suggested by Sonnhammer and Koonin (2002): paralogs derived from an ancestral duplication that do not form orthologous relationships relative to a given speciation event are **outparalogs**, and paralogs derived from a lineage-specific duplication, thus giving rise to co-orthologous relationships, are **inparalogs**.

Ohnologs are paralogous genes that have originated by the process of whole-genome duplication. The name was first given in honor of Susumu Ohno by Wolfe (2000). Ohnologs are interesting for evolutionary analysis because, by definition, all ohnologs in an autopolyploid organism (Chapter 7) have been diverging for the same length of time since their common origin.

García-Moreno and Mindell (2000) suggested the term "gametology" to denote the particular relationships between paralogous genes on nonrecombining, opposite-sex chromosomes. **Gametologs** are markers of genetic sex determination and originate with the erection of barriers to recombination between the sex chromosomes.

Homologs resulting from horizontal gene transfer between two organisms (Chapter 9) are termed **xenologs** (Gray and Fitch 1983).

Types of Data

Molecular data fall into one of two categories: characters and distances. A character provides information about an individual OTU. A distance represents a quantitative statement concerning the dissimilarity between two OTUs.

Character data

A **character** is a well-defined feature that in a taxonomic unit can assume one out of two or more mutually exclusive **character states**. In other words, a character is an independent variable, such as height or the ninety-eighth amino acid position in cytochrome c, and the character state is the value of the character in a particular OTU, such as 1.68 cm or alanine.

Characters are either quantitative or qualitative. The character states of a **quantitative character** (e.g., height) are usually **continuous** and are measured on an interval scale. The states of a **qualitative character** (e.g., the amino acid at a certain positions in a protein) are **discrete**. Discrete characters can be assigned two or more values. A character that can have only one of two character states is referred to as **binary**. When three or more states are possible, the character is referred to as **multistate**.

Genomic data provide many binary characters that are useful in phylogenetic studies, usually taking the form of the presence or absence of a molecular marker. For example, the presence or absence of a retrotransposon at a certain genomic location can be used as a phylogenetic character.

In DNA and protein sequences, the qualitative multistate characters are the positions in the aligned sequences, and the character states are the particular nucleotide or amino acid residues at these positions in each of the taxa under study. For example, if nucleotide A is observed at position 139 of a mitochondrial sequence from pig, then the OTU is pig mitochondrial DNA, the character position is 139, and the character state assigned to the character in this OTU is A.

Assumptions about character evolution

Methods of phylogenetic reconstruction require that we make explicit assumptions about (1) the number of discrete steps required for one character state to change into another, and (2) the probability with which such a change may occur. A character is **unordered** if a change from one character state to another occurs in one step (Fitch 1971). Nucleotides in DNA sequences are usually assumed to be unordered, i.e., a change from any nucleotide to any other nucleotide is possible in one step.

Often a character state cannot change directly into another character state, but must pass through one or more intermediate states. A character is designated as **ordered** if the number of steps from one state to another equals the absolute value of the difference between their state numbers (Farris 1970; Swofford and Maddison 1987). Thus, a change from state 1 to state 5 is assumed to occur in four steps, through the intermediate steps 2, 3, and 4. The process is assumed to be symmetrical, so a change from state 5 to state 1 is also assumed to require four steps. Perfectly ordered characters are rarely encountered in molecular data unless one is willing to make very strong assumptions concerning the mechanism by which a certain character state may change into another. For example, if the number of copies of a certain repetitive sequence in a genome is assumed to increase or decrease in a stepwise fashion, then we may treat the number of repeats as an ordered character such that, for instance, a change from two copies to four copies is assumed to require two steps.

Partially ordered characters are those characters in which the number of steps varies for the different pairwise combinations of character states but for which no definite relationship exists between the number of steps and the character-state number (Swofford and Olsen 1990). Amino acid sequences are the most commonly encountered examples of partially ordered characters in molecular evolution. An amino acid cannot change into all other amino acids in a single step; sometimes two or three steps are required. For example, a tyrosine may only change into a leucine through an intermediate state, i.e., phenylalanine or histidine (see Chapter 1). The number of steps is written down in a matrix format called a **step matrix**, the elements of which indicate the number of steps required between any two character states. Two step matrices are shown in **Figure 5.14**.

Most discrete characters encountered in molecular evolution are **reversible**, i.e., they are assumed to change back and forth with equal probability. However, we sometimes come across characters in which it is reasonable to impose constraints on reversibility. The most common ones are binary characters in which one character state can change into the other quite easily, but the reverse occurs only rarely. For

Figure 5.14 Step matrices. The elements in each matrix represent the number of steps (minimal number of nucleotide substitutions) required for a change between a character state in the column to a character state in the row. (a) A step matrix for nucleotide character states. It is assumed that this case can be suitably represented as a four-state unordered character. (b) A step matrix for amino acids encoded by the standard genetic code. An amino acid position in a protein can be represented as a twenty-state partially ordered character. The diagonals represent the case of no change in character state. (From Graur and Li 2000.)

(a)

	A	C	T	G
A	0	1	1	1
C	1	0	1	1
T	1	1	0	1
G	1	1	1	0

(b)

	A	C	D	E	F	G	H	I	K	L	M	N	P	Q	R	S	T	V	W	Y
A	0	2	1	1	2	1	2	2	2	2	2	2	1	2	2	1	1	1	2	2
C	2	0	2	3	1	1	2	2	3	2	3	2	2	3	1	1	2	2	1	1
D	1	2	0	1	2	1	1	2	2	2	3	1	2	2	2	2	2	1	3	1
E	1	3	1	0	3	1	2	2	1	2	2	2	2	1	2	2	2	1	2	2
F	2	1	2	3	0	2	2	1	3	1	2	2	2	3	2	1	2	1	2	1
G	1	1	1	1	2	0	2	2	2	2	2	2	2	2	1	1	2	1	1	2
H	2	2	1	2	2	2	0	2	2	1	3	1	1	1	1	2	2	2	3	1
I	2	2	2	2	1	2	2	0	1	1	1	1	2	2	1	1	1	1	3	2
K	2	3	2	1	3	2	2	1	0	2	1	1	2	1	1	2	1	2	2	2
L	2	2	2	2	1	2	1	1	2	0	1	2	1	1	1	1	2	1	1	2
M	2	3	3	2	2	2	3	1	1	1	0	2	2	2	1	2	1	1	2	3
N	2	2	1	2	2	2	1	1	1	2	2	0	2	2	2	1	1	2	3	1
P	1	2	2	2	2	2	1	2	2	1	2	2	0	1	1	1	1	2	2	2
Q	2	3	2	1	3	2	1	2	1	1	2	2	1	0	1	2	2	2	2	2
R	2	1	2	2	2	1	1	1	1	1	1	2	1	1	0	1	1	2	1	2
S	1	1	2	2	1	1	2	1	2	1	2	1	1	2	1	0	1	2	1	1
T	1	2	2	2	2	2	2	1	1	2	1	1	1	2	1	1	0	2	2	2
V	1	2	1	1	1	1	2	1	2	1	1	2	2	2	2	2	2	0	2	2
W	2	1	3	2	2	1	3	3	2	1	2	3	2	2	1	1	2	2	0	2
Y	2	1	1	2	1	2	1	2	2	2	3	1	2	2	2	1	2	2	2	0

irreversible characters (Camin and Sokal 1965), it is assumed that changes in character state may occur in only one direction. The presence or absence of a retrosequence (Chapter 9) at a certain location in the genome is one example of an (almost) irreversible character. The reason is that retrosequences frequently insert themselves into the genome but are almost never excised precisely.

In addition to the number of steps between two characters, we may also consider the different probabilities with which different one-step changes occur. For example, we may assign different probabilities of occurrence to transitions and transversions.

Polarity and taxonomic distribution of character states

In terms of temporal appearance during evolution, the character states within a character of interest may be ranked by antiquity (**Figure 5.15**). A primitive or ancestral character state is called a **plesiomorphy** (literally, close to the original form), while the derived state representing an evolutionary novelty relative to the ancestral state is called an **apomorphy** (i.e., away from the original form). An ancestral state that is shared by several taxa is a **symplesiomorphy**. A derived state that is shared by several taxa is a **synapomorphy**. A derived character state unique to a particular taxon is called an **autapomorphy**. A character state that has arisen in several taxa independently (through convergence, parallelism, and reversals) rather than being inherited from a common ancestor is called a **homoplasy**. Some methods of phylogenetic reconstruction rely solely on synapomorphies for the identification of monophyletic clades. Polarized character states, i.e., traits for which we possess knowledge on relative age, are extremely useful in rooting unrooted phylogenetic trees. The rooting can be done

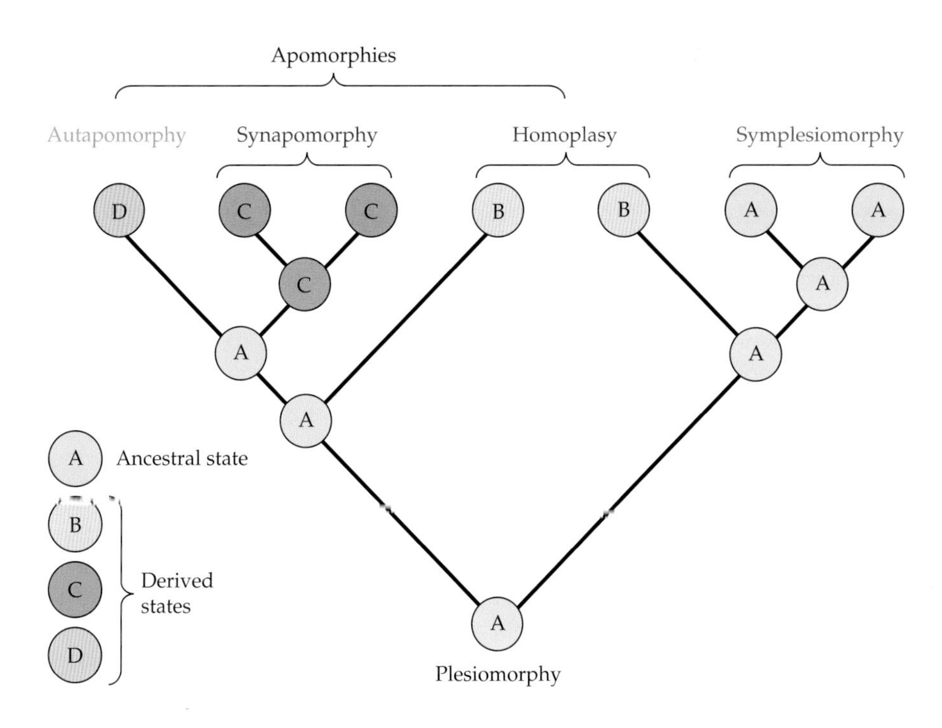

Figure 5.15 Polarity of character states. The character state in the common ancestor of all seven OTUs is A (a plesiomorphy). This character state has been retained in two OTUs, which, therefore, share a symplesiomorphy. Three derived character states (apomorphies) appeared during the evolution of the seven OTUs. The derived character state D appears in a single OTU and is, therefore, an autapomorphy. The derived character C appeared once and is currently shared by two OTUs. Therefore, C is a synapomorphy. The derived character B was acquired independently by two OTUs. It is, therefore, a homoplasy.

in reference to a hypothetical ancestor that possesses plesiomorphies at all the characters used in the analysis. In some cases, such as the presence or absence of a transposable element at a locus, it is easy to determine which character state is ancestral and which is derived. Unfortunately, it is not always possible to polarize every character state in a molecular phylogenetic analysis.

Distance data

Unlike character data, in which values are assigned to individual taxonomic units, **distances** involve pairs of taxa. Some experimental procedures, such as DNA–DNA hybridization, directly yield pairwise distances. Distance data cannot be converted into character data. In such cases, distance methods provide the only means of reconstructing phylogenetic trees. Much of the primary data produced by genomic molecular studies consists of character data. These characters, however, can be transformed into distances, for example, the number of substitutions per site between two nucleotide sequences (Chapter 3).

Swofford and Olsen (1990) outlined three possible reasons for converting characters into distances. First, a long list of character states, such as a DNA sequence, is in itself meaningless in an evolutionary context. On the other hand, if we can say that the similarity between two sequences is 93%, whereas the similarity between one of these sequences and a third one is only 50%, we evoke an intuitive (and often correct) image of a specific evolutionary relationship. Second, as pointed out in Chapter 3, one must take into account multiple substitutions at a site. By making reasonable assumptions about the nature of the evolutionary process, we are able to estimate the number of "unseen" events. Such corrections apply to distances, such as the number of substitutions between two sequences, but not to the sequences themselves. Third, numerous methods exist for inferring phylogenetic trees from distance data. Most of these methods are very fast and efficient and can be used even when the number of OTUs is so large as to preclude the use of many methods that are based on characters (p. 191).

In evolutionary studies, distances (denoted by d_{ij}, where i and j are the two OTUs) are **metric** if they (1) are non-negative ($d_{ij} \geq 0$), (2) are symmetrical ($d_{ij} = d_{ji}$), (3) are distinct ($d_{ij} = 0$ if and only if $i = j$), and (4) abide by the triangle inequality, that is, the distance between OTUs i and j is smaller than the sum of the distances between these two OTUs and a third one, k ($d_{ij} < d_{ik} + d_{jk}$).

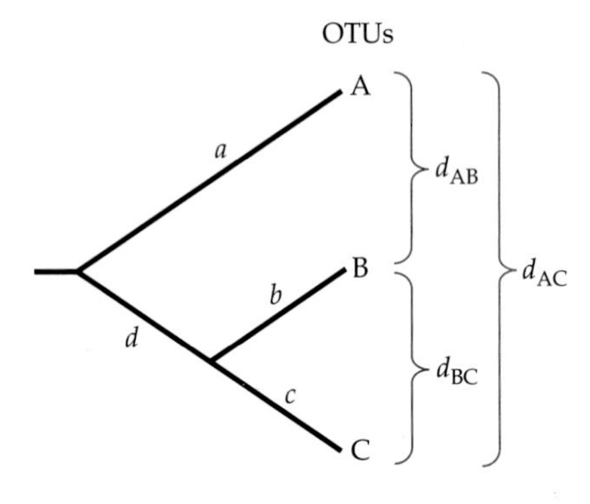

Figure 5.16 Rooted tree with three OTUs illustrating ultrametricity and additivity. Capital letters denote the OTUs. Lowercase italic letters denote the number of changes that have occurred on each branch. Distances are denoted by d_{AB}, d_{AC}, and d_{BC}, where the subscripts denote the OTUs. The tree is ultrametric if $d_{AB} < d_{AC} = d_{BC}$, that is, if $b = c$, $a = d + b$, and $a = d + c$. The tree is additive if $d_{AB} = a + d + b$, $d_{AC} = a + d + c$, and $d_{BC} = b + c$. If the criteria for additivity are not met, the tree is nonadditive.

Distances can be additive, ultrametric, or neither (**Figure 5.16**). Distances are **ultrametric** if all the OTUs are equidistant from the root. This requires all the OTUs under study to evolve at the same rate (Chapter 4). Ultrametric distances are said to abide by the "three-point condition," according to which the two largest distances among three OTUs are equal. Distances are **additive** if the distance between any two OTUs is equal to the sum of the lengths of all the branches connecting them. Ultrametric trees are always additive, but additive trees may not be ultrametric. The distance between two OTUs is calculated directly from molecular data (e.g., DNA sequences), while branch lengths are estimated from the distances between the OTUs according to certain rules (p. 204). Additivity usually does not hold strictly if multiple substitutions have occurred at any of the nucleotide sites (Chapter 3).

Methods of Tree Reconstruction

Inferring a phylogeny is an estimation procedure, in which a "best estimate" of the evolutionary history is made on the basis of incomplete information. Because many different phylogenetic trees can be produced from any given set of OTUs (Table 5.1), we must specify criteria for selecting one or a few trees as representing our best estimate of the true evolutionary history. Most phylogenetic inference methods seek to accomplish this goal by defining a criterion for comparing alternative phylogenies and deciding which tree is better. A phylogenetic reconstruction, therefore, consists of two steps: (1) definition of an **optimality criterion** (or **objective function**), i.e., a value that is assigned to a tree and is subsequently used for comparing one tree to another; and (2) design of specific **algorithms** to compute the value of the objective function and to identify the tree (or set of trees) that has the best values according to this criterion.

Several methods of tree reconstruction employ a specific sequence of steps (an **algorithm**) for constructing the best tree. This class of methods combines tree inference and the definition of the optimality criterion for selecting the preferred tree into a single statement.

Numerous tree-making methods have been proposed in the literature. For a detailed treatment, readers may consult Sneath and Sokal (1973), Nei (1987), Felsenstein (2004), and Nei and Kumar (2000). Here we describe several methods that are frequently used in molecular phylogenetic studies. For simplicity, we consider nucleotide sequence data, but the methods are equally applicable to other types of molecular data, such as amino acid sequences. We divide the methods into **distance matrix**, **character state**, and **maximum likelihood** approaches. Methods belonging to the first approach are based on distance measures, such as the number of nucleotide substitutions or amino acid replacements. They include UPGMA, the neighbors-relation method, and the neighbor-joining method. Methods belonging to the second approach, including the maximum parsimony approach, rely on character states such as the nucleotide or amino acid at a particular site, or the presence or absence of a deletion or an insertion at a certain locus. Methods belonging to the third approach, such as the maximum likelihood method, rely on both character states and distances.

Distance Matrix Methods

In distance matrix methods, evolutionary distances (usually the number of nucleotide substitutions or amino acid replacements between two taxonomic units) are computed for all $n(n-1)/2$ pairs of taxa (where n is the number of OTUs), and a phylogenetic tree is constructed by using an algorithm based on some functional relationships among the distance values.

Unweighted pair-group method with arithmetic means (UPGMA)

UPGMA is the simplest method for tree reconstruction. It was originally developed for constructing taxonomic phenograms, i.e., trees that reflect the phenotypic similarities among OTUs (Sokal and Michener 1958), but it can also be used to construct phylogenetic trees if the rates of evolution are constant. Because of its lack of robustness, UPGMA is seldom used these days, and the reason we present it here is that it is pedagogically useful in illustrating the principle of clustering that is also used by more robust and sophisticated distance methods.

UPGMA employs a sequential clustering algorithm, in which local topological relationships are identified in order of decreasing similarity, and the phylogenetic tree is built in a stepwise manner. In other words, we first identify from among all the OTUs (also referred to as **simple OTUs**) the two that are most similar to each other and treat these as a new single OTU. Such an OTU is referred to as a **composite OTU**. For the new group of OTUs we compute a new distance matrix and identify the pair with the highest similarity. This procedure is repeated until we are left with only two OTUs.

To illustrate the method, let us consider a case of four OTUs A, B, C, and D. The pairwise evolutionary distances are given by the following matrix:

OTU	A	B	C
B	d_{AB}		
C	d_{AC}	d_{BC}	
D	d_{AD}	d_{BD}	d_{CD}

In this matrix, dij stands for the distance between OTUs i and j. The first two OTUs to be clustered are the ones with the smallest distance. Let us assume that d_{AB} is the smallest. Then, OTUs A and B are the first to be clustered, and the branching point, l_{AB}, is positioned at a distance of $d_{AB}/2$ substitutions (**Figure 5.17a**).

Following the first clustering, A and B are considered as the single, composite OTU (AB), and a new distance matrix is computed:

OTU	(AB)	C
C	$d_{(AB)C}$	
D	$d_{(AB)D}$	d_{CD}

In this matrix, $d_{(AB)C} = (d_{AC} + d_{BC})/2$, and $d_{(AB)D} = (d_{AD} + d_{BD})/2$. In other words, the distance between a simple and a composite OTU is the average of the distances between the simple OTU and the constituent simple OTUs of the composite OTU. If $d_{(AB)C}$ turns out to be the smallest distance in the new matrix, then OTU C will be joined to the composite OTU (AB) with a branching node at $l_{(AB)C} = d_{(AB)C}/2$ (**Figure 5.17b**). The final step clusters the last OTU, D, with the new composite OTU, (ABC). The root of the entire tree is positioned at $l_{(ABC)D} = d_{(ABC)D}/2 = [(d_{AD} + d_{BD} + d_{CD})/3]/2$. The final tree inferred by using UPGMA is shown in **Figure 5.17c**.

In UPGMA, the branching point between two simple OTUs, i and j, is positioned at half the distance between them:

$$l_{ij} = \frac{d_{ij}}{2} \tag{5.4}$$

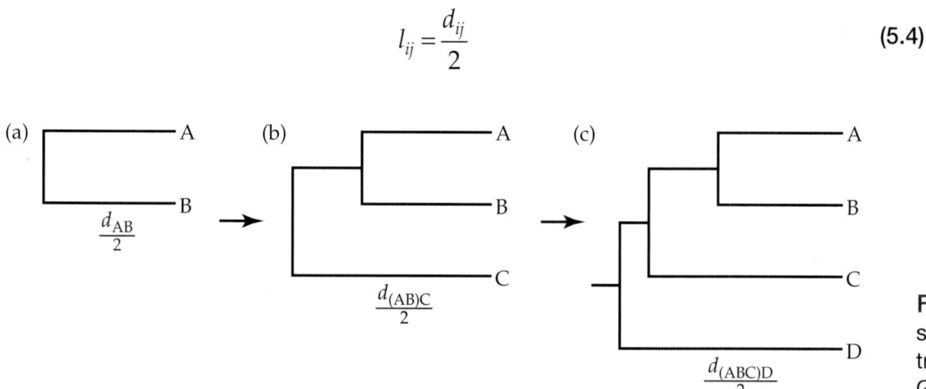

Figure 5.17 Diagram illustrating the stepwise construction of a phylogenetic tree for four OTUs using UPGMA. (From Graur and Li 2000.)

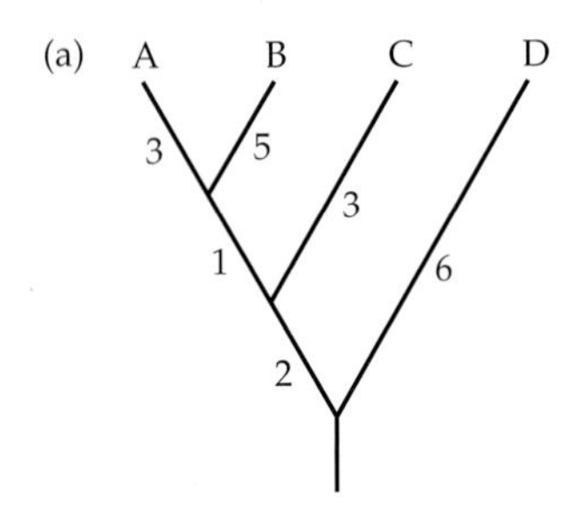

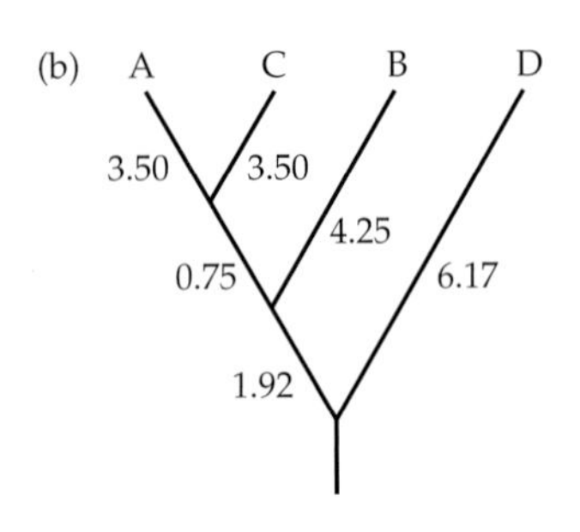

Figure 5.18 Consequences of using UPGMA when the assumption of rate constancy among lineages does not hold. The true phylogenetic tree (a) and the erroneous phylogenetic tree that was inferred by using UPGMA (b). (Modified from Graur and Li 2000.)

The branching point between a simple OTU, i, and a composite OTU, (jm), is positioned at half the arithmetic mean of the distances between the simple OTU and the constituent simple OTUs of the composite OTU:

$$l_{(i)(jm)} = \frac{(d_{ij} + d_{im})/2}{2} \tag{5.5}$$

The branching point between two composite OTUs is positioned at half the arithmetic mean of the distances between the constituent simple OTUs in each composite OTU. For example, the position of the branching point between a composite OTU, (ij), and a composite OTU, (mn), is

$$l_{(ij)(mn)} = \frac{(d_{im} + d_{in} + d_{jm} + d_{jn})/4}{2} \tag{5.6}$$

In the case of a tripartite composite OTU, (ijk), and a bipartite composite OTU, (mn), the position of the branching point is

$$l_{(ijk)(mn)} = \frac{(d_{im} + d_{in} + d_{jm} + d_{jn} + d_{km} + d_{kn})/6}{2} \tag{5.7}$$

UPGMA is one of a small number of methods of phylogenetic reconstruction that yields a rooted tree. Note also that by using UPGMA one obtains the topology of the tree and the branch lengths simultaneously.

We note, however, that if the assumption of rate constancy among lineages does not hold, UPGMA may yield an erroneous topology. For example, suppose that the phylogenetic tree in **Figure 5.18a** is the true tree. By using UPGMA, we obtain an inferred tree (**Figure 5.18b**) that differs in topology from the true tree. For example, OTUs A and C are grouped together in the inferred tree, whereas in the true tree, A and B are sister OTUs. (Note that additivity does not hold in the inferred tree. For example, the true distance between A and B is 8, whereas the sum of the lengths of the estimated branches connecting A and B is 3.50 + 0.75 + 4.25 = 8.50.)

Sattath and Tversky's neighbors-relation method

In an unrooted bifurcating tree, two OTUs are said to be **neighbors** if they are connected through a single internal node. For example, in **Figure 5.19a**, A and B are neighbors, as are C and D. In contrast, A and C are not neighbors, nor are A and D, B and C, or B and D, because they are connected by two nodes. In **Figure 5.19b**, neither A and C nor B and C are neighbors; however, if we combine OTUs A and B, then the composite OTU (AB) and the simple OTU C become a new pair of neighbors. (Note that neighbor taxa may or may not be sister taxa, depending on the position of the root.)

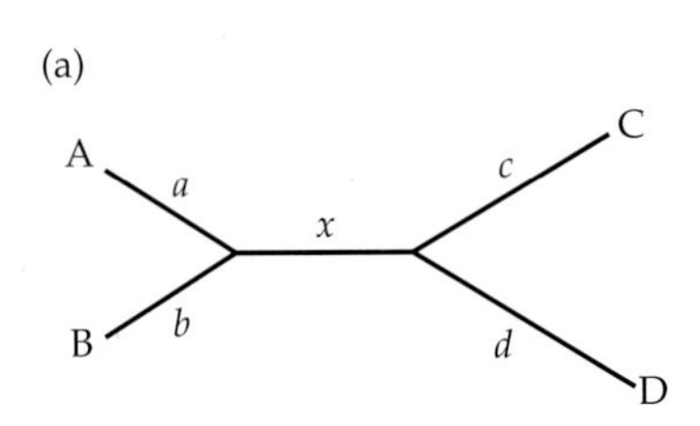

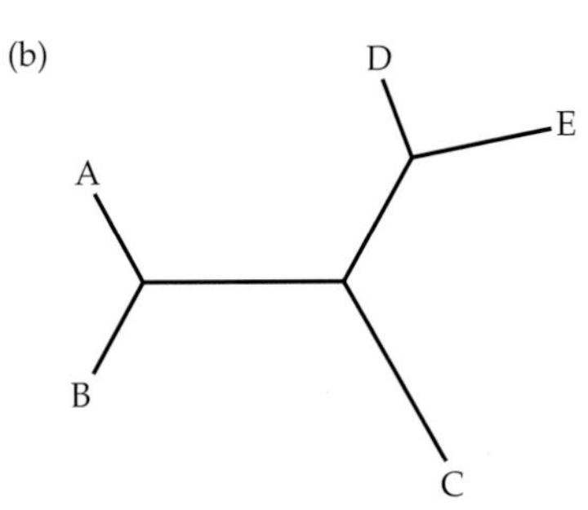

Figure 5.19 Bifurcating unrooted trees with (a) four and (b) five OTUs. The branch lengths in (a) are shown. (From Graur and Li 2000.)

Let us now assume that the tree in Figure 5.19a is the true tree. Then, if additivity holds, we should have

$$d_{AC} + d_{BD} = d_{AD} + d_{BC} = a + b + c + d + 2x = d_{AB} + d_{CD} + 2x \tag{5.8}$$

where a, b, c, and d are the lengths of the terminal branches and x is the length of the central branch. Therefore, the following two conditions hold:

$$d_{AB} + d_{CD} < d_{AC} + d_{BD} \tag{5.9}$$

and

$$d_{AB} + d_{CD} < d_{AD} + d_{BC} \tag{5.10}$$

These two conditions are collectively known as the **four-point condition** (Buneman 1971; Fitch 1981). They may hold even if additivity holds only approximately.

For four OTUs with unknown phylogenetic relationships, the four-point condition provides a simple way for inferring the topology of the tree, because the condition can be used to identify the neighbors (A and B, or C and D), and once a pair of neighbors is identified, the topology of the phylogenetic tree is determined unambiguously. This approach to phylogenetic reconstruction is called the **neighborliness approach**. Note, however, that the four-point condition does not always lead to the true topology, because in practice additivity may not hold. Sattath and Tversky (1977) proposed the following method for dealing with more than four OTUs. First, compute the distance matrix as in UPGMA. For every possible combination of four OTUs, say OTUs *i, j, m,* and *n,* compute the following three values: $d_{ij} + d_{mn}$, $d_{im} + d_{jn}$, and $d_{in} + d_{jm}$. Suppose that the first sum is the smallest; then, assign a score of 1 to both the pair *i* and *j* and the pair *m* and *n*. Pairs *i* and *m*, *j* and *n*, *i* and *n*, and *j* and *m* are each assigned a score of 0. If, on the other hand, $d_{im} + d_{jn}$ has the smallest value, then we assign the pair *i* and *m* and the pair *j* and *n* each a score of 1, and assign the other four possible pairs a score of 0. When all the possible combinations of quadruples are considered, the pair with the highest total score is selected as the first pair of neighbors, and subsequently it is treated as a new single composite OTU. Next, we compute a new distance matrix as in the case of UPGMA and repeat the process to select the second pair of neighbors. The process is repeated until we are left with three OTUs, at which time the topology of the tree is unambiguously inferred.

Sattath and Tversky's method does not provide branch lengths. However, the inferred topology can be used to compute branch lengths separately (p. 204). Other methods based on the neighborliness approach have been proposed by Fitch (1981) and Saitou and Nei (1987). The latter is discussed next.

Saitou and Nei's neighbor-joining method

The **neighbor-joining method** (Saitou and Nei 1987) is also a neighborliness method. It provides an approximate algorithm for finding the shortest (**minimum evolution**) tree. This is accomplished by sequentially finding neighbors that minimize the total length of the tree. The method starts with a starlike tree with *N* OTUs, as shown in **Figure 5.20a**. The starlike tree has no hierarchical structure, i.e., all OTUs are equally distant from all other OTUs.

The length of the starlike tree, i.e., the sum of all the branch lengths, is

$$S_0 = \frac{1}{N-1}\sum_{i<j} d_{ij} \quad \textbf{(5.11)}$$

where *dij* is the distance between OTUs *i* and *j*.

In the next step we consider a tree that is of the form given in **Figure 5.20b**. In this tree, there is only one internal branch connecting nodes X and Y, where X is the node connecting OTUs 1 and 2, and node Y connects all the other OTUs (3, 4, …, *N*). For this tree, the sum of all the branch lengths is

$$S_{12} = \frac{1}{2(N-2)}\sum_{k=3}^{N}(d_{1k}+d_{2k}) + \frac{1}{2}d_{12} + \frac{1}{N-2}\sum_{3\le i<j\le N}^{N} d_{ij} \quad \textbf{(5.12)}$$

Any pair of OTUs can take positions 1 and 2 in the tree, and there are $N(N-1)/2$ ways of choosing the pairs. Among these possible pairs of OTUs (Figure 5.20b), the one that gives the smallest sum of branch lengths is chosen as the first pair of neighbors. This pair of OTUs is then regarded as a single composite OTU, and the arithmetic mean distances between OTUs are computed to form a new distance matrix as in UPGMA. The next pair of OTUs that gives the smallest sum of branch length is then chosen (**Figure 5.20c**). This procedure is continued until all $N-3$ internal branches are found. Saitou and Nei (1987) have shown that in the case of four OTUs, the necessary condition for this method to obtain the correct tree topology is also given by the four-point condition (Equations 5.9 and 5.10).

Maximum Parsimony Methods

The principle of **maximum parsimony** involves the identification of a topology that requires the smallest number of evolutionary changes (e.g., nucleotide substitutions) to explain the observed differences among the OTUs under study. It is often said that the principle of maximum parsimony abides by Ockham's razor, according to

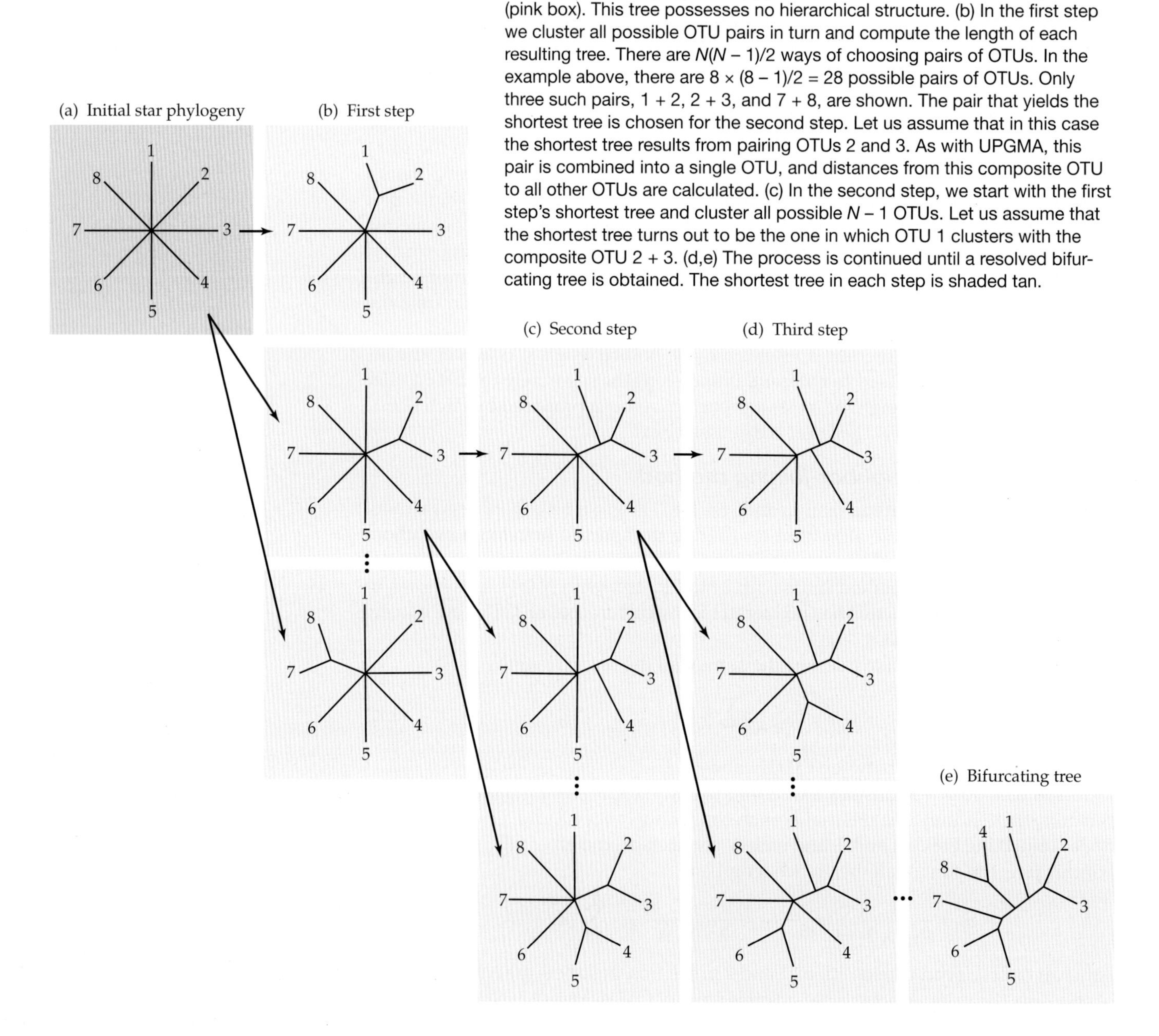

Figure 5.20 Schematic representation of phylogeny reconstruction by the neighbor-joining algorithm. (a) We start with a starlike tree for eight OTUs (pink box). This tree possesses no hierarchical structure. (b) In the first step we cluster all possible OTU pairs in turn and compute the length of each resulting tree. There are $N(N - 1)/2$ ways of choosing pairs of OTUs. In the example above, there are $8 \times (8 - 1)/2 = 28$ possible pairs of OTUs. Only three such pairs, 1 + 2, 2 + 3, and 7 + 8, are shown. The pair that yields the shortest tree is chosen for the second step. Let us assume that in this case the shortest tree results from pairing OTUs 2 and 3. As with UPGMA, this pair is combined into a single OTU, and distances from this composite OTU to all other OTUs are calculated. (c) In the second step, we start with the first step's shortest tree and cluster all possible $N - 1$ OTUs. Let us assume that the shortest tree turns out to be the one in which OTU 1 clusters with the composite OTU 2 + 3. (d,e) The process is continued until a resolved bifurcating tree is obtained. The shortest tree in each step is shaded tan.

which the best hypothesis is the one requiring the smallest number of assumptions. In maximum parsimony methods, we use discrete character states, and the shortest set of pathways leading to these character states is chosen as the best tree. Such a tree is called a **maximum parsimony tree** (sometimes referred to as the "global maximum parsimony tree," for reasons that will become clear shortly). Often two or more trees with the same minimum number of changes are found, so no unique tree can be inferred. Such trees are said to be **equally parsimonious**.

Several different parsimony methods have been developed for treating different types of data (see Felsenstein 1982). The method discussed below was first developed for amino acid sequence data (Eck and Dayhoff 1966) and was later modified for use on nucleotide sequences (Fitch 1977).

We start with the classification of sites. A site is defined as **invariant** if all the OTUs under study possess the same character state at this site. **Variable sites** may be

informative or **uninformative**. A nucleotide site is phylogenetically informative only if it favors a subset of trees over the other possible trees. To illustrate the distinction between informative and uninformative sites, consider the following four hypothetical sequences:

	Site								
OTU	1	2	3	4	5	6	7	8	9
1	A	A	G	A	G	T	T	C	A
2	A	G	C	C	G	T	T	C	T
3	A	G	A	T	A	T	C	C	A
4	A	G	A	G	A	T	C	C	T
Invariant	*					*		*	
Variant		*	*	*	*		*		*
Informative					*		*		*
Uninformative	*	*	*	*		*		*	

There are three possible unrooted trees for each of these four OTUs. Site 1 is uninformative because all sequences at this site have A, so no change is required in any of the three possible trees. At site 2, sequence 1 has A while all other sequences have G, so a simple assumption is that the nucleotide has changed from G to A in the lineage leading to sequence 1. Thus, this site is also uninformative, because each of the three possible trees requires one change. For site 3, each of the three possible trees requires two changes (**Figure 5.21a**), so this site is also uninformative. Note that if we assume for tree I in Figure 5.21a that the nucleotide at the node connecting OTUs 1 and 2 is C instead of G, the number of changes required for the tree remains two. **Figure 5.21b** shows that for site 4, each of the three trees requires three changes; thus site 4 is also uninformative. For site 5, tree I requires only one change, whereas trees II and III require two changes each (**Figure 5.21c**); site 5 is, therefore, informative. The same is true for site 7. Site 9 is also informative, but in contrast to the previous two informative

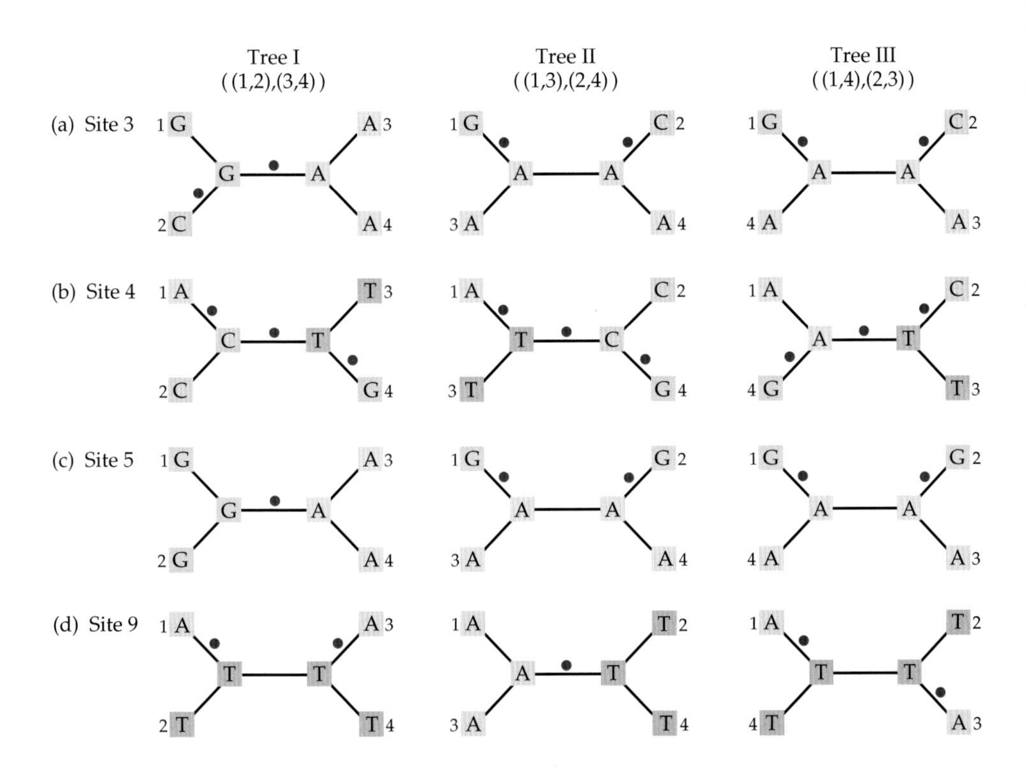

Figure 5.21 Three possible unrooted trees (I, II, and III) for four OTUs (DNA sequences 1, 2, 3, and 4, shown in the table) have been used to choose the most parsimonious tree. Terminal nodes are marked by the sequence number and the nucleotide type at homologous positions in the extant species. Each dot means a substitution is inferred on that branch. Note that the nucleotides at the two internal nodes of each tree represent one possible reconstruction from among several alternatives. For example, the nucleotides at both the internal nodes of tree IIId (bottom right) could be A instead of T, in which case the two substitutions would be positioned on the branches leading to species 2 and 4. Other combinations of nucleotides can be placed at the internal nodes; however, these combinations would require three or more substitutions. Sites 3 and 4 are uninformative, since the number of required substitutions is the same for all three tree topologies. Sites 5 and 9 are informative and favor trees I and II, respectively. (Modified from Graur and Li 2000.)

sites, this site favors tree II, which requires only one change (**Figure 5.21d**), whereas trees I and III require two changes each.

From these examples, we see that a site is informative only when there are at least two different kinds of nucleotides at the site, each of which is represented in at least two of the sequences under study.

To infer a maximum parsimony tree, we first identify all the informative sites. Next, for each possible tree we calculate the minimum number of substitutions at each informative site. In the above example, there are three informative sites. For sites 5, 7, and 9, tree I requires 1, 1, and 2 changes, respectively; tree II requires 2, 2, and 1 changes; and tree III requires 2, 2, and 2 changes. In the final step, we sum the number of changes over all the informative sites for each possible tree and choose the tree associated with the smallest number of changes. In our case, tree I is chosen because it requires only 4 changes at the informative sites, whereas trees II and III require 5 and 6 changes, respectively.

In the case of four OTUs, an informative site can favor only one of the three possible alternative trees. For example, site 5 favors tree I over trees II and III and is thus said to support tree I. It is easy to see that the tree supported by the largest number of informative sites is the most parsimonious tree. For instance, in the above example, tree I is the maximum parsimony tree because it is supported by two sites, whereas tree II is supported by only one site, and tree III by none. In cases where more than four OTUs are involved, an informative site may favor more than one tree, and the maximum parsimony tree may not necessarily be the one supported by the largest number of informative sites.

It is important to note that the informative sites that support the internal branches in the inferred tree are deemed to be synapomorphies, whereas all the other informative sites are deemed to be homoplasies.

When the number of OTUs under study is larger than four, the situation becomes more complicated because there are many more possible trees to consider (Table 5.1), and because inferring the number of substitutions for each alternative tree becomes more tedious. However, the basic goal remains simple: to identify the tree (or set of trees) requiring the minimum number of substitutions.

The inference of the number of substitutions for a given tree can be made by using Fitch's (1971) method. Let us consider a case of six OTUs (1–6) and assume that at a particular nucleotide site the nucleotides in the six sequences are C, T, G, T, A, and A (**Figure 5.22a**). We want to infer the nucleotides at the internal nodes 7, 8, 9, 10, and 11 from the nucleotides at the tips of the tree. The nucleotide at node 7 cannot be determined uniquely, but must be either C or T under the parsimony rule. (If we put A or G at node 7, two nucleotide substitutions rather than one must be assumed to have occurred at this site.) Therefore, the set of candidate nucleotides at node 7 consists of C and T. Similarly, the set of candidate nucleotides at node 8 consists of G and T, and the set at node 9 consists of A, G, and T. However, at node 10, T is chosen because it is shared by the sets at the two descendant nodes, 7 and 9. Finally, the nucleotide at node 11 cannot be determined uniquely, but parsimony requires it be either A or T.

In mathematical terms, the rule used for inferring ancestral states is as follows: The set at an internal node is the intersection (denoted by $\cap$) of the two sets at its immediate descendant nodes if the intersection is not empty (e.g., the nucleotide at node 10 is the intersection of the sets at nodes 7 and 9); otherwise, it is the union (denoted by $\cup$) of the sets at the descendant nodes (e.g., the set at node 9 is the union of the sets at nodes 8 and 5). When a union is required to form a nodal set, a nucleotide substitution must have occurred at some point during the evolution of this position. Thus, the number of unions equals the minimum number of substitutions required to account for the descendant nucleotides from a common ancestor. In the example in Figure 5.22a, this number is 4. The alternative tree in **Figure 5.22b** requires only three unions, i.e., three nucleotide substitutions.

By searching all the possible trees for the six OTUs, we find out that 3 is the minimum number of substitutions required to explain the differences among the nucleo-

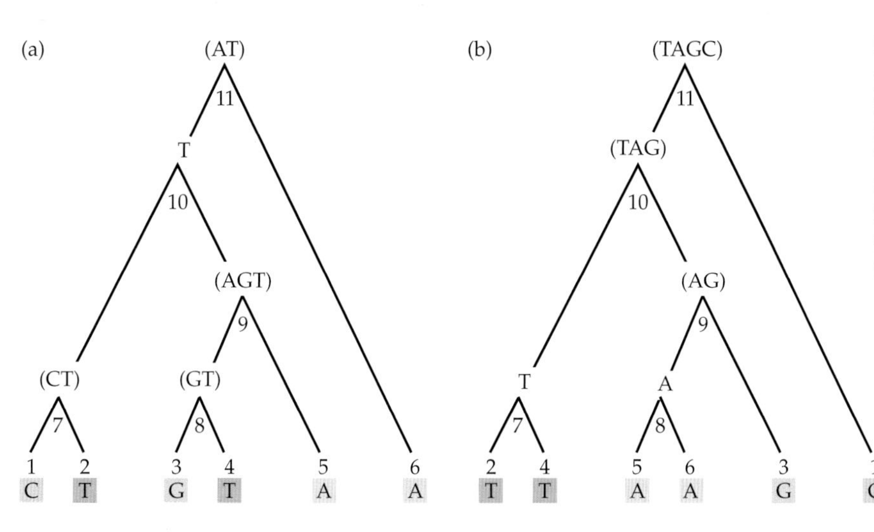

Figure 5.22 Nucleotides at homologous positions in six extant species (1–6) and inferred possible nucleotides in five ancestral species (7–11) according to the method of Fitch (1971). Two different trees, (a) and (b), are depicted. Unions are indicated by parentheses. Note that the inference of an ancestral nucleotide at an internal node is dependent on the topology of the tree. (Modified from Fitch 1971.)

tides at this site. (In addition to the tree shown in Figure 5.22b, there are many other equally parsimonious trees, each requiring three substitutions.)

Although inferring the minimum number of substitutions is straightforward, inferring the evolutionary path (i.e., the true sequence of nucleotide substitutions along each branch) is often difficult (Fitch 1971). Note further that the above procedure neglects all substitutions at uninformative sites. Such substitutions can be easily inferred because the number of substitutions at an uninformative site is equal to the number of different nucleotides present at that site minus 1. For example, if the nucleotides at an uninformative site are A, T, T, C, T, and T in lineages 1, 2, 3, 4, 5, and 6, respectively, then the number of substitutions is 3 – 1 = 2 because regardless of the tree topology, the most parsimonious assumption is that a T ↔ A substitution has occurred in the first lineage and a T ↔ C substitution has occurred in the fourth lineage. The number of substitutions at invariant sites is 1 – 1 = 0.

In computing branch lengths and tree lengths, the total number of substitutions at both informative and uninformative sites should be taken into account.

Weighted and unweighted parsimony

As we have seen, the length of each tree is computed by adding up all the substitutions over all sites. In this computation, all the different nucleotide substitutions have, previously, been given equal weight. This procedure is called **unweighted parsimony**. However, we may wish to give different weights to different types of substitution. For example, we may wish to give a greater weight to transversions, since they usually occur less frequently than transitions (Chapters 1 and 4). Maximum parsimony methods that assign different weights to the various character state changes are called **weighted parsimony**. If transitions are completely ignored and only transversions are used, the method is called **transversion parsimony**.

In the example seen previously (Figure 5.21), we see that the sites supporting tree I (5 and 7) involve transitions. On the other hand, site 9, which supports tree II, involves an A → T transversion. If we give a weight of 1 for each transition and a weight greater than 2 for each transversion, then tree II rather than tree I emerges as the maximum parsimony tree.

Searching for the maximum parsimony tree

When the number of sequences is small, it is possible to look at all the possible trees, determine their lengths, and choose from among them the shortest one (or ones). This type of search for the maximum parsimony tree(s) is called an **exhaustive search**. A simple algorithm can be used for the exhaustive search (**Figure 5.23**). In the first step we connect the first three taxa to form the only possible unrooted tree for three OTUs. In the next step, we add the fourth taxon to each of the three branches of the

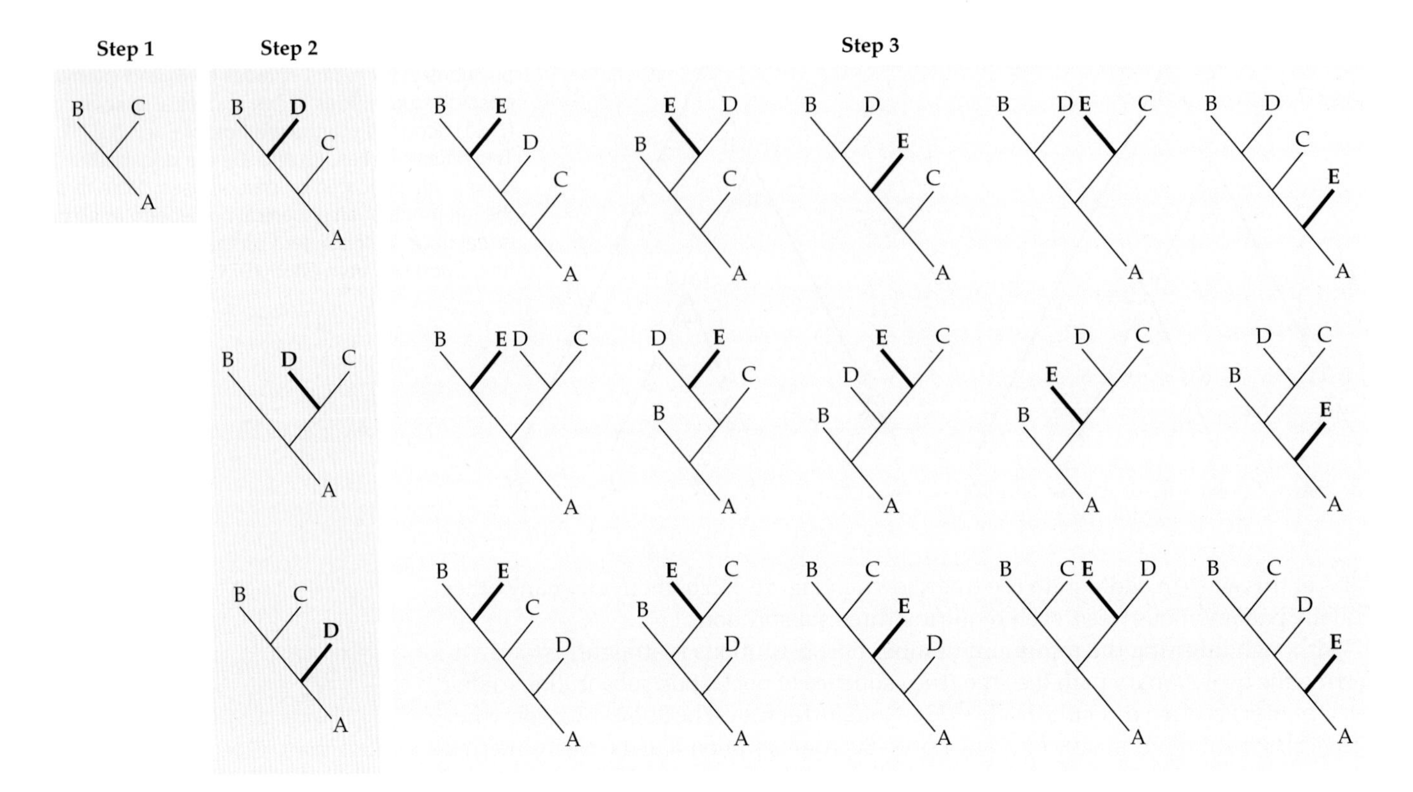

Figure 5.23 Exhaustive stepwise construction of all 15 possible trees for five OTUs. In step 1, we form the only possible unrooted tree for the first three OTUs (A, B, and C). In step 2, we add OTU D to each of the three branches of the tree resulting from step 1, thereby generating three unrooted trees for four OTUs. In step 3, we add OTU E to each of the five branches of the three trees developed in step 2, thereby generating 15 unrooted trees. Additions of OTUs are shown as heavier lines. (Modifed from Swofford et al. 1996.)

three-taxon tree, thereby generating all three possible unrooted trees for four OTUs. In the third step, we add the fifth taxon to each of the five branches of the three four-taxon trees, thereby generating 3 × 5 = 15 unrooted trees. We continue in a similar fashion, adding the next taxon in line to each of the branches in every tree obtained in the previous step.

However, as the number of possible trees increases rapidly with the number of OTUs (Table 5.1), it is virtually impossible to employ an exhaustive search when 12 or more OTUs are studied. Fortunately, there exist shortcut algorithms for identifying all maximum parsimony trees that do not require exhaustive enumeration. One such algorithm is the **branch-and-bound method** (Hendy and Penny 1982). We first consider an arbitrary tree or a tree obtained from a fast method (e.g., the neighbor-joining method) and compute the minimum number of substitutions, L, for the tree. L is then considered as the **upper bound** to which the length of any other tree is compared. The rationale of the upper bound is that the maximum parsimony tree must be either equal in length to L or shorter. The branch-and-bound method works by searching for the maximum parsimony tree by using a procedure similar to that employed for the exhaustive search. In each step of the branch-and-bound algorithm, the length of each tree is compared with the previously determined L value (**Figure 5.24**). If the tree is longer than L, it is no longer used for addition of new taxa in the subsequent steps. The reason is that adding branches to a tree can only increase its length. For example, if a four-taxon tree is longer than L, then all the five-taxon trees descended from it will also be longer than L, and we can therefore ignore them. By dispensing with the evaluation of all the descendant trees from all the partial trees that are longer than L, we may greatly reduce the total number of trees to be considered. Depending on the efficiency of the implementation, the speed of the computer, and the type of data, the branch-and-bound method may be used for finding the maximum parsimony tree for up to 20 OTUs.

Above 20 OTUs, we need to use **heuristic searches**. With heuristic searches, there is no guarantee that the maximum parsimony tree will be found. Nevertheless, it is possible to increase the probability of finding the maximum parsimony tree under certain conditions (Swofford et al. 1996). Numerous heuristics have been suggested

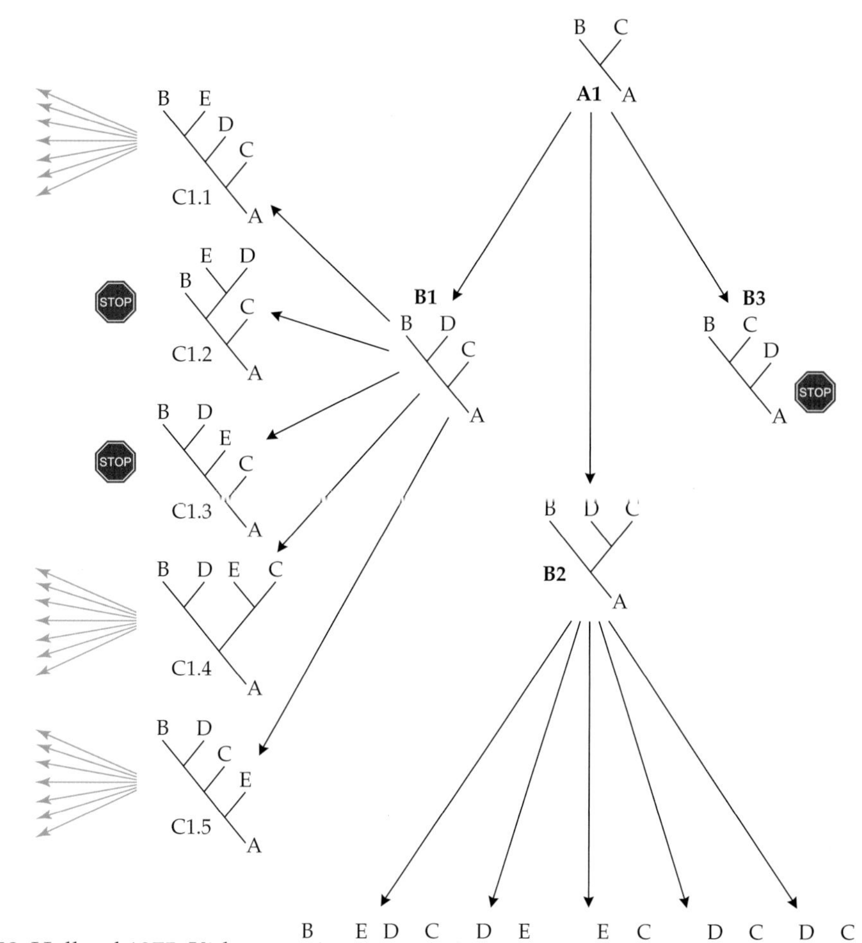

Figure 5.24 An illustration of the branch-and-bound search algorithm for the maximum parsimony tree. We start with the only possible unrooted tree for OTUs A, B, and C (tree A1). Addition of OTU D to each of the three branches in tree A1 results in three unrooted trees (B1–B3). The length of each of these three trees is compared with a previously computed upper-bound value, *L*. Tree B3 is found to be longer than *L*, and branch addition is no longer performed on it (stop sign). In the next step, OTU E is added to each of the five branches of the remaining two trees, B1 and B2, resulting in the formation of 10 trees (C1.1–C1.5 and C2.1–C2.5). Again, each of these trees is compared with the upper bound, and the process of branch addition is only continued for trees shorter than *L*. In the present case, we end up considering only 35 six-taxon trees (green arrows) instead of 105 possible unrooted trees. (Modified from Swofford et al. 1996.)

for use in phylogenetic inference (Croes 1958; Holland 1975; Kirkpatrick et al. 1983; Lundy 1985; Lewis 1998; Salter and Pearl 2001; Barker 2004). Here we present a very simple heuristic method.

The **greedy stepwise-addition search algorithm** (Swofford 1993) is similar to the branch-and-bound algorithm, but because at each step only one path is followed, the search is incomplete and the optimal tree may not be identified at the end of the process. In the first step of the greedy algorithm we connect the first three taxa to form the only possible unrooted tree for three OTUs (A1 in **Figure 5.25**). In the next step, we add the fourth taxon to each of the three branches of the three-taxon tree, thereby generating three possible unrooted trees for four OTUs (B1–B3). The lengths of the three trees are compared with one another, and only the shortest tree is used for addition of new taxa in subsequent steps. Let us assume that tree B2 is the shortest one. In the third step, we add the fifth taxon to each of the five branches of tree B2, thereby generating five unrooted trees. Among these five, we chose the shortest, and at the end of the process proceed to choose one tree as the **local optimum tree**. The adjective "local" is added to emphasize the fact that this tree is the best from among the trees that have been tested, but may not be the **global optimum tree**, i.e., the best of all possible trees.

The problem of the local optimum in heuristic search is illustrated in **Figure 5.26**. Let us imagine that we calculated tree lengths of all the possible unrooted topologies for n OTUs, where n is a very large number. In such a setting, the optimal tree can be identified. We order all the other trees around the optimal tree in order of decreasing similarity to it. Note that the distribution is characterized by many local optima surrounded by trees that are less optimal than they are. Now imagine the realistic scenario in which the identity of the global optimum tree is not known and an exhaustive search is not possible. In such a case, it is likely that the greedy stepwise-addition

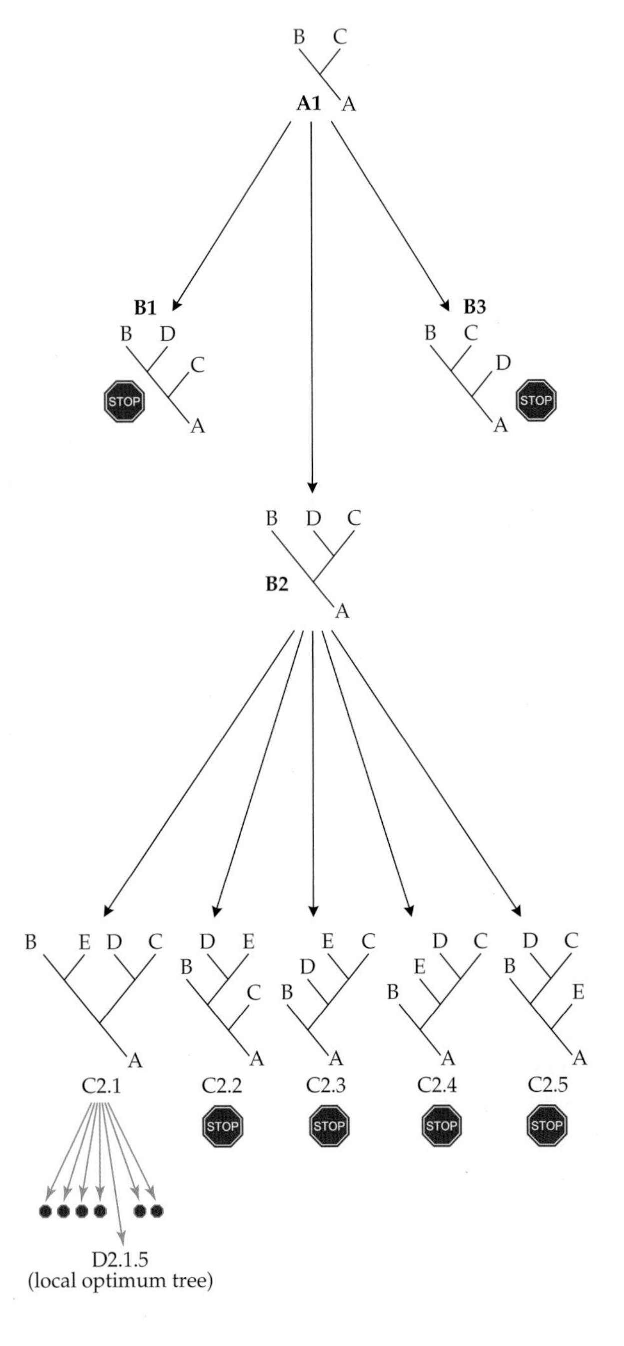

Figure 5.25 An illustration of the greedy stepwise-addition search algorithm for the local maximum parsimony tree. We start with the only possible unrooted tree for OTUs A, B, and C (tree A1). Addition of OTU D to each of the three branches in tree A1 results in 3 unrooted trees (B1–B3). The lengths of these trees are compared with one another. Trees B1 and B3 are found to be longer than B2, and branch addition is no longer performed on them (stop signs). In the next step, OTU E is added to each of the five branches of B2, resulting in the formation of 5 trees (C2.1–C2.5). Again, the tree lengths are compared with one another, and the process of branch addition is only continued for the shortest tree. In the present case, tree D2.1.5 is the local optimum tree. By using a greedy stepwise-addition algorithm, we considered only 7 six-taxon trees (green arrows) instead of the 105 in the exhaustive search or the 35 in the branch-and-bound search.

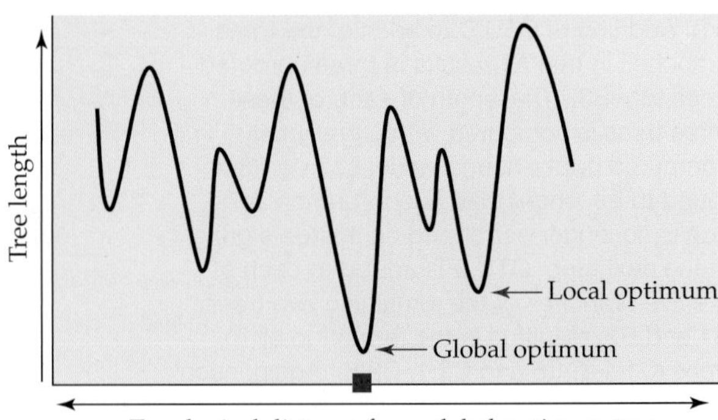

Figure 5.26 Global and local optima in a phylogenetic tree space. The topological distances on the *x*-axis are measured from the global optimum tree (red square).

algorithm merely identifies a local optimum tree that is nowhere near the global optimum. It is possible, however, to improve the search by adding to it a set of predefined rearrangements, a technique commonly referred to as **branch swapping**. Many branch-swapping algorithms are known, and although any one of them is just a "stab in the dark," our chances of identifying the optimal tree increase with the number of rearrangements that are tried.

In **Figure 5.27**, we show one such technique, **subtree pruning and regrafting**. Of course, the global optimum tree may be many rearrangements away from the starting tree. We therefore use an iterative process: If a branch swap succeeds in finding a better tree, we initiate a new round of swapping on this new tree. So long as each round of swapping successfully finds an improved tree, we will eventually arrive at the global optimum. However, if the path to the optimal tree requires us to pass through intermediate trees that are inferior to the best one so far obtained, we will once again find ourselves trapped in a local optimum, albeit a better one.

Maximum Likelihood Methods

The first application of a maximum likelihood method to tree reconstruction was done by Cavalli-Sforza and Edwards (1967) for gene frequency data. Later, Felsenstein (1973, 1981) developed maximum likelihood algorithms for amino acid and nucleotide sequence data. Here we present only the basic principles of the method, without much mathematical detail. For a more detailed presentation, see Nei and Kumar (2000), Felsenstein (2004), and Yang (2006).

The **likelihood**, L, of a phylogenetic tree is the **conditional probability** of observing the data (e.g., the nucleotide sequences), given a phylogenetic tree and a specified model of character state changes (e.g., the pattern of nucleotide substitution). This is usually written as $L = P(\text{data}|\text{tree})$. The aim of the **maximum likelihood method** is to identify the tree with the highest L value from among all the possible trees.

The maximum likelihood method requires a probabilistic model for the process of nucleotide substitution. That is, we must specify the transition probability from one nucleotide state to another nucleotide state in a time interval on each branch. For example, consider the one-parameter model with a rate of substitution λ per site per unit time. Assume that the nucleotide at a given site was i at time 0. Then, from Equations 3.11 and 3.12 (Chapter 3), the probability that the nucleotide at time t is i is given by

$$P_{ii(t)} = \frac{1}{4} + \frac{3}{4}e^{-4\lambda t/3} \quad (5.13)$$

and the probability that the nucleotide at time t is j, $j \neq i$, is given by

$$P_{ij(t)} = \frac{1}{4} - \frac{1}{4}e^{-4\lambda t/3} \quad (5.14)$$

In the next step we compute the probability of observing the nucleotide configuration at a nucleotide site at the tips of a tree. Let us use the case of four sequences (taxa) as an example. For simplicity, we assume a constant rate of substitution and consider the hypothetical rooted tree given in **Figure 5.28a.** The probability for a nucleotide site with nucleotides i, j, k, and l in OTUs 1, 2, 3, and 4, respectively, can be computed as follows. If the nucleotide at the ancestral node (the root) was x, the probability of having nucleotide l in OTU 4 is $P_{xl(t_1+t_2+t_3)}$ because $t_1 + t_2 + t_3$ is the total amount of time between the two nodes. The probability of having nucleotide y at the ancestral node for OTUs 1, 2, and 3 is $P_{xy(t_1)}$, and so on. Therefore, given x, y, and z at the ancestral node and the two other internal nodes, the probability of observing i, j, k, and l at the tips of the tree is equal to $P_{xl(t_1+t_2+t_3)}P_{xy(t_1)}P_{yk(t_2+t_3)}P_{xz(t_2)}P_{zj(t_3)}$.

All the above transition probabilities can be computed by using Equations 5.13 and 5.14. Since in practice we do not know the ancestral nucleotide, we can only assign a probability g_x, which is usually assumed to be the frequency of nucleotide x in the sequence. Noting that x, y, and z can be any of the four nucleotides, we sum over all possibilities and obtain the following probability:

$$h(i,j,k,l) = \sum_x g_x P_{xl(t_1+t_2+t_3)} \sum_y P_{xy(t_1)} P_{yk(t_2+t_3)} \sum_z P_{yz(t_2)} P_{zi(t_3)} P_{zj(t_3)} \quad (5.15)$$

Note that Equation 5.15 represents the probability of observing the configuration of nucleotides i, j, k, and l in OTUs 1, 2, 3, and 4, respectively, for the tree in Figure 5.28a. The probability will be different for any other tree.

A formulation without the assumption of rate constancy can be done in a similar manner. In this case, it is usually more convenient to consider the transition probability in terms of the branch length. For example, we consider $P_{ij(v_a)}$ instead of $P_{ij(t_a)}$, where $v_a = \lambda_a t_a$ and λ_α and t_α are, respectively, the rate of substitution and the time of divergence for the αth branch. If we assume that the internal node connecting OTUs

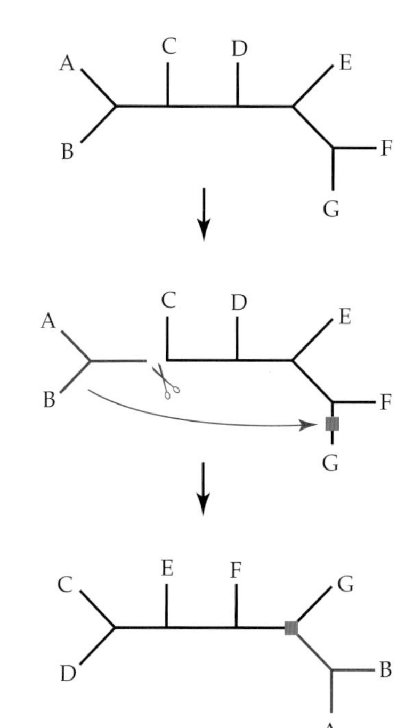

Figure 5.27 Subtree pruning and regrafting. In this method, a subtree (red branches) is pruned from the tree and regrafted to a randomly chosen branch (blue squares). In this example, branch swapping is applied to an unrooted tree with seven OTUs and the subtree consists of OTUs A and B.

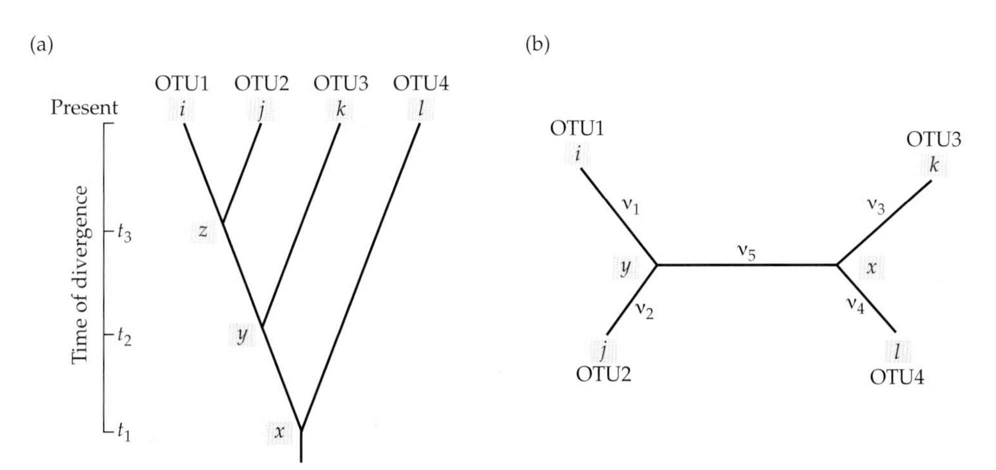

Figure 5.28 Model rooted (a) and unrooted (b) trees for the derivation of the likelihood function. Letters i, j, k, and l represent nucleotides at homologous positions in OTUs. Letters x, y, and z represent nucleotides at homologous positions in HTUs. Branch lengths are represented by $v_1 - v_5$. (From Li and Gouy 1991.)

3 and 4 in the tree in **Figure 5.28b** is the ancestral node, we obtain the following probability (Felsenstein 1981; Saitou 1988):

$$h(i,j,k,l) = \sum_x g_x P_{xl(v_4)} P_{xk(v_3)} \sum_y P_{xy(v_5)} P_{yi(v_1)} P_{yj(v_2)} \qquad \textbf{(5.16)}$$

Let us now discuss briefly the subject of **time reversibility**. In molecular evolution, many convenient substitution models, such as the one- and two-parameter models (Chapter 3), are time-reversible. In other words, the model does not care which sequence is the ancestor and which is the descendant, so long as all other parameters (such as the expected number of substitutions per site between the two sequences) are held constant. When an analysis of real biological data is performed, we usually have no access to the ancestral sequences, only to the sequences of the extant taxa. However, when a model is time-reversible, which species was ancestral is irrelevant; the tree can be rooted using any node as the "ancestral" node. This is because there are no "special" species: all species will eventually derive from one another with the same probability. This is called the **pulley principle** (Felsenstein 1981). Later, the tree can be re-rooted based on new knowledge, or it may be left unrooted. But if the process is *not* time-reversible, the tree must be rooted prior to the analysis.

Previously, we considered a single site. With many nucleotide sites, the likelihood can be computed as follows. Let us consider the simple example in **Figure 5.29a**, which shows a set of aligned nucleotide sequences from four taxa. First, we evaluate

(a)

	1	2	3	4	5	6	7	8	9	... *n*
OTU1	A	A	G	A	C	T	T	C	A	... *N*
OTU2	A	G	C	C	C	T	T	C	T	... *N*
OTU3	A	G	A	T	A	T	C	C	A	... *N*
OTU4	A	G	A	G	G	T	C	C	T	... *N*

(b)

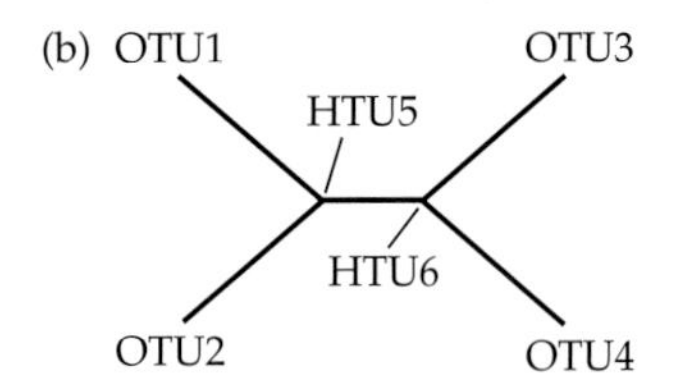

(c)

$L_{(5)}$ = Prob(C, C, A—A, A, G) + Prob(C, C, A—C, A, G) + Prob(C, C, A—T, A, G) + Prob(C, C, A—G, A, G)
+ Prob(C, C, C—A, A, G) + Prob(C, C, C—C, A, G) + Prob(C, C, C—T, A, G) + Prob(C, C, C—G, A, G)
+ Prob(C, C, T—A, A, G) + Prob(C, C, T—C, A, G) + Prob(C, C, T—T, A, G) + Prob(C, C, T—G, A, G)
+ Prob(C, C, G—A, A, G) + Prob(C, C, G—C, A, G) + Prob(C, C, G—T, A, G) + Prob(C, C, G—G, A, G)

(d) $L = L_{(1)} \times L_{(2)} \times L_{(3)} \times \ldots \times L_{(n)} = \prod_{i=1}^{n} L_{(i)}$

(e) $\ln L = \ln L_{(1)} + \ln L_{(2)} + \ln L_{(3)} + \ldots + L_{(n)} = \sum_{i=1}^{n} \ln L_{(i)}$

Figure 5.29 Schematic representation of the calculation of the likelihood of a tree. (a) Data in the form of a DNA sequence alignment of length *n*, where *N* denotes any of the four deoxyribonucleotides (A, C, T, G). (b) One of three possible trees for the four taxa whose sequences are shown in (a). (c) The likelihood of a particular site (in this case, site 5) equals the sum of the 16 probabilities of every possible reconstruction of ancestral states at nodes 5 and 6 in (b). (d) The likelihood of the tree in (b) is the product of the individual likelihoods for all *n* sites. (e) The likelihood is usually evaluated by summing the logarithms of the likelihoods at each site and reported as the log likelihood of the tree. (Modified from Swofford et al. 1996.)

the likelihood of the unrooted tree in **Figure 5.29b**; that is, what is the likelihood for the tree in Figure 5.29b, given the observed data in Figure 5.29a? We assume that all nucleotide sites evolve independently, so we can calculate the likelihood for each site separately and compute the product of the likelihoods at the end.

As an example, let us first compute the likelihood at site 5. At this site, OTUs 1, 2, 3, and 4 have C, C, A, and G, respectively. The unrooted tree in Figure 5.29b has two internal nodes, denoted as HTUs 5 and 6, each of which can have one of the four different nucleotides. Thus, we should consider $4 \times 4 = 16$ possibilities (**Figure 5.29c**). Obviously, some of these possibilities are less plausible than the others, but each has a nonzero probability of generating the pattern of the observed nucleotides at the four tips of the tree. Therefore, the likelihood of observing the nucleotides at site 5 is equal to the sum of 16 independent probabilities. The same procedure is repeated for each site separately, and the likelihood for all the sites is computed as the product of the individual-site likelihoods (**Figure 5.29d**). Note that the final likelihood is no longer a probability, but a product of probabilities. Note also that while the probabilities have to sum up to 1, the likelihoods do not.

For mathematical convenience, the likelihood is usually evaluated by the logarithmic transformation, which transforms multiplication into summation (**Figure 5.29e**). That is, we consider the **log likelihood** ($\ln L$) of the tree. We then proceed to compute the likelihood values for the other possible trees, and the tree with the highest likelihood value is chosen as the maximum likelihood tree.

Note that since the likelihoods depend on the model of nucleotide substitution, a tree with the largest likelihood value under one substitution model may not be the maximum likelihood tree under another model. Moreover, the method is sensitive to the evolutionary model and the estimates of the various parameters. Nonetheless, the maximum likelihood method is regarded by some researchers as the gold standard of molecular phylogenetic reconstruction (e.g., Kumar and Filipski 2008).

We note, however, that exact maximum likelihood estimations can only be performed through exhaustive searches and are extremely time-consuming. Thus, even with fast computers, exact maximum likelihood estimations are only possible with a modest number of taxa. Many heuristic software packages are available in the literature and online; look for current versions of MEGA and PAUP.

Bayesian Phylogenetics

Bayesian statistics started ~250 years ago with a posthumously published paper (Bayes 1763). Unlike the likelihood approach, which considers $L(D|H)$, the likelihood of observing the data D under hypothesis H, the Bayesian approach considers $L(H|D)$—the likelihood that hypothesis H is true, given the observed data D. For a detailed description of the application of this approach to phylogenetics, readers are referred to Felsenstein (2004) and Yang (2006). Here we only provide a brief description of the method.

The Bayesian approach involves the theory of conditional probability (Feller 1968). Consider two events A and B, and let $P(A,B)$ be their joint probability. Then, the probability of occurrence of event A, given that B occurs, is given by

$$P(A \mid B) = \frac{P(A,B)}{P(B)} \tag{5.17}$$

Note that $P(A,B)$ can be written as

$$P(A,B) = P(A)P(B \mid A) \tag{5.18}$$

Therefore, Equation 5.17 can be written as

$$P(A \mid B) = \frac{P(A)P(B \mid A)}{P(B)} \tag{5.19}$$

This is Bayes' theorem in its simplest form.

Let us now compute the probability of "B occurs" by considering two situations—A and "not A"—that denote "A occurs" and "A does not occur," respectively. We can see that the probability that B occurs is given by

$$P(B) = P(A)P(B \mid A) + P(\text{not } A)P(B \mid \text{not } A) \tag{5.20}$$

From Equations 5.19 and 5.20, we obtain

$$P(A \mid B) = \frac{P(A) \times P(B \mid A)}{P(A) \times P(B \mid A) + P(\text{not } A) \times P(B \mid \text{not } A)} \tag{5.21}$$

In statistics, A and "not A" can be two hypotheses—H_1 and "not H_1" (the latter denoted as H_2)—and B can be the observed data D. When there are multiple hypotheses $H_{1,\ldots,}H_n$, the probability that hypothesis H_i is true, given the data D, can be written as

$$P(H_i \mid D) = \frac{P(H_i) \times P(D \mid H_i)}{\sum_{j=1}^{n} P(H_j) \times P(D \mid H_j)} \tag{5.22}$$

In phylogenetics, a hypothesis can be a possible tree, i.e., $H_i = T_i$, in which case Equation 5.22 yields the probability that tree T_i is true, given the data D.

In Equation 5.22, $P(H_i)$ is the **prior probability**, i.e., the probability that H_i is true before the data is analyzed or even seen. In the absence of any information, one usually uses a neutral prior distribution, that is, all n hypotheses are assumed to have the same prior $P(H_i) = 1/n$. Such "priors" are also known as **non-informative** or **flat priors** (Yang 2006). However, if there is evidence that some hypotheses are more likely than others, then unequal priors may be used. For example, if previous studies have suggested that among OTUs 1, 2, and 3, OTUs 1 and 2 are closer to each other than either of them is to OTU 3, then the prior for the hypothesis H_1 = ((1,2),3) can be set higher than the priors for the other two hypotheses: H_2 = ((1,3),2) and H_3 = ((2,3),1). In Equation 5.22, $P(H_i|D)$ is the **posterior probability**, i.e., the estimate of the probability that H_i is true given the data.

The use of priors is extremely controversial. Critics of the Bayesian method, who are sometimes referred to somewhat derisively as "frequentists," contend that assessments of priors are merely licenses for letting personal prejudices influence scientific inferences. However, the effect of priors on the inferences decreases as more and more data are used to make the inference.

Bayesians argue that computing the probability of obtaining a given set of data under a possible hypothesis, i.e., $P(D|H_i)$, is the wrong question because what we really want to know is the probability $P(H_i|D)$, i.e., the probability that a given hypothesis is true given the observed data. Another issue that has generated much discussion in the literature is whether or not the Bayesian approach gives an accurate estimate of the statistical support for a hypothesis. Several simulation studies have indicated that the Bayesian approach tends to overestimate the statistical support of a subtree (e.g., Suzuki et al. 2002). For a more detailed description of the controversy, see Felsenstein (2004) and Yang (2006).

The Bayesian method of phylogenetic inference is extremely tedious computationally, even more so than the maximum likelihood method. However, the developments of heuristics as well as advances in computational hardware have greatly increased its popularity in recent years. In addition to determining the posterior probabilities of phylogenetic topologies, Bayesian inference can be used in a phylogenetic context for such purposes as evaluating uncertainties, computing divergence times, and testing molecular clock hypotheses (Huelsenbeck et al. 2001).

Topological Comparisons

It is sometimes necessary to measure the similarity or dissimilarity among tree topologies. Such a need may arise when dealing with trees that have been inferred from analyses of different sets of data or from different types of analysis of the same data set. Moreover, several methods of tree reconstruction (maximum parsimony, for example) may produce many trees rather than a unique phylogeny. In such cases, it may be advisable to draw a tree that summarizes the points of agreement among all the trees. When two trees derived from different data sets or different methodologies are identical, they are said to be **congruent**. Congruence can sometimes be partial, i.e., limited to some parts of the trees, other parts being **incongruent**.

Topological distance

There are many methods for comparing tree topologies (e.g., Robinson and Foulds 1981). Here we present one commonly used measure of dissimilarity between two tree topologies, Penny and Hendy's (1985) topological distance. The measure is based on tree partitioning and is equal to twice the number of different ways of partitioning the OTUs between two trees,

$$d_T = 2c \qquad (5.23)$$

where d_T is the topological distance and c is the number of partitions resulting in different divisions of the OTUs in the two trees under consideration. (In comparisons between bifurcating trees, d_T is always an even integer.)

Consider, for instance, the trees in **Figure 5.30**. Tree (a) has six OTUs and three internal branches. If we partition this tree at branch 1, we obtain two groups of OTUs: A and B on one side, and C, D, E, and F on the other. Cutting tree (b) at branch 1 results in the same partitioning of the six OTUs. Cutting tree (a) at branch 2 results in the same partitioning of OTUs as the cutting of tree (b) at branch 3, i.e., A, B, E, and F on one side, and C and D on the other. Cutting tree (a) at branch 3 results in a partition of OTUs that cannot be obtained by cutting tree (b) at any of its three internal branches. Therefore, $d_T = 2 \times 1 = 2$.

In comparing trees (a) and (c), we see that none of the partitions in (a) can be found in (c). Therefore, d_T achieves its maximal possible value, i.e., $d_T = 2 \times 3 = 6$. Hence, we conclude that tree (a) is more similar to tree (b) than to tree (c).

Consensus trees

Trees inferred from the analysis of a particular data set are also called **fundamental trees**, i.e., they summarize the phylogenetic information in a data set. **Consensus trees** are trees that have been derived from a set of trees, i.e., they summarize in a single tree the phylogenetic information in a set of fundamental trees. For example, maximum parsimony may sometimes produce many equally parsimonious trees rather than a unique solution. In such cases, it is often difficult to present all the trees, and a consensus tree is usually shown.

In consensus trees the points of agreement among the fundamental trees are shown as bifurcations, whereas the points of disagreement are collapsed into polytomies. There are several different types of consensus trees, but the most commonly used are the **strict consensus** and **majority-rule consensus trees**.

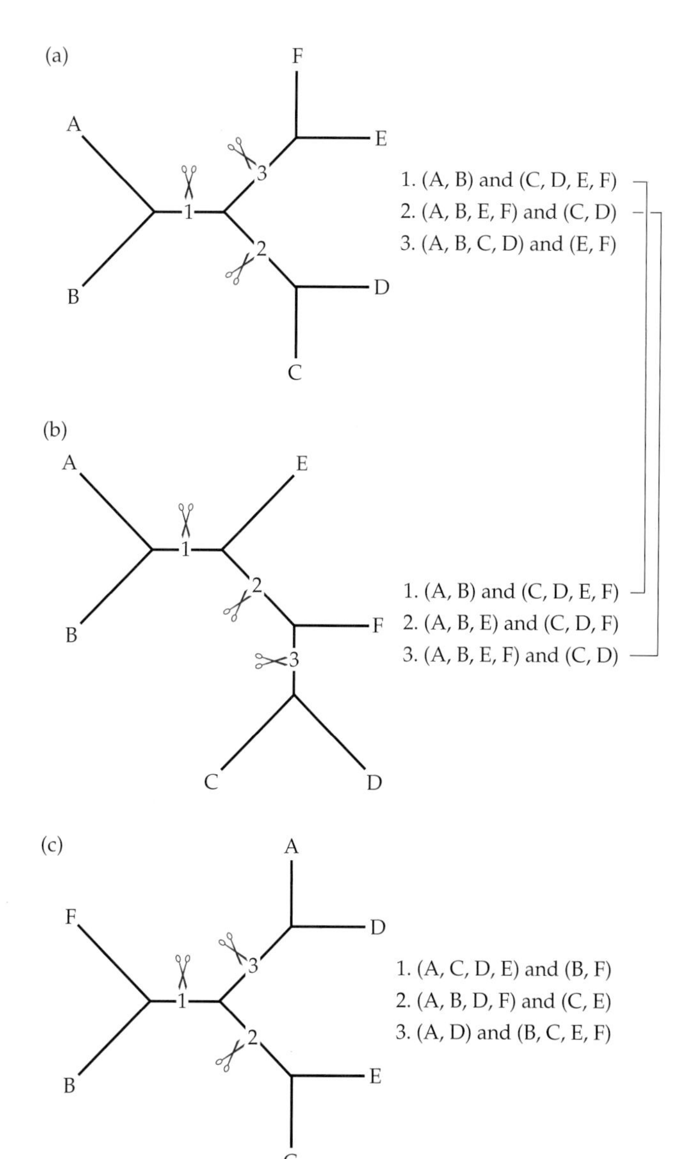

Figure 5.30 Measuring similarity between tree topologies by Penny and Hendy's (1985) method. Each tree can be partitioned three different ways by cutting the internal branches (1–3). The resulting partitions are shown on the right. Note that partitions 1 and 2 in tree (a) are identical, respectively, to partitions 1 and 3 in tree (b). There are no identical partitions between trees (a) and (c). (From Graur and Li 2000.)

Let us assume that we obtained three rooted trees for seven taxa (**Figure 5.31**). In a strict consensus tree, all conflicting branching patterns are collapsed into multifurcations. The resulting consensus tree in this case contains two multifurcations.

Among the majority-rule consensus trees, the most commonly used in the literature is the **50% majority-rule consensus tree**. In this tree, a branching pattern that occurs with a frequency of 50% or more is adopted. In the example in Figure 5.31, the position of taxon A relative to taxa B, C, and D is the same in two out of the three "rival" trees (Figures 5.31b,c), so this pattern is adopted. This tree, therefore, contains only a single multifurcation. It is possible to change the majority-rule percentage to any value; at 100% the result will be identical with the strict consensus tree.

Supertrees

A **supertree** is a phylogenetic tree produced from a set of fundamental trees (also referred to as **source trees**) that possess overlapping sets of taxa. For example, one rooted fundamental tree may be based on a homologous gene set from four taxa, A, B, C, and D, while another may be based on a different homologous gene set from three taxa, C, D, and E. From these trees, a supertree for the five taxa, A, B, C, D, and E, can be build. No fundamental tree can be build from the five taxa, because the set of homologous genes for all the five taxa is empty. In other words, no homologous sequences exist for all these taxa.

There are numerous techniques for building supertrees from source trees (Bininda-Emonds 2004). In **Figure 5.32**, we briefly introduce only one such method, called **matrix representation parsimony**. In this method, the internal nodes of each rooted source tree are identified by a number. In other words, each clade is identified by a number. The taxa that descend from a certain node are given a score of 1; those that do not descend from this node but that are present elsewhere in the source tree are given a score of 0. All other taxa are scored as missing, which is denoted by a question mark (?). We then build a matrix in which each node will contain taxa marked by 1, 0, or ?. If all the source trees are compatible, then a single binary supertree can be built. In the hypothetical case

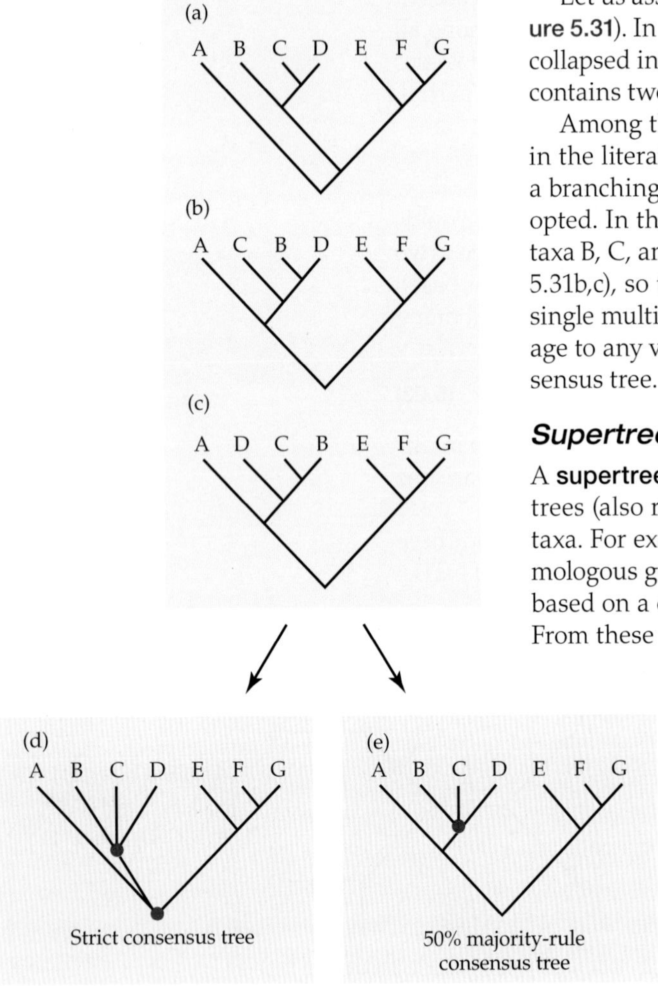

Figure 5.31 Three inferred trees (a, b, and c) can be summarized as a strict consensus tree (d) or as a 50% majority-rule consensus tree (e). Multifurcations are indicated by solid circles.

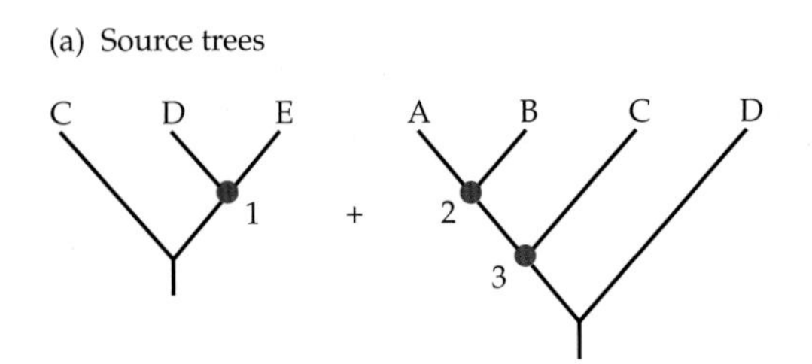

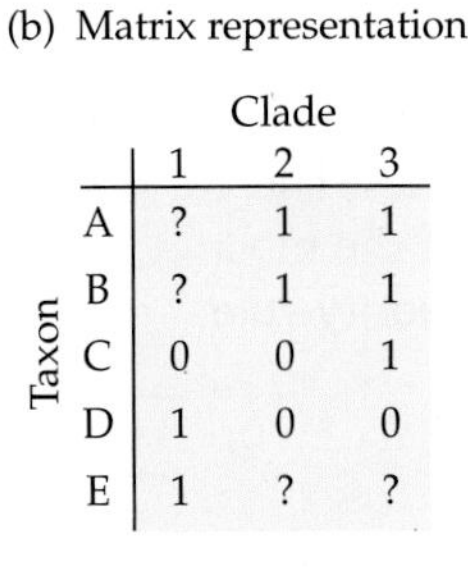

(b) Matrix representation

Taxon	Clade 1	Clade 2	Clade 3
A	?	1	1
B	?	1	1
C	0	0	1
D	1	0	0
E	1	?	?

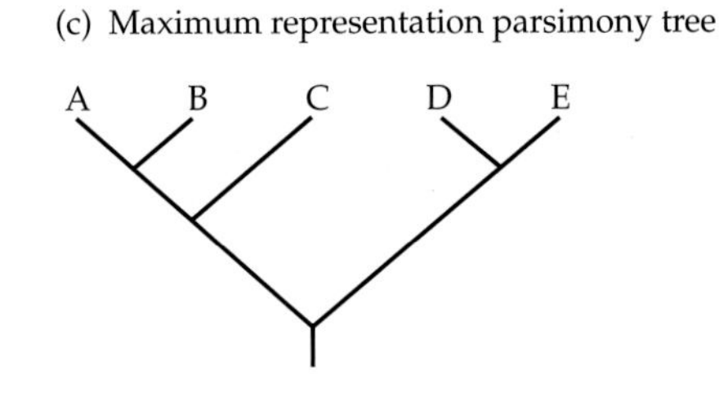

Figure 5.32 Supertree construction by using matrix representation parsimony. (a) Two rooted source trees are shown, one with three taxa and one with four. One internal node is identified in the three-taxon tree; two in the four-taxon tree. (b) The two source trees can be transformed into a matrix. (c) Since the matrix contains no incompatible clades, by using a maximum parsimony algorithm, a single binary matrix representation parsimony supertree can be built.

shown in Figure 5.32, we note that there is no incompatibility of the clades. In the real world, however, some clades may be incompatible with other clades. Interested readers should consult Bininda-Emonds et al. (2002) and references therein for practical solutions to the imcompatibility problem.

To allow use of the maximum number of orthologous loci in the phylogenetic analysis, a subset of supertree methodologies (see Piaggio-Talice et al. 2004) use unrooted source trees consisting of four taxa (**quartets**) as input. With n taxa, the number of possible quartets (N_4) is

$$N_4 = \frac{n!}{24(n-4)!} \tag{5.24}$$

Rooting Unrooted Trees

The majority of tree-making methods yield unrooted trees. As mentioned previously, however, an unrooted tree cannot be said to represent the evolutionary history of divergence among a group of taxa, because it lacks a temporal directionality. In an unrooted tree, one cannot speak of ancestors and descendants, and although it may seem visually counterintuitive, neighboring taxa in an unrooted tree are not necessarily closely related in a phylogenetic sense. The interpretation of common ancestry depends upon rooting, which is thus an important decision in phylogenetic studies.

The rooting of an unrooted tree is equivalent to adding one branch to each of its existing branches. Thus, depending on the placement of the root, the unrooted tree in **Figure 5.33a** can be converted into any of the five rooted trees in **Figure 5.33b**. For example, if the root is placed at the position marked by R_1, OTU B becomes the representative of the most ancient divergence event among the four taxa, whereas if the root is placed at R_5, OTU B comes out as representing a relatively recent divergence event. Note that the direction of the evolutionary time arrow changes with the position of the root. For example, by placing the root at R_1 in Figure 5.33a, we claim that

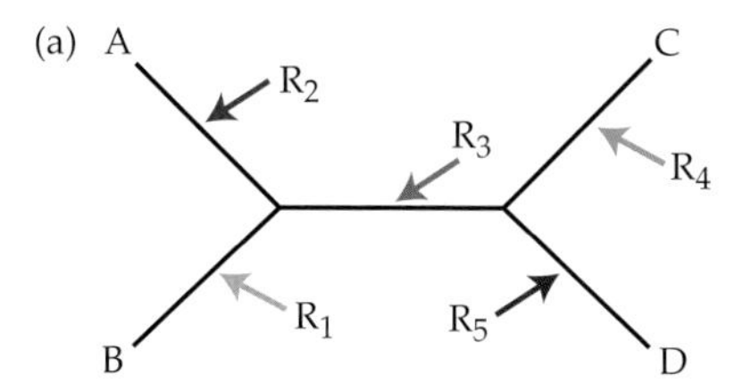

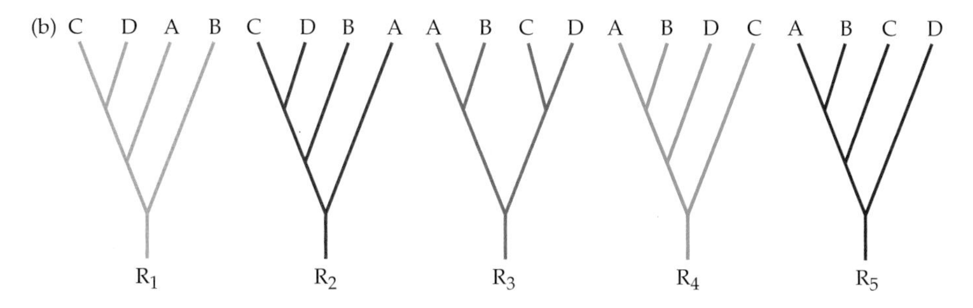

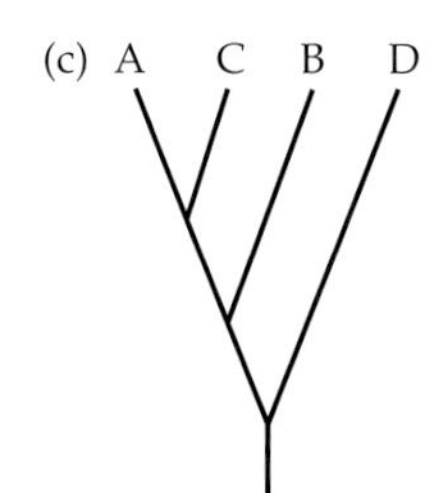

Figure 5.33 An unrooted four-taxon tree (a) can be rooted on each of its branches (R_1–R_5) to obtain five compatible rooted trees (b). The rooted tree in (c) is incompatible with the unrooted tree in (a). (From Graur and Li 2000.)

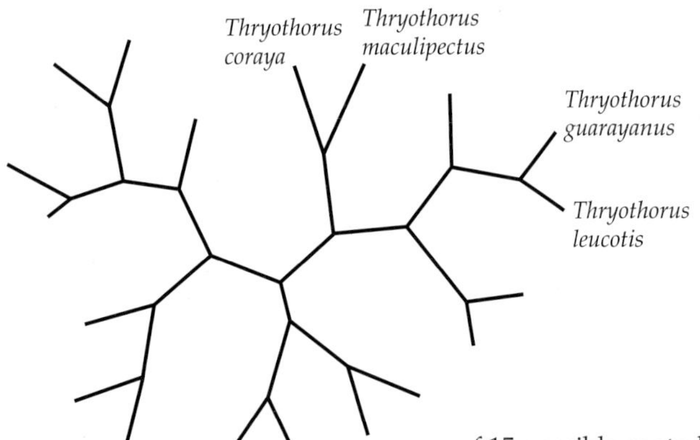

Figure 5.34 An unrooted tree for 19 species belonging to the wren family (Troglodytidae). Only the names of species belonging to the genus *Thryothorus* are shown. The root can in principle be placed on each of the 35 branches of this unrooted tree. It is impossible, however, to root the tree in such a manner as to cluster all the *Thryothorus* species into a monophyletic clade to the exclusion of all other non-*Thryothorus* taxa. Thus, the monophyly of *Thryothorus* can be rejected even without rooting the tree. (Modified from Barker 2004.)

the common ancestor of all the four taxa gave rise to OTU B and the common ancestor of A, C, and D, and thus the existence of the common ancestor of A and B preceded that of the common ancestor of C and D. In comparison, by placing the root at R_5, we obtain the opposite situation, i.e., the existence of the common ancestor of C and D preceded that of the common ancestor of A and B.

Even if we cannot root a tree, an unrooted tree is useful in its own right by (1) reducing the number of rooted phylogenetic trees that need to be considered in subsequent studies and (2) answering specific phylogenetic questions concerning the monophyly or paraphyly of certain OTUs. For example, out of 15 possible rooted trees with four OTUs, only 5 rooted trees are compatible with the unrooted tree in Figure 5.33a. The tree in **Figure 5.33c**, for instance, is incompatible with the unrooted tree in Figure 5.33a. The general rule is that two OTUs cannot form a monophyletic clade in the rooted tree if the path leading from one to the other in the unrooted tree passes through two or more internal nodes. Consequently, even an unrooted tree can provide answers to such questions as whether or not two OTUs constitute a natural clade. One example is shown in **Figure 5.34**. From this unrooted tree it is obvious that, regardless of the ultimate position of the root, the genus *Thryothorus* is not monophyletic. It consists of at least three monophyletic clades, each of which is clustered with non-*Thryothorus* taxa to the exclusion of the other *Thryothorus* clades.

Outgroup rooting

The best method of rooting an unrooted tree is to use an outgroup. An **outgroup** is an OTU (or a group of several OTUs) for which we have incontrovertible external knowledge, such as from taxonomic or paleontological evidence, that they branched off earlier than the other taxa under study (the **ingroup** taxa). The root of the ingroup taxa is placed at the node connecting the ingroup tree to the external branch leading to the outgroup. One such example is shown in **Figure 5.35**.

While we must be certain that the outgroup did indeed diverge prior to the taxa under study, it is not advisable to choose an outgroup that is too distantly related to the ingroup, because in such cases it is difficult to obtain reliable estimates of the distances between the outgroup and ingroup taxa. For example, in reconstructing the phylogenetic relationships among a group of placental mammals, we may use a marsupial as an outgroup. Birds may serve as reliable outgroups, but only if the DNA sequences used have been highly conserved in evolution. Plants and fungi are clearly outgroups, but because they are only very distantly related to the mammals, their use in rooting a mammal tree may result in serious topological errors. On the other hand, the outgroup must not be phylogenetically too close to the other OTUs, because then we cannot be certain that it diverged from the ingroup OTUs prior to their divergence from one another.

The use of more than one outgroup generally improves the estimate of the tree topology, provided again that they are not too distant from the ingroup taxa. If the outgroups are very distant from the ingroup, the use of multiple outgroups may yield worse results than using a single outgroup, because of the long-branch attraction phenomenon.

Midpoint rooting

In the absence of an outgroup, we may position the root by assuming that the rate of evolution has been approximately uniform over all the branches. Under this assumption we put the root at the midpoint of the longest pathway between two OTUs. For example, in the hypothetical unrooted tree in **Figure 5.36**, the longest path is between OTUs B and E. The length of this path is 3 + 6 + 3 + 10 = 22, so we position the root at a distance of 22/2 = 11 from either B or E. Midpoint rooting was found to be a valuable approach with a very high success rate (Hess and De Morales Russo 2007).

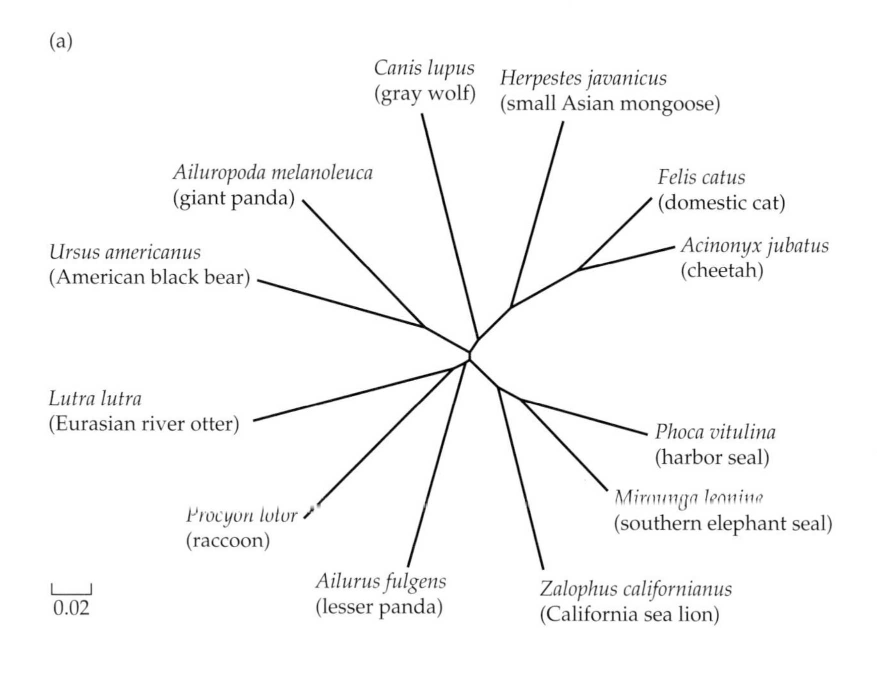

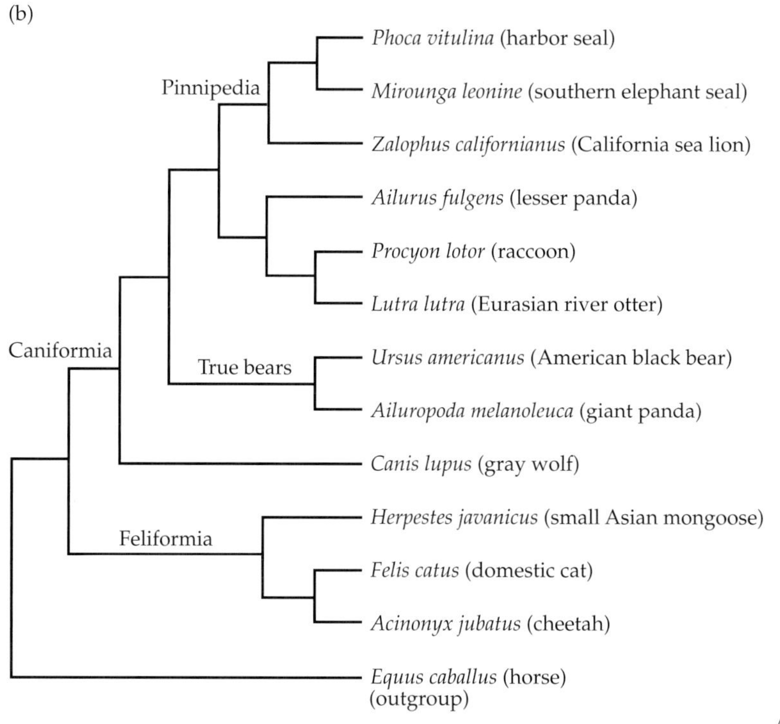

Figure 5.35 Rooting a tree by using an outgroup. (a) An unrooted tree was built by using whole mitochondrial sequences from twelve carnivore species. The gapless length of the alignment was 14,022 nucleotides. In the absence of an outgroup, we cannot decide in many cases which taxonomic groups are monophyletic. However, in some cases, we can decide on the paraphyletic status of a group of species. For example, the two pandas (*Ailuropoda melanoleuca* and *Ailurus fulgens*) are obviously paraphyletic because regardless of the placement of the root, the two taxa are separated by at least four internal nodes. The bar denotes a branch length of 0.02 substitutions per nucleotide site. (b) As outgroup we used the horse (*Equus caballus*), which belongs to a different order of placental mammals (Perissodactyla). The rooted unscaled tree indicates that the most basal divergence event within Carnivora is between suborders Caniformia (doglike) and Feliformia (catlike). The tree also indicates that Pinnipedia (marine fin-footed carnivores) constitutes a monophyletic clade, and that the giant panda is a true bear while the lesser panda is related to raccoons and otters. Finally, the tree indicates that despite many morphological and behavioral idiosyncrasies (e.g., absence of climbing abilities), the cheetah is closely related to cats. (Courtesy of Reed Cartwright and Nicholas Price.)

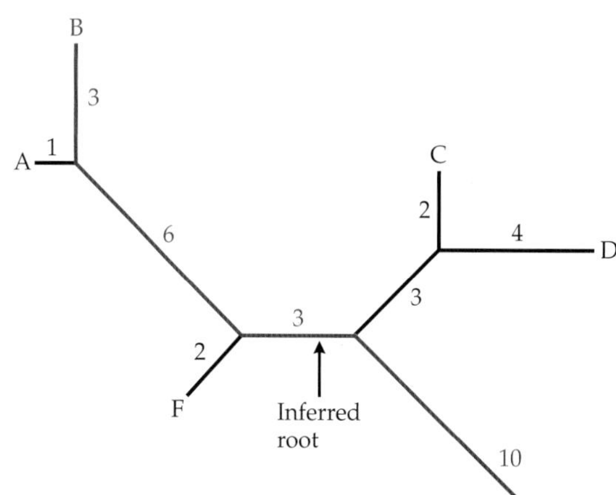

Figure 5.36 A hypothetical unrooted phylogenetic tree with scaled branches that has been rooted at the midpoint of the longest possible pathway (red line) between two OTUs. The numbers of substitutions are marked on the branches. (From Graur and Li 2000.)

Estimating Branch Lengths

So far, we have mainly concerned ourselves with the topology of phylogenetic trees. Here, we consider branch lengths, which provide us with useful information about divergence times and rates of evolution. In UPGMA and maximum likelihood methods, branch lengths are estimated together with the topology. For the maximum parsimony method, the procedure for inferring the minimum number of substitutions on each branch has been described (p. 190). This procedure tends to underestimate the branch lengths, for it is intended to minimize the number of substitutions. The degree of underestimation may not be severe if the tree contains only short branches; otherwise, the underestimation may be quite serious (Saitou 1989; Tateno et al. 1994).

Here, we describe Fitch and Margoliash's (1967) method for estimating branch lengths, assuming that the tree topology has already been inferred by a distance matrix procedure, such as Sattath and Tversky's (1977) neighbors-relation method.

First let us consider the simplest case, i.e., an unrooted tree with three OTUs (X, Y, and Z) and a single node (**Figure 5.37a**). Let x, y, and z be the lengths of the branches leading to X, Y, and Z, respectively. It is easy to see that the following equations hold:

$$d_{XY} = x + y \quad \textbf{(5.25a)}$$

$$d_{XZ} = x + z \quad \textbf{(5.25b)}$$

$$d_{YZ} = y + z \quad \textbf{(5.25c)}$$

From these equations, we obtain the following solutions:

$$x = \frac{d_{XY} + d_{XZ} - d_{YZ}}{2} \quad \textbf{(5.26a)}$$

$$y = \frac{d_{XY} + d_{YZ} - d_{XZ}}{2} \quad \textbf{(5.26b)}$$

$$z = \frac{d_{XZ} + d_{YZ} - d_{XY}}{2} \quad \textbf{(5.26c)}$$

Let us now deal with the case of more than three OTUs. For simplicity, let us assume that there are five OTUs (A, B, C, D, and E) and that the topology and the branch lengths are shown in **Figure 5.37b**. Inferring the branch lengths of this tree require us to proceed in the order in which simple and composite OTUs were clustered in the original distance matrix procedure. At each stage of the process, we solve the branch lengths by analogy to the three-taxon tree in Figure 5.37a.

Let us suppose that OTUs A and B in Figure 5.37b were the first OTUs to be clustered together in the tree reconstruction process. By analogy to the three-taxon tree, we refer to A and B as X and Y, respectively, and put all the other OTUs (C, D, and E) into a composite OTU denoted as Z. By this arrangement, we can apply Equations 5.25a–c and 5.26a–c to estimate the lengths of the branches leading to X, Y, and Z, except that now $d_{XZ} = d_{A(CDE)} = (d_{AC} + d_{AD} + d_{AE})/3$, and $d_{YZ} = d_{B(CDE)} = (d_{BC} + d_{BD} + d_{BE})/3$. Calculating the branch lengths a and b is done by using Equations 5.26a and 5.26b and substituting a for x and b for y. OTUs A and B are subsequently considered as a single composite OTU.

In the next step, suppose that the composite OTU (AB) and the simple OTU C were the next pair to be joined together. By analogy to the three-taxon tree in Figure 5.37a, we denote OTUs (AB) and C by X and Y, respectively, and put the other OTUs (D and E) into a new composite OTU Z. In the same manner as above, we obtain x, y, and z. Note that $y = c$ and $x = f + (a + b)/2$.

The process is continued until all branch lengths are obtained. (Note that sometimes an estimated branch length can be negative. Since the true length can never be negative, the common practice is to replace such an estimate by 0.)

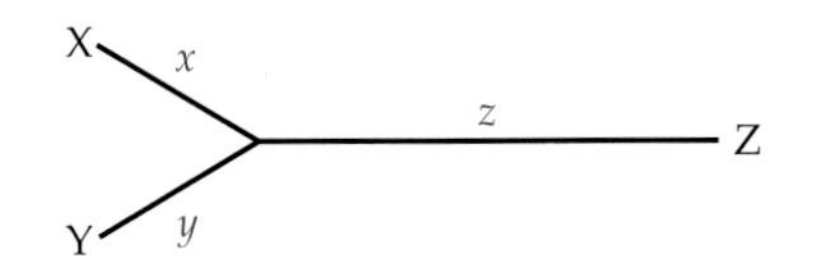

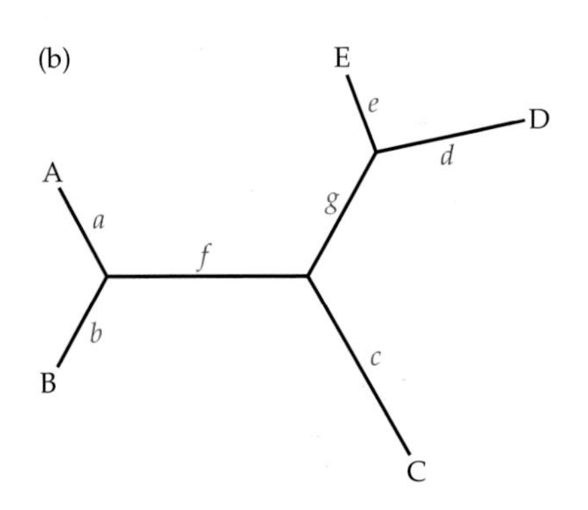

Figure 5.37 Unrooted phylogenetic trees used to compute branch lengths by Fitch and Margoliash's (1967) method. (a) A tree with three OTUs. (b) A tree with five OTUs.

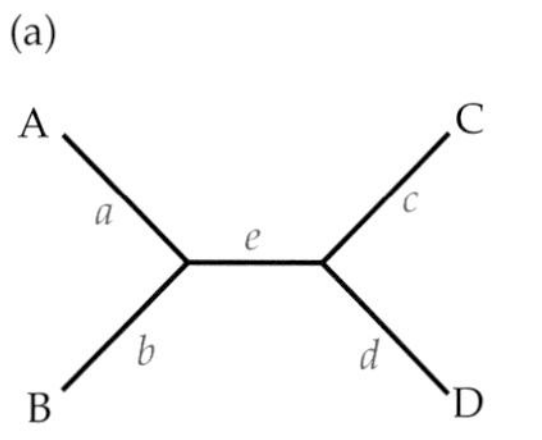

(b)

OTU	A	B	C
B	8		
C	7	9	
D	12	14	11

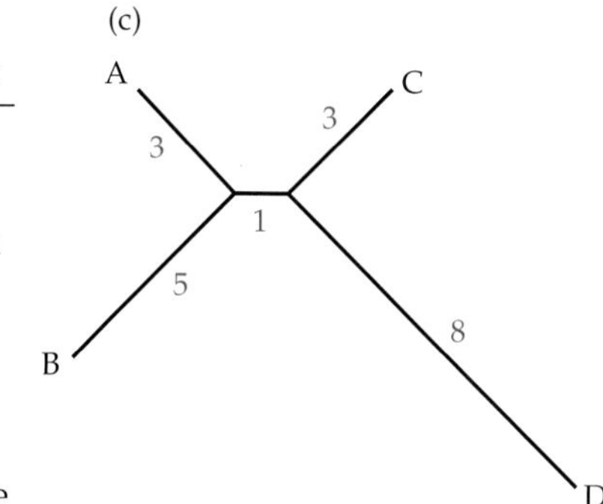

Figure 5.38 Estimating branch length of a phylogenetic tree. The prior information consists of the unrooted topology (a), the distance matrix for the four OTUs (b), and the fact that when the neighbors-relation method was used, OTU A and OTU B clustered first. The inferred branch lengths by the method of Fitch and Margoliash (1967) are shown to scale in (c).

As an example of using the above method, let us compute the branch lengths of the unrooted tree in **Figure 5.38**. We have three pieces of information: (1) the topology (Figure 5.38a), (2) the distance matrix (Figure 5.38b), and (3) the fact that when Sattath and Tversky's neighbors-relation method was used, OTUs A and B clustered first.

We start by putting OTUs C and D into a composite OTU (CD). We then have $d_{AB} = 8$, $d_{A(CD)} = (d_{AC} + d_{AD})/2 = (7 + 12)/2 = 9.5$, and $d_{B(CD)} = (d_{BC} + d_{BD})/2 = 11.5$. From Equations 5.26a–c, we have $a = (8 + 9.5 - 11.5)/2 = 3$, and $b = (8 + 11.5 - 9.5)/2 = 5$. Next we treat OTUs A and B as a composite OTU and denote it by (AB). We then have $d_{(AB)C} = (d_{AC} + d_{BC})/2 = (7 + 9)/2 = 8$, $d_{(AB)D} = (d_{AD} + d_{BD})/2 = (12 + 14)/2 = 13$, and $d_{CD} = 11$. From Equations 5.26a–c we have $c = (8 + 11 - 13)/2 = 3$ and $d = (13 + 11 - 8)/2 = 8$. Given the inferred lengths of the terminal branches, the central branch is inferred as $e = 1$. Branch lengths are shown on the scaled unrooted tree in Figure 5.38c.

Calibrating Phylogenetic Trees and Estimating Divergence Times

Molecular data may be used to date divergence events provided that (1) the rates of evolution for the lineages under study are approximately uniform, and (2) reliable calibration points are available.

Let us first assume that the rate of evolution for a DNA sequence is known to be r substitutions per site per year. To obtain the divergence time, T, between species A and B, we compare the orthologous sequences from the two species and compute the number of substitutions per site, K. As shown in Chapter 4, the rate of substitution is $r = K/2T$. Therefore, T is estimated as

$$T = \frac{K}{2r} \tag{5.27}$$

If the rate of nucleotide substitution is not known, we may use a known divergence time as **calibration**. The known calibration time may be associated with a divergence event between an outgroup taxon (O) and the common ancestor of A and B, e.g., T_O in **Figure 5.39**, or it may be associated with a divergence event between an ingroup taxon (I) and either A or B, e.g., T_I.

Let us assume that we know T_O. Then, the rate of nucleotide substitution, r, is estimated by

$$r = \frac{K_{AO} + K_{BO}}{2(2T_O)} \tag{5.28}$$

The unknown divergence time, T, between species A and B is, then, estimated by

$$T = \frac{K_{AB}}{2r} = \frac{2K_{AB}T_O}{K_{AO} + K_{BO}} \tag{5.29}$$

Conversely, if T_I is known, then

$$r = \frac{K_{AI}}{2T_I} \tag{5.30}$$

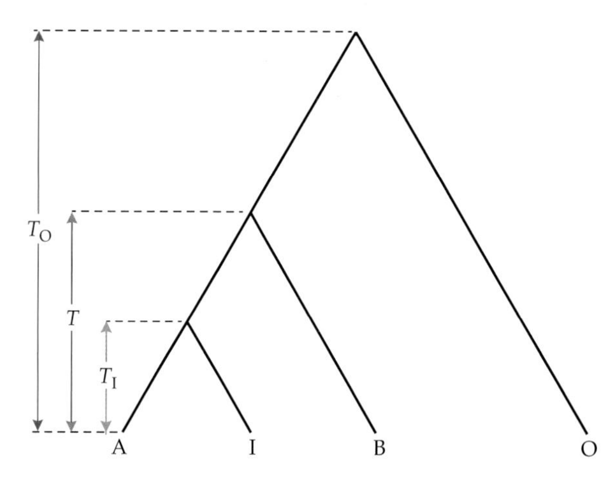

Figure 5.39 Model tree for estimating the divergence time, T, between species A and B. T_I is the divergence time between species A and an ingroup species, I. T_O is the divergence time between the ancestor of species A and B and an outgroup species, O.

In this case, T is estimated as

$$T = \frac{K_{AB}T_I}{K_{AI}} \tag{5.31}$$

As is evident from the above equations, our calculations require calibrations, i.e., "known" divergence times. Ideal calibrations for molecular phylogenetics should be based on data that are not used to build the phylogenetic tree. Typically, calibrations are derived from paleontological data. Here, we describe in a schematic manner how to calibrate a molecular phylogenetic tree by using fossil data. In the case of the phylogeny depicted in **Figure 5.40**, we are interested in estimating a time range for the divergence between A and B. These two taxa delineate an **inclusive clade** to which two other extant taxa (E1 and E2) and two extinct taxa (F1 and F2) belong. We will further assume that we have comprehensive fossil and temporal data from three fossil-yielding strata, 1, 2, and 3. Since it is highly unlikely that paleontologists will ever find the first member of a clade, by definition the oldest fossil belonging to a clade must be younger—sometimes much younger—than the clade to which it belongs. In our case, the oldest fossil belonging to the inclusive clade is F2 from stratum 2, which indicates that the divergence between A and B occurred prior to the time F2 existed.

In actual practice, a fossil's age is customarily defined as a range bounded by the minimum and maximum age estimates of the stratum in which it was found. In our case, the range for F2 is $T_{2(\min)}$ to $T_{2(\max)}$. Thus, $T_{2(\max)}$ may be used as the age of **minimum constraint** or **minimum bound**; in other words, the divergence between A and B must have occurred before $T_{2(\max)}$.

Setting a **maximum constraint** or **maximum bound** for a divergence event is a much more difficult task, because we are looking for absence of fossils belonging to the inclusive clade. In other words, we are looking for "evidence of absence," while the empirical data can only provide us with "absence of evidence." In our hypothetical example, the next oldest stratum is stratum 3. It contains no taxa belonging to the inclusive clade delineated by A and B. $T_{3(\min)}$ can, therefore, be used as a maximum bound. Summing up, we may use the **range calibration** $T_{2(\max)}$ to $T_{3(\min)}$ for the divergence between A and B. We note that if the maximum bound is too old, the range may be too large to be useful in the context of dating divergence events.

The range calibration $T_{2(\max)}$ to $T_{3(\min)}$ may only be used to infer the times of divergence of taxa within the inclusive clade, such as the divergence times between E1 and E2 or E2 and A. One cannot use calibrations outside the inclusive clade to infer

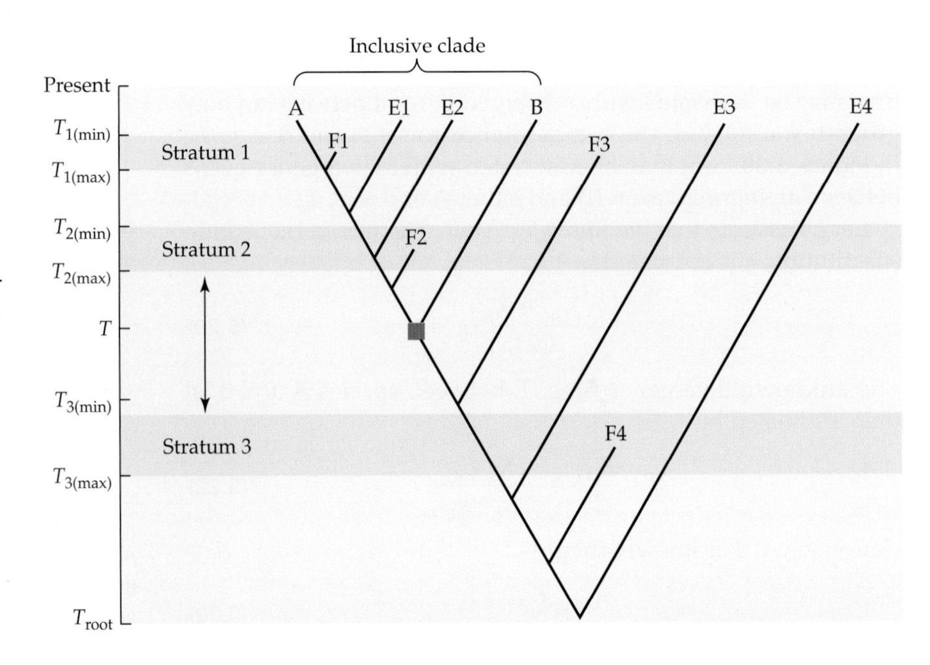

Figure 5.40 Illustration of the principles of calibrating a molecular tree. Taxa A and B delineate an inclusive clade consisting of extant taxa A, E1, E2, and B and extinct taxa F1 and F2. Extant taxa outside the inclusive clade are denoted by E3 and E4. Fossil taxa outside the inclusive clade are denoted by F3 and F4. Well-surveyed fossil-yielding strata for which temporal data exist are shown as shadowed boxes. Each stratum is bound by minimum and maximum dates. For dating divergence events within the inclusive clade, that is, between extant taxa A and E1, we need an estimate of the divergence time, T, between extant taxa A and B (square). The range calibration is marked by the double-pointed arrow. F3 is the closest outgroup taxon to the inclusive clade. Interestingly, this taxon is younger than the oldest ingroup taxon, F2.

times of divergence within the inclusive clade, say between E3 and A. This practice has been shown by Graur and Martin (2004) to yield phantasmagorical estimates. It should be noted, however, that such questionable estimates are still being published (e.g., Hedges et al. 2015).

In the literature, we find different types of calibration (Ho and Phillips 2009). In many cases, a single **point calibration** is used. For example, the divergence between birds and mammals is frequently said to be 310 million years old (e.g., Kumar and Hedges 1998). The use of point calibrations constitutes a logically untenable practice, which fortunately is becoming rarer and rarer in the literature. A **range calibration** is a calibration specified as a time range defined by two bounds. Sometimes, a single-bound **minimum calibration** is given, i.e., that the divergence event occurred before a certain date. Very rarely, a single-bound **maximum calibration** is given, i.e., that the divergence event occurred after a certain date. In the literature, a distinction is sometimes made between **hard bounds** (absolute dates) and **soft bounds** (the divergence event used in the calibration may have occurred with smaller and smaller probabilities after the minimum bound or before the maximum bound).

It is currently agreed that using as many calibration points as possible is better than using one or a few calibration points (Rambaut and Bromham 1998; Weir and Schluter 2008). We note, however, that the many calibrations suggested in the literature may not always agree with one another in all details (Benton and Donoghue 2007; Gandolfo et al. 2008; Parham and Irmis 2008; Wilke et al. 2009; Ksepka et al. 2011).

In addition to calibration, there are many other sources of error that need to be addressed when using molecular data to estimate divergence times. Particular attention should be paid to whether or not the assumption of rate constancy holds. Furthermore, there are errors associated with estimating the number of substitutions, as well as problems associated with among-site variation, ancestral polymorphism, the fact that the divergence among sequences may have predated the divergence among species, and rate variation among lineages. One methodological problem is particularly insidious; it concerns the fact that the confidence intervals around divergence time estimates are asymmetrical. For example, a mean divergence estimate of 23 million years ago may have a 95% confidence interval of 16 to 33 million years (Morrison 2008). Finally, the use of fossils may be problematic, in particular because of uncertainties regarding the , one should pay attention to how certain or uncertain the placement of a fossil is and how certain phylogenetic position of a fossil on a phylogenetic tree. Finally, the incompleteness of the fossil record should be considered when looking for "evidence of absence."

In the literature, methods that can reduce the effects of errors have been proposed (Nei and Li 1979; Li and Tanimura 1987b; Rodriguez et al. 1990; Yang 1993; Lake 1994; Lockhart et al. 1994; Steel 1994; Takezaki et al. 1995; Sanderson 1997; Avise et al. 1998; Edwards and Beerli 2000; Ho et al. 2005; Drummond et al. 2007). Notwithstanding these precautions, estimated divergence times should be treated with plenty of caution.

To date, molecular data have been used not only to date the divergence time among taxonomic groups, but also to determine the impacts of climatic and geological events on diversification (Cooper and Penny 1997; Weir and Schluter 2004; Weir 2006), estimate rates of speciation and extinction (Baldwin and Sanderson 1998; Barraclough and Vogler 2002; Weir and Schluter 2007), determine the timing of dispersal events (Drovetski 2003; Mercer and Roth 2003), and date the origin of gene families (Robinson-Rechavi et al. 2004; Vandepoele et al. 2004). Many of the results, however, are very controversial (Li 1999; Graur and Martin 2004; Penny 2005; Ho 2007; Ho et al. 2008).

Assessing Tree Reliability

Phylogenetic reconstruction is a problem of statistical inference (Edwards and Cavalli-Sforza 1964). Therefore, one should assess the reliability of the inferred phylogeny and its component parts. After inferring a phylogenetic tree, we should ask two questions: (1) How reliable is the tree, or more particularly, which parts of the tree are reliable?

and (2) Is this tree significantly better than another tree? To answer the first question, we need to assess the reliability of the internal branches of the tree. This can be accomplished by several analytical or resampling methods. In phylogenetic studies, one resampling method, the bootstrap, has become very popular and is discussed in the next section. To answer the second question, we need statistical tests for the difference between two phylogenetic trees; in other words, we need to determine whether the difference between two trees is significant or is within the expectation range of random error.

The bootstrap

The **bootstrap** is a computational technique for estimating a statistic for which the underlying distribution is either unknown or difficult to derive analytically (Efron 1979, 1982). Since its introduction into phylogenetic studies by Felsenstein (1985), the bootstrap technique has been frequently used as a means to estimate the confidence level of phylogenetic hypotheses. Here, we will only deal with one type of bootstrapping, called **nonparametric bootstrapping** (Efron 1979). Nonparametric bootstrapping belongs to a class of methods called **resampling techniques** because it estimates the sampling distribution by repeatedly resampling data from the original sample data set.

Figure 5.41 illustrates the bootstrapping procedure in phylogenetics. The data sample consists of five aligned sequences from five OTUs. From these data, a phylogenetic tree was constructed, in this case by the maximum parsimony method. The inferred tree is the null hypothesis to be tested by the bootstrap. Note that this particular null hypothesis consists of two subhypotheses: (1) OTUs 3 and 4 are neighbors, and (2) OTUs 2 and 5 are neighbors.

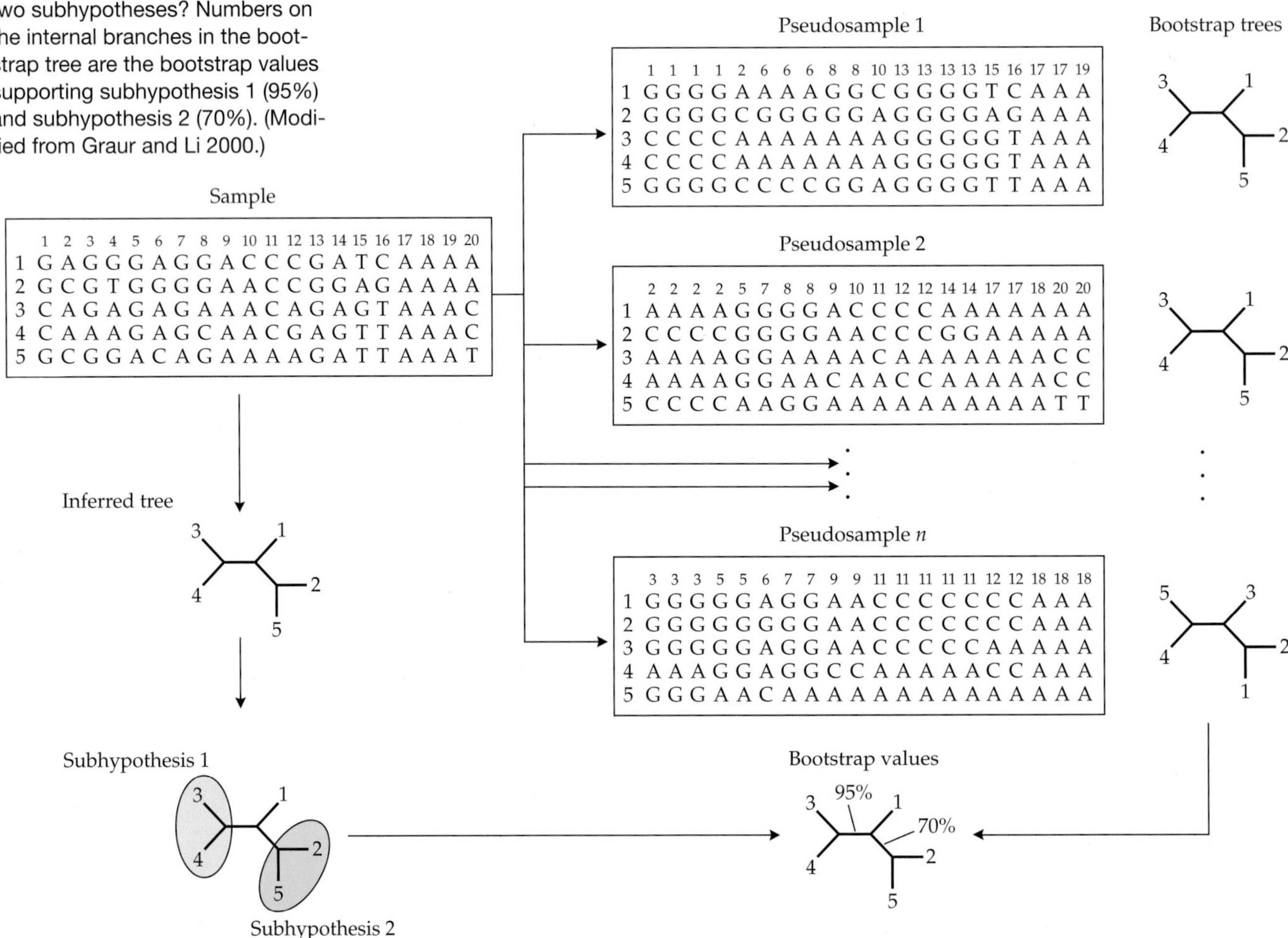

Figure 5.41 The bootstrap technique. Homologous sequences were sampled from five OTUs (1–5). From the data sample an inferred phylogenetic tree was built. The sample was also used to generate *n* pseudosamples by site resampling with replacement. From each of these pseudosamples, a bootstrap tree was generated by using the same method of phylogenetic reconstruction as that employed in the derivation of the inferred tree. The inferred tree was used as a null hypothesis composed of two subhypotheses. According to subhypothesis 1, OTUs 3 and 4 are neighbors. According to subhypothesis 2, OTUs 2 and 5 are neighbors. We ask, What percentage of the *n* bootstrap trees support each of the two subhypotheses? Numbers on the internal branches in the bootstrap tree are the bootstrap values supporting subhypothesis 1 (95%) and subhypothesis 2 (70%). (Modified from Graur and Li 2000.)

To estimate the confidence levels of these subhypotheses, we generate a series of n pseudosamples (usually 500–1,000) by resampling the sites in the sample data with replacement. Sampling with replacement means that a site can be sampled again and again with the same probability as any other site regardless of its resampling history. Consequently, each pseudosample contains sites that are represented several times and sites that are not represented at all. For example, in pseudosample 1, site 1 is represented four times, while sites 3 and 4 are not represented at all. Each pseudosample has the same aligned length as the original sample.

Each pseudosample is, then, used to construct a tree by the same method used for the inferred tree. Subhypothesis 1 is given a score of 1 if OTUs 3 and 4 are neighbors in a bootstrap tree, but a score of 0 otherwise. The score for subhypothesis 2 is similarly decided. The scores for each of the two subhypotheses are added up for all n trees, thus obtaining a **bootstrap value** for each subhypothesis. Bootstrap values are expressed as percentages and are indicated on the internal branches defining the clades. In our particular example, the branch leading to OTUs 3 and 4 is supported by 95% of the bootstrap replicates, while the branch leading to OTUs 2 and 5 is supported by only 70% of the bootstrap replicates.

The statistical properties of bootstrap in the context of phylogenetics are quite complex (Zharkikh and Li 1992a,b, 1995; Felsenstein and Kishino 1993; Hillis and Bull 1993), and there is no consensus on interpreting bootstrap values (Soltis and Soltis 2003). One interpretation due to Felsenstein and Kishino (1993) is that the bootstrap is a conservative estimate of a type I error. According to this approach, if the bootstrap proportion for an internal branch is P, then $1 - P$ equals the probability of a type I error, that is, the probability of falsely accepting an internal branch that is not there. Zharkikh and Li (1992a,b) and Hillis and Bull (1993), however, have shown that bootstrapping tends to underestimate the confidence level at high bootstrap values and overestimate it at low values. Thus, more than 95% of the clades that have a bootstrap value of 85% are correct. Conversely, only about 30% of the clades that have a bootstrap value of 40% are correct. Other researchers view bootstrap values as relative assessments of branch support rather than individual statements of absolute statistical validity (Sanderson 1989; Hillis and Bull 1993). This approach has reached something resembling consensus among practitioners of molecular phylogenetics, although not among statisticians.

There are additional difficulties in the interpretation of results obtained by the bootstrap approach (Felsenstein 1985). First, the bootstrap statements cannot be used for joint confidence statements. That is, for two clades each supported by a bootstrap value of 95%, we might have lower confidence than $(0.95 \times 0.95) \times 100 = 90\%$ in the statement that both clades are present in the true tree.

In the literature, the bootstrap resampling process is often repeated only 100 times, but this number is too low; at least several hundred pseudosamples should be used, particularly when many species are involved. This can be very time-consuming, especially if a computationally time-consuming method such as maximum likelihood is used.

A common practice seen in the literature is to "reduce" the inferred tree by collapsing branches that are associated with bootstrap values that are lower than a certain critical value (**Figure 5.42**). The resulting tree is, of course, multifurcated. By using topological comparisons (pp. 198–200) between a simulated "true" tree and an inferred one, it has been shown that collapsed trees are more similar to the true tree than the original inferred tree (Berry and Gascuel 1996).

We would like to summarize this section with Efron et al.'s (1996) poignant observation that bootstrapping in molecular phylogeny merely provides a "reasonable first approximation to the actual confidence levels of the observed clades."

Tests for two competing trees

Several tests have been devised for testing whether one phylogeny is significantly better than another. Such tests exist for each of the three types of tree reconstruction

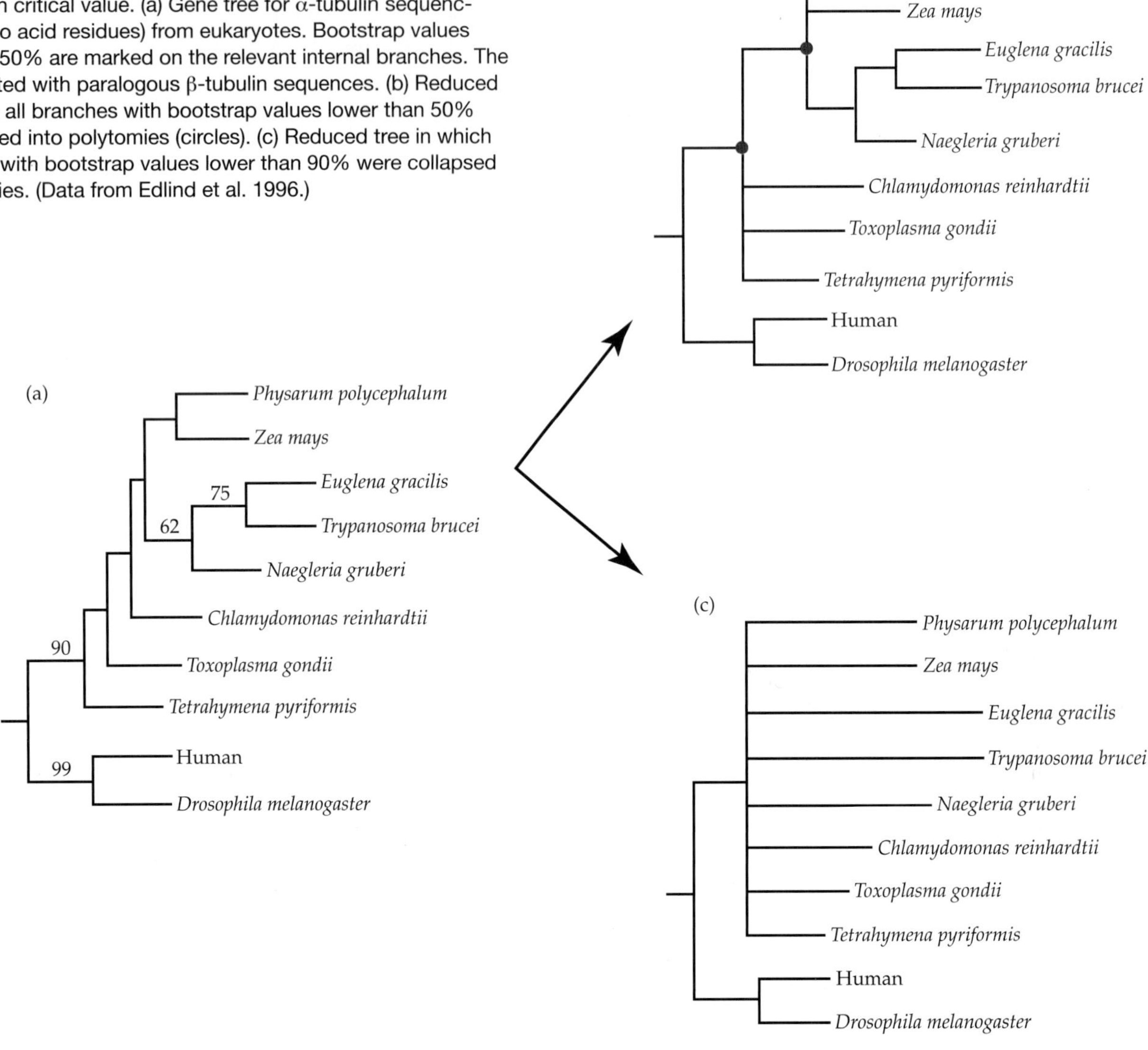

Figure 5.42 Reduction of a phylogenetic tree by the collapsing of internal branches associated with bootstrap values that are lower than a certain critical value. (a) Gene tree for α-tubulin sequences (430 amino acid residues) from eukaryotes. Bootstrap values greater than 50% are marked on the relevant internal branches. The tree was rooted with paralogous β-tubulin sequences. (b) Reduced tree in which all branches with bootstrap values lower than 50% were collapsed into polytomies (circles). (c) Reduced tree in which all branches with bootstrap values lower than 90% were collapsed into polytomies. (Data from Edlind et al. 1996.)

methods (distance matrix, maximum parsimony, and maximum likelihood). Here we present a simple test for testing maximum parsimony trees against alternative phylogenies. For the other methods, readers should consult Rzhetsky and Nei (1992), Tateno et al. (1994), Huelsenbeck and Crandall (1997), and Nye et al. (2006).

Kishino and Hasegawa (1989) devised a parametric test for comparing two trees under the assumption that all nucleotide sites are independent and equivalent. The test uses the difference in the number of nucleotide substitutions at informative sites between the two trees, D, as a test statistic, where $D = \Sigma D_i$, and D_i is the difference in the minimum number of nucleotide substitutions between the two trees at the ith informative site. The sample variance of D is

$$V(D)=\frac{n}{n-1}\sum_{i=1}^{n}\left(D_i-\frac{1}{n}\sum_{k=1}^{n}D_k\right)^2 \qquad (5.32)$$

where n is the number of informative sites. The null hypothesis that $D = 0$ can be tested with the paired t-test with $n - 1$ degrees of freedom, where

$$t=\frac{D/n}{\sqrt{V(D)}\sqrt{n}} \qquad (5.33)$$

Problems Associated with Phylogenetic Reconstruction

No method of phylogenetic reconstruction can be claimed to be better than others under all conditions. Each of the methods of phylogenetic reconstruction has advantages and disadvantages, and each method can succeed or fail depending on the nature of the evolutionary process, which is by and large unknown. Most important is that there are parts of trees that are by their nature difficult to reconstruct. For example, ancient nodes and short branches and, in particular, ancient nodes attached to short branches are extremely difficult to reconstruct (Salichos and Rokas 2013).

Strengths and weaknesses of different methods

UPGMA works well only if rate constancy holds. Its main advantage is the speed of computation. However, speedy algorithms are currently available for other distance matrix methods, and UPGMA is used nowadays for pedagogic purposes only.

The neighbors-relation and the neighbor-joining methods are free of systematic error if the distance data satisfy the four-point condition. The performance of these methods, however, depends on the method used to transform the raw character state data into distances. To the extent that the methods used do not compensate adequately for multiple substitutions at a site, the performance of both methods may be compromised. When the distances are small and the sequences used are long, fairly accurate estimates of distances can be obtained, and these methods may perform well even under nonconstant rates of evolution. Indeed, as noted by Saitou and Nei (1987), when the distances are small, one may even use the Hamming distances, i.e., the observed uncorrected number of differences between two sequences (Chapter 3), and still get the correct tree. Note, however, that if the sequences are short, then the distance estimates are subject to large statistical errors. Moreover, if the rate varies greatly among sites, then the accurate estimation of distances may be unattainable (Chapter 3). Under any of these situations, the performance of the neighbor-joining method may suboptimal. The main advantage of this method is that computation is very fast, and it can be used on enormous numbers of OTUs.

Maximum parsimony methods make no explicit assumptions, except that a tree that requires fewer substitutions is better than one that requires more. Note that a tree that minimizes the number of substitutions also minimizes the number of homoplasies, i.e., parallel, convergent, and back substitutions (Chapter 3). When the degree of divergence between sequences is small and homoplasies are rare, the parsimony criterion usually works well. However, when the degree of divergence is large and homoplasies are common, maximum parsimony methods may yield faulty phylogenetic inferences. In particular, if some sequences have evolved much faster than others, homoplasies are likely to have occurred more often among the branches leading to these sequences than among others, and parsimony may result in erroneous trees. In other words, maximum parsimony methods may perform poorly whenever some branches of the tree are much longer than the other branches, because parsimony will tend to cluster the long branches together (Felsenstein 1978). This phenomenon is called **long-branch attraction** or the **Felsenstein zone** (**Figure 5.43**). Note also that the chance of homoplasy depends on the substitution pattern. For example, if transitions occur more often than transversions, then the chance of homoplasy will be higher than that for the case of equal substitution rates among the four nucleotides. Some of these effects can be remedied by using weighted parsimony (e.g., Swofford 1993), in which the transitional bias is taken into account.

It has been argued that character state methods are more powerful than distance methods because the raw data is a string of character states (e.g., the nucleotide sequence) and that in transforming character state data into distance matrices, some information is lost. We note, however, that while the maximum parsimony method indeed uses the raw data, it usually uses only a small fraction of the available data. For instance, in the example on page 189, only three sites are used while six sites

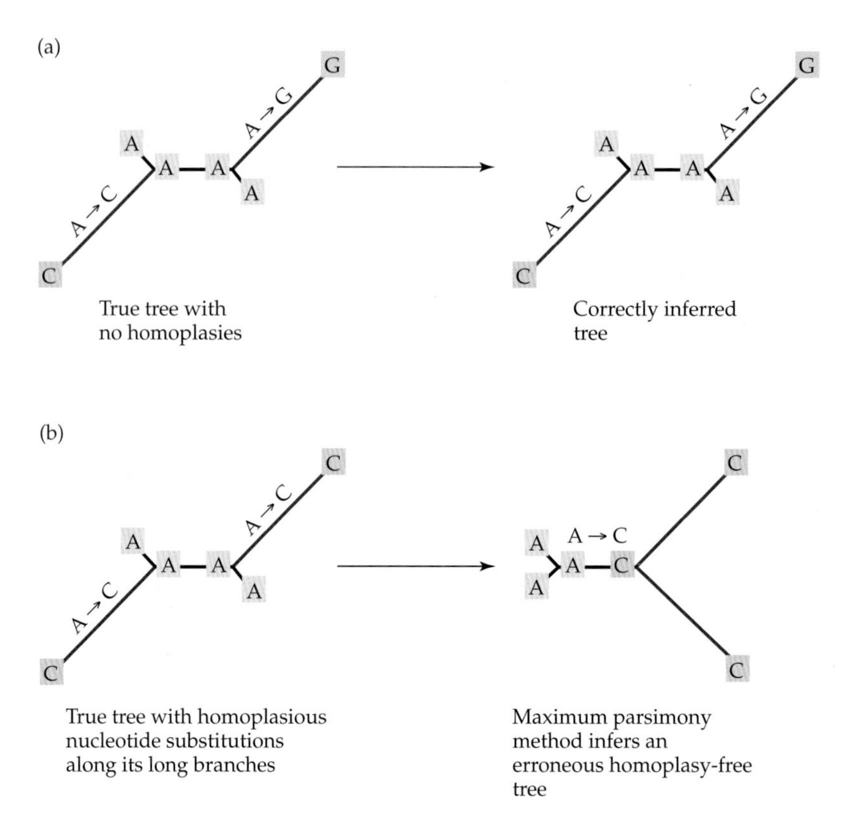

Figure 5.43 The long-branch attraction phenomenon as it affects phylogenetic reconstruction when the true tree has long branches. The phenomenon is illustrated by means of a true four-taxon unrooted tree with two long branches, in which each long branch neighbors a short branch. The letters represent the nucleotides at the terminal and internal nodes. (a) The nucleotide position shown has not experienced homoplasic nucleotide substitutions. Then, by using maximum parsimony we will be able to infer the correct tree. (b) By chance, however, a site may experience homoplasic nucleotide substitutions along the two long branches. As a consequence, the maximum parsimony method will yield an erroneous tree, in which the long branches are inferred to be neighbors. The reason for this error is that the true tree requires two nucleotide substitutions, whereas the erroneous tree requires only a single nucleotide substitution (i.e., it is more parsimonious). In reality, however, branch lengths matter. That is, the probability of nucleotide substitution is proportional to branch length.

are excluded from the analysis. For this reason, this method is often less efficient in identifying clades than some distance matrix methods. Of course, if the number of informative sites is large, the maximum parsimony method is generally very effective.

In maximum parsimony methods, we are required to compare all possible trees. This comparison is feasible only when the number of OTUs is small and the sequences under study are not too long. When the number of sequences and their aligned lengths are large, computation time may become prohibitive for exhaustive search and a heuristic search must be employed. Unfortunately, heuristic approaches do not guarantee obtaining the maximum parsimony tree.

The maximum likelihood method uses the character state information at all sites, and thus can be said to use the "full" information; however, it requires explicit assumptions on the rate and pattern of nucleotide substitution. It has been commonly believed that this method is relatively insensitive to violations of assumptions, but in some studies the method proved to be not very robust (e.g., Tateno et al. 1994). In other words, likelihood methods may perform poorly if the stochastic model used is unrealistic and if some sequences are highly divergent. The main disadvantage of the maximum likelihood method is that its computation is enormously tedious and time-consuming. As in maximum parsimony, the maximum likelihood method requires consideration of all the possible alternative trees, and for each tree it searches for the maximum likelihood length for each branch (Kuhner and Felsenstein 1994). Thus, when the number of OTUs is large, it uses heuristic approaches to reduce the number of trees to be considered (e.g., Felsenstein 1981; Saitou 1988). In such cases, we may not obtain the maximum likelihood tree. Bayesian methods are even more time-consuming, and some studies suggest that they suffer from "overcredibility," i.e., overestimating the confidence of a tree or a clade (Suzuki et al. 2002; Misawa and Nei 2003).

Minimizing error in phylogenetic analysis

Several strategies are available to minimize random and systematic errors in phylogenetic analysis. However, it is not always possible to identify the potential sources of error or bias. Here, we list several do's and don'ts that may increase our chances of recovering the true phylogenetic tree.

The best way to minimize random errors is to use large amounts of data. All other things being equal, a tree based on large amounts of molecular data is almost invariably more reliable than one based on a more limited amount of data. When the sequences do not provide sufficient phylogenetic information (e.g., because they are too short or lacking in variation), no phylogenetic method will produce a sensible result. This said, one should only include reliable data in the analysis. By that we mean that the analysis should be limited to sequences that have been reliably determined, and for which positional homology is certain. (We note, however, that elimination of data deemed "unreliable" may be subjective and arbitrary.) Moreover, we should only use sequences that evolve at an appropriate rate for the phylogenetic question under investigation. Fast-evolving sequences (or parts of sequences, such as third codon positions) should be used for questions regarding close phylogenetic relationships, and slow-evolving sequences should be used for distant phylogenetic relationships. Choosing incorrectly may result in lack of phylogenetic information in the case of slow-evolving sequences, or saturation effects in the case of fast-evolving sequences.

One way to reduce the chance of systematic error leading to inconsistency—i.e., resulting in a wrong inference even when the amount of data is large—is to use more realistic models or more suitable methods of analysis to better match the data. An inferred tree is only as good as the assumptions on which the method of phylogenetic reconstruction is based. For example, base-composition biases are known to have a pronounced effect on phylogenetic reconstruction, and most methods will incorrectly group OTUs with similar base composition (Mooers and Holmes 2000). Some methods seem to be more robust than others to base-composition variability among the taxa (e.g., Lockhart et al. 1994).

Sometimes it is worth examining the assumption of independent evolution among sites. For instance, when dealing with RNA sequences that tend to form internal hairpin structures, nucleotide changes on one side of the stem are frequently matched by complementary changes on the other side of the stem. In these cases, we should consider counting two such changes as one.

Phylogenetic studies often use sequence data from different DNA regions. If all the regions studied have similar rates of nucleotide substitution, then all the data can be combined into a single set. However, if considerable variation in rates exists, regions with different rates should be analyzed separately. In this case, however, it may be difficult to combine the results from the different data sets and to assess the reliability of the clades within the consensus tree.

Phylogenetic errors are expected to be worse with larger distances among the OTUs than with smaller distances. Therefore, having many long distances will tend to compound the problems arising from the long-branch attraction phenomenon. At times it may be advisable to remove long branches from the analysis.

For a character to be useful in a phylogenetic context, it should be informative and reliable. That is, it should provide us with true evolutionary information. Some characters are both informative and reliable. Others are reliable but do not tell us anything useful about the phylogenetic relationships of interest and are thus uninformative. The third category, consisting of "misinformative" characters, is the most problematic. Identification of these unreliable characters is of crucial importance. For example, we know that maximum parsimony methods yield wrong phylogenies when there are many homoplasies in the data. Since fast-evolving characters are likely to result in homoplasies more often than slow-evolving characters, it is advisable to give such characters a lower weight in the analysis. One extreme form of weighting is the elimination of unreliable characters, for example, the elimination of all transitions in transversion parsimony.

Finally, we should realize that, regardless of the precautions taken, inferred trees often contain errors.

Genome Trees

Genome trees are a means of capturing the preponderance of phylogenetic information that is present in genomes. In other words, genome trees attempt to reduce the gene-tree noise due to such processes as incomplete lineage sorting, homoplasic events, and incorrect identification of orthologous sequences, which may obscure the true phylogeny of the species. For example, phylogenetic reconstructions based on 1,070 putatively orthologous genes from 23 yeast genomes resulted in 1,070 distinct gene trees, which were all incongruent with the phylogeny obtained by concatenating the 1,070 genes (Salichos and Rokas 2013). In other words, each gene tree was different from any other gene tree, which is not surprising given that with 23 OTUs, there are approximately 3×10^{23} possible unrooted trees. There are several classes of tree-making methodologies that can be applied to whole genomes (Snel et al. 2005). Some methods, such as those based on nucleotide composition or frequency of particular n-mers (sequences of length n), do not have a solid evolutionary justification and will, hence, be ignored. Some methods, such as those based on mean genetic similarity, i.e., summing up the genetic distances and dividing the sum by the number of genes, or those based on concatenating all genes into a single sequence, are too trivial for further elaboration. Here we present two methods of phylogenetic reconstruction based on shared genomic content. The possibility of using gene-order distances between two genomes in phylogenetics will be discussed in Chapter 10.

Genome trees based on shared gene content

A convenient way to describe a complete genome is by its gene content. As the sharing of genes is such a straightforward and logical approach to comparing genomes, many methods for making gene-content trees have been suggested in the literature. Here, we will describe a very simple approach. First, orthologous genes are identified. Next, one can use the proportion of shared orthologs out of the total number of genes that are found in at least two species to derive a distance measure between a pair of genomes. In the final part, these distances can be used in a combination with a distance-based tree-making method. Alternatively, the presence or absence of an ortholog in a genome can be used as a character state to be used with a phylogenetic method of tree reconstruction such as maximum parsimony.

Genome trees from BLASTology

A fast-and-dirty alternative to computing numerous distances between carefully curated orthologous genes is a class of methods sometimes collectively referred to as BLASTology (BLAST, for *B*asic *L*ocal *A*lignment *S*earch *T*ool, is a database search program; Chapter 3). Here, a distance is calculated on the basis of the average sequence similarity between two genomes or proteomes, explicitly neglecting any knowledge of orthology. Let us describe the most basic approach to BLASTology imaginable (e.g., Henz et al. 2005). In this approach, a local alignment method, such as BLAST, is used to compare two genomes by BLASTing all or a subset of sequences from one genome onto the other. This approach, then, uses the resulting mean BLAST distances to build a distance matrix, and uses this matrix in the phylogenetic reconstruction.

Molecular Phylogenetic Examples

The application of molecular biology techniques and advances in tree reconstruction methodology have led to tremendous progress in phylogenetic studies, resulting in a better understanding of the evolutionary history of almost every taxonomic group. In this section we present several examples where molecular studies have (1) resolved a long-standing issue, (2) led to a drastic revision of the traditional view, (3) pointed to a new direction of research, or (4) solved a societal problem. The field of molecular phylogenetics is progressing very rapidly, however, and some of the views presented here may eventually be revised.

Phylogeny of apes

The issue of the closest living evolutionary relative of humans has always intrigued evolutionary biologists. Interestingly, Darwin was not the first scientist to address the evolutionary origins of humans. In his 1863 *Evidence as to Man's Place in Nature,* Thomas Huxley wrote, "It is quite certain that the Ape which most nearly approaches man, in the totality of its organization, is either the Chimpanzee or the Gorilla." Darwin, in his 1871 *The Descent of Man, and Selection in Relation to Sex,* vacillates among the three great apes (chimpanzee, gorilla, and orangutan) and in the end narrowly and guardedly votes for one of the African apes, with a slight preference for the gorilla. Huxley's and Darwin's choices, however, had little following, and for a very long time taxonomists followed Ernst Haeckel, who in his 1868 *Natürliche Schöpfungsgeschichte* (*The Natural History of Creation*) decreed the orangutan to be our closest relative. The genus *Homo* was therefore assigned to a family of its own, Hominidae, with chimpanzees (*Pan*), gorillas (*Gorilla*), and orangutans (*Pongo*) placed in a separate family, the Pongidae (**Figure 5.44a**). The gibbons (*Hylobates*) were classified either separately (Hylobatidae) or with the Pongidae (**Figure 5.44b**; Simpson 1961). In all these phylogenetic arrangements, humans are placed in one family, while all the other apes are placed in another. This implies that the apes share a more recent common ancestry with one another than with humans. When traditionalists reluctantly put *Homo*

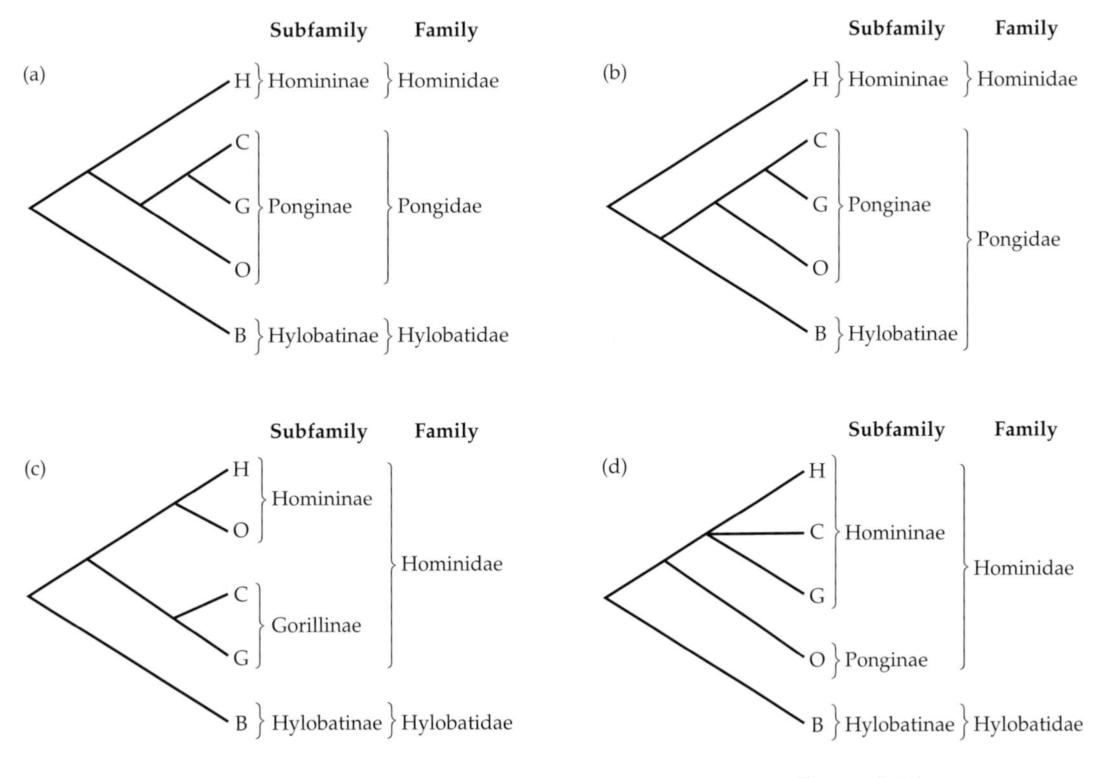

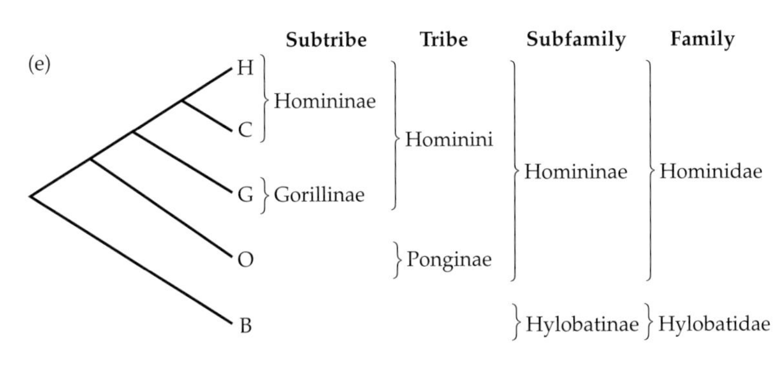

Figure 5.44 Five alternative phylogenies and classifications of extant apes and humans (Hominoidae). Traditional classifications setting humans apart are shown in (a) and (b). The clustering of humans with the orangutan is shown in (c). Cumulative molecular as well as morphological evidence favors the clustering of humans and the African apes (d). A completely resolved phylogenetic tree obtained by using genomic data is shown in (e) Species abbreviations: H, human (*Homo*); C, chimpanzee (*Pan*); G, gorilla (*Gorilla*); O, orangutan (*Pongo*); and B, gibbon (*Hylobates*). The taxonomic nomenclature follows Goodman et al. (1990). (Modified from Graur and Li 2000.)

(a) Rooted

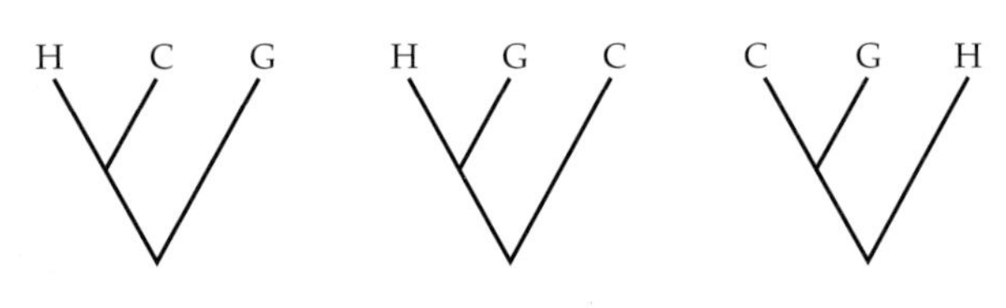

(b) Unrooted

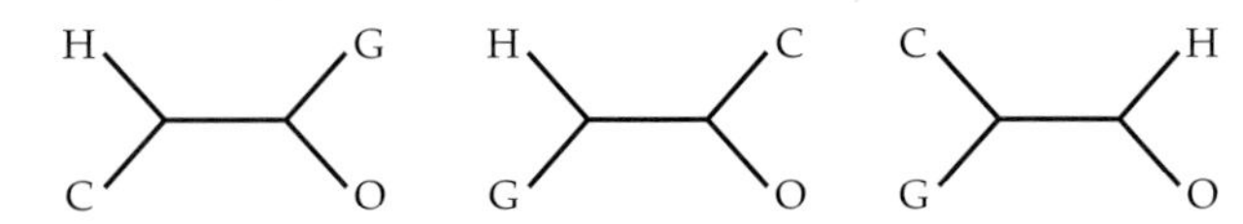

Figure 5.45 (a) Three possible rooted trees for humans (H), chimpanzees (C), and gorillas (G). (b) Comparable unrooted trees with the orangutan (O) as an outgroup. (From Graur and Li 2000.)

in the same clade with an extant ape, it was usually with the Asian orangutan, rather than an African ape (**Figure 5.44c**; Schultz 1963; Schwartz 1984; Grehan and Schwartz 2009). The clustering of humans and orangutans was heavily influenced by the "Aryan thesis," according to which humans originated in Asia (Theunissen 1988).

All these traditional morphological classifications (Simpson 1945; Martin and Martin 1990; Fleagle 1999) were heavily influenced by the Aristotelian principle of progression, and they divided the order Primates into groups that could be ranked on a scale from the primitive to the advanced according to perceived mental abilities. Goodman (1963) correctly recognized anthropocentricity in the systematic arrangements that presuppose humans to represent "a new grade of phylogenetic development, one which is 'higher' than the pongids and all other preceding grades" and suggested using molecular evidence as an objective, non-anthropocentric means of placing humans within the primate phylogenetic tree.

By using a serological precipitation method, Goodman (1962) was the first to demonstrate that humans, chimpanzees, and gorillas constitute a natural clade (**Figure 5.44d**), with orangutans and gibbons having diverged from the other apes at much earlier dates. However, serological, electrophoretic, and DNA-DNA hybridization studies, as well as amino acid sequences, could not resolve the evolutionary relationships among humans and the African apes, and the so-called human-gorilla-chimpanzee trichotomy remained unsolved and controversial, with some data favoring a chimpanzee-gorilla clade (Ferris et al. 1981; Brown et al. 1982; Hixon and Brown 1986), and others supporting a human-chimpanzee clade (Sibley and Ahlquist 1984; Caccone and Powell 1989; Sibley et al. 1990).

In **Figure 5.45** we illustrate three possible rooted trees and corresponding unrooted trees for humans, chimpanzees, and gorillas, with the orangutan as outgroup. Resolving the human-gorilla-chimpanzee trichotomy essentially means favoring one of the three trees over the other two. We will use the DNA sequence data from Miyamoto et al. (1987) and Maeda et al. (1988) to show that the molecular evidence supports the human-chimpanzee clade and, at the same time, to illustrate some of the tree-making methods discussed in the previous sections.

Table 5.2 shows the number of nucleotide substitutions per 100 sites between each pair of the following OTUs: humans (H), chimpanzees (C), gorillas (G), orangutans (O), and rhesus monkeys (R). Let us first apply UPGMA to these distances. The distance between humans and chimpanzees is the shortest ($d_{HC} = 1.45$). Therefore, we

Table 5.2

Mean and standard error[a] of the number of nucleotide substitutions per 100 sites between OTUs

OTU	Human	Chimpanzee	Gorilla	Orangutan	Rhesus monkey
Human		0.17	0.18	0.25	0.41
Chimpanzee	**1.45**		0.18	0.25	0.42
Gorilla	**1.51**	**1.57**		0.26	0.41
Orangutan	**2.98**	**2.94**	**3.04**		0.40
Rhesus monkey	**7.51**	**7.55**	**7.39**	**7.10**	

Source: Sequence data from Koop et al. (1986b) and Maeda et al. (1983, 1988).

[a]Mean values are in bold type, below the diagonal; standard error is shown above the diagonal, in regular type.

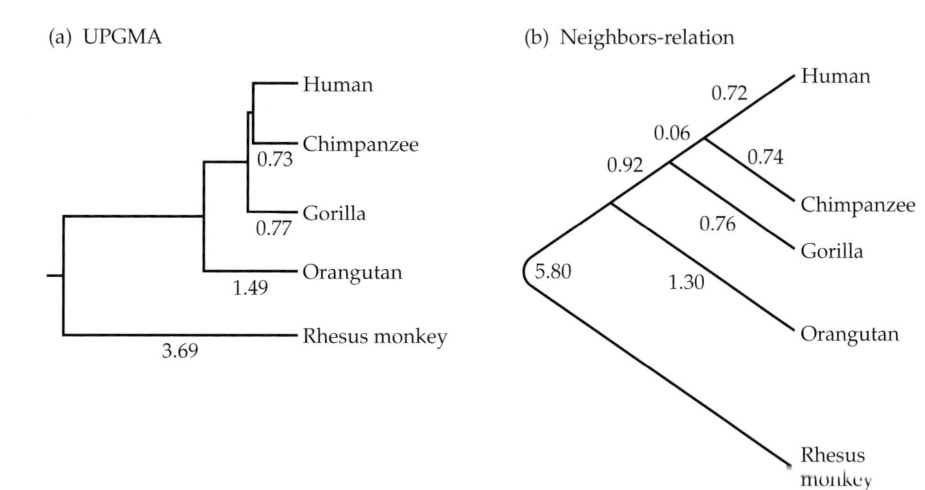

Figure 5.46 Phylogenetic tree for humans, chimpanzees, gorillas, orangutans, and rhesus monkeys inferred by using UPGMA (a) and Sattath and Tversky's neighbors-relation method (b). Note that the tree in (b) is unrooted. (From Graur and Li 2000.)

join these two OTUs first and place the node at 1.45/2 = 0.73 (**Figure 5.46a**). We then compute the distances between the composite OTU (HC) and each of the other species and obtain a new distance matrix:

OTU	HC	G	O
G	1.54		
O	2.96	3.04	
R	7.53	7.39	7.10

Since (HC) and G are now separated by the shortest distance, they are the next to be joined together, and the connecting node is placed at 1.54/2 = 0.77. Continuing the process, we obtain the tree in Figure 5.46a. We note that the estimated branching node for H and C is very close to that for (HC) and G. In fact, the distance between the two nodes is smaller than all the standard errors for the estimates of the pairwise distances among H, C, and G (Table 5.2). Thus, although the data suggest that our closest living relatives are the chimpanzees, the data do not provide a conclusive resolution of the branching order. The position of the orangutan, however, as an outgroup to the human-chimpanzee-gorilla clade is unequivocal.

Next, we use Sattath and Tversky's neighbors-relation method (p. 186). We consider four OTUs at one time. Since there are five OTUs, there are 5!/[4!(5 – 4)!] = 5 possible quadruples. We start with OTUs H, C, G, and O and compute the following sums of distances (data from Table 5.2): $d_{HC} + d_{GO} = 1.45 + 3.04 = 4.49$, $d_{HG} + d_{CO} = 4.45$, and $d_{HO} + d_{CG} = 4.55$. Since the second sum is the smallest, we choose H and G as one pair of neighbors and C and O as the other. Similarly, we consider the four other possible quadruples; the results are shown in **Table 5.3**. Noting from the bottom of that table that (OR) has the highest neighbors-relation score among all neighbor pairs, we choose (OR) as the first pair of neighbors. Treating this pair as a single composite OTU, we obtain the following new distance matrix:

OTU	H	C	G
C	1.45		
G	1.51	1.57	
(OR)	5.25	5.25	5.22

Because only four OTUs are left, it is easy to see that $d_{HC} + d_{G(OR)} = 6.67 < d_{HG} + d_{C(OR)} = 6.76 < d_{H(OR)} + d_{CG} = 6.82$. Therefore, we choose H and C as one pair of neighbors, and G and (OR) as the other. The final tree obtained by this method is shown in **Figure**

Table 5.3

Neighbors-relation scores obtained from the distance matrix in Table 5.2

OTUs compared[a]	Sum of pairwise distances	Neighbor pairs chosen
H,C,G,O	$d_{HC} + d_{GO} = 4.49$	(HG), (CO)
	$d_{HG} + d_{CO} = 4.45$	
	$d_{HO} + d_{CG} = 4.55$	
H,C,G,R	$d_{HC} + d_{GR} = 8.84$	(HC), (GR)
	$d_{HG} + d_{CR} = 9.06$	
	$d_{HR} + d_{CG} = 9.08$	
H,C,O,R	$d_{HC} + d_{OR} = 8.55$	(HC), (OR)
	$d_{HO} + d_{CR} = 10.53$	
	$d_{HR} + d_{CO} = 10.45$	
H,G,O,R	$d_{HG} + d_{OR} = 8.61$	(HG), (OR)
	$d_{HO} + d_{GR} = 10.37$	
	$d_{HR} + d_{GO} = 10.55$	
C,G,O,R	$d_{CG} + d_{OR} = 8.67$	(CG), (OR)
	$d_{CO} + d_{GR} = 10.33$	
	$d_{CR} + d_{GO} = 11.59$	

Total scores: (HC) = 2, (HG) = 2, (HO) = 1, (HR) = 0, (CG) = 1, (CO) = 1, (CR) = 1, (GO) = 1, (GR) = 1, (OR) = 3

[a]H, human; C, chimpanzee; G, gorilla; O, orangutan; R, rhesus monkey.

5.46b. The topology of this tree is identical to that in Figure 5.46a. Note, however, that in this method O and R rather than H and C were the first pair to be joined to each other. This is because in an unrooted tree, O and R are in fact neighbors. The branch lengths in Figure 5.46b were obtained by the Fitch-Margoliash method (p. 204). By using the neighbor-joining method we obtain exactly the same tree as that obtained by the neighbors-relation method (Figure 5.46b).

Finally, let us consider the maximum parsimony method. For simplicity, let us consider only humans, chimpanzees, gorillas, and orangutans, as in Figure 5.45b. **Table 5.4** shows the informative sites for a 10.2 Kb alignment of homologous noncoding sequences from these apes (Koop et al. 1986a; Miyamoto et al. 1987; Maeda et al. 1988; Bailey et al. 1991). For each site, the hypothesis supported is given in the last column. If we consider nucleotide substitutions only, there are 15 informative sites, of which eight support the human-chimpanzee clade (hypothesis I), four support the chimpanzee-gorilla clade (hypothesis II), and three support the human-gorilla clade (hypothesis III). Moreover, there are four informative sites involving a gap, and they all support the human-chimpanzee clade. Therefore, the human-chimpanzee clade is chosen as the best representation of the true phylogeny.

The overall sequence evidence, currently consisting of aligned sequences of about 2 million autosomal base pairs from human, chimpanzee, gorilla, and orangutan, overwhelmingly favors the human-chimpanzee clade (Hobolth et al. 2007). The morphological, physiological, anatomical, and parasitological data yield a more mixed picture: some studies support a clustering of chimpanzee and gorilla into a monophyletic clade to the exclusion of humans (Tuttle 1967; Kluge 1983; Ciochon 1985; Andrews 1987); others support the so-called **Trogloditian hypothesis**, i.e., that *Pan* and *Homo* are sister taxa (Thirnanagama et al. 1991; Retana Salazar 1996; Shoshani et al. 1996; Gibbs et al. 2002). A comprehensive three-dimensional morphometric study on 406 adult specimens belonging to nine hominoid subspecies yielded strong support for the human-chimpanzee clade (Lockwood et al. 2004).

In summary, the closest extant relatives of humans are the two chimpanzee species (*Pan troglodytes* and *P. paniscus*), followed in order of decreasing relatedness by gorillas, orangutans, a clade consisting of gibbons and siamangs, Old World monkeys (Cercopithecidae), and New World monkeys (Platyrrhini). Surprisingly, the phylogenetic and taxonomic picture that emerges from the molecular data is quite close to the one suggested by the father of taxonomy, Carolus Linnaeus, who wrote, "As a naturalist, and following naturalistic methods, I have not been able to discover up to the present a single character which distinguishes Man from the anthropomorphs, since they comprise specimens … that resemble the human species … to such a degree that an inexpert traveler may consider them varieties of Men" (Linnaeus 1758, translated in Chiarelli 1973). In fact, Linnaeus assigned humans, chimpanzees, and orangutans to the same genus, and their original scientific names were *Homo sapiens*, *H. troglodytes*, and *H. sylvestris*, respectively. Linnaeus did not know about the existence of the gorilla, as it was only discovered in 1799 and described in 1847. Later taxonomists decided to isolate humans from the great ape. As a consequence, the orangutan and the chimpanzee were removed from the genus *Homo* by Lacepede in 1799 and by Oken in 1816, respectively.

Table 5.4

Informative sites among human, chimpanzee, gorilla, and orangutan sequences

Site[a]	Sequence[b]				Hypothesis supported[c]
	Human	Chimpanzee	Gorilla	Orangutan	
Data from Miyamoto et al. (1987)					
34	A	G	A	G	III
560	C	C	A	A	I
1287	*	*	T	T	I
1338	G	G	A	A	I
3057–3060	****	****	TAAT	TAAT	I
3272	T	T	*	*	I
4473	C	C	T	T	I
5153	A	C	C	A	II
5156	A	G	G	A	II
5480	G	G	T	T	I
6368	C	T	C	T	III
6808	C	T	T	C	II
6971	G	G	T	T	I
Data from Maeda et al. (1988)					
127–132	******	******	AATATA	AATATA	I
1472	G	G	A	A	I
2131	A	A	G	G	I
2224	A	G	A	G	III
2341	G	C	G	C	III
2635	G	G	A	A	I

Source: Modified from Williams and Goodman (1989).

[a]Site numbers correspond to those given in the original sources. The total length of the sequence used is 10.2 Kb, about twice that used in Table 5.2.

[b]Each asterisk denotes a one-nucleotide indel at the site.

[c]Hypotheses: I, human and chimpanzee are sister taxa; II, chimpanzee and gorilla are sister taxa; and III, human and gorilla are sister taxa.

A proposal was put forward (Goodman et al. 1998, 1999; Wildman et al. 2003) for a partial restoration of Linnaean taxonomy by inclusion within the genus *Homo* of three extant species: human (*Homo sapiens*), common chimpanzee (*Homo troglodytes*), and bonobo (*Homo paniscus*). The reasons for this suggestion are threefold. First, the molecular phylogeny of *Homo* and its relatives indicates conclusively that these three species are evolutionarily closer to one another than any of them is to either gorilla or orangutan. Second, it is currently accepted that taxonomic classification should follow phylogenetics, i.e., a valid taxonomic unit should not be paraphyletic. Third, according to the principle of priority in the International Code of Zoological Nomenclature, the valid name for an animal taxon is the oldest available name that applies to it. Isn't it time to abide by the rules and restore Linnaeus' terminology?

Before the advent of sequence data, paleontologists estimated that the divergence time between humans and the other primates was at least 15 million years ago and probably a great deal earlier. The downward revision started with Sarich and

Wilson (1967), whose microcomplement fixation data yielded divergence times between humans and either chimpanzees or gorillas on the order of 5 million years ago. In the literature, molecular estimates of the human-chimpanzee divergence time vary widely, from less than 3 million years ago (Hasegawa et al. 1985) to almost 14 million years ago (Áranson et al. 1996). A recent estimate in the literature, which is based on massive genomic data and a calibration point of 18 million years ago between human and orangutan, points to an extremely recent divergence date, on the order of 4 million years ago, and the divergence between gorilla and the ancestor of human and chimpanzee at about 6 million years ago (Hobolth et al. 2007). Jensen-Seaman and Hooper-Boyd (2008) noticed that calibration is the main factor affecting estimates of divergence time. Their study of the fossil record indicates that two calibration ranges may be used. The first is for the *Homo-Pongo* split between 23 and 13 million years ago; the second is for the divergence between apes and Old World monkeys between 37 and 23 million years ago. They note that "within these ranges, there is no reason to assume the actual divergence date was near the middle of the range, or for that matter, that any date within the range is more likely than any other date." Under these calibration ranges, the date of split between human and chimpanzee is 6–8 million years ago. Finally, Wilkinson et al. (2011) used both molecular and paleontological data in conjunction with numerous likelihood and Bayesian models to reach a range of 5.3–10.0 million years for the *Homo-Pan* split. Hara et al. (2012) estimated the human-chimpanzee divergence time to be 5.9–7.6 million years, with the gorilla and orangutan lineages diverging 7.6–9.7 and 15–19 million years ago, respectively.

Genomic data from human and chimpanzee support an almost "instantaneous" allopatric model of speciation (speciation by physical isolation) with very little evidence for gene flow following the split between the ancestral populations of *Homo* and *Pan* (Innan and Watanabe 2006; Wakeley 2008; Presgraves and Yi 2009; Webster 2009; Yamamichi et al. 2012).

The utility of polarized character states: Cetartiodactyla and SINE phylogeny

As mentioned previously, some character states can be easily polarized in relationship to one another as ancestral or derived. Let us now discuss one such example involving the insertion of repeated sequences—SINES (short interspersed nuclear elements) and LINES (long interspersed nuclear elements) (Chapter 9). The retroposition process that generates a new SINE or a new LINE insertion is duplicative, i.e., the parent element is retained at its original location while giving rise to copies that insert at new sites throughout the genome. SINE and LINE insertions are thought to be free of the problem of convergence, as the vast number of available target sites within a eukaryotic genome and the nearly random integration pattern of such elements mean that the chance of precisely the same insertion occurring more than once is negligible. SINEs and LINEs are similarly accepted to be free of reversals, as no specific mechanism is known to facilitate their precise deletion. Of course, they can be lost through nonspecific deletion, but the chance that such a loss will exactly match the boundaries of an element is thought to be negligible. As a final bonus, the duplicative nature of retroposition confers an evolutionary directionality onto the characters, from the plesiomorphic (ancestral) "absence" state to the apomorphic (derived) "presence" state. Thus, the presence of a SINE or a LINE at a specific locus in several species may be treated as a synapomorphy. Indeed, SINEs and LINEs have been used to great effect in molecular phylogeny, and here we outline one such case.

The more than 80 species of whales, dolphins, and porpoises, which form the order Cetacea, are among the most fascinating and spectacular of all eutherians (placental mammals). They possess an elaborate communication system indicative of an advanced social structure, and the physical bulk of some cetaceans far exceeds that of the largest dinosaurs. The origin of Cetacea has been an enduring evolutionary mystery since Aristotle, for the transition from terrestriality to an exclusive aquatic lifestyle required an unprecedented number of unique yet coordinated changes in many biological systems. For example, cetaceans have experienced a unique gradual evolutionary reduction of the posterior limbs to a point where the animals have an extremely small pelvis and hindlimbs that do not extend out of the body wall and they are unique in their swimming by dorsoventral oscillations of a heavily muscled tail. In the context of phylogenetics, these unique morphological, anatomical, and behavioral traits constitute autapomorphies for the Cetacea, and they cannot be used to determine the phylogenetic affinities of this order within the eutherian tree.

A link between cetaceans and ungulates (hoofed mammals) was suggested more than a century ago by Flower (1883) and Flower and Garson (1884) on the basis of comparative anatomical information. This view was accepted by Gregory (1910), but two of the most influential paleontologists of the century, Simpson (1945) and Romer (1966), suggested that the cetacean lineage goes back to the very root of the eutherian tree. Flower's view was later endorsed by Van Valen (1966) and Szalay (1969), who argued, mainly on the basis of dental characters, for a connection between cetaceans and mesonychid condylarths, an assemblage of ungulates known primarily from the Paleogene and Eocene epochs. The first paleontological evidence for a connection between cetaceans and artiodactyls (even-hoofed ungulates) was provided by the remains of a middle Eocene (~45-million-year-old) whale exhibiting an artiodactyl-like paraxonic arrangement of the digits on its vestigial hind limbs (Gingerich et al. 1990; Wyss 1990). The discovery by Gingerich et al. (1994) and Thewissen et al. (1994) of the 50-million-year-old fossil cetacean *Ambulocetus natans* in Pakistan provided some insight into the terrestrial–aquatic transition (Novacek 1994).

The first molecular evidence for a close relationship between Cetacea and Artiodactyla was revealed by Boyden and Gemeroy (1950) on the basis of precipitin tests. Sequence evidence started accumulating in the 1980s. Goodman et al. (1982) analyzed seven protein sequences and concluded that Cetacea is a sister taxon of Artiodactyla. This conclusion received further support from studies on mitochondrial DNA sequences (e.g., Irwin et al. 1991; Milinkovitch et al. 1993; Cao et al. 1994).

The order Artiodactyla is traditionally divided into three suborders: Suiformes (pigs and hippopotamuses), Tylopoda (camels and llamas), and Ruminantia (deer, elk, giraffes, pronghorn, cattle, goats, and sheep). Graur and Higgins (1994) inferred the phylogenetic position of Cetacea in relation to the three artiodactyl suborders by using protein and DNA sequence data from cow, camel, pig, several cetacean species, and an outgroup. Their phylogenetic analysis suggested that cetaceans are not only intimately related to the artiodactyls, but deeply nested within the artiodactyl phylogenetic tree, i.e., more closely related to some members of the order Artiodactyla (e.g., Ruminantia) than some artiodactyls are to one another. Thus, the artiodactyls do not constitute a monophyletic clade, unless the cetaceans are included within the order. Cetartiodactyla is currently the preferred name of the clade consisting of artiodactyls and cetaceans (Montgelard et al. 2007).

An unambiguous resolution of cetacean evolutionary affinities has been obtained by Shimamura et al. (1997) and Nikaido et al. (1999), who used the insertion patterns of SINEs and LINEs to resolve the cetartiodactyl phylogenetic tree. **Figure 5.47** illustrates the principles of phylogenetic inference by using SINEs. First, a SINE is identified in a certain species. Then the 5′ and 3′ primers around the SINE unit are used to identify uniquely its genomic location (Figure 5.47a). If the surroundings of the SINE are conserved during evolution, they may be used with the polymerase chain reaction (PCR) to amplify the homologous loci from the genomic DNA of the other species under study (Figure 5.47b). The PCR products are then subjected to electrophoresis, which separates them according to length. A long PCR product indicates the presence of a SINE unit; a short PCR product indicates absence (Figure 5.47c). To ensure that the insertions are indeed homologous (the same SINE at the exact same location), the PCR products may be subsequently sequenced and compared. Because a SINE insertion is essentially an irreversible character state, the presence of a SINE at a specific locus in several species may be treated as a synapomorphy defining a monophyletic clade (Figure 5.47d). For example, the pattern at locus 2 indicates that species A, B, and

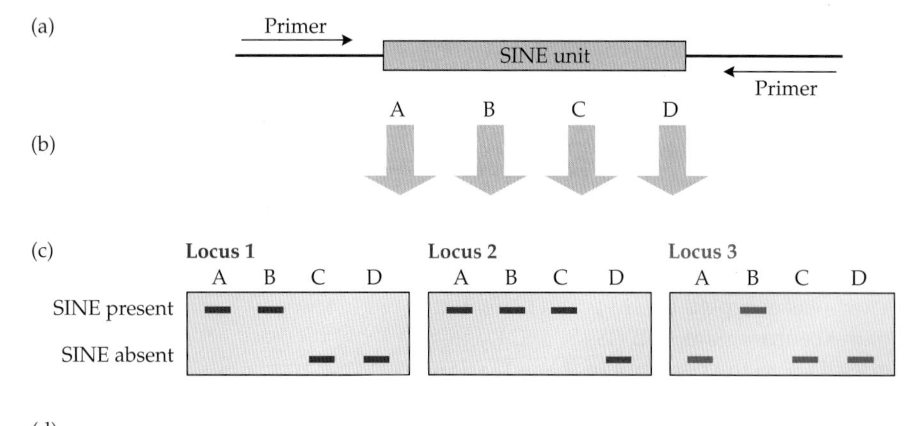

Figure 5.47 Inference of phylogeny from insertion patterns of SINEs. (a) The primers identify the genomic location (locus) of a SINE unit. (b) PCR is used to amplify the homologous loci from the genomic DNA of several species under study (A, B, C, and D). (c) The PCR products are subjected to electrophoresis and separation by length. A long PCR product indicates the presence of a SINE unit; a short PCR product indicates absence. (d) Because SINE insertion is essentially an irreversible character state, the presence of a SINE at a certain locus may be treated as a synapomorphy defining a monophyletic clade (arrowheads 1 and 2) or as an autapomorphy for a single taxon (arrowhead 3). (Courtesy of Norihiro Okada.)

C belong to a monophyletic cluster. SINEs, LINEs, and other transposable elements have become powerful tools in molecular phylogenetics (Shedlock and Okada 2000; Shedlock et al. 2004), and they have been used to untangle many mind-boggling phylogenies, such as that of African cichlid fishes (Terai et al. 2003).

The traditional phylogeny of cetaceans and artiodactyls is shown in **Figure 5.48a**. Shimamura et al. (1997) and Nikaido et al. (1999, 2001) identified synapomorphic and autapomorphic SINE and LINE insertions all over the genome of cetaceans and artiodactyls and used them to reconstruct the phylogeny of cetaceans and artiodactyls (**Figure 5.48b**). For example, they found two *CHR-1* SINEs in Pecora and Tragulina that were not found in any other organisms. One of these SINEs was found in the third intron of the gene for the α subunit of the pituitary glycoprotein hormone; the other was in the third intron of the gene for steroid 21-hydroxylase. These two SINEs demonstrate the monophyly of ruminants.

As shown in Figure 5.48b, four SINE synapomorphies unequivocally indicate that hippopotamuses (Ancodonta) are the closest extant relatives of whales. Incidentally, pigs and peccaries (Suina) were found to be unrelated to hippopotamuses, and the molecular evidence therefore invalidates the monophyly of the suborder Suiformes.

Molecular Phylogenetic Archeology

Ancient DNA is defined as any genetic material recovered from biological samples that have not been preserved specifically for later DNA analysis. In the forensic literature, the term often refers to biological material older than 75 years (Graham 2007). Examples of sources of ancient DNA include archaeological and historical remains; archival, medical, herbarium, and museum specimens; biological material in permafrost; and prefossilized paleontological remains. The field of ancient DNA studies generated much excitement in the 1980s and 1990s, and for a while it seemed that it would revolutionize evolutionary biology. In 1990, Michael Crichton published the novel *Jurassic Park* about the resurrection of dinosaurs by using DNA from nucleated blood cells found in the gastrointestinal tracts of blood-sucking insects, which were trapped in tree sap that later turned into amber. At one point in the novel, a 1200-bp "dinosaur" sequence is shown. On inspection, however, the sequence turned out to be almost identical to pBR322, a ubiquitous genetically engineered cloning vector (Boguski 1992)—a clear contamination that had nothing to do with an animal, let alone a dinosaur. Regrettably, this *Jurassic Park* anecdote turned out to be emblematic of much of the history of the field of ancient DNA studies.

DNA is an unstable molecule that decays spontaneously through hydrolysis, oxidation, and methylation. The chances of unprotected DNA surviving over long periods of time are very low, unless special conditions exist for its preservation. Theoretical calculations suggest that DNA should not survive for more than 10,000–100,000 years, and then only in highly fragmented form (Lindahl 1993a). Informally, potential sources of ancient DNA may be divided into **antediluvian** (literally, before the flood) and **postdiluvian** (Lindahl 1993b). (Obviously, these designations have nothing to do with the mythological flood in the book of Genesis.) It is now universally agreed that antediluvian DNA is not recoverable. The question that remains to be answered is at which age and under what conditions is a potential source of genetic material antediluvian? All reports of DNA recovery from protected and unprotected paleontological sources, such as Miocene plant fossils (Golenberg et al. 1990), dinosaur fossilized bones and eggs (Woodward et al. 1994; An et al. 1995; Li et al. 1995), amber-entombed organisms (Cano et al. 1993a,b), and a 250-million-year-old bacteria (Vreeland et al. 2000), have now been thoroughly discredited (e.g., Austin et al. 1997; Walden and Robertson 1997; Gutiérrez and Marín 1998; Graur and Pupko 2001). One piece of "dinosaur" DNA (Woodward et al. 1994) turned out to be a mitochondrial insertion into the nuclear human genome (Zischler et al. 1995; Bensasson et al. 2001); other "dinosaurs" (An et al. 1995; Li et al. 1995) exhibited uncanny molecular similarity to fungal symbionts of leaf-cutting ants and to plants (Wang et al. 1997). Even sequences from much younger material, such as Neanderthal bones (Green et al. 2006), proved to be of questionable quality (Wall and Kim 2007). There is a very good reason for bookstores to keep *Jurassic Park* in the fiction aisle.

Even postdiluvial samples are difficult to authenticate. For example, the mitochondrial DNA extracted from an Egyptian mummy less than 5,000 years old (Pääbo 1985) turned out to be an artifact (Höss et al. 1994; Marchant 2011). Samples have been shown to be contaminated by old and new DNA—even the reagents used in PCR to amplify DNA have amplifiable animal DNA in them (Leonard et al. 2007).

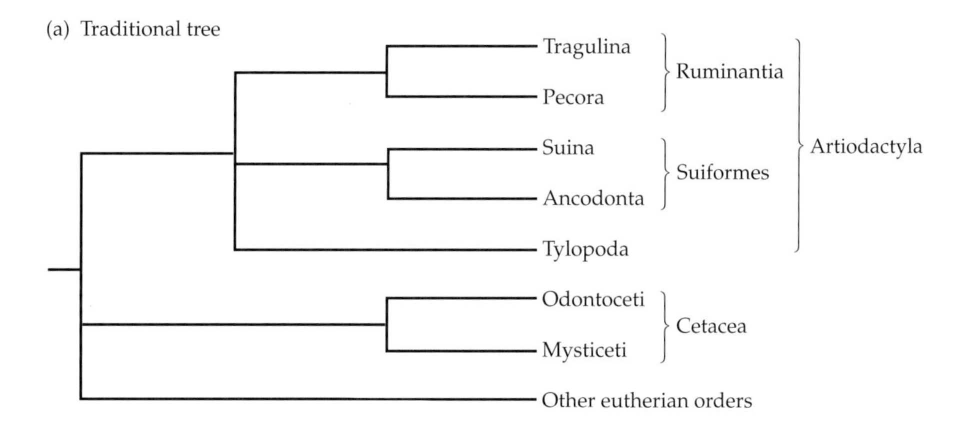

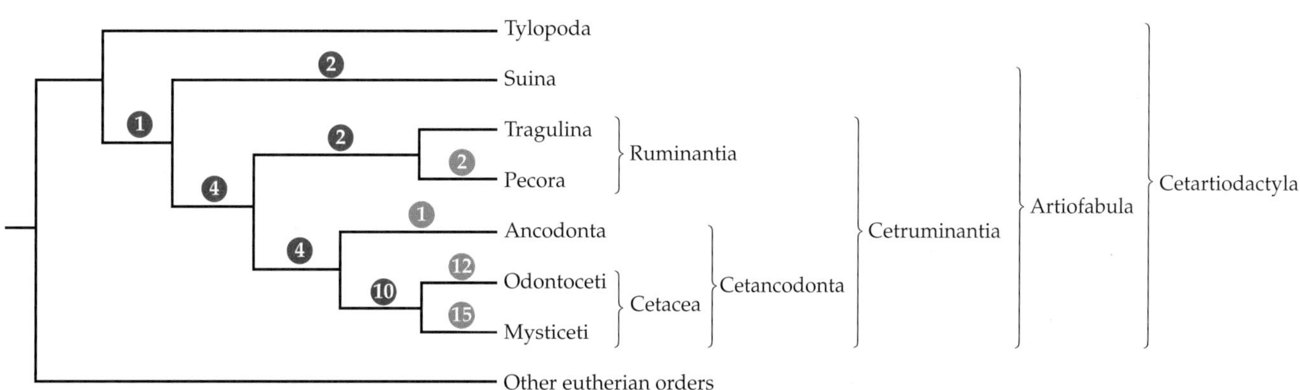

Figure 5.48 (a) Traditional phylogenetic tree and taxonomic nomenclature for cetaceans and artiodactyls. The order Artiodactyla is divided into three suborders: Tylopoda (camels and llamas), Suiformes, and Ruminantia. The phylogenetic relationships among these suborders are morphologically indeterminate. Suborder Suiformes is divided into two infraorders: Suina (pigs and peccaries) and Ancodonta (hippopotamuses). Suborder Ruminantia is divided into two infraorders: Tragulina (mouse deer, or chevrotains) and Pecora (deer, elk, giraffes, pronghorn, cattle, goats, and sheep). The order Cetacea, composed of Odontoceti (toothed whales) and Mysticeti (baleen whales), may or may not be related to Artiodactyla. (b) Molecular phylogenetic tree and revised taxonomic nomenclature for cetartiodactyls. The taxonomic nomenclature follows Montgelard et al. (2007). Numbers of SINE and LINE insertions are listed above the branches. Those within red or green circles represent synapomorphies or autapomorphies, respectively. (Data from Nikaido et al. 1999, 2007.)

When dealing with ancient genetic material, it is important to assess whether or not any postmortem changes in the DNA have occurred. In a concatenated 229 bp sequence from a 140-year-old skin sample, Higuchi et al. (1987) detected two postmortem modifications. Both modifications were transitions that could be attributed to the postmortem deamination of cytosine to uracil. Therefore, some nucleotides in ancient DNA samples may contain postmortem changes. Cooper and Poinar (2000) outlined nine criteria of authenticity for ancient DNA sequences, but sadly their rules are not always followed in practice (Bandelt 2005). These transgressions notwithstanding, there is no doubt that genetic information, albeit in minute quantities, may be preserved in biological material whose age is of the order of a few hundred or a few thousand years. This enables us to use molecular phylogenetic techniques on recently extinct species.

The disextinction of the quagga

When Boer settlers arrived in the Karoo regions of the Cape of Good Hope in the seventeenth century, they found grasslands teeming with herds of zebralike animals, which the natives called quagga and taxonomists labeled accordingly *Equus quagga*. The quagga's coat was distinguished from other zebras by having anterior zebralike stripes, a paucity or absence of stripes over the posterior parts, and a rich brown background color over the upper parts. Due to uncontrolled hunting, the once ubiquitous animal was driven to extinction within a little over 200 years (Hughes 1988). It is uncertain when the last quagga died; the last animal shot in the wild was reported in 1876, and a few specimens seem to have survived in European zoos into the 1880s. The female quagga that died in Amsterdam's Artis Royal Zoo on August 12, 1883 quite possibly marked the extinction of this taxon (Harley 1988).

The phylogenetic affinities of the quagga have always been controversial. Bennett (1980), for instance, placed the quagga in a clade with the domestic horse (*E. caballus*) but apart from the plains zebra (*E. burchelli*), the mountain zebra (*E. zebra*), and Grevy's zebra (*E. grevyi*). In contrast, Eisenmann (1985) clustered the quagga with the zebras, apart from the horse, and Rau (1974) considered the quagga as merely a color variant of the plains zebra, a conclusion supported by radioimmunoassay comparisons (Lowenstein and Ryder 1985).

The first molecular phylogenetic study of the quagga, which was based on a comparison of 229-nucleotide-long mitochondrial sequences, revealed no differences between the quagga and the plains zebra (Higuchi et al. 1987), supporting the view that the quagga is at most a subspecies of *E. burchelli*. More extensive studies were carried out by Leonard et al. (2005) and Lorenzen et al. (2008). These studies showed conclusively that quaggas are not distinct from plains zebras (**Figure 5.49**); they are but one subspecies within one zebra species. This phylogenetic assessment necessitated reconsideration of the morphological evidence previously used to cluster the horse and the quagga (Bennett 1980). For instance, the dental similarities between the quagga and the horse, which were thought to be shared derived character states (synapomorphies), had to be reinterpreted as shared primitive character states (symplesiomorphies). That is, these dental characters have been retained in the horse and quagga lineages but were lost in the other zebras. In addition, the molecular findings necessitated the renaming of the quagga and the plains zebra. According to the rules of taxonomic precedence, the name of the new species that includes both the quagga and the many other plains zebra subspecies should be

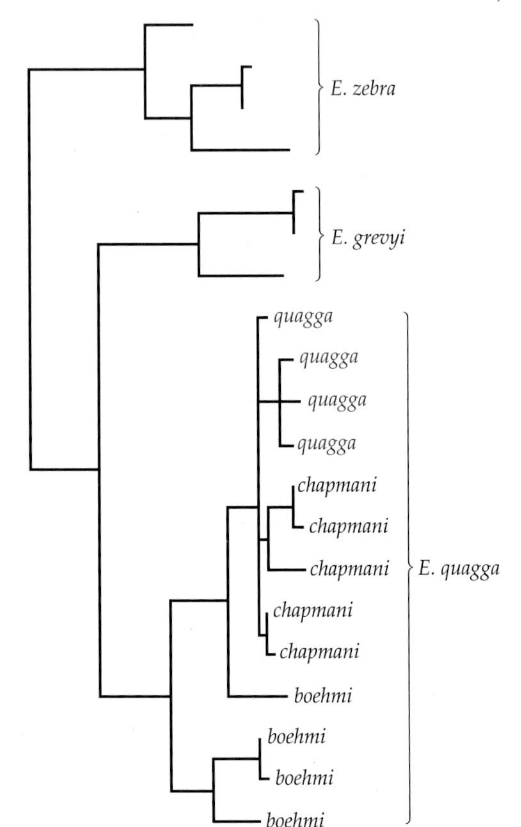

Figure 5.49 A maximum parsimony phylogeny of three zebra species based on 567-bp sequences from the mitochondrial control region. The analysis shows that quagga (*Equus quagga quagga*) is nested within the plains zebra subspecies (*E. q. chapmani* and *E. q. boehmi*). (Modified from Leonard et al. 2005.)

Equus quagga, this term having been coined earlier (1785) than *E. burchelli* (1824). Currently, *E. quagga* is thought to consist of six subspecies: the quagga (*E. q. quagga*), Chapman's zebra (*E. q. chapmani*), Grant's zebra (*E. q. boehmi*), Burchell's zebra (*E. q. burchelli*, the maneless zebra (*E. q. borensis*), and Crawshay's zebra (*E. q. crawshayi*). Incidentally, this revised taxonomy is in agreement with early twentieth century thinking (e.g., Haughton 1932; Dollman 1937).

Interestingly, the newly determined taxonomic status for the quagga provided a rationale for the establishment in 1987 of a quagga "disextinction" project aimed at "reconstructing" the extinct subspecies. The genetic basis of the breeding project relies on the molecular demonstration that the quagga is genetically indistinguishable from other plains zebra subspecies. Given that the main characters that have been used in the past to identify the quagga were its coat patterns, then a breeding program aimed at obtaining the quagga patterns is expected to yield a genuine quagga, rather than a look alike.

The dusky seaside sparrow: A lesson in conservation biology

The last dusky seaside sparrow, called Orange Band, an aging male afflicted with gout and other geriatric diseases, died on June 16, 1987, in a zoo at Walt Disney World, near Orlando, Florida (Mann and Plummer 1992). Dusky seaside sparrows were discovered in 1872, and their melanic spotted appearance led to their classification as a distinct subspecies (*Ammodramus maritimus nigrescens*). The geographic distribution of *A. m. nigrescens* was confined to salt marshes on Merritt Island and the St. Johns River in Brevard County, Florida (**Figure 5.50**). At the time of their discovery, the population of dusky seaside sparrows consisted of about 2,000 individuals. From 1900 onward, the birds were slowly edged out of their range as the salt marshes were flushed with fresh water in an attempt to control mosquitoes. The subspecies was declared endangered in 1967. By 1980, only six individuals, all males, could be found in nature. Obviously, the population was doomed, and an artificial breeding program was launched as a last-ditch attempt to preserve the genes of this subspecies.

In such a case, the conservation program involves the mating of the males from the nearly extinct subspecies with females from the closest subspecies available. The female hybrids of the first generation are then backcrossed to the males, their offspring are again backcrossed to the original males, and the process is continued for as long as the original males live. The crux of such an experiment is to decide from which population to choose the females, i.e., which subspecies is phylogenetically closest to the endangered one.

In the case of *A. maritimus*, there were eight recognized subspecies from which to choose. The geographical ranges of these species are shown in Figure 5.50. On the basis of morphological and behavioral characters, as well as geographic proximity, it was decided that the closest subspecies to *A. m. nigrescens* was Scott's seaside sparrow (*A. m. peninsulae*), which inhabited Florida's Gulf shores. As a consequence of this decision, several *nigrescens* males were mated with *peninsulae* females. Up to three successful backcrosses were accomplished, resulting in one individual that was 87.5% dusky and two that were 75% dusky. These were to be kept inbred with the view of someday releasing the "reconstructed" subspecies into its original habitat.

In order to find out whether the choice of females was correct, Avise and Nelson (1989) compared the restriction enzyme pattern of mitochondrial DNA from the last pure *A. m. nigrescens* specimen with that of 39 individual birds belonging to five of the eight extant

Figure 5.50 Geographic distributions (gold) of the eight taxonomically recognized subspecies of the seaside sparrow, *Ammodramus maritimus*. The former geographic distribution of the dusky seaside sparrow, *A. m. nigrescens*, is emphasized in red. (Modified from Avise and Nelson 1989.)

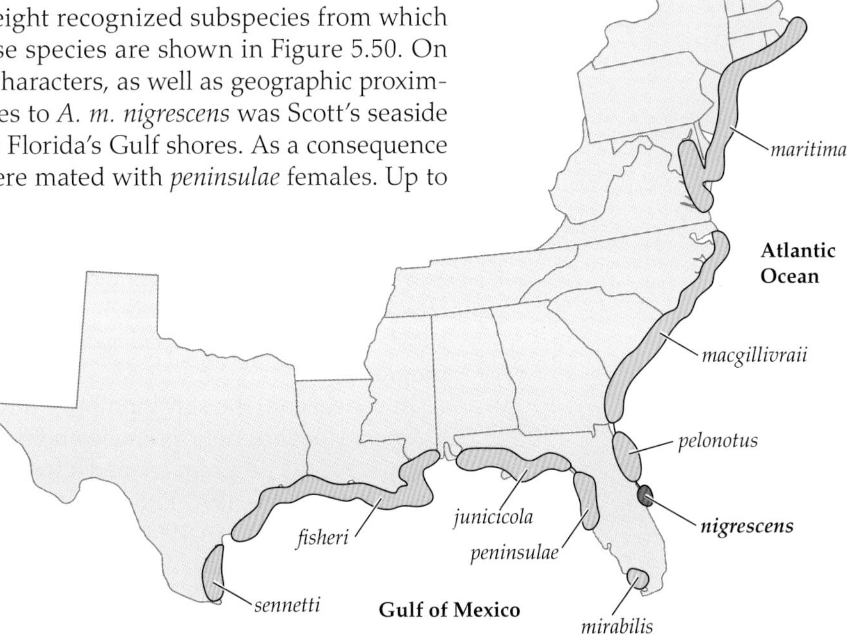

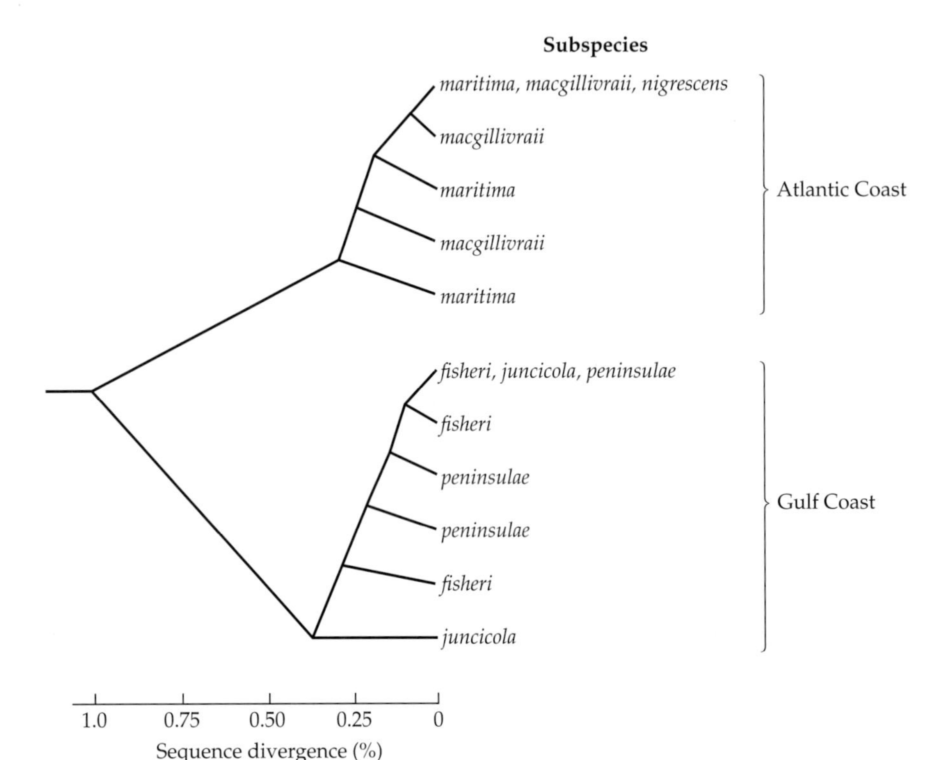

Figure 5.51 UPGMA dendrogram showing the distinction between mitochondrial DNA genotypes of the Atlantic Coast versus Gulf Coast populations of seaside sparrow. Maximum parsimony methods produced many equally parsimonious trees, including one identical in topology to the tree shown. All alternative maximum parsimony trees involved minor branch rearrangements within either the Atlantic clade or the Gulf clade, while the distinction between the two groups remained unaltered. The multiple appearance of the same subspecies name in the dendrogram indicates that different individuals belonging to the same subspecies exhibit different restriction enzyme patterns. Conversely, the appearance of several subspecies names at the end of a single branch indicates that individuals classified as different subspecies on morphological and zoogeographical grounds exhibit identical patterns for the restriction enzymes used. (From Avise and Nelson 1989.)

subspecies of *A. maritimus*. They chose mitochondrial DNA for several reasons. First, mitochondrial DNA in vertebrates is known to evolve very rapidly (Chapter 4), and hence it can provide a high resolution for distinguishing between closely related organisms. Second, mitochondria are maternally inherited, and thus complications due to allelic segregation do not arise.

From the restriction enzyme patterns, Avise and Nelson (1989) reconstructed the evolutionary relationships among several subspecies of *A. maritimus*. As seen from **Figure 5.51**, the Atlantic Coast subspecies, including *A. m. nigrescens*, are nearly indistinguishable from one another. The same is true for the three Gulf Coast subspecies in this study. In comparison, the Atlantic Coast subspecies are quite distinct from the Gulf Coast subspecies. The number of nucleotide substitutions per site between the two groups has been estimated to be about 1%. If mitochondrial DNA in sparrows evolves at about the same rate as mitochondrial DNA in well-studied mammalian and avian lineages (2–4% sequence divergence per million years), then these two groups of subspecies separated from each other some 250,000–500,000 years ago. While these estimates of the time of divergence must remain qualified, given the many sources of uncertainty, not least among which is the problem of calibration, they agree well with the dates obtained for the falling sea level that exposed the Florida peninsula, which serves as a reproductive barrier between the *A. maritimus* subspecies on the two sides of the peninsula.

Most important, Avise and Nelson's (1989) molecular study showed that while the *A. m. nigrescens* subspecies is indistinguishable from the two other Atlantic subspecies (i.e., *A. m. maritima* and *A. m. macgillivraii*), it is quite different from the Gulf subspecies such as *A. m. peninsulae* whose females had been chosen for the breeding program.

In conclusion, the salvation program of the dusky seaside sparrow rested on an erroneous phylogenetic premise and therefore, instead of reconstructing an extinct subspecies, the program created a hybrid. Indeed, the U.S. Department of the Interior ruled that the 1973 Endangered Species Act does not extend to the protection of "mongrels," and in 1990 the dusky seaside sparrow was officially declared extinct (O'Brien and Mayr 1991). The dusky seaside sparrow narrative raises questions about the role of morphological classification in decisions to take heroic measures to save endangered species. Many morphological taxonomies were erected more than a century ago on the basis of very limited and sometimes very questionable data. Clearly, molecular data should supplement morphological taxonomy. Whether the distinctiveness of the dusky subspecies was sufficiently great to have warranted the resources invested in the conservation program is an open question. What is clear, however, is that the strategy was wrong. Knowledge of phylogenetic relationships is essential in making rational decisions for the conservation of biotic diversity. A faulty taxonomy may turn even the most well-intentioned effort into an irreparable fiasco.

We note, however, that even if the breeding plan had been correct, two genomic components of the dusky seaside sparrow would have been lost. Because of the maternal mode of transmission, mitochondrial DNA from the last male dusky seaside sparrow could not have been transmitted to the hybrids in the restorative breeding program. Second, in the ZW sex-determination system of birds, males carry two Z chromosomes and no W chromosome. Thus, the W chromosome of *A. m. nigrescens* would also be extinct.

Molecular Phylogenetics and the Law

Evolutionary biology and criminal investigations have a common goal: reconstructing historical events as accurately as possible (Mindell 2009). To the best of our knowledge, the first encounter between the two occurred in the famous (or infamous) case of the "Florida dentist" (Ou et al. 1992). A group of patients who had undergone invasive procedures in this dentist's office and were subsequently found to be HIV seropositive sued the dentist's estate. A preliminary epidemiologic investigation identified seven HIV-infected patients of the dentist. Because HIV sequences evolve very rapidly, an exact match (such as those expected by DNA fingerprinting in criminal and paternity cases) could not be expected. Rather, forensic analysts needed to allow for the fact that the sequences had evolved and diverged. A phylogenetic analysis of part of the *env* gene of HIV from the dentist, his affected patients, and several HIV-positive controls, i.e., patients who had never come in contact with this particular dentist, indicated significant support for a phylogenetic hypothesis that was consistent with transmission from the dentist to five of the patients (**Figure 5.52a**). Two patients (patients D and F in the figure) seemed to have acquired the virus independently.

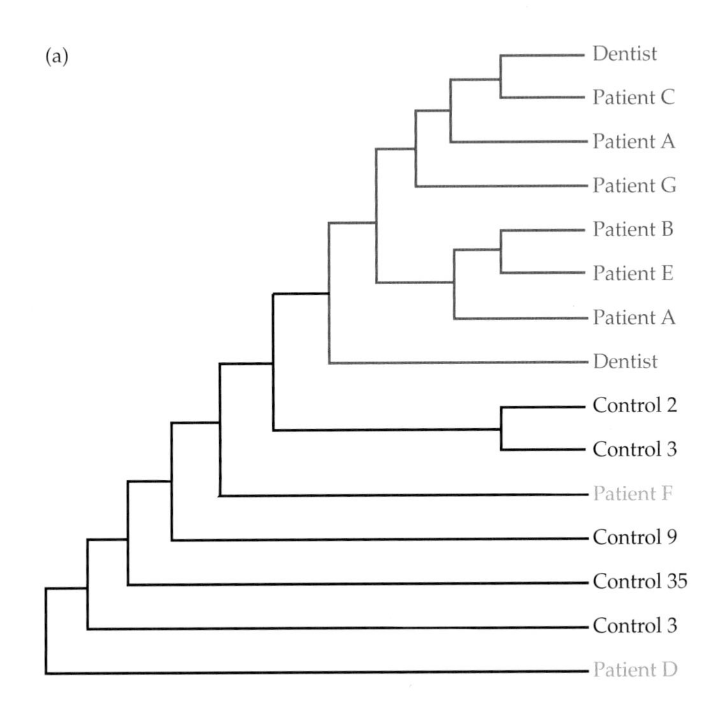

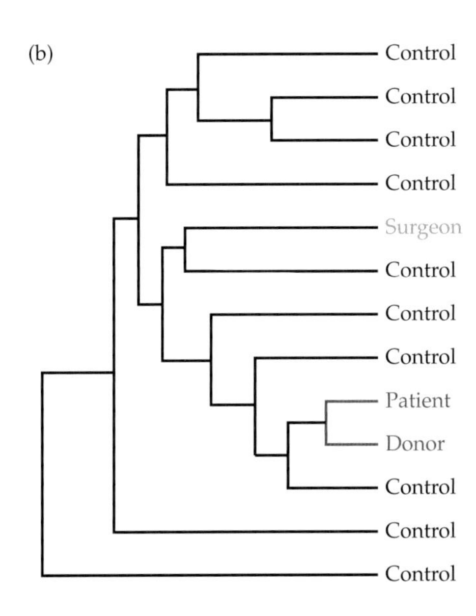

Figure 5.52 Identification of sources of HIV infection by using phylogenetic methodology. (a) The case of the Florida dentist. Simplified unrooted phylogenetic tree of the V3 region of the *env* gene (243 aligned nucleotide positions). The sequences from patients A, B, C, E, and G and the sequences from the dentist constitute a monophyletic clade (red). The sequences from patients D and F (green) are unrelated to either the dentist or to each other, indicating that these two infections occurred independently. Sequences from the same source that clustered together in the phylogenetic reconstruction are shown as a single OTU. The dentist and patient A appear twice each because some of their V3 sequences were so divergent as to assume different phylogenetic positions. Only five controls are shown. (b) The case of the Baltimore surgeon. Simplified unrooted phylogenetic tree based on 343 bp sequences from the *gag* gene of HIV-1. Thirteen different sequences were obtained from the surgeon, 10 from the patient, and 11 from the blood donor. All the sequences from the same source clustered as a monophyletic clade and are, therefore, shown as a single OTU. The sequences from the surgeon and the patient are paraphyletic, and thus the surgeon can be excluded as the source of infection. The patient's infection could be traced back to a unit of transfused blood from an HIV-seropositive donor whose HIV status was unknown at the time of donation.

At about the same time as the Florida case, a patient of an HIV-seropositive surgeon in Baltimore tested positive for HIV. Three possible infection sources were tested: (1) the surgeon, (2) a unit of transfused blood traced back to an HIV-seropositive donor, and (3) independent acquisition. A study by Holmes et al. (1993) ruled out the surgeon as the source of the infection (**Figure 5.52b**). The infection seemed to have originated with the blood donor.

More recently, phylogenetic trees were entered as evidence in the criminal case of *State of Louisiana v. Schmidt*, in which a gastroenterologist injected his former nurse and mistress with solution containing HIV from some of his patients (Metzker et al. 2002). Again, researchers carried out phylogenetic analyses, this time involving enormous amounts of data made possible by the extensive technological progress in sequencing technology between 1992 and the current millennium. The jury found the doctor guilty of attempted murder, and he was sentenced to 50 years in prison, a decision upheld in 2002 by the Supreme Court.

At the Limits of the Tree Metaphor: The Phylogeny of Eukaryotes and the Origins of Organelles

As we will see in the next chapter, the practice of describing evolutionary history as a bifurcating branching tree does not apply to many phylogenetic problems, including the evolutionary relationships among prokaryotes and the origin of eukaryotes from prokaryotic ancestors.

Next we present two phylogenetic problems at the very limits of molecular resolution by phylogenetic trees. We will deal with (1) the phylogenetic relationships within eukaryotes and (2) the origin of the primary organelles.

The phylogeny of eukaryotes

As if to emphasize the unity of creation, in the tenth edition of his *Systema Naturae* (1758), Carolus Linnaeus bunched up the entire natural world within a single taxon, which he called *Imperium Naturae* (the Empire of Nature). Three years later in his *Fauna Suecica* and subsequently in the twelfth edition of *Systema Naturae* (1766–1768), Linnaeus divided *Imperium Naturae* into three kingdoms: *Regnum Animale* (animals), *Regnum Vegetabile* (plants), and *Regnum Lapideum* (stones), a tripartite taxonomy that may have gave rise to the "animal, vegetable, or mineral" game in the early nineteenth century. *Regnum Lapideum*, which was further divided into three classes—*Petrae* (rocks), *Minerae* (minerals), and *Fossilia* (fossils)—soon lost its legitimacy, and for about 50 years the living world, sometimes referred to as Superdomain Biota, was dichotomously divided into plants and animals. In his *Generelle Morphologie der Organismen* (1866), Ernst Haeckel replaced the names *Vegetabile* and *Animale* with **Plantae** and **Animalia**, respectively, and added a new kingdom, **Protista**, consisting of unicellular organisms, which Linnaeus in his tenth edition of *Systema Naturae* placed in class Vermes, within an order called Zoophyta (plantanimals). One of the eight groups within Haeckel's Protista was called *Moneres*. It included all the bacterial groups that were known at the time. Copeland (1938) elevated's Haeckel's *Moneres* to the rank of kingdom (**Monera**), thereby increasing the number of kingdoms to four. This quadripartite division of the living world lasted for about three decades, until Whittaker (1969) replaced it with a five-kingdom system of classification. In this system, the eukaryotes were divided into two multicellular kingdoms—Plantae and Animalia—a unicellular kingdom called Protista, and Whittaker's newest addition, kingdom **Fungi** (formerly regarded as a subkingdom within Plantae), which included both unicellular and multicellular taxa. Whittaker's five-kingdom system became a popular standard, propagated uncritically in countless textbooks and popular science publications.

The five-kingdom system has not withstood the test of time; in fact, it started crumbling almost as soon as molecular data became available (e.g., Sogin et al. 1986). The first thing that became evident was that the diversity of microbial "protists" has been vastly underestimated and underappreciated. As opposed to the propensity of scientists to study "big" organisms, eukaryote biodiversity was found to be dominated by microbial forms. The current molecular view of eukaryotic phylogeny consists of a handful of very large, almost exclusively unicellular taxa, each comprising an enormous diversity of morphological, physiological, ecological, behavioral, and biochemical characteristics (**Figure 5.53**). Most eukaryotic phylogenetic hypotheses are the subject of debate, and the relationships among eukaryotic supergroups are poorly understood.

It is estimated that more than 70% of all recognized eukaryotic taxa are exclusively unicellular, and a sizable minority of taxa include species that are characterized by **simple multicellularity**, i.e., these organisms are made up of many identical cells. *Volvox carteri*, a simple multicellular green alga, provides an example of simple multicellularity. In contrast, organisms characterized by **complex** or **true multicellularity**, i.e., those containing a multitude of differentiated cells, are found in only five clades: brown algae, red algae, green plants, animals, and fungi. With the exception of animals, there is evidence that complex multicellularity arose from simple multicellularity (Knoll 2011).

Interestingly, animals, plants, and fungi (i.e., the kingdoms that have traditionally attracted the most attention in biological studies) turned out to be mere "twigs" at the tips of a few branches in the eukaryotic tree (Olsen and Woese 1996). Moreover, despite the Linnaean tradition of grouping fungi and plants into a "taxon" studied by botanists in contradistinction to the animals that are of interest to zoologists, the kingdoms Animalia and

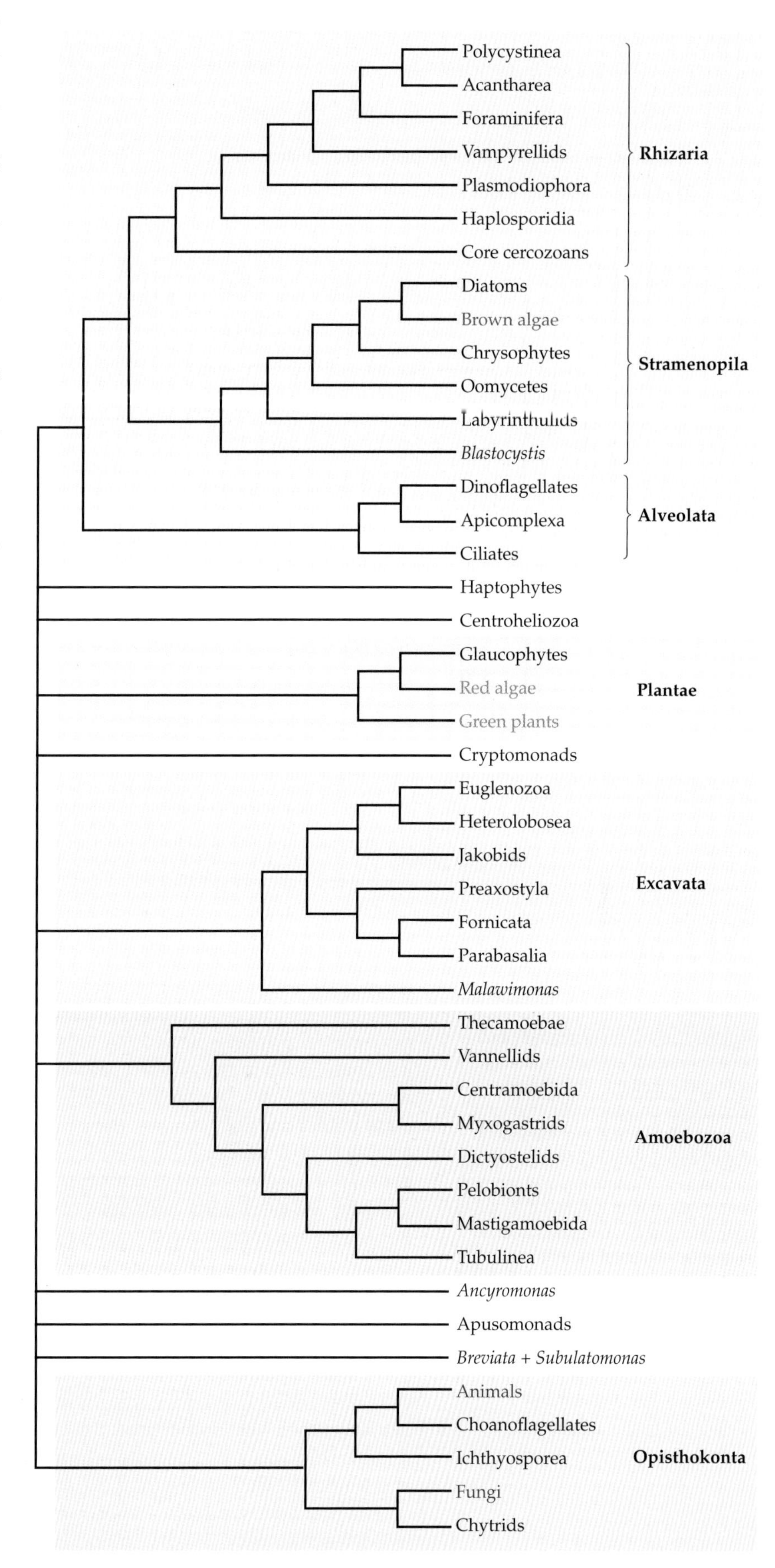

Figure 5.53 An unscaled and mostly unresolved tree for eukaryotes. The vast majority of the higher taxa are either unicellular or contain a few species that exhibit simple multicellularity. Only five clades (colored type) in this tree include species that are characterized by complex multicellularity. Five large "supertaxa" (colored backgrounds) and a handful of smaller taxa are shown. There is little agreement on the monophyly of these taxa, let alone on the branching order among them. Many other phylogenetic arrangements have been proposed in the literature. (Modified from Katz 2012.)

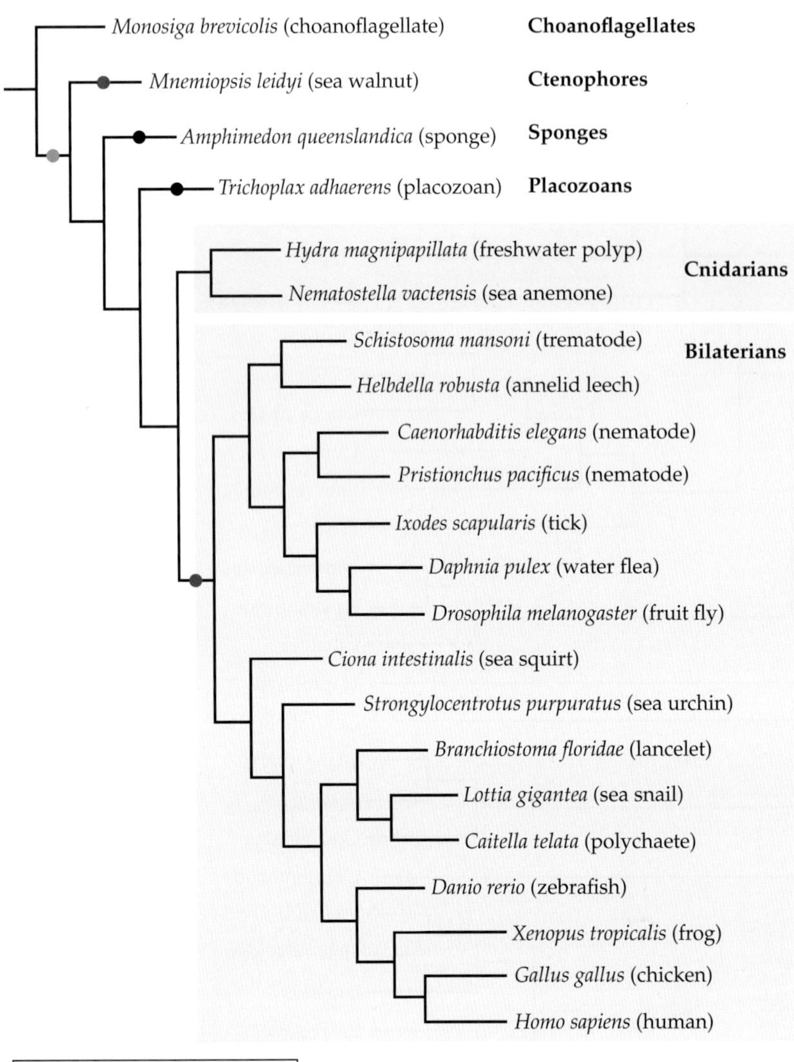

Figure 5.54 An unscaled phylogenetic tree for animals based on their gene content. The genome sequences on which this tree is based indicate that the mesoderm arose twice, once in ctenophorans and once in bilaterians, and the nervous system was lost twice, once in sponges and once in placozoans. (Data from Ryan et al. 2013.)

Fungi turn out to be more closely related to each other than either is to Plantae.

The internal phylogenetic divisions found within Animalia have yielded quite a few surprises. For example, the most basal lineage within the animal kingdom turned out to be the ctenophores (warty comb jellies, or sea walnuts; **Figure 5.54**). Ryan et al. (2013) sequenced the genome of the ctenophore *Mnemiopsis leidyi* and concluded that ctenophores alone—not sponges, not placozoans, and not the clade consisting of both ctenophores and cnidarians—are the most basal extant animals. Their results suggest that, within Animalia, repeated gains and losses of mesodermal cell types must have occurred independently.

Origins of organelles

Mitochondria are double-membrane cellular organelles containing their own genome. They are found in nearly all eukaryotes, and those eukaryotes that lack recognizable mitochondria either have them as relict organelles or had them in the past but subsequently lost them (Embley and Martin 2006; Hampl et al 2009). Mitochondrial genomes vary in length from less than 6 Kb in several alveolates to about 2,900 Kb in muskmelon (*Cucumis melo*). The mitochondrial genome may be circular or linear, segmented or in one piece.

Primary plastids, such as chloroplasts, are also double-membrane organelles containing their own genome. They are extremely diverse in terms of morphology and function and are mainly (but not exclusively) involved in photosynthesis. Like their mitochondrial counterparts, the genomes of primary plastids can also be circular or linear. They vary in size from less than 35 Kb in the poultry pathogen *Eimeria tenella* to about 525 Kb in the green alga *Volvox carteri*. Despite the much smaller variation in size in comparison to that of plant mitochondrial genomes, the genomic content of chloroplasts varies considerably among plants, especially in the numbers of protein-coding genes, tRNA-specifying genes, pseudogenes, and various types of introns and repeats.

There are essentially two types of hypotheses to explain the discrete existence of mitochondrial, plastid, and nuclear genomes within eukaryotic cells. Those in the first category (e.g., Cavalier-Smith 1975) stipulate that the genomes of organelles have **autogenous origins** and are descended from nuclear genes by **filial compartmentalization**, whereby part of the nuclear genome became incorporated into a membrane-enclosed organelle and subsequently assumed a quasi-independent existence. **Endosymbiotic hypotheses** posit that some membrane-enclosed organelles of eukaryotic cells descended from once free-living prokaryotic cells that somehow entered the cytosol of their host and in time, through reduction, became organelles. The symbiotic relationship was subsequently retained because it was mutually beneficial (Martin and Müller 1998). The origin of extranuclear DNA is, therefore, **exogenous**.

The first endosymbiotic hypothesis for the origin of chloroplasts was put forward by Mereschkowsky (1905). An endosymbiotic origin for the mitochondria was first

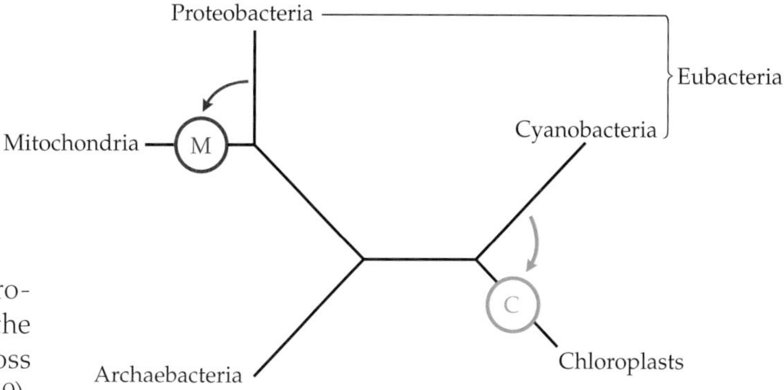

Figure 5.55 Schematic unrooted tree illustrating the phylogenetic affinities of chloroplasts and mitochondria. The mitochondria are derived from a proteobacterium via an endosymbiotic event (M). The chloroplasts are derived from a cyanobacterium via a second endosymbiotic event (C). (Modified from Graur and Li 2000.)

suggested by Portier (1918) and was more explicitly promoted by Wallin (1927). After an endosymbiotic event, the genomes of the endosymbionts were streamlined by loss of genes and by gene transfer to the nucleus (Chapter 9), and they became obligatory symbionts (i.e., incapable of an independent existence outside their hosts). The transformation of the bacterial endosymbiont into an organelle was accompanied by massive genome reduction, transfer of genes to the host nuclear genome, and the acquisition of the ability to reimport proteins encoded by the nuclear genome (Reyes-Prieto et al. 2007; Nowak et al. 2008). In very few cases, host nuclear genes were transferred to the genome of the organelle.

What is the difference between endosymbionts and organelles? **Endosymbionts** are organisms that live within other organisms. Many endosymbionts are obligate—they cannot live outside their hosts, and many obligate endosymbionts are essential for their hosts as well. For example, *Buchnera aphidicola* supplies amino acids to its aphid host. Neither can currently live without the other. **Organelles** have descended from endosymbionts, but they are now double-membrane-bounded compartments of eukaryotic cells. All of the functional proteins in the cytosol of an endosymbiont are encoded by its own genome. By contrast, only a very small fraction of the proteins that function in organelles are encoded by organellar DNA. The majority of organellar proteins are encoded by the nuclear DNA, translated on cytosolic ribosomes, and imported into the organelle with the help of a protein import apparatus. The existence of a protein import apparatus (e.g., the TIM/TOM and the TIC/TOC complexes of mitochondria and plastids, respectively) is what distinguishes organelles from endosymbionts (Cavalier-Smith and Lee 1985; Theissen and Martin 2006).

The molecular evidence overwhelmingly favors the endosymbiotic theory. It indicates that one endosymbiotic event gave rise to all present-day mitochondria, and a second event gave rise to all present-day primary plastids (Keeling 2010). The mitochondria evolved from a proteobacterial ancestor, and all primary plastids were derived from a cyanobacterial ancestor (**Figure 5.55**). It would be naïve to try to identify the "closest living relative" of either mitochondria or plastids, because these two unique events involved prokaryotes with unique collections of genes in their genomes. Because of horizontal gene transfer, divergent evolution, gene loss, and many other evolutionary processes that profoundly affected the makeup of prokaryotic genomes for hundreds of millions of years, identical collections of genes as those possessed by the ancestors of mitochondria and plastids no longer exist. We note that while endosymbiosis is not a treelike process, the endosymbiotic origins of organelles were unraveled by using the methodology of trees.

It is interesting to note that following the initial endosymbiotic events, mitochondria and plastids followed different evolutionary paths. Thus, while mitochondria are a stable fixture of eukaryotic cells that are never completely lost nor shuffled between lineages, plastid evolution has been a complex mix of movement, loss, and replacement (Keeling 2010).

Higher-order endosymbioses are also known (Keeling 2004; Bodył and Moszczyński 2006; Patron et al. 2006; Gould et al. 2008; Archibald 2009). In secondary endosymbiosis, a eukaryote already possessing plastids is engulfed by a second eukaryote. In tertiary endosymbiosis, a eukaryote already possessing a plastid derived from a eukaryote is engulfed by another eukaryote.

Phylogenetic Trees as a Means to an End

So far, we have used molecular phylogenetic trees as a means unto itself, i.e., to reconstruct the sequence of speciation events leading to the creation of a group of taxa under study. Molecular phylogenetic trees, however, can be used for other purposes, for example, as scaffolds on which to map the evolution of molecular and nonmolecular characters. In the next section, we will discuss the use of molecular phylogenetic trees to gain insight into evolutionary principles.

Parallelism and convergence as signifiers of positive selection

Parallelism at the molecular level is defined as the independent occurrence of two or more nucleotide substitutions (or amino acid replacements) of the same type at homologous sites in different evolutionary lineages. Molecular convergence is the occurrence of two or more nucleotide substitutions (or amino acid replacements) at homologous sites in different evolutionary lineages resulting in the same outcome (Chapter 3, Figure 3.6). Given the number of substitutions that have occurred during the evolution of a gene, a certain degree of parallelism and convergence is expected to be observed purely by chance. However, if the numbers of parallel and convergent substitutions significantly exceed the chance expectation, then it is unlikely that they have occurred by random genetic drift, and we must invoke positive selection to explain their existence. Thus, parallel and convergent changes may be taken as evidence for positive adaptive selection. Zhang (2006) proposed four requirements that need to be fulfilled in order to demonstrate adaptive parallel evolution at the protein sequence level. First, similar changes in protein function should occur in independent evolutionary lineages. Second, parallel amino acid substitutions should be observed in these proteins. Third, it must be shown that the parallel substitutions cannot be attributable to chance alone and, therefore, must have been driven by a common selective pressure. Fourth, one must be able to show that the parallel substitutions are responsible for the parallel functional changes. More than a dozen cases of molecular parallel evolution have been reported. Almost all of them satisfy the first and second requirements. A few also satisfy the third requirement, but very few cases satisfy all four criteria. The example of lysozymes in ruminants and colobine monkeys is discussed below. Another example concerns the evolution of digestive RNases in Asian and African colobine monkeys (Zhang 2006).

Lysozyme is a 130-amino-acid-long enzyme whose catalytic function is to cleave the β(1-4) glycosidic bonds between N-acetyl glucosamine and N-acetyl muramic acid in the cell walls of eubacteria, thereby depriving the cells of their protection against osmotic pressures and subsequent lysis. By virtue of its catalytic function and its expression in body fluids such as saliva, serum, tears, avian egg white, and mammalian milk, lysozyme is usually considered a first-line defense against bacterial invasion. In foregut fermenters (i.e., animals in which the anterior part of the stomach functions as a chamber for bacterial fermentation of ingested plant matter), lysozyme is also secreted in the posterior parts of the digestive system and is used to free nutrients from within the bacterial cells.

Foregut fermentation has arisen independently at least three times in the evolution of placental mammals, at least once in marsupials, and at least once in birds. In this section, we will deal with foregut fermentation in two mammalian taxa, the ruminants (e.g., cow, deer, sheep, giraffe) and the colobine monkeys (e.g., langur, proboscis monkey, colobus). In these two cases, lysozyme, which in other animals is not normally secreted in the digestive system, has been recruited to degrade the walls of bacteria that carry on the fermentation in the foregut. Stewart and Wilson (1987) and Kornegay et al. (1994) identified four amino acid positions that have experienced parallel and convergent replacements (**Figure 5.56**). Because is unlikely that the parallel amino acid replacements in the two evolutionary lineages occurred by chance, it is assumed that these changes are adaptations enabling the lysozyme to function in the hostile environment of the stomach.

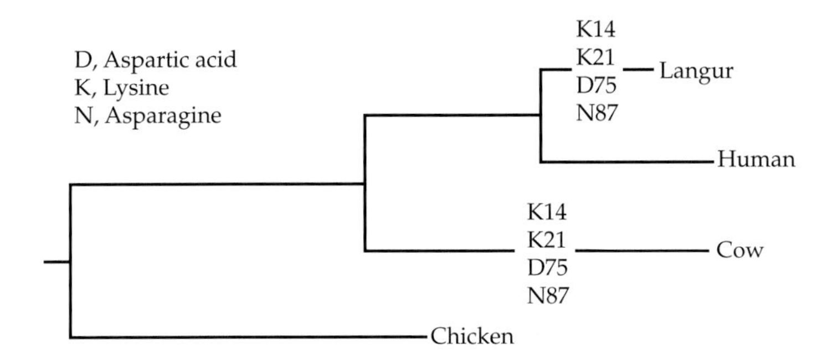

Figure 5.56 Convergent amino acid replacements at four positions (14, 21, 75, and 87) in the foregut lysozymes from cow and langur. The lengths of the branches are proportional to the total numbers of amino acid replacements along them. Only convergent replacements are shown, each denoted by the resultant amino acid followed by the position. (Modified from Kornegay et al. 1994.)

An interesting feature of these parallel and convergent replacements is that they lead to lysozymes with lower isoelectric points (6.0–7.7) than those of lysozymes that are not expressed in the foregut (10.6–11.2). For example, arginine was found to be heavily selected against at the surface of the foregut lysozymes, probably because this amino acid is susceptible to inactivation by digestive products such as diacylglycerol. Moreover, it has been determined that these replacements contribute to a better performance of foregut lysozymes at low pH values and confer protection against the proteolytic activity of pepsin in the stomach.

Clearly, these lysozymes share adaptations to conditions prevalent in the digestive system. Conversely, these lysozymes perform less efficiently at higher pH values than other mammalian lysozymes. In conclusion, it seems safe to deduce that we are dealing here with a case of parallel occurrence of advantageous substitutions in different evolutionary lines resulting in parallel adaptations to similar selective agents. We note, however, that there is some disagreement in the literature on the exact number of advantageous substitutions during the evolution of stomach lysozymes (e.g., Zhang et al. 1997; Yang 1998; Irwin et al. 2011). Schienman et al. (2006) have shown that the recruitment of lysozyme to the foregut in colobine monkeys was aided by several gene duplications and specialization of one of the copies as a digestive enzyme.

Detecting amino acid sites under positive selection

In the previous chapter, we used the ratio of nonsynonymous to synonymous substitutions in the entire sequence to detect positive selection. This approach is equivalent to assuming that all amino acids in a protein are under positive selection. This assumption makes the test too conservative because in reality only a few positions (e.g., functionally and structurally important sites) may have experienced positive selection during their evolution. The ability to infer ancestral nucleotides in a phylogenetic tree, and by extension ancestral codons, allows us to approach the question of selection in a more particular manner, i.e., to examine each site in the coding region separately.

Many methods for detecting selection at single codon sites have been suggested in the literature (Fitch et al. 1997; Nielsen and Yang 1998; Bush et al. 1999; Suzuki and Gojobori 1999; Yang et al. 2000; Suzuki 2004). Here, we describe briefly the method of Suzuki and Gojobori (1999).

The method starts with a set of aligned sequences and a phylogenetic tree. For each codon in the alignment, the ancestral codons in all internal nodes are inferred. Next, the numbers of synonymous and nonsynonymous substitutions at each codon site are determined along the entire phylogenetic tree. Finally, for each codon position, the nonsynonymous to synonymous substitution ratio is calculated. Those sites at which the ratio is significantly greater than 1 (the neutral expectation) are deemed to be under positive selection. We note that all methods for detecting selection at single sites (regardless of degree of sophistication) require large numbers of sequences, completely resolved phylogenetic trees, and branch lengths that are not too long.

Suzuki and Gojobori (1999) applied their method to the case of a variable region (*V3*) in the viral envelope gene (*env*) of HIV-1. The region is 35 amino acids long. In this region, 4 sites (11%) were found to evolve under positive selection, 2 sites (6%)

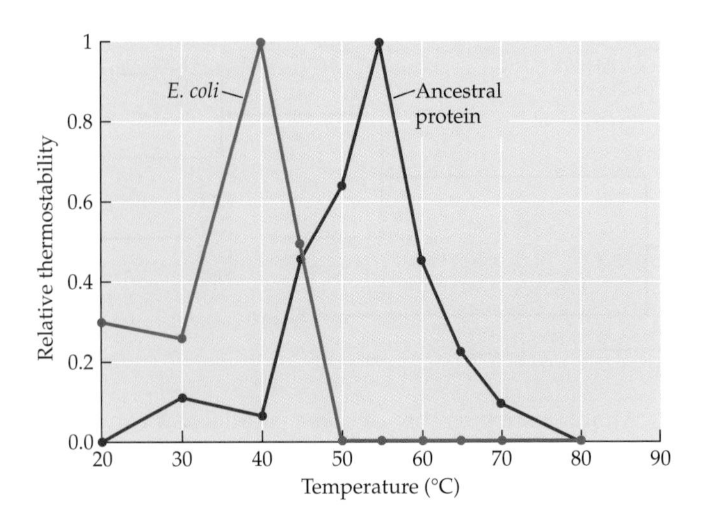

Figure 5.57 Thermostability curves of EF-Tu proteins from an extant mesophilic bacterium, *Escherichia coli*, and a putative ancestor of completely sequenced mesophilic bacteria. Relative thermostability of the protein was defined as the ratio of bound substrate at a certain temperature to the maximum bound substrate at the optimal binding temperature in vitro. The temperature profiles of the ancestral protein and *E. coli* display maximums at approximately 55°C and 40°C, respectively, suggesting that the ancestor of mesophilic bacteria lived at higher temperatures than its descendants. (Modified from Gaucher et al. 2003.)

were under purifying selection, and 29 sites (83%) evolved under no selection (i.e., neutrally).

Reconstructing ancestral proteins and inferring paleoenvironments

The ability to infer in silico the sequence of ancestral proteins, in conjunction with some astounding developments in synthetic biology, allow us to "resurrect" putative ancestral proteins in the laboratory and test their physicochemical properties. These properties, in turn, can be used to test hypotheses concerning the physical environment that the ancestral organism inhabited, i.e., its **paleoenvironment**.

Gaucher et al. (2003, 2008), for instance, used EF-Tu (*E*longation *F*actor *T*hermo*u*nstable) gene sequences from completely sequenced eubacteria to reconstruct candidate ancestral sequences at nodes throughout the bacterial tree. These inferred ancestral proteins were then synthesized in the laboratory, and their activities and thermal stabilities were measured and compared with those of extant organisms. It was reasoned that one might be able to obtain direct evidence on the temperature at which a particular ancestral bacterium lived by the optimal temperature at which its proteins function (its thermostability curve). In a comparison between EF-Tu from *Escherichia coli* and the putative ancestor of all mesophilic bacteria (i.e., bacteria whose growth temperatures range from 20°C to 40°C), the temperature profile of the inferred ancestral protein displayed a maximum at 55°C (**Figure 5.57**), suggesting that the ancestor of modern mesophiles lived at considerably higher temperatures than its descendants.

Similar methodology can be used to infer the properties of ancestral genes and proteins, as well as the crucial amino acid replacement needed to alter protein function (Chang et al. 2002; Thornton et al. 2003).

Mapping nonmolecular characters onto molecular trees

Molecular trees can be used as scaffolds on which to map morphological, physiological, anatomical, and behavioral characters. Insights gained from comparing the evolution of molecular and nonmolecular traits can, in turn, be used to test various evolutionary hypotheses. This approach, which has been advanced numerous times by molecular evolutionists (e.g., Goodman 1989; Graur 1993), was for a long time met with ferocious resistance from some in the paleontological establishment on the grounds that morphological data should not be "relegated to secondary status, in merely providing characters that can be hung like Christmas ornaments on branches defined by molecular data" (Novacek 1993). In the following example, we will illustrate the method of mapping morphological character states onto a molecu-

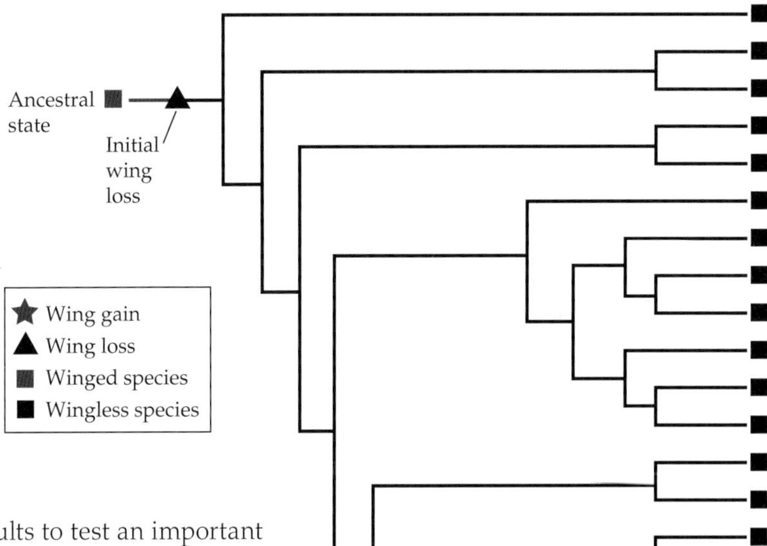

Figure 5.58 Mapping of male wing character states onto a molecular phylogenetic tree based on 18S, 28S, and a partial histone H3 gene sequence from 37 stick insect species. The ingroup tree was rooted by 22 winged outgroups (not shown). Species names were omitted. Two character states were used, winged (red) and wingless (black). Excluding the initial wing loss at the root of the tree, this reconstruction requires six changes in character states, two losses and four wing gains. (Modified from Whiting et al. 2003.)

lar phylogenetic tree, and we will then use the results to test an important nineteenth-century evolutionary hypothesis.

Dollo's law of irreversibility (Dollo 1893) states that evolution is not reversible. This principle most strictly applies to losses of complex morphological characters. Thus, according to this law, once a complex feature is lost, it cannot be regained in the lineage in which it was lost. Some methods of phylogenetic tree reconstruction do not allow reversals to ancestral states and are, hence, called Dollo parsimony methods (e.g., Le Quesne 1974; Farris 1977).

Whiting et al. (2003) tested the law of irreversibility as it applies to the evolution of insect wings, specifically in the order Phasmatodea (stick insects). Thousands of independent transitions from a winged form to winglessness have occurred during the course of insect evolution. Some entire insect orders, such as fleas, lice, and grylloblattids (ice crawlers), are secondarily wingless. However, evolutionary reversals from a flightless to a volant form had never been documented in the Pterygota (winged insects) lineage. Such reversals were considered highly unlikely because wings are complex characters that require intricate interactions between nerves, muscles, sclerites, and wing foils.

Figure 5.58 shows a molecular phylogenetic tree based on 18S and 28S gene sequences and a partial histone H3 gene sequence from 37 species of stick insects (phasmids). When the character states winged and wingless were mapped onto the tree, the winged species neither clustered together nor constituted an outgroup to the wingless taxa. Thus, it is possible to conclude that wings can be restored after being lost for many generations. Moreover, the phylogenetic tree suggests that wings were "resurrected" secondarily on multiple occasions, indicating that wing developmental pathways are somehow conserved in wingless phasmids.

Molecular phylogenetics has uncovered numerous violations of Dollo's law, including the spectacular discovery of the "revolution" of mandibular teeth in frogs after persistence of a toothless character state for more than 200 million years (Wiens 2011).

CHAPTER 6

Reticulate Evolution and Phylogenetic Networks

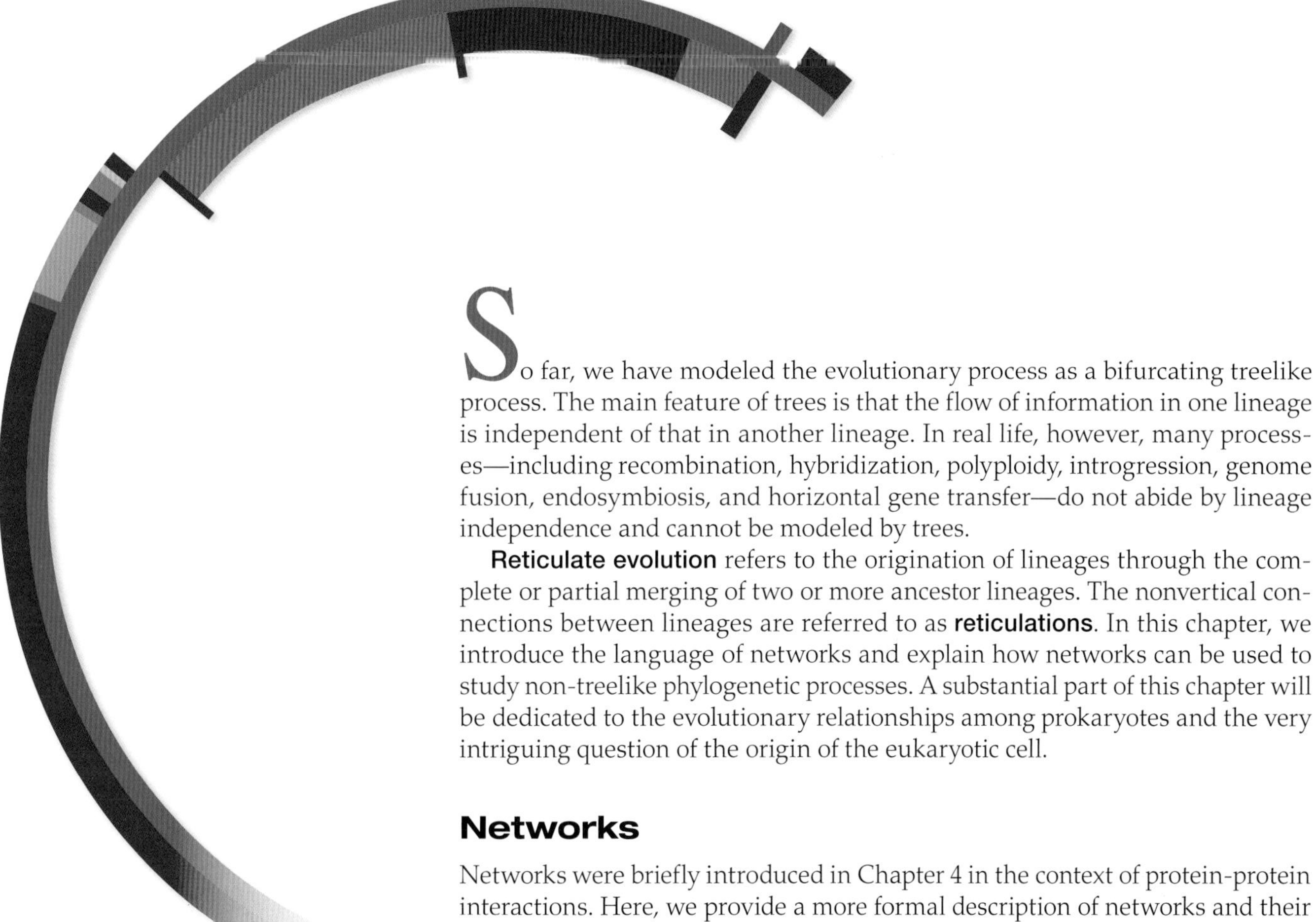

So far, we have modeled the evolutionary process as a bifurcating treelike process. The main feature of trees is that the flow of information in one lineage is independent of that in another lineage. In real life, however, many processes—including recombination, hybridization, polyploidy, introgression, genome fusion, endosymbiosis, and horizontal gene transfer—do not abide by lineage independence and cannot be modeled by trees.

Reticulate evolution refers to the origination of lineages through the complete or partial merging of two or more ancestor lineages. The nonvertical connections between lineages are referred to as **reticulations**. In this chapter, we introduce the language of networks and explain how networks can be used to study non-treelike phylogenetic processes. A substantial part of this chapter will be dedicated to the evolutionary relationships among prokaryotes and the very intriguing question of the origin of the eukaryotic cell.

Networks

Networks were briefly introduced in Chapter 4 in the context of protein-protein interactions. Here, we provide a more formal description of networks and their use in phylogenetics. As mentioned in Chapter 5, trees are connected graphs in which any two nodes are connected by a single path. In phylogenetics, a **network** is a connected graph in which at least two nodes are connected by two or more pathways. In other words, a network is a connected graph that is not a tree. All networks contain at least one **cycle**, i.e., a path of branches that can be followed in one direction from any starting point back to the starting point. Typically, a network is depicted in diagrammatic form as a set of dots or circles for the vertices, joined by lines or curves for the edges (**Figure 6.1, left side**).

The edges may be **directed** or **undirected**. For example, if the vertices represent proteins, and the edges denote participation in the building of a dimer, trimer, or higher order n-mer, then the edges are undirected because if protein A participates in an n-mer with protein B, then protein B also participates in the n-mer with protein A. The networks in Figures 6.1a and 6.1b consist solely of undirected edges.

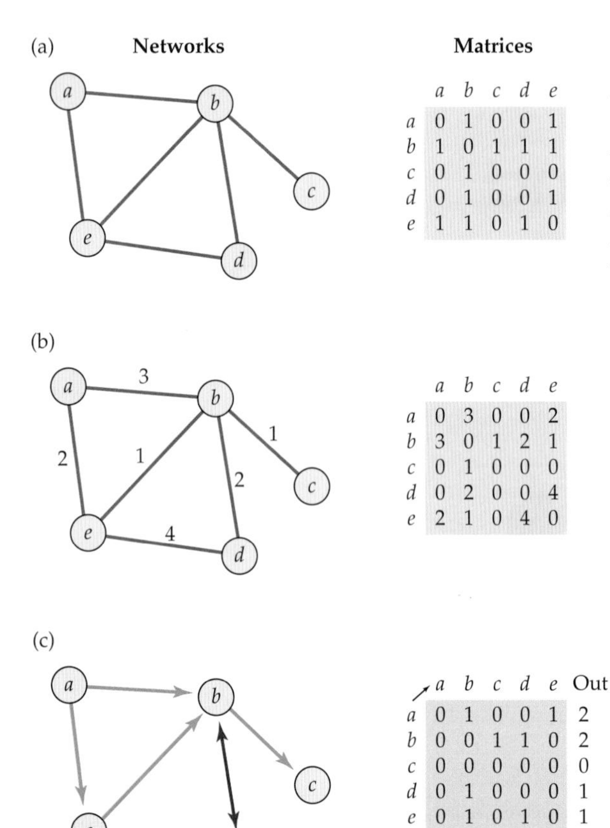

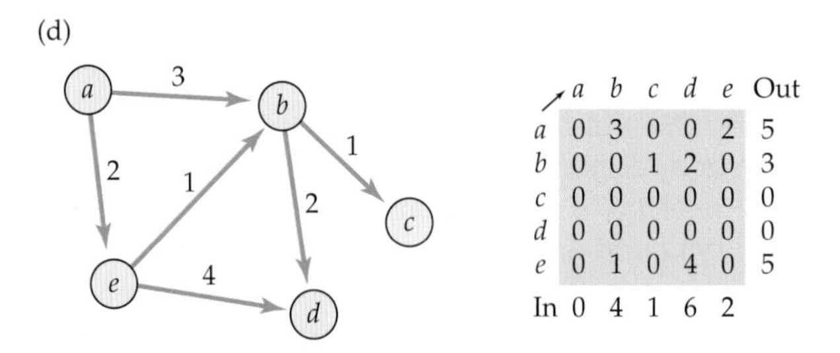

Figure 6.1 Networks, shown on the left, are composed of vertices (circles) and edges (lines or arrows). For each network, the matrix representation is shown on the right. (a) An undirected unweighted network. (b) An undirected weighted network. (c) A directed unweighted network. Unidirectional and bidirectional edges are denoted by green and red arrows, respectively. (d) A weighted directed network. In both (c) and (d), "In" on the matrix denotes the sum of the weights of the vertices pointing toward that vertex, and "Out" denotes the sum of the weights of the vertices pointing away from that vertex.

A directed edge represents an interaction that has at least one direction. For example, the edge may represent the activation of one protein by another. Such an interaction is directed because the fact that protein A activates protein B does not mean that protein B activates protein A. The networks in Figures 6.1c and 6.1d consist solely of directed edges. A directed edge may be **unidirectional**, as is the case when A activates B but B does not activate A, or **bidirectional**, as when A activates B and B activates A (Figure 6.1c). A network is called a **directed network** if all its edges are directed; it is an **undirected network** if all its edges are undirected. In a **mixed network**, some edges are directed and some are undirected.

If the edges of the network are equal in strength of interaction, frequency, likelihood of occurrence, importance, physical distance, and so forth, they are said to be **unweighted**. The network shown in Figure 6.1a contains only undirected, unweighted edges. If the edges of a network are assigned a weight, they are said to be **weighted**. A network containing solely undirected, weighted edges is shown in Figure 6.1b.

Similar to undirected edges, directed edges may also be unweighted (Figure 6.1c) or weighted (Figure 6.1d). Directed, weighted edges are particularly useful when dealing with horizontal gene transfer (Chapter 9), in which the weight of an edge may represent the number of transfers from one taxon to another.

Networks can also be fully represented by matrices (**Figure 6.1, right side**), which are amenable to algebraic manipulations. Networks can be fully defined by an $n \times n$ matrix, where n is the number of vertices. The a_{ij} element in the matrix denotes the edge between vertex i and vertex j. If an edge connects vertex i with vertex j in an undirected unweighted network (Figure 6.1a), then $a_{ij} = 1$, and $a_{ij} = 0$ otherwise. In such a network, $a_{ij} = a_{ji}$. In the matrix representation of an undirected weighted network (Figure 6.1b), a_{ij} equals the weight of the edge connecting vertex i with vertex j, and $a_{ij} = 0$ otherwise. In this type of matrix too, $a_{ij} = a_{ji}$. In the matrix representation of a directed unweighted network (Figure 6.1c), $a_{ij} = 1$ if a directed edge is pointing from vertex i to vertex j, and $a_{ji} = 1$ if a directed edge is pointing from vertex j to vertex i. If the edge is unidirectional (green arrow), then $a_{ij} \neq a_{ji}$. If the edge is bidirectional (red arrow), then $a_{ij} = a_{ji}$. In the matrix representation of a weighted directed network (Figure 6.1d), a_{ij} equals the weight of the directed edge pointing from vertex i to vertex j, and a_{ji} equals the weight of the directed edge pointing from vertex j to vertex i.

Phylogenetic and Phylogenomic Networks

Reticulate (network-like) **evolution** refers to evolutionary processes that violate the independence of evolutionary lineages by allowing some lineages to merge and produce new lineages. Reticulate evolution is modeled by networks instead of by trees (**Figure 6.2**).

While the term "phylogenetic tree" is well defined (Chapter 5), the meaning of "phylogenetic network" is somewhat ambiguous, and in the literature there exist many different usages of the term. Here, we will use a modified form of a definition originally proposed by Huson and Bryant (2005). According to this definition, a **phylogenetic network** is a network in which taxa are represented by nodes and various relationships between two taxa are represented by directed or undirected edges. As in the case of phylogenetic trees, a directed edge may represent ancestry and descent, whereby one taxon or a genomic component is assumed to be derived from another. In contradistinction to phylogenetic trees, however, a directed edge in a phylogenetic network may also represent a partial genetic relationship—that is, one taxon may have only contributed partially to the genome of another taxon. Undirected edges represent relationships between two nodes in which it is impossible with the data at hand to tell which is the ancestor and which is the descendant.

Formally, a phylogenetic network is defined as a graph in which at least one OTU is connected to the common ancestor by two or more paths. This feature distinguishes a phylogenetic network from a phylogenetic tree.

As in the case of phylogenetic trees, in which only rooted ones can explicitly describe the evolutionary process, only **rooted** (or **directed**) **phylogenetic networks** can explicitly describe the evolution of taxa in the presence of reticulate events. Two conditions must be met for a phylogenetic network to be rooted. First, all the branches emanating from all the nodes must have a direction, either **incoming** or **outgoing**, representing the flow of genetic information. Second, there must exist a node within the network with no incoming branches. A network lacking such a node is called a **circuit**. Circuits are seldomly used in phylogenetics.

The methodology for reconstructing phylogenetic networks from empirical data is still in its infancy despite a veritable deluge of publications on the subject (e.g., Nakhleh et al. 2005; Huson and Bryant 2006; Jin et al. 2007a,b; Than et al. 2008; Willson 2008; Huson et al. 2009, 2011; Meng and Kubatko 2009; Nakhleh 2009; Huson and Scornavacca 2011; Morrison 2011). In other words, although many methods for building networks out of comparative molecular data have been proposed in the literature, it would be premature to assess their relative usefulness or the conditions under which they yield trustworthy inferences. Some methods seem to have wider applications; others serve narrower purposes. Some methods employ simple algorithms; others are computationally expensive and time-consuming. Here, we present two simple methods for reconstructing unrooted phylogenetic networks. Those interested in more advanced examinations of network methodology should consult Huson et al. (2011) and Morrison (2011).

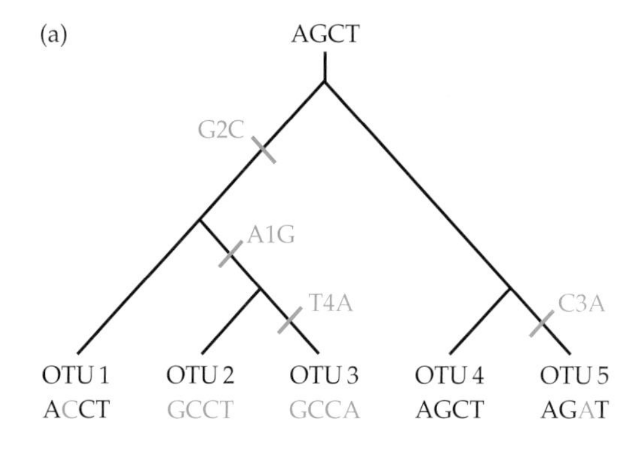

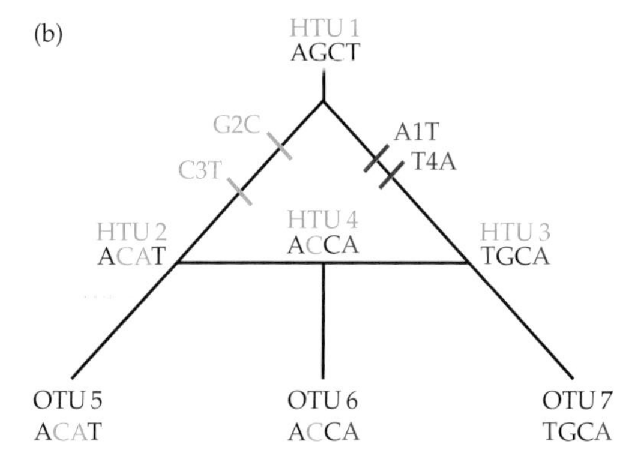

Figure 6.2 Divergent versus reticulate evolution. (a) In a binary bifurcating tree, each lineage is assumed to evolve independently. Thus, the sequences of the extant OTUs (1–5) are expected to trace back to the common ancestor of all OTUs via a single path and to have each accumulated all the substitutions that occurred on their individual paths. (b) An example of reticulate evolution with three OTUs, in which the common ancestor, HTU 1, gave rise to HTUs 2 and 3, and HTU 4 resulted from a reticulate event (recombination) involving HTUs 2 and 3. Reticulate evolution can only be represented by phylogenetic networks, graphs in which two or more paths may connect an OTU to the common ancestor. Nucleotide substitutions are marked on the branches by using the OMIM format (www.ncbi.nlm.nih.gov/omim). For example, "A1G" denotes a change from A to G in nucleotide position 1.

The term **phylogenomic network** refers to a network in which the nodes represent genomes and the edges represent either the presence of homologous genes or instances of horizontal gene transfer. In cases where edges represent homologous genes, an edge may be weighted by the number of genes common to two genomes. In cases where edges represent horizontal gene transfers, an edge may be weighted by the number of such transfers between two genomes. If the donor-recipient relationships are known, the horizontal gene transfers should be represented by directed edges (Dagan 2011).

The median network method

The median network method (Bandelt et al. 1995, 2000) reconstructs phylogenetic networks from binary data. Molecular sequences can be first transformed into binary data, say by using purines (R) versus pyrimidines (Y) instead of the four nucleotides,

Figure 6.3 Median network methodology. The data consist of five OTUs (A–E). There were 42 aligned sites that contained exactly two character states. At the top of the figure, the sequence alignment for these 42 sites is shown in tabular format, with the rows of homologous characters in the order in which they appear in the original sequences. To provide visual clues about the patterns among the sequences, each nucleotide is uniquely colored. We then rearrange the columns so that identical splitting patterns are adjacent to one another. In our example, we have eight different patterns, so there will be eight steps in the process of inferring the phylogenetic network. We note that the order in which the pattern groups are added to the analysis does not affect the result; any order will do. (a) First, we consider the pattern defined by sites 1, 5, 20, 35, and 39. This pattern splits the taxa into two groups, A and B on the one hand, and C, D, and E on the other. The length of the branch between the two groups is 5 character-state differences. (b) Next, we add the pattern defined by sites 17, 18, 27, and 29. This pattern does not contradict the previous split; it merely adds a new split, with C and D on one side and E on the other. (c) In the next step, we add the pattern defined by sites 32 and 41. These sites yield a split that contradicts one of the previous splits. In order to put A and E in a group together, we need to add a pair of branches rather than a single branch. Thus, characters 32 and 41 are represented by two branches rather than one. Note that characters 1, 5, 20, 35, and 39 are now represented by a pair of parallel edges as well. (d) In the next step, we add the group defined by the single site 3. Again, rather than a single branch, a set of parallel branches needs to be added, each originating in a different preexisting group. (e–h) Each of the next four pattern groups splits the data into a single OTU, on the one hand, and the other four OTUs, on the other. Thus, in each case, we need to add a single branch to the growing phylogenetic network. Since the data are binary, every character splits the sequences into two groups (bipartitions), each group being defined by its shared nucleotide pattern. The resulting median network (h) displays all the bipartitions. (i) Some partitions (red dashed lines) are in conflict. Partition I conflicts with partition II and IV; partition II conflicts with I; partition III conflicts with IV; and IV conflicts with I and III. ▶

or by using only nucleotide sites that contain only two nucleotides in all the taxa under study. Alternatively, the method can be used on binary molecular data, such as the presence or absence of a character state. In both cases, constant sites are excluded from the analysis and each binary site is translated into a **split** in which each of the resulting **partitions** contains OTUs that have the same character state at a site. For example, let us assume that the character states at a certain homologous position in OTUs 1, 2, 3, 4, and 5 are R, R, Y, Y, and R, respectively. At this site, the split will result in two partitions (1, 2, 5) and (3, 4).

Let us now consider all the homologous sites in the aligned sequences. Each site that contains two character states can be split into two. We can then cluster all the sites supporting the same split into split patterns. These patterns can be added sequentially in the process of constructing a network, as illustrated in **Figure 6.3**. This method is not much limited by the number of OTUs or the number of patterns (which is, of course, positively correlated with the number of OTUs). However, when the number of patterns increases, it may become very difficult to display the network on a two-dimensional page.

The conditioned-reconstruction method

The **conditioned-reconstruction** method (Lake and Rivera 2004) can be used to infer massive reticulate events, such as genome fusion, rather than events affecting particular genes, such as horizontal gene transfer. The method uses genome-wide statements concerning the presence (P) or absence (A) of orthologous genes. In this method, the distance between two genomes, X and Y, is solely a function of the frequencies of the four possible character state patterns (i.e., PP, PA, AP, and AA). If an orthologous gene is present in both X and Y, then the pattern for that ortholog is PP. Likewise, if the X ortholog is present and the Y ortholog is absent, the pattern is PA, and if the X ortholog is absent and the Y ortholog is present, the pattern is AP. Of course, in dealing with only two genomes, one can only consider the set of all genes present in X or Y—that is, the union of the set of genes present in X and the set of genes present in Y. Hence, the AA set will by necessity be empty, i.e., its frequency will be zero. Because

Informative data

OTU	1	2	3	4	5	6	7	8	9	10	11	12	13	14	15	16	17	18	19	20	21	22	23	24	25	26	27	28	29	30	31	32	33	34	35	36	37	38	39	40	41	42
A	A	C	T	G	A	G	T	C	T	T	C	G	T	T	T	A	T	A	T	T	T	T	C	T	T	C	T	G	T	T	A	C	G	A	T	C	A	T	A	T	T	A
B	A	T	T	G	A	G	C	C	T	T	T	A	C	C	A	G	T	A	C	T	C	T	C	C	C	C	T	C	T	T	A	T	G	C	T	A	G	C	A	C	C	G
C	G	T	T	G	C	A	C	T	G	T	T	A	C	C	A	G	C	G	C	C	T	T	C	T	C	C	C	C	C	T	A	T	G	C	C	A	G	C	G	T	C	G
D	G	T	C	G	C	G	C	C	T	T	T	A	C	C	A	G	C	G	C	C	T	T	C	T	C	C	C	C	C	T	A	T	G	C	C	A	G	C	G	T	C	G
E	G	T	C	T	C	G	C	C	T	G	T	A	C	C	A	G	T	A	C	C	T	C	A	T	C	G	T	C	T	C	G	C	A	C	C	A	G	C	G	T	T	G

Data rearranged into eight nucleotide patterns

OTU	20	35	1	39	5	17	27	29	18	32	41	3	21	24	40	8	9	6	23	26	10	22	30	4	33	31	2	11	36	7	13	14	19	25	38	15	12	28	34	16	37	42
A	T	T	A	A	A	T	T	T	A	C	T	T	T	T	T	C	T	G	C	C	T	T	T	G	G	A	C	C	C	T	T	T	T	T	T	T	G	G	A	A	A	A
B	T	T	A	A	A	T	T	T	A	T	C	T	C	C	C	C	T	G	C	C	T	T	T	G	G	A	T	T	A	C	C	C	C	C	C	A	A	C	C	G	G	G
C	C	C	G	G	C	C	C	C	G	T	C	T	T	T	T	T	G	A	C	C	T	T	T	G	G	A	T	T	A	C	C	C	C	C	C	A	A	C	C	G	G	G
D	C	C	G	G	C	C	C	C	G	T	C	C	T	T	T	C	T	G	C	C	T	T	T	G	G	A	T	T	A	C	C	C	C	C	C	A	A	C	C	G	G	G
E	C	C	G	G	C	T	T	T	A	C	T	C	T	T	T	C	T	G	A	G	G	C	C	T	A	G	T	T	A	C	C	C	C	C	C	A	A	C	C	G	G	G

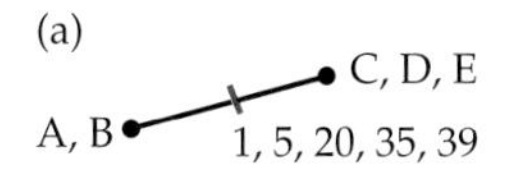

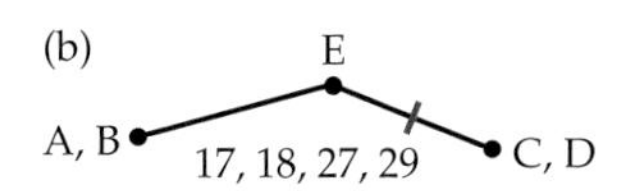

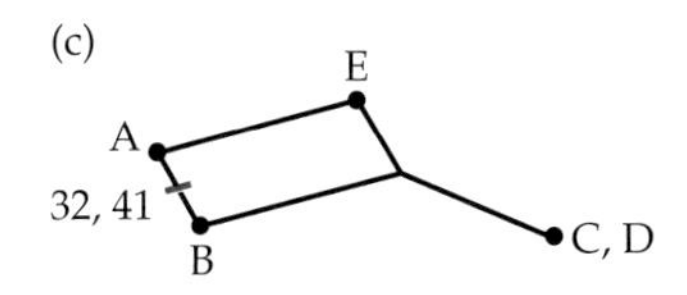

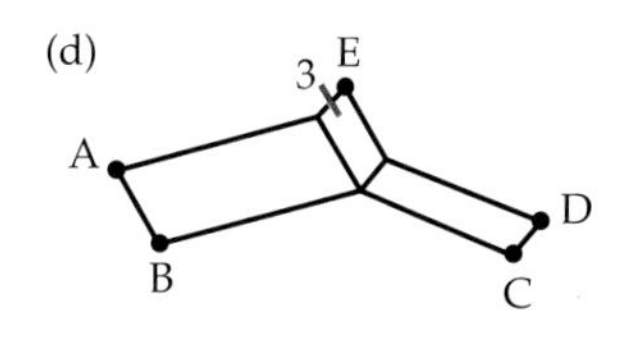

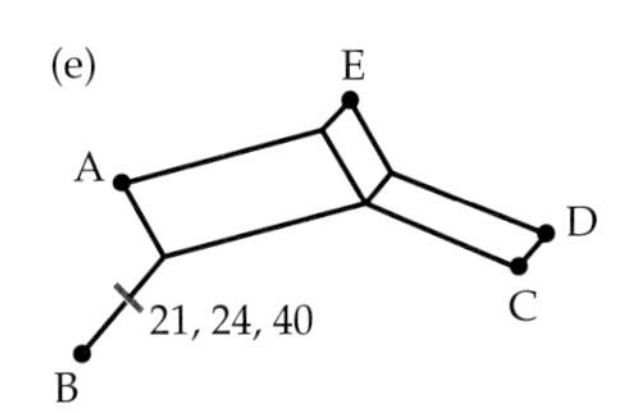

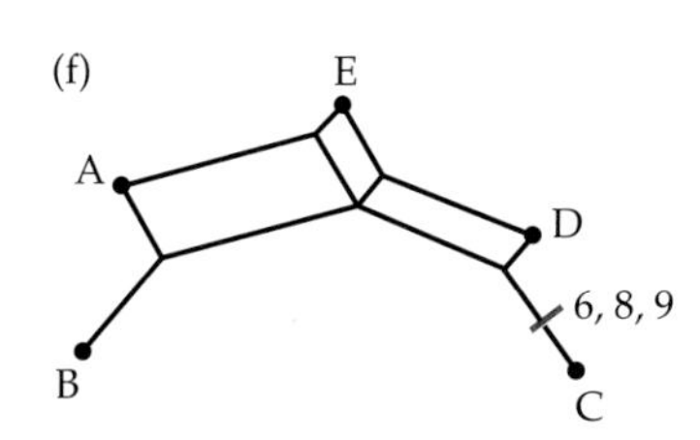

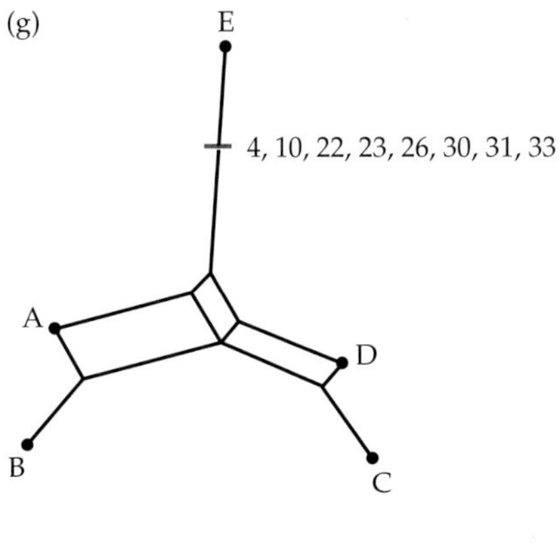

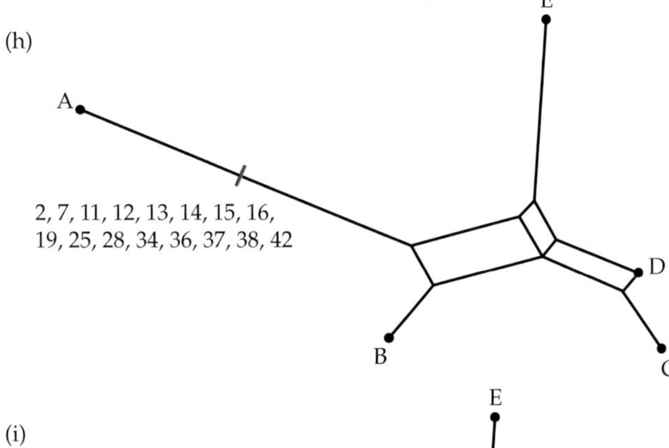

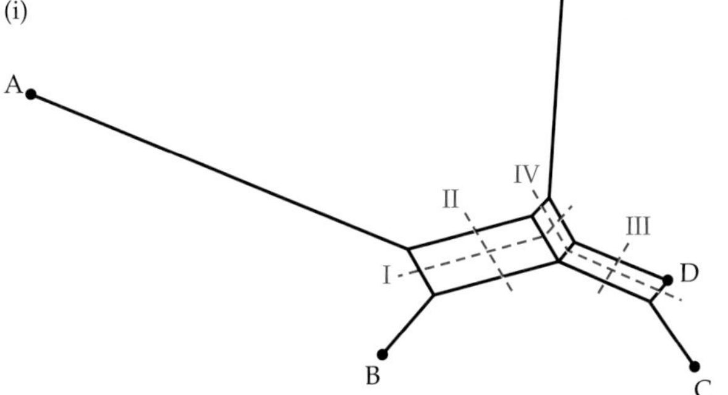

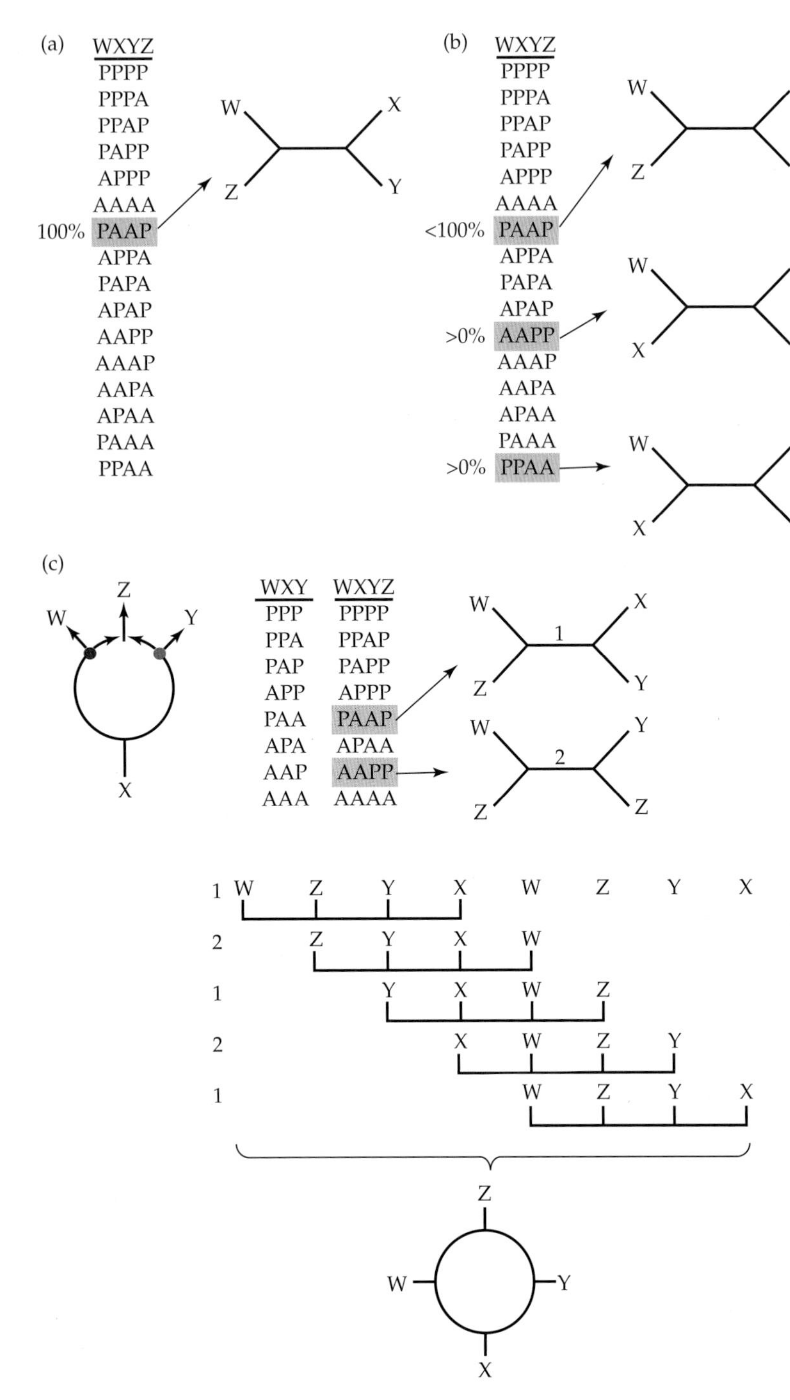

Figure 6.4 The conditioned-reconstruction method. (a) Sixteen possible joint states for four taxa W, X, Y, and Z. In the absence of any reticulate event, we would expect 100% of all genes to support a single unrooted tree. (b) If reticulate evolution is infrequent, we would expect the majority of genes to support one tree but would also expect to find alternative trees that are significantly supported by the data. (c) The fusion of two genomes produces a cycle graph. Z was produced by a fusion of a relative of W that descended from the ancestor marked by the red circle and a relative of Y that descended from the ancestor marked by the blue circle The two four-taxon trees that are simultaneously supported by the four genome patterns are numbered 1 and 2. These four-taxon trees are consistent with a repeated arrangement W-Z-Y-X, W-Z-Y-X, … , which corresponds to the cycle graph shown at the bottom. (Modified from Lake and Rivera 2004.)

of the particular distance model used by Lake and Rivera (2004), the distance between X and Y cannot be computed for cases in which the frequency of AA is zero. To overcome this difficulty, an additional genome, called the **conditioning genome**, is added. Only those genes that are present in the conditioning genome are used. The use of a conditioning genome affects the probabilities of all four joint states in the two-genome example (i.e., it affects the probabilities of PP, PA, AP, and AA). For a gene to be coded as present, it must be present in the conditioning genome and in the genome being coded. For a gene to be coded absent, it must be present in the conditioning genome and absent in the genome being coded. Therefore, if a gene is present in genome X, absent in genome Y, but not present in the conditioning genome, it will not be included in the count of PA patterns. Similar reasoning applies to the PP, AP, and AA categories.

Let us now examine the case of four taxa (W, X, Y, and Z) and one conditioning genome. We assume that there is no horizontal gene transfer affecting subsets of genes. With four taxa and two character states (A and P), there can be 16 possible joint states (**Figure 6.4a**). In the absence of reticulation, we expect all the genes to support a single tree. Of course, sampling and stochastic errors may result in a few genes not supporting the single tree. If reticulate evolution occurs but is infrequent, we expect the majority of genes to support one tree, but we also expect to find alternative trees that are significantly supported by the data (**Figure 6.4b**).

Next, let us examine the case of genome fusion (**Figure 6.4c**). We assume that there are three taxa (W, X, and Y) connected to one another by a traditional unrooted tree, as well as an additional taxon, Z, that was created by a fusion between a relative of W and a relative of Y. We further assume that the respective patterns of presence and absence of genes in W and Y are the same as in their relatives. We note that, with three taxa, there are eight possible joint states.

The rules for assigning P or A to each gene in the fused genome (Z) are simple: every gene present in either genome W or genome Y is present in genome Z. If the same gene ortholog is present in both W and Y, then two copies of that ortholog will exist in genome Z, but only one will be counted. Using these rules, one can deduce all the possible patterns. For illustration, consider a gene that in genomes W, X, and Y has the AAP pattern. Genome Z will contain a copy from the ancestor of Y, and the resulting pattern for W, X, Y, and Z will be AAPP. Thus, a genome fusion between the ancestor of W and the ancestor of Y can only create 8 possible joint states (as opposed to the 16 states that can be created without genome fusion). The sum of the probabilities of these 8 patterns should be 1.

With four genomes, there can be three unrooted trees. Pattern AAPP supports the tree in which genomes W and X are paired on one side of the central branch, and genomes Y and Z are paired on the other side; this is tree (WX)(YZ). Pattern PAAP supports tree (WZ)(XY). The other six patterns are uninformative. Pattern APAP would have supported tree (WY)(XZ); however, pattern APAP cannot appear (i.e., it has a zero probability). If patterns PAAP and AAPP appear with high probability, then two trees are simultaneously supported. After ruling out the possibility of a stochastic artifact, this simultaneous support for contradicting trees can be taken as indicative of reticulation. We can then examine whether or not the trees that have been obtained belong to the subset of trees predicted by the fusion event. Subsequently, we can ascertain whether or not the trees can form a cycle by alternately placing them below each other and successively shifting one taxon to the right, thereby forming a repeating pattern as would be formed if the cycle graph were continuously rolled (Figure 6.4c). If this is the case, then the results support genome fusion.

As we will see later, the conditioned-reconstruction method was used to tackle the origin of the eukaryotic cell.

Inferred reticulations: Are they real?

An **inferred phylogenetic network** is a summary of all the phylogenetic conflicts in the data set, i.e., a list of all the partitions that conflict with other partitions. In other words, an inferred phylogenetic network is a visual representation of all the phylogenetic signals within a data set that cannot be made to conform to a strict treelike phylogeny. Note that even if the true evolutionary process is entirely treelike, phylogenetic conflict may occur because of (1) stochastic error, (2) methodological issues in data collection (e.g., taxon sampling, character sampling, choice of outgroups), (3) homoplasy (e.g., parallelism, convergence, reversal), and (4) inappropriate data analysis (e.g., model misspecification, choice of optimality criterion). An additional source of phylogenetic conflict in otherwise treelike processes may be incomplete lineage sorting due to the ancestral population's polymorphic state at a locus. Of course, scientists are mostly interested in phylogenetic conflicts that uncover real reticulations, such as those arising when the data under study are derived from two or more phylogenetic histories. Unfortunately, deciding whether a phylogenetic conflict represents a real evolutionary reticulation or an artifact due to analytical quirks is very difficult.

It is important to emphasize that any data conflict can create reticulations, irrespective of its source, and factors that have nothing to do with the evolutionary process itself may confound the search for genuine reticulate gene-flow processes (Morrison 2010). Thus, the mere fact that one can build a phylogenetic network with a data set does not mean that the evolutionary processes that gave rise to the data were reticulate. A phylogenetic network may hint at the existence of reticulate processes, or it may merely represent methodological or stochastic flukes in the data. Conversely, the fact that "feeding" sequences to a tree-making program yields a tree does not prove that the evolutionary process that gave rise to the extant sequences under study was exclusively treelike.

Examples of Real-Life Phylogenetic Networks

Many evolutionary processes are known to violate lineage independence and, hence, cannot be modeled by phylogenetic trees. Here, we illustrate the process of reticulation by using as an example the evolution of a common allele in the ABO-blood-group system in humans. The phenomenon of speciation by polyploidy in woodferns will be used to illustrate the conceptual difficulties of dealing with speciation by hybridization, specifically the difficulties arising from the fact that, by definition, species produced by hybridization are paraphyletic. A few other examples of reticulate evolution due to polyploidy will be found in Chapter 7. Later in this chapter, we will deal with the reticulate evolutionary processes in prokaryotes and the genome fusions that may have given rise to the eukaryotic cell.

Reticulate evolution by recombination: A resurrected blood-group allele in humans

In terms of information flow, genetic recombination entails the splitting and rejoining of two unrelated or distantly related DNA sequences to form a new DNA sequence. The recombinant sequence, therefore, is a merger of two evolutionary histories. Understandably, such a process cannot be modeled by trees. Instead, we need to represent it as a network. In **Figure 6.5**, we illustrate the consequence of a recombination event on the structure of the phylogenetic network. We start with four sequences: two sequences that underwent recombination, the resulting recombinant sequence, and an outgroup sequence that is reasonably close to the ancestral sequence that gave rise to the two recombining sequences. We assume that at each stage of the evolutionary process, changes can accumulate independently on each of the branches. The resulting phylogenetic network illustrates what happens when the network is based on data from the recombinant sequence, its parental sequences, and an outgroup. The parental alleles will emerge on opposing branches emanating from the rectangle, and each will have a long external branch. The recombinant allele, on the other hand, will have a short external branch and will be located across from the outgroup. The network will be a reflection of a phylogenetic conflict, whereby one part of the recombinant sequence is closely related to one parent, whereas the other part of the recombinant sequence is related to the other. If the process that gave rise to the network was indeed a recombination event, then the sites supporting one or the other constituent tree will be clustered on the recombinant sequence, rather than being interleaved with one another.

As an example, here we examine the evolution of the human ABO blood group in chromosome 9, which encodes a glycosyltransferase that catalyzes the transfer of carbohydrates to the H antigen (**Figure 6.6**). Three allele groups, *A*, *B*, and *O*, segregate at the ABO glycosyltransferase locus. The human ABO glycosyltransferase gene consists of seven exons, with most of the coding sequence found in exon 7. The *A* and *B* alleles code for glycosyltransferases that

(a)

	Sites														
	1	2	3	4	5	6	7	8	9	10	11	12	13	14	15
out	A	T	T	A	T	C	C	C	G	A	C	C	T	C	T
anc	A	T	C	A	T	C	C	C	G	G	C	C	T	T	T
p1	A	A	C	G	C	C	C	C	C	G	C	C	A	T	A
p2	T	T	C	A	T	T	C	G	G	G	G	G	T	T	T
r1	A	A	C	G	C	C	G	G	G	G	G	C	T	T	T
r2	T	T	C	A	T	T	C	C	C	G	C	C	T	T	A

(b)

	Sites														
	1	2	3	4	5	6	7	8	9	10	11	12	13	14	15
out	A	T	T	A	T	C	C	C	G	A	C	C	T	C	T
~~*anc*~~	~~A~~	~~T~~	~~C~~	~~A~~	~~T~~	~~C~~	~~C~~	~~C~~	~~G~~	~~G~~	~~C~~	~~C~~	~~T~~	~~T~~	~~T~~
p1	A	A	C	G	C	C	C	C	C	G	C	C	A	T	A
p2	T	T	C	A	T	T	C	G	G	G	G	G	T	T	T
r1	A	A	C	G	C	C	G	G	G	G	G	C	T	T	T
~~*r2*~~	~~T~~	~~T~~	~~C~~	~~A~~	~~T~~	~~T~~	~~C~~	~~C~~	~~C~~	~~G~~	~~C~~	~~C~~	~~T~~	~~T~~	~~A~~

(c)

	Patterns					
	I	II	III	IV	V	VI
	2, 4, 5	8, 11	1, 6, 12	7	9, 13, 15	3, 10, 14
out	T A T	C C	A C C	C	G T T	T A C
p1	A G C	C C	A C C	C	C A A	C G T
p2	T A T	G G	T T G	C	G T T	C G T
r1	A G C	G G	A C C	G	G T T	C G T

(d)

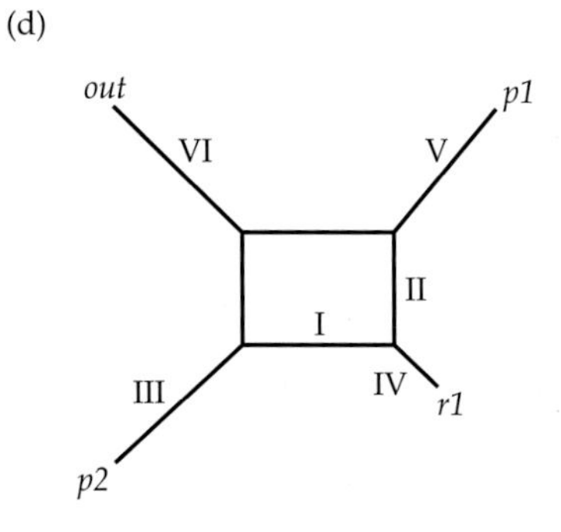

Figure 6.5 A phylogenetic network representation of a recombination event. (a) An alignment of seven sequences derived from a simple evolutionary scenario, in which a 15 bp ancestral sequence (*anc*) gives rise to two descendant sequences (*p1* and *p2*). Subsequently, five nucleotide substitutions are assumed to occur in the lineage leading to *p1* (blue letters), and four substitutions in the lineage leading to *p2* (red letters). Next, a recombination between sites 6 and 7 produces two recombinants, *r1* and *r2*. After the recombination, sequences *p1*, *p2*, and *r1* each accumulate a nucleotide substitution, at site 7, 12, and 13, respectively (green letters). It is also assumed that there are three nucleotide differences (orange letters) between *anc* and an outgroup (*out*), at sites 3, 10, and 14. (b) We assume that we only have access to extant sequences, i.e., we do not have the sequence of *anc*. Moreover, we assume that *r1* and *r2* were produced by a single recombination event and that the transmission of both recombinant alleles to the next generation is highly improbable. Thus, only *r1* is available for the analysis. (c) The 15 aligned sites can be divided into six patterns (I–VI), from which we can build a phylogenetic network. (d) The phylogenetic network represents the relationship of four sequences (*p1*, *p2*, *r1*, and *out*). Patterns determining each split are shown on the branches. The two recombining alleles (*p1* and *p2*) are located on opposing branches emanating from the rectangle. Both have long external branches. The recombinant allele (*r1*), on the other hand, has a short external branch and is located across from the outgroup (*out*). Note that the pattern that supports the short external branch leading to the recombinant allele (*r1*) represents a single nucleotide substitution that occurred after the recombination event. In contrast, the patterns supporting the long external branches of alleles *p1* and *p2* (III and V) represent nucleotide substitutions that occurred both before and after the recombination event.

(a)

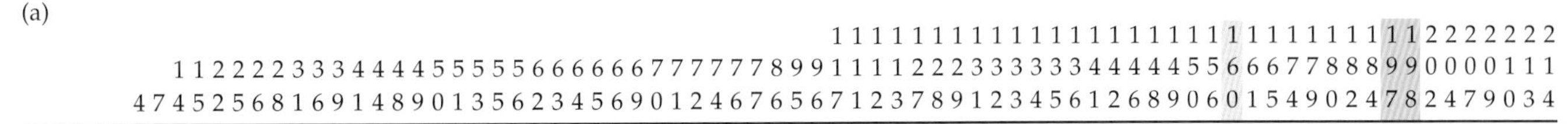


```
Chimpanzee GCTATTAAAGAGGATGACCGACGCATATTTCTGTTCGCCCGTTCTATCGCGACGGGGCGAGGGCGAGCGTCT
O01        GCTGTTAAAACAGTTGGCCGACATATGCTCATAGCAGCTTACCTCGCGAATGTTA-ATAAAAACGAATTCAC
A101       GCCACCGGTACAAACAATTAGTTCTCATATCGATCACTCTACCTCGCGAATGTTGGATAGAAACGGATTCAC
B101       ATCACCGGTACAAACAATTAGTGCTCATATCGATCACTCTACCTCGCGAATGTTGGGCGGGGGACAGTGCAC
```

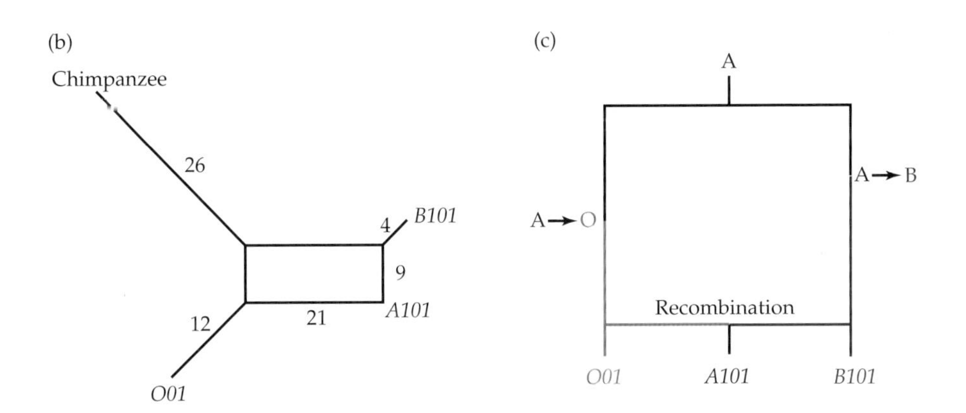

Figure 6.6 Inferring a reticulation from a phylogenetic network. (a) An alignment of sites containing exactly two character states for three major ABO blood-group alleles in humans (O02, A101, and B101) and allele A from chimpanzee. Site numbers follow Kitano et al. (2012). Two sites (197 and 198; red shading) define the difference between A and B alleles in exon 7. Site 160 (blue shading) represents the nucleotide deletion in exon 6 that defines O alleles. Two different splits are shown below the sequences. Red squares denote the split ((chimpanzee, O01),(A101, B101)); blue squares denote the split ((chimpanzee, B101),(O01, A101)). No sites support the third possible split, ((chimpanzee, A101),(O01, B101)). (b) A median phylogenetic network showing the telltale signs of reticulation. (c) A simplified scheme of the evolutionary pathway for three human ABO alleles. The A, B, and O lineages are shown in black, red, and blue, respectively. The two nucleotide substitutions in exon 7 that are crucial for the functional differences between the A and B alleles are shown, accompanied by the resulting amino acid replacements (in parentheses). (Modified from Kitano et al. 2012.)

transfer *N*-acetylgalactosamine or galactose, respectively, to the H antigen. There are two nonsynonymous nucleotide differences in exon 7 that determine the functional distinction between the glycosyltransferases encoded by the *A* and *B* alleles (Figure 6.6c). Allele *O* is similar to allele *A* except for the deletion of a G at position 261 in exon 6, which induces a frameshift that results in a truncated protein devoid of any glycosyltransferase activity.

Each allele group consists of many alleles that differ from one another in sequence. However, all alleles within each allele group exhibit identical enzymatic activities. Within each allele group, some alleles are common, some are rare, and some are private alleles, i.e., they were only found in one or a few individuals.

Kitano et al. (2012) examined the phylogenetic relationships among several common alleles belonging to all three groups. Here, we will only discuss three of these alleles, *O01*, *A101*, and *B101*, as well as an outgroup, the chimpanzee *A* allele. In the aligned sequences, there were 72 sites that consisted of exactly two character states (Figure 6.6a) and could be used with the median network method. The structure of the resulting phylogenetic network (Figure 6.6b) has all the topological features associated with the occurrence of a reticulation event whereby genetic information from both *O01* and *B101* has been transferred to *A101*. We note, for example, that the external branches leading to *O01* and *B101* are much longer than the branch leading to *A101* (which has a length of 0), and that *A101* is located across the rectangle from the outgroup. In the alignment, we note that 20 of the 21 sites supporting the clustering of *A101* and *B101* are clustered in the 5′ part of the gene, while all 9 sites supporting a clustering of *A101* with *O01* are concentrated in the 3′ part. This lack of interleaving is a telltale sign of recombination.

A simplified evolutionary scheme is shown in Figure 6.6c. Based on the fact that chimpanzee populations segregate alleles *A* and *O*, and since *O* is clearly derived from *A*, this allele is assumed to have been the ancestral allele in humans. Approximately 2 million years ago, the *B* allele was derived from an ancestral *A* allele by two nucleotide substitutions. Sometime later, the *O* allele was derived from another ancestral *A* allele by a nucleotide deletion, and because it is nonfunctional the *O* allele has evolved rapidly ever since. Fortuitously, upstream of the nucleotide deletion no other nonsense mutation occurred, and the two positions that define the *A* allele have not been substituted. Thus, by about 1.5 million years ago the original *A* allele became extinct. Approximately 260,000 years ago, a recombination occurred that joined the intact exon 6 from a *B101* and the *A*-defining sites in exon 7 from *O01*. The result was a "resurrected" functional *A* allele. In the period between the extinction of the ancestral *A* allele group and the emergence of *A101*, it is assumed that human populations possessed only two allele groups, *B* and *O*. The resurrection of an *A* allele occurred before the emergence of anatomically modern humans. Nowadays, *A101* has a worldwide distribution and a very high frequency. It is, therefore, reasonable to assume that the increase in *A101* frequency was driven by some kind of very strong positive selection, although the nature of this selection remains a mystery.

Speciation by hybridization: The reticulate evolution of woodferns

Dryopteris is a cosmopolitan fern genus consisting of about 225 species. Of its 13 North American species that reproduce sexually, 7 are diploid, 5 are tetraploid, and one is hexaploid. Using sequence data from nine plastid and two nuclear sequences to reconstruct the phylogenetic relationships among the diploid species revealed that the tetraploid *Dryopteris* species were derived through hybridization of diploid species, while the hexaploid species was a hybrid between a tetraploid mother (the donor of the plastid sequence) and a diploid father (**Figure 6.7**). Sessa et al. (2012) identified four diploid progenitors of four tetraploid and one hexaploid taxa. One hypotheitical diploid taxon, tentatively referred to as *Dryopteris "semicristata,"* has yet to be identified in nature.

Based on divergence time analyses, the earliest *Dryopteris* allopolyploids are thought to have been formed within the last 6 million years. Sessa et al. (2012) found no evidence for recurrent formation of any of the polyploids. In other words, the polyploids are bona fide species rather than accidental hybrids. In addition, the four tetraploids were found to be **transgressive** with respect to geographic range relative to one or both of their diploid "parents." That is, their habitat ranges extend beyond those of the parents, suggesting that ecological advantages in novel habitats may promote long-term regional coexistence of the hybrid taxa with their progenitors.

As is evident from the example in Figure 6.7, it is impossible to define monophyletic clades or sister taxa within phylogenetic networks. Moreover, species concepts that were already problematic in taxa abiding by the tree model become utterly inappropriate in cases in which speciation is driven by reticulate processes such as hybridization.

We note that speciation by hybridization occurs in many plant and animal species. Even the poster child for divergent treelike evolution, the genus *Geospiza* (Darwin's finches), has been shown to experience periodic reticulate gene flow among its constituent species. This gene flow has resulted in a significant reduction in the genetic

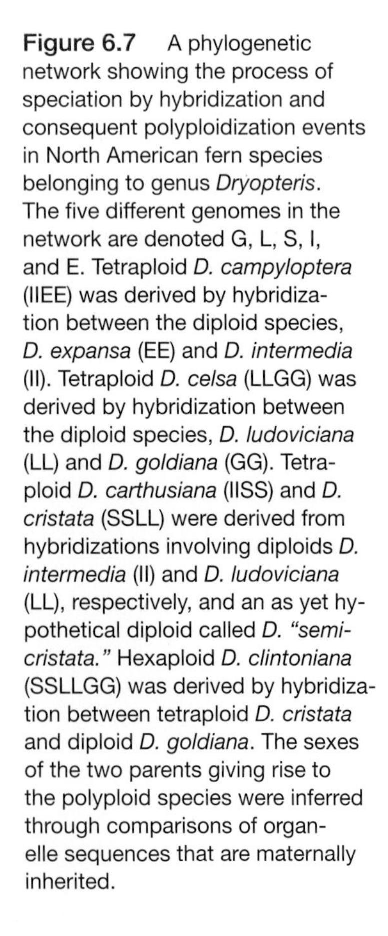

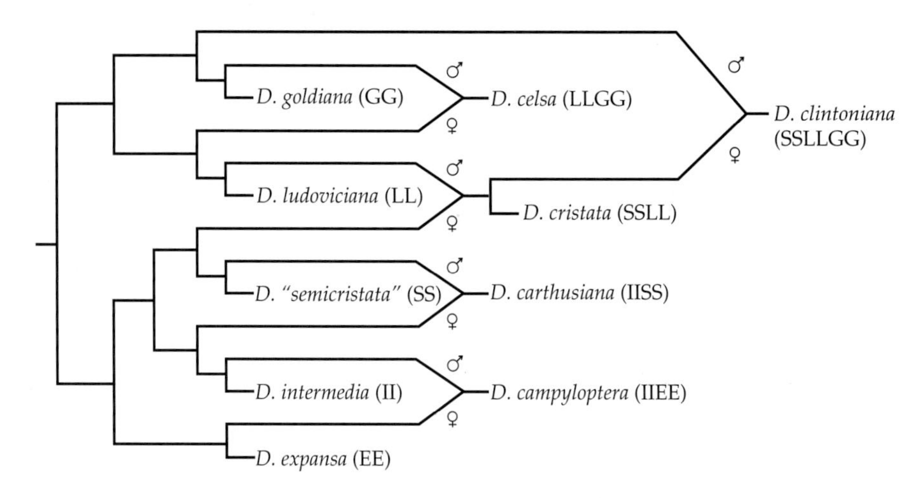

Figure 6.7 A phylogenetic network showing the process of speciation by hybridization and consequent polyploidization events in North American fern species belonging to genus *Dryopteris*. The five different genomes in the network are denoted G, L, S, I, and E. Tetraploid *D. campyloptera* (IIEE) was derived by hybridization between the diploid species, *D. expansa* (EE) and *D. intermedia* (II). Tetraploid *D. celsa* (LLGG) was derived by hybridization between the diploid species, *D. ludoviciana* (LL) and *D. goldiana* (GG). Tetraploid *D. carthusiana* (IISS) and *D. cristata* (SSLL) were derived from hybridizations involving diploids *D. intermedia* (II) and *D. ludoviciana* (LL), respectively, and an as yet hypothetical diploid called *D. "semicristata."* Hexaploid *D. clintoniana* (SSLLGG) was derived by hybridization between tetraploid *D. cristata* and diploid *D. goldiana*. The sexes of the two parents giving rise to the polyploid species were inferred through comparisons of organelle sequences that are maternally inherited.

difference between species as well as the occasional creation of hybrids whose fitness is greater than that of their nonhybrid parents (Arnold and Larson 2004; Grant et al. 2004). Finally, hybrid speciation has also been found in at least one mammalian taxon. The Clymene dolphin, *Stenella clymene*, arose through hybridization between a striped dolphin (*S. coeruleoalba*) father and a spinner dolphin (*S. longirostris*) mother (Amaral et al. 2014). *Stenella clymene* is currently genetically isolated from its parental species, although low levels of introgressive hybridization below the level of detection may be occurring.

The Tree of Life Hypothesis

The "tree of life" has been variously used as a metaphor, an iconographic device, a research tool, and a hypothesis (Mindell 2013). As an iconographic device, the **tree of life** entails the treelike representation of the phylogenetic relationships among all extant and extinct taxa. Of course, it is impossible to include all phylogenetic relationships within a single image, so in practice, the tree of life principally highlights the first divergence events that subsequently gave rise to all the taxa in the world. The **tree of life hypothesis** is the assertion that the evolutionary relationships among all organisms can be accurately described by means of a rooted, bifurcating tree. In this section, we will assess the suitability of the tree of life hypothesis to describe all of evolutionary history.

The tree of life hypothesis is based on two assumptions. The first is the monophyletic view of life, according to which all life forms, be they humans, sponges, onions, algae, or bacteria, are evolutionarily related to one another. This view was first articulated by Charles Darwin in 1859: "All the organic beings which have ever lived on this Earth have descended from one primordial form, into which life was first breathed." The second assumption was put forward by Alfred Wallace a year earlier: "There is a general principle in nature which will cause many varieties to survive the parent species, and to give rise to successive variations departing further and further from the original type." This evolutionary "principle" instituted the practice of describing the evolutionary history of the living world as a branching tree, in which descendant species become isolated almost instantaneously from their ancestors and then become hermetically isolated from one another through progressive divergence. According to the tree of life hypothesis, all of Earth's organisms, both extant and extinct, can be neatly placed at terminal nodes on a gigantic bifurcating tree.

The tree of life hypothesis has had great appeal in the field of molecular evolution, most probably because it made an intuitive and logical connection between a molecular phenomenon (DNA replication) and an evolutionary phenomenon (speciation). As a matter of fact, because DNA replication is binary, at the most fundamental level (i.e., tracing the evolutionary history of individual nucleotide positions), evolution can accurately be described as a bifurcating, rooted, treelike process. As Maddison et al. (2007) put it, "all life on Earth [is] intimately connected in a single tree-like structure of flowing nucleotide sequences, housed in the bodies of organisms they help build. This tree is billions of years in age, with myriad branches, and millions of extant leaves. The existence of this tree, and that each of us is part of one of its leaves, is one of the most profound realizations that we as a species have achieved."

For about 100 years following Wallace's and Darwin's pronouncements, biologists found only a very few exceptions to the "rule" that there is but one tree of life and that this tree is forever bifurcating. The tree of life informed scientists that evolutionary lineages diverge continuously away from one another and that interactions among descendant lineages are rare and, if they occur at all, have only transient and trivial consequences. In the last 50 years, however, numerous reticulate evolutionary events have been discovered, especially among prokaryotes, which have raised doubts about the universality of the tree of life phylogenetic framework.

The initial reaction to the discovery of reticulations was to regard them as "phylogenetic noise," more so since scientists realized quite quickly that methodological artifacts and stochastic errors can give rise to the appearance of reticulation even when

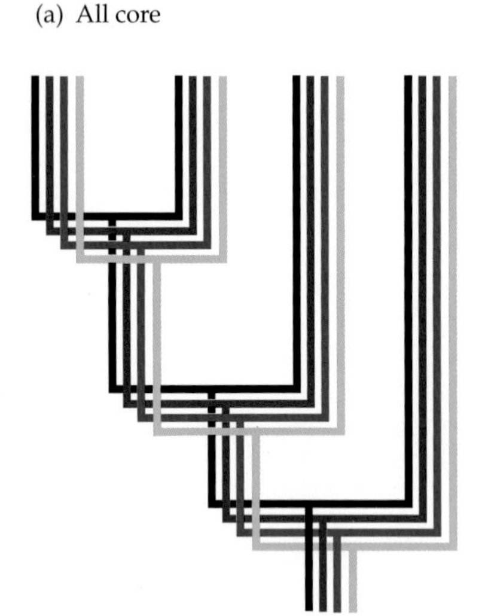

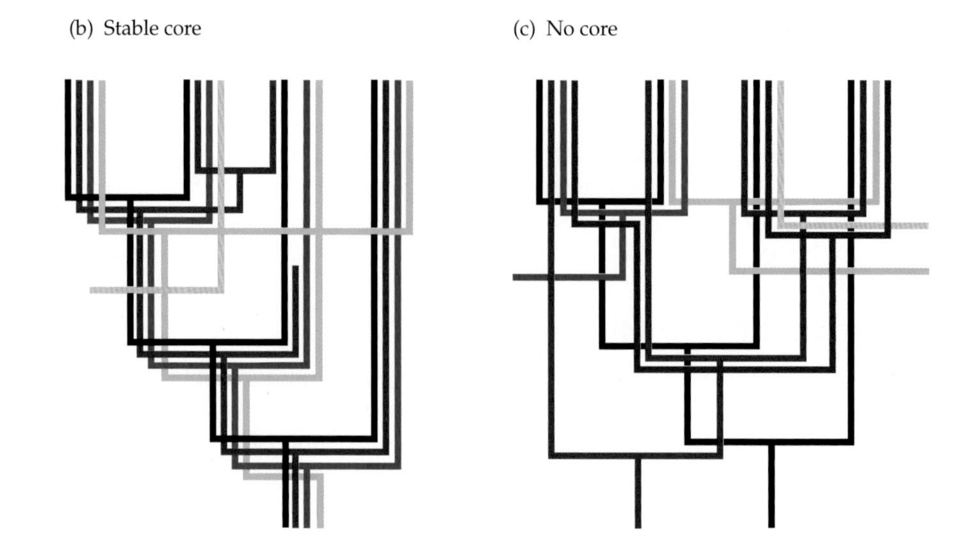

Figure 6.8 Three models for relationships among genome and gene phylogenies for four taxa A, B, C, and D. Each gene lineage is represented by a different color. (a) In the all-core model no genes are exchanged among the genomes, so that each gene in a genome has exactly the same evolutionary history as any other gene in that genome. (b) In the stable-core model, some, possibly even most, genes can be transferred among different genomes over evolutionary time, but a core of genes is immune to this process. Here, the core genome consists of two genes (black and blue lineages) that have identical phylogenetic topologies. These core genes can be used to reconstruct species phylogeny and construct a tree. (c) In the no-core model, no gene is immune from horizontal gene transfer, so in principle each gene can have a different phylogenetic history from any other gene. In the case shown here, no two genes have the same phylogenetic topology.

no reticulation has actually occurred. Under the "noise" assumption, reticulations were assumed to occur either infrequently or not at all and, hence, to have almost no effect on the underlying treelike structure of the evolutionary process. One "quick and dirty" approach was to assume a priori that a phylogenetic process is binary and, hence, that the "true tree" can be "extracted" by putting aside or averaging out the few genes or positions within a gene that have experienced reticulate evolution and then focusing on the "well-behaved" data.

Implicit in the tree of life hypothesis is the assumption that different genes in a genome have identical evolutionary histories (Doolittle 2004). This situation is illustrated in **Figure 6.8a**. We know now that, in its strictest sense, this so-called **standard** or **all-core model** only applies to a subset of macroscopic multicellular eukaryotes. For prokaryotes and for the many unicellular eukaryotes in which horizontal gene transfer is a common occurrence, more realistic models are required to allow for reticulate events.

Figure 6.8b describes the **stable-core model**. In this model, there is a stable core of a few genes that are never horizontally transferred between genomes. If for a group of taxa a stable core of genes can be shown to exist, then these core genes can be used to construct a phylogenetic tree on which reticulate events can subsequently be superimposed. Thus, the validity of a taxon that experiences reticulate evolution but possesses a stable core of genes may be assessed by the tree methodology. If, on the other hand, reticulate evolution predominates, the assumption that a tree exists will yield misleading results.

In **Figure 6.8c**, no gene is exempt from transfer. In this **shifting-core** or **no-core model**, each gene can have an evolutionary history that is different from any other gene. If evolution occurs along the lines of the no-core model, then no phylogenetic tree exists. Paraphrasing Doolittle (1999), scientists frequently fail to find the true tree, not because their methods are inadequate or because they have chosen the wrong genes or the wrong reconstruction method, but because there is no tree to be found in the first place.

Many researchers have expressed the belief that a stable core of genes that are immune to transfer does indeed exist. The belief in a stable core of genes is based on the **complexity hypothesis** (Jain et al. 1999), which states that genes whose functional products (at the DNA, RNA, or protein levels) interact with the products of other genes will coevolve. That is, mutations that will affect the structure and function of one gene product (call it A) will be compensated by mutations that will affect the interacting product of another gene (product B). By means of compensatory mutations, the essential interactions between A and B are preserved throughout evolution. The same is true for a second lineage, in which the homologous gene products A′ and B′

will also coevolve. If gene *B* in one lineage is replaced by gene *B′* from another lineage, then the product of gene *A* may interact less effectively with the product of gene *B′* than with the product of gene *B*. Similarly, product A′ may not be as effective in its interaction with B as it was with B′. Inefficient interactions between gene products may negatively affect the fitnesses of the carriers of these mismatched genes. Such genotypes will subsequently be wiped out by purifying selection. Purifying selection is extremely efficient in prokaryotes, which usually have enormous effective population sizes. The corollary of the complexity hypothesis is that interacting genes would be very difficult to transfer horizontally from taxon to taxon without reducing the fitness of the recipient. Moreover, the more interactions a gene product is involved with, the less likely it is that the gene can be viably transferred horizontally.

One such supposedly "untransferable" gene is the small subunit rRNA-specifying gene that has served as a de facto "universal molecular chronometer" (Woese 1987) for many years. The small subunit rRNA is an important interacting component at the center of an enormously complex structure, the ribosome. It interacts directly and indirectly with two other RNAs and at least 50 proteins in performing a function (translation) that is vital and essential to life.

Because of their numerous interactions, **informational genes**—the genes involved in replication, transcription, ribosome biogenesis, and translation—should be hard to transfer even across very short evolutionary distances. Indeed, these genes are rare among reliably identified horizontally transferred genes. Nonetheless, there are confirmed reports of horizontal transfer involving informational genes (e.g., Green 2005), including small subunit rRNA-specifying genes.

A rigorous test of the stable-core idea for any higher taxon would entail (1) the comparison of all the genomes under study, (2) extracting the set of genes common to all taxa, (3) using these genes to reconstruct phylogenetic trees, and (4) tallying up how many phylogenies are congruent with one another. Efforts in this direction have so far failed (Doolittle 2004); there are fewer than 50 genes shared by most genomes, and few of these genes yield statistically robust trees, let alone trees that can be meaningfully compared with other trees for congruence. Thus, at present it is not possible to prove that there exists a stable core of genes common to all species. Of course, absence of evidence is not evidence of absence.

Let us, however, contemplate the possibility that a stable core of genes as hypothesized in Figure 6.8c does not exist. Let us further imagine that all genes are transferable horizontally, at least in principle, but that the likelihood of transfer varies greatly among genes and among evolutionary lineages. Under such a regime, the success of the transfer would be determined by the nature of the sequence being transferred, the similarities or dissimilarities between the donor and recipient species, and the physical proximity of the donor and the recipient (Chapter 9). Thus, although a stable core of genes may not exist, in the short run at least, the presence of biased horizontal gene transfer may preserve a treelike phylogenetic structure.

The Vertical and Horizontal Components of Prokaryote Evolution

Genome evolution in prokaryotes entails both treelike components generated by vertical inheritance from an ancestor to descendants and network-like components generated by horizontal gene transfer. One of the most important questions in the field of prokaryotic phylogeny is, Can the vertical component of the phylogeny be recovered?

Because of extremely variable rates of evolution among lineages and the great antiquity of many taxonomic groups, reconstructing the evolutionary relationships among prokaryotes would have been a formidable task even in the absence of reticulations. The superimposition of horizontal gene transfers on top of an already difficult-to-reconstruct phylogenetic tree results in a situation in which the reconstruction of evolutionary relationships among prokaryotic taxa is, for all intents and purposes, unattainable. There are two main reasons for this situation.

First, prokaryotic lineages experience a very high rate of **gene turnover** (gene gain and loss), which causes the number of orthologous genes available for phylogenetic analysis to decline precipitously as the number of taxa under study increases. Moreover, in contrast to eukaryotes, in which changes in the gene repertoire occur mostly via DNA duplication, in prokaryotes high rates of horizontal gene transfer determine gene content (Treangen and Rocha 2011). For those of us accustomed to the comforting verticality of vertebrate phylogeny, such radical departures from the Darwinian paradigm seem difficult to swallow. However, if we remember how quickly antibiotic resistance has spread across taxonomic boundaries via horizontal gene transfer, it becomes clear that the process of evolution operates quite differently in prokaryotes than it does in eukaryotes.

Second, it is an underappreciated but very important observation that in prokaryotes, the mechanisms of genetic recombination are identical with the mechanisms of horizontal gene transfer. That is, recombination in prokaryotes is never reciprocal; it is always unidirectional, from donor to recipient, and always entails recombination across small parts of the genome, rather than whole-chromosome pairing and assortment as in eukaryotes. Because prokaryotic growth is clonal, the process of recombination overlaid upon clonal growth generates descendant lineages that have increasingly divergent collections of genes. Over time, "conspecific" cells will have little in common with one another in terms of their genomic content. Thus, the term "genome" in reference to prokaryotic species is meaningless. Instead, we use the term "pangenome" to refer to the collection of all genes found in a "species" (Chapter 10). For example, an analysis of 61 sequenced strains of *Escherichia coli* revealed that there are close to 16,000 gene families distributed among the strains, but only about 4,000 in any individual genome, and slightly less than 1,000 gene families present in all 61 genomes (Lukjancenco et al. 2010). Microbiologists call these 1,000 shared gene families the "core genome" and the remainder the "accessory genome." Concepts like pangenomes, core genomes, and accessory genomes (Chapter 10) are fundamentally foreign to the field of eukaryote phylogenetics, where all individuals belonging to a species have approximately the same number and arrangement of gene loci.

There are currently two extreme schools of thought. One school believes that prokaryotic evolution cannot be dealt with by phylogenetic trees at all (e.g., Bapteste et al. 2009) and that a new metaphor or a new heuristic device is needed. Accepting this view requires a redefinition, or at least a revision, of the term "species" as far as prokaryotes are concerned. The other school of thought believes that, although reticulate evolution is common in prokaryotes, a central treelike signal remains that cannot be realistically explained by a self-reinforcing pattern of biased horizontal gene transfer (Puigbò et al. 2010). In this view, trees are "the natural representation of the histories of individual genes given the fundamentally bifurcating process of gene replication." Some researchers believe that the rate of horizontal gene transfer is too low to obscure an underlying treelike structure. In their opinion, horizontal gene transfers, even when relatively common, should not affect treelike evolutionary history but are seen more as "cobwebs" connecting tree branches (Ge et al. 2005).

Boucher and Bapteste (2009) suggest a distinction between **open lineages** and **closed lineages**. A closed lineage is one in which the majority of evolutionary changes occur vertically. According to their analysis, eubacterial species such as *Staphylococcus aureus* and *Escherichia coli* are closed lineages, while *Streptococcus pyogenes* and *Neisseria gonorrhoeae* represent open lineages. The distinction between open and closed lineages is not absolute, however, and whether a phylogenetic tree may or may not be used in evolutionary reconstruction is not a clear-cut decision.

Prokaryote taxonomy and the meaning of "species" in prokaryotes

Ever since Aristotle, approximately 2,500 years ago, the meaning of the basic biological unit—the "species"—has been debated. Systematics entails the belief that species (e.g., *Homo sapiens* or *Allium cepa*) exist extramentally. That is, species are not arbitrary social constructs; they are real and have verifiable characteristics. In particular, individuals belonging to a species are said to share a combination of traits that unite them as members of that species and distinguish them from members of other species. This approach works quite well as far as multicellular, sexually reproducing eukaryotes are concerned. In prokaryotes, the term "species" is incalculably more abstruse, since in addition to the question "What is a species?" we must also concern ourselves with the question "Do species exist?"

Notwithstanding these difficulties, for pragmatic, clinical, and epidemiological reasons, microbiologists require a taxonomy that is effective, predictive, and stable. As stated in Brenner et al. (2005), "bacterial classifications are devised for microbiologists, not for the entities being classified. Bacteria show little interest in the matter of their classification." How else will microbiologists communicate their science to their peers and the general public? How else can we diagnose the etiological agent of a disease or determine which bacteria to use in mitigating the polluting effects of oil spills? When we have a cough, we need to know whether it's *Legionella*, *Streptococcus*, or something else. The taxonomy determines which particular antibiotic is prescribed, even though in many cases the pathogen may have already acquired, via horizontal gene transfer, the resistance genes that will not only render this particular antibiotic ineffectual, but also cast doubt on its taxonomic status.

Let us first tackle the problem of nomenclature. The taxonomic requirement for describing and naming a new prokaryotic taxon is that one strain be designated as the **type strain** that could subsequently be used as reference for any further taxonomic identification. The designated type strain should be deposited in two international collections and made publicly available. Unfortunately, this system of identification does not seem to work for prokaryotes. First, the vast majority of prokaryotes are not culturable. Put simply, for some prokaryotes we possess the genomic sequence but not its physical embodiment as a living organism. Second, Richter and Rosselló-Móra (2009) found that fewer than 30% of the sequenced genomes in databases actually belonged to the type strain of the species to which they were supposed to belong.

Let us deal with prokaryote classification by noting first that, as opposed to the classification of eukaryotes, prokaryote taxonomy cannot follow phylogenetics because phylogenetic relationships among taxa are mostly unrecoverable. The early classification of prokaryotes was based solely on phenotypic similarities, but in the late 1960s some genome-based methods were developed to evaluate taxonomic relationships. Among them, DNA-DNA hybridization techniques became popular in determining crude genome similarities among organisms. DNA-DNA hybridization tended to be reproducible and was an improvement over methods relying on phenotypes. Over the years, a practice was established whereby strains were deemed to belong to a coherent taxon (or **genospecies**) if they shared DNA-DNA hybridization values with greater than 70% similarity. With the advent of genome sequences, DNA-DNA hybridization techniques became obsolete, and attempts were made to use measures such as the average nucleotide identity between genome pairs as the gold standard for clustering genomes into meaningful taxonomic standards. A value of 95% nucleotide identity between 16S rRNAs was found to be equivalent to the DNA-DNA hybridization critical value of 70% (Richter and Rosselló-Móra 2009). It should be noted that these threshholds (or any others) are by necessity arbitrary. Moreover, if measures of sequence similarity or dissimilarity are applied to commensal and pathogenic strains of *E. coli*, the resulting classification might indicate the existence of many different species rather than the single species recognized so far.

Several comparative measures, which rely on more than a single gene, have been suggested. One such method, **multilocus sequence typing**, in which the sequences of a small number of housekeeping genes (usually seven) are compared, is mainly used in epidemiological studies. Multilocus sequence typing and other comparative sequence methods frequently fail to identify important taxonomic entities, which may have many genes with similar sequences in common with other species, yet differ in gene content. One such example concerns *Burkholderia mallei*, the causative agent of

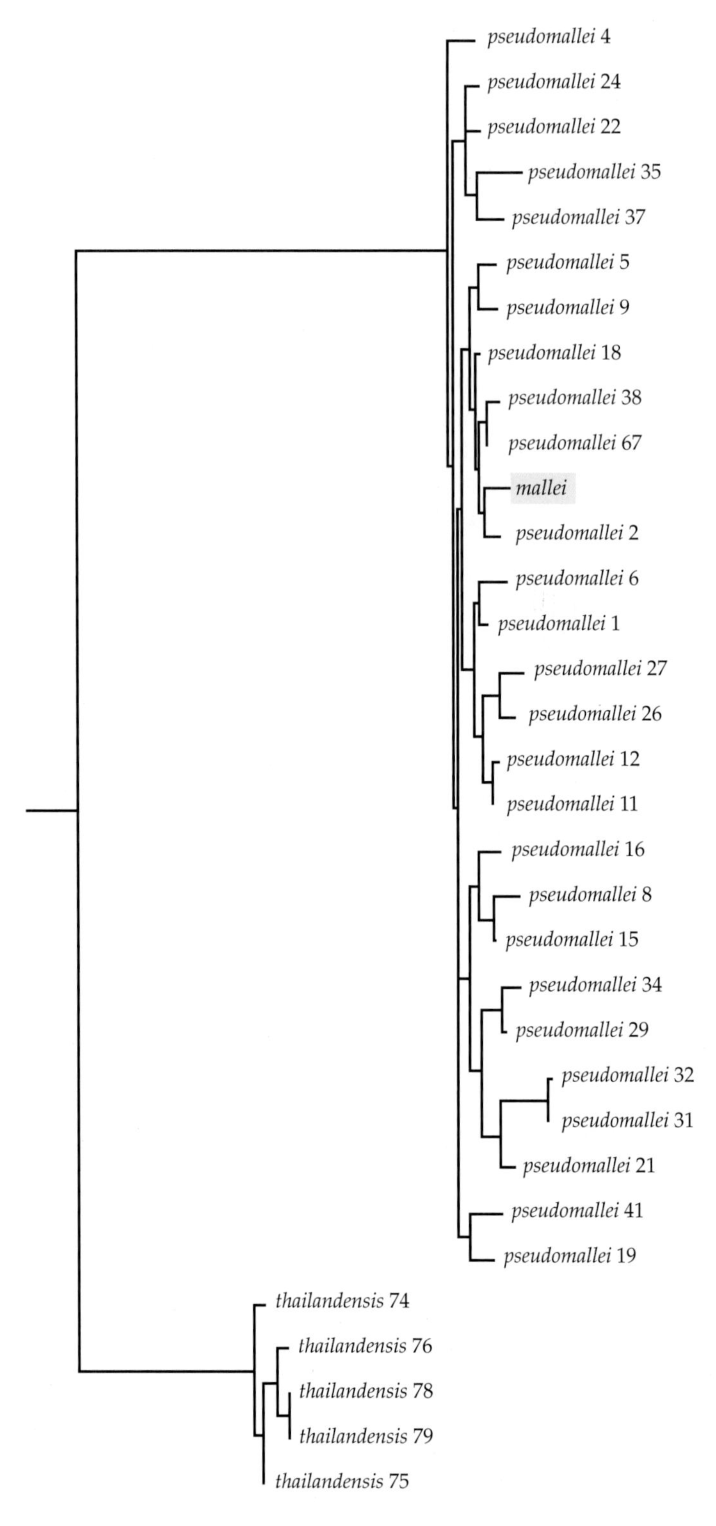

Figure 6.9 A phylogenetic tree for three *Burkholderia* species: *B. mallei*, *B. pseudomallei*, and *B. thailandensis*. The tree is based on sequence similarity of a concatenation of seven housekeeping gene segments (*ace*, *glt*B, *gmh*D, *lep*A, *lip*A, *nark*, and *ndh*) from a large collection of *B. pseudomallei* and *B. thailandensis* isolates, as well as a single isolate of *B. mallei*. (Isolate numbers are listed to the right of the species name.) The total length of the ungapped aligned sequences was 3,399 base pairs. The position of *B. mallei* within the *B. pseudomallei* strains is shaded blue. (Modified from Gevers et al. 2005.)

equine glanders, a disease characterized by pneumonia and necrosis of the tracheobronchial tree if *B. mallei* is inhaled, or by pustular skin lesions, abscesses, and sepsis if the skin is the portal of entry.

Sequence comparisons based on 16S rRNA indicated that *B. mallei* is indistinguishable from the mostly saprophytic *B. pseudomallei* and a *B. pseudomallei*-like bacterium called *B. thailandensis*, which lives on dead or decaying organic matter. In contrast to *B. mallei*, which is an obligate parasite of horses, mules, and donkeys with no other known natural reservoir, *B. pseudomallei* and *B. thailandensis* are free-living organisms. As shown in **Figure 6.9**, a multilocus sequence-typing analysis revealed that all *B. pseudomallei* isolates are tightly clustered and well separated from the *B. thailandensis* isolates. In contrast, *B. mallei* was found to be nested within *B. pseudomallei*. So, while *B. pseudomallei* and *B. thailandensis* are unambiguously distinct species, *B. mallei* is merely a derived clone with a peculiar ecological distribution (an **ecotype**) of *B. pseudomallei*. This means that, on taxonomic and phylogenetic grounds, *B. pseudomallei* and *B. mallei* should not be given separate species names, although the important differences in their biochemical activities and in the clinical symptoms and epidemiology justify their being classified as two species. These differences are due to gene content rather than sequence differences.

The current practice in prokaryote taxonomy is to apply standard phylogenetic tools (Chapter 5) to a set of informational genes, for example, genes that specify ribosomal RNAs or encode proteins for ribosome biogenesis and information-processing functions. These genes are thought to be only very rarely transferred horizontally and are, hence, regarded as suitable markers for reconstructing the vertical evolutionary history of prokaryotes. Each newly discovered or characterized prokaryotic taxon is, thus, defined and assigned to it proper taxonomic grouping on the basis of a very small number of genes, and such classifications may be irrelevant to the traits that are of most interest to clinicians and public health workers.

Basic to any notion of "species" is that in nature they comprise discrete clusters of organisms, defined genomically and phenotypically. That is, genome space "is not uniformly filled by a seamless spectrum of intergrading types" (Helal et al. 2011). Results from **metagenomics**, i.e., the sequencing of genetic material recovered directly from environmental samples consisting of organisms that cannot be grown in culture in the laboratory, suggest that communities of microbes do not exist as discrete sets of species, in which the genomes of different individuals resemble one another to a great extent. Rather, a wide spectrum of organisms displaying disparate levels of sequence dissimilarity may form a species (Huson et al. 2009).

These findings have led some researchers to abandon the strict notion of "species" in prokaryotes. Doolittle and Zhaxybayeva (2009), for instance, came to the conclusion that there is no "principled way" in which questions about prokaryotic species, such as how many species there are, or how large their populations are, or how widely they are distributed, can be answered. Instead, they advocate the acceptance of a fuzzier notion of prokaryotic species. Accordingly, it is possible that some prokaryotes form species either by restrictive horizontal gene transfer or by periodic selection, while others may not form species at all. Thus, "species" is no longer a term describing a category with universal biological application, but merely a useful descriptor of certain types of populations, which we nominally call a "species," although the category itself may not be real. Of course, at some point the degree of fuzziness becomes too extreme for even species nominalism to work.

If we completely abandon the phylogenetic tree metaphor for prokaryotes, we must also be disabused of the notion that there is a uniquely natural hierarchical scheme for prokaryotes, which in turn would mean that the hope of ever unambiguously knowing what species are is a delusion (Doolittle 2009a,b). According to this view, the term "species" in prokaryotes should be recognized as a useless reification that we must and can do without.

The Phylogeny of Everything

In this section, we attempt to trace evolutionary history to its very beginning, thus supplying the foundation on which to build a phylogeny that will encompass all life on Earth. This feat will require us to deal with (1) the eukaryote-prokaryote divide, (2) the eubacterial-archaebacterial divide, (3) the internal taxonomic and phylogenetic divisions within prokaryotes, (4) the origin of the eukaryotic cell, and (5) the identification of the first divergence event in evolutionary history.

The eukaryote-prokaryote divide and the taxonomic validity of Procaryota

The dichotomy between Eucaryota and Procaryota was first proposed by Chatton (1925) and later formalized by Stanier and van Niel (1962) and Ris and Chandler (1963). (In the 1970s, Chatton's original spelling fell out of fashion, and the terms "Eucaryota" and "Procaryota" were gradually replaced by "Eukaryota" and "Prokaryota," respectively.) As opposed to eukaryotes, which were defined by their distinct nucleus and cytoplasm, prokaryotes for a very long time were defined by what they do *not* have: they were said to be those organisms that lack a membrane-enclosed nucleus. Subsequently, other negative character states were added, such as lack of spliceosomal introns and spliceosomes and a primary lack of mitochondria. As a consequence, some claimed that prokaryotes were taxonomically illegitimate because they could only be defined by "a negative and therefore scientifically invalid description" (Woese 1998; Pace 2006, 2009). Pace (2009) even advocated banishing the term "prokaryote" from our textbooks and language. These assessments on the validity of Prokaryota have turned out to be inaccurate, as prokaryotes are now known to share several positive characters in common, including chromosomes attached to membranes (Cavalier-Smith 2007) and cotranscriptional translation, i.e., a process whereby the nascent messenger RNA is translated into a protein as the RNA is still being transcribed (Martin and Koonin 2006; Whitman 2009). This situation is completely different from that in eukaryotes, in which mRNA is produced in the nucleus and then exported to the cytoplasm for translation. Thus, in addition to negative traits, at least two positive traits can be regarded as prokaryotic synapomorphies.

The Eubacteria-Archaebacteria divide

In traditional classification, prokaryotes consisted of a single kingdom, **Bacteria**, that included the cyanobacteria (which once were called blue-green algae and classified within Plantae). The studies of Woese and Fox (1977a) and Fox et al. (1977, 1980) on the rRNA sequences of a few methanogenic bacterial genera (e.g., *Methanobacterium* and *Methanosarcina*) challenged this traditional view. Methanogens are unusual prokaryotes that are obligatory anaerobes in oxygen-free environments such as sewage treatment plants and the intestinal tracts of animals; they generate methane (CH_4) by combining carbon dioxide (CO_2) and molecular hydrogen (H_2). Because of their size, their lack of a nuclear membrane, and their low DNA content, methanogens were considered bacteria, which at the time was synonymous with prokaryotes. According to the traditional view, they were, hence, expected to be more closely related to the other bacteria than to the eukaryotes. However, in terms of rRNA dissimilarity, methanogens turned out to be equidistant from both taxa (Sapp and Fox 2013). On the basis of these findings and the fact that the methanogenic metabolism was thought at the time to be suited to the kind of atmosphere believed to have existed on the primitive Earth (rich in CO_2, but virtually devoid of oxygen), Woese and Fox (1977a) established a new taxon for the methanogens and their relatives, **Archaebacteria** (literally, archaic bacteria). The name "Archaebacteria" implies that this taxon is evolutionarily at least as ancient as the other group of bacteria, which they renamed **Eubacteria** (literally, true bacteria). The term **urkaryotes** (literally, primordial nucleated organisms) was proposed for those ancestors of eukaryotes that existed prior to the endosymbiotic acquisition of mitochondria and chloroplasts from prokaryotes. (This nomenclature gave rise to the Archezoa theory, discussed on page 258.)

As it turned out, Archaebacteria was found to include, in addition to methanogens, many prokaryotes that live in extremely harsh environments (**extremophiles**), such as the **thermophiles** and the **hyperthermophiles**, which live in hot springs at temperatures as high as 110°C, and the **halophiles**, which are highly salt-dependent and grow in such habitats as the Great Salt Lake and the Dead Sea.

What distinguishes Archaebacteria from Eubacteria? Archaebacteria shares a number of features with eukaryotes. For example, the archaebacterial elongation factor EF-2 contains the amino acid diphthamide, the methionyl initiator tRNA is not formylated, the aminoacyl stem of the initiator tRNA terminates with the base pair AU, the DNA polymerases are not inhibited by either aphidicolin or butylphenyl-dGTP, and peptide synthesis is inhibited by anisomycin but not by chloramphenicol as in eubacteria. Archaebacteria also differ from eubacteria in ribosomal protein composition, membrane lipid synthesis, cell wall constituents, and flagellar composition, as well as the enzymes involved in the synthesis of tetrahydrofolate and tetrahydromethanopterin (Sousa and Martin 2014).

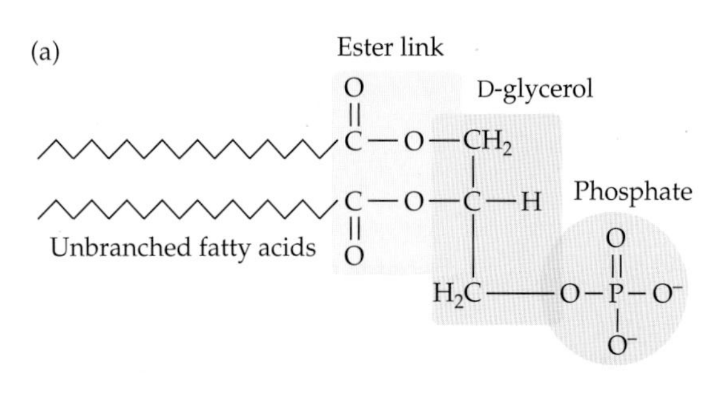

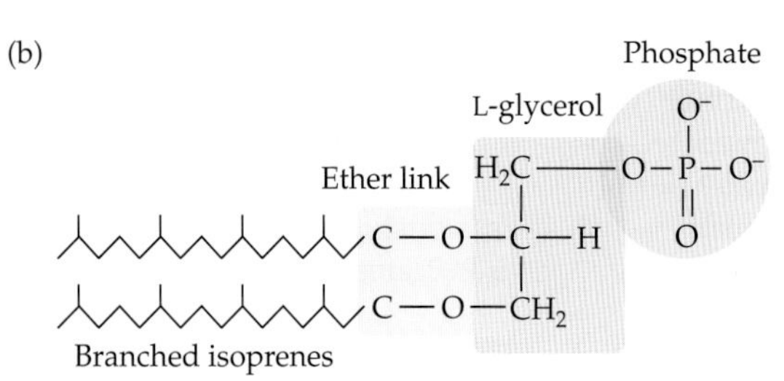

Figure 6.10 The basic chemical structures of membrane phospholipids in eubacteria and eukaryotes (a) and archaebacteria (b).

For taxonomic purposes, archaebacteria are mainly identified by the phospholipids in their membranes (**Figure 6.10**), which are strikingly different from those in eubacteria and eukaryotes. First, their membranes are composed of glycerol-ether phospholipids, whereas eubacteria and eukaryotes have membranes composed mainly of glycerol-ester phospholipids. Second, archaebacterial lipids have a glycerol group with a stereochemistry that is the reverse of that found in other organisms. That is, archaebacterial membranes have L-glycerol, while eubacteria and eukaryotes have D-glycerol. (The L-glycerol ethers constitute a convenient synapomorphy for almost all archaebacterial taxa.) Finally, the lipid tails of archaebacterial phospholipids consist of isoprenoid side chains with multiple side branches. In contrast, the fatty acids in the membranes of other organisms have straight chains with almost no branches. The isoprene side chains of archaebacteria can be joined together. That is, either the two side chains of a single phospholipid can join together, or they can be joined to side chains of another phospholipid on the other side of the membrane. No other group of organisms can form such transmembrane phospholipids.

Another interesting property of the side branches is their ability to form carbon rings. This happens when one of the side branches curls around and bonds with another atom down the chain to make a ring of five carbon atoms. Such rings are thought to provide structural stability to the membrane, since they seem to be more common among species that live at high temperatures. They may work in the same way that cholesterol does in eukaryotic cells to stabilize membranes.

Archaebacteria and Eubacteria are considered the only **superkingdoms** within Prokaryota. The taxonomic rank "superkingdom" is also referred to as **domain** or **urkingdom**.

The tripartite tree of life and its inadequacy

The molecular revelation that prokaryotes are separable into two basic taxa, Archaebacteria and Eubacteria, and that these two taxa are approximately equidistant from eukaryotes prompted Woese et al. (1990) to propose a tripartite taxonomy of all life forms. Accordingly, three **domains** were established: **Bacteria** (formerly Eubacteria), **Archaea** (formerly Archaebacteria), and **Eucarya** (formerly Eukaryota). The reason for the change in nomenclature and, in particular, for the removal of the suffix "bacteria" was to eradicate any hint of kinship between Archaebacteria and Eubacteria, as well as to emphasize that the three taxa were equal in taxonomic rank (Barns et al. 1996; Woese 1996). These taxonomic neologisms were somewhat unfortunate, because *Archaea* is a genus of assassin spiders, *Eucarya* is a genus of quandong trees, and *Bacteria* is a genus of stick insects belonging to order Phasmatodea. The new nomenclature also exhibited a complete disregard for the rules of taxonomy by using a name, Bacteria, that in its original sense included Archaea. Finally, in the literature, the misspelled moniker "Eukarya" has been used about three times as often as the correct term "Eucarya."

The tripartite division of all life forms into Archaea, Eucarya, and Bacteria was customarily illustrated as an unrooted tree (**Figure 6.11**). As with all unrooted trees, turning it into an evolutionary narrative (with a directional temporal arrow) necessitated the rooting of the tree. By definition, however, the evolutionary tree of all the organisms

Figure 6.11 The tripartite division of all life forms, based mainly on ribosomal RNA sequences. The three main lines of descent are Eucarya, Bacteria, and Archaea. (Modified from Badlauf et al. 2004.)

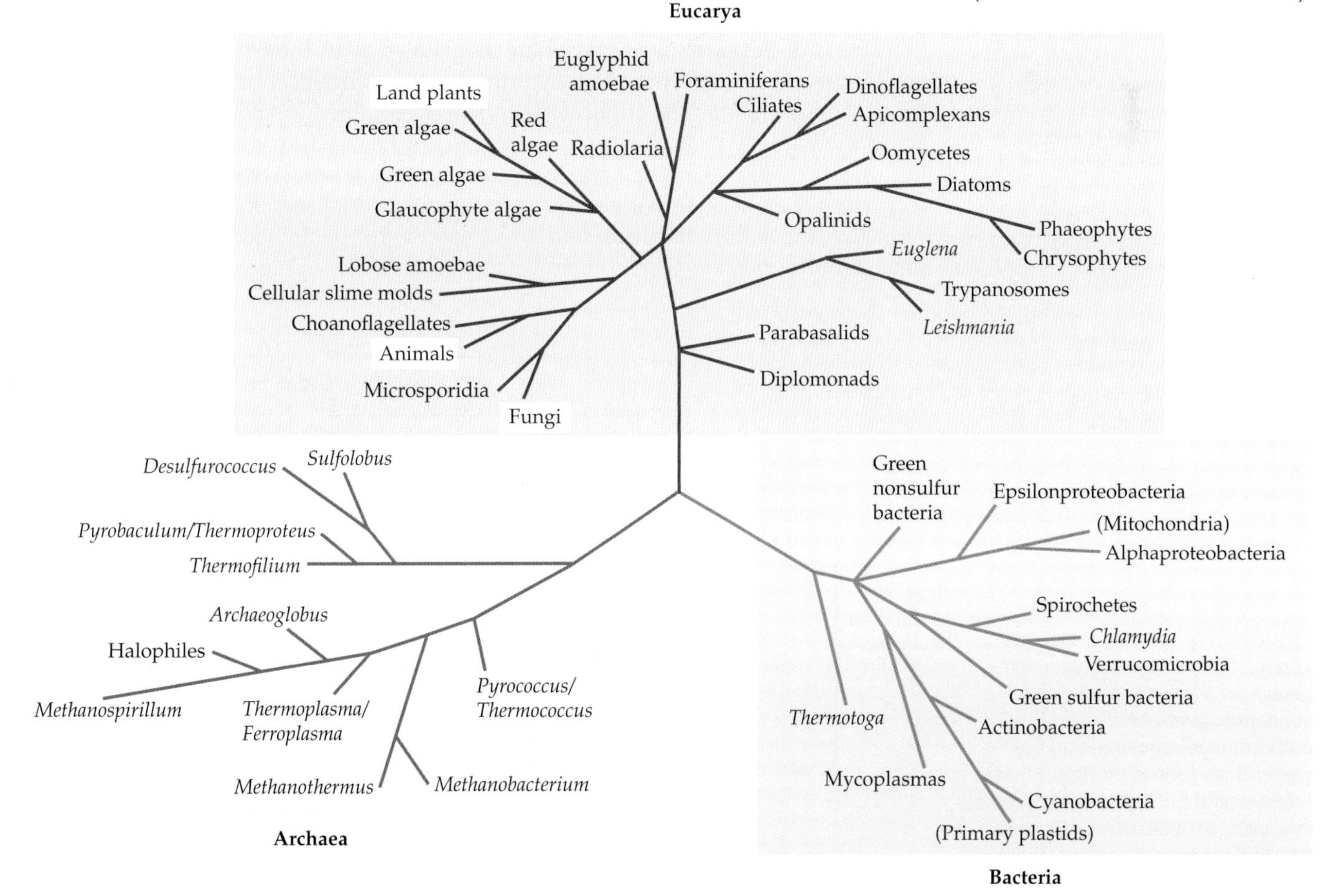

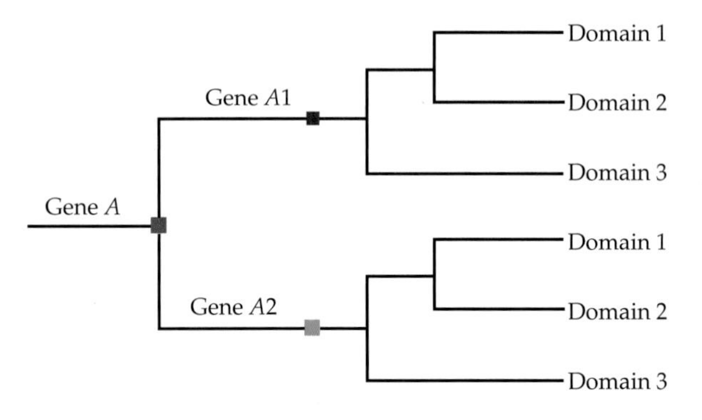

Figure 6.12 Duplication of gene *A* (blue square) into *A1* (red square) and *A2* (green square) prior to the divergence of three domains results in identical topologies for the two subtrees. (Modified from Li 1997.)

in the world has no outgroup. That is, in the "tree of everything," all organisms belong to the ingroup. An ingenious method to infer the root of the tree was suggested by Schwartz and Dayhoff (1978) and put into practice by Gogarten et al. (1989) and Iwabe et al. (1989). The idea was to use a pair of paralogous genes that exist in all organisms and must, therefore, be derived from a gene duplication event that occurred before the divergence of the three domains (**Figure 6.12**). Suppose, for example, that gene *A* gave rise by gene duplication to genes *A1* and *A2* before the divergence of the three lineages. Subsequently, as the three organismic lineages diverged, *A1* and *A2* should also diverge in the same order. Therefore, *A2* sequences may serve as outgroups with which the tree derived from the *A1* sequences can be rooted. Similarly, *A1* sequences can be used to root the tree derived from *A2*.

Iwabe et al. (1989) applied this concept to two paralogous elongation factor genes, *EF-Tu* and *EF-G*, which are present in all prokaryotes and eukaryotes and must, therefore, have been derived from a duplication event that occurred before the three domains diverged from one another. Thus, the *EF-Tu* sequences can be used as outgroups to infer the root of the tree for the *EF-G* sequences, and vice versa (**Figure 6.13**). The *EF-G* subtree indicated that Eucarya (represented by a slime mold and a mammal) is a sister taxon of Archaea (represented by *Methanococcus jannaschii*) to the exclusion of Bacteria (represented by *Micrococcus luteus* and *Escherichia coli*). The *EF-Tu* sequences yielded the same topology. We note that in reconstructing the phylogenetic trees for duplicate genes, we must ensure that our identification of orthologous genes (genes whose homology is due to a speciation event) is correct. This is not always an easy task, especially with distantly related organisms that may have experienced multiple gene acquisitions, losses, and changes in function.

An interesting solution for this problem was suggested by Lawson et al. (1996). In their study of the carbamoyl phosphate synthetase gene, they took advantage of the fact that the gene contains an ancient internal gene duplication that is found in all three domains of life. The duplicated sequences remained linked to each other in the same orientation within the single-copy gene, and hence the orthology identification became a trivial task. When the internal duplication was used to root a gene tree consisting of eight eukaryotic sequences, seven bacterial sequences, and an archaeal sequence, the eukaryotes were found to cluster with the archaeons. For a while, the phylogenetic question seemed comfortably settled: The living world was composed

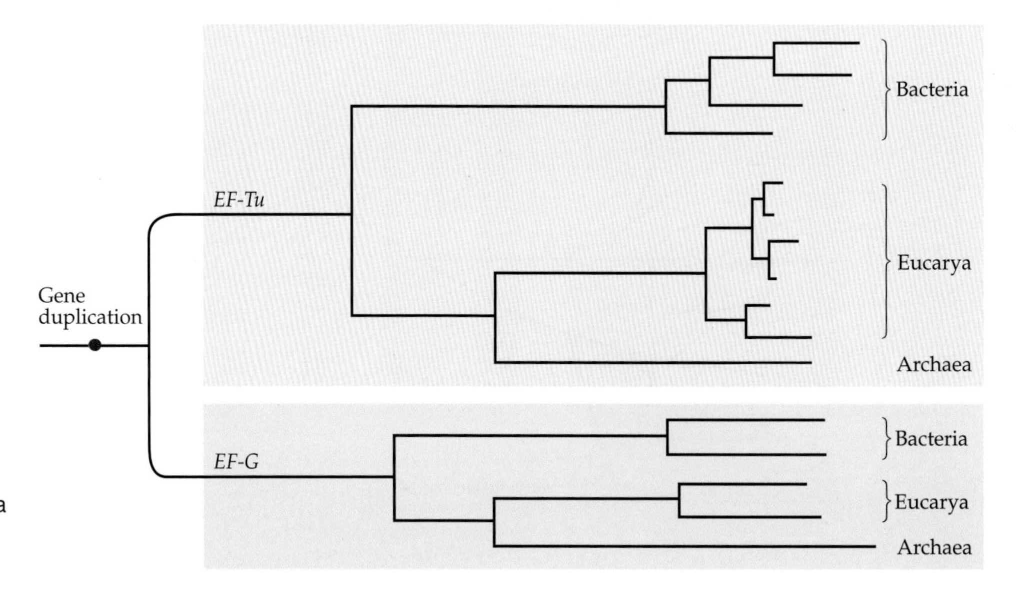

Figure 6.13 Phylogenetic tree inferred from a simultaneous comparison of the duplicated elongation factor genes *EF-Tu* and *EF-G* from Archaea, Bacteria, and Eucarya. Both subtrees indicate that Archaea and Eucarya are sister taxa in a lineage that excludes Bacteria. (Modified from Iwabe et al. 1989.)

of three domains, with Bacteria diverging first from the common ancestor of Eucarya and Archaea. Lamentably, this very simple picture—*Vita est omnis divisa in partes tres*—did not last for long.

The first sign of trouble was the fact that different genes yielded different and conflicting phylogenetic trees. With three monophyletic domains, there are three possible rooted trees (**Figure 6.14a–c**). Moreover, if any of the three domains turns out not to be monophyletic, the number of possible trees increases. With increasingly large genomic databases, the rooting of the "tree of everything" by using molecular data started yielding a very confusing collection of contradictory trees. Each tree in Figure 6.14 has at one time or another been supported by some genes (or some tree-making methodology) and refuted by others. Was the first divergence event a fundamental split between eukaryotes and prokaryotes, as in Figure 6.14a? Or did the first divergence event entail a split between the Bacteria and the common ancestor of Eucarya and Archaea, which later diverged from each other (Figure 6.14b)? Moreover, the suggestion has been made that Archaea may not be monophyletic, and that Eucarya may be a sister taxon of Crenarchaeota (eocytes), a phylum of sulfur-dependent archaeons (Figure 6.14d; Rivera and Lake 1992).

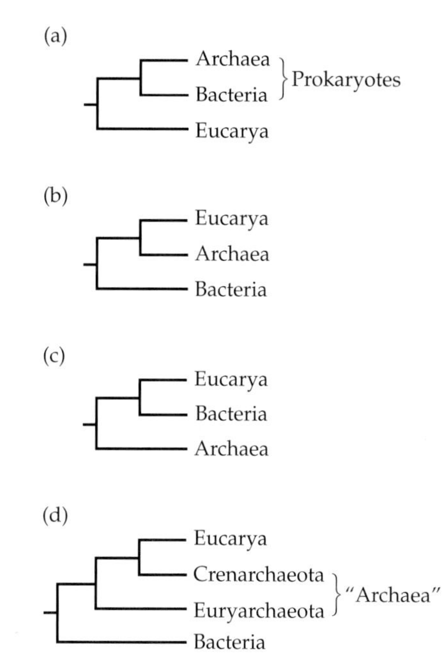

Figure 6.14 Four rooted trees depicting different possible evolutionary relationships of Archaea, Bacteria, and Eucarya. (a) The prokaryote tree, in which Eucarya is the outgroup. (b) The Archaea-Eucarya tree, in which Bacteria is the outgroup. (c) The Eucarya-Bacteria tree, in which Archaea is the outgroup. In all three of these trees, Archaea is monophyletic. (d) The eocyte tree, in which Eucarya is nested within a paraphyletic Archaea (paraphyly is indicated by the use of quotation marks). Crenarchaeota (eocytes) and Euryarchaeota are two purportedly monophyletic subdivisions within Archaea. Each of the four trees is supported by some data sets and contradicted by others. For example, tree (a) is supported by data from DNA topoisomerase II, citrate synthase, and heat shock protein 70; tree (b) is supported by data from elongation factors Tu/G and ribosomal proteins S18, L15, and L5; tree (c) is supported by data from glyceraldehyde 3-phosphate dehydrogenase and malate dehydrogenase; and tree (d) is supported by morphological data pertaining to ribosome structure, as well as by some molecular data sets. (Data from Lake et al. 1984; Saccone et al. 1995; Cox et al. 2008.)

Another problem with the tree of life concerned the inability of researchers to increase the number of genes in phylogenetic studies. The largest number of orthologous protein sequences ever used to study prokaryote and eukaryote relationships by means of strict bifurcating trees was 31 (Ciccareli et al. 2006). These 31 orthologous protein sequences from 191 genomes yielded an alignment in which there were only 1,212 gapless sites that could be used in a phylogenetic analysis. Dagan and Martin (2006) noted that the average prokaryotic proteome consists of about 3,000 protein-coding genes, hence the 31-protein tree represents only about 1% of the average prokaryotic proteome and only 0.1% of a large eukaryotic proteome, such as ours. Thus, although a "tree" was obtained, this tree is not the tree of life, but the "tree of one percent" (Dagan and Martin 2006).

In time, researchers started to think that the conflicting phylogenetic trees obtained by using different genes and proteins tell a true evolutionary story, rather than being artifacts of data, methodology, stochastic error, or a combination of these factors. For example, in a maximum likelihood analysis of 605 protein sequences from Eucarya, Archaea, and Gram-positive and Gram-negative members of Bacteria, Ribeiro and Golding (1998) found 59 trees (10%) significantly supporting the Archaea-Eucarya clade, 14 (2%) significantly supporting the Gram-negative–Eucarya clade, and 3 (0.5%) supporting the Gram-positive–Eucarya clade. The vast majority of the protein trees (529, or 87%) either did not provide significant support for any phylogenetic tree or provided statistically weak support for one of the three possible trees. Ribeiro and Golding (1998) also tested the trees that significantly supported the Gram-negative–Eucarya clade and concluded that it was unlikely for the statistically significant support to be due to convergent evolution or methodological errors. In more recent studies, 15% of all eukaryote genes were traced to cyanobacteria, 26% to proteobacteria, and 10% to archaebacteria (Pisani et al. 2007), indicating that the eukaryotic genome may have multiple origins.

From the late 1990s, it became increasingly clear that the phylogenetic relationships among the three domains of life may not easily be solvable by models employing strictly bifurcating trees. In other words, the tree of life may be impossible to reconstruct, or the tree of life may not be a tree at all.

The Origin of Eukaryotes

More than 20 mutually incompatible evolutionary scenarios have been proposed in the literature to account for eukaryote origins (Pisani et al. 2007). These scenarios can be divided into two broad categories. Scenarios in the first category depict the evolution of eukaryotes from a prokaryote as a gradual process, whereby the prokaryotic lineage destined to become eukaryotic incrementally grew in size and evolved such traits as a membrane-enclosed nucleus, the endoplasmic reticulum, and phagotrophy (the ability to engulf other cells). Only much later did these primitive eukaryotes gain mitochondria through endosymbiosis. The second category of scenarios dispenses with the slow, incremental "progress" from a prokaryotic to a eukaryotic state and instead proposes that eukaryotes were "born" through an "encounter" that resulted in the merger of two prokaryotes.

The origin of eukaryotes: The gradual origin hypothesis

The **gradual origin hypothesis** envisions a slow, gradual process of evolution in which the acquisition of mitochondria occurred after the transition to the eukaryotic state. Adherents of the gradual origin hypothesis consider eukaryotes to be a primary lineage of life, and consider the tree of life to be indeed a bifurcating tree (e.g., Ciccarelli et al. 2006).

Let us examine one corollary of the gradual origin hypothesis. If eukaryotes indeed evolved gradually from a prokaryotic lineage, and if the subsequent evolution of eukaryotes is a run-of-the-mill treelike process, then it is entirely imaginable that descendants of eukaryotic lineages that diverged before the symbiotic event that gave rise to the mitochondria might still exist (**Figure 6.15**). The eukaryotic lineages that diverged before the establishment of the mitochondria would be represented today by organisms devoid of mitochondria. The lack of mitochondria (the amitochondriate condition) is, according to this theory, a plesiomorphy (i.e., an ancestral character state) shared by taxa descended from lineages that diverged at the base of the eukaryote tree. These taxa form a paraphyletic group called "Archezoa" (where the quotation marks denote the lack of monophyly).

Parasitic protists, such as metamonads (e.g., *Giardia* and *Hexamita*), parabasalids (e.g., *Trichomonas*), archamoebae (e.g., *Entamoeba*), and microsporids (e.g., *Encephalitozoon* and *Nosema*), were thought to be the modern descendants of these ancient amitochondriate lineages (Cavalier-Smith 1983, 1989). The mitochondriate clade that diverged into animals, plants, fungi, and the vast majority of protists was called Metakaryota. In other words, the eukaryotes were claimed to have acquired mitochondria as bona fide eukaryotes, i.e., after they had already acquired a membrane-enclosed nucleus.

In time, however, all archezoans were phylogenetically "demoted" from their pre-endosymbiotic status by the findings that each archezoan taxon is nested phylogenetically within a mitochondriate group. In other words, phylogenetic analyses have shown that the lack of mitochondria was always a derived state—a secondary

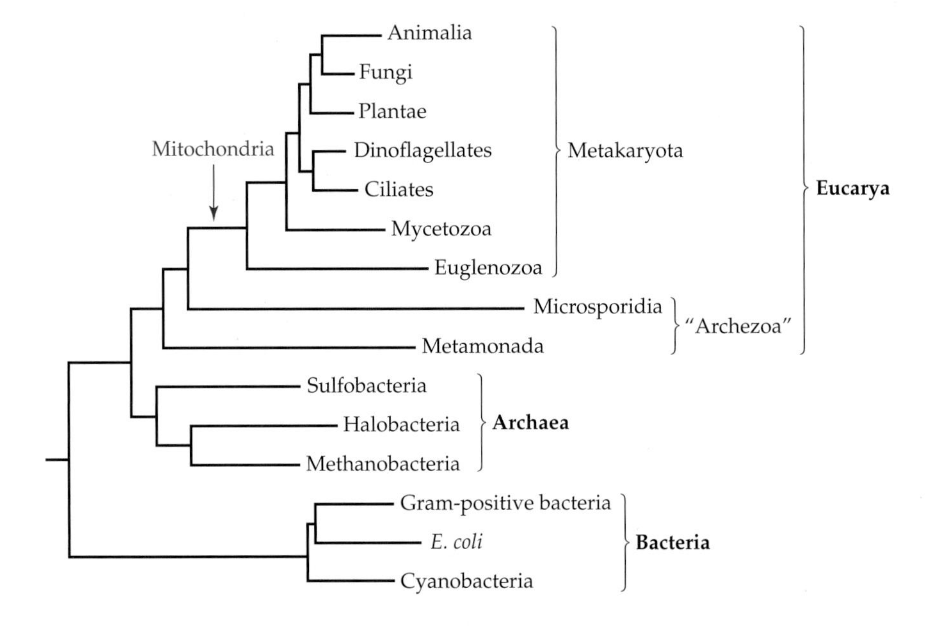

Figure 6.15 The phylogeny of eukaryotes according to the Archezoa theory. The symbiotic event that gave rise to the mitochondria occurred only after several lineages had diverged from the Metakaryota. These early divergent lineages evolved into the paraphyletic "Archezoa," represented here by the Microsporidia and Metamonada. In the scheme presented here, eukaryotes are a sister taxon of Archaea, although this relationship is immaterial to the Archezoa theory.

loss of the organelle from a lineage that once possessed mitochondria (Embley and Hirt 1998; Keeling 1998; Van de Peer 2000; Hampl et al. 2009). Moreover, detailed molecular cell biology and phylogenetic investigations have shown that all so-called amitochondriate eukaryotes examined so far contain a double-membrane-bounded organelle of mitochondrial ancestry, a **mitochondrial organelle**, and thus are not amitochondriate after all (Hjort et al. 2010).

Mitochondrial organelles are functionally and genomically diverse. Müller et al. (2012) classified mitochondria into five functional classes. Classes 1–4 generate ATP; class 5 does not. Class 1 consists of **aerobic mitochondria**, such as the canonical, rat liver–type mitochondrion that is featured in most biology and biochemistry textbooks. Class 1 mitochondria use oxygen as the terminal electron acceptor. Class 2 consists of **anaerobic mitochondria**, which use an endogenously produced electron acceptor, such as nitrate or fumarate, instead of oxygen. As opposed to class 3 and class 4 mitochondria, class 2 do not produce H_2. Class 3 consists of **hydrogen-producing mitochondria**, which, however, possess a proton-pumping electron transport chain as well as a hydrogenase. Class 3 mitochondria can, thus, function aerobically as well as anaerobically as hydrogenosomes—hydrogen-producing organelles. Class 4 mitochondria are the **hydrogenosomes**, i.e., anaerobically functioning ATP-producing mitochondria-derived organelles that can use protons as an electron acceptor, which results in the formation of hydrogen. Class 4 mitochondria do not possess an electron transport chain. Class 5 are the **mitosomes**, which do not produce ATP. Mitochondria belonging to classes 1, 2, and 3 possess a genome; classes 4 and 5 do not.

Interestingly, despite their functional, morphological, and genomic diversity, all mitochondrial organelles can be traced down to a single endosymbiotic event between two prokaryotes. Subsequent evolution gave rise to the different classes of mitochondrial organelles. Some have retained their own genome and translation system; in others, the entire genome has been lost. Irrespective of whether they have a genome or not, however, all types of mitochondrial organelles must import most of their proteome from the cytosol. Moreover, in most eukaryotes, a variable number of essential RNAs are also imported. Thus, the import of macromolecules, both proteins and RNAs, is essential for mitochondrial biogenesis.

The common shared characteristics of all classes of mitochondria are a double membrane and several enzymes with proteobacterial similarities. The genomes of those mitochondria that possess genomes have been revealed through phylogenetic analysis to be vestigial and derived from proteobacteria (Chapter 5). As seen in **Figure 6.16**, each of the classes of mitochondrial organelles may have been derived multiple times independently. More important, however, is the observation that none of the eukaryotic lineages examined so far lacks organelles of mitochondrial origin. The amitochondriate condition does not exist in eukaryotes.

Finally, we note that the nuclear envelope and the nuclear pore complex are made out of proteins of both archaebacterial and eubacterial origins, suggesting that the nucleus arose in a cell that already contained the mitochondrial endosymbiont.

All in all, gradual origin hypotheses have turned out to be untenable and have been almost completely abandoned. For the sake of fairness, however, we should note that some scientists still believe such hypotheses to be capable of explaining the origin of eukaryotes (e.g., Cavalier-Smith 2009; Forterre and Prangishvili 2013).

The origin of eukaryotes: The fateful encounter hypothesis

The alternatives to the gradual origin hypothesis are scenarios that invoke a phylogenetic merger—some say a "freakish" merger—between two prokaryotes. According to these scenarios, collectively referred to as the **sudden origin** or **fateful encounter hypothesis**, the eukaryotic cell is a chimera, and all eukaryotes (including you, ewe, and yew) owe their existence to a fusion between two organisms that occurred more than 2 billion years ago (de Duve 2005; Lane 2009; Yong 2014). In other words, the rise of eukaryotes is assumed to be fundamentally different from the gradual evolutionary transition that gave rise to multicellularity, photosynthesis, terrestriality, and myriad

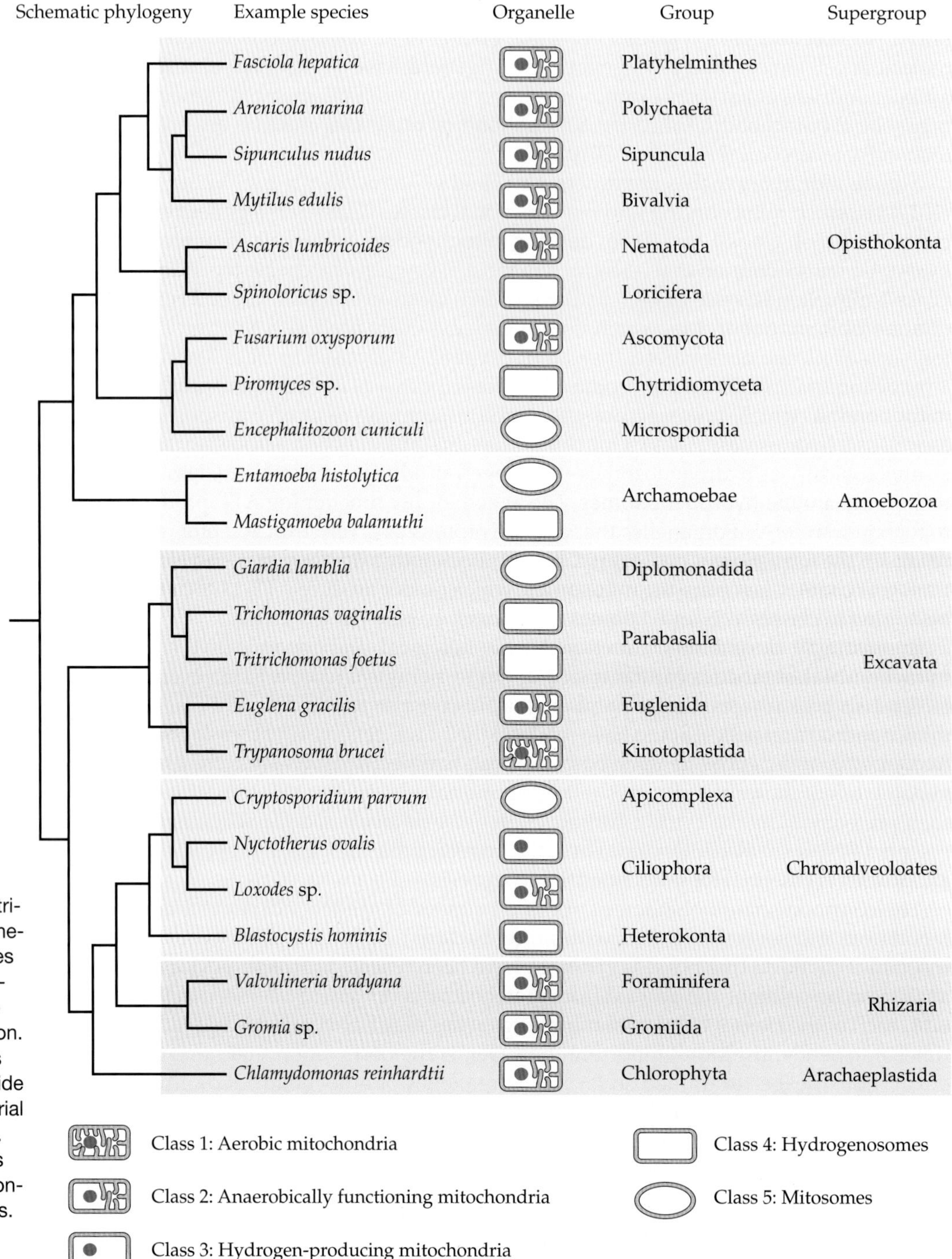

Figure 6.16 Phylogenetic distribution of organelles of mitochondrial origin. The distribution is plotted across a schematic phylogeny for six clades of eukaryotes. The mitochondrion class for each example species is indicated by an icon. The presence of a genome is indicated by a blue circle inside the icon. Sizes of mitochondrial types are not drawn to scale. Note the absence of lineages lacking organelles of mitochondrial origin among eukaryotes. (Modified from Müller et al. 2012.)

other evolutionary novelties. This merger, as far as we know, was a "fluke event of incredible improbability" that only happened once (Yong 2014).

The earliest version of this type of hypothesis was put forward by Martin and Müller (1998), who pointed out that it is very common in nature for eubacteria and archaebacteria to develop a mutually beneficial metabolic association called **mutual syntrophy**, in which one species in a pair lives from the waste products of the other species and vice versa. Mutual syntrophy is known in nature to lead at times to symbiotic associations. A good example of this type of association is the large plasmid that is passed around among members of a group of marine prokaryotes called *Roseobacter*. The plasmid encodes photosynthetic reaction centers and chlorophyll synthesis—everything needed to make its recipient instantaneously photosynthetic (Petersen et al. 2012).

Based on the biochemistry of mitochondria and hydrogenosomes, Martin and Müller (1998) sketched out a scenario for how a merger of two cells may have given rise to eukaryotes. According to this scenario, called the **hydrogen hypothesis**, the merger involved, on the one hand, a methanogenic (methane-producing) archaebacterium that had the capability of deriving energy from the bonding of hydrogen (H_2) and carbon dioxide (CO_2) and, on the other, a eubacterium that fed on organic compounds and produced H_2 and CO_2, which the archaebacterium could use.

Over time, the two organisms involved in the mutual syntrophy that gave rise to eukaryotes became inseparable, i.e., the archaebacteria engulfed the eubacterium; then the archaebacterial component evolved the ability to provide organic compounds to the endosymbiont, which eventually became an organelle—the mitochondrion.

In addition to the hydrogen hypothesis, there are many other variants of the fateful encounter hypothesis, which differ in the reasons for the merger and the exact identities of the archaebacterium and the eubacterium that were involved in the encounter (Searcy 1986; Vellai et al. 1998; Davidov and Jurkevitch 2009). Notwithstanding the differences among the scenarios, all fateful encounter hypotheses are united by one critical feature setting them apart from the gradual origin ideas: they all maintain that the host cell was a bona fide prokaryote. This host prokaryote had not started to grow in size, it did not yet have a nucleus, and it had not embarked on an evolutionary journey on the path of becoming a eukaryote. According to the sudden origin ideas, mitochondria were not merely one of many innovations in the evolution of eukaryotes. The acquisition of mitochondria *was* the origin of eukaryotes; they were one and the same event—eukaryotic inventions came later. The acquisition of mitochondria by a prokaryote was the sine qua non that turned the prokaryote into a eukaryote. Accordingly, all living eukaryotic lineages descended from a common ancestor that had mitochondria (Hampl et al. 2008). Most important in the phylogenetic context is that if eukaryotes are indeed a chimeric lineage, then the tree of life is not a tree, but a network containing a merger of two independent lineages.

Let us now review the evidence pertaining to the fateful encounter hypothesis. Rivera and Lake (2004) used the method of conditioned reconstruction to elucidate the phylogenetic relationships among eukaryotes, eubacteria, and two archaebacterial phyla (**Figure 6.17**). Five unrooted trees with a cumulative posterior probability of 96.3% were found. These trees matched a repeated pattern and could be curled into a ring. In fact, a combinatorial analysis of the genomic fusion of two organisms valid for all possible prefusion trees has shown that the conditioned reconstruction algorithm recovers all permutations of the cycle graph, and only those permutations.

In the next stage of the analysis, Rivera and Lake (2004) set out to identify the "fusion organism." In a tree, each leaf (external node) contacts only one other node in the tree, so eliminating one taxon from the analysis does not otherwise change the tree. Similarly, in a conditioned reconstruction, eliminating a nonfusion organism from the analysis will delete that one leaf from the ring, without affecting the ring. However, eliminating a fusion organism, which by definition contacts two nodes of a ring, will delete the leaf, open the ring, and convert the ring into a tree. When the taxa in Figure 6.17 were systematically removed, the ring opened only when both yeast genomes were simultaneously removed, indicating that the eukaryotic genome is a product of genome fusion between two prokaryotes. Furthermore, the conditioned reconstruction analysis of 24 prokaryotic genomes in the absence of any eukaryotic genomes resulted only in trees with no rings. These results support the conclusion that eukaryotes are the chimerical products of a genome fusion event between two prokaryotes.

By using 5,741 single-gene families distributed across 185 genomes, Pisani et al. (2007) corroborated the chimerical origin of eukaryotic genomes and the derivation of nuclear genes from three main sources: cyanobacteria, proteobacteria, and archaebacteria. These three distinct symbiotic partners gave rise, respectively, to plastids,

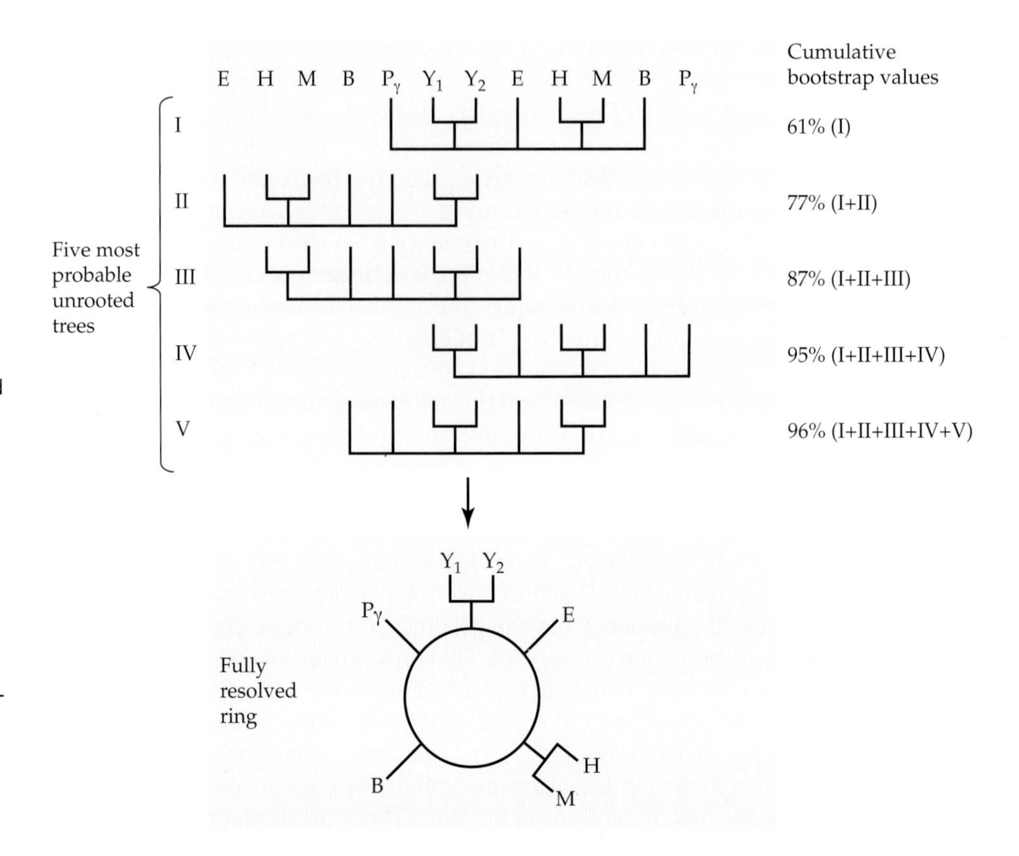

Figure 6.17 Conditioned reconstructions provide evidence for the ring of life. The genomes are from two eukaryotes (*Schizosaccharomyces pombe* and *Saccharomyces cerevisiae*, denoted by Y_1 and Y_2, respectively); an α-proteobacterium (*Xylella fastidiosa*, denoted by P_γ); a bacillus (*Staphylococcus aureus*, denoted by B); and three archaebacteria, two belonging to phylum Euryarchaeota (*Halobacterium* sp. and *Methanosarcina mazei*, denoted by H and M, respectively) and one belonging to phylum Crenarchaeota (the eocyte *Sulfolobus tokodaii*, denoted by E). A euryarchaeote (*Archaeoglobus fulgidus*) was used as the conditioning genome. The analyses are based on 2,408 orthologous gene sets. Of these, 433, 239, 285, 292, 329, 293, 204, and 333 sets contained seven, six, five, four, three, two, one, and no orthologs, respectively. The five most probable unrooted trees (I–V) are shown with leaves pointing upward to emphasize that each is part of a repeating pattern. Cumulative bootstrap scores are shown at the right of each tree. For example, the bootstrap for tree I is 61%, while the cumulative bootstrap value for trees I, II, and III is 87%. Adding additional trees does not significantly improve the cumulative bootstrap values. A fully resolved ring of taxa is shown. (Modified from Rivera and Lake 2004.)

mitochondria, and the nucleus. A schematic depiction of the phylogenetic relationships among eubacteria, archaebacteria, and eukaryotes is shown in **Figure 6.18**.

Eukaryotes as an "organizational upgrade"

The origin of the eukaryotic cell is one of the hardest and most interesting puzzles in evolutionary biology (Lake 2007). Any theory attempting to describe the evolution of eukaryotes must be able to explain the following seven eukaryotic characteristics: (1) The eukaryotic cell is considerably more complex than the prokaryotic cell, possessing, among others, a nucleus with a contiguous endoplasmic reticulum, Golgi bodies, flagella with a 9+2 pattern of microtubule arrangement, and organelles surrounded by double membranes. (2) Only eukaryotes have achieved great size and morphological complexity, whereas prokaryotes have remained small and have not evolved either morphological complexity or multicellularity. (3) The protein-coding genes of eukaryotes are interspersed with introns that need to be removed prior to translation by spliceosomes. (4) The process of transcription is physically and temporally separated from the process of translation. (5) The eukaryote genome consists of components that are archaebacterial and components that are eubacterial. (6) The distribution of the archaebacterial and eubacterial genomic components is not random with respect to function. (7) There are no known precursor structures among prokaryotes from which such attributes could be derived, and no intermediate cell types are known that would point to a gradual evolutionary change of a prokaryote into a eukaryote. For all intents and purposes, the eukaryotic cell represents a sudden "organizational upgrade" or "evolutionary leap." Moreover, any theory on eukaryote evolution must provide a reason why the time it took for prokaryotes to evolve out of inanimate matter is so much shorter than the time it took eukaryotes to evolve out of prokaryotes.

Admittedly, we do not have clear-cut answers to all these conundrums. Testable theories have been suggested, however, and in the following sections, we will deal with several hypotheses that address one or a few of these puzzles.

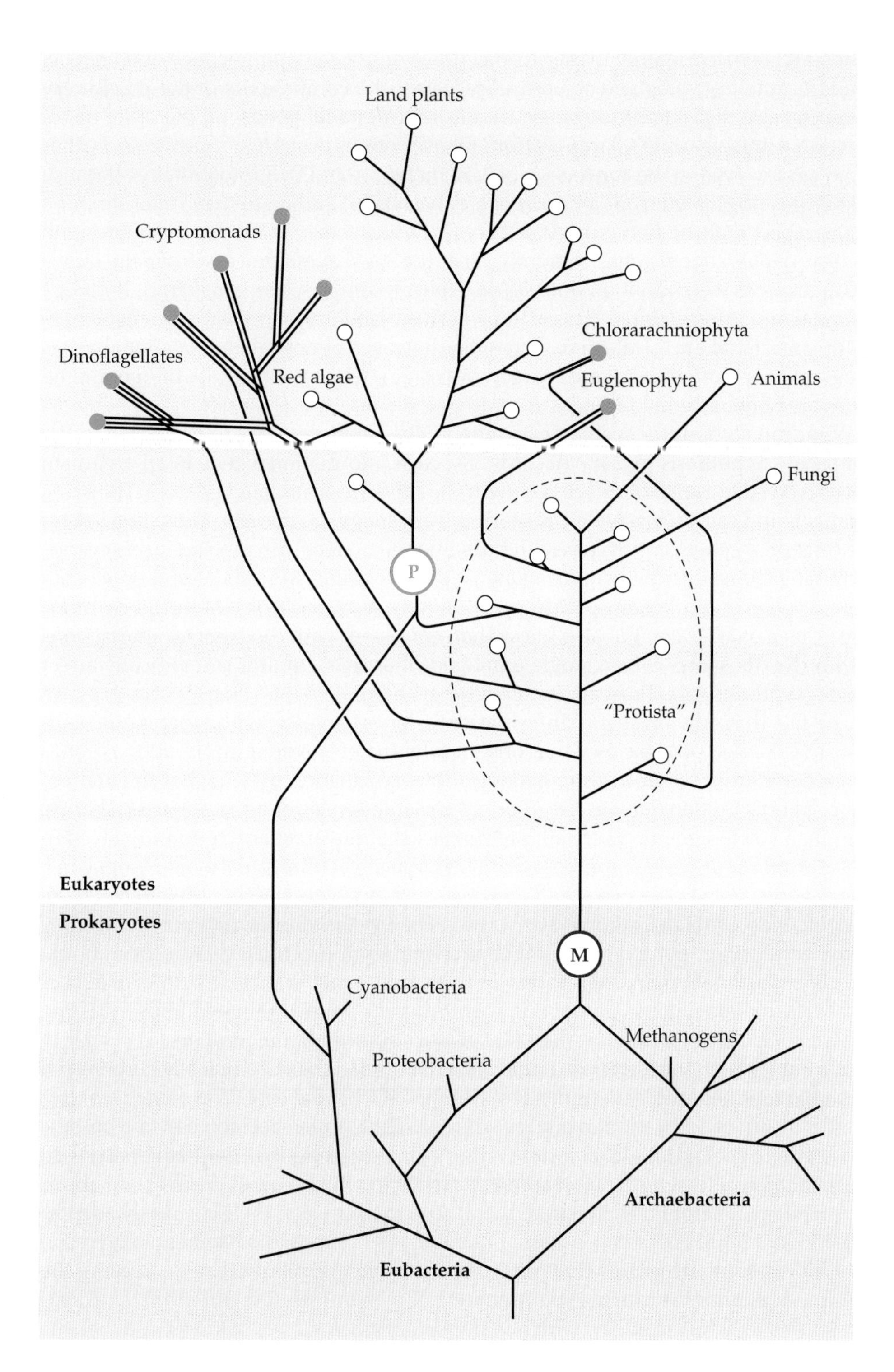

Figure 6.18 Schematic phylogenetic network depicting the evolution of eukaryotes through endosymbiotic events. There are two primary symbiotic events: M (for mitochondria) denotes the endosymbiosis of a proteobacterium into an archaebacterium that gave rise to eukaryotic cells. P (for plastids) denotes the endosymbiosis of a cyanobacterium into a eukaryote that gave rise to the lineage leading to kingdom Plantae. The two primary domains of life are Eubacteria and Archaebacteria, and the root of the network of life is located between the two primary domains. Eukaryotic lineages are marked by circles. Lineages that resulted from primary symbioses are denoted by open circles. Lineages that resulted from secondary and higher-order symbioses are denoted by solid circles. (Modified from Kowallik 2008.)

The nonrandom origin of operational and informational genes in eukaryotes

The origin of the vast majority of eukaryotic genes can be traced to either eubacterial or archaebacterial genes. For example, more than two-thirds of the nuclear genes of the yeast *Saccharomyces cerevisiae* are derived from eubacteria, with the balance from archaebacteria. Intriguingly, the distribution of the archaebacterial and eubacterial genes in the eukaryotic genome is not random with respect to function.

Productive genes (i.e., protein-coding and RNA-specifying genes) can be roughly divided into two major gene classes: operational and informational. **Operational**

genes are involved mainly in day-to-day processes of cell maintenance and encode, in addition to enzymes and structural proteins, the components of the machinery for amino acid and nucleotide biosynthesis. **Informational genes**, on the other hand, function primarily in DNA replication, transcription, protein synthesis, and other processes involved in the conversion of information from DNA into proteins. Because eukaryotes originated from a fusion of a eubacterium and an archaebacterium, one might expect that the first eukaryote contained two sets of informational genes and two sets of operational genes. Assuming that the subsequent processes of gene duplication and loss were random, one would expect to find nuclear genes from both operational and informational classes to be derived randomly from either archaebacterial or eubacterial ancestral genes. Strangely, however, almost all informational genes in eukaryotes seem to be of archaebacterial origin, while the majority of operational genes are of eubacterial origin.

What can account for such a distribution? The solution seems to be related to the complexity hypothesis (p. 248). Accordingly, each informational gene in an organism has coevolved to function effectively with the other informational genes in the same organism and cannot easily be transferred from species to species. Thus, two sets of informational genes cannot coexist in the same cell. Indeed, attempts to fuse a cyanobacterial genome into the genome of a host *Bacillus subtilis* were successful only when the cyanobacterial ribosomal RNA operons were removed from the fused chromosome (Itaya et al. 2005). We also know that, consistent with the complex interactions within the ribosome, even a single damaged ribosomal subunit can radically affect protein synthesis (e.g., Prescott and Dahlberg 1990).

The fact that only one set of informational genes can exist within a genome suggests at least two possible explanations for why the eubacterial informational genes disappeared from the eukaryote genome. One explanation may be that the archaebacterial informational genes are "better" in some sense than the eubacterial ones. One such advantage may be related to the fact that the archaebacterium involved in the fateful encounter was the host, which supplied such informational cytoplasmic components as the ribosomes and the transcription factors, while the eubacterium was the endosymbiont. Alternatively, the archaebacterial informational genes may have been luckier. For example, if a chance mutation inactivated an informational gene of eubacterial origin, then it was probably only a matter of time before all eubacterial informational genes interacting with this gene were eliminated, thus initiating a cascade resulting in the elimination of all such informational proteins.

Why are operational genes of archaebacterial origin almost completely absent from the eukaryotic genome? Valentine (2007) suggested that the evolution of archaebacteria is the story of adaptation to energy stresses. Thus, archaebacteria can outcompete eubacteria in ecological niches where chronic energy stress is a dominant feature. In contrast, many eubacteria are adapted to niches in which energy sources are abundant. The lipid membranes illustrate the differences between the two energy adaptations (Figure 6.10). The archaebacterial membranes, composed of isoprenoid glycerol-ether lipids, form structures that are less permeable to ions, thereby decreasing the amount of energy lost during maintenance of a chemiosmotic potential. Unfortunately, this low permeability, which is important during energy stress, reduces the efficiency of respiration and signal transduction, which is important in eukaryotes. As the protomitochondrion became the "powerhouse" of the eukaryotic cell, adaptations to chronic energy stress became irrelevant (Davidov and Jurkevitch 2007). This evolutionary scenario provides a reasonable explanation not only for the lack of archaebacterial operational genes in eukaryotes, but also for the fact that the eukaryotic membranes, including the cytoplasmic membrane, are eubacteria-like, despite the fact that the host cell was an archaebacterium (Davidov and Jurkevitch 2009).

Why genes in pieces? The origin of the nuclear membrane

The main defining characteristic of a eukaryote, which is the existence of a nuclear envelope separating the nucleus from the cytosol, raises the question of why there should be a need for such compartmentalization. Another characteristic of eukaryotes is the existence of "genes in pieces," in which the coding regions are interrupted by introns that need to be spliced out prior to translation. Martin and Koonin (2006) formulated a hypothesis according to which the separation between transcription and posttranscriptional modifications, on the one hand, and translation, on the other, was needed to deal with the very different kinetic properties of intron splicing and protein synthesis on the ribosome.

According to this hypothesis, the origin of eukaryotes was a prokaryote host that, by definition, lacked a nucleus and acquired another prokaryote that eventually became the mitochondrion. Scientists currently agree that the anucleated host was an archaebacterium and the engulfed prokaryote that evolved into the mitochondrion was a proteobacterium. In the past, a common criticism facing "archaebacterial host" models has been that **phagotrophy** (the engulfment of particulate matter, especially food) is unknown in prokaryotes. According to this argument, prokaryotes can only external materials through **osmotrophy** (the taking in of dissolved nutrients and other molecules from the medium). This argument has lost some of its strength with the discovery of many endosymbioses in which one prokaryote was found to live inside another (nonphagotrophic) prokaryote (Wujec 1979; von Dohlen et al. 2001; Thao et al. 2002).

Let us start the description of the evolution of the eukaryotic cell at the stage in which the proteobacterial ancestor of mitochondrial organelles is engulfed inside an archaebacterial host with one or more cytosolic chromosomes. The first thing that must happen is that the rates of cell division in the host and symbiont become coordinated. Only progeny with synchronized division rates will persist as a consortium of two genomes. If the host archaebacterium lyses, the symbionts are set free, ending the association. However, if a symbiont lyses, a genome's worth of eubacterial DNA is left in the cytosol of the host, free to recombine. As long as there is more than one symbiont per progeny, symbiont lysis can occur repeatedly, resulting in a constant flow of symbiont DNA into the host's chromosomes.

We now know that eukaryotic introns, as well as their cognate spliceosomal small nuclear RNAs (snRNAs), originated from self-splicing group II introns (Cech 1986; Cavalier-Smith 1991). These introns, which behave as mobile elements and are currently found among free-living proteobacteria, are thought to have invaded the intronless archaebacterial genome, which in time became the eukaryotic nuclear genome. Given the antiquity and conservation of many eukaryotic intron positions, it is reasonable to assume that many spliceosomal introns are directly derived from group II introns in the proteobacteria that later became the mitochondrion.

Spreading of group II introns, which are essentially mobile elements, within the chromosomes of the archaebacterial host would presumably pose a hazard to the survival of the early eukaryote. These mobile introns had to be spliced out before translation, to prevent production of frameshifted proteins containing translated noncoding regions, which in all likelihood would be defective. We note, however, that translation is a fast process, on the order of 10 amino acids per second in prokaryotes, whereas splicing is slow, in the range of 0.005–0.01 intron per second. The slow splicing of introns would have made it difficult for them to function in a prokaryote, because the coexistence of DNA and functional ribosomes in the same cell compartment would allow ribosomes to translate unspliced pre-mRNA sequences.

There are three routes for solving the problem of possessing spliceosome-dependent introns in cotranscriptionally translated mRNA. The first solution would be the invention of an extremely fast and efficient spliceosome capable of outrunning the ribosomes. This would entail an unrealistic splicing efficiency in the ancestral spliceosome exceeding that of the modern one. The second solution would be the removal of all introns from the genome. This has not occurred. The third solution would be the invention of a means to physically separate splicing from translation, allowing the former (slow) process to go to completion first, before the latter (fast) process sets in. Physical separation in cells usually entails membranes, so the third solution would involve the invention of a membrane separating splicing from translation, with pores sufficiently large and selective enough to export matured ribosomal subunits, mRNA, and tRNA. The fact that the nuclear membrane consists of proteins of both eubacterial and archaebacterial origin indicates that this was indeed the route pursued by the evolutionary process immediately after the establishment of the endosymbiosis that gave rise to the eukaryote cell. Thus, the incipient function of the nuclear envelope was to allow translation only from postsplicing mRNAs with intact reading frames (Martin and Koonin 2006).

A related problem with having a second genome in such close quarters would be the transfer of mobile elements from the endosymbiont to the host genome. As mobile elements bombard a genome, they can disrupt the proper working of its genes. The selective pressure that forged nucleus-cytosol compartmentalization is assumed to have been the rapid spread of type II introns and other mobile elements following the endosymbiotic event. Thus, the nucleus may have evolved as a defense against mobile element attack. We also note that group II introns and other retrotransposable elements reproduce through RNA-mediated transposition. With the invention of a nucleus, RNA molecules were moved across a barrier outside the nucleus. This barrier reduced the chances of reinsertion of mobile elements back into the DNA genome. Interestingly, according to this view, the nucleus was not important; what was important was a DNA-free cytosol, a dedicated translation compartment that was free of transcriptionally active DNA.

A problem of a much more severe nature arises, however, with the mutational decay of the self-splicing group II introns, resulting in the inactivation of the maturase and the RNA elements that facilitate splicing. Modern examples from prokaryotes and organelles suggest that splicing with the help of maturase and RNA elements provided by intact group II introns in *trans* could have initially rescued gene expression at such loci, although maturase action in *trans* is much less effective than in *cis*. Thus, the mutational decay in the invading introns would create the necessity for a new splicing machinery. We do not currently know how the transition from self-splicing to spliceosome-dependent splicing occurred, as there are no evolutionary grades detectable in the origin of the spliceosome, which apparently was present in its fully fledged state in the common ancestor of all eukaryotic lineages. We note that, in all likelihood, prokaryotes have never possessed spliceosomal introns, so spliceosomes must have originated in eukaryotic cells. Spliceosomes and spliceosomal introns are universal among eukaryotes, although a few lineages, such as microsporidia, have lost them secondarily. Thus, eukaryotic introns must be as old as the eukaryotes themselves, which would suggest that spliceosomal intron origin and spread occurred within a narrow window of evolutionary time: subsequent to the origin of the mitochondrion, but before the diversification of the major eukaryotic lineages.

All complex life is eukaryotic: The energetics of gene expression

Throughout their evolution, prokaryotes have been extraordinarily inventive in terms of their biochemistry and have utilized practically every energy source our planet has to offer. In fact, among microbiologists, a rule of thumb has been that if a chemical reaction yields sufficient energy to support life, a prokaryote that exploits this source must surely exist. However, when judged by their morphological, organismal, and cellular complexity, prokaryotes seem to have advanced little beyond their 4-billion-year-old ancestors. At most, prokaryotes attained a rudimentary level of morphological and behavioral melioration. For example, cyanobacteria and planctomycetes have evolved internal membranes, myxobacteria have evolved a simple form of social multicellularity, and some eubacteria, such as *Epulopiscium fishelsoni* and *Thiomargarita namibiensis*, have attained cell sizes exceeding those of many single-celled eukaryotes. Notwithstanding these sporadic prokaryotic innovations, as a rule, all complex life in the world is eukaryotic. Moreover, complex forms of organization such as multicellularity, sex, and phagocytosis have evolved independently multiple times in eukaryotes.

Enigmatically, almost every "defining" characteristic of eukaryotes is also found in prokaryotes, including nucleus-like structures, recombination, linear chromosomes, internal membranes, multiple replicons, giant cell size, polyploidy, dynamic cytoskeletons, predation, parasitism, introns, intercellular signaling, endocytosis-like processes, and endosymbiosis. It seems as though the prokaryotes made repeated starts up the ladder of complexity, but always fell short. By contrast eukaryotes, despite their meager metabolic repertoire, burst whatever constraints hampered prokaryotes and experimented with the opportunities afforded by greater cell size and more elaborate organization. Why?

The answer, according to Lane and Martin (2010), resides in the bioenergetic changes brought about by the evolution of the mitochondria by endosymbiosis. Broadly speaking, chemical energy in the form of adenosine triphosphate (ATP) is coupled to the transfer of protons (H^+ ions) across a membrane. In prokaryotes, ATP synthesis scales with the surface area of the only available membrane, the plasma membrane. In contrast, protein synthesis scales with cell volume. Consider, for simplicity, a prokaryote that is approximately spherical. If the prokaryote has a typical radius of 1 μm, i.e., one-millionth of a meter, then the ratio of the surface area to the volume is 3 μm^{-1}. If the prokaryotic cell were to have a typical protozoan radius of 50 μm, then the surface-area-to-volume ratio would be 0.06 μm^{-1}, a 50-fold decrease. Thus, larger prokaryotic cells are energetically less efficient than smaller ones.

In eukaryotes, energy production is achieved by hundreds and even thousands of mitochondria, all descended from a singular endosymbiotic event that occurred approximately 1.5 billion years ago. These mitochondria possess highly wrinkled inner membranes that greatly increase the total surface area available for energy-producing oxidative phosphorylation, while at the same time freeing the plasma membrane for other tasks.

Lane and Martin (2010) looked at the amount of power, defined as the amount of energy consumed per unit time, that is available to the cell. In prokaryotes, the power available to the cell is on the order of 0.5 picowatt (pW), or one trillionth (10^{-12}) of a Watt. In comparison, the amount of power available to a eukaryotic cell is on average 2,300 pW, which is approximately 5,000 times more power that that available to a prokaryote.

How does the cell use this energy? The energy cost of DNA replication accounts for a mere 2% of the energy budget of microbial cells during growth. In contrast, protein synthesis accounts for ~75% of the total energy budget. Thus, in dealing with cell energetics, we need to mainly consider gene expression and may, as an approximation, ignore everything else.

Let us now look at the "available power per gene," which was defined as the mean energy available in a cell for expressing one gene per unit time. A prokaryotic gene has on average 0.03 femtowatt (fW) of metabolic power, or one quadrillionth (10^{-15}) of a Watt. In contrast, a eukaryote has on average 57 fW available power per gene, i.e., about 2,000 times more than a prokaryote. Thus, the most important difference between prokaryotes and eukaryotes is the amount of energy available per gene. We note that the energy allocation per gene is greater in eukaryotes than in prokaryotes, despite the fact that eukaryotes have on average 4–6 times more genes than prokaryotes.

What does increased cell complexity entail? The main factors that underlie cell complexity are gene number and, to a lesser extent, genome size. Can a prokaryotic cell increase its genome size and gene number? If the genome size of a prokaryotic cell is increased tenfold, the cost of replicating the genome itself will account for about 20% of its energy budget, which under certain conditions may be sustainable. However, if the number of genes is increased tenfold, a huge energy crisis may ensue, whereby the prokaryote will need to drastically reduce the amount of energy it devotes to the synthesis of each of its proteins. The energy allocation per gene may, thus, reach very low levels, maybe too low for viability.

Can the energy supply be increased in prokaryotes? To do that, a prokaryote would need to grow in size, but as mentioned previously, the increase in plasma membrane surface would be insufficient to offset the greater demand for protein synthesis due to increased cell volume.

Mitochondria bestowed upon eukaryotes abundant energy to expand their genomes by orders of magnitude and to greatly increase their genomic repertoire. Genome size in eukaryotes is on average 500 times larger than the mean DNA content in prokaryotic cells, and some 3,000 new protein families are thought to have originated during the prokaryote-to-eukaryote transition. Moreover, the abundant energy produced by the mitochondria allowed eukaryotes to be "wasteful." Eukaryotic genomes harbor approximately 12 genes per Mb, compared with about 1,000 in prokaryotes. If the average prokaryote had a eukaryotic gene density, it would encode fewer than 100 genes. Prokaryotes must therefore maintain high gene density, around 500–1,000 genes per Mb. They do so by eliminating intergenic and intragenic material (including regulatory elements and microRNAs), by organizing genes into operons, and by restricting the median length of proteins—all of which reduce their energetic costs.

The evolutionary leap from prokaryotes to eukaryotes required orders of magnitude more energy than any prokaryote can provide. For more than 3 billion years prokaryotes have remained simple because of energy constraints. Throughout prokaryote evolution, natural selection has favored small and spare cells with streamlined genomes, rapid reproduction, little superfluous DNA, and tightly disciplined regulation of gene expression. Marching under the banner "Small Is Beautiful," prokaryotes have flourished, multiplied, and in a sense inherited the Earth. From any point of view except that of a eukaryote chauvinist, we live in a prokaryotic world (Harold 2011).

The eukaryotic cell as a one-off innovation and a possible solution to the Fermi paradox

The fossil record does not tell us much about the origin of eukaryotes. Paleontologists have found fossils of prokaryotes dating back 3.5 billion years (Altermann and Kazmierczak 2003). The earliest fossils that have been shown almost certainly to be eukaryotes are approximately 1.7–1.8 billion years old (Rasmussen et al. 2008). Thus, the evolution of life from inanimate matter (abiogenesis) took considerably less time than the evolution of eukaryotes from prokaryotes. Tellingly, no transitional forms from the intervening years have ever been discovered.

The kind of cell merger and subsequent developments that gave rise to the mitochondria are extremely improbable. Indeed, prokaryotes have managed to produce eukaryotes only once in more than 3 billion years, despite prokaryotes' contact with one another all the time. Gradual evolution, on the other hand, is a common occurrence. Why, then, do we believe a less likely scenario over a more likely one? The reason is that the more likely scenario has been empirically refuted, and in the spirit of Sherlock Holmes, we are compelled to hypothesize the "improbable"—a "fateful encounter"—when the only other viable option—the gradual origin of eukaryotes from prokaryotes—has been deemed "impossible."

Interestingly, the improbability of evolving complexity has implications for the search for intelligent alien life. Lane (2010) reasoned that life is certain to emerge on other worlds, as long as the right chemical conditions for life are met. However, without a "fateful merger," such life would forever be simple and microbial.

In the early 1950s, Enrico Fermi noticed a puzzling contradiction between what was thought at the time to be a very high probability that intelligent life exists somewhere among the billions of planets in the Milky Way and our inability to find any signs of such intelligence—a contradiction that has since become known as the Fermi paradox (Jones 1985). The Fermi paradox stemmed from the Drake equation (1962), which is a product of several probabilities used to arrive at an estimate of the number of civilizations in the Milky Way that are capable of interstellar communication. According to the Drake equation, this number, N, was equal to the mathematical product of (1) the average rate of star formation in the galaxy, (2) the fraction of stars that have planets, (3) the average number of planets that can potentially support life per star, (4) the fraction of those planets that actually develop life, (5) the fraction of planets bearing life in which intelligence has developed, (6) the fraction of these civilizations that have developed communication technologies that emit detectable signs into space, and (7) the length of time over which such civilizations release detectable signals. According to Drake's (1962) calculations, *N* ranges from 20 to 50,000,000, which prompted Fermi to puzzle on our inability to encounter intelligent extraterrestrials.

In the original calculations by Drake and his disciples, it was assumed that, from among those planets that can potentially support life, the fraction of planets on which life actually develops is 100%; in other words, life emerges whenever conditions for the emergence of life exist. It was further assumed that the fraction of planets on which intelligence has evolved is equal to the fraction of planets on which life has emerged. Lane (2010) thinks that it is quite likely the first assumption holds, i.e., that extraterrestrial life exists. The second assumption, on the other hand, is a gross overestimation. "The unavoidable conclusion," Lane writes, "is that the universe should be full of bacteria, but more complex life should be rare. And if intelligent aliens do exist, they would probably have something like mitochondria, too."

Archaebacterial Systematics: Clade-Specific Archaebacterial Genes and Clade-Specific Horizontal Gene Imports from Eubacteria

The leading molecular taxonomy database (Federhen 2012) recognizes twelve archaebacterial subdivisions (phyla), of which Crenarchaeota (or Eocyta) and Euryarchaeota are better characterized than the others. Crenarchaeota consists of a single class (Thermoprotei) and five orders, of which Thermoproteales, Desulfurococcales, and Sulfolobales are largely accepted as monophyletic. Euryarchaeota consists of eight classes and thirteen orders, of which nine orders—Haloarchaea, Methanosarcinales, Methanocellales, Methanomicrobiales, Archaeoglobales, Thermoplasmatales, Methanococcales, Methanobacteriales, and Thermococcales—have been shown to be monophyletic. In addition to the above taxa, numerous unclassified, unnamed, and unassigned taxa are listed in the database. Moreover, by using **metagenomics**, which involves sequencing bulk DNA from the environment and assembling more or less complete genomes from such data, numerous new archaebacterial taxa continue to be discovered, resulting in a taxonomy that is in a state of constant flux (Rinke et al. 2013; Raymann et al. 2015; Spang et al. 2015).

The archaebacterial orders were identified by using ribosomal RNA, as well as 30–40 informational protein-coding genes that are found in almost all genomes so far studied. Of course, as mentioned previously, these genes comprise only about 1% of the average prokaryotic genome, so although such trees yield useful taxonomic information, they provide little insight into the remaining 99% of the genome. In particular, such phylogenies do not predict gene content, nor do they reveal the gene innovations underlying the origin of major clades.

Nelson-Sathi et al. (2015) investigated the phylogenetic distribution of 25,762 protein-coding gene families among 134 sequenced archaebacterial genomes and searched for homologs in 1,847 completely sequenced eubacterial genomes. Archaebacterial-specific gene families (i.e., families not found in eubacteria) were found to define 12 traditionally recognized archaebacterial orders: Haloarchaea, Methanosarcinales, Methanocellales, Methanomicrobiales, Archaeoglobales, Thermoplasmatales, Methanococcales, Methanobacteriales, Thermococcales, Sulfolobales, Desulfurococcales, and Thermoproteales (**Figure 6.19**). That is, genes that had no homologs in eubacteria were found to be clade-specific, i.e., to be part of genomes belonging to a single archaebacterial clade and no other. More surprising, however, was the finding that each of these 12 clades had acquired genes from eubacteria, and that 2,264 of these genes occur specifically in only one higher archaebacterial taxon, though at the same time they are ubiquitous among the eubacteria, clearly indicating that they are archaebacterial imports from eubacteria.

Do the origins of the clades coincide with the acquisitions of the imported genes? There can be two explanations for the taxonomic distribution of eubacterial genes in archaebacteria. If an imported gene was acquired at the origination of each

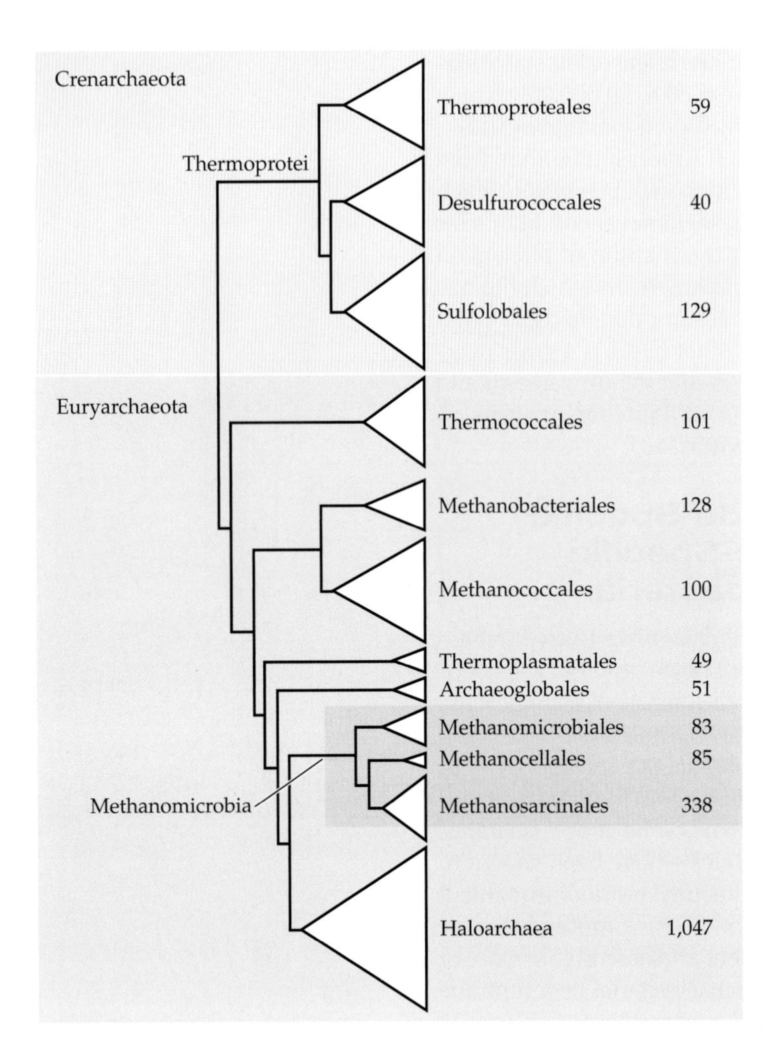

Figure 6.19 Gene acquisitions from eubacteria correspond to major archaebacterial clades. An unrooted tree for archaebacteria was based on 70 informational genes from 134 sequenced genomes. From these data, 12 well-defined archaebacterial orders were identified by phylogenetic analysis. Upon phylogenetic reconstruction, these gene families yielded a monophyletic archaebacterial gene tree that excluded their eubacterial homologs. Each of these clades is characterized by a large number of genes that have homologs in eubacteria but cannot be found in any other archaebacterial clade. Numbers on the right represent gene families found in at least two eubacterial taxa and at least two members of that archaebacterial clade (but in no other archaebacterial clade). For example, there are 129 distinct gene families present in at least two species belonging to Sulfolobales that are not found in any other archaebacterial clade but that are present in at least two eubacterial groups. Maximum likelihood phylogenetic reconstruction of these 129 gene families yields a phylogenetic tree in which all the Sulfolobales genes are monophyletic. The vertical edges of the triangles are proportional to the number of genomes in each group. For example, the shortest vertical edge (Methanocellales) represents three genomes. (Data from Nelson-Sathi et al. 2015.)

archaebacterial clade, then gene phylogenies of the imported genes in an archaebacterial clade should be identical, or at least similar, to the phylogenies for the archaebacterial clade-specific genes from the same clade. Alternatively, an imported gene might have been acquired in one lineage and then spread through the clade, in which case the imported-gene phylogeny should be different from the phylogenies of the archaebacterial clade-specific genes. The null hypothesis—that the import and recipient tree sets were drawn from the same distribution—could not be rejected for six clades: Thermoproteales, Desulfurococcales, Methanobacteriales, Methanococcales, Methanosarcinales, and Haloarchaea. For these six clades, the origin of their group-specific eubacterial genes and the origin of the archaebacterial clade-specific gene were indistinguishable. The eubacteria-to-archaebacteria transfers predominantly consist of genes involved in metabolic functions, with the most frequently represented of these functions being amino acid import, energy production and conversion, inorganic ion transport, and carbohydrate transport.

Have the imported genes been acquired piecemeal by independent horizontal gene transfer events, or have they all been mass transferred at once in a manner similar to that in the origin of eukaryotes? The evidence suggests that, for lineages in which the origin of the eubacterial genes and the origin of the higher archaebacterial taxon are indistinguishable, the latter mechanism seems more likely. The main reason for this assessment is that prokaryotes are known to rapidly eliminate useless genes. Thus, selection is needed to preserve the transferred genes in the recipient. The acquisition of a complete pathway composed of many genes would provide a selectable functional unit. In contrast, one gene from a pathway consisting of many genes would rarely be of any use to its recipient, and would most certainly be eliminated during evolution. This may also be the reason why entire pathways in prokaryotes are organized as operons. Encoded by operons, such pathways may be maintained in the recipient genome more frequently than individual gene transfers. This rationale underlies the **selfish operon theory** that explains why prokaryotes have operons in the first place

(Lawrence and Roth 1996), as well as why operons for the same function can be assembled independently during evolution (Martin and McInerney 2009).

Interestingly, albeit inexplicably, interdomain horizontal gene transfer has been found to be a highly asymmetrical process, with transfers from eubacteria to archaebacteria being more than five times as frequent as transfers from archaebacteria to eubacteria. This extreme asymmetry in interdomain gene transfer probably relates to the specialized lifestyle of methanogens, which served as recipients in 83% of the observed gene transfers.

The Two Primary Domains of Life

Since the eukaryotes are almost certainly chimeric in origin and, therefore, phylogenetically derived, we can ignore them in dealing with the very early stages of cellular evolution. In this section, we discuss attempts to find the root of the tree of life after the reticulation called Eukaryota is removed. The outlines of the very early stages of cellular evolution are fiercely debated in the literature, with little agreement even about the nature of the questions to be asked. One may glean a taste of the ferocity of the disagreements by noting that no universally accepted nomenclature exists. The putative common ancestor of all extant organisms has been referred to as the "progenote" (Woese and Fox 1977b), the "cenancestor" (Fitch and Upper 1987), the "commonote" (Kagawa et al. 1995), the "last universal common ancestor" (Kyripides et al. 1999), the "last universal cellular ancestor" (Philippe and Forterre 1999), and the "universal ancestor" (Doolittle 2000).

Because prokaryote evolution was traditionally viewed as a treelike bifurcating process, efforts to identify the most ancient divergence focused on positioning a root on a phylogenetic tree constructed from one or several genes. Such studies have delivered widely conflicting results on the position of the root, this being mainly due to methodological problems inherent to long divergence times as well as the fact that horizontal gene transfer occurs quite frequently in prokaryotes. Dagan et al. (2010) used a network-based procedure that takes into account both gene presence or absence and the level of sequence similarity among individual gene families. On the basis of 562,321 protein-coding gene families from 191 genomes, they found that the deepest divide in the prokaryotic world is on the branches separating the archaebacteria from the eubacteria. This result takes us back to Fox et al.'s (1977) original division of the prokaryotes into two domains.

Of course, the placement of the root may be erroneous due to unequal rates of sequence change between archaebacteria and eubacteria. Dagan et al. (2010) specifically test for such rate inequalities across all prokaryotic lineages, and neither the archaebacteria as a group nor the eubacteria as a group harbor evidence for elevated evolutionary rates, either in the recent evolutionary past or in their common ancestor. The interdomain prokaryotic position of the root is thus not attributable to lineage-specific rate variation.

These findings suggest that only two primary domains of life exist (Williams et al. 2013). According to the nomenclatorial codes in systematics, these primary domains should be referred to as Archaebacteria and Eubacteria, since these names are senior (i.e., older) synonyms. Archaea and Bacteria are junior synonyms and, hence, invalid.

The Public Goods Hypothesis

Some scientists feel that the failure of the tree metaphor to account for the evolution of prokaryotes necessitates a new metaphor that can (1) accommodate all evolving entities, such as taxa, genes, and genomes, and (2) take into account both horizontal and vertical gene transfer.

McInerney et al. (2011) put forward the **public goods hypothesis**, according to which nucleotide sequences are seen as goods, passed from organism to organism both vertically and horizontally. In the field of economics, from which this hypoth-

Table 6.1

The four categories of goods classified according to rivalry and excludability

	Excludable	Nonexcludable
Rivalrous	Private goods	Common goods
Nonrivalrous	Club goods	Public goods

Source: From McInerney et al. (2011).

esis takes its inspiration, four categories of goods are recognized according to their **rivalry**, on the one hand, and their **excludability**, on the other (**Table 6.1**). A good is **rivalrous** if the consumption of the good by one individual reduces the availability of that good for another individual. It is **nonrivalrous** if the consumption of the good by one individual does not reduce its availability for another individual. A good is **excludable** if it is possible to exclude the good from being available to everybody; it is **nonexcludable** if it is impossible or at least very difficult to exclude the good from being available to others. A good that is both rivalrous and excludable is called a **private good**; a good that is rivalrous and nonexcludable is called a **common good**; a good that is both nonrivalrous and nonexcludable is called a **public good**; and a good that is nonrivalrous and excludable is a **club good**. Most goods with which we are familiar are private goods. Fish stocks, public pasture lands, and fossil energy sources, on the other hand, are common goods. The air that we breathe is an example of a public good. First, the breathing of air by one individual does not greatly reduce the availability of air for other individuals, so this makes air a nonrivalrous good. Second, it is impossible to effectively exclude other individuals from accessing air, so this makes air nonexcludable. (Note, however, that context matters. As far as scuba diving is concerned, air is not a public good, but a private good.) Men-only clubs, private parks, and access to pay-per-view television are examples of club goods.

Since DNA is replicable, it is very difficult to imagine a situation in which DNA is a rivalrous good. In fact, a gene copy can be used almost indefinitely without "using up" the gene. When gene copies are moved either vertically or horizontally, they are mostly moved using a copy-and-paste mechanism, not by a "stealing" mechanism such as cut-and-paste. These features seem to suggest that most genes are nonrivalrous—the use of a gene by one organism does not preclude its use by another. Thus, as far as genes are concerned, we can disregard private and common goods. If we furthermore consider the wide-ranging means of horizontally transferring genes in prokaryotes, along with the experimental demonstration that barriers to gene transfer are few, we must conclude that many genes have the property of being nonexcludable. That is to say, it is very difficult for prokaryotes to completely prevent other prokaryotes from obtaining a particular gene. Thus, many genes are public goods, available for all organisms to integrate into their genomes.

Some DNA sequences, however, are nonrivalrous but excludable, for example, if they are inherited in a strictly vertical fashion, or if their horizontal transfer is restricted by biochemical, taxonomic, and ecological factors. Such sequences might be better described as club goods than public goods. For instance, a protein-coding gene may be excludable if it uses a genetic code that is unique to a particular group of organisms. This DNA sequence could produce a defective protein in an organism that uses a different genetic code. Excludability of the gene in this case would come from an intrinsic characteristic of the molecular sequence and would be independent of function. In another scenario, a protein might only function, say, in the absence of oxygen, and therefore all aerobic organisms would be excluded from using the gene that encodes this protein. In this case, the function of the encoded gene is the feature that would make a gene a club good. Other restrictions leading to excludability may involve toxicity or a function that depends on a highly connected network of proteins. It is currently not possible to state how many genes can be categorized as public goods and how many as club goods.

We note in summary that the public goods hypothesis is currently no more than a conceptual framework; it has yet to yield testable and refutable predictions.

CHAPTER 7

Evolution by DNA Duplication

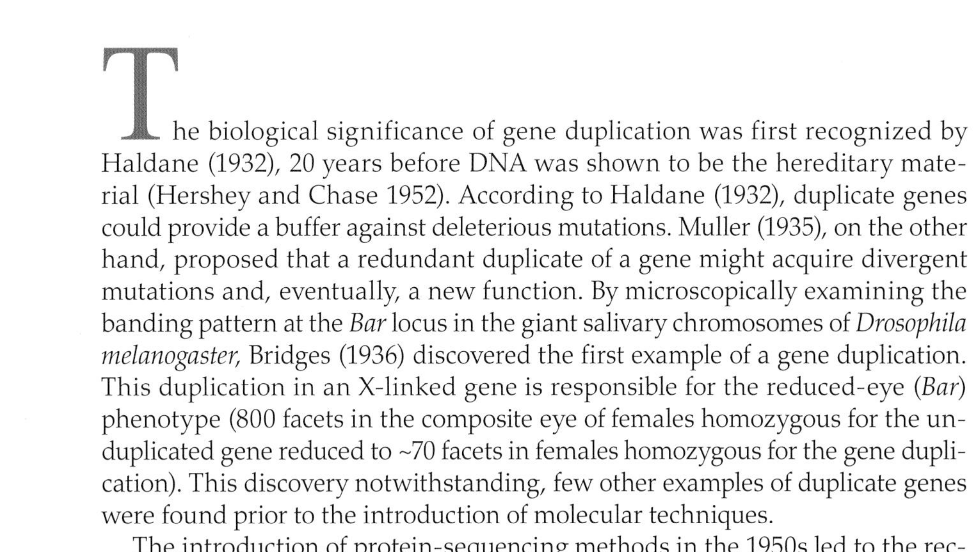

The biological significance of gene duplication was first recognized by Haldane (1932), 20 years before DNA was shown to be the hereditary material (Hershey and Chase 1952). According to Haldane (1932), duplicate genes could provide a buffer against deleterious mutations. Muller (1935), on the other hand, proposed that a redundant duplicate of a gene might acquire divergent mutations and, eventually, a new function. By microscopically examining the banding pattern at the *Bar* locus in the giant salivary chromosomes of *Drosophila melanogaster*, Bridges (1936) discovered the first example of a gene duplication. This duplication in an X-linked gene is responsible for the reduced-eye (*Bar*) phenotype (800 facets in the composite eye of females homozygous for the unduplicated gene reduced to ~70 facets in females homozygous for the gene duplication). This discovery notwithstanding, few other examples of duplicate genes were found prior to the introduction of molecular techniques.

The introduction of protein-sequencing methods in the 1950s led to the recognition that myoglobin and the various chains of hemoglobin have been derived from duplicate genes (Itano 1953, 1957; Rhinesmith et al. 1958; Braunitzer et al. 1961; Ingram 1961). In the 1960s, protein electrophoretic methodology set off an interest in isozymes, which are enzymes mostly encoded by duplicate genes. The study of isozymes provided evidence for the frequent occurrence of gene duplication during evolution, so much so that Ohno (1970) put forward a view that gene duplication was the only means by which a new function could arise: "natural selection merely modified, while redundancy created." Ohno's view was criticized initially, but it gained acceptance in the 1970s and has since stimulated much interest in the study of duplicate genes. Although other means of creating new functions are now known (Chapter 8), Ohno's view remains largely valid. With the advent of gene cloning and sequencing techniques in the 1980s, large quantities of duplicate-gene data started accumulating, culminating in a veritable deluge resulting from whole-genome sequencing.

In this chapter, we discuss the different types of duplication; the evolutionary fates of duplicated genes, including gene death and the acquisition of novel function; and the curious phenomenon of concerted evolution, i.e., the coordinated manner that characterizes the evolution of some repeated DNA sequences.

Types of DNA Duplication

An increase in the number of copies of a DNA segment can be brought about by several types of **DNA duplication** (**Figure 7.1**). These are usually classified according to the size or extent of the genomic region involved. The following types of duplication are recognized: (1) **internal** or **partial gene duplication**, (2) **complete gene duplication**, (3) **partial chromosomal duplication**, (4) **complete chromosomal duplication** (**polysomy**), and (5) **whole-genome duplication** (**polyploidy**). The terms **block duplication**, **regional duplication**, and **segmental duplication** have commonly been used to refer to duplications involving either two or more genes or lengthy genomic segments. Internal gene duplication will be discussed in Chapter 8.

Ohno (1970) argued that whole-genome duplication has generally been more important than regional duplication, because in the latter case only parts of the regulatory system of structural genes may be duplicated and such an imbalance may disrupt the normal function of the duplicated genes. However, as discussed later, regional duplications have apparently played a very important role in evolution.

Mechanisms of DNA Duplication

The mechanisms responsible for DNA duplication fall into two broad categories: (1) unequal crossing over, which is responsible for the generation of tandem repeats (**Figure 7.2**), and (2) transposition, which is responsible for the creation of dispersed sequences (Chapter 9). **Homologous unequal crossing over** refers to a crossing over that is initiated by the presence of highly similar sequences of substantial length, such as repetitive elements. **Nonhomologous crossing over** refers to a process of crossing over that is initiated by **microhomologies**, i.e., two or more short sequences that exhibit similarity to one another. (The term "microhomology" is something of a misnomer, since the similarity between the sequences is mostly due to chance rather than common descent.) Nonhomologous crossing over occurs at much lower rates than homologous events. Both types of crossing over result in two reciprocal chromosomal products: one will contain a duplication; the other, a deletion. It is important to remember that both these events are mutations, i.e., they affect genomic regions without regard to functional constraints and boundaries. The size of the duplicated region can vary from a few base pairs to tens or even hundreds of kilobases, and it can contain no genes, a portion of a gene, an entire gene, a few genes, or many genes.

An important feature associated with gene duplication is that as long as two or more copies of a gene exist in proximity to each other, the process of gene duplication by unequal

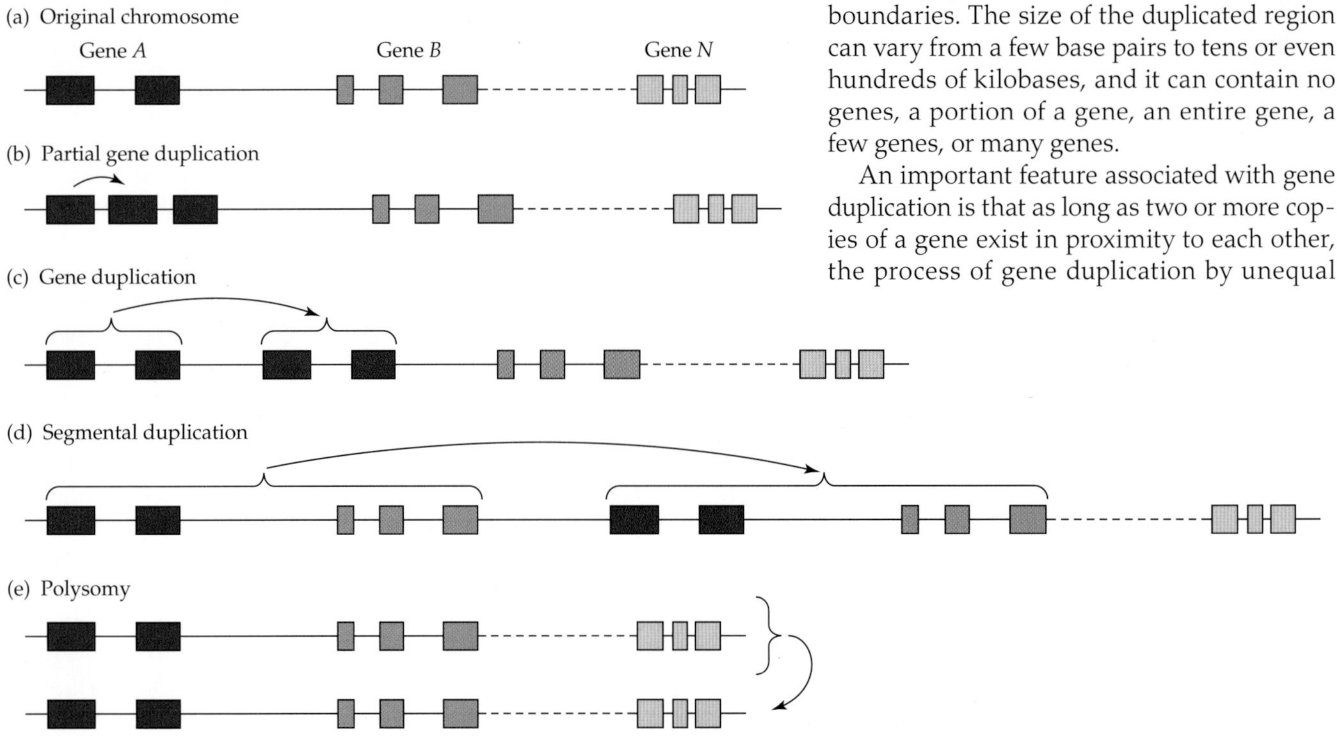

(a) Original chromosome
Gene *A* Gene *B* Gene *N*

(b) Partial gene duplication

(c) Gene duplication

(d) Segmental duplication

(e) Polysomy

Figure 7.1 Types of DNA duplication. (a) A schematic chromosome. Three genes are shown, gene *A* with two exons, *B* with three exons, and *N* with three exons. The dashed line indicates that there are more than three genes on this chromosome. (b) Partial gene duplication results in the duplication of the first exon in gene *A*. (c) Complete duplication of gene *A*. (d) Segmental duplication of adjacent genes *A* and *B*. (e) Complete chromosomal duplication (polysomy).

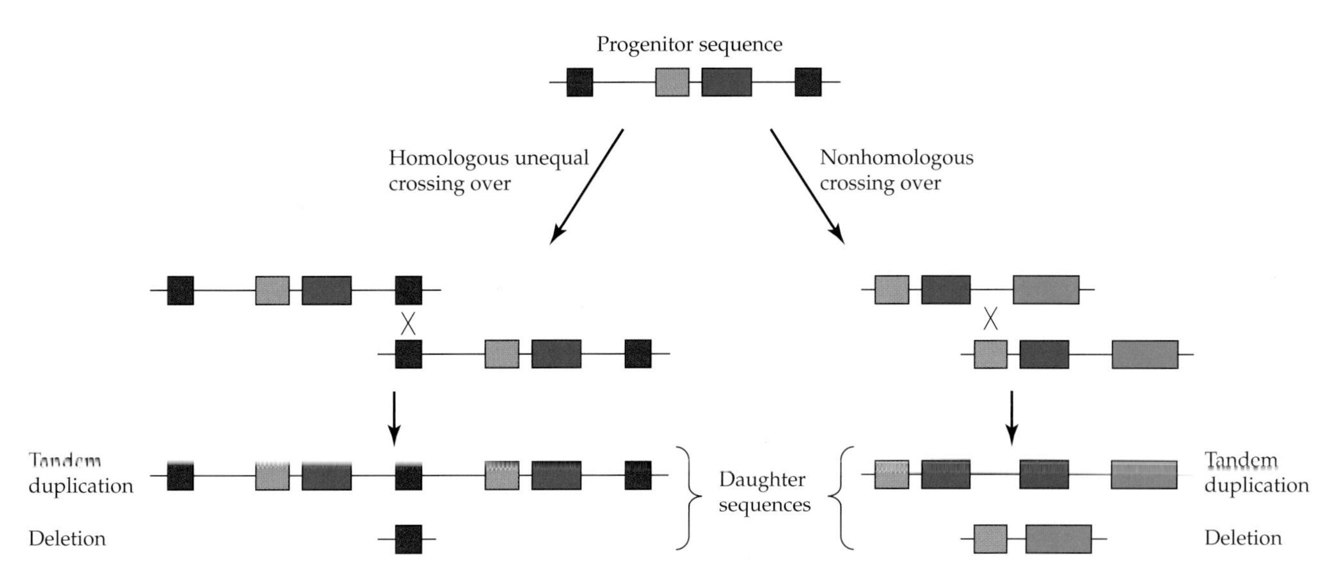

Figure 7.2 Mechanisms for sequence duplication and sequence deletion through crossing over. Single-copy functional sequences are denoted by blue and green rectangles. Repetitive sequences are denoted by red squares. If crossing over is initiated by the presence of similar repetitive sequences, it is referred to as homologous unequal crossing over. If crossing over is initiated by nonhomologous functional sequences, it is referred to as nonhomologous crossing over. In both cases, two reciprocal daughter sequences are produced: one sequence contains a tandem duplication; the other contains a deletion.

crossing over can be greatly accelerated in this region, and numerous copies may be produced.

Gene duplication can also occur by copy-and-paste transposition (Chapter 9). Copy-and-paste transposition can occur through either a DNA or an RNA intermediate. Both DNA-mediated transposition and RNA-mediated retrotransposition can result in the creation of dispersed duplicates.

Sequence homology due to gene duplication is called **paralogy**, in contradistinction with **orthology** and **xenology,** which are due to speciation and horizontal gene transfer, respectively (Chapters 5 and 9).

Dating Duplications

In **Figure 7.3**, genes *Y* and *Z* were derived from the duplication of an ancestral gene *X* and are therefore paralogous, while gene *Y* from species 1 and gene *Y* from species 2 are orthologous, as are genes *Z* from species 1 and gene *Z* from species 2. We can estimate the date of duplication, T_D, from sequence data if we know the rate of substitution in genes *Y* and *Z*. The rate of substitution can be estimated from the number of substitutions between the orthologous genes in conjunction with knowledge of the time of divergence, T_S, between species 1 and 2 (Figure 7.3).

For gene *Y*, let K_Y be the number of substitutions per site between the two species. Then, the rate of substitution in gene *Y*, r_Y, is estimated by

$$r_Y = \frac{K_Y}{2T_S} \tag{7.1}$$

The rate of substitution in gene *Z*, r_Z, can be obtained in a similar manner. The average substitution rate for the two genes is given by

$$r = \frac{r_Y + r_Z}{2} \tag{7.2}$$

To estimate T_D, we need to know the number of substitutions per site between genes *Y* and *Z* (K_{YZ}). This number can be estimated from four pairwise comparisons: (1) gene *Y* from species 1 and

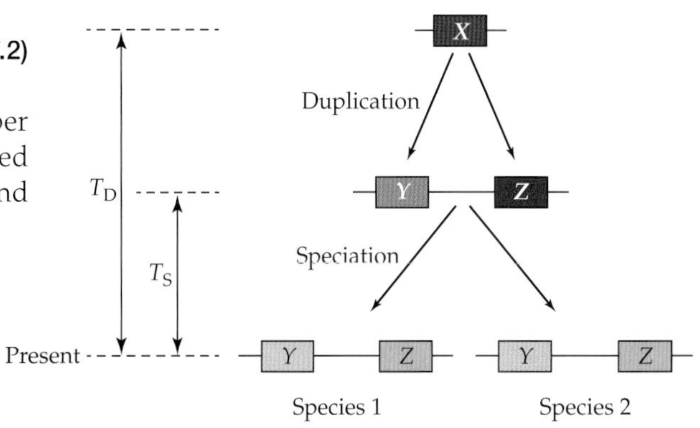

Figure 7.3 Model for estimating the time of a gene duplication event (T_D). Two genes, *Y* and *Z*, were derived from a duplication of gene *X* that occurred T_D time units ago in an ancestral species. The ancestral species split into two species (1 and 2) T_S time units ago.

gene Z from species 2, (2) gene Y from species 2 and gene Z from species 1, (3) gene Y and gene Z from species 1, and (4) gene Y and gene Z from species 2. From these four estimates we can compute the average value for K_{YZ}, i.e., $\overline{K}_{YZ}$, from which we can estimate T_D as

$$T_D = \frac{\overline{K}_{YZ}}{2r} \quad (7.3)$$

Note that, in the case of protein-coding genes, by using the numbers of synonymous and nonsynonymous substitutions separately, we can obtain two independent estimates of T_D. The average of these two estimates may be used as the final estimate of T_D. However, if the number of substitutions per synonymous site between genes Y and Z is large—say, larger than 1—then the number of synonymous substitutions cannot be estimated accurately, and synonymous substitutions may not provide a reliable estimate of T_D. In such cases, only the number of nonsynonymous substitutions should be used. Conversely, if the number of substitutions per nonsynonymous site between the paralogous genes is small, then the estimate of the number of nonsynonymous substitutions is subject to a large sampling error, and in such cases, only the number of synonymous substitutions should be used. Moreover, nonsynonymous sites may experience episodic selection such that their substitution rates may be less clocklike than those at synonymous sites.

In the above discussion, we have assumed rate constancy. Whether this assumption holds can be tested by the four pairwise comparisons mentioned above. If it does not hold, the T_D estimate may be erroneous. As will be discussed later, problems due to concerted evolutionary events may also arise, and these can complicate the estimation of T_D.

Another method for approximating the date of gene duplication events is to consider the phylogenetic distribution of genes in conjunction with paleontological data pertinent to the divergence date of the species in question. For example, all vertebrates with the exception of jawless fishes (hagfishes and lampreys) encode α- and β-globin chains. There are two possible explanations for this observation. One is that the duplication event producing the α- and β-globins occurred in the common ancestor of all vertebrates (Craniata), but the two jawless fish lineages (Myxini and Cephalaspidomorphi) have lost one of the two duplicates. This is possible but not very likely, because such a scenario would require the losses to occur independently in at least two evolutionary lineages. The other explanation is that the duplication event occurred after the divergence of jawless fishes from the ancestor of the jawed vertebrates (Gnathostomata), but before the radiation of the jawed vertebrates from each other. This latter explanation is thought to be more plausible, and the duplication date is commonly taken to be 450–500 million years ago.

Obviously, the above methods can only provide us with rough estimates of duplication dates, so all estimates should be taken with caution. Note that in estimating dates of divergence among species, one can use data from many genes belonging to many gene families. In comparison, in estimating dates of gene duplication, one must rely only on data from genes belonging to a single gene family. Because of the stringent limitations on the sequence data that can be used, estimates of gene duplication are often subject to very large standard errors.

Gene Duplication and Gene Families

The process of complete gene duplication produces paralogous repeats. Depending on their degree of differentiation, repeated genes can be divided into **invariant** and **variant repeats**. Invariant repeats are identical or nearly identical in sequence to one another. Variant repeats are paralogous copies of a gene that, although similar to each other, differ in their sequences to a lesser or greater extent. Surprisingly large numbers of homologous proteins that can perform markedly different functions have been revealed by sequencing projects (Todd et al. 2001). Several such examples are

Table 7.1

Similarity and dissimilarity among pairs of duplicate genes[a]

Gene pair products (organism)	Amino acid sequence similarity (%)	Estimated time of duplication (million years)	Regulation[b]	Chemical attributes[c]	Aggregation properties[d]	Place of expression[e]
Trypsin and chymotrypsin (human)	36	1,500	– – –	– –	+	+
Hemoglobin and myoglobin (human)	23	800	– – –	– –	– – –	– –
Lactate dehydrogenase M and H chains (human)	74	600	– –	– –	+	–
Hemoglobin α and β chains (human)	41	500	– – –	–	–	+
Immunogobulin H and L chains (human)	25	400	– – –	– –	–	+
Lactalbumin and lysozyme (human)	37	350	– – –	– – –	– –	– –
Growth hormone and prolactin (human)	25	330	– –	– –	+	–
Chymotrypsins A and B (human)	79	270	+	+	+	+
Carbonic anhydrases B and C (human)	60	180	–	–	+	+
Insulins I and II (rat)	96	30	–	+	+	+
Growth hormone and lactogen (human)	85	23	– –	–	+	– –
Alcohol dehydrogenase A and S chains (horse)	98	10	–	–	+	+

Source: Modified from Li (1983).

[a] + = similar; – = slightly different; – – = moderately different; – – – = markedly different.

[b] *Regulation* refers to differential expression over tissues or developmental stages, or to the rate of synthesis of the gene product if the two genes are expressed in the same tissue.

[c] *Chemical attributes* include catalytic properties and binding specificities to substrates, inhibitors, antigens, and so on.

[d] *Aggregation* refers to the number of subunits and the types of interactions between them.

[e] *Place of expression* refers to organs of the body or to types of differentiated cells.

listed in **Table 7.1**. They reveal that the rate with which functional differentiation can evolve can be quite variable. For example, lactalbumin, which is a subunit of the enzyme that catalyzes the synthesis of the sugar lactose, and lysozyme, which dissolves certain bacteria by cleaving the polysaccharide component of their cell walls, perform completely unrelated functions, are regulated differently, and are expressed in different tissues. Yet, the divergence time between lactalbumin and lysozyme is estimated to be only a third of that between trypsin and chymotrypsin, which perform very similar digestive functions and differ from each other merely in their recognition sites, with trypsin recognizing arginine and lysine, whereas chymotrypsin recognizes phenylalanine, tryptophan, and tyrosine (Barker and Dayhoff 1980).

Orthologous and paralogous genes, which share a common ancestral gene, are frequently referred to as a **gene family** or **multigene family**. The term **superfamily** was coined by Dayhoff (1978) to distinguish distantly related proteins from closely related ones. Subsequently, many rank terms referring to various degrees of relationships among homologous proteins, such as "subfamily" and "clan," have been suggested in the literature. As with many hierarchical biological terminologies, however, the terms are context-dependent and may sometimes seem quite erratic.

Gene family size, i.e., the number of genes within a gene family, was found to vary considerably among species and among gene families within a species (Li et al. 2001; Gu et al. 2002). Some families consist of **single-copy genes**. Other gene families consist of a small number of paralogous genes repeated within the genome. A few gene families comprise hundreds of duplicated copies, in which case they are referred to as **highly repetitive genes**. The largest protein-coding gene family in *Drosophila melanogaster* is the trypsin family with 111 members, whereas the largest such family in humans is the olfactory receptor family with more than 1,000 members (Gilad et al. 2003; Zhang 2003). From sequence analyses of two yeast genomes (*Saccharomyces cerevisiae* and *Schizosaccharomyces pombe*), *Caenorhabditis elegans, D. melanogaster,* and *Escherichia coli,* Conant and Wagner (2002) inferred that ribosomal proteins and transcription factors generally form smaller gene families than do other protein-coding genes (e.g., those encoding proteins related to controlling cell cycles and metabolism).

The Prevalence of Gene Duplication

Gene duplications arise spontaneously at high rates in bacteria, bacteriophages, insects, and mammals and are generally viable (Fryxell 1996). Thus, the creation of duplications by mutation is not the rate-limiting step in the process of gene duplication and subsequent functional divergence. However, only a small fraction of all duplicated genes are retained, and an even smaller fraction evolve new functions. The reason is the much higher probability of nonfunctionalization in comparison with that of evolving a new function. We note, however, that in large populations, the probability of evolving a new function may be considerable (Walsh 1995; Nadeau and Sankoff 1997).

Genomic studies have shown that large proportions of genes have been generated by gene duplication in all three domains of life. For example, about 40% of all genes in such diverse organisms as *Homo sapiens*, the molicute bacterium *Mycoplasma pneumoniae*, and the fruit fly *Drosophila melanogaster* belong to multigene families. The fraction of genes belonging to multigene families is much higher in some organisms (e.g., 65% in *Arabidopsis thaliana*) and much lower in others (e.g., 17% in the Gram-negative pathogen *Helicobacter pylori*). These figures are almost certainly underestimates because many duplicated genes have diverged so much that virtually no sequence similarity is found.

Lynch and Conery (2000) estimated that in eukaryotes a new duplicate gene arises and is fixed in the population at an approximate rate of one duplication per gene per 100 million years. This rate is the **gene birth rate**, and it was derived from a study of recent duplications. Many fixed duplicated genes later become pseudogenes and may subsequently be deleted from the genome. The rate of duplication that gives rise to stably maintained genes can be calculated by multiplying the gene birth rate by the **retention rate**, which is expected to fluctuate widely with gene function (Zhang 2003).

Modes of Evolution of Multigene Families

Our notions concerning the evolution of multigene families have evolved considerably in the last 50 years. The archetype of evolution of multigene families before 1970 was the globin superfamily of genes (see below). The genes belonging to this family are related by descent and have diverged gradually from one another as the different copies have acquired new gene functions. This mode of evolution may be referred to as **divergent evolution** (**Figure 7.4a**). Evidence for the divergent mode of evolution accumulated rapidly as scientists realized that proteins performing unrelated biological functions may, in fact, be related by descent. Particularly surprising were the findings of Lazure et al. (1981), according to which functionally disparate proteins, such as tonin (a submaxillary angiotensinogen), γ-subunit nerve growth factor, epidermal growth factor-binding protein, and several serine proteinases, are evolutionarily related to one another.

Around 1970, however, a number of groups discovered that ribosomal RNAs in *Xenopus* are encoded by a large number of tandemly repeated genes and that the nucleotide sequences of the intergenic regions of the genes are more similar within a species than between two related species. These observations could not be explained by the model of divergent evolution, and a new model called **concerted evolution**

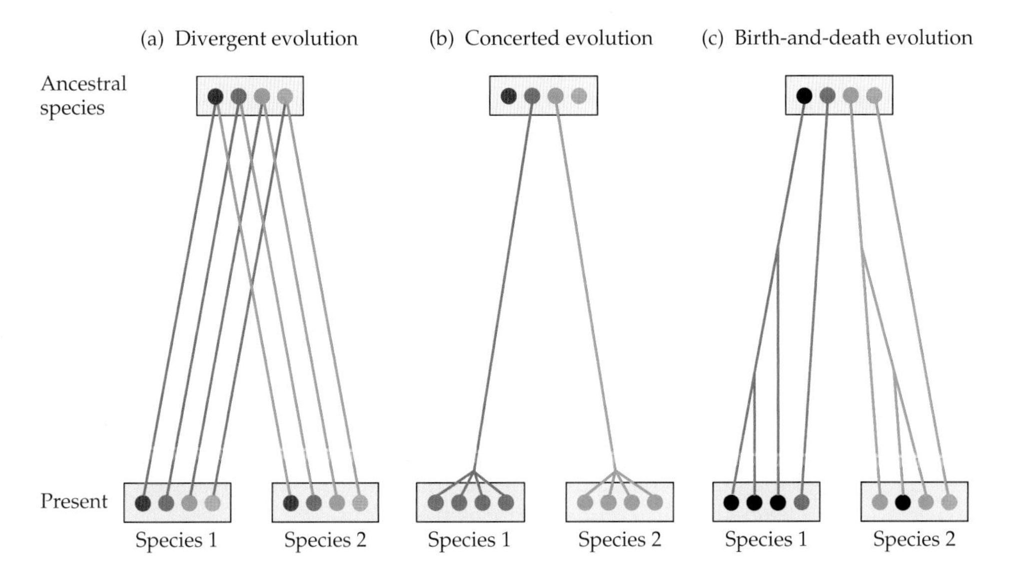

Figure 7.4 Schematic phylogenetic representation of (a) divergent, (b) concerted, and (c) birth-and-death evolution in multigene families. Colored circles represent functional genes; black circles signify pseudogenes.

was proposed (**Figure 7.4b**). In this model all the members of a gene family are assumed to evolve in a concerted manner rather than independently, and a mutation occurring in a repeat may spread through the entire group of member genes by repeated occurrences of unequal crossing over or gene conversion. It was subsequently discovered that the concerted evolution model may not be applicable to a substantial number of multigene families, and a third model, called **birth-and-death evolution**, was proposed. In this model, new genes are "born" by gene duplication, each being maintained in the genome for varying periods of time before "dying" through deletion or nonfunctionalization (**Figure 7.4c**).

Divergent Evolution of Duplicated Genes

Here we deal with the divergent evolution of duplicates created by complete gene duplication only. Divergent evolution refers to the fact that the duplicates evolve independently of one another. Usually such copies diverge in sequence, but in some cases, the copies remain similar to one another because of purifying selection due to functional constraints. How duplicated genes evolve varies from case to case (**Figure 7.5**). In this section, we discuss four general possibilities: (1) **nonfunctionalization**, (2) **retention of original function** or **gene conservation**, (3) **neofunctionalization**, and (4) four models of **subfunctionalization**. For a comprehensive discussion of these and other models pertaining to the evolution of function following gene duplication, see Innan and Kondrashov (2010).

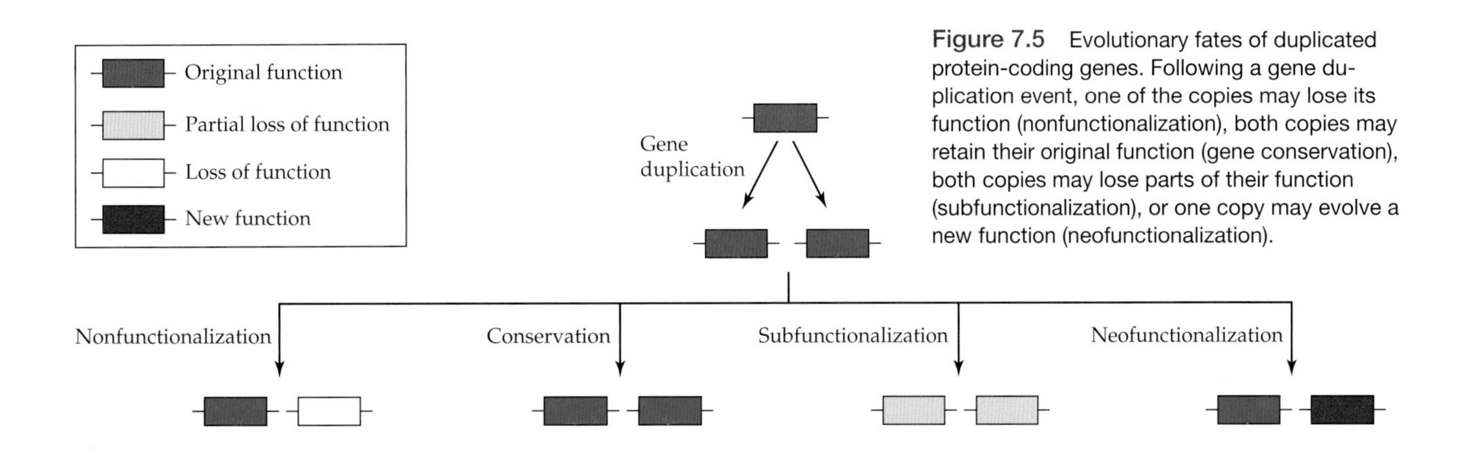

Figure 7.5 Evolutionary fates of duplicated protein-coding genes. Following a gene duplication event, one of the copies may lose its function (nonfunctionalization), both copies may retain their original function (gene conservation), both copies may lose parts of their function (subfunctionalization), or one copy may evolve a new function (neofunctionalization).

Nonfunctionalization and gene loss

The thousands of genetic diseases that have been documented in the medical literature (omim.org) and veterinary literature (omia.org) attest to the fact that mutations can easily destroy the function of a protein-coding gene. The vast majority of such mutations are deleterious and are either eliminated quickly from the population or maintained at very low frequencies by overdominant selection or genetic drift (Chapter 2). As noted by Haldane (1932), however, as long as there are other copies of a gene that function normally, a duplicate gene can accumulate deleterious mutations and become nonfunctional or be lost from the genome by deletion without adversely affecting the fitness of the organism. Indeed, because deleterious mutations occur far more often than advantageous ones, a redundant duplicate gene should be much more likely to become nonfunctional than to either retain its original function or evolve into a new gene (Ohno 1972). Interestingly, while nonfunctionalization is indeed the most likely fate of a redundant duplicate gene, the fraction of duplicated vertebrate genes that have remained functional in extant genomes greatly exceeds the expectation of Ohno's classical model (Prince and Pickett 2002).

The nonfunctionalization or **silencing** of a gene due to deleterious mutations produces a **nonprocessed** or **unprocessed pseudogene**, i.e., a pseudogene that has not gone through RNA processing (Chapter 9). The vast majority of nonprocessed pseudogenes are derived via the nonfunctionalization of duplicate copies of functional genes (duplicated pseudogenes). Some nonprocessed pseudogenes, such as $\psi\beta^X$ and $\psi\beta^Z$ in the goat β-globin multigene family, have been derived from duplication of a preexisting pseudogene (Cleary et al. 1981), and a handful of nonprocessed pseudogenes have been derived from a functional gene without a prior duplication event (Chapter 8).

Table 7.2 lists the structural defects found in several globin pseudogenes. Most of these nonprocessed pseudogenes contain multiple defects such as frameshifts, premature stop codons, and obliteration of splicing sites or regulatory elements, so it has been difficult to identify the mutation that was the direct cause of gene silencing. In a few cases, identification of the "culprit" has been possible. For example, human $\psi\zeta$ contains only a single major defect—a nonsense mutation resulting in the premature termination of translation—that is probably the direct cause of its nonfunctionalization. In some cases, it has been possible to identify the mutation responsible for the nonfunctionalization of a gene through a phylogenetic analysis. For example, human pseudogene $\psi\beta$ in the β-globin family contains numerous defects, each of which could have been sufficient to silence it. The β-globin clusters in chimpanzee and gorilla, our closest relatives, were found to contain the same number of genes and pseudogenes

Table 7.2

Defects in nonprocessed globin pseudogenes[a]

Pseudogene	TATA box	Initiation codon	Frameshift	Premature stop codon	Essential amino acid	Splicing	Stop codon	Polyadenylation signal
Human $\psi\alpha 1$		+	+	+	+	+	+	+
Human $\psi\zeta 1$				+				
Mouse $\psi\alpha 3$	+		+	+		+		
Mouse $\psi\alpha 4$			+		+			
Goat $\psi\beta^x$	+		+	+	+	+	+	+
Goat $\psi\beta^z$	+		+	+	+	+	+	+
Rabbit $\psi\beta^2$			+	+	+	+		

Source: From Li (1983).

[a]A plus sign indicates the existence of a particular type of defect.

as in humans, indicating that the pseudogene was created and silenced before these three species diverged from one another. The three pseudogenes were found to have only three defects in common: either of two nonsynonymous substitutions in the initiation codon (ATG → GTA), a nonsense substitution in the tryptophan codon at position 15 (TGG → TGA), and a deletion in codon 20 resulting in a frameshift in the reading frame and a termination codon in the second exon (Chang and Slightom 1984). Thus, the "list of suspects" was reduced to three. Further studies showed that the same pseudogene exists in all primates and is, therefore, quite ancient. A comparison of the defects among all primate sequences showed that the initial mutation responsible for the nonfunctionalization of ψβ is the one in the initiation codon (Harris et al. 1984). It should be noted that most studies are biased by focusing on changes in the coding region; mutations in promoter regions that silence expression may go undetected.

Nonfunctionalization through missense mutations may not occur very often, because the production of a nonfunctional protein may be deleterious even if there are many functional copies of the gene that produce functional proteins. The reason is that a defective protein may be similar enough to the functional paralog to be incorporated into a biological structure but may interfere with the function of such structure. For example, there are dozens of chorion-coding genes in the genome of the silkworm *Bombyx mori*, yet if even one of them is rendered nonfunctional by a missense mutation, the entire eggshell becomes defective (e.g., Spoerel et al. 1989).

Because they are created by duplication, nonprocessed pseudogenes are usually found in the neighborhood of their paralogous functional genes. There are, exceptions, however; these are cases in which nonprocessed pseudogenes become dispersed as a consequence of genomic rearrangements (Chapter 11). For example, the α-globin cluster in the mouse is located on chromosome 11, and yet a nonprocessed α-globin pseudogene was found on chromosome 17 (Tan and Whitney 1993).

In genomes with large numbers of pseudogenes, nonprocessed pseudogenes are usually less frequent than processed pseudogenes (Chapter 9) but much more abundant than unitary pseudogenes (Chapter 8).

Nonfunctionalization time

The evolutionary history of a nonprocessed pseudogene is assumed to consist of two distinct periods. The first period starts with the gene duplication event and ends when the duplicate copy is rendered nonfunctional. During this period, the would-be pseudogene presumably retains its original function, and the rate of substitution is expected to remain roughly the same as it was before the duplication event. After the loss of function, the pseudogene is freed from all functional constraints and its rate of nucleotide substitution is expected to increase considerably. From the evolutionary point of view, it is interesting to estimate how long a redundant copy of a functional gene may remain functional after the duplication event. To estimate this **nonfunctionalization time**, the following method has been suggested (Li et al. 1981; Miyata and Yasunaga 1981).

Consider the phylogenetic tree in **Figure 7.6**. T denotes the divergence time between species 1 and 2, i.e., the time since the separation between the orthologous functional genes A and B; T_D denotes the time since duplication, i.e., the time of divergence between the functional gene A and its paralogous pseudogene $ψA$; and T_N denotes the time since the nonfunctionalization of pseudogene $ψA$. The numbers of nucleotide substitution per site at the ith position of codons (i = 1, 2, or 3) between $ψA$ and A, between $ψA$ and B, and between A and B are

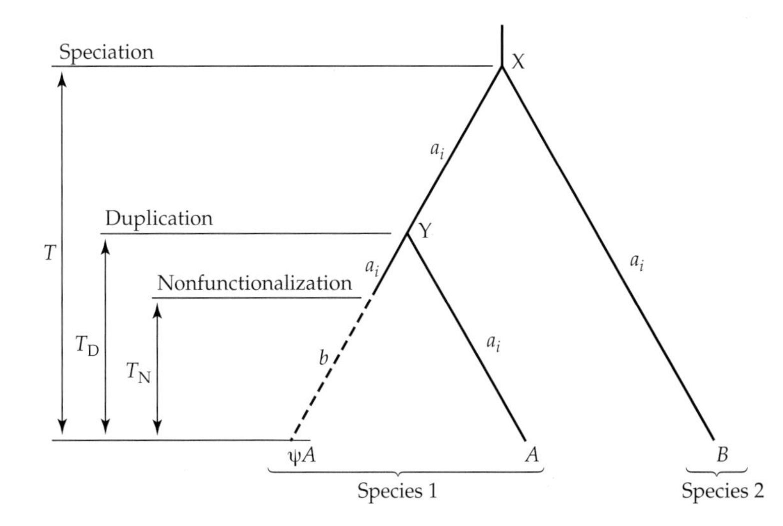

Figure 7.6 Schematic phylogenetic tree used to estimate the nonfunctionalization time of a nonprocessed pseudogene. T denotes the divergence time between species 1 and 2, T_D denotes the time since duplication of gene A, and T_N denotes the time since nonfunctionalization of pseudogene $ψA$. The term a_i denotes the rate of nucleotide substitution per site per year at the ith codon position in the functional genes, while b is the rate of substitution for the pseudogene. The node connecting the orthologous genes is denoted by X; the node connecting the paralogous genes is marked by Y. (Modified from Li et al. 1981.)

denoted as $d_{(\psi AA)i}$, $d_{(\psi AB)i}$, and $d_{(AB)i}$, respectively, and can be calculated directly from the sequence data (Chapter 3).

Let l_i, m_i, and n_i be the numbers of nucleotide substitutions per site at codon position i between points Y and ψA, Y and A, and Y and B, respectively. We then have

$$d_{(\psi AA)i} = l_i + m_i \quad (7.4a)$$

$$d_{(\psi AB)i} = l_i + n_i \quad (7.4b)$$

$$d_{(AB)i} = m_i + n_i \quad (7.4c)$$

Therefore, l_i, m_i, and n_i can be estimated by

$$l_i = \frac{d_{(\psi AA)i} + d_{(\psi AB)i} - d_{(AB)i}}{2} \quad (7.5a)$$

$$m_i = \frac{d_{(\psi AA)i} - d_{(\psi AB)i} + d_{(AB)i}}{2} \quad (7.5b)$$

$$n_i = \frac{-d_{(\psi AA)i} + d_{(\psi AB)i} + d_{(AB)i}}{2} \quad (7.5c)$$

In the following. we assume that the rates of substitution at a given codon position are equal in the functional genes A and B. We denote these rates by a_i, where the subscript i stands for the codon position. We also assume that once ψA became nonfunctional, i.e., once all functional constraints were obliterated, the rate of nucleotide substitution became the same for all three codon positions. We denote this rate by b. A reasonable expectation is that b would turn out to be much larger than a_1 and a_2, and possibly a little larger than a_3. From Figure 7.6, we obtain

$$d_{(\psi AA)i} = 2a_i T_D + (b - a_i) T_N \quad (7.6a)$$

$$d_{(\psi AB)i} = 2a_i T + (b - a_i) T_N \quad (7.6b)$$

$$d_{(AB)i} = 2a_i T \quad (7.6c)$$

If we know T, a_i can be estimated from Equation 7.6c by

$$a_i = \frac{d_{(AB)i}}{2T} \quad (7.7)$$

Let us denote the difference $d_{(\psi AB)i} - d_{(AB)i}$ as y_i. Note that

$$y_i = d_{(\psi AB)i} - d_{(AB)i} = bT_N - a_i T_N \quad (7.8)$$

Therefore, T_D can be estimated from Equations 7.6a and 7.8 by

$$T_D = \frac{\sum d_{(\psi AA)i} - \sum y_i}{2\sum a_i} \quad (7.9)$$

where Σ is the summation over i.

Two simple formulas for estimating T_N and b have been suggested by Li et al. (1981):

$$T_N = \frac{y_{12} - y_3}{a_3 - a_{12}} \quad (7.10)$$

$$b = \frac{a_3}{y_{12} - y_3} \quad (7.11)$$

where $y_{12} = (y_1 + y_2)/2$, and $a_{12} = (a_1 + a_2)/2$.

By setting the time of divergence, T, between mouse, rabbit, and human at about 80 million years ago, Li et al. (1981) estimated that the mouse globin pseudogene $\psi\alpha3$ was created by gene duplication 27 ± 6 million years ago and became nonfunctional

23 ± 19 million years ago. Similarly, the human globin pseudogene $\psi\alpha 1$ was estimated to have been created by gene duplication 49 ± 8 million years ago and to have become nonfunctional 45 ± 37 million years ago. In both cases, it is statistically impossible to determine whether or not the nonfunctionalization time is different from 0. In general, it seems that those redundant duplicates that are ultimately destined to become pseudogenes retain their original function for only very short periods of time following the gene duplication event. Moreover, because gene duplication is a type of mutation that occurs with no regard for functional boundaries, a large fraction of duplicated genes may be "stillborn," i.e., they may be incomplete or may lack the proper regulatory elements necessary for expression, in which case the nonfunctionalization time should be 0.

Retention of original function following gene duplication

Rapid repeated rounds of gene duplication are typically referred to as **gene amplification**. The repeated unit of gene amplification is called an **amplicon**. Gene duplication and amplification produce identical copies of repeated units. With evolutionary time, however, these copies are expected to diverge from one another in sequence, structure, and function. Unexpectedly, in several cases, the copies maintain their sequence similarity to one another for long periods of time and are, therefore, invariant. In some cases, repetition of invariant sequences can be shown to be correlated with the synthesis of increased quantities of a gene product that is required for the normal function of the organism. Such repetitions are referred to as **dose** or **dosage repetitions**.

Selection for gene dosage can involve two different mechanisms. Selection for **increased dosage** involves a positive selection pressure to increase expression from a locus. The **dosage-compensation model**, on the other hand, invokes a negative selection pressure to retain the function and expression levels of both copies in order to preserve the correct stoichiometry—the appropriate amounts of the protein in relation to other proteins.

Selection for increased gene dosage is quite common when there is a metabolic need to produce large quantities of specific RNAs or proteins. For example, a duplication of the acid monophosphatase locus in yeast enables the carrier to produce twice the amount of enzyme, thus exploiting available phosphate more efficiently when phosphate is a limiting factor to growth (Hensche 1975). Perry et al. (2007) studied variation in the number of duplicates of human salivary amylase (*AMY1*) and found that human populations that consume starch-rich diets had on average more copies of *AMY1* per individual than populations relying less on starch in their diet. This translated into higher protein levels and an enhanced ability to break down starches. As these *AMY1* duplicates are very young and extremely similar or even identical in sequence, it is likely that they are maintained by selection, consistent with the dosage model.

There is now good evidence that an increase in gene number by gene amplification can occur quite rapidly under selection pressure for increased amounts of a gene product. For example, the genome of the wild-type strain of the peach-potato aphid (*Myzus persicae*) contains two genes encoding esterases E4 and FE4. The genes are very similar in sequence (98%), indicating that they have been duplicated recently (Field and Devonshire 1998). Following exposure to organophosphorous insecticides, which can be hydrolyzed and sequestered by esterase, resistant strains of *M. persicae* were found to contain multiple copies of *E4* and *FE4*. The gene duplications and subsequent increase in the frequency of the carriers of these duplications within the aphid population are likely to have occurred within the last 50 years, with the introduction of the selective agent. This is consistent with the finding that the individual copies of each duplicated gene, both within and between aphid clones, show no sequence divergence. Gene amplification has been a frequent strategy in the evolutionary response of insect populations to natural and synthetic pesticides (Mouches et al. 1986; Vontas et al. 2000; Li et al. 2007). A similar phenomenon has been observed in bacteria in response to toxic or nutrient-poor environments or antibiotic exposure. In bacteria, gene amplifications may be mediated by transposable-element activity (Chapter 9). Human health may be negatively impacted by amplifications occurring in pathogens. One such example concerns the virulence-correlated amplification of the cholera toxin gene (*ctx*) in epidemic strains of *Vibrio cholerae* (Mekalanos 1983).

Other cases in which selection for increased dosage has been invoked as an evolutionary explanation include the genes specifying rRNAs and tRNAs, which are required for translation (see below), and the genes for histones, which constitute the main protein component of chromosomes and therefore must be synthesized in large quantities, especially during the S phase of the cell cycle, when DNA is replicated.

One prediction of the dosage-compensation model is that following gene duplication, genes for proteins known to form multicomponent complexes or to interact in any other way with other proteins will be preferentially retained in the genome. Some evidence exists that this is indeed the case (Aury et al. 2006; Hughes et al. 2007).

We note that under the increased-dosage model of evolution, the invariant duplicated copies are created rapidly because of stringent positive selection. Under the dosage-compensation model, on the other hand, the invariant copies are actively maintained invariant by stringent selective constraints and purifying selection. Intersequence invariance may also be maintained by two processes that will be dealt with later in this chapter, concerted evolution or birth-and-death evolution

In addition to dosage, Ohno (1970) proposed a second reason for the maintenance of invariant repeats. According to this proposal, a second gene can provide **functional redundancy** in case the original gene is subsequently disabled by mutation. In other words, a second copy can increase **mutational robustness**, i.e., the extent to which a function or a phenotype will remain unchanged in the face of mutations. Many researchers have addressed this model and shown that it is not likely to have played a large role in evolution except in populations with extremely large effective population sizes and very high deleterious mutation rates (Clark 1994; Lynch et al. 2001; O'Hely 2006; Hahn 2009; Price et al. 2011).

Evolution of rRNA-specifying genes

Ribosomal RNA gene copy number varies widely among organisms. Prokopowich et al. (2003) showed that a very strong correlation exists between the number of rRNA-specifying genes and genome size (**Figure 7.7**). This relationship with genome size also holds for the tRNA genes. There may be three reasons for the general positive correlation between genome size and number of copies of RNA-specifying genes. First, large genomes may require large quantities of RNA and the quantity of rRNA may be correlated positively to the number of rRNA-specifying genes. Second, the number of RNA-specifying genes may simply be a passive consequence of genome enlargement by duplication. A third intriguing possibility is that the large number of almost identical copies may be maintained for reasons unrelated to rRNA production. For example, yeast has many copies of rRNA-specifying genes that are not transcribed. Interestingly, when these "extra copies" are lost or deleted, the cells become sensitive to DNA damage induced by mutagens. Apparently, a high density of heavily transcribed genes is toxic to the cells, so in yeast, a gene amplification system has evolved for maintaining large clusters of tandemly repeated copies of mostly untranscribed rRNA-specifying genes (Ide at al. 2010).

As a rule, duplicate copies of rRNA-specifying genes tend to be very similar to one another.

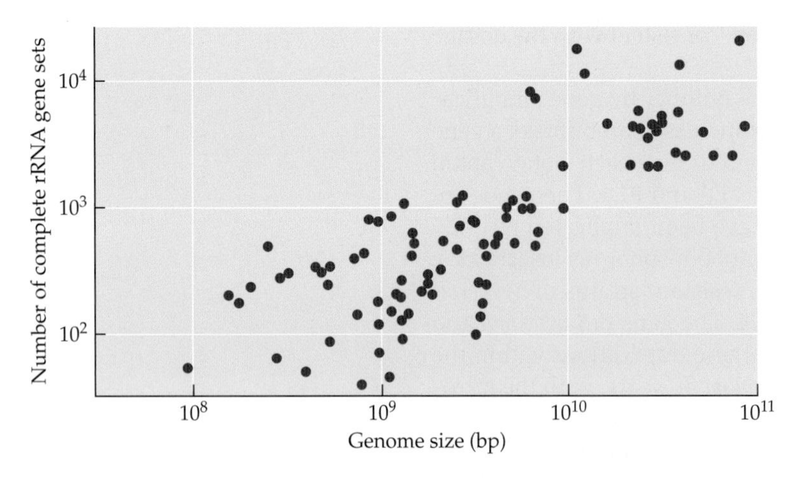

Figure 7.7 Log-log plot illustrating the relationship between genome size and number of complete rRNA gene sets in the genomes of 94 animal species. (Modified from Prokopowich et al. 2003.)

One factor responsible for the homogeneity may be purifying selection, because these genes are required to abide by very specific functional and structural requirements. However, homogeneity often extends to regions devoid of any functional or structural significance, and thus the maintenance of homogeneity requires that other mechanisms be invoked.

Neofunctionalization

Following gene duplication, one of the paralogous genes may undergo neofunctionalization, i.e., it may acquire a new function. In other words, one of the duplicates retains its original function, while the other accumulates molecular changes such that, in time, it can perform a different task (Ohno 1970). Neofunctionalization has also been called "mutation during redundancy" (Hughes 1999). We note, however, that the process of straightforward neofunctionalization, whereby a copy is freed from selective constraint and by chance acquires a different function that is then improved by natural selection, may be extremely infrequent in evolution because (1) functional changes usually require a large number of amino acid replacements, and (2) it is infinitely easier to destroy the function of a gene by mutation than to confer on it a new function by chance. Thus, neofunctionalization should be a minority mechanism as far as the retention of duplicated genes is concerned. The degree to which neofunctionalization can "win out" over nonfunctionalization depends on (1) the number of mutations required to attain the new function and (2) the possible paths within the sequence space to the new function. If the change in function can be achieved by a single amino acid replacement, such as the serine-to-proline replacement that converts the enzyme guanylate kinase into a mitotic spindle-orienting protein (Johnston et al. 2011), then the chances of the change becoming fixed in the population seem considerable. In cases where more than a single change is required, the probability of neofunctionalization decreases rapidly as the number of changes increases. An additional means by which neofunctionalization may be achieved is a single mutation that has big consequences. In the reading frame of protein-coding genes, frameshift mutations resulting from the deletion or insertion of DNA sequences that are not multiples of three nucleotides will have a huge effect on the amino acid sequence. The probability that such a mutation will have a beneficial effect and be fixed in the population is, of course, minuscule. Notwithstanding these almost insurmountable difficulties, neofunctionalization by frameshifting of reading frames does occur in nature (e.g., Okamura et al. 2006; Liu and Adams 2010).

Neofunctionalization may occur through either changes in the coding region or changes in the regulatory elements. Neofunctionalization may refer to minor changes in function, such as quantitative changes in expression pattern and timing, or it may refer to major changes in function, such as changes in substrate and ligand affinity, subunit structure, or subcellular compartmentalization. Some neofunctionalization processes may even turn an enzyme into a structural protein or vice versa.

What are the telltale signs of neofunctionalization in the case of paralogous genes? First, the expectation is that the rates and patterns of evolution of the two duplicate genes will be highly asymmetrical. Second, we expect the ratio of nonsynonymous to synonymous substitution to differ between the two copies. Finally, we expect the faster-evolving paralog to have a higher ratio of nonsynonymous to synonymous substitutions than the other paralog or singleton orthologs from other species. In other words, we expect the fast-evolving paralog to exhibit evidence of an increase in positive Darwinian selection. Comparative studies indicate that neofunctionalization may exert a significant role in the evolution of about 10% of newly duplicated paralogs in primates (Han et al. 2009) as well as in the maintenance of about 6% of the retained genes following the tetraploidization event that gave rise to *Xenopus laevis* (Chain and Evans 2006). In maize, approximately 13% of all the gene pairs that have been created by genome duplication exhibit regulatory neofunctionalization, i.e., are expressed differently in leaves (Hughes et al. 2014). As far as human-specific duplicates are concerned, neofunctionalization may have had an important role in the evolution of genes involved in neuronal and cognitive functions.

In *Drosophila*, neofunctionalization seems to account for the retention of almost two-thirds of duplicate genes (Assis and Bachtrog 2013). Because most gene duplications in *Drosophila* result in the creation of retrogenes, i.e., occur via reverse transcription of mRNA (Chapter 9), it is possible to distinguish between the **parent copy** and the **child** or **daughter copy**. (Such a distinction is not possible if the gene duplication event creates two tandemly arranged copies.) Surprisingly, novel functions nearly always originate in child copies, whereas the function of parent copies persists unaltered. Many young child copies are expressed primarily in the testes, which may be a reflection of the transcriptionally permissive environment in the testes rather than a testament to their functionality (Chapter 8). Nevertheless, the expression breadth, i.e., the number of tissues in which child copies are transcribed, increases over evolutionary time. This finding supports the so-called **out-of-testes hypothesis** (Kaessmann 2010), which posits that testes are a catalyst for the emergence of new genes that ultimately evolve functions in other tissues.

The process of neofunctionalization can be sped up considerably by very strong positive selection, such as that exerted by insecticides. This is particularly true if the acquisition of the novel function can be achieved by very few mutations. One such example was discovered in the sheep blowfly (*Lucilia cuprina*), where a single glycine-to-aspartic acid replacement conferred insecticide resistance by turning a carboxylesterase into an organophosphorous hydrolase (Newcomb et al. 1997).

Here we illustrate the process of neofunctionalization through an example involving a change in substrate specificity of an enzyme.

THE NEOFUNCTIONALIZATION OF MALATE DEHYDROGENASE INTO LACTATE DEHYDROGENASE FOLLOWING GENE DUPLICATION An interesting example of neofunctionalization involves the emergence of a lactate dehydrogenase enzyme following a duplication involving a gene encoding malate dehydrogenase in *Trichomonas vaginalis*, the causative agent of the most common pathogenic protozoan infection of the urogenital tract in industrialized countries (**Figure 7.8**). Malate dehydrogenase catalyzes the interconversion of oxaloacetate and malic acid with the concomitant interconversion of NADH and NAD^+. Lactate dehydrogenase, on the other hand, catalyzes a similar but distinct reaction, the inter-

(a)

L-Malate ($^-OOC-CH_2-CH(OH)-COO^-$) + NAD^+ $\overset{MDH}{\rightleftharpoons}$ Oxaloacetate ($^-OOC-CH_2-C(=O)-COO^-$) + NADH + H^+

L-Lactate ($CH_3-CH(OH)-COO^-$) + NAD^+ $\overset{LDH}{\rightleftharpoons}$ Pyruvate ($CH_3-C(=O)-COO^-$) + NADH + H^+

(b)

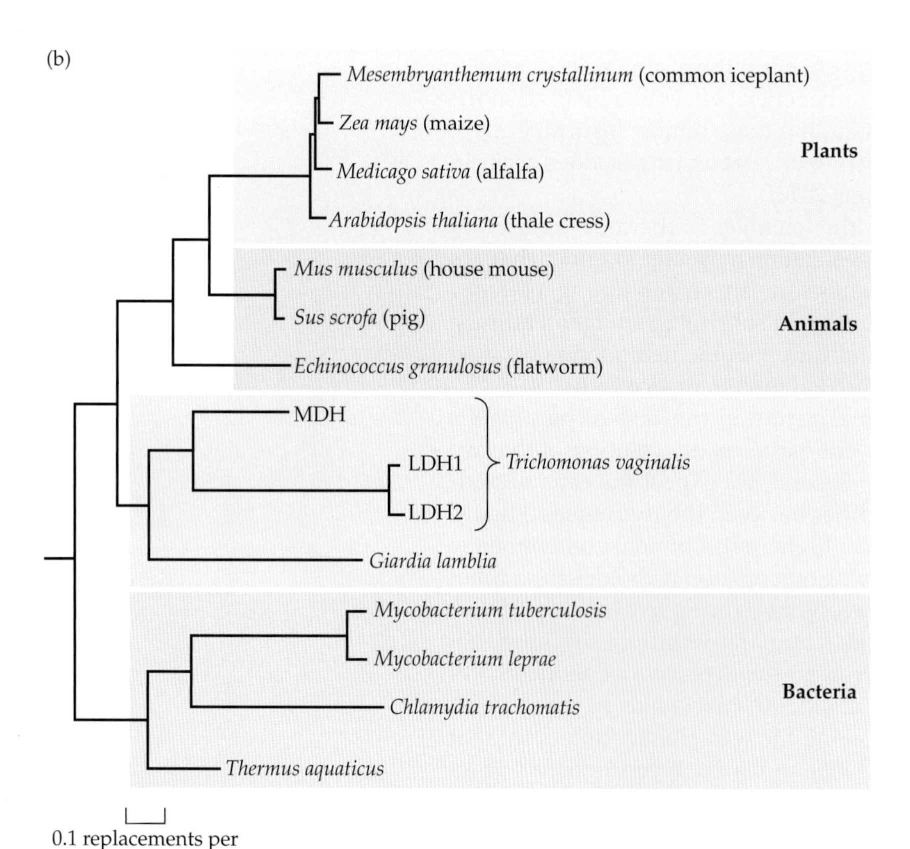

Figure 7.8 (a) The reactions catalyzed by malate dehydrogenase (MDH) and lactate dehydrogenase (LDH). (b) Phylogenetic relationships among malate dehydrogenases from animals, plants, unicellular eukaryotes, and bacteria showing the nested position of the lactate dehydrogenase of *Trichomonas vaginalis* within the malate dehydrogenase family.

conversion of pyruvate and lactic acid with the same concomitant interconversion of NADH and NAD+. A genomic analysis of *T. vaginalis* revealed the existence of two very similar copies of the lactate dehydrogenase gene, each encoding a 333-amino-acid-long enzyme. Surprisingly the two genes, which are clearly the result of a gene duplication, turned out to be closely related to the malate dehydrogenase gene family rather than to the lactate dehydrogenase family. It seems that a recent gene duplication of the original malate dehydrogenase gene created two copies of the gene, one of which was neofunctionalized into a lactate dehydrogenase gene, most probably through a nucleotide substitution that resulted in a leucine-to-arginine replacement at amino acid position 91. Subsequently, the lactate dehydrogenase gene was duplicated again, resulting in two genes, *LDH1* and *LDH2*. The neofunctionalization of malate dehydrogenase to lactate dehydrogenase seems to be quite common in nature, as it has also been recorded in other organisms, such as *Bacillus stearothermophilus* (Wilks et al. 1988).

Multifunctionality and subfunctionalization

During the pre-double helix era (i.e., from the 1910s to the early 1950s) a group of scientists put forward the idea, very controversial at the time, that genes are functionally divisible entities, i.e., that individual genes have not one but a set of functions, that each function is performed by an independently mutable region at the locus, and that each function is separable from the other functions (Emerson 1911; Dubinin 1929; Serebrovsky and Dubinin 1929; Agol 1930; Serebrovsky 1930; Muller 1932; Raffel and Muller 1940; Verderosa and Muller 1954).

Multifunctional proteins, as their name implies, perform more than one function. Multifunctional proteins were thought at first to be exotically rare, but recent evolutionary, biochemical, and genomic studies have revealed that protein multifunctionality is the norm in nature rather than the exception (Khersonsky and Tawfik 2010). **Moonlighting proteins** constitute a subset of multifunctional proteins in which the two or more functions cannot be ascribed to the fusion of protein-coding genes with distinct functions (Copley 2012). In addition, multifunctionality due to mRNA alternative splicing or proteolysis resulting in independently functional proteins is excluded from the definition of "moonlighting." (The evolution of non-moonlighting multifunctional proteins will be discussed in Chapter 8.)

The first examples of moonlighting proteins were recognized in the late 1980s. For example, neuroleukin, a protein found in human skeletal muscles, brain, and bone marrow, which prolongs the life of embryonic nerve cells, was found to function also as phosphoglucose isomerase. Crystallins in the lenses of vertebrates were shown to be identical to various metabolic enzymes, and glyceraldehyde 3-phosphate dehydrogenase in several *Streptococcus* species was found to function as a receptor for fibronectin on the bacterial surface, suggesting that it might play a role in colonization of the pharynx. Moonlighting is not only a fairly common phenomenon (e.g., Hittinger and Carroll 2007; Des Marais and Rausher 2008); it also blurs the traditional distinctions between different types of proteins, such as the difference between enzymes and structural proteins.

Subfunctionalization can most broadly be defined as the partitioning of the set of ancestral functions into different subsets among the duplicated descendants (Ohno 1970; Hahn 2009), and it is predicated on the ancestor gene producing a multifunctional product. Subfunctionalization provides an appealing explanation for the ubiquity of duplicated genes in eukaryote genomes, because the model does not require each duplication event to immediately confer a selective advantage. Because our knowledge of the functions of the ancestral gene is limited, however, it is often difficult to pinpoint different subfunctions of the ancestral gene that have been partitioned among its duplicated descendants.

Many scenarios of subfunctionalization have been described in the literature (Innan and Kondrashov 2010). Here, we will discuss several such examples: subfunctionalization of non-moonlighting multifunctional genes, specialization, gene sharing and escape from adaptive conflict, duplication-degeneration-complementation, and segregation avoidance. We note that the different subfunctionalization scenarios are not mutually exclusive.

SUBFUNCTIONALIZATION OF NON-MOONLIGHTING MULTIFUNCTIONAL GENES Cusack and Wolfe (2007) studied a gene in which the ancestral functions being partitioned among the daughter genes are distinct and readily identifiable because they have nothing in common with each other except the fact that two products of the gene are exported to the chloroplast.

In most flowering plants, the gene for plastid ribosomal protein L32 (*RPL32*) is located in the chloroplast genome. In the lineage leading to the order Malpighiales (a large, heterogeneous plant taxon consisting of such diverse species as poplars, mangroves, cassava, poinsettia, flax, passion fruit, willows, violets, and pansies), however, the plastid gene was transferred to the nuclear genome after the divergence of the lineage from the orders Cucurbitales (which includes the gourd and begonia families) and Fabales (legumes). In these taxa, plastid ribosomal protein L32 is produced in the nucleus and later inported into the chloroplast.

As seen in **Figure 7.9**, in the Burma mangrove (*Bruguiera gymnorrhiza*), which represents the ancestral state, the one-exon *RPL32* gene was inserted in the intron separating exons 7 and 8 of the *SOD* gene. The resulting chimeric gene is alternatively spliced, producing one transcript for *SOD* and one for *RPL32*. In the lineage leading to the poplars (represented in the study by the western balsam poplar, *Populus trichocarpa*, and the silver-leaf poplar, *P. alba*), the *SOD-RPL32* chimeric gene was duplicated twice. The first duplication resulted in one copy that lost the *RPL32* exon and one copy that subsequently lost the ability to encode SOD. Thus, the first duplication resulted in the subfunctionalization of the chimeric gene, producing daughter genes that produce either RPL32 or SOD, but not both.

The RPL32-coding gene (*Poplar1*) has retained a continuous open reading frame derived from the former *SOD* gene. However, the product cannot be a functional SOD protein, as two of its exons have been affected by internal deletions and three exons have disappeared altogether (one was pseudogenized

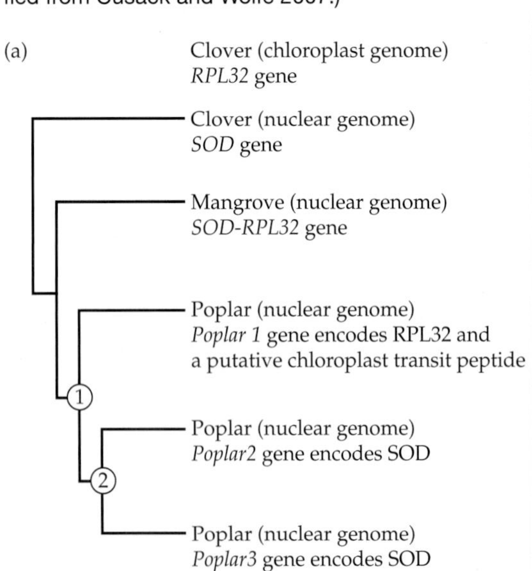

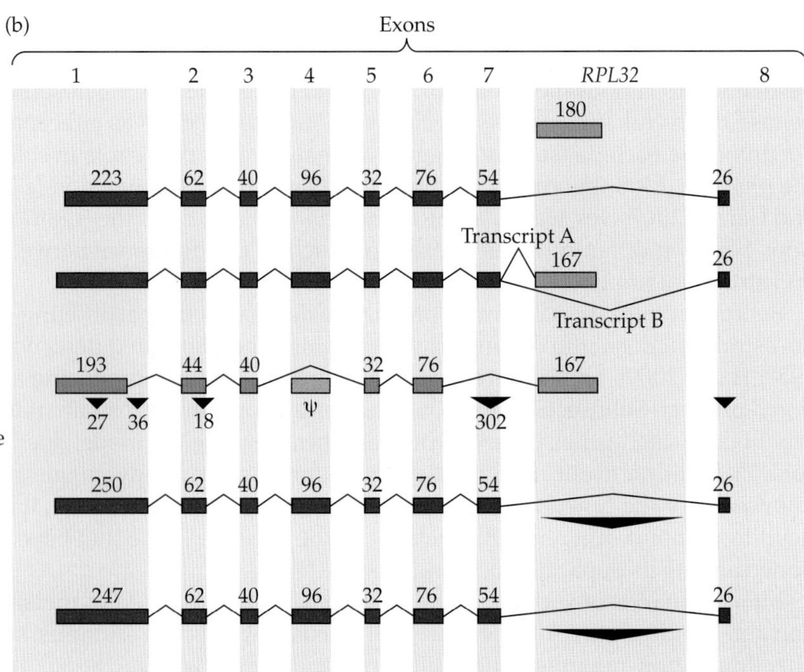

Figure 7.9 Evolution by subfunctionalization in the plant order Malpighiales. (a) The branching order of the nuclear genes is based on pairwise synonymous nucleotide substitution distances. Node 1 represents the first gene duplication in poplar, resulting in *Poplar1* and *Poplar2*. Node 2 is thought to represent a large segmental (or whole-genome) duplication in poplar, because many of the genes neighboring *Poplar2* have homologs neighboring *Poplar3*. (b) Organization of the exons of chloroplast *RPL32* gene and the nuclear *SOD* gene in the barrel clover (*Medicago truncatula*), the *SOD-RPL32* chimeric gene in the Burma mangrove (*Bruguiera gymnorrhiza*), and the *SOD*- and RPL32-coding genes in two species of poplars (*Populus trichocarpa* and *P. alba*). Boxes represent exons and a pseudoexon; the lengths (in bp) of some exons are shown. Introns are not drawn to scale. Red boxes show SOD-coding exons; blue boxes show RPL32-coding exons. Nonframeshifted remnants of five SOD-coding exons are shown in green. The pseudoexon in *Poplar1* (ψ) is shown as a gray box. Triangles indicate sequences deleted in the *poplar* genes (with deletion lengths where known). (Modified from Cusack and Wolfe 2007.)

and two have been deleted). The greatly abbreviated protein is predicted to be a chloroplast transit peptide, i.e., it has retained the ability to transport to the chloroplast while losing its enzymatic activity. Thus, the evolutionary history of *Poplar1* illustrates a second case of subfunctionalization.

The SOD-coding gene later became duplicated a second time, producing two virtually identical genes (*Poplar2* and *Poplar3*). Analyses of gene expression have shown that all three poplar genes are transcribed and that none of them is alternatively spliced anymore. The partitioning process of *SOD-RPL32* in *Populus* can, therefore, be unambiguously categorized as subfunctionalization, because a complementary loss of subfunctions of the ancestral chimeric gene occurred in its first two descendant duplicates.

SPECIALIZATION As a result of specialization, the products of the duplicated genes, although they continue to perform essentially the same function, acquire properties that distinguish them from each other (Ohno 1970; Otto and Yong 2002). For example, while the ancestral gene may have performed its function in all tissues, developmental stages, and environmental conditions, its descendants become specialists, dividing among themselves the different tissues, developmental stages, and environmental conditions (Markert 1964; Ferris and Whitt 1979). Thus, one descendant copy may specialize in catalyzing a certain reaction in the larval liver, whereas the second copy may specialize in catalyzing the same reaction in the adult heart. The specialization model involves positive selection, because tissue, developmental, or any other type of specialization essentially involves the improvement of function.

Good examples of variant repeats created through the process of specialization are families of genes coding for isozymes, such as lactate dehydrogenase, aldolase, creatine kinase, carbonic anhydrase, and pyruvate kinase. **Isozymes** are enzymes that catalyze the same biochemical reaction but may differ from one another in tissue specificity, developmental regulation, electrophoretic mobility, or biochemical properties. Note that isozymes are encoded at different loci, generally by duplicated genes; this is in contrast to **allozymes**, which are distinct forms of the same enzyme encoded by different alleles at a single locus. The study of multilocus isozyme systems has greatly enhanced our understanding of how cells with identical genetic endowment can differentiate into hundreds of different specialized types of cells that constitute the complex body organization of vertebrates. Although all members of an isozyme family serve essentially the same catalytic function, different members may have evolved particular adaptations to different tissues or different developmental stages, thus enhancing the physiological fine-tuning of the cell.

Mammalian lactate dehydrogenase (LDH) is a tetrameric enzyme that catalyzes the reversible interconversion of pyruvate and lactate, a key step in glycolysis and other metabolic pathways. Six paralogous *LDH* genes have been found in the human genome: *LDH-A*, *LDH-C*, and *LDH-6A* on chromosome 11; *LDH-B* and *LDH-6C* on chromosome 12; and *LDH-6B* on chromosome 15. In comparison with the *LDH-A*, *LDH-B*, *LDH-C*, and *LDH-6A* genes, which each possess seven exons, the *LDH-6B* and *LDH-6C* genes are intronless, indicating that these latter two genes may have been created through retroposition (Chapter 9).

Judging by the taxonomic distribution of the various *LDH* genes, the duplication that gave rise to the A and B isozymes occurred in an early vertebrate ancestor subsequent to the divergence of tunicates and lampreys. The duplication that gave rise to the C subunit is much more recent, as the gene for the C subunit was only found in placental and marsupial mammals but is absent from birds and egg-laying monotremes (Stock et al. 1997; Holmes and Goldberg 2009).

The A and B subunits form five tetrameric isozymes, A_4, A_3B, A_2B_2, AB_3, and B_4, all of which catalyze either the conversion of lactate into pyruvate in the presence of the oxidized coenzyme nicotinamide adenine dinucleotide (NAD^+) or the reverse reaction in the presence of the reduced coenzyme (NADH). It has been shown that B_4 and the other isozymes rich in B subunits, which have a high affinity for NAD^+,

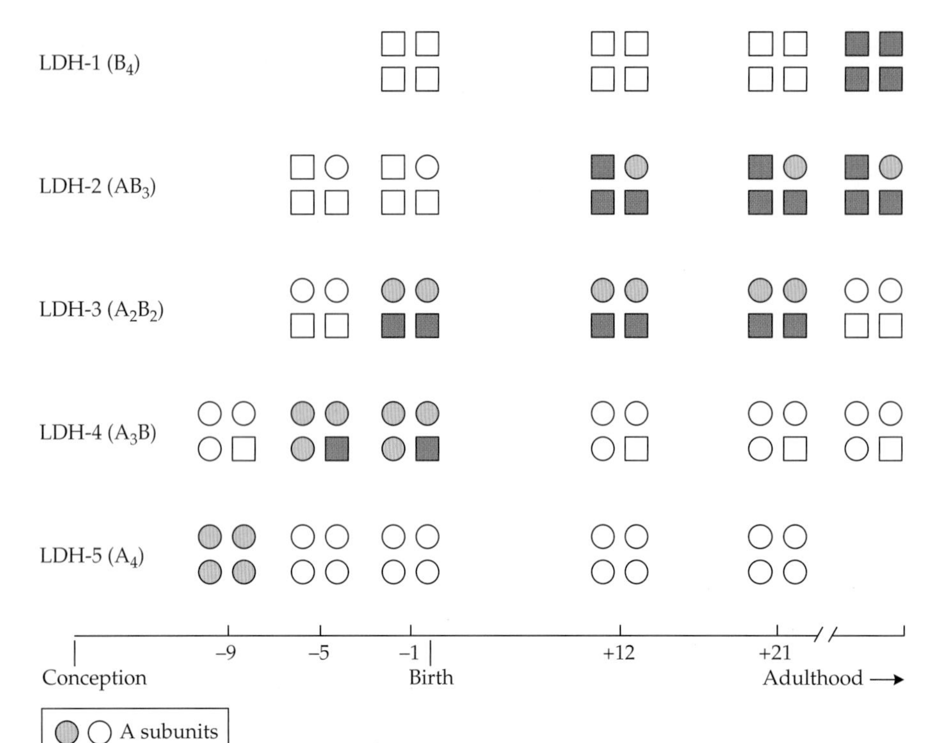

Figure 7.10 Developmental sequences of five lactate dehydrogenase (LDH) isozymes in the rat heart from conception to adulthood. Negative and positive numbers denote days before or after birth, respectively. Colored symbols indicate quantitatively predominant forms. Notice the progressive shift from a preponderance of A subunits (circles) before birth to B subunits (squares) afterward. (Data from Markert and Ursprung 1971.)

function as true lactate dehydrogenase in aerobically metabolizing tissues such as the heart, whereas A_4 and the isozymes rich in A subunits, which have a high affinity for NADH, are especially geared to serve as pyruvate reductases in anaerobically metabolizing tissues such as skeletal muscle. An interesting feature of LDH is that the two subunits can form heteromultimers, thus further increasing the physiological versatility of the enzyme.

Figure 7.10 shows the developmental sequence of LDH production in the heart. We see that the more anaerobic the heart is (specifically, in the early stages of gestation), the higher the proportion of LDH isozymes rich in A subunits will be. Thus, the two duplicate genes have become specialized by different tissues and to different developmental stages by subfunctionalization of expression.

GENE SHARING AND ESCAPE FROM ADAPTIVE CONFLICT A single-copy gene may acquire a new function in addition to its original function. If the two functions become essential for the organism, the gene will be subjected to two or more independent sets of evolutionary pressure. In some cases, the distinct functions may be incapable of further improvement through selection because of detrimental pleiotropic effects of one function on the other function. That is, the multiple functions of the protein cannot be simultaneously optimized by natural selection. This situation has been dubbed **adaptive conflict**. **Gene sharing** (Piatigorsky et al. 1988) refers to the case in which a single gene possesses two or more distinct functions that are not independently selectable.

Gene sharing was first discovered in crystallins, the major water-soluble proteins in the eye lens whose function is to maintain lens transparency and proper light diffraction (Wistow et al.1987; Wistow and Piatigorsky 1987). Crystallins perform their lenticular function by forming a uniform concentration gradient, with the highest protein concentration at the center of the lens. Since the structure and function of eye lenses are almost identical among all vertebrates, it was assumed that all vertebrate crystallins would be correspondingly similar. Comparative studies, however, have revealed an unexpected sequence diversity and taxon specificity among lens crystallins.

Many crystallins turn out to be moonlighting proteins (**Table 7.3**). For example, α-crystallins, which are present in all vertebrate lenses, also act as heat shock proteins and molecular chaperones protecting proteins against physiological stress. Crystallin ε from ducks and crocodiles turns out to be identical in its amino acid sequence to lactate dehydrogenase B and to possess identical enzymatic activity (Wistow et al. 1987). Subsequent work has shown that these "two" proteins are in fact one and the same, encoded by the same single-copy gene (Hendriks et al. 1988). Consistent with the idea that many lens crystallins have additional functions, they are also expressed outside of the lens.

Which function came first? Since the eye is a relatively recent evolutionary invention, whereas metabolic enzymes are ancient proteins, it is presumed that the enzymatic functions came first and the lenticular function later. We also note that enzymes are strictly constrained and fine-tuned by function (e.g., susbstrate recognition and proper catalysis), while crystallins need merely be hydrophobic and maintain a proper

Table 7.3

Examples of gene-sharing and paralogous relationships in taxon-specific eye lens crystallins

Crystallin type	Taxonomic distribution	Gene-sharing enzyme	Paralog
ε	Crocodiles, ducks	Lactate dehydrogenase B	–
ζ	Guinea pigs, degus, rock cavies, camels, llamas	NADPH:quinone oxidoreductase	–
η	Elephant shrews	Aldehyde dehydrogenase I	–
δ	Ducks, chickens	Argininosuccinate lyase	Argininosuccinate lyase
λ	Rabbits, hares	–	Hydroxyacyl CoA dehydrogenase
μ	Kangaroos, quolls	–	Ornithine cyclodeaminase
ρ	Frogs	–	NAPDH-dependent reductase

Source: Modified from Wistow (1993).

size. Thus, crystallins are "borrowed" or "recruited" proteins. As opposed to the case of the mammalian *LDH* genes discussed above, the recruitment of preexisting enzymes as crystallins illustrates a model of molecular evolution in which the evolution of new functions and new patterns of gene expression occur before (rather than after) gene duplication.

Adaptive conflict can be resolved through two main pathways (**Figure 7.11**). First, the conflict can be resolved by reversal, i.e., the loss of one function, usually the new function. This situation is illustrated by the loss of the δ-crystallin function in a bird, the chimney swift (*Chaetura pelagica*). Second, the conflict can be resolved by gene duplication and separation of function by subfunctionalization. This situation is illustrated by the gene duplication and subsequent separation of function between argininosuccinate lyase and δ-crystallin in chicken. Of the two tandemly arranged chicken δ-crystallin genes, *δ1* became specialized for lens expression and produces more than 95% of the lens δ-crystallin mRNA; by contrast, the *δ2* gene, which encodes the enzymatically active argininosuccinate lyase, produces most of the mRNA in nonlens tissues. Interestingly, an intermediate stage has also been found. In duck there are two almost identical δ-crystallin/argininosuccinate lyase genes, which have most probably been derived from a very recent gene duplication event (Piatigorsky 1998a,b). Thus following the emergence of gene sharing, a gene duplication has occurred, but it has not yet been accompanied by functional divergence.

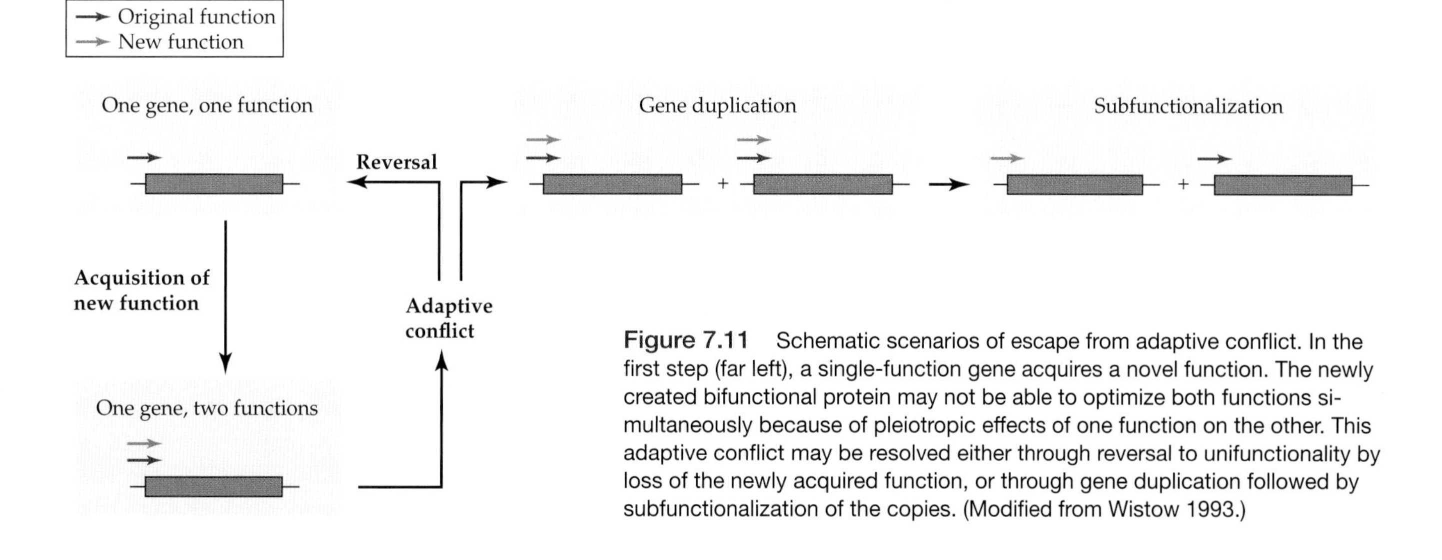

Figure 7.11 Schematic scenarios of escape from adaptive conflict. In the first step (far left), a single-function gene acquires a novel function. The newly created bifunctional protein may not be able to optimize both functions simultaneously because of pleiotropic effects of one function on the other. This adaptive conflict may be resolved either through reversal to unifunctionality by loss of the newly acquired function, or through gene duplication followed by subfunctionalization of the copies. (Modified from Wistow 1993.)

In recent years, the "escape from adaptive conflict" scenario has been validated empirically through the reconstruction and synthesis of ancestral enzymes. In one such case, the ancestral gene encoding a large family of fungal glucosidase genes has been reconstructed in the laboratory (Voordeckers et al. 2012). The original enzyme was primarily active on the disaccharide maltose, which consists of two glucose molecules linked by an $\alpha(1\rightarrow4)$ bond. The enzyme also exhibited some lesser activity on a related disaccharide, isomaltose, in which the two glucose molecules are linked by the slightly different $\alpha(1\rightarrow6)$ bond (Robinson 2012). Structural analyses and activity measurements on resurrected and present-day enzymes suggested that the two activities are incompatible, i.e., they cannot be both fully optimized concomitantly. The "escape from adaptive conflict" was possible through gene duplication followed by the accumulation of substitutions that in some copies optimized the affinity of the enzyme for maltose, while in other copies it optimized the affinity of the enzyme for isomaltose.

Additional evidence for this scenario comes from the crystallins. In humans, the single-gene encoded αB-crystallin performs a chaperone activity in addition to its lenticular function and is expressed in many tissues. In the zebrafish, on the other hand, αB-crystallin is expressed mostly in the lens and its chaperone-like activity is much reduced compared with that of its human ortholog. Interestingly, zebrafish has two nonidentical copies of genes encoding αB-crystallin-like proteins (Smith et al. 2006). The two copies have evolved through gene duplication, after which one paralog maintained its widespread chaperone role while the other adopted a more restricted, nonchaperone role in the lens.

It has been claimed that evolution of most new functions requires a period of gene sharing in which a single protein must serve both its original function and a new function that has become advantageous to the organism. Subsequent gene duplication and subfunctionalization allow one copy to maintain the original function, while the other diverges to optimize the new function. McLoughlin and Copley (2008) have provided some experimental evidence for this hypothesis.

There seems to be a pleasant historical subtext to the story of gene sharing and crystallin evolution, for it has contributed to the partial unraveling of a puzzle that has distressed students of evolution from Darwin onward. It concerns the evolution of the eye and the possibility of evolving a highly complex, totally novel character by natural means. In *The Origin of Species,* in a chapter entitled "Organs of Extreme Perfection and Complication," Charles Darwin wrote, "To suppose that the eye, with all its inimitable contrivances ... could have been formed by natural selection, seems, I freely confess, absurd to the highest possible degree." At least as far as the lens and the cornea of the eye are concerned, it seems that their "inimitability" is not all that inimitable and that some of their constituent proteins can be traced back in evolutionary history to ubiquitous housekeeping enzymes that are found throughout all life-forms on Earth.

DUPLICATION-DEGENERATION-COMPLEMENTATION The duplication-degeneration-complementation process (Force et al. 1999; Lynch et al. 2001) refers to the post-duplication complementary loss of different subfunctions from duplicated genes, after which both copies are needed to maintain the original function. This is illustrated in **Figure 7.12**. In one version of the duplication-degeneration-complementation model, the two duplicate paralogs perform the same function but are expressed at lower levels than that of the original gene, so after duplication both copies are required to produce a sufficient amount of the gene product. Duplication-degeneration-complementation is a selectively neutral process, because the degeneration process merely reflects the accumulation of mutations that would have been deleterious in the original single-copy gene but are neutral because the subfunction destroyed in one copy is complemented by the corresponding intact subfuction in the second copy.

Here, we illustrate the duplication-degeneration-complementation process by using one example of gene duplication in yeast (Sommerhalter et al. 2004; van Hoof 2005). Ribonucleotide reductase is an enzyme that catalyzes the reduction of ribonucleotides to the corresponding deoxyribonucleotides. Ribonucleotide reductase is responsible for maintaining proper levels of DNA precursors for replication and repair and, therefore, plays a crucial and conserved role in all organisms. In eukaryotes the enzyme is a tetramer consisting of two subunits, R1 and R2, with R2 using a diiron (Fe_2) as a prosthetic group. Each of the two subunits is a dimer of the proteins RNR1 and RNR2, respectively.

In the yeast species *Saccharomyces kluyveri*, RNR2 is encoded by a single gene, so the R2 subunit is, by necessity, a homodimer composed of two identical polypeptides. In *S. cerevisiae*, the gene encoding RNR2 has undergone duplication, most probably together with the entire genome. As a result, *S. cerevisiae* has two paralogs, RNR2 and RNR4. Thus, *S. cerevisiae* can produce three R2 subunits, two homodimers (of either RNR2 or RNR4) and a heterodimer composed of one copy of RNR2 and one of RNR4. Interestingly, only the heterodimeric R2 is functional. The reason is that RNR4 lacks the proper residues to accommodate the diiron group, while in RNR2 an invariant phenylalanine has been replaced by threonine, causing structural problems in one of the iron-binding structures. The defects are evidently complementary, since the replacement of *RNR2* gene by *RNR4* or the replacement of *RNR4* by *RNR2* in *S. cerevisiae* result in various degrees of inviability, while the *RNR2* from *S. kluyveri* can replace either and both genes in *S. cerevisiae.*

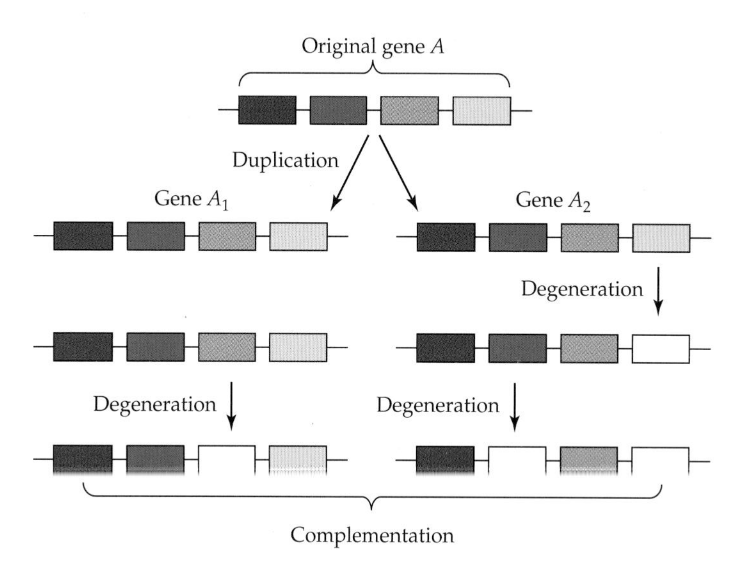

Figure 7.12 Schematic representation of the duplication-degeneration-complementation process. Each box represents a subfunction, with colored boxes representing intact subfunctions and empty boxes representing subfunctions that have been obliterated. The partially functioning duplicated genes A_1 and A_2 are both maintained in the population because the complete set of functions of the original, unduplicated gene can only be met by the two "degenerate" genes together.

SEGREGATION AVOIDANCE The premise of the segregation avoidance model (Spofford 1969) is simple. If overdominant selection occurs at a locus via heterozygous advantage, then lower-fitness homozygotes will be produced each generation regardless of the strength of selection (Chapter 2). This is because it is impossible to maintain populations composed of heterozygotes only. The mean fitness of the population will, therefore, be lower than the highest possible fitness—a phenomenon called **segregational load**. But if an unequal crossing over occurs in a heterozygote that creates a locus duplication containing the two alleles, then individuals can attain a situation akin to permanent heterozygosity, and the population will avoid segregational load (**Figure 7.13**). Hahn (2009) made the astute observation that the segregation avoidance model is not easily classifiable. Is it an instance of retention of ancestral function (gene conservation), because no change to the ancestral sequences was required? Or, is it an instance of subfunctionalization, because the multiple functions of a single locus are subsequently carried out by two loci?

An example of segregation avoidance occurs in the *ace-1* locus of the house mosquito, *Culex pipiens* (Labbé et al. 2007). This locus encodes acetylcholinesterase, the target of organophosphate pesticides. Two alleles are known to segregate at this locus: a susceptible wild-type allele, *ace-1*^{S}, and a resistant allele, *ace-1*^{R}. There is only one amino acid difference between the protein produced by the resistant allele and the protein produced by the wild-type allele, a glycine versus serine at amino acid position 119 of *ace-1*^{R} and *ace-1*^{S}, respectively. Carriers of *ace-1*^{R} have a greatly reduced susceptibility to organophosphate insecticides, thereby increasing their fitness when exposured to the pesticide. Concomitantly, the *ace-1*^{R} allele drastically

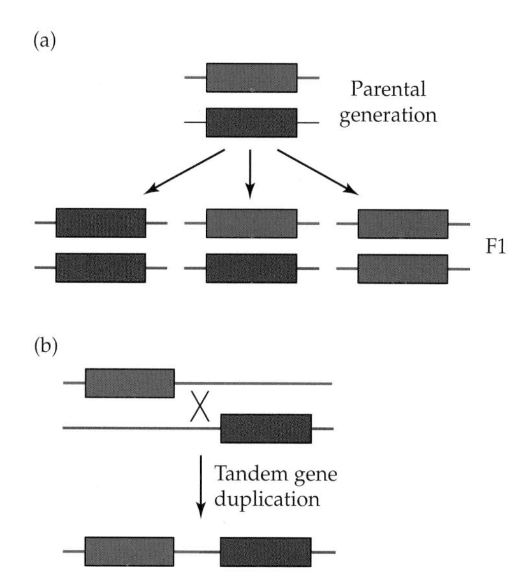

Figure 7.13 A heterozygote at a locus (blue and red rectangles) produces both homozygotes and heterozygotes. (a) If the heterozygote has a higher fitness than either homozygote, a segregational load will be incurred. (b) If, on the other hand, a gene duplication occurs through unequal crossing over, a situation akin to permanent heterozygosity is attained.

reduces the fitness of its carriers in the absence of organophosphate insecticides. Interestingly, a duplication combining resistant and susceptible alleles of the *ace-1* locus (*ace-1*^{D}) has recently appeared in the population. The duplication combines the advantages of the resistant allele at times of pesticide exposure with the advantages of the susceptible allele at all other times, thereby diminishing segregational load.

Segregation avoidance may also explain the evolution of resistance to the acutely toxic organophosphate insecticide monocrotophos in the two-spotted spider mite, *Tetranychus urticae*, through the gene duplication of the acetylcholinesterase gene (Kwon et al. 2009).

Neosubfunctionalization

The subfunctionalization mechanisms described above are not mutually exclusive, although there are other examples that clearly are. Moreover, one can envision hybrid models, such as **neosubfunctionalization**, whereby a new function evolves as a consequence of the subfunctionalization of one of the copies (e.g., He and Zhang 2005; Marcussen et al. 2010).

NEOSUBFUNCTIONALIZATION OF A DUPLICATED PANCREATIC RIBONUCLEASE GENE Colobine monkeys are unique among primates in their use of leafy matter as their primary food source. Similar to ruminants, the colobines have digestive systems in which the leaves are fermented by bacteria in the foregut, and the colobines recover nutrients by breaking up and digesting the bacteria with various enzymes. Here we describe the evolution of pancreatic ribonuclease, which is secreted from the pancreas and transported into the small intestine to degrade RNA. For the colobines, one reason to use pancreatic ribonuclease as a digestive enzyme may be related to the fact that rapidly growing bacteria have huge amounts of nitrogen in their RNA, and high concentrations of ribonuclease are needed to break down bacterial RNA so that nitrogen can be acquired.

Zhang et al. (2002) found that while 15 noncolobine primates have a single pancreatic ribonuclease gene (*RNASE1*), one colobine monkey, the leaf-eating douc langur (*Pygathrix nemaeus*), has a duplicate gene (*RNASE1B*). The unequal crossing over that gave rise to the two genes was inferred to have occurred 2.4–6.4 million years ago. There are 12 nucleotide differences between the coding regions of the two genes, and all 12 nucleotide substitutions giving rise to these differences occurred in the *RNASE1B* lineage. Of the 12 substitutions, 2 were synonymous and 10 were nonsynonymous, a strong additional indication that the *RNASE1B* gene evolved under positive Darwinian selection. Of the 10 nonsynonymous substitutions, one occurred in the signal peptide and 9 occurred in the mature protein. Of these, 7 involved charge changes. Interestingly, all 7 charge-altering substitutions increased the negative charge of the protein (5 from positively charged amino acids to uncharged amino acids, and 2 from uncharged amino acids to negatively charged amino acids).

The charge-altering substitutions reduced the net charge of the *RNASE1B*-encoded protein and the isoelectric point from 9.1 to 7.3. Because RNA (the substrate) is negatively charged, the net charge of the *RNASE1B*-encoded protein should influence its interaction with the substrate and its catalytic performance. Indeed, the optimal pH for the douc langur *RNASE1*-encoded enzyme was 7.4; the optimal pH for the *RNASE1B*-encoded enzyme was 6.3. At the lower pH, the *RNASE1B*-encoded enzyme digested RNA six times as fast as its *RNASE1* counterpart. These results suggest that the amino acid substitutions in one of the duplicates, *RNASE1B*, were driven by positive selection for enhanced enzymatic activity in the much more acidic environment of the colobine small intestine.

Interestingly, the adaptive evolution of RNASE1B came at a price. RNASE1 has the ability to degrade double-stranded RNA (dsRNA) as well. The ability of *RNASE1B* to degrade dsRNA is almost completely lost (merely 0.3% of the activity of RNASE1). As it happens, the nine substitutions that enabled the *RNASE1B*-encoded enzyme to be active in a highly acidic environment have concomitantly reduced its dsRNA enzymatic activity. Of course, *RNASE1B* can afford to loose its dsRNA activity, because its paralog, *RNASE1*, retained it. We thus end up with a case that can be described by a mixed evolutionary model of neofunctionalization and subfunctionalization.

Rates of Evolution in Duplicated Genes

Following a gene duplication event, the resultant duplicates will start to diverge, although occasionally they may be partially or wholly homogenized by gene conversion (p. 302). Since duplication creates redundancy, a commonly held view is that the rate of evolution will be accelerated by relaxation of functional constraint. Lynch and Conery (2000) conducted the first genome-wide analysis of duplicate genes, using data from five animals, two plants, and yeast, and found an elevated rate of nonsynonymous substitution in a large proportion of duplicated genes. This observation was confirmed by subsequent studies (Nembaware et al. 2002; Jordan et al. 2004; Scannell and Wolfe 2008) and has been interpreted to mean that duplicated genes are subject to weaker purifying selection than single-copy genes (Kondrashov et al. 2002).

It has also been observed that duplicated genes often exhibit dissimilar evolutionary rates, i.e., one copy evolves more slowly than the other (Van de Peer et al. 2001; Conant and Wagner 2003; Zhang et al. 2003; Brunet et al. 2006; Scannell and Wolfe 2008). This asymmetry has been taken as evidence supporting Ohno's (1970) model of neofunctionalization, whereby one duplicate (the slow copy) maintains the ancestral functional role and evolves at a rate similar to that of the ancestral gene, while the other copy evolves a novel beneficial function and undergoes rate acceleration due to positive selection (Kellis et al. 2004).

In yeast, Scannell and Wolfe (2008) discovered a sharp rate increase in protein sequence evolution affecting both copies. However, the symmetrical rate acceleration lasted for only a very short time following the duplication event; subsequently a rate-asymmetry pattern was commonly established, with one copy continuing to evolve at an elevated rate while the other reverted to the slower preduplication rate. The generality of these observations, however, remains to be tested in other taxa.

Rates and patterns of expression divergence between duplicated genes

Expression divergence between duplicated genes has long been considered an important step in the functional differentiation between duplicated genes (Markert 1964; Ohno 1970; Ferris and Whitt 1979). Indeed, investigations on tissue expression divergence among enzymes encoded by duplicated genes (isozymes) started soon after the introduction of protein electrophoresis (see Markert 1964). Such studies provided examples of differential tissue expression of duplicated genes and suggested a role for expression divergence in functional refinement and diversification. The availability of sequenced genomes and gene expression data has stimulated numerous studies, and general patterns in the relationship between sequence divergence and expression divergence are beginning to emerge.

Is expression divergence correlated with coding-sequence divergence? Dealing with this question requires two variables: a measure for sequence divergence, and a measure for expression divergence. As far as sequence divergence is concerned, one may use the number of substitutions per site between the two sequences (Chapter 3). For a measure of **expression dissimilarity** or **expression divergence**, one may use $1 - r$, where r is the Pearson correlation coefficient between the expression levels of two genes over different tissues or experimental conditions (Gu et al. 2002b).

In all organisms so far studied, a significant negative correlation was found between the two measures of divergence, indicating that expression divergence increases with sequence divergence and, hence, with time. Studies of duplicate genes in yeast, human, and *Arabidopsis* have revealed a rapid phase of initial expression divergence between duplicates followed by a plateau during which the measure of expression divergence remains relatively unchanged (Gu et al. 2002b; Makova and Li 2003; Blanc and Wolfe

2004; Casneuf et al. 2005; Gu et al. 2005; Yang et al. 2005; Ganko et al. 2008). It is not clear why a rapid divergence in expression between duplicates frequently occurs soon after the duplication event. One possible reason is incomplete duplication of *cis*-regulatory elements in one of the two genes (Katju and Lynch 2003). However, although this may explain the phenomenon as far as dispersed duplications or short tandem duplications are concerned, it is less likely to follow large segmental duplications, because a large duplicated segment would likely include all *cis* regulatory elements. Haberer et al. (2004) noted that in *Arabidopsis*, tandem and segmental duplicate gene pairs have divergent expression even when they share many similar *cis*-regulatory sequences, suggesting that changes in a small fraction of *cis* elements may be sufficient for expression divergence. Most intriguingly, despite the fact that whole-genome duplication should duplicate all regulatory elements, studies have shown that some genes are silenced immediately after allopolyploidization (see Adams and Wendel 2005). One reason why significant expression changes occur so soon after the formation of the polyploid is that transposable elements and other mobile sequences from one genome may invade the other genome in an allopolyploid (Wang et al. 2004).

There are three possible patterns of tissue-specific expression divergence following gene duplication: (1) both copies retain the ancestral pattern of expression, thereby creating redundancy; (2) the two copies diverge asymmetrically, whereby one gene is expressed in only a small subset of the tissues or conditions while its duplicate retains the ancestral pattern of expression; and (3) the two copies diverge complementarily, thus at least one descendant duplicate is expressed in each of the tissues in which the ancestral gene was expressed (**Figure 7.14**).

Data on the pattern of expression divergence in *Arabidopsis thaliana* indicate a preponderance of expression asymmetry over the other two possible patterns of expression divergence (Zhao et al. 2003; Casneuf et al. 2006; Duarte et al. 2006; Ganko et al. 2008). The asymmetric pattern was found in over 70% of gene pairs, whereas the complementary expression pattern was observed in only ~10% of the pairs. Asymmetric expression divergence was observed in many organisms, including yeast, *Drosophila*, rice, and whitefish. Thus, asymmetry in expression divergence between duplicate genes seems to be a universal phenomenon.

One possible reason for asymmetry is that once a *cis*-regulatory element is lost in one copy, leading to a lower expression level, then a similar mutation in the other copy would be deleterious and selected against and, hence, unobservable in natural populations. Papp et al. (2003b) found that the number of *cis*-regulatory elements shared between duplicated genes decreases over evolutionary time, though the total number of element types in the two genes remains relatively constant.

Conant and Wagner (2002) and Kim and Yi (2006) found that the K_A/K_S ratio was related to the degree of expression asymmetry between duplicated genes in yeast. However, the relative roles of positive selection, purifying selection, and relaxation of selective constraints in determining rates and patterns of expression divergence following gene duplication remain to be elucidated.

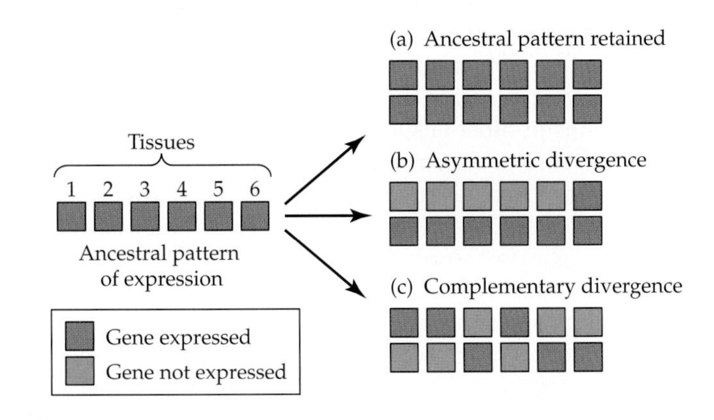

Figure 7.14 Three patterns of tissue-specific expression divergence following duplication of a gene that is expressed in six different tissues. (a) The two copies both retain the ancestral pattern of gene expression. (b) The two genes diverge asymmetrically, whereby one gene is expressed in only a small subset of the tissues, while its duplicate remains expressed in the original six tissues. (c) The two genes diverge complementarily, so no tissue-specific expression is lost. (Modified from Caseneuf et al. 2006.)

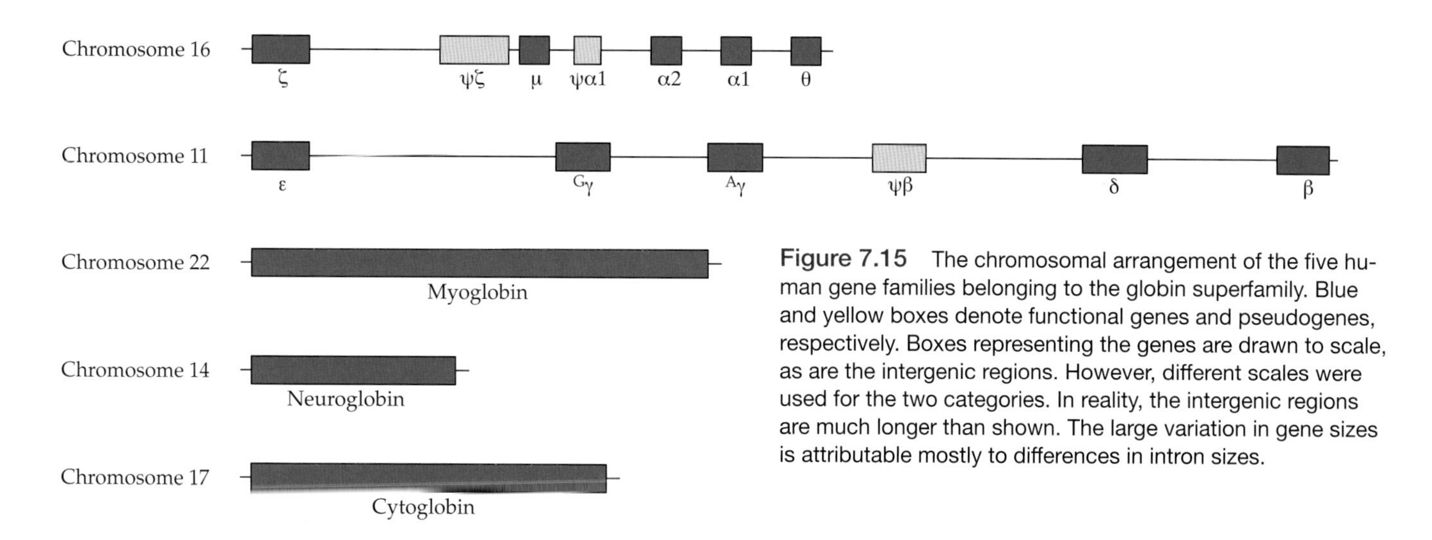

Figure 7.15 The chromosomal arrangement of the five human gene families belonging to the globin superfamily. Blue and yellow boxes denote functional genes and pseudogenes, respectively. Boxes representing the genes are drawn to scale, as are the intergenic regions. However, different scales were used for the two categories. In reality, the intergenic regions are much longer than shown. The large variation in gene sizes is attributable mostly to differences in intron sizes.

Human Globins

Here, we use the evolution of the human globin superfamily of genes as an illustration of the manner in which gene duplication creates working biological systems through the divergent evolutionary pathways described above.

In humans, the globin superfamily of genes consists of five chromosomal clusters, each with at least one functional member: the α-globin family on chromosome 16, the β-globin family on chromosome 11, and three single-gene families: myoglobin on chromosome 22, neuroglobin on chromosome 14, and cytoglobin on chromosome 17 (**Figure 7.15**).

Judging by the fact that globin and globin-like genes exist in both prokaryotes and eukaryotes, the globin superfamily must be very ancient in origin (Lecomte et al. 2005; Vinogradov et al. 2006). Phylogenetic studies indicate that the globin genes represented in the human genome originated more than 800 million years ago, long before mammals existed, probably even preceding the emergence of annelid worms (**Figure 7.16**). The first divergence event gave rise to neuroglobin, which in mammals is predominantly expressed in nerve cells. Neuroglobin is a monomer that reversibly binds oxygen with an affinity higher than that of hemoglobin. It is thought to protect neurons under hypoxic or ischemic conditions, potentially limiting brain

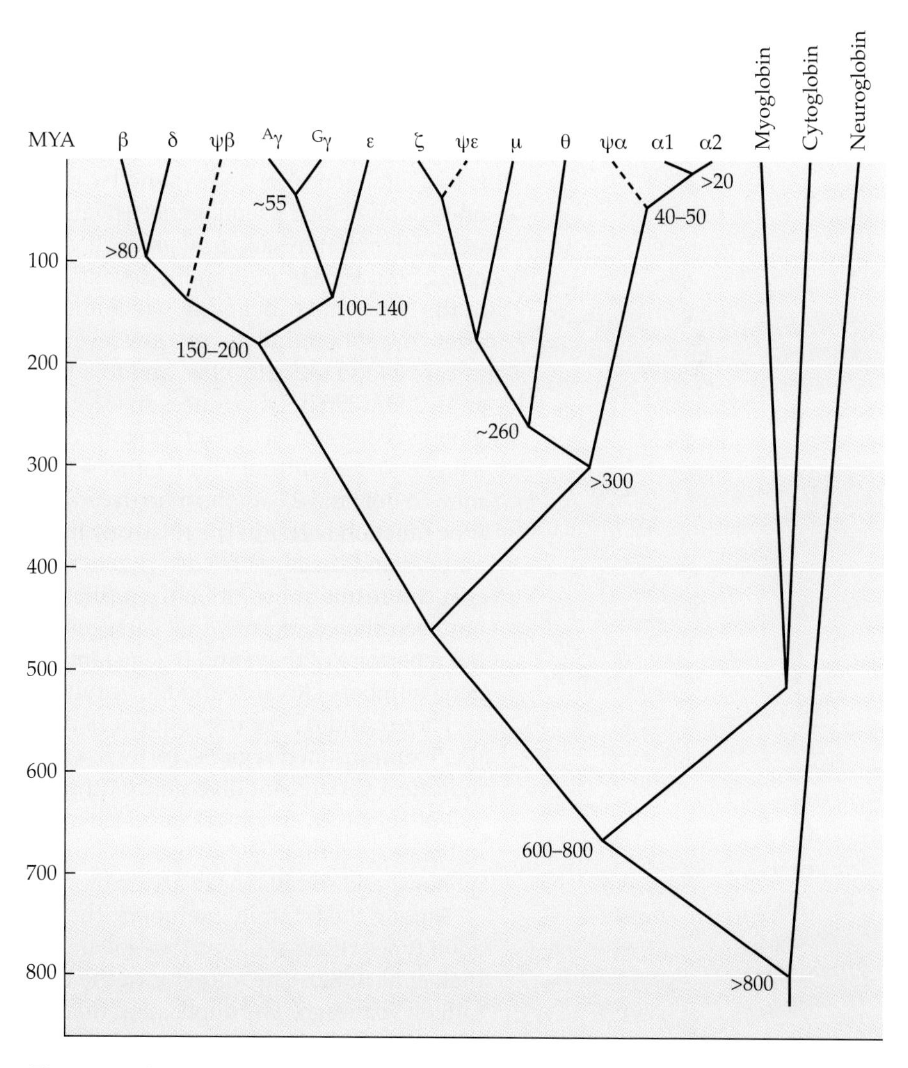

Figure 7.16 Evolutionary history of human globin genes. The broken lines denote pseudogene lineages.

damage. Evolutionarily, it is closely related to the nerve globins of invertebrates (Pesce et al. 2002).

The next divergence event gave rise to the ancestor of myoglobin and cytoglobin, on the one hand, and the ancestor of the α and β gene families on the other. This event occurred in the range of 600 to 800 million years ago. Myoglobin and cytoglobin diverged from each other quite early in vertebrate evolution, and both retained the plesiomorphic monomeric state. Cytoglobin is expressed in many different tissues. Its physiological roles are not completely understood. Although supplying cells with O_2 is a likely function, it is also possible that it acts as an O_2-consuming enzyme, an O_2 sensor, or a scavenger of nitric oxide. Myoglobin is the oxygen-storage protein in muscles, and its affinity for oxygen is higher than that of hemoglobin.

The α and β families diverged following a gene duplication 450–500 million years ago (Figure 7.16). Jawless fishes contain only one type of monomeric hemoglobin, while the vast majority of vertebrates have hemoglobin made up of two types of chains, one encoded by an α family member, the other by a β family member. Thus, the duplication that gave rise to the α- and β-globin families most probably also ushered in the era of tetrameric hemoglobin, the oxygen carrier in blood. This heteromeric structure allowed hemoglobin to acquire several novel capabilities that are absent in the monomeric globins. Among these are (1) binding of four oxygen molecules cooperatively, (2) responding to the acidity and carbon dioxide concentration inside red blood cells (the Bohr effect), and (3) regulating its own oxygen affinity through the level of organic phosphate in the blood.

In humans, the α-globin gene cluster is located on chromosome 16. It spans about 30 Kb and includes seven loci, for five functional genes and two pseudogenes: ζ, ψζ, μ, *ψα1*, *α2*, *α1*, and θ. The β cluster consists of five functional genes: the embryonic gene ε; two fetal genes, $^{G}\gamma$ and $^{A}\gamma$; and two adult genes, β and δ. The family also contains one nonprocessed pseudogene, ψβ (also referred to as ψη). The two families have diverged in both physiological properties and ontological regulation. In fact, distinct hemoglobins appear at different developmental stages: $\zeta_2\varepsilon_2$ and $\alpha_2\varepsilon_2$ in the embryo, $\alpha_2\gamma_2$ in the fetus, and $\alpha_2\beta_2$ and $\alpha_2\delta_2$ in adults. The θ gene is mainly transcribed 5–8 weeks after conception, but at very low levels. The μ gene is transcribed quite abundantly in cord-blood reticulocytes, and to a lesser extent in adult-blood reticulocytes (Goh et al. 2005, 2007). Differences in oxygen-binding affinity have evolved among these globins. For example, the embryonic and fetal hemoglobins ($\zeta_2\varepsilon_2$, $\alpha_2\varepsilon_2$, and $\alpha_2\gamma_2$) have a higher oxygen affinity than either adult hemoglobin ($\alpha_2\beta_2$ and $\alpha_2\delta_2$), mainly because they do not bind 2,3-diphosphoglycerate as strongly as the adult forms. Consequently, they function better in the relatively hypoxic (low oxygen) environment in which the embryo and the fetus reside. This example illustrates once again how gene duplication can result in evolutionary refinements of physiological systems. In addition, it has been shown recently that each gene may produce more than one transcript; thus, the repertoire of the α and β gene families may be even greater than implied by mere gene numbers (Alvarez and Ballantyne 2009).

The *α1* and *α2* coding sequences are identical, but there are some differences in the 5′ untranslated regions, introns, and 3′ untranslated regions. This would seem to indicate a very recent divergence time. However, the similarity could also be the result of gene conversion or concerted evolution, a phenomenon that will be discussed in the next section. The two genes are present in humans and all the apes (including gibbons) and so must have arisen more than 20 million years ago.

Among the β family members, the adult types (β and δ) diverged from the nonadult types (γ and ε) about 150–200 million years ago (Efstratiadis et al. 1980; Czelusniak et al. 1982). The ancestor of the two γ genes diverged from the ε gene 100–140 million years ago. The duplication that created $^{G}\gamma$ and $^{A}\gamma$ occurred after the separation of the simian lineage (Anthropoidea) from the prosimians about 55 million years ago (Hayasaka et al. 1992). Soon after that, the ancestor of the two γ genes, which was originally an embryonically expressed gene, became a fetal gene. The change in temporal ontogenetic expression was brought about by sequence changes within a 4 Kb region surrounding the γ gene (TomHon et al. 1997). The divergence between the δ and β genes is estimated to have occurred before the eutherian radiation, more than 80 million years ago (Goodman et al. 1984; Hardison and Margot 1984).

We emphasize that the description above only applies to the evolution of globins currently found in the human genome. The evolution of the globin superfamily of genes in vertebrates constitutes a wonderful, but as yet only partially deciphered, narrative of gene and genome duplications, gene birth-and-death processes, convergence and divergence, and horizontal movements of globin genes from one locus to another (see Goh et al. 2005; Hardison 2007; Opazo et al. 2008a, 2008b; Patel et al. 2008; Hoffman et al. 2010; Patel and Deakin 2010).

Concerted Evolution

From the mid-1960s to the mid-1970s, a large number of DNA reannealing and hybridization studies were conducted to explore the structure and organization of eukaryotic genomes. These studies revealed that the genome of multicellular organisms is composed of highly and moderately repeated sequences as well as single-copy sequences (Chapter 11). They also revealed an unexpected evolutionary phenomenon, namely that the members of a repeated-sequence gene family are sometimes very similar to each other within one species, although members of the family from even fairly closely related species may differ greatly from each other. A fine illustration of this phenomenon was first provided by Brown et al. (1972) in a comparison of the ribosomal RNA genes from the African frogs *Xenopus laevis* and *X. borealis* (the latter being misidentified at the time as *X. mulleri*).

In these two *Xenopus* species, as well as in the vast majority of multicellular eukaryotes, the genes specifying the 18S and 28S ribosomal RNAs (rRNAs) are present in hundreds of copies and are arranged in one or a few tandem arrays (Long and Dawid 1980). Each repeated unit consists of the 18S, 5.8S, and 28S RNA-specifying genes, two **external transcribed spacers** (ETS1 and ETS2), two **internal transcribed spacers** (ITS1 and ITS2), and an **intergenic nontranscribed spacer** (IGS) (**Figure 7.17**). The IGS and the two external transcribed spacers are sometimes referred to as the **intergenic region** (IGR). The transcribed segment produces a 45S RNA precursor from which the functional ribosomal RNAs are produced by means of enzymatic cleavage. The transcribed repeats are separated from each other by the IGS.

In a comparison of the ribosomal RNA genes of *X. laevis* and *X. borealis*, Brown et al. (1972) found that, while the rRNA-specifying sequences of the two species were virtually identical to each other, the nucleotide sequences of their IGS regions differed by about 10%. In contrast, the IGS regions were very similar within each individual and among individuals within each species. Thus, it appears that the IGS regions in each species evolved together, although they diverged rapidly between species. This observation could not be explained by the divergent evolution model, according to which the differences in nucleotide sequence between different repeats of the same species are expected to be as large as those between repeats of different species (**Figure 7.18**). One simple explanation for the intraspecific homogeneity may be that the function of the repeats depends strongly upon their specific nucleotide sequence, so new mutations have been either eliminated by purifying selection or fixed by positive selection (**Figure 7.19a**). This explanation requires that the same advantageous

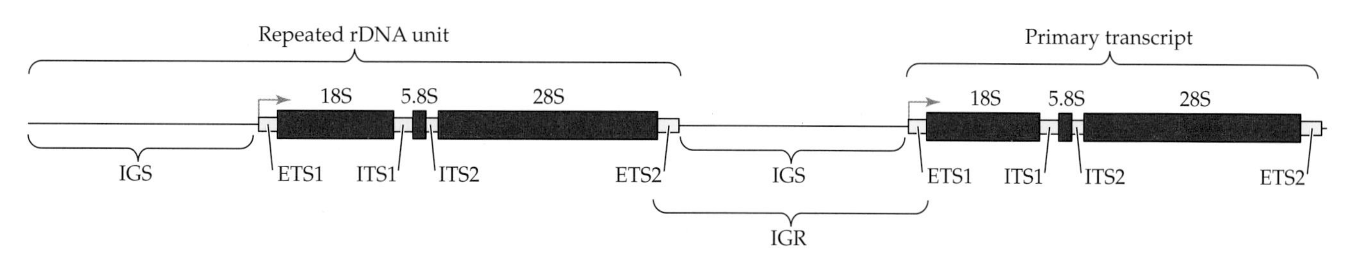

Figure 7.17 Two repeated rDNA units of *Xenopus laevis*. The 18S, 5.8S, and 28S rRNA sequences are transcribed together with the external transcribed sequences (ETS1 and ETS2) and the internal transcribed sequences (ITS1 and ITS2) as a single primary RNA precursor that is posttranscriptionally processed to produce the mature rRNA molecules. Green arrows denote transcription initiation sites. IGS, intergenic spacer; IGR, intergenic region.

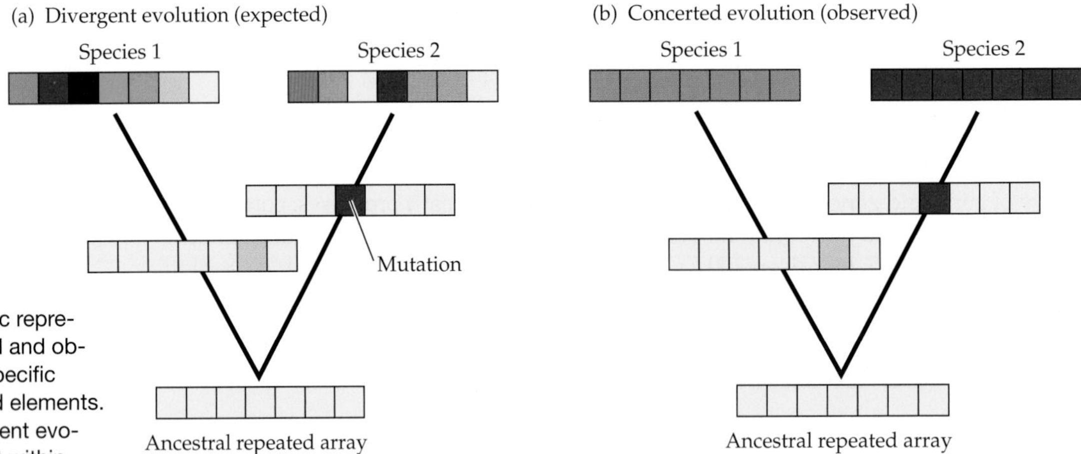

Figure 7.18 Schematic representation of the expected and observed patterns of intraspecific variation among repeated elements. (a) Under classical divergent evolution, each repeated unit within an array evolves independently. As a consequence, the similarity between any two randomly chosen units within a species is expected to be the same as that between two units randomly chosen from different species. (b) The observed patterns of intraspecific variation reveal a high degree of within-species homogeneity among repeated units. Differences from the ancestral sequence are shown in different colors.

mutation occur repeatedly in all copies of the gene. Moreover, the IGS regions have no known function and do not appear to be subject to stringent selective constraints, and yet are conserved within the species. Another simple explanation may be that the family has arisen from a recent amplification of a single unit (**Figure 7.19b**). In this case, the homogeneity would simply reflect the fact that there has not been enough time for the members of the gene family to diverge from one another. If this is the case, it is expected that the homogeneity of the family would gradually decrease because over evolutionary time mutations would accumulate in the family members through genetic drift, particularly in regions that are not subject to stringent structural constraints. Under this model of independent evolution, the degree of intraspecific variation among the repeated elements is expected to be approximately equal to the degree of interspecific variation.

The empirical data from *Xenopus* supports neither model. Rather, the intraspecific homogeneity among the repeated units seems to be maintained by a mechanism through which mutations can spread horizontally to all members in a multigene family. Brown et al. (1972) concluded that a "correction" mechanism must have operated to spread a mutation from one sequence to another faster than new changes can arise by mutation in these sequences (**Figure 7.19c**). They called this phenomenon, which

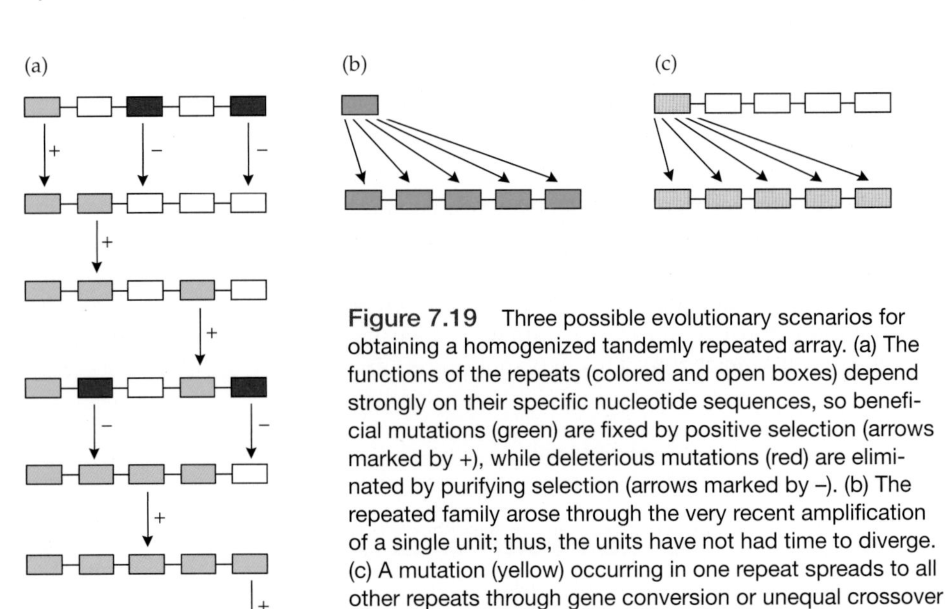

Figure 7.19 Three possible evolutionary scenarios for obtaining a homogenized tandemly repeated array. (a) The functions of the repeats (colored and open boxes) depend strongly on their specific nucleotide sequences, so beneficial mutations (green) are fixed by positive selection (arrows marked by +), while deleterious mutations (red) are eliminated by purifying selection (arrows marked by –). (b) The repeated family arose through the very recent amplification of a single unit; thus, the units have not had time to diverge. (c) A mutation (yellow) occurring in one repeat spreads to all other repeats through gene conversion or unequal crossover regardless of its selective advantage or disadvantage. (Modified from Graur and Li 2000.)

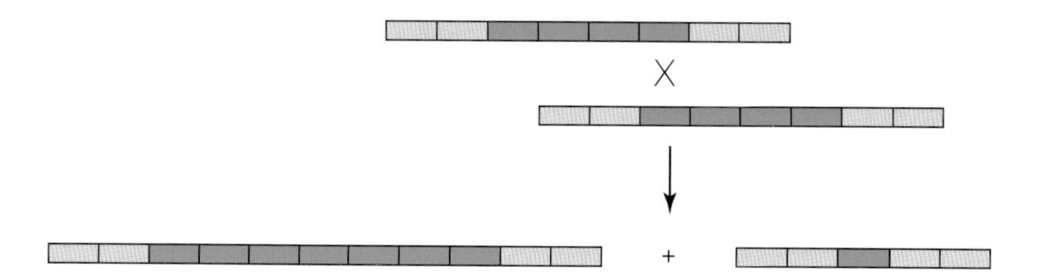

Figure 7.20 Model of the homogenizing process of repeated units in an array through unequal crossing over (×).The two repeat types have equal frequencies (50%) in the parental arrays. As a result of unequal crossing over, the repeat makeups of the two daughter arrays are more homogeneous than those of the parents (64% and 80%).

must originate in a single individual and later become fixed in the population, horizontal evolution, in contrast to vertical evolution, which refers to the spread of a mutation in a breeding population from one generation to the next. Many other terms have since been suggested in the literature; the term most commonly used nowadays is **concerted evolution** (Zimmer et al. 1980).

The term "concerted evolution" implies that an individual member of a gene family does not evolve independently of the other members of the family. Rather, repeats in a family exchange sequence information with each other, either reciprocally or nonreciprocally, so a high degree of intrafamilial sequence homogeneity is maintained. Through genetic interactions among its members, a multigene family evolves as a unit in a concerted fashion. The result of concerted evolution is a homogenized array of nonallelic homologous sequences. It is important to emphasize that concerted evolution requires not only the horizontal transfer of mutations among the members of the family (homogenization), but also the spread of the homogenized array of repeats in the population (fixation).

Gene conversion and unequal crossover are considered to be the two most important mechanisms responsible for concerted evolution. Although these two mechanisms have received the most extensive quantitative coverage in the literature and will be explained in some detail below, there are other mechanisms, such as slipped-strand mispairing and duplicative transposition, that can result in the creation of homogeneous families of repeated sequences.

Unequal crossing over

Unequal crossing over may occur either between the two sister chromatids of a chromosome during mitosis in a germline cell, or between two homologous chromosomes at meiosis. Unequal crossing over is a reciprocal recombination process that creates a sequence duplication in one chromatid or chromosome and a corresponding deletion in the other (Chapter 1). A hypothetical example in which an unequal crossover has led to the duplication of three repeats in one daughter chromosome and the deletion of three repeats in the other is shown in **Figure 7.20**. As a result of this exchange, the repeated arrays in both daughter chromosomes have become more homogeneous than those in the parental chromosomes. If this process is repeated, the numbers of each variant repeat on a chromosome will fluctuate with time, and eventually one type will become dominant in the array. **Figure 7.21** illustrates how one type of repeat may spread throughout a gene family through repeated rounds of unequal crossing over. The process of concerted evolution by unequal crossing over has been investigated mathematically in detail and has received considerable experimental support (Smith 1976; Ohta 1984; Nei and Rooney 2005).

In the example of the rRNA genes in *Xenopus laevis* and *X. borealis*, the family size is large and it is conceivable that the number of genes may fluctuate with time without adverse consequences. Thus, unequal crossing over, rather than gene conversion, was apparently the driving force in the homogenization of the repeats.

Like unequal crossing over, slipped-strand mispairing is an expansion-contraction process that leads to the homogenization of the members of a tandem

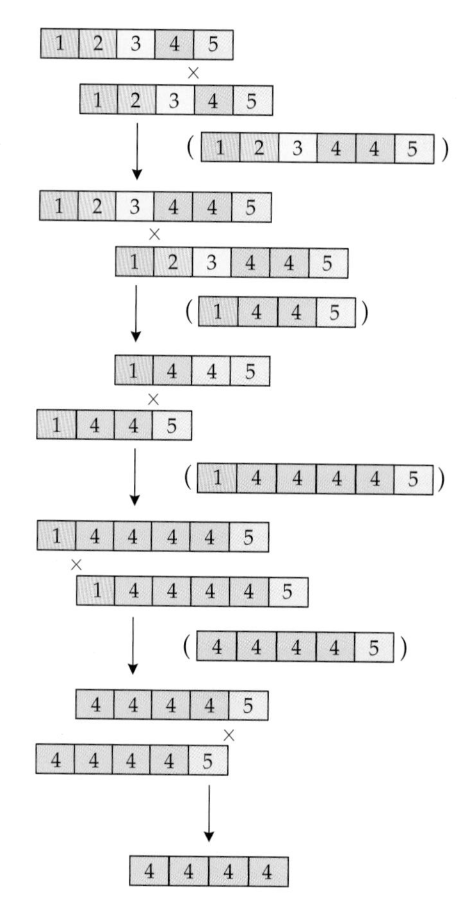

Figure 7.21 Concerted evolution by unequal crossing over (×). Repeated cycles of unequal crossover events cause the duplicated sequences on each chromosome to become progressively more homogenized. Different sequences are numbered and differently colored. The sequences in parentheses on the right are the ones selected for the next round of unequal crossover. The end result is a takeover by the "green" repeat (sequence 4). Note that unequal crossing over affects the number of repeated sequences on each chromosome. (Modified from Ohta 1980.)

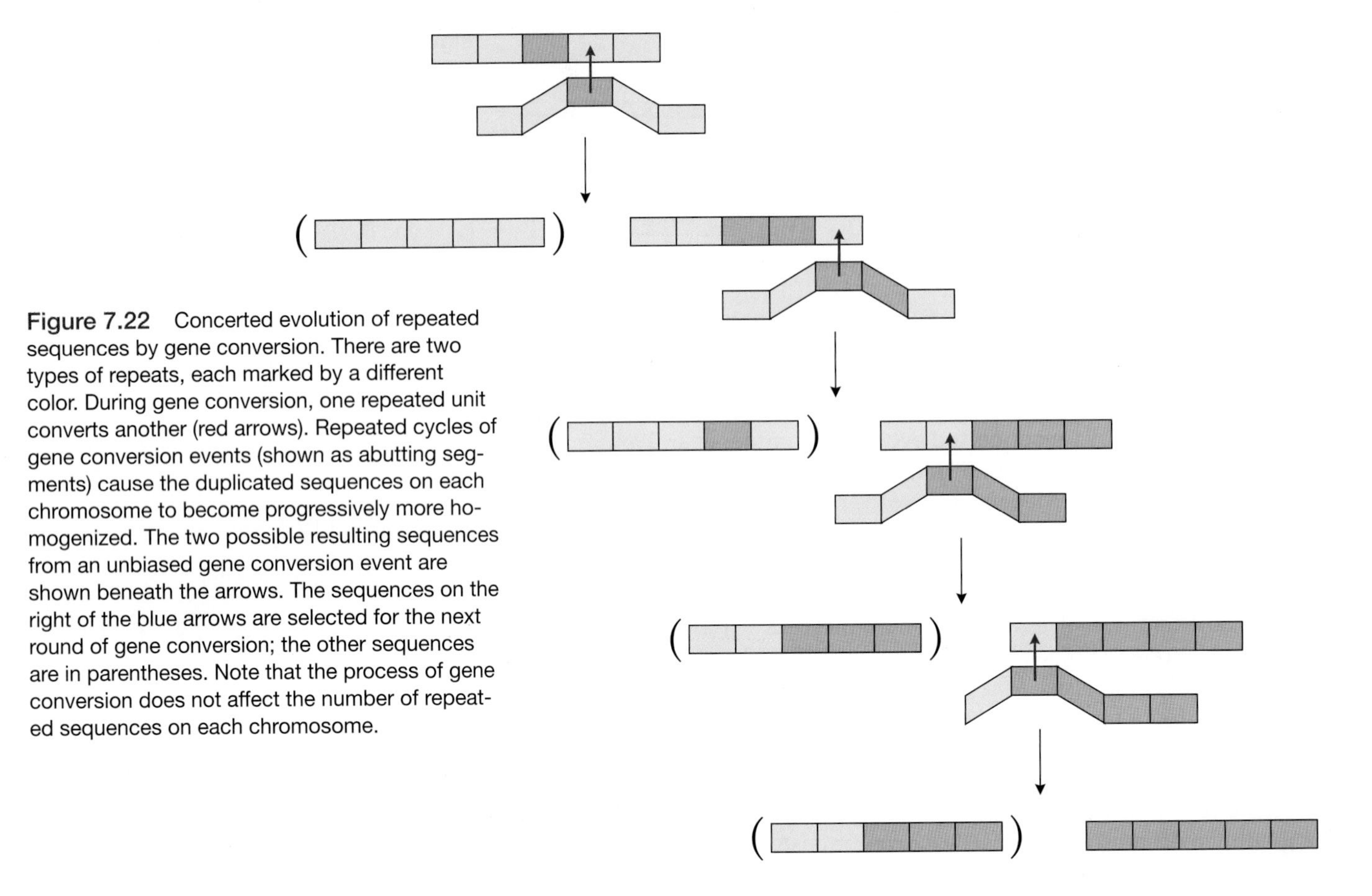

Figure 7.22 Concerted evolution of repeated sequences by gene conversion. There are two types of repeats, each marked by a different color. During gene conversion, one repeated unit converts another (red arrows). Repeated cycles of gene conversion events (shown as abutting segments) cause the duplicated sequences on each chromosome to become progressively more homogenized. The two possible resulting sequences from an unbiased gene conversion event are shown beneath the arrows. The sequences on the right of the blue arrows are selected for the next round of gene conversion; the other sequences are in parentheses. Note that the process of gene conversion does not affect the number of repeated sequences on each chromosome.

repeat family. As will be addressed in Chapter 11, while unequal crossover usually affects large tracts of DNA, slipped-strand mispairing is involved in the generation of tandem arrays of short repeats.

Gene conversion

Gene conversion is a nonreciprocal recombination process in which two sequences interact in such a way that one is converted by the other (Chapter 1). Although the mechanisms of gene conversion have not yet entirely been elucidated, several models of gene conversion have gathered support in the literature (see Chen et al. 2007). Theoretical studies (e.g., Ohta 1990) have shown that gene conversion can produce concerted evolution (**Figure 7.22**).

Based on the chromatids involved in the process, gene conversion can be divided into several types (**Figure 7.23**). When the exchange occurs between two paralogous sequences on the same chromatid, the process is called **intrachromatid conversion**. An exchange between two paralogous sequences from complementary chromatids is called **sister chromatid conversion**. **Classical conversion** involves exchanges between two alleles at the same locus. **Semiclassical conversion** involves an exchange between two paralogous genes from two homologous chromosomes. If the exchange occurs between paralogous sequences located on two nonhomologous chromosomes, the process is called **ectopic conversion**. From the viewpoint of the evolution of duplicated genes, the most important types of gene conversion are the **nonallelic conversions** (i.e., conversions between genes located at different loci and not between allelic forms).

Gene conversion may be biased or unbiased. **Unbiased gene conversion** means that sequence A has as much chance of converting sequence B as sequence B has of converting sequence A. **Biased gene conversion** means that the two possible direc-

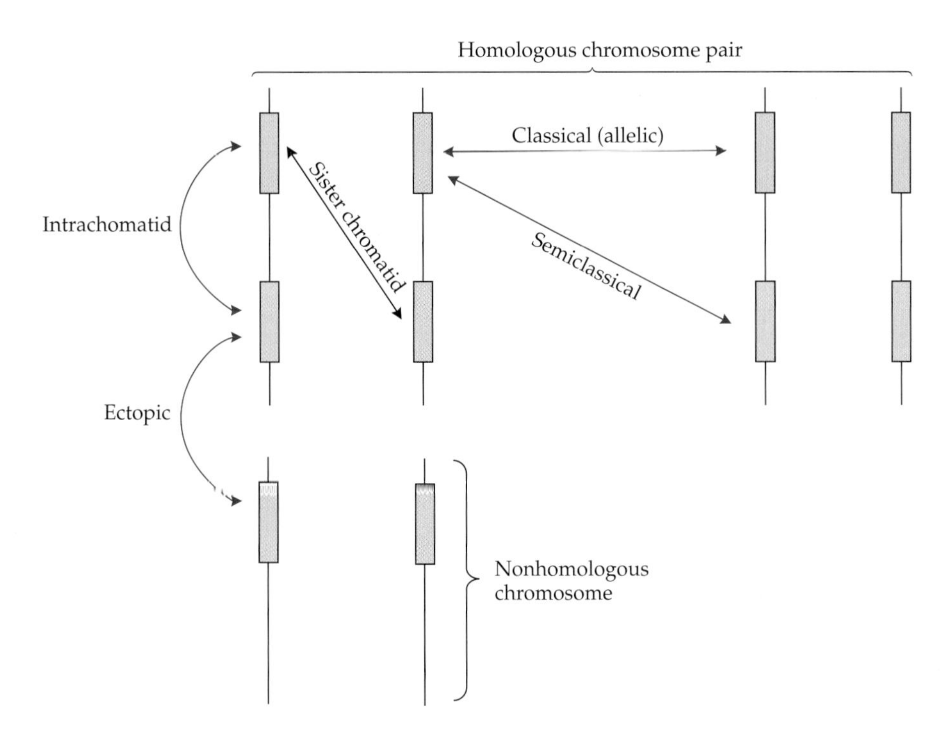

Figure 7.23 Types of gene conversion (red double-headed arrows) between repeated sequences (green boxes). The two chromatids of a chromosome are shown as neighboring pairs. Homologous double-stranded chromosomes are shown at the top; a nonhomologous chromosome is shown at the bottom. (Modified from Graur and Li 2000.)

tions of gene conversion occur with unequal probabilities. If deviation from parity occurs, we may speak of **conversional advantage** or **disadvantage** of one sequence over the other. Available data indicate that biased gene conversion is more common than the unbiased type. One particularly important type of biased gene conversion is **GC-biased gene conversion**, in which GC-rich sequences are favored over AT-rich ones (p. 304).

The amount of DNA involved in a gene conversion event varies from a few base pairs to a few thousand base pairs. Finally, the rate and the probability of occurrence of gene conversion vary with location; some locations are more prone to conversions than others.

DETECTING GENE CONVERSION EVENTS Many procedures for detecting putative gene conversion events have been put forward in the literature (see Drouin et al. 1999). In this section, we will briefly describe the method of Sawyer (1999), which asks whether or not a run of consecutive identical sites from among all polymorphic sites between two paralogous sequences is longer than would be expected by chance.

Consider the set of six aligned DNA sequences in **Figure 7.24a**. The length of the alignment is 50. We first identify the longest run of matched sites between any two sequences in the alignment. In our case, we find that sequences 2 and 4 have a run of 27 consecutive sites in common. Because we aim to identify identities due to gene conversion rather than extreme conservation due to stringent selective constraint or insufficient time for divergence subsequent to the duplication event, we next discard the 26 sites that are monomorphic in the six sequences. The condensed alignment is shown in **Figure 7.24b**. Now, the longest run of matched sites is 14 in an alignment of size 24. One way to calculate the statistical significance of this run is to use a runs test, which is a statistical procedure that examines whether or not a string of data occurs randomly given a specific distribution. The number of differences between sequences 2 and 4 in the condensed alignment is 6 out of 24, i.e., $p = 0.25$. Let us now use the runs test to calculate the probability of obtaining a run of matched sites of size 14 or longer starting at position 3. This probability is $0.25 \times (1 - 0.25)^{14} = 0.0045$. We note that there are 11 possible starting points for runs of length 14 or more in an alignment of length 24. Thus, the corrected probability of obtaining a run of length 14 or more is $0.0045 \times 11 = 0.05$. We now need to concern ourselves with the fact that

(a) Length = 50

```
  ↓↓ ↓  ↓↓  ↓↓↓↓       ↓↓ ↓↓  ↓↓  ↓ ↓↓↓    ↓↓ ↓↓↓↓ ↓
1 GCAGAGTGCTATAACAAGAACGAGTACCGGTGTATCTAATGTAAATGTAT
2 GCAGAGGGCTTTAACTTCTACGAGTACCGGTGTTTCTAATGTAAATGTTT
3 ACACAACGCTGTAACTAGAAAGATTAGGGGTGTGTCTGGCGTACATGTAT
4 GCACAGGGCTTTAACTTCTACGAGTACCGGTATTTCTGGCGTACATGTTT
5 ACACACAGCTATAACATCTAAGAGTACCGGTGTATCTGGCGTAAATGTAT
6 GCACATGGCGGTAACAAGTTCGAGTATTGGAATCTCTTACATATATGTAT
```

(b) Length = 24

```
1 GGGTTAAAGAACGCCTGAAATGAA
2 GGGGTTTTCTACGCCTGTAATGAT
3 ACACTGTAGAAATGGTGGGGCGCA
4 GCGGTTTTCTACGCCTATGGCGCT
5 ACCATAATCTAAGCCTGAGGCGAA
6 GCTGGGAAGTTCGTTAACTACATA
```

Figure 7.24 Sawyer's (1999) statistical test for detecting gene conversion. (a) An alignment of six paralogous DNA sequences. The length of the alignment is 50. The longest run of matched sites between any two sequences in the alignment is 27, between sequences 2 and 4 (shaded). Red arrows mark monomorphic sites. (b) A condensed alignment after removing monomorphic sites. The length of the condensed alignment is 24. The longest run of matched sites is 14 (shaded). The probability of finding a run of 14 matches between two sequences from among 15 nonindependent comparisons is much larger than 5%, hence we cannot reject the null hypothesis of no gene conversion.

with six sequences there are 15 sequence-pair nonindependent comparisons. Thus, the true probability of obtaining such a run of matched sites by chance becomes larger than 5%. Thus, despite the fact that sequences 2 and 4 display a large run of sequence identity, we cannot rule out chance in explaining this identity, and there is no support for the gene conversion hypothesis.

PREVALENCE OF GENE CONVERSION How often do duplicate genes undergo gene conversion? Ezawa et al. (2006) examined 2,641 gene pairs in the mouse and rat genomes that were duplicated after the human-rodent split but before the mouse-rat split. They found strong evidence of gene conversion in 488 pairs (18%), the vast majority of which (407/488 = 83%) were linked on the same chromosome. Since detection of gene conversion has a low power, this represents an underestimate. Thus, gene conversion seems to occur with a high frequency between young duplicate genes.

GC-BIASED GENE CONVERSION There is evidence that in the vast majority of eukaryotes, gene conversion is biased, in that G and C nucleotides are favored over A and T, a phenomenon called **GC-biased gene conversion** (Galtier et al. 2001). GC-biased gene conversion is expected to increase the GC content of recombining DNA segments over evolutionary time, and it is considered a major contributor to the variation in GC content within and between genomes. GC-biased gene conversion has major methodological implications, for example, in detecting positive selection. Interestingly, GC-biased gene conversion can often be an antiadaptive force, by favoring GC-rich alleles (or paralogs) even if the AT-rich ones happen to be superior in fitness (Galtier et al. 2009).

Examples of gene conversion

In the examples described below, only two duplicate genes are involved, so concerted evolution must have occurred through gene conversion alone. In general, when the family size is small, unequal crossover events resulting in a number of repeats below a critical threshold may have a low fitness, so unequal crossover most probably does not play an important role in the concerted evolution of such gene families.

THE $^{A}\gamma$- AND $^{G}\gamma$-GLOBIN GENES IN THE GREAT APES An interesting case of concerted evolution by gene conversion involves the $^{A}\gamma$- and $^{G}\gamma$-globin genes, which were created by a duplication approximately 55 million years ago, after the divergence between prosimians and simians. Since the African great apes (humans, chimpanzees, and gorillas) diverged from one another at a much later date, we would expect the $^{G}\gamma$ orthologous genes from apes to be much more similar to one another than to any of the $^{A}\gamma$ paralogs. However, as shown in **Figure 7.25a**, this is only true for the 3′ part of the gene, which contains exon 3. The 5′ part, which contains exons 1 and 2 (Figure 7.25b), exhibits a different phylogenetic pattern, i.e., paralogous exons within each species resemble each other more than they resemble their orthologous counterparts in other apes (Slightom et al. 1985). This discrepancy is obvious when counting the nucleotide differences between two paralogous genes from the same species. In humans, for example, the 5′ parts of $^{A}\gamma$ and $^{G}\gamma$ differ from each other at only 7 out of 1,550 nucleotide positions (0.5%). In contrast, the 3′ part shows a difference that is 20 times larger, 145 out of 1,550 nucleotides (9.4%). Assuming that the 5′ and 3′ parts are subject to similar functional constraints, we may conclude that the 5′ end of the gene underwent gene conversion. This conclusion is strengthened by the fact that the second intron in both genes in all apes contains a stretch of the simple repeated DNA sequence $(TG)_n$ that can serve as a hotspot for the recombination events involved in the process of gene conversion.

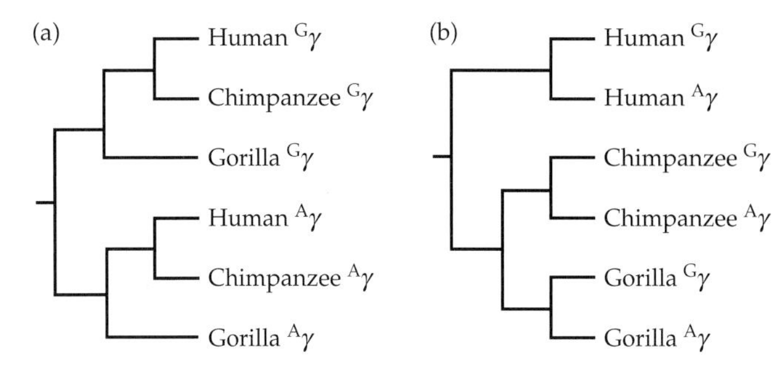

Figure 7.25 Phylogenetic trees for (a) exon 3 and (b) exons 1 and 2 of the $^{G}\gamma$- and $^{A}\gamma$-globin genes from human, chimpanzee, and gorilla. (Modified from Graur and Li 2000.)

Assuming that both the converted and unconverted parts of the genes evolve at equal rates, it is possible to date the last gene conversion event by using the degrees of similarity between the two sequences in conjunction with the date for the gene duplication event. The last conversion event that has been fixed in the human lineage occurred about 1–2 million years ago, i.e., after the divergence between human and chimpanzee, which indicates that the conversions in the chimpanzee and gorilla lineages occurred independently. Indeed, there are indications that gene conversion events between the 5′ parts of $^{A}\gamma$ and $^{G}\gamma$ genes occur quite frequently in human populations (e.g., Kalamaras et al. 2008).

GENE CONVERSION OF GENES AND PSEUDOGENES: WHEN DEATH IS NOT FINAL, LIFE IS PRECARIOUS, AND DISTINGUISHING BETWEEN THE TWO IS DIFFICULT

Pancreatic ribonuclease is a ubiquitous protein secreted by the pancreas in all vertebrates. In mammals, this protein is usually encoded by a single-copy gene. In the ancestor of true ruminants (suborder Pecora), the gene underwent two rounds of duplication, from which emerged three paralogous genes encoding pancreatic, seminal, and cerebral ribonucleases. Interestingly, a functional gene for seminal ribonuclease was only found in the closely related bovine species *Bos taurus* (cattle), *Bubalus bubalis* (Asian water buffalo), and *Syncerus caffer* (Cape buffalo), whereas in all other pecorans, such as deer and giraffe, the orthologous sequence was found to be a pseudogene. Even in *Tragelaphus imberbis* (the lesser kudu), which belongs to the same subfamily (Bovinae) as *Bos*, *Bubalus*, and *Syncerus*, the orthologous sequence is a pseudogene (Confalone et al. 1995; Breukelman et al. 1998).

One explanation for the data is that the gene was nonfunctionalized repeatedly and independently in numerous ruminant lineages within families Bovidae and Giraffidae (Sassi et al. 2007). The other most parsimonious explanation is that the original seminal ribonuclease gene in the ancestor of the true ruminants was a pseudogene that was subsequently "resurrected" in one lineage and became expressed in the seminal fluid, while in the other lineages, it stayed "dead" (**Figure 7.26**). Because of the taxonomic distribution of the seminal ribonuclease genes and pseudogenes, it is possible to date this resurrection at between 5 and 10 million years ago, after the divergence of the lesser kudu, but before the divergence of the Asian water buffalo.

A detailed analysis of the ribonuclease sequences indicated that the resurrection may have involved a gene conversion event, i.e., a transfer of information from the gene for the pancreatic enzyme to that for the seminal ribonuclease (Trabesinger-Ruef

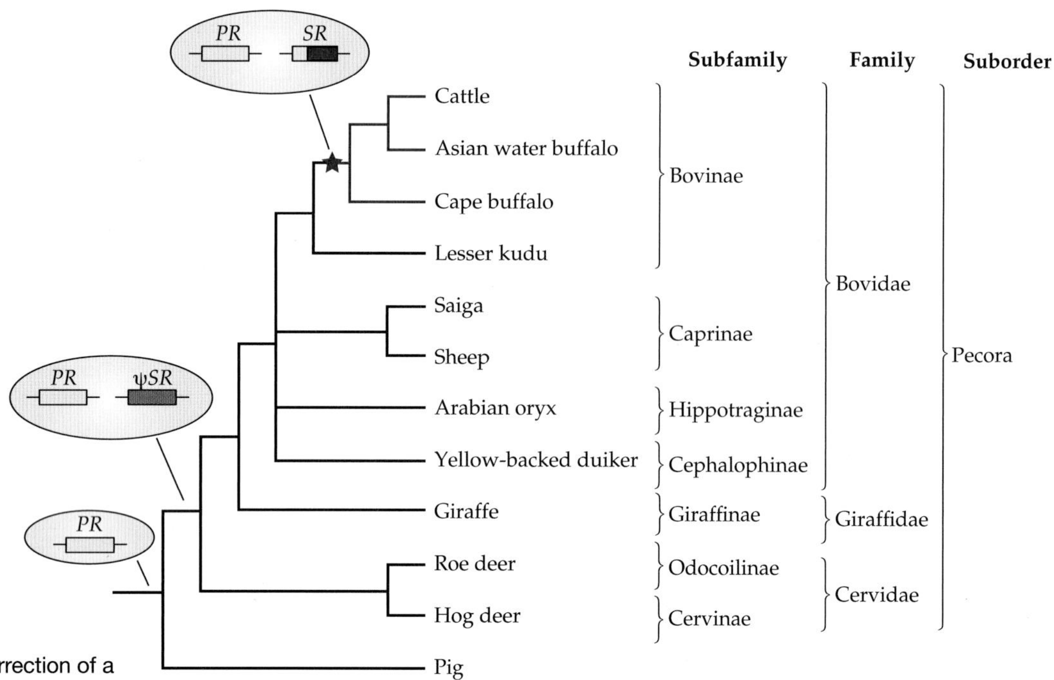

Figure 7.26 The resurrection of a ribonuclease pseudogene by gene conversion in a bovine lineage. Species in which the seminal ribonuclease protein is functional are marked with a plus sign (+). In the ancestor of the pecorans, the pancreatic ribonuclease (*PR*) gene (empty box) was triplicated to yield a nonfunctional seminal ribonuclease (ψ*SR*) pseudogene (green box) and a functional cerebral ribonuclease gene (not shown). In the ancestor of cattle and buffaloes, the 5′ end of the pseudogene was converted by the functional gene and became a functional gene (red star) encoding seminal ribonuclease (*SR*). Species: cattle, *Bos taurus*; Asian water buffalo, *Bubalus bubalis*; Cape buffalo, *Syncerus caffer*; lesser kudu, *Tragelaphus imberbis*; saiga, *Saiga tatarica*; sheep, *Ovis aries*; Arabian oryx, *Oryx leucoryx*; yellow-backed duiker, *Cephalophus sylvicultor*; giraffe, *Giraffa camelopardalis*; roe deer, *Capreolus capreolus*; hog deer, *Axis porcinus*; pig, *Sus scrofa*. (Modified from Graur and Li 2000.)

et al. 1996). The gene conversion event involved only a small region at the 5′ end of the gene, including about 70 nucleotides in the untranslated region. Interestingly, this conversion not only removed a deletion that caused a frameshift in the reading frame, but also restored two amino acid residues that are vital for the proper functioning of ribonuclease: histidine at position 12 and cysteine at position 31.

Thanks to gene conversion, death (as far as genes are concerned) does not by necessity connote finality. Thus, while pseudogenes are nonfunctional by definition, genomic extinction is not their only fate. They may, albeit very rarely, take part in the evolution of functional genes through gene conversion, unequal recombination, and transposition (Chapter 9).

We note that gene conversion is a mutational process, i.e., the proximity of a gene to a pseudogene may not only spell rebirth for the pseudogene, but can also spell death for the gene. One such example of gene death by gene conversion concerns *CYP21A2* (cytochrome P450 gene, family 21, subfamily A, polypeptide 2). In humans, *CYP21A2* is a 10-exon gene that is located on chromosome 6 in a region in which many major histocompatibility and complement genes are interspersed with one another. The gene has a paralogous nonprocessed pseudogene in the vicinity. Interestingly, many mammals have one functional copy and one nonfunctional copy of the *CYP21A2* gene; however, the nonfunctionalization event giving rise to the pseudogene occurred independently in many of the various lineages. Thus, for instance, the ortholog of the human functional gene is a pseudogene in mouse, and the ortholog of the human pseudogene is a functional gene in mouse.

Hundreds of mutations in *CYP21A2* have been characterized in the clinical literature (mostly causing a disease called congenital adrenal hyperplasia), and about 75% of them have been found to be due to gene conversion (Mornet et al. 1991). In at least one case, by using a de novo mutation in an individual, it was possible to pinpoint an intrachromatid gene conversion event involving 390 nucleotides in the maternal chromosome as the cause of nonfunctionalization (Collier et al. 1993).

The relative roles of gene conversion and unequal crossing over

As a mechanism for concerted evolution, gene conversion appears to have several advantages over unequal crossover. First, unequal crossover generates changes in the number of repeated genes within a family, which may sometimes cause a significant dosage imbalance. For example, the deletion of one of the two α-globin genes following an unequal crossover gives rise to a mild form of α-thalassemia in homozygotes (Lupski 1998). Gene conversion, on the other hand, causes no change in gene number.

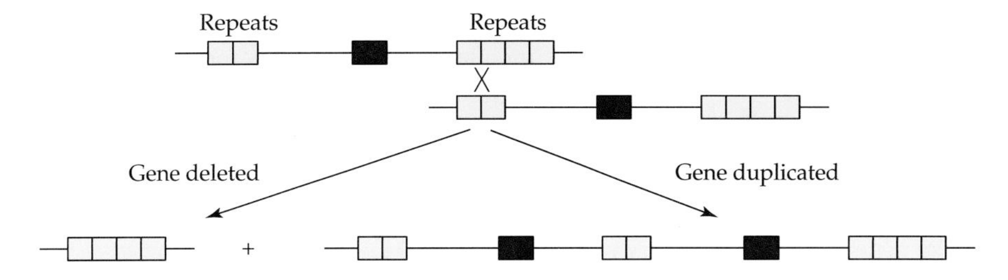

Figure 7.27 Crossover involving dispersed repeats (yellow boxes). The purple box denotes a unique gene. In the crossover event, the unique gene is deleted in one chromosome and duplicated in the other. (Modified from Graur and Li 2000.)

Second, gene conversion can act as a correction mechanism not only on tandem repeats but also on dispersed repeats within a chromosome (Jackson and Fink 1981; Klein and Petes 1981), between homologous chromosomes (Fogel et al. 1978), or between nonhomologous chromosomes (Scherer and Davis 1980; Ernst et al. 1982). In contrast, unequal crossover is severely restricted when repeats dispersed on nonhomologous chromosomes are involved. It can probably act effectively on nonhomologous chromosomes only if the repeated genes are located on the telomeric parts of the chromosome (the ends of the chromosome arms), as in the case of rRNA genes in humans and apes, but will be greatly restricted if the dispersed repeats are located in the middles of chromosomes, as in the case of rRNA genes in mice, lizards, and *Drosophila melanogaster.* If the repeats are dispersed on a chromosome, unequal crossover can result in the deletion or duplication of the genes that are located between the repeats. For example, **Figure 7.27** shows a hypothetical case of unequal crossover between two repeated clusters, resulting in the deletion of a unique gene in one chromosome and a corresponding duplication in the other. Either one or both chromosomes could have a deleterious effect on their carriers.

Third, gene conversion can be biased, i.e., have a preferred direction. Experimental data from fungi have shown that bias in the direction of gene conversion is common and often strong (Lamb and Helmi 1982), and theoretical studies have shown that even a small bias can have a large effect on the probability of fixation of repeated mutants (Nagylaki and Petes 1982; Walsh 1985).

For the above reasons, some authors (Baltimore 1981; Dover 1982; Nagylaki and Petes 1982) have proposed that gene conversion plays a more important role in concerted evolution than unequal crossover. This is probably true for dispersed repeats, because in this case gene conversion can act more effectively than unequal crossover. It is also probably true for small multigene families (e.g., the duplicated α-globin genes in humans), because in such families unequal crossover may cause severe adverse effects. In large families of tandemly repeated sequences, however, unequal crossover may be as acceptable a process as gene conversion. Indeed, in such cases, unequal crossover may be faster and more efficient than gene conversion in bringing about concerted evolution, for several reasons.

First, in such families, the number of repeats apparently can fluctuate greatly without causing significant adverse effects. This is suggested by the observations that the number of RNA-specifying genes in *Drosophila* varies widely among individuals of the same species and among species (Ritossa et al. 1966; Brown and Sugimoto 1973). Moreover, in humans, several families of tandem repeats that exhibit extraordinary degrees of variation in copy number have been found (Nakamura et al. 1987). Second, in a gene conversion event, usually only a small region (the heteroduplex region) is involved, whereas in unequal crossover the number of repeats that are exchanged between the chromosomes can be very large. For example, in yeast, a single unequal crossover event was shown to involve on average seven repeats of rRNA genes, i.e., ~20,000 bp (Szostak and Wu 1980), whereas a gene conversion track may not exceed 1,500 bp (Curtis and Bender 1991). Obviously, the larger the number of repeats exchanged, the higher the rate of concerted evolution will be (Ohta 1983). In some cases, this advantage of unequal crossover may be large enough to offset those of gene conversion. Finally, the empirical data show that in some organisms (e.g., yeast), unequal crossover occurs more frequently than nonallelic gene conversion. Of course, the observed lower rate of gene conversion might have been due to a detection bias, for it is generally much easier to detect unequal crossover than gene conversion.

Factors Affecting Concerted Evolution

How cohesively the members of a repeated-sequence family evolve together depends on several factors, including the number of repeats (i.e., the size of the gene family), the arrangement of the repeats, the structure of the repeated unit, the functional constraints imposed on the repeated unit, the mechanisms of concerted evolution, and the selective and nonselective processes at the population level.

Number of repeats

It is quite easy to see that the rate of concerted evolution is dependent on the number of repeats. For example, if there are only two repeats on a chromosome, a single intrachromosomal gene conversion will lead to homogeneity of the repeats on the chromosome. On the other hand, when there are more than two repeats on the chromosome, more than a single conversion may be required to homogenize the sequences.

Smith (1974, 1976) seems to be the first author to have conducted a quantitative study of the effect of family size on the rate of homogenization in a multigene family. His simulation study indicated that the number of unequal crossover events required for the fixation of a variant repeat in a single chromosomal lineage increases roughly with n^2, where n is the number of repeats on the chromosome.

Arrangement of repeats

There are, roughly speaking, two types of arrangement of repeated units. In some gene families, the members are highly dispersed all over the genome. One example is the human *Alu* family, whose approximately one million members are interspersed with single-copy sequences throughout the genome (Chapter 11). This type of arrangement is the least favorable for concerted evolution, because it greatly reduces the chance of unequal crossover and gene conversion and because unequal crossover will often lead to disastrous genetic consequences. Thus, the high similarities among *Alu* sequences are most probably due to relatively recent amplification events of source sequences (Chapter 9) rather than to concerted evolution.

In the second type of arrangement, all members of a family are clustered either in a single tandem array or in a small number of tandem arrays located on different chromosomes. This arrangement is the most favorable for unequal crossover and gene conversion. If the repeats are located on more than one chromosome, the rate of unequal crossover is greatly reduced, unless the clusters occur at the ends of chromosome arms (as in the case of the rDNA family in humans). Moreover, the rate of gene conversion would also be reduced. However, Ohta and Dover (1983) have shown that such a reduction in gene conversion rate has only a minor effect on the extent of identity between genes, unless the conversion rate between genes on nonhomologous chromosomes becomes very low, or unless the number of nonhomologous chromosomes on which gene family members reside is large.

Structure of the repeat unit

The structure of the repeat unit refers to the numbers and sizes of coding regions (i.e., exons) and noncoding regions (i.e., introns and spacers) within the repeat unit. As noncoding regions generally evolve rapidly, it is difficult to maintain a high degree of similarity among the repeats if each repeat contains large or numerous noncoding regions. We note that homogeneity and concerted evolution go hand in hand, since both unequal crossover and gene conversion depend on sequence similarity for misalignment of repeats. Thus, the higher the homogeneity among the repeats in a family, the higher the rates of unequal crossover and gene conversion.

Zimmer et al. (1980) estimated that in the great apes, the rate of concerted evolution in the α-globin gene region is 50 times higher than that in the β-globin gene region. They suggested that the rate in the β region has been greatly reduced because the introns and flanking sequences are highly divergent between the two β genes. It is interesting to note that the β genes have introns that are several times longer than those of the α genes, and that the intergenic region between the two β genes is 2,400 bases longer than that between the two α genes. Indeed, Zimmer et al. (1980) suggested that the larger introns and intergenic region in the β genes arose as a response to selection against unequal crossover, which may produce a single homogenized gene out of the β- and δ-globin genes (called hemoglobin Lepore), whose expression is under the control of the δ promoter and is deleterious in the homozygous state.

We note, however, that qualitative (as opposed to quantitative) arguments concerning putative advantages associated with protection against mutational events (e.g., avoidance of pretermination codons, prevention of crossover events) are usually highly exaggerated, because the selective advantage for a reduction in the rate of a mutational event would at most be as large as the mutation rate itself. Assuming that mutational events occur at rates of 10^{-5} to 10^{-9}, the selective advantage would be insignificant except in taxa with enormous effective population sizes. Thus, the large introns and intergenic regions might have arisen by chance rather than by selection. It is possible that the introns and the intergenic regions were already large before the divergence of the apes, and that this promoted the divergence between the two β genes rather than vice versa.

Functional requirements and selection

Let us first consider two extreme situations. One is that the function has an extremely stringent structural requirement, often requiring large amounts of the same gene product (dose repetitions). The rRNA genes and the histone genes are well-known examples. The other extreme is that the function requires a large amount of diversity. The immunoglobulin and histocompatibility genes belong to this category.

In general, the rate of concerted evolution is expected to be higher in the former type than in the latter type of families. In rRNA genes, purifying selection will tend to eliminate new variants and promote homogeneity, which in turn will facilitate unequal crossover and gene conversion among members of multigene families, thus accelerating the process of concerted evolution. In immunoglobulin genes, on the other hand, an individual with many identical copies owing to concerted evolution would be at a severe disadvantage, as its arsenal of immunoglobulins against pathological antigens would be limited. Thus, concerted evolution may not play an important role in the evolution of such families.

A general feature of concerted evolution is that it cannot persist for long periods of time in the absence of purifying selection. This is because both gene conversion and unequal crossing over require a high degree of sequence similarity between the duplicate genes so that pairing between two paralogous sequences can occur. Without selection, sequence similarity tends to decrease with time because of the accumulation of mutations. Teshima and Innan (2004) studied concerted evolution under a mathematical model that allows mutation and gene conversion, but no selection. They found that concerted evolution of duplicate genes can persist for only short periods of time. As seen in **Figure 7.28**, the average sequence difference (d) between two duplicated sequences is 0 at the time of the gene duplication event, but it will increase with time to reach the equilibrium value (d_0) that is determined by the opposing effects of mutation and gene conversion (phase I). Then, d will fluctuate around the equilibrium value because of stochastic effects (phase II) until chance carries it over a threshold value (d_t), above which gene conversion can no longer occur. From this point onward, d will increase monotonically with time (phase III). In this model, a clear-cut threshold value is assumed, such that the probability of gene conversion suddenly becomes 0 when d exceeds the threshold. In reality the probability is likely to decrease gradually as d increases. Under this situation, the transition from phase II to phase III is less

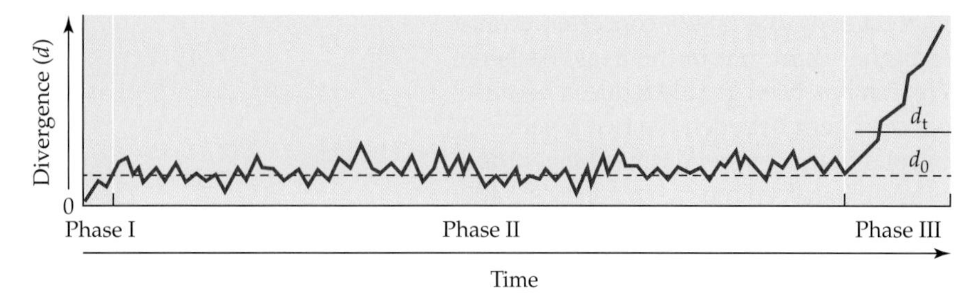

Figure 7.28 A schematic example of the changes in average sequence divergence (d) between two paralogs after a gene duplication event at time 0. In phase I, d tends to increase until it reaches an equilibrium value (d_0; dashed line). In phase II, d fluctuates around the equilibrium value. In phase III, d happens to exceed a threshold (d_t), above which neither gene conversion can occur, so d increases monotonically with time. The time duration of phase II follows an exponential distribution that depends on the rate of gene conversion, the length of the conversion tract, and the rate of mutation. (Modified from Teshima and Innan 2004.)

sharp but will eventually occur. Thus, without purifying selection to maintain sequence similarity, concerted evolution between duplicated genes cannot continue for a very long time.

For protein-coding genes, functional constraints at the protein level will retard the rate of divergence between sequences, so concerted evolution can persist for a longer time than for sequences without function. This is particularly true for genes coding for highly constrained proteins such as histones. However, even for protein-coding genes, sequence divergence can occur at degenerate sites where synonymous substitution can occur. In this case, the strength of codon usage bias can make a difference in the persistence of concerted evolution, with genes that exhibit highly biased codon usage being more disposed to concerted evolution than genes that exhibit no codon preference (Gao and Innan 2004).

Lin et al. (2006) pointed out that purifying selection is a constant force that operates in each and every generation, whereas gene conversion is unlikely to occur with any constancy or regularity. Moreover, purifying selection can occur without gene conversion, whereas concerted evolution conversion cannot continue for long without purifying selection. Thus, purifying selection is more potent in maintaining sequence homogeneity between duplicated genes than either gene conversion or unequal crossover.

Finally, we note that concerted evolution by unequal crossing over may be affected by centripetal selection, i.e., purifying selection against too many or too few repeats. Specifically, unequal crossing over is known to create a large variation in the number of repeats among individuals in a population. If a certain number of copies is required, i.e., if an optimum copy-number range exists, then centripetal selection may become an important force shaping concerted evolution patterns.

Population size

Population size affects the rate of concerted evolution because concerted evolution requires not only the horizontal spread of genetic variation among members of a gene family, but also the fixation of such homogeneous variants within the population. Obviously, the time required for a variant to be eliminated from a population or to become fixed in a population is dependent on the population size (Chapter 2).

Positive natural selection will accelerate the process of concerted evolution because the rate and probability of fixation for a variant favored by natural selection will be larger than those for selectively neutral variants. The effect of biased gene conversion on the evolution of multigene families would be similar to that of positive selection, albeit somewhat weaker. In addition, biased gene conversion will be more effective when the number of repeats is large (Walsh 1985). Both natural selection and biased gene conversion work more effectively in large populations than in small ones, because the effect of random genetic drift decreases with population size.

Evolutionary Implications of Concerted Evolution

Concerted evolution allows the spread of a variant repeat to all gene family members. This capability has profound evolutionary consequences. In this section we discuss the effects of concerted evolution on the spread of advantageous mutations, the rate of divergence between duplicate genes, and the generation of genic variation.

Spread of advantageous mutations

Through concerted evolution, an advantageous mutant can spread rapidly and replace all other repeats within a gene family. We note that the selective advantage that a single variant can confer on an organism is usually very small. The advantage would, however, be greatly amplified if the mutation were to spread within the genome. Thus, through concerted evolution, a small selective advantage can become a great advantage. In this respect, concerted evolution surpasses independent evolution of individual gene family members (Arnheim 1983; Walsh 1985).

Arnheim (1983) compared the evolution of RNA polymerase I transcriptional control signals with that of RNA polymerase II transcriptional control signals. RNA polymerase I transcribes rRNA genes, whereas RNA polymerase II transcribes protein-coding genes (Chapter 1). RNA polymerase I transcriptional control signals appear to have evolved much faster than the signals for RNA polymerase II. For example, in cell free transcription systems, a mouse rDNA clone does not work in a human cell extract, but clones of protein-coding genes from astonishingly diverse species can be transcribed in heterologous systems (e.g., silkworm genes in human cell extracts, and mammalian genes in yeast). Arnheim (1983) argues that in the case of transcription units for RNA polymerase I, mutations that favorably affect transcription initiation were propagated throughout the rDNA multigene family as a consequence of concerted evolution, and they could become species-specific. On the other hand, in the case of transcription units for RNA polymerase II, advantageous mutations affecting transcription initiation that occur in any one gene would not be expected to be propagated throughout all genes, for they belong to many different families.

Retardation of paralogous gene divergence

The traditional view concerning the creation of a new function is that a gene duplication event occurs, and one of the two resultant genes gradually diverges and becomes a new gene. It is now clear that the process may not be as simple as previously assumed. As long as the degree of divergence between the two genes is not large (as is the case immediately after the duplication event), one copy may be deleted by unequal crossover or converted by the sequence of the second copy by gene conversion. In the former case, an additional duplication would be required to create a new redundant copy, while in the latter case divergence must start again from scratch. Thus, divergence of duplicate genes may proceed much more slowly than traditionally thought.

Generation of genic variation

From an evolutionary point of view, there is an analogy between the evolution of multigene families and the evolution of subdivided populations. We may regard each repeat in a multigene family as a deme in a subdivided population. The transfer of information between repeats is then equivalent to the migration of genes or individuals between demes. It is well known that migration reduces the amount of genetic difference between demes but increases the amount of genic variation (i.e., the number of alleles) in a deme. Similarly, transfer of information between repeats will reduce the genetic difference between repeats but will increase the amount of genic variation at a locus (Ohta 1983, 1984; Nagylaki 1984). Indeed, some loci in the mouse major histocompatibility complex are highly polymorphic, with as many as 50 alleles being observed at a locus, and it has been suggested that the high polymorphism is due to concerted evolution (Weiss et al. 1983). An alternative explanation is that the alleles have persisted in the population for very long periods of time (Figueroa et al. 1988), probably being maintained by overdominant selection (Hughes and Nei 1989). Of course, these two mechanisms are not mutually exclusive, and they may both operate at these loci.

Methodological pitfalls due to concerted evolution

It has been customary to assume that, following a gene duplication, the two resultant genes diverge monotonically with time. Under this assumption, as we have previously shown, it is rather simple to infer the time of the duplication event (p. 275). An unfortunate feature of concerted evolution is that it erases the record of molecular divergence during the evolution of paralogous sequences. Thus, when dealing with very similar paralogous sequences from a species, it is usually impossible to distinguish between two possible alternatives: (1) the sequences have only recently diverged from one another by duplication, or (2) the sequences have evolved in concert. One way to distinguish between the alternatives is to use a phylogenetic approach. For example, the two α-globin genes in humans are almost identical to one another. Initially they were thought to have duplicated quite recently and that there had not been sufficient time for them to diverge in sequence. However, duplicated α-globin genes were also discovered in distantly related species, and so one had to assume either that multiple gene duplication events occurred independently in a great number of evolutionary lineages, or that the two genes are quite ancient, having been duplicated once in the common ancestor of these organisms, their antiquity subsequently obscured by concerted evolution.

Under concerted evolution, gene duplications appear younger than they really are. Phylogenetic reconstructions based on sequence comparisons can only go back to the last erasure of the evolutionary history. We must therefore use taxonomic information concerning the distribution of duplicated genes versus unduplicated ones to infer the time of gene duplication. In large multigene families, gene correction events are expected to occur frequently, and in such cases it will be even more difficult to trace the evolutionary relationships among the family members. Thus, concerted evolution should be taken into account when attempting to reconstruct the evolutionary history of paralogous genes. Failure to consider this possibility may result in faulty phylogenetic reconstructions.

Positive selection or biased gene conversion? The curious histories of HAR1 and FXY

A particularly intriguing methodological pitfall concerns the confusion between GC-biased gene conversion and selection (see Galtier and Duret 2007; Galtier et al. 2009). GC-biased gene conversion (Marais 2003) is a recombination-associated segregation distortion favoring GC-rich sequences over AT-rich sequences. A trivially expected effect of GC-biased gene conversion is an increase in the GC content of those DNA sequences undergoing conversion. This feature of GC-biased gene conversion has the unfortunate result of causing scientists to identify genomic regions under positive selection where none exist. Moreover, it serves to illustrate a particularly "unintelligent" characteristic of the evolutionary process, i.e., that a mutational process (gene conversion) not only does not promote adaptation but may also lead to the fixation of deleterious traits.

Pollard et al. (2006) proposed an elegant approach for detecting lineage-specific instances of positive selection. They were particularly interested in human sequences that experienced positive selection as candidates for adaptations involved in what makes us human. They first sought orthologous regions that exhibit high levels of conservation in all vertebrates but were very different in humans. Among these regions, they discovered 49 regions in which the substitution rate was significantly elevated in humans in comparison to the rates in other organisms. These regions were called "human accelerated regions" (HARs) and were christened *HAR1–HAR49*. The 118 bp *HAR1*, for instance, showed only two differences between chimpanzees and chickens (>300 million years of divergence) but 18 between chimpanzees and humans (~6 million years of divergence). The strong conservation of HARs among nonhuman vertebrates clearly indicates that they are under functional constraints. The sudden acceleration in substitution rates in the human lineage cannot be explained simply by loss of function and relaxation of selection, because the amount of change in the human lineage is much higher than expected for a neutrally evolving sequence—the average human-chimpanzee divergence in noncoding sequences is less than 2%. HARs are, therefore, good candidates for having evolved under positive selection in humans.

We note, however, that forces other than natural selection can lead to an increased nucleotide substitution rate. First, the rate of mutation can vary along genomes; however, the intragenomic variation in mutation rates that has been reported so far is relatively small (about twofold) compared with the 20-fold increase in the substitution rate in *HAR1*. The second possible explanation is biased gene conversion, in particular GC-biased gene conversion. From an evolutionary point of view, GC-biased gene conversion will yield results that are almost indistinguishable from positive selection (Galtier 2004). A GC-biased gene conversion event will lead to the transmission and eventual fixation of AT → GC mutations. The result will mimic the consequences of an episode of adaptive evolution, including an increased substitution rate.

How can we distinguish between positive selection and GC-biased gene conversion? This question is particularly pertinent to the study of regions that do not code for proteins, for in protein-coding regions we may examine the ratios of nonsynonymous to synonymous substitution as indicators of selection regimes. First, we note that although under GC-biased gene conversion, GC-rich sequences are favored, under positive selection there is no a priori reason why advantageous alleles should systematically be GC-rich. Notably, all 18 nucleotide substitutions observed in *HAR1* are AT → GC changes. Second, *HAR1* is located in a highly recombining region. This is a necessary condition for GC-biased gene conversion, but not for any known type of selection. Third, a mutational process such as GC-biased gene conversion has no regard for function. Indeed, the region of rapid evolution in *HAR1* is not restricted to the 118 bp region, which is conserved in all vertebrates with the exception of humans, but extends to flanking sequences.

A spectacular "experiment of nature" illustrating the effects of biased-gene conversion can be seen in the case of the evolution of the *FXY* gene in mammals (**Figure 7.29**). This gene is X-linked in human, rat, and short-tailed mouse (*Mus spretus*) but was recently translocated to another location on the X chromosome in the house mouse, *M. musculus*. In the house mouse, it now overlaps the boundary between the X-specific region and the pseudoautosomal region, with the 5′ end of the gene (exons 1–3) being located in the X-specific region, which does not recombine in males, and the 3′ end (exons 4–10) in the pseudoautosomal region, a short segment of homology between the X and Y chromosomes, which is a highly recombining region in both males and females. As a consequence of the move, the 3′ *FXY* sequence experienced a sudden increase in recombination rate, followed by a dramatic increase in GC content and in inferred substitution rates. No such acceleration was observed in the 5′ region.

At the protein level, the consequences of the translocation are equally dramatic. The 667-amino-acid protein is highly conserved between mammals (excluding *M. musculus*). The human and *M. spretus* sequences, which diverged more than 80 million years ago, differ by just 6 amino acid replacements. The rat and *M. spretus* sequences, which diverged 10–12 million years ago, differ by 5 amino acid replacements. The house mouse sequence, however, has accumulated as many as 28 amino acid replacements since the divergence from *M. spretus*, roughly 1–3 million years ago (Kurzweil et al. 2009). This corresponds to more than a 100-fold increase in the rate of amino acid replacements in the *M. musculus* lineage. Interestingly, all of these amino acid replacements have occurred in the regions encoded by the 3′ exons and were caused by AT → GC substitutions. Obviously, this elevated substitution rate has nothing to do with selection or adaptation. If directional selection had been acting on the protein sequence, then silent sites (most third-codon positions and introns) should have remained unaffected, but they did not. The estimated numbers of synonymous substitutions on the

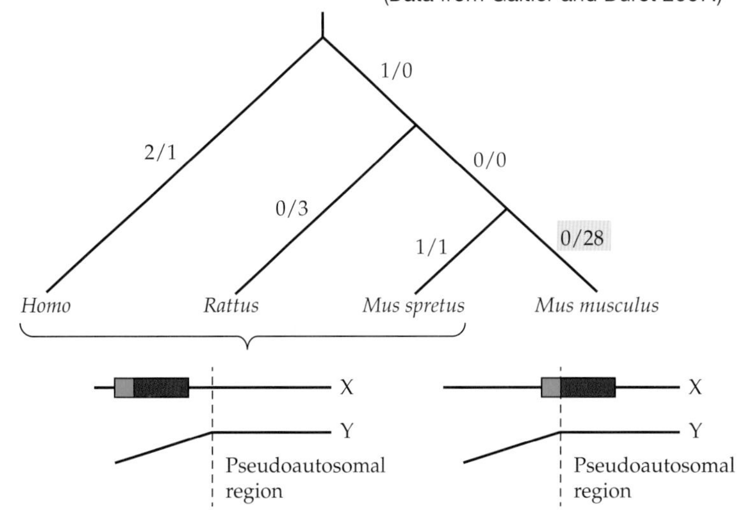

Figure 7.29 Evolutionary history of *FXY* genes in four mammalian species. In most mammals the *FXY* gene is located in the X-specific region, which cannot recombine in males (diagram below tree). In the house mouse, *Mus musculus*, the gene was translocated to a new position on the X chromosome. This position overlaps the pseudoautosomal boundary (vertical dashed line), with exons 1–3 (blue) located in the X-specific region and exons 4–10 (red) located in the pseudoautosomal region. For each branch, the numbers of nonsynonymous substitutions that have occurred in the 5′ and 3′ ends of the gene are given on the left and right sides of a forward slash, respectively. A huge increase in the rate of amino acid replacement can be seen to have occurred in the 3′ end of the *FXY* gene in the *M. musculus* lineage. (Data from Galtier and Duret 2007.)

M. musculus branch are 1 for the 5′ end of the gene and 163 for the 3′ end. Moreover, if selective pressures were at work, they should have affected the entire gene, and not only its 3′ end. Finally, there is no reason to assume that all adaptive changes will be caused by AT → GC substitutions.

An important lesson from *HAR1* and *FXY* is that evolution is neither intelligent nor monotonically directional with respect to adaptation. GC-biased gene conversion can theoretically overcome purifying selection and lead to the fixation of deleterious AT → GC mutations. Conversely, it can overcome positive selection and erase an advantageous GC → AT mutation. Despite initial claims, it is unlikely that *HAR1* and *FXY* contribute to adaptation. Rather, they represent imperfections, whose functions were somehow preserved in spite of many "undesired" effects of gene conversion.

Birth-and-Death Evolution

The birth-and-death model for the evolution of gene families was proposed by Hughes and Nei (1989). In this model, new gene copies are produced by gene duplication. Some of the duplicate genes diverge functionally; others become pseudogenes owing to deleterious mutations or are deleted from the genome. The end result of this mode of evolution is a multigene family with a mixture of functional genes exhibiting varying degrees of similarity to one another plus a substantial number of pseudogenes interspersed in the mixture.

The birth-and-death model was initially put forward to explain the unusual pattern of evolution of the major histocompatibility genes in mammals (Hughes and Nei 1989; Klein et al. 1993; Nei et al. 1997). Subsequently, the model was used to explicate the evolution of such large and heterogeneous multigene families as the immunoglobulins and olfactory receptors, as well as that of highly conserved gene families such as the genes for histones and ubiquitins (Nei and Rooney 2005). Interestingly, even gene families consisting of only a few paralogs, such as the opsin and hemoglobin families, exhibit telltale signs of birth-and-death evolution (below).

Expansion and contraction of gene families

An important prediction of the birth-and-death process is that gene family size will vary among taxa as a result of differential birth and death of genes among different evolutionary lineages. Thus, an understanding of the evolutionary forces governing the birth-and-death process is predicated upon an accurate accounting of the number of births (duplications) and deaths in each lineage. This "bookkeeping" turns out to be anything but a trivial undertaking. Two computational methods have been employed for this type of analysis. One method requires a well-supported species tree and a detailed gene tree for each member of the multigene family. By reconciling the gene tree with the species tree, one can infer the number of gene gains and losses on each branch of the species phylogeny (Zmasek and Eddy 2001). The second method uses maximum likelihood to infer family size at each internal node in the species phylogeny. It requires a priori knowledge of the number of gene copies in each family for each OTU, as well as estimates of divergence time between each pair of taxa (De Bie et al. 2006).

In **Figure 7.30**, a completely resolved phylogenetic tree for five *Saccharomyces* species is shown. There were 3,517 gene families shared by the five species. Of these, 1,254 (~37%) have changed in size across the tree. By inferring the most likely ancestral gene family sizes for all of these gene families, it was possible to deduce the number of changes in gene family size on all eight branches of the tree. Note that on each branch in the tree, the vast majority of gene family sizes remain static. Expansions outnumbered contractions on four of the eight branches, and contractions outnumbered expansions on the other four. Such evolutionary reconstructions can also be used to test for differences among lineages. For example, let us compare the numbers of expansions and contractions on the branches leading to *S. mikatae* and *S. cerevisiae* from their common ancestor, approximately 22 million years ago. On

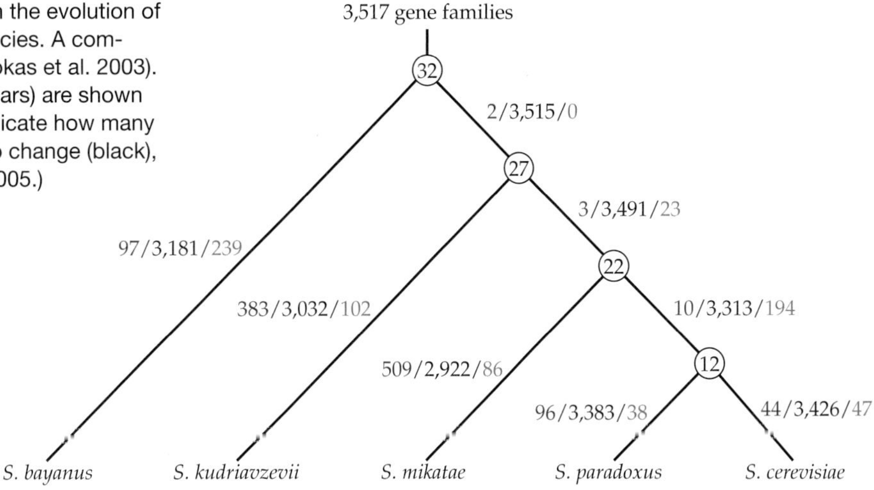

Figure 7.30 Expansions and contractions in the evolution of 3,517 gene families in five *Saccharomyces* species. A completely resolved phylogenetic tree is shown (Rokas et al. 2003). Estimates of divergence times (in millions of years) are shown in circles. The numbers along the branches indicate how many gene families experienced expansions (red), no change (black), or contractions (blue). (Data from Hahn et al. 2005.)

the lineage leading to *S. mikatae* there were 509 families that expanded and 86 families that contracted—a ratio of 6:1. In contradistinction, on the lineage leading to *S. cerevisiae* a smaller number of families changed their size, and the ratio of expanded families (54) to contracted ones (241) was inverted, 1:5. This lineage specificity of change in families and functions implies that adaptation via copy number change is not a peculiarity of specific gene families: rather, it is a general mechanism that affects many different gene families depending on lineage-specific evolutionary pressures.

Rates of gene gain and loss are determined by an often difficult to disentangle interplay of mutation, fixation, and retention probabilities (Demuth and Hahn 2009). Analyses of gene family evolution in mammals revealed highly dissimilar rates of gene turnover across taxa. For example, the gene turnover rate in primates is nearly twice that in nonprimate mammals (0.0024 versus 0.0014 gains and losses per gene per million years). A further acceleration must have occurred in the great-ape lineage, such that humans and chimps gain and lose genes almost three times faster (0.0039 gains and losses per gene per million years) than the other mammals. In eukaryotes, new duplicates are "born" at the rate of 0.001–0.016 per gene per million years (Gu et al. 2002; Lynch and Conery 2003).

Rapid gene family expansion (amplification) in functionally important genes suggests adaptive scenarios in which natural selection favors additional copies either for increased dosage or for an increased arsenal of molecular weaponry. Particular cases illustrating this phenomenon have already been presented (p. 283). Here we look at general trends. In eukaryotes, gene amplification appears not to be common, as opposed to the situation in prokaryotes, where it is ubiquitous. The rarity of gene amplification in eukaryotes may be due to the ineffectiveness of selection in small populations rather than actual differences in mutational input. The population size effect is reinforced by the fact that among eukaryotes, adaptive amplification appears most frequently in yeast, followed by insects, and is rare or absent in vertebrates. Gene amplifications in yeast are responsible for, among other effects, resistance to copper toxicity and growth under resource-limited conditions. As mentioned on page 293, independent amplifications of certain esterase genes are responsible for resistance to organophosphate pesticides (Field et al. 1988; Vontas et al. 2000), the most dramatic case being a 250-fold copy number increase in resistant strains of the mosquito *Culex pipiens* (Mouches et al. 1986).

In vertebrates, the evidence suggests a very limited role for gene amplification under positive selection. In mammals, less than 2% of all families display telltale signs of selective amplification. Gene families involved in immune defense, metabolism, cell signaling, chemoreception, and reproduction tend to amplify more frequently than would be expected by chance (see Demuth and Hahn 2009).

Examples of birth-and-death evolution

The birth-and-death process will be illustrated by the evolution of three gene families. Each was chosen to illuminate a different property or consequence of the process.

EVOLUTION OF THE OLFACTORY RECEPTOR GENE REPERTOIRE Olfactory receptors are G-coupled proteins that have seven α-helical transmembrane regions. Olfactory receptor genes are chiefly expressed in sensory neurons of the main olfactory epithelium in the nasal cavity. Vertebrates use different olfactory receptors and different combinations of olfactory receptors to detect many types of volatile or water-soluble chemicals. The number of functional olfactory receptor sequences in sequenced chordate genomes ranges from one in elephant shark to about 2,000 in the African elephant. The number of functional olfactory receptors is small in comparison to the number of known recognizable odorants, but olfactory receptors are known to function in a combinatorial manner, whereby a single receptor may detect multiple odorants, and a single odorant may be detected by multiple receptors (Malnik et al. 1999). Moreover, it was demonstrated that some olfactory receptors are "generalists" that bind to a variety of ligands, whereas others are "specialists" that are narrowly tuned to a small number of ligands.

In addition to intact olfactory receptor genes, the genomes of vertebrates contain numerous truncated genes and pseudogenes, both of which are presumably devoid of function. The percentage of nonfunctional olfactory receptor sequences ranges from 12% in zebrafish to 73% in tree shrew. The numbers of functional and nonfunctional olfactory receptor sequences in chordates are shown in **Figure 7.31**. (Olfactory receptors in invertebrates, including insects, nematodes, echinoderms, and mollusks, are not homologous to vertebrate olfactory genes. It therefore appears that genes encoding chemosensory receptors have evolved many times independently throughout animal evolution.)

As with other genes for G-coupled proteins, olfactory receptor genes do not have any introns in their coding regions. However, they often have introns and exons upstream of their coding regions. The noncoding exons can be alternatively spliced to generate multiple mRNA isoforms; however, this results in the same protein. Thus, it is not clear whether or not the alternative splicing of olfactory receptor genes has any biological significance. Olfactory receptor-coding genes constitute one of the largest gene families in tetrapods. For example, approximately 5% of the human proteome consists of olfactory receptors. Olfactory receptor genes form genomic clusters and are dispersed on many chromosomes. In the human genome, for instance, they are located on all chromosomes except 20 and Y.

Olfactory receptor gene loci make up one of the most genetically polymorphic regions in the human genome. Copy number variation is especially common, with many olfactory receptor loci being present in some individuals but not in others. Olfactory receptor gene coding regions also harbor a large number of single-nucleotide polymorphisms, some of which lead to inactivation of the functional gene and the creation of a segregating pseudogene. At least 35% of all human olfactory receptor genes exhibit copy number polymorphism or segregating pseudogenes. Therefore, the numbers of olfactory receptor genes vary enormously among individuals. This variation, in turn, causes olfactory perception to differ widely among individuals. "Specific anosmia" refers to the inability to perceive the odor of a specific substance in an individual having a generally good sense of smell. Specific anosmias are quite common (e.g., Gross 2007). In the United States, for instance, one in ten people cannot smell the extremely poisonous gas hydrogen cyanide, which for most people smells faintly like almonds; 12% cannot detect musky odors, which are common perfume ingredients; and one in 1,000 people cannot smell butyl mercaptan, the rancid issue of skunks, which is used as an additive to natural gas to enable its detection when it escapes or leaks from pipes.

The evolution of the olfactory receptor multigene family is characterized by an extremely rapid process of birth-and-death evolution, whereby genes are frequently added to the repertoire through gene duplication and neofunctionalization, and at the same time frequent gene losses occur by pseudogenization and deletion. The rapidity of the process can be deduced, for instance, by comparing the olfactory receptor repertoire of humans and chimpanzees, which have diverged from each other quite

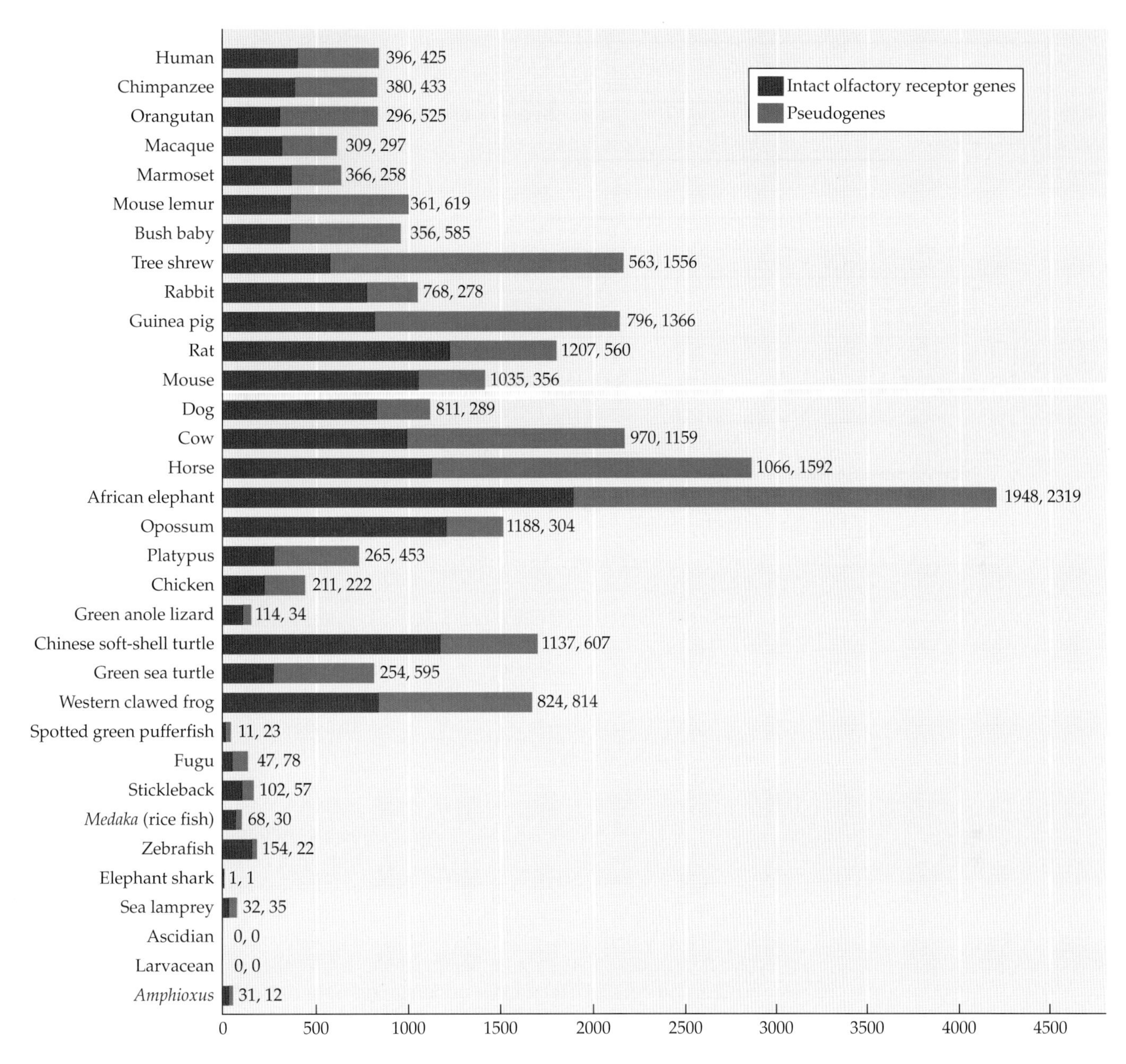

Figure 7.31 Numbers of intact olfactory receptor genes (red) and pseudogenes (blue) in chordates (vertebrates, cephalochordates, and urochordates). Truncated olfactory receptor genes, which are presumably nonfunctional, were included within the count for pseudogenes. The numbers next to the bars represent intact genes (left) and pseudogenes (right). (Modified from Niimura 2012.)

recently. Despite the fact that these two species have similar numbers of functional olfactory receptor genes and a similar fraction of pseudogenes, about 25% of the olfactory receptor genes are species-specific. As seen in **Figure 7.32**, it is estimated that since chimpanzee-human divergence, chimpanzees have gained 8 new species-specific olfactory receptor genes, while humans have gained 18. The rate of gene loss is faster, with chimpanzees and humans losing, respectively, 20% and 19% of the olfactory receptor repertoire of their common ancestor. If we choose as point of reference the most recent common ancestor of all placental mammals, then the largest gain of olfactory receptors occurred in the lineage leading to elephant, resulting in a doubling of the olfactory receptor repertoire. The most impressive loss of olfactory receptors

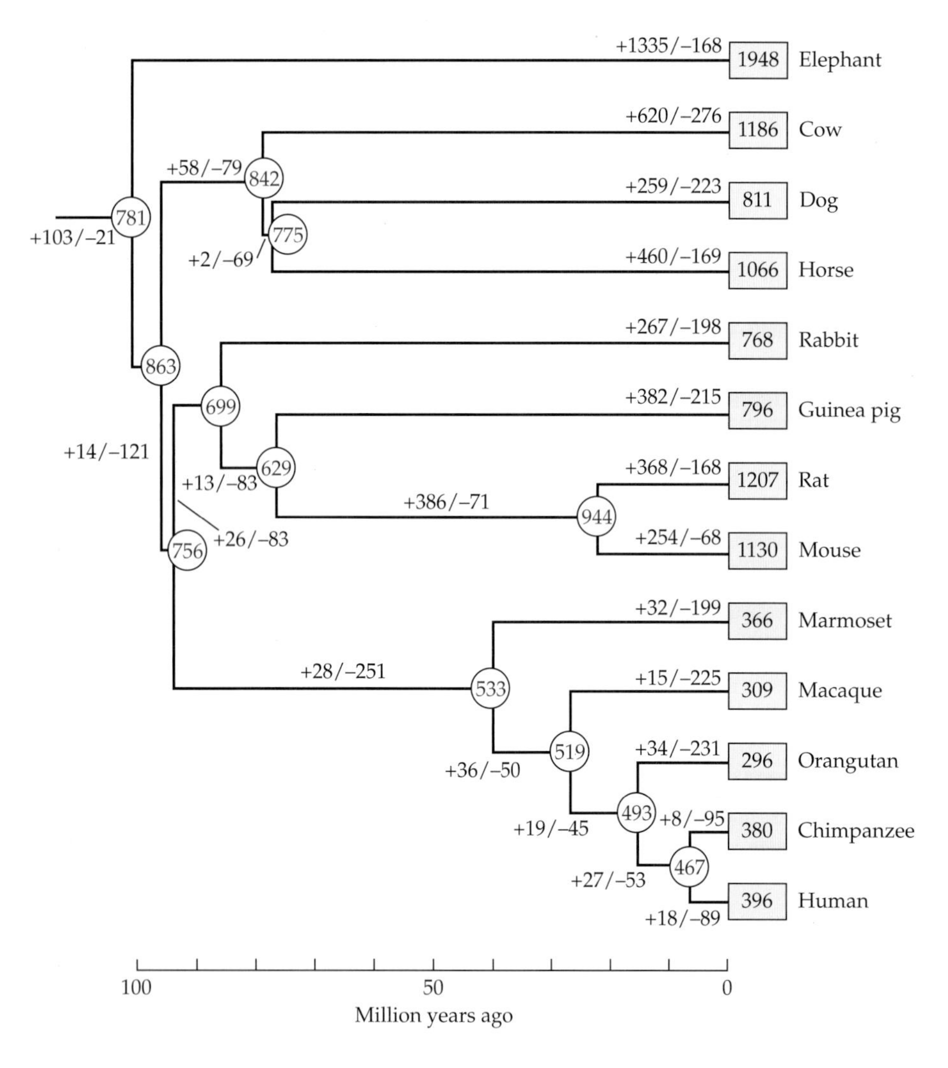

Figure 7.32 Changes in the number of olfactory receptor sequences during the evolution of placental mammals. Numbers in yellow boxes indicate intact olfactory receptor genes in extant species. Numbers in red represent functional olfactory receptor genes in ancestral nodes estimated by the tree reconciliation method (Chapter 11). Estimated numbers of gene gains and gene losses in each branch are shown as + and −, respectively. Speculative divergence times at the nodes of the tree were taken from www.timetree.org. (Modified from Niimura et al. 2014.)

occurred in the lineage leading to primates, in particular orangutans, which lost approximately 70% of the olfactory receptors found in the placental common ancestor. The mean birth and death rates of olfactory receptor genes during placental evolution are 6.2×10^{-3} and 5.9×10^{-3} per gene per million years, respectively.

The extreme rapidity of the birth-and-death gene process characterizing the evolution of the olfactory receptor gene family looks at first glance to be an exceptional case. A deeper understanding of the process, however, reveals that the evolutionary processes affecting this gene family abide by all the evolutionary rules we have discussed previously. For example, olfactory receptor gene subfamiles that have undergone many gene duplications tend to evolve faster than those subfamilies in which the number of duplications has been smaller (Niimura et al. 2014). This is understandable because gene duplication creates redundancy and a relaxation of functional constraint.

The vertebrate olfactory receptor multigene family originated most probably more than 550 million years ago in chordates. Phylum Chordata consists of three subphyla: vertebrates, urochordates (e.g., tunicates), and cephalochordates (e.g., lancelets). Among the three subphyla, cephalochordates are the most basal clade. The amphioxus, a member of Cephalochordata, lacks any distinctive olfactory apparatus. Nevertheless, more than 30 vertebrate-type olfactory receptor genes were identified within its genome. Amphioxus olfactory receptor genes are highly divergent from the vertebrate counterparts, but they are clearly olfactory receptor genes rather than G-coupled proteins unrelated to olfaction. No olfactory receptor-like genes were found in any of the two sequenced urochordate genomes. The absence of vertebrate-type olfactory receptor genes from urochordate genomes suggests that all olfactory receptor genes were lost in this lineage.

Phylogenetic analysis has shown that vertebrate olfactory receptor genes can be classified into at least seven subfamilies (α, β, γ, δ, ε, ζ, and η), each of which originated from one or a few ancestral genes in the most recent common ancestor of vertebrates (Nei et al. 2008). The taxonomic distribution of these subfamilies suggests that they can be further divided into three functional categories (**Table 7.4**). One category contains α and γ, which are present in tetrapods but absent in fish (with the exception

Table 7.4

Relative abundance of olfactory receptor sequences[a]

Organism	Subfamily[a] α	β	γ	δ	ε	ζ	η
Zebrafish	—	4/3	1/0	62/7	12/1	37/4	38/7
Medaka	—	3/0	0/1	33/15	3/1	9/3	20/10
Stickleback	—	1/0	0/3	71/45	4/0	18/5	8/4
Fugu	—	1/0	—	30/50	2/1	4/6	10/21
Spotted green pufferfish	—	—	—	4/14	2/0	2/0	3/9
Western clawed frog	8/4	14/8	752/780	27/16	13/4	—	10/2
Green anole lizard	1/0	—	111/34	—	—	—	—
Chicken	9/4	—	202/218	—	—	—	—
Platypus	31/39	—	234/414	—	—	—	—
Opossum	216/37	5/4	967/263	—	—	—	—
Cow	140/132	2/0	828/1,027	—	—	—	—
Dog	159/55	1/1	651/233	—	—	—	—
Mouse	110/50	3/0	922/306	—	—	—	—
Rat	132/31	2/1	1,073/528	—	—	—	—
Macaque	36/76	0/1	273/220	—	—	—	—
Chimpanzee	64/39	0/1	316/393	—	—	—	—
Human	58/43	0/1	329/371	—	—	—	—

Source: Data from Niimura (2012).

[a] Functional olfactory receptor genes and presumed nonfunctional (truncated genes and pseudogenes) olfactory receptor sequences are listed to the left and right of the slashes, respectively.

of a single intact γ gene in zebrafish). The second category consists of δ, ε, ζ, and η genes, which are found in teleost fishes and amphibians, whereas reptiles, birds, and mammals completely lack them. Interestingly, amphibians retain both categories of genes. This observation suggests that the primary function of α and γ receptors is to detect airborne odorants, while the primary function of δ, ε, ζ, and η receptors is to detect water-soluble odorants. The third category consists of the genes for β olfactory receptors, which are present in both aquatic and terrestrial vertebrates. It has been speculated that β olfactory receptors may detect odorants, such as alcohol, that are both water-soluble and airborne (Niimura 2009), although at present experimental evidence for this hypothesis is lacking.

Water-soluble odorants disperse more slowly than airborne odorants. Nevertheless, smell is an important source of information also for aquatic animals. Fish, for instance, use olfactory information for finding food, avoiding predators, and identifying potential mates and as identifiers of geographical locations. For example, the salmon's remarkable homing ability relies on olfaction and odorant memory. Most fish have four nostrils through which water flows in one direction carrying odorants. This one-way flow allows fish to constantly access new odor information. Fish detect mainly four groups of water-soluble molecules as odorants: amino acids, gonadal steroids, bile acids, and prostaglandins. These are nonvolatile chemicals, and thus humans cannot smell them. As shown in Table 7.4, teleost fishes (zebrafish, medaka, stickleback, fugu, and spotted green pufferfish) generally have much smaller numbers of olfactory receptor genes than mammals, but a much larger representation of gene subfamilies.

In nonamphibian tetrapods, an enormous expansion in the number of α and γ genes occurred. As shown in Figure 7.31, primates tend to have smaller numbers of olfactory receptor genes than other mammals, possibly because primates rely on vision more heavily than on olfaction in sensing the world around them. Indeed, the relative sizes of the olfactory bulb in the brain and olfactory epithelium in the nasal cavity are smaller in primates than in most other mammals. On the basis of their nostril morphology, the order Primates is classified into two suborders, Strepsirrhini and Haplorhini. Strepsirrhines and haplorhines are characterized, respectively, by the presence or absence of the rhinarium, the moist and hairless surface at the tip of the nose, so familiar to us on cats and dogs, that is used to detect the direction of odorants. Haplorhines also have a smaller olfactory epithelium than strepsirrhines. Moreover, most strepsirrhines are nocturnal while most haplorhines are diurnal, and color vision is well developed in haplorhines. Therefore, the reliance on olfaction is decreased in haplorhines compared with that in strepsirrhines. Consistent with these observations, the number of putatively functional olfactory receptor genes is smaller in haplorhines than in strepsirrhines.

Old World monkeys generally have a smaller number of functional olfactory receptor genes than rodents and a higher proportion of pseudogenes. For example, mice have approximately 600 more functional genes than humans and a much lower proportion of pseudogenes (~24% compared with 52% in the human genome). Why do rodents and Old World monkeys differ so conspicuously in the number of functional genes and proportion of pseudogenes? A popular explanation is that Old World monkeys are equipped with trichromatic color vision (see below) and, therefore, olfaction is no longer of vital importance. Thus, it was originally thought that olfactory receptor genes were lost concomitantly with the acquisition of trichromatic vision. The "vision priority hypothesis" (Gilad et al. 2004) states that the evolution of color vision in primates may have decreased primate reliance on olfaction, which resulted in the relaxation of selective constraint and the consequent accumulation of pseudogenes. A phylogenetic analysis by Niimura (2012) indicated that there was no sudden loss of olfactory receptor genes coincident with the emergence of trichromacy. Neither was the accumulation of pseudogenes sudden. Rather, olfactory receptor genes were lost gradually in all primate lineages, not only in those that possess trichromatic vision. In humans, the process of pseudogenization in olfactory receptor genes seems to be ongoing, as many loci seem to be polymorphic and segregate a functional allele and a pseudogene, whereas their corresponding homologous loci in chimpanzees have only functional alleles, thus indicating that the pseudogene is a derived character state in humans (Menashe et al. 2003).

The vision priority hypothesis was based on the assumption that the number of functional olfactory receptor genes is correlated to olfactory versatility. In this view, a decrease in the number of functional olfactory receptor genes would cause a decrease in smell ability. This assumption is most probably flawed. For example, dogs, which are famous for their keen olfactory sense, nevertheless possess an unremarkable number of olfactory receptor genes. While the number of olfactory receptor genes in a species may indeed be positively correlated with the number of odorants among which it can discriminate, the sensitivity to a specific odorant may be determined by levels of expression of particular olfactory receptor genes.

What happens to the olfactory apparatus of terrestrial mammals that become secondarily aquatic? Can the ability to sense water-soluble odorants be reacquired after it has been lost? Can olfactory receptors suitable for smelling airborne odorants evolve into receptors for water-soluble ones? The answers seem to be negative. For example, dolphins and other toothed whales, which secondarily adapted to the marine habitat, have completely lost the olfactory apparatus. A toothed whale has no olfactory bulbs, and its nose (blowhole) is located at the top of its head where its sole function is breathing. Apparently the olfactory system that had been used in the terrestrial ancestors neither works in water nor can evolve into one that does. Indeed, the number of intact olfactory receptor sequences in the dolphin genome is extremely small, and almost all of them are pseudogenes. As it takes time to accumulate disruptive mutations in a coding region of a gene after the gene is no longer subject to functional constraint, even intact-looking genes in the dolphin genome are likely to be nonfunctional.

What determines the size and the makeup of the olfactory receptor gene repertoire in a species? They are determined by the mutational process of gene birth and death as well as by historical contingency. Whether a new gene is retained and whether a functional gene is allowed to die seem to reflect selective constraints determined by the reliance of the organism on olfaction. At present, we have no strong evidence for positive selection, nor can we rule out positive selection. We cannot tell whether the numbers and types of olfactory receptor genes in vertebrates are affected more by selection or by chance. Similarly, we are clueless about the factors determining the numbers of pseudogenes. A case in point is the extraordinary repertoire of olfactory receptors found in African elephants. Are the olfactory capabilities of African elephants really exceptional? As yet, no systematic studies assessing olfactory capabilities in African elephants have been published. However, anecdotal evidence indicates that African elephants do indeed have an extraordinarily keen sense of smell. For example, Bates et al. (2007) reported that African elephants can distinguish between two ethnic groups—one with a tradition of elephant hunting and another with no history of direct threat to elephants—by using a mixture of olfactory and visual cues. Additionally, African elephants are said to be able to recognize up to 30 individual family members from olfactory cues in mixtures of urine and earth (Bates et al. 2008).

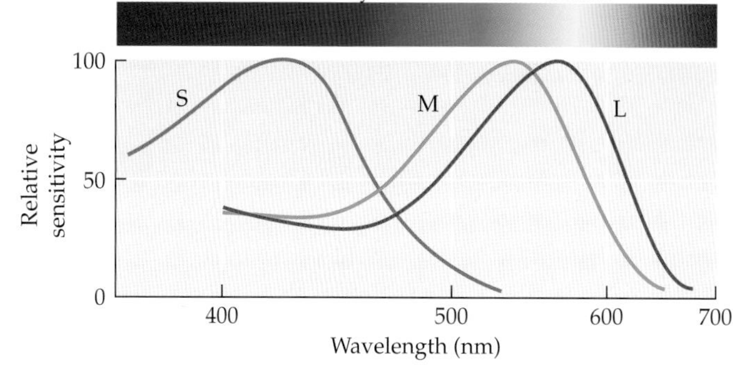

Figure 7.33 Sensitivity spectra for the three photoreceptors in humans. S, short wavelength; M, medium wavelength; L, long wavelength. (Modified from Bowmaker and Dartnall 1980.)

PRIMATE OPSINS Color vision in primates is mediated in the eye by up to three types of photoreceptor cells (cones), which transduce photic energy into electrical potentials. Each type of color-sensitive cone expresses one type of color-sensitive pigment (photopigment). Each photopigment consists of two components: a transmembrane protein called opsin, and either of two lipid derivatives of vitamin A, 11-*cis*-retinal or 11-*cis*-3,4-dehydroretinal. Variation in spectral sensitivity, i.e., color specificity, is determined by the sensitivity maximum of the opsins (**Figure 7.33**). In humans, there are three opsins, S, M, and L, which are maximally sensitive at approximately 430, 530, and 560 nanometers (nm), respectively. S, M, and L opsins are sometimes referred to as "blue," "green," and "red" opsins, respectively, although this nomenclature is somewhat confusing because the sensitivity maxima of the receptors do not coincide with the wavelengths ordinarily recognized as corresponding to these color names.

Each color stimulates one or more kinds of cones. Red light, for example, stimulates L cones much more strongly than M cones, and S cones hardly at all. Blue light, on the other hand, stimulates S cones, but L and M cones more weakly. The brain combines the information from each type of receptor to give rise to different perceptions of different wavelengths of light.

The S opsin is encoded by an autosomal gene, while the L and M opsins are encoded by X-linked genes. In humans, the 364-amino-acid sequences of the L and M opsins are 96% identical, but they share only 43% amino acid identity with the S opsin, which is shorter. The *S* gene and the ancestor of the *L* and *M* genes diverged about 500 million years ago (Yokoyama and Yokoyama 1989). In contrast, the close linkage and high sequence similarity between the *L* and *M* genes, as well as their restricted taxonomic distribution, point to a relatively recent duplication event (~25–35 million years ago). Three amino acid replacements account for the differences in spectral tuning between the red and the green opsins (**Figure 7.34**).

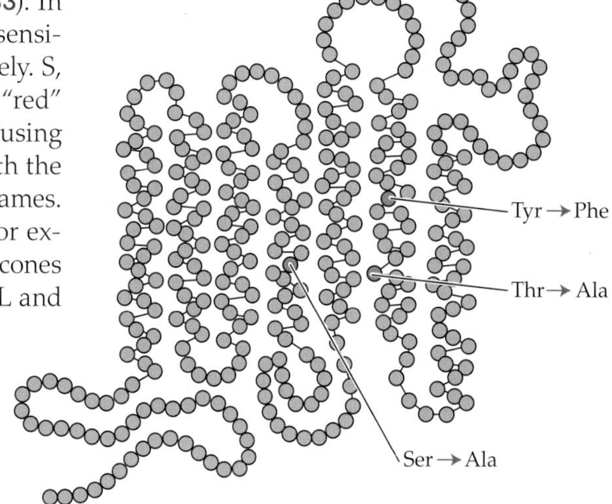

Figure 7.34 Three amino acid replacements in the red opsin protein account for the spectral tuning of the green opsin. At position 180, swapping serine for alanine produces a 6 nm shift of the absorption spectrum; tyrosine to phenylalanine at position 277 provides a 9 nm shift; and changing a threonine to an alanine at position 285 confers another 15 nm shift in maximum absorption. Together, these three changes produce the 30 nm gap between the maximum absorptions of the red and green opsins. (From Grens 2014.)

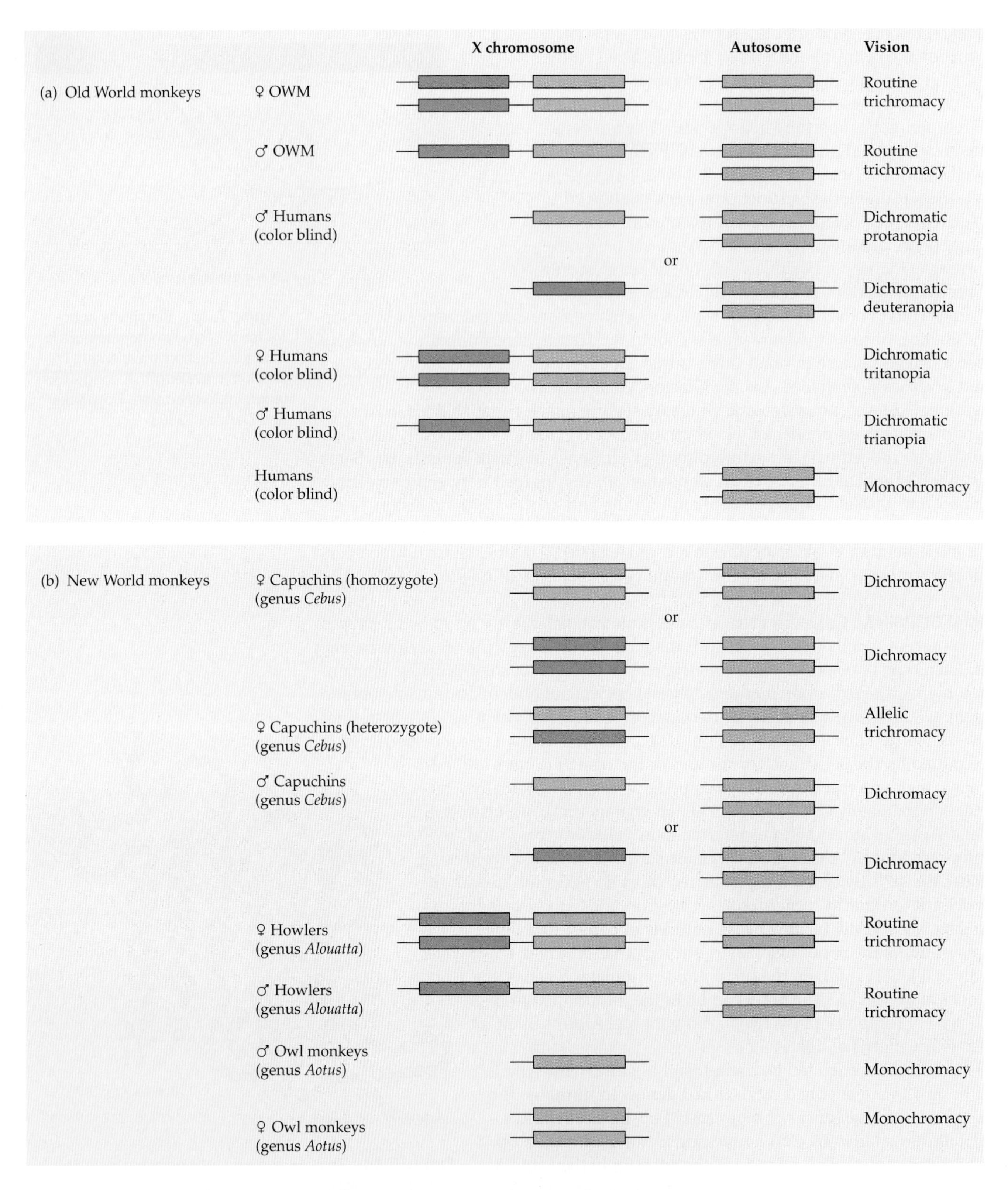

Figure 7.35 Molecular basis of trichromatic, dichromatic, and monochromatic vision in males and females of humans and some Old World monkeys, New World monkeys, and strepsirrhines. Note the distinction between routine and allelic trichromacy. L, M, and S opsins are depicted as red, green, and bue boxes, respectively.

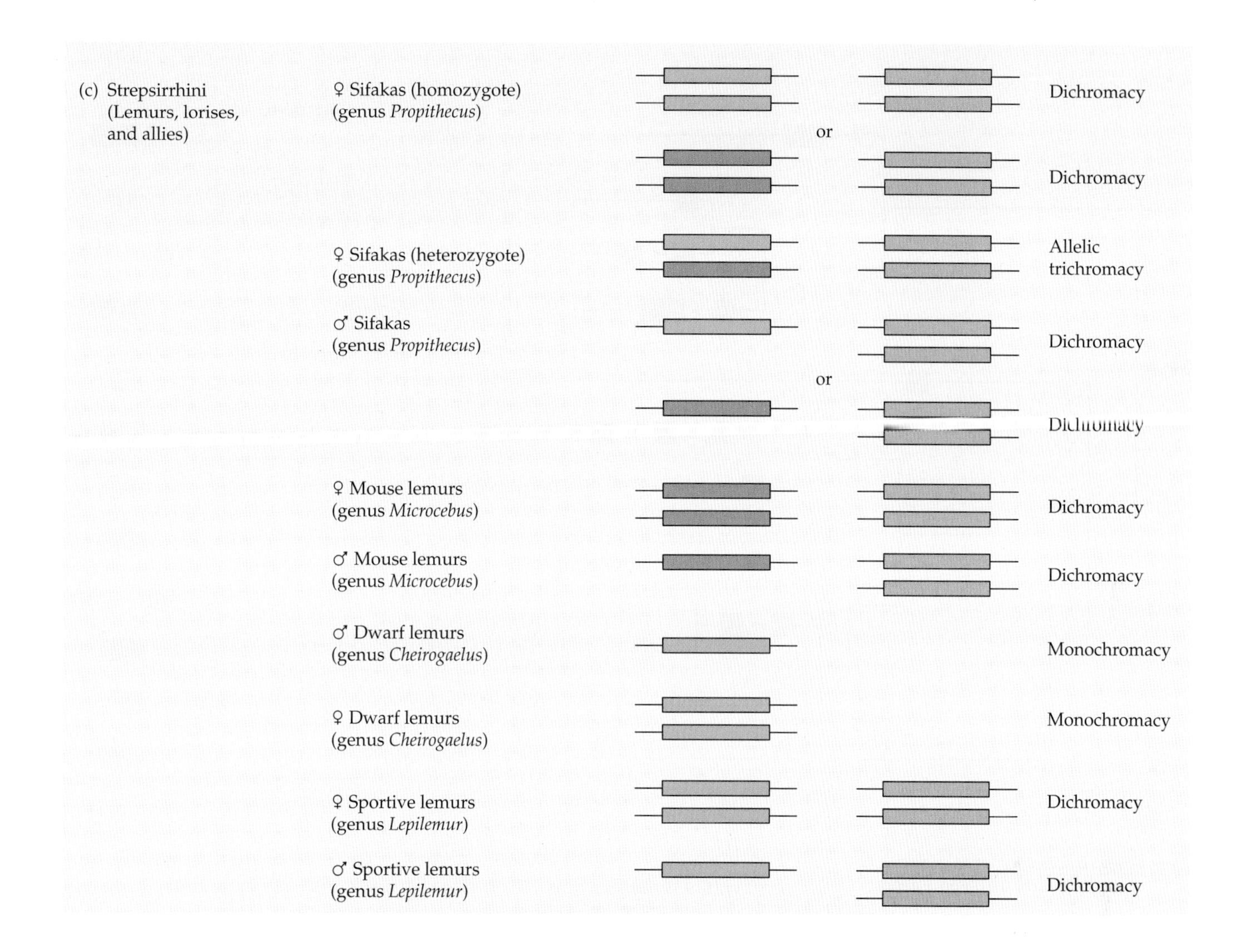

All Old World monkeys (including humans) possess trichromatic vision that is achieved through three distinct opsins encoded by three separate loci. This type of trichromacy is called **routine trichromacy** because all individuals, regardless of sex, can achieve it (**Figure 7.35**). Genetic defects in any one of the opsin genes can lead to dichromacy (which in humans is referred to as color blindness). There are three types of dichromacy: protanopia due to L photopigment deficiency, deuteranopia due to M photopigment deficiency, and the extremely rare tritanopia due to S photopigment deficiency. Because of X linkage, protanopia and deuteranopia are considerably more common in males than in females. Monochromacy can occur if both L and M photopigments are faulty.

Most prosimians (Strepsirrhini) and New World monkeys (Platyrrhini) carry only one X-linked pigment gene and are therefore dichromatic. The ancestral X-linked opsin is thought to resemble the M opsin, and indeed most prosimians and New World monkeys are protanopic. However, because shifts in the maximal sensitivity of opsins can be achieved quite easily by nonsynonymous substitutions in as few as 3–5 codons, in a few diurnal taxa of prosimians, *L* alleles have been produced. In some lineages (e.g., woolly lemurs), the *L* allele became fixed in the population at the expense of the *M* allele. In consequence, these taxa are deuteranopic. In other cases (e.g., sifakas and capuchins), a polymorphic state consisting of two or more alleles is maintained in the population. As an example, in white-faced capuchin monkeys (*Cebus capucinus*), there exist two alleles at the X-linked opsin locus, the maximal-sensitivity peaks of which are similar to those of human L and M opsins, respectively.

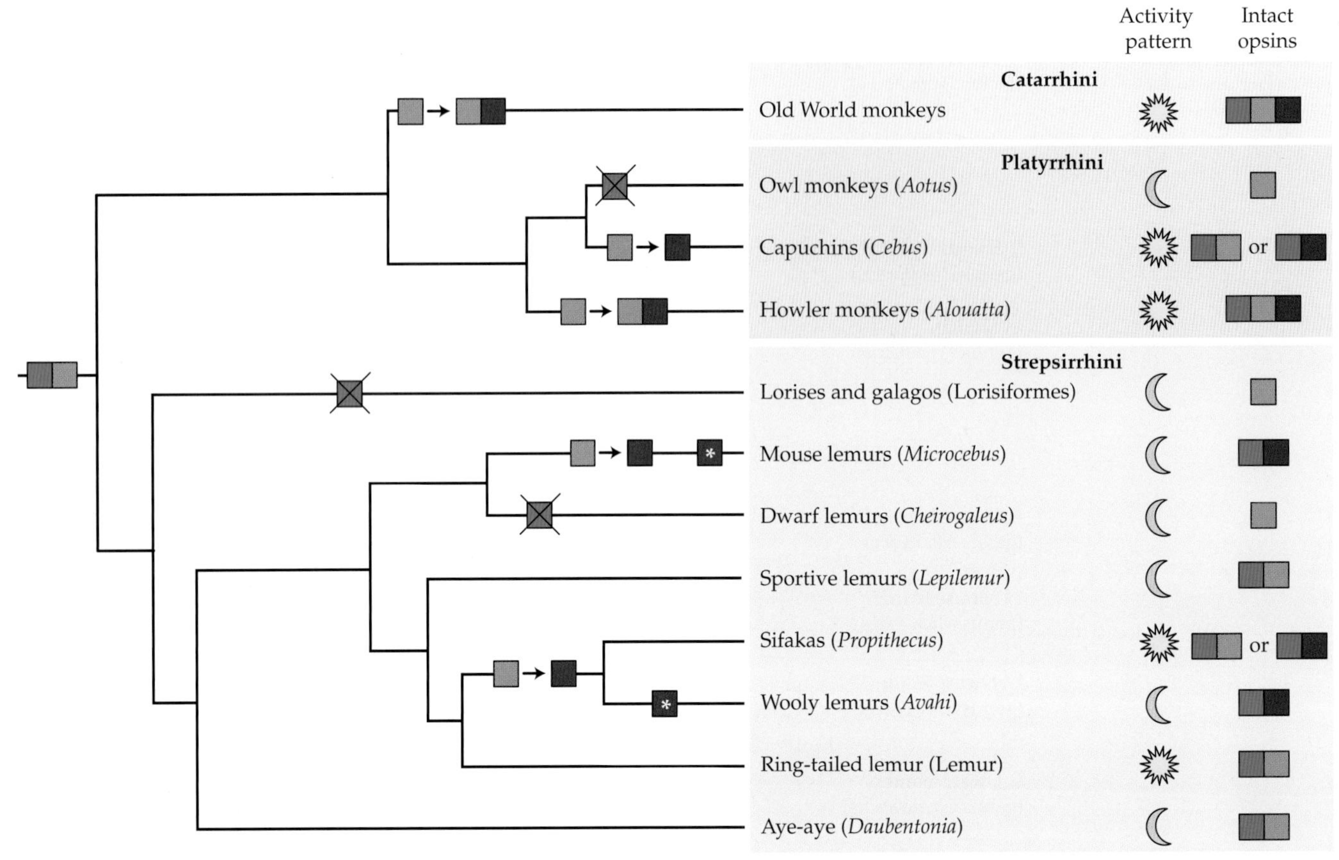

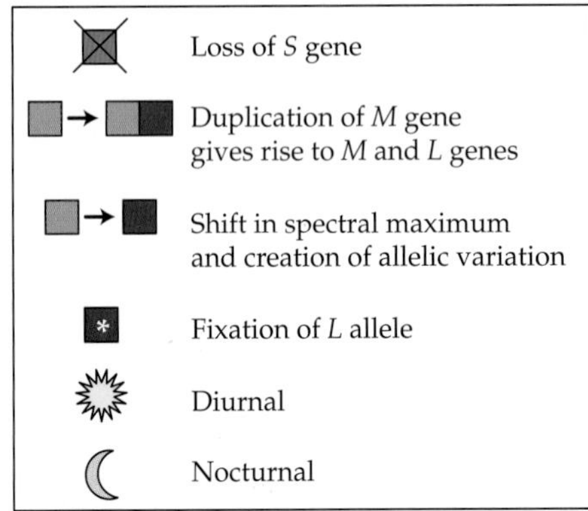

Figure 7.36 The birth-and-death process of primate color vision genes in relation to activity patterns (diurnal or nocturnal). Phylogenetic relationships are based on *SINE* presence/absence as well as on mitochondrial DNA sequences (Roos 2003; Roos et al. 2004). S, M, and L opsins are depicted as blue, green, and red boxes, respectively. Asterisks indicate fixations of the *L* alleles. In *Cebus* and *Propithecus*, an X-linked balanced polymorphism for M and L opsin alleles is maintained, such that heterozygous females may achieve allelic trichromacy. Note the absence of either routine or allelic trichromacy in nocturnal animals. (Data from Perry et al. 2007.)

For this reason, while males and homozygous females are dichromatic, heterozygous females are trichromatic (Figure 7.35). This type of trichromacy is called **allelic trichromacy**.

The long-term maintenance of a high level of polymorphism at the X-linked opsin locus most probably requires some strong form of balancing selection, for example, overdominance or frequency-dependent selection. One possible selective advantage of trichromatic vision is thought to be the ability to detect ripe fruits against a background of dense green foliage.

Interestingly, the evolution of opsin genes in primates can be characterized as a birth-and-death process. As seen in **Figure 7.36**, trichromacy arose twice through independent gene duplications, once in the lineage leading to Catarrhini, the other in the lineage leading to howler monkeys (genus *Alouatta*). Similarly, independent gene deaths of the S opsin in some nocturnal lineages gave rise to independent instances of monochromacy.

BIRTH-AND-DEATH EVOLUTION WITH STRONG PURIFYING SELECTION IN HISTONE MULTIGENE FAMILIES Histones are small basic nuclear proteins in eukaryotes. They are involved in the packaging of DNA and the regulation of gene expression. There are five major histone classes, which can be classified into two groups according to their functional and structural features: core histones (H2A, H2B, H3, H4) and linker histones (H1). Histone genes are extremely conserved in evolution, although *H1* genes evolve faster, on average, than the genes encoding core histones.

How can we distinguish between concerted evolution and birth-and-death evolution when the genes are under extreme functional constraints? A number of groups studied the evolution of several histone gene families (Piontkivska et al. 2002; Rooney et al. 2002; Eirín-López et al. 2004). They reasoned that if concerted evolution is the main driving force in the evolution of histone multigene families, then the number of synonymous substitutions per synonymous site and the number of nonsynonymous substitutions per nonsynonymous site must both be close to zero because gene conversion affects both synonymous and nonsynonymous sites in the same manner. In contrast, if protein similarity is caused by strong purifying selection but every member gene in the family evolves independently, then the number of synonymous substitutions per synonymous site is expected to be greater than the number of nonsynonymous substitutions per nonsynonymous site because synonymous substitutions will be subject to very weak or no purifying selection.

When this approach was applied to histone *H1*, *H3*, and *H4* genes from diverse groups of organisms, the synonymous substitution rate was higher than the nonsynonymous rate in the vast majority of comparisons. These results, therefore, indicate that the members of the histone gene families are mainly subject to strong purifying selection, rather than evolving in unison by gene conversion.

There were some exceptions to these observations. For example, chicken *H1* genes showed similarly low levels of synonymous and nonsynonymous divergence. There are three possible explanations. First, a very recent series of gene duplications produced the duplicates and, hence, there was simply not enough time to accumulate nucleotide substitutions. Second, there is a high degree of codon bias in the genes, which means that synonymous similarity is maintained either by selection at synonymous sites or by biased mutation patterns resulting in high GC contents at synonymous positions. Finally, the results could be explained by gene conversion.

The death of gene families

If a single gene is left in a family, its loss will result in the extinction of that gene family. The death of gene families is simply the continuation of the loss of genes and does not involve any special evolutionary processes. Naively, we might expect that the complete loss of a biochemical function via loss of the last member of a gene family would be deleterious, and consequently rare. Therefore, the loss of gene families might serve as an indicator of shifts in the physiological constraints of an organism. Genome-wide analyses in animals suggest that gene families producing metabolic enzymes frequently undergo independent extinction in multiple lineages (Hughes and Friedman 2004). This may indicate that shifts in nutrient availability or acquisition are most often responsible for conditions that permit gene family extinction.

Mixed Concerted Evolution and Birth-and-Death Evolution

It should be noted that concerted evolution and birth-and-death evolution are not mutually exclusive processes. Indeed, in a gene family some members may undergo concerted evolution, while others may undergo birth-and-death evolution. One such example is the β-globin family (Figure 7.15), which consists of five functional genes (ε, $^{G}\gamma$, $^{A}\gamma$, δ and β) and one pseudogene (ψβ). Within this family, divergent evolution has created four functionally distinct subfamilies (ε, γ, δ, and β), the genes $^{G}\gamma$ and $^{A}\gamma$ have experienced concerted evolution by gene conversion, and one duplicate has died (ψβ). Another example is the α-globin family, which consists of five functional genes (α_1, α_2, μ, θ, and ζ) and two pseudogenes (ψα and ψζ). The α_1 and α_2 genes are virtually identical within each of the genomes of the great apes studied to date, likely because of gene conversion. However, the other genes differentiated considerably, and two underwent nonfunctionalization.

In general, in a small gene family, there is usually a strong constraint on family size (gene number), so the number of genes in the family cannot fluctuate much. As a consequence, unequal crossover does not play a major role in the evolution of such a gene family, and if there is evidence of concerted evolution, it is likely due to gene conversion. In contrast, in a large gene family, some fluctuations in family size may occur because of unequal crossing over or nonfunctionalization.

Polysomy

Euploidy refers to a chromosome number that is an exact multiple of the haploid chromosome number. Thus, haploids, diploids, triploids, tetraploids, pentaploids, and any other multiple of the haploid state are euploids. **Aneuploidy** refers to the condition in which the number of chromosomes in a cell is not an integer multiple of the typical haploid set for the species. The duplication of a complete chromosome is called **polysomy**; the duplication of a segment of a chromosome is called **partial polysomy**.

Polysomy is almost invariably deleterious or lethal in animals. In mammals, for instance, it is frequently associated with lethality or infertility. In humans, only three trisomies are not lethal at birth—Down syndrome (trisomy 21), Patau syndrome (trisomy 13), and Edwards syndrome (trisomy 18)—but the evolutionary fitness of individuals with these conditions is zero. Autosomal monosomy, i.e., the presence of a single autosome, is invariably lethal. Sex chromosomes, on the other hand, can routinely withstand monosomy. In fact, the presence of a lone X chromosome without a counterpart (X or Y), such as is the karyotype in Turner syndrome, is a nonlethal condition in humans. Severe deleterious manifestations are often associated with partial polysomy, such as partial trisomy and partial tetrasomy (e.g., cat eye syndrome, which involves a small segment of chromosome 22). Therefore, chromosomal duplication or deletion—either complete or partial—is not expected to contribute significantly to animal evolution.

Even in plants, in which polyploidy is common and polysomic individuals are produced with great regularity, permanent polysomy is hardly ever tolerated. One of the earliest studies on plant trisomy, and thus far the most thorough, was carried out by Blakeslee and colleagues (e.g., Blakeslee and Avery 1919; Blakeslee et al. 1920). Jimsonweed (*Datura stramonium*) has 12 pairs of chromosomes, and all 12 trisomics have been identified in nature. Each one was distinguishable by specific morphological features, and they all were "of feeble growth."

Polyploidization

Polyploidy entails the addition of one or more complete sets of chromosomes to the original set. An organism whose cells contain two copies of each autosome is a diploid, an organism with four copies is a tetraploid, one with six copies is a hexaploid, and so on. The gametes of diploid organisms are haploid, those of tertraploids are diploid, those of hexaploids are triploid, and so on. Organisms with an odd number of autosomes, such as the triploid domestic banana plant (*Musa acuminata*), can neither undergo meiosis nor reproduce sexually. Since its discovery in the early 1900s, polyploidy has been recognized and an important evolutionary phenomenon, particularly in vascular plants (de Vries 1904; Lutz 1907). Paralogs created en masse after polyploidization are sometimes referred to as either **ohnologs**, an eponym coined in honor of Susumu Ohno by Leveugle et al. (2003), or **homeologs**, a homophone that is sometimes difficult to distinguish aurally from "homologs."

There are two main types of polyploidy. **Autopolyploidy**, or **genome doubling**, entails the multiplication of one basic set of chromosomes. **Allopolyploidy** is the condition arising from the combination of genetically distinct, but similar, chromosome sets. Thus, autopolyploids are derived from within a single species, whereas allopolyploids arise via hybridization between two species.

Hybridization may not always result in polyploidy. **Homoploidy** is a process of hybridization between two species that results in the creation of a distinct new species without a change in ploidy level (Gross 2012). In homoploids, one chromosome in each homeologous pair is derived from one parental species while the other chromosome is from the other parental species. Thus, the result of homoploid hybridization between two diploid species is a diploid. *Senecio squalidus* is a homoploid species that has been studied extensively (e.g., Brennan et al. 2012). The species originated roughly 300 years ago, and both parental species (*S. aethnensis* and *S. chrysanthemifolius*) as well as hybrids are still present in their native habitat on the slopes of Mount Etna, Sicily (James and Abbott 2005). *Senecio squalidus* was introduced by humans to the British Isles and is currently classified as an invasive species.

AUTOPOLYPLOIDY Autopolyploidy may be common in plants, although its prevalence may be quite underestimated in the taxonomic literature (Soltis et al. 2007). One species that is doubtlessly a true autopolyploid, rather than an allopolyploid derived from two very similar diploids, is the potato, *Solanum tuberosum* (Potato Genome Sequencing Consortium 2011).

The main selective advantage of autopolyploidy in nature may be the fact that populations of autopolyploids can maintain much higher levels of heterozygosity than their diploid progenitors can, because of their polysomic inheritance; that is, the number of allelic combinations at each locus is larger than the three possible combinations in diploids (Muller 1914; Haldane 1930; Moody et al. 1993). Let us consider, for example, an autotetraploid (*aabb*) derived from a heterozygous diploid (*ab*). Assuming simple tetrasomic inheritance, the genotype *aabb* is expected to produce diploid gametes in the ratio 1*aa*:4*ab*:1*bb*. In the progeny, the ratio of the genotypes will be 1*aaaa*:8*aaab*:18*aabb*:8*abbb*:1*bbbb*. That is, heterozygotes (*aaab, aabb, abbb*) are expected to outnumber homozygotes (*aaaa, bbbb*) 17 to 1. In comparison, in diploids the heterozygote-to-homozygote ratio is 1:1. Moreover, autopolyploids can maintain more than two alleles per locus, allowing them to produce a larger variety of allozymes than diploids, which, in principle, may allow them to achieve higher fitness values (Parisod et al. 2010). Finally, by reason of their polysomal inheritance, autopolyploid populations have larger effective population sizes than diploids, suggesting that selective processes are much more effective relative to random genetic drift (Ronfort 1998).

Autotetraploidy is a common mutational occurrence in nature. Indeed, somatic autotetraploidy is found in almost all organisms, including algae, plants, mollusks, insects, and vertebrates (Nagl 1990). However, during evolutionary history autotetraploids seem to have survived only rarely. The reason is that, in many cases, autotetraploidy may be deleterious and will be strongly selected against. Deleterious effects include (1) prolongation of cell division time, (2) increase in the volume of the nucleus, (3) increase in the number of chromosome disjunctions during meiosis, (4) genetic imbalances, and (5) interference with sexual differentiation when the sex of the organisms is determined either by the ratio between the number of sex chromosomes and the number of autosomes (as in *Drosophila*) or by the degree of ploidy (as in Hymenoptera).

ALLOPOLYPLOIDY Allopolyploidy is much more common in nature than autopolyploidy. It is extremely common in plants; about 80% of all land plants may be allopolyploids (**Figure 7.37**). In animals, allopolyploidy is not nearly as prevalent; however, it is by no means nonexistent. Allopolyploidy has been found in both parthenogenetic and sexually reproducing species of insects, fish, reptiles, and amphibians. For example, *Xenopus laevis*, the African clawed frog of laboratory fame, is an allotetraploid. No cases of polyploidy have ever been found in birds. And, although two mammalian species—the red vizcacha rat (*Tympanoctomys barrerae*) and the golden vizcacha rat (*Pipanacoctomys aureus*)—were suspected to be tetraploids, much disagreement exists in the literature about their ploidy status (Gallardo et al. 1999, 2004; Svartman et al. 2005; Suárez-Villota et al. 2012).

In the last 15,000 years, the domestication of plants, but not of animals, frequently involved allopolyploid species. The most famous example is, of course, bread wheat (*Triticum aestivum*), which is an allohexaploid containing three distinct sets of chromosomes derived from three different diploid species through tetraploid intermediaries (**Figure 7.38**).

Many microbes too were domesticated, and in the case of the lager-brewing yeast, *Saccharomyces pastorianus*, the domestication involved an allopolyploid species. There are two broad categories of beer: ale, which is produced by the top-fermenting yeast

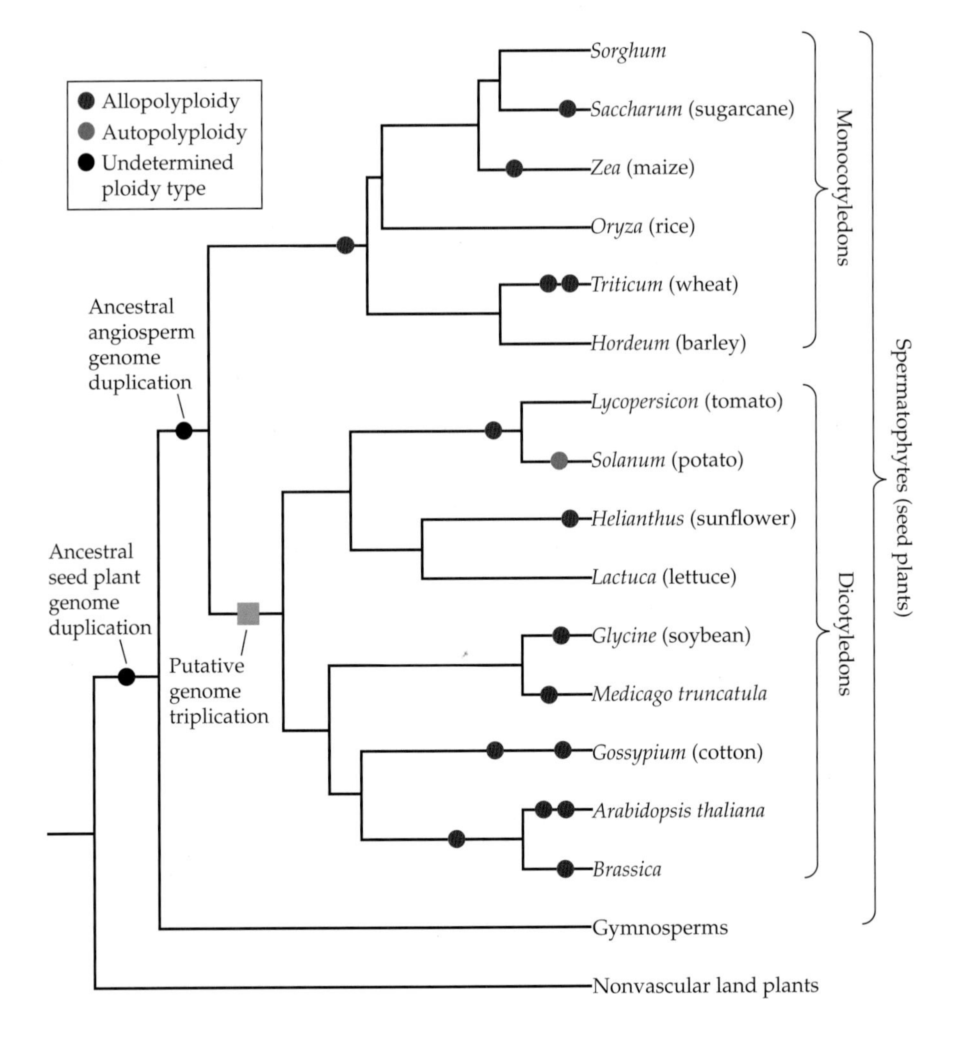

Figure 7.37 Inferred polyploidy events during the evolution of angiosperms. The ancestral angiosperm genome duplication is estimated to have occurred 190–230 million years ago. The ancestral seed plant duplication is estimated to have occurred 320–350 million years ago. (Data from Adams and Wendel 2005 and Jiao et al. 2011.)

S. cerevisiae; and lager, made with the bottom-fermenting yeast *S. pastorianus*. Ale is the older of the two, probably having been produced as early as 7,500 BCE in the Euphrates valley of modern Syria and Iraq (Bamforth 1998; Hornsey 2003). Lager brewing arose in early fifteenth-century Bavaria, gained broad acceptance by the late nineteenth century, and has since become the most popular technique for producing alcoholic beverages. Unlike ales and wines, lagers require slow, low-temperature fermentation that is carried out by cryotolerant *S. pastorianus* strains. Libkind et al. (2011) have shown that the domesticated species *S. pastorianus* was created by the fusion of an *S. cerevisiae* strain used to produce ale and a newly discovered cryotolerant species called *Saccharomyces eubayanus* found in Patagonian forests. The draft genome sequence of *S. eubayanus* is 99.5% identical to the non-*S. cerevisiae* portion of the *S. pastorianus* genome. The 0.5% difference is due to several changes in the genome of *S. eubayanus* following the tetraploidization event, changes almost certainly driven by the stringent artificial selection in the brewing environment.

CONSEQUENCES OF POLYPLOIDY At a phenotypic level, the effects of polyploidization are often mild and idiosyncratic (Otto 2007). In some cases, moreover, polyploidy seems to have almost no effect on the phenotype. For example, diploid, autopolyploid, and allopolyploid *Chrysanthemum* species vary in chromosome number from 18 to 198, yet they are almost indistinguishable from one another. Similar observations have been made in roses (*Rosa*), leptodactylid toads (*Odontophrynus*), and goldfish (*Carassius*).

Cell volume generally rises with increasing genome size (Cavalier-Smith 1978; Gregory 2001), although the exact relationship between ploidy and cell volume var-

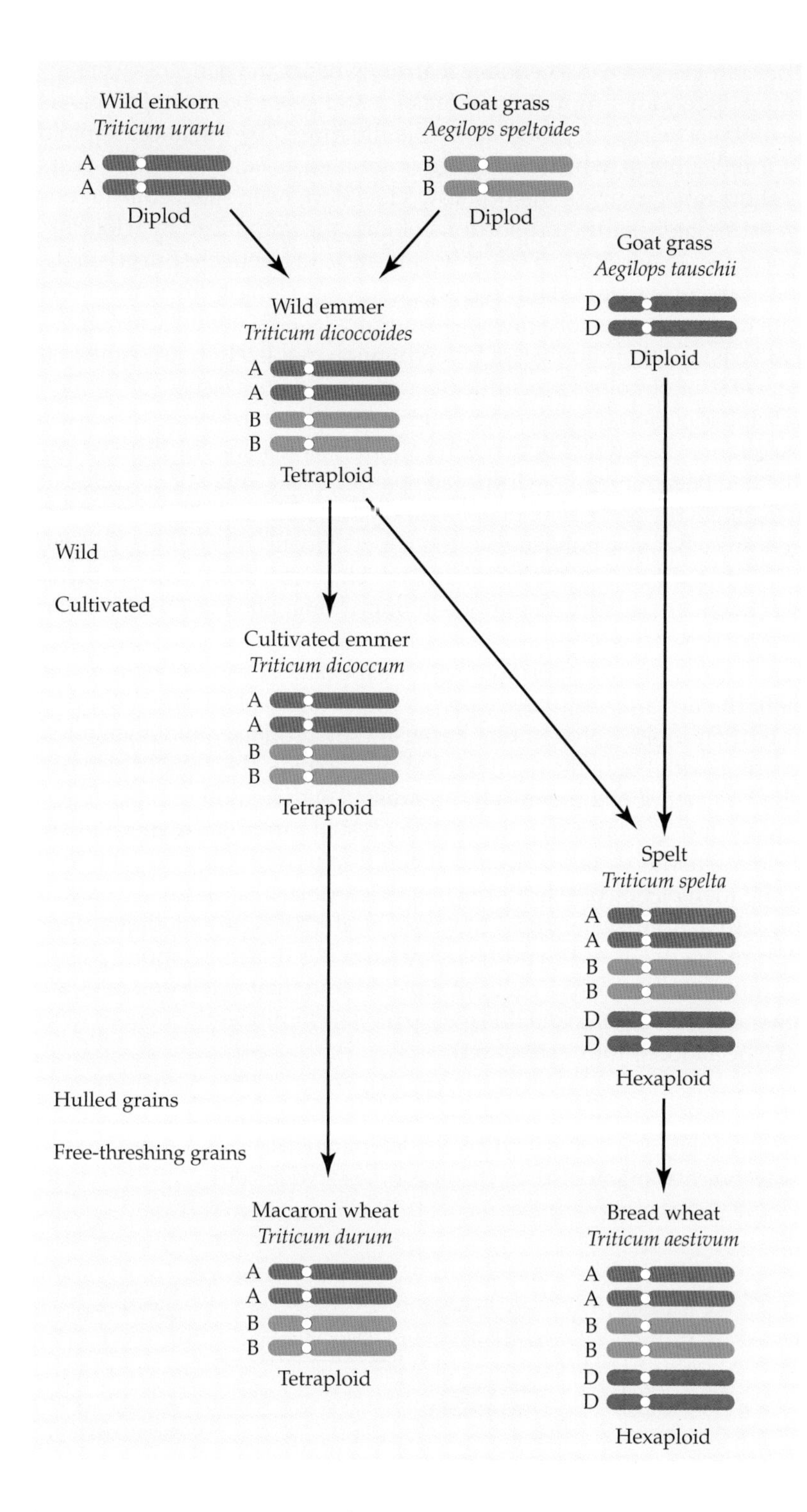

Figure 7.38 The evolution of wheat from diploid grasses to the free-threshing tetraploid macaroni (or durum) wheat and the hexaploid bread wheat. Molecular data and molecular clock considerations indicate that *Triticum urartu* and *Aegilops speltoides* diverged from a common ancestor ~6.5 million years ago. *Aegilops tauschii* turned out to be a homoploid derived through a hybridization between a relative of *T. urartu* and a relative of *Ae. speltoides* ~5.5 million years ago. The tetraploidization event giving rise to wild emmer, *Triticum dicoccoides*, occurred ~800,000 years ago. *Triticum dicoccum* was cultivated as early as 17,000 years ago. Cultivated emmer has larger grains than wild emmer, a fact attributed to selection. The hybridization of *T. dicoccoides* with *Ae. tauschii* that gave rise to the hexaploid *Triticum spelta* occurred less than 400,000 years ago, and *T. spelta* was cultivated ~10,000 years ago. Cultivated emmer and spelt separately acquired a mutation that changed how grains grew in the ears, from being enclosed by hard shells to being looser and easier to thresh. This yielded bread wheat and macaroni wheat, free-threshing wheats that are about 8,500 years old. (Data from Dvorak and Akhunov 2005 and Marcussen et al. 2014.)

ies among environments and among taxa. Interestingly, although average cell size is larger in polyploids, the size of the adult polyploidy organism may or may not be altered; as a rough generalization, polyploidization is more likely to increase adult body size in plants and invertebrates than in vertebrates (Otto and Whitton 2000; Gregory and Mable 2005). The poor correlation between cell size and organismal size was even remarked upon by Albert Einstein, who stated, "Most peculiar for me is the fact that in spite of the enlarged single cell, the size of the animal is not correspondingly increased" (Fankhauser 1972).

An important feature of many newly formed polyploids is that their genomes are unstable and undergo rapid repatterning and segmental loss (Feldman et al. 1997;

Wendel 2000). For example, extensive genomic rearrangements and gene loss were recorded almost immediately after formation of *Brassica* and *Arabidopsis* allopolyploids (e.g., Song et al. 1995). The rapidity of gene loss is illustrated by the allohexaploid bread wheat, *Triticum aestivum*, which may have originated as early as 10,000 years ago. In this very short time, many of the triplicated loci have been silenced. Aragoncillo et al. (1978) estimated that the proportions of enzymes produced by triplicate, duplicate, and single loci in wheat are 57%, 25%, and 18%, respectively. Surprisingly, however, ohnologs can persist for millions of years, allowing a long-term contribution of these gene pairs to the evolution of a lineage. For example, ~8% of duplicated genes have remained in yeast ~100 million years following polyploidization (Seoighe and Wolfe 1999), and ~77% of ohnologs are still detectable in *Xenopus laevis* ~30 million years after allotetraploidization (Hughes and Hughes 1993).

Another consequence of polyploidy concerns allopolyploids, in which transposable elements that had been repressed within each parent lineage may be activated in hybrids and can facilitate the movement of genes and promote unequal crossing over. In addition to genomic changes, polyploids often show evidence of changes in gene expression. This is especially true of allopolyploids, which exhibit changes in methylation, disruption of heterochromatin, alterations in imprinting, and biased expression of genes by species of origin.

Finally, polyploidy may be an important factor in speciation, especially in plants (Wood et al. 2009). In particular, sexually reproducing autotetraploids are automatically isolated from their diploid progenitors because they produce diploid gametes; were these to combine with the haploid gametes of the diploids, they would give rise to triploid progeny. As mentioned previously, organisms with an odd number of autosomes cannot reproduce sexually, so polyploidy represents an effective mechanism of reproductive isolation. Stebbins (1971) postulated that polyploids represent dead ends because of the inefficiency of selection when deleterious alleles can be masked by multiple copies. In a large-scale study, Mayrose et al. (2011) provided quantitative corroboration of the dead-end hypothesis by showing that speciation rates of polyploids are significantly lower than those of diploids, and their extinction rates are significantly higher. Together, polyploid lineages exhibited significantly reduced rates of diversification (speciation minus extinction).

Diploidization

Diploidization is the evolutionary process whereby a tetraploid species "decays" to become a diploid with twice as many distinct chromosomes (Wolfe 2001). A newly creat-

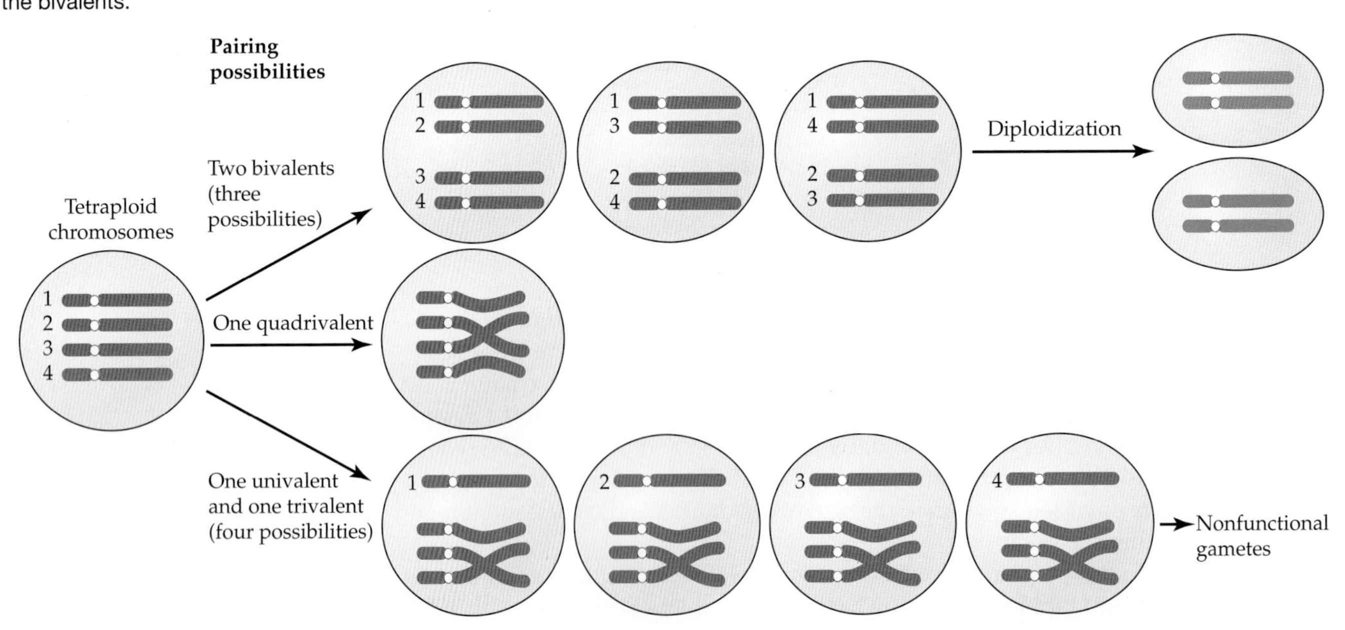

Figure 7.39 Immediately after tetraploidization, the homologous and homeologous chromosomes can exchange genetic material through recombination. The four homologous and homeologous chromosomes (numbered) may pair as two bivalents (three possible combinations) or a quadrivalent. Both pairing types can yield functional gametes. The four chromosomes may also pair as a univalent and a trivalent (four possible combinations), yielding nonfunctional gametes. As the homeologous chromosomes diverge from each other, they can no longer form a quadrivalent, and recombination is restricted to the homologous chromosomes in the bivalents.

Figure 7.40 Diploidization and its phylogenetic consequences. Following allotetraploidization, one locus may become fixed for alleles that originate from one parent, whereas some other locus might retain alleles from both parents, so a molecular clock analysis of the duplicated *A′* and *A″* loci in the paleopolyploid descendant may point to a recent time of divergence, corresponding to the diploidization date. The locus for *B* remains polymorphic for the two parental alleles during the tetrasomic phase, so molecular estimates of the divergence time between its diploidized daughter *B′* and *B″* loci might correspond to the speciation date between the two parent species. (Modified from Wolfe 2001.)

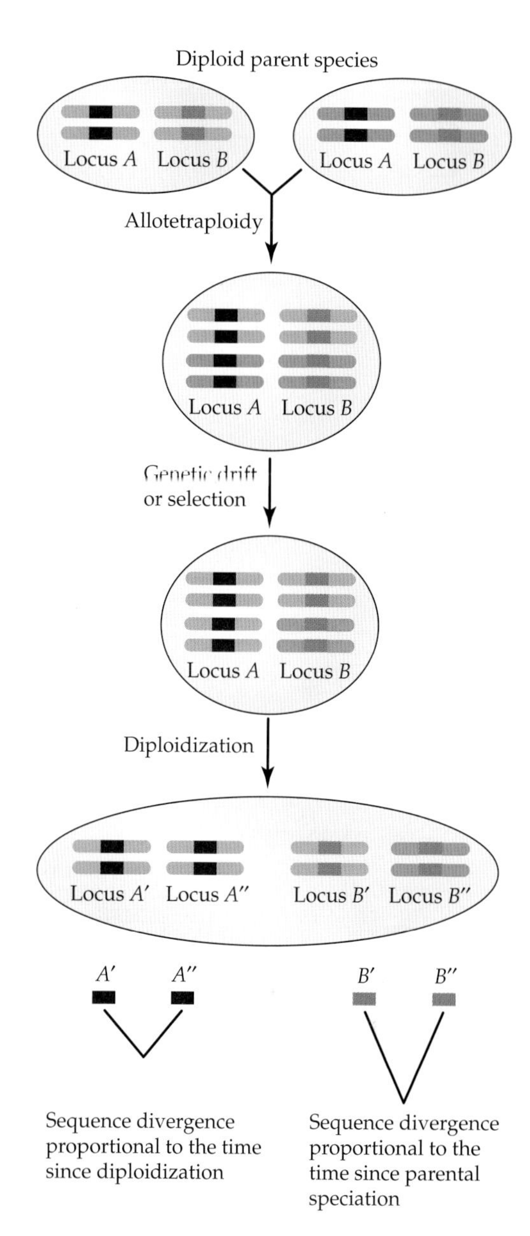

ed polyploid is sometimes referred to as **neopolyploid**; after diploidization it is called a **paleopolyploid**. In the literature, a distinction is sometimes made between paleopolyploids, whose diploid ancestors are unknown or extinct (e.g., *Xenopus laevis*), and **mesopolyploids**, whose diploid ancestors are known or extant (e.g., tetraploid and hexaploid wheat species) (see Mangenot and Mangenot 1962; Guerra 2008).

The molecular basis of diploidization is not well understood, but it presumably occurs through the accumulation of DNA differences among the homeologous chromosomes. That is, as the two genomes undergo mutations, translocations, and chromosomal rearrangements, as well as unbalanced changes in chromosome number, they will eventually become a single new genome, a situation that has been dubbed **cryptopolyploidy** (literally, a hidden polyploidy). In other words, an ancient polyploid will no longer be distinguishable from a diploid (Cavalier-Smith 1985a).

The key event in diploidization is the switch from having four chromosomes that may form a **quadrivalent** or four types of trivalents at meiosis, to having two pairs of chromosomes each of which forms a **bivalent** (**Figure 7.39**). In population genetics terms, this is the switch from having four alleles at a single locus (**tetrasomic inheritance**) to having two alleles at each of two distinct loci (**disomic inheritance**).

In an allotetraploid, each locus will initially have four alleles, two from each parent. In time, one locus may become fixed for alleles that originate from one parent, whereas some other locus might retain alleles from both parents (**Figure 7.40**). After diploidization and further sequence evolution, the amount of sequence divergence between some paralogous loci is expected to be proportional to the time elapsed since diploidization, whereas at other loci it will correspond to the time since the speciation between the parents of the allopolyploid. This situation is called **segmental allotetraploidy**, and the consequence is that when phylogenetic trees are drawn, some pairs of paralogous loci will point to one divergence date, whereas other pairs of loci will point to a different date. The maize genome seems to have this structure (Gaut and Doebley 1997).

In both plants and animals (such as in the case of salmonid taxa), a single species can harbor a mixture of tetraploid and diploidized loci. In other words, diploidization does not necessarily happen simultaneously for all chromosomes or even for all loci on a particular chromosome. If this is commonplace, tree-based analyses of paleopolyploids may yield very confusing results. The consequence of independent diploidization dates for each locus would be a continuum of divergence dates for duplicated loci, ranging from the very recent back to the parental speciation date.

Distinguishing between gene duplication and genome duplication

Most genomes contain gene duplications. They can be the result of either (1) gene or segmental duplications or (2) whole-genome duplications. How can one distinguish between the two mechanisms? We note that we are mainly concerned with ancient polyploidizations rather than recent ones (such as that in wheat), in which the

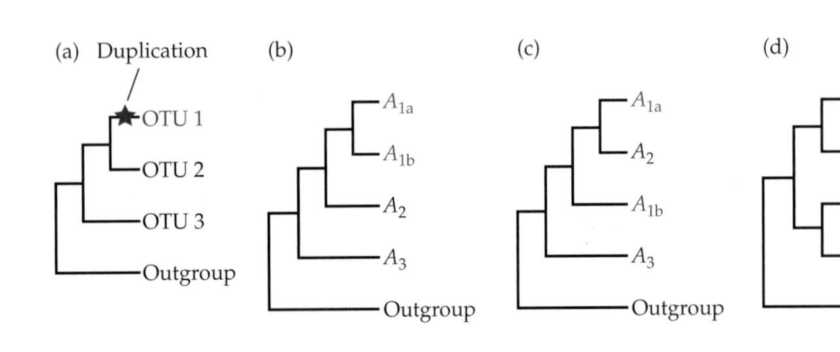

Figure 7.41 (a) Phylogenetic tree for three OTUs and an outgroup, in which a genome duplication (red star) occurred in the lineage leading to OTU 1. (b) The expected phylogenetic positions for duplicated genes A_{1a} and A_{1b} from OTU 1 if indeed they are ohnologs. If, however, A_{1a} and A_{1b} from OTU 1 are not ohnologs, they may be found in different phylogenetic positions, such as in trees (c) or (d).

differences are trivially simple. Several telltale signs have been proposed as evidence of ancient polyploidization.

A good indication of polyploidy would be if all gene trees exhibited the expected topology. For illustration, let us assume that the species tree is as in **Figure 7.41a** and that a genome duplication occurred in the lineage leading to OTU 1, leading to the duplication of all the genes in the genome. In such a case, each of the paralogous genes should yield the same tree (**Figure 7.41b**). If, however, the paralogous genes yield alternative trees (**Figure 7.41c,d**), then it is unlikely that all the gene duplications occurred at the same time. We need, however, to add a caveat, to the effect that the

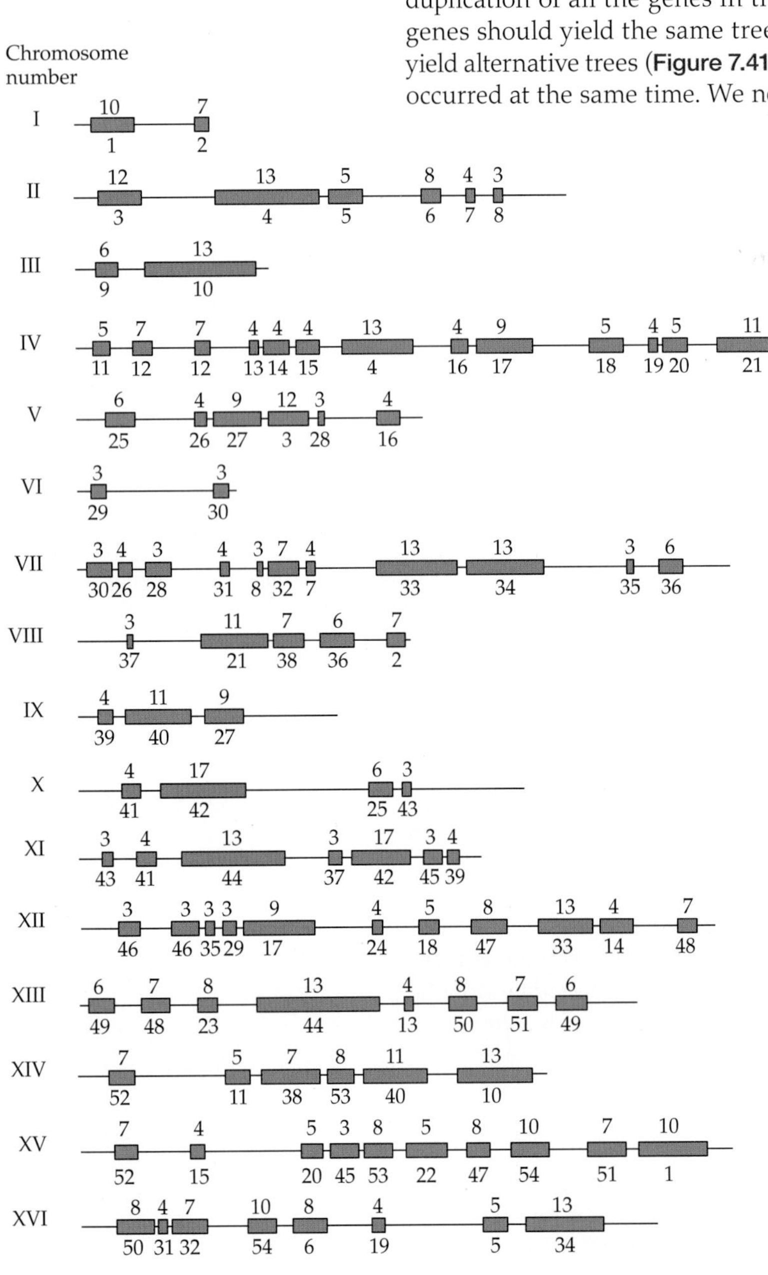

Figure 7.42 Locations of 54 nonoverlapping blocks of doubly conserved synteny (blue boxes) in the yeast genome. (In more recent studies, a greater number of blocks of doubly conserved synteny were found.) The two copies of each duplicated region are given the same number, below their respective boxes. Numbers are listed in order of chromosomal occurrence. The number of homologous genes in each duplicated region is listed above its box. Chromosome numbers are given in roman numerals. (Modified from Wolfe and Shields 1997.)

expectation that all gene pairs yield the same phylogenetic topology assumes that diploidization occurred at the same time in all the genes. Moreover, gene loss and processes of recombination, such as unequal crossing over and gene conversion, as well as gene duplication that occurs subsequent to genome duplication, may hinder the expected result.

Another piece of information that is typically adduced as evidence of genome duplication concerns regions of double synteny, i.e., two or more genomic regions containing paralogous arrays of genes. Here, we present two cases in which the hypothesis of whole-genome duplication was investigated.

THE YEAST GENOME: TETRAPLOIDY OR REGIONAL DUPLICATIONS? *Saccharomyces cerevisiae* has long been suspected of being a cryptotetraploid (Smith 1987). Wolfe and Shields (1997) systematically searched the complete yeast proteome for regions of double synteny. The criteria used for defining two regions as duplicated were (1) a sequence similarity between the two regions associated with a probability of less than 10^{-18} that it was fortuitous; (2) at least three protein-coding genes or open-reading frames in common, with intergenic distances of less than 50 Kb; and (3) conservation of gene order and relative orientation of the genes. According to these criteria, Wolfe and Shields (1997) identified 54 nonoverlapping pairs of duplicated regions spanning about 50% of the yeast genome (**Figure 7.42**). Of the approximately 5,800 genes in the yeast genome, about 900 were paralogs located in duplicated chromosomal regions (also called **blocks of doubly conserved synteny**, or **paralogons**).

There are two possible explanations for these observations. Either (1) the duplicated regions formed independently by many regional duplications occurring at different times during the evolution of *S. cerevisiae*, or (2) the duplicated regions were produced simultaneously by a single tetraploidization event, followed by massive rearrangements of the genome and loss of many redundant duplicate genes. There are two reasons to favor the latter model. First, 50 of the duplicated regions have maintained the same orientation with respect to the centromere. Second, based on a Poisson distribution, 54 independent regional duplications would be expected to result in about seven triplicated regions (i.e., duplicates of duplicates), but none was observed.

Wolfe and Shields (1997) proposed that *S. cerevisiae* is an ancient tetraploid, formed through the fusion of two ancestral diploid yeast genomes, each containing about 5,000 genes. They estimated the tetraploidization event to have occurred approximately 100 million years ago in the ancestor of four *Saccharomyces* species. The new species then became a cryptotetraploid, and about 90% of the duplicate gene copies were lost through sequence decay or deletion. Some 70–100 subsequent map disruptions (i.e., regional translocations) were inferred to have been required to explain the current chromosomal distribution of the duplicate genes (Seoighe and Wolfe 1998).

These conclusions were almost immediately challenged because they were based on a small set of yeast genes (less than 10%) and could be explained almost as easily by independent duplications of chromosomal segments (e.g., Koszul et al. 2004; Martin et al. 2007).

The clearest way to prove the existence of an ancient polyploidization event would be to find a species that descended directly from a common ancestor along a lineage that diverged before the whole-genome duplication event. Phylogenetic data indicate that *Kluyveromyces waltii* and its paraphyletic congeneric *K. lactis* diverged from the lineage leading to *S. cerevisiae* and the related yeast *S. bayanus* before the polyploidization event (**Figure 7.43**).

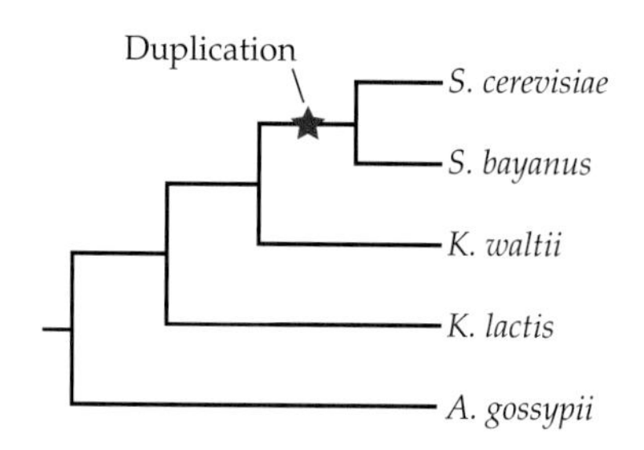

Figure 7.43 Schematic phylogenetic tree for five yeast species: *Saccharomyces cerevisiae*, *S. bayanus*, *Kluyveromyces waltii*, *K. lactis*, and *Ashbya gossypii*. The whole-genome duplication event is marked with a red star. The consequencies of such an event are illustrated in Figure 7.44.

The expected genomic signature of whole-genome duplication is illustrated in **Figure 7.44**. Following duplication, sister regions would undergo gene loss by deletion; one or the other of the two paralogous copies of each gene would be lost in most cases, with both paralogs being retained only very rarely. Eventually, the only residual signature that two regions arose from ancestral duplication would be the presence of a few paralogous genes in the same order and orientation scattered amidst a multitude

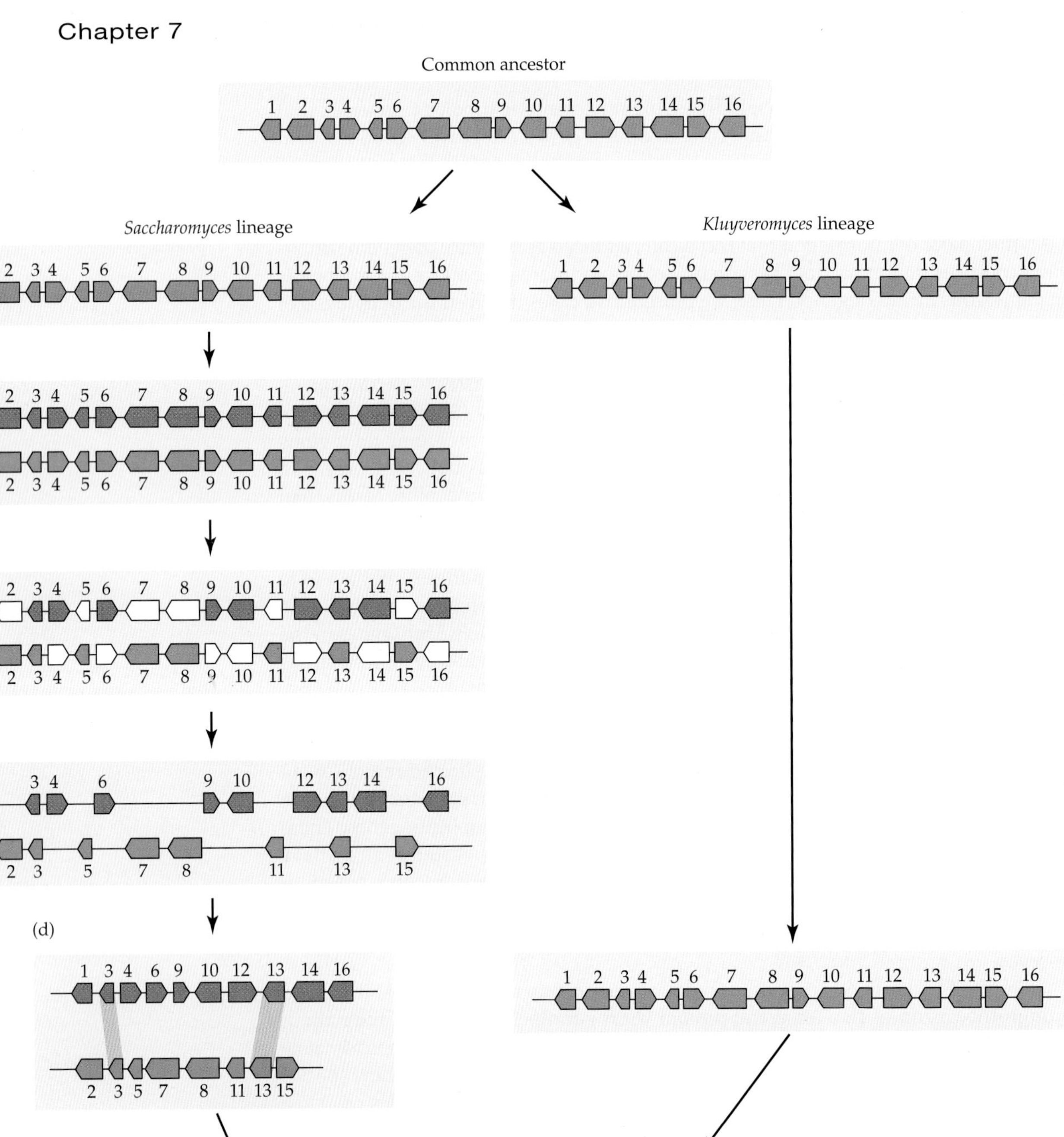

Figure 7.44 Consequences of genome duplication and subsequent gene loss in a comparison between a species that underwent duplication and one that diverged before the duplication event. (a) After divergence from *Kluyveromyces waltii*, the *Saccharomyces* lineage underwent a genome duplication event, creating two copies of every gene and chromosome. (b) The vast majority of duplicated genes underwent mutation and gene loss. (c) Sister segments retained different subsets of the original gene set, keeping two copies for only a small minority of duplicated genes. (d) Within *S. cerevisiae*, the only evidence comes from the conserved order of duplicated genes (numbered 3 and 13) across different chromosomal segments; the intervening genes are unrelated. (e) Comparison with *K. waltii* reveals the duplicated nature of the *S. cerevisiae* genome, interleaving genes from sister segments on the basis of the gene order in *K. waltii*, which is supposed to reflect the ancestral gene order. Spacing between *S. cerevisiae* genes is set to match *K. waltii* chromosomal positions. The pointed side of each gene denotes the direction of its transcription. In the *Saccharomyces* lineages, duplicated blocks are in blue and orange, and nonfunctionalized genes about to decay into pseudogenes or be deleted are in white. (From Kellis et al. 2004.)

of unrelated genes. Paired regions containing such signatures would be relatively short because chromosomal rearrangements would have disrupted global gene order over time, leaving only weak evidence of ancestral duplication.

As *K. waltii* diverged before the whole-genome duplication, it would be expected to display a 1:2 mapping pattern with *S. cerevisiae*. This synteny map should have the following properties: (1) nearly every region in *K. waltii* should correspond to two sister regions in *S. cerevisiae*; (2) nearly every region of *S. cerevisiae* should correspond to one region of *K. waltii*; and (3) the two sister regions in *S. cerevisiae* should each contain an ordered subsequence of the genes in the corresponding region of *K. waltii*, resulting in an interleaving pattern as in Figure 7.44.

Kellis et al. (2004) identified a total of 253 blocks of doubly conserved synteny containing 75% of *K. waltii* genes and 81% of *S. cerevisiae* genes. A detailed view of a region of chromosome 1 of *K. waltii* and the cross-species mapping of genes from two blocks in *S. cerevisiae*, one from chromosome 4 and the other from chromosome 12, are shown in **Figure 7.45**. This picture repeated itself with virtually every block of doubly conserved synteny.

The picture that emerges is that a genome duplication event doubled the number of chromosomes in the *Saccharomyces* lineage, but subsequent gene events led to the current *S. cerevisiae* genome, which is only 13% larger than that of *K. waltii* and contains only 10% more genes. The polyploid genome returned to functional normal ploidy through a large number of deletion events. In principle, gene loss can occur by large segmental deletions or individual gene deletions, and it can either be balanced between the two sisters or act primarily on one of them. Analysis of duplicated blocks revealed that gene loss occurred by many small deletions (the average size of a lost segment is two genes) and was typically balanced between the two sister regions (average balance 57% to 43%). A similar pattern of gene loss was observed in the evolution of the ciliate *Paramecium tetraurelia*, whose nearly 40,000 genes arose through at least three successive whole-genome duplications (Aury et al. 2006).

VERTEBRATE POLYPLOIDY? THE 2R HYPOTHESIS Based on the then common belief that vertebrates possess more genes than invertebrates and his belief that gene duplications are invariably maladaptive, Ohno (1970) suggested that one or more genome duplications occurred in the lineage leading to vertebrates. He was not very explicit about the number and timing of the genome duplications. Following the discovery

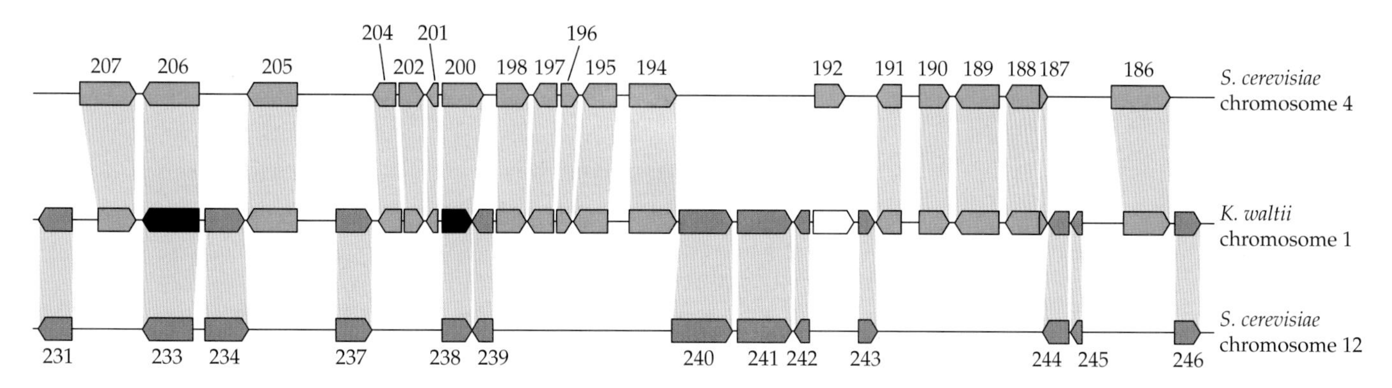

Figure 7.45 A comparison of a portion of chromosome 1 of *Kluyveromyces waltii* with two blocks of doubly conserved synteny in *Saccharomyces cerevisiae*, one from chromosome 4 (orange) and one from chromosome 12 (blue). Genes are arbitrarily numbered as in the *Saccharomyces* Genome Database (Cherry et al. 1997), with the pointed sides indicating the direction of transcription. Each gene of *K. waltii* is shown in red or blue if it has a match in *S. cerevisiae* chromosomes 4 or 12, respectively; in black if it has a match in both *S. cerevisiae* chromosomes; and in white if there is no match in any of the two syntenic blocks. Spacing between *S. cerevisiae* genes is set to match *K. waltii* chromosomal positions. (Modified from Kellis et al. 2004.)

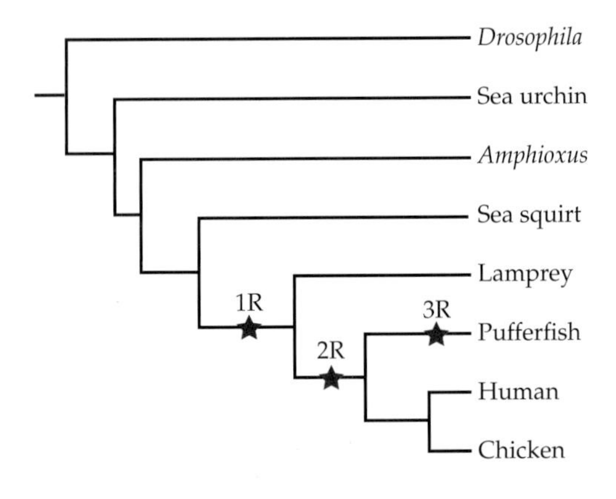

Figure 7.46 The timing of three putative whole-genome duplications in the evolution of deuterostomes, with a protostome (*Drosophila*) as outgroup. Genome duplications 1R and 2R mark the two duplications assumed by the 2R hypothesis. The 3R marks a genome duplication assumed to have occurred in teleost fishes.

that the four Hox gene clusters in mammals had evolved from an ancestral cluster by quadruplication, Kappen et al. (1989) suggested that two polyploidization events occurred in the vertebrate lineage. Holland et al. (1989) placed the first putative duplication event on the lineage leading to the vertebrates after the divergence of the amphioxus (*Branchiostoma lanceolatum*) and sea squirts (genus *Ciona*) (**Figure 7.46**). The second putative duplication was placed on the lineage leading to the jawed vertebrates (Gnathostomata) after the divergence of hagfish and lampreys (Cyclostomata). In time, the assumption of two rounds of tetraploidization on the evolutionary branch leading to jawed vertebrates came to be known as the "2R hypothesis" (Hughes 1999), leading to some memorable puns, including the Hamletian "2R or not 2R" (Hughes and Friedman 2003). A 2R scenario is shown in **Figure 7.47**. In this particular scenario, deletions of paralogs may occur subsequent to genome duplication. Within vertebrates, an additional genome duplication may have occurred in bony fishes (Figure 7.46; Jaillon et al. 2004).

Let us now examine a simple expectation of the 2R hypothesis. In the absence of any gene duplications prior to, in between, or subsequent to the two rounds of genome duplication, the expectation is an (*AB*)(*CD*) topology, where *A*, *B*, *C*, and *D* are paralogous genes (**Figure 7.48a**). Any other topology (e.g., **Figure 7.48b,c**) may be interpreted as a refutation of the 2R hypothesis. Hughes and Friedman (2003) looked for four-member paralogous gene families in the human genome that have a single homolog in *Drosophila* and reconstructed their phylogenetic relationships. Of the 92 resolved phylogenetic topologies, only 22 (24%) supported the 2R hypothesis. Out of the 53 phylogenies in which all internal branches received statistically significant support, only 11 topologies (21%) supported the 2R hypothesis, leading the authors to "decisively" reject the hypothesis. We remind the reader, however, of the very strict assumptions of this test. If one selects only those human paralogs that duplicated before divergence of tetrapods from the bony fishes, the 2R hypothesis cannot

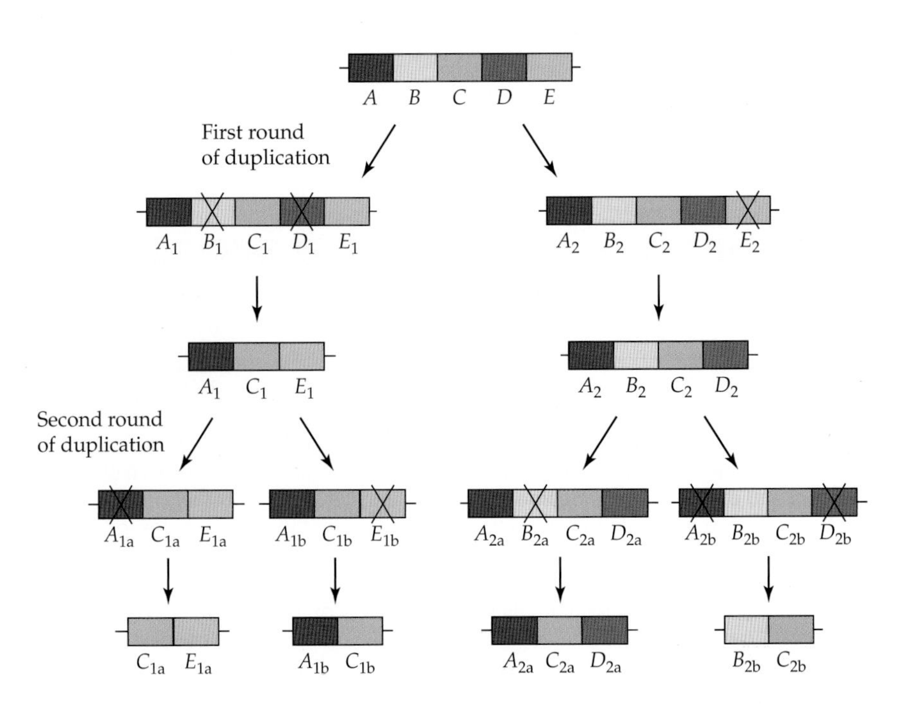

Figure 7.47 Model of the 2R hypothesis (with deletions) and its phylogenetic implications. Starting with five linked genes (*A–E*; top row), two rounds of gene duplication result in nine genes (bottom row). At least one copy of each ancestral gene is present in the genome at all times. Only the descendants of gene *C* retained all four resultant copies. (Modified from Wolfe 2001.)

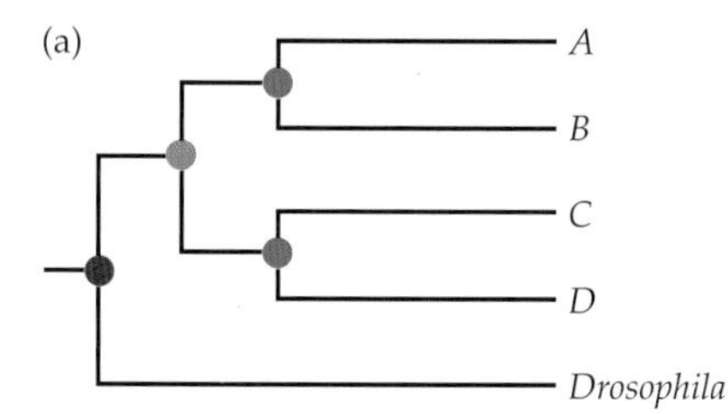

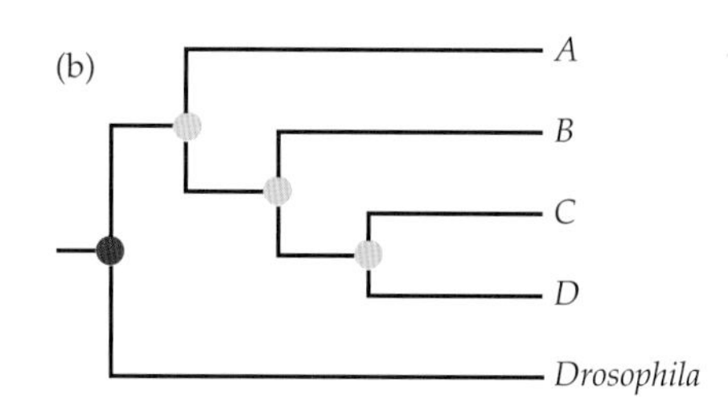

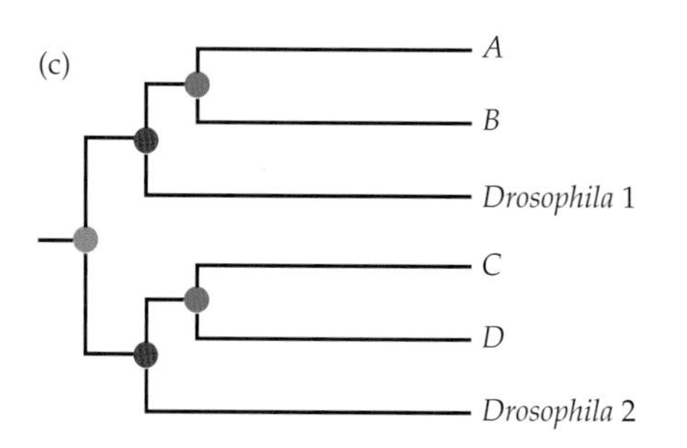

Figure 7.48 (a) Expected phylogenetic topology for a four-member (*A–D*) gene family in humans and a *Drosophila* ortholog under the 2R hypothesis. (b) A phylogeny that is inconsistent with the 2R hypothesis and is most parsimoniously explained by three single-gene duplications. (c) The expected phylogeny if one genome duplication (green circle) predates the deuterostome-protostome split. The number of four-member gene families in the human genome yielding topologies (a), (b), and (c) are 22, 38, and 32, respectively. (From Hughes and Friedman 2003.)

be rejected (Dehal and Boore 2005). McLysaght et al. (2002) and Hokamp et al. (2003) found paralogons covering over 44% of the human genome, many more than would be expected by chance. Moreover, molecular clock analyses of all protein families in humans that have orthologs in the fruit fly *Drosophila melanogaster* and the nematode *Caenorhabditis elegans* indicated that a burst of gene duplication activity took place 350–650 million years ago. These findings together lend support to the hypothesis that one or more polyploidization events occurred early in the vertebrate lineage.

Thus, although the 2R hypothesis remains highly contentious, and may not be solvable solely by phylogenetics, the possibility remains that vertebrates (including the readers of this book) may in fact be cryptooctoploids, and the fish they consume cryptohexadecaploids!

CHAPTER 8

Evolution by Molecular Tinkering

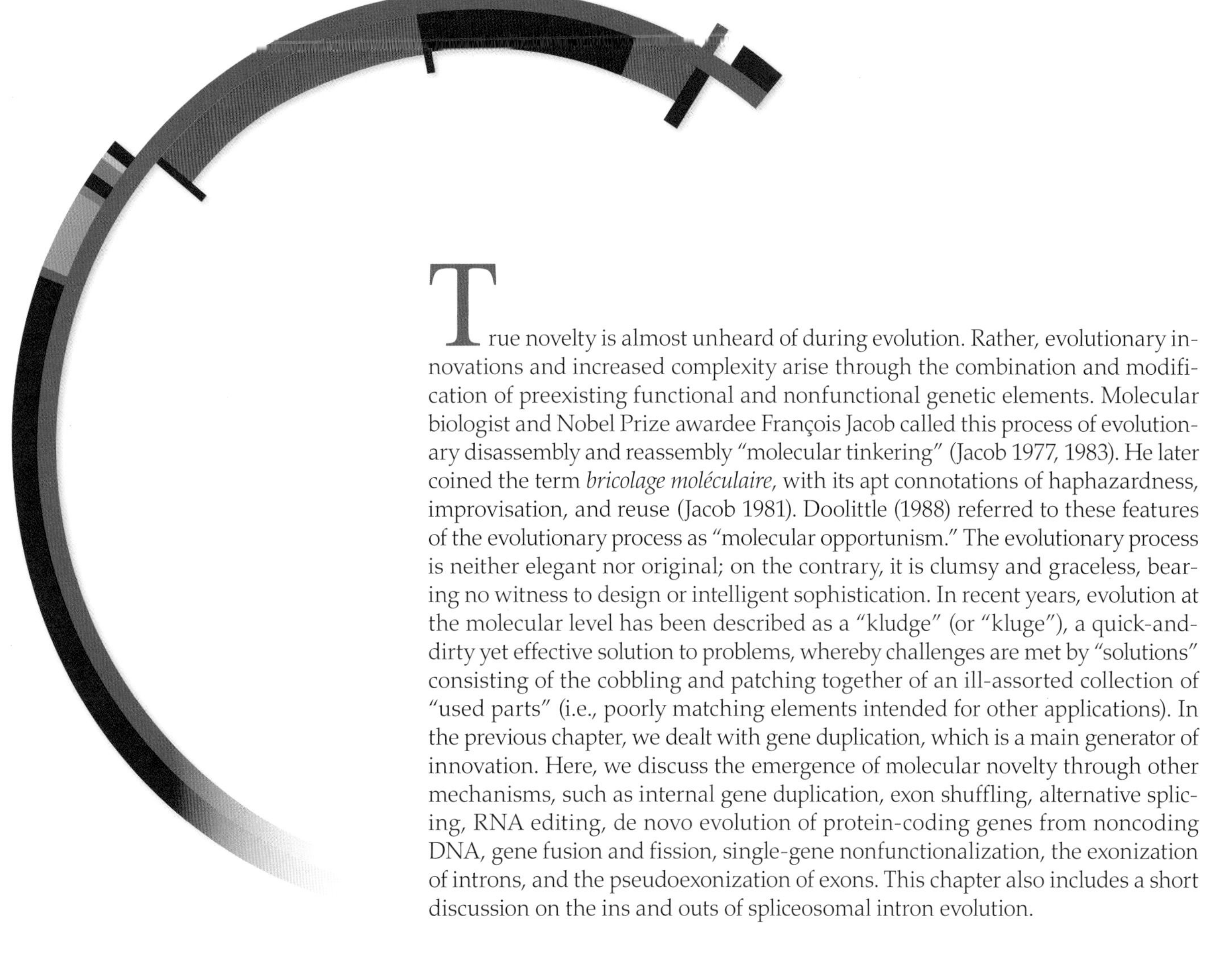

True novelty is almost unheard of during evolution. Rather, evolutionary innovations and increased complexity arise through the combination and modification of preexisting functional and nonfunctional genetic elements. Molecular biologist and Nobel Prize awardee François Jacob called this process of evolutionary disassembly and reassembly "molecular tinkering" (Jacob 1977, 1983). He later coined the term *bricolage moléculaire,* with its apt connotations of haphazardness, improvisation, and reuse (Jacob 1981). Doolittle (1988) referred to these features of the evolutionary process as "molecular opportunism." The evolutionary process is neither elegant nor original; on the contrary, it is clumsy and graceless, bearing no witness to design or intelligent sophistication. In recent years, evolution at the molecular level has been described as a "kludge" (or "kluge"), a quick-and-dirty yet effective solution to problems, whereby challenges are met by "solutions" consisting of the cobbling and patching together of an ill-assorted collection of "used parts" (i.e., poorly matching elements intended for other applications). In the previous chapter, we dealt with gene duplication, which is a main generator of innovation. Here, we discuss the emergence of molecular novelty through other mechanisms, such as internal gene duplication, exon shuffling, alternative splicing, RNA editing, de novo evolution of protein-coding genes from noncoding DNA, gene fusion and fission, single-gene nonfunctionalization, the exonization of introns, and the pseudoexonization of exons. This chapter also includes a short discussion on the ins and outs of spliceosomal intron evolution.

Protein Domains

A **domain** is a well-defined region within a protein that constitutes a compact, stable, independently folding unit with a unique three-dimensional structure. A protein domain consists of a continuous stretch of amino acids. Domains have size limits; the domains described so far vary in size from ~40 amino acids to ~700 amino acids. The mean size of domains of known structures is ~174 residues (Gerstein 1997). Often, protein domains convey a distinct function, such as substrate binding. A protein structural domain that can function and evolve independently is sometimes referred to as a **module** (Gō and Nosaka

1987). We note that some functions are conferred by amino acid residues that are scattered throughout the polypeptide; that is, a function may not be colinear with the amino acid sequence. This caveat is important when considering possible evolutionary mechanisms by which **multidomain proteins**, i.e., proteins possessing two or more different domains, have come into existence. If a function is encoded by a contiguous module, then the duplication of this module will result in an increase in the number of amino acid sequences performing this function. In contrast, if a function is performed by scattered amino acid residues, then the effects of duplicating a contiguous stretch of DNA may not be functionally desirable. Indeed, the internal repeats found in many proteins are frequently the result of domain duplication (e.g., Barker et al. 1978; Staub et al. 2002a,b,c; Garb and Hayashi 2005).

Domain architecture, or the order of domains in a protein, is frequently considered a fundamental level of a protein's functional complexity. A protein can consist of a single domain or multiple domains. The majority of proteins in all genomes studied to date are multidomain proteins; two-thirds of the proteins in prokaryotes and about four-fifths of the proteins in eukaryotes have two or more domains (Chothia et al. 2003). The domains of a multidomain protein can belong either to the same domain type or to different types. Repeats of a single domain type attest to a history of internal gene duplications. The presence of multiple domain types may hint at processes such as exon shuffling, gene fusion, and transposition. The prevalence of proteins with more than two domains and the recurrent appearance of the same domain in otherwise nonhomologous proteins indicate that domains are routinely "reused" in the evolution of new proteins. Domains have, in fact, been likened to Lego blocks that can be combined in various ways to build proteins with completely new functions (Das and Smith 2000). Over 10,000 protein domains have been collected in databases (Buljan and Bateman 2009). Using these domains along with species phylogenies, one can investigate the gain and loss of protein domains during evolution. In animals, most gains and losses of domains seem to occur at protein termini. The bias toward protein termini is largely due to the fact that insertion and deletion of domains at most other positions in a protein are likely to disrupt the structure and function of the protein.

A protein structural domain is usually a compact, mostly globular substructure that has more interactions within itself than with the rest of the protein. One can thus identify a structural domain by using two characteristics related to the protein's three-dimensional structure: its compactness and the extent of its isolation from other parts of the protein or from other domains (e.g., Islam et al. 1995; Taylor 1999). Assigning functionality requires, in addition, various biochemical tests, such as cleaving and isolating the domain under study and testing its activity. Gō (1981) exploited the fact that interdomain distances are normally larger than intradomain distances and designed a simple graphical method (the **Gō plot**) for identifying domains in globular proteins. In this method, the amino acid residues of a protein are listed consecutively on the two axes of a two-dimensional matrix. With the tertiary structure of a protein as a given (**Figure 8.1a**), a plus sign is entered in the matrix if the distance between two corresponding residues is greater than a certain predetermined critical value (**Figure 8.1b**). For globular proteins, this value is usually the radius of the sphere containing the globular protein, denoted as R. In the ideal case, domains may be identified unambiguously as empty, nonoverlapping, right-angle triangles whose hypotenuses are on the diagonal of the Gō plot and whose sides are defined by distinct rectangles containing clusters of plus signs. In less than ideal cases, complex statistical methods are required to identify the most likely locations of the boundaries between modules. **Figure 8.1c** shows a Gō plot for the β subunit of human hemoglobin.

Internal Gene Duplication

Internal gene duplication is a process by which tandem repeated sequences within a gene are created by sequence duplication. A trivial consequence of internal gene duplication is **gene elongation**. Gene elongation is an important step in the evolution

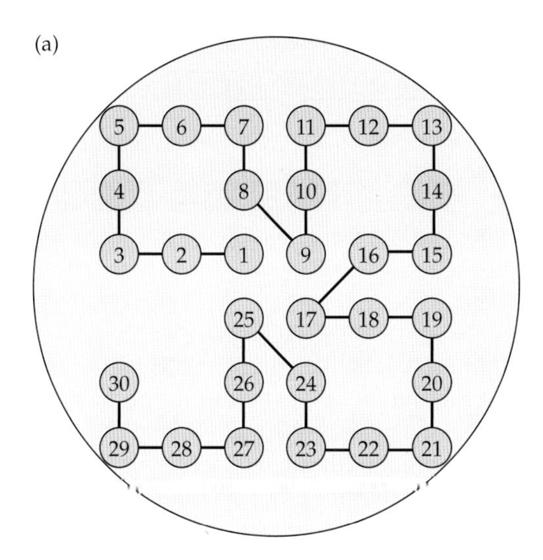

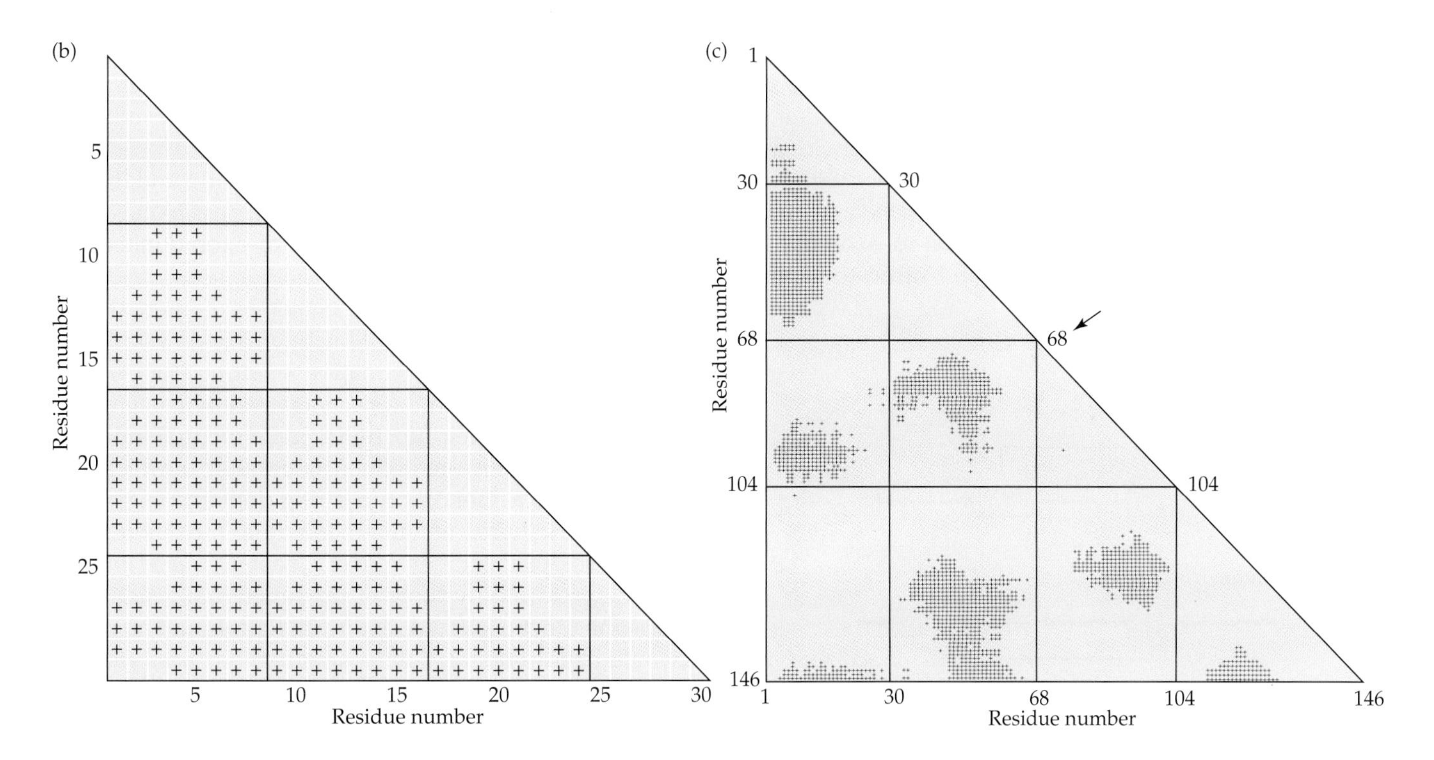

Figure 8.1 Gō plots for identifying structural modules. (a) A simplified two-dimensional protein consisting of 30 amino acids. The distance between two adjacent amino acids either horizontally (e.g., 6 and 7) or vertically (e.g., 17 and 24) is one unit. The radius of the sphere containing this protein is 3.5 units. (b) A plus sign (+) is entered in the matrix if the distance between two corresponding amino acid residues is greater than the radius of the sphere containing this schematic protein. Four modules were identified: 1-8, 9-16, 17-24, and 25-30. (c) Gō plot for an actual protein, the human β chain of hemoglobin. The radius of the protein is 27 Å. Four protein modules were identified: 1-30, 31-68, 69-104, and 105-146. In contrast, the human β-globin gene consists of only three exons, corresponding to amino acid positions 1-30, 31-104, and 105-146. Based on the one-module, one-exon hypothesis, Gō (1981) postulated that a merger had occurred between two exons as a result of the loss of a central intron. The predicted positions of the introns are numbered on the hypotenuse. An arrow marks the predicted position of the central intron. This prediction was borne out by the finding of just such an intron in plant leghemoglobin genes, as well as in several invertebrate globin genes.

of complex genes from simple ones. Of course, the elongation of genes can also occur by means other than internal gene duplication. For example, a mutational change converting a stop codon into a sense codon can elongate the gene (Chapter 1). Other mechanisms for gene elongation include insertion of a foreign DNA segment into an exon, exonization of intronic sequences, and the occurrence of mutations obliterating or creating splicing sites. These types of molecular changes, however, would most probably disrupt the function of the elongated gene, because the added regions would almost certainly encode an irrelevant array of amino acids. Indeed, in the vast majority of cases, such molecular changes have been found to be associated with pathological manifestations. For instance, the α-hemoglobinopathies (i.e., α-hemoglobin abnormalities resulting in pathological manifestations) Constant Spring, Icaria, Koya Dora, and Seal Rock resulted from mutations turning the stop codon into codons for glutamine, lysine, serine, and glutamic acid, respectively, thus adding 30 additional residues to the α chains of these variants (Weatherall and Clegg 1979).

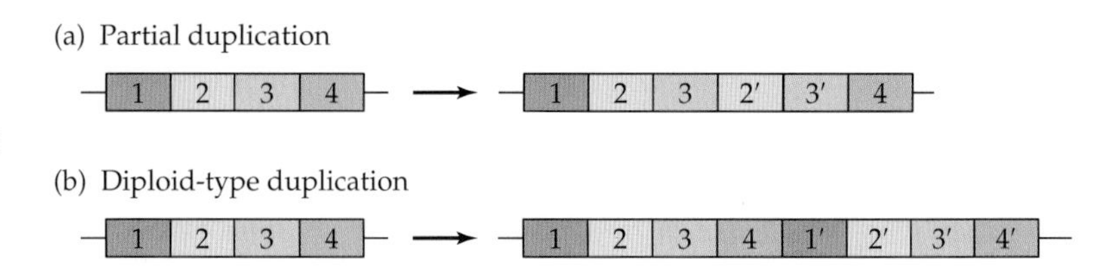

Figure 8.2 Two types of internal gene duplication. Domains are numbered; duplicated domains are marked by the number of the duplicated domain with a prime. (a) In partial duplication, one or a few adjacent domains are duplicated. The number of duplicated domains is smaller than the total number of domains in the protein. (b) In diploid-type duplication, all the domains in the protein are duplicated, resulting in a protein twice the size of the original one.

By contrast, duplication of a structural domain is less likely to be problematic. Indeed, such a duplication can sometimes even enhance the function of the protein produced, for example by increasing the number of active sites (a quantitative change), thus enabling the gene to perform its functions more rapidly and efficiently, or by having a synergistic effect yielding a new function (a qualitative change). Of course, as in all mutational processes, the duplication of a structural domain can also cause trouble.

There are two types of internal gene duplications, partial gene duplication and diploid-type duplication (**Figure 8.2**; Shimizu et al. 2004). **Partial gene duplication** involves less than the entire gene sequence. A **diploid-type gene duplication** involves the entire gene, but as opposed to the process of whole-gene duplication that yields two copies of the gene (Chapter 7), the diploid-type gene duplication results in a single gene twice as long as the original one.

In genes with introns, partial gene duplications may affect splicing and, hence, may create introns out of exons and exons out of introns. Four such possibilities exist: **exon-to-exon duplication**, **exon-to-intron duplication**, **intron-to-exon duplication**, and **intron-to-intron duplication** (**Figure 8.3**).

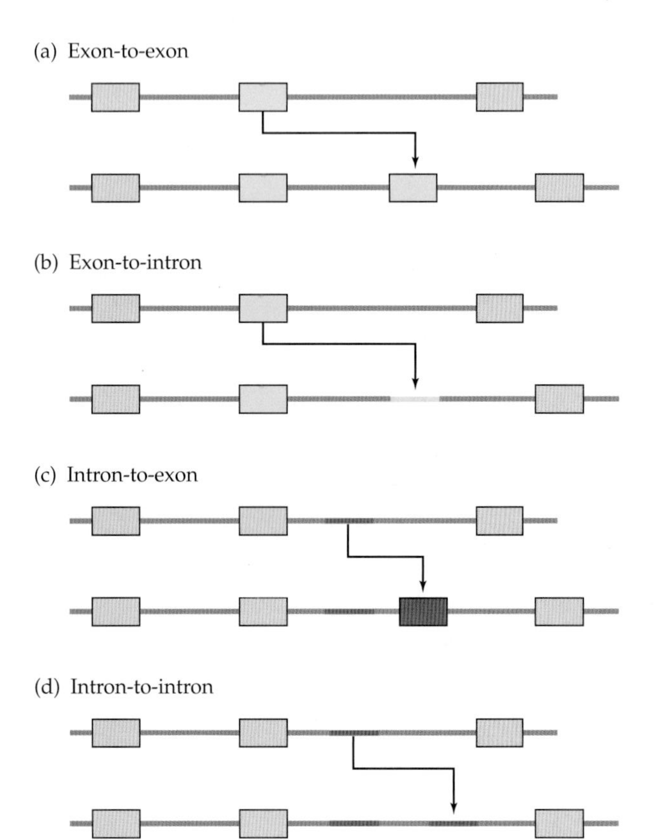

Figure 8.3 Types of partial gene duplication. Exons are shown as boxes.

Several structural and functional outcomes of internal gene duplication have been described (Taylor and Sadowski 2010). In a manner analogous to the situation resulting from whole-gene duplication, the internal repeats created by partial gene duplication may retain their functional and structural independence with no significant interdomain association, in which case they are referred to as **beads-on-a-string domains** (e.g., Phan et al. 2007). Alternatively, in a manner akin to the duplication-degeneration-complementation model (Chapter 7), each internal duplicate may retain only some of the functional residues, in which case the function would require all repeats and no repeat alone would be functional. Such a situation has been dubbed **inseparable domains**.

A diploid-type gene duplication of a protein-coding gene that used to produce single units for **homodimers** (i.e., dimers made of identical subunits) may result in a protein that folds as a **pseudodimer**, i.e., a single polypeptide mimicking the structure of the dimer (**Figure 8.4**). There are many examples of proteins with two or more domains of nearly identical structure that are likely the result of diploid-type duplication. The *Escherichia coli* thioester dehydrase is a homodimer. Each monomer has the same fold as the domains of another protein, thioesterase. Interestingly, thioesterase is a monomer made up of two equal-sized domains of almost identical structure. Thus, thioesterase must have undergone a diploid-type duplication that resulted in the creation of a pseudodimer.

Finally, in a manner akin to neofunctionalization (Chapter 7), the internal copies produced by internal gene duplication may diverge in sequence, ultimately resulting in each of the repeats performing a different function. For instance, the variable and constant regions of immunoglobulin genes were most probably derived from a common primordial domain but have since acquired distinct properties (Leder 1982). Thus, despite common molecular ancestry, the variable region of immunoglobulins binds antigens, while the constant region mediates nonantigenic functions. Many complex genes might have arisen in this manner.

Even if an internal duplication does not involve any active sites, it may still be beneficial. For example, a duplication event involving a structural domain engaged in conferring spatial stability to the protein, or protecting its active parts from degradation, or altering its cellular interactivity may affect the function and longevity of the protein. For example, PEST motifs—amino acid sequences rich in proline (P), glutamic acid (E), serine (S), and threonine (T)—were found to be cleaved and subsequently degraded rapidly inside eukaryotic cells (Rogers et al. 1986; Belizario et al. 2008). In some cases, duplicated PEST domains were found in proteins, thus ensuring their very rapid degradation, which is evidently important in such proteins as regulatory nuclear factors and calmodulin-binding proteins, which have specific and transient functions (Chevaillier 1993). It has also been suggested that, as long as they do not interfere with normal function, redundant duplicate domains may be maintained indefinitely within the genome, and may in time serve as raw materials for creating new functions.

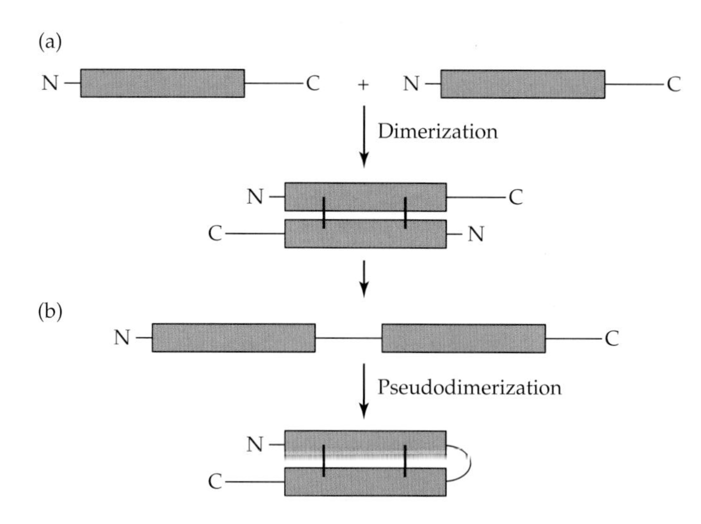

Figure 8.4 Schematic depiction of protein homodimerization (a) and pseudodimerization (b) as a result of a diploid-type duplication of a protein-coding gene. The vertical bars indicate bonds (e.g., cystine bonds) between the two subunits. Amino (N) and carboxyl (C) termini are indicated.

Properties and prevalence of internal gene duplication

Internal gene duplications have occurred frequently in the evolution of eukaryotic genes (Letunic et al. 2002). The fraction of protein-coding genes in the human genome that contain one or more duplicated regions is ~18% (Björklund et al. 2006; Gao and Lynch 2009), and the fraction containing an exon duplication is 11% (Letunic et al. 2002). In comparison, internal gene duplications are infrequent in prokaryotes. The mean fraction of proteins containing domain repeats in eubacteria and archaebacteria is less than 5% (Björklund et al. 2006), and some of them may in fact have been acquired from a eukaryote by horizontal gene transfer (Chapter 9). For instance, a domain called ankyrin was found repeated in tandem 19 times in the syphilis bacterium *Treponema pallidium*. This domain is found in other bacteria, but never consecutively repeated so many times. In contrast, this domain is commonly repeated numerous times in animal genes. Hence, a likely scenario is that this repeat has been horizontally transferred from a eukaryotic host to the bacterium, rather than expanded in the bacterial lineage (Björklund et al. 2006).

In the *Caenorhabditis elegans* genome, the median size of duplications was estimated at 1.4 Kb, i.e., shorter than the average gene length (2.5 Kb). This indicates that partial gene duplications are more frequent than complete gene duplications (Katju and Lynch 2003). An examination of six completed sequenced genomes (human, mouse, fruit fly, nematode, rice, and *Arabidopsis thaliana*) revealed that internal gene duplications occur with a frequency of 0.001–0.013 duplications per gene per million years (Gao and Lynch 2009).

Domain repeats are often expanded through duplications of several domains at a time, while the duplication of one domain is less common. Long repeats containing several domains in tandem have been observed to be particularly common in multicellular species. Some domain families show distinct duplication patterns. For example, the muscle protein nebulin in vertebrates contains a large number of repeated actin-binding domains. The domains have mainly been expanded through duplications of either single domains or a seven-domain super-repeat (Björklund et al. 2010).

Table 8.1

Examples of human proteins with internal domain duplications taking up 75% or more of the total length

Protein	Length of protein[a]	Mean length of repeat[a]	Number of repeats[b]	Percent repetition[c]
α1β-Glycoprotein	474	91	5	96
Calbindin	260	43	6	99
Calcium-dependent regulator protein	148	74	2	100
Hemopexin	439	207	2	94
Hexokinase	917	447	2	97
Lactase-phlorizin hydrolase	1,927	480	3	79
Ribonuclease/angiogenin inhibitor	461	57	8	99
Serum albumin	584	195	3	100
Tropomyosin α chain	284	42	7	100
Villin	826	360	2	87

[a] Length is measured in number of amino acid residues.
[b] Some repeats may be truncated in comparison with the duplicated consensus unit.
[c] Percent of the total length occupied by repeated sequences.

The fraction of internal gene duplications observed in the middles of proteins is somewhat higher than expected by chance. Hence, internal gene duplications exhibit a different pattern from that of domain shuffling (p. 349), which mainly involves the addition and deletion of domains at the termini of proteins.

In **Table 8.1** we present a few examples of human genes for which there is evidence of internal duplication during their evolutionary histories. All involve one or more domain duplications, and some of the sequences (e.g., tropomyosin α chain) were derived from serial duplications of a primordial sequence, resulting in a repetitive structure that takes up the entire length of the protein. In each of the examples in Table 8.1, the duplication event can easily be inferred from protein or DNA sequence similarity. We note, however, that not all internally duplicated regions may be identifiable. Sometimes, internally duplicated regions may have diverged from one another to such an extent that the sequence similarity between them is no longer discernible. In some cases, such as in the case of the constant and variable regions of immunoglobulin genes, a common ancestry due to internal gene duplication was inferred by comparing the secondary and tertiary structures of the protein domains, because these structures tend to be conserved better than amino acid sequences (Hood et al. 1975; Worth et al. 2009).

Here, we present a few examples of evolution by internal gene duplication.

THE DINUCLEOTIDE-BINDING REGIONS OF GLYCERALDEHYDE 3-PHOSPHATE DEHYDROGENASE AND ALCOHOL DEHYDROGENASE One example of neofunctionalization by internal gene duplication involves the dinucleotide-binding regions of glyceraldehyde 3-phosphate dehydrogenase and alcohol dehydrogenase. In many species, these binding regions consist of two domains (**Figure 8.5**). Each of these domains consists of three homologous β-folding units, denoted βA, βB, and βC in the N-terminal domain and βD, βE, and βF in the C-terminal domain. The entire dinucleotide-binding region is encoded by five exons, three for the N-terminal domain and two for the C-terminal domain. Each coenzyme-binding domain can bind only a mononucleotide. The entire duplicated region, however, can bind not only two mono-

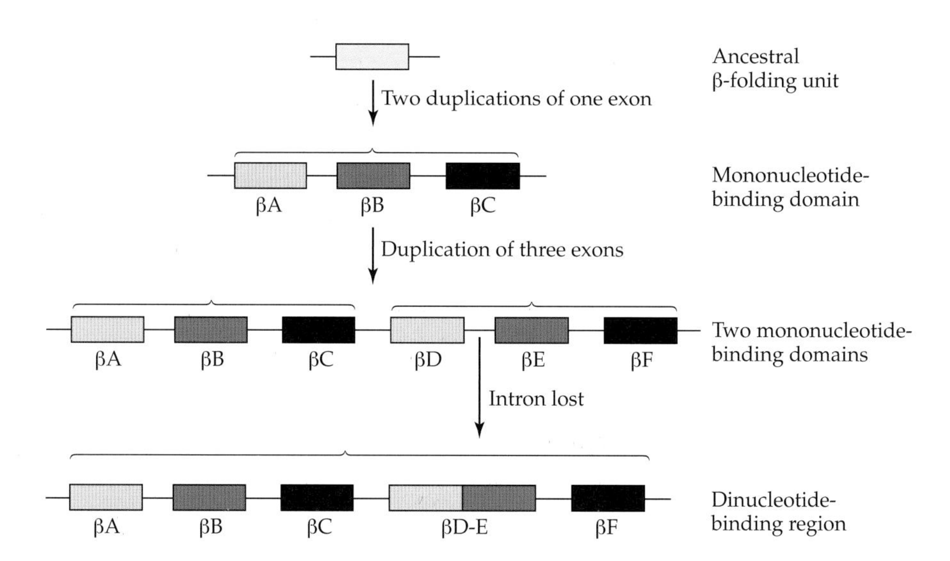

Figure 8.5 The evolution of a dinucleotide-binding domain. Exons are shown as boxes connected by introns, shown as lines. An ancestral β-folding unit encoded by an exon underwent two duplication events to produce a triexonic mononucleotide-binding domain. The duplication of the three exons resulted in the creation of two mononucleotide-binding domains. Subsequent modifications of the primary sequence gave rise to a single dinucleotide-binding region. The present pentaexonic arrangement is explained by the loss of an intron. (From Graur and Li 2000.)

nucleotides, but also a dinucleotide. Thus, in comparison with the unduplicated domain, the duplicated domain can also recognize a novel molecular entity.

THE OVOMUCOID GENE Ovomucoid is an inhibitor of trypsin, a serine proteinase inhibitor that catalyzes the digestion of proteins. In birds, it is present in the albumen (egg white). The ovomucoid polypeptide consists of three structural and functional domains (**Figure 8.6**). Each domain is capable of binding one molecule of either trypsin or another serine proteinase. The DNA regions coding for the three functional domains clearly share a common origin and are separated from each other by phase 0 introns (Stein et al. 1980; Sharp 1981). Domains I and II, I and III, and II and III exhibit 27%, 38%, and 35% amino acid sequence identity, respectively, if the gaps in the alignment are ignored. Each of the three regions consists of two exons interrupted by a phase 2 intron, and the two exons encoding the two parts of each domain exhibit no similarity between them. Thus, the ovomucoid gene appears to have been derived from a primordial single-domain gene by two internal duplications, each of which involved two neighboring exons.

ENHANCEMENT OF FUNCTION IN THE α2 ALLELE OF HAPTOGLOBIN A well-known example of an enhancement in function as a consequence of an internal gene duplication is that of the haptoglobin *α2* allele in humans (Smithies et al. 1962). Haptoglobin is a tetrameric protein made up of two α and two β chains. Both chains are produced by the same gene as a single polypeptide, which is subsequently cleaved at an arginine residue to generate the α and β subunits. Haptoglobin is found in the blood serum, where it functions as a transport glycoprotein that removes free hemoglobin from the circulation of vertebrates. In

Figure 8.6 Schematic secondary structure of the secreted ovomucoid protein from chicken. The protein sequence is shown using the one-letter abbreviation for amino acids (Chapter 1, Table 1.3). Each protein domain is encoded by two exons. Introns B, D, and F are phase 0 introns. Introns C, E, and G are phase 2 introns, i.e., they are located between second- and third-codon positions. Intron A interrupts the 5′ noncoding region and is not shown.

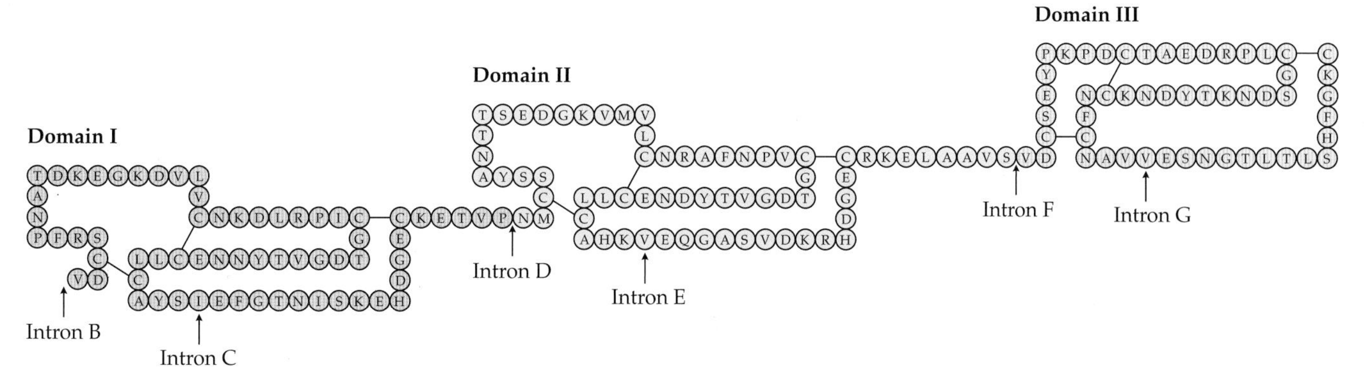

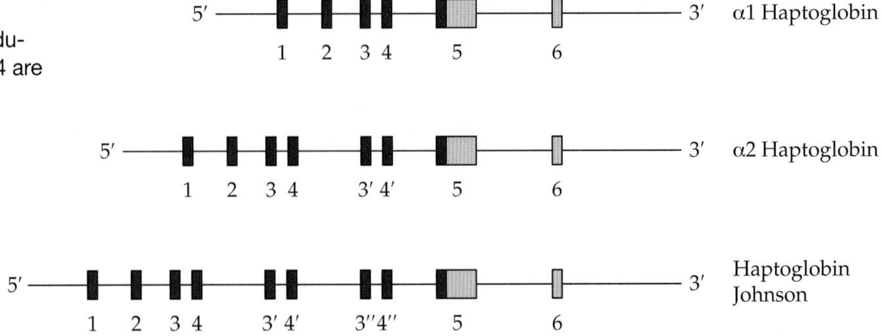

Figure 8.7 Three alleles of haptoglobin α in humans. In α2 haptoglobin, exons 3 and 4 are duplicated; in haptoglobin Johnson, exons 3 and 4 are triplicated.

human populations, haptoglobin α is polymorphic because of the existence of three common alleles: there is slow α1 (due to the *α1S* allele), fast α1 (due to *α1F*), and α2 (due to *α2*). The *α2* allele was most probably created by a nonhomologous crossover within different introns of two *α1* alleles within a heterozygous individual carrying both the *α1S* and *α1F* electrophoretic variants. The length of the internal duplication was about 1.7 Kb, of which 177 bp were derived from exons 3 and 4. The duplication nearly doubled the length of the encoded polypeptide (from 84 to 143 amino acids). As a consequence, the stability of the haptoglobin–hemoglobin complex and the efficiency of rendering the heme group of hemoglobin susceptible to degradation increased considerably (Black and Dixon 1968).

The *α2* allele is probably of recent origin; it appeared after the human-chimpanzee split, most probably as recently as 100,000 years ago, and yet it attained fairly high frequencies, especially in Southeast Asia (Mourant et al. 1976; Langlois and Delanghe 1996). Such a rapid increase in frequency indicates a significant selective advantage for individuals carrying the *α2* allele. If, indeed, individuals carrying *α2* alleles enjoy a selective advantage over carriers of *α1* alleles, it is likely that in the future the *α2* allele will become fixed in the human population at the expense of the *α1* variants. Support for this hypothesis has been obtained by studying the bacteriostatic (growth-inhibiting) properties of haptoglobin, which rest on the ability of haptoglobin to restrict the access of bacteria to hemoglobin-derived iron that is critical for bacterial growth. As has been demonstrated empirically, the *α2* allele is much more efficient against streptococcal infections than the *α1* allele (Wasserzug et al. 2007).

Interestingly, an even longer allele, *α3* (for haptoglobin Johnson), has been found in human populations (**Figure 8.7**). This allele contains a threefold tandem repeat of the same 1.7 Kb segment implicated in the *α2* duplication (Oliviero et al. 1985).

ORIGIN OF AN ANTIFREEZE GLYCOPROTEIN GENE: A "COOL" TALE OF INTERNAL GENE DUPLICATIONS The body fluids of most teleosts (ray-finned fishes) freeze at temperatures ranging from –1.0°C to –0.7°C. Therefore, most fish cannot survive the freezing temperatures of approximately –2°C of the Arctic and Antarctic Oceans. Freezing resistance in such fishes is frequently due to the existence in the blood of a protein that lowers the freezing temperature by adsorbing to small ice crystals and inhibiting their growth and recrystallization, which might otherwise puncture the cell membranes with fatal consequences. There are several such proteins, termed antifreeze protein types I, II, and III, and antifreeze glycoprotein. Where did these proteins come from? In the case of the antifreeze glycoprotein gene in the Antarctic toothfish (*Dissostichus mawsoni*), the answer provides an interesting example of acquisition of a novel function through internal gene duplication.

There are many different antifreeze glycoproteins in this species, each of which is composed mostly of two simple tripeptide repeats: Thr-Ala-Ala and Thr-Pro-Ala. (Note that the proline codon family differs from the alanine codon family by a single nucleotide.) The antifreeze glycoproteins are encoded by a large gene family, in which each gene encodes a large polyprotein precursor that is cleaved posttranslationally to yield multiple antifreeze glycoprotein molecules.

Chen et al. (1997) characterized one antifreeze glycoprotein gene from the Antarctic toothfish and discovered that it derives from a gene encoding a trypsinogen-like serine proteinase (**Figure 8.8**). The evolutionary history of the antifreeze glycoprotein gene could be reconstructed with a great degree of confidence, mainly because the gene was inferred to have arisen only 5–14 million years ago, an estimate that agrees well with the presumed date of the freezing of the Antarctic Ocean (10–14 million years ago). Its relative newness essentially means that its evolutionary history has not yet been obscured by superimposed molecular changes.

As shown in Figure 8.8, an initial deletion in the ancestral trypsinogen-like serine proteinase gene, starting at nucleotide 6 in the second exon and ending one nucleotide before the start of exon 6, created a new gene with two exons. Four nucleotides from the first intron, five nucleotides from the second exon, one nucleotide from the fifth intron, and five nucleotides from the sixth exon became the coding region of a newly created exon, whose short frameshifted reading frame encoded a tetrapeptide (Thr-Ala-Ala-Gly) before reaching a new in-frame stop codon (TGA). The 12 bp sequence encoding the first three amino acids (Thr-Ala-Ala) was duplicated twice to create a fourfold repetition of this tripeptide. Following this initial round of duplication, a

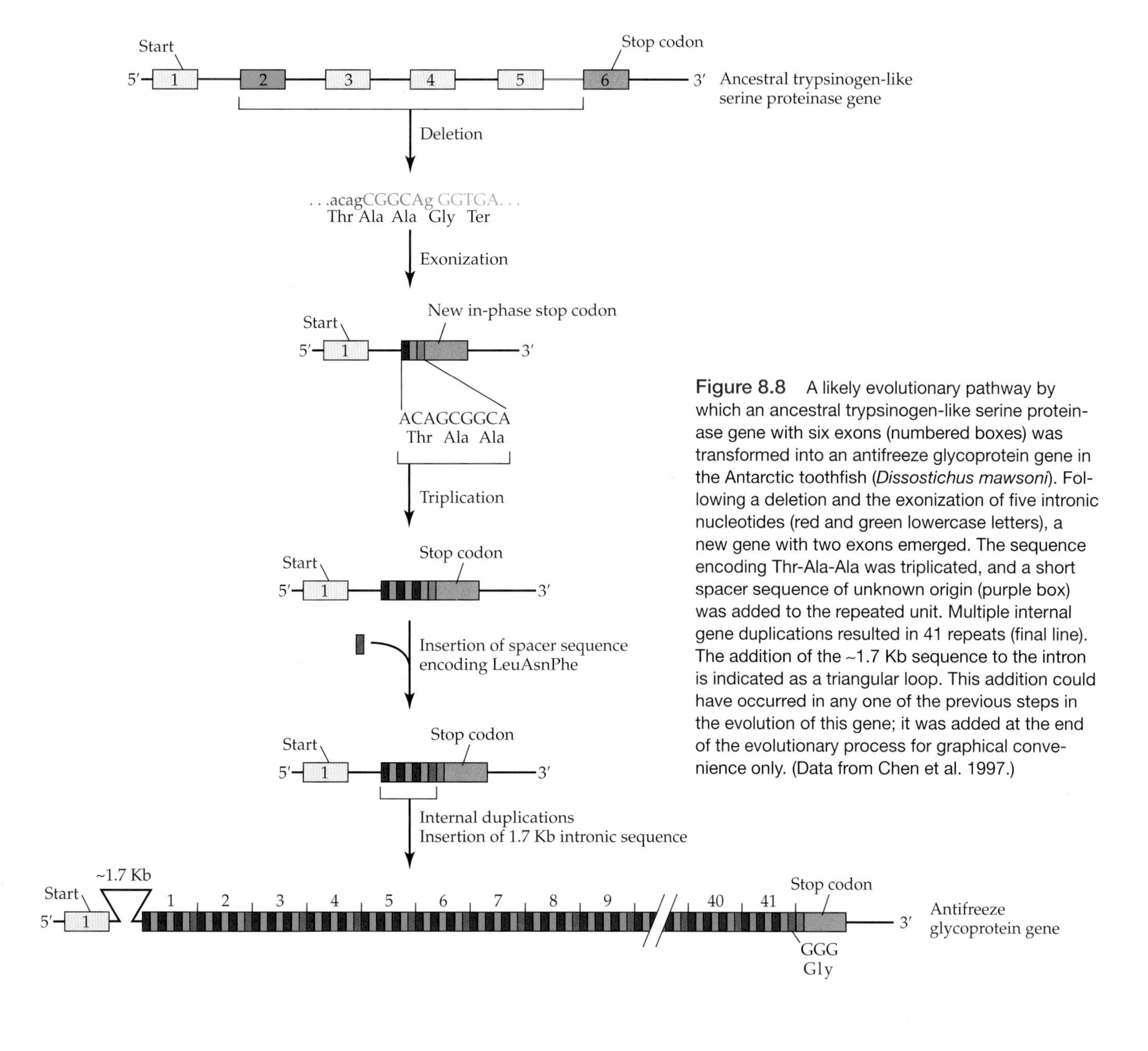

Figure 8.8 A likely evolutionary pathway by which an ancestral trypsinogen-like serine proteinase gene with six exons (numbered boxes) was transformed into an antifreeze glycoprotein gene in the Antarctic toothfish (*Dissostichus mawsoni*). Following a deletion and the exonization of five intronic nucleotides (red and green lowercase letters), a new gene with two exons emerged. The sequence encoding Thr-Ala-Ala was triplicated, and a short spacer sequence of unknown origin (purple box) was added to the repeated unit. Multiple internal gene duplications resulted in 41 repeats (final line). The addition of the ~1.7 Kb sequence to the intron is indicated as a triangular loop. This addition could have occurred in any one of the previous steps in the evolution of this gene; it was added at the end of the evolutionary process for graphical convenience only. (Data from Chen et al. 1997.)

short spacer sequence of unknown origin was added to the 3′ end of the repeated unit. The resulting sequence was then duplicated multiple times to yield 41 repeats followed by the Gly codon from the original tetrapeptide and a stop codon. Some of the spacer sequences encode peptide motifs that serve as signals for the cleavage of the antifreeze glycoprotein polypeptide into the active proteins. Sometime along the line from the ancestral trypsinogen-like proteinase gene to the antifreeze glycoprotein gene, a sequence approximately 1.7 Kb long was added to the intron. Most probably, this insertion had no functional consequences.

We note that a completely new function has been created by multiple mutational events (mostly internal gene duplications) in a very short time span. Thus, the new gene must have been subjected to intense positive selection, most probably due to an abrupt shift in environmental conditions (Logsdon and Doolittle 1997). Interestingly, close evolutionary relatives of *D. mawsoni*, which presumably migrated to inhabit temperate waters at much lower latitudes, have retained functional antifreeze glycoprotein genes (Cheng et al. 2003).

Convergent processes of internal gene duplication leading to the creation of antifreeze proteins and glycoproteins have occurred many times in the evolution of polar (Arctic and Antarctic) fishes (Chen et al. 1997; Patarnello et al. 2011). For example, in the case of the Antarctic eelpout, *Lycodichthys dearborni*, a type III antifreeze protein was evolved from a copy of a sialic acid synthase gene following a single internal gene duplication (Deng et al. 2010).

Exon-Domain Correspondence

Several possible relationships may be theoretically envisioned between the structural domains of a protein and the arrangements of the exons in the gene encoding it. Gō (1981) and Liu and Grigoriev (2004) found that in many globular proteins for which the internal modular division had been determined, a more or less exact correspondence existed between the exons of the gene and the structural domains of the protein product (**Figure 8.9a**). In many cases, an exon was found to encode several adjacent modules (**Figure 8.9b**). The domains encoded by the exons in Figures 8.9a and b are **exon-bordering domains** (i.e., domain borders coinciding precisely or approximately with exon borders). In some cases, a single module was found to be encoded by more than one exon (**Figure 8.9c**). A complete discordance between the modular structure of a protein and the division of its gene into exons (**Figure 8.9d**) was not found by Gō (1981) in the original study, but cases of complete or partial discordance were subsequently discovered in many protein-coding genes. Notwithstanding the large number of exceptions noted above, the correlation between domains and exons was found to be not only statistically significant (Vivek et al. 2003), but also biologically intriguing.

When an exon encodes a domain (Figure 8.9a), the exon may be regarded as an independent unit that can be duplicated, deleted, or shuffled. The same is true for the case of two or more domains encoded by an exon (Figure 8.9b). These exons may be duplicated or shuffled, and it is possible to envision cases in which the addition of a domain to a protein may be beneficial. In comparison, the duplication, deletion, and shuffling of random sequences are

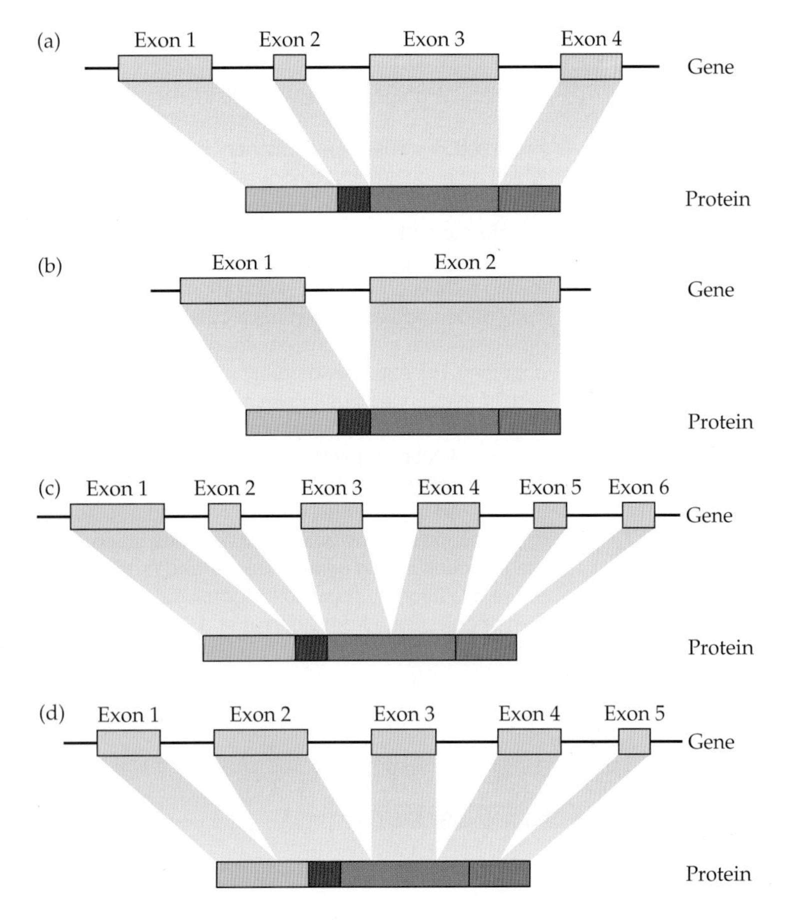

Figure 8.9 Four possible relationships between the arrangement of exons in a gene and the structural domains of its protein. (a) Each exon corresponds exactly to a structural domain. (b) An exon encodes two or more domains. (c) A single structural domain is encoded by two or more exons. (d) Lack of correspondence between exons and domains. The structural domains of the protein are designated by different-colored boxes. The domains in (a) and (b) are exon-bordering domains. (Modified from Graur and Li 2000.)

likely to be useless or even deleterious. A survey of nine vertebrate genomes revealed that duplicated domain-containing exons tend to be retained by proteins more often than other exons (Liu et al. 2005). Moreover, such exons also tend to have a higher mobility and be more widely distributed among proteins.

Exon-domain correspondence features prominently in discussions concerning the origin of spliceosomal introns (p. 383).

Mosaic Proteins

A **mosaic** (or **chimeric**) **gene** is a gene that contains sequences that are also found in nonhomologous genes. Thus, one part of a mosaic gene may be homologous to gene *A*, while another part of the mosaic gene may be homologous to gene *B*, where genes *A* and *B* are unrelated. This type of mixed kinship is sometimes referred to as **partial paralogy**. Proteins encoded by mosaic genes are referred to as **mosaic** (or **chimeric**) **proteins**. (In the literature, a distinction is sometimes made between naturally evolving mosaic proteins and artificially designed **combinatorial proteins**.) The general means by which mosaic genes are produced is called **domain shuffling**. So far, several molecular mechanisms for domain shuffling have been identified: (1) exon shuffling, (2) gene fusion, (3) gene fission, and (4) fragment joining through transposable-element recruitment of small chromosome fragments. (The last mechanism will be discussed in Chapter 9.)

One of the first described mosaic proteins was tissue plasminogen activator (**Figure 8.10**), an important component of the blood-clotting system in vertebrates. Tissue plasminogen activator is activated by coagulation factor XIIa to yield the active form, which converts plasminogen into its active form, plasmin, which in turn dissolves

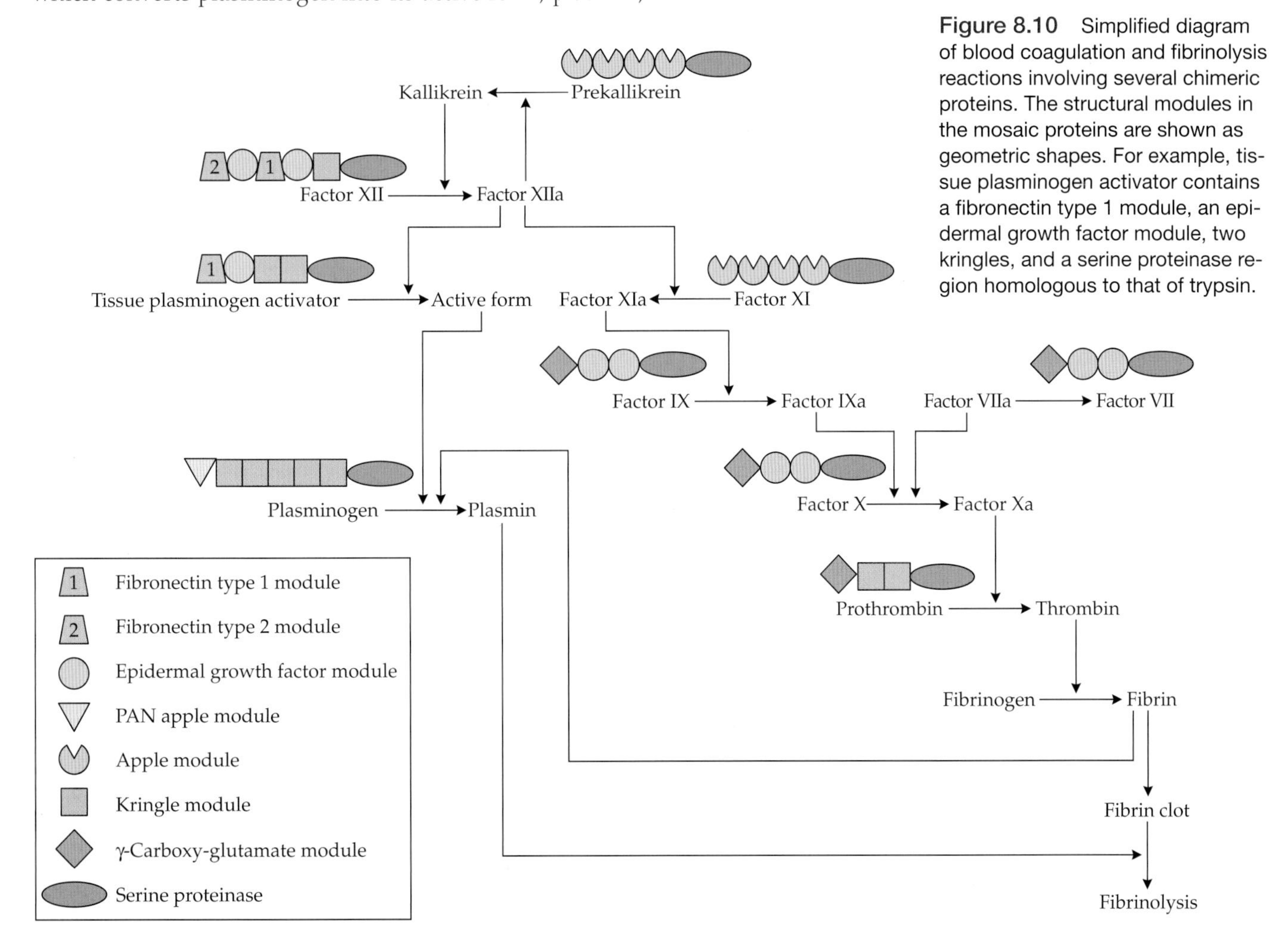

Figure 8.10 Simplified diagram of blood coagulation and fibrinolysis reactions involving several chimeric proteins. The structural modules in the mosaic proteins are shown as geometric shapes. For example, tissue plasminogen activator contains a fibronectin type 1 module, an epidermal growth factor module, two kringles, and a serine proteinase region homologous to that of trypsin.

fibrin, a soluble fibrous protein in blood clots. The conversion of plasminogen into plasmin is greatly accelerated by the presence of fibrin, the substrate of plasmin. Fibrin polymers bind both plasminogen and tissue plasminogen activator, thus aligning them for catalysis. This mode of molecular alignment allows plasmin production only in the proximity of fibrin, thus conferring fibrin specificity to plasmin. The physiological significance of this molecular mechanism is that it ensures that plasminogen activation takes place predominantly on the surface of fibrin, thus restricting plasmin action to its proper substrate.

Interestingly, tissue plasminogen activator and prourokinase—the precursor of the urinary plasminogen activator—have similar catalytic activities; however, prourokinase lacks fibrin specificity. A comparison of the amino acid sequences of tissue plasminogen activator and prourokinase showed that the former contains a 43-residue sequence at its amino-terminal end that has no counterpart in prourokinase (**Figure 8.11a**). This segment can form a finger-like structure and is homologous to one of the three finger domains responsible for the fibrin affinity of fibronectin—a large glycoprotein present in the plasma and on cell surfaces that promotes cellular adhesion (**Figure 8.11b**). Deletion of this segment in tissue plasminogen activator leads to a loss of its fibrin affinity. The sequence similarity of tissue plasminogen activator with fibronectin is restricted to this domain (denoted as fibronectin type 1 domain). Thus, this domain must have been acquired by tissue plasminogen activator from either fibronectin or a similar protein.

Tissue plasminogen activator also contains a segment homologous to portions of the epidermal growth factor precursor and the growth factor-like regions of other proteins, such as the coagulation factors VII, IX, X, and XII (Figures 8.10 and 8.11). In

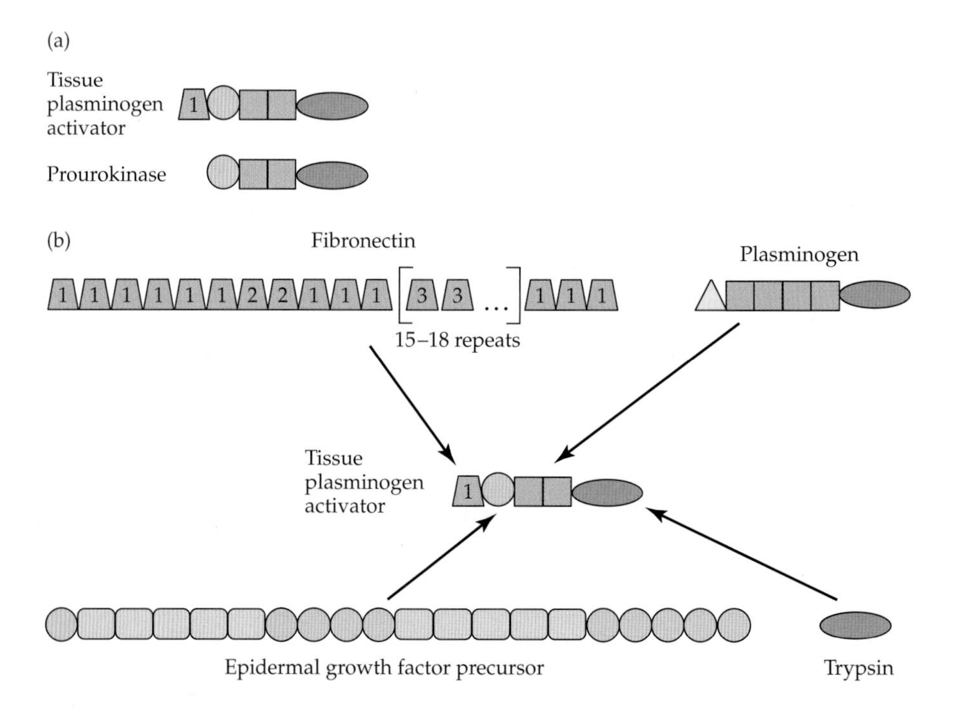

Figure 8.11 (a) Comparison of the modular structures of tissue plasminogen activator and prourokinase. (b) Possible origin of the modules acquired through exon insertion in the tissue plasminogen activator protein.

addition, the carboxy-terminal regions of tissue plasminogen activator are homologous to the protease parts of trypsin and other trypsin-like serine proteinases, such as prothrombin and plasminogen, which are enzymes that hydrolyze proteins into peptide fragments. Finally, the nonproteinase part of tissue plasminogen activator contains two structures similar to the kringles of plasminogen. (A kringle is a cysteine-rich sequence that contains three internal disulfide bridges and forms a pretzel-like structure resembling an eponymous Danish cake.)

Thus, during its evolution, tissue plasminogen activator acquired at least five DNA segments from at least four other genes: plasminogen, epidermal growth factor, fibronectin, and trypsin (Figure 8.11b). Moreover, the junctions of these acquired units coincide precisely with the borders between exons and introns (Ny et al. 1984), lending further credibility to the idea that exons can be shuffled from one gene to another.

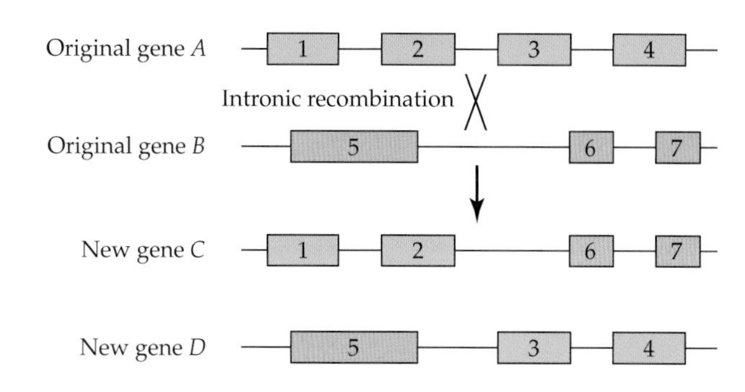

Figure 8.12 An example of an intronic recombination event between two nonhomologous genes resulting in exon shuffling. Exons are shown as numbered boxes.

Exon Shuffling

Exon shuffling is a process by which exons from different sources are mixed, duplicated, or deleted. The mixing of exons that encode one or more protein domains results in the creation of mosaic proteins.

Several molecular mechanisms have been observed to recombine different genic and nongenic regions to generate mosaic genes (Arguello et al. 2007). Unequal crossing over between nonhomologous genes or **nonhomologous recombination** is the most important mechanism in exon shuffling (**Figure 8.12**). This type of recombination is also referred to as **illegitimate recombination**, i.e., a recombination between two nonhomologous sequences that share either no sequence similarity or very little similarity between them. Many nonhomologous unequal recombination events are not really nonhomologous, as they occur through crossovers between dispersed repetitive sequences residing in introns. By being similar in sequence to one another and derived from a common source (i.e., homologous), these repetitive sequences increase the chances of a crossing-over event. In primates, about 30% of such recombination events occur through crossovers between *Alu* elements (Chapter 9). Another mechanism by which exon shuffling may occur is **nonallelic homologous recombination**, i.e., recombination between divergent paralogous genes.

There are three types of shuffling of protein-coding exon: exon duplication, exon insertion, and exon deletion (**Figure 8.13**). In most cases, **exon duplication** refers to the tandem duplication of an exon. **Exon insertion** is the process by which exons are inserted within an intron of a different protein-coding gene (**ectopic exon insertion**) or within a different intron of the same gene (**intragenic exon insertion**). (In the literature, the term "exon shuffling" is mostly reserved for ectopic exon insertion.) An **exon deletion** may result in the truncation of the reading frame of the gene, or it may cause a frameshift in the reading frame. As shown in Chapter 1, frameshifts may either elongate or shorten the encoded protein. All these types of exon shuffling are known to have occurred in the

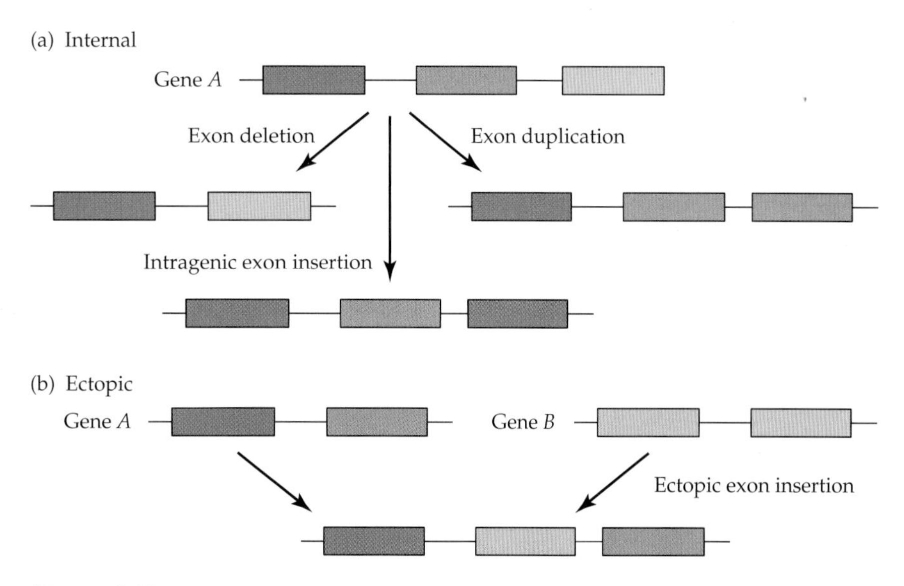

Figure 8.13 Internal (a) and ectopic (b) exon shuffling.

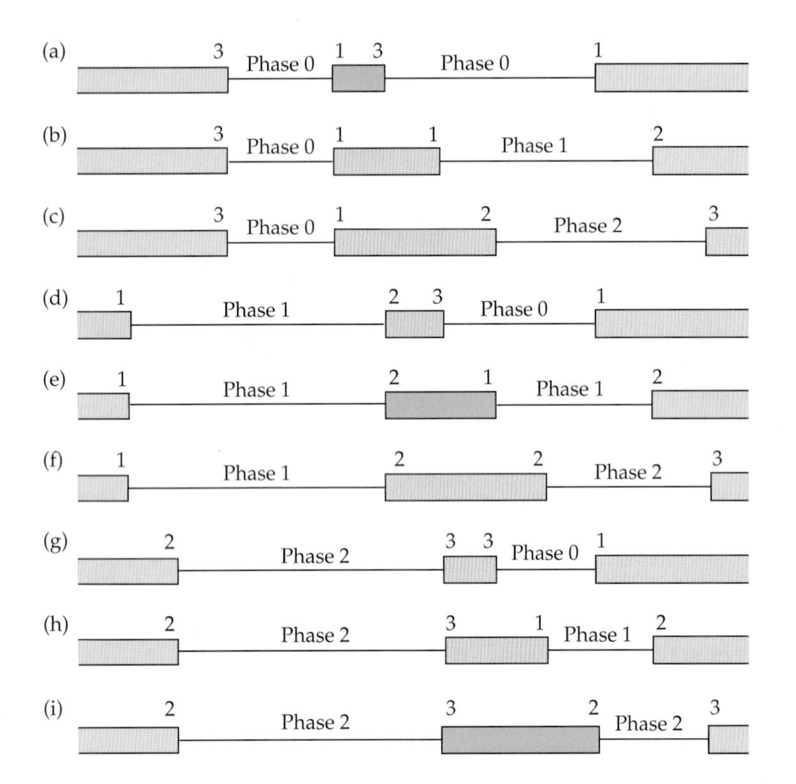

Figure 8.14 Phases of introns and classes of exons. Exons are represented by boxes. A number at an exon-intron junction indicates the codon position of the last nucleotide of the exon, while a number at an intron-exon junction indicates the codon position of the first nucleotide of the exon. Green and blue boxes represent symmetrical and asymmetrical exons, respectively. Yellow boxes represent exons of unknown class. (Modified from Graur and Li 2000.)

evolutionary process of creating new proteins and new functions. Evolutionarily successful exon shuffling events almost always involve exon-bordering domains (Nagy and Patthy 2011).

Phase limitations on exon shuffling

For an exon to be inserted, deleted, or duplicated without causing a frameshift in the reading frame, certain phase limitations of the exonic structure of the gene must be respected. In order to understand this phase limitation constraint, let us consider the different types of introns in terms of their possible positions relative to the coding regions. Introns residing between coding regions are classified into three types according to the way in which the coding region is interrupted. An intron is of **phase 0** if it lies between two codons, of **phase 1** if it lies between the first and second nucleotides of a codon, and of **phase 2** if it lies between the second and third nucleotides of a codon (**Figure 8.14**).

Exons are grouped into classes according to the phases of their flanking introns. For example, the middle exon in Figure 8.14b is flanked by a phase 0 intron at its 5′ end and by a phase 1 intron at its 3′ end; it is said to be a class 0-1 exon. An exon that is flanked by introns of the same phase at both ends is called a **symmetrical exon**; otherwise it is **asymmetrical**. For example, the middle exon in Figure 8.14a is symmetrical. Of the nine possible classes of exons, three are symmetrical (0-0, 1-1, and 2-2), and six are asymmetrical (0-1, 0-2, 1-0, 1-2, 2-0, and 2-1). The length of a symmetrical exon is always a multiple of three nucleotides.

Only symmetrical exons can be duplicated in tandem or deleted without affecting the reading frame (**Figure 8.15**). Duplication or deletion of asymmetrical exons would disrupt the reading frame downstream. Similarly, only symmetrical exons can be inserted into introns. For example, in **Figure 8.16a** the insertion of a 0-1 exon into a phase 0 intron causes a frameshift in all the subsequent exons. Insertion of a 0-2 asymmetrical exon into a phase 1 intron causes a

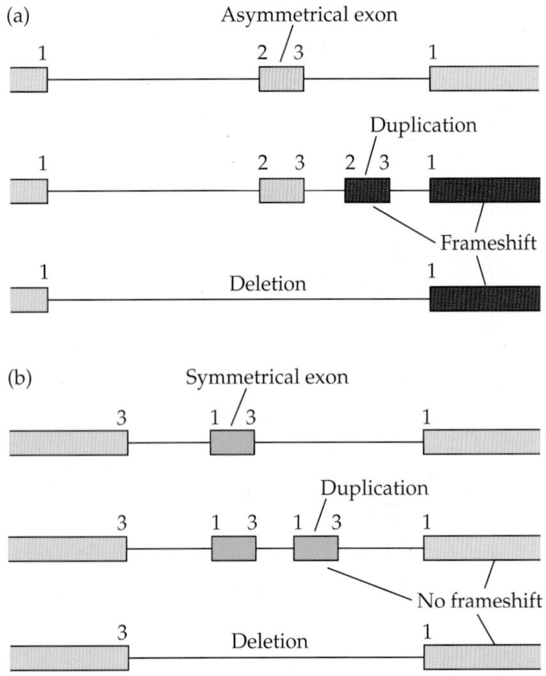

(a) Asymmetrical exon
1 2 / 3 1
Duplication
1 2 3 2 / 3 1
Frameshift
1 Deletion 1

(b) Symmetrical exon
3 1 / 3 1
Duplication
3 1 3 1 / 3 1
No frameshift
3 Deletion 1

Figure 8.15 Consequences of exon duplication or deletion. Numbers and colors are explained in Figure 8.14. (a) A gene containing a 1-2 asymmetrical exon. The duplication or deletion of this exon causes frameshifts (red) in the downstream exons. In addition, the duplication of the asymmetrical exon results in frameshifts in the second copy of the duplicated exon itself. (b) A gene containing a 0-0 symmetrical exon. Duplication or deletion of this exon causes no frameshifts in the downstream exons. (Modified from Graur and Li 2000.)

frameshift not only in all downstream exons but also in the inserted exon itself (**Figure 8.16b**).

Insertion of symmetrical exons is also restricted; a 0-0 exon can be inserted only into phase 0 introns, a 1-1 exon can be inserted only into phase 1 introns, and a 2-2 exon can be inserted only into phase 2 introns. For example, **Figure 8.16c** shows that the insertion of a 0-0 exon into an intron of phase 1 causes a frameshift in the inserted exon and at the 5′ end of the nearest downstream exon, while **Figure 8.16d** shows that the insertion of a 0-0 exon into a phase 0 intron causes no frameshift.

Not surprisingly, a plurality of exons coding for the modules of mosaic proteins are symmetrical (Patthy 1987; Kawashima et al. 2009; Nagy and Patthy 2011). For example, all the exons that encode the noncatalytic regions of the proteases taking part in the fibrinolytic, blood coagulation, and complement cascades are class 1-1 exons. The reason for this particular class choice in the evolution of these proteases is not clear. A plausible explanation is that we are dealing with a "frozen accident," in which an ancestral phase 1 intron could only accept class 1-1 exons. Consequently, all subsequent insertions and duplications in this region could only involve symmetrical 1-1 exons. Thus, an initial bias may have led to a predominance of phase 1 introns.

It has been suggested that the proliferation of 1-1 symmetrical exons is associated with the rise of multicellularity in animals (Kaessmann et al. 2002). In contrast, 0-0 symmetrical exons tend to be overrepresented among protein-coding genes found in both eukaryotes and prokaryotes, which is compatible with the suggestion of primordial domain shuffling in the progenote (the last common ancestor of all cellular life-forms).

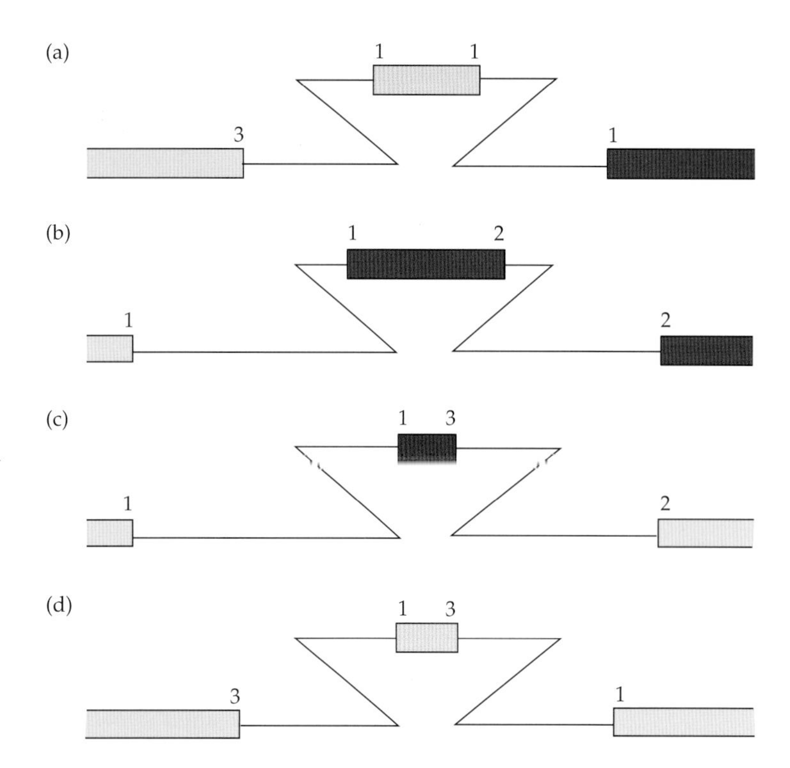

Figure 8.16 Consequences of exon insertion into introns. Numbers and colors are as explained in Figure 8.14, with frameshifted exons in red. (a) Insertion of a 0-1 asymmetrical exon into a phase 0 intron. (b) Insertion of a 0-2 asymmetrical exon into a phase 1 intron. (c) Insertion of a 0-0 symmetrical exon into a phase 1 intron. (d) Insertion of a 0-0 symmetrical exon into a phase 0 intron. The insertions in (a) and (b) cause frameshifts in the exons downstream of the insertion. The inserted exons in (b) and (c) are frameshifted. Only the insertion in (d) does not cause frameshifts in either the inserted exons or the downstream exons. (Modified from Graur and Li 2000.)

Since nonrandom intron phase usage is a necessary consequence of exon duplication or insertion, this property may be used as a diagnostic feature of gene assembly through exon shuffling. For example, the genes coding for type III collagen, β-casein, and the precursor of growth hormone have predominantly class 0-0 exons, consistent with the suggestion that these proteins have evolved by exon shuffling. On the other hand, phosphoglycerate kinase, glyceraldehyde 3-phosphate dehydrogenase, and triosephosphate isomerase genes contain a mixture of intron types, and consequently exon shuffling might not have played an important role in the formation of these genes.

Prevalence of domain shuffling and the evolutionary mobility of protein domains

Many examples of domain shuffling, exon shuffling, mosaic proteins, and frequently shuffled domains have been discovered. Protein domains are thought of as the common currency of protein structure and function (Buljan and Bateman 2009). In particular, domain-shuffling events are thought to have contributed to the evolution of many vertebrate-specific characteristics, such as cartilage and the inner ear (Kawashima et al. 2009). As seen in **Table 8.2**, different taxa exhibit differences in the propensity to form multidomain proteins. The proportion of multidomain proteins is highest in metazoans (39%) and lowest in archaebacteria (23%). Furthermore, the multidomain proteins of metazoans tend to be longer than those in archaebacteria,

Table 8.2

Mean numbers of domains and multidomain proteins in some major taxonomic groups

Taxon	Proteins	Domains	Multidomain proteins	Mobile domains
Bacteria	273,859	4,079	73,076 (27%)	1,974 (48%)
Archaea	23,728	1,725	5,529 (23%)	776 (45%)
"Protozoa"	16,756	1,967	5,298 (32%)	932 (47%)
Plants	57,620	2,562	20,359 (35%)	1,305 (51%)
Fungi	20,371	2,249	6,434 (32%)	1,102 (49%)
Metazoa	129,881	3,272	51,085 (39%)	1,748 (53%)

Source: Modified from Tordai et al. (2005).

such that, for instance, multidomain proteins with more than five domains are five times more frequent in metazoans than in archaebacteria, and proteins with more than ten domains are nine times more frequent (Tordai et al. 2005).

Smaller domains are more likely to take part in the making of multidomain proteins, while larger domains are found mostly in single-domain proteins. It is noteworthy in this respect that the most versatile mobile domains, for example, the epidermal growth factor–like domain (EGF-like domain), Sushi, and kringles, are less than 100 amino acid residues in length. A possible explanation for this phenomenon is that smaller, compact domains are more likely to satisfy the folding-autonomy criterion that is crucial for their structural integrity in multidomain proteins (Tordai et al. 2005).

It has long been known that the propensity of individual domains to form multidomain architectures differs widely among domains. Some domains are never observed in multidomain proteins, and these are referred to as **static domains**; others are extremely widely used and are referred to as **mobile** (or **promiscuous**) **domains**. Quantifying mobility, however, requires a precise definition of the term. In principle, "mobility" may refer to (1) the number of proteins in which a domain is present, (2) the number of copies of the domain in the sample or the entire proteome, (3) the number of other domains with which the domain co-occurs, or (4) the number of multidomain architectures (linear sequences of domains) in which the domain is found. In **Figure 8.17**, we show an example with which each of these measures of domain mobility can be calculated.

(a)

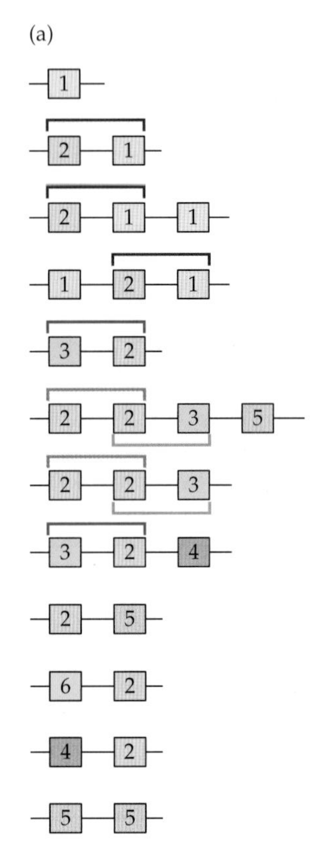

(b)

Domain type	Number of proteins in which the domain is present	Number of domain copies	Number of co-occurring domains	Number of two-domain local architectures in which the domain occurs
1	4	6	1	3
2	10	12	5	9
3	4	4	3	3
4	2	2	2	2
5	3	4	2	3
6	1	1	1	1

Figure 8.17 Measures of domain mobility in a sample of proteins. (a) A sample consisting of 12 proteins (one protein per line) with a total of 29 domains (boxes) and six different domain types (differently colored and numbered 1–6). There is a single one-domain protein, six two-domain proteins, four three-domain proteins, and one four-domain protein. Ten of the proteins contain more than one domain type. Of the two that contain a single domain type (the first and last in this list), one is a one-domain protein, the other a two-domain protein. Of the six different domains, one (6) is static, i.e., found in a single protein. The most promiscuous domain type is 2. "Local architecture" is defined as the order of two domain types (square brackets). With 6 domain types, 36 two-domain architectures are possible; only 12 of these are found in our sample (1-1, 1-2, 2-1, 2-2, 2-3, 2-4, 2-5, 3-2, 3-5, 4-2, 5-5, and 6-2). Four two-domain architectures (2-1, 3-2, 2-2, and 2-3) appear in more than one protein. In two proteins, one domain type takes part in two overlapping domain architectures (brackets both above and below). (b) The table compiles the measures of domain mobility for the six domain types.

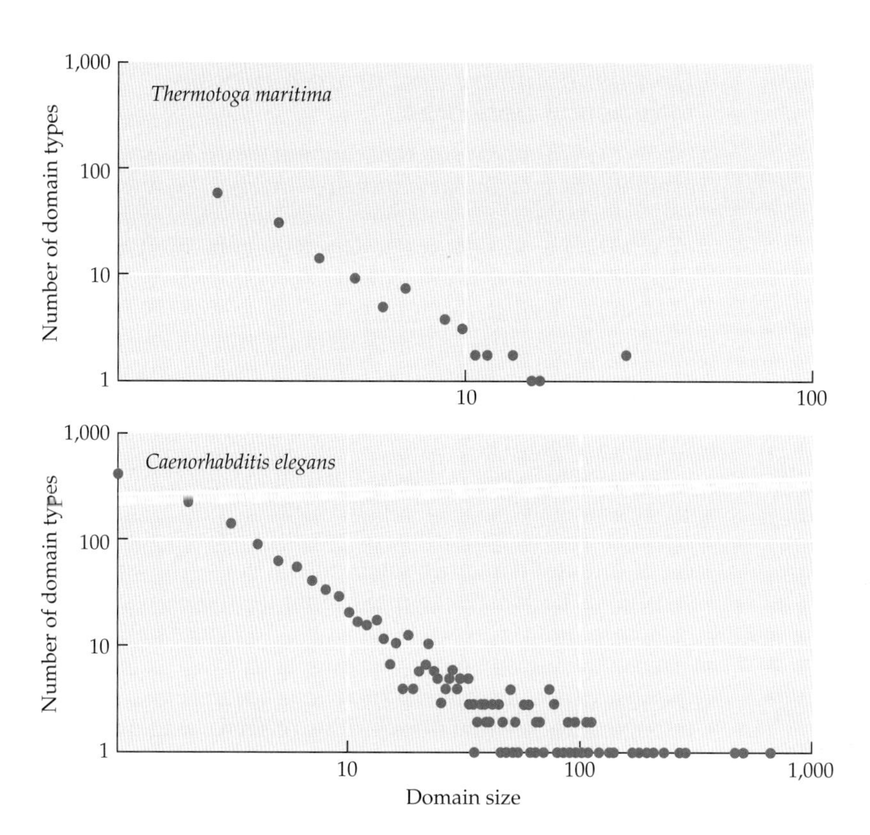

Figure 8.18 Size distribution of protein domain types in hyperthermophilic bacterium *Thermotoga maritima* and the nematode *Caenorhabditis elegans*. Note the log-log plot. (Modified from Koonin et al. 2002.)

The vast majority of domains in all organisms so far examined occur in a relatively small number of local architectures, whereas a small minority of domains serve as versatile building blocks of multidomain proteins, appearing hundreds of times in the proteome. A large proportion of mobile domains are involved in protein-protein interactions. This, together with the fact that they show evidence of being affected by strong purifying selection, implies that these domains were able to become promiscuous in the first place because they had a potential to be useful in various contexts. In **Figure 8.18** we present the frequency distribution of domains by domain type in a prokaryote and a eukaryote. In animals, some of the top-ranking mobile domains are the protein kinase domain, the epidermal growth factor domain, the ankyrin repeat, the immunoglobulin and the immunoglobulin I-set domains, the fibronectin types 1 and 3 domains, and the pleckstrin homology domain (Tordai et al. 2005). Of course, the exact relative rank of each domain type depends on the particular measure of mobility.

Sometimes, two- or three-domain combinations tend to recur in the same orientation in different protein contexts (**Figure 8.19**). These units are termed **supra-domains**. Vogel et al. (2004) identified close to 1,500 two- and three-domain combinations that were significantly over-represented relative to individual domains. Over one-third of all multidomain proteins in both eukaryotes and prokaryotes contain supra-domains.

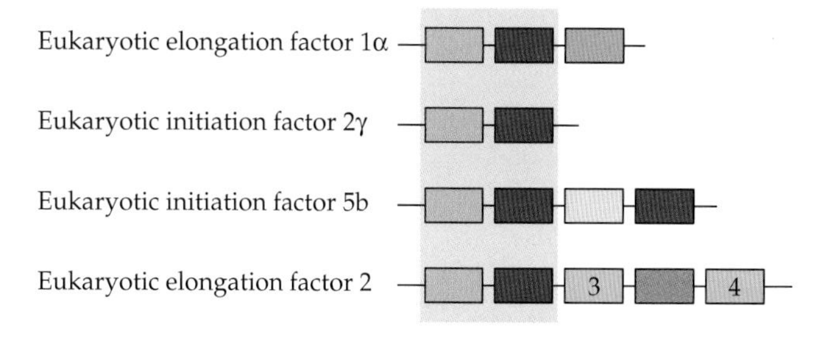

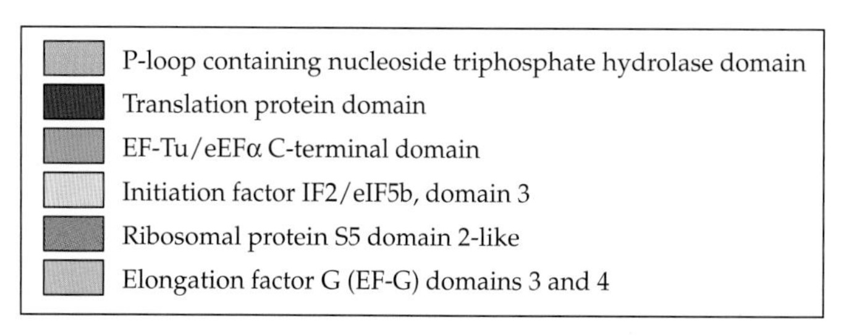

Figure 8.19 An example of a recurring supra-domain (gray shading). The P-loop containing the nucleoside triphosphate hydrolase domain and the translation protein domain occur as one unit in at least 35 different domain architectures, of which four are shown. Each of these four domain architectures is found in one or more proteins, of which only one is listed. (Modified from Vogel et al. 2004.)

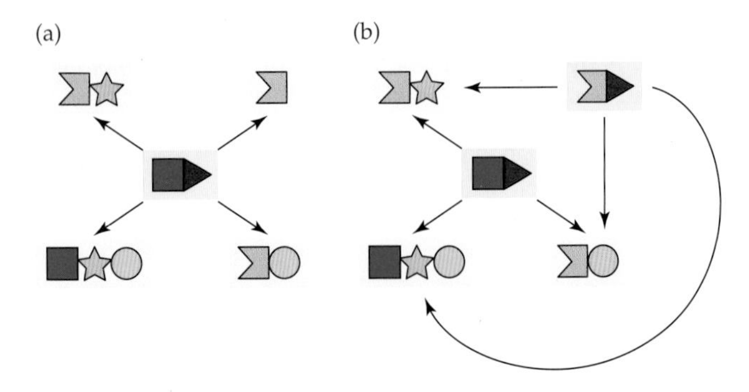

Figure 8.20 Effects of domain shuffling on network structure. (a) A network consisting of five proteins (boxes) made out of five modules (geometrical shapes). Four proteins are multidomain proteins, of which three contain two domains each, and one contains three domains. One domain (the red triangle) interacts with all four other domains (arrows) but does not interact with proteins that contain the same domain. This domain is solely responsible for all protein-protein interactions in the network. The network consists of five nodes and four edges. The network has a single hub, i.e., a node with more than one interaction. (b) Domain shuffling resulted in the addition of one domain (the red triangle) to a protein. This event dramatically changed the topology of the network, generating an additional hub, reducing the number of interactions for each of the two hubs from four to three, and increasing the total number of interactions in the network to six, thereby creating a more interconnected network.

Domain shuffling and protein-protein interaction networks

Gilbert (1978) proposed that domain shuffling speeds up evolution by creating new proteins. Indeed, the notion that new proteins (including chimeric ones) are solely responsible for phenotypic novelty was prevalent until the late 1980s. Although this notion is still valid for specific cases, many new phenotypes have been found to emerge through changes in patterns of gene expression, as well as through restructuring of protein-protein interaction networks. Changes in protein-protein interaction networks can occur through modifications to the topology, the creation of new hubs (nodes with a high degree of interactivity), and increases in the interconnectivity among the components of the network.

A significant proportion of protein-protein interactions occur through domain-domain interactions. Thus, domain shuffling should have an important role in the evolution of protein-protein interaction networks. In **Figure 8.20**, we illustrate the effects of the shuffling of a protein domain on the structure of a protein-protein interaction network. From the figure, we see that the shuffling of even a single domain can dramatically alter the structure of a network.

In the last two decades, we have witnessed an increasing interest in the role of domain shuffling in the origin and evolution of biological networks. For example, by mapping domain-domain interactions onto protein-protein interaction networks, Itzhaki et al. (2006) showed that interacting domains are repeatedly used throughout evolution to mediate protein-protein interactions. The relationship between domain shuffling and protein-protein interaction networks was also made clear by the finding of Cancherini et al. (2010) that the average number of interactions was significantly higher in proteins with evidence of at least one domain shuffling event when compared with proteins that have experienced no domain shuffling in their evolutionary history. The set of proteins with at least one shuffled domain that are self-interacting was found to be much greater than expected by chance. This reinforces the notion that self-interacting domains are important for the evolution of protein-protein interaction networks. Moreover, there is evidence that multi-interacting domains are shuffled more frequently than other domains. Thus, domain shuffling seems to be a major mechanism in the evolution of hubs. Indeed, domain-shuffling events are associated with the emergence of biological novelty, such as multicellularity. Most of the proteins that compose the extracellular matrix of multicellular organisms were built by domain shuffling. The same is true for cell surface receptors, whose expansion in multicellular animals is clearly associated with domain shuffling. In vertebrates, Kawashima et al. (2009) found several vertebrate-specific transcription factors to be the products of domain shuffling.

Gene Fusion and Fission

Gene fusion refers to the creation of a new gene by the joining of the 3′ region of an upstream gene to the 5′ region of a downstream gene. Fused protein-coding genes encode mosaic proteins consisting of two or more domains. **Gene fission** (or **gene fragmentation**) refers to the converse process, whereby a multidomain gene is split into two or more separate transcription units. Because prokaryotic genes are devoid of introns, the modular assembly of domains into multidomain proteins in these taxa is accomplished in the majority of cases by gene fusion and fission (Pasek et al. 2006). Numerous cases of fusion and fission have been described in prokaryotes (**Table 8.3**).

Table 8.3

Number of inferred fissions and fusions in the genomes of eubacteria and archaebacteria

Species	Number of genes in genome	Number of inferred fusions	Number of inferred fissions
Eubacteria			
Mycoplasma genitalium	468	2	1
Mycoplasma pneumoniae	677	2	0
Rickettsia prowazekii	834	6	0
Borrelia burgdorferi	850	3	1
Chlamydia trachomatis	876	8	0
Treponema pallidum	1,031	6	0
Aquifex aeolicus	1,522	12	8
Helicobacter pylori	1,590	9	0
Haemophilus influenzae	1,717	18	3
Synechocystis PCC6803	3,168	24	4
Mycobacterium tuberculosis	3,924	36	1
Bacillus subtilis	4,100	19	1
Escherichia coli	4,290	33	2
Archaebacteria			
Methanococcus jannaschii	1,735	12	5
Methanobacterium thermoautotrophicum	1,871	16	5
Pyrococcus horikoshii	2,061	4	3
Archeoglobus fulgidus	2,407	19	8

Source: Modified from Snel et al. (2000).

Gene fusion and fission, however, are not restricted to prokaryotes; many examples have also been documented in unicellular and multicellular eukaryotes (e.g., Nakamura et al. 2007; Salim et al. 2011).

Successful fusion requires that both halves of the new gene function properly despite the loss of the controlling elements from the downstream gene which, following fusion, becomes regulated by the promoter of the upstream constituent gene. Therefore, only fusions in which the two linked genes can function in the same compartment of the cell, or at the same developmental stage, or in response to the same stimuli will be tolerated. While it has been hypothesized that two genes with unrelated functions may merge and be retained in the genome (Miozzari and Yanofsky 1979; Galperin and Koonin 2000), almost all bifunctional fusion genes discovered to date show a functional relationship between the proteins that comprise the fusion.

The number of genes resulting from fusion was found to increase with genome size, which is to be expected if fusion occurs randomly, such that a larger number of genes by chance will contain a larger number of fused genes. Gene fusion occurs four to five times more frequently than fission (Snell et al. 2000; Kummerfeld and Teichmann 2005). The prevalence of fusion may be due to the fact that fusion allows for the potentially advantageous physical coupling of functions that are biologically coupled (e.g., Marcotte et al. 1999), whereas fission involves a separation between genes whose structures and functions may have coevolved (and possibly coadapted) over long periods of evolutionary time. Additionally, gene fusion only involves the loss of the termini of the genes being fused, whereas gene fission requires that the resulting downstream gene somehow obtain promoter and terminator sequences, a start codon, a stop codon, and possibly other regulatory elements. These requirements add up to fission being an extremely remote long shot.

It has been suggested that gene fission occurs more frequently in thermophiles (Snel et al. 2000), possibly because larger thermal fluctuations can lead to increased error rates in replication. According to this view, if we assume that the number of errors that occur in the process of creating a functional protein from DNA (e.g., errors in transcription, translation, or folding) is proportional to sequence length, then with an error rate of 10% per 100 codons, 81% of proteins with a length of 200 amino acids will be functional (90% × 90%) and 19% will be defective. Under the same conditions, the error rate for two separate proteins with 100 amino acids each will be only 10%. Since at higher temperatures the error rate increases, the above difference may become more pronounced in conferring an adaptive advantage on fissioned genes over fused genes in thermophiles and hyperthermophiles (organisms growing at temperatures above 45°C and 80°C, respectively). This finding has not, however, been independently replicated (Kummerfeld and Teichmann 2005; Sung 2011).

In a study of a family of Gram-positive, heterotrophic, endospore-producing bacteria called Bacillaceae, Sung (2011) found an accelerated rate of gene fragmentation in the lineage leading to the etiological agent of anthrax, *Bacillus anthracis*, which is a common evolutionary trend found in lineages that have recently become host-restricted. In this bacillus, gene fission is not associated with gene duplication, and fissioned genes seem to evolve under relaxed selective constraints. This raises the possibility that the products of gene fission are nonfunctional (i.e., pseudogenes).

So far, a plausible mechanism for functional gene fission suggested in the literature involves two steps. First, a multidomain gene is duplicated; then, in each of the duplicates a complementary degenerative process occurs. This chain of events resembles the previously discussed process of subfunctionalization and specialization (Chapter 7). One such example of duplication and degeneration involves a family of genes in *Drosophila melanogaster* called *monkey king*. This gene family originated 1–2 million years ago as retrotransposed duplicates of a multidomain ancestor, and fission was achieved through gene duplication and subsequent partial degeneration. All other proposed fission mechanisms in the literature have been found wanting, and none could yield active, positively selected genes.

Here, we present two examples of gene fusion, each illustrating a different facet of this process.

THE EVOLUTION OF A SHORTCUT IN THE METHIONINE SALVAGE PATHWAY: WHEN 1 + 1 = 3 The methionine salvage pathway is universally used to regenerate L-methionine from methylthioadenosine, a by-product of certain reactions involving S-adenosylmethionine. In prokaryotes, the enzymes that convert methylthioadenosine to methionine are: mtnA (S-methyl-5-thioribose-1-phosphate isomerase), mtnB (methylthioribulose-1-phosphate dehydratase), mtnC (2,3-diketo-5-methylthio-1-phosphopentane phosphatase), mtnD (1,2-dihydroxy-3-keto-5-methylthiopentene dioxygenase), mtnE (methionine aminotransferase), mtnK (5-methylthioribose kinase), mtnN (5′-methylthioadenosine/S-adenosylhomocysteine nucleosidase), mtnP (methylthioadenosine phosphorylase), mtnW (2,3-diketo-5-methylthiopentyl-1-phosphate enolase), and mtnX (2-hydroxy-3-keto-5-methylthiopentenyl-1-phosphate phosphatase).

Part of the methionine salvage pathway is shown in **Figure 8.21**. In the ciliate *Tetrahymena*, a mosaic gene, *mtnBD*, was created from a fusion of *mtnB* and *mtnD*. The N-terminus of *mtnBD* shares a strong sequence similarity with *mtnB* from other species, while the C-terminus shares a strong similarity with *mtnD*. This fusion seems to be unique to *Tetrahymena* and its closest relatives, as it is not present in the genome of any other fully sequenced ciliate. Surprisingly, the *Tetrahymena* genome

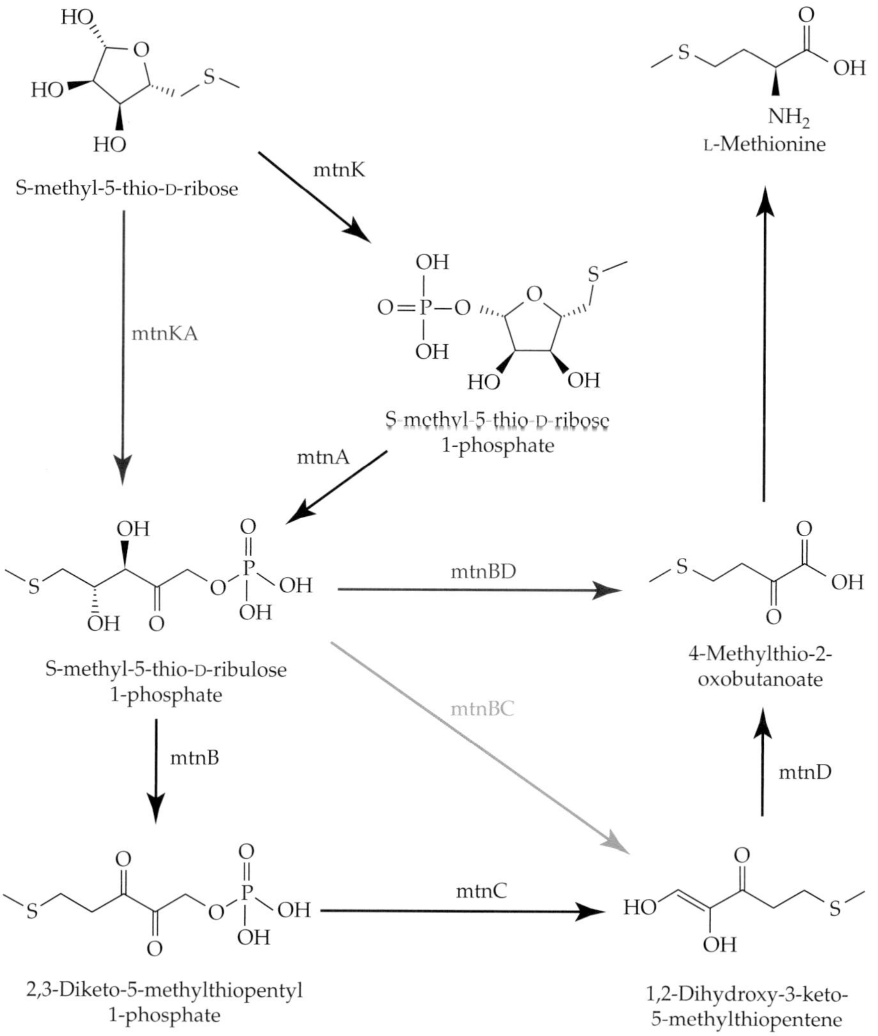

Figure 8.21 The methionine salvage pathway. Enzyme names (see text) are shown beside the arrows. The genes coding for mtnB and mtnC are fused in *Arabidopsis thaliana*, resulting in a gene called *mtnBC*. In *Tetrahymena thermophila*, the genes for mtnB and mtnD are fused into *mtnBD*, and the genes for mtnK and mtnA are fused into *mtnKA*. Reactions encoded by fused proteins are shown in green and red for *A. thaliana* and *T. thermophila*, respectively. (Data from Salim et al. 2009.)

lacks *mtnC*, whose product catalyzes the conversion of 2,3-diketo-5-methylthiopentyl-1-phosphate to 1,2-dihydroxy-3-keto-5-methylthiopentene. It has been shown experimentally that the product of the newly fused gene can catalyze this reaction too, in addition to the two expected reactions. This indicates a gain of function as a result of the fusion.

THE *PIPSL* GENE IN APES: EMERGENCE BY GENE FUSION AND REVERSE TRANSCRIPTION In the human genome, the 15-exon *PIP5K1A* gene is located on chromosome 1. It encodes an enzyme called phosphatidylinositol-4-phosphate 5-kinase type 1α. The enzyme catalyzes the phosphorylation of phosphatidylinositol-4-phosphate to form phosphatidylinositol-4,5-biphosphate. This reaction is involved in a variety of cellular processes, such as actin cytoskeleton organization, cell adhesion, migration, and phagocytosis. The gene *PSAMD4* (or *S5a*) is located 5,167 nucleotides downstream of the *PIP5K1A* gene (**Figure 8.22**). It is a 10-exon gene that encodes the 26S proteasome non-ATPase regulatory subunit 4, which binds and presumably selects ubiquitin conjugates for destruction.

The two neighboring genes produce two separate mRNAs. Rarely, however, the two genes may be cotranscribed, and a fused transcript called *PIP5K1A-PSAMD4-TIC* may be produced. This mRNA contains the first 13 exons from *PIP5K1A* and the last 9 exons from *PSAMD4*.

By a rare coincidence, the splicing between *PIP5K1A* and *PSAMD4* occurs between same-phase introns, thus allowing translation to proceed without frameshifts

(a) Chromosome 1

PIP5K1A *PSAMD4*

1 2 3 4 5 6 7 8 9 10 11 12 13 14 15 1 2 3 4 5 6 7 8 9 10

PIP5K1A-PSAMD4-TIC

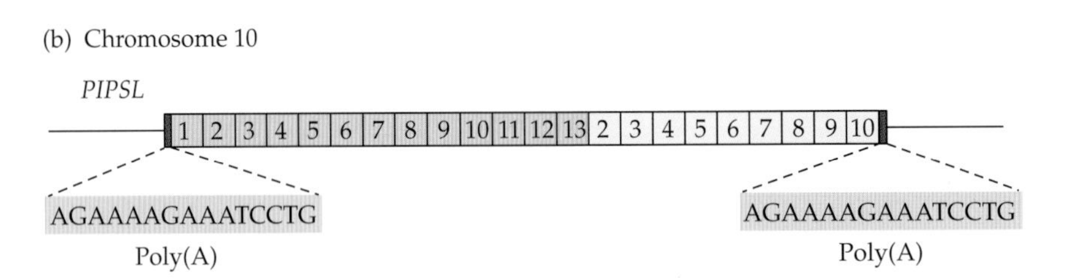

Figure 8.22 Gene fusion through reverse transcription. (a) Neighboring 15-exon *PIP5K1A* and 10-exon *PSAMD4* genes on human chromosome 1 are spliced to form two mRNAs (green pattern, above the gene). Rarely, a fused transcript called *PIP5K1A-PSAMD4-TIC* is produced. This mRNA contains the first 13 exons from *PIP5K1A* and the last 9 exons from *PSAMD4* (black pattern, below the gene). (b) The *PIP5K1A-PSAMD4-TIC* mRNA has been retrotransposed (reverse transcribed and integrated into a different location in the genome) by a common human retrotransposable element called *L1*. The resulting intronless gene on chromosome 10 is called *PIPSL*. The gold-boxed sequences denote duplications due to the insertion of the cDNA into the genome (target site duplication). The poly(A)-rich region is shown. (Data from Babushok et al. 2007.)

or premature stop codons. In the ancestor of the great apes (humans, chimpanzees, gorillas, and orangutans), a *PIP5K1A-PSAMD4-TIC* mRNA molecule was retrotransposed (reverse transcribed and integrated into a different location in the genome) by the ubiquitous human retrotransposable element *L1*. The resulting intronless gene on chromosome 10 is called *PIPSL*. By another "stroke of luck" the entire *PIP5K1A-PSAMD4-TIC* was retrotransposed without being truncated or rearranged, as happens to the majority of *L1* integrants. The fact that the reverse-transcribed *PIPSL* is transcribed is another extraordinarily rare occurrence, for it requires the retrotransposed gene to land downstream of a promoter.

Interestingly, the fused intronless gene has accumulated more nonsynonymous substitutions than synonymous ones, a phenomenon indicative of positive Darwinian selection. In particular, the ratio of nonsynonymous to synonymous substitutions in the *PIP5K1A*-derived part of the *PIPSL* gene in the chimpanzee lineage is 4:1. The *PIPSL* gene is mainly transcribed in the testes, although this tissue specificity may be incidental rather than functional because of the transcriptionally permissive environment in the testis, where RNA polymerase II is known to increase up to 1,000 times, allowing for high transcription levels from nonpromoter and weak promoter sequences (Babushok et al. 2011).

Domain Accretion

Domain accretion is the addition of one or more domains to the existing domain architecture of a protein (Koonin et al. 2000; **Figure 8.23**). Domain accretion increases the number of domains and alters the functionality of the protein. One way the presence of additional domains may affect function is by allowing the protein to interact with more proteins, nucleic acids, and cofactors. Thus, domain accretion is frequently detected in regulatory proteins or proteins involved in complex interactions, such as signal transduction. By its involvement in cell-to-cell signaling, domain accretion is found, as expected, to be more widespread in eukaryotes than in prokaryotes. In some lineages, protein complexity appears to be gained by gene duplication accompanied by domain accretion, rather than by gene duplication and domain accretion independently (Cohen-Gihon et al. 2011).

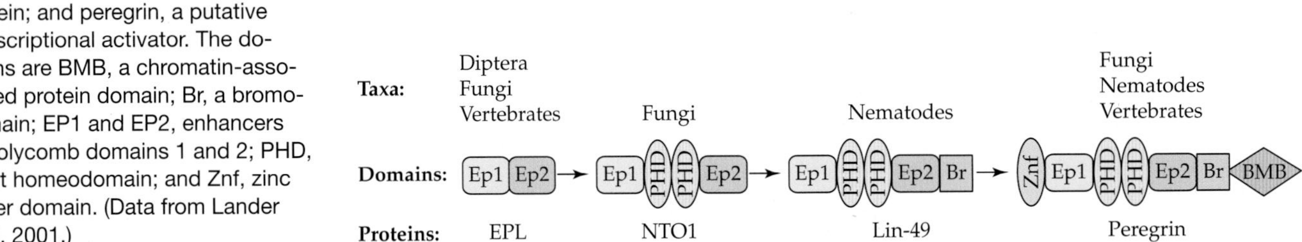

Figure 8.23 An example of domain accretion. The proteins are EPL, enhancer of polycomb-like protein; NTO1, a subunit of the NuA3 histone acetyltransferase complex; Lin-49, a bromodomain protein; and peregrin, a putative transcriptional activator. The domains are BMB, a chromatin-associated protein domain; Br, a bromodomain; EP1 and EP2, enhancers of polycomb domains 1 and 2; PHD, plant homeodomain; and Znf, zinc finger domain. (Data from Lander et al. 2001.)

Domain gain and domain loss occur preferentially at the termini of the proteins rather than in the middles of proteins (Buljan and Bateman 2009). The reason for this preferential localization may be the fact that protein termini are normally charged, flexible, and found at the surface of proteins, so it is easy to imagine that additions (or deletions) of domains there are less likely to disrupt the rest of the structure, especially if the concerned domains are independent structural units. On the other hand, addition of domains to connector regions between domains, even if these regions are unstructured and do not have functional roles, is more likely to disrupt the rest of the structure. Therefore, changes at the termini may be more tolerable than those that occur between domains.

Strategies of Multidomain Gene Assembly

A corollary of domain shuffling, as well as of fusion, fission, and domain accretion, is that some complex biological functions that require several enzymes may be specified by genes encoding different combinations of protein modules. We may find single-module proteins in some species, while in others we may find different combinations of multimodular proteins. One such instance is seen in the genes involved in the synthesis of fatty acids from acetyl CoA. This multistep process requires seven enzymatic activities and an acyl-carrier protein (**Table 8.4**). In most bacteria, these functions are carried on by discrete monofunctional proteins. However, in fungi, these activities are distributed between two nonidentical polypeptides encoded by two unlinked intronless genes, *FAS1* and *FAS2*. *FAS1* encodes two of the seven enzymatic activities (β-ketoacyl synthase and β-ketoacyl reductase) as well as the acyl-carrier protein. *FAS2* encodes the rest of the five enzymatic activities (Chirala et al. 1987; Mohamed et al. 1988).

In animals, all the functions are integrated into a single polypeptide chain called fatty acid synthase (or megasynthase). Characterization of the fatty acid synthase gene in the rat (Amy et al. 1992) revealed that the gene product contains eight modules, including one module that performs a dual function (as acetyl transferase and malonyl transferase), and another whose function is most probably unrelated to fatty acid synthesis but may have a role in determining the tertiary structure of this multimodular protein (**Figure 8.24**). The gene is composed of 43 exons separated by 42 introns. With only one exception, boundaries between adjacent modules coincide with the locations of introns. Thus, the fatty acid synthase genes in fungi and mammals are most probably mosaic proteins that have assembled from single-domain proteins like the ones found in bacteria, most probably by exon shuffling. The fact that the arrangement of domains is different in fungi from that in mammals indicates not only that

Table 8.4

Principal biochemical reactions in the synthesis of fatty acids from acetyl CoA in eukaryotes and eubacteria

Reaction	Enzyme
1. Acetyl CoA + condensing enzyme domain ↔ acetyl-condensing enzyme	Acetyl transferase
2. Malonyl CoA + acyl-carrier peptide ↔ malonylacyl-carrier peptide	Malonyl transferase
3. Acetyl-condensing enzyme + malonylacyl-carrier peptide ↔ β-ketoacyl-carrier peptide	β-Ketoacyl synthase
4. β-Ketoacyl-carrier peptide + NADPH + H^+ ↔ β-hydroxyacyl-carrier peptide + $NADP^+$	β-Ketoacyl reductase
5. β-Hydroxyacyl-carrier peptide ↔ 2-butenoylacyl-carrier peptide + H_2O	β-Hydroxyacyl dehydratase
6. 2-Butenoylacyl-carrier peptide + NADPH + H^+ ↔ butyrylacyl-carrier peptide + $NADP^+$	Enoyl reductase
7. Butyrylacyl-carrier peptide + condensing enzyme domain ↔ butyryl-condensing enzyme + acyl-carrier peptide	Thioesterase

Source: From Graur and Li (2000).

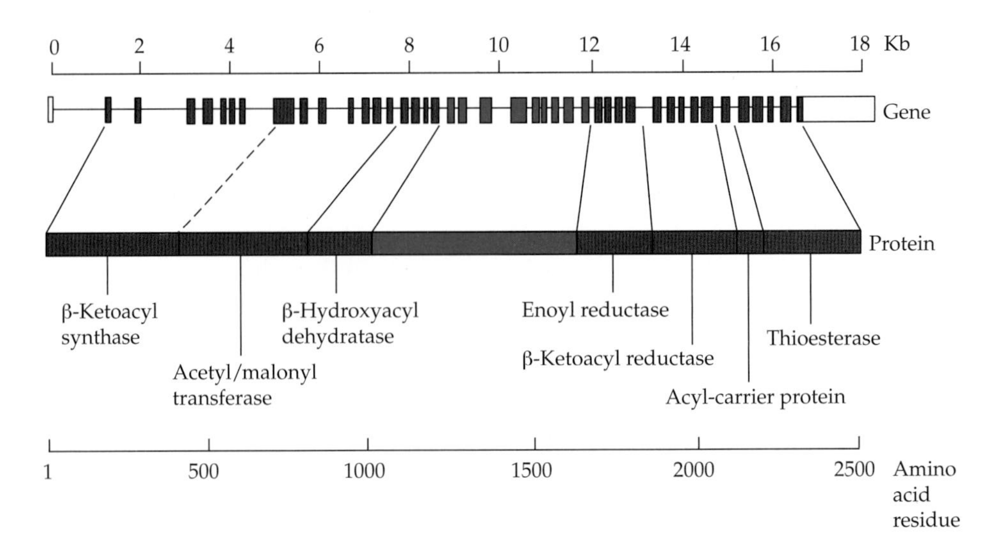

Figure 8.24 The fatty acid synthase gene in rat encodes a multifunctional protein. Exons are shown as boxes separated by introns. Colored boxes indicate protein-coding regions. Seven domains are marked in blue on the amino acid sequence. The function of this interdomain region (blue box) is unknown at this time, but its purpose may be to hold two fatty acid synthase subunits together and generate a dimer with two centers of fatty acid synthesis. With the exception of the boundary between the β-ketoacyl synthase and the acetyl/malonyl transferase domains (dashed line), all boundaries between adjacent modules coincide with the locations of introns. (Modified from Graur and Li 2000.)

the two lineages evolved multimodularity independently, but also that different strategies may be employed in the assembly of genes encoding multimodular proteins.

Evolution by Exonization and Pseudoexonization

Because donor and acceptor splicing sites can be obliterated or created de novo quite easily by mutation, exons may sometimes appear or disappear by processes other than exon shuffling. **Exonization** is the process through which an intronic sequence becomes an exon. Most exonized intronic sequences are deleterious, and reports of exonization events resulting in serious clinical manifestations abound in the medical literature (e.g., Raponi et al. 2006).

For an exonization event to survive during evolution, first it must abide by the same rules as exon insertion, and second, the added polypeptide should not interfere with function of the protein to which it was added. Not surprisingly, there are only very few cases in which one can state with any degree of confidence that a certain amino acid sequence was derived through the exonization of an intronic sequence An interesting example of exonization involves the evolution of sugar receptors in two moth species, the tobacco budworm, *Heliothis virescens*, and the domestic silkworm, *Bombyx mori* (Kent and Robertson 2009). In the ancestor of four paralogous sugar-receptor genes in *B. mori*, a novel symmetrical exon (class 2-2) emerged within a phase 2 intron. This novel exon must have been quite short, as its extant descendants encode 15- to 20-amino-acid-long polypeptides. Interestingly, the rates of evolution of the genes that contain the exonized sequence seem to be similar to the rates of evolution of the paralogs that lack the novel exon, indicating that the selective constraints have neither lessened nor increased after the addition of the exon.

Pseudoexonization refers to the nonfunctionalization of an exon. For pseudoexonization to occur without disruption of the reading frame, the rules pertaining to exon deletion must be respected. The evolution of the enamelin gene in amniotes will serve to illustrate the process of pseudoexonization. Enamelin is a crucial protein for

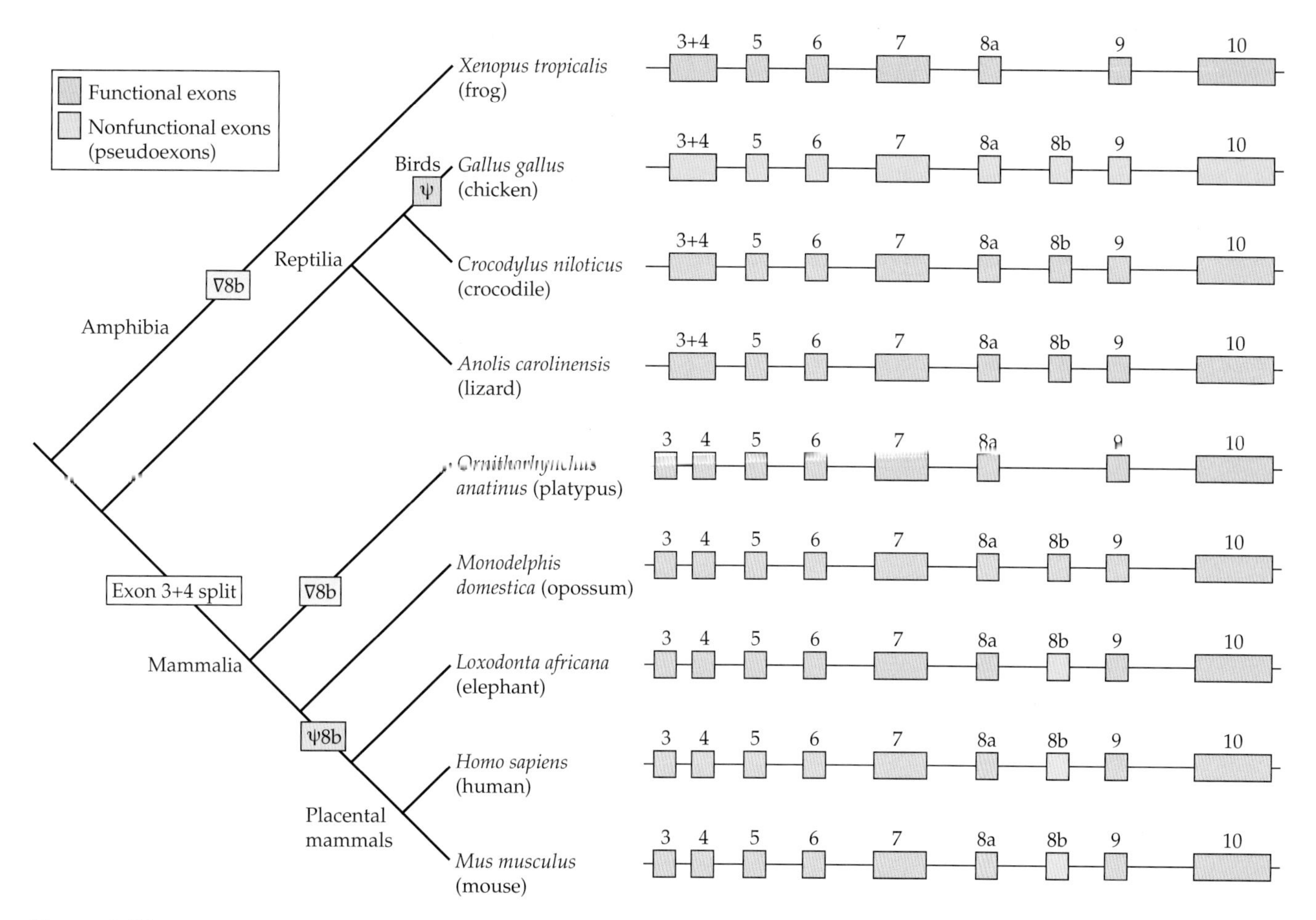

Figure 8.25 The independent loss and pseudoexonization of exon 8b of the enamelin-coding genes in tetrapod lineages. A schematic phylogenetic tree for nine tetrapods is shown on the left. Exons and pseudoexons are denoted by blue and yellow rectangles, respectively, and are numbered. In the lineage leading to birds, the enamelin-coding gene underwent total nonfunctionalization (Ψ) and became a unitary pseudogene. We note, however, that the vast majority of its pseudoexons, including 8b, are still identifiable. An independent pseudoexonization event affecting exon 8b occurred in placental mammals (ψ8b). Pseudoexon 8b is still detectable in many, but not all, placental taxa. The designation ∇8b marks independent losses of exon 8b in the platypus and frog lineages; it is not known whether these deletions took place before or after pseudoexonization. A split of exon 3+4 into two exons (3 and 4), possibly due to an intron insertion, is also shown. (Modified from Al-Hashimi et al. 2010.)

enamel formation and mineralization. Homozygous mice lacking both copies of the enamelin gene fail to produce true enamel on their dentin, and their teeth crumble easily. As seen in **Figure 8.25**, the small ancestral exon 8b evolved into a pseudoexon at least three times independently. In some cases, for example, frog and platypus, the pseudoexon is no longer recognizable; in other cases, for example, African elephant, human, and mouse, the pseudoexon retained sufficient similarity to its functional orthologs to be detectable.

As with nucleotide substitutions, deletions and insertions, and fusion and fission events, it is not possible to infer the polarity of exonization and pseudoexonization events in pairwise comparisons (**Figure 8.26**). We need rooted phylogenetic trees to make such decisions. Figure 8.26 also illustrates the possibility that exonization events may sometimes affect entire introns, thereby causing exon fusion, as well as the possibility that pseudoexonization may be partial, i.e., affect only part of an exon, thereby shortening it.

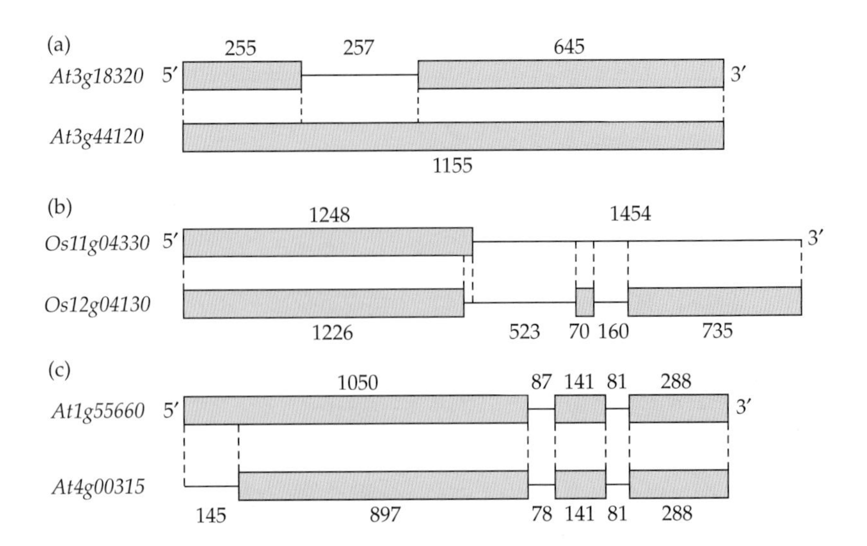

Figure 8.26 The exon-intron structures of three pairs of plant paralogs show telltale signs of exonization or pseudoexonization. Exons and introns are denoted by blue rectangles and horizontal lines, respectively. The length of each region is given in base pairs. Dashed vertical lines indicate sequence homology between exonic and intronic regions. Gene names follow a format of chromosome-based nomenclature. The first two letters denote species (*At*, *Arabidopsis thaliana*; *Os*, *Oryza sativa*); these are followed by the chromosome number, the functional designation of the sequence (*g*, for protein-coding gene), and a five-digit code for the location of the sequence, counted from the tip of the short arm of the chromosome toward the tip of the long arm. In (a), either two exons were joined into one by the exonization of an intronic sequence or an exon was split into two by the pseudoexonization of its middle part. In (b), two straightforward exonization/pseudoexonization events affected the second and third exons. A change in sequence at the 3′ end of the first exon either elongated the exon or shortened it. In (c), either a sequence at the 5′ end of the exon became a pseudoexon or a sequence at the 3′ end of the intron became exonized. (Modified from Xu et al. 2012.)

Evolution of Overlapping Genes

By its very nature, a double-stranded DNA sequence has six possible reading frames (**Figure 8.27**). If a DNA segment codes for a protein, then in the vast majority of cases a single reading frame is used. In a few cases, however, more than one frame is used, resulting in the production of two or more proteins from the same DNA segment. Two or more genes wholly or partially encoded by the same sequence are referred to as **overlapping genes**. Overlapping genes make use of different reading frames and different initiation and termination codons. In this section we only discuss protein-coding genes, while noting that gene overlaps are also known for RNA-specifying genes. [For example, the genes specifying tRNA^{Ile} and tRNA^{Gln} in the human mitochondrial genome are located on different strands and there is a three-nucleotide overlap between them that reads 5′—CTA—3′ in the former and 5′—TAG—3′ in the latter (Anderson et al. 1981). We will not discuss here RNA-specifying genes overlapping protein-coding genes (e.g., Nagano and Fraser 2011). We also restrict our discussion to protein-coding regions overlapping other protein-coding regions. Cases such as that of an exon from one gene overlapping an intron from another gene will be discussed in the section "Nested and Interleaved Genes" (p. 375).]

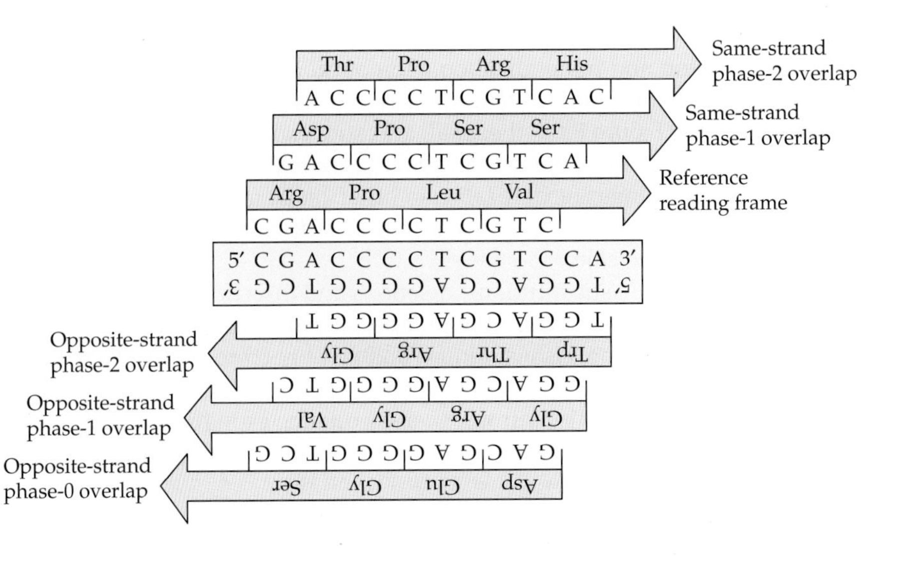

Figure 8.27 Each double-stranded DNA sequence (central blue box) contains six reading frames (tan arrows) that may be used by same-strand or opposite-strand overlapping genes.

(a)

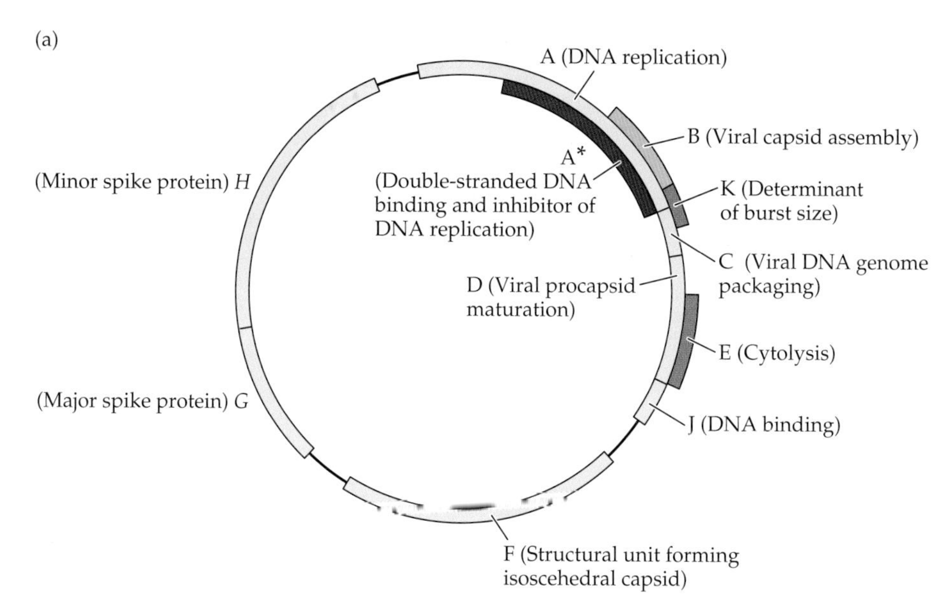

(b)

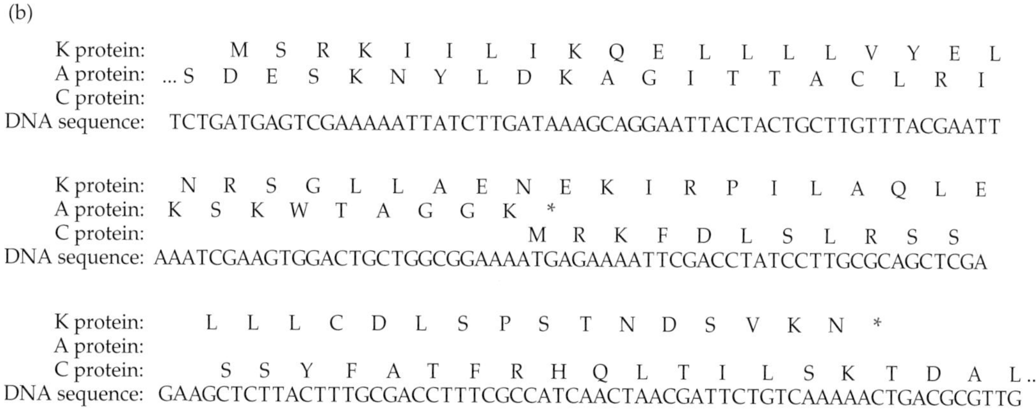

Figure 8.28 (a) Map of the circular genome of the single-stranded DNA bacteriophage ΦX174. Protein-coding genes are shown in blue. Same-strand phase 1 and phase 2 overlapping genes are in red and green, respectively. The genes encoding protein A and A* use the same reading frame. Gene *A** is an abbreviated form of gene *A* that uses an initiation codon 516 nucleotides upstream of the one used by *A*, and hence yields a shorter protein. (b) Sequence of the *K* gene, showing overlap with the 5′ end of the *A* gene and the 3′ end of the *C* gene. Red asterisks indicate stop codons. One-letter abbreviations are used for the amino acids.

Overlapping genes are widespread in DNA and RNA viruses, as well as in organelles and prokaryotes, but are much rarer in nuclear eukaryotic genes. **Figure 8.28a** shows the genetic map of the single-stranded DNA bacteriophage ΦX174. Several overlapping genes are observed. Because the bacteriophage is single-stranded, only same-strand overlaps are possible. Relative to genes *A* and *D*, respectively, genes *B* and *E* are phase 1 overlaps. Gene *K*, on the other hand, is a phase 2 overlap of both genes *A* and *C*. A more detailed illustration of the overlap between gene *K*, on the one hand, and genes *A* and *C*, on the other, is given in **Figure 8.28b**.

The overlap between two protein-coding genes may be internal or terminal (**Figure 8.29a**). In **internal overlapping genes**, the coding region of one of the genes is completely contained within the coding region of the second gene. In **terminal** (or **partially**) **overlapping genes**, the overlap only spans parts of the coding regions of the two genes. Overlapping genes can reside on the same strand or on opposite strands. In **same-strand overlap**, the two genes are transcribed from the same strand. In **opposite-strand overlap**, each of the overlapping genes uses one of the two complementary DNA strands. Same-strand overlaps are also referred to as **head-to-tail** or **codirectional overlaps**, because the 3′ end of one gene overlaps the 5′ end of the other. There are two types of terminal opposite-strand overlaps. In **head-to-head** or **convergent overlap**, the termination codon of one gene is located on the complementary strand of the reading frame of the other gene. In **tail-to-tail** or **divergent overlap**, the initiation codon of one gene is located on the complementary strand of the reading frame of the other gene. As mentioned previously, every double-stranded DNA sequence has six reading frames. In this discussion, we designate the reading frame that is used in

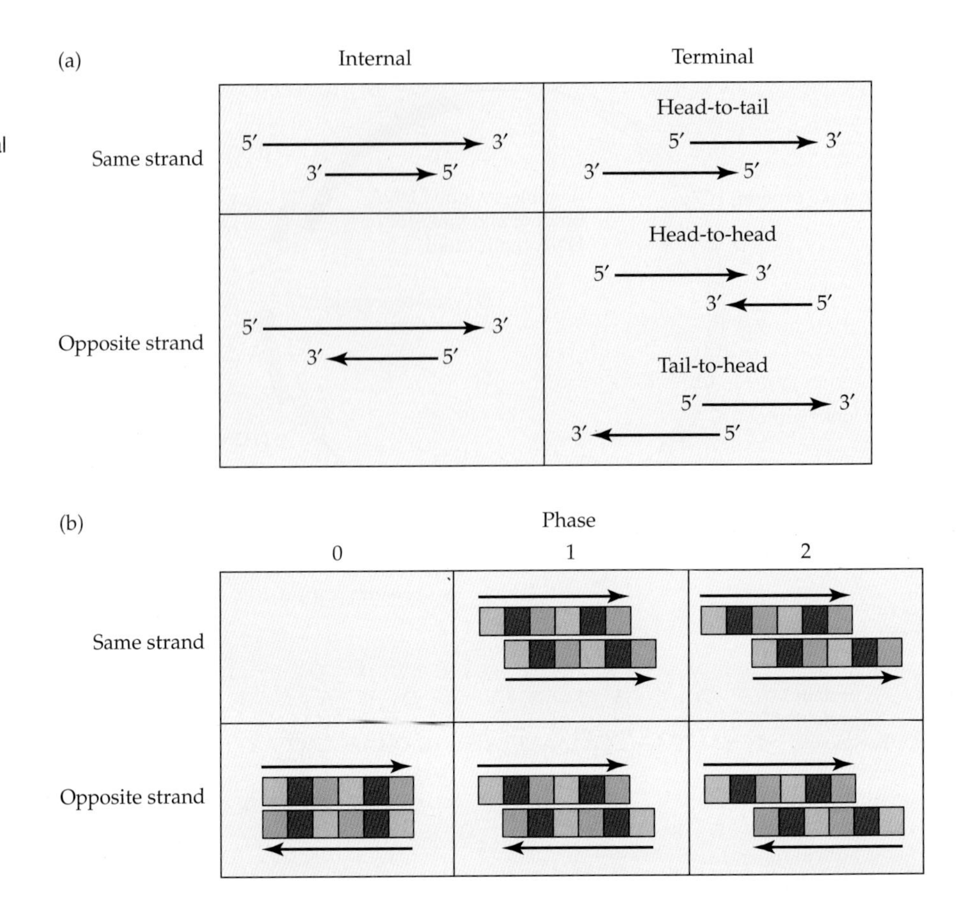

Figure 8.29 Orientations and phases of internal and terminal same-strand and opposite-strand gene overlaps. (a) Same-strand and opposite-strand internal gene overlaps. Genes are represented by arrows (5′ → 3′). (b) Orientations and phases of gene overlap in protein-coding genes. Genes can overlap on the same strand or on the opposite strand. The reference gene in a pair of overlapping genes is called phase 0. Same-strand overlaps can be of two phases (1 and 2); opposite-strand overlaps can be of three phases (0, 1, and 2). First and second codon positions, in which ~5% and 0%, respectively, of the changes are synonymous, are in light and dark red, respectively. Third codon positions, in which ~70% of the changes are synonymous, are shown in blue.

translating the reference gene as phase 0. A phase 0 gene may have five overlap phases, two (phase 1 and phase 2) on the same strand as the reference sequence, and three (phase 0, phase 1, and phase 2) on the opposite strand (**Figure 8.29b**).

The rate of evolution is expected to be slower in stretches of DNA encoding overlapping genes than in similar DNA sequences that only use one reading frame. The reason is that the proportion of nondegenerate sites is much higher in overlapping genes than in nonoverlapping genes, thus immensely reducing the proportion of synonymous mutations out of the total number of mutations (Miyata and Yasunaga 1978). For example, as far as phase 1 and phase 2 same-strand overlaps as well as phase 0 and phase 1 opposite-strand overlaps are concerned, there can be only very few synonymous changes at any site, because no site in the sequence is completely synonymous in both reading frames. In the case of phase 2 opposite-strand overlaps, there exist sites that serve as third codon positions of both genes, and hence a few of the sites can sometimes be fourfold synonymous in both genes, but even in this case the number of synonymous sites is vastly reduced relative to DNA sequences using a single reading frame. The sequence interdependence imposed by gene overlap also adds quite a bit of complexity to almost all evolutionary analyses. For example, if one estimates selection as the ratio of nonsynonymous to synonymous substitutions and assumes that the overlapping genes evolve independently of one another, chances are the results will indicate that one gene is subject to purifying selection and the other is under positive selection (e.g., Hughes et al. 2001). On the other hand, if the overlap is taken into account (Sabath et al. 2008), then the appearance of positive selection disappears.

We note that since gene duplication is a widespread phenomenon in evolution, the coupling between two genes due to overlap can easily be severed by complementary subfunctionalization. Thus, the maintenance of overlapping genes, as opposed to two nonoverlapping copies, most probably requires quite strong selective pressures (say, against increasing the genome size). Overlapping genes enable the production of more proteins from a given region of DNA than would be possible if the genes were arranged sequentially. Indeed, for the bacteriophage ΦX174, the overlapping of genes is a necessity. The amount of DNA present in the circular single-stranded DNA genome of this virus would not be nearly sufficient to encode the 11 bacteriophage proteins if transcription occurred in a linear fashion, nonoverlapping gene after nonoverlapping gene.

A survey of nearly 13,000 overlapping gene pairs from 58 completely sequenced prokaryotic genomes that employ the standard genetic code reveals that the vast majority of overlaps (85%) are same-strand overlaps (Lillo and Krakauer 2007). In approximately 50% of the cases, the overlap is five nucleotides or shorter in length. The rules governing these short overlaps are different from those applying to gene pairs with longer overlaps, because they are stringently constrained by the requirement for an initiation codon to overlap another initiation codon, or for an initiation codon to overlap a stop codon, or for a stop codon to overlap another stop codon. As far as long same-strand overlaps are concerned, the most frequent are phase 1 overlaps. The most frequent long opposite-strand overlaps are phase 2 and phase 0, for head-to-head and tail-to-tail overlaps, respectively.

The question arises of how overlapping genes may have come into existence during evolution. One possibility may involve an overlapping reading frame being turned into a bona fide new gene. We note that open reading frames of considerable lengths abound on either the same strand or the complementary strand of a protein-coding gene. Because only 3 of 64 possible codons are termination codons, even a random DNA sequence might contain open reading frames hundreds of nucleotides in length. Interestingly, codon usage biases (Chapter 4) are especially conducive to the creation of long, nonstop reading frames on the complementary strands of protein-coding genes (Silke 1997). Of course, the chances that a completely overlapping reading frame will actually contain the prerequisites of transcriptional and translational functionality are quite remote, as are the chances that the product will be useful or at least nondeleterious. Moreover, if de novo creation of overlapping genes were common, we would expect the overlap regions between genes to be fairly long, as opposed to the empirical fact that overlaps are very short in the vast majority of cases. Indeed, de novo origination of internal overlapping genes, whereby one protein-coding gene completely nests within the same strand or the opposite strand of another gene, is exceedingly rare. One such example was found in yeast, in which one protein was found not only to be encoded by the opposite strand of another gene, but also to play a role in its regulation (Li et al. 2010).

In most cases, the creation of an overlap seems to be a by-product of gene elongation, whereby one gene extends its coding region into that of a neighboring gene. In **Figure 8.30** we illustrate two possibilities. In the first scenario, a stop codon is obliterated by mutation such that translation proceeds until a downstream in-frame stop codon is encountered. In the process, an overlap is created. The second possibility involves the obliteration of an initiation codon, so translation starts upstream of the original initiation codon, again "invading" a neighboring gene. Fukuda et al. (1999, 2003) have shown that most overlapping genes are generated by the loss of a stop codon in a gene by deletion, point mutation, or frameshift, causing the gene to elongate to the next stop codon. In their studies, they also found a positive correlation between the number of overlapping genes and the total number of genes. These findings imply that the creation of overlapping genes is mainly driven by mutation. Support for this mutationist hypothesis is provided by the fact that most of the overlapping genes are derived from gene pairs that have very short intergenic regions, especially genes residing within the

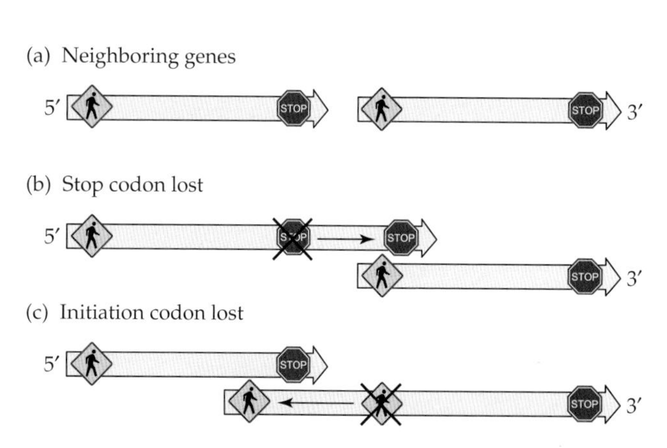

Figure 8.30 (a) Two same-strand neighboring genes. (b,c) Two scenarios for the creation of same-strand overlapping genes from the genes in (a). (b) The stop codon of the upstream gene is obliterated by mutation, causing translation to continue until it encounters the next stop codon. (c) The initiation codon of the downstream genes is obliterated by mutation. Translation, therefore, starts at an upstream site.

same operon. This reflects the ease with which mutation can lead to gene elongation and the creation of an overlapping structure. These results seem also to indicate that overlapping genes in bacteria have not evolved so that the genome size is minimized; rather, overlapping genes seem to be generated by random events.

Alternative Splicing

Alternative splicing of a primary RNA transcript results in the production of different mRNAs from the same DNA segment, which in turn may be translated into different polypeptides. The different mRNAs and proteins produced from a single gene locus by alternative splicing are called **isoforms**. Because of alternative splicing, the distinction between exons and introns is not absolute but depends on the mRNA of reference. There are two types of exons: **constitutive** (i.e., exons that are included within all the mRNAs transcribed from a gene) and **facultative** (i.e., exons that are sometimes **spliced in** and sometimes **spliced out** of the mature transcript). Exons that contain stop codons, which prematurely terminate the translation of the mRNA containing them, are called **poison exons** (Hollins et al. 2005). The truncated proteins produced due to the splicing in of poison exons are usually nonfunctional. Thus, poison exons can only exist in conjunction with alternative splicing. Many cases of alternative RNA processing have been found in eukaryotes, as well as in eukaryotic transposable elements and viruses.

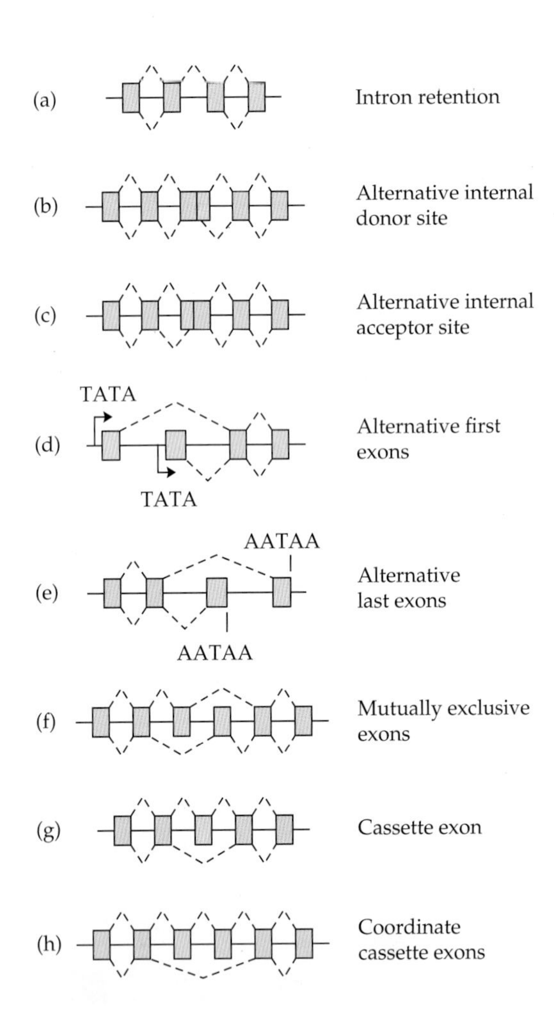

Figure 8.31 Types of alternative splicing. Constitutive and facultative exons are shown as blue and yellow boxes, respectively. Introns are depicted as horizontal lines. Alternative splicing pathways are shown by the diagonal dashed lines above and below the gene. Transcription initiation and polyadenylation sites are denoted TATA and AATAA, respectively. (Modified from Graur and Li 2000.)

There are different types of alternative splicing. Perhaps the simplest form of alternative splicing is **intron retention** (**Figure 8.31a**). An unspliced intron can result in the addition of a peptide segment if the reading frame is maintained. Of course, intron retention may also result in the premature termination of translation due to frameshifts. One such example is the periaxin gene in mouse (Dytrich et al. 1998). The gene contains seven exons and through intron retention encodes two proteins involved in the initiation of myelin deposition in peripheral nerves. One mRNA retains the intron separating exons 6 and 7 and, as a consequence, a translation stop is brought into the reading frame 21 codons after the 3′ end of exon 6. The inclusion of the intron results in a large mRNA (5.2 Kb), but because of the premature stop codon, a short protein isoform is produced (148 amino acids). The other alternatively spliced mRNA is shorter (4.6 Kb), but it encodes a much longer isoform (1,391 amino acids). Intron retention accounts for less than 5% of all alternative splicing events in multicellular eukaryotes.

Alternative splicing sometimes involves the use of **alternative donor sites** or **alternative acceptor sites**, also referred to as alternative 5′ splicing sites and alternative 3′ splicing sites, respectively (**Figure 8.31b,c**). In these types of splicing, an exonic sequence may be partially exonic in one isoform and partially intronic in another. Similarly, an intronic sequence may be partially exonic in one isoform and partially intronic in another. Such use of competing splice sites was found in several transcription units of adenoviruses, as well as in eukaryotic cells such as with the *transformer* gene in *Drosophila melanogaster* (see below). The use of alternative donor and acceptor sites accounts for more than a quarter of all alternative splicing events in multicellular eukaryotes.

In some instances, different mRNAs that are produced from the same gene differ from one another only at their 5′ or 3′ ends. This is usually the result of **alternative first exons** (also referred to as **alternative transcription initiation**) or **alternative last exons** (also referred to as **alternative transcription termination**), respectively (**Figure 8.31d,e**). The α and β isoforms of the mouse caveolin-1 gene are produced by alternative transcription initiation sites (Kogo and Fujimoto 2000).

Alternative polyadenylation sites are quite common in eukaryotic nuclear genes (e.g., Thekkumkara et al. 1992). In some cases, the alternative polyadenylation sites are located in internal introns or exons and therefore the use of such alternative sites will result in the production of different protein isoforms. Alternative polyadenylation sites that are located in the 3′ untranslated region (3′ UTR) will result in transcripts with 3′ UTRs of different length but encoding the same protein. Alternative polyadenylation sites within internal introns or exons have the potential to affect gene expression qualitatively by producing distinct protein isoforms. In comparison, alternative polyadenylation sites within the 3′ UTR have the potential to affect gene expression quantitatively (Di Giamartino et al. 2011).

Some cases of alternative splicing involve the use of **mutually exclusive exons**, i.e., two exons are never spliced out together, nor are both retained in the same mRNA (**Figure 8.31f**). One such instance is the M1 and M2 forms of pyruvate kinase, which are produced from a single gene by mutually exclusive use of exons 9 and 10.

Finally, we have alternative splicing involving **cassette exons** (**Figure 8.31g**), in which a particular exon (the cassette) is either spliced in or spliced out (**exon skipping**). Exon skipping is quite common, accounting for nearly 40% of all alternative splicing events in multicellular eukaryotes. Sometimes, two or more exons can be skipped together, a phenomenon referred to as **coordinate cassette exons** (**Figure 8.31h**). Usually the reading frame is maintained whether or not the cassette exons are included within the mature mRNA, although cases are known in which the inclusion or exclusion of a cassette exon may cause frameshifts and/or bring into frame a termination codon.

Within a single gene, we may find different types of alternative splicing. For example, in the rat gene for fast skeletal muscle troponin-T, the presence of cassette exons in conjunction with mutually exclusive exons theoretically allows the production of at least 128 different proteins (Morgan et al. 1993). A huge increase in the protein-coding repertoire is seen in human neurexin genes. There are three paralogous neurexin genes in the human genome, *NRXN1*, *NRXN2*, and *NRXN3*. Each of these genes can be transcribed using either an upstream promoter or a promoter downstream of exon 17, thus producing a total of six primary transcripts. These transcripts are subsequently alternatively spliced, yielding in excess of 800 different proteins (Rowen et al. 2002).

The most striking example of alternative splicing complexity may be the single pre-mRNA for a *Drosophila melanogaster* axon guidance receptor gene, *Dscam*, which can be processed to generate potentially 38,016 different mature transcripts (Schmucker et al. 2000). Even if only a subset of the potential protein isoforms is ever produced in vivo, this combinatorial use of alternative exons undoubtedly represents an extravagant source of diversity for a single gene, especially given that the entire *Drosophila* genome consists of only ~14,000 protein-coding genes.

Sex determination and alternative splicing

Alternative splicing is often involved in developmental regulation. An intriguing situation is seen in several genes involved in the process of sex determination in *Drosophila melanogaster*. At least four genes, *Sexlethal* (*sxl*), *transformer* (*tra*), *doublesex* (*dsx*), and *fruitless* (*fru*), are spliced differently in males and females (**Figure 8.32**). The presence of two X chromosomes in diploid flies signals female identity and triggers the expression of the *sxl* gene early in development. The alternatively spliced transcript in females skips over exon 3 and is translated into the Sxl protein. In males, exon 3, which contains an in-frame termination codon, is spliced in, and no functional protein is produced. Sxl is an RNA-binding protein that controls several alternative splicing decisions. Here, we shall mention two of these. First, Sxl represses the inclusion of exon 3 in its own pre-mRNA, thus initiating a positive feedback loop that maintains functional Sxl expression in females. Second, Sxl blocks the more upstream of two competing 3′ splice sites in *tra* pre-mRNA, thus allowing expression of the Tra protein in females. In males, an in-frame stop codon prematurely terminates translation. Tra is also a splicing regulator that takes part in controlling two splicing decisions

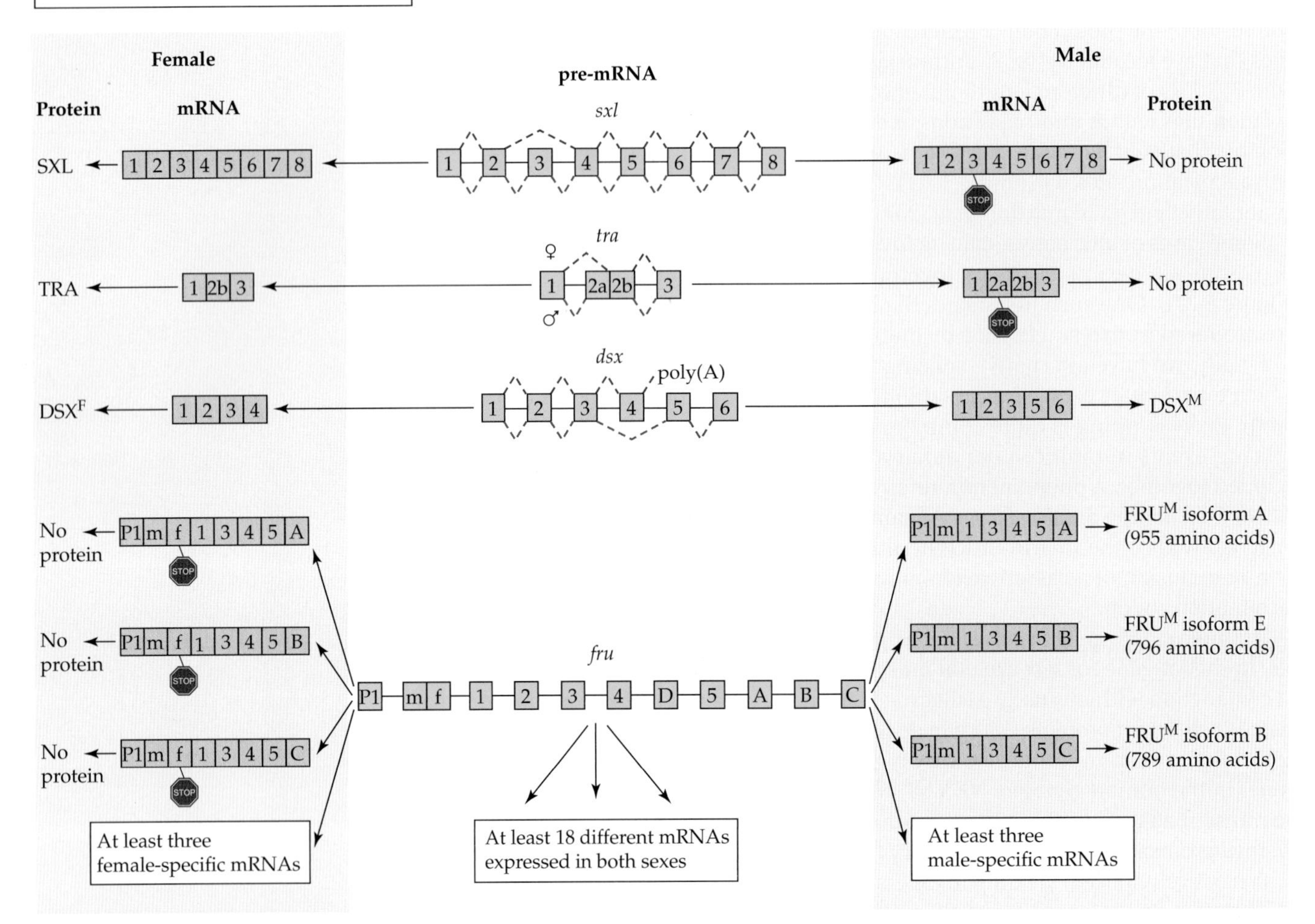

Figure 8.32 Sex-specific patterns of splicing and translation of the *Sexlethal* (*sxl*), *transformer* (*tra*), *doublesex* (*dsx*), and *fruitless* (*fru*) genes in *Drosophila melanogaster* females (left) and males (right). Termination codons in poison exons are indicated (red stop signs). Mature mRNAs that contain poison exons are translated into nonfunctional proteins. (Data from Lopez 1998 and Salvemini et al. 2010.)

required for somatic sexual differentiation. First, Tra activates the more upstream of two competing 3′ splice sites in the *dsx* pre-mRNA. This results in the incorporation of a female-specific terminal exon rather than the default exon used in males. The two isoforms, Dsx^F in females and Dsx^M in males, are responsible for gender-appropriate gene expression patterns for most somatic sexual characters. Tra also activates the more downstream of two competing 5′ splice sites in the *fru* pre-mRNA, thus preventing the expression of male-specific Fru proteins, which are transcriptional regulators required for male courtship behavior, as well as for the differentiation of a male-specific abdominal muscle.

The genetic basis of sex determination varies widely among insects. Despite these differences, comparative studies have shown that alternative splicing has been frequently used in insect evolution to control expression of key sex-determining genes. Whether an individual ends up as a male or a female frequently involves a binary "decision" based on alternative splicing. A comparison of alternative splicing in the *tra* gene of several insect taxa is shown in **Figure 8.33**.

Evolution of alternative splicing

The evolution of alternative splicing requires that an alternative splice junction site be created de novo. Since splicing signals are usually 5–10 nucleotides long, it is possible that such sites are created with an appreciable frequency by mutation. Indeed, many such examples are known in the literature. In particular, many synonymous and non-

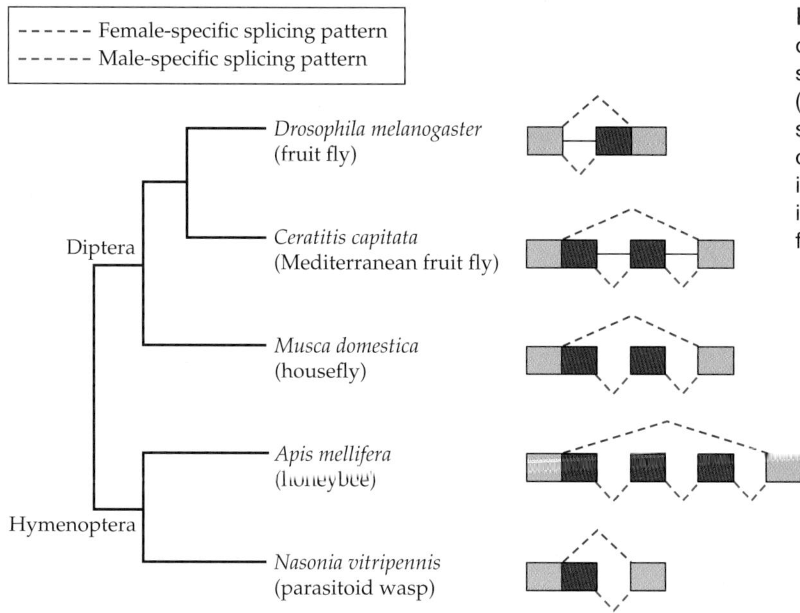

Figure 8.33 Phylogenetic tree and comparisons of the portions of the *tra* genes that are alternatively spliced to give rise to sex-specific products in flies (Diptera) and hymenopterans. The male-specific poison exons are shown as red boxes. Note that the *tra* orthlogs exhibit different types of alternative splicing, including alternative internal donor sites, alternative internal acceptor sites, and cassette exons. (Modified from Salz 2011.)

synonymous substitutions are known to affect splicing (Cartegni et al. 2002). In cases in which the newly created splice site is stronger than the old splice site, pathological manifestations may ensue. Such a mutation will obviously have deleterious effects and is not ever expected to become fixed in the population. However, if the newly created splice site is much weaker, then most mRNAs will be of the original type, and only small quantities of the new mRNA will be made. Such a change will not obliterate the old function, and yet will create an opportunity to produce and, most importantly, put to the test a new protein, possibly one with a new useful function.

The strength of the splicing sites of an alternatively spliced exon can be measured by means of a **retention ratio**, which is calculated as the number of mRNA sequences that contain the alternatively spliced exon divided by the total number of mRNA sequences produced from the gene (Sorek et al. 2002). A constitutive exon will have a retention ratio of 1. Splicing, whether constitutive or alternative, entails the removal of introns from the pre-mRNA by the spliceosome, which is composed of several splicing factors. Precise recognition of introns is required for correct splicing. This recognition is achieved by the binding of the splicing factors to signals of varying specificity that are located in both the intron and its flanking exons. In vertebrates, three signals are known to direct splicing: the 5′ splice site at the 5′ end of the intron, the 3′ splice site at the 3′ end of the intron, and a branch site upstream of the 3′ splice site. Combinations of coadapted splicing factors and strong splicing signals will result in constitutive splicing, while weak signals and not-so-well-adapted factors will result in alternative splicing. The weaker the signal is, the smaller the retention ratio will be. What constitutes a weak or a strong splicing site depends on the interaction between the splicing factor and the splicing signal, and the strength of a splicing site, and this interaction may be taxon-specific. So, while the entire "rule book" is not yet known, some rules have been deciphered (Ast 2004). For example, the intronic sequence 5′-G-T-A-3′ immediately following a G on the 3′ end of an exon is considered a strong signal resulting in constitutive splicing, while 5′-G-T-G-3′ will mostly result in alternative splicing, and 5′-G-T-C-3′ will result in exon skipping (no splicing).

Evolutionary reconstructions have shown that the splicing signals of the early eukaryotes were most probably weak, which implies that alternatively spliced exons, rather than constitutive exons, were the norm in the common ancestor of eukaryotes.

Based on huge compilations of partial and complete mRNAs, at least 75% of human protein-coding genes have been found to possess two or more isoforms. Most of them have very low retention ratios. This indicates that most mRNA isoforms are produced by aberrant rather than regulated splicing and that they are almost certainly not functional (Keren et al. 2010). The production of superfluous RNA molecules has been dubbed "transcriptional noise" (Bird 1993). Of course, transcriptional noise has its evolutionary uses, as a testing ground for new potentially functional products.

Novel alternatively spliced isoforms are often created by the exonization of intronic sequences or by the pseudoexonization of exonic sequences (Kondrashov and Koonin 2003; Modrek and Lee 2003). Newly exonized sequences and newly inserted exons are initially alternatively spliced at harmlessly low retention ratios, and in the exceedingly rare cases in which they are found advantageous, there will be selective pressures to increase the splicing strength toward increasing the retention ratio. *Alu* sequences, which constitute more than 10% of the human genome (Chapter 9), can be used to illustrate this process. At least 5% of all human alternative exons are derived from *Alu* elements. In contrast, no human constitutively spliced exon was ever found to contain an *Alu* sequence, despite the small number of substitutions required to transform an alternatively spliced exon into a constitutively spliced one. The reason must, therefore, be strong negative selective pressure. Indeed, all constitutive *Alu*-containing exons found to date are associated with disease-causing alleles (Kreahling and Graveley 2004).

Increasing proteome diversity: Alternative splicing or gene duplication?

New gene functions are generally acquired through redundancy. This may occur through gene duplication (Chapter 7) or an increase in transcript diversity. One mechanism for increasing transcript diversity is alternative splicing. Kopelman et al. (2005) discovered that there is an inverse correlation between the size of a gene family and its propensity to produce alternative splice variants. **Singletons**, i.e., genes that comprise gene families of size 1, are more likely to use alternative splicing than genes that have undergone duplication. An extreme example of this phenomenon can be seen in the ten largest gene families in the human genome. Of the 415 genes in these families for which we have splicing data, fewer than 3% use alternative splicing, compared with more than 50% for singletons.

The rate of alternative splicing was roughly three times higher in conserved singletons in human and mouse than in genes that have duplicated in both lineages. This relationship could reflect an inherent property of certain gene families to prefer proliferation through gene duplication as opposed to alternative splicing. But singletons in one mammalian species whose orthologs in the other have duplicated since divergence had a substantially higher rate of alternative splicing than their duplicated orthologs had. This finding implies that inherent family inclinations do not underlie the inverse correlation and suggests instead that demand for proliferation may be fulfilled independently by either mechanism.

Two evolutionary models are consistent with these results: either duplicates lose splice variants, or singletons acquire them. If splice variant loss prevails, we should expect anciently duplicated genes (that have had more time to lose splice variants) to have little or no alternative splicing. In reality, a positive correlation between the age of a duplication event and the fraction of alternative splicing was found. This suggests that splice variant acquisition, rather than loss, better explains the observations.

Further support for the notion of evolutionary gain of splice variants is provided by the observation that alternative splicing frequently evolves through exon duplication. This suggests that acquiring splice variants is concomitant with an increase in the number of exons; in fact, singletons possess on average more exons than genes with conspecific paralogs.

Thus, gene duplication and alternative splicing may not be independent evolutionary properties of a gene. Rather, an evolutionary necessity for diversification may be "satisfied" through either gene duplication or alternative splicing, whichever happens to occur first. This phenomenon is consistent with the suggestion that alternatively spliced exons are "internal paralogs" (Modrek and Lee 2002).

De Novo Origination of Genes

A protein-coding sequence that is only found in a single evolutionary lineage is called an **ORFan** (a homophone for "orphan" that includes the acronym for "open reading frame"). Several explanations have been proposed for the appearance of ORFans. These include such unexciting possibilities as rapid evolution resulting in the loss of similarity to homologs or horizontal gene transfer from an as yet uncharacterized source. ORFans may also be trivial artifacts of gene loss in lineages other than the ones in which the ORFans appear. Here, we are interested in **de novo genes**, i.e., protein-coding genes that were derived from noncoding regions of the genome, such as intergenic regions, introns, or the complementary strand of protein-coding genes.

From an evolutionary viewpoint, the derivation of a protein-coding gene from noncoding DNA is the least likely route of producing a new function. In fact, the emergence of complete functional genes—with promoters, open reading frames, and protein products that perform a useful function—has been likened to the "transmutation of lead into gold" (Siepel 2009). The reason for this implausibility is that the conversion of a noncoding sequence to a coding sequence requires a large number of steps, each with a very low probability of occurrence. The first step is the creation of an open reading frame (ORF). Given the structure of the genetic code, the creation of an ORF of a length comparable to a short exon is quite likely. Indeed, spurious ORFs abound in all the genomes so far studied. Conversion of such an ORF into a coding sequence, however, requires the sequence in question to be transcribed, escape degradation at the nuclear exosome, associate with ribosomes, be translated, and escape degradation by proteases. Finally, the protein must not have deleterious effects on the cell in which it is produced, for example, creating toxic conformations such as insoluble amyloids. These conditions are unlikely to be met in nature, and indeed the de novo origination of genes is quite a rare event.

For a gene to be defined as a de novo gene, it must have no functional orthologs or paralogs, and it must not be derived from transposable elements. Its open reading frame as well as its functional promoters must be forged out of noncoding DNA "like a blacksmith shaping a new tool from raw iron" (Siepel 2009). We note that the validation of de novo protein-coding genes and their products is complicated because it requires showing not only that the gene is functional, but also that its evolutionary antecedents (i.e., the sequences from which it was derived) are nonfunctional.

The most likely model for the de novo evolution of a functional protein-coding gene from a nongenic sequence is one involving a transitory **protogene** phase (Carvunis et al. 2012). According to this model, there is a progression between nongenic DNA and functional protein-coding gene that is the mirror image of the degeneration of a functional gene into nongenic DNA through a pseudogene as the intermediary stage (**Figure 8.34a**). This idea is based on the fact that many noncoding sequences are frequently transcribed, and some are even translated at low rates (Wilson and Masel 2011). The protein products of such protogenes can essentially be tested in vivo, with the most strongly deleterious polypeptides being purged by purifying selection. The fact that random polypeptides are tested and can be improved by natural selection makes de novo origination of genes much more plausible.

In recent years, a handful of de novo genes were reported in the literature. Zhou et al. (2008) and Chen et al. (2010), for instance, estimated that as many as 12% of the newly emerged genes in *Drosophila melanogaster* since its divergence from *D. willistoni* approximately 35 million years ago may have arisen de novo from noncoding DNA. The total number of de novo genes in the *D. melanogaster* genome, however, is minuscule; only 18 out of a total of approximately 13,000 protein-coding genes are so far suspected of being de novo genes. The situation in yeast is similar, confirming that de novo creation of coding sequences is indeed rare (Ekman and Elofsson 2010). In humans, less than 60 de novo protein-coding genes have been discovered so far, with slightly more than a dozen genes appearing after the divergence between human and chimpanzee (Li et al. 2010; Knowles and McLysaght 2011; Wu et al. 2011).

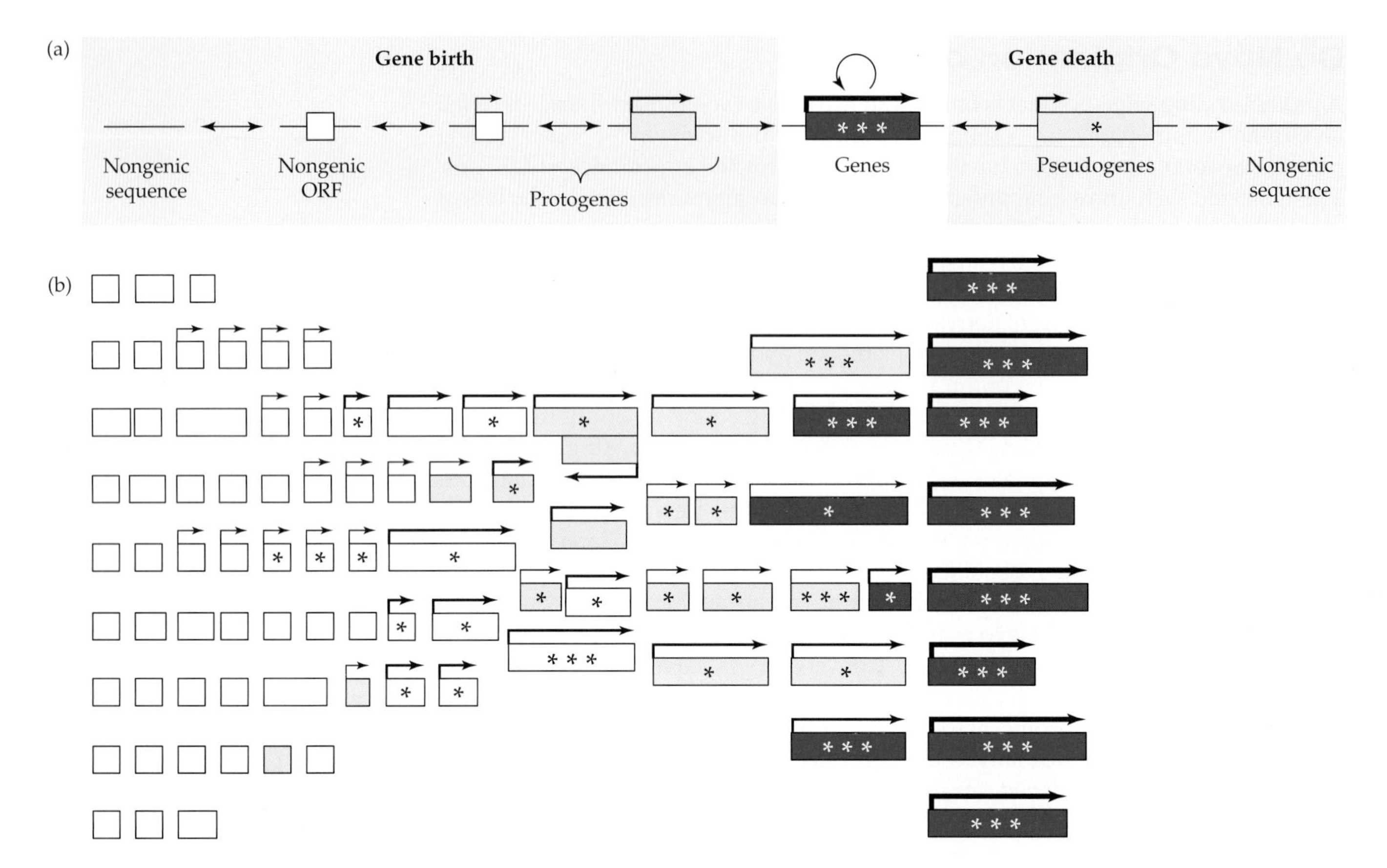

Figure 8.34 Origination of genes from nongenic sequences through a protogene stage. (a) Schematic representation of de novo gene birth and gene death as mirror images; protogenes are for gene birth what pseudogenes are for gene death. An open reading frame (ORF) may appear by chance in a nongenic region. This process is reversible (bidirectional arrows). This nongenic ORF may be transcribed and translated, at which stage its products are tested by natural selection. If the protogene does not pass the test of natural selection, it will irreversibly revert to a nongenic sequence. Protogenes that pass the test of selection may become longer and more conserved evolutionarily, and their sequences may acquire functionally appropriate codon usage, amino acid abundances, and structural features. A protogene may either revert to nongenic status or evolve into a gene. The circular arrow indicates gene origination from preexisting genes, such as through gene duplication. There can be no reversion of genes to protogenes (unidirectional arrow), because gene decay produces pseudogenes. Pseudogenes are highly related in sequence to existing genes but have accumulated disabling mutations such that translation of functional proteins is no longer possible. The premise that pseudogene formation represents irreversible gene death has been challenged by reports of pseudogene resurrection (bidirectional arrow). After enough evolutionary time, mutational decay renders the pseudogenes indistinguishable from nongenic sequences. We note that whereas pseudogenes resemble known genes, protogenes resemble no known genes. (b) A continuum model for the de novo emergence of genes, showing populations of nongenic ORFs, protogenes and genes and the impossibility to delineate exact boundaries between these three categories. ORFs are shown as rectangles. The length of a rectangle is proportional to the ORF length. Dark blue indicates a high degree of sequence conservation; lesser conservations are shown in light blue and white. The widths of the arrows above the rectangles indicate expression levels. Asterisks denote sequence features conducive to expression and function, such as codon usage and amino acid abundances.

As expected, de novo genes are mostly intronless and rather small. The mean length of proteins encoded by de novo genes in yeast (71 amino acids) is only about 14% of the mean length of more conventionally derived proteins. In humans, the putative de novo genes encode proteins that are about one-third of the length of the average annotated protein. Little is known about the function of any de novo gene, and there is no evidence of positive selection for any of these genes described so far. In fact, the vast majority of them evolve very fast, indicating an almost total lack of functional constraints. In humans and *Drosophila*, de novo genes show a strong tendency for testis-specific expression (Levine et al. 2006; Begun et al. 2007; Chen et al. 2007; Wu et al. 2011), which as mentioned previously may be a reflection of the transcriptionally permissive environment in the testis, rather than a functional imperative.

Quantitatively, de novo genes do not seem to contribute significantly to the expansion of the protein repertoire. However, their contribution to the evolutionary process may be to add novel elements to the proteome that can subsequently be used in conjunction with other elements to enlarge the proteome. In at least one case, it is possible to estimate the relative prevalence of different mechanisms in creating new genes. The lineages leading to *Drosophila melanogaster* and *D. simulans* diverged from each other approximately 3 million years ago (Russo et al. 1995), after which at least 72 new genes emerged in the *D. melanogaster* lineage. As expected, the vast majority of these genes—more than 90%—were created by gene duplication, mostly tandem gene duplication. Retroposition (Chapter 9) contributed close to 7% of the new genes, while de novo origination accounted for only two genes (Zhou et al. 2008). Close to 30% of the new genes created by either duplication or retroposition displayed evidence of chimerism by either domain shuffling or fusion.

Nested and Interleaved Genes

An intron may sometimes contain an open reading frame that encodes a protein or part of a protein that is completely different in function from the one encoded by the flanking exons. Such a gene is referred to as a **nested gene**, and its product is sometimes called an **intron-encoded protein**. One such example is shown in **Figure 8.35a**. From a mechanistic point of view, a nested gene, which by definition is transcribed from the same strand as the neighboring exons, which belong to a different gene altogether, may be regarded as a special instance of alternative splicing. However, the manner in which the transcription and splicing machinery maintain the individual identities of the nested gene and the gene into which it is nested is something of a mystery.

Assis et al. (2008) envisioned four scenarios for the origination of nested gene structure: (1) insertion of a gene into an intron of another gene, (2) de novo gene emergence from an intronic sequence, (3) internalization following the acquisition of a new exon by a neighboring gene, and (4) fusion of two flanking genes (**Figure 8.36**). The vast majority of nested genes were found to have emerged by insertion of a gene into an intron.

At least three hypotheses could explain the maintenance of nested gene structures. First, a nested structure might confer a selective advantage because of a functional or coregulatory relationship between the nested gene and the gene into which it is

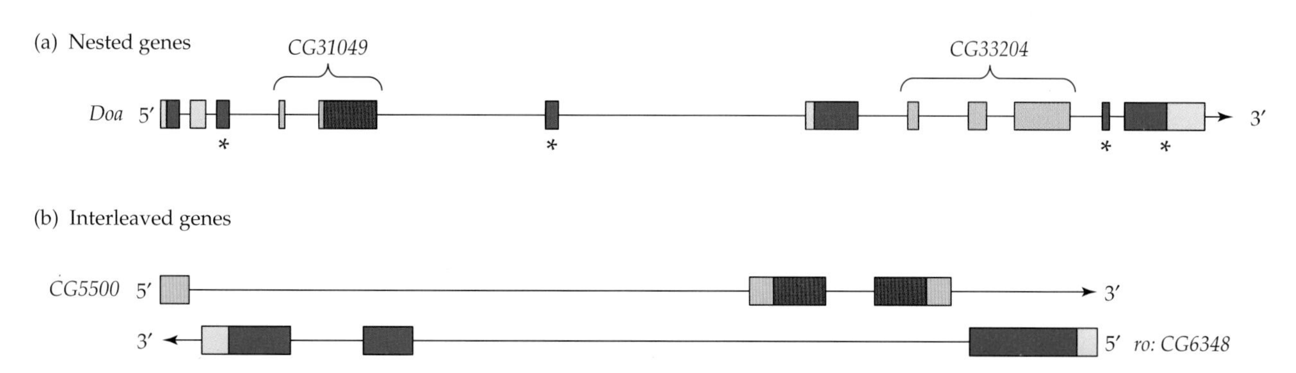

Figure 8.35 Nested and interleaved genes in *Drosophila melanogaster*. (a) In this example of nested genes, two genes, *CG31049* and *CG33204* (red and green, respectively), map to the second and third introns, respectively, of the *Darkener of apricot* (*Doa*) gene. *Doa* is an alternatively spliced gene with four constitutive exons (each marked with an asterisk) and three alternatively spliced exons. (b) Interleaved genes are transcribed on opposite strands from the same genomic region. Their exons do not overlap but map to introns of the gene on the complementary strand. In this example, the last two exons of *CG5500* map to the first intron of *rough* (*ro:CG6348*), and the last two exons of *ro:CG6348* map to the first intron of *CG5500*. Exons are shown as solid rectangles, with the protein-coding regions in dark color and the untranslated regions in light color; introns are represented by horizontal lines. (Modified from Celniker and Rubin 2003.)

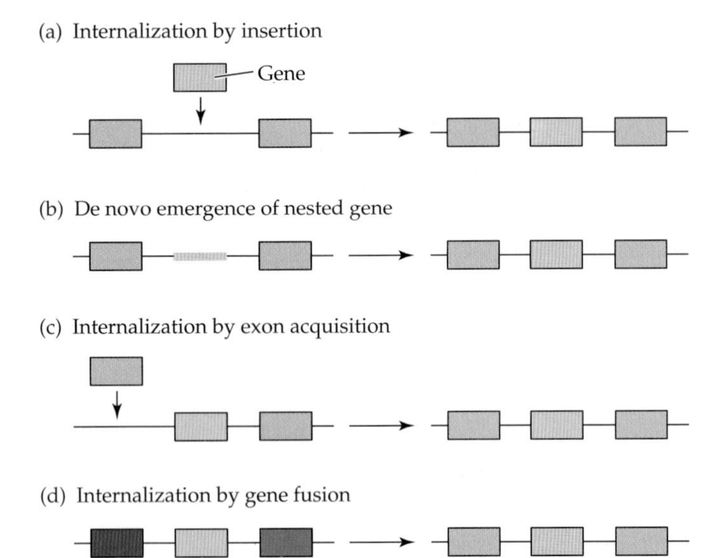

Figure 8.36 Scenarios for the origin of a nested gene structure. (a) Internalization following the insertion of a gene into an intron of another gene. (b) De novo evolution of a gene from an intronic sequence (yellow line). (c) Internalization of a gene after exon acquisition by an adjacent gene. (d) Internalization of a gene by the gene fusion of its upstream and downstream neighboring genes. Yellow, nested genes; green, exons of the surrounding gene; red and blue, genes involved in gene fusion. (Modified from Assis et al. 2008.)

nested. Second, according to the transcriptional collision model, members of a nested gene structure could interfere with each other's transcription, resulting in alternative expression of these genes in different tissues or during different development stages (Da Lage et al. 2003; Crampton et al. 2006; Osato et al. 2007). Finally, acquisition of a nested gene structure could be a neutral process, driven by the presence of numerous long introns that provide convenient "niches" for the insertion of genes. Each of these hypotheses leads to a distinct prediction about the relationship between the expression of the two genes in a nested pair. The functional coregulation hypothesis predicts a positive correlation between levels of expression, the transcriptional collision hypothesis predicts a negative correlation, and the neutral hypothesis predicts no correlation. Assis et al. (2008) analyzed gene expression data from human and *D. melanogaster* genomes and concluded that the patterns of expression were consistent with the neutral hypothesis. The observation that nested genes reside in relatively long introns is also compatible with the neutral hypothesis.

Analyses of nested protein-coding genes in vertebrates, fruit flies, and nematodes revealed substantially higher rates of evolutionary gains than losses of nested structures. Since the accumulation of nested gene structures could not be attributed to any obvious functional relationships between the genes involved, nested genes seem to represent one more means of increasing genomic organizational complexity via a neutral process.

Interleaved genes are transcribed from opposite strands of the same genomic region. Their exons do not overlap; instead, one exon is located on one strand while the other exon is located on the complementary strand (**Figure 8.35b**). In the *D. melanogaster* genome, there are close to 900 nested genes, but only 26 interleaved genes have been identified (Celniker and Rubin 2003). The origination of interleaved genes most probably follows the same pathway as that described for de novo genes.

Gene Loss and Unitary Pseudogenes: A Molecular Revisiting of the "Law of Use and Disuse"

One of the main components of Jean-Baptiste Lamarck's theory of evolution was the law of "use and disuse," the idea that frequently used body parts become stronger and larger from generation to generation, while parts not used waste away and disappear. Interestingly, Darwin admitted "use and disuse" as an important evolutionary mechanism (in addition to natural selection), and his pre-molecular intuition was quite correct in emphasizing "disuse" over "use." For instance, he made the observation that the reduced size of the eyes in moles and other burrowing mammals is "probably due to gradual reduction from disuse, but aided perhaps by natural selection." In the case of cave animals, when referring to the loss of eyes, he stated, "I attribute their loss wholly to disuse."

In hindsight, we know that (1) mutations are random in respect to their effect on the fitness of the organism carrying them, (2) deleterious mutations are considerably more frequent than advantageous mutations, and (3) purifying selection is considerably more prevalent than positive selection. As a consequence, "use" cannot in itself bring about desirable evolutionary change, while "disuse," i.e., the elimination or reduction of selective constraint, can lead to the loss of genes and their resultant functions.

Gene loss can occur through either of two processes: nonfunctionalization or deletion of a single-copy gene from the genome. (From comparative studies, we now know that in many cases deletion follows nonfunctionalization.) The nonfunctionalization of a single-copy gene results in the creation of a **unitary pseudogene**, i.e., a pseudogene that has no functional paralogs in the genome. Unitary pseudogenes are the genetic analogs of vestigial anatomical structures, such as the wings of ostriches and the eyes of subterranean moles, which have lost their functions and their structural integrity, but are recognizable as eyes through comparisons with functional structures in other organisms. The nonfunctionalization of a single-copy gene is typically deleterious and unlikely to be fixed in a population. This fact notwithstanding, unitary pseudogenes have been found in both eukaryotes and prokaryotes. As expected, unitary pseudogenes are rare; the human genome, for instance, contains fewer than 100 unitary pseudogenes (Zhang et al. 2010), in comparison with more than 20,000 pseudogenes that possess functional paralogs. Interestingly, there are 11 "unitary" pseudogenes that are polymorphic, i.e., they have both functional and nonfunctional alleles that are currently segregating in the population.

How can a nonfunctional pseudogene lacking a functional counterpart in the genome become fixed in the population? This can happen either by random genetic drift, if the selection against the loss of the gene product no longer operates or is very weak, or through positive selection, if the loss of the gene entails a selective advantage on the organism. The possibility that gene loss is driven by positive selection is as intriguing as it is counterintuitive. In particular, Olson (1999) made the case that gene loss is "an engine of evolutionary change" in his "less is more" hypothesis. His argument relies on the well-known observations that (1) mutations that lead to loss of function are considerably more numerous than gain-of-function mutations, and (2) only a very minute fraction of all mutations (including loss-of-function ones) are advantageous. Thus, loss-of-function mutations may predominate among all adaptive mutations. When a novel selective force starts acting on a population, the fixation of a loss-of-function mutation may be the most likely outcome.

We note that, in most cases, it is not possible to state with any degree of confidence whether the loss of a gene was fixed in the population by drift, following the elimination of selective constraint (disuse), or by positive selection. It is also very difficult to state whether the pseudogenization was the cause or the consequence. Did the giant panda became the only member of Carnivora to subside on an exclusively vegan diet because of the pseudogenization of the umami receptor gene (Zhao et al. 2010), or did the receptor gene became dispensable after the panda adopted a vegan way of life? Was the increase in cranial capacity in humans caused by the pseudogenization of a sarcomeric myosin heavy chain gene *MYH16* expressed in the masticatory muscles of other mammals? Or was the pseudogenization of *MYH16* merely made possible following a change in the consumption of hard and gritty victuals? The verdict is not yet in (Stedman et al. 2004; Perry et al. 2005; McCollum et al. 2006). Notwithstanding these difficulties, sometimes the anatomical, biochemical, and physiological consequences of gene loss can be spectacular, as for instance in the case of blackfin icefish (*Chaenocephalus aceratus*), one of 16 recognized species of Antarctic icefish that lack hemoglobin and red blood cells. This species lost the linked α- and β-globin genes some 8.5 million years ago, and as a result its blood is colorless (Ruud 1954; di Prisco et al. 2002).

Sometimes it is possible to establish a tentative order for pseudogenization events, as in the case of tooth-specific genes in baleen whales (Mysticeti). Whales belonging to suborder Mysticeti are filter feeders that use baleen (bristles made of keratin) to sift zooplankton and small fish from ocean waters. Adult mysticetes lack teeth, although tooth buds may be present in fetal stages. But which tooth-specific gene was the first to be lost, thereby rendering the other ones superfluous? Paleontological data indicate that enamel-capped teeth were lost in the Oligocene. The cause for the loss seems to be the pseudogenization of the gene for enamelysin in the common ancestor of living baleen whales (Meredith et al. 2011). The "culprit" for the pseudogenization event seems to be a retrotransposable element that became inserted into the second exon of the gene. As a consequence, other genes, such as those encoding ameloblastin, enamelin, and amelogenin, became free of all selective constraints and were sporadically pseudogenized in different lineages.

Here, four examples of gene loss will be presented in some detail. They aptly illustrate the fact that, contrary to intuition, loss of function may not always be deleterious. Some losses may be selectively neutral, and some may even confer a fitness advantage. These examples were chosen not only because they illustrate different evolutionary pathways to gene loss, but also because they highlight the methodological difficulties in establishing the root causes of each loss. A "forward genomics" approach that links phenotype changes to gene loss has been proposed by Hiller et al. (2012).

THE PSEUDOGENIZATION OF L-GULONO-γ-LACTONE OXIDASE, OR WHY DO YOU NEED TO EAT FRUIT AND GREEN VEGETABLES EVERY DAY? While the majority of mammals possess the ability to synthesize L-ascorbic acid, a number of taxa, including Simiiformes (Old World and New World monkeys), Hystricognathi (e.g., guinea pigs and porcupines), as well as most bat genera (Chiroptera) have lost the the ability to produce L-ascorbic acid (i.e., are hypoascorbemic) and are, therefore, susceptible to the disease scurvy unless they obtain L-ascorbic acid from their diets. In addition to these mammals, hypoascorbemia has been found in all members of the very large infraclass Teleostei (ray-finned bony fishes) and in some passeriform birds. For these organisms, ascorbic acid is a vitamin (vitamin C).

The reason these animals cannot manufacture their own ascorbic acid is that they lack a protein called L-gulono-γ-lactone oxidase, an enzyme that catalyzes the reaction L-gulono-1,4-lactone + O_2 → L-xylo-hex-2-ulono-1,4-lactone + H_2O_2 → L-ascorbic acid. In animals that are not prone to scurvy, this protein is produced by a single-copy gene (Koshizaka et al. 1988). In humans and other Old World monkeys, L-gulono-γ-lactone oxidase is a pseudogene, containing such molecular defects as the deletion of 7 noncontiguous exons (out of 12), three deletions and an insertion (each capable of frameshifting the reading frame), and obliterations of intron-exon boundaries (Nishikimi et al. 1994; Inai et al. 2003; Lachapelle and Drouin 2011). In addition to these changes, the pseudogene locus in humans accumulated six *L1* and 13 *Alu* retrotransposable-element insertions (Chapter 9).

The unitary pseudogene for L-gulono-γ-lactone oxidase in the guinea pig contains different defects from those in the human pseudogene (**Figure 8.37**). Only exons 1 and 5 and a 3′ portion of exon 6 were deleted. These differences indicate that nonfunctionalization of the L-gulono-γ-lactone oxidase-coding gene occurred independently in hystricomorphs and catarrhines (Nishikimi et al. 1992). Indeed, the inactivation of the enzyme in Old World monkeys is estimated to have occurred more than 60 million years ago, whereas the inactivation in the guinea pig lineage occurred less than 15 million years ago (Lachapelle and Drouin 2011).

Given the high daily requirement and important functions of ascorbic acid, the widespread loss among vertebrates of the capacity to synthesize it raises two interconnected questions. First, why are all known losses of the capacity to synthesize ascorbic

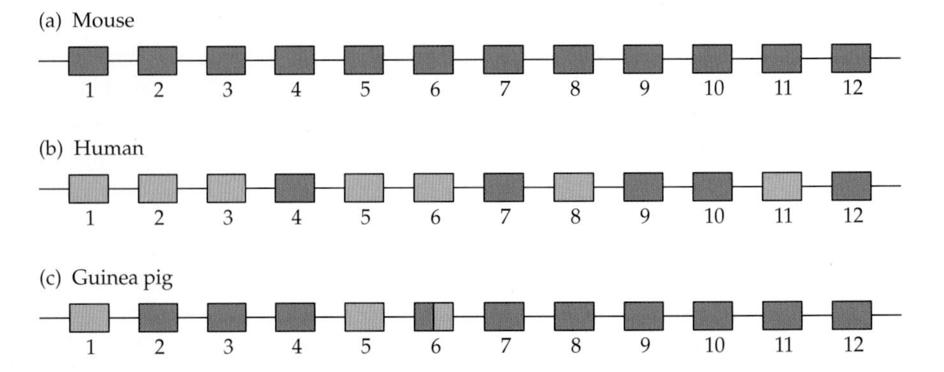

Figure 8.37 Schematic representation of the L-gulono-γ-lactone oxidase gene structure in a murid (mouse) and the structures of the L-gulono-γ-lactone oxidase pseudogenes in an anthropoid primate (human) and a hystricomorph (guinea pig). Functional exons (blue) and deleted exons (yellow) are numbered as in the mouse functional gene. A partial deletion of exon 6 in guinea pig is shown.

acid the result of mutations in the L-gulono-γ-lactone oxidase gene? Second, was the loss of the enzyme driven by positive selection or random genetic drift?

The first question is easy to address. L-gulono-γ-lactone oxidase catalyzes the final step of ascorbic acid biosynthesis. Therefore, losing this enzyme will only impact the ability to make ascorbic acid and nothing else. In contrast, losing genes for other enzymes in the synthetic pathway that starts with D-glucose and ends with ascorbic acid would affect the production of many other molecules. For example, losing the gene coding for gluconolactonase would also affect caprolactam degradation and the pentose phosphate pathway.

Explaining the frequent loss of L-gulono-γ-lactone oxidase genes by saying that this enzyme only affects the production of a single compound implies that losing this capacity is not selected against, i.e., that such a loss does not constitute a selective disadvantage. The observation that all species that have lost the capacity to synthesize ascorbic acid have a vitamin C-rich diet seems to support this claim. This explanation is consistent with the fact that all wild anthropoid primates consume much more vitamin C than their presumed daily allowance. The same is true for bats, such as *Artibeus jamaicensis*, that do not produce ascorbic acid. The converse is also true: no species lacking L-gulono-γ-lactone oxidase activity has ever been found to have a diet poor in vitamin C, such as a diet composed uniquely of seeds.

An alternative hypothesis is that losing the capacity to synthesize ascorbic acid is evolutionarily advantageous because ascorbic acid leads to the formation of hydrogen peroxide and the depletion of glutathione. However, if this hypothesis were true, then as far as the L-gulono-γ-lactone oxidase gene is concerned, we should never observe gene resurrections. Since L-gulono-γ-lactone oxidase gene resurrections have been documented in bats and passerine birds (Drouin et al. 2011), this hypothesis is not tenable.

THE HUMAN-SPECIFIC PSEUDOGENIZATION OF THE GENE ENCODING CYTIDINE MONOPHOSPHATE N-ACETYLNEURAMINIC ACID HYDROXYLASE All nucleated cells in nature are covered with a dense and complex coating of sugar chains (glycans) that have numerous biological roles. In vertebrates, as well as in some invertebrate phyla, the outer ends of glycan chains are often capped by sialic acids. Sialic acids are essential for embryonic development and many critical endogenous functions. For instance, they are used by complement factor H and by sialic acid-binding immunoglobulin-like lectins (siglecs) as signals for "self" recognition by the innate immune system. By virtue of their abundance on each cell (dozens to hundreds of millions of copies), numerous pathogens have evolved the ability to use sialic acids as receptors that can facilitate their penetrating the cell.

Some microbes that interact with vertebrates have convergently "reinvented" sialic acids, i.e., they have positioned sialic acids on their outer membrane or surface, and this "molecular mimicry" allows them not only to camouflage themselves from the adaptive immune systems, but also to engage siglecs in the dampening of the innate immune response. For all these reasons, sialic acids seem perfectly positioned at the nexus of the evolutionary arms race between vertebrate hosts and their pathogens.

Two major sialic acids are found on most cell types in mammals: N-acetylneuraminic acid (Neu5Ac) and N-glycolylneuraminic acid (Neu5Gc). These molecules differ by a single oxygen atom (**Figure 8.38**), which is added to Neu5Ac in the cytosol in a reaction catalyzed by the enzyme cytidine monophosphate N-acetylneuraminic acid hydroxylase (CMAH). Most mammalian tissues contain both these sialic acids. Humans are the only known exception; Neu5Gc is missing from normal human tissues (Varki 2010). This human-specific difference was caused by an *Alu*-mediated deletion of a 92 bp exon in the *CMAH* gene. The timing of the mutation is estimated to have been ~2–3 million years ago, which coincides with the emergence of *Homo erectus*. The fact that the inactivated allele quickly became fixed in all human populations indicates that the driving force was positive selection of considerable strength (Hayakawa et al. 2001).

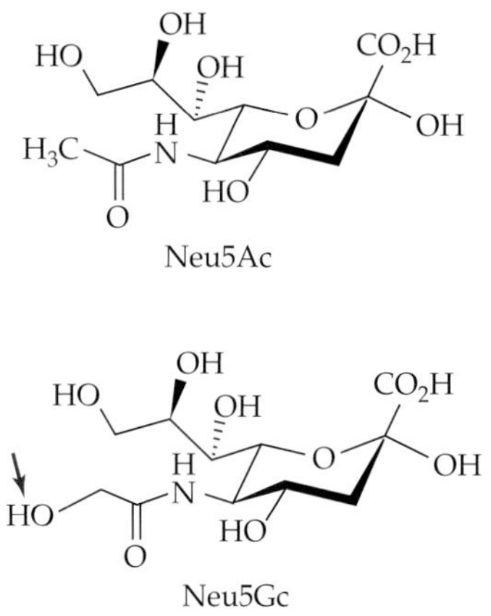

Figure 8.38 The two major sialic acids found in mammals, N-acetylneuraminic acid (Neu5Ac; top) and N-glycolylneuraminic acid (Neu5Gc; bottom). These molecules differ from each other by a single oxygen atom (red arrow), which is added to Neu5Ac by the enzyme cytidine monophosphate N-acetylneuraminic acid hydroxylase. This enzyme is not functional in humans, which makes humans unique among mammals in their inability to produce Neu5Gc.

Because Neu5Gc is produced from Neu5Ac by an enzymatic reaction, the pseudogenization of the enzyme resulted in the loss of Neu5Gc and an excess of Neu5Ac. This change affected the relative efficacy of various pathogens to interact with humans. Humans became resistant to pathogens binding Neu5Gc and more susceptible to pathogens preferring to bind Neu5Ac. Particularly interesting is a difference in the binding preference between malarial parasites of humans and other apes. For example, while chimpanzees succumb to the malarial parasite *Plasmodium reichenowi*, humans cannot be infected with it because this species has an almost exclusive affinity for Neu5Gc. Indeed, it has been suggested that because the prevailing malarial parasite used to be *P. reichenowi*, the loss of Neu5Gc might have been extremely beneficial, explaining the rapid spread of this "null" allele in the population. According to this hypothesis, *Plasmodium falciparum*, the current human malarial parasite, arose only later, when a strain of *Plasmodium* evolved the ability to bind preferentially to Neu5Ac-rich erythrocytes of humans.

Of course, the inactivation of a gene, any gene, may have pleiotropic effects. To test these effects, Hedlund et al. (2007) genetically engineered *CMAH*-null mice and studied the in vivo consequences. First, no evidence for an alternate pathway to produce Neu5Ac was found. As with humans, there was an accumulation of the precursor Neu5Ac. Interestingly, *CMAH*-null mice exhibited a few abnormalities reminiscent of the human condition. Adult mice exhibited age-related abnormalities of the inner ear, resulting in a diminished acoustic startle response. Adult animals also showed slower wound healing, as do humans compared with nonhuman primates. We note that these "deleterious" effects may not have had any evolutionary impact, as they occurred in postreproductive individuals.

Finally, we note that although humans do not produce Neu5Gc, this sialic acid may be acquired through consumption of food containing animal products, such as meat and milk. Since Neu5Gc is no longer recognized as self, it may elicit immune reactions that might contribute to human-specific diseases, such as rheumatoid arthritis, bronchial asthma, and type 2 diabetes (e.g., Kavaler et al. 2011). Again, such diseases may not have any evolutionary impact, as they tend to occur late in life. A particularly interesting twist to the story of sialic acids was discovered by Byres et al. (2008). It seems that a toxin from a particularly aggressive strain of *Escherichia coli* that causes serious gastrointestinal disease in humans has a strong preference for Neu5Gc. Assimilation of dietary Neu5Gc creates high-affinity receptors on human gut epithelia and kidney vasculature. Thus, a receptor of a bacterial toxin can be generated by metabolic incorporation of an exogenous factor derived from food, allowing the toxin to ravage human cells. Ironically, foods rich in Neu5Gc are also the most common source of toxin-secreting bacterial contamination.

CASPASE-12: FIGHTING SEPSIS BY GENE INACTIVATION Caspase-12 is a cysteine protease involved in the processing and activating of inflammatory cytokines, such as interleukins 1 and 18. In humans, the gene encoding caspase-12 consists of eight exons and is located on the long arm of chromosome 11. Due to a C → T substitution that transformed a glutamine codon (CAA) into a termination codon (TAA), human populations are polymorphic for the presence or absence of a stop codon in exon 4. The C allele produces a full-length active caspase protein; the T allele produces a truncated inactive form. Unsurprisingly, the full-length caspase-12 was shown to be ancestral, while the inactive enzyme is derived. The highest frequencies of the active allele were found in South Africa (~60%). Outside of Africa, the active allele was detected at very low frequencies (on average less than 1%), with most human populations in Europe, Asia, North and South America, and Oceania being fixed for the inactive allele. Xue et al. (2006) and Wang et al. (2006) determined that the C → T mutation most probably arose in Africa less than 150,000 years ago, long before the out-of-Africa migrations of modern humans. Initially, the inactive form may have been a neutral or almost neutral allele. Subsequently, strong positive selection beginning 50,000–60,000 years ago drove the inactive allele to near fixation.

The strong selection pressure seems to have been sepsis, i.e., the severe pathology in which the bloodstream is overwhelmed by bacteria as a consequence of an otherwise local infection. Sepsis is one of the most common causes of death in infants and children around the world; indeed, deaths ascribed to the four major infectious killers—pneumonia, diarrhea, malaria, and measles—are often the result of fatal sepsis. In modern hospitals, individuals with two copies of the inactive caspase-12 gene are both more likely to escape severe sepsis and more likely to survive if they do develop it, whereas heterozygotes show an intermediate level of protection (Saleh et al. 2004; Scott and Saleh 2007). The degree of protection against sepsis due to caspase-12 deficiency has been experimentally demonstrated in genetically engineered mice (Saleh et al. 2006). Therefore, avoidance and survival of sepsis may have been the selective force that led to the spread of the inactive form of the caspase-12 gene.

Many infectious diseases that are known to cause sepsis require large host population sizes to maintain themselves and, thus, would have been rare or absent in archaic humans, when population sizes were small. Consequently, in small populations, there would have been no advantage associated with the inactive gene, and the evolutionary conservation of the gene in most mammals suggests that there may even have been a disadvantage. Thus, selection for the inactive gene probably started in earnest only when the human population size became large. Indeed, as mentioned above, selection is thought to have strengthened about 50,000–60,000 years ago, as human populations increased in size and density.

Analyses of DNA sequence variation among African and non-African populations strongly suggest that the spread of the T allele has been driven by strong positive selection and that the selective advantage was directly conferred by the C → T nonsense mutation (Wang et al. 2006).

THE TEMPTATION OF CATS AND DOGS: THE INDEPENDENT PSEUDOGENIZATION OF TASTE RECEPTOR GENES In a quarterly magazine dedicated to the lifestyles of "modern dogs," one can find many recipes for sweet treats, indicating that our canine friends have a "sweet tooth." In comparison, try to tempt a cat with a sweet treat and you will be met with contemptuous indifference. Why the difference?

Most mammals perceive five basic taste qualities: sweet, umami, bitter, salty, and sour. The receptors for sweet and umami tastes belong to a class of proteins called type 1 taste receptors (Tas1r). The sweet taste is mediated by two type 1 taste receptors: Tas1r2 and Tas1r3. Umami (the characteristic taste imparted by monosodium glutamate and 5′ nucleotides such as inosinate and guanylate) is similarly mediated by two receptors: Tas1r1 and Tas1r3. A survey of 18 species belonging to Carnivora revealed that those species that are exclusive meat eaters have lost the ability to produce Tas1r2 (**Figure 8.39**). The pseudogenization of the *Tas1r2* gene, which encodes the Tas1r2 receptor, has occurred independently at least six times in this eutherian order (Li et al. 2005; Jiang et al. 2012). In some cases, it is possible to pinpoint the mutation that incapacitated the gene; in others, many reading-frame-disturbing mutations accumulated and it is not possible to state which was first. In the lineage leading to sea lion and fur seal, the initiation codon ATG mutated to ATA, thus preventing the gene from being translated. This change has been found in both species, indicating that the change occurred in their common ancestor. After the divergence between sea lion and fur seal, several other reading-frame-disturbing mutations accumulated independently in the two lineages. In the Pacific harbor seal, a nonsense mutation and a 2 bp deletion, both in exon 6, likely rendered the gene nonfunctional. An insertion of a T in exon 3 was most probably responsible for the inactivation of the gene in the Asian small-clawed otter, and a 1 bp deletion in exon 2 rendered the gene inactive in the spotted hyena. In the fossa, a nonsense mutation in exon 3 and a single nucleotide insertion in exon 4 were found. Because the insertion is fixed in the population, while the nonsense mutation exists in a polymorphic state, it is likely that the insertion was the incapacitating mutation. *Tas1r2* from banded linsang carries eight reading-frame-disturbing mutations: an insertion of one nucleotide and a deletion of ten nucleotides

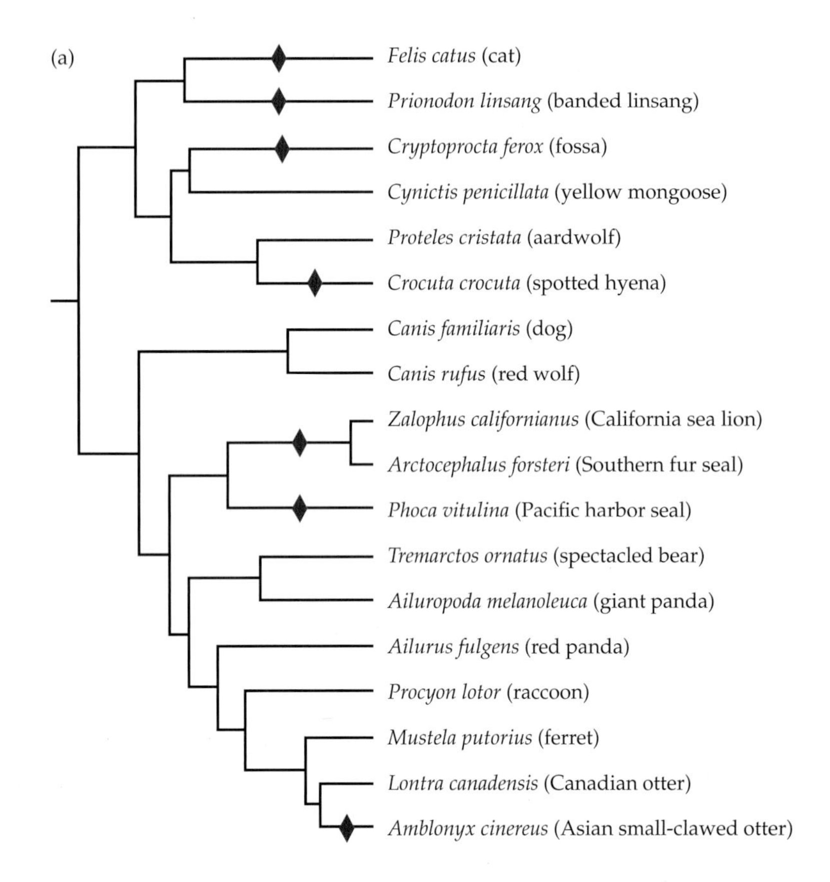

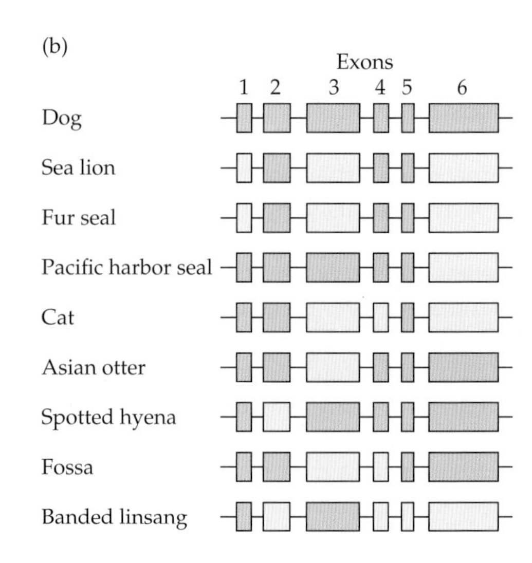

Figure 8.39 Independent losses of the sweet-receptor gene in different lineages of the order Carnivora. (a) A phylogenetic tree of the *Tas1r2* gene from 18 species within Carnivora. Pseudogenization events are marked with a red diamond. (b) Schematic diagram showing the exons affected by ORF-disrupting mutations (light tan rectangles). The intact exon structure of the dog *Tas1r2* gene is shown as a reference. (Modified from Jiang et al. 2012.)

in exon 2, a 14 bp insertion in exon 4, a 20 bp deletion and a 2 bp deletion in exon 5, and two 1 bp deletions and a 28 bp insertion in exon 6. In cat, a 247 bp deletion was observed in exon 3, as well as five nonsense mutations in exons 4, 5, and 6.

Are these independent gene losses an indication that the losses were driven by positive selection, or are they evidence that loss by disuse (i.e., diminished selective constraint) is prevalent? Unfortunately, the losses occurred too long ago for us to be able to identify selective sweeps. However, no one has come up with a convincing reason for a selective advantage associated with the inability to taste sweet flavors, so the independent losses of functional Tas1r2 genes may be due to the accumulation of gene-disabling mutations that nonetheless were neutral with respect to fitness. Interestingly, in the sea lion, not only Tas1r2 but also Tas1r1 and Tas1r3 became pseudogenized, consistent with their feeding behavior, which entails swallowing food whole without chewing, thereby rendering the gustatory apparatus obsolete.

Functional Convergence

In molecular evolution, the term "convergence" has many and sometimes contradictory meanings (Doolittle 1994). The term always refers to the situation in which a similar result is reached from disparate starting points. However, in one usage "convergence" may refer to the molecular changes causing two sequences to look more similar rather than more dissimilar to one another, while in another usage it may refer to the evolutionary mechanism (usually adaptive evolution) that causes entities to appear more related than they are. Here, we will use the term only in the sense of **functional convergence**, whereby the same molecular functionality arises independently on more than one occasion. Numerous examples of functional convergence have been found in nature—not surprising, since the function of a protein is frequently determined by only a few of its amino acids. For instance, the ability to catalyze the hydrolysis of peptide bonds has evolved independently many times; there are sulfhydryl proteases, metalloproteases, aspartyl proteases, serine proteases, threonine proteases, cysteine proteases, glutamic acid proteases, and many others. Moreover, the serine proteases have evolved at least three different times.

An interesting case of functional convergence is that of the myoglobin in the red-muscle tissue of the buccal mass in the abalone *Sulculus diversicolor* and related prosobranchian mollusks (Suzuki et al. 1996). *Sulculus* myoglobin consists of 377 amino acids, which makes it 2.5 times larger than myoglobins belonging to the globin superfamily. The protein carries a heme group and can bind oxygen reversibly, but its oxygen affinity is somewhat lower than those of other vertebrate and invertebrate myoglobins. Intriguingly, the amino acid sequence exhibits no similarity to any other myoglobin or hemoglobin but was shown to be homologous with the enzyme indoleamine 2,3-dioxygenase, which degrades tryptophan and other indole derivatives into kynurenines. In mammals, indoleamine dioxygenase performs a very important function during pregnancy by preventing the immunological rejection of the fetus by the mother (Munn et al. 1998).

The taxonomic distribution of indoleamine dioxygenase-derived myoglobins in conjunction with a molecular phylogeny inferred from 18S rRNA sequences (Winnepenninckx et al. 1998) indicates that the recruitment of indoleamine dioxygenase as a myoglobin occurred once, about 270 million years ago, in the ancestor of *Sulculus* and its relatives *Nordotis, Battilus, Omphalius,* and *Chlorostoma.*

Origin and Evolution of Spliceosomal Introns

The existence of spliceosomal introns, which cannot splice themselves out of a transcript independently, raises the question of their relative antiquity or modernity, which in turn raises the question of whether it is evolutionarily more plausible for introns to be lost or gained.

Since mechanisms are known for both intron excision and intron insertion, we need to approach these questions evolutionarily rather than mechanistically. Let us first examine the case of vertebrate hemoglobin α and β chains. From Gō plots, we know that these proteins can be divided into four domains. At the gene level, however, the genes consist of only three exons, the second of which encodes two adjacent domains. Gō (1981) postulated that during evolution a merger between two exons had occurred as a result of the loss of a central intron. Indeed, the genes for homologous globins in plants (leghemoglobins) were subsequently found to contain an additional intron at or very near the position predicted by the domain structure of vertebrate globins (Landsman et al. 1986). A similar intron was found in the globin genes of the codworm *Pseudoterranova decipiens* (Dixon et al. 1991). Interestingly, the globin gene of the free-living nematode *Caenorhabditis elegans* was found to contain a single intron corresponding in position to the central intron in leghemoglobins (Kloek et al. 1993). Thus, during the evolution of the globin gene family from a four-exon ancestral gene, several lineages lost some or all of their three introns, thereby generating a panoply of exon-intron permutations (**Figure 8.40**). The evolution of the globin family seems to indicate that gene loss predominates.

Following the discovery of a possible correspondence between protein domains and exons, Gilbert (1978) suggested that this correspondence might be traced back to the original proteins, which had a single domain and were encoded by monoexonic (i.e., one-exon) genes. Subsequently, it has been hypothesized that introns are a primitive feature of genes—a view that became known as the introns-early hypothesis or the exon theory of genes (Darnell 1978; Doolittle 1978; Gilbert 1987). According to this view, ancient genes possessed introns, which were subsequently retained by eukaryotes but lost in prokaryotes (**Figure 8.41a**).

An opposing introns-late hypothesis was proposed by Cavalier-Smith (1985a, 1991), according to which early genes had no introns (**Figure 8.41b**). According to this hypothesis, introns arose in the eukaryotic cell before the endosymbiotic process that gave rise to the mitochondria.

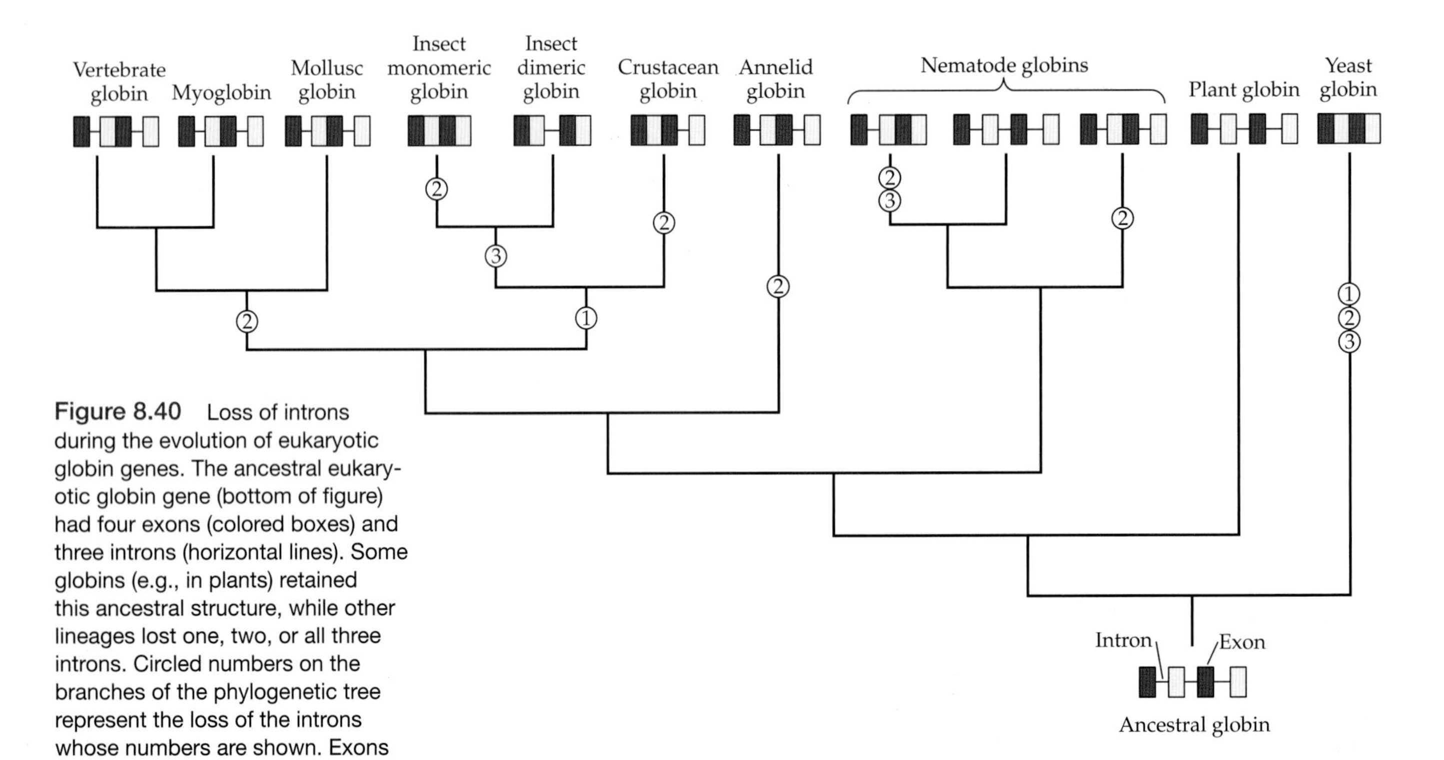

Figure 8.40 Loss of introns during the evolution of eukaryotic globin genes. The ancestral eukaryotic globin gene (bottom of figure) had four exons (colored boxes) and three introns (horizontal lines). Some globins (e.g., in plants) retained this ancestral structure, while other lineages lost one, two, or all three introns. Circled numbers on the branches of the phylogenetic tree represent the loss of the introns whose numbers are shown. Exons and introns are not drawn to scale. (Modified from Graur and Li 2000.)

As detailed in Chapter 6, the eukaryotic cell almost certainly arose as a reticulate event involving the endosymbiosis of a eubacterium, which became the mitochondrion, into an archaebacterium. Thus, the emergence of the eukaryotic cell and the endosymbiotic process that gave rise to the mitochondria are but two facets of the same event. According to this hypothesis (Martin and Koonin 2006), mobile eubacterial elements, such as group II introns, that were previously held in check in the eubacterial genome invaded the archaebacterial genome, resulting in huge mortality. The transmogrification of group II introns into spliceosomal introns required many evolutionary innovations, not least among them the physical separation between the process of mRNA splicing and the process of translation. This separation was

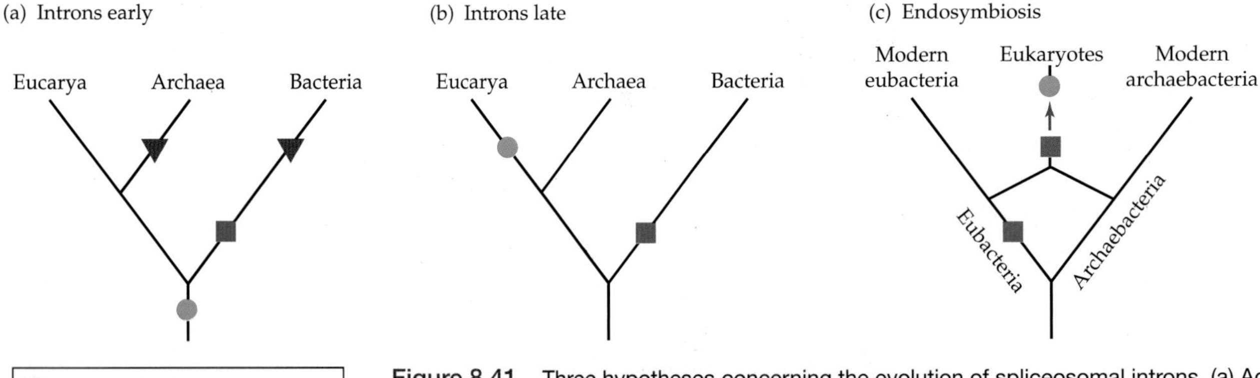

Figure 8.41 Three hypotheses concerning the evolution of spliceosomal introns. (a) According to the introns-early hypothesis, introns originated in the ancestor of all life-forms. They were retained by eukaryotes but lost independently twice in the two prokaryotic lineages. (b) According to the introns-late hypothesis, early genes had no introns. Spliceosomal introns arose in eukaryotes before they acquired mitochondria through endosymbiosis. (c) A more recent hypothesis envisions the emergence of the eukaryotic cell through the endosymbiosis of a eubacterium into an archaebacterium, i.e., the emergence of the eukaryotic cell and the endosymbiotic process that gave rise to the mitochondria are but two facets of the same event. According to this hypothesis, group II introns that originated in eubacteria evolved into spliceosomal introns in the eukaryotic nuclear genome.

achieved through the invention of the nuclear membrane. The evolution of introns according to this scenario is shown in **Figure 8.41c**. The exon theory of genes was subjected to an investigation by Stoltzfus et al. (1994). No significant correspondence between exons and units of protein structure was detected, suggesting that the exon theory of genes is untenable.

The predominant current view in the literature is that spliceosomal introns are a eukaryotic invention and the vast majority of the components of the spliceosome, whether proteins or RNAs, had already existed in the ancestor of extant eukaryotes (Collins and Penny 2005). The dynamics of intron gain and loss in eukaryotes seems to be lineage-specific, with different lineages exhibiting a wide range of gain and loss rates (Carmel et al. 2007). In most eukaryotic lineages, the rate of intron loss greatly exceeds that of intron gain, with vertebrates exhibiting a ratio of 3:1 between loss and gain. The ratio between intron loss and gain seems to be even more pronounced in mammals, with many studies reporting numerous intron losses but no intron gains (e.g., Coulombe-Huntington and Majewski 2007; but see Kordiš 2011). The main exception to the general rule that intron losses outnumber intron gains is the Magnoliophyta (flowering plants) lineage, in which the ratio between loss and gain was found to be 3:4. The Ascomycota (sac fungi) lineage experienced rates of intron gain and loss that were significantly above the respective mean for the other eukaryotes. Evolutionary reconstructions by Koonin (2009) indicated that the last common ancestor of all extant eukaryotes had an unexpectedly high number of introns in protein-coding genes. Combined with the evidence of the origin of spliceosomal introns from invading group II self-splicing introns, Koonin's results suggest that up to 80% of early ancestral eukaryotic genomes may have consisted of sequences derived from group II introns.

A population study by Li et al. (2009) revealed parallel intron gains often occurring at exactly the same location in different alleles. More than half of the gains were found to be associated with short sequence repeats, indicating that intron gain may be mediated by repair of staggered double-strand breaks. These findings also suggest that hotspots for intron gain may exist in the genome, thereby weakening the claim that intron gain is an irreversible synapomorphy.

A Grand View of Molecular Tinkering: Suboptimality and Gratuitous Complexity

Nature has frequently tempted observers to describe its beauty and grandeur in terms of perfection, optimality, efficiency, and good design, overlooking the fact that nature has never invented anything as simple and as obvious as a true wheel or a flame-resistant moth. Despite ample eons to perfect its products, evolution mostly achieves functions that are more or less satisfactory, and by and large organisms make due with either suboptimal structures or gratuitously complex adaptations. Evolution seldom accomplishes anything that is better than "good enough," and many of its results would have benefited from a second opinion. From the inside-out eye, in which the nerves and the blood vessels lie on the surface of the retina instead of behind it (thereby creating a blind spot), to the recurrent laryngeal nerve that travels from the brain to the larynx via a ridiculous detour around the aortic arch (which in the giraffe takes about 6 meters instead of the 30 centimeters the nerve could take on the shortest route), nature is replete with bad, ugly, and nonsentient design. In his famous paper "Evolution and Tinkering," François Jacob (1977) wrote:

> *Natural selection does not work as an engineer works. It works like a tinkerer—a tinkerer who does not know exactly what he is going to produce but uses whatever he finds around him whether it be pieces of string, fragments of wood, or old cardboards. ... The tinkerer ... always manages with odds and ends. What he ultimately produces is generally related to no special project, and it results from a series of contingent events, of all the opportunities he had to enrich his stock with leftovers.*

In the following sections, we will discuss two consequences of tinkering in evolution: suboptimal design and gratuitous complexity.

Tinkering in action: The patchwork approach to the evolution of novel metabolic pathways

The way in which a population copes with a radical change in its environment is fairly simple: it either adapts or becomes extinct. The process of adaptation, however, can use only what is available, and what is available may be quite substandard. There are many reasons why evolution does not usually produce optimal solutions. One cause is the lack of necessary genetic variation; an increase in fitness is only possible if a better allele exists in the population at exactly the right evolutionary time. A second concerns pleiotropic trade-offs, whereby changing one feature for the better might cause another feature to be changed for the worse. Thus, even if the right variability for one trait is available in the population, it may not be the right variability for another trait. And, notwithstanding these two reasons, optimality is difficult to attain mostly because of historical constraints. Evolution can only use the raw materials that have been inherited in a population from previous-generation ancestors. There is no design from scratch in evolution, no rethinking; there is only tinkering with whatever Lego blocks have survived intact through preceding generations. Bats would have much better wings if top-down engineering were allowed in evolution. Alas, evolution relies on bottom-up design in which the wing must be made with minimally modified preexisting components. Here, we will describe the very rapid evolution of a novel metabolic pathway out of old metabolic components, and the inescapable suboptimal performance resulting from such evolutionary tinkering.

A xenobiotic is a substance that is foreign to a biological system, and it usually adversely affects it. Exposure to xenobiotics requires organisms (mostly microorganisms) to adapt very quickly, especially if the xenobiotic compound is lethal. Recent human perturbations of the environment have included the introduction of a great number of xenobiotics and, from an evolutionary point of view, have created new opportunities for microorganisms to evolve metabolic pathways that either exploit new carbon sources or detoxify toxic compounds. Because most extant metabolic pathways evolved many millions of years ago and anthropogenic compounds have been introduced, at most, mere hundreds of years ago, xenobiotics allow us to study the initial stages in the creation of novel metabolic pathways.

Pentachlorophenol (PCP) is an organochlorine compound that has been extensively used as a wood preservative, pesticide, disinfectant, and precursor of herbicides, leading to widespread contamination of groundwater and soil. PCP is expected to be recalcitrant to biodegradation for several reasons. First, it is a xenobiotic compound, and therefore microorganisms are unlikely to have ever encountered it, let alone have the enzymes required for its catabolism. Second, it is highly chlorinated, and the resistance of chlorinated aromatic compounds to biodegradation is known to generally increase with degree of chlorination. Finally, it is highly toxic because it uncouples oxidative phosphorylation and perturbs membrane properties.

In spite of these unpropitious characteristics, it has been discovered that several soil bacteria, such as *Sphingomonas chlorophenolica*, can degrade PCP. Because PCP was introduced into the environment recently (in the 1930s) and heavy use only started in the 1970s, it is likely that the PCP degradation pathway has been assembled in the past few decades. The pathway consists of at least three enzymes recruited from two other catabolic pathways. Two of these enzymes, pentachlorophenol hydroxylase and 2,6-dichlorohydroquinone dioxygenase, may have originated from a pathway involved in the degradation of naturally occurring chlorinated phenols. The third enzyme, a reductive dehalogenase, may have evolved from a maleylacetoacetate isomerase normally involved in the degradation of tyrosine.

This recently assembled pathway seems not to function very well. Copley (2000) lists three aspects of the pathway that are clearly suboptimal. The first deficit is the poor catalytic effectiveness of pentachlorophenol hydroxylase, which appears to limit the ability of the bacterium to funnel carbon through the pathway. The slow rate of catalysis is undoubtedly related to the fact that the enzyme is not specific to the substrate, whereas in nature rate acceleration is usually achieved by the narrowing of substrate specificity. The second problem is that the dehalogenase is inhibited by its substrates, most probably because one of the two glutathione molecules required in the reaction binds to a site that partly overlaps with the binding site for the substrate. Finally, the regulation of the three enzymes is out of sync. Ideally, all of the enzymes in a pathway should be coordinately regulated and induced by the first substrate. However, the expression of dehalogenase is not induced by PCP.

The PCP degradation pathway illustrates a great truth about the evolutionary process in general, and the historical constraints affecting the evolution of novel traits in particular. Products of evolution are far from ideal. No self-respecting watchmaker would dare peddle such cumbersome and ugly ware as the PCP-degrading pathway. However, even at its present limited level of effectiveness, the newly evolved PCP degradation pathway manages to achieve something wonderful: it allows *S. chlorophenolica* to survive and multiply in the presence of PCP. The PCP degradation pathway is, thus, a perfect example of a kludge, whereby an extraordinary environmental challenge is met with a barely working solution cobbled together of poorly matched parts originally intended for other purposes.

Another example of suboptimal performance, that of Rubisco, highlights the fact that adaptations are mostly changes from a generalized to a specialized state, and specializations are particular to a certain time, a certain place, and certain conditions. With the change in conditions, it is not always possible for a specialized macromolecule to acquire a new specialization. Rubisco (ribulose 1,5-bisphosphate carboxylase/oxygenase) is in all likelihood the most important enzyme in the world, as virtually all organic carbon derives ultimately from the fixation of CO_2 through a carboxylation reaction catalyzed by this enzyme. Rubisco evolved when the atmospheric composition was very different from what it is today, and it has not adapted to the modern atmosphere. First, the enzyme reacts with either CO_2 or O_2. The reaction with oxygen competes with the reaction in which CO_2 is fixed, and it feeds a pathway called photorespiration. Photorespiration causes the loss of up to 25% of the carbon that is fixed by the carboxylation reaction. Second, Rubisco is one of the most inefficient enzymes on the planet, with a catalytic rate of only 3–10 molecules of CO_2 fixed per second per molecule of enzyme, which incidentally explains why Rubisco is the most abundant protein in the world. The enzyme performs very badly at current levels of atmospheric CO_2, which is why some agricultural practices entail elevating CO_2 concentrations inside greenhouses. Of course, none of these deficiencies mattered when Rubisco first evolved, as there was no oxygen in the atmosphere and the level of CO_2 was much higher than it is today. Rubisco is, hence, a superb example of unintelligent design (Ellis 2010), a relic of *temps perdu*, incapable of improvement without the aid of intelligent scientific intervention.

Another example of "history as destiny" and the potentially dire consequences of suboptimal performance is the case of the cancer suppressor p53. Cancers are rare because their development is actively restrained by a range of tumor suppressors, of which p53 is an extremely important example. As evidenced from its wide phylogenetic distribution, p53 is an evolutionarily ancient protein, a transcription factor whose original function was to serve as a coordinator of metazoan stress responses. Many types of stress and damage activate p53. It also performs many physiological functions in its "inactivated" state. Its role in tumor suppression is a relatively recent adaptation and has only become beneficial since large, long-lived organisms acquired sufficient size and somatic regenerative capacity to necessitate specific mechanisms to rein in rogue proliferating cells. However, the evolutionary "appropriation" of this protein from one function to another entails compromises that restrict its efficacy as a tumor suppressor. In fact, p53 has a fundamental flaw as far as cancer suppression is concerned: it does not sense the oncogenetic signal itself. In vertebrates, p53 has acquired a function additional to the many it already had and has become a "jack of all trades, master of none" (Junttila and Evan 2009).

Irremediable complexity by constructive neutral evolution

Many macromolecular machines appear gratuitously complex, comprising more components than their basic functions seem to demand. The most gratuitously complex macromolecular machine in nature may be the eukaryotic spliceosome, which removes introns from pre-mRNA in a process called splicing. The spliceosome uses hundreds of proteins and five small nuclear RNAs to do a job that self-splicing introns can do alone. The mitochondrial RNA editing system of trypanosomes, in which hundreds of guide RNAs and several large protein complexes insert and delete uridine residues to restore mRNAs to their "ancestral" state, is similarly baroque.

When faced with molecular complexity, we can find several types of explanations in the literature. The selectionist account invariably invokes selection for improved function. Thus, a molecular machinery comprising many proteins may be said to be the result of gradual evolutionary gene accretion, whereby proteins were progressively added to the machinery for faster, more stable, and more efficient performance of whatever the task of the machinery was in the first place. As an alternative to such selectionist thinking, Stoltzfus (1999) invoked fixation of neutral or slightly deleterious features as a general and unavoidable source of complexity in taxa with small effective population sizes. Such nonselective processes could account, for instance, for the origins and spread of transposable elements and other contributors to the high DNA content of many eukaryotes. Neutrally fixed complexity, however, could be neutrally "unfixed," say, through an increase in population size and its exposure to purifying selection. We know, however, that most molecular complexity cannot be simplified, which is why it is referred to as **irremediable complexity**.

Constructive neutral evolution is a ratchet-like (unidirectional) process that can increase complexity regardless of whether it is necessary or not (Covello and Gray 1993; Stoltzfus 1999; Lynch 2007; Lukeš et al. 2011). In this process, interactions between the components arise before the interactions are needed, or indeed used. The interactions become obligatory when the function of the added component is no longer performed by the first component. In **Figure 8.42**, we illustrate a stepwise process of increasing biochemical complexity by constructive neutral evolution. Let us start with a function performed by a single gene, denoted by *A*. Gene *A* carries on its function in a self-regulated manner, but the self-regulation is vulnerable to mutation. In the next step, *A* becomes regulated by *B*, either because gene *A* has acquired a binding site for *B*, or because *B* acquires a mutation that enables it to serve as an activator of *A*. At this point, gene *A* has two redundant activation pathways and is therefore insensitive to loss of one of them. Thus, the functional relationship between *A* and *B* can be regarded as **presuppressor mutation**, i.e., *B* can suppress the negative effects of future mutations in the regulation of *A*. In principle, this process could be iterated ad nauseam as *B* connected to a further upstream gene *C*, and *C* connected to *D*, and so forth. The result would be the molecular equivalent of one of Rube Goldberg's famously overengineered machines, in which the simple task of window cleaning necessitates a pedestrian, a banana peel, a rake, a horseshoe, a rope, a sprinkling can, a mop, a dog, a folding sign, and an ashtray (Wolf 2000).

Let us examine a simple case of irremediable complexity in the mitochondrial genome of the ascomycete fungus *Neurospora crassa* (Akins and Lambowitz 1987). The *Neurospora* mitochondrial genome contains several group I introns; however, while most of them self-splice, others require the enzyme tyrosyl-tRNA synthetase to help with the process. As a consequence, the *N. crassa* mitochondrial tyrosyl-tRNA synthetase performs two functions: an old one, in which it aminoacylates mitochondrial $tRNA^{Tyr}$, and a new one, in which it acts as a cofactor in the splicing of several group I introns. Given the narrow phylogenetic distribution of this trait, the bifunctionality of tyrosyl-tRNA synthetase is almost certainly a derived trait.

The driving force for the emergence of the new interaction between the synthetase and the introns is typically stated to be "the need to compensate for structural defects acquired by group I introns" (Paukstelis and Lambowitz 2008). However, this order of events puts the cart before the horse. Specifically, introns bearing defects would be at an immediate selective disadvantage and would be likely to be eliminated from the population long before the gene encoding tyrosyl-tRNA synthetase acquired, through mutations, the ability to suppress such deleterious mutations. If the order of events were reversed, then there would be no deleterious intermediate. That is, if the binding interaction arose first—either fortuitously or for some reason unrelated to splicing—and its existence could compensate for splicing-inactivating mutations, the loss of independence by the intron would be selectively neutral (or at most slightly disadvantageous).

Immediately after the intron loses its splicing independence, there may exist a time interval when the intron might, at least in principle, revert to independence, but the intron is much more likely to accumulate more and more splicing-disabling mutations, and so the intron will ratchet toward greater and greater dependency on the synthetase over time.

Thus, constructive neutral evolution is a directional force that drives increasing complexity. There is no need for positive selection. So far, the constructive neutral evolution model has been most successful in explaining molecular systems of peripheral importance or limited distribution (such as splicing or RNA editing). There is no reason why constructive neutral evolution might not contribute to the generation of such complex cellular systems as the ribosome, the mitochondrial respiratory complexes, protein import, and the cytoskeleton. Much of the bewildering intricacy of cells could consist of originally fortuitous molecular interactions that have become more or less fixed by constructive neutral evolution. Indeed, although complexity in biology is generally regarded as evidence of "fine-tuning" and "sophistication," large biological conglomerates might be better interpreted as the consequences of a "runaway bureaucracy" (Gray et al. 2010).

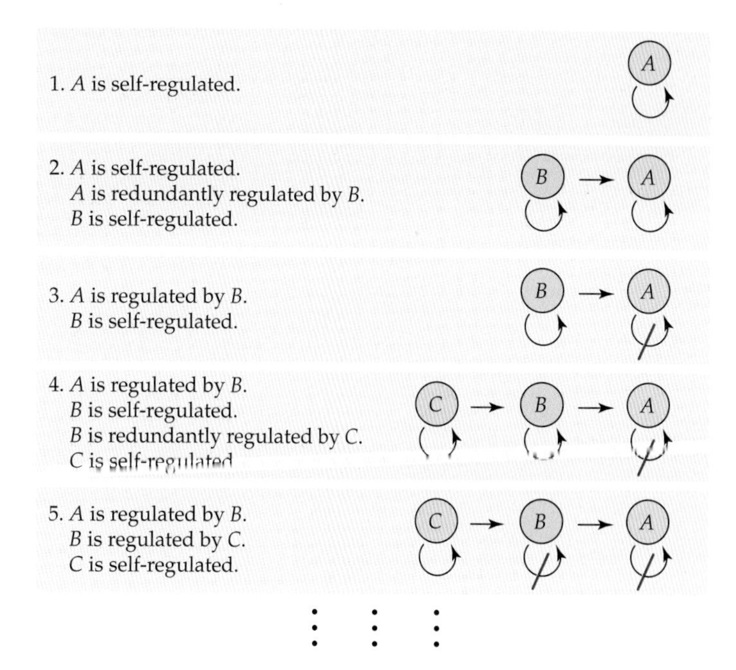

Figure 8.42 Schematic depiction of the stepwise process of increasing biochemical complexity by constructive neutral evolution. Gene *A* carries on its function in a self-regulated manner (semicircular arrow). Its regulation is vulnerable to mutation. In the next step, *A* becomes regulated by the product of gene *B*, either because gene *A* has acquired a binding site for protein B or because gene *B* acquires a mutation that enables it to serve as an activator of *A*. At this point, gene *A* has two redundant activation pathways and is therefore insensitive to the loss of either. Should a mutation cause a redundantly regulated allele *A* to lose its ability to self-regulate (crossed semicircular arrow), *B* has already been established as an activator. In principle, this process could be iterated as *B* acquires sensitivity to a further upstream gene *C* and, subsequently, loses its ability to constitutively express. Note that the process of increased complexity is irreversible because loss-of-function mutations are abundant, while reversals to previous activity are almost unheard of. (Modified from Lynch 2007.)

What do the scientific term "irremediable complexity" and the creationist term "irreducible complexity" have in common? Both terms refer to the situation in which a biological system is composed of several parts that contribute to its function, and in which the removal of one or more of the parts effectively causes the system to cease functioning. The difference is one of functional antecedent. That is, while creationist theologians assume that each component of an irreducibly complex system is devoid of function and has been fashioned for the sole purpose of being part of a system, the components of an irremediably complex system are hypothesized to have discoverable functional antecedents, albeit in contexts other than the system in question. Apologists for religious creationism frequently use the bacterial flagellum as an example of irreducible complexity. However, as shown by Pallen and Matzke (2006), with two exceptions, all the ~50 proteins of the bacterial flagellum are either nonessential for flagellar function or have nonflagellar homologs. The two exceptions—the genes encoding axial proteins FliE and FlgD—are probably derived from genes encoding other axial proteins, but homologous relationships cannot be confirmed with currently available data.

CHAPTER

9

Mobile Elements in Evolution

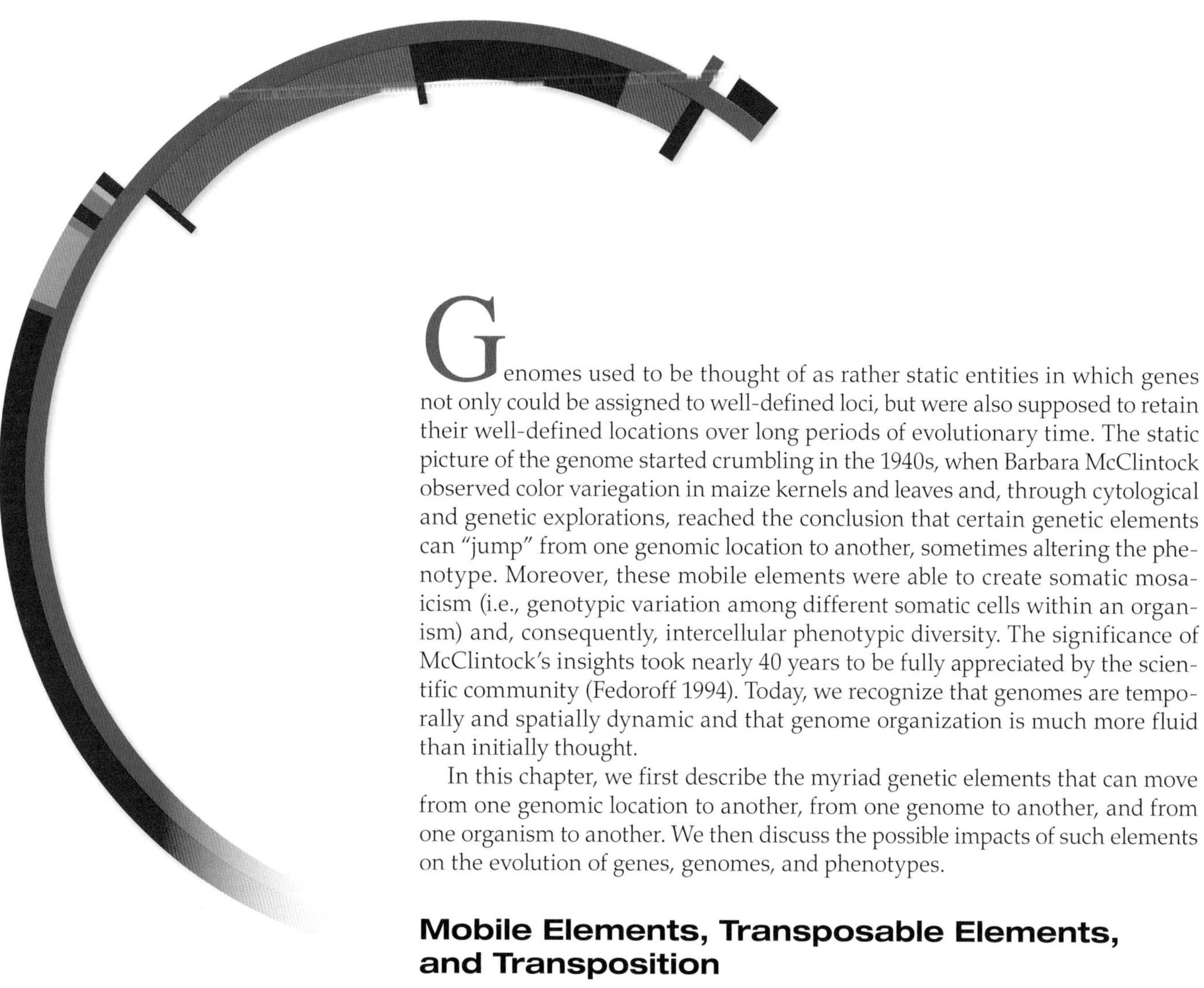

Genomes used to be thought of as rather static entities in which genes not only could be assigned to well-defined loci, but were also supposed to retain their well-defined locations over long periods of evolutionary time. The static picture of the genome started crumbling in the 1940s, when Barbara McClintock observed color variegation in maize kernels and leaves and, through cytological and genetic explorations, reached the conclusion that certain genetic elements can "jump" from one genomic location to another, sometimes altering the phenotype. Moreover, these mobile elements were able to create somatic mosaicism (i.e., genotypic variation among different somatic cells within an organism) and, consequently, intercellular phenotypic diversity. The significance of McClintock's insights took nearly 40 years to be fully appreciated by the scientific community (Fedoroff 1994). Today, we recognize that genomes are temporally and spatially dynamic and that genome organization is much more fluid than initially thought.

In this chapter, we first describe the myriad genetic elements that can move from one genomic location to another, from one genome to another, and from one organism to another. We then discuss the possible impacts of such elements on the evolution of genes, genomes, and phenotypes.

Mobile Elements, Transposable Elements, and Transposition

In this book, we distinguish between two terms that are sometimes used interchangeably in the literature. By **mobile element** (or **mobile DNA**), we refer to any sequence that is able to integrate in the genome. The term **transposable element** is reserved for the subset of mobile elements that are able to move or propagate intragenomically (within a genome). Thus, all transposable elements are mobile elements, but not all mobile elements are transposable elements. We note that some transposable elements can move both intragenomically and intergenomically (between genomes).

Transposition is defined as the movement of genetic material from one genomic location, the **donor site**, to another, the **target site**, within the same genome (**Figure 9.1**). The process of transposition is, ipso facto, a recombination event in that it involves the breaking and reforming of phosphodiester bonds. Transposition, however, requires no extended sequence similarity between the transposable element and its sites of insertion. In many cases, when a transposable element is inserted into the target site, a small segment of the host DNA is duplicated at the insertion site, a phenomenon referred to as **target-site duplication**. This occurs because the double-stranded target site is cleaved in a staggered manner prior to the insertion of the transposable element. The single-stranded flanks are then repaired, and two repeats in the same orientation (**direct repeats**) are created on both sides of the integrated transposable element. These repeats are frequently regarded as the hallmarks of transposition. Nonetheless, a few DNA-mediated transposons (e.g., *helitron* and *crypton*) and a few retrotransposons (e.g., *DIRS*, *ngaro*, and *viper*) do not generate target-site duplications.

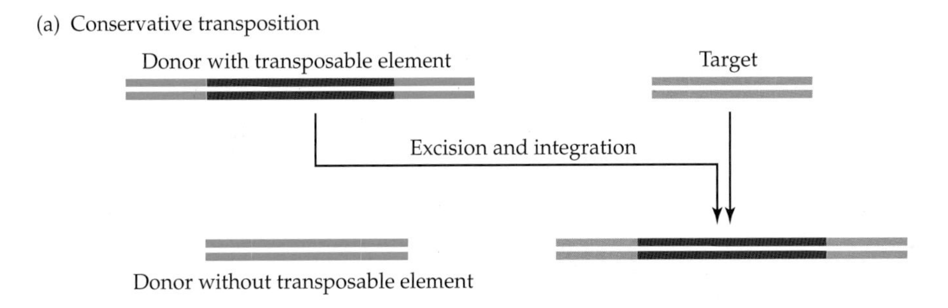

(b) Replicative transposition
Donor with transposable element
Target
Replication
Integration
DNA
Donor with transposable element

(c) Retrotransposition

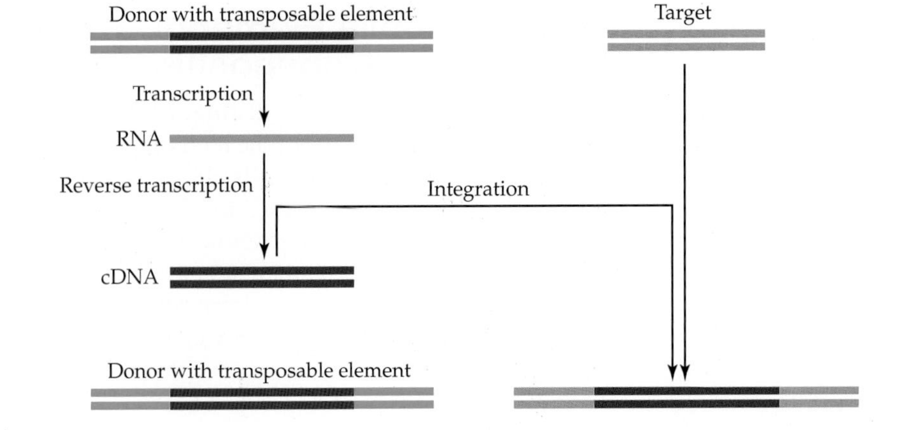

Figure 9.1 (a) Conservative DNA-mediated transposition. Following transposition to the target site, the transposable element is lost at the donor site. (b) Replicative DNA-mediated transposition. The element is replicated, and one copy is either inserted or copied into the target site while the other copy remains at the donor site. (c) Retrotransposition. The element is transcribed into RNA, which is then reverse-transcribed into cDNA. The cDNA copy is either inserted or copied into the target site. The black box represents the target site being duplicated following transposition. We note, however, that not all transposition events result in target-site duplication.

Donor site
Transposable element
Target site
cDNA
RNA

Classification of Transposable Elements

The literature dealing with transposable elements teems with classification schemes that are not always compatible with one another or applicable to all types of transposable elements. In particular, different classification schemes are used for prokaryotic and eukaryotic transposable elements. In the following sections, we will introduce some of the more general, intuitive, and useful classification schemes.

Conservative and replicative transposition

Transposition may be conservative or replicative (Figure 9.1a,b). In **conservative transposition**, the transposable element is excised from the donor site and either inserts itself into the target site (the **cut-out, paste-in model**) or is copied into the target site (the **cut-out, copy-in model**). The excised element may be linear or circular, single- or double-stranded (**Figure 9.2**). Conservative transposition in itself does not increase the number of copies of the transposable element; however, copy number increase may occur through other means, such as DNA duplication.

In **replicative** (or **duplicative**) **transposition**, the transposable element is duplicated, so the transposing entity is a copy (or a copy once removed) of the original element. This process results in two copies of the transposable element, one at the donor site and one at the target site. In replicative transposition, the transposable element at the donor site is copied, and the copy either inserts itself at the target site (the **copy-out, paste-in model**) or is copied into the target site (the **copy-out, copy-in model**). The copied element may be DNA or RNA. RNA elements are always linear and single-stranded. DNA elements can be linear or circular, single- or double-stranded. In addition to the two-step modes of transposition, replicative transposition may also occur through the direct copying of the transposable element from the donor site to the target site (the **copy-in model**). This type of transposition has not been studied as extensively as the other models. Retroposition (see below) is a type of replicative transposition.

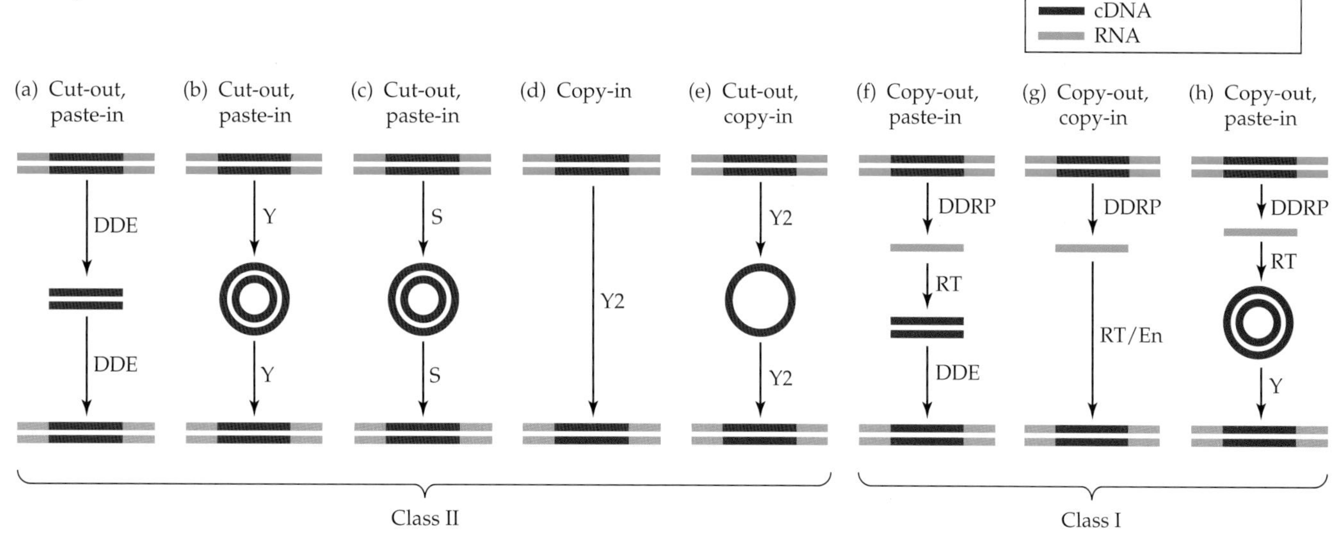

Figure 9.2 Classification of transposable elements by mode of transposition and the enzymes catalyzing different steps of the transposition process. (a) Cut-out, paste-in, class II transposon using a DDE transposase. (b) Cut-out, paste-in, class II transposon using a Y transposase. (c) Cut-out, paste-in, class II transposon using an S transposase. Note the double-stranded circular DNA intermediates in (b) and (c). (d) Copy-in, class II transposon using a Y2 transposase. (e) Cut-out, copy-in, class II transposon using a Y2 transposase. Note the single-stranded circular DNA intermediate. (f) Copy-out, paste-in, class I transposon using a DDE transposase. (g) Copy-out, copy-in, class I transposon using a DDE transposase. Note the double-stranded cDNA intermediate. (g) Copy-out, copy-in, class I transposon that does not have a DNA intermediate. (h) Copy-out, paste-in, class I transposon using a Y transposase. Note the double-stranded circular cDNA intermediate. *Abbreviations:* DDRP, DNA-dependent RNA polymerase; RT, reverse transcriptase; RT/En, reverse transcriptase/endonuclease; DDE, transposases defined by a triad of highly conserved amino acids at the catalytic site, comprising two aspartic acids (D) and a glutamic acid (E); Y, tyrosine transposase; Y2, rolling-circle transposase; S, serine transposase. (Modified from Curcio and Derbyshire 2003.)

Most transposable elements use only one type of transposition; others may utilize multiple modes of transposition. In some cases, the switch between modes of transposition requires few changes. For example, in the bacterial transposon *Tn7*, the switch is accomplished by a single nonsynonymous substitution (e.g., May and Craig 1996).

DNA- and RNA-mediated transposition

Finnegan (1989) distinguished between classes of transposition according to whether the transposition intermediate is RNA or DNA. To distinguish between the DNA- and the RNA-mediated modes of transposition, the latter has been termed **retrotransposition** (or **retroposition**; Figure 9.1c). Transposition and retroposition are found in both eukaryotic and prokaryotic organisms. Transposable elements that use the RNA-mediated mode of transposition are referred to as **class I transposable elements** or **retrotransposable elements**. Transposable elements that use DNA-mediated transposition are called **class II transposable elements**.

Enzymatic classification of transposable elements

A classification of transposable elements based on the enzymes involved in transposition was suggested by Curcio and Derbyshire (2003). It is based on the fact that the number of protein families that mediate transposition singly or in combination with DNA-mediated RNA polymerase and/or reverse transcriptase seems to be rather small (Figure 9.2). We note that the same enzyme type can be involved in both DNA- and RNA-mediated transposition. Of course, given that this field of scientific inquiry is in its infancy, there may be many more protein families involved in transposition to be discovered.

Autonomous and nonautonomous transposable elements

Transposable elements may or may not encode all the components of the transposition machinery. The former are referred to as **autonomous transposable elements**, the latter as **nonautonomous transposable elements**. Nonautonomous transposable elements can only transpose from one genomic location to another by appropriating the necessary enzymes, either from autonomous transposable elements or from the genome in which they reside. For example, the autonomous *Ac* (*Activator*) elements in maize can transpose regardless of the genetic background. In contrast, the nonautonomous *Ds* (*Dissociation*) elements cannot transpose unless one or more copies of *Ac* are also present in the genome. The evolution of nonautonomous transposable elements and their dependence on autonomous transposable elements will be discussed separately for DNA- and RNA-mediated transposition.

Active and fossil transposable elements

A transposable element is defined as **active** if it can be transposed either autonomously or nonautonomously. Active transposable elements may be rendered immobile by mutation, in which case they are referred to as **fossil transposable elements**. A transposable-element family may contain different combinations of active autonomous, active nonautonomous, fossil autonomous, and fossil nonautonomous transposable elements. For example, the human genome contains approximately 50,000 fossil autonomous and 200,000 fossil nonautonomous DNA-mediated transposable elements. Intriguingly, however, the human genome seems to contain no active DNA-mediated elements, although a gene derived from a transposable element, *THAP9*, has been shown to retain DNA transposase activity and the ability to facilitate transposition in other species (Majumdar et al. 2013).

Taxonomic, developmental, and target-site specificity of transposition

Some transposable elements can transpose themselves in all cells; others are highly specific. For example, *Tc1* elements in the nematode *Caenorhabditis elegans* and *P* elements in *Drosophila melanogaster* are usually mobile only in germ cells. In maize, the transposition of many elements was found to be regulated by the host developmental stage. From an evolutionary point of view, the developmental timing of transposition is particularly important because it affects the probability that a transposable element will be passed onto future generations. One pertinent example involves the *LINE-1* transposable elements in mammals. Branciforte and Martin (1994) found that these elements are mainly active during two stages of meiosis (leptotene and zygotene), at which times DNA-strand breakages occur. This timing increases the probability that the transposable elements will insert themselves into new sites as well as the probability that the genome with the increased number of elements will be passed on to the next generation.

The genomic locations of the target sites for transposition are extremely variable among transposable elements. Most transposable elements show little or no target-site specificity; some show an exclusive preference for a specific genomic location, and a few elements are known to have intermediate specificities. An extreme case of specificity is exemplified by *IS4*, which incorporates itself exactly and always at the same location in the galactosidase operon of *E. coli*, and thus each individual bacterium can contain only one copy of the element. Similarly, transposon *Tn554* in *Staphylococcus aureus* primarily transposes into a single site in one orientation only. Others, such as bacteriophage *Mu*, can transpose themselves at random into almost any genomic location. Many transposable elements show intermediate degrees of genomic preference. For example, 40% of all *Tn10* transposons in *E. coli* are found in the *lacZ* gene, which constitutes a minute fraction of the host genome. Some transposable elements exhibit affinities for a particular type of nucleotide composition. For example, *IS1* favors AT-rich insertion sites (Devos et al. 1979). Many elements exhibit a special affinity for palindromic sequences. *IS630*, for example, prefers 5′—CTAG—3′ sequences (Tenzen and Ohtsubo 1991), while the retrotransposons *Gypsy* and *297* in *D. melanogaster* prefer 5′—TACATATGTA—3′ sequences (Linhero and Bergman 2012). Chromosomal preferences have also been found. A *Drosophila miranda* retrotransposable element called TRIM exhibits a preference for the Y chromosome (Steinemann and Steinemann 1991), while *Tn502*, a transposable element in Gram-negative bacteria, is notable for its site-specific insertion preference into plasmids rather than chromosomes (Petrovski and Stanisich 2010). One of the most peculiar biases in transposition is that of the *DIRS1* transposable element of the slime mold *Dictyostelium discoideum*, which preferentially inserts itself into other *DIRS1* sequences (Cappello et al. 1984), thereby disabling the elements at the target site and creating an onionlike structure with the active element at the "core" surrounded on both sides by progressively older elements.

Finally, some transposable elements are known to be species-specific, while others seem to thrive in many different hosts. In the case of the *mariner* transposable element, it can maintain efficient transposition in many different species, even if the species belong to different taxonomic kingdoms (Gueiros-Filho and Beverley 1997). It moves easily from species to species through horizontal gene transfer.

DNA-Mediated Transposable Elements

DNA-mediated transposable elements are divided into insertion sequences and transposons, according to the numbers and kinds of genes they contain.

Insertion sequences

Insertion sequences are the simplest transposable elements. They carry no genetic information except that which is necessary for transposition. Insertion sequences in prokaryotes range in size from 700–2,700 bp in length; those in eukaryotes tend to be longer. They contain short inverted repeats at both ends (9–40 bp), with the 5′ end and the 3′ end sometimes differing slightly from each other. Over 1,500 different insertion sequences have been described in the literature (Siguier et al. 2006).

Prokaryotic insertion sequences are denoted by the prefix *IS* followed by the type number. Additional letters are sometimes added to indicate the taxonomic source of the insertion element (e.g., *ISRm2* is the second insertion element discovered in the α-proteobacterium *Sinorhizobium meliloti*).

The structure of the insertion sequence *IS1* from the intestinal bacteria *E. coli* and *Shigella dysenteriae* is shown schematically in **Figure 9.3a**. *IS1* is approximately 770 nucleotides in length, including two inverted, nonidentical terminal repeats of 23 bp each. It contains two out-of-phase reading frames, *insA* and *insB*, from which a single protein is produced by translational frameshifting at a run of adenines. The N-terminal part of the protein encoded by *insA* is an inhibitor of transposition; the C-terminal part is a DDE transposase, which catalyzes the insertion of the transposable element into the target site.

Although the term "insertion sequence" is usually mentioned in the context of prokaryotic genomes, many eukaryotic DNA-mediated transposable elements contain a single open reading frame encoding a transposase and are, hence, by definition insertion sequences. As a matter of fact, the *Ac* element in maize, originally described by Barbara McClintock, is a eukaryotic insertion sequence.

Transposons

Transposons are mobile elements, usually about 2,500–7,000 bp long, that exist mostly as families of dispersed repetitive sequences in the genome. Transposons are distinguished from insertion sequences by also carrying so-called **exogenous genes**, i.e., genes that encode functions other than those related to transposition itself. (Note that the nomenclature is muddled in the literature, and the term "transposon" is sometimes used to denote any transposable element, including insertion sequences and retrotransposons.)

In bacteria, transposons are denoted by the prefix *Tn* followed by the type number. Transposons in bacteria often carry genes that confer antibiotic resistance (e.g., *Tn554*), heavy-metal resistance (e.g., *Tn21*), or heat resistance (e.g., *Tn1681*) on their carriers. **Plasmids**, which are autonomously replicating extrachromosomal molecules distinct from the bacterial genome, can carry such transposons from cell to cell, and as a consequence, resistance can quickly spread throughout the population of bacteria exposed to such environmental factors.

Several bacteriophages (viruses of bacteria) are in fact transposons or **transposable bacteriophages**. For example, bacteriophage *Mu* (short for *Mutator*) is a very large transposon (~37,000 bp) that encodes not only the enzymes that regulate its transposition, but also approximately 55 additional proteins (Morgan et al. 2002). Of these, some perform regulatory functions, some are enzymes, and some are structural proteins that construct the packaging of the virion.

Some bacterial transposons are **composite** (or **compound**) **transposons**, so named because two complete, independently transposable insertion sequences in either orientation flank one or more exogenous genes (**Figure 9.3b**). Interestingly, in composite transposons, not only can the entire transposon move as a unit, but also one or both of the flanking sequences can transpose independently. Since the transposition functions are encoded by the insertion sequences, composite transposons do not usually contain additional transposase genes. Composite transposons can be either symmetrical or asymmetrical. **Symmetrical composite transposons** contain the same insertion sequence on both sides. **Asymmetrical composite transposons** contain different insertion sequences. For instance, *Tn1547* from *Enterococcus faecalis*, which

Figure 9.3 Schematic representation of four transposable elements in bacteria. Black triangles denote inverted repeats. (a) Insertion sequence *IS1* from *Escherichia coli* contains two out-of-phase reading frames (*insA* and *insB*) flanked by two imperfect inverted terminal repeats. (b) Symmetrical composite transposon *Tn9* from *E. coli* contains two copies of *IS1* flanking the *cat* gene, which encodes a protein conferring chloramphenicol resistance. (c) Symmetrical simple transposon *Tn3* from *E. coli*, which confers streptomycin resistance, contains three genes, two of which (*tnpR* and *bla*) are transcribed on one strand, and the third (*tnpA*) on the other. *Tn3* is flanked by two perfect inverted repeats, 38 bp long. (d) Asymmetrical simple transposon *Tn554* from *Staphylococcus aureus* lacks terminal repeats and contains seven well characterized protein-coding genes plus an open reading frame (ORF), which is abundantly transcribed and translated, but its function is not known. Three of the genes encode transposases (*tnpA*, *tnpB*, and *tnpC*) and are transcribed as a unit. The *spc* and *ermA* genes confer spectinomycin and erythromycin resistance, respectively. The *spc* gene, which encodes an S-adenosylmethionine-dependent methylase, as well as the *peptide L* and *peptide 1* genes, are transcribed on a different strand from the other genes. Intergenic regions (lines) contain no open reading frames. (Modified from Graur and Li 2000.)

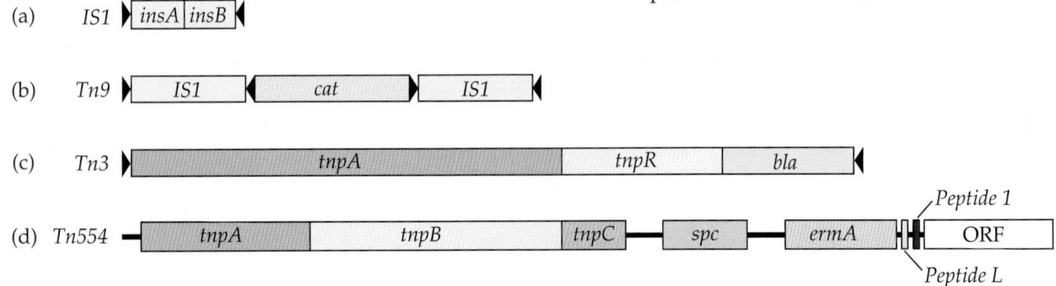

confers vancomycin resistance, is flanked by an *IS16* element on one side and an *IS256* element on the other.

Transposons that are not composite transposons (i.e., do not contain insertion sequences) are sometimes referred to as **simple** or **unit transposons**. Simple transposons also come in two varieties, symmetrical and asymmetrical. Those flanked by short repeated sequences are called **symmetrical** (or **complex**) **transposons** (**Figure 9.3c**); those that are not are called **asymmetrical** (**Figure 9.3d**).

Hypercomposite transposons are made up of two or more transposons. For instance, *Tn5253* in *Streptococcus pneumoniae* conveys resistance to both tetracyclines and chloramphenicol because it is made of two transposons (*Tn5251* and *Tn5252*), each carrying one set of resistance genes.

The coding regions of some transposons in animals (e.g., *P* elements in *Drosophila*) are interrupted by spliceosomal introns. In bacterial transposons (e.g., *Tn5397* in the bacterium *Clostridium difficile*), self-splicing introns have been found (Mullany et al. 1996). Some transposable elements employ alternative splicing. For example, alternative splicing regulates transposition of the *P* elements in *Drosophila*.

An interesting subclass of transposons comprises the *helitrons*, which are distinguished from other DNA transposons by the fact that they encode a rolling-circle replication protein and a helicase. They replicate intragenomically by using a copy-in mechanism, i.e., they do not generate transposition intermediates.

Some transposons not only transpose, but also promote their own transfer from one cell to another. These **conjugative transposons** differ from plasmids in that they do not replicate autonomously, but are replicated passively, as part of the genome in which they are inserted. One of the most extensively studied conjugative transposons is the 18 Kb *Tn916*, which encodes tetracycline resistance. *Tn916* was originally identified on the chromosome of *E. faecalis* strain DS16. Its nucleotide sequence is flanked by imperfect inverted repeats and contains 24 open reading frames (ORFs). The *TN1545* conjugative transposon from *S. pneumoniae* confers resistance to three antibiotic compounds (tetracycline, erythromycin, and kanamycin). Interestingly, in one strain of *S. pneumoniae*, an insertion sequence (*IS1239*) became integrated inside transposon *TN1545*.

Nonautonomous DNA-mediated transposable elements

Most nonautonomous DNA-mediated transposable elements in eukaryotes are derived from autonomous elements through deletion of internal segments. Such elements have lost their capability to encode some or all of the components of the transposition machinery but have retained the ability to be recognized by the transposases. Of particular interest are transposable elements flanked by **t**erminal **i**nverted **r**epeats (TIRs), also referred to as **TIR-transposable elements**. In TIR-transposable elements, the transposition is initiated by the transposase recognizing the TIRs. Thus, in the presence of an active autonomous transposase-producing element, a nonautonomous element will be transposed if and only if it has intact TIRs. Nothing else seems to matter. Interestingly, in rare cases a nonautonomous TIR element may also arise when two closely spaced genomic sequences mutate into active TIRs (usually very short ones), thereby conferring mobility on the sequence bounded by the two TIRs.

A few examples of autonomous and nonautonomous DNA-mediated transposable elements belonging to the CACTA family are shown schematically in **Figure 9.4**. CACTA DNA-mediated transposable elements are so named because they are flanked by inverted repeats that terminate in a conserved CACTA motif. CACTA elements are present in thousands of copies in many plant genomes, most conspicuously wheat, rice, sorghum, and other grasses. Some autonomous CACTA elements (e.g., *Balduin*

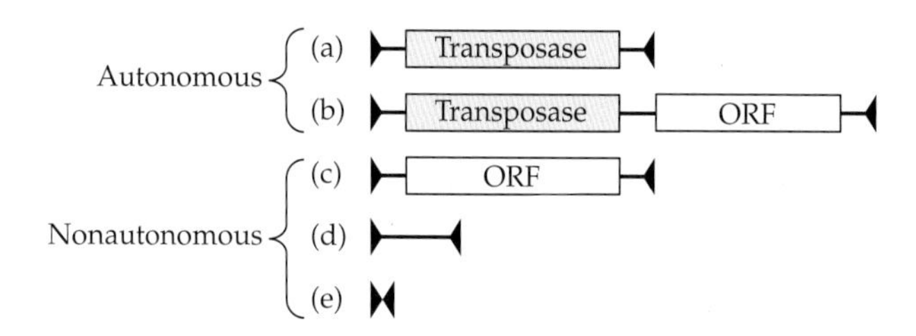

Figure 9.4 Schematic representation of autonomous and nonautonomous DNA-mediated transposable elements in eukaryotes. Boxes represent coding regions; triangles represent terminal inverted repeats (TIRs). (a) An autonomous insertion sequence containing a single protein-coding gene for transposase. (b) An autonomous transposon containing a transposase gene and an additional open reading frame (ORF). (c) A nonautonomous transposable element containing an ORF that is not related to transposition. (d) A nonautonomous transposable element devoid of intact open reading frames. The two TIR regions are separated by a small internal sequence. (e) A minimal transposable element, consisting solely of two terminal repeats.

and *Isaac*) contain a single transposase-coding gene and may, therefore, be properly referred to as insertion sequences. Other autonomous CACTA elements (e.g., *Caspar*) contain two ORFs, one for transposase and one for a protein with an unknown function. Nonautonomous CACTA elements either contain only the reading frame for the protein with the unknown function, or are devoid of intact ORFs altogether. One family of CACTA transposable elements (*Caspar* in wheat) has been studied in some detail, and it has been inferred that the initial autonomous *Caspar* transposon had a size of approximately 10 Kb. During the evolution of these elements, a large number of deletion derivatives were established, which further evolved and diverged. A large number of elements became nonautonomous by losing their transposase gene; others lost both genes and were reduced basically to their two TIR regions separated by a small internal domain. A possible final product of this size reduction is the *SNAC_AP003446-1* transposon in rice that does not even contain an internal domain but consists exclusively of two TIR sequences. At approximately 60 bp in length, *SNAC_AP003446-1* may represent the "minimal transposable element."

For reasons that are as yet unknown, some nonautonomous DNA-mediated transposable elements have immensely increased their numbers within some genomes. For example, there are thousands of **miniature inverted-repeat transposable elements** (**MITEs**) in the genomes of plants and animals, and the vast majority of them are inactive; only very few active ones are known (e.g., Jiang et al. 2003).

Retroelements

Retroelements are DNA or RNA sequences that contain a gene encoding the enzyme **reverse transcriptase**, which catalyzes the synthesis of a DNA molecule from an RNA template. A DNA synthesized from an RNA template by means of a reaction catalyzed by reverse transcriptase is called **complementary DNA** (**cDNA**). Not all retroelements are transposable intragenomically or even mobile intergenomically. Retroelements can be divided into three categories: (1) transposable elements that move within a genome by replicative RNA-mediated transposition (but may also move intergenomically), (2) mobile nontransposable elements that only move intergenomically, and (3) nonmobile elements. Here, several kinds of retroelements are presented in a somewhat subjective order of increased complexity. We note that simple retroelements may be ancestral, i.e., their simplicity is primitive, or they may be derived from more complex retroelements, i.e., their simplicity may be the result of reductive evolution. In the following survey, we have made no attempt at exhaustiveness, especially since new and curious retroelements are being discovered with great regularity (e.g., Doulatov et al. 2004; Gladyshev and Arkhipova 2011).

Retrons

Retrons are essentially reverse transcriptase–encoding genes. They are widely distributed among eubacterial species, but within each species retrons tend to be rare (Lampson et al. 2005). Retrons do not possess the capability to either transpose or excise. So far, natural populations of retron-carrying genomes have been found to possess only a single retron copy, either in the intergenic part of the genome or inside a prophage (a viral genome that has became integrated into the bacterial chromosome). Interestingly, with few exceptions, retrons take part in the production of a mysterious DNA-RNA hybrid molecule called **multicopy single-stranded DNA** (**msDNA**). An msDNA is composed of a small, single-stranded DNA linked to a small, single-stranded RNA molecule. The 5′ end of the DNA molecule is joined to an internal guanosine residue of the RNA molecule by a unique 2′—5′ phosphodiester bond (**Figure 9.5**). The msDNAs are produced in many hundreds of copies per cell and, in all likelihood, form large-molecular-weight nucleoprotein complexes composed of multiple msDNA copies bound to protein components.

The natural function of msDNA has remained elusive despite 30 years of study. Since knockout mutations that do not express msDNA are viable, the production of

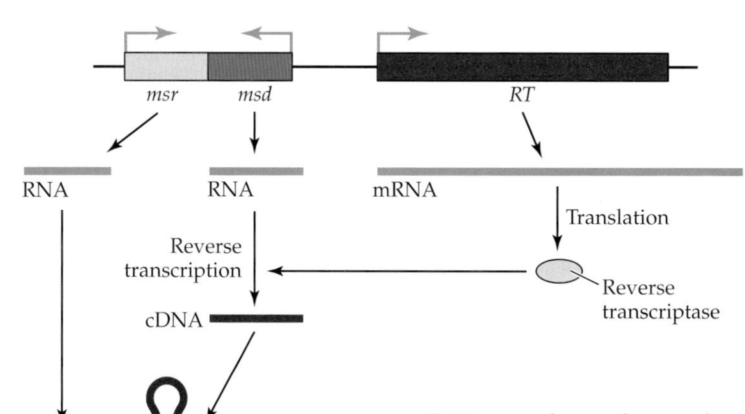

Figure 9.5 Structure of a retron locus and the synthesis of multicopy single-stranded DNA (msDNA). Three RNA sequences are independently transcribed: *msr*, *msd*, and *RT* (a reverse transcriptase mRNA). The *msd* transcript is reverse transcribed into cDNA by the reverse transcriptase (gray oval). The green arrows denote transcription initiation sites and the direction of transcription. The branched msDNA hybrid consists of the *msr* transcript and the *msd* reverse transcript joined together via a 2′—5′ phosphodiester bond. The two nucleic acids in msDNA form intrastrand loop-and-stem structures. (Modified from Graur and Li 2000.)

msDNA was deemed not to be essential for life under laboratory conditions. Moreover, overexpression of msDNA was found to be mutagenic, most probably by the mismatched base pairs in msDNA attracting repair proteins and rendering them unavailable whenever and wherever they are really needed (Lampson et al. 2005).

Elfenbein et al. (2015) discovered that msDNAs of zoonotic *Salmonella* pathogens and enteropathogenic *E. coli* strains are necessary for colonization of the intestine. Using mutant strains of *Salmonella* incapable of producing msDNA, they showed that msDNA is needed for growth in the absence of oxygen. In particular, mutants that lack msDNA exhibit abnormal patterns of protein composition when grown in oxygen-deficient conditions. These findings indicate that msDNA regulates protein production during anaerobic metabolism.

TERT genes

Telomerases are nucleoproteins whose function is to add DNA sequence repeats to the 3′ end of the telomeres at the ends of linear eukaryotic chromosomes. Human telomeres, for instance, consist of the sequence TTAGGG tandemly repeated many thousand times. Because of incomplete DNA replication, a few of these repeats are lost from the tips of the chromosomes in each replication cycle, but de novo addition of TTAGGG repeats by the enzyme telomerase partially or wholly compensates for this loss. Telomerases in all eukaryotic species share at least two components essential for their catalytic activity: a **TERT (te**lomerase **r**everse **t**ranscriptase) protein and a telomerase RNA. The TERT-encoding gene is, therefore, a retroelement. In humans, this retroelement is located on chromosome 5.

Mitochondrial retroplasmids

Extragenomic DNA and RNA molecules (plasmids) are frequently detected in fungal mitochondria. There are two main groups of plasmids. The first group consists of plasmids that have been derived from the mitochondrial genome via intra- or intergenomic recombination. These plasmids seem to be associated with fitness-reducing manifestations such as senescence and growth defects, and may be unimportant evolutionarily. The second group consists of autonomously replicating **true mitochondrial plasmids** (or **true plasmids**) that exhibit little or no sequence similarity with the host mitochondrial genome. True mitochondrial plasmids fall into specific groups based on their mode of replication and can be further divided by structure into linear and circular plasmids. All RNA mitochondrial plasmids described so far are linear and encode an RNA-dependent RNA polymerase. DNA plasmids can be linear or circular. Linear DNA plasmids usually encode a DNA-dependent RNA polymerase as well as a DNA-dependent DNA polymerase. Circular DNA plasmids generally have a single ORF encoding a DNA-dependent DNA polymerase. Finally, there is a group of double-stranded DNA plasmids that encode reverse transcriptase and, hence, are

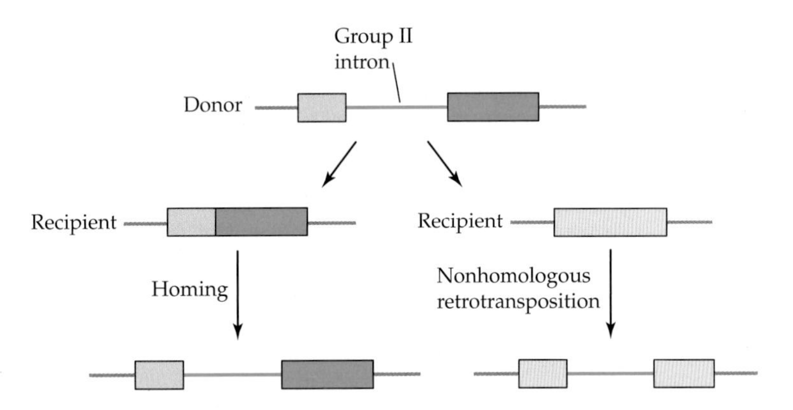

Figure 9.6 A group II intron (blue line) residing between two exons may retrotranspose into the homologous position of an intronless allele of the same gene (homing), or retrotranspose into a nonhomologous genetic position of either the same gene or a different gene.

referred to as **retroplasmids**. There are circular retroplasmids (e.g., *Mauriceville* and *Varkud* in the mitochondria of the red bread mold *Neurospora crassa*) and linear ones (e.g., *FOXC* in the mitochondria of the filamentous fungus *Fusarium oxysporum*). The reverse transcriptase of circular retroplasmids does not need a primer. In contrast, reverse transcription of linear retroplasmids is protein-primed. The fact that the reverse transcriptase of *FOXC* utilizes protein primers is very unusual; only hepadnaviruses (below) are know to employ protein primers. This similarity indicates that *FOXC* plasmids may be related evolutionarily to hepadnaviruses (Galligan et al. 2011).

Group II introns and twintrons

Self-splicing introns are introns that can fold into catalytic structures capable of removing their own sequences from the RNA transcripts of protein-coding and RNA-specifying genes. Self-splicing introns are classified into three groups—group I, group II, and group III—according to the biochemical reaction they use to excise themselves out of primary RNA transcripts. Some **group II introns** qualify as retroelements because they contain a reverse transcriptase-coding gene. Group II introns were first discovered in the genomes of fungal and plant organelles (Lambowitz and Belfort 1993). **Group II twintrons** are self-splicing introns located within the sequence of another intron. Some group II introns may also act as retrotransposable genetic elements. As seen in **Figure 9.6**, they can integrate into the homologous position of an intronless allele of the same gene (homing) and, at much lower frequencies, into other sites (retroposition).

Group III introns are short introns that are related by descent to group II introns (Doetsch et al. 1998) but are devoid of open reading frames. Because they do not contain reverse transcriptase-coding genes (or any other protein-coding genes), they do not qualify as retroelements. Group III introns are found in a small number of protist eukaryotes (e.g., *Euglena gracilis*).

Retrotransposons

Retrotransposons (or **retroposons**) are transposable elements that use RNA-mediated transposition but do not construct virion particles, i.e., they lack the *env* (*envelope*) gene, and so, unlike retroviruses, cannot independently transport themselves across cells. Initially, the retrotransposons were divided into **LTR retrotransposons** and **non-LTR retrotransposons** according to whether or not their coding sequences were flanked by **long terminal repeats** (**LTRs**). Evolutionary studies indicate that most non-LTR retrotransposons constitute a monophyletic group. The LTR retrotransposons, on the other hand, are paraphyletic and possess structures and mechanisms similar to those of the vertebrate retroviruses. Some LTR retrotransposons have even acquired *env*-like reading frames that may enable them to move from cell to cell (i.e., they are in practice retroviruses).

The *G3A* element and the *I* factor in *D. melanogaster* are typical non-LTR retrotransposons (**Figure 9.7a**). They contain two open reading frames. ORF2 in *G3A* consists of seven exons separated by very short introns. Autonomous non-LTR retrotransposons are also referred to as LINEs (see below). *LINE-1* (*L1*) retrotransposons represent the most abundant family of non-LTR retrotransposons in many mammalian taxa. The consensus *L1* structure has a poly(A) tail at one end and contains two large open reading frames, ORF1 and ORF2, with about 375 and 1,300 codons, respectively. *L1* encodes two proteins, ORF1p and ORF2p, both of which are required for retrotransposition. ORF2p contains amino acid sequence motifs characteristic of reverse transcriptases.

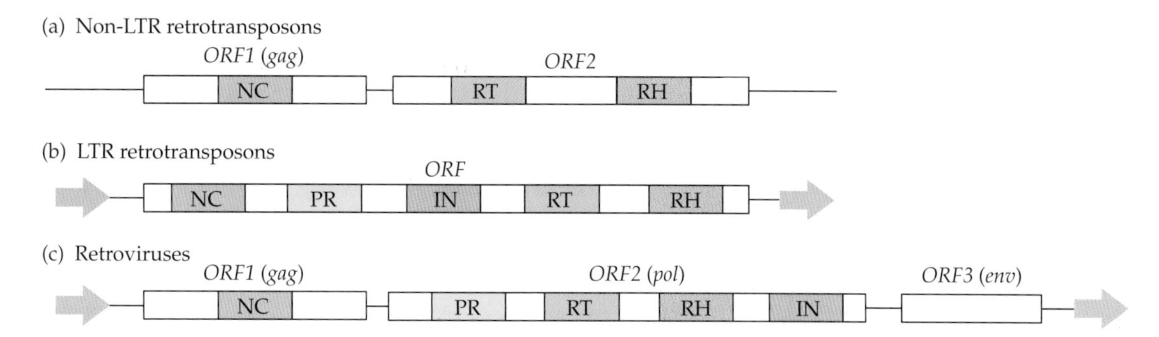

Figure 9.7 Structural comparison of transposable retroelements. Open reading frames separated by termination codons are shown as white boxes. Long terminal repeats are represented by arrows. (a) An I factor non-LTR-retrotransposon from *Drosophila melanogaster*. (b) A *copia* LTR-retrotransposon from *Drosophila melanogaster*. (c) A consensus retrovirus containing only the *gag*, *pol*, and *env* genes. Identified domains are indicated by solid boxes. NC, nucleocapsid protein; PR, aspartate protease; RT, reverse transcriptase; RH, RNase H; IN, integrase. Neither ORFs nor domains are drawn to scale. Areas of overlap between neighboring ORFs, e.g., between *gag* and *pol*, are not shown. The internal order of the ORFs and the domains may differ among transposable retroelements. Different transposable retroelements may also differ in their ORF and domain complements.

The *Ty* elements in yeast and the *copia* elements in *Drosophila* are examples of LTR retrotransposons. They contain LTRs at both ends and a single long open reading frame with regions of similarity to the *pol* gene of retroviruses (**Figure 9.7b**).

Retroviruses

Retroviruses are RNA viruses that reproduce in the host cell by using reverse transcriptase to produce DNA from their RNA genomes. Retroviruses are the most complex of all retroelements. They are similar in structure to retrotransposons but contain at least three protein-coding genes: *gag*, *pol*, and *env* (**Figure 9.7c**). The *env* gene encodes a precursor of two or more envelope proteins, *gag* encodes several core structural proteins, and *pol* encodes several enzymes (including a reverse transcriptase). Many retroviruses possess additional genes. HIV (the AIDS virus), for example, possesses at least six genes. The coding region of a retrovirus is flanked by LTRs. The LTRs contain promoters for transcription (in the proviral stage) and reverse transcription (in the viral stage), as well as an enhancer and a polyadenylation signal. The presence of a functional *env* gene is correlated with the infectivity of retroviruses, and acquisition of *env* is thought to be a key step in the evolution of retroviruses from LTR retrotransposons (Kim et al. 2004).

The life cycle of a retrovirus is shown in **Figure 9.8**. After the retroviral particle, called the **virion**, invades a host cell, its genomic RNA is reverse transcribed into viral cDNA. This DNA can integrate into the host genome and become a **provirus**. Next, the proviral DNA is transcribed into RNAs, which can serve both as mRNAs for synthesizing viral proteins and as viral genomes that can be packaged into infectious virion particles. Once a virion is formed, the cycle can start again.

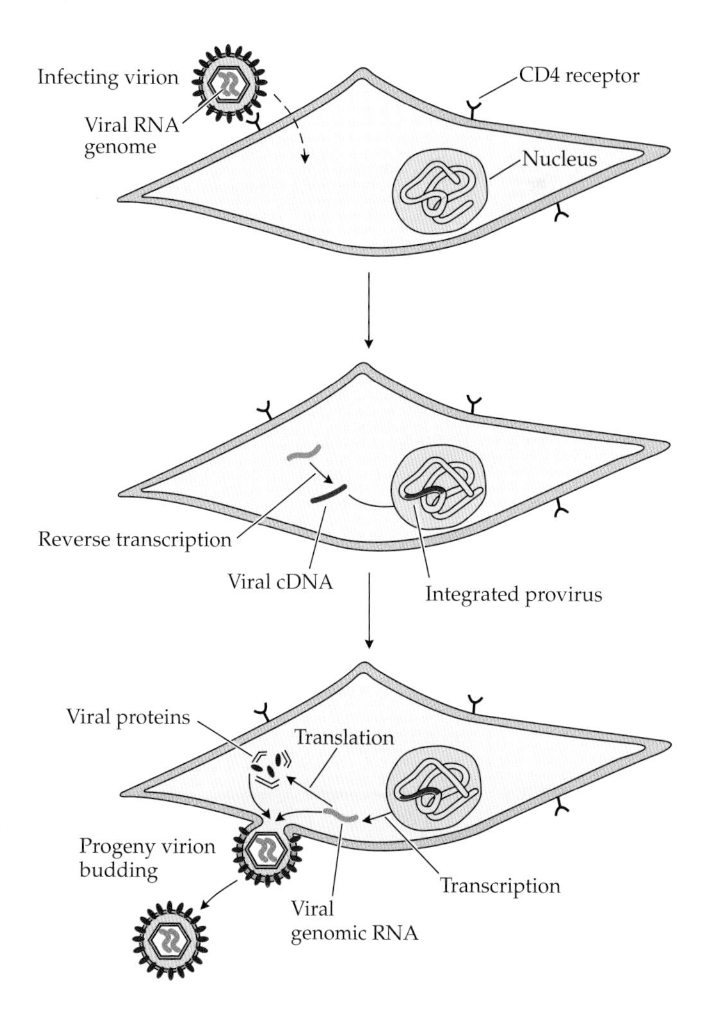

Figure 9.8 Life cycle of a retrovirus. The virion attaches to a receptor on the surface of the cell. The two copies of genomic RNA are injected into the cytoplasm, where they are reverse-transcribed by the enzyme reverse transcriptase. The cDNA penetrates the nucleus and becomes integrated within the genome of the host cell. The integrated provirus is transcribed into RNA that serves as both a genomic RNA and as mRNA for the synthesis of viral proteins. The genomic RNA and the structural and enzymatic viral proteins assemble into a new infectious virion particle that buds off the cell membrane. (Modified from Graur and Li 2000.)

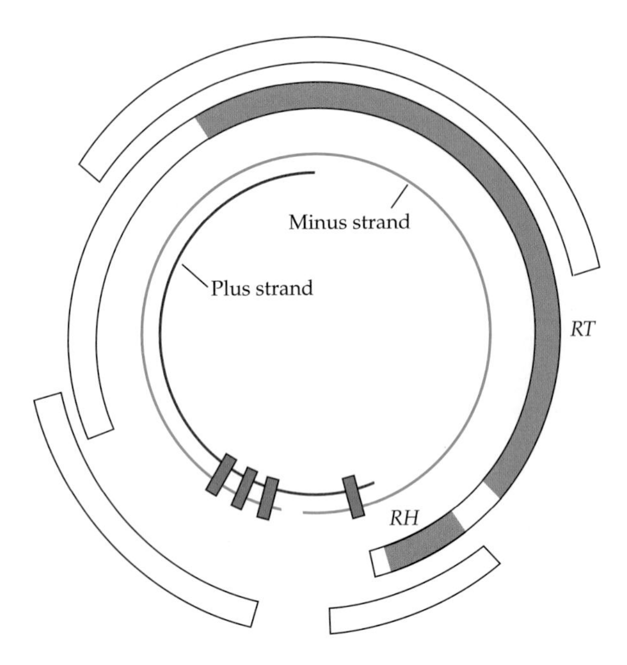

Figure 9.9 Schematic representation of the partially double-stranded genome of the hepadnavirus. The lengths of the minus and plus strands are 3.5 and 2.1 Kb, respectively. Bars spanning the two strands indicate sequence motifs involved in interstrand interactions. Four overlapping open reading frames are shown as arcs. Identified domains are indicated by solid arcs: *RT*, reverse transcriptase gene; *RH*, RNase H gene. (Modified from Graur and Li 2000.)

Most retroviruses are mobile elements, i.e., they have the ability to integrate within the genome; a few are also transposable elements that can move from locus to locus within a genome.

Pararetroviruses

Pararetroviruses, such as the hepatitis B virus (a **hepadnavirus**) and the cauliflower mosaic virus (a **caulimovirus**), are double-stranded or partially double-stranded circular DNA viruses. The structure of the hepatitis B virus is shown in **Figure 9.9**. The uniqueness of pararetroviruses lies in their mode of replication. Rather than replicating the two strands, as do many double-stranded DNA viruses, pararetroviruses use only one strand for synthesizing both daughter strands (**Figure 9.10**). Let us refer to this strand as the plus strand and to the complementary one as the minus strand. First, the plus strand is replicated into a minus DNA strand. Second, the plus strand is transcribed into a minus RNA strand. Finally, the minus RNA strand is reverse transcribed into a plus DNA strand, which combines with the minus DNA strand created by DNA replication, and together they form the double-stranded DNA genome of the pararetrovirus.

The amino acid sequence encoded by the longest reading frame in pararetroviral genomes was found to be homologous to the N-terminal region of the *pol* gene product of the retroviruses (Toh et al. 1983). Therefore, pararetroviruses are similar to retroviruses in their endogenous capability to produce reverse transcriptase, but they lack the ability to insert themselves into the host genome. This disqualifies them as transposable elements, although they clearly share a common evolutionary origin with the retroviruses.

Evolutionary origin of retroelements

The fact that the reverse transcriptases of all retroelements have some amino acid identity with one another suggests a common evolutionary origin. Because of the simplicity of retrons as opposed to the complexity of retroviruses, and because of the antiquity of bacteria, Temin (1986, 1989) suggested that the path of evolution went from retrons to non-LTR retrotransposons to LTR retrotransposons to retroviruses. This evolutionary scheme is characterized by a progressive increase in the structural complexity of the retroelements. Of course, one can as easily imagine the opposite evolutionary trend, whereby simple elements are derived from more complex elements through loss of function, or a mixed trend, whereby periods of progressive and reductive evolution are temporally interspersed with one another.

Many authors have attempted to study the evolutionary relationships among retroelements, and many contradicting phylo-

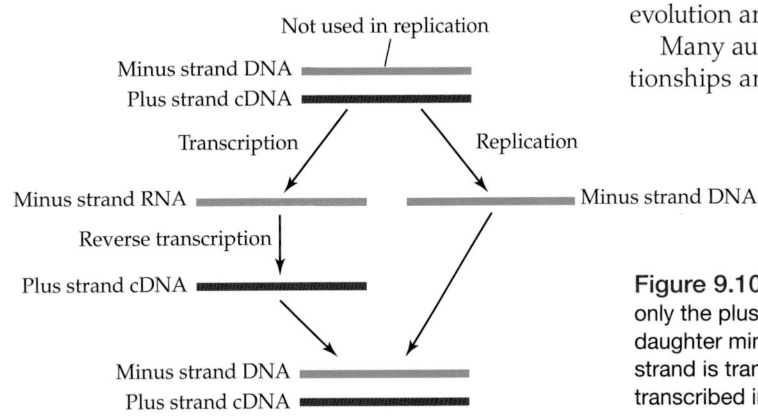

Figure 9.10 DNA replication in pararetroviruses. Of the two strands, only the plus strand is used. The replication of the plus strand yields a daughter minus strand. To generate a daughter plus strand, the plus strand is transcribed into a minus RNA strand, which in turn is reverse transcribed into a plus DNA strand.

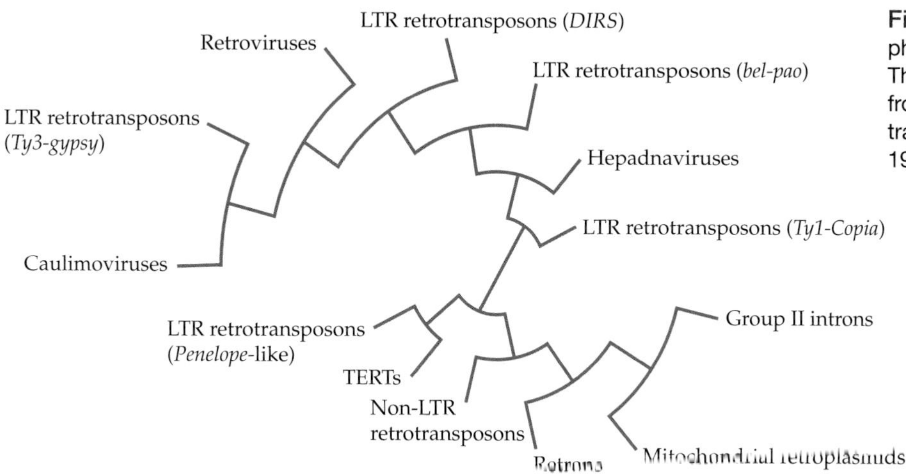

Figure 9.11 Schematic representation of phylogenetic affinities among retroelements. The unrooted, unscaled tree was derived from the amino acid sequences of the reverse transcriptase domains. (Data from Eikbush 1994 and Chang et al. 2008.)

genetic trees (rooted and unrooted) have been published. There are two main reasons for the daunting difficulty in reconstructing retroelement phylogeny. First, the alignable parts of retroelements, which mostly consist of relatively short sequences of reverse transcriptase genes, have diverged considerably from element to element. In many cases, the pairwise alignment between two sequences yields lower than 25% identity levels, which in phylogenetic studies is referred to as the "twilight zone." Such distantly related sequences are difficult to align reliably, and it is almost impossible to safely estimate phylogenetic relationships (Chang et al. 2008). The second difficulty is the very high frequency of reticulate events in the evolution of retroelements, which may render meaningless whatever trees we do manage to obtain.

Be that as it may, **Figure 9.11** shows a schematic unrooted phylogeny of retroelements derived from reverse transcriptase sequence data, as well as additional data pertaining to their three-dimensional structure. Two main conclusions can be made from the tree. First, it seems almost certain that LTR retrotransposons are paraphyletic, as are the pararetroviruses, the group of simple retroelements (retrons, group II introns, TERT genes), and the group of LTR retroelements (LTR retrotransposons, retroviruses, and pararetroviruses). Second, there is no possible position for the root that will yield either a purely progressive evolution scheme toward greater and greater complexity, or a purely reductive evolution scheme through gene loss toward greater simplicity.

Nonautonomous and fossil retrotransposable elements

Nonautonomous retrotransposable elements—i.e., transposable elements that do not encode the proteins necessary for their retroposition but instead rely on proteins provided by retroelements—have been found in many eukaryotic genomes. Nonautonomous retrotransposable elements can be classified into **nonautonomous LTR retrotransposable elements** and **nonautonomous non-LTR retrotransposons**. The reason for using the term "retrotransposable element" in the former category is that we have so far been unable to distinguish between nonautonomous elements derived from LTR retrotransposons and elements derived from retroviruses, which also contain LTRs. Typically, nonautonomous LTR retrotransposable elements lack all coding capacity but have retained the LTRs, as well as a primer-binding site and a polypurine tract. These are the minimal features required for replication, because the LTRs contain the promoter needed to produce a template RNA, and the primer-binding site and the polypurine tract are needed to prime reverse transcription. The success of some nonautonomous LTR retrotransposable elements is quite remarkable; for example, there are about a thousand copies of the nonautonomous *Dasheng* element in the maize genome.

For most nonautonomous LTR retrotransposable elements, it is unclear which autonomous element is involved in the mobilization of which nonautonomous element.

Striking sequence similarities within and adjacent to the LTR sequence between the nonautonomous *Dasheng* element and the autonomous *RIRE2* element, however, make it very probable that *RIRE2* provides the proteins needed to retrotranspose *Dasheng*.

The nonautonomous *Dasheng* element is large, ranging in size from 5.5 to 8.5 Kb. Such large nonautonomous LTR elements are sometimes referred to as large retrotransposon derivatives (LARDs). Two other types of nonautonomous LTR retrotransposable elements have been identified in plants. Terminal-repeat retrotransposons in miniature, or TRIMs, also lack an internal coding domain but, in contrast to LARDs, are very small—usually less than 540 bp in length. At less than 300 bp in length, small LTR retrotransposons, or SMARTs, are even smaller. So far, no autonomous partners for either TRIMs or SMARTs have been found.

Nonautonomous non-LTR retrotransposons seem to be much more numerous (or at least better characterized). There are two types of nonautonomous non-LTR retrotransposons—those transcribed by RNA polymerase II (as is the case with autonomous non-LTR retrotransposons), and those transcribed by RNA polymerase III, also referred to as SINEs. *HAL1* (half-*L1*) retrotransposons in placental and marsupial genomes are an example of nonautonomous non-LTR retrotransposons that are transcribed by RNA polymerase II (Bao and Jurka 2010). They contain only one open reading frame, ORF1, i.e., they do not encode reverse transcriptase.

The vast majority of retrotransposable elements found in sequenced genomes no longer transpose; they constitute a "fossil record" of past retroelement activity (e.g., Katzourakis et al. 2009). Of particular interest are the integrated **endogenous retroviruses** (**ERVs**) or **virogenes**. ERVs are derived from ancient viral infections, which somehow became integrated into the germline and are currently inherited vertically from parents to offspring. Vertebrate genomes contain a substantial number of former proretroviral sequences in various states of inactivation since their integration. Ancient ERVs commonly contain many point and segmental mutations (mainly deletions). They may even be reduced to single long terminal repeats by homologous recombination between two LTRs. The more recent integrations may have retained some coding capacity. Some recent ERV elements have been co-opted and are used advantageously by the host as a means of protection against new retrovirus infections. This mechanism, called **superinfection resistance**, works best against closely related exogenous retroviruses and uses the simple mechanisms of **receptor occupancy**—i.e., the ERV produces a protein that binds to the receptor on the cell, thereby depriving the exogenous retrovirus of appropriate cell receptors (Nethe et al. 2005).

RNA and proteins encoded by ERVs are frequently found in immunodeficient mice, and through recombination and spontaneous association, these RNAs and proteins can combine to produce complete retroviral genomes encapsulated in infective virion particles (Young et al. 2012). Of particular interest are reactivated viruses that can only infect other species, a phenomenon called **xenotropism**. Examples are retroviruses of gibbons and koalas that are derived from a rodent ERV and that cause leukemia in their new hosts (Weiss and Stoye 2013). Interestingly, the newly acquired koala virus is currently colonizing the germline as a new ERV (Tarlinton et al. 2006). Thus, a virus embedded in the chromosomes of one host species without causing apparent harm can trigger a potentially devastating epidemic in another.

Around 8% of the human genome is of retroviral origin (van der Kuyl 2012). **Human ERVs** (**HERVs**) are used to mainly populate works of apocalyptic fiction, such as *Darwin's Radio* by Greg Bear, in which pandemonium follows their revivification from the dead. In the nonfiction world, however, no replication-competent HERV has been identified so far, although HERV RNA sequences have been found in the blood of patients with HIV-1 infection (Contreras-Galindo et al. 2013), and the env protein of one human endovirus, HERV-K, has been implicated in the pathogenesis of sporadic amyotrophic lateral sclerosis by causing malformations and malfunctions of the axons and dendrites of neurons (Li et al. 2015).

LINEs and SINEs

The genomes of all multicellular eukaryotes contain several types of highly repetitive interspersed sequences (Chapter 11). For example, over one-third and possibly as much as two-thirds of the human genome consists of interspersed repetitive sequences. These sequences, originally detected as rapidly reannealing components of genomic DNA, were initially divided according to length and abundance into two major classes, referred to as **long interspersed repetitive elements (LINEs)** and **short interspersed repetitive elements (SINEs)**. LINEs and SINEs were initially differentiated from one another by length and number of genomic repeats (Singer 1982). Following sequencing studies and evolutionary analyses, LINEs and SINEs were redefined by structure, type of transposition, and autonomy of intragenomic movement.

LINEs are non-LTR retrotransposons that typically range in length from 3 to 7 Kb. Sequencing studies have indicated that most LINEs in the genome are fossil non-LTR retrotransposons, which are unable to move or multiply. The number of active LINEs within any well-characterized genome is very small. A functional LINE contains one or two open reading frames that encode domains possessing endonuclease and reverse transcriptase activity. Some reverse transcriptase domains exhibit LINE specificity, i.e., the reverse transcriptase from one LINE will only recognize the 3′ end of the LINE that produces it, or a similar sequence, and will be much less efficient at recognizing and subsequently reverse transcribing other LINEs. Fossil LINEs are mostly partial LINE sequences that have lost the ability to transpose either autonomously or not autonomously. In a sense, they are pseudogenes of LINEs.

SINEs typically range in length from 75 to 500 bp. They do not code for proteins required for retroposition; in fact, they do not possess any open reading frame of significant length. Thus, SINEs are nonautonomous elements that must be aided in the process of retroposition by autonomous genetic elements. Interestingly, as opposed to nonautonomous non-LTR retrotransposons, whose transcription into RNA is catalyzed by RNA polymerase II, SINEs use RNA polymerase III, which is otherwise used to transcribe ribosomal 5S rRNA, tRNA, and other small RNAs. Indeed, SINEs usually contain sequences derived from polymerase III-transcribed RNA (tRNA, 7SL RNA, 5S RNA, snRNA), or they contain a mosaic of different polymerase III promoters.

SINEs derived from 7SL RNA

Alu sequences, so named because they contain a characteristic restriction site for an *Arthrobacter luteus*-derived endonuclease called *Alu*I, are approximately 280 bp long. There are over one million *Alu* elements interspersed throughout the human genome, i.e., about 11% of the human genome consists of *Alu* sequences.

Ullu and Tschudi (1984) found that *Alu* sequences are in fact derivatives of a gene specifying 7SL RNA, an abundant cytoplasmic component of the signal recognition particle that is essential in the process of removal of signal peptides from secreted proteins (Walter and Blobel 1982). The active gene is highly constrained, and its sequence is conserved among such divergent taxa as humans, *Xenopus*, and *Drosophila*. *Alu* sequences have been derived from a functional 7SL RNA-specifying gene by a series of steps (**Figure 9.12**). First, a so-called fossil *Alu* monomer (FAM) was derived from 7SL RNA by a large central deletion. A further deletion resulted in the creation of the free right *Alu* monomer (FRAM). The free left *Alu* monomer (FLAM) was derived independently from 7SL RNA. Finally, a fusion of FRAM to FLAM gave rise to the predominating dimeric *Alu* family in primates. The human genome also contains a number of monomeric *Alu* elements, which descended without duplication from ancestral FAM, FRAM, and FLAM sequences, as well as a handful of tetrameric *Alu* sequences.

Kriegs et al. (2007) suggested that the 7SL-derived SINEs originated in the ancestor of Supraprimates (also called Euarchontoglires), a higher taxon within mammals

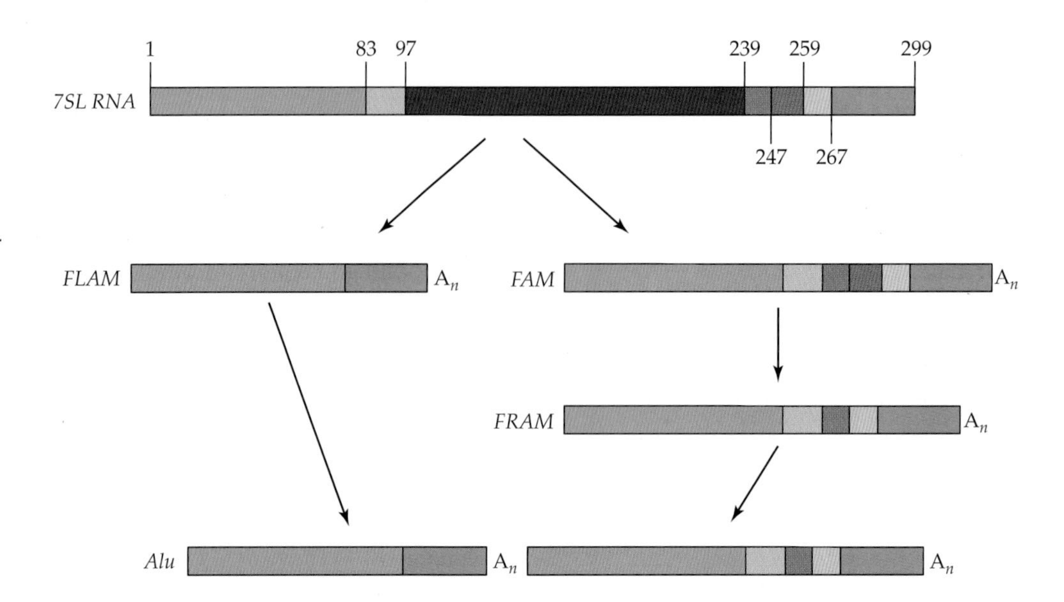

Figure 9.12 Origin of *Alu* sequences. Different regions of the *7SL RNA* gene are different colors to emphasize deletions and rearrangements. The functional 7SL RNA-specifying gene gave rise to *FAM* and *FLAM* by independent deletions. A further deletion in *FAM* gave rise to *FRAM*. The fusion of *FRAM* and *FLAM* gave rise to the dimeric *Alu* family in primates.

that includes primates, scandentians (tree shrews), rodents, dermopterans (flying lemurs), and lagomorphs (rabbits). So far, however, 7SL-derived SINEs have not been found in either dermopterans or lagomorphs. The Supraprimates hypothesis has been weakened considerably by the finding that 7SL-derived SINEs are also founds in pigs (Yu at al. 2015).

The 7SL RNA gene is transcribed by RNA polymerase III, which does not require promoters outside of the transcribed region. The success of *Alu* sequences should in part be attributed to the fact that they have retained the proper promoters for transcription. However, this is not the entire story. To produce hundreds of thousands of copies, *Alu* sequences must also multiply by duplicative retroposition, and they must do that despite lacking autonomous transposable capabilities. It has been shown that LINEs can generate cDNA copies of any mRNA transcript by means of a process involving the LINE-encoded endonuclease and reverse transcriptase. There are, however, features in the *Alu* sequence (mostly its 3′ region) that increase its rate of retrotransposition by about three orders of magnitude relative to regular mRNAs (Dewannieux et al. 2003).

Large-scale sequencing studies of primate genomes have provided a great deal of detail on the evolution of *Alu* elements. Because there is no specific mechanism for removal of *Alu* insertions, *Alu* evolution is dominated by the accumulation of new *Alu* inserts. These new *Alu* inserts accumulate mutations over time and are rendered inactive.

Different periods of evolutionary history have given rise to different families and subfamilies of *Alu* elements, each containing a small number of active *Alu* elements that serve as the source of subsequent families. The earliest dimeric *Alu* elements are the J family, of which about 160,000 copies are still identifiable in the human genome. The J family gave rise to the very active S family, of which approximately 650,000 copies are identifiable in the human genome. The dominant S subfamilies are Sx, Sq, Sp, and Sc. More recently, *Alu* amplifications in Old World monkey and ape lineages have been almost exclusively from members of the Y family, with Ya5 and Yb8 subfamilies dominating in humans. In particular, members of the Y family are responsible for all genetic diseases attributed to *Alu* insertion (Deininger 2011).

As mentioned previously, the *Alu* amplification rate peaked with the S family ~40 million years ago, when the majority of *Alu* elements currently in the human genome were generated. The current rate of *Alu* retroposition in humans is estimated to be on the order of one new *Alu* insertion every ~20 births (Cordaux et al. 2006). *Alu* retrotransposition rates vary in a lineage-specific manner. For example, the rate of *Alu* accumulation in the orangutan lineage is much lower than that in humans, while the rate in the macaque lineage is much higher.

Because we are dealing with evolution, we are, of course, focused on the germline. We note that there may be ~150 full-length retroelements in the human genome and ~3,000 in the mouse genome. These intact elements are potentially active in both germline and somatic cells. Retroelements that are active in somatic cells have the potential to create somatic mosaicism, causing variation among cells and altering the transcriptome of individual cells. For example, *LINE-1* non-LTR retrotransposons are known to be mobilized early in development of the nervous system and later during adult neurogenesis. As a consequence, neuron-to-neuron genomic variation in genomic DNA content has been observed in both rodents and humans (Singer et al. 2010). It has also been noted that the rate of *Alu* transposition in somatic cells is many orders of magnitude higher than the rate in the germline, as well as being tissue-specific. Of particular interest may be the fact that approximately 15,000 somatic SINE and LINE insertions were observed in the human brain (Baillie et al. 2011). Such insertions have the potential to cause disease (Hancks and Kazazian 2012).

SINEs derived from tRNAs and SINEs containing 5S rRNA

The vast majority of SINEs in nature are tRNA-derived. That is, *Alu* elements are not in any way representative of SINEs in general. A typical tRNA-derived SINE is a chimerical molecule consisting of three parts: a tRNA-derived region, a region whose origin remains to be elucidated, and a LINE-derived region (**Figure 9.13**).

Because of the rapid rate of SINE evolution, the identification of the tRNA species from which the tRNA-related region of a SINE had been derived is extremely difficult in the majority of cases, and errors in identification may have crept into the literature. Several tRNA species have been identified unambiguously as origins of SINEs. In particular, many SINEs derived from tRNA^{Lys} (the most commonly used tRNA species) have been identified. At present, we do not know whether or not all tRNA types can form SINEs.

SINEs containing 5S rRNA have a structure very similar to those derived from tRNA (Figure 9.13). The tRNA-derived SINEs are substantially more numerous than those possessing a 5S rRNA–derived sequence at their 5′ end. There may be several reasons for this numerical inequality. First, the difference may be due to the greater availability of tRNA molecules in the cell. Another possibility involves template switching, i.e., the process in which the strand of DNA being synthesized switches from its current template to a new template. According to this hypothesis, all 5S rRNA-containing SINEs have descended from tRNA-derived SINEs. That is, template switching is envisioned to add a 5S rRNA region to the 5′ terminus of the tRNA-derived SINE. Since the presence of two promoters for RNA polymerase III may be deleterious for the SINE, there is selective pressure for the promoters residing within the tRNA-derived region to be deleted. After the deletion of promoters, some remaining tRNA region may still be discernable in 5S rRNA-containing SINES. Sometimes, the tRNA part has been deleted completely and is no longer identifiable. The evidence so far indicates that all 5S rRNA–containing SINEs may have been so derived.

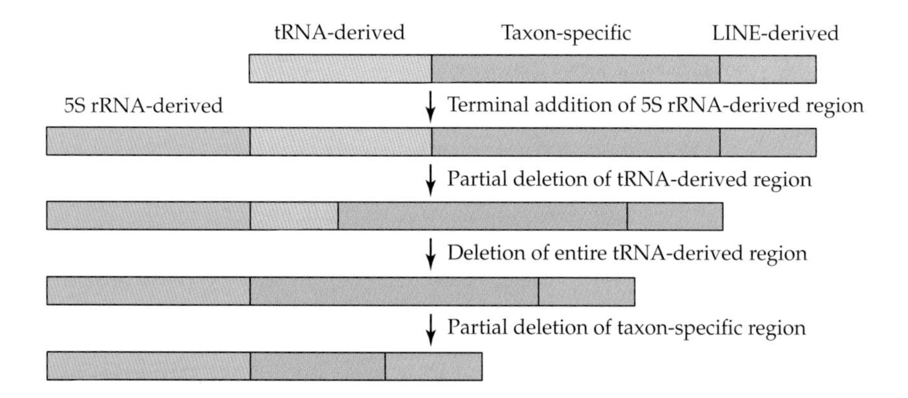

Figure 9.13 A plausible evolutionary scenario for the evolution of 5S rRNA-containing SINEs from tRNA-derived SINEs. The taxon-specific regions are usually of unidentified origin. In this scenario, a 5S rRNA sequence becomes joined to a 5′ end of a tRNA-derived SINE. Subsequently, a partial or complete deletion of the tRNA-derived sequence gives rise to a 5S rRNA-containing SINE. The 5S rRNA-containing SINE may become shorter by the partial deletion of the taxon-specific region. (Modified from Nishihara et al. 2006.)

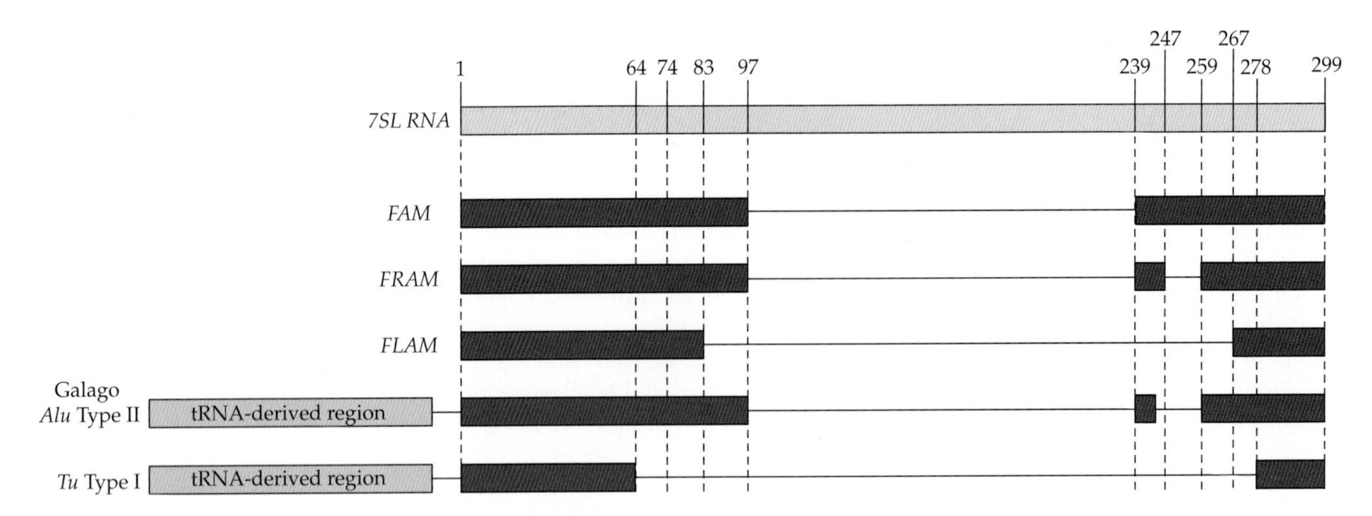

Figure 9.14 A comparison of the *7SL* gene and monomeric 7SL-derived SINE with two monomeric mosaic SINEs. Solid lines that connect blocks indicate large deletions. Common deletion breakpoints are indicated by vertical dotted lines that indicate corresponding positions in the 7SL RNA. The tRNA-derived regions of galago *Alu* type II and *Tu* types I and II are boxed. (Modified from Nishihara et al. 2002.)

SINEs containing snRNA

Kojima (2015) has recently discovered a group of SINEs whose 5′ ends originated from either the U1 or U2 small nuclear RNA (snRNA) genes. These SINEs, named *SINEU*, are distributed among crocodylians and classified into three families and numerous subfamilies. The structures of the SINEU families and subfamilies indicate the recurrent addition of a U1- or U2-derived sequence onto the 5′ end of SINEU-1 elements. Some SINEU elements are ancient, as they are shared among alligators, crocodiles, and gharials, which presumably diverged from one another more than 100 million years ago.

Mosaic SINEs

Mosaic or **composite SINEs** are nonautonomous retrotransposable elements that combine sequences derived from different SINE types. For example, the *Alu* type II SINE from *Galago crassicaudatus* (a prosimian) most probably originated by the joining of a tRNA-derived region and 7SL-derived sequence, such as FRAM (**Figure 9.14**). The *Tu* type I SINE from the tree shrew (*Tupaia belangeri*) combines a tRNA-derived region and a highly abbreviated 7SL-derived sequence. The tRNA-derived region of the *Tu-type-I* SINE includes internal RNA polymerase III promoters, which have been deleted from the 7SL-derived part.

The human genome contains 4,000–5,000 copies of a nonautonomous mosaic retroelement called *SVA* (so called after its three main components, *SINE-R*, *VNTR*, and *Alu*). Interestingly, over half of the *SVA* elements in primate genomes are full-length. *SVA* elements are relative newcomers to primate genomes. They arose as a family of repeats after the divergence of hominids from Old World monkeys, and their distribution is restricted to hominids. Similar to *Alu* and *L1* elements, *SVA* elements also appear to have followed a model of relatively limited proliferation, which has given rise to a hierarchical subfamily structure of elements with shared diagnostic sequence characters.

Where there's a SINE, there's a LINE

An enduring mystery in molecular evolution has been the amazing efficiency with which SINEs, which are in themselves devoid of self-replicating mechanisms, have produced hundreds of thousands of genomic copies. The discovery that the 3′ end of a tRNA-derived SINE exhibits sequence similarity with the 3′ end of a LINE (Ohshima et al. 1996) provided the key for solving this mystery. **Figure 9.15** shows one such example, the alignment of the 3′ end of a SINE and a LINE from the silkworm *Bombyx mori*. About 80 nucleotides at the 3′ ends of both sequences are remarkably similar, while no discernible similarity is observed upstream of the 3′ region. Since this region is known to be recognized by the LINE-encoded reverse transcriptase,

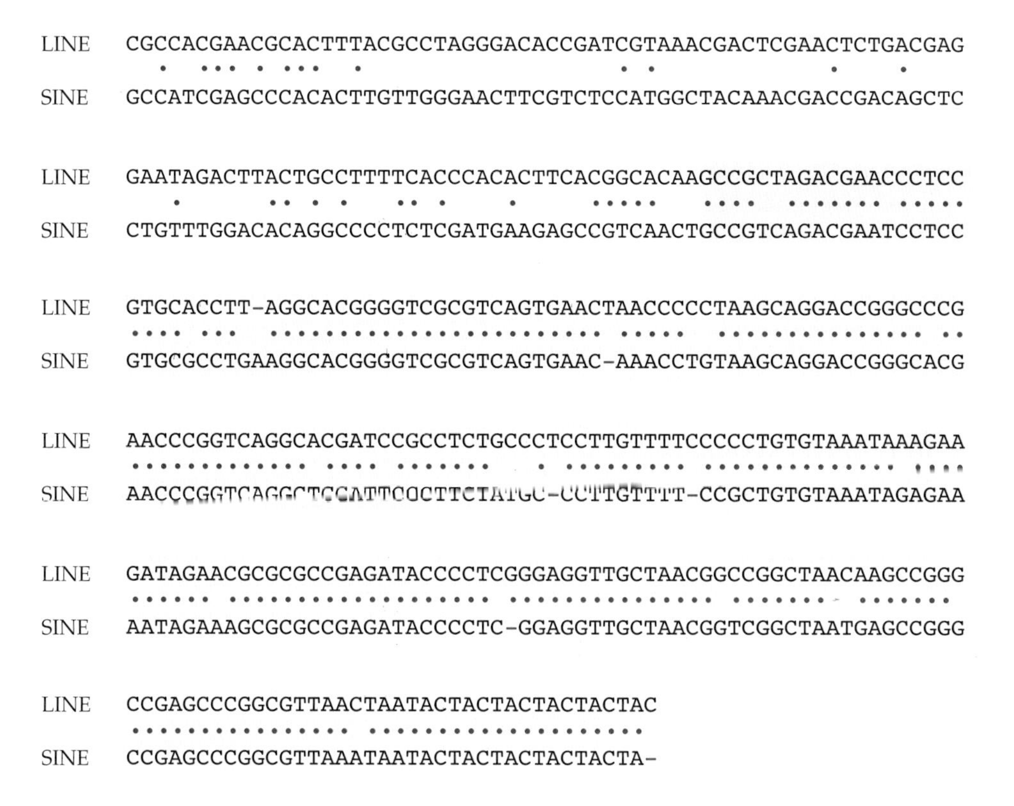

Figure 9.15 Alignment of the Mg SINE from the rice blast fungus *Magnaporthe grisea*, with the 3′ end of the Mg LINE from the same species. Red dots indicate identical nucleotides in the two sequences. Gaps are indicated by red dashes. The shaded regions indicate significant sequence similarity between the SINE and LINE sequences. (Data from Ohshima and Okada 2005.)

Ohshima et al. (1996) proposed that each SINE is propagated within the genome by using the enzyme from a corresponding LINE (**Figure 9.16**). In some cases, two or more SINEs were found to have the same 3′ tail as a single LINE. Many such **LINE/SINE couples** have been identified in many diverse taxa, such as ruminants, reptiles, salmonid fishes, sharks, silkworms, tobacco plants, and fungi (Okada et al. 1997; Platt and Ray 2012).

The above model makes three interesting predictions. First, if there is a tRNA-derived SINE, a LINE partner must be present in the genome of the same organism. LINEs can exist on their own; SINEs cannot. Second, the model predicts that once a LINE partner becomes inactive, the SINE partner will lose its ability to retrotranspose. Finally, a LINE partner should have a longer evolutionary history, and hence a broader phylogenetic distribution, than its SINE partner. All these predictions have been validated (Terai et al. 1998).

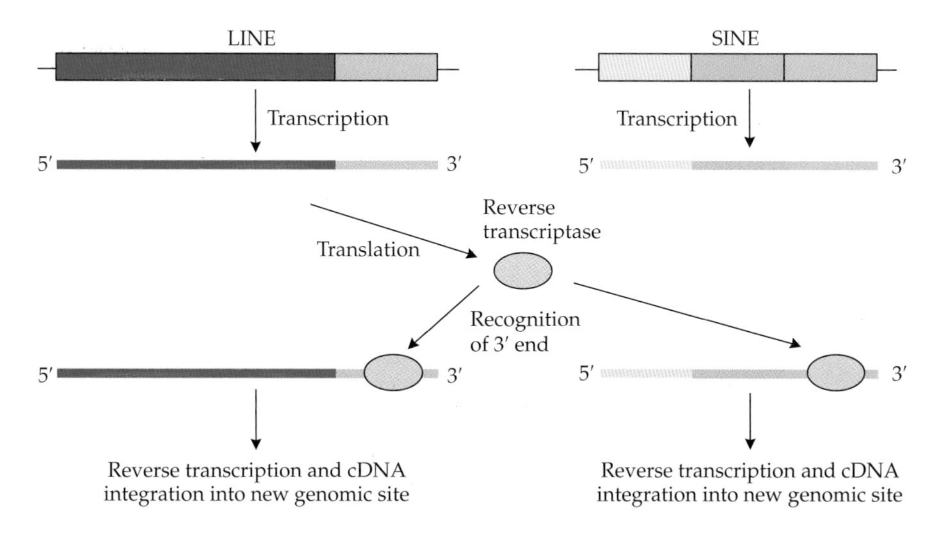

Figure 9.16 The reverse transcriptase (shaded oval) required for the retroposition of tRNA-derived and 5S rRNA-containing SINEs is provided by the partner LINE. SINE transcripts are recognized by the enzyme at their 3′ LINE-derived end (light blue) and are subsequently reverse transcribed and integrated into a new genomic target site. The 3′ sequences of the LINE and the SINE are shown in the same shade of light blue to emphasize their similarity to each other. The dark blue box represents a LINE-specific sequence. The taxon-specific and tRNA-derived sequences in the SINE are shown in green and yellow, respectively. (Courtesy of Norihiro Okada.)

In terms of their ability to recognize SINEs, there are two types of LINEs, **stringent LINEs** and **relaxed LINEs**. A stringent LINE can only mobilize SINEs if they have a 3′ end derived from the stringent LINE. A relaxed LINE does not have such a stringent requirement for the SINEs it amplifies. It is thought that all LINEs were originally of the stringent type. However, during evolution, some LINEs lost their stringency. This occurred independently in some plant and mammalian lineages. In mouse and rat, for example, the B2 SINEs, which originated from $tRNA^{Lys}$, have no *LINE-1*-derived tail at their 3′ end, yet they are amplified by *LINE-1*. A relaxed LINE, therefore, can genomically amplify a SINE regardless of its origin, whereas a stringent LINE cannot. If a genome contains both stringent and relaxed LINEs, then two different SINE structures may coexist, one with a LINE-related 3′ tail, and one with only poly(A) structure.

Rate of SINE evolution

Because of their sheer number, SINEs, especially *Alu* sequences, were initially thought to have a biological function. We now know that only an infinitesimal fraction of SINEs have been co-opted into functionality. The rest have become fossil elements subject to no form of selection. Indeed, their rate of evolution seems to be determined solely by the rate of mutation, and they evolve at the same rate as well-characterized functionless pseudogenes.

Retrosequences

Retrosequences (also referred to as **retrotranscripts** or **retrocopies**) are genomic sequences that have been produced through the reverse transcription of RNA and the subsequent integration of the resulting cDNA into the genome. Retrosequences may be thought of as "accidental retroelements" in that they have been produced with the aid of a reverse transcriptase from a retroelement but exhibit no special adaptation for retroposition such as a binding site for the enzyme. Indeed, each retrosequence type in a genome tends to be either unique or repeated only a few times, whereas nonautonomous retroelements are usually highly repetitive. Here, we describe retrosequences derived from either mRNA or transcripts of RNA-specifying genes.

A process of producing retrosequences is shown in **Figure 9.17**. If a gene is not transcribed within a germline cell, but only in specialized somatic cells, the creation of a retrosequence requires the RNA to cross cell barriers from the somatic cell to the germline cell. This can happen when an RNA molecule becomes encapsulated within the virion particle of a retrovirus and is then transported to the germline cell where it is reverse transcribed (Linial 1987). This process is referred to as **retrofection**. Retrofection appears to be very common in some taxa, such as mammals, but not in others.

Since retrosequences originate from RNA sequences, they bear marks of RNA processing and are hence also referred to as **processed sequences**. The diagnostic features of retrosequences include (1) the lack of introns; (2) precise boundaries coinciding with the transcribed regions of the gene from which they were derived; (3) stretches of poly(A) at the 3′ end; (4) short direct repeats at both ends (indicating that transposition may have been involved); (5) various posttranscriptional modifications, such as the addition

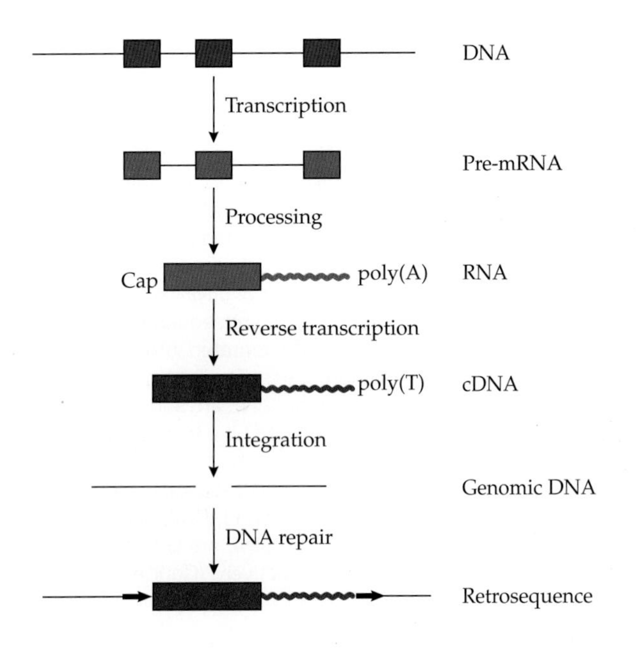

Figure 9.17 Creation of a processed retrosequence. The solid boxes represent exons. The wavy lines indicate a poly(A) tail in mRNA and the complementary poly(T) in cDNA. The DNA is transcribed into pre-mRNA, then processed into mRNA. The mRNA is reverse transcribed into cDNA, which becomes integrated into the genomic DNA. The gaps are repaired, so two direct short repeats flanking the inserted retrosequence are created (horizontal arrows).

or removal of short stretches of nucleotides; and (6) chromosomal positions different from the locus of the original gene from which the RNA was transcribed.

Retrogenes

A **retrogene** or **processed gene** is a functional retrosequence that produces a protein that is identical or nearly identical to the one produced by the gene from which the retrogene has been derived. There are three main reasons why it is highly unlikely that a reverse-transcribed sequence will retain its functionality. First, the process of reverse transcription is very inaccurate, and many differences (mutations) between the RNA template and the cDNA may occur. Second, unless a processed gene has been derived from a gene transcribed by RNA polymerase III, it usually does not contain the necessary regulatory sequences that reside in the nontranscribed regions, so it will likely be inactive even if its coding region remains intact after retroposition. Two processed metallothionein-I pseudogenes, one in human and one in rat, appear to represent such a case (Karin and Richards 1984; Anderson et al. 1986). Their coding regions are intact, and in the case of the rat pseudogene even the 5′ regulatory sequences are intact. Nevertheless, these sequences are not transcribed, presumably because of the lack of some regulatory sequences or the formation of unstable transcripts. Third, a processed gene may be inserted at a genomic location that may not be adequate for its proper expression. Indeed, in the vast majority of cases, a processed sequence is "dead on arrival."

How many functional retrogenes are there? Sakharkar et al. (2002) showed that 20% of mammalian protein-coding genes lack introns, but are they all retrogenes? The identification of retrogenes has been traditionally based on the assumption that both the parental gene and its descendant retrogene are present in the genome. Therefore, only genomic sequence loci that were homologous to introned genes were considered to be retrogenes. However, one should consider the possibility that the parental gene was lost or pseudogenized after duplication, and that the retrogene, which took over its function, does not have a multi-exon homolog. Ciomborowska et al. (2013) conducted a comparative analysis of human, chicken, and nematode genes and identified 25 "orphan retrogenes" in the human genome. These orphan retrogenes had introned homologs in other species, but none in the human genome. Interestingly, seven of these retrogenes are known to be associated with human disease. A compilation of retrocopies from animal genomes has been published by Kabza et al. (2014), raising the possibility that the fraction of restrosequences that is functional may be larger than previously thought. We note that the mere existence of retrogenes illustrates how opportunistic and how inventive the evolutionary process can be.

The human phosphoglycerate kinase (PGK) multifamily consists of an active X-linked gene, a processed X-linked pseudogene, and an additional autosomal gene. The X-linked gene contains 11 exons and 10 introns. Its autosomal homolog, on the other hand, is unusual in that it has no introns and is flanked at its 3′ end by remnants of a poly(A) tail, i.e., two diagnostic features of a processed sequence. Initially, a great deal was made of the fact that the autosomal PGK gene has not only maintained an intact reading frame and the ability to transcribe and produce a functional polypeptide, but has also acquired a novel tissue specificity: it is produced only in the male testis (McCarrey and Thomas 1987). However, we now know that transcription in the testis may not be indicative of function, because the testis possesses a transcriptionally permissive environment, in which there is as much as a thousandfold increase in RNA polymerase II, thus allowing for high transcription levels from even nonpromoter or weak promoter sequences (Soumillon et al. 2013). Many other putative retrogenes have been described in many species, but for most of them the only pertinent evidence we have is that they are intronless, that their reading frames are intact, and for a minority of them, that they are transcribed and translated. In most cases, we have no clue about the evolutionary forces driving their evolution subsequent to the integration of the retrosequence into the genome.

In a few cases, we have evidence that the retrogene not only maintained an intact reading frame, but also evolved under positive selection. Such is the case of a retrogene called *c1orf37-dup* (Yu et al. 2006). This intronless gene on human chromosome 3 was derived from an 8-exon gene called *c1orf37* on chromosome 5. Retrogene *c1orf37-dup* contains all the hallmarks of retroposition, a lack of introns, a remnant of a poly(A) tail at the 3′ end, and 14 bp direct repeats at both ends. It has retained the full reading frame of *c1orf37* (1,128 bp) without any insertion or deletion and is predicted to encode a 376-amino-acid peptide. The retrogene has no orthologs in chimpanzee, gorilla, or orangutan, indicating that it arose in the human lineage after its separation from chimpanzees. Two lines of evidence lead us not only to ascribe functionality to this retrosequence but also to believe that it has undergone adaptive evolution. First, only a single polymorphism (a nonsynonymous change from ATG to GTG) was discovered in a sample of 122 chromosomes from different continents. Nucleotide diversity was very low even for the 5′ and 3′ flanking regions. Second, the frequency of the derived character state (GTG) was very high (54%) for an allele so young. These findings suggest that the evolution of *c1orf37-dup* is being shaped by a selective sweep, and that we are in the middle of it. The second reason for believing that *c1orf37-dup* has evolved under positive selection concerns its pattern of expression. While *c1orf37* has a ubiquitous expression in all tissues, *c1orf37-dup* was only found in lung, pancreas, thymus, intestine, and blood; it was undetectable in heart, placenta, liver, muscle, kidney, spleen, prostate, testis, ovary, or colon. These findings indicate that *c1orf37-dup* may have acquired a new regulatory system.

In at least one case, it has been demonstrated that the expression of a recently acquired retrogene is strongly associated with and most probably the cause of a large phenotypic change (Parker et al. 2009). Fibroblast growth factor 4 is a protein that in mammals is encoded by the three-exon gene *Fgf4*. Most dogs have a single copy of this gene on chromosome 18. In several breeds, however, an intronless retrogene of *Fgf4* is found upstream of the original gene. All three exonic sequences are present in the retrogene, and the protein-coding regions are identical to those of the original gene, which is strongly indicative of the retrogene's newness. The retrogene is only found in short-legged breeds displaying limb morphology characteristic of chondrodysplasia, also known as disproportional dwarfism (**Figure 9.18**). The chondrodysplasic phenotype (shortened, curved bones) is due to the growth plates of the long bones calcifying too early in development. Chondrodysplasia is a primary require-

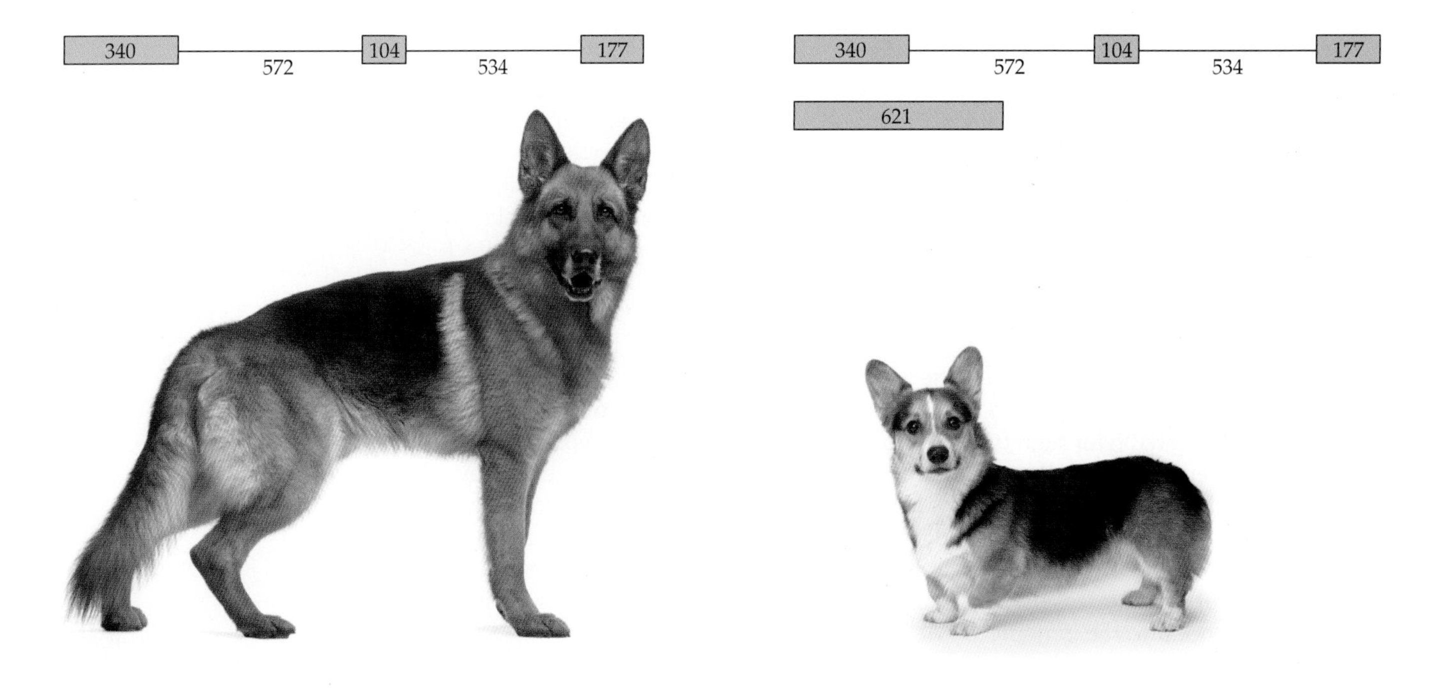

Figure 9.18 Long-legged dogs, such as the German shepherd (left), have a single three-exon gene encoding fibroblast growth factor 4 (*Fgf4*) on chromosome 18. Short-legged dogs, such as the Welsh corgi (right) have in addition an intronless *Fgf4* retrogene that is identical in its coding regions to the gene from which it was derived. The retrogene is almost certainly responsible for the chondrodysplasic phenotype. Exons are represented by green boxes, introns by horizontal lines. Numbers are exon lengths in nucleotides. (Left, photograph © iStock.com/GlobalP; right, photograph courtesy of David McIntyre.)

ment of the American Kennel Club for about a dozen official dog breeds, including dachshund, corgi, basset hound, and Pekingese. Because the retrogene was found in the identical genomic position in 19 short-legged breeds, it is likely that the chondrodysplasic phenotype arose only once, most probably before the divergence of early dogs into different breeds.

Retrogenes can, in principle, originate from and be integrated into both autosomes and sex chromosomes. However, there is a statistically significant excess of retrogenes that originated from the X chromosome and were retrotransposed into autosomes (Ding et al. 2006). Most of these X chromosome-derived autosomal retrogenes have evolved a male-specific expression profile, being expressed predominantly in germ cells, while their progenitor genes are usually housekeeping genes that are widely expressed. One explanation for this phenomenon, at least as far as mammals are concerned, is the "compensation hypothesis" (McCarrey and Thomas 1987). In many mammals, the sex chromosomes are condensed into an "XY body" during the meiotic pachytene period of spermatogenesis, and the sex-linked genes become transcriptionally silent, whereas the autosomal genes are active. The evolution of autosomal, testis-specific retrogenes that have originated in the X chromosome has been hypothesized to compensate for the inactivation of the X-linked genes that are required for meiotic germ cells. These retrogenes might play important roles in spermatogenesis.

Semiprocessed retrogenes

The vast majority of mammals contain a single preproinsulin gene. In contrast, myomorph (mouselike) rodents contain two. The rat preproinsulin I gene (and its ortholog in mouse) represents an instance of a **semiprocessed retrogene**. The gene contains a single 119 bp intron in the 5′ untranslated region. In comparison, its paralog, preproinsulin II, contains an additional 499 bp intron within the region coding for the C-peptide (Lomedico et al. 1979). All preproinsulin genes from other mammals, including other rodents, contain two introns as well. Moreover, the preproinsulin I gene is flanked by short repeats, and the polyadenylation signal is followed by a short poly(A) tract (Soares et al. 1985). These features suggest that the preproinsulin I gene has been derived from a partially processed preproinsulin II pre-mRNA, from which only one intron has been excised. Preproinsulin I appears to have been derived from an aberrant pre-mRNA transcript that initiated 500 bp upstream of the normal cap site. This retrogene might have maintained its function following integration into a new genomic location precisely because the aberrant transcript contained 5′ regulatory sequences not normally transcribed. The preproinsulin I gene arose in the common ancestor of the subfamily Murinae, which includes Old World mice and rats, approximately 20 million years ago (Shiao et al. 2008).

Some retrogenes may have acquired introns after their integration into the genome, and hence they only look like semiprocessed retrogenes. Most intron gains in retrons have occurred within untranslated exons in the 5′ flanking regions of the retrogene, and only few retrogenes acquired introns in their coding regions or their 3′ untranslated regions. Generally, retrogenes with 5′ introns display high transcription levels and show broader spatial expression patterns than intronless retrogenes. In contrast, 3′ introns seem to be down-regulated (Fablet et al. 2009). Distinguishing between semiprocessed retrogenes and retrogenes that have experienced intron gain can be accomplished by comparison of the location of the intron to the location of introns in the gene that gave rise to the retrogene.

Retropseudogenes

A **retropseudogene** or **processed pseudogene** is a retrosequence that has lost its function. It bears all the hallmarks of a functional retrosequence but has molecular defects that prevent it from being expressed. A comparison between a functional gene and a processed pseudogene is shown in **Figure 9.19**. Many processed pseudogenes are truncated during reverse transcription and retrofection; both 5′ and 3′ truncations are very common, and some retropseudogenes are truncated on both sides (Ophir

```
        Met Ala Thr Lys Ala Val Cys Val Leu Lys Gly Asp Gly Pro Val
SOD-1   ATG GCG ACG AAG GCC GTG TGC GTG CTG AAG GGC GAC GGC CCA GTG
          • ••   •       •  •    •  •   •           •  • •    •
ψ69.1   ATA ATG ATG AAG GTC ATG TAC ATG TTG AAG GGC CAG AGC CCG GTG
        Ile Met Met Lys Val Met Tyr Met Leu Lys Gly Gln Ser Pro Val

        Gln Gly Ile Ile Asn Phe Glu Gln Lys          Glu Ser Asn Gly
SOD-1   CAG GGC ATC ATC AAT TTC GAG CAG AAG  Intron  GAA AGT AAT GGA
             ••          •       •                        •       •
ψ69.1   CAG GCG A-C ATC CAT TT- GAG CAG AAG          GAA AAT --- GAA
        Gln Ala   Thr  Ser  Ile * **
                              STOP

        Pro Val Lys Val Trp Gly Ser Ile Lys Gly Leu Thr Glu Gly Leu
SOD-1   CCA GTG AAG GTG TGG GGA AGC ATT AAA GGA CTG ACT GAA GGC CTG
            • •  •           •           •       •           •   •
ψ69.1   CCA TTT ATG GTG T-C AGA AGC ATT ACA GGA TTG ACT GAA CGC CAG

        His Gly Phe His Val His Glu Phe Gly Asp Asn Thy Ala
SOD-1   CAT GGA TTC CAT GTT CAT GAG TTT GGA GAT AAT ACA GCA G Intron
          • •                                        •   •
ψ69.1   CAC AGA TTC CAT GTT CAT CAG TTT GGA G-T A-T AAC ACA G
```

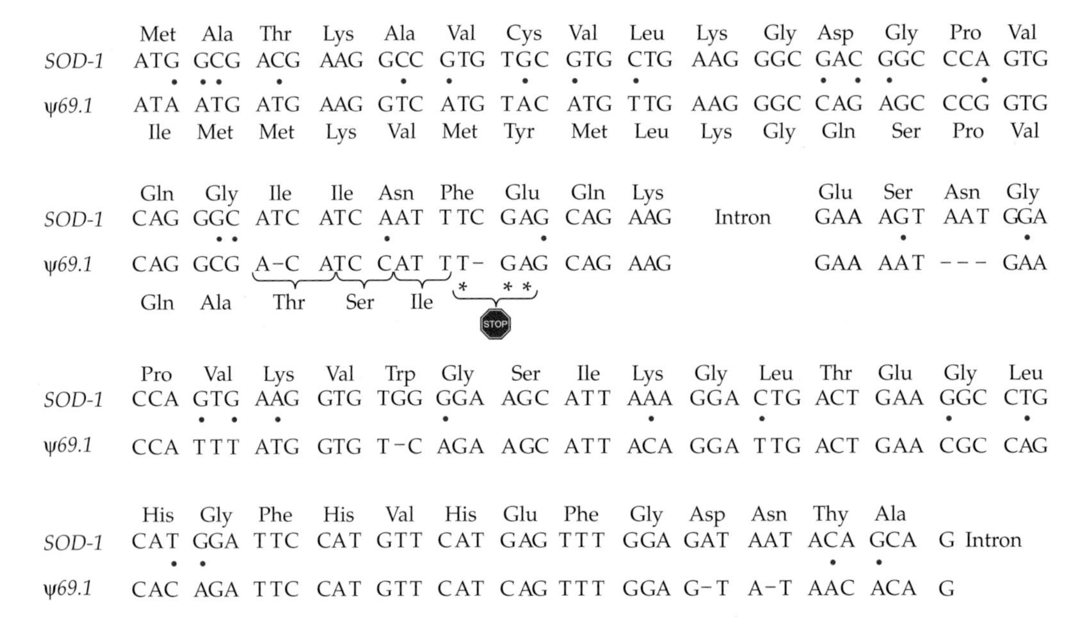

Figure 9.19 Comparison between the first two exons of the human Cu/Zn superoxide dismutase gene (*SOD-1*) and the homologous parts of a processed pseudogene (ψ*69.1*). Red dots denote substitutions; red minus signs denote deletions. Note the absence of introns and the premature termination codon. (Modified from Graur and Li 2000.)

and Graur 1997). Truncation can occur during (1) transcription initiation or termination (upstream or downstream of the normal sites), (2) RNA processing (e.g., faulty splicing), or (3) reverse transcription (e.g., failure of the enzyme to start properly or to complete the reverse transcription of the RNA molecule).

Most retropseudogenes are derived from processed RNA transcripts, though in rare cases retropseudogenes have been found that have been derived from unprocessed or semiprocessed RNA transcripts (Weiner et al. 1986). Retropseudogenes derived from all types of RNA transcripts are known. However, the vast majority of retropseudogenes are derived from RNA polymerase II and III transcripts (i.e., mRNA, snRNA, tRNA, 7SL RNA, and 5S rRNA). Only very few retropseudogenes derived from RNA polymerase I transcripts (i.e., 18S, 28S, and 5.8S rRNAs) are known (e.g., Wang et al. 1997).

Transfer RNA retropseudogenes are particularly interesting because they provide one of the most compelling pieces of evidence that processed pseudogenes are indeed derived through the reverse transcription of RNA. All nuclear tRNA molecules possess a CCA sequence at their 3′ terminus (**Figure 9.20**). This sequence is not encoded by the tRNA-specifying gene, but it is added enzymatically after transcription. In contrast, genomic tRNA retropseudogenes often contain the CCA sequence at their 3′ end (Reilly et al. 1982).

The number of retropseudogenes per gene is highly uneven. In the human genome, only 10% of protein-coding genes have retropseudogenes. The 79 ribosomal protein-coding genes in the human genome account for nearly 20% of the entire retropseudogene population. Other genes that have multiple retrotransposed pseudogenes include housekeeping genes—genes that code for structural proteins and metabolic enzymes. In general, genes that have multiple retropseudogenes tend to be highly expressed, have short transcripts, and have lower GC content (Zhang and Gerenstein 2004). Mouse orthologs of human genes that have multiple retrotransposed pseudogenes tend to have many retropseudogenes too.

```
tRNA^Phe-specifying gene   5′ TCTAAAGGTCCCTGGTTCGATCCCGGGTTTCGGCA--- 3′
tRNA^Phe                   5′ UCUAAAGGUCCCUGGUUCGAUCCCGGGUUUCGGCACCA 3′
ψtRNA^Phe                  5′ TCTAAAGGTCCCTGGTTCGATCCCGGGTTTCGGCACCA 3′
```

Figure 9.20 Comparison of $tRNA^{Phe}$-coding gene, $tRNA^{Phe}$, and a $tRNA^{Phe}$ pseudogene ($\psi tRNA^{Phe}$). The sequence CCA at the 3′ end of the functional $tRNA^{Phe}$ molecule is added posttranscriptionally. The fact that the CCA sequence frequently appears in the genomic sequences of tRNA retropseudogenes attests to its RNA origin.

Processed pseudogenes have been found in animals, plants, and even bacteria and viruses. However, although processed pseudogenes are abundant in mammals, they are relatively rare in other animals, including chicken, amphibians, and *Drosophila*. The rate of emergence of processed pseudogenes in the human and mouse lineages has been estimated to be about 1–2% per gene per million years (Sakai et al. 2007). Estimates of the number of processed pseudogenes in the human genome range from approximately 8,000 to 20,000 (Torrents et al. 2003; Zhang et al. 2003), but given the difficulty in identifying old functionless and rapidly evolving sequences, the true number of processed pseudogenes must be much higher. In comparison, *Drosophila* has only 34 processed pseudogenes, and the nematode *Caenorhabditis elegans* has a little over 200. These disparities cannot be attributed to differences in genome sizes or numbers of protein-coding genes. There are several possible explanations for the differences in relative abundance. First, different organisms may experience different rates of DNA loss, which mostly affects nonessential DNA. For example, the rate of DNA loss in *Drosophila* is approximately 75 times faster than that in mammalian pseudogenes (Petrov and Hartl 1997). This high rate of DNA loss may be in part responsible for the dearth of pseudogenes in *Drosophila* species.

The differences in retropseudogene number may also be due to the substrate specificity of the reverse transcriptases that generate processed pseudogenes. In humans, most reverse transcriptase is produced by *LINE-1* elements that are not very specific and can initiate reverse transcription of almost all RNA transcript. In contrast, the enzyme produced by the predominant retrotransposon in the chicken genome, CR1, exhibits high specificity (International Chicken Genome Sequencing Consortium 2004). Finally, it has been hypothesized that the differences in retropseudogene number are related to the characteristics of gametogenesis (Weiner et al. 1986). While spermatogenesis is very similar among mammals, chicken, amphibians, and *Drosophila*, mammalian oogenesis differs from that in the other animals by a prolonged lampbrush stage that lasts from birth to ovulation. This state of relatively suspended animation may last for up to 40 years in humans, but for only a few months in amphibians and for less than three weeks in birds, and it is virtually absent in *Drosophila*. If this hypothesis is correct, and retroposition in mammals does occur predominantly in the female germline, then it is expected that mammals should possess a greater number of retrosequences than other animals.

An additional prediction derived from this hypothesis is that retrosequences should be found in high numbers on the X chromosome and in intermediate numbers on autosomes but should be rare on the Y chromosome, which is not carried by females. As in the case of male-driven evolution (Chapter 4), the reason for this expectation is that autosomes spend half their time in males and females, X chromosomes spend two-thirds of their time in females and only one-third of their time in males, and Y chromosomes are always in males (Miyata et al. 1987). Thus, according to the oogenesis hypothesis, processed pseudogenes should be roughly 33% more abundant in X chromosomes than in autosomes and not be present in Y chromosomes. Indeed, Drouin (2006) found a significant excess of processed pseudogenes in human and mouse X chromosomes. The excess was in the range of 26–32%, i.e., very close to the 33% expectation. The oogenesis hypothesis could also be further tested by comparing the distribution of processed pseudogenes to that of unprocessed pseudogenes. There was no significant excess of unprocessed pseudogenes on the X chromosome. Processed pseudogenes were also found to be significantly less abundant in Y chromosomes than in autosomes, while the situation was reversed as far as unprocessed pseudogenes were concerned. We note that processed pseudogenes are found on Y, whereas the oogenesis hypothesis predicts they should be absent from this chromosome. Most retropseudogenes on Y are located within the pseudoautosomal region that pairs with the X chromosome during meiosis, and they may have been transferred from the X chromosome to the Y chromosome through recombination (Skaletsky et al. 2003).

How does the number of processed pseudogenes compare with the number of functional genes from which they were derived? The number varies immensely from gene to gene, with some genes being especially adept at producing large numbers of processed pseudogenes. For example, in the genome of the mouse, about 200 processed pseudogenes have been produced from a single gene for glyceraldehyde 3-phosphate dehydrogenase. In fact, the number of processed pseudogenes may even be underestimated in many cases, because old pseudogenes may have diverged in sequence from their parental functional genes to such an extent that they are no longer detectable by molecular probes derived from their functional homologs. It will not be at all surprising to find out that the total number of processed pseudogenes (detectable and undetectable) exceeds the number of functional genes. What determines the variation in retropseudogene number per functional paralog? Genes that give rise to many retropseudogenes were found to be (1) widely expressed, (2) highly conserved evolutionarily, (3) short, and (4) AT-rich (Gonçalves et al. 2000; McDonell and Drouin 2012).

Due to the ubiquity of reverse transcription, the genomes of mammals are literally bombarded with copies of reverse-transcribed sequences. The vast majority of these copies have been nonfunctional from the moment they were integrated into the genome. Moreover, such sequences cannot be easily rescued by gene conversion, since they are mostly located at great chromosomal distances from the parental functional gene (Chapter 7). The phenomenon of a functional locus pumping out defective copies of itself and dispersing them all over the genome has been likened to a volcano generating lava, and the process has been termed the **Vesuvian mode of evolution** (Lewin 1981).

The vast majority of retropseudogenes become nonfunctional and, hence, free from all selective constraint as soon as they are integrated as chromosomal sequences inside the genome. Because of their lack of function, pseudogenes are affected by two evolutionary processes (Graur et al. 1989b). The first involves the rapid accumulation of point mutations. This accumulation eventually obliterates the sequence similarity between the pseudogene and its functional homolog, which evolves much more slowly. The nucleotide composition of the pseudogene will come to resemble its nonfunctional surroundings more and more, eventually "blending" into them. This process has been called **compositional assimilation**. The second evolutionary process is characterized by pseudogenes becoming increasingly shorter compared with the functional gene. This **length abridgment** is caused by the excess of deletions over insertions. This process is very slow, however. It has been estimated that it would take about 400 million years for a mammalian processed pseudogene to lose half of its length. Of course, mammals are only about 200 million years old, so the mammalian genome is expected to contain major chunks of very ancient pseudogenic DNA. Obviously, these ancient pseudogenes have by now lost almost all similarity to the functional genes from which they have been derived. In other words, mammalian processed pseudogenes are created at a much faster rate than the rate by which they are obliterated by deletion. It has therefore been concluded that abridgment is too slow a process to offset the steady Vesuvian increase in genome size (Graur et al. 1989b).

Endogenous non-retroviral fossils

As mentioned previously, the vast majority of viral sequences found in sequenced genomes are fossil retroelements. More recently, endogenous sequences derived from viruses that neither transpose nor produce reverse transcriptase have been identified in many animal and plant genomes. Particularly puzzling was the discovery of genomic sequences derived from RNA viruses that neither have a DNA stage in their life cycle nor usually enter the host cell's nucleus. That said, **endogenous viral elements** (**EVEs**) derived from all groups of non-retroviruses (i.e., double-stranded DNA, single-stranded DNA, double-stranded RNA, and the two types of single-stranded RNA viruses) have been identified (Cock et al. 2010; Horie et al. 2010; Holmes 2011; Cui and Holmes 2012).

The "Ecology" of Transposable Elements

There are different ways to look at transposable elements from an evolutionary point of view. One way is to regard transposition as a mutation and, as such, most likely to be either deleterious or neutral. Of course, some transposable elements may have a positive effect on the host (see below); indeed, it would be strange if mutations never resulted in advantageous changes. Unfortunately, some researchers latch onto sporadic cases in which a transposable element has been shown to have acquired a function to claim that all transposable elements have a beneficial function (e.g., Biémont and Vieira 2006). This is equivalent to claiming that all mutations are beneficial based on the fact that a handful of mutations are.

Another viewpoint for dealing with transposable elements was put forward by Brookfield (2005). He suggested that we view genomes, particularly eukaryotic genomes, as ecological niches inhabited by a vast number of genetic elements whose primary function is to multiply in numbers within an essentially "passive" genome. In this view, the families of transposable elements correspond to species—competing with one another, cooperating with one another, and taking advantage of one another. Such an ecosystem may only rarely reach a stable and sustainable equilibrium in terms of the relative abundances of the different families of transposable elements. Usually, perturbations occur because of extrinsic or intrinsic factors. For example, the ecological equilibrium may be disturbed if one transposable-element family suddenly experiences a burst of replication or if it suddenly ceases to be active. Similarly, an invasive transposable element may change the absolute and relative distributions of the different transposable-element families.

A third way of dealing with transposable elements is to regard them as very successful genomic parasites (as well as parasites of genomic parasites). In this view, their success is due to a very simple reason: they cheat (Lisch and Slotkin 2011). That is, the spread of transposable elements is mostly explained by their ability to replicate faster than the host genome. Indeed, Hickey (1982) has shown that a high rate of transposition can overcome the effects of even the most stringent purifying selection regime against the spread of transposable elements.

In the following sections, we will briefly enumerate the mechanisms with which the host genome attempts to control the reproduction of transposable elements and the mechanisms employed by these "genomic parasites" to evade control. Later, we will present a simple model to predict transposable-element copy number.

Transposable elements and the host genome: An evolutionary tug-of-war

Transposable elements have been widely successful over the course of genome evolution, in some cases achieving "majority status" within a genome. For example, approximately two-thirds of the human genome consists of transposable elements, while the genome of maize (*Zea mays*) is composed of roughly 85% transposable elements. Is this a sign that the activity of transposable elements is unrestrained? Not at all. All multicellular organisms have evolved a suite of mechanisms whose primary function is to restrain transposable-element activity (Levin and Moran 2011). One class of such mechanisms involves small RNAs, i.e., endogenous small interfering RNAs (endo-siRNAs) and Piwi-interacting RNAs (piRNAs). Another mechanism involves the methylation of the C residues in the integrated transposable elements, thereby suppressing their expression. Finally, histone posttranslational modifications are employed to repress chromosomal activity of regions containing transposable elements. One of the biggest mysteries is how a host cell distinguishes between sequences belonging to the transposable element and other genomic sequences. This is a difficult question, as there is no sequence or motif that defines all transposable elements, nor do they have a characteristic size.

Of course, transposable elements are not passive participants in this tug-of-war. Judging by their sheer numbers, they have evolved means to escape the restraints imposed by the host system. Various strategies of evasion employed by transposable elements have been described by Lisch and Slotkin (2011). One such strategy involves the preferential insertion into regions transcribed by RNA polymerase II. Successful silencing of a transposable element at such a site may also result in the silencing of the nearby gene as well. Another evasive strategy consists of blurring the line between host and parasite by adopting parts of the host genome. Retrotransposable elements that contain tRNA or rRNA segments may be difficult to detect, or silencing by the host may be costly because it also silences bona fide RNA-specifying genes. Some transposable elements are also very small and AT-rich, making silencing by DNA or histone methylation impractical. Finally, some transposable elements minimize the harm to the host by targeting integration to regions devoid of genes, such as heterochromatin and telomeric regions, and some even "earn an honest living" by playing a vital role for the host, such as telomere maintenance (p. 427).

A system composed of transposable elements and their host should sooner or later reach a state of equilibrium between the regulatory mechanisms exerted by the host and the evasion strategies employed by the transposable elements. However, stress—induced by environmental changes, inbreeding, or the introduction of a novel transposable element into the system—may disrupt the equilibrium, usually by unleashing massive bursts of transposition. One such burst has been documented experimentally in *Arabidopsis* (Tsukahara et al. 2009). With their capacity to drive nonadaptive host evolution, mobilized transposable elements can rapidly restructure the genome. This process has been likened to a punctuated equilibrium-like escape from stasis and a prelude for diversification and speciation (Zeh et al. 2009).

Transposable elements and segregation distortion

One of the most puzzling phenomena in transposable-element evolution is not so much their ability to spread within a genome, but their ability to increase their frequency in the population. The reason that this is puzzling is that on average each transposition event is deleterious. While transposing within their host genome, transposable elements may disrupt essential gene functions (e.g., by insertion within the coding region of a gene) and thus decrease host fitness. Rarely, a transposition event causes an increase in fitness, but such instances are exceedingly rare and, hence, negligible.

Let us consider, for example, the spread of *P* elements within *Drosophila melanogaster* populations that has occurred in a very short period of time (Struchiner et al. 2005). On average 1% to 30% of *P* element transposition events result in a lethal mutation. Moreover, studies indicate that mean homozygous fitness, viability, and fertility can be reduced by as much as 55% in *Drosophila* strains invaded by *P* elements. Yet, despite decreasing their hosts' fitness, such elements fix in sexual populations. The reason for their increase in population frequency must be their ability to increase their representation in their hosts' gametes to a greater extent than they decrease their hosts' fitness by transposition. Through replicative transposition to homologous chromosomes, some transposable elements engage in a kind of segregation distortion (Chapter 2). For example, the F1 generation of a cross between an individual whose gametes all contain transposable elements and one not containing any transposable elements would be expected to have transposable elements in 50% of its gametes. In reality, approximately 70% of the gametes have been found to contain one or more copies of the *P* element. What happened is that the *P* element behaved like a contagious particle and "infected" some or all of the chromosomes that did not previously carry it. This "segregation distortion" would confer a 20% (70% – 50%) selective advantage on *P*-carrying gametes. Empirical studies indicate that the insertion of a *P* element reduces the fitness of its carriers by 19%. Even so, gametes carrying the element would be found in 56.7% (81% × 70%) of the gametes and thus have a 6.7% advantage over noncarriers. Successful transposable elements "cheat" the meiotic segregation laws more than they harm their hosts, and such elements can easily fix in the host population.

Evolutionary dynamics of transposable-element copy number

The simplest models dealing with transposable-element dynamics attempt to explain the main features of the distribution and abundance of transposable elements within the host population as functions of basic genetic mechanisms, such as transposition and excision rates, selection on host individuals in relation to number of copies of elements, and mechanisms of self-regulation by the transposable elements. Other models emphasize the "community ecology" aspects of the interaction between transposable elements and their hosts (e.g., Leonardo and Nuzhdin 2002; Brookfield 2005; Venner et al. 2009). Finally, there are models that incorporate elements from both approaches (e.g., Struchiner et al. 2005) and models that also incorporate competition between different families of transposable elements within a genome (e.g., Le Rouzic and Capy 2006).

Here, we consider a simple deterministic model of describing the dynamics of transposable-element copy number (Charlesworth 1985). In this model, the number of transposable elements within a genome is determined by three factors: (1) u, the probability that a transposable element produces a new genomic copy by replicative transposition, (2) v, the probability that the element is excised, and (3) the intensity of selection against increased numbers of transposable elements within the genome. The values of u and v have been determined experimentally for many transposable elements. In laboratory populations of *Drosophila*, transposition rates were found to vary among transposable elements and among species from 10^{-3} to 10^{-5} per element per generation (e.g., Charlesworth et al. 1992; Vieira and Biémont 1997). Excisions are much rarer, although not unheard of (Nuzhdin 1999). Thus, in the absence of selection against the transposable element, the number of copies in the genome is expected to increase indefinitely.

If the number of transposable elements is maintained at an equilibrium—an assumption that may or may not hold in nature—then selection must operate against an increase in copy number. Indeed, there are indications that viability decreases with copy number. Let us assume that the fitness of an individual, w, decreases with copy number, n. The justification for this assumption is that the insertion of a transposable element frequently has deleterious effects, say due to its insertion into a coding region, and that with increasing numbers, the probability of a deleterious effect increases. It can be shown that as long as w decreases with n in a monotonic fashion (i.e., when the addition of each transposable element always decreases fitness), then regardless of the exact relationship between n and w, the mean fitness of a population at equilibrium, $\bar{W}$, relative to an individual lacking transposable elements is

$$\bar{W} = e^{-n(u-v)} \tag{9.1}$$

In the case of the genome of an inbred laboratory strain of *D. melanogaster*, there are 5,390 annotated elements belonging to 121 families (Kaminker et al. 2002; Bergman et al. 2006). Since v is much smaller than u, $u - v \approx u$. Assuming that $u = 10^{-5}$, we obtain $\bar{W} = 0.95$. The reduction in fitness is $s = 1 - 0.95 = 0.05$. If we assume that the effect on fitness of each transposition event is independent of that of another transposition event, then the reduction in fitness with each additional transposon is approximately $0.05/5{,}390 \approx 10^{-5}$. Such a small selection coefficient indicates that (1) the selection coefficient needed to control copy number need not be large, and (2) the number of copies of transposable elements within a genome is strongly influenced by random genetic drift.

An alternative to selection against increase in copy number would be a mechanism of self-regulated transposition, i.e., a rate of transposition that decreases with copy number or a rate of excision that increases with copy number (Charlesworth 1988; Charlesworth and Langley 1989). Evidence for direct control of transposition rate by transposable-element copy number is provided by the *I* retrotransposon responsible for the *I-R* system of hybrid dysgenesis. The activity of the *I* element promoter has been found to be sensitive to the number of copies of the first 186 nucleotides of the element, which constitute the 5′ untranslated region (Chaboissier et al. 1998). At

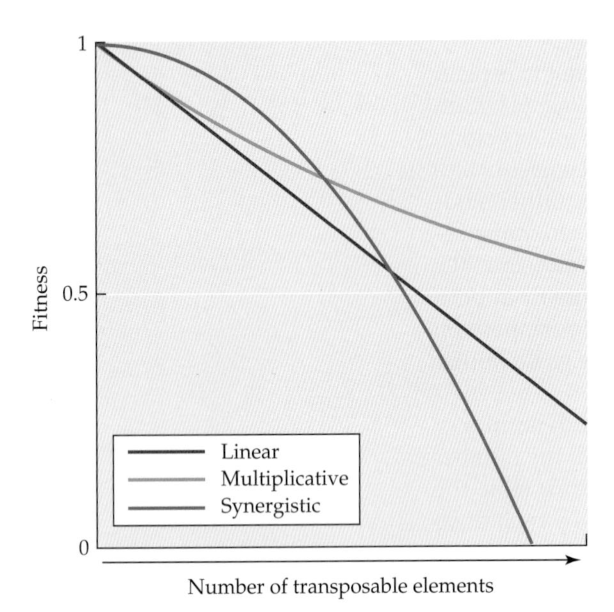

Figure 9.21 Three possible interactions among replicative transposition events and their effects on the fitness of the host. In the linear (or additive) model, the fitness decline due to the insertion of transposable elements is the sum of the individual effects (a straight line). Under the synergistic epistasis model, as the number of transposable elements increases, the deleterious interactions grow stronger, resulting in an ever faster decline of fitness (a concave curve). In the multiplicative epistasis model, the deleterious effects of adding a new transposable element get weaker and weaker as the number of transposable elements increases (a convex curve).

low copy number, the *I* element transposes frequently, but this rate drops as its copy number increases, falling to a very low level when the number of transposable elements within the genome reaches approximately 15 copies per haploid genome.

Finally, it is possible that copy number is determined by the effective population size. That is, the addition of each transposable element may be regarded as a neutral or slightly deleterious mutation. In taxa with small effective population sizes, such mutations may be fixed because of the ineffectuality of purifying selection (Le Rouzic et al. 2007; Lynch 2007). Such a nonselective process could account for the spread of transposable elements in multicellular eukaryotes, which typically have much smaller population sizes relative to prokaryotes. Of course, if copy number is mainly determined by stochastic effects, we need to take into account the dynamics of fitness reduction as a function of number of transposable elements.

In **Figure 9.21**, we show three possible forms of interaction among transposition events (Lee and Langley 2010). In the **linear** (or **additive**) **model**, the fitness decline due to the insertion of n transposable elements equals the sum of the fitness effects of each of these n elements independently. If this is the case, then the equilibrium copy number of transposable elements will be determined by the interaction between the degree of fitness reduction caused by the addition of each transposable element and the effective population size. Thus, for a given fitness reduction per transposable element, the number of transposable elements tolerated by the genome will be higher for species with low effective population sizes than for species with large effective population sizes. If the fitness effects can be described by the **multiplicative epistasis model**, i.e., if the deleterious effects of adding a new transposable element get weaker and weaker as the number of transposable elements increases, then the equilibrium number of transposable elements is expected to be larger than that expected under the linear model. If the fitness effects can be described by the **synergistic epistasis model**, i.e., if the deleterious effects of adding a new transposable element get stronger and stronger as the number of transposable elements increases, then the equilibrium number of transposable elements in the genome is expected to be much lower than that expected under the linear model.

Genetic and Evolutionary Effects of Transposition

With very rare exceptions (e.g., the unicellular parasites *Encephalitozoon cuniculi* and *Bathycoccus prasinos*) all eukaryotic genomes possess transposable elements, sometimes in huge numbers. The same is true for the majority of prokaryotic genomes. Here we will deal with the genetic and evolutionary effects of transposition. We will deal with a range of effects, from the more obvious ones, such as the fact that the insertion of a transposable element into the genome constitutes a mutation that may have phenotypic effects, to the most unexpected ones, such as the fact that transposition may create reproductive barriers within a species. The contribution of transposable elements to genome size will be discussed in Chapter 11.

Transposable elements as mutagens

As proven by countless human genetic diseases (Deininger 2011), transposable elements have an enormous mutagenic potential. Deletions, insertions, frameshifts, inversions, duplications, translocations, genome rearrangements, and the generation of intron-like sequences have all been noted as consequences of transposition events. In *Drosophila*, for instance, transposable elements are responsible for 50–80% of the mutations that result in visible phenotypic changes (Ashburner et al. 2005). In humans, active retrotransposable elements (LINEs, SINEs, and their relatives) are responsible for many de novo genetic disease-causing mutations, many polymorphic genetic diseases, and many cancers (e.g., Wallace et al. 1991; Lamprecht et al. 2010).

Transposable elements can generate a wide spectrum of mutations, from subtle regulatory changes to gross genomic rearrangements. Here, we will distinguish between direct effects of transposition, i.e., the consequences of the insertion (or excision) of a transposable element into (or from) the genome, and indirect effects, i.e., the consequences of the presence of transposable elements in the genome.

The insertion of a transposable element into the genome will have no phenotypic consequences if the element is inserted in the vast nongenic regions. If, on the other hand, the transposable element is inserted within a gene or adjacent to it, it may alter gene expression. Similarly, excision of transposable elements may be imprecise, resulting in the addition or deletion of bases.

In the simplest case, the insertion of a transposable element into the coding region of a protein-coding gene will almost certainly obliterate the reading frame, which in turn may result in drastic phenotypic and evolutionary consequences. For example, the loss of the gelatinous copulatory plug in *Caenorhabditis elegans* was due to the insertion of a retrotransposon into an exon of a mucin-like coding gene. The reduced male-male competition associated with the origin of hermaphroditism in *C. elegans* may have permitted the global spread of this loss-of-function mutation (Palopoli et al. 2008).

Another example concerns the loss of teeth in baleen whales, a condition that may have started with the inactivation of one of the genes involved in enamel production through the insertion of a transposable element. Meredith et al. (2011) discovered that the *enamelysin* gene, which encodes a tooth-specific metalloproteinase that is expressed in the early stages of enamel development, was inactivated in the common ancestor of living baleen whales by the insertion of a SINE called *CHR-2* into its second exon.

Another example of transposable-element mutagenesis is of particular historical interest because it concerns the wrinkled phenotype in the pea seeds that feature prominently in Gregor Mendel's 1865 paper. Bhattacharyya et al. (1990, 1993) discovered that the wrinkled variant arose through the insertion of a 0.8 Kb transposable-element sequence into the reading frame of a gene encoding a starch-branching enzyme (**Figure 9.22**). The pea sequence was found to be very similar to the *Ac/Ds* family of transposable elements from maize. Due to the inactivation of the gene, the total amount of amylose (starch) and the proportion of amylopectin (branched starch) were greatly reduced in homozygotes, while the amount of sucrose was increased. Increased sucrose in seeds causes a greater uptake of water, thereby increasing seed size. During seed desiccation, these seeds lose more water than seeds from plants possessing a functional starch-branching enzyme, resulting in a wrinkled pea.

Transposition may also affect gene expression without directly affecting the reading frame. For example, a transposable element may contain regulatory elements, such as promoters, which may influence the mode and rate of transcription of nearby genes. For example, the presence of an insertion sequence in the promoter region of the *gal* operon of *E. coli* results in the operon being expressed constitutively, i.e., the regulation of the operon is disrupted (Shapiro 1983). Another example involves retroviruses. The long terminal repeats of retroviruses often contain strong enhancers, which greatly influence the expression of nearby genes (e.g., Lamprecht et al. 2010). Similarly, *Ty* (which stands for "transposon yeast") elements in *Saccharomyces cerevisiae* are known to increase the expression of downstream genes. This may be beneficial in some specific circumstances, although in most cases the metabolic imbalance produced by such a change is probably detrimental.

One case in which the overtranscription of a gene due to the presence of a transposable element turned out to be beneficial for its carriers concerns insecticide resistance. One type of DDT resistance in *Drosophila melanogaster* was found to be caused by the overexpression of a cytochrome P450 gene called *Cyp6g1*. The overexpression

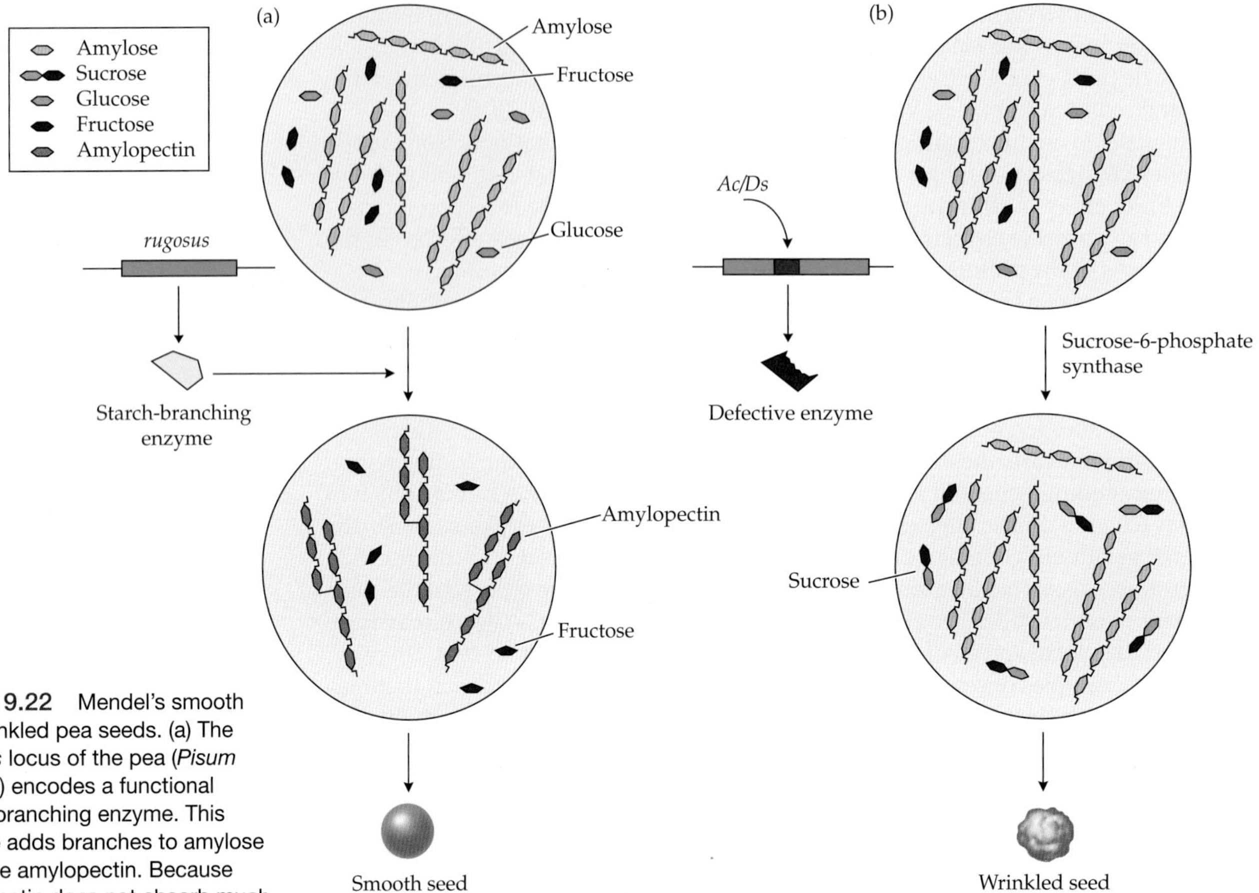

Figure 9.22 Mendel's smooth and wrinkled pea seeds. (a) The *rugosus* locus of the pea (*Pisum sativum*) encodes a functional starch-branching enzyme. This enzyme adds branches to amylose to create amylopectin. Because amylopectin does not absorb much water, the seeds do not enlarge much when hydrated and therefore remain smooth following desiccation. Because amylose and amylopectin are made of glucose units, less glucose joins fructose to make sucrose in smooth seeds than in wrinkled seeds. (b) Insertion of an *Ac/Ds* transposable element into *rugosus* disrupts the reading frame and an inactive starch-branching enzyme is produced. A greater water uptake increases the size of hydrated seeds, and water loss through desiccation causes the seeds to become wrinkled. Since less glucose is converted to amylopectin, glucose may combine with fructose to form sucrose (thus wrinkled seeds taste sweeter than smooth ones). (Modified from Graur and Li 2000.)

was triggered by the upstream insertion of an *Accord* transposable element (Daborn et al. 2002).

In a study of the transcriptome of a rice strain called EG4 that contains 1,000 *mPing* DNA transposons, in comparison with only 50 copies in the Nippobare strain, Naito et al. (2009) discovered that gene expression of 82% of 31,439 alleles was indistinguishable between the two strains. The other 18% of the alleles were either upregulated or downregulated in the presence of the DNA transposon.

Transposable elements that contain splice donor or acceptor sites may affect the processing of the primary RNA transcript even if the element has incorporated itself within a noncoding region of a gene, such as an intron. For example, a *MIR* element was found to determine the site of alternative splicing in the human acetylcholine receptor gene (Murnane and Morales 1995).

Many transposable elements promote large segmental changes, for example by moving a DNA sequence from one genomic location to another. In particular, there is evidence that *helitrons*, which most probably multiply by rolling-circle replication, can produce transcripts containing segments from different genes, giving rise to the hypothesis that these transposable elements have a role in exon shuffling. Indeed, Morgante et al. (2005) and Kapitonov and Jurka (2007) found evidence that autonomous *helitron* elements seem to continuously produce new nonautonomous elements responsible for the replicative insertion of gene segments into new locations, thus bringing into close contact gene fragments from otherwise distantly located loci. Many of these chimerical gene candidates are transcribed (Bennetzen 2005), but a proven case of functional gene creation by rolling-circle activity remains to be demonstrated.

Finally, as mentioned above, transposons in bacteria often carry genes that confer antibiotic or other forms of resistance on their carriers. Plasmids can carry such transposons from cell to cell, either within a species or between different bacterial species, so resistance can spread quickly throughout a population or an entire bacterial ecosystem. Thus, transposons may help species to survive adverse environments.

So far, we have enumerated a number of direct effects due to the transposition event itself. The indirect effects of transposition can be equally dramatic. The mere presence in the genome of transposable elements—especially interspersed repeated copies of a transposable element—can promote gross genomic rearrangements, such as inversions, deletions, translocations, and duplications. These rearrangements can take place as an indirect result of transposition if, as a result of transposition, two sequences that previously had little similarity with each other are now sharing a similar transposable element, making unequal crossing over between them possible. This may occasionally produce a beneficial gene duplication or gene rearrangement. For example, a duplication of the entire growth hormone gene early in human evolution might have occurred via an *Alu-Alu* recombination event (Barsh et al. 1983). Similarly, the duplication that gave rise to the $^{G}\gamma$- and $^{A}\gamma$-immunoglobin genes may have resulted from recombination between two *LINE-1* sequences in an early ancestor of simians (Maeda and Smithies 1986; Fitch et al. 1991).

In the majority of cases, however, unequal crossing over constitutes a deleterious mutation. In the human genome, for instance, recombination between distant *Alu* sequences has been found to be responsible for many hereditary afflictions. Take, for instance, the gene encoding the low-density lipoprotein (LDL) receptor. Numerous unequal crossing-over events, facilitated by the presence of multiple *Alu* sequences within its introns, have given rise to numerous exon deletions and duplications in this gene. In most cases these changes result in a hereditary disorder called familial hypercholesterolemia, which is characterized by very high levels of low-density ("bad") cholesterol and extreme risks of heart attack at an early age. One such unequal crossing-over event between two *Alu* sequences in introns 4 and 5 resulted in the deletion of exon 5 (Hobbs et al. 1986). Exon 5 normally encodes the sixth repeat of the ligand-binding domain of the LDL receptor, and its deletion resulted in a shortened protein that reaches the cell surface and reacts with antireceptor antibodies but does not bind LDL. A deletion of exon 14 by the same mechanism has also been observed (Lehrman et al. 1986). Lehrman et al. (1987) have shown that the duplication of seven exons in the LDL receptor gene caused by an *Alu-Alu* recombination also resulted in familial hypercholesterolemia.

The insertion of transposable elements into members of a multigene family can reduce the rate or limit the extent of gene conversion between the members of the family, thereby increasing the rate of divergence between duplicate genes (Hess et al. 1983; Schimenti and Duncan 1984). This may happen because the process of gene conversion involves pairing of homologous DNA (Chapter 7), and nonhomologous regions created by the uneven insertion of transposable elements might reduce the frequency of such pairings and consequently the chance of gene conversion between duplicate genes.

There is also evidence that some transposable elements may cause an increase in the rate of mutation. For example, strains of *E. coli* that contain the transposable element *Tn10* were found to have elevated rates of insertion, while the insertion sequence *IS1* increases the probability of deletion in adjacent regions. Under most conditions, this trait will be deleterious to the carrier. However, under severe environmental stress it is possible that an elevated rate of mutation might be advantageous because some mutations may be better suited to the new circumstances and their carriers will have a higher fitness than noncarriers.

Finally, the presence of a transposable element can turn an otherwise immobile piece of DNA into a mobile one. For example, the presence of *IS1* elements on both sides of a gene may turn a gene into a composite transposon-like entity and may allow the gene to transpose into plasmids, by which it can be transferred to other cells. An additional method by which a transposable element confers mobility on other sequences is exemplified by autonomous retroelements, which encode reverse transcriptases that confer mobility on SINEs and retrosequences. Some retrotransposons, such as the *LINE-1* elements in the human genome, can comobilize a 3′ flanking region and move it to a new location. The co-mobilization is accomplished by the aberrant overtranscription of the element in the 3′ direction, followed by its reverse transcription and integration back into the genome at a different location (Moran et al. 1999; Pickeral et al. 2000). Some transposable elements that multiply by replicative transposition may incorporate functional regulatory sequences, such as binding sites, and consequently facilitate the expansion of these elements within a genome (e.g., Schmidt et al. 2012).

Transposable elements are part of the genetic heritage of all organisms, and as long as they continue to move and multiply within and between genomes, they will produce mutations. The vast majority of these mutations will occur in unconstrained functionless regions of the genome and will behave like neutrall mutations. Some of them will be deleterious, and depending on the degree of fitness reduction and the effective population size, they will be selected against. A small minority of them will evolve a useful function and will be co-opted by the host genome.

Transposable elements and somatic mosaicism

In evolutionary studies we are mostly interested in changes occurring in the germline. We note, however, that there are ~150 full-length retroelements in the human genome and ~3,000 in the mouse genome. These intact elements are potentially active, not only in the germline but also in somatic cells. In the case of *Alu* elements, it has been noted that the rate of transposition in somatic cells is many orders of magnitude higher than in germline cells. Of course, it is unlikely that transposition will occur simultaneously and equally in all somatic cells, and hence the differential transpositional activity in different somatic cells will create **somatic mosaicism**, a situation in which the somatic cells of the body are of more than one genotype. Somatic mosaicism due to transposable elements occurs when individual somatic cells in a multicellular organism harbor different transposable elements at different insertion sites.

Coufal et al. (2009) discovered de novo *L1* retroposition in the human brain, thereby illustrating the fact that transposable elements contribute to somatic mosaicism within an individual. They found that each brain cell contained approximately 80 additional copies of the *L1* retrotransposon in comparison with heart and liver cells. So far, approximately 15,000 new somatic SINE and LINE insertions have been identified in the human brain (Baillie et al. 2011), along with considerable neuron-to-neuron genomic variation (Singer et al. 2010; Evrony et al. 2012). In the *Drosophila* brain, somatic mosaicism is considerable and is driven by both DNA transposition and retroposition (Perrat et al. 2013). This "jumping-gene roulette" (Martin 2009) has fueled all manner of unfounded speculation on its effects, particularly on the human brain. Sadly, as with all other mutations, the effects of a random insertion of transposable elements into the genome of a brain cell are hard to predict.

The molecular domestication of transposable elements

The vast majority of transposable elements neither add to nor detract from the fitness of their host. In fact, most of them are not even active as transposable elements. Most transposition events are neutral in terms of their effect on fitness; a few are deleterious. Rarely, transposable elements are co-opted by the organism to perform a useful function. As stated by Orgel and Crick (1980), "It would be surprising if the host genome did not occasionally find some use for particular selfish DNA sequences, especially if there were so many different sequences widely distributed over the chromosomes."

The term **exaptation** was coined by Gould and Vrba (1982) to denote the co-option of a sequence or a structural feature for a function other than that for which it was originally shaped by natural selection. As "selfish DNA," transposable elements have been shaped by natural selection to multiply and evade the host machinery that functions to keep them in check. In the course of evolution, however, some transposable elements or parts of transposable elements have become established as useful components of the host genome; they have been exapted for functions unrelated to their being transposable elements. Miller et al. (1996) coined the term **molecular domestication** in reference to the process of transposable-element exaptation. One of the most important contributions of transposable elements to host genome evolution is as a source of raw material that can be used for the assembly of new genes or functions. Transposable elements have numerous properties that predispose them for molecular domestication. For example, some transposable elements contain palindromic structures that may be co-opted by microRNA genes (Feschotte and Pritham 2007), while others have been reported to contribute useful domains in the origination of chimerical protein-coding genes (Wang et al. 2006; Feschotte and Pritham 2007; Jurka et al. 2007; Elrouby and Bureau 2010; Schmitz and Brosius 2011).

In the following sections, we describe several instances of molecular domestication. They were chosen mainly for their suitability to illustrate the importance and range of exaptations by transposable elements. Many other examples can be found in Volff (2006) and Alzohairy et al. (2012). We note, however, that the evidence for transposable element exaptation derived from genome-wide studies should be treated with caution (de Souza et al. 2013).

THE BIRTH OF NEW EXONS THROUGH EXONIZATION AND NEOFUNCTIONALIZATION OF TRANSPOSABLE ELEMENTS The number of exons in a genome is not static; rather, it changes continuously through evolutionary processes of exon gain and loss. Here we concern ourselves with the process of exonization by exaptation or neofunctionalization of transposable elements. The particular example that will be used concerns the molecular mechanisms leading to the exonization of *Alu* elements (Sorek 2007).

Many *Alu* sequences contain sequences that are similar to splice sites and are sometimes referred to as **pseudosplice sites**. Specifically, the antisense orientation of many *Alu* sequences contains a poly(T) stretch that can serve as the polypyrimidine tract that is necessary upstream of the 3′ splice site. This poly(T) is the complement to the terminal poly(A) of many retrotransposons. Following insertion of an *Alu* into an intron in the antisense orientation, mutations that change the pseudosplice site into a real splice site may occur, after which the splicing machinery will recognize part of the *Alu* sequence as a bona fide exon (**Figure 9.23**).

Several studies have characterized the series of mutations needed to occur within an *Alu* element in order for it to be exonized. In fact, by comparison of exonized and nonexonized *Alus*, it has been shown that the number of mutations leading to exonization can be surprisingly small. Such understandings

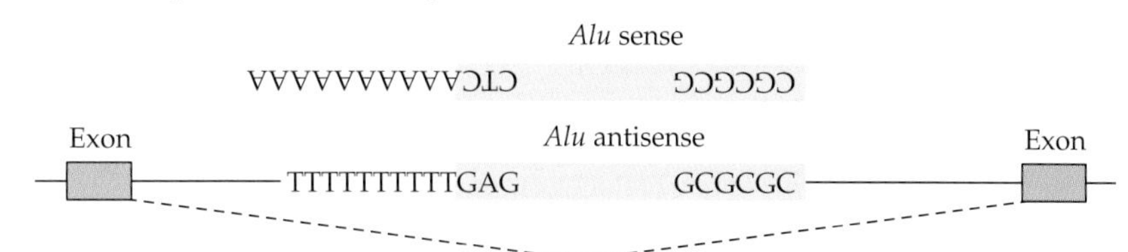

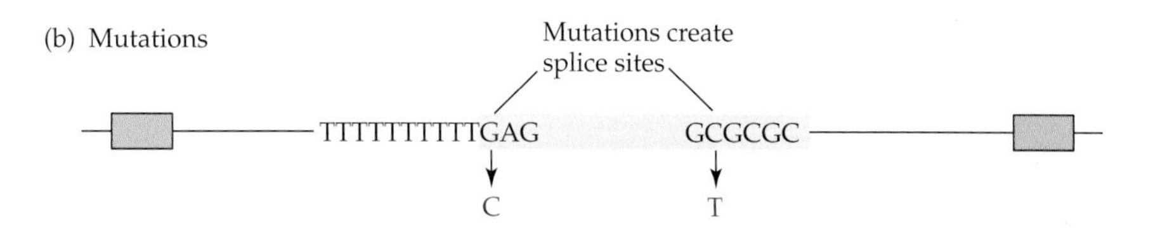

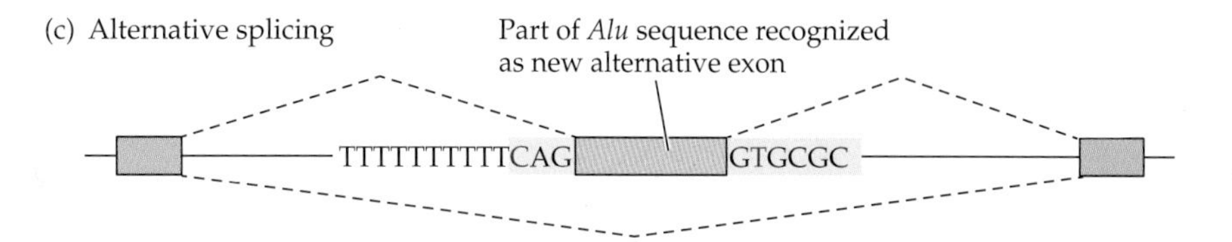

Figure 9.23 Schematic description of the partial exonization of an *Alu* element. (a) An *Alu* is inserted into an intron by retrotransposition. The antisense orientation of many *Alu* elements contain pseudosplice sites, i.e., sequences that are similar to splice sites. (b) Subsequently, mutations (red letters) turn the pseudosplice sites into splice sites. (c) These mutations cause part of the *Alu* sequence to be recognized as a new alternatively spliced exon (yellow box). The two alternative splicing patterns are shown above and below the pre-mRNA. Typically, the *Alu*-containing transcript is a minor splice form, as in most cases the created splice sites are weak. (Modified from Sorek 2007.)

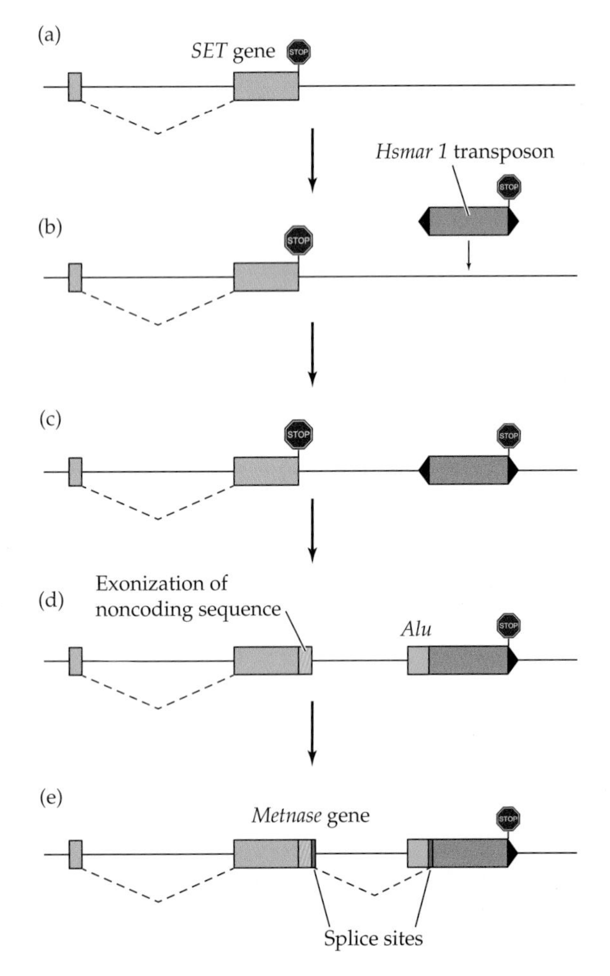

Figure 9.24 A simplified chronology of the origination of the *Metnase* gene. (a) The *SET* protein-coding gene consists of two exons (green rectangles) separated by an intron (black line). The structure of the gene is conserved throughout all mammalian genomes examined to date and, with few exceptions, is terminated by a stop codon. (b) The *Hsmar1* transposon is shown with terminal inverted repeats (black triangles) and the transposase coding sequence (pink rectangle). (c) The transposon was inserted downstream of the SET-coding gene in the primate lineage, after the split between tarsier and the anthropoids but before the divergence of extant anthropoid lineages (58–40 million years ago). (d) The insertion of an *Alu* sequence (blue square) probably obliterated the 5′ terminal inverted repeat. A deletion or a sense mutation obliterated the stop codon in the second exon of the SET-coding gene. These changes most probably resulted in the exonization of a noncoding sequence (yellow). Ultimately, two splice sites (heavy dark blue lines) completed the creation of the *Metnase* gene. (Modified from Cordaux et al. 2006.)

have enabled researchers to synthetically exonize intronic *Alu* elements by introducing the appropriate mutations into its sequence.

In principle, the insertion of a ~100 bp exonized *Alu* sequence in the middle of a coding region should be deleterious, since in the vast majority of cases the new exon would either frameshift the reading frame or introduce a premature stop codon into it. How, then, can such exonizations be tolerated? The answer lies in alternative splicing. Newly born exons are alternatively spliced at a much higher frequency than old exons. Specifically, nearly all exonized *Alu* elements are alternatively spliced. Moreover, newly created exons are spliced into the RNA transcript at very low frequencies, i.e., only a minute fraction of the transcripts will contain the new exon. Thus, the vast majority of transcripts will retain their old sequence, which essentially means that the exonization of an *Alu* element constitutes a neutral or at most a slightly deleterious change. Under these conditions, exonized sequences can increase the size of the transcriptome while at the same time maintaining the intactness of the original proteome. This situation allows evolution to "test" the new exons. Successful exons may become more prominently expressed, and some may even become constitutively spliced. Indeed, it has been shown that inclusion levels of *Alu*-derived exons become higher with time, and that this increase is associated with mutations creating stronger splice sites.

We note, however, that the vast majority of exonized *Alu* sequences do not add a great deal of functionality to the proteome. Indeed, more than 79% of all *Alu*-derived exons create frameshifts or introduce premature stop codons. Moreover, although exonized transposable elements have been detected in 4% of RNA transcripts, they are found in only 0.1% of proteins. This indicates that most exonized sequences are nonfunctional and that they are "tolerated" in transcripts only because they are alternatively spliced and the frequency of their inclusion into RNA is low.

THE CHIMERIC *METNASE* GENE The *Metnase* (or *SETMAR*) protein-coding gene has three exons that extend over 13.8 Kb. In the human genome it is located on the short arm of chromosome 3. Structural studies have indicated that the Metnase protein exists as a dimer. The first two exons of the *Metnase* gene encode a protein with significant similarity to a domain called SET that is found among others in the enzyme histone methyl transferase. The third exon is a transposase sequence derived from a *mariner*-like transposon called *Hsmar1*.

A simplified chronology of the birth of the *Metnase* gene is shown in **Figure 9.24**. The *Hsmar1* transposon and the *Metnase* gene are found in all anthropoid lineages (humans, apes, Old World monkeys, and New World monkeys) but not in the tarsier, indicating that the gene was created 58–65 million years ago. Comparison of the rates of nonsynonymous and synonymous substitution in the transposase region indicates that the third exon evolved under purifying selection in all anthropoid lineages. However, when the analysis was performed separately for the 5′ and the 3′ halves of the exon, the 5′ half displayed a very strong signal of purifying selection, whereas the 3′ half seems to have evolved neutrally. Thus, the transposable element might have been domesticated for a function located in the N-terminal region of the transposase protein. All eukaryotic transposases studied thus far contain two major functional domains: an N-terminal region that is responsible for DNA binding to the terminal inverted repeats, and a C-terminal region containing the catalytic domain responsible for the cut-and-paste transposition reaction. Therefore, the transposon might have been recruited for its DNA-binding capabilities rather than its catalytic activities (Cordaux et al. 2006).

DROSOPHILA TELOMERES Telomeres serve two essential functions: (1) they prevent the recognition of natural chromosome ends as DNA breaks (the end-capping function), and (2) they counteract incomplete end replication by adding DNA to the ends of chromosomes (the end elongation function). In most organisms studied, telomerase fulfills the end elongation function. Telomeres of most animals, plants, and unicellular eukaryotes are made up of tandem arrays of repeated DNA sequences produced by the enzyme telomerase. The repeats that telomerase reverse transcribes onto the ends of chromosomes are surprisingly similar from species to species: in vertebrates the repeated sequence is TTAGGG, in nematodes it is TTAGGC, and in *Arabidopsis* it is TTTAGGG. In contrast to the slight variations in repeat size, the length and chromatin structure of the arrays can differ greatly between species and within species. In humans, for instance, telomeres are 5,000–15,000 bp in length and vary with age and cell type. *Drosophila* species have an unusual variation on this theme; telomeres consist of tandem arrays of sequences produced by successive transpositions of three non-LTR retrotransposons, *HeT-A*, *TAHRE*, and *TART* (**Figure 9.25**).

HeT-A is the most abundant element. It contains a single open reading frame that encodes a protein that may form an association with RNA within a ribonucleoprotein complex. *HeT-A* lacks a second ORF that is present in the other two non-LTR retrotransposons. Interestingly, the promoter for *HeT-A* transcription does not reside within the 5′ region, but within the 3′ UTR. Therefore, transcription of a *HeT-A* element originates from the 3′ UTR of the element closest to the chromosome end. This mechanism is thought to protect the 5′ region of the retrotransposon.

Full-length *TART* elements have two ORFs. The first is similar to the one in *HeT-A*. The second ORF encodes a reverse transcriptase that possesses an endonuclease domain. Both 5′ and 3′ UTRs have promoter activities. *TAHRE* is the least abundant element. Although like *TART* it encodes two proteins, *TAHRE* shares greater sequence similarity with *HeT-A*. This led to the suggestion that *HeT-A* evolved from *TAHRE* by losing one ORF. Promoter activities in *TAHRE* have been detected in the 3′ UTR, representing another similarity between *HeT-A* and *TAHRE*. It is interesting to note that *HeT-A*, the most abundant element, does not encode a reverse transcriptase. This implies that the reverse transcriptase for *HeT-A* transposition must be supplied from *TAHRE*, *TART*, or another retroelement.

THE ORIGIN OF THE ADAPTIVE IMMUNE SYSTEM OF JAWED VERTEBRATES V(D)J recombination is the process by which immunoglobulins and T-cell receptors are assembled by somatic rearrangement of the variable (V), diversity (D), and joining (J) gene segments. V(D)J recombination combines V, D, and J gene segments nearly randomly, and because of this randomness it is able to generate the almost unlimited variety of antibodies that defines the adaptive immune system. The two essential components of V(D)J recombination are the RAG1 and RAG2 proteins. In humans, these two proteins are encoded by neighboring genes on chromosome 11.

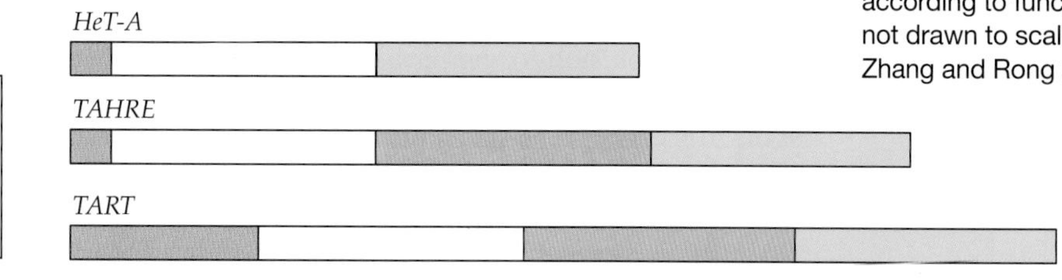

Figure 9.25 The telomeres of *Drosophila* species consist of tandem arrays of sequences produced by successive transpositions of three non-LTR retrotransposons, *HeT-A*, *TAHRE*, and *TART*. Each element with the retrotransposan is represented as a rectangular box, colored according to function. Elements are not drawn to scale. (Modified from Zhang and Rong 2012.)

The analogy of the process of V(D)J recombination to a transposition reaction is striking. First, the recombination signal sequences adjacent to the V(D)J gene segments were found to be similar to signals found at the termini of integrated transposable elements, such as the *Transib* DNA transposons in echinoderms, insects, nematodes, and fungi. Second, *RAG1*, which provides the catalytic core for the reaction, was found to be closely related in sequence to transposases encoded by *Transib* elements. Subsequently, transposition activity has been empirically demonstrated for the *RAG1/RAG2* complex. Finally, it has been observed that several eukaryotic transposases utilize a cleavage chemistry similar to that seen in V(D)J recombination.

RAG genes are present in all jawed vertebrates, from cartilaginous fishes to mammals, but are absent from lamprey and hagfish. The phylogenetic distribution suggests that *RAG* genes became established in a vertebrate ancestor sometime during the interval between the divergence of jawless fishes and cartilaginous fishes (470–550 million years ago).

The acquisition of V(D)J recombination is often regarded as a key step in the evolution of jawed vertebrates. There is currently little doubt that V(D)J recombination is the product of a fortuitous event of DNA transposon domestication, whereby a *RAG* transposon became integrated into a V-like gene in the genome of an ancestral vertebrate. The integration of the transposon split the gene into two components that could subsequently be rejoined after *RAG*-induced double-strand breakage and removal of the intervening DNA.

HYBRID DYSGENESIS Among *Drosophila*, **hybrid dysgenesis** is a syndrome of correlated abnormal genetic traits that is spontaneously induced in one type of hybrid between certain mutually interactive strains, but usually not in the reciprocal hybrid (Sved 1976; Kidwell and Kidwell 1976). Hybrid dysgenesis has attracted much attention from molecular and evolutionary biologists because its main manifestation is the creation of a barrier against hybridization between strains or populations, which has been speculated to be a cause of speciation.

There are several dysgenic systems in different *Drosophila* species, some caused by a single transposable element, and others, like the *Helena-Paris-Penelope-Ulysses-Telemac* system in *D. virilis*, caused by "coalitions" of transposable elements (Vieira et al. 1998). Here we will deal with just one system, the *P-M* system in *D. melanogaster.*

The asymmetry of hybrid dysgenesis is shown in **Figure 9.26**. When a male from a *P* strain mates with a female from an *M* strain, the offspring are dysgenic; in the reciprocal mating, the offspring are normal. The dysgenic traits of the *P-M* system include (1) failure of the gonads to develop if the first instar is exposed to temperatures above 27°C, (2) recombination in males (an unnatural occurrence in *Drosophila,* in which recombination is usually restricted to females), (3) chromosomal breakages and rearrangements, (4) distortion of Mendelian transmission proportions (including sex ratios), and (5) high frequencies of lethal and nonlethal mutations, mostly due to chromosomal nondisjunction or the insertion and excision of the mobile elements.

The cause of the *P-M* dysgenesis is a family of transposable elements called *P* elements (**Figure 9.27**). *M* strains do not carry *P* elements, while in *P* strains there are 30–50 *P* elements per genome. Not all the *P* elements in a genome are autonomous; some contain deletions and are therefore either nonautonomous or completely inactive. *P* elements transpose by a cut-and-paste mechanism that does not inherently increase copy number. Indeed, for a long time, the means by which *P* elements increase their numbers in the genome was something of a puzzle, especially given the fact that *P* elements entered *D. melanogaster* by horizontal transfer merely 80 years ago (p. 444). A solution to the problem was provided by the discovery that *Drosophila P* elements preferentially transpose to replication origins on the chromosome (Spradling et al. 2011). Integrating at replication origins might help a cut-and-paste transposon increase its genomic copy number by allowing the element to "sense" the activation of the origin of replication and time its transposition so that each departing element would leave a copy on the sister strand to act as a template for repair of the resulting

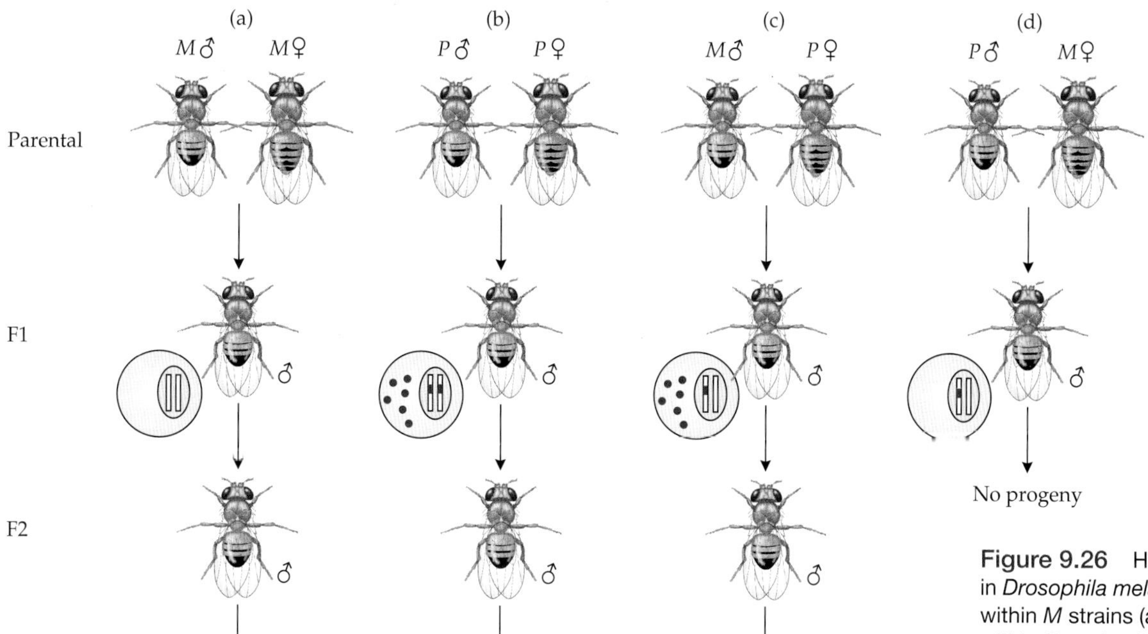

Figure 9.26 Hybrid dysgenesis in *Drosophila melanogaster*. Matings within *M* strains (a) and matings within *P* strains (b) produce normal progeny. (c) Crosses between *M* males and *P* females also produce normal progeny. (d) Crosses between *P* males and *M* females produce dysgenic offspring, many of which are sterile. The cytotypes are shown schematically on the left sides of the F1 flies. *P*-carrying chromosomes are denoted by a red box. The maternally inherited cytoplasm may or may not carry *P*-encoded repressors (red circles). In the absence of such repressors, *P* elements tend to transpose uncontrollably, resulting in dysgenic traits. Such a situation can only occur in *P* male × *M* female matings.

double-strand break. Thus, by locating themselves at origins of replication and timing their transposition to the S phase of the cell cycle, cut-and-paste transposable elements can multiply intragenomically.

The transposase-coding region in the *P* element consists of four exons separated from one another by three short introns (each less than 100 bp in length). In germline-cell mRNA, these three introns are excised and a functional transposase is made. In somatic-cell mRNA, the third intron is retained, and as a consequence a truncated protein is produced. This truncated protein acts as a repressor of transposition, and hence transposition of *P* elements only takes place in germline cells (Roche et al. 1995).

An interesting observation has been made by Kidwell (1983) concerning the distribution of *P*-carrying strains. *P* characteristics were not found in any *D. melanogaster*

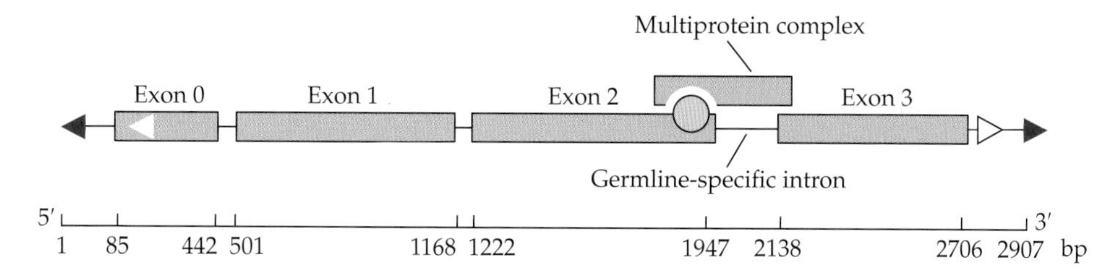

Figure 9.27 Schematic structure of a complete 2,907 bp *P* element in *Drosophila melanogaster*. The element is flanked by 31 bp terminal inverted repeats (red triangles) and by 11 bp subterminal inverted repeats (white triangles). The element contains a single gene encoding a transposase (766 amino acids). The coding region contains four exons (blue boxes) interrupted by three short spliceosomal introns (lines). Production of the functional protein depends on the splicing of the third intron from the pre-mRNA. The splicing of this intron is prevented in somatic cells by the binding of a multiprotein inhibitory complex (green box) to a site located in exon 2 (yellow circle). One component of this multiprotein complex (*P* element somatic inhibitory protein, or PSI) has a very low abundance in germline cells, effectively limiting the production of functional transposase to these cells only (Adams et al. 1997). In somatic cells, a shorter and inactive polypeptide is encoded by the mRNA that retains the third intron (Rio et al. 1986). (From Graur and Li 2000.)

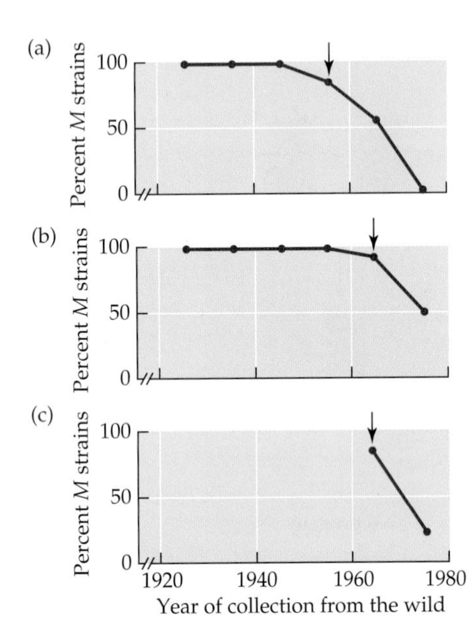

Figure 9.28 Temporal changes in the frequencies of *M* strains (i.e., strains devoid of *P* elements) in natural populations of *Drosophila melanogaster* in (a) North and South America; (b) Europe, Africa, and the Middle East; and (c) Australia and the Far East. Arrows denote the first appearance of *P* strains. Note that the proportion of *M* strains decreased first in the American continents and only later in the other continents. (Modified from Kidwell 1983.)

strains collected before 1950, and collections made subsequently showed increasing frequencies of *P* with decreasing age (**Figure 9.28**). A similar observation was made with another dysgenic system, *I-R*.

Two hypotheses have been suggested to explain this distribution of *P* elements. Engels (1981b) proposed that most strains in nature are of the *P* type but that they tend to lose the transposable elements in the laboratory. The second hypothesis postulated a recent introduction of *P* transposable elements into *D. melanogaster* populations followed by their rapid spread in formerly *M* populations (Kidwell 1979). There are several reasons why the latter hypothesis is favored over the first one. First, *P*-carrying laboratory strains that have been monitored for dozens of years were never observed to lose their *P* characteristics. Second, there seems to be a geographical cline in the distribution of *P* strains, with American populations showing earlier signs of carrying *P* elements than European, African, and Middle Eastern strains, which in turn seem to have acquired *P* elements before Australasian populations. Last, there is now evidence that *P* elements have been recently acquired from a distantly related *Drosophila* species (p. 444).

Transposition and Speciation

Speciation or **cladogenesis** (i.e., the creation of two or more species from one parent species) is one of the most important evolutionary processes. Unfortunately, at the molecular level, it is also one of the least understood processes. We know little of the means by which new species arise from old ones. What we do know is that the process of speciation requires the creation of a **reproductive barrier** between two populations belonging to the same (parent) species, after which they can no longer interbreed. Hybrid dysgenesis has been thought for a while to represent an early stage in the process of speciation, by acting as a postmating reproductive isolation mechanism between different populations of the same species (e.g., Periquet et al. 1989). Indeed, the sterility of hybrids produced by crosses between the sibling species *D. melanogaster* and *D. simulans* is very similar to that due to postzygotic reproductive isolation (e.g., rudimentary gonads, segregation distortion). There are problems with this view, however. First, although hybrids exhibit reduced fitness and are therefore partially isolated in terms of reproduction, the transposition of *P* elements in the germline virtually ensures that most of the chromosomes transmitted to the hybrids will bear *P* elements, and the cytotype will eventually change to the *P* type as well. Thus, provided that the reduction in fitness in hybrids is not too great, the *P* element will quickly spread through the entire population. Indeed, theoretical considerations indicate that the hybrid must be almost completely sterile for an effective reproduction isolation to last. Second, *P* elements have the ability to transpose themselves horizontally (p. 444) as an infectious agent from individual to individual. Thus, an entire population will be rapidly taken over by *P*, so hybrid dysgenesis is likely to last for very short periods of time in nature. Indeed, many species belonging to the genus *Drosophila* are known in which all individuals and all populations carry *P* or *P*-like elements, and consequently hybrid dysgenesis does not occur in any of these species.

Since the discovery of transposable elements, numerous other mechanisms for speciation by transposition have been proposed. For example, it has been suggested that mass replicative transposition of elements containing regulatory sequences in one population may cause a so-called **genetic resetting** of the genome, whereby many genes will be subject to a novel mode of gene regulation. Such a population would obviously become reproductively isolated from a population that retained the old form of gene regulation. The discovery that some *Alu* repeats contain functional retinoic acid response elements (Vansant and Reynolds 1995), which are known to function as transcription factors, indicates that it is possible to alter the expression of numerous genes through the genomic dispersion of transposable elements.

Another suggestion invokes a mechanism of **mechanical incompatibility**, also caused by mass replicative transposition. In this case, it is assumed that in one population the transposable elements may expand their numbers to such an extent that they cause a significant increase in the size of the chromosomes. A hybrid organism that inherited a set of large chromosomes from one parent and a set of small chromosomes from the other would experience difficulties in chromosome pairing during meiosis, and would most probably be sterile. The evidence for this mechanism is at present somewhat sparse (e.g., Stelzer et al. 2011).

The most promising hypothesis linking transposable elements and speciation invokes rapid genome remodeling by transposable elements multiplying within a genome and creating new regulatory gene networks. Since transposable elements are often regulated by the host, one can envision the evolution of a control system in one population and its absence from another. The mating between an individual that exerts control over its transposable elements and one that doesn't may result in the amplification of transposable elements and the lowering of the fitness of the hybrid. For example, hybrids of *Helianthus annuus* (sunflower) and *H. petiolaris* (prairie sunflower) have genomes 50% larger than parental individuals, the result of transposable amplification (Rebollo et al. 2010). Other tentative lines of evidence indicate that bursts of transposition coincide with speciation events (Böhne et al. 2008; Rebollo et al. 2010; Symonová et al. 2013).

Finally, some transposable elements may affect function and, hence, produce reproductive barriers. For at least one retrotransposable element, *AmnSINE1*, there is evidence of its involvement in generating mammalian-specific morphologies and, hence, of its importance in mammalian radiation (Okada et al. 2010).

Horizontal Gene Transfer

Horizontal gene transfer (also called **lateral gene transfer**) is defined as the movement of genetic information from one genome to another, specifically between two species. The term "horizontal gene transfer" was coined in contradistinction to the more familiar "vertical gene transfer," in which a parent passes genetic information on to its progeny. An organism that has incorporated genetic information from a different organism as a stable part of its genome is called a **transgenic organism**. In the not so distant past, transgenic organisms were considered Frankensteinian curiosities of the laboratory. More recently, we have come to realize that horizontal gene transfer is extremely prevalent in evolution, such that all organisms are in effect transgenics. Sequence homology due to horizontal gene transfer is called **xenology**, as opposed to paralogy and orthology, which are due to gene duplication and speciation, respectively.

Telltale signs of horizontal gene transfer

When should we suspect that a horizontal gene transfer event may have occurred? First, horizontal gene transfer may be suspected in cases in which an outstanding discontinuity in the phylogenetic distribution of a certain DNA sequence is discovered. In the past, such so-called patchy distributions were frequently invoked to infer horizontal gene transfer. It is important to keep in mind, however, that taxonomic discontinuity may also result from other evolutionary processes, such as stochastic loss of genes in different lineages, differential fixation of ancestral polymorphisms, differential purifying selection acting to preserve sequences in some taxa but not others, and variable rates and modes of evolution (Schaack et al. 2010).

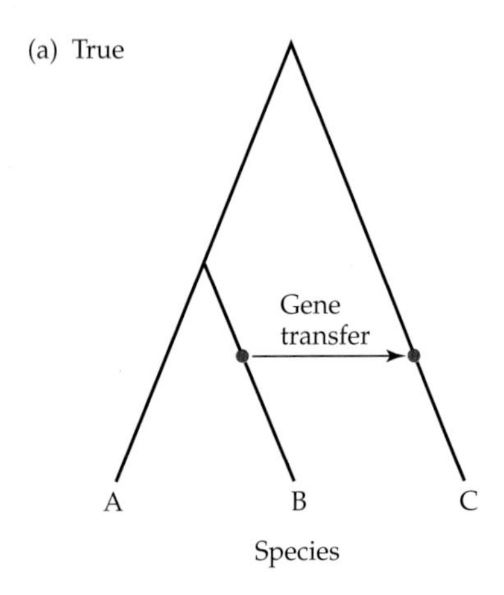

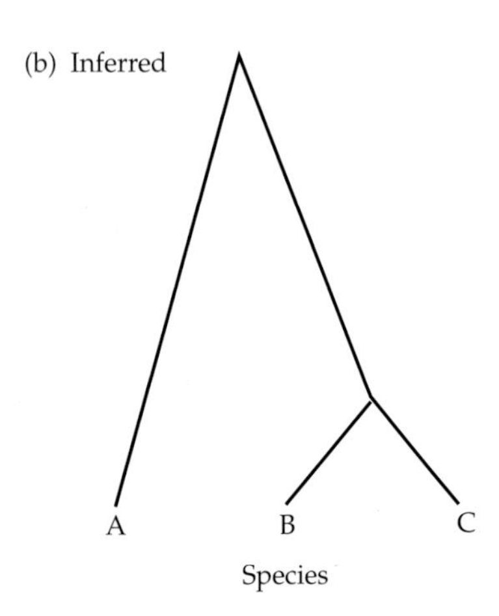

Figure 9.29 Erroneous phylogenetic tree reconstruction due to horizontal gene transfer (arrow). (a) The true phylogenetic network. (b) The inferred tree. (From Graur and Li 2000.)

Horizontal gene transfer may also be suspected when a notable discrepancy between gene phylogeny and species phylogeny is uncovered, in particular when sequence similarity seems to reflect geographical proximity or ecological affinities rather than phylogenetic kinship. Consider, for example, the phylogenetic tree in **Figure 9.29a**. Let us assume that a piece of DNA was transferred from species B to species C after the divergence of species A and B. On the basis of sequence comparisons involving any gene other than the horizontally transferred one, we expect to be able to reconstruct the correct phylogenetic relationships among the species. In contrast, if we use the horizontally transferred piece of DNA, we obtain the erroneous tree in **Figure 9.29b**. Again, we note that many factors other than horizontal gene transfer may conspire to yield a discrepancy between the species tree and the gene tree (Chapter 5), so additional evidence is required to infer a horizontal transfer event.

Because gene order and orientation are rarely conserved in evolution (Chapter 11), horizontal gene transfer may be suspected if strict synteny (i.e., the same order and orientation of genes) is observed between distantly related taxa. One such example involves the genes responsible for the production of sterigmatocystin, a highly toxic secondary metabolite and a precursor to the aflatoxins, in the fungi *Aspergillus nidulans* and *Podospora anserine*. The order and relative orientation of the 23 genes in the gene cluster are almost perfectly conserved between these two unrelated species. In contrast, very little similarity of order and relative orientation of these genes exists between the closely related species *A. nidulas* and *A. flavus* (Slot and Rokas 2011).

Finally, as we saw in Chapter 4, there are compositional features that characterize genes within a genome, reflecting the fact that all vertically inherited genes within a species share similar mutation patterns and similar codon adaptations. In other words, at the time of the integration into the recipient genome, horizontally transferred genes reflect compositional patterns of the donor genome, which may be very different from those of the recipient genome in such characteristics as GC content and codon usage. (We note that some incongruent compositional features may arise for reasons that have nothing to do with xenology, such as gene-specific patterns of expression.) Of course, the compositional idiosyncrasies of a horizontally transferred sequence will only last for a short evolutionary while, as mutation and selection will "naturalize" the imported DNA sequence and erase all signs of its "exotic" origin. This process is called **amelioration**. Because of amelioration, compositional methods usually fail to detect ancient horizontal transfer events. By modeling the kinetics of the amelioration process, one can estimate the time of acquisition of horizontally transferred sequences. Lawrence and Ochman (1997) estimated that horizontally transferred sequences have accumulated in *E. coli* at a rate of 31 Kb per million years.

Following Koonin et al. (2001), horizontal transfer events of functional genes can be classified into three distinct categories with respect to the horizontally acquired gene and homologous genes in the recipient lineage: (1) acquisition of a new gene that does not have a homolog in the recipient genome, (2) acquisition of a new homolog of a preexisting gene, and (3) xenologous gene displacement, i.e., the replacement of a gene by a distant homolog.

Mechanisms of horizontal gene transfer among prokaryotes

Many mechanisms for horizontal gene transfer among prokaryotes have been described. Some have only been found in a handful of taxa, while others have wider taxonomic distributions. Here we describe several mechanisms that are either proven or suspected to promote horizontal gene transfer (**Figure 9.30**). Other prokaryotic means for horizontal gene transfer are known to exist (e.g., Aas et al. 2002; Klieve et al. 2005).

TRANSFORMATION **Transformation** involves the uptake of **histone-free** or **naked DNA** from the environment. The phenomenon of transformation was first observed by Frederick Griffith (1928) in his studies of the conditions responsible for the acquisition of a capsule by unencapsulated strains of *Streptococcus pneumoniae*. However, the mo-

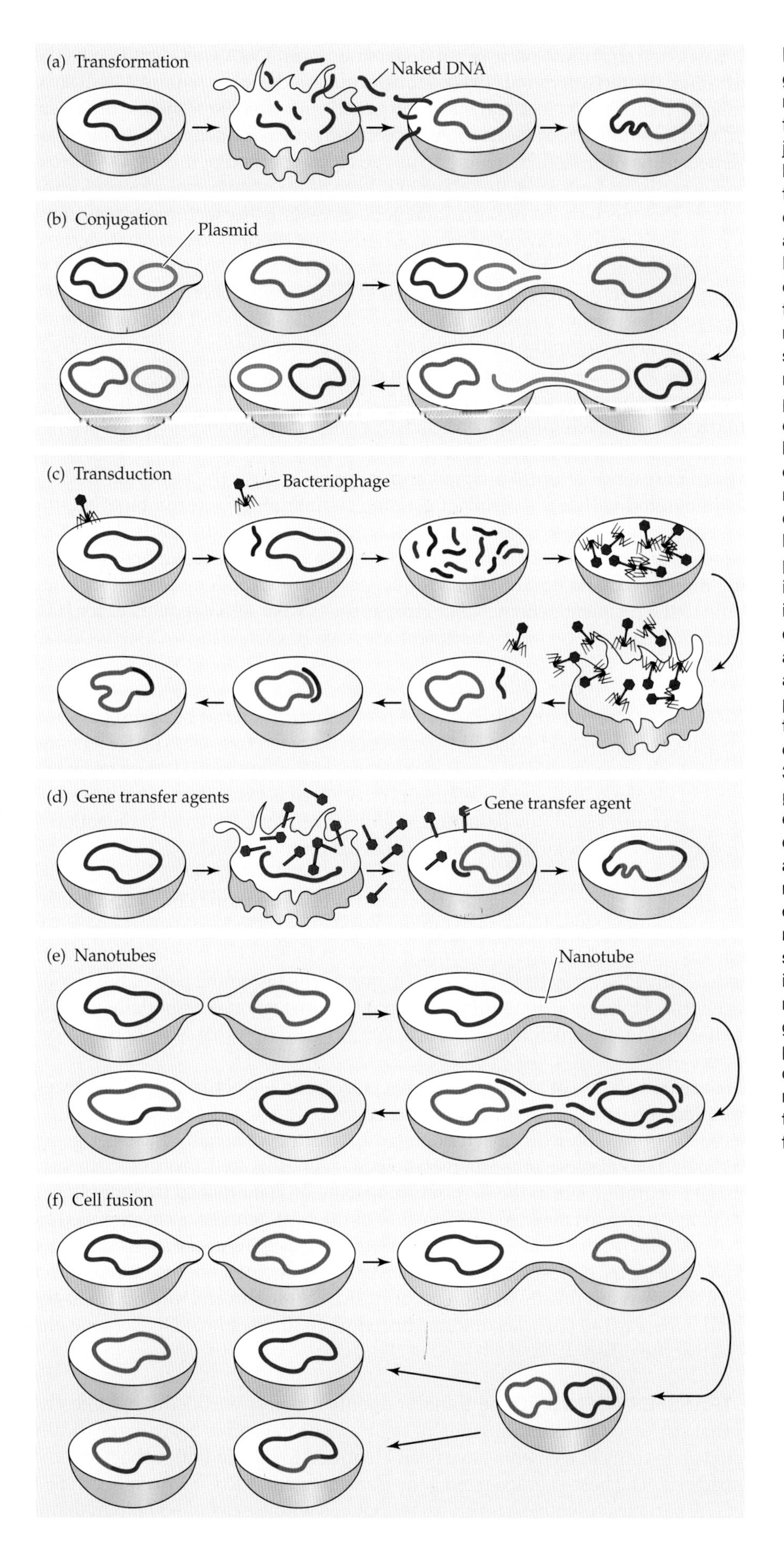

Figure 9.30 Mechanisms of horizontal gene transfer in prokaryotes. (a) Transformation entails the uptake of naked DNA fragments from the environment. (b) Conjugation is the transfer of genetic material by a direct cell-to-cell bridge-like connection between two bacterial cells by means of a conjugative genetic element, such as a plasmid or a transposable element. Plasmids can integrate into the recipient chromosomes by homologous recombination. (c) In transduction, bacteriophages recognize possible hosts by specific cell-surface receptors. Instead of its own genome, a bacteriophage may accidentally package genomic chunks from the previous host. These are transferred to the new host during the integration of the genome of the bacteriophage into the host chromosome. DNA integration into the host chromosome is generally mediated by bacteriophage-encoded enzymes. (d) DNA packaged in gene transfer agent particles is imported into the recipient in a generalized transduction process mediated by a cellular recombination. Lysis is most probably the only way by which gene transfer agents can escape a cell, although the process has not been observed. (e) The transfer of DNA through nanotubes is still conjectural. The nanotubes are between 30 and 130 nm (10^{-9} m) wide and up to 1 μm (10^{-6} m) long. The tube's dimensions correlate with the distance between the cells it connects. Cells in close proximity are commonly connected by several small nanotubes, while thicker tubes connect distant cells. The rate of transfer via the nanotubes correlates negatively with the size of the transferred substance. Cellular interconnection mediated by nanotubes is not species-specific. However, morphology and diameter of the tubes seem to depend on characteristics of the connected cells. (f) In archaebacteria, cell fusion and recombination may result in the reciprocal transfer of DNA between two taxa. (Modified from Popa and Dagan 2011.)

lecular basis of transformation was not understood until DNA was identified as the transforming agent (Avery et al. 1944). Transformation does not require a dedicated molecular or cellular vehicle to transport the genetic information from one organism to another. The simplicity of transformation is commonly regarded as a sign of antiquity. Hence, transformation is thought of as the oldest method of mixing DNA from different organisms, long predating the emergence of sex. Additional support for the antiquity of this mechanism of horizontal gene transfer is the fact that transformation is nonspecific; any stray bit of DNA, whether synthetic DNA or DNA derived from a long-dead animal, such as from a mammoth bone, can be incorporated into a live bacterium genome (Overballe-Petersen et al. 2013). Transformation is enabled by 20–50 proteins during a stage in the life cycle of the cell called the **competence state**. In bacteria grown in the laboratory, the competence state exists for a very short period of time in the late exponential growth phase. Competence is rapidly lost in the early stationary phase.

While transfer of DNA into the cell is mostly nonspecific, the incorporation of DNA into the genome is by no means random. In some species, effective transformation requires the presence of **uptake signal sequences**. Because the imported DNA is usually integrated within the recipient genome via homologous recombination, the donor's uptake signal sequences are recipient genome-specific. That is, DNA that has sequence motifs that are also found in the recipient genome has a better chance of being integrated. The uptake signal sequences are 8–12 bp long and are defined according to the recipient genome. For example, the uptake signal sequences for *Haemophilus influenzae* and *Neisseria gonorrhoeae* are AAGTGCGGT and GCCGTCTGAA, respectively.

CONJUGATION **Conjugation** is the transfer of genetic material between bacterial cells by a direct cell-to-cell bridge-like connection between two cells (Lederberg and Tatum 1946). The transfer of DNA is accomplished from a donor cell to a recipient cell by means of a conjugative genetic element, most often a plasmid or a transposable element. The transfer is highly regulated and unidirectional. Bacterial conjugation is often regarded as the bacterial equivalent of sexual reproduction.

TRANSDUCTION The cross-cellular transport for **transduction** (Zinder and Lederberg 1952) is provided by a bacteriophage (bacterial virus). The bacteriophage particle may encapsulate a DNA sequence from a host, and when the particle attaches to another cell, the DNA may be injected into it, eventually becoming part of the second host's genome. Horizontal gene transfer through transduction is essentially an error in viral packaging, in which the viral particle transfers DNA that isn't its own from cell to cell.

There are two classes of transduction: specialized and generalized. In **specialized transduction**, a particular bacteriophage packages only a specific region of the host genome. This is exemplified by bacteriophage λ, which occasionally packages genes involved in either galactose or biotin metabolism—genes located adjacent to the bacteriophage integration site in the *E. coli* chromosome. In **generalized transduction**, a bacteriophage packages a nonspecific portion of the host genome, as exemplified by the transposable bacteriophages *Mu* in *E. coli* and *RcapMu* in *Rhodobacter capsulatus*, which package pieces of host DNA at the ends of their genomes. For the viral particles of *Mu*, this packaging results in ~100 bp of host DNA at one end and ~2 Kb at the other end, whereas the particles of *RcapMu* contain ~30 bp of host DNA at one end and ~3 Kb at the other.

GENE TRANSFER AGENTS **Gene transfer agents** are host-encoded virus-like elements that package random fragments of the host chromosome. Several genetically unrelated gene transfer elements have been identified to date in both eubacteria and archaebacteria. The best-characterized gene transfer agent is RcGTA from *R. capsulatus*. RcGTA particles resemble a tailed bacteriophage with a head of ~30 nm in diameter and a tail of ~50 nm in length. Each RcGTA particle contains ~4 Kb of DNA from the host. The particle is encoded by a ~14 Kb cluster of 15–17 cotranscribed genes. Thus, each RcGTA particle can carry at most about 30% of the sequence needed to encode it—an indication that RcGTA cannot use the system for its own propagation. Of course, in the vast majority of cases, the RcGTA particle carries random DNA stretches that have nothing to do with the genes encoding it.

The question of how RcGTA genes are maintained in the population is a difficult one. Gene transfer agents escape the cell by lysis, i.e., by killing the donor cell. The death of the donor cell also signifies the death of the genes encoding the gene transfer agents. So, the question arises about how a "suicidal" agent can be maintained in the population. One answer invokes a mechanism similar to kin selection, whereby the cell releasing the gene transfer agents dies so that the recipient cell can benefit from a molecular trait acquired through horizontal gene transfer (West et al. 2006, 2007). From an evolutionary viewpoint, such explanations require a Herculean suspension of disbelief.

NANOTUBES **Nanotubes** are tubular protrusions composed of membrane components that can bridge neighboring cells and may conduct the transfer of DNA and proteins from one cell to another. Nanotubes have only been discovered recently (Dubey and Ben-Yehuda 2011), and their role and relative importance in horizontal gene transfer is not yet clear.

CELL FUSION In archaebacteria, horizontal gene transfer is sometimes enabled by **cell fusion**, i.e., a process by which cells establish physical contact through the creation of cytoplasmic bridges. The expansion of these bridges leads to the creation of fused cells, which will subsequently divide. Cell fusion can lead to either segregation of the chromosomes that will result in reversal of the original state or through recombination to the creation of hybrid cells harboring genetic material from the two cells (Naor and Gophna 2013).

Prevalence and limitations of horizontal gene transfer in prokaryotes

In a study of 190 prokaryotic genomes, Dagan and Martin (2007) estimated the minimum horizontal gene transfer rate to be on the order of 1.1 events per gene family. Indeed, at least 65% (and probably 100%) of gene families have experienced horizontal gene transfer during the course of their evolution. Garcia-Vallvé et al. (2013) estimated that at least 2–15% of the genes in prokaryote genomes have been horizontally transferred.

Are there barriers to gene transfer among prokaryotes? Here we discuss several types of barriers: (1) functional, (2) compositional, (3) phylogenetic, and (4) ecological. Within the category of functional barriers we include toxicity and usefulness. For an imported sequence to be integrated successfully into a host genome, the first condition that should be met is that the sequence should not be toxic to the host. Sorek et al. (2007) approached the question of toxicity by looking at genome sequencing projects that employ shotgun methodology, whereby multiple copies of the genome that is being sequenced are randomly sheared into overlapping fragments of DNA, and plasmids containing the cloned fragments are transformed into an *E. coli* cell. The cloned fragments are then sequenced, and the overlapping sequences are used for genome assembly. Because cloned fragments contain the full set of genes belonging to the sequenced organism, genome sequencing can be viewed as a large-scale experiment in horizontal gene transfer to *E. coli*, where each gene in a genome undergoes multiple transfer attempts to the *E. coli* host. Sorek et al. (2007) noted that in the course of nearly all sequencing projects, a small fraction of the genome being sequenced fails to clone in *E. coli*, resulting in sequence gaps. These gaps are subsequently sequenced without cloning in a process called finishing. From among the unclonable (or untransferable) genomic regions of 79 prokaryotic genomes, the researchers found 61 genes that could not be cloned from 5 or more different genomes. This high failure rate of transformation into *E. coli* suggests that specific genes, rather than experimental protocol or random error, have caused horizontal gene transfer to fail. Many of the genes that could not be transferred into *E. coli* turned out to encode ribosomal proteins and other proteins involved in translation. This observation is consistent with computational analyses in which genes involved in translation were found to be underrepresented in horizontal gene transfer events. Indeed, experimental evidence suggests that some of these sequences are toxic, i.e., they drastically reduce the fitness of the *E. coli* host. Interestingly, although some specific genes were found to be transfer-resistant when taken from a wide range of prokaryotes, no single gene was found to be untransferable among all genomes examined.

Because prokaryotes tend to delete nonfunctional and unneeded DNA, a horizontally transferred sequence needs to be both functional and retained by the host genome. For example, transferring a gene for a component of the photosynthetic apparatus is likely to be successful only if the recipient species already performs photosynthesis. If the transferred gene is of no use to the recipient prokaryote, it is unlikely to be kept and will soon be lost; we will never even know that such a transfer occurred. Similarly, if the transfer occurs between species that inhabit different niches, the transfer is unlikely to be successful. For example, genes from thermophiles, which perform best at high temperatures, are unlikely to perform adequately in psychrophiles, which grow and reproduce best in cold temperatures.

Another feature related to functionality that determines horizontal transfer success seems to be gene dosage, with xenologous gene displacement occurring more frequently than the addition of paralogs, which in turn occurs more frequently than the acquisition of genes devoid of homologs (Koonin et al. 2001).

In terms of compositional barriers, we note that genes employing suboptimal codons that do not fit the tRNA pool of the recipient cell were initially thought to be untransferable, but more recent studies have shown that the impact of codon usage might have been a little exaggerated. Thus, although gene variants with suboptimal codon usage are indeed less efficient in producing the protein encoded by the gene (Kudla et al. 2009), this shortcoming quickly disappears through substitutions in the promoter region or elsewhere in the genome (Amorós-Moya et al. 2010).

Another compositional factor affecting horizontal gene transfer is overall similarity between donor and recipient in terms of GC content. In the vast majority of cases, the difference in GC content between donor and recipient is less than 5%. Indeed, genome sequence similarity accounts for about 25% of the variation in horizontal gene transfer frequency (Popa et al. 2011).

Several studies have suggested that phylogenetic barriers are common and that horizontal gene transfer is quite frequent for short and intermediate evolutionary distances, but uncommon between distantly related taxa. Wagner and de la Chaux (2008) analyzed 2,091 horizontal gene transfers in 438 completely sequenced prokaryotes and found only 30 cases (~1.5%) among distantly related clades. Although rare, instances of gene transfer between archaebacteria and eubacteria have been described (Rest and Mindell 2003; Gophna et al. 2004).

Finally, physical proximity and ecological conditions are important in facilitating horizontal gene transfer. In transformation, the distance between the donor and recipient depends upon the stability of DNA within the environment. Conjugation requires that the donor and recipient be close enough for the formation of the conjugation tunnel. Because bacteriophages are usually mobile, transduction facilitates horizontal gene transfers over larger distances. All in all, the vast majority of horizontal gene transfers occur within the same habitat, and interhabitat gene transfers are rare (Popa and Dagan 2011). Interestingly, the same is true for the human microbiome, where horizontal gene transfer occurs mostly among bacteria that inhabit the same body site (Smillie et al. 2011).

The importance of proximity is aptly illustrated by the acquisition of carbohydrate-hydrolyzing enzymes by a gut bacterium. Porphyran is a complex sulfated carbohydrate found in abundance in seaweed species, such as those belonging to the genus *Porphyra* from which the sushi ingredient nori is made. The backbone of the molecule porphyran can be hydrolyzed by a group of enzymes called porphyranases. In nature, porphyranase-coding genes are found in the genomes of a group of Gram-negative anaerobic bacteria called Bacteroidetes. One such species, *Zobellia galactanivorans*, lives and feeds on the seaweeds from which nori is prepared. In Japanese individuals, who consume on average 14 grams of seaweeds per day, a gut bacterium, *Bacteroides plebeius*, was found into which genes encoding porphyranases and associated proteins from *Z. galactanivorans* have been horizontally transferred (Hehemann et al. 2010). Comparative studies failed to detect porphyranase-coding genes in gut bacteria from North American individuals, who either do not consume seaweed at all or consume pasteurized nori.

Genomic consequences of gene transfer among prokaryotes

The main evolutionary significance of horizontal gene transfer in prokaryotes is the acquisition of new genes. It has been estimated that about 90% of all expansions of protein families in prokaryotes originated as horizontal transfer of genes. That is, horizontal gene transfer, rather than gene duplication, drives the expansion of the genomic repertoire in prokaryotes. There are reasons to believe that, at least initially, xenologs are less important and, hence, less constrained than paralogs. Indeed, xenologs have been shown to evolve faster than paralogs and to share fewer protein-protein interactions or regulatory mechanisms (Treangen and Rocha 2011).

When different individuals belonging to the same eukaryotic species are compared with one another, a certain degree of dissimilarity is expected; however, their complements of genes will generally be identical or almost identical. In contrast, when the first three completely sequenced *E. coli* genomes were compared, only 39% of their protein-coding genes were found in all three genomes (Welch et al. 2002). This finding not only illustrates the ubiquity of horizontal gene transfer, but also has made it necessary to coin a new genetic nomenclature and abolish the term "genome" in reference to prokaryotic species (Chapter 10). More importantly, however, the fact that due to horizontal gene transfer, genomes from individuals belonging to the same prokaryotic species have little in common in terms of gene content calls into question the very notion of "species," and it limits the usefulness of the tree metaphor in reconstructing phylogenetic relationships among prokaryotes (Chapter 6).

Clinical consequences of gene transfer among prokaryotes

In some bacteria, genes of clinical importance have turned out to be xenologs. For example, the genomes of many *Staphylococcus aureus* strains contain xenologs that confer extreme virulence as well as resistance to antibiotics. The protein-coding genes that encode toxins responsible for toxic shock syndrome, food poisoning, necrotizing pneumonia, and scalded-skin syndrome are recent acquisitions through horizontal gene transfer, as are the many genes conferring antiseptic and antibiotic resistance (Lindsay and Holden 2004; Holden et al. 2010).

An extreme case of evolution by horizontal gene transfer is seen in *Yersinia pestis*, the causative agent of plague. Three global pandemics now generally attributed to *Y. pestis* have been recorded. The first was the so-called Plague of Justinian (541–767), which is thought to have originated in East or Central Africa and spread via Egypt to countries surrounding the Mediterranean. The second and most infamous pandemic was the Black Death, which starting in 1346 spread from the Caspian Sea to devastate all of Europe and which spawned subsequent epidemic outbreaks into the early nineteenth century; the virus may have been imported from central Asia. The third pandemic began in the Yunnan region of China in the mid-nineteenth century and spread globally via marine shipping from Hong Kong in 1894—three years before *Yersinia pestis* was first described in the literature (Yersin 1897).

A comparison of housekeeping genes from *Yersinia pestis* with orthologs from its closest relative, the mildly pathogenic *Y. pseudotuberculosis*, revealed almost no differences between the two species. Thus, the divergence event that gave rise to *Y. pestis* must be very recent, 1,500–22,000 years ago. During this time, *Y. pestis* acquired 32 new genes and two unique plasmids, pPCP1 (9.6 Kb) and pMT1 (102 Kb); the plasmids play roles in tissue invasion and capsule formation, as well as infection of the flea vector. Other changes that have occurred in *Y. pestis* since its divergence from a *Y. pseudotuberculosis* strain are most probably related to the fact that *Y. pestis* is a parasite: relative to the *Y. pseudotuberculosis* genome, *Y. pestis* has completely lost 317 genes, while 149 other genes have become nonfunctional as unitary pseudogenes (Chain et al. 2004).

Horizontal Gene Transfer Involving Eukaryotes

In discussing horizontal gene transfer events in which a eukaryote is the donor or the recipient or both, we need to distinguish between unicellular and multicellular eukaryotes. At least in principle, the transfer of DNA in unicellar eukaryotes seems less fraught with obstacles than that involving multicellular eukaryotes, where the passage of DNA from one organism to another may be difficult. Within multicellular eukaryotes, we need to distinguish between those taxa in which a strict partition exists between somatic cells and germline cells, and other taxa, such as some plants and planarians, in which the distinction between soma and germ is hazier. If a strict division exists between somatic cells and germline cells, the horizontal gene transfer must occur in the germline for it to be established in the evolutionary lineage. At present, horizontal gene transfer involving eukaryotes seems to be not nearly as rampant as that among prokaryotes, although we know that horizontal gene transfer is important in the evolution of eukaryotes (e.g., Andersson 2005; Gladyshev et al. 2008).

The mechanisms by which horizontal gene transfer into eukaryotes takes place are largely a matter of speculation; agents that have been implied are viruses, bacteria, fungi, and mechanical cell piercing. From an ecological point of view, all indications are that symbionts—especially endosymbionts, parasites, predators, and transmittable viruses—play a major role in facilitating horizontal gene transfer in eukaryotes.

Horizontal gene transfer from eukaryotes to prokaryotes

Because prokaryotes can take up free DNA from the environment in a manner that does not require a specific molecular vector, one could imagine that horizontal gene transfer from eukaryotes to prokaryotes should be a common occurrence. This is not so, however, most probably because the expression of eukaryotic genes requires the involvement of different pieces of molecular machinery that are not available in prokaryotes. So far, only sporadic reports of eukaryote-to-prokaryote gene transfer exist in the literature (Jenkins et al. 2002; Schlieper et al. 2005; Arias et al. 2012).

Horizontal gene transfer from prokaryotes to eukaryotes

Horizontal gene transfer from prokaryotes to eukaryotes should, in principle, be fraught with almost insurmountable difficulties. First, the foreign genetic material must enter the right cell, i.e., a germline cell. Second, once inside the cell, the DNA must reach the nucleus, become incorporated into the nuclear genome, and be expressed properly. These requirements constitute a very tall order, and hence gene transfer into eukaryotes, particularly multicellular ones, is expected to be quite rare. We note, moreover, that discovering prokaryote-to-eukaryote gene transfer is difficult, as eukaryote whole-genome sequencing projects routinely exclude bacterial sequences on the assumption that these represent contaminations. For example, the publicly available nuclear genome sequence of *Drosophila ananassae* does not contain any horizontally transferred bacterial sequences, despite the ample evidence that such sequences exist (see below). Thus, the frequent claim in the literature that horizontal gene transfer from prokaryotes to eukaryotes is very rare may be an artifact of the deliberate "eviction" of prokaryotic sequences from eukaryotic genome assemblies. We also note that it is very difficult to distinguish between cases in which a prokaryotic gene was acquired directly from a prokaryote and cases in which a "prokaryotic" gene originated in an organelle.

In the following sections, we describe three classes of prokaryote-to-eukaryote gene transfer: endosymbiotic gene transfer, gene transfer under extreme selective pressures, and gene transfer involving antibacterial products.

ENDOSYMBIOTIC GENE TRANSFER The existence of intracellular prokaryotic endosymbionts within the germline of animals gives rise to the possibility that a gene originating in the endosymbiont may be acquired by the host genome. Such a transfer is called **endosymbiotic gene transfer**.

Wolbachia pipientis is a maternally inherited endosymbiont that infects a wide range of arthropods, including at least 20% of insect species. It is present in developing gametes and so provides circumstances conducive to heritable transfer of bacterial genes to the eukaryotic hosts. Hotopp et al. (2007) discovered a great number of *Wolbachia* sequences that have been horizontally transferred to host genomes. These host genomes include insects, crustaceans, arachnids, and nematodes. Of particular interest was the discovery of numerous associations between *Drosophila* retrotransposons and *Wolbachia* genes, indicating that retrotransposons may play a role in the *Wolbachia*-to-host horizontal gene transfer.

The horizontally transferred segments exhibited a wide range of lengths, with the record being held by an almost complete *Wolbachia* genome that was transferred to chromosome 2L of the Hawaiian species *Drosophila ananassae*. Two lines of evidence indicate that the *Wolbachia* DNA sequences do not represent bacterial contamination but have indeed been integrated within the *D. ananassae* genome. First, by using fluorescence microscopy and *Wolbachia*-specific probes, it could be shown that the *Wolbachia* sequence has indeed been integrated into a chromosome (**Figure 9.31**). Second, *Drosophila* lines containing the *Wolbachia* inserts continued to exhibit characteristic *Wolbachia* sequences even after treatment with antibiotics, whereas *Drosophila* lines lacking *Wolbachia* inserts in their genome did not exhibit any bacterial sequences after being cured with antibiotics.

Are the transferred genes functional? So far, there is little evidence that the transferred prokaryotic genes are functional in the eukaryotic host. For example, only very low expression levels have been found for some transferred genes, and this may represent no more than background noise. Their dynamics seem to be similar to those of mitochondrial sequences that have recently transferred to the nucleus (see below). That is, a balance may exist between new nonfunctional gene copies being inserted in the host genome and old copies being deleted or mutated beyond recognition.

There are, however, cases in which transferred prokaryotic genes are expressed in the eukaryotic recipient, a necessary but not sufficient condition for demonstrating functionality of the horizontally acquired genes. For instance, the pea aphid *Acyrthosiphon pisum* seems to have acquired about a dozen genes and gene fragments from bacteria (Nikoh and Nakabachi 2009; International Aphid Genomics Consortium 2010). These sequences have probably been acquired independently from facultative symbionts such as *Regiella*, *Spiroplasma*, *Rickettsia*, and *Wolbachia*, but interestingly the aphid's obligate primary symbiont, *Buchnera aphidicola*, has not transferred any expressed gene to the aphid host. Apparent transferred genes include those encoding carboxypeptidases, amidases, acetylmuramidases, and lipoproteins. Some of these genes are highly expressed in bacteriocytes, specialized cells that harbor *B. aphidicola*, which has a reduced genome and lacks these genes, whereas most other bacteria, including *Buchnera*'s close free-living relatives, possess them. These genes may, thus, be functionally essential to maintain *Buchnera*. In addition, functionality is implied by the observation that the

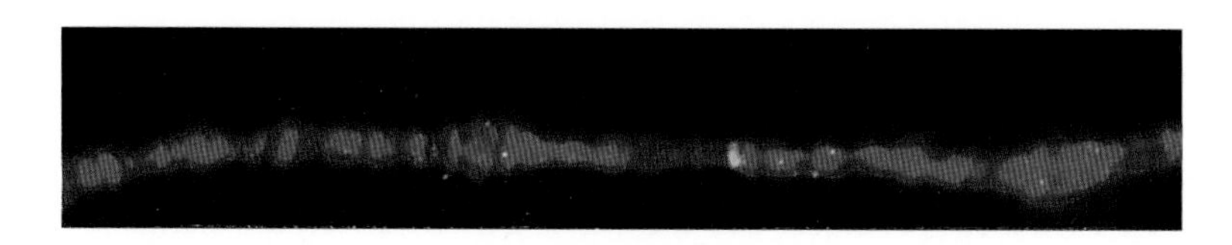

Figure 9.31 A *Wolbachia pipientis* genome (green) inserted into chromosome 2L (red) of the Hawaiian fruit fly, *Drosophila ananassae*. (From Hotopp et al. 2007; © AAAS and used with permission.)

bacterial sources of these genes are not currently present in the aphid, implying that the transfer has not occurred recently. Finally, pseudogenization of these genes would be expected in the absence of positive selection for function (Ros and Hurst 2009).

NICHE ADAPTATION TO EXTREME ENVIRONMENTAL CONDITIONS THROUGH HORIZONTAL GENE TRANSFER Prokaryotes have extremely broad niche ranges. Some may grow in extremely acid environments (pH ≤ 3); others thrive in highly alkaline conditions (pH ≥ 9). Some prokaryotes inhabit hydrothermal vents and reproduce at temperatures and hydrostatic pressures as high 122°C and 300 atmospheres, respectively, while others are capable of growth at subfreezing temperatures of –15°C. Still other prokaryotes are radioresistant (i.e., capable of sustaining high levels of ionizing radiation) or osmophilic (i.e., capable of surviving huge osmotic pressures) or metallotolerant (i.e., capable of tolerating high levels of dissolved heavy metals). In contrast, eukaryotes occupy a much narrower range of niches. Interestingly, in at least one case, the ability of a eukaryote to adapt to life in an extreme environment has been helped by the acquisition of genes from prokaryotes.

The microbial eukaryote *Galdieria sulphuraria* is a red alga that inhabits hot, metal-rich, and acidic environments. It can grow at pH values ranging from 0 to 4 and temperatures up to 56°C, close to the upper temperature limit for eukaryotic life. In addition, *G. sulphuraria* displays high salt tolerance and inhabits ecological niches with high concentrations of arsenic, aluminum, cadmium, mercury, and other toxic metals.

The *G. sulphuraria* genome was found to contain 6,623 protein-coding genes, of which at least 5% were acquired horizontally through 75 independent gene transfer events (Schönknecht et al. 2013). Some of these proteins are involved in important adaptive processes. For instance, the two largest protein families in *G. sulphuraria* are most probably derived from adenosine triphosphatases (ATPases) from archaebacteria. These ATPases have not been observed in other eukaryotes. Following the acquisition of an ancestral ATPase gene from archaebacteria, two large gene families evolved by duplications and divergence. A correlation is known to exist in archaebacteria between ATPase gene copy number and optimal growth temperature. Thus, *G. sulphuraria*'s adaptation to heat may have been facilitated by the acquisition and subsequent expansion of an archaebacterial ATPase gene. In fact, the vast majority of *G. sulphuraria*'s adaptations, including its salinity tolerance, its coping with a millionfold H^+ gradient across its plasma membrane, its heavy-metal detoxification abilities, and its glycerol uptake and metabolism, are due to horizontal gene transfer.

THE ACQUISITION OF ANTIBACTERIAL TOXINS One potential benefit that genes of bacterial origin could provide to eukaryotes is the capacity to produce antibacterials, which have evolved in prokaryotes as the result of more than a billion years of interbacterial competition. One such case, involving type VI secretion amidase effector proteins, was discovered by Chou et al. (2014). Type VI secretion amidase effectors are potent bacteriocidal enzymes that degrade bacterial cell walls. Genes encoding these proteins have been transferred to eukaryotes on at least six occasions, and the resulting domesticated genes have been preserved for hundreds of millions of years through purifying selection. Chou et al. showed that these genes have acquired eukaryotic secretion signals, are expressed within the recipient organisms, and encode active antibacterial toxins that possess a substrate specificity matching extant proteins in prokaryotes. Finally, they showed that the protein in the deer tick *Ixodes scapularis* limits proliferation of *Borrelia burgdorferi*, the etiologic agent of Lyme disease. Thus, a family of horizontally acquired toxins honed to mediate interbacterial competition now confer antibacterial capacity on eukaryotes.

Horizontal transfer among eukaryotes

It is not clear how horizontal gene transfer between eukaryotes is accomplished. Transformation and conjugation are almost certainly out of the question, because it is difficult to imagine how these processes could transfer genetic information into the germline. Transduction is imaginable, although the evidence for such a process in eukaryotes is nonexistent. Most probably, horizontal gene transfer between eukaryotes is accomplished through various viruses, which are capable of both incorporating foreign genetic information into their genomes and crossing species boundaries. There are tentative clues that both retroviruses and DNA viruses devoid of an RNA stage may be involved (e.g., Benveniste and Todaro 1976; Bishop 1981; Piskurek and Okada 2007).

The vast majority of eukaryote-to-eukaryote horizontal transfers involve transposable elements. There may be no more than a couple of well-established instances of a genuine protein-coding gene being transferred from one eukaryote to another. Indeed, the history of the field abounds with putative examples of eukaryote-to-eukaryote gene transfer that have turned out on further examination to be anything but horizontal gene transfer. Horizontal gene transfer of functional genomic sequences among eukaryotes seems to be something of a rarity. In the following sections we will discuss several examples of horizontal transfers among eukaryotes involving transposable elements, and one example in which an active and functional protein-coding gene has been successfully transferred from one eukaryote to another.

Horizontal gene transfer among plants

Horizontal transfer of functional genes between multicellular eukaryotes is exceedingly rate. One such case involves the giant witchweed, *Striga hermonthica*, a parasitic dicot plant that only infects monocots, including major crops such as sorghum (*Sorghum bicolor*) and rice (*Oryza sativa*). *Striga hermonthica* has acquired a gene from *Sorghum* via horizontal gene transfer (Yoshida et al. 2010). The gene, *ShContig9483*, exhibits significant similarity to genes from monocots, but no similarity whatsoever to any known dicot gene. Its pattern of evolution strongly suggests that it is a functional gene, although its function is presently unknown.

Horizontal transfer of a functional gene from fungi to aphids

Animal body color is an ecologically important trait, often involved in mimicry, aposematism (warning coloration), or crypsis (camouflage). In the pea aphid *Acyrthosiphon pisum*, an insect that feeds on the phloem sap of plants, red- and green-colored individuals frequently coexist in natural populations. Body color exhibits Mendelian inheritance, with red being dominant to green. Aphids have two major natural enemies: lady beetles, which preferentially attack red aphids, and parasitoid wasps, which mostly parasitize green aphids. The color polymorphism in aphid populations is most probably maintained through a frequency-dependent selection regime driven by opposite predation pressures (Fukatsu 2010).

Moran and Jarvik (2010) determined that green aphids contain only yellow carotenoids, whereas red aphids possess red carotenoids in addition to the yellow ones. Because animals are generally devoid of genes required for carotenoid biosynthesis, the carotenoids identified in aphids were expected to originate from either food or microbial associates of the aphid. However, carotenoids are lipid-soluble compounds and are unlikely to occur in plant phloem sap. Moreover, carotenoids in aphids differ from those in their host plants. Finally, the genome sequences of the bacterial symbionts of *A. pisum* contained no carotenoid biosynthetic genes. Where, then, do the carotenoids come from?

Within the *A. pisum* genome, several genes related to carotenoid biosynthesis were identified. Surprisingly, phylogenetic analyses revealed that these biosynthetic genes are nested within fungal homologs (**Figure 9.32**). The most parsimonious explanation is that these genes were transferred from a fungus to an aphid ancestor. Because these genes were also found in the green peach aphid *Myzus persicae*, the horizontal gene transfer event must have occurred before these two species diverged. Interestingly, the phylogenetic analysis revealed a second independent horizontal gene transfer event, one from a cyanobacterial ancestor to the ancestor of plants.

Although regarded as idiosyncratic in the past, horizontal gene transfers have now been shown to occur in diverse organisms. In most cases, however, the roles of the

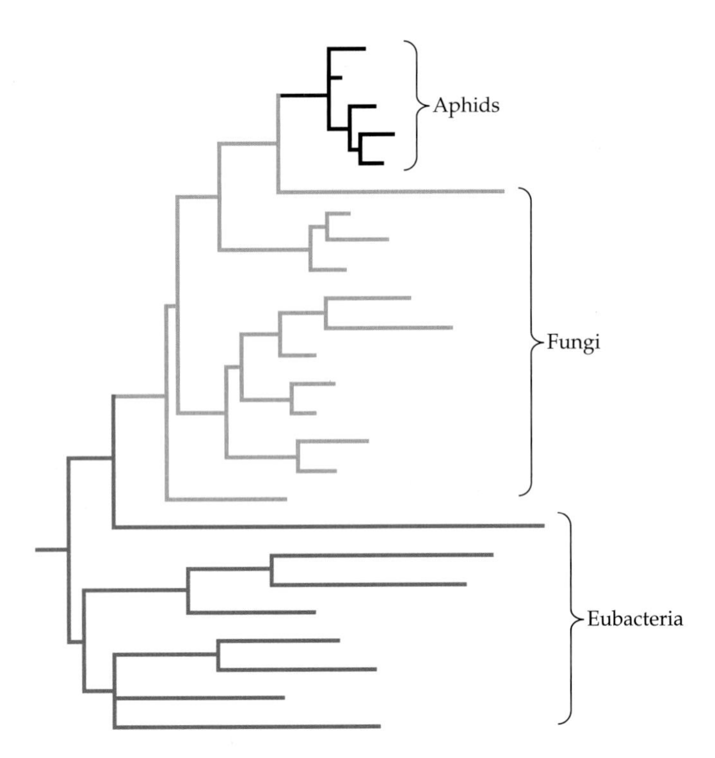

Figure 9.32 Phylogenetic tree for carotenoid desaturase proteins from aphids, fungi, and eubacteria. No homologs were detected in any animal genomes other than aphids. (Data from Moran and Jarvik 2010.)

horizontally transferred sequences are unknown. The case of the aphid carotenoid genes is remarkable in that the horizontal gene transfer not only established a prominent external trait (body color), but may have also contributed to the ecological adaptation of the host.

Horizontal transfer of transposable elements among animals

How can genetic material be transferred between two non-mating animals? Transposable elements, with their inherent mobility within genomes, may be candidates for such transfer vectors. Indeed, in several studies, transposable elements were found to cross species and even phylum barriers with what seems like relative ease. Moreover, judging by the sequence similarity among horizontally transferred elements, one can show that extraordinary feats of dispersal among widely divergent taxa can occur in a relatively short period of time (Pace et al. 2008). In the three cases described next, we have evidence for the transfer of transposable elements and for the propagation of the elements in the recipient species, but we lack evidence for any useful function performed by the newly acquired DNA sequences for the recipient.

HOST-PARASITE INTERACTIONS FACILITATE TRANSFER OF TRANSPOSONS ACROSS PHYLA A study by Gilbert et al. (2010) has shown that host-parasite interactions are extremely important in facilitating the movement of transposons from one animal species to another. Their study dealt with the kissing bug, *Rhodnius prolixus* (also called triatomine bug or assassin bug), a hemipteran that feeds on the blood of various tetrapods, such as opossums, squirrel monkeys, and bats, and serves as vector for *Trypanosoma cruzi*, the etiological agent of Chagas disease. The exchange of large quantities of blood and saliva between the bugs and their hosts during feeding is known to facilitate the spread of trypanosomes and could also provide a route for the horizontal transfer of transposons.

Rhodnius prolixus carries in its genome four distinct transposon families: *SPACE INVADERS*, *hAT1*, *OposCharlie1*, and *ExtraTerrestrial*. In **Figure 9.33**, we see these four families of transposable elements exhibiting an extreme form of disjointed taxonomic distribution among 52 completely sequenced vertebrate genomes. The distribution of the four transposon families makes absolutely no phylogenetic sense. Moreover, the degree of sequence similarity is also peculiar and does not correlate with phylogenetic relationships. This phylogenetic discontinuity and the extraordinary sequence similarity among distantly related species cannot be explained by processes other than horizontal gene transfer. Moreover, judging by the fact that the transposable elements from the different animals are quite similar in sequence, the multiple horizontal gene transfers must have occurred quite recently.

Some taxa carry more than a single element; however, in addition to *R. prolixus*, only one other organism carries all four transposon families—the little brown bat *Myotis lucifugus*. Another puzzling aspect of the taxonomic distribution is the extensive overlap among the transposon families. Mapping the occurrence of these transposons onto the phylogeny of 102 animals for which whole genome assemblies are currently available reveals a strikingly nonrandom distribution. The probability of observing this distribution by chance alone, i.e., the probability that the transposons have been acquired independently, is approximately 5×10^{-8} (Gilbert et al. 2010), which strongly suggests that those hosts that possess two or more transposon families acquired them together at one time. Unfortunately, the exact routes of horizontal gene transfer in this case are not easy to reconstruct, but there seems little doubt that host-parasite interactions play a major role in the transfer.

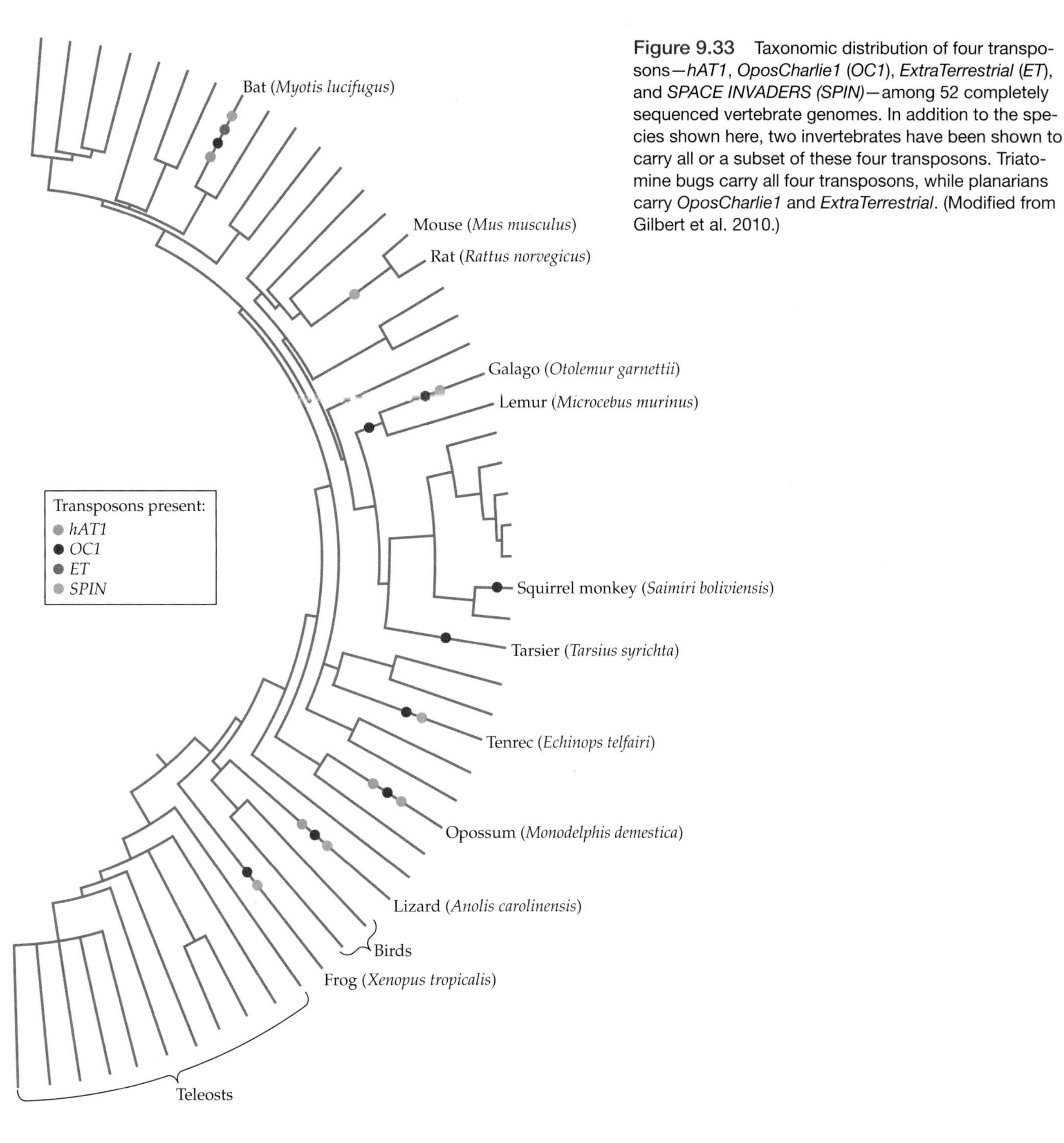

Figure 9.33 Taxonomic distribution of four transposons—*hAT1*, *OposCharlie1* (*OC1*), *ExtraTerrestrial* (*ET*), and *SPACE INVADERS* (*SPIN*)—among 52 completely sequenced vertebrate genomes. In addition to the species shown here, two invertebrates have been shown to carry all or a subset of these four transposons. Triatomine bugs carry all four transposons, while planarians carry *OposCharlie1* and *ExtraTerrestrial*. (Modified from Gilbert et al. 2010.)

Although the evolutionary consequences of the horizontal transfer of transposons are not yet clear, the sheer amount of DNA generated by the amplification of the transposons (e.g., approximately 100,000 copies of *SPACE INVADERS* in tenrec), as well as the myriad ways through which mobile elements can alter the structure and function of genes and genomes, supports the idea that the exchange of genetic material between animal hosts and parasites could have important evolutionary ramifications.

HORIZONTAL TRANSFER OF ENDOGENOUS RETROVIRUSES FROM BABOONS TO CATS As mentioned previously, the vertebrate genome contains many sequences that are homologous to retroviruses. Endogenous retroviruses (ERVs or virogenes) are normal constituents of the nuclear DNA of many multicellular eukaryotes. There

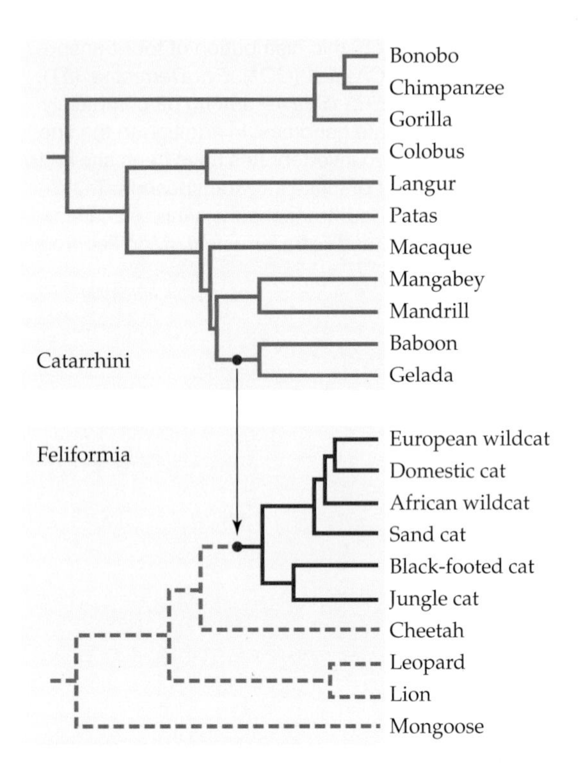

Figure 9.34 Phylogenetic trees for Old World monkeys (Catarrhini) and catlike carnivores (Feliformia). The branches leading to taxa that contain the type C virogene are shown as solid red and blue lines. The virogenes from the baboon and the gelada exhibit the highest similarity to the feline virogene. Therefore, a horizontal gene transfer probably occurred about 10 million years ago, from the ancestor of the baboon and the gelada to the ancestor of the domestic cat group after the divergence of lion, leopard, and cheetah lineages. (Modified from Graur and Li 2000.)

are several examples of endogenous retroviral sequences being transferred between vertebrate species (Benveniste 1985), one of which involves a type C retrovirus from baboons (**Figure 9.34**).

Sequences homologous to the baboon ERV have been detected in the cellular DNA of all Old World monkeys. The sequence similarity between them is closely correlated with the phylogenetic relationship among the species. Thus, the ERV must have existed for at least 30 million years in the genomes of primates. Less expected was the finding that six cat species closely related to the domestic cat (*Felis catus*) also contain this sequence, although it is present in neither more distantly related Felidae, such as lions, leopards, and cheetahs, nor in any other carnivore. It is thus highly probable that this sequence was horizontally transferred between species sometime in the past.

The date and direction of the horizontal transmission can be deduced from two types of data: sequence similarity and paleogeographical information. All cat species that contain the baboon ERV are from the Mediterranean area, while Felidae from Southeast Asia, the New World, and Africa lack the sequence. Therefore, the transfer must have occurred after the major radiation of the Felidae and been limited to one zoogeographical area. This conclusion points to a date of about 5–10 million years ago for the horizontal gene transfer event.

The direction of transmission can be deduced by considering the distribution of the sequence among primates on the one hand and cats on the other. Since all Old World monkeys possess the endogenous retrovirus, while only a few felid species possess it, it is reasonable to assume that the cats acquired the sequence from the baboons and not vice versa. This conclusion is strengthened by considering that the virogene in cats is more similar to that in three species of baboons (*Papio cynocephalus*, *P. papio*, and *P. hamadryas*) and the closely related gelada (*Theropithecus gelada*) than to that of any other primate sequence. Therefore, the sequence must have been transferred to cats from the ancestor of the baboons and the gelada shortly after their divergence from the mandrill-mangabey clade (Figure 9.34). The date derived from the Old World monkey species tree agrees well with the date derived from the cats.

HORIZONTAL TRANSFER OF *P* ELEMENTS BETWEEN *DROSOPHILA* SPECIES Another example of horizontal transfer involves the *P* transposable elements in *D. melanogaster*. As mentioned previously, *P* elements have spread rapidly through natural populations of *D. melanogaster* in less than a hundred years. *P* elements have not been detected in any of the hundreds of closely related *melanogaster* group species, such as *D. mauritiana*, *D. sechellia*, *D. simulans*, and *D. yakuba* (Daniels et al. 1990). Where, then, did these elements come from?

Interestingly, all species of the distantly related *D. willistoni* and *saltans* groups were found to contain *P* and *P*-like elements. Moreover, drosophilid genera such as *Scaptomyza*, and even nondrosophilid genera such as *Lucilia*, were also found to contain *P*-like elements. However, the *P* element from *D. willistoni* was found to be identical to the ones in *D. melanogaster* with the exception of a single nucleotide substitution, indicating that *D. willistoni* may have indeed served as the donor species in the horizontal gene transfer of *P* elements to *D. melanogaster*.

There are several reasons to suspect that this horizontal gene transfer occurred quite recently. First, the near identity of the *P* sequences from *D. melanogaster* and *D. willistoni* suggests a very short time of divergence. Second, the near absence of genetic variability in the *P* sequences from *D. melanogaster* from even very distant geographic locations indicates that the time since the introduction of *P* elements into *D. melanogaster* is too short for genetic variability to have accumulated. Finally, the geographical

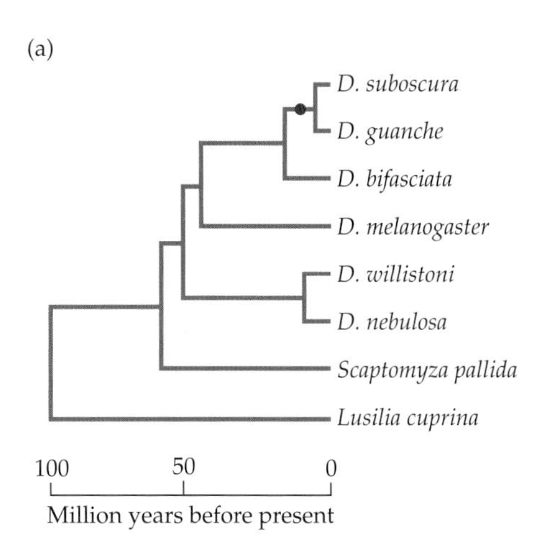

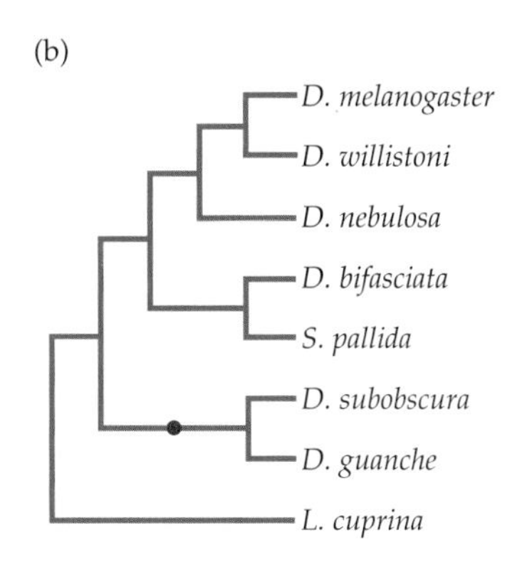

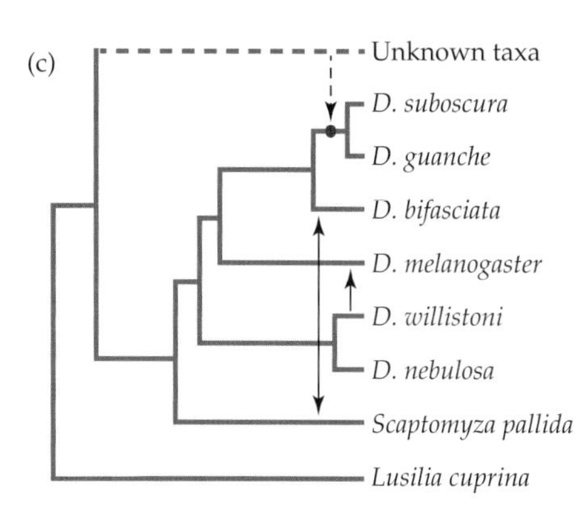

Figure 9.35 Comparison of a species tree for *Drosophila* and related taxa (a) with a gene tree derived from *P* element sequences (b). With the exception of the *subobscura-guanche* clade (red dots), the two trees are completely incongruent. (c) To resolve the incongruence, several horizontal gene transfers have to be assumed. Only a few of these transfers could be localized on the branches of the species tree. The only horizontal transfer whose direction could be unambiguously inferred, i.e., from *D. willistoni* to *D. melanogaster*, is marked with a single-headed arrow. The direction of the horizontal gene transfer event involving *D. bifasciata* and *Scaptomyza pallida* could not be determined (double-headed arrow). Some putative transfers proposed by Clark et al. (1994) necessitate assuming the existence or former existence of an unknown species or group of species (dashed lines) from which *P* elements were transferred to the ancestor of *D. subobscura* and *D. guanche* (dashed arrow). (Data from Clark et al. 1994.)

pattern of appearance of *P* elements in *D. melanogaster*, with populations in the American continents acquiring it first (Figure 9.28), suggests a very recent invasion, probably within the last 100 years (Kidwell 1983; Anxolabéhère et al. 1988; Gamo et al. 1990).

Clark et al. (1994) and Clark and Kidwell (1997) conducted a phylogenetic analysis of *P* elements from *Drosophila* and other dipteran species. The phylogenetic trees obtained from these analyses were found to be incongruent with the species tree that has been reliably inferred from molecular, morphological, and paleontological data (**Figure 9.35**). Numerous instances of horizontal gene transfer had to be postulated to restore congruence between the two trees. However, in many cases, the horizontal transfers must have been quite ancient, such that they could not be exactly localized to particular internal branches in the species tree. Furthermore, in some cases the existence of unknown extant or extinct taxa had to be assumed. From such an unknown species or group of species, *P* elements are assumed to have been transferred to the ancestor of *D. subobscura* and *D. guanche*.

Houck et al. (1991) identified the semiparasitic mite *Proctolaelaps regalis* as a potential suspect in the transfer of *P* elements from *D. willistoni* to *D. melanogaster*. Although *P. regalis* is not the most common mite co-occurring with these two *Drosophila* species, several of its anatomical, behavioral, and ecological attributes (such as its peripatetic mode of fluid-feeding on the immature stages of flies) and its geographic overlap with North American *Drosophila* species make it a very suitable putative vector in the transfer. Indeed, *P. regalis* possesses the capacity to acquire *P* elements from *P*-carrying strains of *Drosophila*.

Houck et al. (1991) have also defined a minimum number of conditions that have to be satisfied in order for this horizontal gene transfer to take place. These are (1) the two *Drosophila* females from the donor and the recipient species must lay their eggs in proximity to one another; (2) the recipient egg must be less than 3 hours old, i.e., it must not have reached the 512-cell stage; (3) the germline of the recipient embryo must incorporate a complete copy of *P* before the element degrades in the cytoplasm; and (4) the receiving embryo must survive the injuries inflicted by the mite. If, as seems likely, each of these events has a low independent probability in nature, then the combined multiplicative probability must be extremely low, and the interspecific horizontal gene transfer of *P* elements probably occurs only very rarely.

BDELLOID ROTIFERS: CHAMPIONS OF INDISCRIMINATE HORIZONTAL GENE TRANSFER Bdelloid rotifers are tiny freshwater invertebrates with a set of unique and bizarre characteristics among animals: they can withstand desiccation at any life stage, they are impervious to levels of ionizing radiation that would kill other animals, and their lineage has survived over many millions of years without standard sexual reproduction, i.e., through meiotic processes that do not require paired homologous chromosomes (Signorovitch et al. 2015). On top of these peculiarities, bdelloids have an uncanny ability to acquire genes from other organisms through horizontal gene transfer. Boschetti et al. (2012) have shown that horizontal gene transfer occurs at an unprecedented scale in bdelloids. Approximately 10% of their active protein-coding genes are "foreign," mostly originating from bacteria and unicellular eukaryotes, but now functioning as part of the bdelloid repertoire. About 80% of the horizontally transferred and active genes code for enzymes, and these make a major contribution to bdelloid biochemistry; 39% of enzyme activities have a foreign contribution, and in 23% of cases the activity in question is uniquely specified by a foreign gene. This indicates that some of the idiosyncratic biochemistry of bdelloids, such as toxin degradation and antioxidant generation, which are unknown in other animals and which are expected to provide the bdelloid with a selective advantage, have been acquired from other species. It has been suggested that the bdelloids' massive gene transfer may represent a possible mechanism for survival without sex, by diversification of functional capacity and even replacement of defective genes by functional xenologs (Gladyshev et al. 2008; Boschetti et al. 2012).

Promiscuous DNA

With the exception of some secondary losses, all eukaryotes have at least two separate genomes, a large nuclear genome (**nDNA**) and a much smaller mitochondrial genome (**mtDNA**). Plants and some protists have an additional plastid genome (**ptDNA**). (A plastid containing the green photosynthetic pigment chlorophyll is called the **chloroplast**.) The plastids and the mitochondria ultimately derive from free-living prokaryotes (Chapter 5). Collectively, mtDNA and ptDNA are referred to as organelle DNA (**orgDNA**).

Organelle genomes were suspected for a long time to have lost many of their original genes or to have had them transferred to the nuclear DNA during the long evolutionary time since the beginning of the endosymbiotic relationship. However, until the 1980s, evidence for such transfer was lacking, and the three genetic systems were thought to be largely discrete. The first instance of a DNA transfer from the mitochondria to the nucleus was discovered by Dubuy and Riley (1967), who used DNA-DNA hybridization to detect mitochondrial DNA in the nuclear genome of mouse and the parasitic flagellate *Leishmania enriettii*. Subsequently, Stern and Lonsdale (1982) reported the first instance of an organelle-to-organelle (chloroplast-to-mitochondrion) transfer, and Timmis and Scott (1983) reported the first DNA transfer from the chloroplast to the nucleus. This type of "disrespect" for organellar genomic barriers has been dubbed **promiscuous DNA** by two science commentators (Ellis 1982; Lewin 1982).

Promiscuous DNA provides evidence for intercompartmental DNA flow between organelles and between organelles and the nucleus. To date, examples have been found for five out of the six possible types of DNA transfer among genomes (**Figure 9.36**). Plastid genomes seem to be almost impenetrable to DNA transfer, and so far, no evidence exists for nucleus-to-plastid transfer, while mitochondrion-to-plastid transfer seems to be exceedingly rare (Goremykin et al. 2009; Iorizzo et al. 2012). On the other hand, the plastid genome seems to be the most common donor of promiscuous DNA (Thorsness and Weber 1996; Nakazono and Hirai 2008), as plastid-derived sequences have been found in both nuclear and mitochondrial genomes. Moreover, there is evidence for plastid sequences migrating to the mitochondrial genome and subsequently being transferred to the nucleus (Notsu et al. 2002).

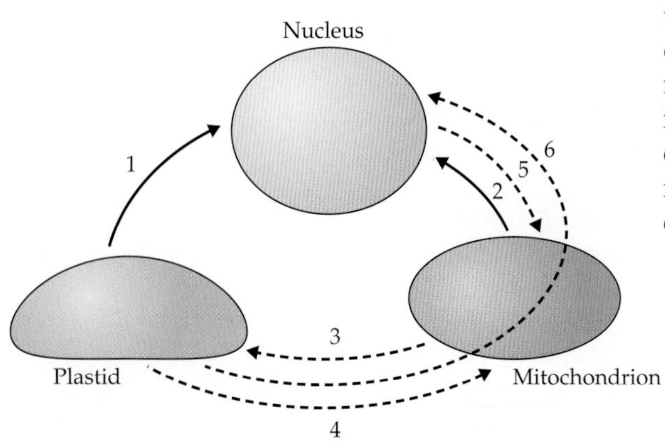

Figure 9.36 Known types of intracellular DNA transfer: (1) plastid-to-nucleus, (2) mitochondrion-to-nucleus, (3) mitochondrion-to-plastid, (4) plastid-to-mitochondrion, (5) nucleus-to-mitochondrion, and (6) plastid-to-nucleus via the mitochondrion. Solid arrows indicate frequent events, dashed arrows indicate rare events.

Transfer of intact functional genes to the nucleus

Mitochondria and plastids are descendants of proteobacterium-like and nitrogen-fixing cyanobacterium-like progenitors, respectively. Although the organelles have retained much of their prokaryotic biochemistry, their genomes currently encode only a small fraction of the organelles' proteins. In mitochondria the current numbers of protein-coding genes range from 3 in several *Plasmodium* species to 163 in a subspecies of *Zea mays*. The numbers of protein-coding genes in plastid genomes range from 23 in the parasitic western underground orchid, *Rhizanthella gardneri*, to 209 in the Korean pine, *Pinus karaiensis*. In comparison, extant cyanobacteria and proteobacteria have 1,500–7,500 genes and 3,500–8,500 genes, respectively (Timmis et al. 2004). The huge reduction in coding capacity of organelles is partly due to gene loss and partly due to the transfer of genes to the nucleus, a process associated with the loss of the organelles' genetic autonomy. Most functional genes of organellar origin are thought to have been transferred to the nucleus soon after the endosymbiotic event, whereas subsequent transfers of functional genes are thought to have been rare. These transfers have contributed significantly to the nuclear genome. For example, 18% (or 4,500 out of 25,000) of all nuclear genes in *Arabidopsis thaliana* are estimated to be of cyanobacterial origin. The contribution of protomitochondrial genes to the nuclear genome may be even higher (Esser et al. 2004).

In animals, functional-gene transfer from mitochondria to the nucleus has almost certainly ceased (Kleine et al. 2009). In contrast, in flowering plants, transfer of some functional organelle genes to the nucleus seems to be an ongoing process. Massive ongoing transfers of organellar genes to the nucleus are sometimes seen in parasitic plants (e.g., Xi et al. 2013).

In principle, until an organelle-derived gene is successfully established in the nucleus, two copies of the gene may be retained—one in the organelle and one in the nuclear genome. Indeed, the nucleus- and plastid-encoded genes for the ribosomal protein 16S in *Arabidopsis thaliana* and its relatives (*A. arenosa*, *A. lyrata*, and *Crucihimalaya lasiocarpa*) are known to have coexisted for at least 125 million years (Roy et al. 2010).

Transfer of nonfunctional DNA segments from organelles to the nucleus: Numts and nupts

While at present the DNA transfer from plastids and mitochondria to the nucleus that results in functional nuclear genes has mostly ceased, the transfer of DNA from organelles to the nucleus that results in nonfunctional DNA segments continues to occur quite frequently. As a result, the nuclear genome of eukaryotes has become replete with pseudogenes of mitochondrial origin. Both mtDNA and ptDNA are known to give rise to nonfunctional sequences in the nuclear genome. While mtDNA gives rise to **nuclear mitochondrial DNA sequences**, abbreviated as **numts** (pronounced "new mights"), ptDNA produces **nuclear plastid DNA sequences**, or **nupts** (pronounced "new peats"). Collectively *numt*s and *nupt*s are referred to as **nuclear organelle DNA (norgDNA)**. As complete nuclear genome sequences continue to accumulate, numerous norgDNA sequences have been identified.

*Numt*s have been found in the vast majority of eukaryotes so far studied, while *nupt*s are abundant in plants. The largest number of confirmed *numt*s is found in the western honeybee (*Apis mellifera*), which has close to 2,000 *numt*s in its nuclear genome (Behura 2007). The largest number of *nupt*s so far is found in the common grape vine (*Vitis vinifera*), which has almost 4,000 *nupt*s in its nuclear genome.

The fraction of the nuclear genome represented by *numts* or *nupts* is usually less than 0.1%, but it can be higher in flowering plants and some fungi. The abundance of *numt*s in eukaryotic genomes ranges from undetectable (e.g., in the mosquito *Anopheles gambiae* and the zebrafish *Danio rerio*) to about 2.1 Mb in the opossum *Monodelphis domestica*. Similarly, the abundance of *nupt*s in nuclear genomes varies from undetectable (e.g., in the brown alga *Aureococcus anophagefferens*) to more than 1 Mb in the short-grained *japonica* variety of rice (*Oryza sativa*). Because mutation and deletion rapidly render *numt* and *nupt* sequences undetectable, the observed value represents a steady state, reflecting the balance between organelle-DNA insertion, deletion, duplication, and mutational decay.

Not only do *numt*s and *nupt*s vary in total length within the nuclear genome, the sizes of individual sequences and the frequency distributions of sequence length vary as well. The largest organelle DNA insertion reported so far is a 620 Kb partially duplicated *numt* in *Arabidopsis thaliana*. The largest unduplicated *numt* is a 262 Kb *numt* in *A. thaliana*, and the largest unduplicated *nupt* is a 131 Kb *nupt* in rice. Since *numt*s and *nupt*s are devoid of function, the largest *numt*s and *nupt*s are of more recent origin; because of deletion and recombination, *numt*s and *nupt*s become shorter with evolutionary time.

For *numt*s and *nupt*s to become part of the nuclear genome, organellar DNA must physically reach the nucleus and integrate into the nuclear chromosome (Hazkani-Covo et al. 2010). In theory, there are two pathways for organellar DNA to reach the nucleus. It can reach it either as DNA fragments following organelle death and disintegration (e.g., Campbell and Thorsness 1998), or as cDNA following reverse transcription (Nugent and Palmer 1991). We now know that the incorporation of organellar DNA into the nucleus occurs mostly by nonhomologous end-joining of DNA segments at double-strand breaks (Blanchard and Schmidt 1996; Ricchetti et al. 1999, 2004; Hazkani-Covo and Covo 2008; Hazkani-Covo et al. 2010). Thus, transcription and reverse transcription are not factors in *numt* and *nupt* formation. This, in conjunction with the fact that *numt*s and *nupt*s lack function, yields three testable predictions: (1) organisms with multiple organelles in their cells will have more *numt*s and *nupt*s than organisms with a single organelle, (2) all parts of the organelle genome will be represented in the nuclear genome regardless of function or lack thereof, and (3) large nuclear genomes will have more *numt*s and *nupt*s than small nuclear genomes. All these predictions are borne out by empirical observations (Huang et al. 2005; Hazkani-Covo et al. 2010; Smith et al. 2011).

The *numt*s have been referred to as "poltergeists" (German for "noisy ghosts") because of their tendency to throw research results into confusion (Hazkani-Covo et al. 2010). Due to their sequence similarity to mitochondrial DNA, *numt*s are responsible for many instances of misidentification. For example, *numt*s sequences are frequently misreported as mitochondrial variations, and at least one *numt* was erroneously implicated in causing disease (Thangaraj et al. 2003). In addition, *numt*s are known to interfere with DNA barcoding, a technique that uses short mitochondrial segments to identify organisms. Because of the existence of *numt*s, it has been suggested that the DNA barcoding approach is unreliable, at least in primates and arthropods (Thalmann et al. 2004). As mentioned in Chapter 5, *numt*s have also been responsible for "fooling" scientists into believing that 80-million-year-old dinosaur bones harbored DNA (Woodward et al. 1994; Zischler et al. 1995). Finally, *numt*s may be responsible for erroneous reports of horizontal gene transfer among plants (Goremykin et al. 2009).

Rates and evolutionary impacts of norgDNA insertion

The rate of norgDNA insertion has been assessed in several species. In yeast, the probability of a *numt* insertion into the nucleus is about 2×10^{-5} per cell per generation. In humans, the rate of *numt* integration in the germline has been estimated to be $5–6 \times 10^{-6}$ per germline cell per generation. Thus, on average, any two haploid nuclear human genomes should differ by the presence or absence of orgDNA in at least two loci. In male gametes of the tobacco plant, the chloroplast-to-nucleus DNA transfer frequency is about 2×10^{-5}.

The rate of *numt* insertion differs among lineages and at different times within an evolutionary lineage. For example, since the human-chimpanzee divergence, the number of *numt*s that have integrated in the chimpanzee nuclear genome is about twice that in the human genome. Moreover, by using a phylogenetic reconstruction of *numt* insertions in Old World monkeys, it was determined that a burst of insertions occurred in the ancestor of the great apes (human, chimpanzee, gorilla, orangutan) after their divergence from gibbons (Jensen-Seaman et al. 2009).

From an evolutionary viewpoint, *numt* and *nupt* insertions can essentially be regarded as nuclear mutations. By their very nature, mutations can be neutral, deleterious, or advantageous. Most *numt* insertions are neither deleterious nor beneficial. Rarely, however, a *numt* insertion may interfere with the proper function of a genomic element. Indeed, several familial diseases caused by the de novo insertions of *numt*s into genes are known (Turner et al. 2003; Hazkani-Covo et al. 2010). One such case involves a germline insertion of a 41 bp *numt* sequence into the breakpoint junction of a chromosomal translocation; another concerns a severe inherited bleeding disorder where a 251 bp *numt* insertion has introduced a novel splicing site into the gene for coagulation factor VII (Borensztajn et al. 2002). Other cases include a de novo 72 bp mtDNA insertion into a gene for a zinc finger transcription factor that created a premature stop codon and was responsible for a sporadic case of a developmental disease called Pallister-Hall syndrome, and a recent intronic *numt* insertion in *MADH2*, a tumor suppressor gene that is often mutated in colorectal carcinomas. Because about 25% of *nupt*s and *numt*s are found within exons or introns in flowering plants, it can be concluded that norgDNA insertions represent a considerable challenge to the functional integrity of the nuclear genomes of plants (Leister 2005).

All in all, the vast majority of *numt*s and *nupt*s behave like selectively neutral mutations, and once *numt*s and *nupt*s are in the nucleus, decaying processes are expected to determine their fates. Indeed, in a study of the rice genome, Matsuo et al. (2005) found that following their integration into the nuclear genome, *nupt*s are rapidly fragmented and shuffled around, and about 80% of them are eliminated or become unrecognizable within a million years or so.

Are all norgDNA-sequence insertions neutral or deleterious? There are at least two tentative reports on norgDNA sequences being recruited and adapted as functional sequences in the nuclear genome (Noutsos et al. 2007; Rousseau-Gueutin et al. 2011). While many more cases will undoubtedly be found, beneficial norgDNA insertions are expected to be the exceedingly rare exceptions rather than the rule.

CHAPTER

10

Prokaryotic Genome Evolution

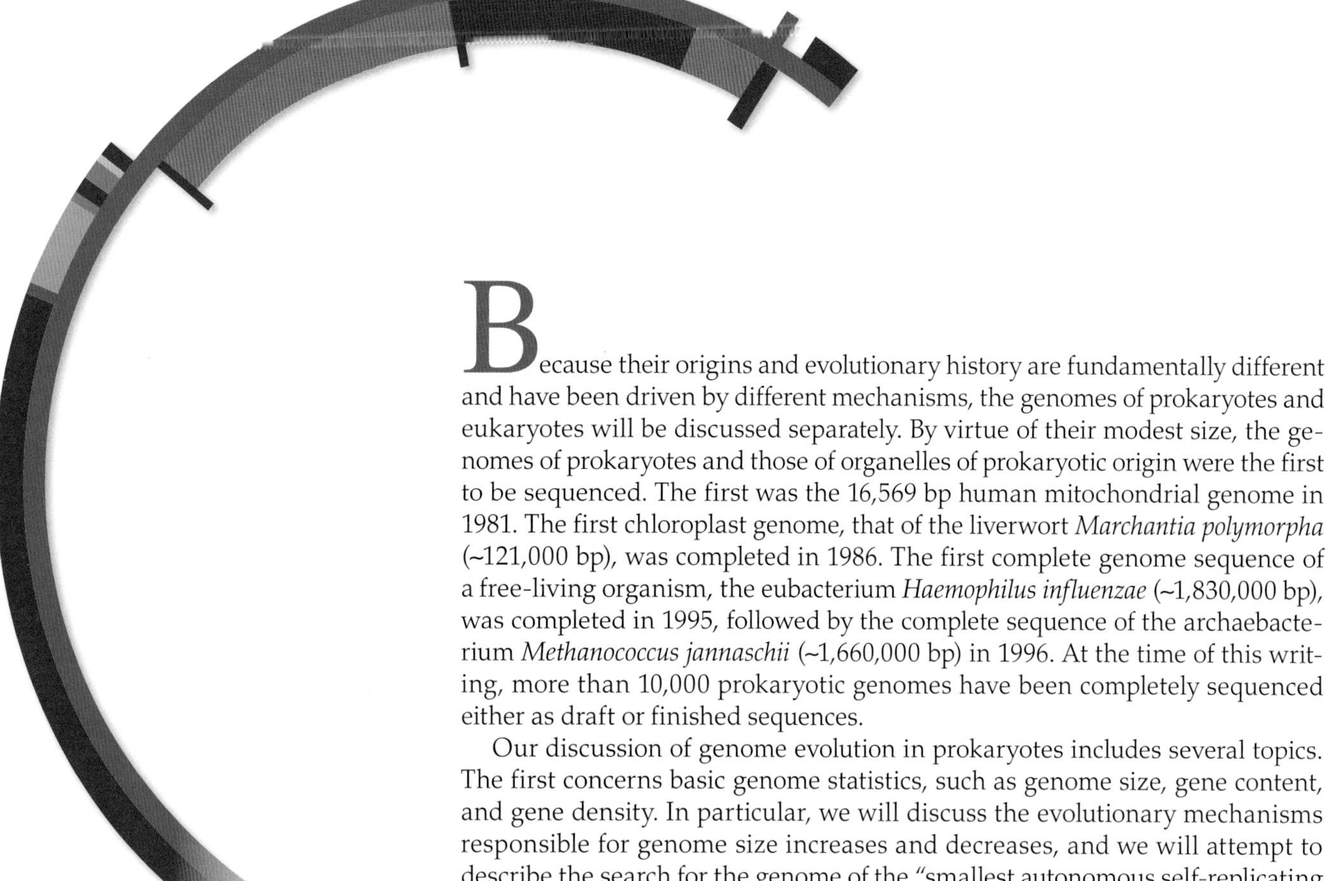

Because their origins and evolutionary history are fundamentally different and have been driven by different mechanisms, the genomes of prokaryotes and eukaryotes will be discussed separately. By virtue of their modest size, the genomes of prokaryotes and those of organelles of prokaryotic origin were the first to be sequenced. The first was the 16,569 bp human mitochondrial genome in 1981. The first chloroplast genome, that of the liverwort *Marchantia polymorpha* (~121,000 bp), was completed in 1986. The first complete genome sequence of a free-living organism, the eubacterium *Haemophilus influenzae* (~1,830,000 bp), was completed in 1995, followed by the complete sequence of the archaebacterium *Methanococcus jannaschii* (~1,660,000 bp) in 1996. At the time of this writing, more than 10,000 prokaryotic genomes have been completely sequenced either as draft or finished sequences.

Our discussion of genome evolution in prokaryotes includes several topics. The first concerns basic genome statistics, such as genome size, gene content, and gene density. In particular, we will discuss the evolutionary mechanisms responsible for genome size increases and decreases, and we will attempt to describe the search for the genome of the "smallest autonomous self-replicating entity," i.e., address the "minimal genome" question. The second topic to be dealt with in this chapter concerns gene order and the dynamics of evolutionary change in gene order. How are genes distributed along and among the chromosomes, and what mechanisms are responsible for the reshaping of gene order during evolution? The third topic concerns nucleotide composition, in particular addressing the relative importance of mutation versus selection in determining genomic GC content.

At the end of this chapter, we will discuss the evolution of the standard genetic code and its alternatives. Specifically, we will ask, How can the rules of translation, which affect all proteins in a cell, be changed without deleterious effects, and under what conditions?

Table 10.1

Genome size ranges in some prokaryotic taxa

Taxon	Genome size range (Kb)	Ratio (highest/lowest)
Eubacteria	**112–14,780**	**132**
Actinobacteria	930–11,950	13
Cyanobacteria	1,440–9,060	6
Firmicutes	1,380–8,820	6
Proteobacteria	112–14,780	132
Spirochaetes	900–4,710	5
Tenericutes	510–1,810	4
Archaebacteria	**490–5,750**	**12**
Crenarchaeota	1,300–2,990	2
Euryarchaeota	1,240–5,750	5
Thaumarchaeota	1,640–2,830	2

Genome Size in Prokaryotes

Bacterial genome sizes vary over approximately two orders of magnitude (**Table 10.1**). The average eubacterial genome size is approximately one-third larger than that of the average archaebacterium. The variation in genome size in eubacteria is approximately ten times larger than that in archaebacteria (**Figure 10.1**). The smaller genome-size variation in archaebacteria either may reflect the true state of affairs in nature or it may be due to the fact that considerably fewer archaebacteria than eubacteria have been sequenced. The fact that the mean genome size of archaebacteria is also smaller than that of eubacteria supports the claim that the size variation in archaebacteria is not an artifact of sampling. Genome size variation in prokaryotes is much smaller than that in eukaryotes.

The smallest sequenced eubacterial genome is that of the endosymbiotic β-proteobacterium *Nasuia deltocephalinicola* (~112 Kb), which is found within the cells of the phloem-feeding leafhopper *Macrosteles quadrilineatus*. The genome of *N. deltocephalinicola* contains 137 protein-coding genes, 2 rRNA-specifying genes, and 29 tRNA-specifying genes. The largest eubacterial genome belongs to a strain of the soil-dwelling δ-proteobacterium *Sorangium cellulosum* (~14.8 Mb). It contains 10,400 protein-coding genes and 103 RNA-specifying genes. The ratio of the largest to the smallest eubacterial genome is approximately 132. Eubacterial genome sizes were once thought to exhibit a bimodal frequency distribution (Koonin and Wolf 2008), but this bimodality was found to be due to biases in sampling (Graur 2014).

The smallest sequenced archaebacterial genome belongs to the hyperthermophilic *Nanoarchaeum equitans* (491 Kb), a marine species that appears to be an obligate ectoparasite on another archaebacterium, *Ignicoccus hospitalis*. It contains 540 protein-coding genes and 45 RNA-specifying genes. *N. equitans* cannot synthesize lipids but obtains them from its host; it, hence, must be in contact with the host to survive. The largest archaebacterial genome that has been sequenced so far is that of the methane-producing *Methanosarcina acetivorans* (5.8 Mb), which contains 4,540 protein-coding genes and 181 RNA-specifying genes. The ratio of the largest to the smallest archaebacterial genome is slightly less than 12.

The numbers of protein-coding genes in eubacteria range from 116 to 10,400 (approximately a 90-fold range). In other words, the variation in the number of genes is only slightly smaller than the variation in C values. The numbers of protein-coding genes in archaebacteria range from 540 to 4,540 (approximately an 8-fold

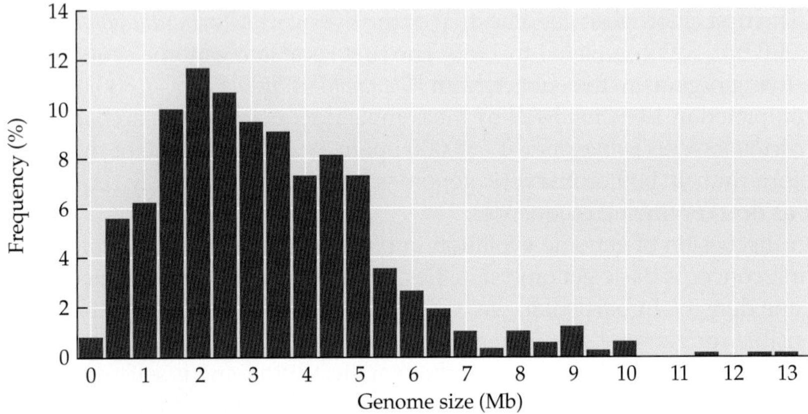

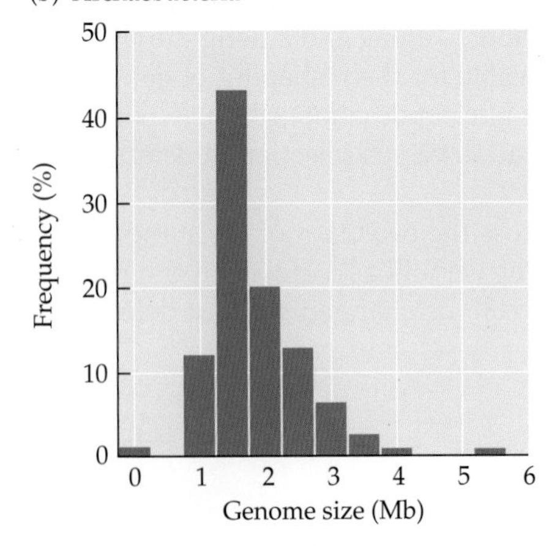

Figure 10.1 Frequency distribution of eubacterial (a) and archaebacterial (b) genome sizes. The figure is based on data from 998 completely sequenced eubacterial genomes and 109 archaebacterial ones. (Courtesy of Yichen Zheng.)

range) in comparison with a 12-fold range for genome sizes. As in eubacteria, the variation in the number of genes in archaebacteria is only slightly smaller than the variation in C values. (The relative rankings of genome sizes, on the one hand, and protein-coding gene contents, on the other, are not completely congruent. For example, at 139 Kb, the genome of strain PCVAL of the citrus mealybug endosymbiont *Tremblaya princeps* is larger than the genome of *N. deltocephalinicola* by about 27 Kb, yet it contains 21 fewer protein-coding genes.)

In reviewing gene content, we note that some annotated protein-coding genes may not actually be genes, but rather open reading frames that occur by chance. Skovgaard et al. (2001) discovered this overestimation by comparing the length distribution of annotated genes with the length distribution of known proteins, and finding too many short open reading frames being annotated as protein-coding genes. According to their calculations, the genome of *Escherichia coli*, for example, contains ~3,800 protein-coding genes, whereas ~4,300 such genes are annotated in databases. A similar discrepancy was found in other prokaryotic genomes.

In the vast majority of cases, the productive fraction (protein-coding genes and RNA-specifying genes) of the prokaryote genome is estimated to be around 90%. Let us consider the model prokaryote, *E. coli*, as an example. The genome size of *E. coli* (substrain MB1655, strain K12) is 4,639,675 bp. There are ~4,300 protein-coding genes that account for ~88% of the genome. In addition to the protein-coding genes, there are about 200 RNA-specifying genes that account for 0.8% of the genome. Nontranscribed functional genes serving regulatory functions constitute less than 0.01% of the genome. The rest of the genome is presumably nonfunctional; it is made of intergenic regions, various types of tandem and dispersed repeated sequences, about 100 pseudogenes, and variable numbers of transposable elements, mainly insertion sequences. In total, the amount of junk DNA in *E. coli* is expected to account for approximately 10% of the genome. It is thought that the situation of *E. coli* is representative of many free-living prokaryote genomes.

In archaebacteria and nonpathogenic eubacteria, 1–8% of all gene-like sequences turned out to be pseudogenes (Liu et al. 2004; Lerat and Ochman 2005). Obligate pathogenic eubacteria have on average a higher proportion of pseudogenes in their genomes that their nonpathogenic brethren. In some extreme cases, such as that of, the intracellular parasites *Rickettsia prowazekii* (the etiological agent of epidemic typhus) and *Mycobacterium leprae* (the etiological agent of Hansen's disease—formerly called leprosy), the ratio of pseudogenes to functional genes can be as high as 80% (Singh and Cole 2011). In archaebacteria and nonpathogenic eubacteria, most pseudogenes arise from failed horizontal gene transfer. In pathogenic eubacteria, on the other hand, many pseudogenes are unitary pseudogenes that arose from the nonfunctionalization of single-copy genes. This phenomenon is consistent with the process of genome size reduction experienced by some parasitic organisms (see below).

For both eubacteria and archaebacteria, an almost perfect correlation exists between genome size and gene number (**Figure 10.2**). That is, the variation in prokaryotic genome size can be wholly explained by gene number. Because of this almost perfect correlation between genome size and number of protein-coding genes, there is only a modest variation in gene density among prokaryotes. In eubacteria, the density ranges from 0.38 genes per Kb in the aphid symbiont *Serratia symbiotica*, to 1.35 genes per Kb in the causative agent of feline infectious anemia, *Mycoplasma haemofelis*. In archaebacteria, the density ranges from 0.76 genes per Kb in the rumen methanogen *Methanobrevibacter ruminantium*, to 1.26 genes per Kb in the ammonia-oxidizing archaebacterium *Nitrososphaera gargensis*.

The pangenome, the core genome, and the accessory genome

Prokaryotes exhibit large intraspecific variation in gene content. For example, only about 67% of all protein-coding genes in *Legionella pneumophila*, the causative agent of Legionnaires' disease, were found in five completely sequenced strains of this bacterium (**Figure 10.3**).

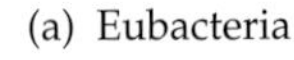

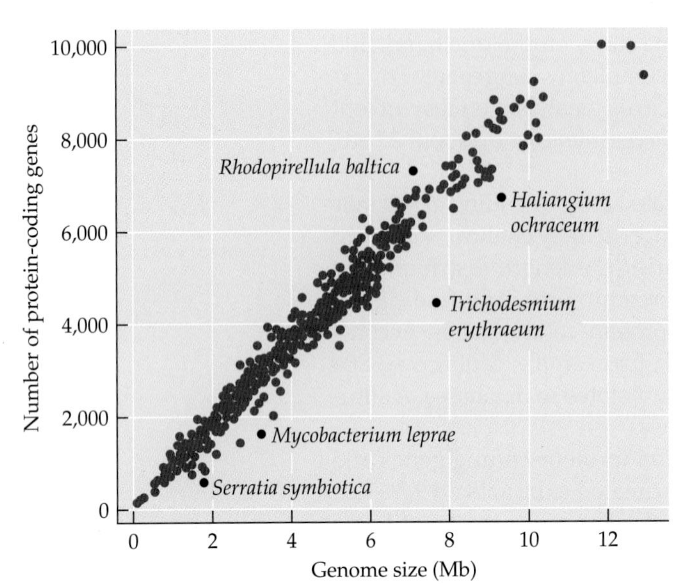

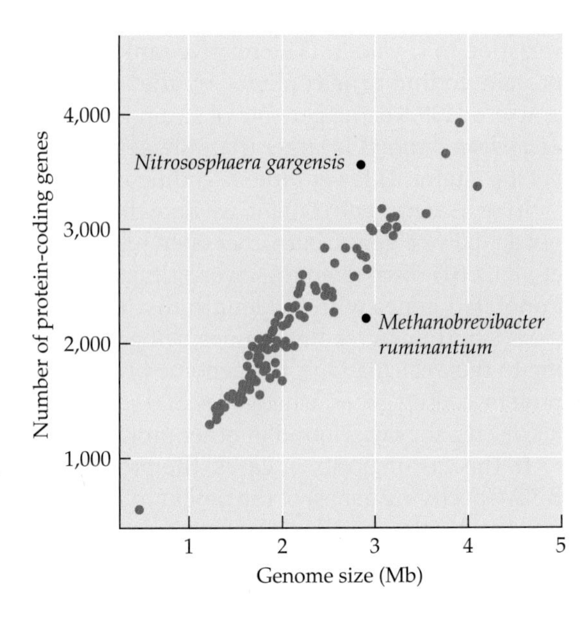

Figure 10.2 Relationship between gene number and genome size in 998 completely sequenced eubacterial species (a) and 109 archaebacterial ones (b). Some notable taxa deviating from the almost perfect linear correlation between the two variables are noted. (Courtesy of Yichen Zheng.)

Because of this variation, as far as prokaryotes are concerned, we need to distinguish among several genome categories. The **pangenome** (or **supragenome**) describes the full complement of genes in a species. In other words, a species' pangenome is the union of the gene sets of all of its strains. The pangenome consists of the **core genome**, i.e., all the genes that are found in all the strains, and the **accessory genome** (also **dispensable**, **auxiliary**, or **distributed genome**), i.e., genes present in one or more strains. Genes that are found in a single genome or a single strain are referred to as **unique genes** (also **character genes** or **strain-specific genes**).

We note that the sizes of the pangenome, the core genome, and the accessory genome depend on the sample size. With the increase in the number of sequenced genomes, the sizes of the pangenome and the accessory genomes are expected to increase, while the core genome is expected to decrease (**Figure 10.4**). Indeed, with three sequenced genomes, the core genome of *E. coli* was found to contain almost 3,000 genes while the pangenome consisted of more than 7,500 genes (Welch 2002). From a comparison of 17 genomes, Rasko et al. (2008) predicted through extrapolation to infinity that the core genome of *E. coli* would plateau at approximately 2,200 genes, while the pangenome would consist of about 13,000 genes. These predictions turned out to be quite off the mark. Using 20 sequenced genomes (a mere 3 genomes more

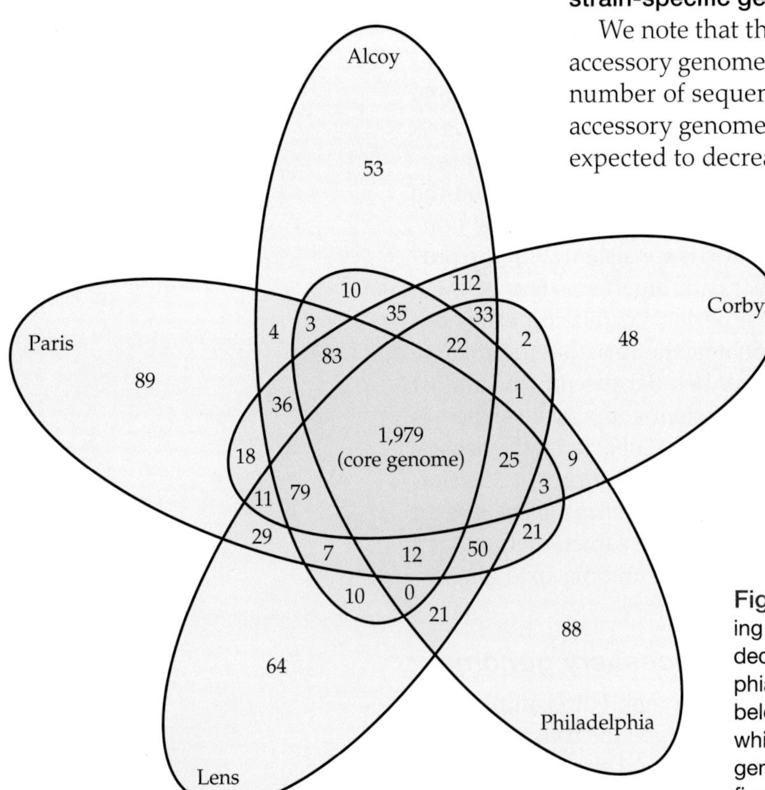

Figure 10.3 Venn diagram showing the 2,957 protein-coding genes in the pangenome of *Legionella pneumophila* as deduced from a comparison of five strains: Alcoy, Philadelphia, Lens, Paris, and Corby. Of these, 1,979 genes (66.9%) belong to the core genome that is common to all five strains, while 978 belong to the dispensable genome. A total of 342 genes were found to be specific to a single strain out of the five. (From D'Auria et al. 2010.)

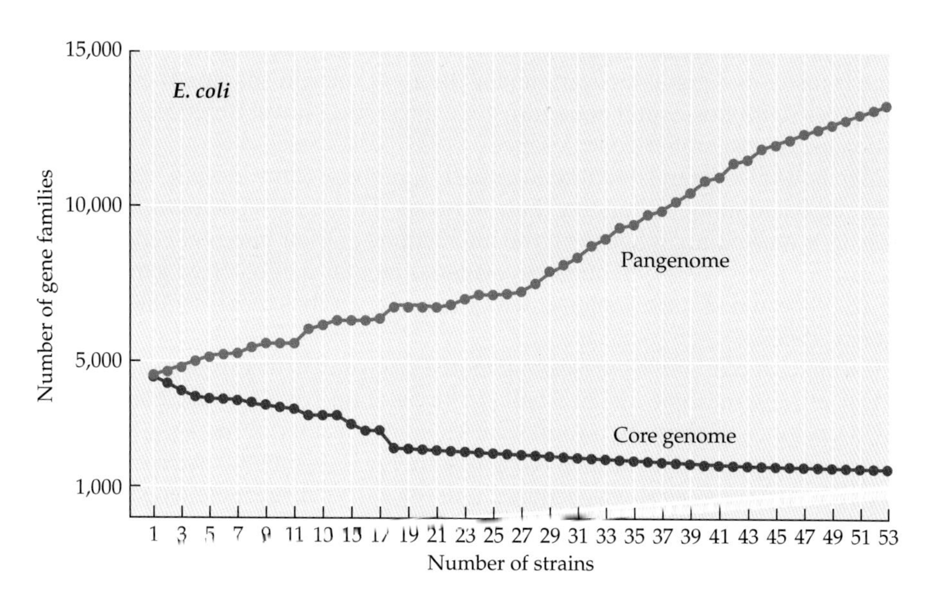

Figure 10.4 Number of gene families in the pangenome and core genome of *Escherichia coli* as a function of the number of sequenced genomes. Blue, pangenome; red, core genome. (Modified from Lukjancenko et al. 2010.)

than Rasko et al.), Tenaillon et al. (2010) found that the core genome decreased to less than 2,000 genes, while the pangenome ballooned to almost 18,000 genes. It is currently estimated that the core genome of *E. coli* consists of approximately 11% of the gene families found in the pangenome (Lukjancenko et al. 2010).

The core genome is usually interpreted as encoding the basic functions of the species, while the specialized features that are necessary in one niche but not in another are specified by the accessory genome. The pangenome is, then, taken to represent the genes that are necessary to colonize the totality of habitats within which the species is found.

The concepts of pangenome and core genome can be applied not only to species, but also to higher taxa. Of course, when the number of genomes increases to hundreds and even thousands, the definition of the core genome needs to be relaxed somewhat to include genes that are missing in only a small fraction of the genomes. Lapierre and Gogarten (2009) defined an **extended core genome**, which includes genes present in at least 99% of the sampled genomes. By extrapolating from a sample of 573 sequenced eubacterial genomes, Lapierre and Gogarten (2009) concluded that the eubacterial pangenome is of infinite size, i.e., the number of genes belonging to the pangenome increases with no sign of plateauing with increased sample size. Complementarily, the eubacterial core genome (sensu stricto) was found to be empty, but the extended core genome was found to consist of ~250 genes.

Increases and decreases in prokaryotic genome sizes

As with all study of evolution, discussing the evolution of genome size in prokaryotes requires us to distinguish between relevant mutational input, on the one hand, and the two evolutionary forces that can fix mutational changes in the population: selection and random genetic drift. Because of the almost perfect correlation between genome size and protein-coding gene number, and because of the fact that the bulk of the prokaryotic genome is taken by protein-coding genes, changes in genome size must be attributed to expansions and contractions in the protein-coding repertoire.

At the mutational level, expansions of the protein-coding gene repertoire can, in principle, occur by genome duplication, gene duplication, operon duplication, duplicative transposition, or horizontal gene transfer. So far, no evidence has been found for genome duplication in prokaryotes, while the contributions of operon duplication and duplicative transposition to genome size in prokaryotes are negligible. Treangen and Rocha (2011) have shown that 88% of all expansions of protein families are due to horizontal gene transfer. Moreover, horizontally transferred genes—xenologs—tend to persist longer in prokaryotic lineages than duplicated genes, while duplicated genes—paralogs—are more highly expressed and evolve more slowly. This suggests that gene transfer and gene duplication play different roles in shaping the evolution of prokaryotes, with horizontal gene transfer contributing novel functions and gene duplication leading to higher gene dosages.

Paralogs have also been shown to take part in more protein-protein interactions and to be regulated more elaborately than xenologs. This is as expected, since copies of a gene are expected to "inherit" all the interactions of their parental gene and the vast majority of its regulatory sequences, while horizontally transferred genes represent a blank slate and may only interact with molecular entities that resemble the ones in the donor species.

At the level of mutation, decreases in genome size can occur through either large deletions spanning multiple loci or small deletions of one or a few nucleotides. The relative importance of these two routes varies among bacterial lineages as well as within lineages. For example, from a comparison of endosymbiotic *Buchnera* species to their free-living relatives, it was deduced that the early stages of genome reduction involved large deletions spanning as many as 50 genes (Mira et al. 2001). By contrast, the rate of more recent reductions in the genome size of *Buchnera* species is quite slow. We note that gene death—pseudogenization—while extremely important biochemically and metabolically, does not in itself alter genome size; it merely removes all functional and selective constraints from the affected gene.

If we exclude horizontal gene transfer, there is an inherent mutational bias toward deletions over insertions in bacterial genomes (e.g., Mira et al. 2001). Moreover, deletions tend to be larger than insertions. Thus, in the absence of selection or when selection is overwhelmed by random genetic drift due to a decrease in effective population size, genomes are expected to decrease in size. Unfortunately, testing this prediction in a straightforward manner requires information on effective population size (N_e), which is usually difficult to come by, especially in microbes.

An ingenious solution was found by Kuo et al. (2009), who used an indirect proxy for N_e. Their reasoning was as follows: Because point mutations causing amino acid replacements are often deleterious, the rate of nonsynonymous substitution per nonsynonymous site (K_A) is usually much smaller than the rate of synonymous substitution per synonymous site (K_S). As the relative importance of genetic drift increases in species with low effective population sizes, many slightly deleterious amino acid replacements would escape purifying selection, causing a genome-wide increase in K_A/K_S ratios. Of course, K_A can also increase as a result of positive selection; however, positive selection is unlikely to affect all genes at once. Based on genomic K_A/K_S ratios in 42 pairs of bacterial species, Kuo et al. (2009) discovered that genome size exhibits a strong negative correlation with the degree of genetic drift. Support for the validity and reliability of the K_A/K_S ratio as a proxy for N_e can be gleaned from the fact that many species with high K_A/K_S ratios ($K_A/K_S > 0.06$) have lifestyles indicative of reduced effective population sizes. Such species include endosymbionts, extremophiles, vector-borne pathogens, and pathogens with limited transmission routes. Thus, lifestyles expected to result in an increase in the degree of genetic drift appear to be associated with genome reduction.

A commonly held view of bacterial genome size evolution argues that selection for rapid and efficient replication is a major force that drives the streamlining of bacterial genomes (e.g., Maniloff 1996; Vellai and Vida 1999). For this hypothesis to hold, a negative correlation should exist between genome size and generation time. No such correlation has been found (Mira et al. 2001). Moreover, bacteria harboring the smallest genomes are often obligate endosymbionts, whose lifestyle precludes selection for rapid cell division. Thus, the strong association between elevated levels of drift and small genome sizes suggests that genome reduction in bacteria is predominately a nonadaptive process.

In their extremely influential paper, Lynch and Conery (2003) argued that genome size is determined "passively" by random genetic drift in response to long-term population-size reductions. Interestingly, nonadaptive processes may lead to opposite outcomes as far as genome size is concerned. That is, random genetic drift is thought to lead to increases in genome size in many (but not all) eukaryotes, while in prokaryotes it leads to genome size diminution. The disparity is due to fundamentally different patterns of mutational input. Whereas gene and genome duplication as well as the proliferation of transposable elements are widespread in many eukaryotic genomes, these processes are inconsequential as far as prokaryotic genome size is concerned. In the absence of selection, the expectations concerning genome size may, therefore, be antipodal.

So far, we have only dealt with decreases in genome size. Do the same principles apply to genome increases? The genomes of some prokaryotes, such as the cyanobacterium *Scytonema hofmanni*, the planctomycete *Schlesneria paludicola*, and the filamentous soil bacterium *Ktedonobacter racemifer*, are quite large and contain more than 12,000 protein-coding genes. For a prokaryote to possess so many genes is remarkable, since such a gene content exceeds not only that of many unicellular eukaryotes, but also that of many animals, for example, honeybee (*Apis mellifera*) and body louse (*Pediculus humanus*). These findings raise the question, Can the genome of a prokaryote increase in size indefinitely? While genome reduction seems to be driven by random genetic drift, the upper limit of prokaryote genome size seems to be constrained by selection due to limitations on bacterial cell volume and the fact that most bacterial chromosomes have a single origin of replication.

In principle, large genome sizes in free-living bacteria may reflect more frequent acquisition of new genes, a greater need for metabolic versatility, or a larger effective population size, allowing for the evolutionary maintenance of genes that are only weakly beneficial. Konstantinidis and Tiedje (2004) hypothesized that metabolic versatility is key and that large genomes evolve in environments where resources are scarce but diverse and where there is little penalty for slow growth. Such conditions are characteristic of soil, where many nonsymbiotic soil bacteria are known to possess very large genomes (6–8 Mb), and where generation time is long (three generations per year).

Genome Miniaturization

The question of "use and disuse" in evolution is as old as the discipline itself. As a matter of fact, this is one of the very few elements that Darwin admitted to having adopted from Lamarck. At the genome level, prokaryotes present the most compelling evidence for this rule in that a drastic reduction in genome size (**genome miniaturization**) is invariably associated with loss of function. In particular, parasitic or endosymbiotic modes of life were found to affect genome size profoundly.

Obligate prokaryotic endosymbionts have smaller genomes than obligate parasites, which in turn have smaller genomes than free-living prokaryotes (Gregory and DeSalle 2005). Because the genome size in prokaryotes is primarily determined by gene number, the question to be addressed is why endosymbionts, and to a lesser extent parasites, have fewer genes in their genomes.

In the following sections we will discuss genome size reduction due to endosymbiosis and parasitisim separately.

Genome size reduction in intracellular symbionts and parasites

Intracellular symbionts and parasites, especially those residing within eukaryotic cells, are affected by three main factors that can each result in either genome size reduction or the inability to increase genome size. First, by being sequestered in hosts, intracellular prokaryotes do not regularly encounter other bacteria, so the opportunity for gene uptake through horizontal gene transfer is greatly reduced. Moreover, in time such prokaryotes might lose the pathways required to incorporate exogenous DNA into their genomes. Thus, a major mutational source of genome size increase is severely restricted.

Second, the effective population size of intracellular prokaryotes may be reduced to such an extent as to shift the evolutionary balance from selection to random genetic drift. When selection is not strong enough to maintain them, genes are either lost through large deletions or inactivated by deleterious mutations and subsequently eroded by small deletions. When a bacterial lineage adapts to a lifestyle that reduces its long-term N_e, such as obligate endosymbiosis or intracellular parasitism, new and otherwise deleterious mutations may become fixed by random genetic drift. As mentioned previously, deletions outnumber insertions at the mutation level (Mira et al. 2001); hence, mutation and drift will result in genome size reduction.

The third factor that affects genome size in intracellular prokaryotes is the fact that these organisms no longer have to use their own genes for many functions. By becoming reliant on metabolites provided by the host, the selective constraints on many endosymbiont genes are greatly reduced. Over many generations, endosymbiotic bacteria will tend to lose the genes they do not use, as well as quite a few genes that they do use but that are only mildly constrained functionally. The combination of these three factors means that the dominant processes affecting genome size in endosymbionts are deletion bias and random genetic drift.

The evolution of intracellular symbionts and intracellular parasites is similar in most respects, with the exception that it is advantageous for the host to evolve mechanisms to ward off the parasite, and it is advantageous for the parasite to evolve countermeasures. Thus, although parasitism is to a large extent accompanied by genome miniaturization and gene loss, an intracellular parasite must also add genes to its repertoire. In *Mycoplasma* species, for instance, a significant number of genes devoted to coding adhesins (adhesive proteins), attachment organelles, and variable membrane-surface antigens directed toward evasion of the immune system have been acquired (Razin 1997a,b).

Let us now discuss the temporal dynamics of genome miniaturization in intracellular prokaryotes. Because the bulk of the prokaryotic genome is made of coding regions, the fixation of deleterious mutations due to relaxation of selection and the decrease in effective population size is also expected to lead to gene death and the creation of pseudogenes, with genome diminution expected to lag behind gene nonfunctionalization. Bacterial lineages that have only recently became host restricted, such as the human pathogen *Mycobacterium leprae* or the secondary endosymbiont of the tsetse fly *Sodalis glossinidius*, illustrate this initial stage of genome reduction, in which the genome still contains a substantial number of pseudogenes and a proliferation of transposable elements and repetitive sequences within the genome is usually observed (Moran and Plague 2004). *Mycoplasma genitalium* and *Rickettsia prowazekii*, which have fewer pseudogenes, might represent a more advanced stage in this process of genome size reduction. Finally, as the pseudogenes are being deleted, the tight correlation between genome size and number of genes is restored. Indeed, in cases in which the decrease in effective population size occurred long ago, such as in the obligate endosymbionts *Carsonella ruddii* and *Sulcia muelleri*, the genomes can become very small (less than 250,000 bp) but contain little except protein-coding genes. The difference between two endosymbionts as far as their pseudogene contents are concerned is, thus, an indication of (1) the time that has elapsed since the lifestyle shift occurred, i.e., how long ago the functional constraints were lessened or abolished; and (2) the rate, size, and frequency of deletion events, i.e., how long it takes for mutations to delete a pseudogene.

Is genome miniaturization in intracellular prokaryotes a gradual process? Frank et al. (2002) put forward a scenario according to which the adoption of a host-restricted intracellular lifestyle initially causes a proliferation of transposable and mobile elements as well as an increase in the number of repetitive sequences. This process initially leads to a short-term increase in nonfunctional sequences within the genome. Subsequently, the repetitive sequences that have accumulated in the initial stages of host restriction facilitate homologous recombination events, leading to loss of large blocks of DNA as well as to a reduction in the number of the repeated sequences themselves. Thus, initial decreases in genome size are expected, according to this scenario, to be much greater than subsequent DNA losses. Dagan et al. (2006) examined the evolutionary sequence of gene-death events during the process of genome miniaturization in three bacterial species that have experienced extensive genome reduction: *Mycobacterium leprae, Shigella flexneri,* and *Salmonella typhi.* Based on an analysis of evolutionary distances, they proposed a two-step "domino effect" model for reductive genome evolution. The process starts with a gradual, gene-by-gene-death sequence of events. Eventually, a crucial gene within a complex pathway or a central place within a network of functional genes is rendered nonfunctional, triggering a "mass gene extinction" of the dependent genes.

The miniaturization of organelle genomes

Primary organelles are of unquestionable prokaryotic ancestry, with mitochondria and chloroplasts being evolutionarily related to proteobacteria and cyanobacteria, respectively (Chapter 5). We note, however, that while free-living proteobacteria have in excess of 1,000 protein-coding genes and often many more, the protein-coding gene repertoire in mitochondria and mitochondria-derived organelles has mostly shrunk to single- or low-double-digit values. The situation in plastids is similar; genome size and gene content of chloroplasts are greatly reduced in comparison with those of cyanobacteria from which they are derived.

Interestingly, while their genomes have shrunk by orders of magnitude, from full-fledged eubacterial genomes to sizes comparable to the genomes of plasmids, mitochondria and plastids have retained the bulk of their prokaryotic biochemistry. It is important, therefore, to note that, as opposed to the situation in endosymbiotic prokaryotes, genome size reduction in organelles did not significantly diminish the need for the gene products encoded by genes that are no longer in the organelle genome. Such a situation is only made possible through gene transfer to the nuclear genome (Chapter 9). Thus, while the genome of the organelles has shrunk, the eukaryotic nuclear genome has been the recipient of mitochondrial and plastid DNA and has, consequently, ballooned to enormous size and complexity.

As we have seen, prokaryotic endosymbionts and intracellular parasites also have reduced genomes, albeit not to the same extent as organelle genomes. The mechanism of genome reduction in intracellular prokaryotes, however, differs fundamentally from that in organelles. In organelles, genome miniaturization is facilitated not only by gene loss, but also by gene transfer. Similar to the situation in endosymbionts, some biological functions of organelles (e.g., biotin synthesis) may have been rendered unnecessary, leading to the elimination of some genes. However, as opposed to the situation in endosymbionts, organelles continue to require the products of many of the genes that have been transferred to the chromosomes of their eukaryotic hosts. Thus, organelle-genome reduction is not simply an extreme form of intracellular genome reduction. In intracellular prokaryotes, genome reduction is driven by specialization to a nutrient-rich intracellular environment; in the case of organelles, the export of essential genes to the host genome is accompanied by the import of thousands of essential proteins and RNAs from the host into the organelle. The yeast nuclear genome, for instance, contains about 300 protein-coding genes that function exclusively in the mitochondria. Its mitochondrial genome, however, contains only eight protein-coding genes. Even the mitochondrial genomes with the largest coding capacities contain far fewer genes than the number required for independent existence.

For the products of the organelle genes that have been transferred to the nucleus to be "reimported" into the organelle, the genes must first be expressed properly in the nuclear genome. This in many cases requires the genes to be inserted into preexisting promoters. In addition, proteins encoded by nuclear genes that are exported to a particular organelle often contain a type of signal peptide called a **transit peptide** that marks the protein for transfer to the appropriate organelle. Transit peptides are generally cleaved from the mature protein upon arrival.

Many of the nuclear genes acquired from the organelles encode components of the organelles themselves, such as transporters, building blocks of their inner membranes, metabolic enzymes, and proteins of the oxidative phosphorylation and protein synthesis machinery. Some, however, have been co-opted into other functions. A search for nuclear genes in the *Arabidopsis thaliana* genome that branch with cyanobacterial homologs in phylogenetic trees, or that have homologs in cyanobacteria only, and that are sufficiently conserved in sequence to allow phylogenetic analysis indicates that roughly 18% of the protein-coding genes in *Arabidopsis* are acquisitions from the plastid (Timmis et al. 2004). Many of these genes have been co-opted into functions that are unrelated to the chloroplast. Among eukaryotes, ~630 nuclear-encoded proteins have been identified that originated from mitochondria, of which less than 30% were predicted to be targeted to mitochondria in either yeast or human. So, the proteins encoded by many nuclear genes that are derived from organelle DNA ultimately take on new functions.

The deletion and functional replacement of mitochondrial genes by nuclear copies has effectively stopped in animals, but the process still actively continues in plants, which have a larger number of mitochondrially encoded proteins. Studies of flowering plants have uncovered many cases of gene transfer that result in a nuclearly expressed gene. For example, the mitochondrial *rps10* gene has been independently transferred to the nucleus of various plant species many times (Adams and Palmer 2003). Contemporary losses and transfers from the organelle to the nucleus are more common in plastids than in mitochondria. Independent nuclear relocations of the chloroplast translation-initiation factor 1 gene (*infA*) were found to be accompanied by mutational decay and/or deletion of the original chloroplast sequence.

The fact that the ability to transfer functional organelle genes to the nucleus still exists raises an important question: Why have most organelles retained genomes at all? Several hypotheses have attempted to address this question. The principal reason that organelles retain a genome may have to do with the ability of the organelle to perform membrane-associated ATP synthesis (Allen 1993). According to this hypothesis, chloroplasts and mitochondria need to be able to maintain close control over the expression of protein-coding genes involved in electron flow through their bioenergetic membranes so that they can synthesize proteins selectively in response to altered energy needs. In the presence of molecular oxygen, the organelle needs to be able to adjust protein production to decrease production of toxic reactive oxygen species (Lane et al. 2013). In support of this hypothesis, it is worth noting that organelles that lose this bioenergetic function entirely also lose their genomes. Similarly, when membrane-bound electron transport becomes lost during evolution, for instance because of the adoption of a parasitic lifestyle, the organelle genome becomes expendable and may become reduced or lost (e.g., Molina et al. 2014; Smith and Lee 2014).

In the following sections, we discuss the evolution of genome sizes separately for mitochondria and plastids.

The evolution of mitochondrial genome sizes

Some mitochondrially derived organelles have lost their genomes altogether, while in mitochondria that have retained a genome, the numbers of protein-coding genes can be as low as 3, such as in the protozoan parasite *Babesia bovis*, or close to 70 in some protozoans. With very few exceptions, animal mitochondrial genomes contain 12–14 protein-coding genes. The most gene-rich mitochondrial genomes are found in fungi (e.g., pink crust fungus, *Phlebia radiata*); in plants (e.g., beet, *Beta macrocarpa*; common tobacco, *Nicotiana tabacum*; and maize, *Zea mays*); and in protozoa (e.g., *Reclinomonas americana* and *Andalucia godoyi*). We note, however, that even the most gene-rich mitochondrial genomes have not retained much more than 0.1–1% of the gene content in contemporary proteobacteria.

Given the numbers of genes in free-living proteobacteria versus the numbers in mitochondria, one must conclude that the bulk miniaturization of the mitochondrial genome must have occurred immediately following the endosymbiotic event that gave rise to eukaryotes, which is estimated to have happened more than 1.5 billion years ago (Dyall et al. 2004). Only few subsequent losses occurred later in evolution, sometimes independently and convergently in different lineages.

From an evolutionary point of view, gene-rich mitochondrial genomes are especially interesting, as they may represent a construct similar to that in the ancestral organelle. Of course, one needs to take into account the possibility that some gene-rich mitochondrial genomes represent a derived state following some recent acquisitions of genes and other elements by horizontal gene transfer.

Let us now consider the gene-rich mitochondrial genome of the placozoan *Trichoplax adhaerens*, a metazoan with the simplest known body plan of any animal, possessing no organs, no basal membranes, and only four distinct somatic cell types. Mitochondrial genomes of animals (metazoans) are typically 15–24 Kb circular molecules that contain nearly identical sets of 12–14 protein-coding genes and 24–25 rRNA- and tRNA-specifying genes. Animal mitochondrial genomes lack significant intragenic spacers and are generally without introns. In comparison, the mitochondrial genome of *T. adhaerens* was found to be more than twice the size of the typical metazoan mitochondrial genome; to contain large intragenic spacers, several introns, and 17 open reading frames, including 12 for known protein-coding genes, and to have protein-coding regions that are generally larger than those in other animal mitochondrial genomes. In addition, the mitochondrion of *T. adhaerens* contains 27 RNA-specifying genes. Does this situation represent an ancestral or a derived state? If the large mitochondrial genome of *T. adhaerens* is an ancestral state, then one must deduce that while most gene loss and gene transfer from the mitochondrial to the nuclear genome occurred during the emergence of multicellular animals, the final compaction of the genome occurred after the divergence of placozoans from the rest of the metazoans (Dellaporta et al. 2006). The problem with this hypothesis is that sponges, which are more basal phylogenetically than placozoans, possess the same mitochondrial gene set as all other animals, so the situation in *T. adhaerens* seems to be an autapomorphy. Moreover, the existence of introns and open reading frames unrelated to any mitochondrial function would seem to indicate that the increase in the mitochondrial genome size and gene content in *T. adhaerens* represents a relatively recent derived character state. One may, therefore, conclude that the gene content of animal mitochondria remained, with few exceptions, relatively unchanged relative to that in the common ancestor of animals.

The gene-rich mitochondrial genomes of the freshwater jakobid protozoans *Reclinomonas americana* and *Andalucia godoyi* may, however, represent an ancestral state or a state as close to the ancestral state as we are likely to ever find (Lang et al. 1997; Burger et al. 2013). Characterized jakobid mitochondrial genomes are of intermediate size, ranging from 65 to 100 Kb. The AT content is moderate (64–74%), and the proportion of coding regions (including introns) versus intergenic sequence is high (80–93%). Jakobids stand out from other eukaryotes by reason of the elevated gene complement of their mitochondrial genomes, which specify molecular functions not encoded by any other mitochondria. Jakobid mitochondria are also exceptional in their possessing many bacteria-like features, such as bacterial regulatory elements that are presumably used to control gene expression.

The mitochondrial genome of *A. godoyi* is one of the most slowly evolving mitochondrial genomes, so we expect its nuclear genome to have preserved primitive features. The inferred ancestral state and the sequence of gene deletions in jakobid mitochondria are shown in **Figure 10.5**.

Similar to the situation in eukaryotes (Chapter 11), mitochondrial genomes can sometimes increase enormously in size without any increase in gene content. For example, the mitochondria of two species belonging to genus *Silene*, the clammy cockle (*S. noctiflora*) and the sand catchfly (*S. conica*), have experienced a huge recent acceleration in their mutation rates accompanied by a massive proliferation of nonfunctional DNA (Sloan et al. 2012a,b). As a consequence, the genomes of *S. noctiflora* and *S. conica* are 6.7 and 11.3 Mb, respectively, i.e., larger than most prokaryotic genomes and even some eukaryotic ones. This proliferation of nonfunctional DNA coincided with many architectural changes, including the emergence of multichromosomal structures (59 and 128 circular chromosomes) ranging in size from 44 to 192 Kb.

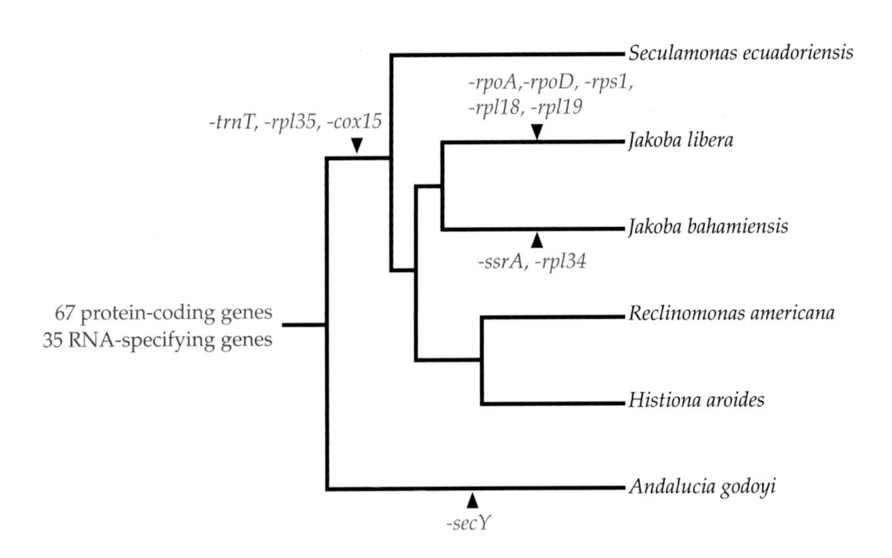

Figure 10.5 Protein-coding (red) and RNA-specifying (blue) gene deletions during the evolutionary history of six jakobid mitochondrial genomes. Deletion events are denoted by minus signs and mapped onto a maximum parsimony phylogenetic tree. The deleted genes include a tRNA^{Thr}-specifying gene (*TrnT*), a tmRNA-specifying gene (*ssrA*), four large-subunit ribosomal protein-coding genes (*rpl*), a small-subunit ribosomal protein-coding gene (*rps*), two genes encoding core RNA polymerases (*rpo*), a gene encoding a cytochrome c oxidase subunit (*cox*), and a gene encoding a transporter (*secY*). (Modified from Burger et al. 2013.)

In comparison, their close relatives, the bladder campion (*S. vulgaris*) and the white campion (*S. latifolia*) retained mitochondrial genomes with run-of-the-mill sizes (427 and 253 Kb, respectively). Interestingly, each of the four *Silene* mitochondrial genomes contains exactly 25 protein-coding genes. The only difference seems to be that the smallest genome contains 9 tRNA-specifying genes, while the largest contains only 2.

The evolution of plastid genome sizes

The oldest fossil red algae are 1.2 billion years old. Thus, plastids must have originated prior to this date, most probably 1.2–1.5 billion years ago (Dyall et al. 2004). The evidence so far points to a single origin for all the plastids in the taxon Plantae, which consists of three clades: Glaucophyta, Rhodophyta (red algae), and Viridiplantae (green algae and land plants).

The ancestor of primary plastids was a free-living cyanobacterium and, therefore, must have possessed several thousand genes, as do its contemporaries. Subsequent to the emergence of a plastid protein import apparatus (a prerequisite for relocating genes that encode proteins required by the organelle to the nucleus), plastids relinquished most of their genes to the genomes of their host cells. Given that even the largest extant plastid genomes contain not many more than 200 genes, the mass gene loss and relocation must have occurred very close to the onset of endosymbiosis. Gene loss and transfer to the nucleus continued at a slower pace in parallel during algal diversification, yet the same core set of genes (for photosynthesis and translation) has been retained in all lineages. Plotting the process of chloroplast genome reduction onto a chloroplast genome phylogeny (**Figure 10.6**) shows that parallel losses in independent lineages outnumber unique losses that are shared by descendant lineages by a ratio of more than 10 to 1. Therefore, the similarity in gene content among contemporary plastid genomes is the result of rampant convergent evolution.

Parasitism involves an intimate association between two organisms: provision of many metabolic and physiological requirements by one, the host, for the other, the parasite. Parasitism invariably entails loss of genetic functions in the parasite and a consequent reduction in genome size. Parasitic plants that no longer photosynthesize are known to have small chloroplast genomes. For example, the beechdrop *Epifagus virginiana*, a nonphotosynthetic parasitic relative of lavender, basil, and catnip, has a very small chloroplast genome (~70,000 bp) that contains only 42 genes. Understandably, all genes for photosynthesis and chlororespiration are absent. It is not clear, however, why all chloroplast-encoded RNA polymerase genes, as well as many ribosomal protein-coding genes and tRNA-specifying genes, have also been lost (Wolfe et al. 1992a,b).

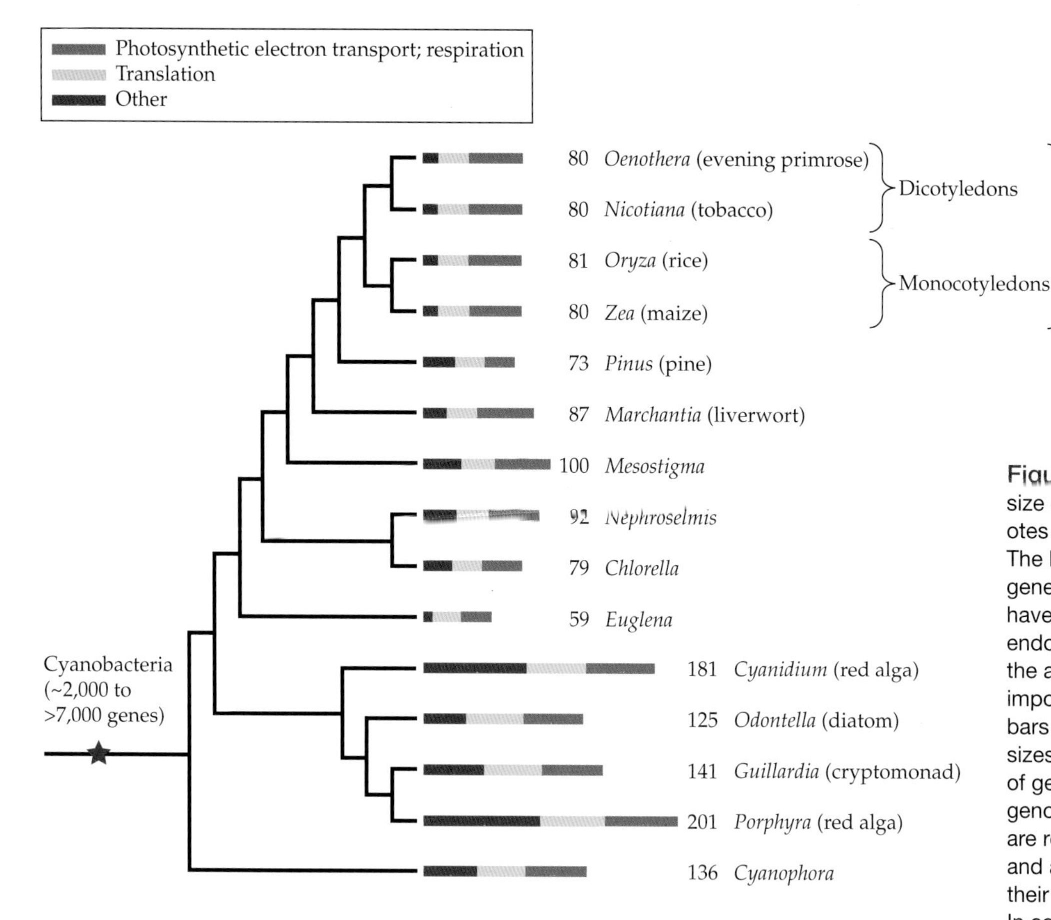

Figure 10.6 Reduction in the size of plastid genomes in eukaryotes over the last 1.2 billion years. The biggest reduction in plastid gene content (star at left) must have occurred close to the onset of endosymbiosis but subsequent to the appearance of a plastid protein-import apparatus. The sizes of the bars are proportional to genome sizes. Numbers indicate the number of genes that are present in each genome. Functional gene categories are represented by different colors and are drawn to scale according to their relative fraction in the genome. In comparison to cyanobacteria, not only do the chloroplasts contain a smaller number of protein-coding genes, but the fraction of the genome dedicated to functions other than electron transport, respiration, ATPase, and translation is much smaller than in free-living cyanobacteria. The glaucocystophyte *Cyanophora* most probably represents the most ancient divergence within photosynthetic eukaryotes. *Mesostigma* represents one of the earliest divergence events within green plants (Viridiplantae). (Data from Timmis et al. 2004.)

The Minimal Genome

A cell that contains only components that are absolutely essential for life is called a **minimal cell**. A minimal cell has a **minimal genome** that contains only genes that are required for survival. In other words, the minimal genome is made of the smallest set of genes an organism needs to live. If the term "life" is comprehended solely from the perspective of the genetic material, then the minimal genome is the answer to the question "what is life?" In reality, a minimal genome is only a theoretical construct; it is not something we can find in nature. First, the "minimal genome" depends on the environment in which the organism lives. With one extraordinary possible exception discovered so far—*Desulforudis audaxviator*, a motile, sporulating eubacterium capable of surviving at 60°C and pH 9.3 without organic compounds, light, or oxygen (Chivian et al. 2008)—no organism lives by itself. There is no self-sufficiency in nature. Thus, a minimal genome for life in brackish waters off the coast of Georgia would be different from a minimal genome for existence as an endocellular symbiont in a leafhopper species. Moreover, a minimal cell would require an "ideal" environment, i.e., one that is free of all selective pressures, and its minimal genome, unlike any other genome in nature, would lack all redundant and superfluous DNA.

The minimal genome can be understood either mechanistically, as a biological model with the smallest number of variables that is capable of sustaining cellular life, or evolutionarily, as a reconstruction of the first and presumably simplest cellular organism. From a mechanistic (or synthetic) point of view, there are two approaches to the minimal genome: the bottom-up and the top-down approaches (Szathmáry 2005; Forster and Church 2006). Traditionally, scientists working in the area of prebiotic evolution have pursued the so-called **bottom-up approach**, which is based on the hypothesis that a continuous increase in molecular complexity can transform inanimate matter into a self-reproducing cellular entity that may be considered to be a living organism. A simplified description of this approach (Hayden 1999) reads as follows: "One day a scientist will drop gene number 297 into a test tube, then number 298, then 299 … and presto: what was not alive a moment ago will be alive now. The creature will be as simple as life can be. But it will still be life. And humans will have made it, in an ordinary glass tube, from off-the-shelf chemicals." For a number of reasons, this approach has not yet been successful, although great strides have been made in the field of genome synthesis (e.g., Gibson et al. 2010).

The **top-down approach** essentially starts with the genome of an existing bacterium from which DNA segments are gradually subtracted, and the process is continued for as long as the cells retain their viability. In *E. coli*, reduced genomes have been constructed by diverse genomic deletion methods, such as suicide plasmid–mediated genomic deletion, linear DNA-mediated genomic deletion, site-specific recombination-mediated genomic deletion, and transposon-mediated random deletion (Nilsson et al. 2005). A principal economic motivation for in vivo genome-reduction experiments is the construction of efficient "minimal-genome factories" for the production of valuable primary gene products (RNAs and proteins) as well as secondary metabolites. Reduced-genome *E. coli* strains generally have genomes that are up to 30% smaller than those of their wild-type ancestors (e.g., Mizoguchi et al. 2007). The genomes of other microorganisms, such as *Bacillus subtilis* and *Corynebacterium glutamicum,* have also been reduced for industrial synthetic purposes. By design, prokaryotes bred to become minimal-genome factories are also selected for faster and more efficient growth patterns than their nonreduced ancestors. Therefore, the fact that these reduced-genome bacteria also grow very fast should not be taken as support for the hypothesis that genome size constrains growth rates. Nonetheless, the fact that such reduced-genome bacteria can be bred is consistent with the hypothesis that bacterial genomes contain redundant or nonessential segments, albeit not as many as eukaryotes (Chapter 11).

The evolutionary search for the genome of the "smallest autonomous self-replicating entity," which presumably would resemble the first cellular organism, was begun in the late 1950s (Morowitz 1984). The search led to studies of Mollicutes, a eubacterial taxon consisting of mycoplasmas, most of which are distinguished by the absence of a cell wall. Some mollicutes have very small genomes, and most of these are symbionts or parasites.

One such mollicute, *Mycoplasma genitalium* G37, has a genome that is approximately 0.58 Mb in length and a GC content of 32% and contains 475 protein-coding genes, 43 RNA-specifying genes, and 6 pseudogenes. In nature, *M. genitalium* is an obligatory parasite, yet it can be grown in pure culture in the lab, where it has become notorious as a contaminant of cell cultures. Because of its ability to grow in the lab, albeit very slowly and only when its growth medium is supplemented by many extracts and complex nutrient sources (Peterson and Fraser 2001), it is considered the free-living organism with the smallest genome. *Mycoplasma genitalium* has a reduced metabolism and little genomic redundancy. Consequently, for many years its genome was expected to be a close approximation to the minimal set of genes needed to sustain bacterial life. There is no evidence, however, that the protein-coding and RNA-specifying capacities of *M. genitalium* actually represent the minimal requirement for sustaining independent life. It is possible that a certain degree of genetic redundancy exists even in this extremely streamlined genome.

In the following sections, we describe three evolutionary approaches for inferring the minimal genome for cellular life.

The comparative genomics approach: Identifying the core genome of all life forms

Koonin and Mushegian (1996) and Mushegian and Koonin (1996a) were the first to employ a comparative genomics approach to estimate the gene content of the minimal genome. This approach is based on the hypothesis that the genes that are conserved in distantly related species are almost certainly essential for cellular function and may, hence, approximate the minimal gene set. In principle, the estimate of the minimal gene complement should be made by identifying the set of all orthologous protein-coding genes that are common to all organisms. In practice, however, an estimate of the minimal gene complement is made by using a sample of organisms. Viewed this way, the comparative genomics approach is merely an extension of the core genome concept as applied to all life forms.

From a comparison of the proteomes of *E. coli*, *H. influenzae*, and *M. genitalium*, the initial minimal set of protein-coding genes was inferred by Mushegian and Koonin (1996a) to include 239 protein-coding genes (**Figure 10.7**). However, in addition to these protein-coding genes, other vital genes must be included in the minimal set. These other genes cannot be identified in the first step of the analysis, because of the phenomenon of **nonorthologous gene displacement**, in which proteins that are dissimilar in sequence perform the same function. Nonorthologous gene displacement can be caused by many evolutionary processes, such as gene birth and death and functional convergence through neofunctionalization. For example, the function of the glycolytic enzyme phosphoglycerate mutase is performed in different bacteria by two proteins that are unrelated to each other. One is encoded by the *gpm* gene and is 2,3-bisphosphoglycerate-dependent; the other is encoded by *yibO* and is 2,3-bisphosphoglycerate-independent. In *M. genitalium* the phosphoglycerate mutase function is performed by the *yibO* gene product, whereas in *H. influenzae* the same function is performed by the protein encoded by the *gpm* gene. Because the two phosphoglycerate mutases are nonhomologous, the intersection of the two proteome sets would contain neither of them, despite the fact that their common catalytic function is most probably indispensable for life. About two dozen genes involved in such nonorthologous gene displacement were discovered, and these were added to the initial minimal set. Finally, genes that appear to be specific for parasitic bacteria or to represent functional redundancy were removed, resulting in a bacterial version of a minimal gene set of 256 genes. Under the assumption that the average protein-coding gene is 750 bp in length, one can calculate that the minimal genome size inferred above is only about a third the genome size of *M. genitalium*.

Based on the comparative approach, the minimal gene set was found to include (1) a nearly complete system of translation, including aminoacyl-tRNA synthesis, tRNA maturation and modification, ribosomal proteins, ribosome function, maturation and modification, and translation factors; (2) a nearly complete set of genes for DNA metabolism, including DNA replication, DNA repair, restriction, and modification; (3) a rudimentary set of genes for DNA recombination; (4) a simple transcription apparatus consisting of four RNA polymerase units, a single σ factor for the initiation of RNA synthesis, three transcription factors, and proteins involved in RNA degradation; (5) a large set of genes involved in protein processing, such as posttranslational modification, folding, secretion, translocation, and turnover; (6) a limited number of protein-coding genes involved in energy and anaerobic metabolism; (7) several genes encoding enzymes for lipid and cofactor biosynthesis; (8) several transmembrane transport proteins; and (9) a few protein-coding genes that are poorly characterized functionally (Luisi et al. 2006). This minimal set is notable in that it does not contain the necessary machinery for the biosynthesis of amino acids and nucleotides, which presumably must have been procured "ready-made" from the environment.

Figure 10.7 Venn diagram of common orthologous protein-coding genes among three species. *Mycoplasma genitalium* and *Haemophilus influenzae* have 240 orthologs in common, *M. genitalium* and *E. coli* have 257, and *H. influenzae* and *E. coli* have 1,128. There are 239 orthologs common to all three species. (Modified from Graur and Li 2000.)

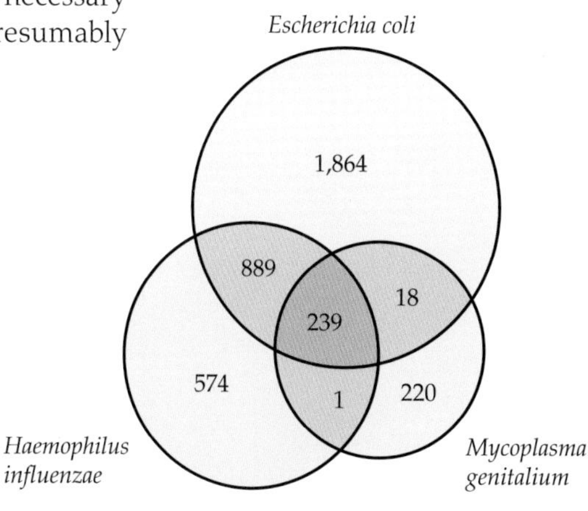

We note that *M. genitalium* has a genome that is almost twice the size of this estimated minimal genome (Maniloff 1996). We further note that by increasing the number of bacterial species in the analysis, the estimate for the number of genes in the minimal genome will approach zero. As mentioned previously, it is possible to relax the core-genome notion and rely on the notion of an extended core genome (p. 455). By using a sample of 573 eubacterial genomes, Lapierre and Gogarten (2009) concluded that the eubacterial extended core genome consists of about 250 genes.

As also mentioned previously, estimates of minimal genome sizes are dependent on the environment. A similar comparative analysis can be conducted on other groups of organisms that are not free-living, thereby addressing the question of the absolutely minimal complement of genes required for cellular life regardless of lifestyle. The smallest prokaryotic genomes—including the record holder, the genome of *Nasuia deltocephalinicola*—are found in maternally transmitted obligate symbionts of hemipterans, such as mealybugs, leafhoppers, cicadas, and sharpshooters. Inside those hosts, these eubacterial symbionts provide their hosts with amino acids and other essentials that the hosts are unable to synthesize by themselves. Bennett and Moran (2013) estimated the symbiotic relationship between the bacteria and the insects to have originated over 260 million years ago. The ancestors of *N. deltocephalinicola* and its relatives started out as free-living microbes that had genome sizes on a par with *E. coli*. But once they got inside a host, they were able to lose DNA without a decrease in fitness. The insects provided the bacteria with a stable home, even to the extent of evolving special cells (bacteriocytes) and special organs (bacteriomes) to shelter them, and their survival was ensured by their being passed down from mother insect to its offspring (Moran and Bennett 2014). The bacteria could, therefore, cast aside many genes that were previously essential, such as genes involved in generating energy. All they needed to do was to provide a service—synthesizing one or more amino acids for the host.

At the time of this writing, *N. deltocephalinicola* holds the record for the smallest genome in a cellular organism. There are, however, millions of other species of insects left to investigate, so it is conceivable that smaller genomes are yet to be discovered. It is, thus, interesting to use the comparative method to estimate how much smaller the genomes of such symbionts can get. All known insect symbionts share a core genome of 82 genes; thus, 82 genes are most probably necessary to survive as a symbiont. However, a symbiont also needs to provide some benefit to its host, or the host will get rid of it. Since it takes at least 11 genes to synthesize a single amino acid, a minimal symbiont will have 82 + 11 = 93 genes. Assuming a mean gene length of 750 bp, these 93 genes could fit in a genome as small as 70,000 bp. Thus, the minimal genome size for an endosymbiont may be about 63% of the genome size of *Nasuia* (Zimmer 2013).

Probabilistic reconstruction of gene content in the last universal ancestor of life

In the comparative genomics approach, we assume that the minimal genome mostly consists of genes that have not been lost in any descendant lineages. This assumption is, of course, unrealistic, and we use nonhomologous gene displacement to compensate for this lack of realism.

Another approach that can be used is the backward-in-time approach, in which the genomic makeup of the last universal common ancestor of all life (LUCA) is "reconstructed" by making certain assumptions on the probabilities of gene loss and gain. Many such studies have been described in the literature (Mirkin et al. 2003; Ouzounis et al. 2006; Dagan and Martin 2007; Mushegian 2008; Kannan et al. 2013). Gene estimates derived from this method are generally larger than those derived from the straightforward comparative method. Usually, the LUCA genome is inferred to possess 500–1,000 protein-coding gene families. Regardless of the method used to infer the size and the content of the LUCA genome, not a single contemporary genome has ever been found to contain representatives of all the ancestral gene families. The largest number of inferred ancestral gene families in an organism has been 537 in *Salmonella enterica*.

Mushegian (2008) defined **LUCA-likeness** as the proportion of the ancestral gene families out of all the gene families in a genome. The highest LUCA-likeness values were observed in two thermophilic eubacteria, *Aquifex aeolicus* and *Thermotoga maritima*.

The experimental gene inactivation approach: Gene essentiality

An **essential gene** is one that is critical for the survival of the organism carrying it and whose individual inactivation causes inviability. Understandably, gene essentiality depends on the conditions in which an organism lives; a gene may be essential under one set of conditions but not another. A gene that is not essential is referred to as a **dispensable gene**. The experimental method of estimating the minimal genome size through gene inactivation was pioneered by Itaya (1995), who knocked out by mutagenesis 79 randomly selected protein-coding loci in the Gram-positive bacterium *Bacillus subtilis* (**Figure 10.8**). Knockouts at only 6 of these loci rendered *B. subtilis* unable to grow and form colonies, while bacteria carrying incapacitating mutations at the rest of the 73 loci retained their ability to multiply. The functions of only 3 of these 6 essential protein-coding genes are known unambiguously. They are *dnaA* and *dnaB*, which are involved in the initiation of DNA replication, and *rpoD*, whose protein product takes part in RNA synthesis.

Of course, genes that are individually dispensable may not be simultaneously dispensable, because of the presence of alternative cellular pathways, functionally redundant gene copies created by gene duplication, or compensation by other genes in the genome. To make sure that this is not the case, Itaya (1995) also constructed bacteria with multiple knockouts. Interestingly, even when 33 loci were incapacitated simultaneously, the bacterium and its progeny retained their ability to form colonies. No combination of genes that were deemed individually to be dispensable resulted in inviability. Thus, 73 out of 79 genes (92%) were inferred to be truly dispensable. Given that the length of the genome of *B. subtilis* is 4.2×10^6 bp, and assuming that the genomic ratio of indispensable to dispensable genes is the same as that in the sample, the length of the indispensable genome was estimated to be 3.2×10^5 bp. Using 1.25 Kb as the average size of a protein-coding gene, Itaya (1995) obtained an estimate of the minimal gene set of 320,000/1,250 = 254 genes. Given that the comparative and the experimental approaches used unrelated methodologies and data, the agreement between the two results is astounding.

In more recent studies, genome-scale identifications of essential genes have been performed in different groups of organisms by employing many different experimental strategies. Gene inactivation is generally accomplished by one of three main experimental approaches: massive transposon mutagenesis, the use of antisense RNA to inhibit gene expression, or the systematic and directed inactivation of each individual gene present in a genome. The results indicate that percentage of essential genes in the whole genome is usually very low (**Table 10.2**). Even the highly reduced genome of *M. genitalium* contains a substantial fraction of genes that are dispensable (Glass et al. 2006). We note, however, that the results may vary a lot.

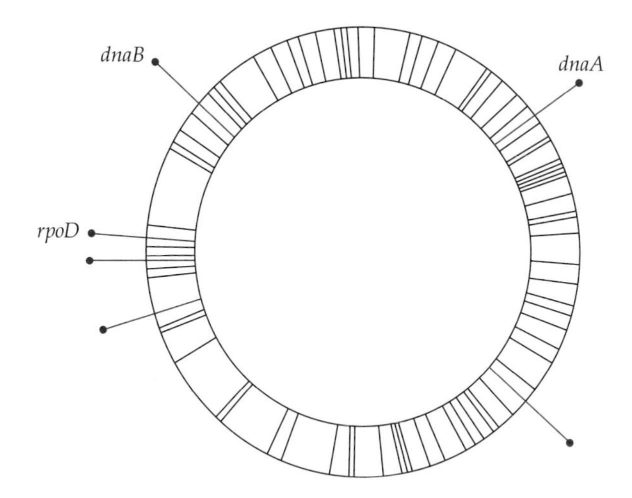

Figure 10.8 Genomic locations of the 79 randomly chosen loci (lines) in *Bacillus subtilis* that have been knocked out by mutagenesis. Six indispensable loci are shown as longer, red lines. Only three named genes are indispensable. (From Graur and Li 2000.)

Table 10.2

Percentage of essential genes as inferred from large-scale gene inactivation experiments in prokaryotes

Organism	Essential genes as percentage of genome
Bacillus subtilis	6–7
Caulobacter crescentus	13
Corynebacterium glutamicum	22
Escherichia coli	7–14
Francisella novicida	23
Francisella tularensis	22
Haemophilus influenzae	39
Helicobacter pylori	21
Mycobacterium tuberculosis	7–20
Mycoplasma genitalium	55–79
Mycoplasma pulmonis	38–40
Neisseria meningitidis	27
Pseudomonas aeruginosa	6–12
Salmonella typhimurium	6–12
Staphylococcus aureus	6–25
Streptococcus pneumonia	5–6
Vibrio cholerae	20

GC Content in Prokaryotes

Among prokaryotic genomes, the mean percentage of guanine and cytosine, or **GC content**, varies from 13.5% in *Zinderia insecticola*, a β-proteobacterial endosymbiont of the spittlebug *Clastoptera arizonana*, to 74.7% in the cellulolytic actinobacterium *Cellulomonas fimi* (**Figure 10.9**). The GC content range in archaebacteria is narrower, from 27.6% to 66.6%, possibly because the taxonomic sampling is not as extensive as that in eubacteria. Intriguingly, the distribution of GC content seems to be independent of

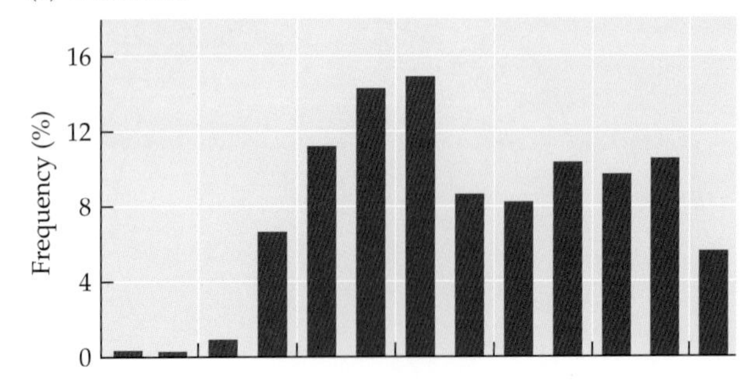

(b) Archaebacteria

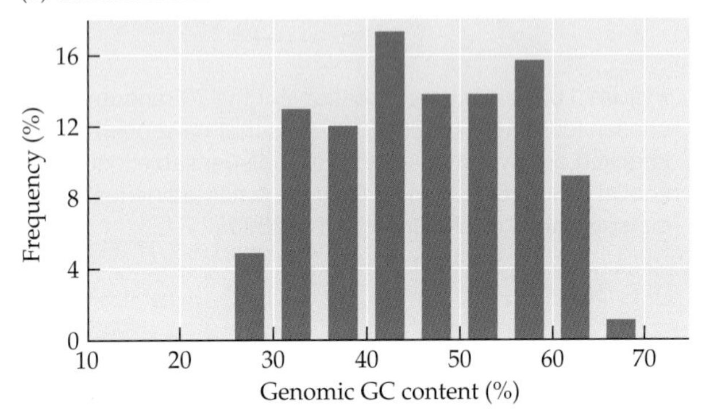

Figure 10.9 Frequency distribution of eubacterial (a) and archaebacterial (b) GC contents. The figure is based on data from 998 completely sequenced eubacterial genomes and 109 archaebacterial ones. (Courtesy of Yichen Zheng.)

phylogenetic relationships, indicating that changes in GC content occur frequently and rapidly. There are essentially two types of hypotheses to explain the variation in GC content in bacteria, the selectionist view and the mutationist view.

Possible explanations for variation in GC content

The **selectionist view** regards genomic GC content as a property determined by selective forces, i.e., a form of adaptation to extrinsic or intrinsic conditions. Numerous such selective factors have been proposed (Wu et al. 2012). Here, we will only discuss three of them.

The first selectionist scenario in the literature invoked ultraviolet radiation as the main factor affecting GC content (Singer and Ames 1970). Since UV radiation induces the formation of thymidine dimers, a higher GC content would presumably convey selective advantage on organisms living in niches that are susceptible to direct and intense sunlight. Indeed, the genomes of bacteria in the upper layers of the soil were initially found to possess higher GC content than bacteria that are not exposed, such as intestinal bacteria. This hypothesis, however, was refuted by the discovery of many GC-rich bacteria that inhabit the colon—a place where the sun literally does not shine (Schell et al. 2002).

Another selectionist scenario that was popular for a time was one based on the fact that G:C pairs are thermally more stable than A:T pairs because of their additional hydrogen bond as well as their different stacking pattern (Chapter 1). It was, therefore, reasoned that there may be a relationship between GC content and the temperatures to which bacteria are exposed. Indeed, initial studies by Argos et al. (1979), Kagawa et al. (1984), and Kushiro et al. (1987) seemed to indicate that in thermophilic eubacteria, which inhabit very hot niches, there may be a preferential usage of amino acids encoded by GC-rich codons (e.g., alanine and arginine) and an avoidance of amino acids encoded by GC-poor codons (e.g., serine and lysine). However, in an extensive study of 764 prokaryotic species, Galtier and Lobry (1997) found no correlation between GC content and optimal growth temperatures.

A third selectionist hypothesis relates aerobiosis and anaerobiosis to GC content, claiming that aerobic prokaryotes have on average higher GC contents than anaerobic taxa (Naya et al. 2002). Two main selective factors have been suggested to explain this difference. The first was the pattern of amino acid utilization, which differs between aerobes and anaerobes. This difference may necessitate more or less GC in the genome. Of course, as with all correlative relationships, the causality may be reversed, so one may claim that the reason for the differences in the amino acid composition may be due to differences in genomic GC content rather than vice versa. A second selective factor that has been deemed responsible for the difference in GC composition between aerobes and anaerobes is thermodynamic stability. It has been claimed that the stacking of GC-rich DNA would diminish the accessibility of oxygen radicals to the nucleotides and would, therefore, be advantageous in aerobic lifestyles.

So far, most selectionist mechanisms in the literature have been rejected, either because they were found to be valid in only a few taxa and not in others, or because support for them diminished with increase in taxonomic sampling. Of course, the possibility remains that GC content is indeed determined mainly by selection, but the selective agent or agents remain to be identified.

The mutationist view invokes biases in the mutation pattern to explain the variation in GC content (Sueoka 1964; Muto and Osawa 1987). According to this view, the GC content of a given bacterial species is determined by the balance between (1) the rate of substitution from G or C to T or A, denoted as u; and (2) the rate of substitution from A or T to G or C, denoted as v. At equilibrium, the GC content is expected to be

$$P_{GC} = \frac{v}{v+u} \tag{10.1}$$

Therefore,

$$\frac{u}{v} = \frac{1-P_{GC}}{P_{GC}} \tag{10.2}$$

The ratio u/v is also called the **GC mutational pressure**. When u/v is 3.0, the GC content at equilibrium will be 25%. Such is the situation in *Mycoplasma capricolum*. When the ratio is 1, the GC content will be 50%, as in *E. coli*. When it is 0.33, the GC content will be 75%, as in *Micrococcus luteus*. We note, however, that including sites at which selective constraints are strong may yield erroneous values of u/v. To estimate the GC mutational pressure, it is advisable to use DNA sites at which selective constraints are absent or very weak, such as fourfold degenerate sites. For example, the GC content at fourfold degenerate sites in *M. capricolum* protein-coding genes is lower than 10%. Consequently, u/v must be higher than 9.0. Similarly, in *M. luteus*, P_{GC} at fourfold degenerate sites is larger than 0.9; therefore, $u/v \approx 0.11$.

The majority of prokaryotic genomes—some 90%—are composed chiefly of protein-coding genes. Thus, in addition to GC mutational pressure, the genomic GC content should also be affected by selective constraints. The weaker the selective constraint is in a particular region, the stronger the effect the GC mutational pressure will have on the GC composition. **Figure 10.10a** shows the correlation between the GC content in coding regions and the GC content at the three codon positions for 1,038 prokaryotic genomes. All three correlations are positive and highly significant statistically, indicating that mutation plays a major role in determining genomic GC content. If we consider the analysis as a linear regression analysis with GC content in coding regions as the independent variable and the GC contents at all the codon positions as the dependent variables, we observe that the slopes of the regression line are different. The steepest slope is observed for the third position of codons. The slopes for the first and second positions, while positive, are more moderate. These observations can be easily explained by the fact that selection at the mostly degenerate third position of codons is expected to be much less stringent than that at the first and second positions (Chapter 4), so mutation pressure is more important in determining the GC level for the third

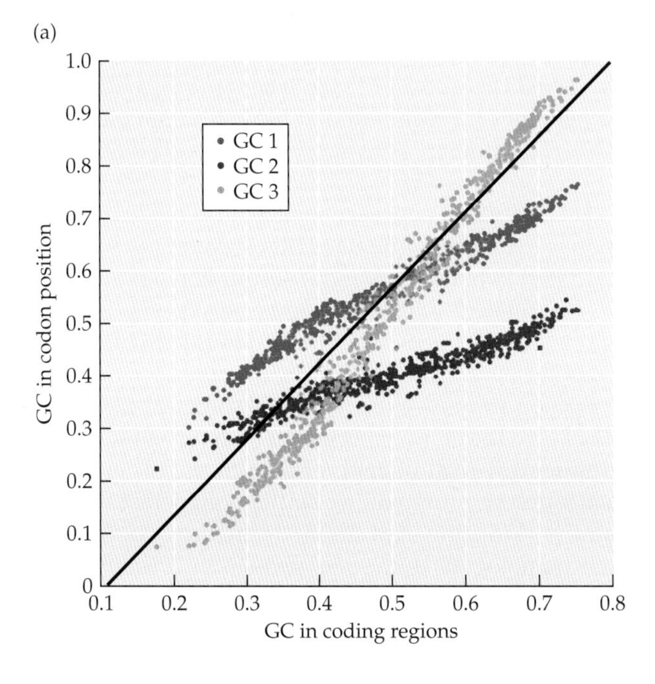

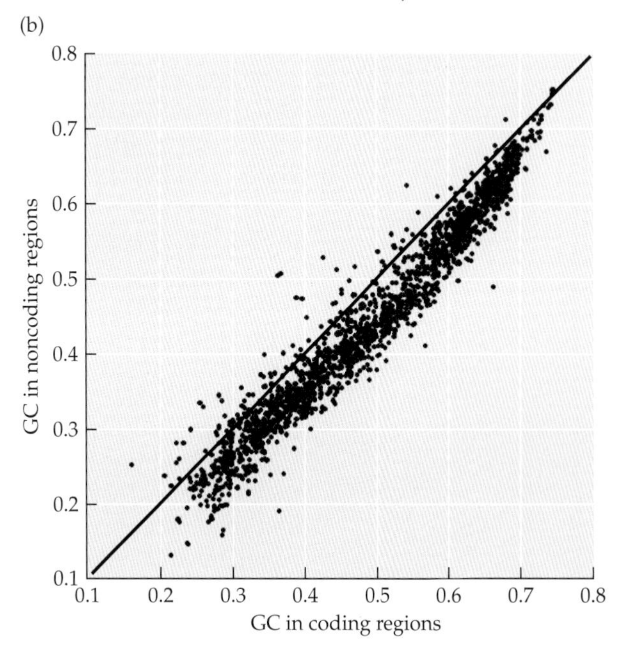

Figure 10.10 (a) Correlation between the GC content in coding regions and the GC content at first (blue), second (red), and third (green) codon positions based on 1,038 prokaryotic genomes. (b) Correlation between the GC content in coding regions and that in noncoding regions. The diagonal lines represent a perfect correspondence between the GC contents on the two axes. (Modified from Brocchieri 2014.)

codon position than for the other two positions. In Figure 10.10b, we see a comparison between the GC content of coding regions and the GC content of noncoding (mostly intergenic) regions. In the vast majority of cases, the GC content in the coding regions is somewhat higher than that in the noncoding regions, indicating that different driving forces or a combination of different driving forces affect the two GC contents.

Hildebrand et al. (2010) examined patterns of synonymous polymorphism in eubacteria and discovered a large excess of G:C → A:T mutations over A:T → G:C mutations in all but the most AT-rich bacteria. Such a mutational bias implies that GC-rich bacterial genomes should experience a rapid decline in GC content. The mere fact that GC-rich genomes exist indicates that mutation alone cannot explain the GC content in GC-rich species. Possibly, selection is an important factor in determining GC content in prokaryotes with extremely high and extremely low GC contents, whereas the GC content of the vast majority of bacteria is determined by mutational biases only.

An intriguing finding in support of mutation as a major determinant of GC content was obtained by Akashi and Yoshikawa (2013). They looked at the mutational bias in a wild-type strain of *Bacillus subtilis* and found that it is AT-biased. However, after deletion of the *mutS* gene, which encodes a protein that is involved in the repair of mismatches in DNA, the pattern of mutation becomes GC-biased.

Chargaff's parity rules

More than half a century ago, Erwin Chargaff noted certain regularities in the base composition of nucleic acids, which he thought "reflected the existence in all DNA preparations of certain structural principles" (Chargaff 1950, 1951). In particular, for double-stranded DNA, he identified an invariant component of base composition, i.e., that the frequency of A equals that of T and the frequency of C equals that of G. A few years later, the **first parity rule** was used by Watson and Crick (1953) in support of their double-helix model for the structure of DNA. Notwithstanding this seminal contribution, Chargaff made three other fundamental observations concerning regularities in the nucleotide composition of DNA. The first was the **cluster rule**, which states that individual bases are clustered in runs to a greater extent than expected at random (Chargaff 1963). The second was the **GC rule**, which states that the GC ratio (G + C)/(G + C + A + T) or GC content tends to be constant within a species but varies among species (Chargaff 1951, 1979). The third was the **second parity rule**, which states that to a close approximation, the first parity rule also applies to single-stranded DNA derived from a double-stranded DNA molecule (Rudner et al. 1968).

The validity of the second parity rule became clear when full double-stranded genome sequences became available. For example, the top strand of the double-stranded DNA vaccinia virus, where "top" denotes the strand that is listed in the database, has 64,902 As and 64,899 Ts, as well as 32,535 Cs and 32,375 Gs. Chargaff's second parity rule does not apply to organellar genomes (mitochondria and plastids) smaller than ~20–30 Kb, single-stranded DNA viral genomes, artificial double-stranded DNA, or any type of RNA genome.

It has been shown that the second parity rule applies to double-stranded sequences in which the pattern of substitution is the same in the two DNA strands. Such a situation arises when the patterns of mutation and the patterns of selection do not differ between the two strands (Lobry 1995; Sueoka 1995; Lobry and Lobry 1999).

In addition to Chargaff's rules, other regularities in double-stranded DNA have been documented. For example, purines (A and G) are almost always more frequent in the leading strand than in the lagging strand—a phenomenon sometimes referred to as **Szybalski's rule** (Lao and Forsdyke 2000). In particular, a preference for G over C in the leading strand is an almost universal phenomenon. On the other hand, the preference for A over T depends on the proofreading by the polymerase-α subunit that replicates the leading strand (Worning et al. 2005). There are two forms of polymerase-α subunit: one form is homologous to the *polC* gene from *B. subtilis*, and the other is homologous to the *dnaE* gene from *E. coli*. In some bacteria the DNA polymerase machinery contains two identical *dnaE*-like subunits; in others, it contains one *dnaE*-like subunit and one *polC*-like subunit. Genomes with more A on the leading strand than on the lagging strand contain both a *dnaE* homolog and a *polC* homolog. Genomes that only possess a *dnaE* homolog have more T on the leading strand than on the lagging strand.

GC Skew and Gene-Density Asymmetries Are Related to DNA Replication Biases

While the second parity rule usually applies to whole genomes, there are local deviations from it. These deviations are not random. Rather they seem to be related to the manner in which the DNA is replicated. Here, we discuss GC-skew and gene-density asymmetries.

Replichores and chirochores

A **replichore** is a double-stranded DNA in which one strand is replicated in its entirety as a leading strand while the other is replicated in its entirety as a lagging strand. In other words, a replichore is a DNA fragment that does not contain within it either an origin of replication or a terminus of replication. **Figure 10.11** depicts an idealized circular genome with one origin of replication and one terminus of replication. From the origin of replication, one half of the genome is replicated clockwise, while the other half is replicated counterclockwise. This genome can, therefore, be divided into two replichores. In the vast majority of eubacterial genomes, there exist two replichores of approximately equal length. A small number of eubacterial genomes have been described in which the length of one of the replichores is less than 40% of the total length of the circular chromosome (Morton and Morton 2007).

The differences in the way the leading and lagging strands of DNA are replicated (Chapter 1) may result in **strand-dependent mutation patterns**, also called **asymmetric directional mutational pressure** (Lobry 1996; Lobry and Sueoka 2002). That is, certain types of point mutations will occur at different frequencies in the leading and the lagging strands. The mutational asymmetry may lead to strand-dependent substitution patterns, which in turn may cause the nucleotide frequencies of each of the two DNA strands to deviate from the second parity rule. Put another way, within each strand $f_A \neq f_T$ and $f_C \neq f_G$.

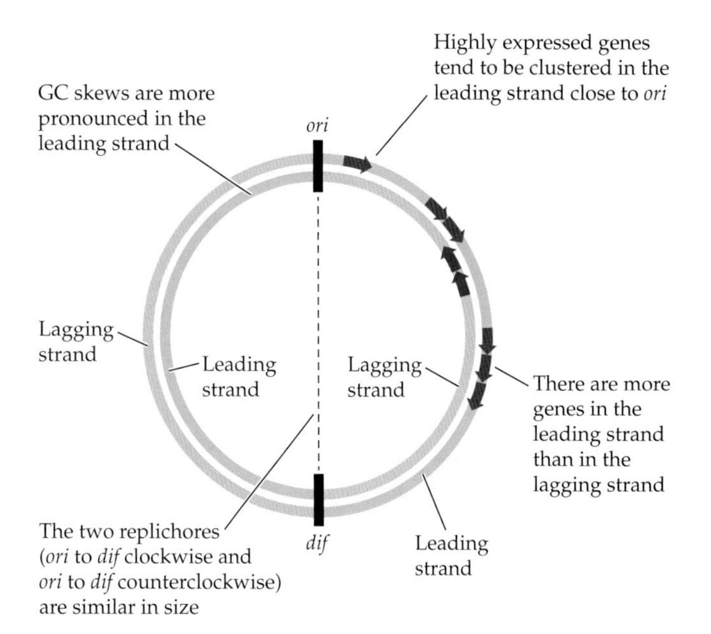

Figure 10.11 A schematic double-stranded circular genome with one origin of replication (*ori*) and one terminus of replication (*dif*). From the origin of replication, one half of the genome is replicated clockwise, while the other half is replicated counterclockwise. This genome can, therefore, be divided into two replichores that are roughly equal in size. Leading and lagging strands are in blue and green, respectively. The dark blue arrows represent protein-coding genes.

Deviations from equal nucleotide frequencies on a strand are usually quantified by using a variable called the **skew**, $S_{X=Y}$, which is a measure of inequality between the frequency of nucleotide X (f_X) and the frequency of nucleotide Y (f_Y) on a strand. It is calculated as

$$S_{X=Y} = \frac{f_x - f_y}{f_x + f_y} \quad \textbf{(10.3a)}$$

For example, the AT skew ($S_{A=T}$), which is the deviation from $f_A = f_T$, is calculated as

$$S_{A=T} = \frac{f_A - f_T}{f_A + f_T} \quad \textbf{(10.3b)}$$

and the GC skew ($S_{G=C}$), which is the deviation from $f_G = f_C$, is

$$S_{G=C} = \frac{f_G - f_C}{f_G + f_C} \quad \textbf{(10.3c)}$$

Mathematically, there are six possible skews: $S_{G=C}$, $S_{A=T}$, $S_{A=C}$, $S_{A=G}$, $S_{C=T}$, and $S_{G=T}$. Under the second parity rule, $S_{G=C}$ and $S_{A=T}$ are expected to be 0. The other four skews do not have a biological meaning. Skews can be expressed as percentages by multiplying the value in Equation 10.3c by 100.

Skew values can be calculated for sliding windows of predetermined lengths and can be plotted on a **skew diagram** (**Figure 10.12**). The existence of local skews on one strand indicates that the proportions of complementary nucleotides (G:C and A:T) between the two strands deviate from racemic (1:1) proportions. That is, an excess of G over C or A over T indicates that the strand on which the skew is calculated is heavier than the complementary strand.

Lobry (1996c) examined three bacterial genomes and found considerable local deviations from $f_G = f_C$. That in itself was not particularly surprising. What was surprising, however, was the spatial distribution of the skews. The skews abruptly switched sign at the origin and the terminus of replication. Thus, the origin and terminus of replication could be identified as the points of inflection from positive to negative and from negative to positive skew, respectively.

There are two useful ways to visualize the abruptness of the switch in skew values. One way is to plot the cumulative skew over the length of the genome (**Figure 10.13a**). The second is by means of a vectorial representation (Mizraji and Ninio 1985; Lobry 1996a; Roten et al. 2002). In this representation the sequence is shown as a trajectory composed of equal-length steps, in which A, C, T, and G dictate a step leftward, downward, rightward, and upward, respectively (Figure 10.13b). The switch in $S_{C=G}$ values at the origin and terminus of replication proved to be of such general pertinence in eubacteria that it has been used with great success to identify these sites computationally in many species for which empirical information was lacking (e.g., Lobry 1996b).

Interestingly, the GC skew was used to rectify a general error in the literature concerning the location of the terminus of replication. In eubacteria, the *Ter* loci, which bind Tus/Rtp proteins, were thought for many years to be the termini of replication. However, several lines of evidence indicated that *Ter* sites have nothing to do with replication termination. First, numerous copies of *Ter* were found to be distributed over a large portion of the chromosome, while the termination of replication is usually restricted to one region. Second, *Ter* sequences were found not to be conserved among eubacteria. Finally, *Ter* mutants lacked any measurable phenotype. GC-skew calculations indicated that the replication termination is most likely located at or near a site called *dif*, which is the site of action of the XerCD site-specific recombinase (Hendrickson and Lawrence 2007).

A **chirochore** is a single-stranded segment of DNA with a relatively constant positive or negative GC or AT skew. In *E. coli* and other bacteria for which it has been determined that the main cause of GC and AT skews is replication asymmetry, a chirochore is a replichore (if one omits the single-stranded versus double-stranded DNA distinction). There are, however, cases in which the terms "chirochore" and "replichore" cannot be applied to the same segment.

Is mutation sufficient to explain the existence of chirochores? Many selectionist hypotheses have been put forward, but no selective factor has ever been found to be

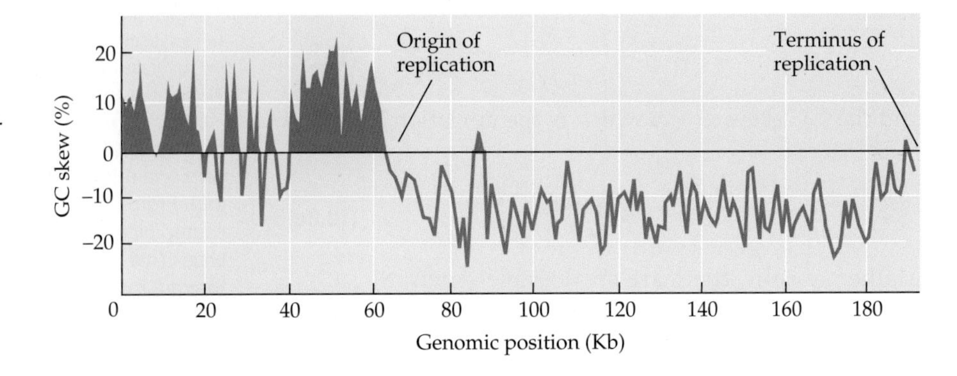

Figure 10.12 Skew diagram for C and G in a 200 Kb genomic sequence of *Bacillus subtilis*. Note that the GC skew downstream of the terminus of replication site but upstream of the origin of replication site is mostly positive, while the GC skew downstream of the origin of replication site but upstream of the terminus of replication site is mostly negative. (Modified from Lobry 1996b.)

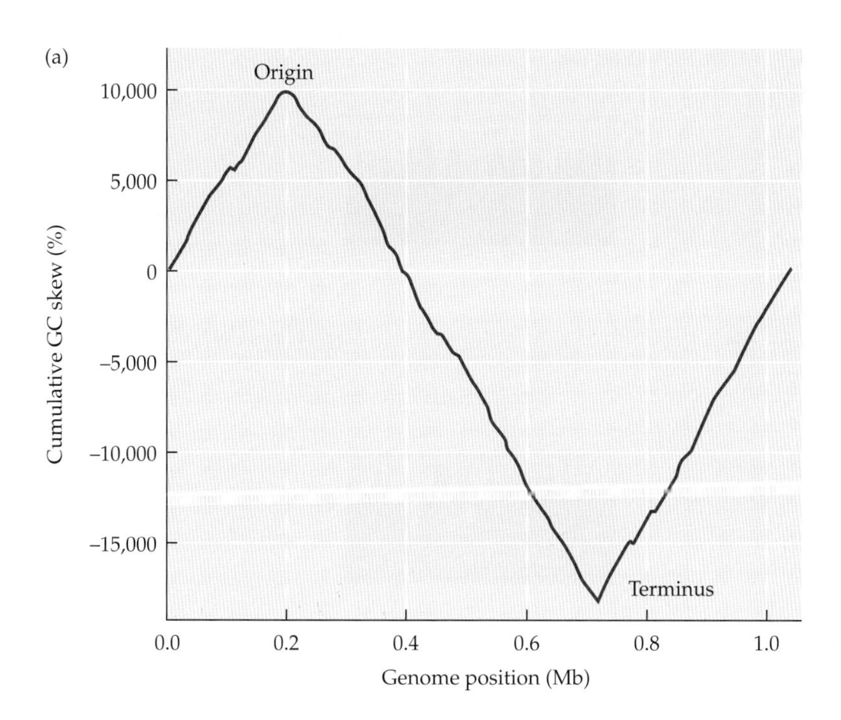

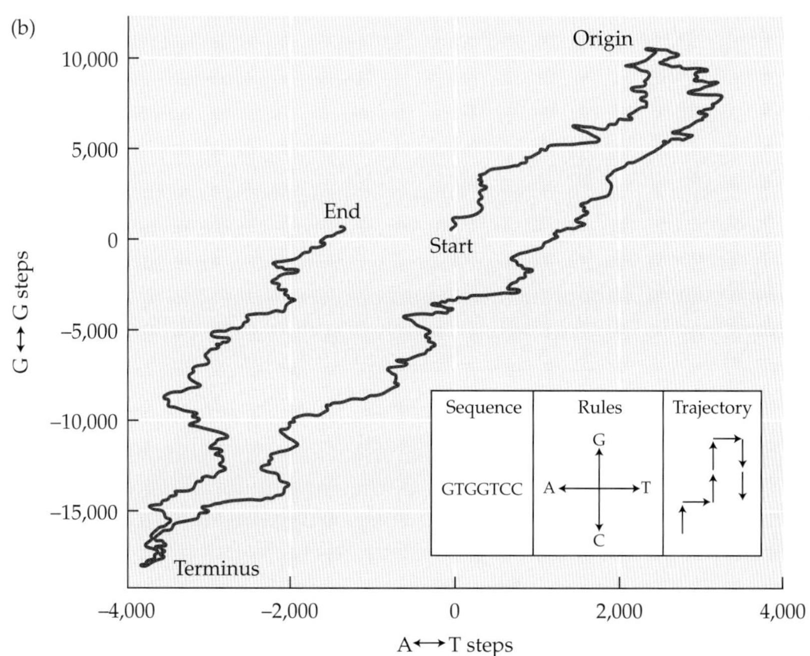

Figure 10.13 Two methods of visualizing chirochoric structures and identifying the origin and terminus of replication. (a) Cumulative GC skew diagram for *Chlamydia trachomatis*. (b) Vectorial representation of the genomic sequence of *C. trachomatis*. The sequence is drawn as a trajectory of equal-length steps in which the direction of each step is dictated by the type of nucleotide at the position (inset). Since the genome is circular, the trajectory starts and ends at arbitrary points in the genome. Plots were drawn by using Comparative Genometrics (Roten et al. 2002).

nearly as important as mutational biases in explaining the existence of chirochores (Frank and Lobry 1999). There are two main lines of evidence for mutation being the driving force in the evolution of chirochores. First, in many cases the boundaries between chirochores coincide with the boundaries between replichores, and this immediately suggests a link with the processes of replication and DNA repair. Second, the relative bias from $S_{C=G} = 0$ is larger for intergenic regions and third codon positions than for first and second codon positions, as expected if the bias is mainly mutational (**Figure 10.14**). Indeed, to avoid complications due to selection at the amino acid level, GC skew is customarily measured at third codon positions or at third codon positions of fourfold degenerate codons.

So far we have dealt with the simple case of circular eubacterial genomes with a single origin of replication and a single terminus of replication. Some eubacteria and

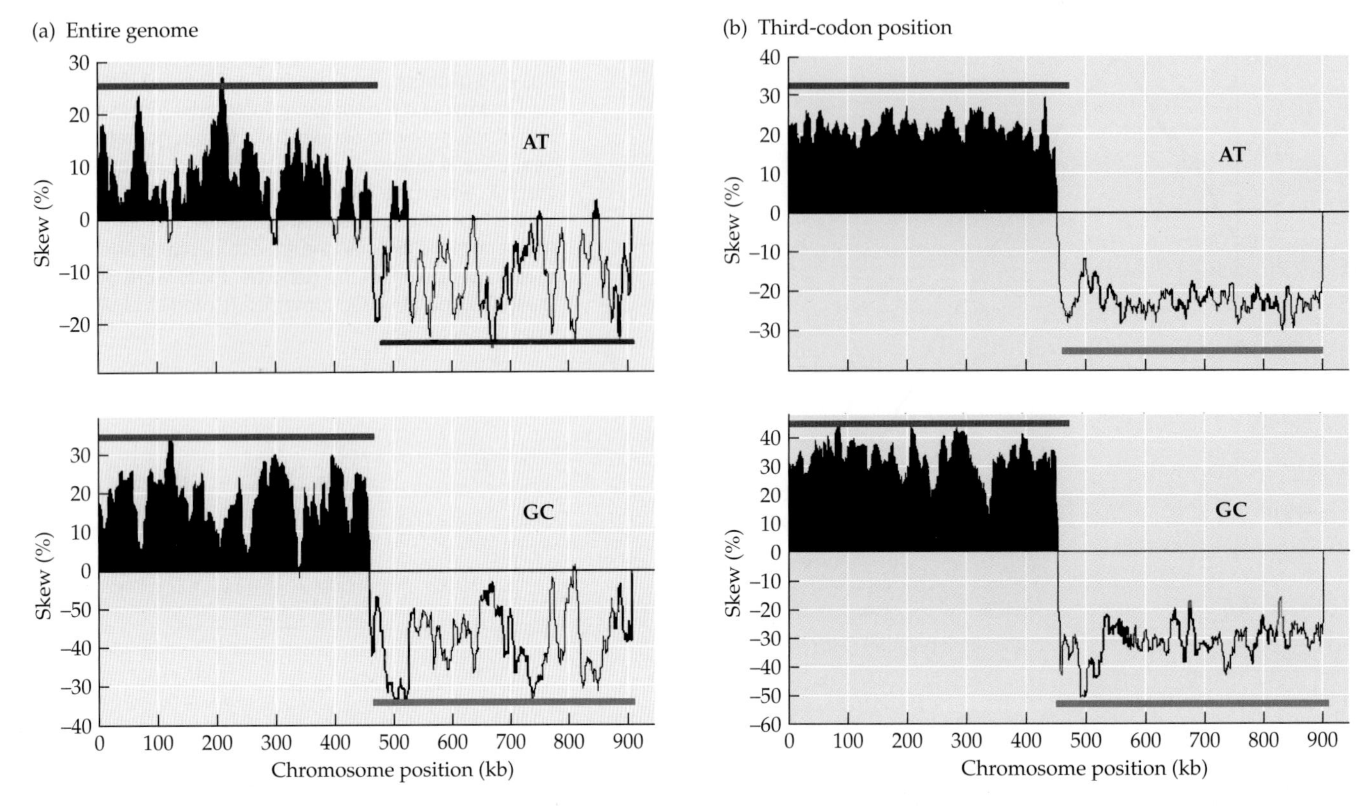

Figure 10.14 Skew diagrams for the 910,724 bp genomic sequence of *Borrelia burgdorferi*, the causative agent of Lyme disease. Positive deviations from 0 are indicated in black. The two chirochores are shown as blue and red lines. (a) AT and GC skews for the entire genome. (b) AT and GC skews calculated by using only third positions of codons. Note that the skews are more evident when only positions under little or no selection, such as third codon positions, are used. (Courtesy of Jean Lobry.)

most archaebacteria, however, deviate from the asymmetrical pattern of nucleotide distribution. One of the principal reasons for this deviation is the existence of multiple origins and/or multiple termini of replication within a genome. **Figure 10.15** presents cumulative GC skews for two archaebacteria, the halophilic *Halobacterium salinarum* and the hyperthermophilic neutrophile *Hyperthermus butylicus*. In both cases, the GC skew evidence indicates that multiple origins of replication exist within the genome.

Despite several studies by one group of scientists indicating that strand compositional asymmetries may also be associated with replication origin in eukaryotes (Brodie of Brodie et al. 2005; Touchon et al. 2005; Audit et al. 2007), the verdict so far is that the process of replication in eukaryotes, particularly multicellular eukaryotes, is different from that in prokaryotes and, hence, compositional asymmetries between the two strands cannot predict either origins or termini of replication (Méchali 2010).

The location of genes in leading and lagging strands

The number of eubacterial protein-coding genes is higher in the leading strand than on the lagging strand. In a sample of 725 eubacterial genomes, Mao et al. (2012) found that the proportion of genes in the leading strand varied between 49% and 92%, with a mean of 62% (**Figure 10.16**). The largest proportion of genes on the leading strand was found in the extremely thermophilic Gram-positive anaerobe *Carboxydothermus hydrogenoformans*, which has the interesting property of producing hydrogen as a waste product while feeding on carbon monoxide and water. The lowest proportion of genes on the leading strand was found in *Prochlorococcus marinus*. It and its congeners are speculated to be the most abundant photosynthetic taxon on Earth. The number of genes on the lagging strand exceeded 50% in only 2% of the taxa in the Mao et al. (2012) sample, and never by more than 1%.

Why should there be a preference for the leading strand over the lagging strand? DNA is used as template for both replication and transcription. When a gene is transcribed from the leading strand, the replication machinery (**replisome**) and the RNA

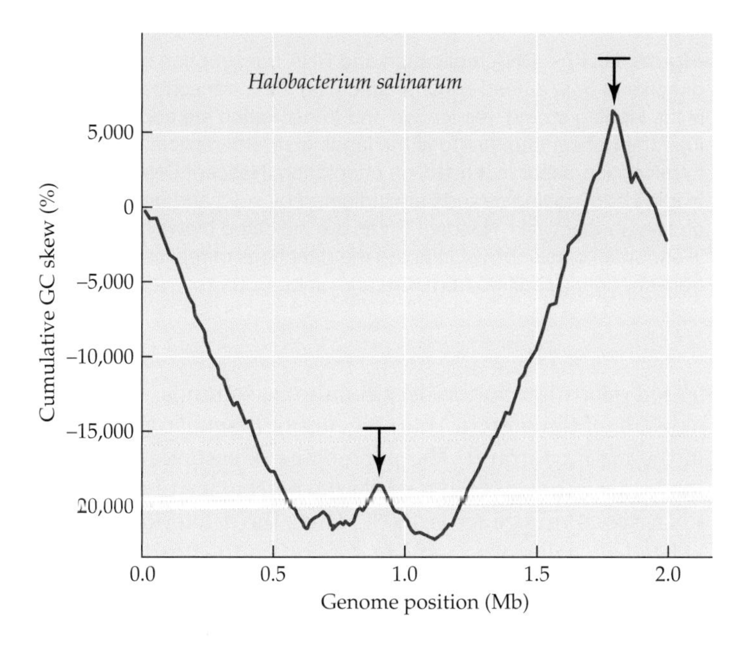

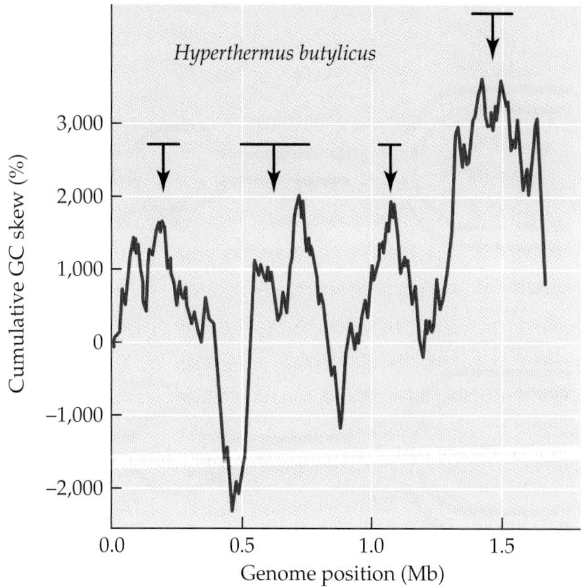

Figure 10.15 Cumulative GC skews in two archaebacteria, *Halobacterium salinarum* and *Hyperthermus butylicus*. The skews strongly suggest that *Halobacterium* has two replication origins and *Hyperthermus* has four (downward arrows). The horizontal bars represent the approximate range within which the replication origins were located through computational analysis. Plots were drawn by using Comparative Genometrics (Roten et al. 2002).

polymerase move in the same direction (**Figure 10.17a**). Because in bacteria DNA polymerase proceeds 10–20 times faster than RNA polymerase (Rocha and Danchin 2003), the two machineries may occasionally collide codirectionally. By contrast, when a gene is transcribed from the lagging strand, DNA and RNA polymerases may frequently collide head on (Figure 10.17b). **Head-on collisions** are thought to result in transcriptional abortion, replication delay, and mutagenesis and be more deleterious than **codirectional collisions** (Merrikh et al. 2012; Chen and Zhang 2013). It should, thus, be advantageous for the antisense strands of protein-coding genes to be preferentially located on the leading strand as a means of avoiding head-on collisions. This is indeed the case in the vast majority of eubacterial genomes.

If the coorientation of replication and transcription is indeed advantageous, then this phenomenon should be more pronounced in important genes than in less important ones. Two measures of importance, expressiveness and essentiality, have been studied (Rocha and Danchin 2003), and in two species it has been shown that coorientation is not driven by expressiveness. That is, coorientation is only slightly more prevalent in highly expressed genes than in lowly expressed ones. Rather, in both *Bacillus subtilis* and *E. coli*, essentiality of the transcript product was inferred to drive

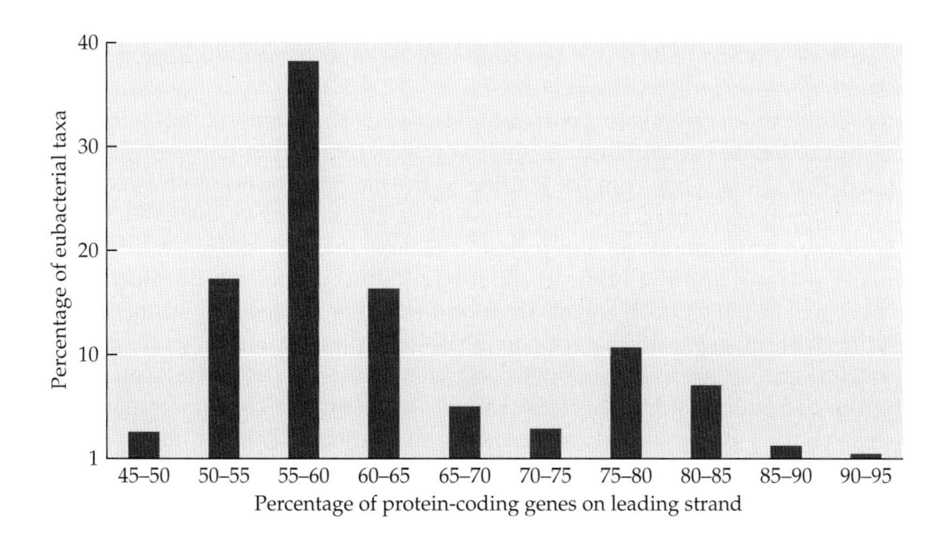

Figure 10.16 Frequency distribution of the percentage of protein-coding genes transcribed from the leading strand, based on a sample of 725 eubacterial genomes. (Data from Mao et al. 2012, courtesy of Yichen Zheng.)

(a)

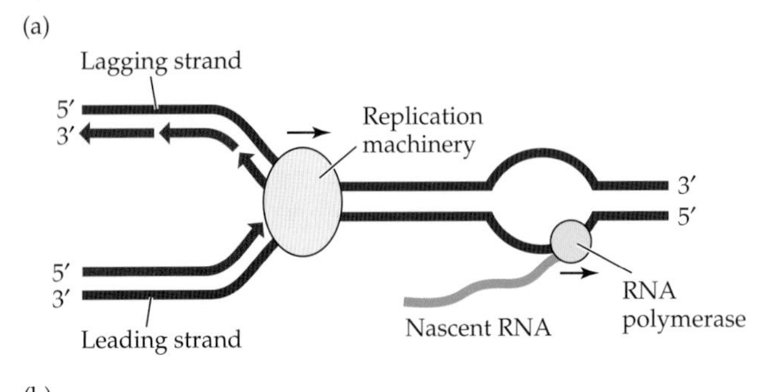

(b)

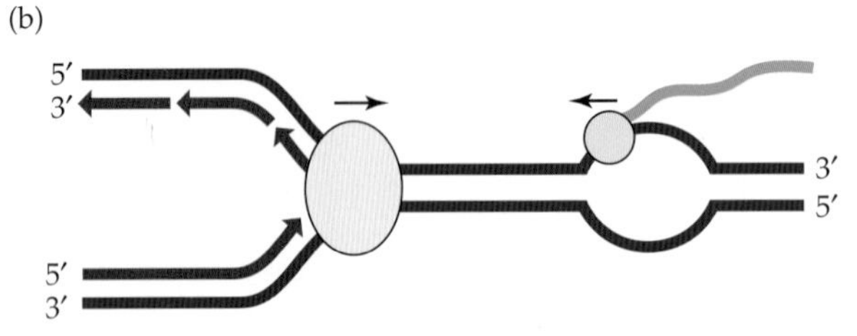

Figure 10.17 DNA replication and RNA transcription can be codirectional or in a head-on orientation. (a) If the transcribed strand is the leading strand, replication and transcription are codirectional. (b) If the transcribed strand is the lagging strand, replication and transcription occur in a head-on orientation. Nascent DNAs on the leading and lagging strands are indicated by solid and broken red arrows, respectively. Nascent RNAs are indicated by wavy green lines. Small black arrows indicate the directions of replication and transcription. (Modified from Kim and Jinks-Robertson 2012.)

biased gene distribution between the two strands. For example, 75% of the protein-coding genes of *B. subtilis* are located in the leading strand. The percentage of essential genes in the leading strand is 94%, while the percentage of ribosomal RNA-specifying genes is 100%. Thus, the deleterious nature of head-on collisions seems to depend on the function of the genes being transcribed, with collisions involving essential genes having a more deleterious effect on fitness than collisions involving highly expressed genes.

Gene-density asymmetry and nucleotide-composition asymmetry between the leading and the lagging strands are highly and positively correlated with each other, indicating that nucleotide content and coding density are consequences of the same cause, **asymmetric directional mutational pressure**. As noted by Rocha (2004) and Chen and Chen (2007), the mutational asymmetry can arise either during the process of DNA replication (**replication-associated mutational pressure**) or during transcription (**transcription-associated mutational pressure**).

The mutational asymmetry in replication-associated mutational pressure is related to the distance from the origin of replication and the terminus of replication and is independent of gene content. Indeed, highly expressed genes tend to be located in the leading strand in proximity to the origin of replication, *ori* (Rocha 2008). Transcription-associated mutational pressure, on the other hand, is higher in areas of high densities of highly transcribed genes. The higher the density of transcribed genes on a strand and the more transcripts produced per gene, the longer the two DNA strands will be unpaired and, hence, prone to mutation. This mechanism should cause compositional bias between the two strands only in genomes with strong bias in density of transcribed genes, while the contribution of replication-associated mutational pressure to asymmetry in nucleotide usage should be low in genomes with weak bias in gene density. According to a study by Chen and Chen (2007), transcription-associated mutational pressure is a universal feature of all prokaryotic genomes, while replication-associated mutational pressure is absent or extremely weak in many prokaryotic genomes.

Because of the observed positive correlation between the fraction of protein-coding genes in the leading strand and the GC skew, it is interesting to investigate the fate of genes that translocate from the leading to the lagging strand or vice versa. In an analysis of paralogs from 50 eubacterial genomes, Mackiewicz et al. (2003) distinguished between **trans-paralogs**, i.e., pairs of paralogs in which one copy is located in the leading strand and the other in the lagging strand, and **cis-paralogs**, i.e., pairs of paralogs in which both copies are found in either the leading strand or the lagging strand. Because paralogs are the result of gene duplication, they are initially located in the same strand. In *trans*-paralogs, then, one copy must have changed position from one strand to the other, for example, by inversion. Thus, *trans*-paralogs can serve as proxies for gene rearrangements in the genome. The number of *trans*-paralogs was found to vary from 6% in *Chlamydia pneumoniae* to 52% in *Helicobacter pylori*. A highly significant negative correlation was found between GC skew and fraction of *trans*-paralogs (**Figure 10.18**).

Let us now consider the evolutionary fate of *trans*-paralogs. An inversion inside a replichore will turn a part of the leading strand into a lagging strand and a part of the lagging strand into a leading strand. Such an inversion is thus expected to lower the GC skew while at the same time decreasing the gene-density asymmetry between the leading and the lagging strands (Esnault et al. 2007). Genomic rearrangements involving movements from one strand to the other are expected initially to reduce both compositional and gene-density asymmetries. Subsequently, however, mutation is expected to restore compositional asymmetries relatively quickly, whereas the repositioning of genes is expected to take considerably longer.

What determines the proportion of protein-coding genes that are transcribed from the leading strand, and is there ever an advantage in being transcribed from the lagging strand? Paul et al. (2013) suggested that some genes are adapted to the higher mutation rate on the lagging strand. Thus, genes under strong negative selection against amino-acid-changing mutations tend to be co-oriented with replication in the leading strand, while genes under positive selection for amino-acid-changing mutations are more commonly found on the lagging strand, indicating faster adaptive evolution in many genes in the head-on orientation. Chen and Zhang (2013) refuted this adaptationist scenario and showed that a better explanation for the existence of genes in the lagging strand is balance between deleterious mutations that move genes from the leading strand to the lagging strand and purifying selection against such changes.

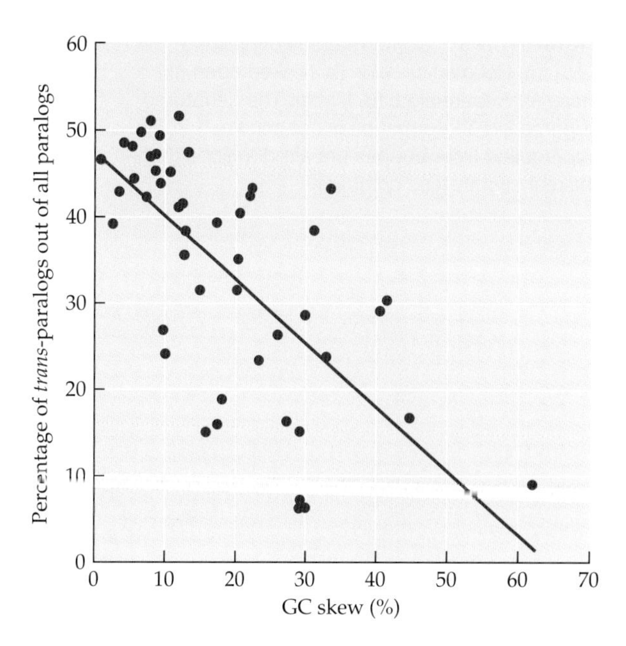

Figure 10.18 Relation between the fraction of all paralogs that are *trans*-paralogs and compositional asymmetry as measured by the GC skew at the third-position codons of protein-coding genes from 50 genomes. (Modified from Mackiewicz et al. 2003.)

Chromosomal Evolution in Prokaryotes

Organisms and organelles contain two types of genetic material: chromosomes and extrachromosomal elements. **Chromosomes** contain one or more unconditionally essential genes, i.e., genes that are needed under all possible circumstances. **Extrachromosomal elements**, on the other hand, contain genetic information that may be important under certain circumstance but is not necessary under all conditions.

The main classes of extrachromosomal elements are plasmids and episomes. **Plasmids** are autonomously replicating extrachromosomal molecules distinct from the chromosomal genome. They exist solely in an autonomous state and are inherited independently of chromosomes. Their replication rate may be considerably higher than that of the chromosomal DNA, although in many instances there seems to be a 1:1 relationship between plasmid and chromosomal replication rates. Plasmid DNA can assume one of five conformations: (1) relaxed circular, (2) nicked open-circular, (3) supercoiled, (4) denatured supercoiled, and (5) linear (**Figure 10.19**). Plasmids range in size from about 500 nucleotides to ~2 Mb. Plasmids for which no function has been described and which have no discernable phenotypic effect are referred to as **cryptic plasmids**. Plasmids have been discovered in both prokaryotes and eukaryotes, as well as in mitochondria and plastids. A nucleus or an organelle may carry two or more plasmid types simultaneously.

Episomes, too, contain only nonessential genetic information, but they are capable of alternating between two states: independently replicating within a cell, or integrated into the chromosome. Viral prophages are examples of episomes. An episome that loses its ability to become attached to the chromosome becomes a plasmid; one that loses its ability to detach itself from the chromosome becomes part of the chromosome.

The best-known phenotypic effects of extrachromosomal elements in bacteria are (1) antibiotic, heavy-metal, and heat resistance; (2) virulence and pathogenicity; (3) autotrophy, i.e., the ability to produce energy and nutrients from inorganic

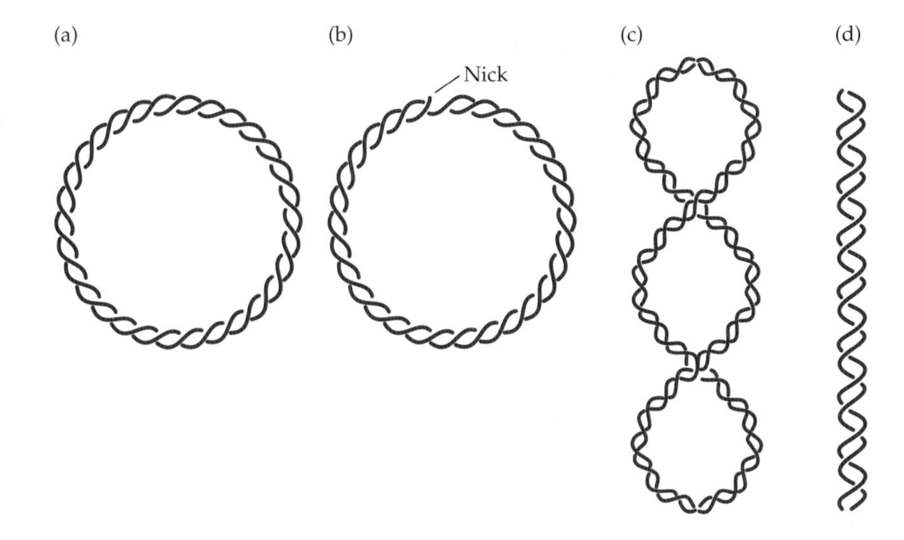

Figure 10.19 Conformations of plasmid DNA. (a) Relaxed circular. (b) Nicked open circular. (c) Supercoiled. (d) Linear. The denatured supercoiled conformation is not shown. It is like supercoiled DNA but has unpaired regions that make it slightly less compact.

compounds by using such mechanisms as photosynthesis or chemosynthesis; and (4) antigenic plasticity, i.e., the ability to evade the immune system of the host.

Evolution of chromosome number in prokaryotes

The vast majority of prokaryotes contain a single chromosome. There are, however, exceptions. Prokaryotes having two or more chromosomes are said to possess a **multipartite genome**. In prokaryotes with multipartite genomes, one chromosome usually contains the majority of genes; this chromosome is referred to as the **primary chromosome**. The other chromosomes are referred to as **secondary** or **accessory chromosomes**. The primary chromosome is designated CI; the secondary ones are designated CII, CIII, and so on. At present, we know of no prokaryote species with four or more chromosomes. In the absence of complete information on the essentiality of the genes they carry, it is impossible to tell a plasmid from an accessory chromosome, and designations may change as knowledge accumulates. For example, in *Sinorhizobium meliloti*, former plasmids pSymA and pSymB are now designated CII and CIII, respectively.

At present, 50 bacterial species—mostly proteobacteria—have been reported to contain multiple chromosomes (Choudhary et al. 2012). While the majority of these taxa are free-living, a substantial minority, including *Agrobacterium*, *Brucella*, *Sinorhizobium*, and *Vibrio*, are plant and animal pathogens.

In species belonging to the same genus, the sizes of the CI chromosomes do not vary much, while a significant size difference is seen among the CII chromosomes. For example, in *Burkholderia* species with two chromosomes, the ratio of the largest to the smallest CI chromosome is less than 2. In comparison, the ratio of the largest to the smallest CII chromosome is close to 30. That said, gene density in accessory chromosomes is indistinguishable from that in primary ones. The large variation in CII sizes in closely related groups of bacteria together with the disjoined taxonomic distribution of multipartite genomes suggests that accessory chromosomes have originated independently multiple times throughout bacterial evolution.

The mechanisms by which accessory chromosomes evolved in prokaryotes are not fully understood. Three main hypotheses have been proposed in the literature to explain the existence of multipartite genomes in prokaryotes (Egan et al. 2004). The first is the **plasmid hypothesis**, which states that the accessory chromosome evolved from a plasmid through the acquisition of essential genes from the primary chromosome. The second is the **schism hypothesis**, which states that the accessory chromosome evolved from a split of the primary chromosome into two chromosomes of unequal sizes. The third hypothesis is the **lateral transfer hypothesis**, which states that the accessory chromosome was captured from another bacterium (possibly another species). This third hypothesis can be combined with the first hypothesis by envisioning a small, plasmid-derived chromosome being transferred from one cell to another or from one species to another by conjugation.

Chromosomes differ from plasmids in that they are indispensable for survival under all conditions. While plasmids may give the cell a competitive advantage under one set of conditions, they are dispensable under most growth conditions. As such, an accessory chromosome needs to be replicated in synchrony with the primary chromosome once per cell cycle. How does a bacterium coordinate chromosomal replication and segregation of its divided genome so that each daughter cell receives a full genome complement at cell division? This coordination may be achieved by (1) a common replication machinery, (2) distinct chromosome-specific replication factors, or (3) some combination of the two. In *Vibrio cholerae*, it has been shown that the replication of the two chromosomes requires (1) a set of proteins that are common to both chromosomes, and (2) two sets of chromosome-specific proteins.

Do accessory chromosomes confer an evolutionary advantage? One idea is that replicating two small chromosomes simultaneously would be speedier than the process of replicating the sum of the two chromosomes. Some weak support for this hypothesis is obtained from experiments in which the three chromosomes of *S. meliloti* were integrated into a single chromosome and a decrease in growth rate was recorded as a consequence. However, several bacterial species, such as *Myxococcus xanthus*, possess extremely large genomes with only one primary chromosome and without any obvious evolutionary disadvantage.

Another possibility may be that multiple chromosomes benefit the organism via chromosome-specific gene regulation, whereby in some environments a higher expression of genes located in either CI or CII is needed. Classifying genes into functional categories lead to the discovery that genes encoding proteins involved in translation, DNA replication, and cellular processes are overrepresented on primary chromosomes, while genes encoding proteins involved in transcription, signal transduction, and metabolic functions are overrepresented on secondary chromosomes. The overrepresentation of certain functions on either CI or CII is likely indicative of adaptation of certain functions to the accessory chromosome.

In support of this hypothesis, it has also been shown that in *V. cholerae*, genes involved in metabolism, nutrient starvation, quorum sensing, DNA repair, and pathogenicity, which are overrepresented in CII, are overexpressed under pathogenic conditions, while genes on CI maintain similar expression levels under both free-living and pathogenic conditions.

In a few species, it has been shown that genes in accessory chromosomes evolve faster, on average, than genes in primary chromosomes (Cooper et al. 2010). Two possible factors can account for this difference: (1) either the rate of mutation is greater in the accessory chromosomes than in the primary ones, or (2) purifying selection is weaker. At present it is not possible to unambiguously decide which of these factors is responsible for the rate difference, although analyses of codon usage studies indicate that purifying selection may be weaker in accessory chromosomes.

Within the genus *Brucella*, a Gram-negative bacterial group pathogenic to animals and humans, we find species with either a single chromosome or two. For example, *B. melitensis*, a pathogen of sheep and goats, has two circular chromosomes, 2,100 Kb and 1,150 Kb in size. In other *Brucella* species with two chromosomes, the sizes of the chromosomes may be different, for example, 1,850 Kb and 1,350 Kb in the porcine pathogen *B. suis*. Interestingly, the size of the chromosome in single-chromosome strains is about the same as the total chromosome size for two-chromosome strains. Jumas-Bilak et al. (1998) have shown unambiguously that these are bona fide chromosomes rather than extrachromosomal elements, by mapping all known genes of the two chromosomes on the one chromosome of a single-chromosome *Brucella* species. Moreover, they could explain all the chromosomal variation in *Brucella* by paralogous recombination among three rRNA-specifying loci.

Using an elaborate laboratory protocol, Itaya and Tanaka (1997) succeeded in dividing the chromosome of *Bacillus subtilis* into two independently replicating subgenomes.

The resulting two-chromosome bacteria were viable. This finding indicates that the evolution of chromosome number in bacteria may be restricted by mutational input rather than by selection against multichromosomality. That is, the generation of two viable chromosomes out of one chromosome requires many low-probability steps in a particular order. We note, however, that given the long evolutionary history and the diversity of lineages in prokaryotes, even a low rate of mutational input should have led to a significant increase in chromosome number. The rarity of multichromosomal prokaryotic species indicates that selection against multichromosomality must be quite powerful.

Estimating the number of gene order rearrangement events

To be able to study the evolution of gene order rearrangements, we must first estimate the number of events (e.g., inversions, transpositions, deletions, and insertions) that are necessary to change the gene order of one extant genome into another. This will give us an estimate of the number of gene order rearrangement events that have occurred since the divergence of two genomes from each other. In the following sections we present two such methods.

THE ALIGNMENT REDUCTION METHOD A simple method called the **alignment reduction method** has been suggested by Sankoff et al. (1992). In this method, we compute a so-called **evolutionary edit distance** (E) between two genomes, say 1 and 2. Evolutionary edit distance has two components: the **deletion distance** (D), which is the minimal number of deletions and insertions necessary for the two genomes to have identical sets of genes, albeit in different orders, and the **rearrangement distance** (R), i.e., the minimal number of inversions and transpositions necessary to convert the gene order of genome 1 into the gene order of genome 2.

$$E = D + R \tag{10.4}$$

To estimate E, we employ three simple geometrical procedures: **deletion**, **bundling**, and **inversion** (**Figure 10.20**). We start by connecting homologous genes by lines. At this stage, we distinguish between homologous pairs that have the same genomic orientation and those that are inverted relative to one another. A different type of line

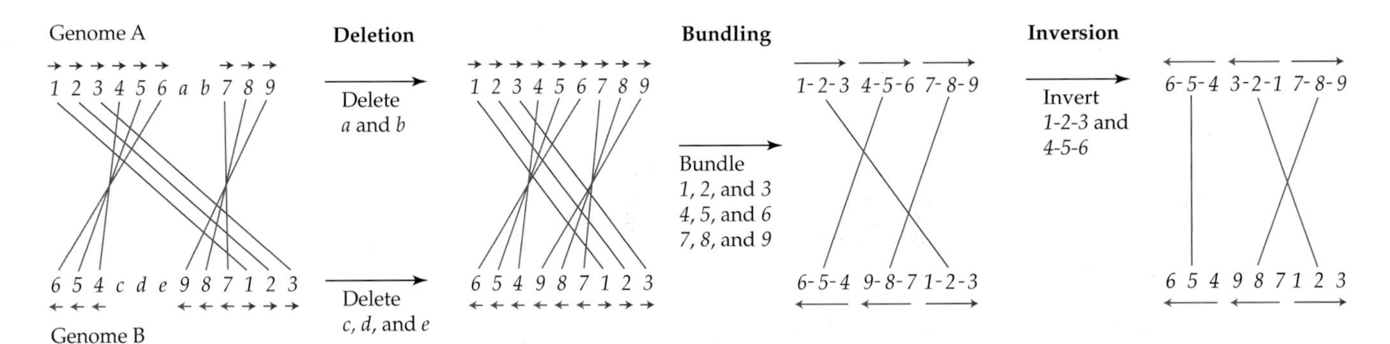

Figure 10.20 The three basic geometrical procedures employed by the alignment reduction method to infer the number of gene order rearrangement events between two genomes. The orientation of the genes (*1–6* and *a–e*) is marked by red and blue arrows. Homologous genes are connected by lines. Red lines indicate the same orientation; blue lines indicate reverse orientation. The first procedure is deletion, where genes that are absent in either of the two genomes are deleted. Genes *a* and *b* were deleted from genome A. Genes *c*, *d*, and *e* were removed from genome B. Although five genes were deleted, the number of deleted segments (i.e., the deletion distance, *D*) is two. The second procedure is bundling. Bundling involves the clustering of adjacent genes if the genes have the same order relative to each other and the same relative orientation in the two genomes. Following bundling, the bundled genes are marked by hyphens. Bundling carries no weight in computing the evolutionary edit distance. The third procedure is inversion. The inversion of segments *1-2-3* and *4-5-6* in genome A causes a change in their orientation relative to genome B. The delete, bundle, and invert arrows are located at the top if the procedure applies to genome A only; at the bottom if the procedure applies to genome B only; and in the middle if the procedure applies to both genomes. (Modified from Sankoff et al. 1992.)

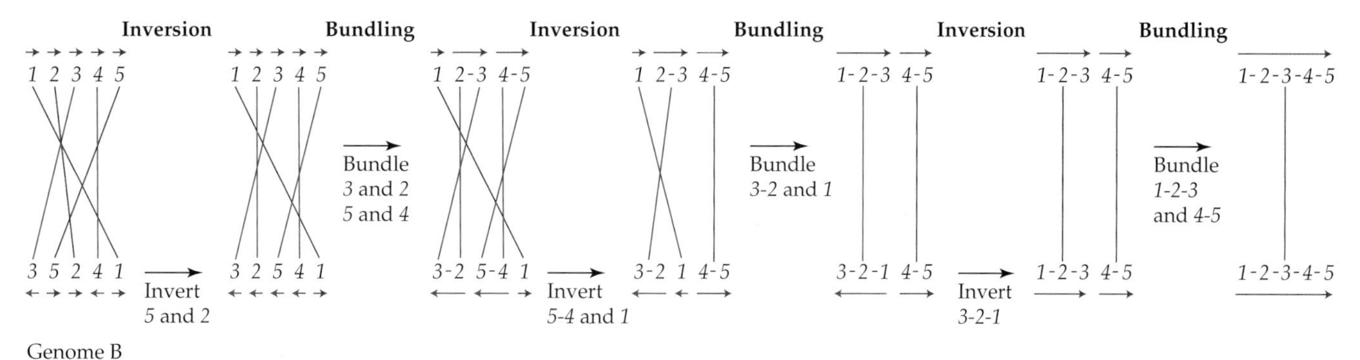

Figure 10.21 Using the alignment reduction method to infer the rearrangement distance for two genomes with five genes each. The orientation of the genes is marked by red and blue arrows. Homologous genes are connected by lines. Red lines and arrows indicate the same orientation in the two genomes; blue lines and arrows indicate that the gene in genome B is in the reverse orientation relative to its homolog in genome A. Bundled genes are indicated by hyphens. Bundling affects both genomes; hence, the bundle arrows are placed in the middle. Inversion procedures are always performed on genome B. The total number of inversions it takes to convert the gene order in genome B to that in genome A is three. (Modified from Sankoff et al. 1992.)

marks each kind of pair. The estimation of D is the easiest: all genes that are absent in either of the two genomes are removed from the analysis. D is equal to the number of segments that were removed. Note that in the case illustrated in Figure 10.20, we removed five genes but only two segments, so $D = 2$.

If certain genes are adjacent to one another in both genomes, and if all these genes have the same relative order, whether in the same or in the reverse orientation, we may combine them into one bundle. We note that bundling is just an algorithmic procedure that carries no weight in computing the rearrangement distance.

The third procedure is inversion, i.e., the conceptual rotation of a segment 180° without changing its genomic location. An inversion will, of course, change the relative orientation of the homologous segments, and we must keep track of such changes. In the alignment reduction method, we invert and bundle until the alignment of the two genomes reduces to a single segment in the same orientation in both genomes. R is the sum of all inversions along the way.

A hypothetical case of inferring the number of events that have occurred since the divergence of two (very small) genomes from each other is shown in **Figure 10.21**. In this case, we do not bother with D, since the two genomes have the same set of genes, so $D = 0$. The minimal solution to the problem in Figure 10.21 turns out to be $R = 3$.

Sankoff et al. (1992) have applied the method to infer the number of indels and segmental changes that have occurred in the evolution of seven animal mitochondrial genomes (**Table 10.3**). From this table, we can deduce that gene number has been quite a conserved trait throughout animal evolution, whereas rearrangements occurred quite frequently. For example, while mitochondrial gene content has not changed at all from sea urchins to humans, at least 16 rearrangements of the 30 common genes are inferred to have occurred since they last shared a common ancestor.

The alignment reduction method has two main computational problems. First, finding the smallest possible R is what computer scientists call a computationally hard problem—a problem requiring computing times that increase exponentially with the number of genes in the two genomes. Thus, the alignment reduction method is only applicable to small and evolutionarily conserved genomes. It is barely applicable to seven circular genomes with only 30 genes in common. With larger genomes and more taxa, computing time becomes prohibitive. Some progress has been made in developing algorithms for computing rearrangement distances if only one type of operation, for example, inversion, is allowed (Bernt et al. 2013). Nevertheless, there are still major unsolved problems, including the lack of an efficient, exact algorithm

TABLE 10.3

Evolutionary edit distance between pairs of animal mitochondrial genomes[a]

OTUs[b]	*Hs*	*Gg*	*Sp*	*Ap*	*Po*	*Dy*	*As*
Hs		1	18	16	19	13	25
Gg	0		19	17	17	12	26
Sp	0	0		2	1	26	27
Ap	4	4	4		1	22	24
Po	1	1	1	5		23	24
Dy	0	0	0	4	1		28
As	1	1	1	5	2	1	

Source: Modified from Sankoff et al. (1992).

[a]Deletion distances and rearrangement distances are below and above the diagonal line, respectively.

[b]*Hs*, *Homo sapiens* (human); *Gg*, *Gallus gallus* (chicken); *Sp*, *Strongylocentrotus purpuratus* (sea urchin); *Ap*, *Asterina pectinifera* (starfish); *Po*, *Pisaster ochraceus* (starfish); *Dy*, *Drosophila yakuba* (fruit fly); *As*, *Ascaris suum* (pig roundworm).

for rearrangement problems that allow transpositions. To circumvent the algorithmic complications caused by transpositions, a **double cut-and-join rearrangement** operation was proposed (Yancopoulos et al. 2005; Bergeron et al. 2006; Friedberg et al. 2008). In this method, a certain gene order is cut at two places, and the resulting segments are subsequently rejoined in a different order.

The second problem with the alignment reduction method is that no exact algorithm is known for applying it to three or more genomes.

THE BREAKPOINT DISTANCE METHOD In this method, we start with two genomes that possess the same set of genes. (If the set of genes differs between the two genomes, then we need first to purge the unique genes from each genome.) The purpose of the analysis is to calculate how many breakpoints one needs to make in one genome to be able to assemble the second genome. An application of the method to a simple case of two circular genomes is shown in **Figure 10.22**. As opposed to the evolutionary edit distance, the **number of breakpoints** is a distance measure that does not require assumptions about the mechanisms involved in gene order evolution. The measure has the added advantage that it is easy and fast to calculate.

If we use the number of breakpoints as a genetic distance between two genomes, we need to realize that its value is correlated with the number of homologous genes in the two genomes. To compare breakpoint distances between different pairs of genomes, a normalized measure is required. Suyama and Bork (2001) introduced

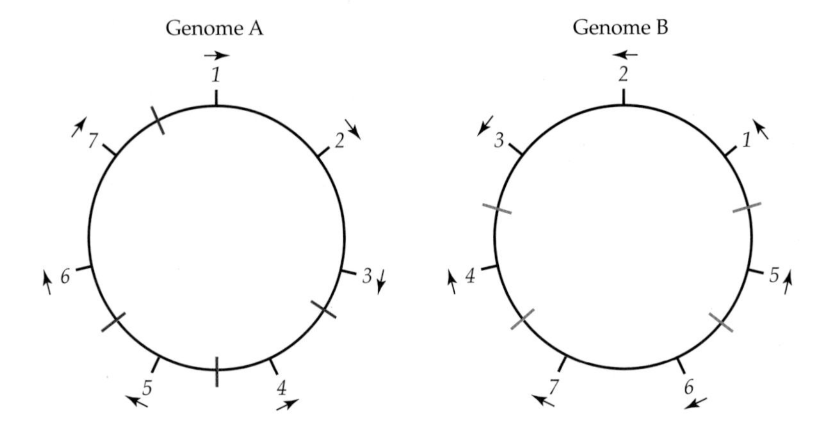

Figure 10.22 Identifying breakpoints between circular genomes A and B. Arrows denote the direction of transcription. We note that gene *4* and gene *5* have different neighbors on both sides in genome A than they do in genome B. Genes *6* and *7* and genes *1*, *2*, and *3* appear in the same order in both genomes. If we cut genome A at four breakpoints (red lines), we can use the resulting segments to assemble the gene order in genome B. Alternatively, we can cut genome B at four breakpoints (blue lines) to assemble the gene order in genome A. The neighboring disruption frequency is 4/7 = 0.57.

a measure called the **neighborhood disruption frequency**, which is the number of breakpoints of gene neighbors divided by the number of shared homologous genes between the two genomes. Neighborhood disruption frequencies can range between 0 (complete conservation of gene order) and 1 (complete shuffling of genes).

Gene order evolution

Because prokaryotes are severely restricted in recombination and because some of their genes function as units (operons), it was thought the evolution of prokaryotes would be characterized by spatial and temporal stability (Ochman and Wilson 1987). With the complete sequencing of the first eubacterial genomes, however, it became clear that eubacterial gene order was anything but conserved (Mushegian and Koonin 1996b).

To illustrate this lack of conservation, Watanabe et al. (1997) compared gene order between *Haemophilus influenzae* and *Mycoplasma genitalium*. They identified 184 orthologous genes. Gene order conservation was found to be rare. In fact, the genes in the two genomes seem to have been reshuffled so frequently as to appear to be randomly ordered. The situation was even worse in a comparison between *H. influenzae* and *E. coli*, despite the fact that these two species are more closely related to each other than *H. influenzae* and *M. genitalium*. It seems that gene order colinearity is almost never conserved in bacterial evolution. Only very short regions of gene order conservation were found, and these disappeared as more species were included. Similarities in gene order between distantly related taxa invariably turned out to be due to recent horizontal gene transfer (Coenye and Vandamme 2005). Gene order turned out to be quite an unremarkable molecular marker; with few exceptions gene order is similar among closely related species, but the similarity rapidly decreases with genetic distance (Tamames 2001). The decrease in similarity is slower in some gene clusters than in others, indicating that some degree of purifying selection for maintaining gene order does indeed operate in some cases. This selection, however, seems to be very weak. The reasons for the maintenance of gene order by selection are not yet understood.

Suyama and Bork (2001) found an almost perfect correlation between neighborhood disruption frequencies and mean numbers of amino acid replacements per site for the orthologous genes shared by the two genomes. The exception involved comparisons involving *Chlamydia* and *Mycoplasma* species, whose genome orders seem to have evolved much more slowly than those of other species, i.e., fewer rearrangements have occurred in comparison to the expectations derived from distances based on amino acid sequences. There may be two reasons for the unique behavior of these two taxa. First, the two lineages seem to lack a number of enzymes involved in cutting and rearranging the genome. Second, it is possible that these two lineages evolve much faster than the other bacteria in terms of amino acid replacements, while the genomic rearrangement rate is unexceptional. Hence, large amino acid distances may be indicative of faster evolutionary rates rather than divergence time.

Repeated sequences within a genome seem to constitute a risk factor for gene order rearrangements (Achaz et al. 2003; Rocha 2003).

Operon evolution

As mentioned previously, some protein-coding genes in prokaryotes are arranged consecutively within an operon, which is transcribed into a single polycistronic mRNA molecule from which the different proteins are subsequently translated (Chapter 1). An operon may consist of genes with related functions, genes that are needed in successive steps of a biochemical pathway, or genes that interact with one another to form multisubunit protein complexes. Beside the fact that genes within operons are transcribed as a unit, operons may contain additional genetic elements that control their coordinated expression. Operons tend to be quite compact; in most taxa, genes in the same operon are usually separated by fewer than 20 bp of DNA. Furthermore, the stop codon of an upstream gene in an operon often overlaps the start codon of its downstream gene (Price et al. 2006).

Because operons are ubiquitous, and because so few operons are conserved among bacteria, it has been inferred that operons are born and die frequently during evolution (Itoh et al. 1999; Osborn and Field 2009). The prevailing view in the past was that operons exist because they facilitate coregulation. However, genes that are in the same operon in one bacterium are frequently found in different operons in other bacteria. Do changes in operon structure lead to changes in gene expression patterns, or can genes be cotranscribed in some organisms and coregulated from distinct promoters in other organisms without functional consequences? Are changes in operon structure neutral or even slightly deleterious, as suggested by the frequent loss of operons, or are changes in operon structure the result of positive selective pressures? Here, we will discuss the genomic mechanisms responsible for changes in operon structure. In particular, we will investigate how operons form, how they are maintained, and how they die.

In discussing the birth of operons, we need to mention the fact that many operons have been acquired, hook, line, and sinker, through horizontal gene transfer. Of course, this pattern of operon acquisition does not tell us how new operons are born; however, the ubiquity of horizontally transferred operons as opposed to the rarity of single-gene transfers indicates that the horizontal transfer of entire metabolic pathways or entire protein complexes has a better chance of being advantageous than the transfer of a single gene.

For reasons of parsimony, we will discuss only the birth of two-gene operons; the creation of multigene operons in a single evolutionary step is unlikely. Novel operons can be created either through the addition of a neighboring gene in close proximity to an existing gene or through a deletion reducing the distance between two genes. In both cases, it is assumed that the resulting proximity of two genes will cause them to be transcribed as a unit. As far as operon creation by addition is concerned, there can be two sources of neighboring genes. The first source is extrinsic, i.e., horizontal gene transfer occurs. In some cases, the source of the new neighboring gene can be ascertained through sequence similarity with unrelated taxa; in other cases, the newly acquired neighboring gene is classified as an **ORFan**, i.e., a gene that lacks homologs outside of a group of closely related taxa and whose origin is unknown (Chapter 8). Many ORFans are thought to be derived from bacteriophages (Daubin and Ochman 2004). The second source is intrinsic. Two genes may be brought in close proximity through genome rearrangements. Finally, two genes can become close neighbors through the deletion of intervening genes or through the deletion of intergenic regions.

Modifications to preexisting operons seem to be less common than the formation of new operons. In *E. coli*, Price et al. (2006) identified 455 new operons but only 81 modification events. In **Figure 10.23**, we detail the various mechanisms by which an operon can be modified. By far the most common mechanisms of modifying an operon involve the addition of a gene. The gene can be appended to the 3′ end of the operon or prepended to the 5′ end. If a gene is appended to an existing operon, the operon will retain its original promoter. If a gene is prepended to an operon, the operon will fall under the control of a new promoter. Insertions of genes into an operon or the replacement of one gene by another are known to occur, albeit very rarely. The joining of two operons has also been observed, albeit infrequently. One interesting example of an operon that evolved through the joining of three smaller operons is the *his* operon in proteobacteria (Fani et al. 2005).

An operon can be lost by the deletion of one or more genes, by replacement by another operon, say one derived through horizontal gene transfer, or else by the splitting apart of the constituent genes (**Figure 10.24**). The splitting of the constituent genes can occur by gene rearrangement or by insertion of either another gene with its own promoter or an intergenic sequence that will increase the physical distance between the genes. For the constituent genes to remain active after the splitting of the operon, it is necessary that each of them evolve a promoter before the splitting occurs.

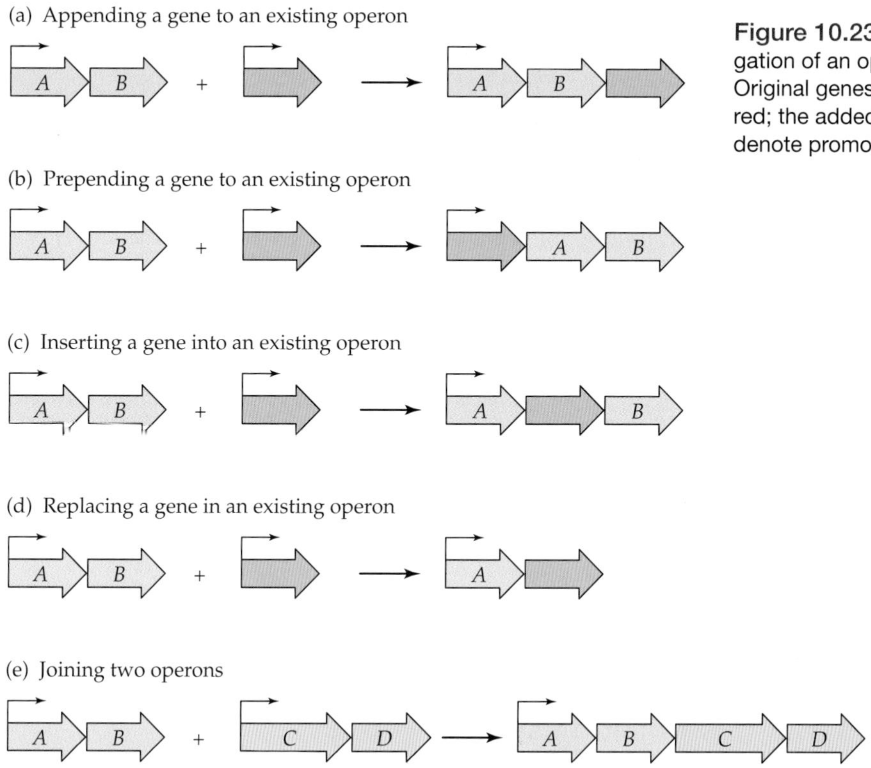

(a) Appending a gene to an existing operon

A B + → A B

(b) Prepending a gene to an existing operon

A B + → A B

(c) Inserting a gene into an existing operon

A B + → A B

(d) Replacing a gene in an existing operon

A B + → A

(e) Joining two operons

A B + C D → A B C D

Figure 10.23 Mechanisms of operon elongation. The elongation of an operon entails the loss of at least one promoter. Original genes in the operon are in blue; added genes are in red; the added operon is in yellow. Arrows above the genes denote promoters.

What drives the evolution of operons? At present, it is not possible to provide a very definite answer. Some findings suggest that the evolution of operons may be driven by selection. First, both the birth and the death of operons are known to lead to large changes in expression patterns. Since gene expression is thought to be under strong selection, then the integrity of the operons that determine expression to a great extent should also be under selection. Second, some operons are known to acquire new genes in a relatively short period of evolutionary time, a phenomenon suggestive of positive Darwinian selection. On the other hand, operons are born and die frequently and are seldom conserved evolutionarily. Moreover, many operons contain genes with functions that are unrelated to one another, raising the possibility that some operons are put together haphazardly by mutational processes and random genetic drift (de Daruvar et al. 2002). These and other findings suggest that selection for maintaining operon structure may be either very weak or nonexistent (Itoh et al. 1999; Hazkani-Covo and Graur 2005).

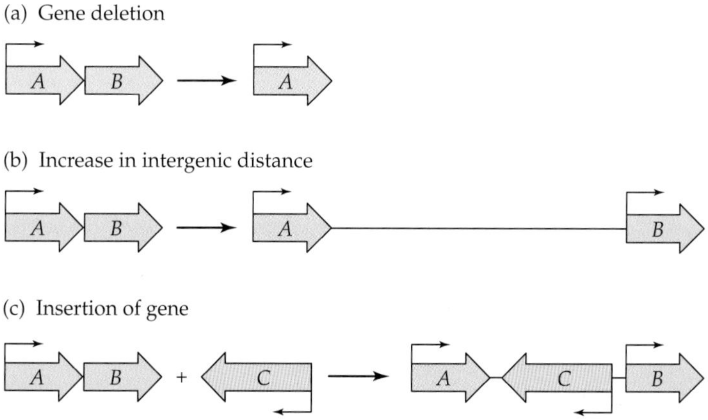

(a) Gene deletion

A B → A

(b) Increase in intergenic distance

A B → A B

(c) Insertion of gene

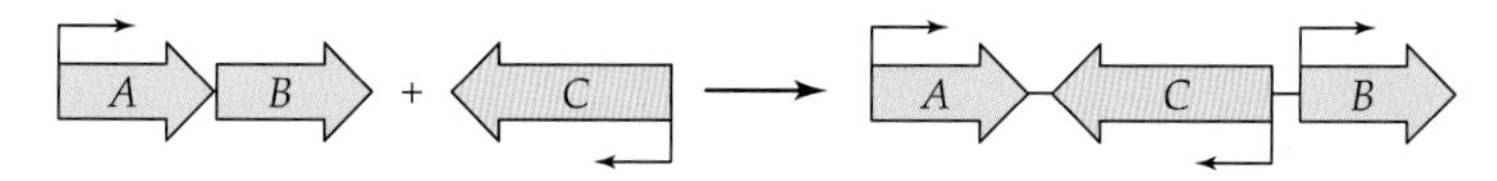

Figure 10.24 Mechanisms of operon death. In addition to the deletion of genes (a) or the entire operon, operon death may be due to increase in physical distance between the constituent genes (b) or the insertion of a disrupting gene with its own promoter (c). Original genes are in blue; disrupting genes are in yellow. Arrows above and below the genes denote promoters.

The Emergence of Alternative Genetic Codes

The vast majority of prokaryotic and eukaryotic genomes use the standard genetic code (Chapter 1). Two main kinds of explanations have been proposed for this phenomenon. The **amino acid–codon interaction theory** (Woese 1969) states that the specific codon assignments originated in direct chemical interactions between nucleic acids and amino acids, and therefore the standard genetic code is a relic of a world in which translation did not require intermediates, such as tRNA and aminoacyl-tRNA synthetases. The **frozen accident theory** (Crick 1968) states that the code is universal not because of any chemical or physical imperative, but because the genetic code happened to evolve to a certain point by either chance or selective optimization (e.g., Figureau 1989), and when the genome grew to such an extent that it specified the production of many proteins, the rules of translation could not be further altered without affecting many proteins at once. Such a drastic change would not stand a chance of being benign, let alone beneficial. A useful nonmolecular example of a "frozen accident" is the arrangement of letters on a keyboard. The QWERTY arrangement, which is used in the vast majority of languages written in the Latin alphabet, proved very difficult to

TABLE 10.4

Fifteen codons that have been reassigned in at least one alternative genetic code

Codon	Standard genetic code	Alternative genetic code[a]																
		2	3	4	5	6	9	10	12	13	14	15	16	21	22	23	24	25
AAA	Lys						Trp				Asn			Asn				
AGA	Arg	Stop[b]			Ser		Ser			Gly	Ser			Ser			Ser	
AGG	Arg	Stop[b]			Ser		Ser			Gly	Ser			Ser			Lys	
AUA	Ile	Met	Met		Met					Met				Met				
CGA	Arg		Absent															
CGC	Arg		Absent															
CUA	Leu		Thr															
CUC	Leu		Thr															
CUG	Leu		Thr						Ser									
CUU	Leu	Thr	Thr															
UAA	Stop					Gln					Tyr							
UAG	Stop					Gln						Gln	Leu		Leu			
UCA	Ser														Stop			
UGA	Stop	Trp	Trp	Trp	Trp		Trp	Cys		Trp	Trp			Trp			Trp	Gly
UUA	Leu															Stop		

[a]Alternative genetic codes that differ from the standard genetic code only by having unassigned codons are not shown. Similarly, partially characterized alternative genetic codes are not shown. The alternative genetic codes are numbered according to Elzanowski and Ostell (2013): 2, vertebrate mitochondrial code; 3, yeast mitochondrial code; 4, mold, protozoan, and coelenterate mitochondrial code and *Mycoplasma/Spiroplasma* nuclear code; 5, invertebrate mitochondrial code; 6, ciliate, dasycladacean, and *Hexamita* nuclear code; 9, echinoderm and flatworm mitochondrial code; 10, euplotid nuclear code; 12, alternative yeast nuclear code; 13, ascidian mitochondrial code; 14, alternative flatworm mitochondrial code; 15, *Blepharisma* nuclear code; 16, chlorophycean mitochondrial code; 21, trematode mitochondrial code; 22, *Scenedesmus obliquus* mitochondrial code; 23, *Thraustochytrium* mitochondrial code; 24, *Pterobranchia* mitochondrial code; 25, candidate division SR1 and Gracilibacteria code.

[b]The AGA and AGG codons in the vertebrate mitochondrial code are unassigned, but because of ribosomal –1 frameshifting and the fact that they are always preceded by U they are read as the stop codon UAG (Chapter 1).

change without causing havoc for the users. Indeed, only few minor deviations from this arrangement are known (e.g., QUERTZ in German and AZERTY in French).

The first alternative genetic code was deduced from the protein-coding sequences within the human mitochondrial genome (Barrell et al. 1979). It is estimated that alternative genetic codes have arisen independently at least 24 times in mitochondrial genomes and at least 10 times in nuclear genomes (Ring and Cavalcanti 2008). A list of codons that have been reassigned in at least one alternative genetic code, in comparison with the standard genetic code, is shown in **Table 10.4**. Since the amino acid–codon interaction theory precludes the existence of nonuniversal genetic codes, the existence of such codes is widely construed as a refutation of this theory. The frozen accident theory, on the other hand, does not preclude the existence of nonuniversal genetic codes; however, it does not provide an obvious mechanism for changing the rules of translation without causing massive deleterious effects on proteins. How, then, might an alternative genetic code evolve? How can codons be reassigned?

Jukes (1985) and Osawa and Jukes (1989, 1995) noticed that alternative genetic codes are frequently associated with small genomes—many of them mitochondrial ones—as well as with biased GC content. It was, then, concluded that all alternative genetic codes are derived from the standard genetic code, i.e., the standard genetic code is a plesiomorphy. Osawa and Jukes (1989) came up with the first satisfactory explanation for how changes in the genetic code might occur without a massive reduction in the fitness of the organism. Their **codon-capture hypothesis** (later also referred to as the **codon disappearance mechanism**) postulates that a codon may disappear from the genome—say, because of biased GC mutational pressures or random genetic drift—and may later reappear (as a rare codon) with a different amino acid assignment.

To illustrate this mechanism, let us consider codon AAA, which codes for lysine in the standard genetic code, whereas in the mitochondria of echinoderms (e.g., starfish and sea urchins) it is used for asparagine, and in the mitochondria of hemichordates (e.g., acorn worms) it is unassigned (Castresana et al. 1998). Let us start with the situation in the standard genetic code, where lysine is coded by two codons (AAA and AAG) that are recognized by a tRNA^{Lys} with the anticodon UUU (**Figure 10.25a**). GC mutational pressure may change many AAA codons into AAG without affecting amino acid sequences. In small genomes, AAA

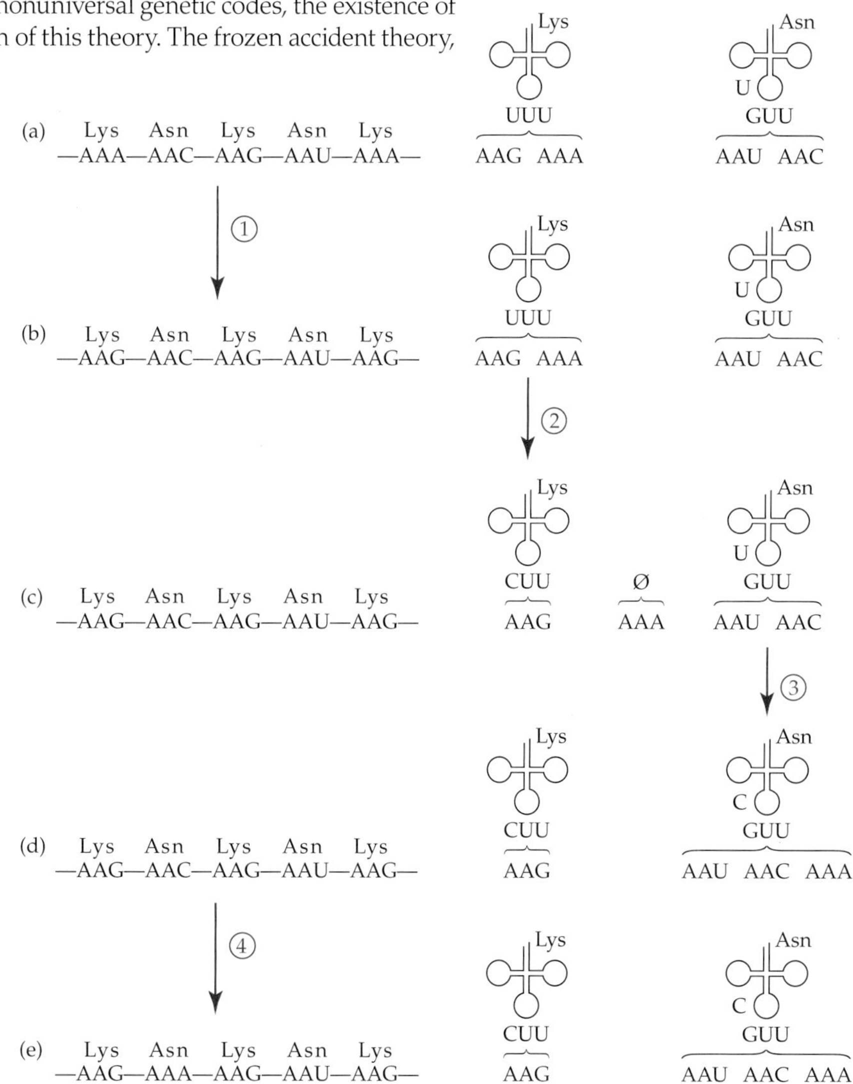

Figure 10.25 A four-step evolutionary scenario for the reassignment of codon AAA from lysine (as in the standard genetic code) to asparagine (as in the mitochondrial genetic code of echinoderms) by the codon-capture hypothesis. (a) In the standard genetic code, lysine is coded by two codons (AAA and AAG) that are recognized by tRNA^{Lys} with anticodon UUU (cloverleaf). Asparagine is coded by two codons (AAU and AAC) that are recognized by tRNA^{Asn} with anticodon UUU and a uridine (U) at position 33. (b) In step 1, codons AAA disappear from the genome. (c) A mutation changes anticodon UUU into CUU, thereby restricting its recognition to AAG only. AAA becomes an unassigned codon (Ø). (d) A mutation changes position 33 in tRNA^{Asn} from U to C, thereby enabling it to recognize AAA as well. (e) An AAA codon reappears in the genome and is translated into asparagine. The situation in (c) represents the mitochondrial genome of hemichordates. The situation in (e) represents the mitochondrial genome of echinoderms. Note that the amino acid sequence remained unchanged throughout this entire evolutionary scenario. (Modified from Castresana et al. 1998.)

may disappear altogether (**Figure 10.25b**). Anticodon UUU does not pair strongly with AAG, so a mutation in the $tRNA^{Lys}$-specifying gene that changes its anticodon to CUU will be advantageous (**Figure 10.25c**). Anticodon GUU of $tRNA^{Asn}$ is used in the standard genetic code to pair with codons AAU and AAC (asparagine). According to the wobble rules (Chapter 1; Table 1.5), $tRNA^{Asn}$ should have also recognized codon AAA. However, a uridine (U) upstream of the anticodon at position 33 restricts the codon recognition of $tRNA^{Asn}$ to AAU and AAC only. In the absence of AAA codons, however, there will be no selective constraint on position 33, and it may change by chance to C. This change would allow $tRNA^{Asn}$ to recognize AAA codons, should such codons reappear in the genome (**Figure 10.25d**). If an AAA codon reappears in the genome, it will be translated into asparagine (**Figure 10.25e**).

The codon-capture hypothesis, which assumes a stage in which a codon entirely disappears from the genome, works well for small genomes. However, this model cannot explain codon reassignments in large genomes, such as nuclear genomes. In the **ambiguous intermediate mechanism** (Schultz and Yarus 1994, 1996), a codon does not need to disappear in order to be reassigned. Rather, a transient period is postulated during which a codon is ambiguously translated into two distinct amino acids. In other words, a codon that was previously recognized by one tRNA acquires an affinity for another tRNA. Subsequently the first tRNA is lost and the codon is recognized exclusively by the second tRNA.

A third mechanism, the **unassigned-codon mechanism**, was proposed by Sengupta and Higgs (2005). Under this scenario, a transient period is postulated during which no tRNA can translate the codon, i.e., the codon is unassigned. The codon becomes reassigned when a new tRNA starts recognizing this codon. We note that truly unassigned codons, i.e., codons unrecognized by any tRNA, should be lethal to the organism and, hence, unlikely to persist in the population even for short periods of time. However, several cases are known where a different tRNA was able to translate such a codon—albeit inefficiently—after the tRNA-specifying gene that was specific to that codon had been deleted from the genome (Yokobori et al. 2001). Deletion of tRNA-specifying genes appears to be frequent in mitochondrial genomes, and many such events are thought to instigate codon reassignments (Sengupta et al. 2007).

An interesting subcategory of codon reassignment that mixes elements from the unassigned-codon mechanism and the ambiguous intermediate mechanism was created in the lab by Mukai et al. (2010). It involves the elimination of a release factor that recognizes the amber termination codon, UAG. In the *E. coli* genome, UAG is found at the ends of about 300 open reading frames. Release factor 1, encoded by the *prfA* gene, is the only molecule that recognizes UAG and terminates protein synthesis. Amber suppressor tRNAs, however, occur naturally in *E. coli* and may occasionally translate UAG to amino acids. As a consequence of amber suppression, UAG is recognized ambiguously, as a sense codon and a stop signal. The reassignment of UAG from a stop codon to a sense codon should have been, in principle, a one-step process, i.e., the elimination of *prfA*. Unfortunately, such a deletion is known to be lethal. For the *prfA* deletion not to be lethal, two conditions must be met: (1) the cell should contain suppressor tRNAs that recognize UAG, such as $tRNA^{Gln}$ or $tRNA^{Tyr}$; and (2) seven instances of UAG out of the 300 or so in the genome should be mutated to a different codon. These results are unexpected, since reassignment of a stop codon results in aberrant proteins with long peptide tails. Surprisingly, unnecessary peptide tails in many proteins do not seem to impede cell growth and reproduction. More generally, however, these findings indicate that codon reassignment does not always necessitate the disappearance of all affected codons or even a majority of them.

In **Figure 10.26**, we illustrate the sites on a tRNA molecule that affect its specific bindings to the anticodon and the aminoacyl-tRNA synthetase. The reassignment of a codon can, in principle, be effectuated either by changes in the pairing between the codon and the anticodon or by changes in the aminoacyl-tRNA synthetase such that it recognizes a different tRNA than its original cognate. Changes in the tRNA/aminoacyl-tRNA synthetase recognition can occur through either changes in the enzyme or

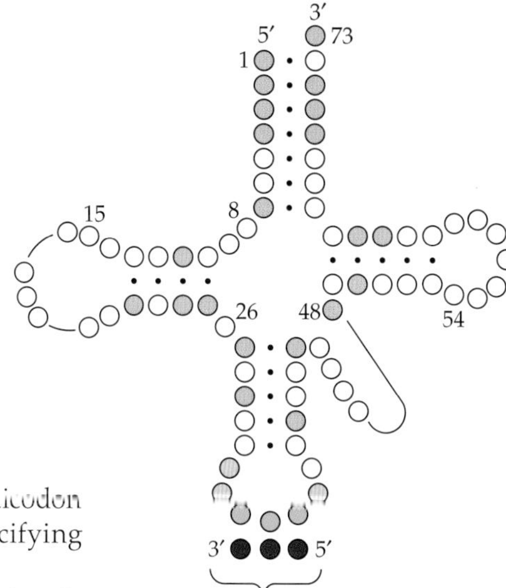

Figure 10.26 The translation of a codon (red) into an amino acid is determined by recognition of the codon (red) by a specific anticodon (green), as well as by aminoacyl-tRNA synthetase (not shown) recognizing its correct cognate tRNA. In the cloverleaf depiction of the tRNA molecule, specific nucleotide sites (circles) that determine the specificity of the recognitions between codon and anticodon and between the aminoacyl-tRNA synthetase and its cognate tRNA are shown. In addition to the anticodon (green), the aminoacyl-tRNA synthetase uses several nucleotide positions (blue) as principal signifiers for recognizing its cognate tRNA. (Only those sites most strongly associated with aminoacyl-tRNA synthetase recognition are shown.) In addition to the nucleotides in the anticodon, the specificity of the binding of the anticodon to the codon may be determined by the nucleotides upstream and downstream of the anticodon (yellow).

changes in the tRNA. Changes in the binding between the codon and the anticodon can be caused by either changes in the primary sequence of the tRNA-specifying gene or posttranscriptional modifications of the tRNA molecule.

In the lab, codon reassignment has been frequently achieved through the alteration of the activity and specificity of aminoacyl-tRNA synthetases (e.g., Link and Tirrell 2005). So far, however, no alternative genetic code in nature has been found that can be attributed to changes in the amino acid sequence of an aminoacyl-tRNA synthetase; all alternative genetic codes in nature have been traced to changes in the tRNA.

Interestingly, in the majority of cases in which the codon reassignment was found to be due to changes in the codon/anticodon recognition, these changes could be attributed to posttranscriptional modifications rather than mutations in the anticodon (Knight et al. 2001). For example, squid and starfish mitochondria translate AGR, where R is any purine, as serine instead of arginine, as in the standard genetic code. The gene for the single tRNA that decodes the AGN block, where N is any nucleotide, has a GCT anticodon, which according to the wobble rules should have only paired with AGY, where Y is any pyrimidine. However, conversion of the first position of the anticodon from G to 7-methylguanosine allowed it to decode any codon that has AG in the first two codon positions and any nucleotide in the third codon position.

Codon reassignment can also occur as a result of changes in the tRNA that affect its recognition by aminoacyl-tRNA synthetases, in particular, changes in the acceptor stem of the tRNA. For instance, such a change is responsible for the translation of CUU and CUA codons into alanine in the mitochondrial genetic code of *Ashbya gossypii*, a filamentous plant pathogen related to the yeast genus *Saccharomyces*. In contrast, in the yeast mitochondria, CUU and CUA codons are translated into threonine and leucine, respectively, according to the standard genetic code (Ling et al. 2014). A phylogenetic analysis indicates that $tRNA^{Ala}_{UAG}$ in the *A. gossypii* mitochondria is derived from $tRNA^{Thr}_{UAG}$ in yeast. The fact that the former is recognized by alanyl-tRNA synthetase, while the latter is recognized by threonyl-tRNA synthetase, is due to a difference in the acceptor stem. While in $tRNA^{Ala}_{UAG}$ *A. gossypii* has G and U at positions 3 and 70 in the acceptor stem, yeast $tRNA^{Thr}_{UAG}$ has A and U.

Alternative genetic codes have only been found in small genomes that encode small numbers of proteins. This is understandable given the potential of codon reassignment to affect the entire proteome. Originally, it was thought that extreme GC contents are necessary for the establishment of alternative genetic codes; however, the finding of alternative codes in small genomes of intermediate GC composition (e.g., the α-proteobacterial symbiont *Hodgkinia cicadicola*, whose GC content is 58%) indicates that base composition may not be as important as genome size as a driver of codon reassignment (McCutcheon et al. 2009).

Noncanonical genetic codes have been found to have a patchy phylogenetic distribution (e.g., Cocquyt et al. 2010). Thus, codon reassignment may not be as rare as intuitively thought. Indeed, stop codon reassignment seems to be a common event in environmental samples of eubacteria, eukaryotes, and viruses, but so far absent in archaebacteria (Ivanova et al. 2014).

What are the consequences of codon reassignment on protein composition? Massey et al. (2003) compared the genome of *Saccharomyces cerevisiae*, which uses the standard genetic code, with that of *Candida albicans*, a closely related yeast in which the CTG codon has been reassigned from leucine to serine. They showed that coding positions occupied by CTG in *C. albicans* align with positions that contain a serine codon in *S. cerevisiae*. Similarly, positions occupied by CTG in *S. cerevisiae* align with positions occupied by leucine codons in *C. albicans*. In other words, the codon reassignment seems to have had almost no effect on the protein sequences.

Codon reassignments and alternative genetic codes can be used to test the relationship between number of codons and amino acid frequencies. King and Jukes (1969) showed that the frequency of amino acids in proteins correlates with the number of synonymous codons that code for each amino acid. They proposed that this correlation occurs because most amino acids in proteins are the result of neutral replacements that do not affect protein structure or function. No correlation between number of codons and amino acid frequencies would be expected if amino acid composition is determined by selection. Gilis et al. (2001) suggested an alternative explanation, hypothesizing that the genetic code evolved to take into account preexisting amino acid frequencies. For example, arginine, leucine, and serine are each coded by six codons and are common amino acids in proteins. According to King and Jukes' hypothesis, they are common in proteins because most amino acids arise by random mutations in the genome, and so the number of codons for an amino acid should determine its frequency. If Gilis et al.'s hypothesis is correct, however, then these three amino acids have six codons each because all three were common when the code was established, and their frequency acted as an evolutionary pressure determining their number of synonymous codons.

Ring and Cavalcanti (2008) tested these theories by taking advantage of the fact that glutamine is coded by four codons in the ciliates *Tetrahymena thermophila* and *Paramecium aurelia*, but by only two codons in organisms that use the standard genetic code. If the number of codons impacts the frequency with which an organism uses a given amino acid through neutral processes, glutamine should be more common in *Tetrahymena* and *Paramecium* than in organisms that use the standard code. Indeed, the frequencies of glutamine residues in *Tetrahymena* and *Paramecium* were 9.8% and 9.1%, respectively, whereas in eukaryotes that use the standard genetic code the frequencies of glutamine residues ranged from 2.8% to 5.3%. When the GC content of the genomes was taken into account, the observed frequencies of glutamine were not significantly different from the expected frequencies under no selective constraint. These results suggest that the number of codons coding for an amino acid and the GC content determine the frequency of amino acids in proteins, supporting the King and Jukes (1969) hypothesis. Moreover, these results indicate that only a negligible fraction of amino acid sites within the entire proteome is subject to selection.

CHAPTER 11

Eukaryotic Genome Evolution

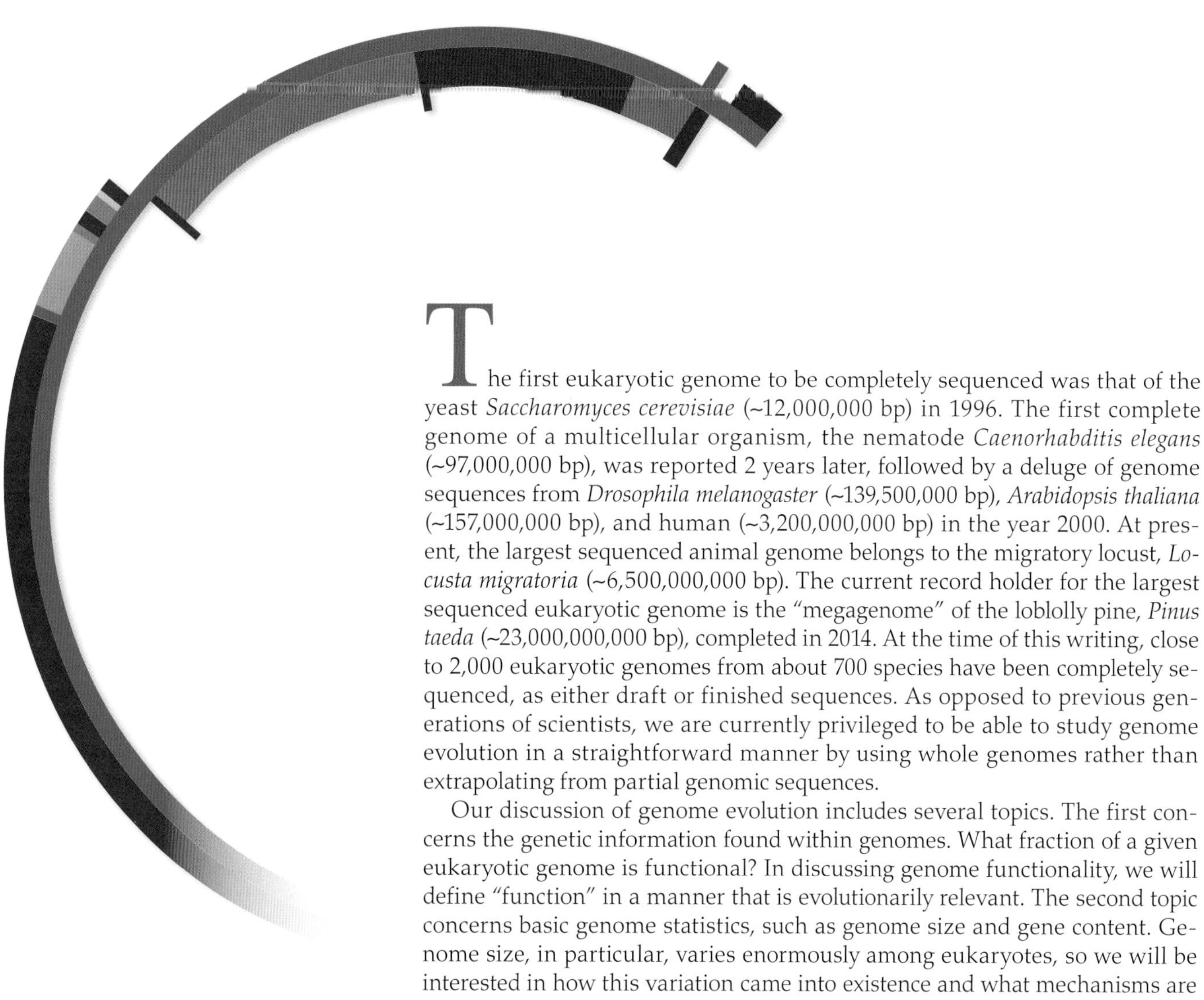

The first eukaryotic genome to be completely sequenced was that of the yeast *Saccharomyces cerevisiae* (~12,000,000 bp) in 1996. The first complete genome of a multicellular organism, the nematode *Caenorhabditis elegans* (~97,000,000 bp), was reported 2 years later, followed by a deluge of genome sequences from *Drosophila melanogaster* (~139,500,000 bp), *Arabidopsis thaliana* (~157,000,000 bp), and human (~3,200,000,000 bp) in the year 2000. At present, the largest sequenced animal genome belongs to the migratory locust, *Locusta migratoria* (~6,500,000,000 bp). The current record holder for the largest sequenced eukaryotic genome is the "megagenome" of the loblolly pine, *Pinus taeda* (~23,000,000,000 bp), completed in 2014. At the time of this writing, close to 2,000 eukaryotic genomes from about 700 species have been completely sequenced, as either draft or finished sequences. As opposed to previous generations of scientists, we are currently privileged to be able to study genome evolution in a straightforward manner by using whole genomes rather than extrapolating from partial genomic sequences.

Our discussion of genome evolution includes several topics. The first concerns the genetic information found within genomes. What fraction of a given eukaryotic genome is functional? In discussing genome functionality, we will define "function" in a manner that is evolutionarily relevant. The second topic concerns basic genome statistics, such as genome size and gene content. Genome size, in particular, varies enormously among eukaryotes, so we will be interested in how this variation came into existence and what mechanisms are responsible for producing such variation. The third topic to be dealt with in this chapter concerns gene order and the dynamics of evolutionary change in gene order. How are genes distributed along and among the chromosomes? What mechanisms are responsible for the reshaping of gene order during evolution? We will also be interested in the fraction of the genome made of repetitive sequences, their origin, and the pattern of their chromosomal distribution. The fourth topic concerns the nucleotide composition of genomes. Is there heterogeneity in composition among different regions of the genome? What mechanisms can give rise to localized differences in nucleotide composition?

Functionality and Nonfunctionality in Eukaryotic Genomes

Because genomes are products of evolution rather than "intelligent design," all genomes contain functional and nonfunctional parts. Among people unversed in evolutionary biology, a misconception exists that natural processes can produce a genome that is wholly functional. In fact, evolution can produce such a genome if and only if (1) effective population size is enormous—infinite, to be precise; (2) the deleterious effects of increasing genome size by even a single nucleotide are considerable; and (3) generation time is short. Not even in the commonest of bacterial species populations are these conditions met, let alone in species with very small effective population sizes (such as humans) or extremely long life spans (such as the palm *Eutrepe globoso*, whose generation time is ~100 years; Hempe and Petit 2006). A genome that is 100% functional is a logical impossibility. Thus, in discussing genome evolution—in particular, in attempting to estimate the functional and functionless fractions of a genome—it is first necessary to provide a rigorous definition of "function."

What does "function" mean in the context of evolution?

Like many words in the English language, the word "function" has numerous meanings. What meaning, then, should we use in an evolutionary context? In biology, there are two main concepts of function: the selected-effect and causal-role functions.

The **selected-effect function**, also referred to as the **proper biological function**, is a historical concept. In other words, it explains the origin, the cause (etiology), and the subsequent evolution of the trait (Millikan 1989; Neander 1991). Accordingly, for a trait, *T*, to have a selected-effect function, *F*, it is necessary and (almost) sufficient that the following two conditions hold: (1) *T* originated as a "reproduction" (a copy or a copy of a copy) of some prior trait that performed *F* (or some function similar to *F*, say *F'*) in the past, and (2) *T* exists because of *F* (Millikan 1989). In other words, the selected-effect function of a trait is the effect for which the trait was selected and/or by which it is maintained. The selected-effect function answers the question, Why does *T* exist? (The parenthetical "almost" was added to the definition above to account for extremely recent and extremely rare de novo functions.)

The **causal-role function** is an ahistorical and nonevolutionary concept (Cummins 1975; Amundson and Lauder 1994). That is, for a trait, *Q*, to have a causal-role function, *G*, it is necessary and sufficient that *Q* performs *G*. The causal-role function answers the question, What does *Q* do?

To illustrate the difference between these two concepts of function, let us consider an example (modified from Griffiths 2009). There are two identical or almost identical sequences in the genome. The first, TATAAA, has been maintained at a certain genomic position by natural selection for the purpose of binding a transcription factor; hence, its selected-effect function is to bind this transcription factor. A second sequence has arisen by mutation and, purely by chance, has come to resemble the first sequence; therefore, it also binds the transcription factor. However, the transcription factor binding to the second sequence has neither an adaptive nor a maladaptive consequence. Thus, the second sequence has no selected-effect function, but its causal-role function is to bind a transcription factor.

Using the causal-role concept of function in the biological sciences can lead to bizarre outcomes. For example, while the selected-effect function of the heart can be stated unambiguously to be the pumping of blood, the heart may be assigned many additional causal-role functions, such as adding 300 grams to body weight, producing sounds, preventing the pericardium from deflating onto itself, and providing an inspiration for love songs and Hallmark cards (Graur et al. 2013). The thumping noise made by the heart is a favorite of philosophers of science; it is a valuable aid in medical diagnosis, but it is not the evolutionary reason we have a heart. An even greater absurdity of using the causal-role concept of function arises when we realize that every nucleotide in a genome has a causal role—it is replicated! Does that mean that every nucleotide in the genome has evolved for the purpose of being replicated?

Distinguishing what a genomic element does (its causal-role activity) from why it exists (its selected-effect function) is very important in biology (Huneman 2013; Brunet and Doolittle 2014). Ignoring this distinction, and assuming that all genomic sites that exhibit a certain biochemical activity are functional, as was done by the ENCODE Project Consortium (2012) and Sundaram et al. (2014), is equivalent to claiming that following a collision between a car and a pedestrian, a car's hood would be ascribed the "function" of harming the pedestrian, while the pedestrian would have the "function" of denting the car's hood (Hurst 2012).

Most biologists follow the Dobzhanskyan dictum, according to which biological sense can only be derived from evolutionary context. Hence, with few exceptions, they use the selected-effect concept of function. We note, however, that the causal-role concept may sometimes be useful, for example, as an ad hoc device for traits whose evolutionary history and underlying biology are obscure. Furthermore, we note that all selected-effect functions have a causal role, while the vast majority of causal-role functions do not have a selected-effect function. It is, thus, wrong to assume that all causal-role functions are biologically relevant. Doolittle et al. (2014), for instance, prefers to restrict the term "function" to selected-effect function, and to refer to causal-role functions as "activities" or "effects."

The main advantage of the selected-effect definition of function is that it suggests a clear and conservative method for inferring function in a DNA sequence—only sequences that can be shown to be under selection can be claimed with any degree of confidence to be functional. The selected-effect definition of function has led not only to the discovery of many new functional genes, for example, microRNA-specifying genes (Lee et al. 1993) or functional "pseudogenes" (Svensson et al. 2006), but also to the rejection of hypotheses concerning putative functions for such genomic elements as *numt*s (Hazkani-Covo et al. 2010).

From an evolutionary viewpoint, a function can be assigned to a DNA sequence if and only if it is possible to destroy it (Graur et al. 2013). All functional entities in the universe can be rendered nonfunctional by the ravages of time, entropy, mutation, and what have you. Unless a genomic functionality is actively protected by selection, it will accumulate deleterious mutations and will cease to be functional. The absurd alternative is to assume that function can be assessed independently of selection, i.e., that no deleterious mutation can ever occur in the region that is deemed to be functional. Such an assumption is akin to claiming that a television set left on and unattended will still be in working condition after a million years because no natural events, such as rust, erosion, static electricity, or the gnawing activity of rodents can affect it (Graur et al. 2103). A convoluted "rationale" for discarding natural selection as the arbiter of functionality was put forward by Stamatoyannopoulos (2012). This paper should be read as a cautionary tale of how biology can be corrupted by the uncritical use of the causal-role concept of function.

Function should always be defined in the present tense. In the absence of prophetic powers, one cannot use the potential for creating a new function as the basis for claiming that a certain genomic element is functional. For example, the fact that a handful of transposable elements have been co-opted into function (Chapter 9) cannot be taken as support for the hypothesis that all transposable elements are functional. In this respect, the Aristotelian difference between "potentiality" and "actuality" is crucial.

Finally, we should discuss the proper manner in which null hypotheses concerning the functionality or nonfunctionality of a particular genomic element should be phrased. Most science practitioners adhere to the Popperian system of demarcation, according to which scientific progress is achieved through the falsification of hypotheses that do not withstand logical or empirical tests. Thus, a null hypothesis should be phrased in such a manner as to spell out the conditions for its own refutation.

Should one assume lack of functionality as the null hypothesis, or should one assume functionality? Let us consider both cases. A statement to the effect that a genomic element is devoid of a selected-effect function can be easily rejected by showing that the element evolves in a manner that is inconsistent with neutrality. If, on the other hand, one assumes as the null hypothesis that an element is functional, then failing to find telltale indicators of selection cannot be interpreted as a rejection of the hypothesis, but merely as a sign that we have not searched thoroughly enough or that the telltale signs of selection have been erased by subsequent evolutionary events. There exists a fundamental asymmetry between verifiability and falsifiability in science: scientific hypotheses can never be proven right; they can only be proven wrong. The hypothesis that a certain genomic element is functional can never be rejected and is, hence, unscientific. According to physicist Wolfgang Pauli (quoted in Peierls 1960), a hypothesis that cannot be refuted "is not only not right, it is not even wrong."

What do genomes do? An evolutionary classification of genomic function

Genomic sequences are frequently categorized according to biochemical activity, regardless of whether or not such activity is biologically meaningful. The need for a rigorous evolutionary classification of genomic elements by selected-effect function arises from two erroneous and sometimes deliberately disingenuous equivalencies that are frequently found in the literature. The first equivalency, usually espoused in the medical and popular literature (e.g., Krams and Bromberg 2013; Mehta et al. 2013), misleadingly uses "noncoding DNA"—i.e., all regions in the genome that do not encode proteins—synonymously with "junk DNA"—i.e., all regions in the genome that are neither functional nor deleterious. The second, even more pernicious equivalency transmutes every biochemical activity into a function (e.g., ENCODE Project Consortium 2012; Kellis et al. 2014; Sundaram et al. 2014).

The classification scheme presented here is based on Graur et al. (2015). It starts with the premise that all genomes are the products of natural evolutionary processes, rather than intelligent design and, hence, that they consist of functional and nonfunctional parts, where function is understood as selected-effect function.

We first divide the genome into **functional DNA** and **rubbish DNA** (**Figure 11.1**). Functional DNA refers to any segment in the genome whose selected-effect function is that for which it was selected and/or by which it is maintained. Most functional sequences in the genome are maintained by purifying selection. Less frequently, functional sequences exhibit telltale signs of either positive or balancing selection. A causal-role activity, such as "low-level noncoding RNA transcription" (e.g., Kellis et al. 2014), is an insufficient attribute of functionality.

Functional DNA is further divided into **literal DNA** and **indifferent DNA**. In literal DNA, the order of nucleotides is under selection. Strictly, a DNA element of length l is defined as literal DNA if its function can be performed by a very small subset of the 4^l possible sequences of length l. For example, there are three possible sequences of length 3 that can encode isoleucine according to the standard genetic code, as opposed to the much larger number, $3^4 = 64$, of possible three-nucleotide sequences. Functional protein-coding genes, RNA-specifying genes, and untranscribed control elements are included within this category.

Indifferent DNA includes genomic segments that are functional and needed, but the order of nucleotides in their sequences is of little consequence. These sequences serve as spacers or fillers; they can act as protectors against frameshifts or have nucleotypic functions such as determining nucleus size (p. 519).

The third position in fourfold degenerate codons may be regarded as a simple example of indifferent DNA; the nucleotide that resides at this position is unimportant, but the

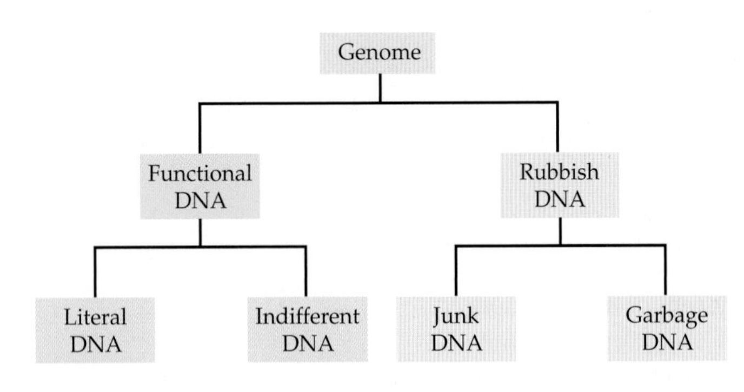

Figure 11.1 An evolutionary classification of genomic elements according to their selected-effect function.

position itself needs to be occupied. Thus, indifferent DNA should show no evidence of selection for or against point mutations, but deletions and insertions should be under selection. For example, Nóbrega et al. (2004) deleted 2,356 Kb from the mouse genome, yet mice homozygous for the deletions were indistinguishable from wild-type littermates with regard to morphology, reproductive fitness, growth, longevity, and a variety of parameters assaying general homeostasis. Thus, these sequences should be considered junk DNA rather than indifferent DNA.

Rubbish DNA (Brenner 1998) refers to genomic segments that have no selected-effect function. Rubbish DNA can be further subdivided into **junk DNA** and **garbage DNA**. We have written evidence that the term "junk DNA" was already in use in the early 1960s (e.g., Aronson et al. 1960; Ehret and de Haller 1963); however, it was Susumu Ohno (1972, 1973) who formalized its meaning and provided an evolutionary rationale for its existence. "Junk DNA" refers to a genomic segment on which selection does not operate, and hence it evolves neutrally. Of course, some junk DNA may acquire a useful function in the future, although such an even is expected to occur only very rarely. Thus, the "junk" in "junk DNA" is identical in its meaning to the colloquial "junk," such as when a person mentions a "garage full of junk," in which the implications are that (1) the garage is still serviceable despite being cluttered with useless objects, and (2) some of the useless objects may become useful in the future. Of course, like the junk in the garage, the majority of junk DNA will never acquire a function. Junk DNA and the junk in one's garage are also similar in that "they may be kept for years and years and, then, thrown out a day before becoming useful" (David Wool, personal communication).

The term "junk DNA" has generated a lot of controversy. First, because of linguistic prudery and the fact that "junk" is used euphemistically in off-color contexts, some biologists find the term "junk DNA" to be "derogatory" and "disrespectful" (e.g., Brosius and Gould 1992). An additional opposition to the term "junk DNA" stems from false teleological reasoning. Many researchers (e.g., Makalowski 2003; Wen et al. 2012) use the term "junk DNA" to denote a piece of DNA that can never, under any evolutionary circumstance, be selected for or against. Since every piece of DNA may become advantageous or deleterious by gain-of-function mutations, this type of reasoning is faulty. A piece of junk DNA may indeed be co-opted into function, but that does not mean that it will be, let alone that it currently has a function. Finally, some opposition to the term is related to the antiscientific practice of assuming functionality as the null hypothesis (Petsko 2003).

"Garbage DNA" refers to sequences that exist in the genome despite being actively selected against. The reason that detrimental sequences are observable is that selection is neither omnipotent nor rapid. At any slice of evolutionary time, segments of garbage DNA (presumably on their way to becoming extinct) may be found in the genome. The distinction between junk DNA and garbage DNA was suggested by Brenner (1998):

> *Some years ago I noticed that there are two kinds of rubbish in the world and that most languages have different words to distinguish them. There is the rubbish we keep, which is junk, and the rubbish we throw away, which is garbage. The excess DNA in our genomes is junk, and it is there because it is harmless, as well as being useless, and because the molecular processes generating extra DNA outpace those getting rid of it. Were the extra DNA to become disadvantageous, it would become subject to selection, just as junk that takes up too much space, or is beginning to smell, is instantly converted to garbage by one's wife, that excellent Darwinian instrument.*

Each of the four functional categories described above can be (1) transcribed and translated, (2) transcribed but not translated, or (3) not transcribed. Hence, we may encounter, for instance, junk DNA, literal RNA, and garbage proteins.

Later in this chapter, we will discuss the relative amounts of literal, indifferent, junk, and garbage DNA in different genomes.

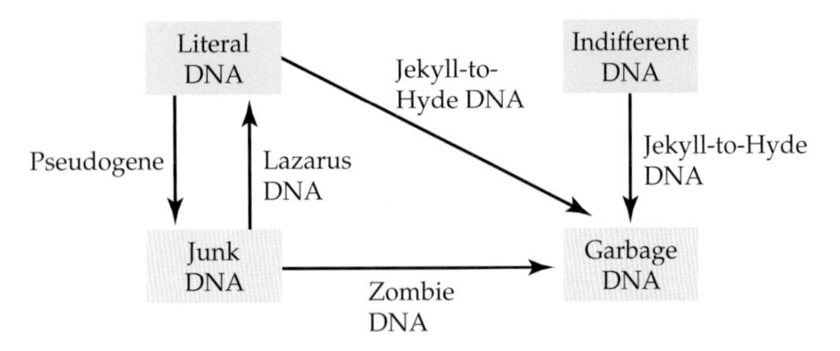

Figure 11.2 Some possible changes in functional affiliation of genomic elements.

Changes in functional affiliation

The affiliation of a DNA segment to a particular functional category may change during evolution. With four functional categories, there are twelve possible affiliation changes. Several such changes are known to occur quite frequently (**Figure 11.2**). For example, junk DNA may become garbage DNA if the effective population size increases; the opposite will occur if the effective population size decreases (Ohta 1973). Many of the 12 possible changes have been documented in the literature. Pseudogenes, for instance, represent a change in functional status from literal DNA to junk DNA, while some diseases are caused by either a change from functional DNA to garbage DNA (e.g., Chen 2003) or from junk DNA to garbage DNA (Cho and Brant 2011).

Rubbish DNA mutating to functional DNA may be referred to as **Lazarus DNA**, named after Lazarus of Bethany, the second most famous resurrected corpse in fiction (John 11:38-44; 12:1; 12:9; 12:17). Similarly, functional DNA may mutate to garbage DNA, in which case the term **Jekyll-to-Hyde DNA** has been suggested—the term being derived from Robert Louis Stevenson's 1886 novel on the transformation of a benevolent entity into a malicious one. Garbage DNA may also be derived from junk DNA, for which the name **zombie DNA** was suggested by Kolata (2010).

Detecting Functionality at the Genome Level

The availability of intraspecific and interspecific genomic sequences has made it possible not only to test whether or not a certain genomic region is subject to selection, but also to exhaustively scan the genome for regions likely to have been the target of selection and that are, hence, of functional importance. For a given species, such scans allow us to estimate the proportion of the genome that is functional. In the literature, numerous approaches for detecting selection through comparisons of DNA sequences have been proposed (e.g., Nielsen 2005; Andrés et al. 2009; Li 2011; Grossman et al. 2013; Vitti et al. 2013; Lawrie and Petrov 2014). A few approaches for detecting selection from sequence comparisons have been described in previous chapters of this book. The main difference between these tests and tests involving genome comparisons is that the latter involve multiple tests at multiple loci. Thus, many of the sites that are identified through such methods as having been subjected to selection are false positives. The statistics must therefore be adjusted for the number of tests performed, using standard techniques for multiple comparisons and adjusting significance levels to account for false discovery rates (e.g., Massingham and Goldman 2005).

Some methods for detecting selection at the genomic level require comparisons among species, some rely on intraspecific comparisons, and yet others require both types of data. Some methods are applicable to protein-coding genes only; some are applicable to all sequences. Some are based on comparisons of allele frequencies, some are based on linkage disequilibrium measures, and some rely on population-differentiation measures, such as genetic distances. Some are suitable for detecting purifying selection, and some are suitable for detecting positive or balancing selection. A straightforward method of estimating the **functional fraction** of a genome is to add up the genomic fractions that are under (1) positive, (2) negative, and (3) balancing selection.

Detecting functional regions subject to purifying selection is relatively straightforward. In interspecific comparisons, homologous genomic regions under purifying selection are expected to be more similar to one another than unselected regions. The rationale for this expectation is that many mutations in functional sequences are deleterious and, hence, weeded out of the population. Thus, purifying selection is primarily observable as highly conserved regions. We note, however, that evolutionary processes other than selection, such as mutational coldspots and gene conversion, can result in sequence conservation (Ahituv et al. 2007).

Balancing selection at the level of the genome is one of the least studied areas in evolutionary genomics. This lack of interest is somewhat unexpected, given that a strong association between balancing selection and pathology has been hypothesized for a large number of human diseases, such as sickle-cell anemia, cystic fibrosis, and phenylketonuria. Andrés et al. (2009) devised a method for identifying regions that concomitantly show excessive levels of nucleotide polymorphism (as measured by the number of polymorphic sites in a gene), as well as an excess of alleles at intermediate allele frequencies.

All in all, the main emphasis in the last two decades has been on methods intended to detect positive selection (**Table 11.1**). The principal reason for the emphasis

TABLE 11.1

Some approaches for detecting positive selection

Type of data and approach used	Assumptions, rationale, caveats, and references
Interspecific data	
Ratio of nonsynonymous to synonymous substitution (K_A/K_S) test	Synonymous mutations are assumed to be selectively neutral. Nonsynonymous mutations that reach fixation are assumed to be advantageous. If the rate of nonsynonymous substitution (K_A) to synonymous substitution (K_S) is significantly larger than 1, positive selection is inferred. The test is only applicable to protein-coding regions.
Identification of accelerated regions	Regions that undergo accelerated change in one lineage but are conserved in related lineages are candidates for positive selection. Caution should be exercised, however, as loss of function or accelerated mutation rates also lead to acceleration of evolutionary rates.
Inter- and intraspecific data	
McDonald-Kreitman test	Synonymous mutations are assumed to be neutral. Only nonsynonymous mutations may be adaptive. Selectively advantageous mutations will be fixed in the population much more rapidly than neutral mutations and, hence, are less likely to be found in a polymorphic state. Under neutrality, the expectation is that the ratio of fixed nonsynonymous differences to fixed synonymous differences will be the same as the ratio of nonsynonymous polymorphisms to synonymous polymorphisms. A significant difference between the two ratios can therefore be used to reject the neutral mutation hypothesis. The test is only applicable to protein-coding regions.
Hudson-Kreitman-Aguadé test	Levels of polymorphism and divergence should be correlated because both are primarily functions of the mutation rate, unless selection causes one to exceed the other.
Intraspecific data	
Frequency-based methods	In a selective sweep, a genetic variant reaches high prevalence together with nearby linked variants (high-frequency derived alleles). From this homogenous background, new alleles arise but are initially at low frequency (surplus of rare alleles).
Linkage disequilibrium-based methods	Selective sweeps bring a genetic region to high prevalence in a population, including the causal variant and its neighbors. The associations between these alleles define a haplotype, which persists in the population until recombination breaks these associations down.
Population differentiation-based methods	Selection acting on an allele in one population but not in another creates a marked difference in the frequencies of that allele between the two populations. This effect of differentiation stands out against the differentiation between populations with respect to neutral (i.e., nonselected) alleles.

Source: Modified from Vitti et al. (2013), in which references to the various methods may be found.

on positive selection at the expense of purifying and balancing selection is that positive Darwinian selection is presupposed to be the primary mechanism of adaptation.

Detecting genomic regions that have experienced positive selection requires more nuanced procedures and significantly more data than are required by methods for detecting purifying selection. Moreover, the various methods for detecting positive selection at the level of the genome are known to yield many false positives. Although each method has its own particular strengths and limitations, there are several challenges that are shared among all tests. First, deviations from neutrality expectations may be explained by factors other than selection. Demographic events, such as migration, and population expansions and bottlenecks, can often yield signals that mimic selection. This recognition has led some researchers to adopt approaches that explicitly attempt to separate demographic effects from selection effects (Li and Stephan 2006; Excoffier et al. 2009). Second, even when confounding effects are dealt with, the interpretation of selection may not be straightforward. For example, rate-based tests identify as "functional" all regions in which evolutionary rates have been accelerated. Such regions may indeed be subject to positive selection, but the acceleration may also be due to the relaxation of selective constraint due to total or partial nonfunctionalization or to an increase in the rate of mutation. Distinguishing among these possibilities requires a case-by-case analysis—a proposition that is antithetic to the ethos of Big Science genomics and bioinformatics.

Because positive selection leaves a number of footprints on the genome, and each test is designed to pick up on a slightly different signal, researchers sometimes combine multiple metrics into composite tests toward the goal of providing greater power of detection and a finer spatial resolution. Scores of such tests are typically referred to as **composite scores**. Tests employing composite scores come in two distinct forms. First, some methods calculate a composite score for a contiguous genetic region rather than a single nucleotide site by combining individual scores at all the sites within the region. The motivation for such an approach is that, although false positives may occur at any one nucleotide site by chance, a contiguous region of positive markers is unlikely to be spurious and most likely represents a bona fide signature of selection (e.g., Carlson et al. 2005). In the other type of method, composite scores are calculated by combining the results of many tests at a single site. The purpose of these methods is to utilize complementary information from different tests in order to provide better spatial resolution and pinpoint selection to the root cause (e.g., Zeng et al. 2006; Grossman et al. 2010).

It is very important to note that regardless of method or combination of methods, there are both factors that conspire to underestimate the functional fraction of the genome and factors that conspire to overestimate the functional fraction of the genome. That is, some of the genomic segments identified as functional through telltale signs of positive selections may be false positives, while others may elude detection. For example, functional sequences may be under selection regimes that are difficult to detect, such as positive selection or weak purifying selection. In addition, selection may be difficult to detect as far as recently evolved species-specific elements (e.g., Smith et al. 2004) or very short genetic elements are concerned (e.g., De Gobbi et al. 2006). These factors would cause the fraction of the genome that is under selection and, hence, functional to be underestimated. On the other hand, selective sweeps, background selection, and significant reductions in population size (bottlenecks) would cause an overestimation of the fraction of the genome that is under selection (e.g., Williamson et al. 2007).

THE DOMESTICATION SYNDROME AND SIGNATURES OF POSITIVE SELECTION IN THE GENOME OF THE DOMESTIC CAT Charles Darwin (1868) was the first to notice that domesticated species not only are tamer than their wild forebears, but also tend to display a suite of behavioral, physiological, and morphological characteristic features, including piebald coat color (large depigmented areas), floppy and reduced ears, shortened muzzles, small teeth, small cranial capacity, shortened tails, fewer vertebrae, more frequent estrous cycles, curly tails, wavy hair, and the persistence of juvenile behavior into adulthood (Trut 1999). This combination of traits is referred to as the **domestication syndrome** (Wilkins et al. 2014). What do these traits have in common? How can one explain in a straightforward and parsimonious fashion the evolution of such a collection of seemingly disparate traits? Wilkins et al. (2014) began with the assumption that the primary selective pressure during the initial stages of domestication is on behavior, in particular tameness or the lack of fearful or aggressive responses to human caretakers. Such a reduction in acute fear and long-term stress is a prerequisite to successful breeding in captivity.

Wilkins et al. (2014) noticed that the diverse traits that characterize the domestication syndrome are all linked to neural crest cells, a class of stem cells that first appear during early embryogenesis at the dorsal edge of the neural tube and then migrate ventrally throughout the body, giving rise to the cellular precursors of many cell and tissue types and indirectly promoting the development of others. Tissues derived from neural crest cells include much of the skull, the sympathetic ganglia, the adrenal gland, pigment-related melanoblasts in both head and trunk, and tooth precursors. Cranial neural crest cells are also crucial precursors of bony, cartilaginous, and nervous components of the craniofacial region, including the jaws, hyoid, larynx, and external and middle ears. Wilkins et al. (2014) suggest that the initial selection for tameness led to mild neural crest cell deficits during embryonic development, which in turn produced as unselected by-products the morphological changes that are typical of the domestication syndrome. The authors listed three routes by which neural crest cell deficits could be produced: reduced numbers of original neural crest cells, lesser migratory capabilities of neural crest cells and consequently lower numbers at the final sites, and decreased proliferation of neural crest cells at the final sites. Because the characteristic domestication syndrome phenotypes are located in parts of the body that are relatively distant from the sites of origination of neural crest cells, such as the face, limb extremities, tail, and belly midline, the authors suspect that migration defects are particularly important. Slower neural crest migration, for example, results in decreased size and function of neural crest-derived organs, such as the adrenal gland, which is responsible for the fight-or-flight hormone response. As a result, domesticated animals have decreased fear and stress responses that allow them to live alongside humans.

To test this hypothesis, Montague et al. (2014) compared the genomes of 22 domestic cats (*Felis silvestris catus*) from six breeds with the genomes of four individuals belonging to either of two wildcat subspecies: the European wildcat (*F. s. silvestris*) and the African wildcat (*F. s. lybica*). Domestic cats, of which there may be 600 million worldwide, have been part of the human household for ~9,500 years, and their association with humans presumably started by their attraction to the rodent-infested grain stores of early farming societies. It is thought that cats are more self-domesticated or semidomesticated than domesticated, i.e., humans have not engaged in extensive selective breeding of cats as they have with dogs. Unlike the many domesticated mammals bred for food, herding, hunting, or security, most of the 40 or so cat breeds currently recognized by cat fancier associations originated extremely recently, within the past 150 years, and largely due to selection for aesthetic rather than functional traits.

Montague et al. (2014) discovered five chromosomal regions containing 13 protein-coding genes that exhibit convincing signatures of positive selection. Interestingly, each of these five regions contained at least one gene involved in neural crest cell migration and maintenance, in support of Wilkins et al.'s (2014) hypothesis.

Phenotypic validation of positive selection

Selection acts directly on the portion of the phenotypic variation that influences fitness, and only indirectly on DNA sequence variation. In contrast, inferences of positive selection are made directly from patterns of genetic variation and linkage disequilibrium without taking phenotypic data into account. To turn the tentative evidence concerning positive selection at the DNA level into a convincing adaptive scenario, we need to link genotypes to phenotypes (Akey 2009). Advances in genomic technology have made possible the identification of hundreds of candidate genetic variants with evidence of recent positive natural selection. Many of these candidate variants are false positives; others represent real adaptations. Unfortunately, many genes have unknown effects, making their adaptive advantages difficult to uncover.

In the following sections, we discuss two human loci that were found to have experienced recent positive selection. In one case, the connection to the phenotype was made through the study of a homologous gene in a distantly related organism; in the second, the effects of the candidate adaptive alleles were discovered by replacing the homologous allele in a laboratory animal with the selected human allele and then studying the phenotypic effects in vivo.

SKIN PIGMENTATION LIGHTENING IN EUROPEANS Body coloration in vertebrates is controlled by specialized pigment cells called melanocytes in birds and mammals and melanophores in other vertebrates. These cells produce an insoluble polymeric pigment called melanin. Melanin plays two very important roles: (1) the protection of DNA in skin cells from the ravages of ultraviolet radiation, and (2) the enhancement of visual acuity by controlling light scatter in the retina.

A protein-coding gene called *slc24a5,* which encodes a protein called solute carrier family 24 member 5, serves as an example of how phenotypic context, comparative biology, and the pattern of selection at a locus can facilitate an understanding of adaptation (Lamason et al. 2005). In zebrafish, *Danio rerio,* there exists a phenotype called *golden* that is characterized by lower than normal pigmentation of the skin and the retinal epithelium. Linkage analysis revealed that the *golden* phenotype was due to a C → A transversion in the *slc24a5* gene on chromosome 18, which caused a TAC codon (tyrosine) to turn into the TAA stop codon.

The homologous *slc24a5* gene in humans was found to be located on the long arm of chromosome 15. Looking at human population data from different geographic regions uncovered a strong signature of positive selection at this locus in European populations. Zebrafish data and evidence for positive selection in humans guided demonstration that a G → A nonsynonymous transition in codon 111 of the *slc24a5* gene leading to an alanine-to-threonine replacement was a major determinant of pigmentation levels. This single change in *slc24a5* accounts for 25–40% of the average difference in skin tone between Europeans and West Africans.

The allele frequency for the derived threonine variant ranges from 99% to 100% among populations of European descent, whereas the ancestral alanine allele has a frequency of 93–100% in African, indigenous American, and East Asian populations (Beleza et al. 2013a). The threonine variant is not only associated with light skin pigmentation in Europeans, but is also associated with lighter skin pigmentation among admixed African Americans and African Caribbeans.

According to Lamason et al. (2005), the threonine-encoding allele may have been the subject of the strongest positive selection ever observed in European populations. So strong, in fact, that the allele reached fixation in a very short time within the last 11,000–19,000 years (or less than 1,000 generations). It is theorized that selection for the derived allele was based on the need for sunlight to produce cholecalciferol (vitamin D). In northerly latitudes, where there is less sunlight, a greater requirement for body coverage due to the cold climate, and a diet that is frequently poor in vitamin D, lighter skin, which can generate more vitamin D, confers a large selective advantage.

IN VIVO VALIDATION OF POSITIVE SELECTION: THE *EDARV370A* ALLELE IN EAST ASIAN AND NATIVE AMERICAN HUMAN POPULATIONS In genome-wide scans, a derived allele of the ectodysplasin A receptor gene, *EDAR,* was found to exhibit strong signs of recent positive selection in East Asian and Native American populations. The protein encoded by the ancestral *EDAR* allele, which is common in African and European populations, has a valine at amino acid position 370. In the derived allele, a T → C nonsynonymous substitution resulted in a valine-to-alanine (V-to-A) replacement, which gave the allele its name, *EDARV370A*. The alanine-encoding allele originated in an area currently encompassing northeastern China and eastern Siberia approximately 35,000–38,000 years ago, and its frequency increased rapidly in East Asian populations due to strong positive selection. It is estimated that the *EDARV370A* allele conferred a 7–8% selective advantage on its carriers (Kamberov et al. 2013). This is one of the highest selective coefficients ever measured in human populations.

EDARV370A is found in very high frequencies in East Asians (e.g., 88% in Chinese and Japanese samples), whereas its frequencies in African, Middle Eastern, European, and Central and South Asian populations are close to zero. High *EDARV370A* allele frequencies are also found in the indigenous peoples of the Americas, who descended from northeastern Asians who had migrated to the New World via Beringia (the region surrounding the Bering Strait, the Chukchi Sea, and the Bering Sea) about 15,000 years ago. For example, the *EDARV370A* allele frequency in the Pima people of Arizona and Mexico and the Karitiana people of Brazil is 100%.

The Val-to-Ala replacement affects a region within the EDAR protein called the "death domain," which has been shown to interact with a protein called TNF-receptor-associated-factor 2. Not much was known about the function or phenotypic consequences of EDAR, except that the *EDARV370A* allele was thought to influence the thickness of head hair in a codominant fashion, with homozygotes for the *EDARV370A* allele having hair follicles with an average cross-sectional area that was slightly larger than those of heterozygotes, who in turn had hair follicles slightly larger that those of homozygotes for the ancestral allele (Fujimoto et al. 2008).

Kamberov et al. (2013) tested the biological consequences of the Val-to-Ala replacement in the human *EDARV370A* allele by introducing the same mutation in the mouse orthologous gene. They took advantage of the fact that while the human and mouse EDAR proteins differ from each other at 8% of their 448 amino acids, their 74-amino-acid death domains are identical (**Figure 11.3**). By using homologous recombination

```
Human    MAHVGDCTQTPWLPVLVVSLMCSARAEYSNCGENEYYNQTTGLCQECPPCGPGEEPYLSC
Mouse    MAHVGDCKWMSWLPVLVVSLMCSAKAEDSNCGENEYHNQTTGLCQQCPPCRPGEEPYMSC
         *******    ************* ** ******** ******** **** ****** **

Human    GYGTKDEDYGCVPCPAEKFSKGGYQICRRHKDCEGFFRATVLTPGDMENDAECGPCLPGY
Mouse    GYGTKDDDYGCVPCPAEKFSKGGYQICRRHKDCEGFFRATVLTPGDMENDAECGPCLPGY
         ****** *****************************************************

Human    YMLENRPRNIYGMVCYSCLLAPPNTKECVGATSGASANFPGTSGSSTLSPFQHAHKELSG
Mouse    YMLENRPRNIYGMVCYSCLLAPPNTKECVGATSGVSAHSSSTSGGSTLSPFQHAHKELSG
         ********************************** **    *** ***************

Human    QGHLATALIIAMSTIFIMAIAIVLIIMFYILKTKPSAPACCTSHPGKSVEAQVSKDEEKK
Mouse    QGHLATALIIAMSTIFIMAIAIVLIIMFYIMKTKPSAPACCSSPPGKSAEAPANTHEEKK
         ****************************** ********** * **** **     ****

Human    EAPDNVVMFSEKDEFEKLTATPAKPTKSENDASSENEQLLSRSVDSDEEPAPDKQGSPEL
Mouse    EAPDSVVTFPENGEFQKLTATPTKTPKSENDASSENEQLLSRSVDSDEEPAPDKQGSPEL
         **** ** * *  ** ****** *  **********************************

Human    CLLSLVHLAREKSATSNKSAGIQSRRKKILDVYANVCGVVEGLSPTELPFDCLEKTSRML
Mouse    CLLSLVHLAREKSVTSNKSAGIQSRRKKILDVYANVCGVVEGLSPTELPFDCLEKTSRML
         ************* **********************************************

                 A
Human    SSTYNSEKAVVKTWRHLAESFGLKRDEIGGMTDGMQLFDRISTAGYSIPELLTKLVQIER
Mouse    SSTYNSEKAVVKTWRHLAESFGLKRDEIGGMTDGMQLFDRISTAGYSIPELLTKLVQIER
         ************************************************************

Human    LDAVESLCADILEWAGVVPPASQPHAAS
Mouse    LDAVESLCADILEWAGVVPPASPPPAAS
         ********************** * ***
```

Figure 11.3 Pairwise amino acid alignment between the human and mouse EDAR proteins. Asterisks indicate identical amino acids in the two sequences. The identical death domains are shown in bold letters. The V-to-A replacement is shown in red.

in embryonic stem cells, Kamberov et al. (2013) introduced a T → C point mutation into the murine gene, resulting in the Val-to-Ala replacement in the protein. Thus, the death domain of the mouse protein became identical to its counterpart in East Asian human populations, while the rest of the protein was unaltered. These so-called **knockin** mice were, then, used to derive mutant homozygotes and heterozygotes. Finally, the phenotypes of the three genotypes were compared.

The *EDARV370A* allele turned out to have many **pleiotropic** effects, i.e., to affect a great number of seemingly unrelated traits. First, the mutated allele affected hair shaft thickness. The mouse coat contains a mix of four different types of hair: guard, awl, auchene, and zigzag (Duverger and Morasso 2009). Hair thickness is determined by the number of medulla cells (1, 2, 3, or 4) involved in the development of the hair follicle. Guard and zigzag hairs invariably consist of a single medulla cell; awl and auchene hairs are variable, each follicle consisting of 1–4 medulla cells. Homozygous mutant mice had more of the thickest, four-cell awl hairs and fewer three-cell hairs than wild-type homozygotes. Similarly, mutant homozygotes had a higher proportion of thicker auchene four- and three-cell hairs than either heterozygotes or wild-type homozygotes. This finding supports the hypothesis that *EDARV370A* was responsible for the thicker scalp hair of East Asians.

Second, the mutated allele affected mammary gland branch density and mammary fat-pad size. Mammary gland branches were found to be considerably denser in mutant homozygotes relative to the other two genotypes. The mammary fat-pad size (or breast size) of *EDARV370A* homozygotes was found to be ~15% smaller than that of wild-type homozygotes, with heterozygotes having intermediate fat-pad sizes. Thus, the human *EDARV370A* allele is most probably responsible for the rarity of DD-cup bras in East Asian lingerie shops.

The third finding concerned eccrine (sweat) glands. The finding that knockin homozygotes had significantly more eccrine glands than either heterozygotes or wild-type homozygotes was largely unexpected, as human *EDARV370A* has not previously been suspected of having a connection with this trait. Following the findings in the mice, the association between *EDARV370A* and number of active eccrine glands was validated in humans by comparing 2,226 *EDARV370A* homozygous and 340 heterozygous ethnic Han individuals from the eastern coastal city of Taizhou, China. Despite 34 centuries of searching for and documenting racial differences (Gossett 1997), the higher density of sweat glands in East Asians than in Europeans and Africans had not been described before 2013.

Of course, all model organisms have their limitations. Because the teeth of rodents are so different from those of humans, the hypothesis that a causal connection exists between *EDARV370A* and incisor single- and double-shoveling could not be validated.

The many traits affected by *EDARV370A* make it very difficult to identify any one trait that has been favored by selection. One possibility is that selection favored an increased number of eccrine glands because such individuals have the ability to evaporate water more efficiently. Such a trait may be advantageous for East Asian hunter-gatherers in warm and humid climates such as the one that existed in China between 32,000 and 40,000 years ago. Another possibility is that *EDARV370A* increased in frequency by sexual selection. Joshua Akey (cited in Wade 2013) hypothesized that since hair and breasts are visible sexual signals, the *EDARV370A* allele might have spread through the population by men preferring women with thick hair and small breasts. If this is the case, eccrine gland density and incisor shoveling are incidental traits that confer no advantage. The sexual selection hypothesis is supported by analogy with comparable traits in European populations, in which the alleles underlying blue eyes and blond hair exhibit strong signals of selection, despite the fact that no intrinsic fitness advantage exists for these two traits. The reader may, of course, object to this statement on account of the correlation that supposedly exists between eye and hair color, on the one hand, and skin color on the other. After all, as mentioned above, lighter skin in northern latitudes has been shown to be subject to selection in connection with the synthesis of vitamin D.

The problem is that eye and hair color correlate only very weakly with skin color. The correlation that we believe exists between skin color and eye color is an artifact of observing mainly inbred European populations or inbred African populations. In mixed European-African populations, the correlation between skin and eye color was observed to be 0.38, so only 14% of the variation of eye color could be explained by the variance of skin color (Beleza et al. 2013b). This observation points to different genetic causes for skin and eye color, and thus to different selection regimes.

What proportion of the human genome is functional?

While it is undisputed that many functional regions within genomes have evolved under complex selective regimes, such as selective sweeps, balancing selection, and recent positive selection, it is widely accepted that purifying selection persisting over long evolutionary times is the most common mode of evolution (Rands et al. 2014). Studies that identify functional sites by using the degree of conservation between sequences from two (or more) species have estimated the proportion of functional nucleotides in the human genome to be 3–15% (Ponting and Hardison 2011; Ward and Kellis 2012). We note, however, that each lineage gains and loses functional elements over time, so the proportion of nucleotides under selection needs to be understood in the context of divergence between species. For example, estimates of constraint between any two species will only include sequences that were present in their common ancestor and that have not been lost, replaced, or nonfunctionalized in the lineages leading up to the genomes of the extant species under study. **Functional element turnover** is defined as the loss or gain of purifying selection at a particular locus of the genome when changes in the physical or genetic environment cause a locus to switch from being functional to being nonfunctional or vice versa.

By using genomic data from 12 mammalian species and an estimation model that takes into account functional element turnover, Rands et al. (2014) estimated that 8.2% of the human genome is functional, with a 95% confidence interval of 7.1–9.2%. Because of the difficulties in estimating the functional fraction of a genome, evolutionary biologists treat such estimates as somewhat underestimated. Thus, a claim that 10% or even 15% of the human genome is functional would be tolerable. On the other hand, a claim that 80% of the human genome is functional (e.g., ENCODE Project Consortium 2012) is misleading in the extreme and logically risible.

Unsurprisingly, in Rands et al.'s (2014) study, constrained coding sequences turned out to be much more evolutionarily stable, i.e., experienced less functional element turnover, than constrained noncoding sequences. From among noncoding sequences, the sequences that were most likely to be functional were enhancers and DNase 1 hypersensitivity sites. Transcription factor binding sites, promoters, untranslated regions, and long noncoding RNAs (lncRNAs) contributed little to the functional fraction of the human genome, with lncRNAs exhibiting the most rapid rate of functional element turnover of all the noncoding element types.

How much garbage DNA is in the human genome?

Because humans are diploid organisms and because natural selection is notoriously slow and inefficient in ridding populations of recessive deleterious alleles, the human genome is expected to contain garbage DNA. The amount of garbage DNA, however, should be quite small—many orders of magnitude smaller than the amount of junk DNA.

Deleterious alleles should exhibit a few telltale signs. First, they should be maintained in the population at very low frequencies. The reason for the rarity of deleterious alleles is that at low frequencies, the vast majority of such alleles will be found in the heterozygous state, unexposed to purifying selection. Second, deleterious alleles should only very rarely become fixed in populations. Thus, one can identify them by using the ratio of polymorphic alleles to fixed alleles (Chapter 2). Third, as shown by Maruyama (1974), a deleterious or slightly deleterious allele should, on average, be younger than a neutral allele segregating at the same frequency. The young age of deleterious alleles is due to the fact that although purifying selection is not very efficient, it does eventually eliminate deleterious alleles from the population.

Many studies have shown that human genomes consist of measurable quantities of garbage DNA. Tennessen et al. (2012), for instance, sequenced 15,585 protein-coding genes from 2,440 individuals of European and African ancestry and found that out of the more than 500,000 single-nucleotide variants, the majority were rare (86% with a minor allele frequency less than 0.5%). Of the average 13,595 single-nucleotide variants that an individual carried, ~43% were missense, nonsense, or affected splicing—i.e., they affected protein sequence. The rest of the mutations were synonymous. About 47% of all variants (74% of nonsynonymous and 6% of synonymous variants) were predicted by at least one of several computational methods to be deleterious, and almost all of these deleterious variants (~97%) had very low frequencies. Fu et al. (2013) analyzed 15,336 protein-coding genes from 6,515 individuals and estimated that ~73% of all single-nucleotide variants and ~86% of the variants predicted to be deleterious arose in the past 5,000–10,000 years. Sunyaev et al. (2001) estimated that the average human genotype carries approximately 1,000 deleterious nonsynonymous single-nucleotide variants that together cause a substantial reduction in fitness.

Chun and Fay (2009) approached the problem of distinguishing deleterious mutations from the massive number of nonfunctional variants that occur within a single genome by using a comparative genomics data set of 32 vertebrate species. They first identified protein-coding sites that were highly conserved. Next they identified amino acid variants in humans at protein sites that are evolutionarily conserved. These amino acids variants are likely to be deleterious. Application of Chun and Fay's method to human genomes revealed close to 1,000 deleterious variants per individual, approximately 40% of which were estimated to be at allele frequencies smaller than 5%. This method also indicated that only a small subset of deleterious mutations can be reliably identified.

So far, we have discussed population genetics and evolutionary methods for predicting the deleteriousness of protein-altering variants based on population properties. There are, however, methods that combine evolutionary and biochemical information to make such inferences (Cooper and Shendure 2011). Nonsense and frameshift mutations are the most obvious candidates for being deleterious, as they often result in genetic diseases due to loss of protein function . However, this class of variation is not unambiguously deleterious, as in some cases nonsense and frameshift mutations do not interfere with functional protein production. Moreover, even if the mutation results in the loss of function, it may not always be harmful.

Considering nonsynonymous variants, the simplest and earliest approach to estimate deleteriousness was to use discrete biochemical categorizations such as "radical" versus "conservative" amino acid changes. However, there are now numerous more sophisticated approaches to classifying nonsynonymous variants on both quantitative and discrete scales. These methods can be divided into "first-principles approaches" and "trained classifiers."

First-principles approaches explicitly define a biological property of deleterious variants and make predictions on the basis of similarity or dissimilarity to that property. For example, first-principles approaches may use the presence of frameshifts in the coding regions as identifiers of deleteriousness (e.g., Sulem et al. 2015). By contrast, **trained classifiers** generate prediction rules by identifying heuristic combinations of many potentially relevant properties that optimally differentiate a set of true positives from a set of true negatives. First-principles approaches have the advantage of greater interpretability; for example, radical and conservative annotations of amino acid substitutions have a straightforward biochemical interpretation. However, first-principles methods are only as good as the assumptions that they make and do not model all of the relevant factors. Conversely, a trained-classifier approach effectively yields a "black box" prediction and will be prone to the biases and errors in the data. However, trained classifiers have the advantage of being specifically tunable to the desired task (e.g., predicting disease causality) and are capable of incorporating many sources of information without requiring a detailed understanding of how that information is relevant.

We note that all methods for predicting deleteriousness are prone to estimation error. For example, all methods use multiple sequence alignments and phylogenetic reconstructions. Low qualities of alignment or erroneous phylogenetic trees may result in low-quality inferences. Moreover, the sampling of species is crucial. A sample consisting of sequences that are very similar to one another offers less power of detection, thus increasing the number of false negatives. Conversely, inclusion of distant sequences may increase the number of false positives.

Finally, we note that many methods exploit biochemical data, including amino acid properties (such as charge), sequence information (such as presence of a binding site), and structural information (such as the presence of a β-sheet). The integration of these data with comparative sequence analysis significantly improves predictions of deleteriousness

Genome Size, DNA Content, and C value

Following Winkler (1920), early usages of the term "genome" most probably referred to either the monoploid genome, such as that in prokaryotes, or to the haploid genome, such as that in the nucleus of animal sperm cells. In time, however, the term acquired a more flexible or inclusive meaning, so that in the literature, "genome" may refer to a property of a cell, an individual, a population, a species, or a higher taxonomic unit. In eukaryotes, it is customary to deal with the genomes of organelles separately from the nuclear genome. Thus, humans are said to have two genomes, nuclear and mitochondrial, while chrysanthemums have three genomes: nuclear, mitochondrial, and plastid. (Because of their prokaryotic ancestry, mitochondrial and plastid genomes were discussed in Chapter 10.)

The terms "genome size" and "DNA content" are used quite inconsistently in the literature (Greilhuber et al. 2005). They may refer to the sizes of monoploid or haploid genomes, or they may refer to the total amount of DNA in a cell. It is, therefore, advisable to use the unambiguous term **C value** (Swift 1950) whenever referring to monoploid or haploid genomes. The "C" in "C value" stands for "constant" (Greilhuber et al. 2005), to denote the fact that the intraspecific variability in haploid genome size is substantially smaller than the interspecific variability (Vendrely and Vendrely 1948).

Genome size variation and genomic content in eukaryotes

As a rule, eukaryotes have much larger genomes than prokaryotes, but there are exceptions. For instance, the yeast *Saccharomyces cerevisiae* has a genome that is smaller than that of the proteobacterium *Sorangium cellulosum*. Eukaryotic endosymbionts and intracellular parasites possess greatly reduced genomes. C values in eukaryotes range from less than 4×10^5 bp in the secondary plastid of the chlorarachniophyte *Bigelowiella natans* (Douglas et al. 2001; Gilson et al. 2006) to approximately 1.5×10^{11} bp in the canopy plant *Paris japonica* (Pellicier et al. 2010), close to a 400,000-fold range (**Table 11.2**). Green plants, in particular, exhibit a huge variation in C values, while genome size variation in mammals is negligible.

What can explain the huge variation in eukaryotic genome size? Let us first investigate whether any genomic compartment correlates with genome size. First, we note that in contradistinction to the situation in prokaryotes, only a miniscule fraction of the eukaryotic genome is occupied by protein-coding sequences. Moreover, the number of protein-coding genes does not correlate with genome size. To illustrate this fact, let us compare the human genome with the genome of a teleost fish called *Takifugu rubripes* (formerly *Fugu rubripes*). At 400 Mb, the *T. rubripes* genome is one of the smallest vertebrate genomes (Aparicio et al. 2002; Noleto et al. 2009). Interestingly, although the length of the *Takifugu* genome is less than about one-eighth that of the human genome, it contains a comparable number of protein-coding genes (Aparicio et al. 2002). It is for this reason that Sydney Brenner regarded the *Takifugu*

TABLE 11.2

Range of C values in a sample of monophyletic eukaryotic higher taxa[a]

Taxon	Genome size range (Kb)	Ratio (highest/ lowest)
Dinophyceae (dinoflagellates)	1,470,000–220,000,000	150
Fungi (fungi)	2,220–893,000	402
Metazoa (animals)	19,600–130,000,000	6,642
Porifera (sponges)	39,200–1,760,000	45
Cnidaria (cnidarians)	225,000–1,810,000	8
Rotifera (rotifers)	245,000–1,200,000	5
Tardigrada (water bears)	78,400–804,000	10
Nematoda (roundworms)	19,600–2,450,000	125
Platyhelminthes (flatworms)	58,000–20,100,000	342
Annelida (annelid worms)	58,800–7,490,000	127
Mollusca (mollusks)	421,000–7,690,000	18
Crustacea (crustaceans)	137,000–63,300,000	462
Chelicerata (chelicerates)	78,400–7,350,000	94
Hexapoda (insects)	88,200–16,600,000	188
Myriapoda (myriapods)	274,000–2,100,000	8
Echinodermata (echinoderms)	529,000–4,310,000	8
Bryozoa (bryozoans)	196,000–1,570,000	8
Onychophora (velvet worms)	4,340,000–19,500,000	4
Gastrotricha (gastrotrichs)	49,000–617,000	13
Cyclostomata (jawless fishes)	1,260,000–4,500,000	4
Chondrichthyes (cartilaginous fishes)	1,480,000–16,700,000	11
Osteichthyes (bony fishes)	343,000–130,000,000	380
Amphibia (amphibians)	931,000–118,000,000	127
Crocodylia (crocodiles)	2,440,000–3,380,000	1
Squamata (squamates)	1,030,000–3,850,000	4
Testudines (turtles)	1,750,000–5,330,000	3
Aves (birds)	891,000–2,120,000	2
Mammalia (mammals)	1,600,000–8,230,000	5
Viridiplantae (green plants)	9,800–149,000,000	15,204
Pteridophyta (pteridophytes)	88,200–71,200,000	807
Bryophyta (mosses)	167,000–7,810,000	47
Spermatophyta (seed plants)	63,400–149,000,000	2,350
Eukaryota (all eukaryotes)	**373–149,000,000**	**399,000**

Source: Data from www.genomesize.com, http://data.kew.org/cvalues, www.zbi.ee/fungal-genomesize, LaJeunesse et al. (2005), Hackett and Bhattacharya (2006), and Bennett and Leitch (2011).

[a]Only monophyletic taxa that contain multicellular organisms were included.

genome as "the *Reader's Digest* version" of the human genome (quoted in Purves et al. 2004). The small genome size of *Takifugu* can be attributed to a reduction in intron and intergenic lengths, a lack of significant amounts of repetitive sequences, and a very small number of pseudogenes—the genome of *T. rubripes* has no mitochondrial pseudogenes and only 162 nuclear pseudogenes versus at least 15,000 pseudogenes in the human genome (Hazkani-Covo et al. 2010).

The situation in *Takifugu* is not by any means universal. Some small-genome organisms have *Takifugu* characteristics; others do not. For example, the carnivorous bladderwort plant *Utricularia gibba* has a tiny genome, yet it accommodates a typical number of protein-coding genes (~30,000) for a plant (Ibarra-Lachlette et al. 2013). In contrast, the miniature genome of another carnivorous plant, *Genlisea aurea*, has a smaller number of protein-coding genes than its large-genome relatives (Leushkin et al. 2013).

Can genome size variation be explained by other variables related to protein-coding genes? While there are small differences in the mean mRNA lengths among different organisms, no correlation exists between this variable and genome size. For instance, mRNAs are slightly longer in multicellular organisms than in protists (1,400–2,200 bp versus 1,200–1,500 bp), whereas genome sizes in protists can be much larger than those of multicellular organisms. Similarly, organisms with larger genomes do not always produce larger proteins, nor do they have longer introns.

Finally, while there exists a significant positive correlation between the degree of repetition of several RNA-specifying genes and genome size (Chapter 7), these genes constitute only a negligible fraction of the eukaryotic genome, such that the variation in the number of RNA-specifying genes cannot explain the variation in genome size.

Intraspecific variation in genome size

Intraspecific variability is the sine qua non of evolution. No trait can evolve unless it exhibits genetic variability within the population. As mentioned previously, the "C" in "C value" stands for "constant," implying that genome size lacks intraspecific variability. This is not strictly true, although the variation within species is many orders of magnitude smaller that that among species.

Information on intraspecific genome size variation is quite scarce, and the data is mostly limited to laboratory model organisms. Ellis et al. (2014), for instance, studied 211 laboratory lines of *D. melanogaster* and found that the average genome size ranged from 169.5 to 192.8 Mb, a range of about 1.1-fold. Furthermore, the distribution of genome sizes was skewed toward large genome sizes, possibly indicating that increases in genome size are more frequent than decreases.

Studies of intraspecific genome size variation in natural populations are even scarcer. In one such study, a 1.2-fold intraspecific variation in genome size was discovered in the grass species *Festuca pallens* (Šmarda et al. 2008). This study also revealed that genome size is heritable, as the genome sizes of seedlings were found to be highly correlated with those of their mothers.

Mutations That Increase or Decrease Genome Size

In this section we discuss mutational processes that can increase or decrease genome size. In particular, we ask ourselves which among these processes can concomitantly increase or decrease the fraction of nonfunctional DNA in the genome. Increases in genome size can be caused by genome duplication, various types of subgenomic duplications, mononucleotide and oligonucleotide insertions, and replicative transposition. Decreases in genome size can occur either through unequal crossing over or deletion. Genome size increases can occur either gradually or in a punctuated manner, in which large increases in genome size occur in a relatively short period of time. There are no analogous processes that can cause large and sudden decreases in genome size. Thus, as opposed to genome size increases, which can occur in fits and starts, decreases in genome size can only occur in a gradual manner.

Insertions and subgenomic duplications can increase genome size; however, these process are expected to increase the size of a genome only very slowly, such that their contributions to genome size are thought to be negligible. Moreover, neither of these two processes is expected to alter the fraction of nonfunctional elements in the

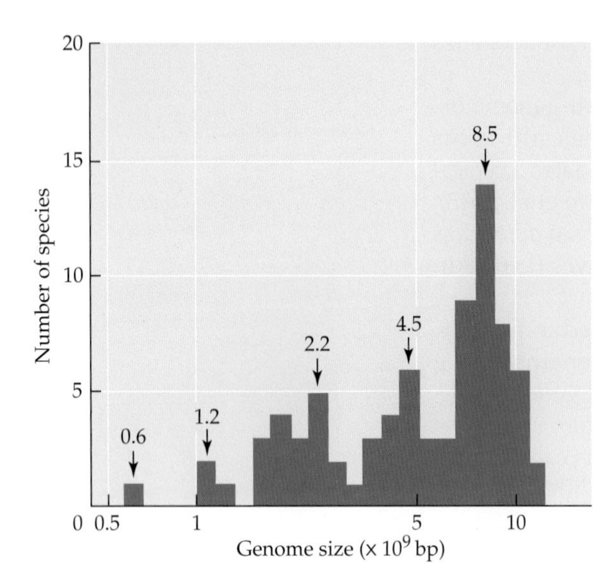

Figure 11.4 Frequency distribution of genome sizes in 80 grass species (family Poaceae). Peaks in the multimodal distribution are marked with arrows. Note that the abscissa is in logarithmic scale. (Modified from Sparrow and Nauman 1976.)

genome. For example, the probabilities of duplicating a functional sequence or a nonfunctional sequence are proportional to the prevalence of such sequences in the genome, so following subgenomic duplication, the fraction of junk DNA in the genome neither increases nor decreases.

The contribution of genome duplication to genome size

Genome duplication is the fastest route to genome size increase; in one fell swoop, genome size is doubled. Genome duplication, however, does not increase the fraction of junk DNA in the genome unless the newly created functional redundancy is quickly obliterated by massive gene nonfunctionalization. Of course, in recently formed polyploids, one cannot speak of an increase in the C value, since this value refers to the size of the haploid genome and does not depend on the degree of ploidy. The contribution of gene duplication to the variation in C values only comes into play after the polyploid species has been diploidized and become a cryptopolyploid (Chapter 7).

Let us now examine the genomic consequences of polyploidy on the distribution of genome sizes. A polymodal distribution of genome sizes has been registered in many groups of eukaryotes (Otto 2007). This is particularly evident in grasses (family Poacea), where genome sizes exhibit a polymodal distribution with peaks at 0.60, 1.18, 4.51, and 8.53 Gb (**Figure 11.4**). Similar distributions have been observed in echinoderms, insects, and fungi, and to a lesser extent in amphibians and bony fishes. Thus, genome duplication seems to be an important mechanism in the evolution of genome size in eukaryotes. Interestingly, each round of genome duplication appears to have involved small losses of DNA, such that the amount of DNA after each round of duplication increases by a factor of slightly less than 2 (Sparrow and Nauman 1976).

If polyploidization is indeed an important determinant of genome size evolution, than we should observe an excess of even numbers over odd numbers of chromosomes. In a sample of more than 18,000 plants (ferns, monocotyledons, and dicotyledons), the percentage of species with even numbers of chromosomes was 57%, while only 43% of the plants had odd numbers of chromosomes (Otto 2007). One such distribution is shown in **Figure 11.5**.

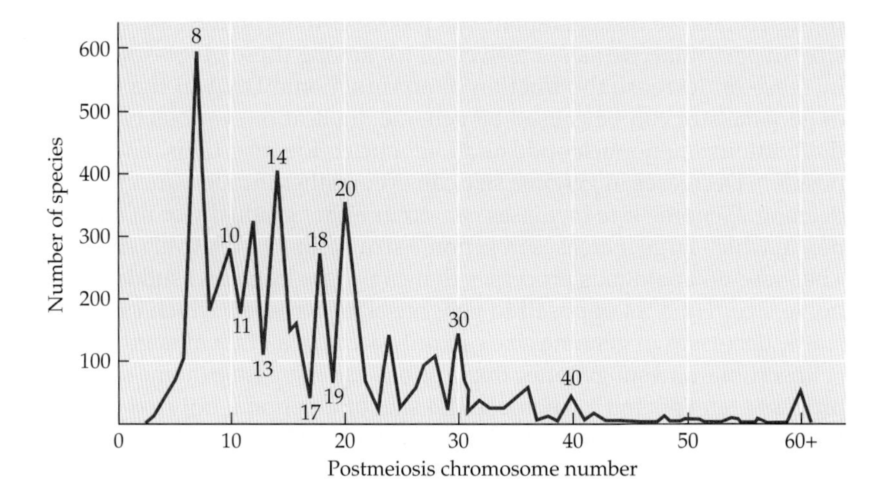

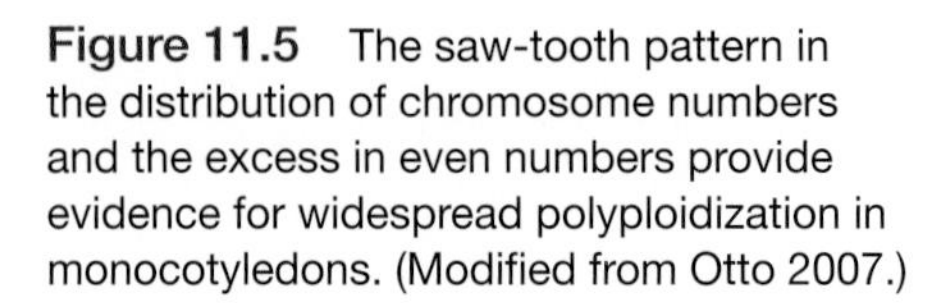

Figure 11.5 The saw-tooth pattern in the distribution of chromosome numbers and the excess in even numbers provide evidence for widespread polyploidization in monocotyledons. (Modified from Otto 2007.)

The contribution of transposable elements to genome size

Because of their ability to increase in number, many transposable elements can have profound effects on the size and structure of genomes. Indeed, replicative transposable elements, especially retrotransposons, have the potential to increase their copy number while active, and many are responsible for huge genome size increases during relatively short periods of evolutionary time. One such extreme example concerns the genome of *Oryza australiensis*, a wild relative of the cultivated rice *O. sativa*, which has experienced a recent burst of replicative transposition involving three LTR retrotransposon families. In the last 3 million years, the genome of *O. australiensis* has accumulated more than 90,000 new retrotransposon copies, leading to more than a twofold increase in its size (Piegu et al. 2006).

The vast majority of eukaryotic genomes studied to date contain large numbers of transposable elements, and in many species, such as maize (*Zea mays*), the edible frog (*Pelophylax esculentus*, formerly known as *Rana esculenta*), and the largest genome sequenced so far, the loblolly pine (*Pinus taeda*), transposable elements constitute the bulk of the genome (e.g., Kovach et al. 2010). Organisms devoid of transposable elements are extremely rare. The only organisms known to lack transposable elements altogether or to possess only two to three copies of transposable elements are a group of yeast species that includes *Ashbya gossipii*, *Kluyveromyces lactis*, *Zygosaccharomyces rouxii*, and *Schizosaccharomyces octosporus* (Dietrich et al. 2004; Rhind et al. 2011). We note, however, that most of the DNA that originated as "selfish DNA" (p. 523) is no longer selfish. Much of it is currently in a degenerate state—dead and moribund transposable elements that are no longer capable of transposition.

Replicative transposition is the only mutational process that can greatly and rapidly increase genome size and at the same time significantly increase the fraction of junk DNA in the genome. The reason for this is that transcription and reverse transcription are very inaccurate methods of copying genetic information (Chapter 9), and hence most of the increase in genome size due to transposable elements will consist of nonfunctional elements, i.e., junk DNA.

The claim that most junk DNA is made of transposable elements and their incapacitated descendants yields a quantitative prediction—it is expected that a positive correlation should exist between the genomic fraction inhabited by transposable elements and genome size. We note, however, that estimating the numbers and kinds of transposable elements in a genome is far from trivial or routine. First, different sequenced genomes differ from one another in the quality of the sequences. Because repetitive elements are difficult to sequence and problematic to use in genomic assemblies, the fraction of repetitive DNA is frequently underrepresented in low-quality genome sequences. Second, available algorithms for detecting repeated elements are known to perform with varying degrees of success in different species. The reason is that algorithms use a database of known transposable sequences as their reference, so the repertoire of transposable elements in one species may be better characterized than that in another species. Third, most transposable elements have neither been co-opted into a function by the host nor retained their ability to transpose. Thus, they evolve in an unconstrained fashion, losing their similarity to other members of their transposable-element family very rapidly. Algorithms that rely on similarity measures are, hence, unable to identify a significant proportion of dead transposable elements. Finally, most algorithms for detecting repeated transposable elements fail to discover short elements. As a consequence, the fraction of a genome that is taken up by transposable elements is more often than not extremely underestimated. For example, about 50% of the human genome has been identified as derived from transposable elements by using algorithms that rely on a database of consensus element sequences. By using a method that identifies oligonucleotides that are related in sequence space to one another, de Koning et al. (2011) found out that 66–69% of the human genome is derived from repetitive sequences.

Despite the difficulties described above, a positive correlation between total sequence length of transposable elements and genome size is seen in many groups

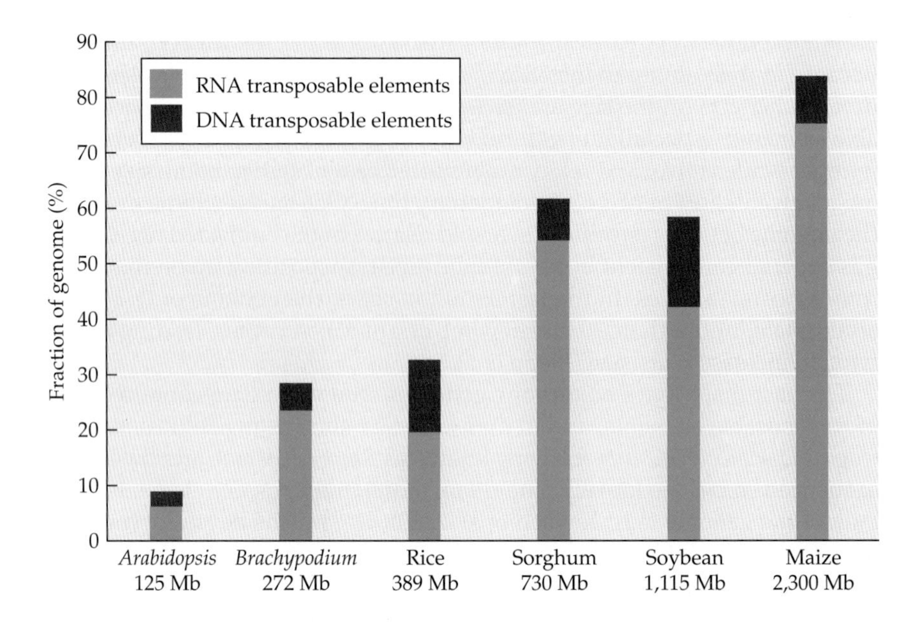

Figure 11.6 The genomic fraction occupied by RNA and DNA transposable elements in six plant species. Genome sizes are indicated along the abscissa. (Modified from Slotkin et al. 2012.)

(**Figure 11.6**). For example, the Pearson correlation coefficient in a sample of 66 vertebrate genomes was 0.77 (Tang 2015). In other words, ~60% of the variation in genome size was explained by the variation in total length of transposable elements. In a sample of 16 diploid angiosperm species, the correlation coefficient was 0.99 (Tenaillon et al. 2010). Many other researchers have reported positive correlations between the number of transposable-element copies per genome and genome size (Kidwell 2002; Biémont and Vieira 2006; Lynch et al. 2011; Chénais et al. 2012; Hawkins et al. 2012; Chalopin et al. 2015). The rule is simple: large genomes have huge numbers of transposable elements; small genomes have very few (e.g., Roest Crollius et al. 2000; Kovach et al. 2010; Ibarra-Laclette et al. 2013; Kelley et al. 2013).

What is more important in determining genome size: transposable-element copy number or transposable-element length? A large part of the genomes of *Drosophila* species is a graveyard for dead members of about 30 families of transposable elements. Inactivated mobile elements can, however, multiply locally through such processes as unequal crossing over. Comparison of dispersed repeats between closely related species suggests that transposition can quickly increase the fraction of dispersed repeats in the genome. For example, the sibling species *D. melanogaster* and *D. simulans* diverged from each other less than 2.5 million years ago. During this relatively short period of time, their genomes have become very different in terms of the fraction of dispersed repeats (21% in *D. melanogaster* and 3% in *D. simulans*). This difference is most likely explained by an increase in copy number in *D. melanogaster* rather than a decrease in *D. simulans*, because loss of genomic transposable elements is known to be a very slow process. We notice, however, that length abridgment is faster in *D. simulans* than in *D. melanogaster* (p. 564), so we must conclude that the contribution of copy number to genome size is greater than that of repeat length.

Deletions and genome size

The difference in genome size among organisms can be explained by differences in either the rates of accumulation of nonfunctional DNA or the rates with which organisms get rid of junk DNA. For quite a long time it has been known that the genomes of *Drosophila* species contain very few transposable elements (Vanin 1985; Weiner et al. 1986; Wilde 1986). Petrov et al. (1996) and Petrov and Hartl (1998) investigated the reason for this phenomenon and found that dead *Helena* transposable elements lose DNA at unusually high rates. They put two and two together and suggested that DNA regions that are not subject to selective constraints are deleted at "rampant" rates. They further extrapolated that different deletion rates, rather than accumulation rates, may contribute to the divergence in genome size among taxa. Their assumption was that high rates of deletion are not confined to *Helena* alone, but that the phenomenon is of general application to all selectively unconstrained regions.

To test this assumption, they compared intron sizes between two *Drosophila* species. *Drosophila virilis* has a genome twice as large as that of *D. melanogaster* (Moriyama et al. 1998). Admittedly, part of the difference could be attributed to heterochromatin, but even if this factor is taken into account, the genome of *D. virilis* is still about 36% larger than that of *D. melanogaster.* In their comparison of 115 complete introns collected from 42 orthologous genes, they found that the difference in intron length between the two species was statistically significant. The difference in the mean length of the introns between *D. virilis* and *D. melanogaster* (394 and 283 bp, respectively) was 39%, which was surprisingly close to the size difference in the nonrepetitive fraction between the genomes. Thus, it seems that some organisms are simply more efficient at "throwing out the trash" than others (Petrov and Hartl 1997).

A follow-up study by Gregory (2004) revealed that DNA loss by small deletions is too slow to affect differences between genomes. Thus, small-deletion bias is far too weak to be accepted as the primary determinant of genome size. Moreover, more recent studies in *Drosophila* revealed that while the deletion rate may be higher than the insertion rate at the mutation level, purifying selection is stronger on deletions than on insertions, resulting in a net increase in genome size (Huang et al. 2014).

Genomic Paradoxes in Eukaryotes

A **genomic paradox** refers to the lack of correspondence between a measure of genome size or genome content and the presumed amount of genetic information "needed" by the organism, i.e., its **complexity** or **organismal complexity**. Different measures of complexity have been put forward in different scientific fields. However, because they ask similar questions about complexity, ecologists, evolutionary biologists, computer scientists, and economists have come up with similar approaches to measuring complexity (Lloyd 2001). All the different measures of complexity of an entity attempt to answer three basic questions: (1) How hard is the entity to describe? (2) How hard is it to create (or evolve)? and (3) What is its degree of organization? Not all measures of complexity are applicable to the living world, and some suggested measures are so vaguely and impressionistically defined as to defy application. For example, Szathmáry and Maynard Smith (1995) defined the complexity of an organism as "the richness and variety of morphology and behavior."

The simplest measures or proxies for organismal complexity are the **number of cells** and the **number of cell types** (Kaufman 1971). As mentioned in Chapter 5, the vast majority of eukaryotic taxa are unicellular, i.e., they have a single cell and a single cell type. In animals, the number of cells varies from 10^3 in some nematodes to 10^{12} in some mammals; the number of cell types varies from 4 and 11 in placozoans and sponges, respectively, to approximately 200 in mammals (Klotzko 2001; Srivastava et al. 2008; Goldberg 2013). The number of cell types in plants is about one order of magnitude smaller than that in mammals (Burgess 1985).

The C-value paradox

While organismal complexity as measured by the number of cell types ranges from 1 to approximately 200, C values range over a 400,000-fold range, as mentioned above. Numerous lines of evidence indicate that there is no relationship between genome size and organismal complexity. For example, many unicellular protozoans, which by definition have a single cell and a single cell type, possess much larger genomes than mammals, which possess trillions of cells and 200 or more cell types and are presumably infinitely more complex than protists. The same goes for the comparison between plants and mammals. Many plants have much larger genomes than mammals while at the same time possessing fewer cell types. Furthermore, organisms that are similar in morphological and anatomical complexity (e.g., flies and locusts, onion and lily,

Paramecium aurelia and *P. caudatum*) exhibit vastly different C values (**Table 11.3**). This lack of a positive correlation between organismal complexity and genome size is also evident in comparisons of sibling species (i.e., species that are so similar to each other morphologically as to be indistinguishable phenotypically). In insects, protists, bony fishes, amphibians, and flowering plants, many sibling species differ greatly in their C values, even though by definition no difference in organismal complexity exists. The example of sibling species is extremely illuminating, since it tells that the same level of complexity can be achieved by vastly different amounts of genomic DNA.

This lack of correspondence between C values and the presumed complexity of the organisms has become known in the literature as the **C-value paradox** (Thomas 1971) or the **C-value enigma** (Gregory 2001a). There are two facets to the C-value paradox. The first concerns the undisputable fact that genome size cannot be used as a predictor of organismal complexity. That is, from knowledge of the C value it is impossible to say whether an organism is unicellular or multicellular, whether the genome contains few or many genes, or whether the organism is made of a few or a couple of

TABLE 11.3

C values of a few eukaryotic organisms ranked by genome size

Species	C value (Mb)	Species	C value (Mb)
Ostreococcus tauri (unicellular alga)	10	*Carcharias obscurus* (dusky shark)	2,700
Saccharomyces cerevisiae (baker's yeast)	13	*Canis familiaris* (dog)	2,900
Pratylenchus coffeae (plant-parasitic nematode)	20	*Rattus norvegicus* (rat)	2,900
Neurospora crassa (red bread mold)	39	*Xenopus laevis* (African clawed frog)	3,100
Tethya actinia (sponge)	39	***Homo sapiens* (human)**	**3,300**
Genlisea margaretae (carnivorous plant)	63	*Nicotiana tabacum* (tobacco plant)	3,800
Caenorhabditis elegans (nematode)	78	*Locusta migratoria* (migratory locust)	6,600
Chlorella ellipsoidea (green alga)	80	*Spirogyra setiformis* (desmid alga)	7,000
Selaginella apoda (meadow spike-moss)	84	*Paramecium caudatum* (ciliate)	8,600
Mayetiola destructor (Hessian fly)	88	*Schistocerca gregaria* (desert locust)	9,300
Ascidia atra (sea squirt)	160	*Allium cepa* (onion)	15,000
Drosophila melanogaster (fruit fly)	180	*Podisma pedestris* (short-horned grasshopper)	17,000
Paramecium aurelia (ciliate)	190	*Chara contraria* (multicellular alga)	19,000
Oryza sativa (rice)	590	*Triturus cristatus* (warty newt)	19,000
Mycale laevis (sponge)	620	*Thuja occidentalis* (western white cedar)	19,000
Strongylocentrotus purpuratus (sea urchin)	870	*Coscinodiscus asteromphalus* (centric diatom)	25,000
Gymnosporangium confusum (rust fungus)	893	*Pinus ayacahuite* (Mexican white pine)	35,000
Limnodynastes ornatus (ornate burrowing frog)	930	*Lilium formosanum* (lily)	36,000
Scomber scombrus (mackerel)	950	*Psilotum nudum* (skeleton fork fern)	71,000
Gallus domesticus (chicken)	1,200	*Amphiuma means* (two-toed salamander)	84,000
Erysiphe cichoracearum (powdery mildew)	1,500	*Pinus resinosa* (Canadian red pine)	68,000
Cyprinus carpio (common carp)	1,700	*Necturus lewisi* (Neuse River waterdog)	120,000
Lampetra planeri (brook lamprey)	1,900	*Lepidosiren paradoxa* (South American lungfish)	120,000
Boa constrictor (snake)	2,100	*Protopterus aethiopicus* (marbled lungfish)	130,000
Gnetum ula (gnetum)	2,200	*Paris japonica* (monocotyledon)	150,000
Parascaris univalens (horse roundworm)	2,400		

Source: Data from www.genomesize.com, http://data.kew.org/cvalues, www.zbi.ee/fungal-genomesize, LaJeunesse et al. (2005), Hackett and Bhattacharya (2006), Tavares et al. (2014), and other sources.

hundred cell types. The second facet reflects an unmistakably anthropocentric bias. For example, some salamanders have genome sizes that are almost 40 times bigger than that of humans. So, although there is no definition of organismal complexity that shows that salamanders are objectively less complex than humans (Brookfield 2000), it is still very difficult for us humans to accept the fact that our species, from the viewpoint of genome size, does not look at all like the "pinnacle of creation" and the "paragon of animals." Realizing that our genome is so much smaller than those of frogs, sturgeons, shrimps, squids, flatworms, mosses, onions, daffodils, and amoebas can be quite humbling, if not outright insulting. Moran (2007) referred to the anthropocentric difficulties of some people to accept their reduced genomic status as the "deflated-ego problem."

The C-value paradox requires us to explain why some very complex organisms have so much less DNA than less complex ones. Why do humans have less than half the DNA in the genome of a unicellular ciliate? Since it would be illogical to assume that an organism possesses less DNA than the amount required for its vital functions, the logical inference is that many organisms have vast excesses of DNA over their needs. Hence, large genomes must contain unneeded and presumably functionless DNA. This point of view, incidentally, is not very popular with people who regard organisms as epitomes of perfection, either because they believe in creation directed by an omniscient intelligent designer or because they lack training in evolutionary biology and erroneously believe that natural selection is omnipotent and can produce perfection.

Possible solutions to the C-value paradox

Three main hypotheses have been proposed to explain the C-value paradox. In the first, the selectionist hypothesis, large eukaryotic genomes are assumed to be entirely or almost entirely composed of literal DNA. In the second, the nucleotypic hypothesis, eukaryotic genomes are assumed to be almost entirely functional, but to contain mostly indifferent DNA and very little literal DNA. In the third, the neutralist hypothesis, genomes are assumed to contain large amounts of junk DNA—with bigger genomes containing larger amounts of junk DNA than smaller genomes. A few different albeit overlapping versions of the neutralist hypothesis will be presented.

THE SELECTIONIST HYPOTHESIS The first **selectionist hypothesis** at the level of the genome was put forward by Zuckerkandl (1976a), who asserted that there is very little nonfunctional DNA in the genome and that the vast majority of the genome performs essential functions, such as gene regulation, protection against mutations, maintenance of chromosome structure, and binding of proteins. Consequently, the excess DNA in large genomes is is not really excessive. Over the many years since the publication of Zuckerkandl's article, this theory was rejected multiple times. The ENCODE Project Consortium (2012) resurrected the selectionist hypothesis (without acknowledging its originator), but their conclusion that the human genome is almost 100% functional was reached through various pseudoscientific means, such as by employing the causal role definition of biological function and then applying it inconsistently to different biochemical properties, by committing a logical fallacy known as "affirming the consequent," by failing to appreciate the crucial difference between junk DNA and garbage DNA, by using analytical methods that yield false positives and inflate estimates of functionality, by favoring statistical sensitivity over specificity, and by emphasizing statistical significance over the magnitude of the effect (Eddy 2012; Doolittle 2013; Graur et al. 2013; Niu and Jiang 2013; Brunet and Doolittle 2014; Doolittle et al. 2014; Palazzo and Gregory 2014). There are numerous arguments against selectionist hypotheses. Here, we present two such arguments: the "onion test" and the concept of genetic load.

Selectionist claims to the effect that genomes are entirely or almost entirely functional are usually made in the context of the human genome. For wholly unscientific reasons, humans are frequently assumed to occupy a privileged position, against which all other creatures are measured. In the literature, it frequently seems that, as far as humans are concerned, the equations of population genetics are suspended and evolution abides by a different set of rules. Thus, while no one will ever insist that ferns, salamanders, and lungfish—all of which have vastly larger genomes than humans— are devoid of junk DNA, one can often encounter National Institutes of Health bureaucrats denying that human junk DNA exists (Zimmer 2015). It is in the human context, therefore, that we will examine whether or not selectionist claims hold water.

THE ONION TEST By focusing on the human genome, selectionist hypotheses fail to account not only for the immense diversity of genome sizes among eukaryotes, but also for the lack of correlation between genome size, on the one hand, and the organismal complexity or the presumed number of protein-coding genes, on the other. It should be noted that organisms that have much larger genomes than humans are neither rare nor exceptional. For example, the genome sizes of more than 200 salamander species have been determined thus far, and without exception they have been found to be 4–35 times larger than the human genome. As noted by Orgel and Crick (1980), if one assumes that genomes are 100% functional, then one must conclude that the number of genes needed by a salamander is 20 times larger than the number needed by humans.

These observations pose an insurmountable challenge to any claim that most eukaryotic DNA is functional. The challenge is beautifully illustrated by the **onion test** (Palazzo and Gregory 2014). The domestic onion, *Allium cepa*, is a diploid plant ($2n = 16$) with a haploid genome size of roughly 16 Gb, i.e., approximately five times larger than that of a human. (The example of the onion was chosen arbitrarily, presumably for its shock value—any other species with a large genome could have been chosen for this comparison.) The onion test simply asks, "If most DNA is functional, then why does an onion require five times more DNA than a human?" (We note that the comparison does not have to involve humans. It could as easily be asked for pufferfish versus lungfish, which have a ~350-fold difference, or for other species belonging to the genus *Allium*, which display a more than fourfold range in C values.)

A selectionist solution to the onion test would need to cast the human genome in the role of Goldilocks—not too small for the "pinnacle of creation" and "paragon of animals," yet not big enough for it to be littered by junk DNA. The onion genome, on the other hand, would of necessity be assumed to possess junk DNA lest we humans regard onions as our superiors.

GENETIC LOAD: CAN LARGE GENOMES BE 100% FUNCTIONAL? Many evolutionary processes can cause a population to have a mean fitness lower than its theoretical maximum. For example, deleterious mutations may occur faster than selection can get rid of them; recombination may break apart favorable combinations of alleles, thus creating less fit combinations; and genetic drift may cause allele frequencies to change in a manner that is antagonistic to the effects of natural selection. **Genetic load** is defined as the reduction in the mean fitness of a population relative to a population composed entirely of individuals having the maximal fitness. The basic idea of the genetic load was first discussed by Haldane (1937) and later by Muller (1950).

If w_{max} is the fitness of the fittest possible genotype and $\bar{w}$ is the mean fitness of the actual population, then the genetic load (L) is the proportional reduction in mean fitness relative to highest possible fitness:

$$L = \frac{w_{max} - \bar{w}}{w_{max}} \quad (11.1)$$

Because we are interested in the fraction of the genome that is functional, the only genetic load that we will consider is the **mutational genetic load**, i.e., the reduction in mean population fitness due to deleterious mutations. Haldane (1937) showed that the decrease of fitness in a species as a consequence of recurrent mutation is approximately equal to the total mutation rate multiplied by a factor that ranges from 1 to 2 depending on dominance, inbreeding, and possible sex linkage. For example, for a recessive mutation, it has been shown that as long as the selective disadvantage of the mutant is larger than the mutation rate and the heterozygote fitness is neither larger nor smaller that the fitness values of the homozygotes, the mutational load is approximately equal to the mutation rate (Kimura 1961; Kimura and Maruyama 1966). Thus,

$$L = \mu \tag{11.2}$$

and

$$\overline{w} = (1 - \mu)^n \tag{11.3}$$

where n is the number of loci. (Here we will only deal with recessive mutations but will note that under the same conditions, the mutational load for a dominant mutation is approximately twice the mutation rate.)

Interestingly, in randomly mating populations at equilibrium, the mutational load does not depend on the strength of the selection against the mutation. This surprising result comes from the fact that alleles under strong selection are relatively rare, but their effects on mean fitness are large, while the alleles under weak purifying selection are common, but their effects on mean fitness are small; the effects of these two types of mutation neatly cancel out. As a result, in order to understand the magnitude of mutational load in randomly mating populations, we conveniently need only know the deleterious mutation rate, not the distribution of fitness effects.

Let us now consider the connection between mutational genetic load and fertility. The **mean fertility of a population** ($\overline{F}$) is the mean number of offspring born per individual. If the mortality rate before reproduction age is 0 and mean fertility is 1, then the population will remain constant in size from generation to generation. In real populations, however, the mortality rate before reproduction is greater than 0 and, hence, mean fertility needs to be larger than 1 to maintain a constant population size. In the general case, for a population to maintain constant size, its mean fertility should be

$$\overline{F} = \frac{1}{\overline{w}} \tag{11.4}$$

Let us assume that there are 10,000 loci in the genome and that the mutation rate is $\mu = 10^{-5}$ per locus per generation (Nei 2013). Under the assumption that all mutations are deleterious and recessive, the mutational load is $L = 10{,}000 \times 10^{-5} = 0.1$ and the mean fitness of the population is $\overline{w} = (1 - 10^{-5})^{10{,}000} \approx 0.90$. Therefore, the average fertility is $\overline{F} = 1/0.90 = 1.11$. That is, each individual should have, on average, 1.11 descendants for the population to remain constant in size. For mammals, such a load is easily bearable.

Let us now assume that the entire diploid human genome ($2 \times 3 \times 10^9$ bp) is functional. If the length of each functional element is the same as that of a bacterial protein-coding gene, i.e., 1,000 nucleotides, then the human genome should consist of 3 million functional loci. With a mutation rate of 10^5 per locus per generation, the total mutational load would be $L = 30$ and the mean population fitness would be 9×10^{-14}. The average individual fertility required to maintain such a population would be $\overline{F} = 1.1 \times 10^{13}$. That is, to maintain an approximately constant population size, each man and woman in the population would have to beget 11,000,000,000,000 children and all but one would have die before reproductive age. The absurdity of such numbers was realized by Muller (1950), who suggested that genetic load values cannot exceed $L = 1$. As a matter of fact, he believed that the human genome has no more than 30,000 genes, i.e., a genetic load of $L = 0.3$, an average fitness of $\overline{w} = 0.72$, and an average fertility per individual of $\overline{F} = 1.39$. More recent estimates by Keightley and Eyre-Walker (2000) suggest that humans have a genetic load of between 0.78 and 0.95 depending on whether most of our mutations are recessive or dominant, respectively.

Above, we assumed that deleterious mutations have an additive effect on fitness. Any factor that increases the number of deleterious mutations removed from the population, such as negative epistasis or inbreeding, will reduce the mutational load. On the other hand, any factor that decreases the efficacy of selection, such as positive epistasis or reduction in effective population size, will increase the mutational genetic load. Be that as it may, let us now consider the implications of the mutational genetic load on the fraction of the genome that is functional.

Studies have shown that the genome of each human newborn carries between 56 and 103 point mutations that are not found in either parental genome (Xue et al. 2009; Roach et al. 2010; Conrad et al. 2011; Kong et al. 2012). If 80% of the genome is functional, as implied by the ENCODE Project Consortium (2012), then 45–82 deleterious mutations arise per generation. For the human population to maintain its current size under these conditions, each of us should have on average 3×10^{19} to 5×10^{35} (30,000,000,000,000,000,000 to 500,000,000,000,000,000,000,000,000,000,000,000) children, which is clearly bonkers. This absurd situation does not arise if the vast majority of point mutations are neutral, i.e., if the human genome consists mostly of junk and indifferent DNA.

Why so much of the genome is transcribed—or is it?

The ENCODE Project Consortium (2012) consisted of approximately 500 researchers and cost in excess of $300 million. Its purpose was to identify all functional elements in the human genome. Its main findings were that 75% of the genome is transcribed, 56% is associated with modified histones, 15% is found in open-chromatin areas, 9% binds transcription factors, and 5% consists of methylated CpG dinucleotides. Do these findings indicate that the human genome is almost entirely functional? Let us dissect one finding, the claim that much of the genome is transcribed—a phenomenon that was originally described by Comings (1972). The interested reader can find detailed refutations of all the ENCODE Project Consortium's other findings in Graur et al. (2013) and Palazzo and Gregory (2014).

ENCODE systematically catalogued every transcribed piece of DNA as functional. In real life, whether a transcript has a function depends on many additional factors. For example, ENCODE ignores the fact that transcription is fundamentally a stochastic process that is inherently noisy (Raj and van Oudenaarden 2008). Some studies even indicate that 90% of the transcripts generated by RNA polymerase II may represent transcriptional noise (Struhl 2007). In fact, many transcripts generated by transcriptional noise may even associate with ribosomes and be translated (Wilson and Masel 2011). Moreover, ENCODE did not pay attention to the number of transcripts produced by a DNA sequence. The vast majority of their newly "discovered" polyadenylated and unpolyadenylated RNAs are present at levels below one copy per cell and are found exclusively in the nucleus, never in the cytoplasm (Palazzo and Gregory 2014). In ENCODE's method, a DNA segment that produces 1,000 transcripts per cell is counted as equivalent to a segment that produces a single RNA transcript once in a blue moon.

We also note that ENCODE used almost exclusively pluripotent stem cells and cancer cells, which are known as transcriptionally permissive environments. In these cells, the components of the RNA polymerase II enzyme complex can increase up to 1,000 times, allowing for high transcription levels from random sequences. In other words, in these cells transcription of nonfunctional sequences, that is, DNA sequences that lack bona fide promoters, occurs at high rates (Babushok et al. 2007). The use of HeLa cells is particularly suspect, as these cells ceased long ago to be representative of human cells. For example, as opposed to humans who have a diploid number of 46, HeLa cells have a "hypertriploid" chromosome number, i.e., 76–80 regular-size chromosomes, of which 22–25 no longer resemble human chromosomes, as well as a variable number of "tiny" chromosomal fragments (Adey et al. 2013). Indeed, HeLa has been recognized as a bona fide biological species called *Helacyton gartleri* (Van Valen and Maiorana 1991).

The human genome contains many classes of sequences that are known to be abundantly transcribed, but are typically devoid of function. Pseudogenes, for instance, have been shown to evolve very rapidly and are mostly subject to no functional constraint. Yet up to one-tenth of all known pseudogenes are transcribed (Pei et al. 2012), and some are even translated, chiefly in tumor cells (Kandouz et al. 2004). Pseudogene transcription is especially prevalent in pluripotent stem, testicular, germline, and cancer cells (Babushok et al. 2007). Unfortunately, because "functional genomics" is a recognized discipline within molecular biology, while "nonfunctional genomics" is only practiced by a handful of "genomic clochards" (Makalowski 2003), pseudogenes have always been looked upon with suspicion and wished away. Gene prediction algorithms, for instance, tend to zombify pseudogenes in silico by annotating many of them as functional genes.

Another category of sequences that are devoid of function yet transcribed are introns. When a human protein-coding gene is transcribed, its primary transcript contains not only functional reading frames but also introns and exonic sequences devoid of reading frames. In fact, only 4% of pre-mRNA sequences is devoted to the coding of proteins; the other 96% is mostly made of noncoding regions. Because introns are transcribed, ENCODE concluded that they are functional. But, are they? Some introns do indeed evolve slightly more slowly than pseudogenes, although this rate difference can be explained by a minute fraction of intronic sites involved in splicing and other functions. There is a longstanding debate over whether or not introns are indispensable components of eukaryotic genomes. In one study (Parenteau et al. 2008), 96 introns from 87 yeast genes were deleted. Only three of these deletions (~3%) seemed to have a negative effect on growth. Thus, in the majority of cases, introns evolve neutrally, whereas a small fraction of introns may be under selective constraint (Ponjavic et al. 2007). Of course, we recognize that some human introns harbor regulatory sequences (Tishkoff et al. 2006), as well as sequences that produce small RNA molecules (Hirose et al. 2003; Zhou et al. 2004). We note, however, that even those few introns under selection are not constrained over their entire length. Hare and Palumbi (2003) compared nine introns from three mammalian species (whale, seal, and human) and found that only about a quarter of their nucleotides exhibit telltale signs of functional constraint. A study of intron 2 of the human *BRCA1* gene revealed that only 300 bp (3% of the length of the intron) may be constrained (Wardrop et al. 2005). Thus, the practice of summing up all the lengths of all the introns and adding them to the pile marked "functional" is misleading.

The human genome is also populated by a very large number of transposable elements. Transposable elements, such as LINEs, SINEs, retroviruses, and DNA transposons, may, in fact, account for up to two-thirds of the human genome (de Koning et al. 2011) and for more than 31% of the transcriptome (Faulkner et al. 2009). Both human and mouse have been shown to transcribe SINEs (Oler et al. 2012). The phenomenon of SINE transcription is particularly evident in carcinoma cell lines, in which multiple copies of *Alu* sequences are detected in the transcriptome (Umylny et al. 2007). Moreover, retrotransposons can initiate transcription on both strands (Denoeud et al. 2007). These transcription initiation sites are subject to almost no evolutionary constraint, casting doubt on their "functionality." Thus, while a handful of transposons have been domesticated into functionality, one cannot assign a "universal utility" to all retrotransposons (Faulkner et al. 2009).

Whether transcribed or not, the majority of transposons in the human genome are merely parasites, parasites of parasites, or dead parasites, whose main "functions" appear to be causing frameshifts in reading frames, disabling RNA-specifying sequences, and simply littering the genome.

Life History and Cellular Correlates of Genome Size

A variety of cellular, organismal, and life history parameters have been reported to correlate with haploid genome sizes (C values). Unfortunately, these relationships are never universal; they are only apparent in limited taxonomic contexts. For example, although genome size is inversely correlated with metabolic rate in both mammals and birds, no such relationship is found in amphibians.

Many correlates of genome size were described in plants (Greilhuber and Leitch 2013) but have no applicability outside this kingdom. For example, species with large genomes flower early in the spring, while later-flowering species have progressively smaller genomes. Additional examples concern weediness (the ability to invade arable lands) and invasiveness (the ability to colonize new environments), which were both found to be negatively correlated with genome size.

An intriguing hypothesis for explaining genome size in plants casts phosphorus as a key player. Phosphorus is an important ingredient in the synthesis of nucleic acids (DNA and RNA), as well as membrane lipids and many enzymes. Yet despite its being the twelfth most abundant element in the environment, it is not readily accessible to plants and is often present in such low amounts that it may be considered a limiting nutrient for DNA synthesis. Indeed, some plants have evolved specific mechanisms to cope with limited levels of phosphorus. For example, it has been observed that unicellular algae under chronic phosphorus shortage have the capacity to restructure their membranes by replacing phosphorus-rich lipids with sulfur- or nitrogen-containing lipids. Because large genomes require increased supplies of phosphorus, it has been hypothesized that polyploids and plants with large genomes are at a selective disadvantage in phosphorus-limited environments (Sterner and Elser 2002). Phosphorus-depleted soils should, accordingly, be populated by species with small genome sizes. Indeed, plants that live in mineral-poor environments seem to have particularly small genomes. Moreover, the smallest plant genome reported so far was found in a family of carnivorous plants that grow in nutrient-poor environments (Greilhuber et al. 2006). Some experimental support for the phosphorus hypothesis has been obtained by Šmarda et al. (2013) in their long-term fertilization experiment with 74 vascular plant species.

The nucleocytoplasmic ratio

For over 100 years, cytologists have been aware of a positive correlation between the volume of the nucleus and the volume of the cytoplasm. These observations led to the concept of the **nucleocytoplasmic** or **cytonuclear ratio**, according to which the ratio of the nuclear volume to the cytoplasmic volume is constant, reflecting the need to balance nuclear and cytoplasmic processes. The nucleocytoplasmic ratio was determined to be approximately 15–25% in plant and 5–10% in animal cells (Price et al. 1973; Ledda et al. 2000). Of course, this ratio is not a precise absolute constant; it varies during ontogenetic development, and the different methods of determining the ratio yield somewhat different results (Trombetta 1942; Ledda et al. 2000). Notwithstanding these caveats, the nucleocytoplasmic ratio turned out to be mostly impervious to alterations by such processes as increasing or decreasing nuclear DNA content, varying the growth conditions, or subjecting the organism to various drug treatments. Indeed, deviations from the nucleocytoplasmic ratio are frequently associated with disease and abnormal developmental patterns (Jorgensen et al. 2007; Neumann and Nurse 2007).

Given that a positive correlation exists between genome size and nuclear volume and that cytoplasm volume and cell volume are similarly correlated, the nucleocytoplasmic ratio is frequently presented as a positive correlation between genome size and cell size. Indeed, a rough correlation between C value and cell size has been noted in some of the earliest studies of genome size evolution (Mirsky and Ris 1951) and has since been confirmed in many groups of animals, plants, and protists (**Figure 11.7**). Cavalier-Smith (1993) has claimed that the relationship between C value and cell size ranks among the most fundamental rules of eukaryote cell biology. We note, however, that the best correlations are found among closely related taxa. For example, while a significant correlation between pollen size and genome size was found in a sample of 16 wind-pollinated grass species, a large-scale analysis of 464 angiosperms failed to confirm the correlation (Greilhuber and Leitch 2013). These findings make it clear that

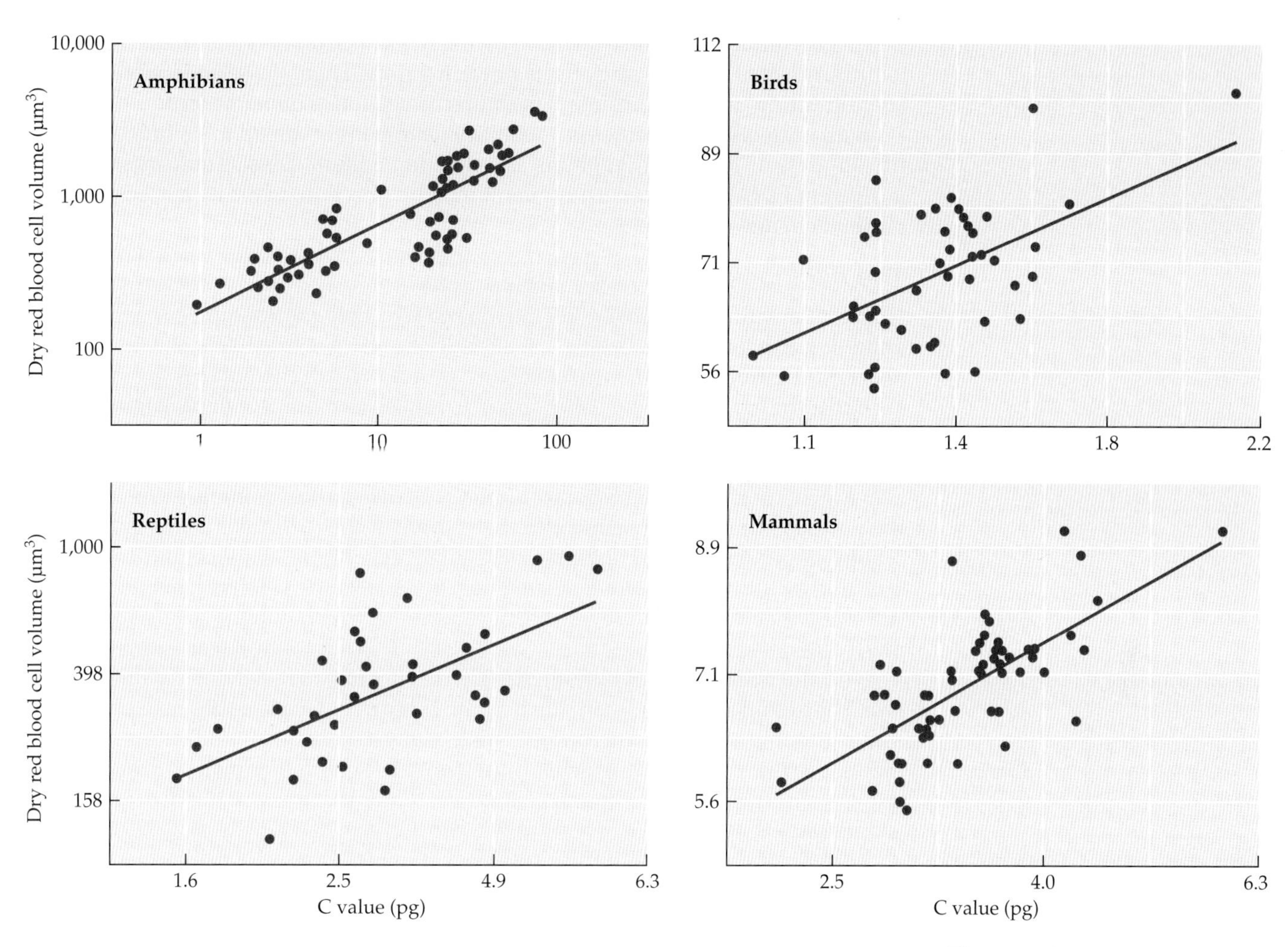

Figure 11.7 A $\log_{10}$-$\log_{10}$ plot illustrating the positive correlation between red blood cell volume and haploid nuclear DNA content (C value) in vertebrates. (Modified from Gregory 2001b, 2005b.)

genome size evolution cannot be understood without reference to the particular biology of the organisms under study. We should also note here that while many studies on the relationship between C values and cell size in plants rely on pollen size, pollen itself is not a single cell. Each pollen grain consists of many vegetative (somatic) cells and a single generative (germline) cell. Finally, we should note that, in many taxa, cell size and genome size do not seem to be correlated (e.g., Starostova et al. 2008, 2009).

The coincidence hypothesis

Several explanations for the correlation between genome size and cell size have been put forward in the literature. Under the **coincidence hypothesis**, most DNA is assumed to be junk and genome size is assumed to increase through mutation pressure. The increase in the amount of DNA in the genome is not, however, a boundless process. At a certain point, the genome becomes too large and too costly to maintain, and any further increases in genome size will be deleterious and selected against. In the coincidence hypothesis, it is assumed that bigger cells can tolerate the accumulation of more DNA (Pagel and Johnstone 1992). In other words, the correlation between genome size and cell size is purely coincidental. The cell and the nucleus are envisioned as finite containers to be filled with DNA.

Nucleotypic hypotheses

Nucleotypic hypotheses assume that the genome has a **nucleotypic function** in addition to its function as a carrier of genetic information. In other words, the genome is assumed to affect the phenotype and the fitness of the organism in a manner that is dependent on its length but independent of its sequence. As a consequence, genome size may be under secondary selection owing to its nucleotypic effects. Let us assume, for instance, that genome size affects flowering time. Then, the selection for earlier or later flowering times will result in an indirect selection on genome size.

In all nucleotypic hypotheses, the entire genome is assumed to be functional, although only a small portion of it is assumed to be literal DNA. Thus, most DNA is indifferent DNA, whose length is maintained by selection, while its nucleotide composition changes at random.

The nucleoskeletal hypothesis

A particularly popular nucleotypic explanation for the relationship between C value and cell size was put forward by Cavalier-Smith (1978, 1982, 1985a, 2005), who envisioned the DNA to be a nucleoskeleton around which the nucleus is assembled, such that the total amount of DNA in a cell, as well as its degree of packaging, exerts a strong effect on nucleus size and subsequently on cell size. According to this **nucleoskeletal hypothesis**, the DNA is not only a carrier of genetic information, but also a structural material element—a nucleoskeleton that maintains the volume of the nucleus at a size proportional to the volume of the cytoplasm. Since larger cells require larger nuclei, selection for a particular cell volume will secondarily result in selection for a particular genome size. Thus, the correlation between genome size and cell size arises through a process of coevolution in which nuclear size is adjusted to match alterations in cell size.

With the nucleoskeletal hypothesis, the driving force is selection for optimal cell size. For example, if a large cell size became adaptively favorable because of changes in the environment, then there would also be positive selection for a corresponding increase in nuclear volume, which in turn would be achieved primarily through either increases in the amount of indifferent DNA or modifications in the degree of DNA condensation.

At this point, we need to raise two questions: (1) Does cell size matter? and (2) Does genome size affect cell size? The answer to the first question is that cells do have an optimal size, i.e., they need to be not too big and not too small. Based on studies in *C. elegans* and other systems, it has been argued that a cell's size is limited by the physical properties of its components (Marshall et al. 2012). For example, in order to proliferate, a cell has to divide, and for faithful cell division, the molecular machinery, such as the centrosome (the organelle that serves as the main microtubule organizing center) and the mitotic spindle, must be constructed at the right position and with the correct size. This may not be accomplished in extremely large or extremely small cells due to the physical properties of microtubules and chromosomes. If a cell exceeds the upper size limit, its centrosome and mitotic spindle may not be able to position themselves at the center of the cell, leading to nonsymmetrical cell division. Moreover, in such a case, microtubules may not reach the cell cortex, potentially leading to insufficient spindle elongation. If a cell falls below a lower size limit, its centrosome may not be able to position itself in a stable manner at the center of the cell due to the excess elastic forces of the microtubules. In addition, there may not be sufficient space for accurate chromosome segregation.

We do not know at present what the upper or lower limits of cell size are. We do, however, know that some cells are so extremely large that they most certainly exceed whatever the theoretical upper limit for cell size may be. Three such examples are the shelled amoeba, *Gromia sphaerica*, which can reach a diameter of 3 cm; ostrich eggs, which can reach 15 cm in diameter and can weigh more than 1 kg; and the record holders, unicellular seaweeds belonging to genus *Caulerpa*, whose tubular stolons may extend to a length of three or more meters. These enormous examples cannot, however, be taken as evidence that an upper limit for cell size does not exist. What these examples show is that there exist molecular and cellular devices for escaping the consequences of large cell size during cell division. None of these enormous cells undergoes regular binary fission; they divide either by cleaving only a small portion of their mass (e.g., bird eggs) or by becoming multinucleate for at least part of their life cycle and producing large numbers of diminutive progeny or gametes (e.g., *Gromia* and *Caulerpa*).

As to the second question—does genome size affect cell size?—the evidence is quite thin. First, the correlation between cell size and genome size is imperfect at best. Second, there is evidence for contributors to cell size other than DNA. Levy and Heald (2007) studied the regulation of nuclear size in two related frog species: *Xenopus laevis* and *X. tropicalis*. *Xenopus laevis* is a larger animal than *X. tropicalis*, and its cells are tetraploid while the cells of *X. tropicalis* are diploid. The two species also differ in another aspect: the cells and nuclei of *X. laevis* are larger. Because *Xenopus* nuclei can be assembled in a test tube using chromatin (a DNA-protein complex) and extracts of egg cytoplasm, one can test whether or not DNA has a role in determining cell size. Levy and Heald (2007) added sperm chromatin from either *X. laevis* or *X. tropicalis* to egg extracts from either *X. laevis* or *X. tropicalis*. They found that, although both extracts can trigger assembly of the nuclear envelope around the chromatin, the cytoplasmic extract from *X. laevis* forms larger nuclei than the *X. tropicalis* extract, regardless of the DNA used. This indicates that one or more cytoplasmic factors determine nuclear size, while DNA content seems to have no discernable effect. Can we, therefore, state that the nucleoskeletal hypothesis has been invalidated? Given that what we have here is one laboratory experiment involving a single amphibian genus, it may be too early to discard the nucleoskeletal hypothesis, although its universality has certainly been called into question.

Is small genome size an adaptation to flight?

Cavalier-Smith (1995) hypothesized that high metabolic rates require relatively small cells because such cells have a larger surface-to-volume ratio and a higher rate of gas exchange per unit volume than larger cells. Because of the correlation between cell size and genome size postulated by his nucleoskeletal hypothesis, he predicted high metabolic rates to be associated with small genome sizes. Indeed, in amniotes (reptiles, birds, and mammals), genome size was found to roughly correlate negatively with metabolic rate. The observations that birds and bats have the smallest genomes among vertebrates and that, among birds, flightless species tend to have the largest genomes led to the idea that the metabolically intense demands of powered flight resulted in strict selective constraints on cell size and indirectly on genome size (Hughes and Hughes 1995).

Organ et al. (2007) estimated the sizes of osteocytes (bone cells) in 31 extinct species of dinosaurs by measuring the cavities in fossilized bone in which cells once resided. By calibrating the relationship between osteocyte size and genome size using data from extant species, they estimated dinosaur genome sizes. Their analysis indicated that small genomes evolved before the divergence among saurischian taxa (**Figure 11.8**). Thus, such unmistakable nonflying species as *Brontosaurus* and *Tyrannosaurus rex* were found to have smaller cells (and presumably lesser quantities of DNA) than some flying species. Most species from the other primary clade of dinosaurs, the ornithischians, had larger cell sizes, comparable to those of most extant nonavian reptiles, although some lineages exhibited reduced cell sizes. Finally, some extant nonavian reptiles—for example, monitor lizards (genus *Varanus*)—possess genomes as small as those of flying birds.

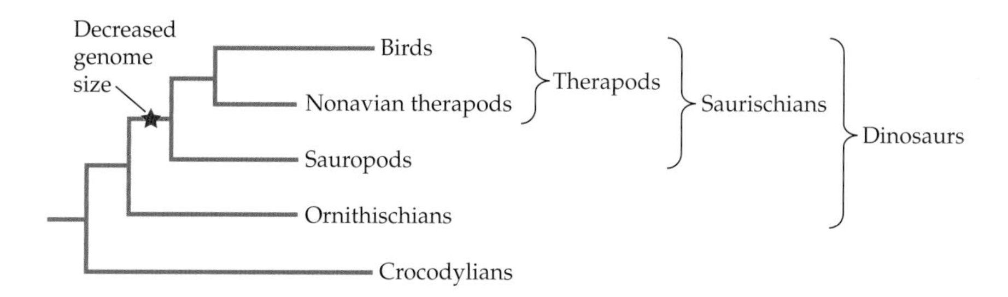

Figure 11.8 A major decrease in genome size (as inferred from measurements of cell size) occurred approximately 250 million years ago in the saurischians, one of the two main dinosaur branches (red star). With the exception of birds, none of the other saurischian lineages evolved flying capabilities.

In summary, the decrease in genome size occurred approximately 250 million years ago, considerably earlier than the emergence of the first birds about 150 million years ago, invalidating the theory that small genomes constitute an adaptation to the metabolic costs associated with flight.

The C-Value Paradox: The Neutralist Hypothesis

As explained previously, a scientific hypothesis should spell out the conditions for its refutation. As far as the C-value paradox is concerned, the simplest scientific hypothesis is that the fraction of DNA that looks superfluous is, in fact, superfluous. The assumption that a vast fraction of the genome evolves in a neutral fashion means that this DNA does not tax the metabolic system of eukaryotes to any great extent and that the cost (e.g., in energy, time, and nutrients) of maintaining and replicating large amounts of nonfunctional DNA is negligible.

The first researcher to suggest that part of the genome may lack function was Darlington (1937), who recognized the difficulties in getting rid of redundant DNA, because of its linkage to functional genes: "It must be recognized that the shedding of redundant DNA within a chromosome is under one particularly severe restriction, a restriction imposed by its contiguity, its linkage with DNA whose information is anything but dispensable." The theme of redundancy was later adopted by Rees and Jones (1972) and by Ohno (1972), the latter being credited with popularizing the notion that much of the genome in eukaryotes consists of junk DNA. In particular, Ohno (1972) emphasized the interconnected themes of gene duplication and trial and error in genome evolution. Gene duplication can alleviate the constraints imposed by natural selection by allowing one copy to maintain its original function while the other accumulates mutations. Only very rarely will these mutations turn out to be beneficial. Most of the time, however, as Ohno (1972) pointed out, one copy will be degraded into a pseudogene: "The creation of every new gene must have been accompanied by many other redundant copies joining the ranks of silent DNA base sequences." Genomes are not *arcs de triomphe* celebrating successful evolutionary conquests. "Triumphs as well as failures of nature's past experiments appear to be contained in our genome" (Ohno 1972).

The discovery of transposable elements, and more importantly the observation that the vast majority of transposable elements are nonfunctional and highly degraded by mutation, added support for the junk DNA hypothesis, as did the discovery of other nonfunctional genomic elements, such as pseudogenes, introns, and highly repetitive DNA.

Under the neutralist hypothesis, we need to address three issues. First, we need to ask what mutational processes can create junk DNA, i.e., what does junk DNA consist of? Second, we need to identify the evolutionary driving forces that can maintain junk DNA. Third, we need to explain why different organisms possess vastly different quantities of junk DNA. The first question is relatively easy to answer. Notwithstanding the fact that gene and genome duplications can create redundancies that may ultimately result in junk DNA, and that satellite DNA can too contribute to the nonfunctional DNA fraction, there is currently no doubt whatsoever that the bulk of junk DNA is derived from transposable elements. By estimating the relative contributions of the major types of transposable elements, genomes can be classified into four main categories: (1) genomes in which DNA transposons predominate (e.g., *Amphioxus*, *Ciona*, most teleost fishes, *Xenopus*), (2) genomes in which LINEs and SINEs predominate (e.g., lamprey, elephant shark, *Takifugu*, coelacanth, chicken, mammals), (3) genomes with a predominance of LTR retrotransposons (e.g., the tunicate *Oikopleura*), and (4) genomes in which no particular transposable-element type predominates (e.g., *Tetraodon*, stickleback, reptiles, zebra finch). Some genomes are particularly poor in DNA transposable elements and contain almost exclusively retroelements (e.g., elephant shark, coelacanth, birds, mammals); however, there are no genomes that contain almost exclusively DNA transposable elements (Chalopin et al. 2015).

In the following sections, we discuss the **selfish DNA hypothesis**, which attempts to explain how superfluous and even deleterious elements can multiply within genomes and spread within populations, as well as the **mutational hazard hypothesis**, which provides an explanation for the fact that different species harbor vastly different quantities of junk DNA.

Selfish DNA

Selfish DNA (Doolittle and Sapienza 1980; Orgel and Crick 1980) is a term that applies to DNA sequences that have two distinct properties: (1) the DNA sequences form additional copies of themselves within the genome, and (2) the DNA sequences either do not make any specific contribution to the fitness of the host organism or are actually detrimental. Some selfish DNA also engages in transmission ratio distortion and horizontal gene transfer—two processes through which it can further increase its frequency in the population.

The vast majority of selfish DNA comprises class I and class II transposable elements that (1) are active, i.e., have the ability to produce copies of themselves by replicative transposition, and (2) have not been domesticated, i.e., co-opted into function (Chapter 9). A minor fraction of selfish DNA consists of promiscuous DNA as well as tandemly repeated sequences. As an approximation, here we will use the terms "selfish DNA," "selfish DNA elements," and "transposable elements" interchangeably.

Because of their ability to increase in number, selfish DNA elements can have profound effects on the size and structure of genomes. Two main classes of hypotheses have been put forward to explain the long-term persistence of selfish DNA in the genome. One class of hypotheses proposes that the process reflects two independent equilibria. At the genome level, a balance is achieved between transposition or retroposition, on the one hand, and the mechanisms utilized by the host to restrain transposable-element activity, on the other. At the population level, a balance is achieved between the rate with which new copies of transposable elements are created and the efficiency with which negative selection gets rid of genotypes carrying transposable elements. The efficiency of selection, in turn, depends on population genetic parameters such as effective population size (Brookfield and Badge 1997; Le Rouzic et al. 2007; Levin and Moran 2011). Under this class of hypotheses, a genome is assumed to be a closed system, in which the activity of transposable elements is counteracted by such intrinsic entities as small interfering RNAs, Piwi-interacting RNAs (piRNAs), DNA methylation, and histone modifications (Chapter 9). In this closed system, transposable elements employ various evasive strategies, such as preferential insertion into regions transcribed by RNA polymerase II. Sooner or later, the system comprising the host genome and the transposable elements reaches a stable equilibrium, after which nothing much happens until internal or external perturbations, such as mutations or environmental stress, disrupt the equilibrium, at which point either a burst of transposition is unleashed or the transposable elements become forever incapacitated.

The discovery that some transposable elements, notably the *P* element of *Drosophila*, are able to colonize new genomes by means of horizontal transfer (Daniels et al. 1990) unveiled an additional way for transposable elements to persist over evolutionary time. Horizontal escape of an active transposable element into a new genome would allow the element to evade a seemingly inevitable extinction in its original host lineage resulting from host elimination or inactivation due to mutational decay (Hartl et al. 1997).

Although the inherent ability of transposable elements to integrate into the genome suggested a proclivity for horizontal transfer (Kidwell 1992), the extent to which such processes affected a broad range of transposable elements and their hosts was not clear. Schaack et al. (2010) revealed more than 200 cases of horizontal transfer involving all known types of transposable elements, which may mean that virtually all transposable elements can horizontally transfer.

From a genome-wide study across *Drosophila* species, it was estimated that approximately one transfer event per transposable-element family occurs every 20 million years (Bartolome et al. 2009). In several instances, we have evidence for horizontal transfer of transposable elements among extremely distant taxa, including at least 12 movements across phyla. So far, all these "long jumps" were found to involve DNA transposons, suggesting one of two possibilities: either DNA elements are better adapted to invade genomes than RNA elements, or the preponderance of DNA elements represents a case of ascertainment bias due to the fact that DNA elements are studied more intensively in an evolutionary context, while the research on RNA elements is almost exclusively focused on a narrow taxonomic range of so-called model organisms.

It is becoming increasingly clear that the life cycle of a transposable-element family is akin to a birth-and-death process in that it starts when an active copy colonizes a novel host genome and ends when all copies of the transposable-element family are lost or inactivated—by chance, through the accumulation of disabling mutations; or by negative selection, in a process that may be driven by host defense mechanisms; or by the fact that each transposable element contributes negatively to the fitness of the organism.

There are two major ways for transposable elements to escape extinction: the first is to horizontally transfer to a new host genome prior to extinction; the second is to inflict minimal fitness harm. As with other parasites, it is possible that transposable elements will make use of different strategies at different times. Each strategy has a phylogenetic signature. In cases in which horizontal transfer is frequent, there should be dramatic incongruence between the phylogenies of the transposable-element family and those of their various host species. In these cases, horizontal transfer might allow the transposable element to colonize a new genome in which host suppression mechanisms are inefficient.

In cases where a transposable-element family has persisted for long periods in a host lineage, the reduced frequency of horizontal transfer can be inferred from the similarity between the phylogenies of the transposable element and the hosts. Persistence could be achieved, for instance, through self-regulatory mechanisms that limit copy number or by evolving targeting preference for insertion into "safe havens" in the genome, such as through preferential transposition into high-copy-number genes or heterochromatin. The *LINE-1* element of mammals provides an exceptional example of endurance, having persisted and diversified over the past 100 million years with virtually no evidence of horizontal transfer.

The mutational hazard hypothesis

So far, we have provided plausible explanations for the persistence of junk DNA within genomes. We have not, however, addressed the question of genome size disparity: Why do certain organisms possess minute quantities of junk DNA, whereas the genomes of other eukaryotic organisms are made almost entirely of nonfunctional DNA? A possible explanation may be that some genomes have not been invaded by transposable elements while others have been invaded many times. Another explanation may be that some genomes have been invaded by very inefficient transposable elements, and others by very prolific ones. A third explanation may be that some genomes are extremely efficient at warding off selfish DNA while others are more permissive. Since these three explanations yield no predictions concerning the relationship between the amount of junk DNA and population genetic variables, to the best of our knowledge they have not been tested. Here, we present a hypothesis that uses differences in selection intensity and selection efficacy to explain the observed difference between organisms with much junk DNA and those with little junk DNA.

The mutational hazard theory (Lynch and Conery 2003; Lynch 2006) postulates that virtually all increases in genome size in eukaryotes impose a slight fitness reduction. Thus, eukaryotic genomes are assumed to contain varying amounts of excess DNA that behaves like junk DNA in species with large effective population sizes and like garbage DNA in species with small effective population sizes.

As explained in Chapter 2, the fate of a slightly deleterious allele is determined by the interplay between purifying selection and random genetic drift. Purifying selection acts by decreasing the frequency of the slightly deleterious allele at a rate that depends upon its selective disadvantage, s, where $s < 0$. Random genetic drift changes the allele frequencies in a nondirectional manner at a mean rate that is proportional to $1/N_e$, where N_e is the effective population size. Thus, whether a mutation that increases genome size is selected against or has the same probability of becoming fixed in the population as a neutral mutation is determined by its selective disadvantage relative to the effective size of the population into which it is introduced (**Figure 11.9**). Lynch and Conery (2003) argued that the ineffectiveness of selection in species with low N_e is the key to understanding the evolution of genome size, and the main prediction of the mutational-hazard theory is that large genomes will be found in species with small effective population sizes. How can we test this hypothesis?

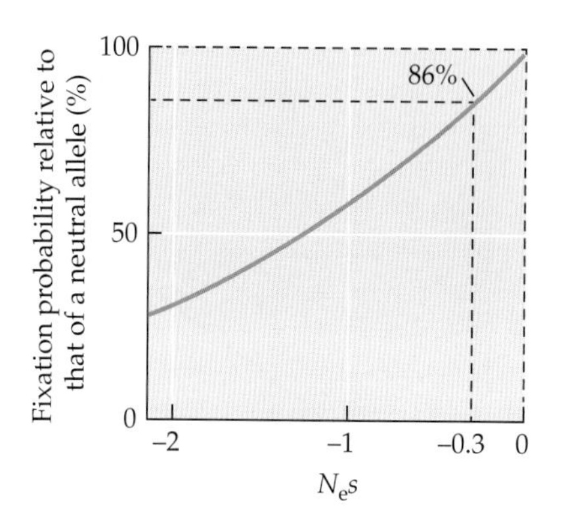

Figure 11.9 The relative fixation probability of a deleterious allele relative to that of a strictly neutral allele ($N_e s = 0$) as a function of $N_e s$. The relative fixation probability of the strictly neutral is set at 1. As an example, a mutation with $N_e s = 0.3$ has a fixation probability that is 86% of that of a neutral allele. (Modified from Yi 2006.)

First, we must ascertain that two mutational requirements are met. The first mutational requirement is that mutations resulting in genome increases will outnumber mutations resulting in genome diminution. In many prokaryotes, for instance, this condition is not met. As a consequence, random genetic drift causes the genomes of many prokaryotes with small effective population sizes to get smaller rather than larger (Chapter 10). In most eukaryotes, the proliferation of transposable elements overwhelms all other mutations and, hence, at the mutation level, the condition that genome increases outnumber genome decreases is met.

The second mutational requirement is that the addition of each transposable element to the genome should on average decrease the fitness of its carrier. Two types of empirical data support this assumption. First, it has been shown that in *D. melanogaster*, each transposable-element insertion decreases the fitness of an individual by 0.4% (Pasyukova et al. 2004). Second, it has been shown that bottlenecks affect the diversity and activity level of transposable elements, with populations that experienced size reductions having an increased level of both transposable-element activity and diversity (Lockton et al. 2008; Picot et al. 2008).

After determining that the mutational requirements of the mutational hazard hypothesis are met, we need to ascertain that the postulated relationship between genome size and effective population size (N_e) is supported by empirical data. In principle, N_e can be estimated directly by monitoring the variance of allele frequency changes across generations. In practice, however, as the expected changes in allele frequency from generation to generation are extremely small unless N_e is tiny, this approach is difficult to put into practice, because errors in estimating allele frequencies will overwhelm the true change in allele frequencies unless the sample size is enormous. As a consequence, most attempts to estimate N_e have taken a circuitous route, the most popular being to indirectly infer effective population size from the levels of within-population variation at nucleotide sites assumed to evolve in a neutral fashion. The logic underlying this approach is that if μ is the rate of neutral point mutations per generation per site and if N_e is roughly constant, then an equilibrium level of variation will be reached in which the mutational input to variation, 2μ, is matched by the mutational loss via genetic drift, $1/(2N_e)$. At equilibrium, the nucleotide site diversity (Chapter 2) is expected to be approximately equal to the ratio of these two forces, i.e., $4N_e\mu$ for a diploid population and $2N_e\mu$ for haploids. Thus, to estimate effective population size, one needs to empirically determine the degree of nucleotide site diversity, which is a relatively straightforward process, and the rate of mutation, μ, which is a slightly more difficult thing to do.

After measuring nucleotide site variation at synonymous sites and taking into account the contribution from mutation, average N_e estimates turned out to be ~10^5 for vertebrates, ~10^6 for invertebrates and land plants, ~10^7 for unicellular eukaryotes, and more than 10^8 for free-living prokaryotes (Lynch et al. 2011). Although crude, these estimates imply that the power of random genetic drift is substantially elevated in eukaryotes—e.g., at least three orders of magnitude in large multicellular species relative to prokaryotes. It is also clear that the genetic effective sizes of populations are generally far below actual population sizes.

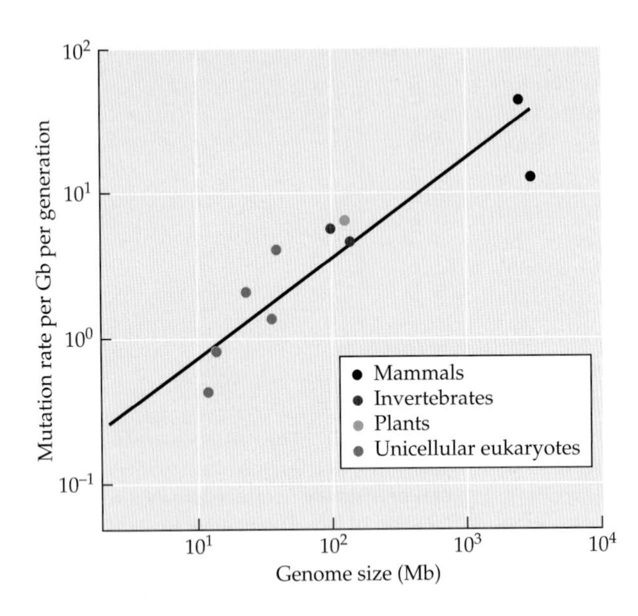

Figure 11.10 A positive correlation between mutation rate and genome size in eukaryotes. (Modified from Lynch 2010.)

At present, there is insufficient data on effective population sizes from a sufficiently diverse sample of taxa to test the mutational hazard hypothesis directly. There is, however, indirect evidence supporting it. As mentioned in Chapter 2, the mutation rate per generation is expected to be higher in species with small effective population sizes than in species with large effective population sizes. The mutational hazard hypothesis asserts that organisms with small effective population sizes should have larger genomes than organisms with large effective population sizes. Thus, support for the mutational hazard hypothesis can be obtained indirectly by showing that a positive correlation exists between mutation rate and genome size. Such a correlation has indeed been obtained by (**Figure 11.10**).

GENOME SIZE AND BOTTLENECKS: THE SIMULTANEOUS ACCUMULATION OF *ALUS*, PSEUDOGENES, AND *NUMTS* WITHIN PRIMATE GENOMES As far as slightly deleterious mutations are concerned, the importance of random genetic drift relative to selection becomes especially pronounced during profound reductions of population size (bottlenecks). If genome size can be shown to have increased concomitantly with reductions in population size, then the mutational hazard theory will gain evidential support.

Gherman et al. (2007) compared the evolution of *numt* insertions in primate genomes and compared them with the insertions of two other classes of nonfunctional elements, *Alus* and pseudogenes (Britten 1994; Bailey et al. 2003; Ohshima et al. 2003). As mentioned in previous chapters, these elements are unlikely to be functional, since their rate of evolution indicates a complete lack of selective constraint and they possess no positional, transcriptional, or translational features that might indicate a beneficial function subsequent to their integration into the nuclear genome. Using sequence analysis and fossil dating, Gherman et al. (2007) showed that a probable burst of integration of *Alus*, pseudogenes, and *numts* in the primate lineage occurred close to the prosimian-anthropoid split, which coincided with a major climatic event called the Paleocene-Eocene Thermal Maximum (~56 million years ago). During this event, which lasted for about 70,000 years, average global temperatures increased by approximately 6°C, massive amounts of carbon were added to the atmosphere, the climate became much wetter, sea levels rose, and many taxa went extinct or suffered decreases in population size. Thus, the increase in primate genomes can be largely accounted for by a population bottleneck and the subsequent neutral fixation of slightly deleterious nonfunctional insertions. The fact that three classes of nonfunctional elements that use vastly different mechanisms of multiplying in the genome increased their numbers simultaneously effectively rules out selectionist explanations. These findings suggest that human and primate genomic architecture, with its abundance of repetitive elements, arose primarily by evolutionary happenstance.

Is it junk DNA or is it indifferent DNA?

Distinguishing between neutralist and nucleoskeletal explanations has been quite difficult. Pagel and Johnstone (1992) hypothesized that a major cost of junk DNA is the time required to replicate it. Organisms that develop at a slower pace may therefore be able to "tolerate" greater amounts of junk DNA, and thus a negative correlation across species between genome size and developmental rate is predicted. In contrast, the prediction of the nucleoskeletal hypothesis is for a positive correlation between genome size and cell size. Unfortunately, organisms with large cells also tend to develop slowly, whereas faster-growing organisms typically have smaller cells. Thus, according to the skeletal DNA hypothesis, a negative correlation between developmental rate and the C value is also expected. However, according to the nucleotypic hypothesis, the relation between developmental rate and genome size occurs only secondarily, as a result of the relationship between developmental rate and cell size.

Pagel and Johnstone (1992) studied 24 salamander species. The size of the nuclear genome was found to be negatively correlated with developmental rate, even after the effects of nuclear and cytoplasmic volume had been removed. However, the correlations between genome size, on the one hand, and nuclear and cytoplasmic volumes, on the other, became statistically insignificant once the effects of developmental rates had been removed. These results support the hypothesis that most of the DNA in salamanders is junk rather than indifferent DNA. Whether Pagel and Johnstone's results represent a true phenomenon or one restricted to *Salamandra* is still a controversial subject (Gregory 2003), especially since in eukaryotes, the "cost" of replicating DNA may not be correlated perfectly with genome size.

Trends in Genome Size Evolution

Here, we are interested in what happens to genome size during evolution. Do genomes, as a rule, become larger during evolution? Or, are changes in genome size haphazard, with genome size increases and decreases distributed randomly along the branches of phylogenetic trees? Several algorithms for inferring ancestral genome sizes (and other continuous-value variables) on rooted phylogenetic trees are available (e.g., Swofford and Maddison 1987; Maddison 1991), and these methods allow us to infer the direction of change in genome size on each branch. Although the taxonomic representation is far from comprehensive, it seems that, in general, genomes tend to increase in size during evolution.

For example, in a study of eight salamander families, which have some of the largest genomes among vertebrates and also some of the lowest reported levels of genetic diversity, Herrick and Sclavi (2014) found that genome size is correlated with phylogenetic age, indicating that genome sizes in salamanders tend to increase with time.

In plants the general direction of the trends is similar to that described above. Johnston et al. (2005) studied genome size evolution of 28 species belonging to Brassicaceae—a large family of flowering plants informally known as mustards. A rooted tree with 28 OTUs has 54 branches (Chapter 5). Increases in genome size were observed along 51 branches (94%), while decreases in genome size were only seen in three branches. These three branches were terminal ones, indicating that the reductions in genome size occurred only very recently. Similar results were obtained in diploid grass species as well as other taxa. Bennetzen and Kellogg (1997) rightfully characterized eukaryotic genome evolution is a "one-way ticket to genomic obesity."

Is there an upper limit to genome size?

Are there selective disadvantages associated with the accumulation of large amounts of nonfunctional DNA? Or is this phenomenon, also referred to as **genomic bloating** (Plague et al. 2008), of no noticeable selective consequence? We first note that very large genomes, that is, those above 20 Gb, are rare. They are, however, distributed widely across the eukaryotic lineages (Metcalfe and Casane 2013). So far, such genomes have been found in diatoms, dinoflagellates, animals, and green plants.

There are sporadic reports of drawbacks in maintaining large amounts of nonfunctional DNA. It has been claimed, for example, that large genomes exhibit a greater sensitivity to mutagens than small genomes (Heddle and Athanasiou 1975). However, large-scale studies addressing the question of whether genomes can grow ad infinitum or whether there are upper limits to the genomic fraction of good-for-nothing DNA have been carried out only in plants.

Within the flowering plants, the distribution of genome sizes is significantly skewed, with decreasing numbers of species in the large-genome categories. One way to explain this skew is to suggest that increases in genome size are rare and that only a few lineages have experienced them. This explanation cannot be true, however, since both transposable elements and polyploidy are exceedingly common in plants, and both of these evolutionary processes are known to rapidly increase genome size. Thus, it appears that there has been enough time and ample means for all plant genomes to become large. In itself, the fact that not all plant genomes are large can already serve as an argument against the possibility of unchecked genomic enlargements.

Vinogradov (2003) and Knight et al. (2005) found that species with large genomes are less likely to generate progenitor species, through either decreased speciation rates or increased rates of extinction. However, the relationships between genome size and speciation rates were nonlinear. For example, while lineages with small genomes may be either speciose or nonspeciose, lineages with large genomes are almost invariably nonspeciose.

A similar relationship was evident when trends across environmental gradients were examined. Species with small genomes were found in widely varying habitats; however, lineages with very large genomes seem to be excluded from the most extreme habitats. Again, it does not appear that genome size is generating a consistent causative effect across the whole genome-size range; rather it appears that lineages with the largest genome sizes are unable to survive in extreme environments. Finally, large-genome plants produce larger seeds and have significantly reduced photosynthetic rates.

From an evolutionary viewpoint, invasive species constitute a great success. Interestingly, in a study of the European reed canarygrass (*Phalaris arundinacea*) that has invaded North America, Lavergne et al. (2010) found that the invading plants had smaller genomes than the native European ones. The authors argue that the smaller genome size had phenotypic effects, such as higher growth rate, and that these traits enhanced the invasive potential of the species.

Knight et al. (2005) proposed the **large-genome constraint hypothesis**, according to which plant species with large genomes are restricted to less stressful environments with longer growing seasons. This restricted ecological tolerance, combined with the fact that their heavier seeds have restricted dispersion abilities, most probably means that species with large genomes have small effective population sizes. This in turn may increase the probability of extinction due to "mutational meltdown" (Chapter 2). Moreover, the inability to colonize extreme environments may decrease the chances of long-term isolation and allopatric speciation of large-genome lineages, thus explaining why large-genome plants are not speciose.

All these findings suggest that large genomes constitute a maladaptive trait and that increases in genome size have an evolutionary cost. Unfortunately, it is impossible to state at present whether or not these findings are applicable to other taxa or to flowering plants only.

Genome miniaturization in eukaryotes

As is the case in prokaryotes (Chapter 10), genome reduction is also observed in intracellular symbiotic and parasitic eukaryotes, most prominently in secondary organelles that have been derived from free-living eukaryotes. In the following sections we discuss the extreme genome reduction in microsporidia and the case of organelles derived from eukaryotic cells.

GENOME REDUCTION IN MICROSPORIDIA Microsporidia are obligate, spore-forming, intracellular parasites that invade animal cells. Most microsporidia infect insects, but they are also responsible for diseases in crustaceans and fish, as well as for emerging and opportunistic zoonoses in humans. Their phylogenetic notoriety stems from the fact that they lack mitochondria and were once thought to represent one of the most basal divergence events within Eukaryota (Chapter 6). They have since been recognized as fungi, although due to the fact that they evolve much faster than the other fungi, it has not been possible as yet to assign them to a specific fungal lineage (James et al. 2006).

The genomes of some microsporidia are examples of extreme reduction. At approximately 2.3 and 2.9 Mb in length, the smallest sequenced microsporidian genomes are those of *Encephalitozoon intestinalis* and *E. cuniculi*, respectively. Thus, the smallest microsporidian genome has a length that is only ~18% of that of the brewer's yeast *Saccharomyces cerevisae*. The genomes of *E. intestinalis* and *E. cuniculi* contain 1,833 and 1,997 protein-coding genes, respectively, i.e., about a third of the number of genes in yeast. The proteins of the two *Encephalitozoon* species are approximately 15% shorter than those of yeast, and the vast majority of their protein-coding genes are intronless (Katinka et al. 2001). For example, there are only 14 very short spliceosomal introns (23–47 bp) in the protein-coding genes of *E. intestinalis*. In the two *Encephalitozoon* species, genes are tightly packed, with intergenic spaces averaging just over 100 bp, and several protein-coding sequences physically overlap with those of their neighbors. Interestingly, there are 25 rRNA-specifying genes in *E. intestinalis*, the same as in *S. cerevisae*, but only 16 tRNA-specifying genes versus 275 in yeast (17%). Finally, *E. intestinalis* has shed all its pseudogenes—*S. cerevisae* carries 19 in its genome—and all transposable elements seem to have been completely eradicated from its genome.

Is the extreme genomic petiteness of the two *Encephalitozoon* species a consequence of parasitism, or are the diminutive sizes of their genomes explainable by other factors? Corradi et al. (2009) determined a draft genome sequence of *Octosporea bayeri*, a microsporidian intracellular pathogen of the crustacean *Daphnia magna* (water flea), whose genome, at 24 Mb, is ten times bigger than that of *E. intestinalis*. This huge size difference between *O. bayeri* and *E. intestinalis* could be due to several factors. The proximate reasons may be (1) whole or partial genome duplications, (2) an increase in the number or lengths of introns, (3) expansions of gene families, (4) expansion of the functional diversity of the genes, (5) an increase in the lengths of intergenic regions, (6) an increase in the numbers of tandem repetitive sequences, or (7) an increase in the number of dispersed repetitive sequences due to transposable-element activity. Ultimately, the reasons for the existence of an intracellular parasite with a large genome may be either selective or nonselective. Selection may come into play if for some reason *O. bayeri* is less biochemically dependent on its host than *E. cuniculi* and, hence, requires more genes and a greater variety of functionalities. The second possibility is that the greater size of the *O. bayeri* genome has nothing to do with function, degree of host dependency, or selection.

The large size of the *O. bayeri* genome does not exhibit telltale signs of either whole or extensive partial genome duplication. Nor does *O. bayeri* possess more numerous and longer introns than the *Encephalitozoon* species. The genome of *O. bayeri* was found to contain only 19% more protein-coding genes that the *E. cuniculi* genome, an excess that is too small to explain the one-order-of-magnitude difference in genome size. Corradi et al. (2009) assigned all identified protein-coding genes in *O. bayeri* and *E. cuniculi* to one of 11 functional categories. The differences between the two species were quite small: in five categories, there were more proteins in *O. bayeri* than in *E. cuniculi*; in two categories the number of proteins in *O. bayeri* and *E. cuniculi* was exactly the same; while in four categories there were fewer proteins in *O. bayeri* than in *E. cuniculi* (**Figure 11.11**). It seems, therefore, that *O. bayeri* does not have a greater variety of functionalities, merely a little more of the same or a little less of the same as *E. cuniculi*—not nearly enough to explain the genome size differences.

The cause for the genome difference must be found elsewhere. First, *O. bayeri* has a fivefold lower gene density than species with smaller genomes. As a consequence the mean length of the intergenic regions is longer—some reaching lengths that exceed 5.5 Kb. Second, the genome of *O. bayeri* seems to contain many highly repeated sequences. Because the genome sequence of *O. bayeri* has not yet been assembled into chromosomes, it is difficult to tell whether or not the tandemly repeated sequences are telomeric. The vast majority of the dispersed repeated sequences could not be identified; however, 74 contigs were found to contain DNA segments that were highly similar in sequence to known fungal transposable elements. It, therefore, seems that

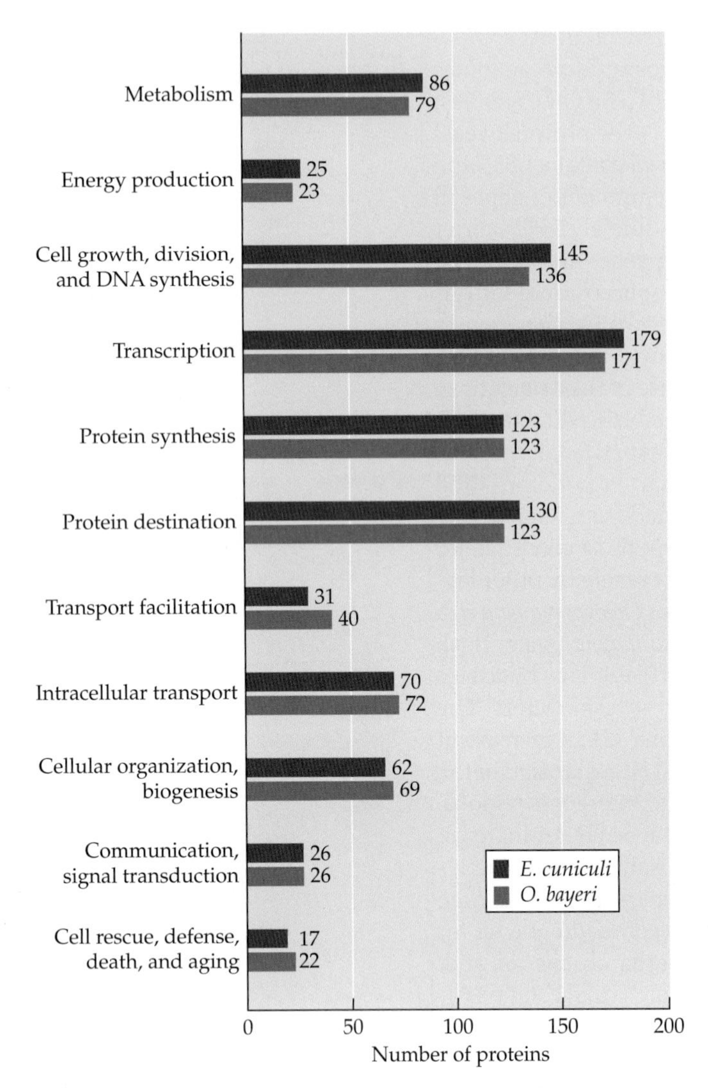

Figure 11.11 Distribution of *Octosporea bayeri* and *Encephalitozoon cuniculi* proteins among 11 functional categories. (Modified from Corradi et al. 2009.)

a proliferation of transposable elements may be responsible for the genome size increase in *O. bayeri.*

Outside of microsporidia, genomic knowledge is still in its infancy. For medical and economic reasons, the vast majority of unicellular organisms whose genomes have been sequenced belong to pathogenic lineages, while only few free-living species have been sequenced. This biased taxonomic sampling means that comparative genomic information is meager. Thus, for the majority of intracellular parasitic taxa, such as *Plasmodium, Trypanosoma,* and *Entamoeba,* it is impossible at present to determine whether or not they have experienced genome reduction in comparison with their free-living relatives.

GENOME REDUCTION IN EUKARYOTICALLY DERIVED PLASTIDS **Primary endosymbiosis** refers to the uptake and retention of a cyanobacterium by a eukaryote and the subsequent evolution of a **primary plastid** bounded by two membranes. All primary plastids in land plants, green algae, red algae, and glaucocystophyte algae derive from a single primary endosymbiosis event (Chapter 6). Some plastids in nature, however, are not primary plastids. Lee (1977) and Gibbs (1978) noticed that some plastids are surrounded by three or more membranes. Subsequently, it was hypothesized that such plastids originated through endosymbiosis involving a photosynthetic eukaryote carrying a primary plastid. This process was dubbed **secondary endosymbiosis**, and the resulting organelles were called **secondary plastids** (**Figure 11.12**). As a consequence of the secondary endosymbiotic event, secondary plastids are initially enveloped by four membranes: two from the primary plastid, one from the plasma membrane of the engulfed alga, and one from the host. In at least two lineages, the dinoflagellates and the euglenids, one membrane was subsequently lost—most likely the one derived from the plasma membrane of the algal endosymbiont. Secondary endosymbioses involving both green and red algae are known.

TABLE 11.4

Features of nucleomorph genome sequences

Taxon	Genome size (Kb)	Number of protein-coding genes	Average intergenic spacer (bp)	Number of spliceosomal introns
Chlorarachniophytes				
Lotharella oceanica	612	610	147	1,011
Bigellowiella natans	373	284	82	951
Cryptophytes				
Chroomonas mesostigmatica	703	505	200	24
Hemiselmis andersenii	571	472	132	0
Guillardia theta	55	487	93	17
Cryptomonas paramecium	486	466	102	2

Source: Data from Tanifuji et al. (2014) and Moore et al. (2012).

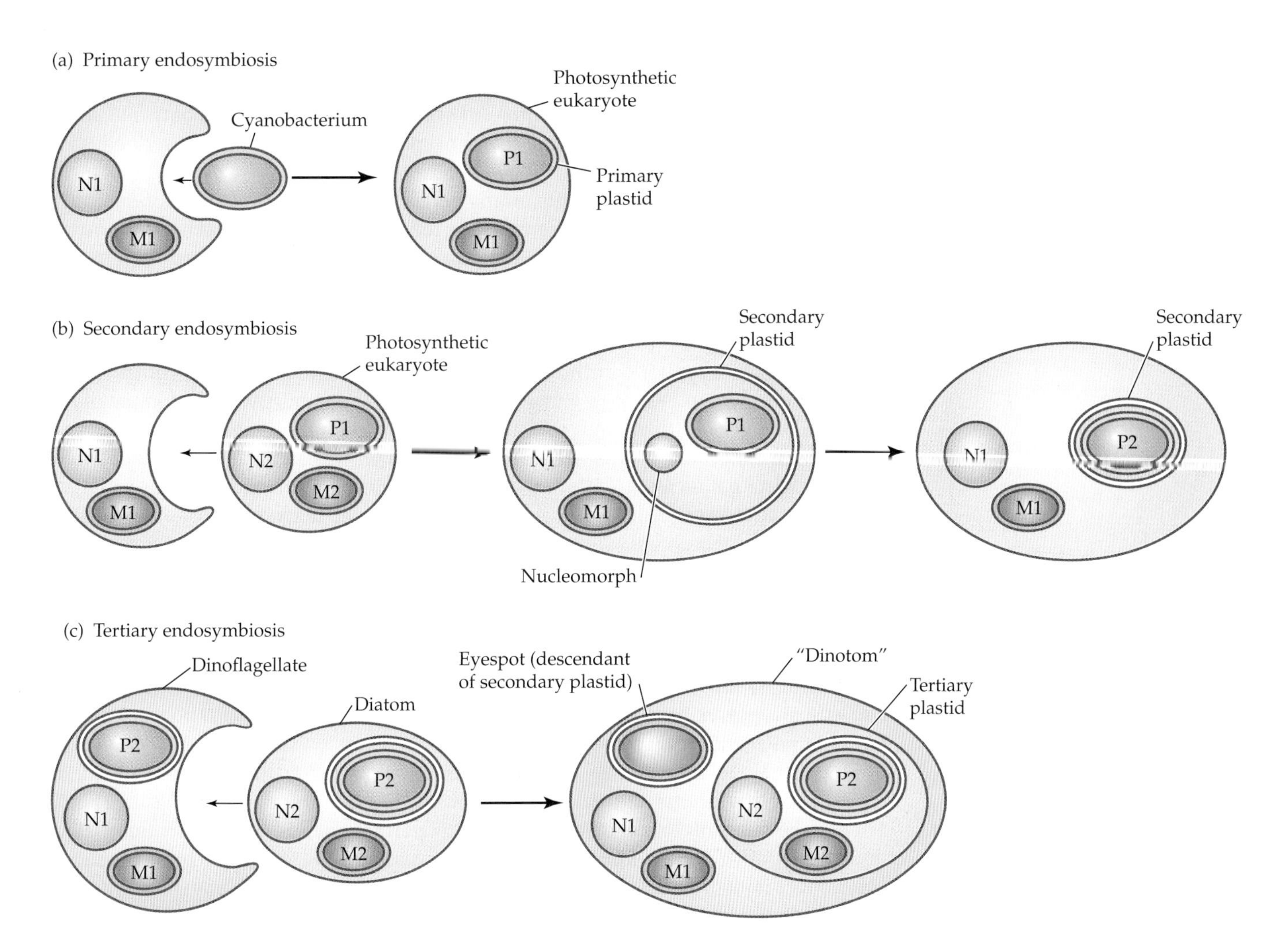

Figure 11.12 Schematic representation of plastid evolution through primary, secondary, and tertiary endosymbioses. (a) In primary endosymbiosis, a eukaryote containing a nucleus (N1) and a double-membraned mitochondrion (M1) engulfs a double-membraned cyanobacterium. Through shedding of its genes by either transfer to the host nucleus or deletion, and by acquiring the ability to import gene products from the host, the cyanobacterium becomes a primary plastid (P1). Such is the situation, for instance, in land plants. (b) In secondary endosymbiosis, a eukaryote engulfs a photosynthetic eukaryote such as a green or red alga that possesses a nucleus (N2), mitochondria (M2), and primary plastids (P1). In time, the four-membrane secondary endosymbiont (P2) looses its mitochondria, and its nuclear genome, now called a nucleomorph (NM), is drastically reduced in size and content. Such is the situation in cryptomonads and chlorarachniophytes. Subsequently the nucleomorph may be lost, as is the situation in euglenids. (c) In tertiary endosymbiosis, a dinoflagellate with a nucleus (N1), mitochondria (M1), and three-membrane secondary plastids engulfs a diatom, which possesses a nucleus (N2), mitochondria (M2), and four-membrane secondary plastids (P2). The resulting organism—a "dinotom," such as *Durinskia*—contains the host nucleus (N1) and mitochondria (M1), as well as a nonphotosynthetic descendant of the host's secondary plastid, which is referred to as the eyespot. The tertiary plastid of the dinotom contains a nonreduced nucleus (N2), whose genome may be ten times larger than that of the human genome, as well as mitochondria (M2). The quote marks around "Dinotom" indicate that this taxon is paraphyletic (Courtesy of Patrick Keeling.)

In all cases described so far, the secondary endosymbiont has lost its mitochondria; in the vast majority of cases, the secondary endosymbiont has also lost its nucleus. A typical example is *Euglena gracilis*, a freshwater photosynthetic flagellate. Its secondary plastid, which was almost certainly derived from a green alga, is bounded by three membranes, and the nucleus of the organelle was lost.

The nucleus of the secondary plastid was retained, albeit in a highly vestigial form called a **nucleomorph**, in only two lineages: the chlorarachniophytes and the cryptomonads, in which the secondary plastids originated from a green alga and a red alga, respectively. Nucleomorph genomes are highly reduced (**Table 11.4**), on the order of 373–703 Kb. Interestingly, despite their polyphyletic origins, the nucleomorphs of both chlorarachniophytes and cryptomonads are divided into three linear chromosomes. Nucleomorph genomes are the quintessence of compactness: the number of protein-coding genes is the smallest for eukaryotes, the average space between adjacent genes is a mere 80–200 bp, some genes overlap and others are cotranscribed, and the spliceosomal introns are tiny.

As with primary endosymbionts, many genes have been lost from secondary endosymbionts, while others have been transferred to the host nucleus. The reimport of nucleus-encoded plastid-targeted proteins is, however, more complicated than in the case of primary plastids, because of the existence of additional membranes around the secondary plastid. Hence, a bipartite targeting system is necessary, with one component targeting the proteins to the inner membranes, and the second targeting it to the plastid.

According to Cavalier-Smith (2005), the optimal size for the nucleomorph's genome is zero, which is indeed the genome size for the majority of secondary plastids. The only reason that some cells retained a genome was to produce ~40 chloroplast proteins. Because of this need, the nucleomorph was saddled with 300 or more housekeeping genes just to allow these 40 or so essential proteins to be produced and imported into the chloroplast, as well as to allow the nucleomorph genome to be replicated and segregated.

Higher-order symbioses, i.e., tertiary and serial secondary endosymbiosis, are only known in dinoflagellates. **Serial secondary endosymbiosis** is thought to have occurred only once, in the lineage leading to the dinoflagellate genus *Lepidodinium*, where an ancestral secondary plastid of red algal origin has been replaced by a green algal one. The *Lepidodinium* plastid genome confirms its ancestry from a green alga, whereas analysis of its nucleus-encoded plastid-targeted genes shows that it had acquired many genes from a red alga.

Species belonging to genus *Durinskia* represent a case of **tertiary endosymbiosis** in which a photosynthetic dinoflagellate acquired a plastid from diatoms (Figure 11.12). There is no question that the endosymbiont is permanent and completely integrated into the dinoflagellate host cell cycle, but the diatom retained its own nucleus and even intact mitochondria, so it is unclear whether there has been any gene transfer to the host and whether host-encoded proteins are targeted to the symbiont. In other words, we do not know whether this tertiary symbiont is a bona fide organelle or not (Gagat et al. 2014; Hehenberger et al. 2014).

Protein-Coding Gene Number Variation and the G-Value Paradox

Before discussing the variation in protein-coding gene number in eukaryotes, we need to mention some estimation difficulties. First, it is important to realize that, owing to technical and fiscal constraints, the data set of sequenced eukaryotic genomes consists almost entirely of species with relatively small genomes, and therefore most of the variation in genome size found in eukaryotes is either unrepresented or severely underrepresented in genomic databases (Gregory 2005a). We note, moreover, that determining the number of protein-coding genes is a very difficult task even in completely sequenced, completely assembled, and completely annotated eukaryotic genomes. The main reasons for the difficulties are (1) the existence of very large introns, (2) the huge variation in the length of the first and last exons, (3) the uneven distribution of protein-coding genes along the genome, and (4) the fact that spurious long open reading frames that are not in fact coding proteins are ubiquitous within the genome.

The enormity of any genomic annotation task enforces our dependence on automated tools for assembling genomes and detecting genes. In an ideal world, the automated part would be followed by manual curation and experimental verification. In the real world, however, researchers rely almost exclusively on gene prediction algorithms whose performances are mediocre at best. For example, these algorithms have the peculiar and exasperating tendency to "resurrect" pseudogenes in silico, turning dead genes into fictional doppelgangers of protein-coding genes (Nelson 2004; Ezkurdia et al. 2014). In other words, the automated annotation of genomes tends to overestimate the number of protein-coding genes. For instance, the number of protein-coding genes in the yeast *Saccharomyces cerevisiae* was estimated by two-dimensional electrophoresis to be about 3,000. The number of protein-coding genes actually identified in the genomic sequence is more than double that (about 6,200 putative genes). Furthermore, estimates of the number of genes in a genome tend to change with time and additional scholarship. For example, in 2001, with the publication of the initial sequences of the human genome, the number of protein-coding genes was estimated to be 26,000–30,000. In 2004, the number was reduced to ~24,500, and in 2007 it was further decreased to ~20,500. The newest estimate for the number of protein-coding genes in the human nuclear genome is approximately 19,000 (Ezkurdia et al. 2014).

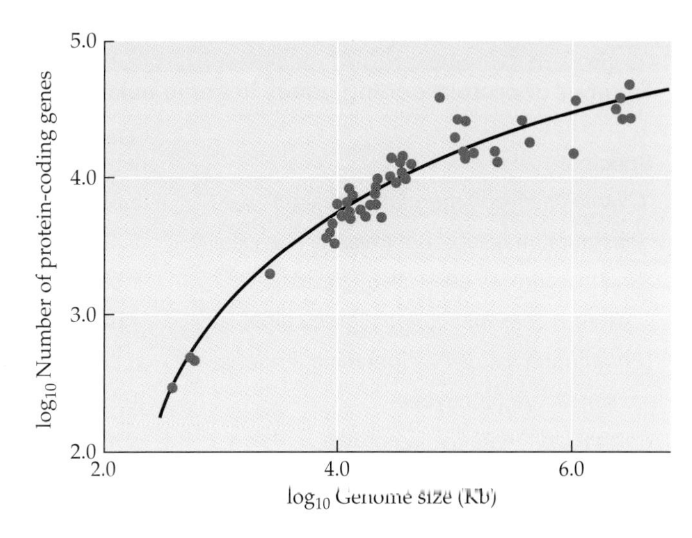

Figure 11.13 The relationship between the number of protein-coding genes and genome size in eukaryotes can be described by a logarithmic regression model. Note that even when the two variables are $\log_{10}$ transformed, the relationship between the two is not linear. This indicates that percentagewise, genome size increases much faster than the number of protein-coding genes. (Modified from Hou and Lin 2009.)

Notwithstanding the uncertainties in assessing protein-coding gene number, one can state quite confidently that the number of protein-coding genes in eukaryotes varies roughly from about 300 in the nucleomorph of the chlorarachniophyte *Bigelowiella natans* to about 100,000 in wheat (*Triticum aestivum*), a mere 300-fold range, as opposed to the 400,000-fold range for genome size. A weak and nonlinear correlation exists between gene number and genome size in eukaryotes. That is, as genome size increases, the number of genes increases too, but disproportionately more slowly than genome size (**Figure 11.13**). Thus, the protein-coding fraction of the genome decreases very rapidly with genome size, from 82% in the smallest eukaryotic genomes to less than 1% in many large genomes, such as those of some plants (Hou and Lin 2009). Moreover, as opposed to the situation in prokaryotes (Chapter 10), genome size variation in eukaryotes cannot be explained by the variation in number of protein-coding genes. That is, some relatively small genomes contain a relatively large number of protein-coding genes. For example, at 200 Mb in length, the genome of the microcrustacean *Daphnia pulex* is about one-fifteenth the size of the human genome, yet it contains approximately 50% more protein-coding genes (Colbourne et al. 2011).

As with the C-value paradox, there does not seem to be any relationship between the number of genes and organismal complexity (**Table 11.5**). This lack of correlation has been referred to as the **N-value paradox** (Claverie 2001) or the **G-value paradox** (Hahn and Wray 2002). For unknown reasons, the term "G-value paradox" became more popular in the literature than the earlier term. As with the C-value paradox, there are two facets to the G-value paradox. The first concerns the undisputable fact that number of protein-coding genes cannot be used as a predictor of organismal complexity. For example, there are single-celled species, such as *Trichomonas vaginalis* and *Emiliana huxleyi*, that have vastly more protein-coding genes than animals and plants, which are presumably more complex. Let us consider, for instance, the case of two well-studied model organisms: *Caenorhabditis elegans* and *Drosophila melanogaster*. The number of cells in the body of individual *C. elegans* nematodes is known with precision: 1,031 and 959 cells, for males and hermaphrodites, respectively. In contrast, the body of *D. melanogaster* flies has about 10^8 cells. Yet despite the five orders of magnitude difference in number of cells, the genome of *C. elegans* contains approximately 6,500 more protein-coding genes than that of *D. melanogaster*. Conversely, some organisms that are so similar to one another they have been assigned to the same genus—for example, the sea squirts *Ciona savignyi* and *C. intestinalis*—have very different gene numbers in their genomes.

The second aspect of the G-value paradox is unavoidably anthropocentric. That is, it is very difficult for humans to accept the fact that *Homo sapiens* does not possess the

TABLE 11.5

Number of protein-coding genes in some eukaryotic species

Species[a]	Gene number	Species[a]	Gene number
**Cyanidioschyzon merolae* (red alga)	5,009	**Bigelowiella natans* (chlorarachniophyte alga)	21,706
**Plasmodium falciparum* (malaria parasite)	5,429	*Mus musculus* (mouse)	22,606
**Saccharomyces cerevisiae* (yeast)	6,692	*Poecilia formosa* (Amazon molly)	23,615
**Ostreococcus lucimarinus* (green alga)	7,603	*Hordeum vulgare* (barley)	24,287
**Entamoeba histolytica* (parasitic amoeba)	8,283	**Tetrahymena thermophila* (ciliate)	24,725
Apis mellifera (honeybee)	10,157	**Guillardia theta* (cryptomonad alga)	24,945
Petromyzon marinus (lamprey)	10,415	*Danio rerio* (zebrafish)	26,459
Ciona savignyi (sea squirt)	11,616	*Amborella trichopoda* (plant)	27,313
**Dictyostelium discoideum* (slime mold)	13,212	*Arabidopsis thaliana* (mustard weed)	27,416
Drosophila melanogaster (fruit fly)	13,937	*Theobroma cacao* (cacao)	29,188
Gallus gallus (chicken)	15,508	*Vitis vinifera* (grape)	29,971
Ciona intestinalis (sea squirt)	16,671	*Physcomitrella patens* (moss)	32,273
**Phytophthora infestans* (potato blight)	17,785	*Solanum lycopersicum* (tomato)	34,675
Xenopus tropicalis (western clawed frog)	18,442	*Oryza sativa* (rice)	35,679
Takifugu rubripes (fugu)	18,523	*Solanum tuberosum* (potato)	39,021
Pan troglodytes (chimpanzee)	18,759	*Populus trichocarpa* (poplar)	41,377
Latimeria chalumnae (coelacanth)	19,569	*Manihot esculenta* (cassava)	30,666
Tetraodon nigroviridis (pufferfish)	19,602	*Selaginella moellendorffii* (lycophyte)	34,799
Canis familiaris (dog)	19,856	**Emiliania huxleyi* (haplophyte)	38,544
Bos taurus (cow)	19,994	*Zea mays* (maize)	39,469
Caenorhabditis elegans (nematode)	20,447	*Brassica rapa* (rape)	41,018
Monodelphis domestica (opossum)	21,327	*Pinus taeda* (loblolly pine)	50,172
***Homo sapiens* (human)**	**20,364**	*Malus × domestica* (hybrid apple)	57,386
Oreochromis niloticus (tilapia)	21,437	**Trichomonas vaginalis* (flagellated protist)	60,815
Ornithorhynchus anatinus (platypus)	21,698	*Triticum aestivum* (bread wheat)	99,504

Source: Data from www.ensembl.org, fungi.ensembl.org, protists.ensembl.org, plants.ensembl.org, and www.ncbi.nlm.nih.gov/genome/browse.
[a]Organisms marked with an asterisk (*) are unicellular.

genome with the most protein-coding genes. Having a genome that contains fewer genes than apple, zebrafish, unicellular algae, cassava, and a protozoan responsible for vaginal infections can be hard to swallow. This is doubly hard, since gene number estimates in humans have plummeted precipitously from approximately 6.7 million in 1964 to 100,000 in the 1990s, to 30,000–40,000 in 2001, and to 19,000 in 2014 (Vogel 1964; Pertea and Salzberg 2010; Ezkurdia et al. 2014).

Possible solutions to the G-value paradox

The realization that the genomes of multicellular organisms contain considerable amounts of junk DNA seemed for a while to have resolved the C-value paradox. Implicitly, however, this resolution rested on the assumption that once the junk DNA was taken into account, the total number of functional elements, especially protein-coding genes, would be found to correlate with organismal complexity. However, fully sequenced eukaryotic genomes have made it clear that we have not yet resolved the C-value paradox—it has merely been replaced by the G-value paradox (Hahn and Wray 2002).

Several solutions to the G-value paradox have been proposed in the literature. According to one class of solutions, the number of protein-coding genes underestimates the number of functions required for achieving complexity. First, gene regulation can increase the repertoire of functions carried by a single gene. Second, alternative splicing and posttranslational modification can greatly increase the number of gene products relative to the number of genes. For example, if the number of splice variants that affect protein-coding regions is considered, then the number of proteins encoded by the human genome is about 69,000, i.e., 400% more than the number of protein-coding genes. In comparison, *C. elegans* contains a smaller proportion of alternatively spliced genes, resulting in only a 33% increase in number of proteins over the number of genes. Third, some functions are not carried out by proteins, but by RNA or even by untranscribed DNA. Fourth, the existence of multifunctional proteins can greatly increase the number of functions relative to the number of genes. For example, the proportion of multifunctional proteins encoded by the human genome seems to be greater than those in *Drosophila* and *C. elegans*. This type of solution to the G-value paradox has been called the "Swiss Army knife" explanation. Finally, as the number of genes increases, the number of interactions among them increases exponentially.

A second class of solutions invokes "G-value inflation." That is, the number of protein-coding genes may sometimes be an overestimate of the number of unique (nonredundant) genes. For example, fully 40% of the protein-coding genes in *C. elegans* are the result of gene duplication, which may account for the large number of genes in this nematode, relative to *Drosophila*.

A third class of solutions to the G-value paradox claims that one should not expect a correlation between genomic complexity and organismal complexity. According to this view, the genome is a Rube Goldberg machine (Chapter 8) that is overly complex. Each task, according to this view, can be done more efficiently with fewer genes. Indeed the G-value paradox is only a paradox under the assumption that a system's complexity corresponds to the number of its components. Complexity theory has taught us that a simple relationship between the complexity of instructions (genes) and the complexity of the product (organism) does not exist (Kauffman 1993). For example, biosynthesis of the amino acid proline requires three different enzymes in enterobacteria, two enzymes in fungi and higher plants, but only a single, unrelated enzyme in *Clostridium* and other Firmicutes (Herrmann and Somerville 1983). In fact, the number of genes necessary to synthesize any given amino acid is known to vary across the living world (Herrmann and Somerville 1983). This is not functional redundancy, but simply different recipes for making the same product. The historical necessity of working with existing materials also means that the number of biological instructions does not always correlate with the complexity of the biological products. During the development of *C. elegans*, for instance, 131 cells are produced that subsequently undergo apoptosis (programmed cell death). Thus, approximately one-eighth of the total cell production is unnecessary (Sulston and Horvitz 1977). Like the human appendix and other vestigial organs, these cells may have evolved in ancestors in which they had a function.

The I value

Hahn and Wray (2002) tentatively suggested the term **I value** for the amount of functional information contained in a genome. This as yet unmeasurable variable incorporates the effective number of genes as well as the complexity added by gene expression and the number of interactions among the genes and the gene products. The effective number of genes will equal the number of genes plus a certain addition due to alternative splicing, posttranslational modifications, and other processes that cause a single gene to produce more than one functional product. From this sum one will need to subtract the degree of gene redundancy and other processes that decrease the number of functional products relative to the number of genes. At present we do not know how to estimate the I value.

Gene Number Evolution

Studying changes in gene number during genome evolution requires that we reconstruct the sequence of gene gains and gene losses along the branches of a phylogenetic tree. For simplicity, let us assume that all gene gains are due to gene duplication, thereby ignoring the tiny minority of gains due to horizontal gene transfer and de novo genes, and that all the gene losses are equivalent, thereby ignoring the difference between nonfunctionalization and deletion. Here we estimate the number of gains and losses by considering each gene family separately. The method outlined below is called the **tree reconciliation** method. It was proposed by Goodman et al. (1979) and subsequently modified and extended by numerous other authors (Hahn 2007).

We start with two rooted trees, a gene tree and a species tree (**Figure 11.14**). If the gene tree and the species tree are identical to each other, no tree reconciliation is needed, and the most parsimonious explanation is that there were neither gains nor losses in the evolution of the gene family under study. If, on the other hand, the two trees differ from each other, then we need to perform tree reconciliation. The purpose of tree reconciliation is to infer the minimum number of gene gains and losses, as well as their placements on the species tree, that will suffice to explain the gene tree.

In the first step of the procedure, we duplicate as many subtrees as necessary in the species tree to get a new subtree that has the same topology as the gene tree. This tree is referred to as the **reconciled gene tree**. The number of subtrees that need to be duplicated in order to get a subtree that is identical to the gene tree is equal to the number of gene duplications necessary to explain the gene tree. In the next step, we prune as many branches as necessary for the reconciled gene tree to become identical with the species tree. The number of pruned branches equals the number of gene losses.

If both the species tree and the gene tree are inferred without error, and if the reconciliation algorithm is appropriate, then the correct history of duplication and loss should be recovered, albeit with varying degrees of computational efficiency. If, however, one of the trees is not correct, then it becomes necessary to assume additional gene gains and losses in order to reconcile the two trees. These inferred gains and losses are artifacts of the incorrect topologies. Let us consider the example in Figure 11.14a. We will assume that the gene tree is erroneous and that the true tree is actually identical to the species tree. By using the wrong gene tree we came to the conclusion that four gains and losses occurred in this gene family, whereas in reality none occurred. Moreover, tree reconciliation methods assume that there are no missing data. If some data are missing, then further overestimations of gene gains and losses may occur.

What have we learned so far about gene number increases and decreases with evolutionary time? First, we learned that gene gains and losses occur quite frequently. In mammals, yeast, and *Drosophila*, approximately 17 genes are estimated to be duplicated and fixed in the population every million years. In many cases, extreme expansions and contractions of gene families can occur in relatively short periods of time. We also found that rates of gene gain and loss do not follow the same patterns as rates of nucleotide substitution. For example, as far as nucleotide substitution is concerned, the rates in primates are lower than those in other mammals, and the rates in apes are higher than those in Old World monkeys. In contrast, the rates of gene gain and gene loss are higher in apes than

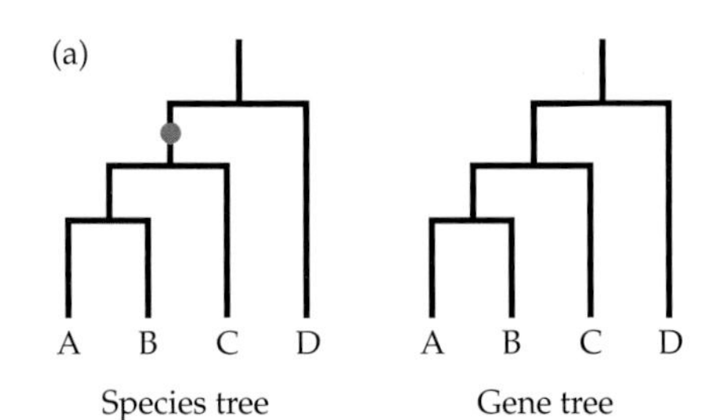

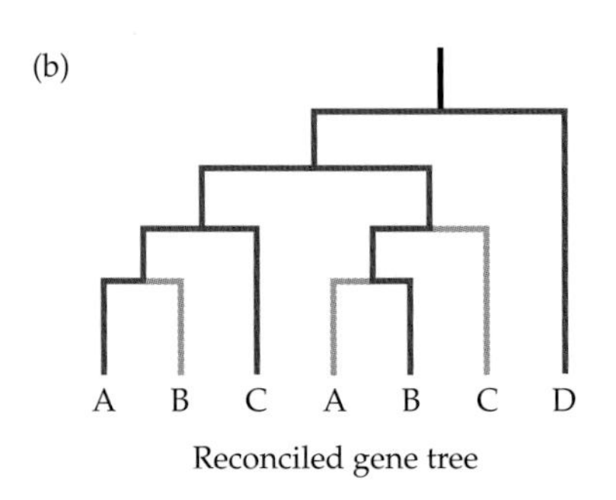

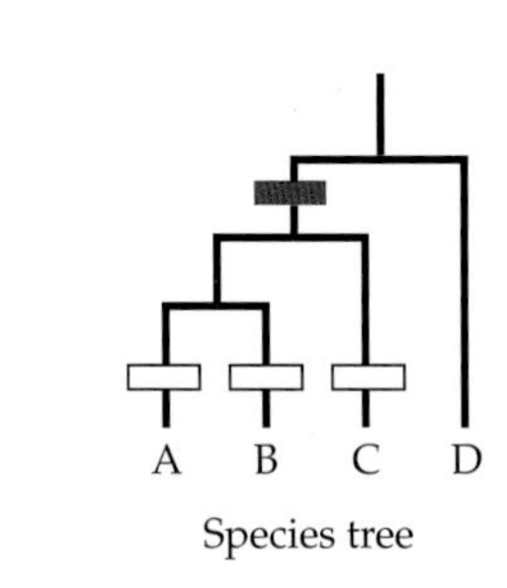

Figure 11.14 Inferring the numbers of gene duplications and losses by the tree reconciliation method. (a) The four-species rooted tree differs in its topology from the gene tree. To reconcile the two trees, a duplication of the subtree descending from the branch marked with a blue circle on the species tree is required. (b) In the resulting reconciled gene tree, there exists a subtree (red) that is identical in its topology to the gene tree in (a). The reconciled gene tree, however, contains more OTUs than the gene tree. Hence, three branches (gray) must be pruned. (c) In the final species tree, the subtree duplication is interpreted as a gene gain by duplication (red rectangle), and the three branches that were pruned are interpreted as gene losses (empty rectangles).

in Old World monkeys, which in turn are higher than those in rodents and carnivores (Hahn et al. 2007).

Taking gene gains and losses into account can frequently alter our perception of the degree of similarity or dissimilarity among species. For example, after comparing ~2.4 Gb of high-quality orthologous sequences from human and chimpanzee, the Chimpanzee Sequencing and Analysis Consortium (2005) concluded that the sequence divergence between these two species is approximately 1%. Much has been made about the near identity between the two species. The patterns of gene gain and loss tell a different story: along the lineages leading to humans and chimpanzees, there occurred at least 715 gene gains and at least 815 gene losses (Demuth et al. 2006). Thus, humans and chimpanzees differ from each other by about 6% of their complement of genes. This result illustrates that the mythological "near identity" between the human and the chimpanzee genomes is just that—mythological.

Szathmáry and Maynard Smith (1995) and Bird and Tweedie (1995) suggested that gene number increases do not occur continuously during evolution, but rise in discrete steps. They also thought that two of the biggest steps in gene number increase occurred at the transition from prokaryotes to eukaryotes and the transition from invertebrates to vertebrates. The latter claim is only partially true, as the endosymbiotic event that gave rise to eukaryotes seems to have indeed resulted in a large increase in gene number, but the empirical evidence does not support a huge increase in gene number at the transition between vertebrates and invertebrates. In fact, multicellularity, which has evolved independently in dozens of lineages, may be responsible for increases in gene numbers far larger than those associated with the evolution of vertebrates.

What determines the probability that a duplicated gene will be retained in the population? Observations indicate that duplicate genes are lost much more slowly in multicellular than in unicellular eukaryotes, and there is a clear tendency for the half-life of duplicate genes to increase with genome size (Lynch and Force 2000).

Preservation of both members of a duplicate pair can be promoted either by one of the copies acquiring a beneficial mutation (neofunctionalization), while the other copy retains its original function, or by the two copies becoming subfunctionalized to the extent that their joint expression is necessary to fulfill the essential functions of the ancestral locus. The probability of preservation by neofunctionalizing mutations under positive selection should not exhibit any correlation with either population size or genome size. In contrast, the probability of fixation of the two subfunctionalized copies should be larger in small populations. This is exactly what has been observed, leading Lynch and Force (2000) to conclude that gene number evolution is driven by the fixation of slightly deleterious mutations is species with small effective population sizes.

Methodologies for Studying Gene Repertoire Evolution

Genomes differ from one another not only in the number of constituent genes, but also in the types of genes they contain, i.e., their **gene repertoire**. Two genomes may have similar gene numbers, but the types of genes present in each may be very different. In eukaryotes, differences in gene repertoire may be due to such evolutionary processes as gene duplication, nonfunctionalization, neofunctionalization, and, to a lesser extent, the acquisition or loss of xenologous sequences. Differences in gene repertoire may also result from the loss of detectable sequence similarity among homologs.

It is typically very difficult to identify the reasons for the presence or absence of functionality in a genome. For example, a causal relationship has been postulated between carnivory and the presence of genes encoding umami (savory) taste receptors. Zhao et al. (2010) raised this possibility following their discovery that the genome of the giant panda—an animal that has lost its dietary reliance on meat despite being phylogenetically a carnivore—does not have a functional umami-encoding gene. What is left in the genome of the giant panda is a pseudogene indicating that a nonfunctionalization event occurred between 1.3 and 10 million years ago. The hypothesis of a connection between the absence of a functional umami gene and the evolution of herbivory in the panda is extremely alluring but for the fact that the genomes of some mammals (such as cow and horse) that are much stricter in their herbivory than the giant panda still retain intact genes for the umami taste receptor.

The study of gene repertoire evolution is still in its infancy, and we know very little about the effectors of gene-repertoire change. In fact, every time a new genome is published, one reads about the idiosyncracies of that genome and the importance of this or that gene for adaptation. For example, much attention is given to autapomorphies. (An autapomorphy is a singular derived character state found in a single taxon; Chapter 5.) In particular, there exists a large body of literature dealing with human autapomorphies, or what "make us human" (e.g., Mikkelsen 2004; Newton 2007; Pollard 2009).

Because we do not know much about the evolutionary forces driving gene repertoire, genome sequence publications frequently promise but rarely deliver coherent hypotheses concerning adaptation. For example, Jia et al. (2013) promise in their title that the "*Aegilops tauschii* draft genome sequence reveals a gene repertoire for wheat adaptation." Sadly, the article bearing this title mentions no gene involved in wheat adaptation, nor for that matter what this adaptation consists of. Unfortunately, such dissonance between promises made and promises kept is quite common in the genomic literature.

In this section, we will mostly pay attention to the emerging methodologies for comparing gene repertoires among species. Because most studies on gene repertoire in the literature concern themselves with protein-coding genes only, we have not discussed the handful of studies dealing with the evolution of RNA-specifying gene repertoires (e.g., Brameier 2010).

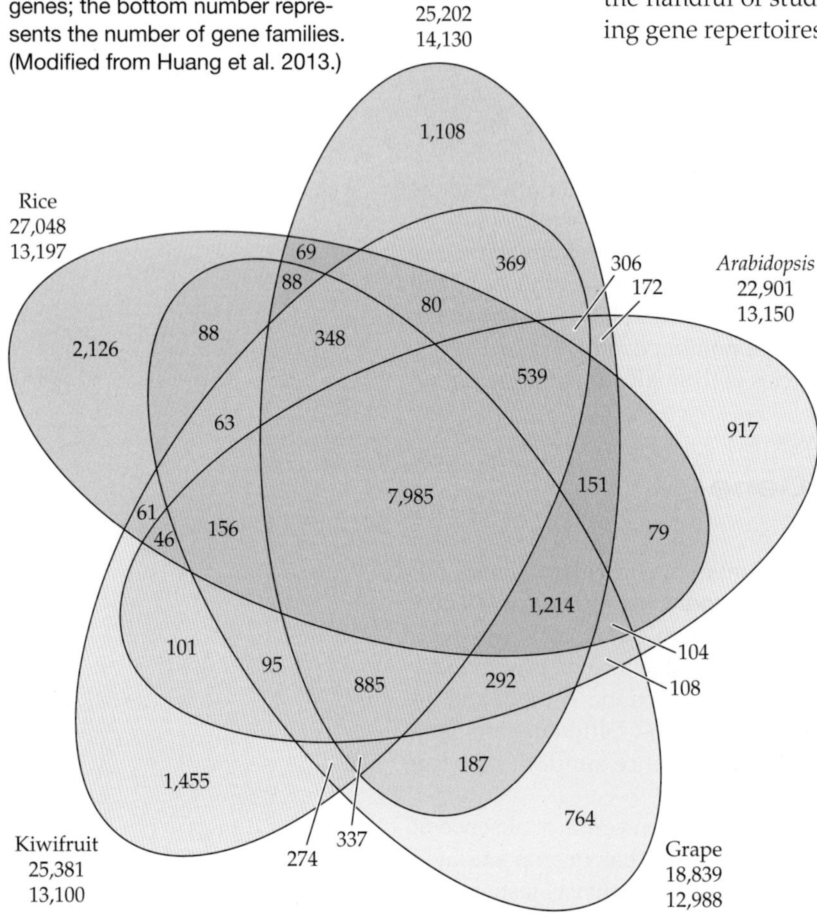

Figure 11.15 Venn diagram of homologous gene families among five plant species. The top number under the name of the species denotes number of protein-coding genes; the bottom number represents the number of gene families. (Modified from Huang et al. 2013.)

Gene-family cluster analysis

In **gene-family cluster analysis**, the first step is to collect all the protein sequences encoded by a genome. For a gene that encodes multiple proteins by alternative splicing, only one variant—usually the longest—is used. In the next stage of the analysis, the proteins are clustered by family, each consisting of all the recognizable paralogs of a protein. The number of protein families usually turns out to be approximately half the number of proteins (e.g., Huang et al. 2013). The process is repeated for all the genomes in the comparison. In the final stage of the analysis, reciprocal searches for sequence similarity are used to identify homologous protein families from the different taxa. The results of gene-family cluster analyses are usually presented in the form of a Venn diagram.

In the example in **Figure 11.15**, we present a comparison of five plants. The numbers of gene families vary between close to 13,000 in grape to more than 14,000 in tomato. Only about 8,000 gene families are found in all five taxa. The numbers of autapomorphous gene families, i.e., gene families that are found in one species but not in the other four, ranges from 764 in grape to 2,126 in rice.

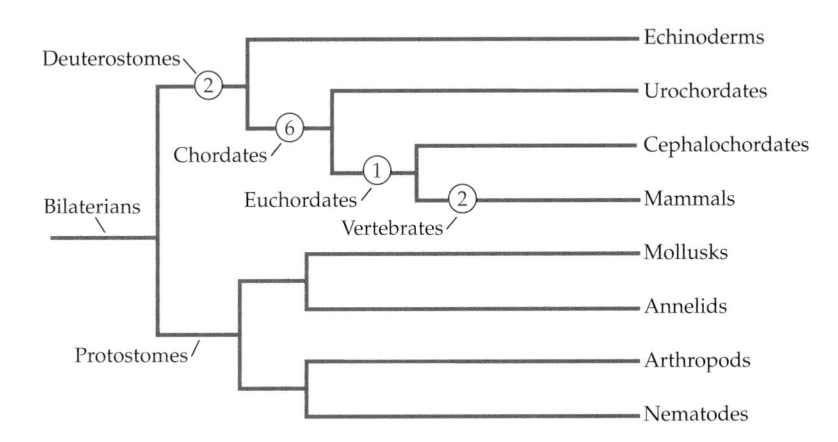

Figure 11.16 Phylogeny of bilaterian animals showing the gene loss pattern that led to mammals lacking 11 gene families. The number of gene families lost in each branch is indicated in red. Due to lack of data from vertebrates other than mammals, it is impossible to say whether the loss of two gene families (uracil nucleosidase and short-chain dehydrogenase) occurred in all vertebrates or in mammals or whether one loss occurred in vertebrates and the other in mammals. The phylogeny shown here conforms to the Ecdysozoa hypothesis, which posits that nematodes are closer to arthropods than either taxon is to annelids. The pattern of gene family loss would be unchanged under the Coelomata hypothesis, according to which arthropods are an outgroup of deuterostomes.

Gene-family cluster analyses in conjunction with phylogenetic distributions can be used to trace gene gains and losses. For example, by using this method, Danchin et al. (2006) discovered that eleven gene families that were present in the last common ancestor of opisthokonts, which consists of choanoflagellates, fungi, metazoans, and four additional genera (*Amoeboaphelidium*, *Aphelidium*, *Fonticula*, and *Nuclearia*), are not found in mammals. By using the phylogenetic distribution of these genes, it was possible to infer that of these eleven gene families, two were lost in deuterostomes, six in chordates, one in euchordates, and two on the branch leading to mammals (**Figure 11.16**).

Functional clustering of proteins

Instead of dealing with tens of thousands of protein-coding gene families, one can try to cluster the protein families by functionality. Of course, since the function of the vast majority of proteins has not been experimentally deduced, one needs to use a computational method for assigning putative function to proteins. One such method, for instance, infers the putative function of a protein whose function is not known by comparing its sequence with all proteins of known functions, and assuming that its function is similar to that of proteins with similar sequences whose functions are known. Tatusov et al. (1997) put forward a method that classifies all protein-coding genes in a genome according to sequence similarity.

In their initial report, Tatusov et al. (1997) studied proteins encoded by seven taxa: two Gram-negative eubacteria (*E. coli* and *Haemophilus influenzae*), two Gram-positive eubacteria (*Mycoplasma genitalium* and *M. pneumoniae*), a cyanobacterium (*Synechocystis* sp.), an archaebacterium (*Methanococcus jannaschii*), and a eukaryote (*Saccharomyces cerevisiae*). They found that they could cluster all proteins into 720 **COGs** (**C**lusters of **O**rthologous **G**roup**s**). (The use of the term "orthologous" is not really justified, since it is extremely difficult to ascertain orthology in such distantly related taxa. However, using "homologous" or "similar" instead of "orthologous" would have resulted in acronyms that are not as euphonious as "COGs.")

Tatusov et al. (2003) improved the classification system by introducing two major changes. First, they separated the prokaryotic clusters, which retained the name COGs, from the eukaryotic clusters, which were named **KOGs** (for eu**K**aryotic **O**rthologous **G**roup**s**). Second, they reduced the number of clusters to 25 (denoted by the letters of the English alphabet from A to Z, skipping the letter X). These COGs and KOGs were in turn clustered into four **groups of functions**: (1) cellular processes and signaling, (2) information storage and processing, (3) metabolism, and (4) poorly characterized. COGs and COFs are collectively referred to as **functional clusters**. A comparison of COGs and KOGs is shown in **Figure 11.17**. The most obvious difference between prokaryotes and eukaryotes concerns proteins that are only found in eukaryotes, i.e., proteins related to the structure of the nucleus, proteins in the

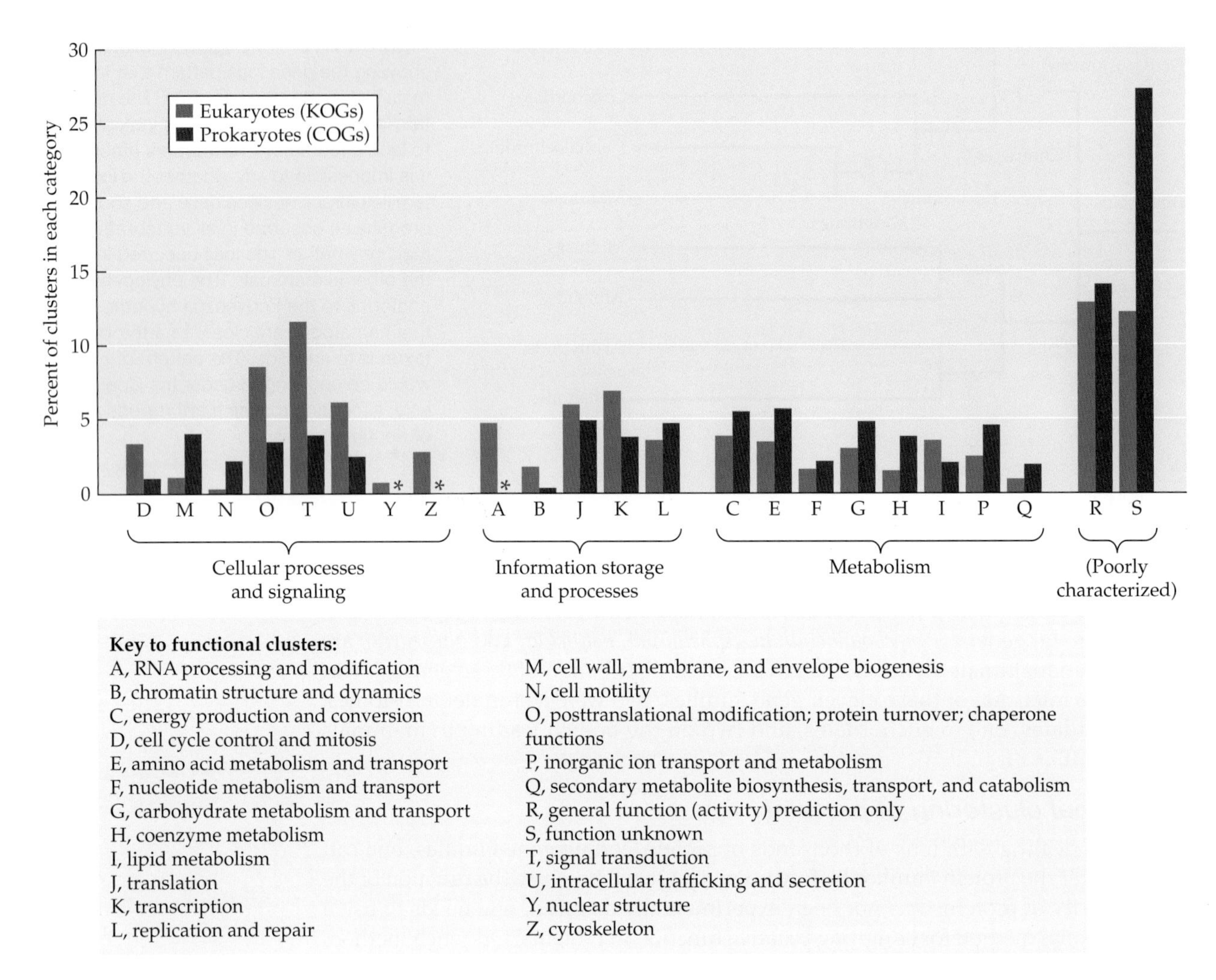

Figure 11.17 Classification and relative abundance of functional categories in prokaryotes (COGs) and eukaryotes (KOGs). Four groups of functional clusters are shown below brackets. Red asterisks indicate functional categories that are absent from prokaryotes. (Modified from Tatusov et al. 2003.)

cytoskeleton, and proteins involved in RNA processing. Another difference that is very evident from the figure is the fact that a higher percentage of prokaryotic genes have poorly characterized functions. This difference may be due to the fact that as a group, prokaryotes have a much bigger and much more varied gene repertoire than eukaryotes.

Figure 11.18 shows a comparison of KOGs between unicellular eukaryotes (*Saccharomyces cerevisiae*, *Schizosaccharomyces pombe*, and *Encephalitozoon cuniculi*) and multicellular eukaryotes (*Arabidopsis thaliana*, *D. melanogaster*, *C. elegans*, and *Homo sapiens*). The main difference seems to be that a much higher percentage of the multicellular proteome than the unicellular proteome is devoted to signal transduction. This increase is compensated for by a relative decrease in the fraction of the multicellular proteome that is devoted to translation, intracellular trafficking and secretion, RNA processing and modification, DNA replication and repair, cell cycle control, energy production and conversion, and amino acid metabolism and transport.

In studying protein-coding gene repertoire, it is sometimes useful to consider domains rather than all proteins. As mentioned in Chapter 8, a domain is an evolutionary unit whose coding sequence can be duplicated or shuffled. Small proteins contain just one domain; large proteins may be formed by combinations of domains. Domains are typically 100–250 amino acid residues in length, though smaller and larger domains are known. In individual genomes, the number of members in different domain families fits a power-law distribution. That is, a few families have many members; many families have few members. In a comparison among animals, plants, and fungi, only about 10–20% of all protein domains were found to be kingdom-specific.

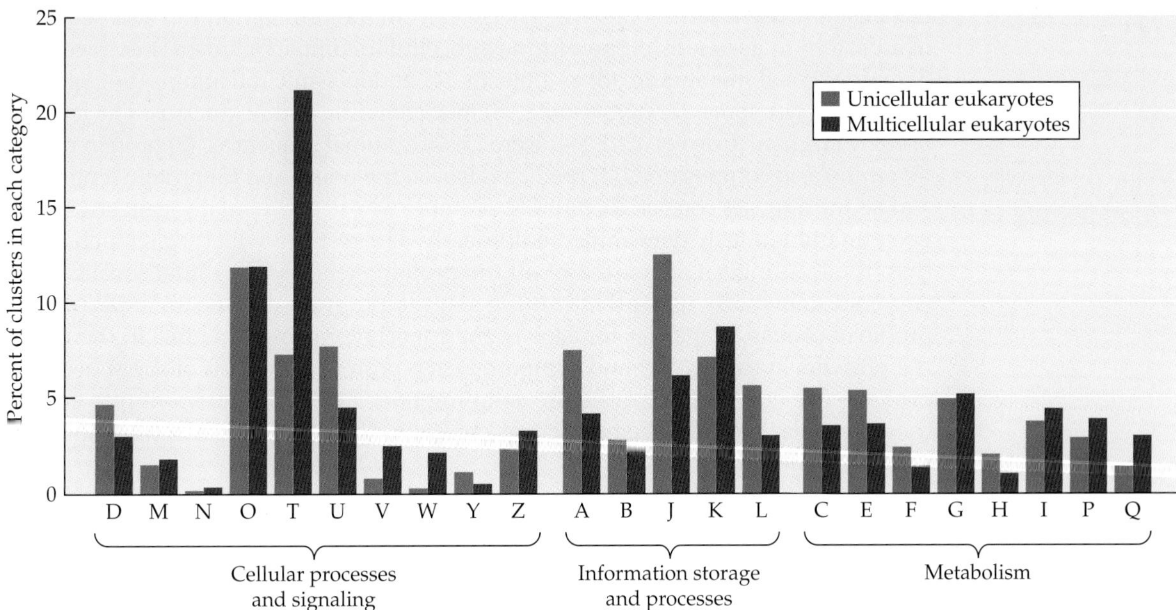

Figure 11.18 The difference in protein-coding gene repertoires of unicellular and multicellular eukaryotes. The classification of functional categories and clusters (KOGs) is the same as in Figure 11.17, except that the poorly characterized clusters (R, S) are not shown. KOG numbers for unicellular organisms are based on the genomes of the yeasts *Saccharomyces cerevisiae* and *Schizosaccharomyces pombe* and the microsporidian *Encephalitozoon cuniculi*. The numbers for multicellular organisms are based on the genomes of the thale cress *Arabidopsis thaliana*, the fruit fly *Drosophila melanogaster*, the nematode *Caenorhabditis elegans*, and *Homo sapiens*. (Data from Eugene Koonin and Yuri Wolf; courtesy of Fei Yuan.)

This implies that all but a small proportion of the protein repertoire originated before the divergence among eukaryotic taxa (Chothia et al. 2003).

Supervised machine learning and the subcellular localization of proteins

Machine learning is a scientific discipline that concerns itself with algorithms that can learn from data (Flach 2012). Rather than following strictly static program instructions, machine-learning algorithms operate by building models from example inputs that can, then, be used to make decisions. Machine learning is often synonymized with **statistical learning**, a discipline whose aim is to design algorithms for implementing statistical methods, especially in cases in which analytical solutions are intractable.

A machine-learning algorithm performs a **task**. The most common task in biology is the **classification** of **objects** into a finite number of **categories**. (In the field of machine learning, objects are also referred to as **examples** or **instances**, and categories are known as **labels**.) **Features** are the set of attributes associated with the objects. Ideally, the number of features should be the same for each object. The algorithm uses the features to classify the objects into categories, or assign labels to the examples. For example, the objects may be people; the features may be height, weight, and favorite color; and the classification task may be dividing the people into males and females. In other words, the algorithm uses three gender-neutral features for gender prediction.

In machine learning, there are two types of data sets. The **training set** is used by the algorithm to devise models that can, then, be used to classify the objects in the **test set**. There are two main learning scenarios. In **supervised learning**, the training data are labeled. In **unsupervised learning**, the training data are not labeled. Both supervised and unsupervised learning are used in biology. Here, however, we only present an example of the use of supervised learning.

The goal of supervised machine learning is to build a concise **model** or **classifier** of the distribution of labels in terms of the features of the labeled objects in the training set. The resulting classifier is, then, used to assign labels to the objects in the test set based on features that were found to be predictive in the training set. Hazkani-Covo et al. (2004) used supervised learning to predict the subcellular localization of proteins and to compare the subcellular assignments of proteins of a unicellular organism, *Saccharomyces cerevisiae*, and two multicellular ones, *D. melanogaster* and *C. elegans*.

Several computational tools for predicting the subcellular localization of eukaryotic proteins have been published (e.g., Huang et al. 2007, 2008; Zhang et al. 2014). Such tools usually use a series of features derived from the primary amino acid sequence of a protein to assign it to one of nine subcellular compartments: (1) extracellular domain, (2) cell membrane, (3) cytoplasm, (4) endoplasmic reticulum, (5) Golgi apparatus, (6) lysosome, (7) peroxisome, (8) mitochondria, and (9) nucleus. The features used by Hazkani-Covo et al. (2004) were (1) N-terminal sequences, (2) protein motifs, (3) amino acid composition, (4) predicted isoelectric point, and (5) protein length.

The training set consisted of all the proteins whose subcellular localizations had been unambiguously determined empirically. The test set, which was much larger than the training set, consisted of all the proteins whose subcellular localizations were not known. By using the training set, the program derived an experimental profile of protein properties for each of the subcellular compartments. It, then, took the proteins in the test set and compared each protein with each of the subcellular profiles. The end product was a list detailing how well each protein fit into each of the subcellular profiles. The best fit was chosen as the subcellular assignment for the protein in question.

Hazkani-Covo et al. (2004) found that the transition to multicellularity is characterized by an increase in the total number of proteins encoded by the genome. Interestingly, this increase was found to be distributed unevenly among the subcellular compartments. That is, a disproportionate increase in the number of proteins in the extracellular domain, the cell membrane, and the cytoplasm is observed in multicellular organisms, while no such increase is seen in other subcellular compartments. A possible explanation involves signal transduction. In terms of protein numbers, signal transduction pathways may be roughly described as a pyramid with an expansive base in the extracellular domain (the numerous extracellular signal proteins) that progressively narrows at the cell membrane and cytoplasmic levels and ends in a narrow tip consisting of only a handful of transcription modulators in the nucleus. These observations suggest that extracellular signaling interactions among metazoan cells account for the uneven increase in the numbers of proteins among subcellular compartments during the transition to multicellularity.

Gene ontology

In philosophy, **ontology** is the study of the nature of being, becoming, existence, and reality. In computer and information science, an ontology concerns the formal naming and definition of types, properties, and interrelationships of the entities that exist for a particular domain of discourse. **Gene ontology** is an ongoing bioinformatic initiative that attempts to unify the representation of genes and gene products across all species. More specifically, the project aims to produce a "structured, precisely defined, common, controlled vocabulary for describing the roles of genes and gene products in any organism" (Gene Ontology Consortium 2000).

Gene ontology was originally constructed by a consortium of researchers studying the genomes of three model organisms: *Drosophila melanogaster, Mus musculus,* and *Saccharomyces cerevisiae.* Many other organism databases have since joined the consortium, contributing not only to the collection of annotation data, but also to the development of the ontologies themselves and the tools to view them and apply them to the data.

From a practical view, an ontology consists of a representation of gene properties that are detectable or observable, and of the relationships among those properties. The basic categories of gene ontology are (1) biological process, (2) molecular function, and (3) cellular components. Examples of the three categories of gene ontology are shown in **Figure 11.19**.

Biological process refers to a biological objective to which the gene product contributes. A process is accomplished via one or more ordered assemblies of molecular functions. Processes often involve a chemical or physical transformation, in the sense that something goes into a process and something different comes out of it. The example in Figure 11.19a is of DNA metabolism. Other broad examples include cell growth and maintenance and signal transduction. Examples of more specific or

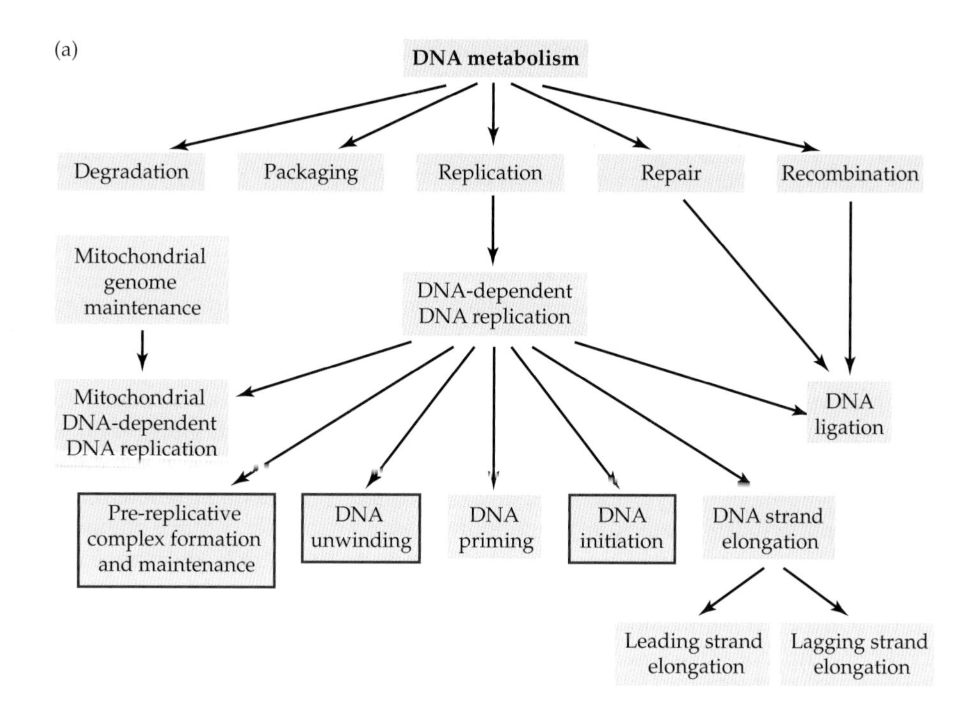

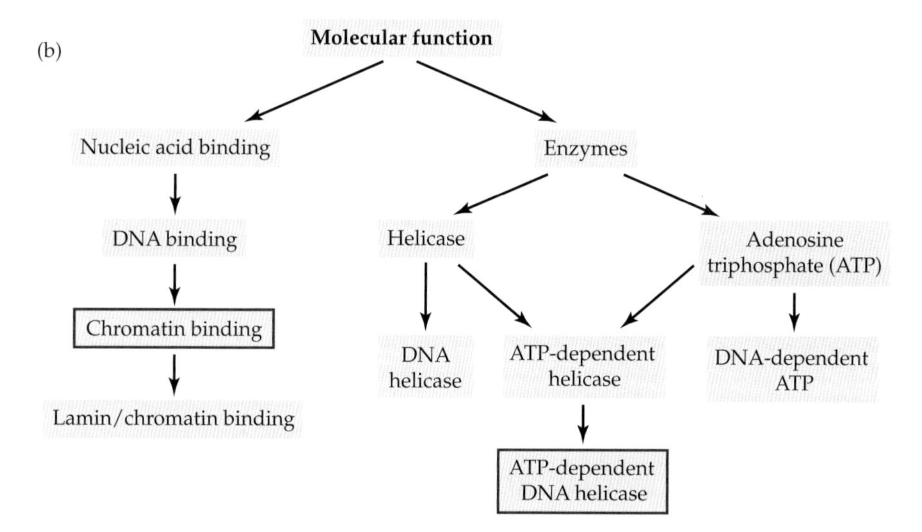

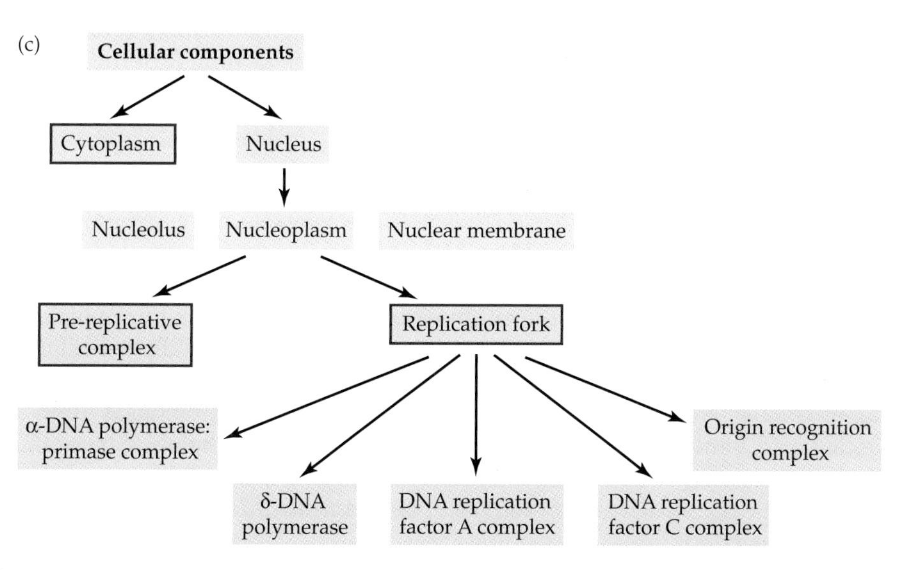

Figure 11.19 Three examples of the structure and style used to represent gene ontologies and to associate genes with nodes within an ontology. The ontologies are built from a structured, controlled vocabulary. They are designed for a generic eukaryotic cell and are flexible enough to represent differences among organisms. The multiple positions of one gene, *MCM2*, which encodes a protein called DNA replication licensing factor MCM2, are shown as red-bordered boxes. (a) Biological process ontology. This example illustrates a portion of the biological process ontology pertaining to DNA metabolism. Note that a node may have more than one parent. For example, DNA ligation has three parents: DNA-dependent DNA replication, DNA repair, and DNA recombination. (b) Molecular function ontology. This ontology is not intended to represent a reaction pathway, but instead reflects conceptual categories of gene product function. A gene product can be associated with more than one node within an ontology, as illustrated by the MCM2 proteins, which have been shown to bind chromatin and to possess ATP-dependent DNA helicase activity. (c) Cellular component ontology. Here we are interested in the physical location of a gene product. (Note that these illustrations are the products of work in progress and are subject to change as new evidence becomes available.) (Modified from Gene Ontology Consortium 2000.)

lower-level process terms are translation, pyrimidine metabolism, and DNA-dependent DNA replication.

Molecular function is defined as the biochemical activity (including specific binding) of a gene product. It describes only what is done without specifying where or when the event actually occurs. Examples of broad molecular function terms are "enzyme," "transporter," and "ligand." Examples of narrower functional terms are "adenylate cyclase" and "Toll receptor ligand."

Cellular component refers to the place in the cell where a gene product is active. These terms reflect our understanding of eukaryotic cell structure. As is true for the other ontologies, not all terms are applicable to all organisms. Cellular component terms include "ribosome" and "Golgi apparatus."

Biological process, molecular function, and cellular component are all attributes of gene products, and each attribute may be assigned independently to the gene product. The relationships between a gene product and biological process, molecular function, and cellular component may be one-to-one or one-to-many, reflecting the biological reality that a particular protein may function in several processes, contain domains that carry out diverse molecular functions, and participate in multiple alternative interactions with other proteins, organelles, or locations in the cell.

A gene ontology is a directed acyclic graph, i.e., a network containing a collection of vertices and directed edges such that by starting at a particular vertex and following any sequence of directed edges emanating from this vertex, one cannot loop back to the initial vertex again. Gene ontology terms are the nodes (vertices).

The ontologies are dynamic, in the sense that they exist as networks that change as more information accumulates, but have sufficient uniqueness and precision so that databases based on existing ontologies can automatically be updated as the ontologies mature. The ontologies are flexible in another way; they are species neutral and can be used regardless of the many differences in the biology of diverse organisms. In this way, gene ontology can be understood and used by a wide biological community.

Only few evolutionary studies have so far used gene ontology. There are, however, signs that gene ontology may find uses in evolutionary research (Primmer et al. 2013). For instance, when a particular molecular function is suspected to be of importance, gene ontology can be queried to retrieve a list of functionally relevant candidate genes for further investigation or to test specific evolutionary hypotheses. Alternatively, gene ontology can be used to identify differences in the abundance of ontologies among species.

Chromosome Number and Structure

Cytogenetics is a branch of genetics that is concerned with the study of the number, structure, and function of chromosomes. In the following sections, we introduce the vocabulary of cytogenetics that will serve us subsequently in dealing with the evolution of chromosome number and the dynamics of gene order rearrangements.

Chromosome number variation

The term **karyotype** refers to the number and types of chromosomes in the nucleus of a eukaryotic cell. The term is also used to denote the complete set of chromosomes in a species or an individual. Through meiosis and sexual reproduction, almost all eukaryotes switch between a haploid and a diploid state. The oscillation between diploid and haploid stages is called the **alternation of generations**, where the diploid to haploid state is accomplished through meiosis, and the haploid to diploid state is accomplished through gamete fusion. Only very few eukaryotic taxa, such as bdelloid rotifers (Flot et al. 2013), do not go through alternation of generations. Within a eukaryotic taxon, either the haploid state predominates over the diploid state or the diploid state predominates over the haploid state. Most fungi, algae, and mosses are haploid during the principal stage of their life cycles, while in animals and seed plants the diploid state predominates.

The **haploid number** of chromosomes is designated by n; for example, the haploid number of chromosomes in humans is $n = 23$. By extension, the **diploid number** of chromosomes is $2n$; thus, in humans, $2n = 46$. In some cases there is evidence that the n chromosomes in the haploid set have resulted from genome duplications of an originally smaller set of chromosomes. This "base" number—the number of unique chromosomes in a haploid set—is called the **monoploid number** and is denoted by x. The monoploid number is also known as the **basic**, **cardinal**, or **fundamental number**. In diploid species, as well as in polyploid species that have completed the diploidization process, $x = n$. Thus, in humans $x = n = 23$. In polyploid species, n is a multiple of x. For example, bread wheat, *Triticum aestivum*, has been derived from three different ancestral diploids (Chapter 7), each of which had 7 chromosomes in its haploid gametes. The monoploid number of bread wheat is, thus, $x = 7$, and the haploid number is $n = 3 \times 7 = 21$.

Chromosome number is usually constant within species but hugely variable among taxa. The lowest possible haploid number, $n = 1$, was observed in the nematode *Parascaris univalens* (Goday and Pimpinelli 1984) and in male (haploid) *Myrmecia croslandi* ants (Crosland and Crozier 1986). The high record for haploid number belongs to a fern species, *Ophioglossum reticulatum*, with $n = 720$ (Khandelwal 1990). The highest number of chromosomes in animals was found in the shortnose sturgeon *Acipenser brevirostrum* with $n = 186$.

Surprisingly, chromosome number does not correlate with DNA content. In jawless fishes, for instance, the lamprey *Eptatretus stoutii* (Pacific hagfish) with $n = 24$ has twice the amount of DNA of *Lampetra planeri* (brook lamprey) with $n = 73$. Similarly, the yeast *Saccharomyces cerevisiae* has 16 chromosomes and about 1.2×10^4 Kb of DNA, while the lily *Lilium longiflorum* has 10,000 times as much DNA and only 12 chromosomes. The lack of correlation between chromosome number and DNA content is nicely illustrated by pea species belonging to the genus *Lathyrus*. Despite the fact that nuclear DNA content varies from 3.4 Gb to 14.3 Gb, all *Lathyrus* species have the same number of chromosomes (Narayan and Rees 1976). Despite the fact that mammals have quite a narrow range of genome sizes (Table 11.2), haploid chromosome number varies a great deal, from $n = 3$ in the ova of the Indian muntjac, *Muntiacus muntjak vaginalis*, to $n = 102$ in the plains viscacha rat, *Tympanoctomys barrerae*.

The number of chromosomes was also found to bear no relationship to any measure of organismal complexity. Thus, closely related and morphologically similar species frequently exhibit a large difference in chromosome number. For instance, within the single plant family Asteraceae we find that n varies from 2 in *Xanthisma gracile* and *Brachyscome dichromosomatica* to 131 in *Cotula scariosa*. A spectacular example of chromosome number variability between closely related and morphologically similar taxa exists in species of gall-forming scale insects belonging to the genus *Apiomorpha*, in which chromosome number varies from $n = 4$ in *A. baeuerleni* to $n = 192$ in *A. macqueeni* (Cook 2000).

Chromosome morphology and chromosome types

Eukaryotic chromosomes have two main longitudinal differentiations, the centromere and the telomere, which are both responsible for maintaining the integrity of the chromosome and for conserving and transmitting the genetic material. The **centromere** is the location on each chromosome that maintains sister chromatid cohesion and regulates accurate chromosome segregation during cell division. During cell division, the **spindle fibers** that pull sister chromatids apart attach to the centromere via a proteinaceous structure called the **kinetochore**. In microscopic images, centromeres are characterized by a constricted region of the chromosome (often referred to as the **primary constriction**) where the two **sister chromatids** (the identical copies of a chromosome) are most closely in contact. In most eukaryotes, the centromere consists of large arrays of repetitive DNA in which individual repeat units are similar but not identical to one another. In humans, the primary centromeric repeat unit is called **α-satellite** (or **alphoid sequence**). **Telomeres** are unique structures at the ends of linear chromosomes that allow for the lagging strand to be replicated.

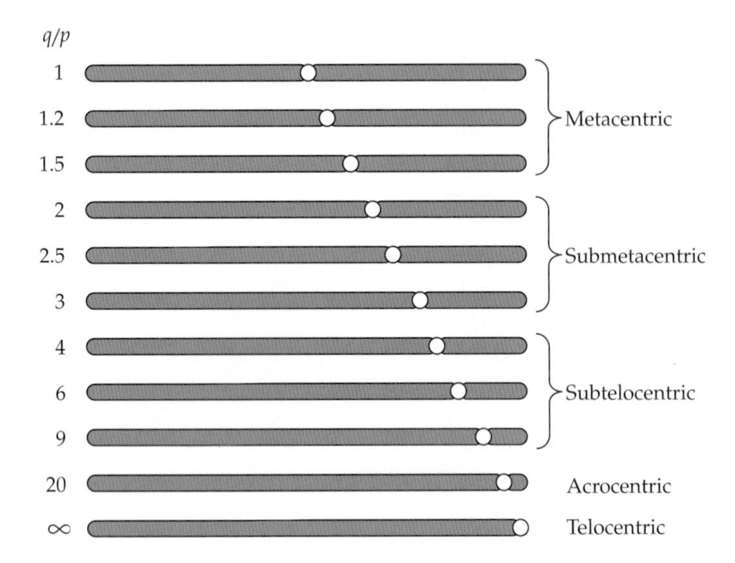

Figure 11.20 Classification of eukaryotic chromosomes by centromere position. The ratio of the long arm (*q*) to the short arm (*p*) is shown on the left for each chromosome type.

Unless the centromere is located at one of the terminal ends of the chromosome, it divides the chromosome into two **arms**, labeled *p* (the shorter of the two) and *q* (the longer one). Depending on the position of the centromere, eukaryotic chromosomes can be divided into five types: **metacentric**, **submetacentric**, **subtelocentric**, **acrocentric**, and **telocentric** (**Figure 11.20**). In metacentric chromosomes the centromere is located in the middle of the chromosome such that the two arms are approximately equal in length. In humans, two chromosomes, 1 and 3, are metacentric. If the *p* arm is present but is so short that it is hard to observe, then the chromosome is acrocentric. The human genome includes five acrocentric chromosomes: 13, 14, 15, 21, and 22. The centromere of a telocentric chromosome is located at the terminal end of the chromosome. The house mouse (*Mus musculus*) has only telocentric chromosomes. Humans, on the other hand, do not possess telocentric chromosomes. In submetacentric and subtelocentric chromosomes, the two arms are of unequal length. If the centromere is located closer to an end of the chromosome than to its middle, it is subtelocentric; if the centromere is located closer to the middle of the chromosome than to its end, it is submetacentric (Levan et al. 1964). Chromosomes that are not telocentric are referred to as **atelocentric**.

In most eukaryotes, the kinetochore protein complex assembles at a single locus termed the "centromere" to attach chromosomes to spindle microtubules. In some organisms, the centromere is diffuse, i.e., the entire length of the chromosome can act as the centromere. These so-called **holocentric chromosomes** have the unusual property of attaching to spindle microtubules along their entire lengths. Holocentric chromosomes have been found scattered throughout the plant and animal kingdoms, as well as in Rhizaria—a clade of protists characterized by the presence of needlelike pseudopodia. Holocentric chromosomes are particularly common in arthropods and nematodes (e.g., *C. elegans*).

The **fundamental number** (*FN*) of a karyotype is the number of visible chromosomal arms per set of chromosomes. Because some chromosomes are single-armed (acrocentric or telocentric), $FN \leq 2 \times 2n$. Due to the presence of five acrocentric chromosome pairs (13, 14, 15, 21, and 22), the fundamental number in humans is $FN = 82$. The **fundamental autosomal number** of a karyotype is the number of visible chromosomal arms per set of autosomes (non-sex chromosomes).

Chromosome size variation

Chromosome sizes in eukaryotes vary between 17 Kb in chromosome LG5 of the zebra finch (*Taeniopygia guttata*) to ~750 Mb in chromosome 1 of the opossum *Monodelphis domestica*. Notwithstanding the large interspecific variability in chromosome size, within diploid eukaryotic species, there seems to be a limit to intraspecific chromosome-size variation. That is, the vast majority of the chromosomes in a species have lengths that range between 0.4 and 1.9 times the mean chromosome length (Li et al. 2011). The evolutionary mechanisms for constraining intraspecific chromosome-size variation across a wide taxonomic range are not known. Mean chromosome size does not correlate with genome size. That is, a certain genome size may be attained via a large number of small chromosomes or a small number of large chromosomes. Interestingly, there is no correlation between mean chromosome size and chromosome number.

The chromosomes within a species may be similar in size to one another, in which case the karyotype is called a **symmetrical karyotype**; if the chromosomes differ markedly in size, the karyotype is an **asymmetrical karyotype**.

Euchromatin and heterochromatin

In eukaryotic cells, genomic DNA is folded with histone and nonhistone proteins to form **chromatin**. Each chromatin unit, or **nucleosome**, contains 146 bp of DNA, which is wrapped around an octamer of histones. Histone-modifying enzymes, chromatin-remodeling complexes, and DNA methylation help compact and organize genomes into discrete chromatin domains. This organization underlies many aspects of chromosome behavior, such as transcription, recombination, and DNA repair (Grewal and Jia 2007).

Chromosomes are usually not visible in the nuclei of nondividing cells. However, some parts of chromatin may become visible after staining at interphase, giving rise to the distinction between heterochromatin and euchromatin (Heitz 1928). **Euchromatin** refers to chromosomal segments that decondense during parts of the cell cycle to the extent that they become invisible during late telophase and the subsequent interphase. In contrast, **heterochromatin** refers to chromosomal segments that remain compact and can be stained and rendered visible during interphase. From a functional point of view, heterochromatin represents chromosomal segments that are highly condensed, transcriptionally inaccessible, and highly ordered in nucleosomal arrays. Euchromatin, on the other hand, is accessible and easily transcribed. From a genomic point of view, heterochromatin is dominated by repetitive DNA sequences.

Heterochromatin can be categorized as **constitutive heterochromatin** if it maintains a compact state in all cells at all times. Constitutive heterochromatin is usually encountered around the centromeres and near the telomeres. For example, constitutive heterochromatin can be observed in human cells in the vicinity of the centromeres of chromosomes 1, 9, and 16. In addition, male cells have a large block of constitutive heterochromatin on the long arm of the Y chromosome. Heterochromatin can form in chromosomal regions that contain a high density of repetitive DNA elements, such as satellite sequences and transposable elements. These regions remain condensed throughout the cell cycle.

In contrast to constitutive heterochromatin, which is identical in all cells of the body, **facultative heterochromatin** refers to sequences that may be densely packaged as heterochromatin in one cell or at one developmental stage, but packaged lightly as euchromatin in other cells or developmental stages. Facultative heterochromatin may be found at developmentally regulated loci, where the chromatin state can change in response to cellular signals and gene activity. The best known example of facultative heterochromatin is the product of X-chromosome inactivation in female cells, in which one X chromosome is packaged as heterochromatin and to a large extent, but not completely, inactivated, while the other X chromosome is packaged as euchromatin.

A key feature of heterochromatin is its ability to spread from a specific **nucleation site** to neighboring chromosomal domains. Heterochromatin can, therefore, influence gene expression in a region-specific, sequence-independent manner. When heterochromatin spreads, it generally silences the expression of nearby sequences. In female mammalian X-chromosome inactivation, heterochromatin spreads from a specific nucleation site, ultimately silencing most of the X chromosome, thereby regulating gene dosage. Heterochromatin can also repress recombination, which protects genome integrity by prohibiting illegitimate recombination between dispersed repetitive DNA elements. In fact, control of "parasitic" transposable elements has been suggested as the original evolutionary benefit of heterochromatic silencing.

Different organisms contain varying amounts of heterochromatin. The human and *D. melanogaster* genomes, for instance, are approximately 20% and 33% heterochromatic, respectively. Because of their highly repetitive nature, heterochromatic regions of the genome are extremely difficult to sequence. Only few sequencing projects have attempted to deal with heterochromatin. In *D. melanogaster*, 24 Mb of heterochromatic sequence have been annotated (Smith et al. 2007). Unsurprisingly, more than 77% of the heterochromatin was found to be composed of fragmented and nested transposable elements and other repeated DNA sequences. What was surprising, however, was the fact that *Drosophila* heterochromatin was found to contain "islands" of highly conserved protein-coding genes embedded in "oceans" of repeated sequences. Approximately 250 protein-coding genes were found, and they are almost certainly functional, since they have conserved homologs in other insects. Similar results were obtained in an analysis of 16.6 Mb of heterochromatin in the malaria vector *Anopheles gambiae* (Sharakova et al. 2010). It is still unknown how genes embedded within heterochromatin can be transcribed.

Chromosomal Evolution

In the following discussion we deal with evolutionary processes driven by **chromosomal** or **microscopic mutations**, i.e., mutations that affect the karyotype and can be detected through optical microscopy. A tentative classification of chromosomal mutations and their effects on chromosome number and gene number is shown in **Table 11.6**. Evolution at the chromosome level concerns itself with chromosome numbers, gene numbers, and gene order within and between chromosomes.

Chromosome number evolution

Many processes can bring about changes in chromosome number. Polyploidy and polysomy, which increase chromosome number, were discussed in Chapter 7. Here we present three other processes that affect chromosome number: chromosome fusion, chromosome fission, and chromosome elimination. An increase in chromosome number due to chromosome fission is an **agmatoploidy**. A decrease in chromosome number due to fusion is a **symploidy**.

Chromosome fusion, which reduces chromosome number, can occur through three processes (**Figure 11.21**). **Centric fusion** (also referred to as **centromeric fusion**) is the fusion of two chromosomes at their centromeres. If the chromosomes involved in the fusion are atelocentric, one arm from each of the chromosomes involved in the translocation will be lost. If a chromosome is acrocentric and the arm that is lost is the *p* arm, which contains very little functional genetic material, then the centromeric fusion can occur without significant harm. A centric fusion involving two metacentric chromosomes is also referred to as a **Robertsonian translocation**.

Telomeric fusion (also referred to as **tandem fusion**) is the fusion of two chromosomes at one of their telomeres. Subsequently, one centromere and the fused telomeres will become nonfunctional. In telomeric fusion there is no loss of genetic information.

Centromere-telomere fusion (also called **jumping translocation**) is the fusion of two chromosomes at the centromere of one and

TABLE 11.6

A classification of chromosomal mutations and their effects on gene and chromosome numbers

Mutation	Effect on gene number	Effect on chromosome number
Pericentric inversion	None	None
Paracentric inversion	None	None
Reciprocal translocation	None	None
Nonreciprocal translocation	None	None
Telomeric fusion	None	Decrease
Centric fusion	Decrease	Decrease
Centromere-telomere fusion	Decrease	Decrease
Fission	None	Increase
Conservative transposition	None	None
Replicative transposition	Increase	None
Block interchange	None	None
Duplication	Increase	None
Insertion	Increase	None
Deletion	Decrease	None
Polyploidy	Increase	Increase
Polysomy	Increase	Increase
Chromosome elimination (monosomy)	Decrease	Decrease

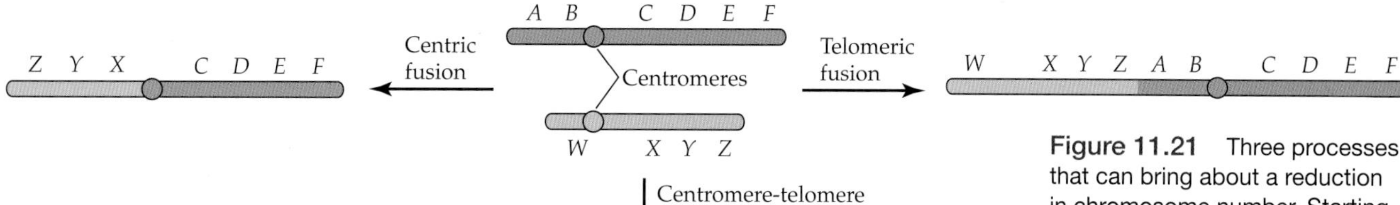

Figure 11.21 Three processes that can bring about a reduction in chromosome number. Starting with two metacentric chromosomes (middle), centric fusion (left) results in gene loss and a decrease in chromosome number. In telomeric fusion (right) one centromere and the fused telomeres become nonfunctional, but there is no change in either gene content or gene order. In centromere-telomere fusion (bottom), the genes on one arm of one chromosome are lost, and one centromere becomes nonfunctional, but there is no change in gene order. Genes are denoted by letters. (Modified from Graur and Li 2000.)

a telomere of the other. Such fusions have been described in pathological cases, such as Turner syndrome, often accompanied by the creation of an **isochromosome**, i.e., a chromosome that has lost one arm and is made of two exact copies of the other arm. Quantitatively, centromere-telomere fusions have played somewhat of a lesser evolutionary role than the other two types of chromosomal fusion. However, centromere-telomere fusion seems to have had a role in the evolution of the ancestor of the domestic pig (*Sus scrofa*), which as recently as 12 million years ago experienced two centromere-telomere fusions, creating two large chromosomes out of four smaller ones (Thomsen et al. 1996). The ancestral state is still retained in the closely related babirusa (*Babyrousa babyrussa*).

Chromosome fission, i.e., the split of one chromosome into two, should in principle be much more difficult to accomplish than fusion. The reason is that a functional chromosome needs to possess one centromere for proper chromosomal segregation and inheritance of genetic information and two telomeres for proper DNA replication. The split of one chromosome into two will result in two chromosomes, both devoid of one telomere and one devoid of a centromere. Thus, while it is not hard to understand how two chromosomes can fuse and subsequently lose a centromere and two telomeres between them, it is difficult to imagine how a centromere and two telomeres can be acquired de novo. We note, however, that **neocentromeres**, i.e., new centromeres, can occasionally appear in chromosomal locations other than that of the original centromere (Amor and Choo 2002). In humans, neocentromeres do not have alphoid DNA sequences, but are able to function as regular centromeres, i.e., form primary constrictions and assemble functional kinetochores. Unfortunately, the many human neocentromeres that have been detected so far are invariably associated with chromosomal rearrangements as well as developmental and congenital abnormalities. Admittedly, we have no evidence linking neocentromeres and chromosome fission; however, we have evidence that neocentromere formation is important in evolution, particularly in the process of repositioning the centromere within a chromosome.

Ventura et al. (2001) studied the X chromosomes from the black lemur (*Eulemur macaco*) and the ring-tailed lemur (*Lemur catta*). The X chromosome in the black lemur is telocentric, while that of the ring-tailed lemur is almost perfectly metacentric. The orders of the genetic markers on these chromosomes, however, were perfectly collinear with each other, unequivocally pointing to neocentromere emergence as the most likely explanation of centromere repositioning, rather than to translocation of an existing centromere into this region. According to one model, a rearrangement or reduction in size of the alphoid DNA array may reduce kinetochore-binding capacity and impair centromere function. A latent neocentromere can then be activated (Hsu et al. 1975). The neocentromere is initially imperfect, but, in subsequent generations, selection pressure will improve kinetochore maturation through duplication of existing sequences or accumulation of repetitive DNA. The original centromeric alphoid DNA would subsequently contract in the absence of selection pressure and would ultimately disappear. In evolutionary terms, centromere relocation would rapidly lead to the reproductive isolation of emerging species, providing a formidable mechanism for speciation. Thus, although we lack direct evidence, it is reasonable to assume that the neocentromeres play a role in chromosome fission.

Following chromosome fission, the two resulting chromosomes will each lack one telomere. Deletion or absence of a telomere from a chromosome necessitates a compensatory mechanism to stabilize chromosome ends. Two main mechanisms were proposed to stabilize chromosome ends following terminal deletions: **telomere healing**, i.e., the restoration of telomere sequences, and **telomere capture**, i.e., the formation of a derivative chromosome adopting through fusion a telomere from another chromosome. With the exception of some cancer cells, which employ telomere capture (Labib et al. 2014), most experimental data indicate that telomere healing by direct addition of telomeric repeats, also called **neotelomeres**, is the main mechanism of dealing with terminal chromosome deletions (Kulikowski et al. 2010.). In *Drosophila*, for instance, neotelomeres are created with relative ease whenever terminally deleted chromosomes arise by mutation.

Given the inherent mechanistic difficulties of chromosomal fission, do we know for certain that fissions actually occur during evolution? We do, and phenomenologically we can also state that fissions occur at much lower frequency than fusions. For example, within odd-hoofed ungulates (Perissodactyla), the ratio of fusions to fissions is about 18 to 1 (Trifonov et al. 2008). Are we as knowledgeable of the process of chromosome fission as we are of fusion? Not yet.

The standard chromosomal complement of a cell—say, the 46 chromosomes in a human diploid cell—is referred to as the set of **A chromosomes**. Extra chromosomes, which may exist in some cells or in some individuals from a species but not in others, are referred to as **B chromosomes** (also referred to as **supernumerary** or **accessory chromosomes**). As far as A chromosomes are concerned, **chromosome loss** or **elimination** is a process that has been mainly documented in somatic cells and is, therefore, of limited evolutionary interest. It is, however, an important mechanism in the evolution of B chromosomes (Camacho et al. 2000). B chromosomes have probably arisen from A chromosomes but follow their own evolutionary pathway. Their irregular mitotic and meiotic behavior allows them to increase or decrease their numbers in the germline, enabling non-Mendelian inheritance with transmission rates either above or below the 0.5 that is standard in A chromosomes. Occasionally, they may be eliminated from the genome. B chromosomes have been found in all major groups of animals and plants.

Chromosome fission is extremely common in holocentric chromosomes, i.e., chromosomes with diffuse centromeres. The reason is that fragments of holocentric chromosomes can be inherited in a Mendelian fashion because they retain the capability to attach to the spindle apparatus (Hipp et al. 2013).

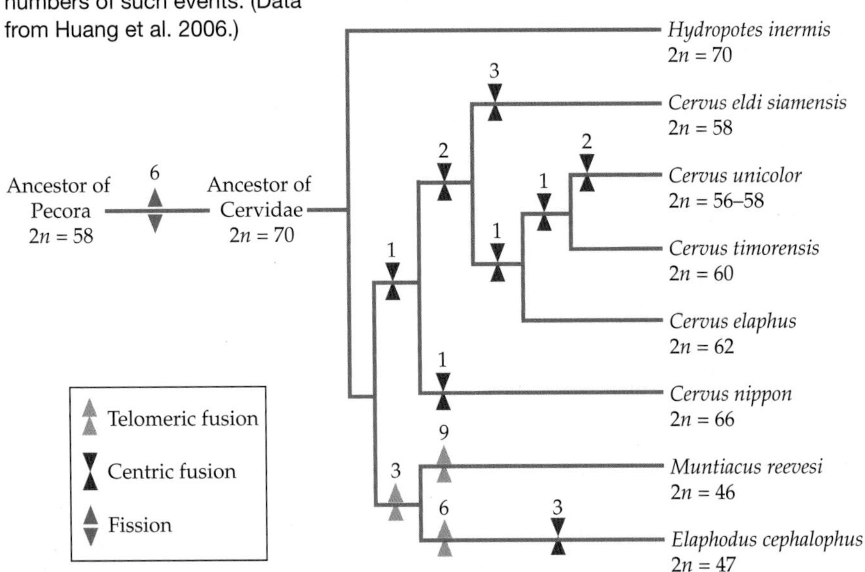

Figure 11.22 Chromosomal fusions and fissions and changes in diploid chromosome number in the evolution of eight cervid (deer) species. The ancestral cervid karyotype of $2n = 70$ was retained in a few species (e.g., *Hydropotes inermis*), while telomeric and centric fusions resulted in a wide range of chromosome number reductions in many cervid taxa. In some species, such as the sambar (*Cervus unicolor*), intraspecific variation in chromosome number has been observed. Interestingly, while evolution within the Cervidae entailed only fusions, six chromosomal fissions were inferred to have occurred from the $2n = 58$ karyotype of the ancestor of all Pecora (an infraorder that includes all kosher even-toed mammals with ruminant digestion) to the $2n = 70$ ancestral cervid karyotype. The numbers above the fission and fusion symbols indicate inferred numbers of such events. (Data from Huang et al. 2006.)

CHROMOSOME NUMBER EVOLUTION IN DEER Some lineages seem to experience frequent increases and decreases in chromosome numbers by chromosome fission and fusion, respectively. The deer family (Cervidae) is one such example. Deer species have diverse karyotypes, their diploid chromosome numbers ranging from $2n = 6$ in female Indian muntjacs (*Muntiacus muntjak vaginalis*) to $2n = 80$ in the Siberian roe deer (*Capreolus capreolus pygargus*). Huang et al. (2006) found that the chromosomal evolution within Cervidae entailed only fusions, resulting in great reductions in chromosome numbers. In contrast, the ancestral cervid karyotype of $2n = 70$ evolved through fissions only (**Figure 11.22**).

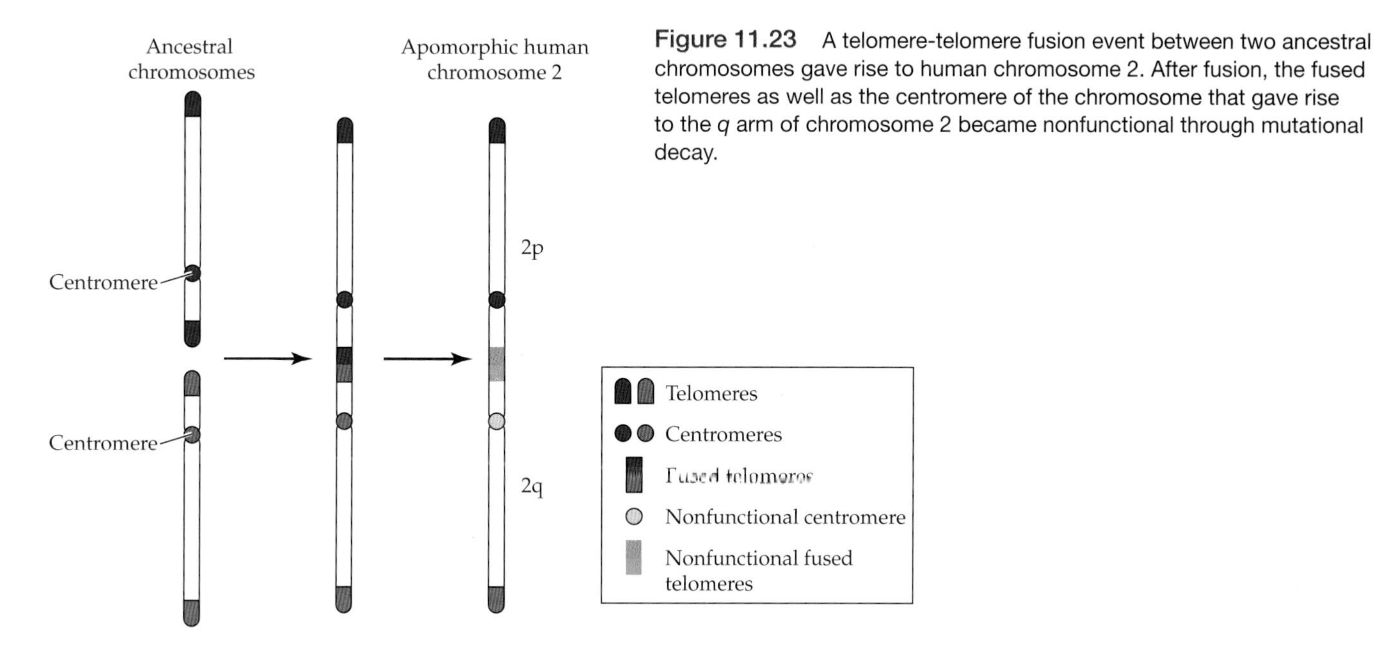

Figure 11.23 A telomere-telomere fusion event between two ancestral chromosomes gave rise to human chromosome 2. After fusion, the fused telomeres as well as the centromere of the chromosome that gave rise to the *q* arm of chromosome 2 became nonfunctional through mutational decay.

CHROMOSOME NUMBER REDUCTION IN THE HUMAN LINEAGE With the exception of humans, all extant members of Hominidae (great apes) have 24 pairs of chromosomes. Humans have only 23 pairs. Human chromosome 2 is thought to have resulted from an end-to-end fusion of two ancestral (plesiomorphic) chromosomes (**Figure 11.23**). Thus, human chromosome 2 represents a derived (apomorphic) state, while chimpanzees, gorillas, and orangutans have retained the plesiomorphic state.

Several lines of evidence support the fusion hypothesis. First, in the closest extant relative of humans, the chimpanzee, two chromosomes, 2p and 2q, are nearly identical in their DNA sequences to different contiguous parts of the two arms of human chromosome 2. The same is true of the chromosomes of the more evolutionarily distant gorilla and orangutan (Yunis and Prakash 1982).

Second, human chromosome 2 contains vestigial telomeric sequences (IJdo et al. 1991). Two inverted arrays of telomeric repeats, 5′–(TTAGGG)*n*–3′and 5′–(CCCTAA) *n*–3′ in a head-to-head arrangement, were found on the long arm of human chromosome 2. The inverted repeats seem to be relics of a telomere-telomere fusion that gave rise to human chromosome 2. Such interstitial telomeric sequences have frequently been used in evidence for chromosome fusion events (e.g., Lee et al. 1993).

Third, human chromosome 2 contains vestigial centromeric sequences (Avarello et al. 1992). Two centromeric alphoid-like sequences were detected on the long arm of chromosome 2. This finding, too, supports the hypothesis that human chromosome 2 originated through telomeric fusion.

Chromosomal rearrangements

Chromosomal rearrangements are mutations involving long DNA sequences that result in changes in the location and/or order of genes. Chromosomal rearrangements may sometimes involve multiple mutational processes; for instance, a DNA sequence may be excised from one location and incorporated in reverse orientation in a different location.

Different processes may bring about chromosomal rearrangements. Here, we distinguish between **intrachromosomal rearrangements** and **interchromosomal rearrangements**. Processes that can result in intrachromosomal changes in gene order are shown in **Figure 11.24**. **Chromosomal inversions** involve the 180° rotation of a DNA segment, resulting in a reversed gene order with respect to the original one. There are two types of inversions: **pericentric** and **paracentric**. In the former, the inverted segment includes the centromere; in the latter it does not. **Chromosomal deletions** may

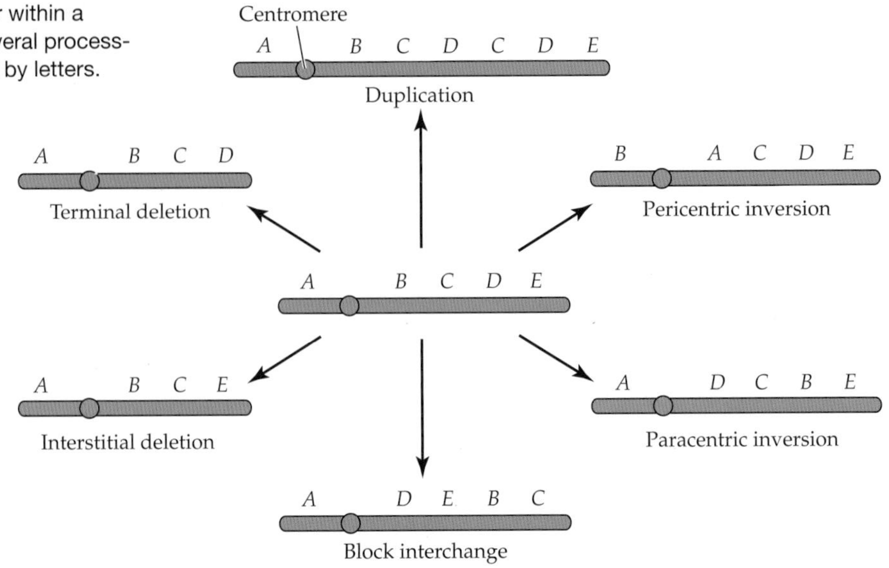

Figure 11.24 Gene content and gene order within a chromosome (center) may change through several processes of gene rearrangement. Genes are denoted by letters. (Modified from Graur and Li 2000.)

Centromere
A B C D C D E
Duplication
A B C D
Terminal deletion
B A C D E
Pericentric inversion
A B C D E
A B C E
Interstitial deletion
A D C B E
Paracentric inversion
A D E B C
Block interchange

be **terminal** or **interstitial**. Alternatively, parts of a chromosome may be duplicated in a process we referred to as partial polysomy in Chapter 7. A **block interchange** is a rearrangement event that exchanges two, not necessarily consecutive, contiguous regions in a chromosome, maintaining the original orientation. Block interchange is a very useful operation in computer algorithms designed to deal with chromosomal rearrangements; it is unclear, however, whether such a change ever occurs in nature. In **Figure 11.25**, we illustrate two processes involving interchromosomal rearrangements that can affect gene order: **reciprocal** and **nonreciprocal translocations**.

Studying gene order rearrangements in eukaryotes is much more complicated than in bacteria for several reasons. First, the eukaryotic genome contains many repeated genes, and deciding whether two genes from two organisms are orthologous or paralogous may be quite complicated. Second, gene order rearrangements may involve the movement and exchange of genetic information between chromosomes as well as within chromosomes, whereas the vast majority of prokaryotes are unichromosomal. For example, genes from human chromosome 1 are found on nine different chromosomes in mice. Finally, eukaryotes have much larger numbers of genes than prokaryotes, a serious impediment for all computer algorithms designed to deal with chromosomal rearrangements (Zhao and Bourque 2010).

Comparisons among multichromosomal organisms necessitate the introduction of terms that will help us (1) characterize positional relationships among genes, (2) describe positional changes within and between chromosomes, and (3) quantify the number and type of genome rearrangements between taxa. The definitions used here have been taken from Renwick (1972), Nadeau and Taylor (1984), Nadeau (1989), and Ehrlich et al. (1997).

Synteny ("same thread" in Greek) refers to the occurrence of two or more genes on the same chromosome (**Figure 11.26**). Genes are defined as syntenic whether or not they are genetically linked. **Conserved** or **shared synteny** refers to two or more syntenic genes in one species whose orthologs are syntenic in another species. **Conserved linkage** pertains to the conservation of both synteny and linkage in two or

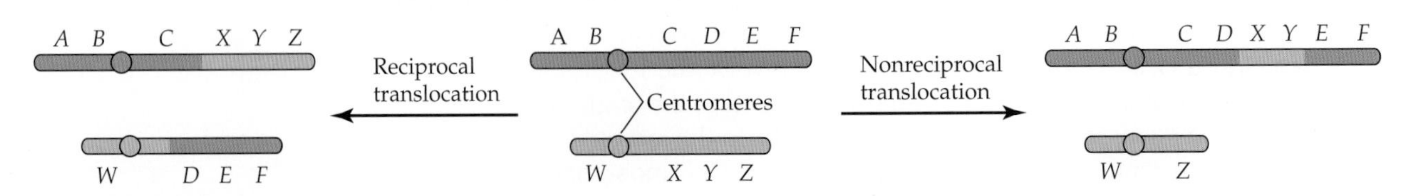

Figure 11.25 Changes in gene order and gene distribution among chromosomes through chromosomal exchanges.

more pairs of orthologous genes regardless of their order. In comparing two directed linear orders of biological entities (e.g., codons in a coding region of a gene, amino acids in a protein, genes on a chromosome, temporal gene expression), "collinearity" refers to the conservation of the exact order. In the context of positional relationships among genes, **collinearity** refers to the conservation of synteny, linkage, and gene order between two species. (Despite its frequent use in the literature, we advise against the use of the tautological term "conserved collinearity.") Chromosomal segments exhibiting collinearity between two or more species are frequently referred to as **synteny blocks**.

A **disrupted synteny** refers to cases where two or more genes are located on the same chromosome in one species but their orthologs are located on different chromosomes in the second species. A **disrupted linkage** refers to a difference in the relative proximity of two genes between two species. Finally, a **disrupted collinearity** refers to a difference in gene order between two species. We note that a disrupted synteny also counts as a disrupted linkage, and a disrupted linkage also counts as a disrupted collinearity.

We emphasize that all the above terms are qualitative, hence the use of quantitative modifiers, such as a "high degree of collinearity" (e.g., Tang et al. 2008), is as logical as claiming that a woman is "slightly pregnant." We also note that the use of these terms in the literature is frequently inconsistent with the original meaning of the terms (Passarge et al. 1999). For example, the terms "synteny" and "conserved synteny" are frequently confused with each other.

From the comparison of any two multichromosomal genome maps, it is possible to obtain information on the following variables: (1) the number of conserved syntenies, (2) the distribution of number of genes among conserved syntenies, (3) the number of conserved linkages, (4) the distribution of number of genes among conserved linkages, (5) the number of collinear segments, and (6) the distribution of number of genes among collinear segments. By assuming a uniform distribution of genes over the genome, and by using a maximum likelihood approach with gene order data from nine well-mapped mammalian genomes (cow, chimpanzee, human, baboon, hamster, mouse, rat, mink, and cat), Ehrlich et al. (1997) estimated the number of breakpoints or synteny disruptions required to explain the differences in karyotypes between any two genomes. Their main findings were that (1) gene order rearrangements occur with amazing rapidity; (2) the rates of synteny disruption vary widely among mammalian lineages, with the mouse lineage experiencing a rate of synteny disruptions 25 times higher than that of the cat lineage; and (3) despite a priori theoretical considerations implying that interchromosomal rearrangements should be strongly selected against, they were found to occur approximately four times more frequently than intrachromosomal rearrangements.

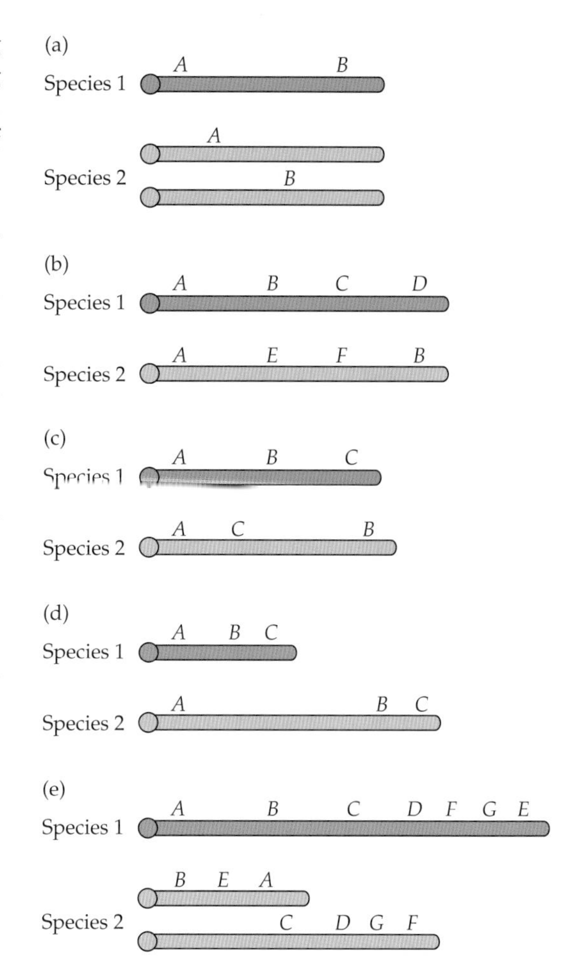

Figure 11.26 Examples of positional conservations and disruptions of syntenic orthologous genes. Genes are shown in relation to the centromere (circle). (a) In species 1, genes *A* and *B* are syntenic; in species 2, they are not. (b) Shared synteny between orthologous gene pairs *A* and *B* in species 1 and 2. (c) Shared linkage of genes *A*, *B*, and *C* in species 1 and 2. (d) Genes *A*, *B*, and *C* in species 1 are collinear with their respective orthologs in species 2. (e) Comparison of seven orthologous gene pairs between species 1 and 2 reveals cases of conserved synteny (e.g., *A* and *B*), conserved linkage (e.g., *G* and *F*), collinearity (e.g., *C* and *D*), disrupted synteny (e.g., *A* and *C*), disrupted linkage (e.g., *A* and *B*), and disrupted collinearity (e.g., *G* and *F*).

Many methods have been suggested in the literature to quantify the number of chromosomal rearrangements required to "transform" one genomic order into another (e.g., Housworth and Postlethwait 2002). The simplest quantitative approach to the problem is to count how many gene neighbors in one species are also neighbors in the other species (Keogh et al. 1998). Another method is to use highly similar short sequences from two taxa to identify syntenic regions. This method was pioneered by the Mouse Genome Sequencing Consortium (2002) in their comparison of the human and mouse genomes. Their analysis started with the identification of regions having a minimum of a 40 bp perfect match, with the additional condition that each such mouse sequence had a unique match in the human genome and that each such human sequence had a unique match in the mouse genome. Such regions,

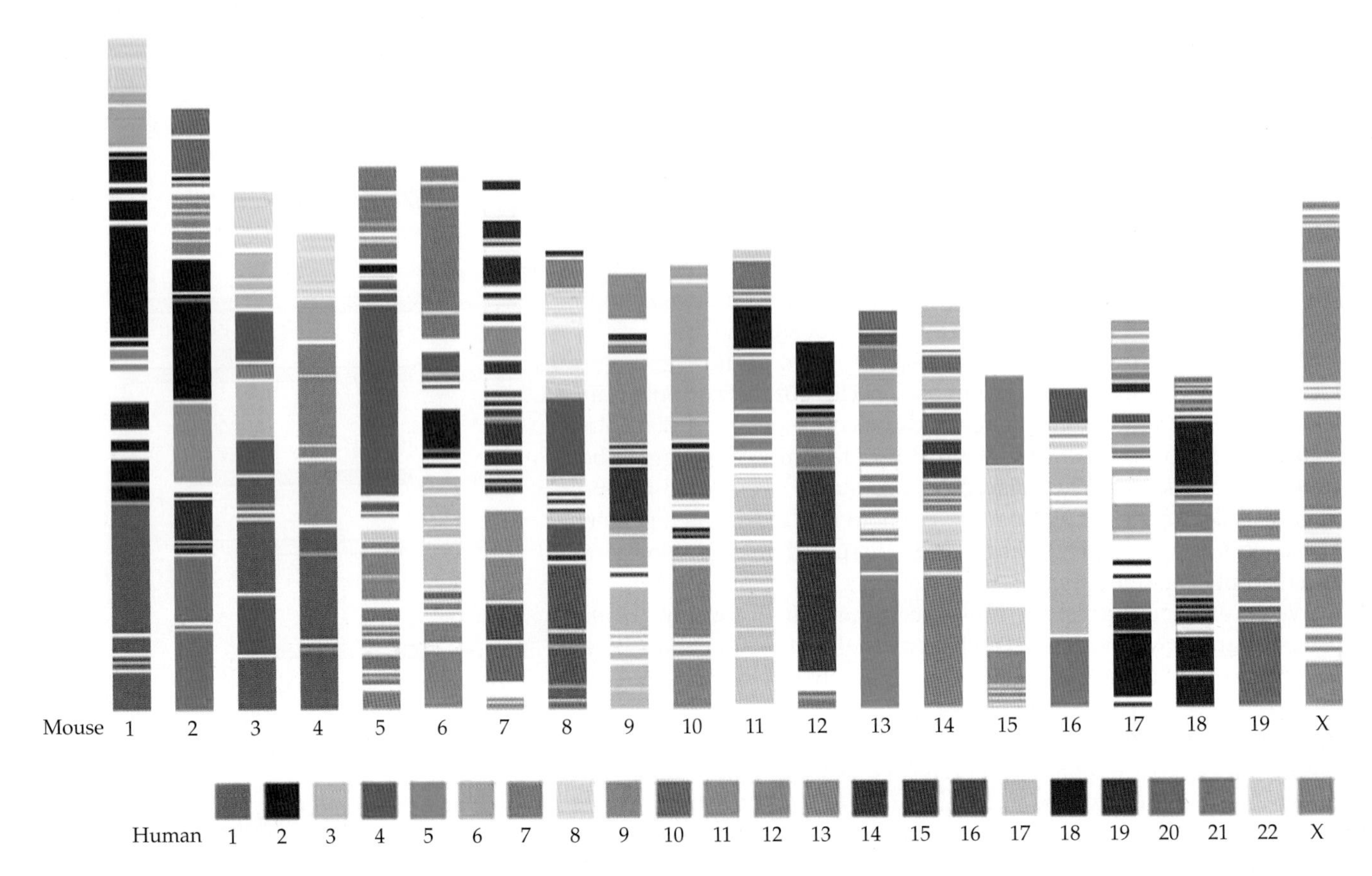

Figure 11.27 The 342 conserved syntenic segments between human and mouse chromosomes. Each color corresponds to a particular human chromosome, as shown along the bottom of the figure. (From Mouse Genome Sequencing Consortium 2002.)

which the authors called "landmarks," were deemed to represent orthologous sequence pairs derived from the same ancestral sequence. About 558,000 orthologous landmarks were identified. The landmarks had a total length of roughly 188 Mb, comprising about 7.5% of the mouse genome. The locations of the landmarks in the two genomes were then used to identify regions exceeding 300 Kb in length, in which a series of landmarks were found in the same order on a chromosome in each species. The human and mouse genomes yielded a total of 342 such segments (**Figure 11.27**), each containing on average 1,600 landmarks. The segments varied greatly in length, from 303 Kb to 64.9 Mb, with a mean of 6.9 Mb. The segment sizes were broadly consistent with a random breakage model of genome evolution. Thus, at this level of resolution, there was no evidence for selection on gene order across the genome. In total, about 90.2% of the human genome and 93.3% of the mouse genome were found to reside within conserved segments.

The nature and extent of synteny conservation differed substantially among chromosomes. The X chromosomes are represented as single, reciprocal syntenic segments. Human chromosome 20 corresponds entirely to a portion of mouse chromosome 2, with nearly perfect conservation of order along almost the entire length. Human chromosome 17 corresponds entirely to a portion of mouse chromosome 11, but extensive rearrangements have divided it into at least 16 segments. Other chromosomes, however, show evidence of much more extensive interchromosomal rearrangements.

With a map of conserved syntenic segments between the human and mouse genomes, it was possible to calculate the minimal number of rearrangements needed to "transform" one genome into the other. When applied to the 342 syntenic segments, the most parsimonious path consisted of 295 rearrangements. This analysis also suggested that chromosomal breaks have a tendency to reoccur in certain regions. Kemkemer et al. (2009) noted that specific chromosomal regions appear to be prone to recurring rearrangements in different mammalian lineages. They referred to the phenomenon of hotspots of chromosomal rearrangement as **breakpoint reuse**.

It has also been suggested that chromosomal rearrangements and segmental duplications tend to occur in the same regions (e.g., Bailey et al. 2004; Armengol et al. 2005; Bailey and Eichler 2006; Mehan et al. 2006; Giannuzzi et al. 2013). However, as pointed out by Bailey et al. (2004), segmental duplications are not necessarily the cause of the chromosomal rearrangements, nor are the chromosomal rearrangements the cause of the segmental duplications. Rather, the analyses supported a nonrandom model of chromosomal evolution that implicates specific regions within the mammalian genome as being predisposed to both recurrent segmental duplication and chromosomal rearrangement.

Evolutionary patterns of chromosomal rearrangements

An almost universal finding in studies dealing with chromosomal rearrangements is that gene order conservation decreases monotonically with evolutionary distance (e.g., Keogh et al. 1998). The monotonic decay in synteny conservation suggests that gene order changes constitute neutral mutations rather than adaptive processes

In animals, the rate of synteny conservation decay has been shown to be roughly proportional to the rate of amino-acid-sequence divergence in orthologous proteins, so at ~50% mean amino-acid-sequence divergence, all traces of ancestral gene order are lost (Zdobnov et al. 2005; Zdobnov and Bork 2007). These observations indicate that, compared with prokaryotes, eukaryotes show a much slower decay of synteny conservation—even at ~90% amino acid sequence identity, prokaryotes lose all gene order conservation beyond conserved operons. In general, measures of gene order similarity become useless at large genetic distances. Thus, in comparisons between *Saccharomyces cerevisiae* and *Schizosaccharomyces pombe*, which diverged from each other approximately 420 million years ago, not even one common pair of gene neighbors was found (Keogh et al. 1998).

By assuming a uniform distribution of genes over the genome, and by using a maximum likelihood approach with gene order data from nine mammalian genomes (cow, chimpanzee, human, baboon, hamster, mouse, rat, mink, and cat), Ehrlich et al. (1997) estimated the number of rearrangements required to explain the differences in karyotypes between any two genomes. Their main findings were that (1) rearrangement rates vary widely among mammalian lineages, with the mouse lineage, for instance, experiencing a rate of synteny disruptions 25 times higher than that of the cat lineage, and (2) despite a priori expectations that interchromosomal rearrangements should be strongly selected against, they were found to occur approximately four times more frequently than intrachromosomal rearrangements.

Is gene order conserved?

Why do some groups of genes stay linked to one another during evolution? Poyatos and Hurst (2007) looked at many factors that had been previously suspected to drive gene order conservation, such as metabolic relationships; gene coexpression; physical proximity; density of lethal mutations, i.e., essential gene density; recombination rates; gene coregulation, i.e., number of common regulatory motifs between two genes; and the distance between the gene products in the protein-protein interaction network. They found that the physical distance between neighboring genes was the best predictor of gene order conservation. That is, no selection was needed to explain gene order conservation. The linkage between any two genes was found to decay as a simple consequence of the accumulation of random mutational events that disrupted linkage.

As a rule, the size of synteny blocks decreases with evolutionary time. If disrupted collinearity occurs randomly, then the probability of observing synteny blocks of a certain length should vary exponentially with length of the synteny block. In a comparison of 12 insect genomes, Zbodonov and Bork (2007) found that this probability can be fitted much better to a power function, suggesting a more complex pattern of genomic rearrangements. They also found that as a rule there is no selection for conservation of gene order; however, they have identified many pairs of genes whose proximity to each other seems to be under selective constraint. The most prominent example of gene order conservation was a large number of genes encoding glucose-methanol-choline oxidoreductases nested within a 80 Kb intron of the *D. melanogaster Flo-2* gene on the X chromosome. Although this arrangement might have been frozen in evolution as a result of selection pressure to keep the intron and its parent gene intact, there is no evidence for similar constraint in other nested genes.

Gene Distribution Between and Within Chromosomes

In this section, we deal with two aspects of gene distribution. First, we address the question of gene density: Is the density of protein-coding genes more or less the same throughout the genome, or are there regions within the genome that exhibit a significantly higher or lower gene density than other regions? Second, we tackle the possibility of functional clustering of genes. That is, we ask, are genes randomly strewn across the genome, or do some genes cluster by function or pattern and timing of expression or affiliation to a certain biochemical pathway?

Gene density

In both prokaryotes and eukaryotes, gene density is inversely correlated with genome size. Thus, gene density in most eukaryotes is much lower than in most prokaryotes. For example, gene density in *Mycoplasma genitalium* is 0.8 genes/Kb. The density drops to 0.6 genes/Kb in *Escherichia coli*, which has a genome 8 times larger. In eukaryotes, the density is approximately 0.6 gene/Kb in the yeast, 0.2 gene/Kb in *Caenorhabditis elegans* and *Arabidopsis thaliana*, and 0.1 genes/Kb in *Drosophila melanogaster*. In species with larger genomes, such as zebrafish, wheat, and maize, the density drops to 0.02 genes/Kb. In humans protein-coding gene density is 0.007 gene/Kb, and in the largest genome sequenced so far, *Pinus taeda*, the density is 0.002 gene/Kb.

In some organisms, protein-coding genes are distributed about evenly among chromosomes, i.e., the number of genes on each chromosome is proportional to its length. Such is the case for the distribution of the 6,692 protein-coding genes among the 16 chromosomes of the yeast *S. cerevisiae* (**Figure 11.28**). The distribution of yeast genes within chromosomes is not even. There are regions with a high gene density and regions with a low gene density (**Figure 11.29**).

In the nematode *C. elegans*, there are over 19,000 genes distributed among six chromosomes with a total of about 97 Mb in length. The chromosomal distribution is less uniform than that in the yeast, with the X chromosome having a lower gene density than that of the other chromosomes, but the departure from uniformity is not very large. In some organisms, such as protists, plants, fungi, insects, urichordates, birds, and marsupials, chromosome length explains at least 83% of the variation in gene number per chromosome (Salzburger et al. 2009). The situation in other taxa is less clear. For example, in bony fishes, some species (e.g., *Tetraodon nigroviridis*) exhibit a more or less uniform distribution of genes, while in others (e.g., *Danio rerio*) some genomic regions are gene-rich and some are gene-poor.

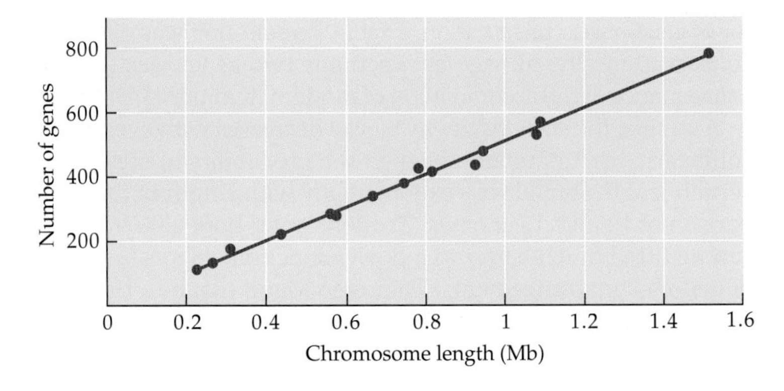

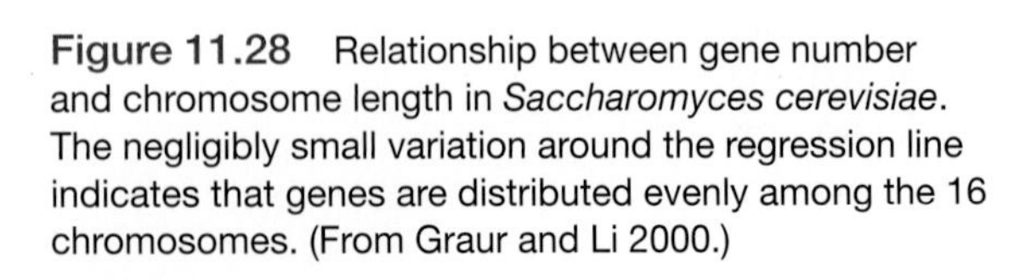
Figure 11.28 Relationship between gene number and chromosome length in *Saccharomyces cerevisiae*. The negligibly small variation around the regression line indicates that genes are distributed evenly among the 16 chromosomes. (From Graur and Li 2000.)

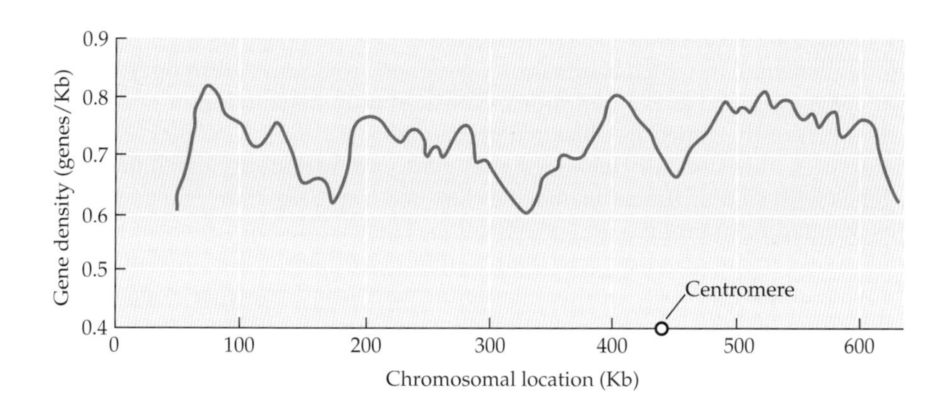

Figure 11.29 Periodicity in gene density along chromosome XI of *Saccharomyces cerevisiae*. (Modified from Sharp and Matassi 1994.)

Placental mammals are exceptional in that chromosome length explains very little of the variation in gene number per chromosome (from 28% in cattle, through 39% in humans, to 70% in rats). In humans, the highest gene densities are seen in chromosomes 19 and 22, with approximately 19.7 and 13.7 protein-coding genes per million nucleotides, respectively. The chromosomes with the lowest gene densities in human are Y and 13, with approximately 2.9 and 3.4 protein-coding genes per million nucleotides. Thus, gene density distribution among human chromosomes exhibits a sevenfold range.

In species in which gene density is low and the intragenomic variation in gene density is large, it is possible to distinguish between **gene deserts**, i.e., long chromosomal regions that are devoid of protein-coding genes (Ovarchenko et al. 2005; Taylor 2005) and clusters of protein-coding genes in high-density regions, collectively called the **gene space** (Carels et al. 1995). There is, unfortunately, no universally agreed upon definition of "gene desert." In some publications, gene deserts are defined as intergenic regions larger than a certain threshold. Frequently, this threshold is set at 500,000 bp (e.g., Salzburger et al. 2009). In other studies, gene deserts are defined as a certain top percentile (e.g., 3%) from among the length distribution of intergenic regions (e.g, Hellsten et al. 2010). The vast majority of placental genomes contain numerous gene deserts. For example, the fractions of placental genomes taken up by gene deserts in dogs and humans are 20% and 35%, respectively. The only exception to this rule so far is the genome of *Bos taurus*, only a minute fraction of which (~4%) is occupied by gene deserts.

In large plant genomes, the gene space represents only a small fraction (12–24%) of the nuclear genome, while some gene deserts (e.g., those in maize) seem to be enormous in size (Barakat et al. 1998).

Based on comparison of the conservation levels of the genomes of human and chicken, gene deserts were classified into two categories, slowly evolving **stable gene deserts** and rapidly evolving **variable gene deserts** (Ovarchenko et al. 2005). Stable gene deserts are suspected to contain long-range *cis*-regulatory sequences acting on neighboring genes (Nobrega et al. 2003; Uchikawa et al. 2003; Kimura-Yoshida et al. 2004). Gene ontology categories of genes neighboring stable gene deserts show a strong bias toward transcriptional and developmental regulators, suggesting that the presence of long-range enhancers may play a role in the evolutionary maintenance of the stable gene deserts. In contrast, gene deserts of the variable class are most probably nonessential to function, since they can be deleted without obvious phenotypic consequences (Nobrega et al. 2004).

Do genes cluster by function?

As opposed to prokaryotes, in which operon organization governs to some extent the arrangement of genes in the genome, the vast majority of eukaryotes seem to have no such simple organizing principle. Nearly all eukaryotic mRNAs are monocistronic, so there is no single dominant mechanistic basis for clustering of genes. There are, however, eukaryotic exceptions, i.e., (1) the genomes of kinetoplastids (e.g., *Trypanosoma* and *Leishmania*), in which the majority of genes are organized in operon-like units that are transcribed as polycistronic mRNAs; and (2) nematodes, in which approximately 15% of the genes are clustered in operons.

In eukaryotes, it is typically assumed that genes are randomly distributed. Of course, genes with similar functions may be clustered in proximity to one another if the genes resulted from tandem gene duplication. This is, however, a trivial type of clustering. A less trivial question is to ask whether or not genes sharing a property other than sequence paralogy tend to cluster in the genome. Lercher et al. (2002) looked at a group of 5,112 human genes that excluded gene duplicates. They divided the genes according to their expression in 14 different tissues. They first tested the hypothesis that the genome is organized into subregions, each specializing in genes needed in a certain tissue. They found no support for this hypothesis. Next, they classified the genes into (1) tissue-specific genes, i.e., genes expressed in one or two tissues; (2) genes with intermediate tissue specificity, i.e., genes expressed in three to eight tissues; and (3) housekeeping genes, i.e., genes expressed in nine or more tissues. These three categories of genes turned out to be nonrandomly distributed, but the nonrandomness could be entirely explained by housekeeping genes that showed strong clustering in GC-rich regions of the genome.

In the literature, however, one may find claims of nonrandom gene order organization in eukaryotic genomes. That is, statistically significant clustering has been observed among genes that are considered related by some criterion or another (Koonin 2009). In particular, many reports have documented statistically significant clustering of coexpressed genes. Thus, it has been shown that approximately 25% of the yeast genes that are expressed in the same stage of the mitotic cell cycle are clustered on chromosomes. Similar findings have been reported for the nematode *C. elegans*, even after the contribution of operons was subtracted, as well as for *D. melanogaster.* In many cases, however, such reports suffer from a well-known problem that afflicts quantitative analyses involving large data sets—confusing statistical significance with significance and ignoring effect size (Carver 1993; Snyder and Lawson 1993). Thus, a comparison of clusters of coexpressed genes in human and mouse genomes revealed statistically significant conservation, purportedly in support of the functional significance of these clusters; however, the clusters encompassed less than 5% of the genes in each of the two genomes (Sémon and Duret 2006).

The Repetitive Structure of the Eukaryotic Genome

As opposed to prokaryotic genomes, which consist mostly of single-copy DNA (i.e., DNA sequences that are found only once in the genome), eukaryotic genomes can be divided into single-copy DNA, i.e., nucleotide sequence that appear once in the genome, and **repetitive DNA**, i.e., nucleotide sequences of various lengths and compositions that occur several times in the genome. Repetitive DNA was discovered by Britten and Kohne (1968) in their classic study of DNA **denaturation and annealing**. (The term "denaturation" is also referred to as **disassociation**, while the term "annealing" is also called **reassociation**, **renaturation**, or **reannealing**.) In denaturation and annealing experiments, a DNA sample is (1) sheared into small pieces approximately 400 nucleotides long, (2) thermally melted into single strands, then (3) allowed to anneal through gradual cooling.

The kinetics of the reannealing process are measured by a parameter called C_0t (pronounced "cot"), in which C_0 is the initial molar concentration of the complementary strands of DNA, and t is the incubation time required for completing half of the reannealing reaction under controlled conditions. If the sequences of all initial DNA fragments are identical, complementary fragments will encounter each other readily and the mixture will reanneal quickly. At the other extreme, reannealing is very slow if the original DNA contains many different sequences, such that the concen-

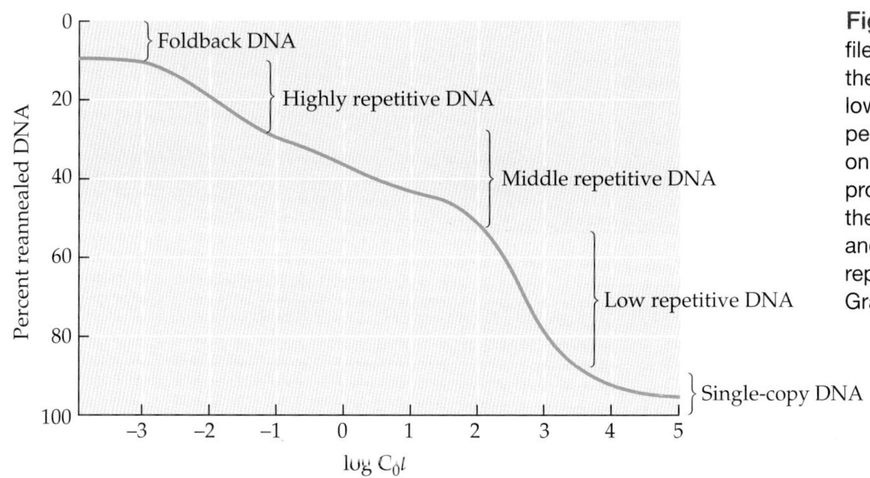

Figure 11.30 A schematic reassociation profile of eukaryotic DNA. DNA is purified, sheared, thermally melted into single strands, and then allowed to reassociate through gradual cooling. The percentage of reassociated double-stranded DNA on the vertical axis is shown as a function of the product of DNA concentration and time (C_0t) on the horizontal axis. With the exception of foldback and single-copy DNA, the division into different repetitive categories is arbitrary. (Modified from Graur and Li 2000.)

trations of complementary strands are very low and they encounter each other only infrequently.

Two extreme C_0t value ranges are encountered (**Figure 11.30**). The first corresponds to a genomic fraction that reanneals almost instantaneously. This fraction, which is called **foldback DNA**, consists of palindromic sequences that can form hairpin double-stranded structures. Each sequence in this fraction can reanneal with itself, and hence reannealing occurs as soon as the denatured DNA is allowed to renature. The foldback DNA fraction is usually very small, although in some organisms it may reach values in excess of 10%. The other extreme C_0t value range corresponds to the fraction of **single-copy** or **unique DNA**, which only reanneals very slowly. In between these two unambiguously defined fractions, we find the bulk of the genome, which consists of sequences that are repeated to a greater or lesser extent in the genome. Repetitive DNA can be arbitrarily divided into three fractions called **highly repetitive DNA**, **middle repetitive DNA**, and **low** or **lowly repetitive DNA**. The highly repetitive fraction is made up of sequences that are repeated 10^5 to 10^7 times in the genome. The middle repetitive fraction consists of sequences that appear in the genome 10 to 10^5 times. The lowly repetitive fraction consists of sequences that are repeated 2 to 10 times in the genome. We note that the division into highly, middle, and lowly repetitive DNA is arbitrary; there is a continuum of numbers of repeats in the genome. Hence, these terms are terms of convenience; they do not represent truly distinct DNA classes.

In recent years it has become possible to infer the proportion of repetitive DNA from completely sequenced genome sequences. We note, however, that for the reasons listed below, the inferred repetitive fraction of the genome is almost always underestimated. Let us consider the reasons for this situation. First, the bulk of repeated DNA is nonfunctional and as such it evolves very rapidly. As a result, many algorithms designed to detect repeated sequences fail to identify highly diverged repeats as well as short repeats. Second, algorithms for identifying repeated sequences perform well as far as known repeated sequences are concerned, but they fail to identify novel repeated sequences. Third, different algorithms for detecting repeated DNA frequently produce reciprocally nonoverlapping inferences. That is, some sequences that are identified as repetitive by one algorithm are not identified by another algorithm, and those identified by the other algorithm are not identified by the first one. This is particularly true as far as tandem repeats are concerned. As a consequence of these difficulties, much repeated DNA goes unreported or underreported. Of course, progress is being made. Thus, by using traditional methodology, approximately one-third of the human genome was found to consist of repeated sequences; with the use of modern methods, the fraction increased to 66–69% (de Koning et al. 2011).

The proportion of the genome that is taken up by repetitive sequences varies widely among taxa. In yeast, this proportion amounts to about 20% of the genome. In

Figure 11.31 Eukaryotic genomes can be roughly divided into unique and repetitive DNA. The two main categories of repetitive DNA (tandem repeats and dispersed repeats) along with their constituent subcategories are shown. Satellite DNA constitutes the bulk of tandemly repeated DNA. The majority of dispersed repeated sequences are transposable elements.

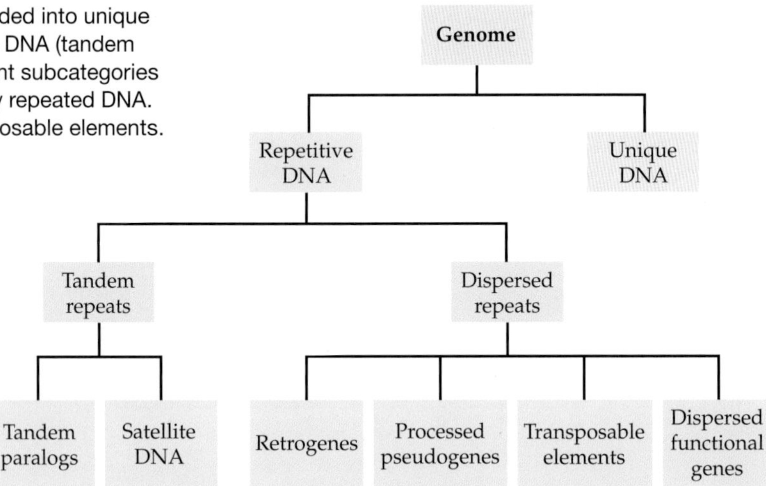

animals, the proportion ranges from about 5% in the nonbiting midge *Chironomus tetans* to close to 90% in the newt *Necturus maculosus*. In mammals, the genomic fraction of repetitive DNA may exceed 70%. In plants, the proportion varies from 3% in the carnivorous bladderwort plant, *Utricularia gibba*, to approximately 80% in angiosperm genomes with haploid DNA contents larger than 5 Gb. The fraction of repetitive DNA in the largest sequenced genome to date, the loblolly pine (*Pinus taeda*), is estimated to be at least 82%.

On the basis of the pattern of dispersion of repeats, the repetitive fraction was found to consist of two major types of repeated families: **tandemly repeated sequences** and **dispersed repeated sequences** (**Figure 11.31**). Here, we will only discuss tandemly repeated sequences, as dispersed repeated sequences, which consist of transposable and mobile elements, have been dealt with extensively in Chapter 9 and earlier in this chapter.

Tandemly repeated sequences

Satellite DNAs, which were discovered by Kit (1961) and Sueoka (1961), are DNA sequences that can be separated from the main DNA component of the genome by physical methods, such as centrifugation in density gradients of cesium salts. The reason for the separation is that satellite DNAs are made of highly repeated sequences that happen to have extremely low or extremely high GC contents. Thus, upon fractionalization of whole-genome DNA and separation by density gradient, they will form one or more bands that are clearly distinguishable from the smear created by the other DNA fragments, which have a more heterogeneous composition. GC content in satellites ranges from as low as 1% in the crabs *Cancer gracilis* and *C. antennarius*, to as high as 73% in the trypanosomal pathogen *Leishmania infantum* and the midge *Chironomus plumosus*.

Tandemly repeated DNA sequences are repetitive elements that form arrays of **repeat units** within the genome. Because some tandem repeats have GC contents that are similar to the GC content of the genome in which they reside, upon centrifugation they will cluster with the bulk of heterogeneous sequences. Thus, not all tandemly repeated sequences qualify as satellite DNA, but all satellite DNAs qualify as tandemly repeated DNA. In recent years, however, the terms "satellite DNA" and "tandemly repeated DNA" have frequently been used interchangeably.

Tandemly repeated DNA can be divided into several categories according to the sizes of the individual repeats, though the specific classification schemes tend to vary widely and confusedly in the literature. **Microsatellites** (Litt and Luty 1989) are defined as arrays made of repeat units that are eight nucleotides or less in length (Richard et al. 2008; López-Flores and Garrido-Ramos 2012). They were found in every organism studied so far, and are usually clustered within dispersed arrays that are shorter than 1Kb. Microsatellites are also known as **simple sequence repeats (SSRs)** or **short tandem repeats (STRs)**. Some authors do not consider mononucleotide repeats, such as poly(A), to be microsatellites.

Some microsatellites are characterized by a low degree of repetition at a particular locus but may be found at thousands of genomic loci. The vast majority of such **dispersed microsatellites** occur at high frequencies in noncoding regions but are rare in coding regions. This finding indicates that the existence of microsatellites within coding regions constitutes a deleterious character state. Exceptions to this rule are trimers and hexamers that are nearly two times more prevalent in protein-coding exons compared with introns and intergenic regions. Their high frequency in coding regions may be explained by the fact that they do not cause frameshifts and may, thus, be much better tolerated than other tandemly repeated arrays.

The repeat units in **minisatellites** (Jeffreys et al. 1985) are 9–100 bp in length. Richard et al. (2008) noted that in yeast, microsatellites and minisatellites are frequently found within the introns of genes. Remarkably, they are not found in the same type of genes, with microsatellites being found mainly in genes encoding nuclear transcription factors, and minisatellites being found mainly in genes encoding cell wall proteins. Since the shortest minisatellite repeat unit found in genes encoding cell wall proteins was nine nucleotides long, Richard et al. (2008) proposed that tandem repeated sequences consisting of mono- to octanucleotide repeats should be called microsatellites, whereas those consisting of nonanucleotide and longer repeats should be called minisatellites. Minisatellites are abundant in centromeres and in subtelomeric regions (near the telomere) and pericentromeric regions (i.e., regions situated on either side of the centromere). In humans, the nematode *C. elegans*, and the pufferfish *Tetraodon nigroviridis*, minisatellites are abundant in subtelomeric regions of the chromosomes, whereas in *Arabidopsis thaliana*, these tandem repeats are abundant in pericentromeric regions (Roest-Crollius et al. 2000; Vergnaud and Denoeud 2000). No discernable pattern of microsatellite distribution was found in *Saccharomyces cerevisiae*.

Tandemly repeated DNA arrays whose constituent repeat units have a length on the order of 10^2–10^3 bp are called **midisatellites** (Nakamura et al. 1987), **macrosatellites** (Fowler et al. 1988), **megasatellites** (Gondo et al. 1998), or simply **satellites**.

For added terminological confusion, some textbooks (e.g., Strachan and Read 2011) classify satellites "on the basis of total array length, not the size of the repeat unit." Given this nomenclatorial discombobulation, it is advisable to use precise descriptions when dealing with tandemly repeated DNAs. The sequence of the repeat unit as it appears on the lagging strand should be clearly stated, as should the repeat size, the sequence variation among repeat units, the total array length (measured in number of repeats), and the chromosomal location.

The vast majority of tandemly repeated sequences contain very short repeat units (1–5 bp), which are also referred to as **simple repeats**. Dinucleotide repeats are the commonest microsatellites in nature. In the human genome, the most common single repeat is CA. Other common repeats in the human genome are A, AT, AAT, AAC, AAG, and AGC (Subramanian et al. 2003). Not all tandemly repeated DNA consists of short simple repeats. For example, the killer whale, *Orcinus orca*, contains about half a million copies of a sequence 1,579 bp long, accounting for approximately 15% of its genome (Widegren et al. 1985).

The repeat units that make each satellite DNA array are not strictly identical; a repeat unit may differ to a certain extent from the consensus sequence. For example, in the kangaroo rat *Dipodomys ordii*, there exists a satellite whose **consensus repeat unit** is TTAGGG (Salser et al. 1976). In fact, less than one-quarter of the repeats are TTAGGG. There are at least 11 other variants, of which TTAGAG, TTAGGT, and TGAGGG are the runners-up in frequency (Miklos and Gill 1982).

The vast majority of eukaryotic genomes contain considerable amounts of tandemly arrayed repetitive DNA sequences, and in some species, they can account for more than half of the genome (Sumner 2003). Even small genomes may contain a huge proportion of tandemly repetitive sequences. For example, 44% of the *Drosophila virilis*

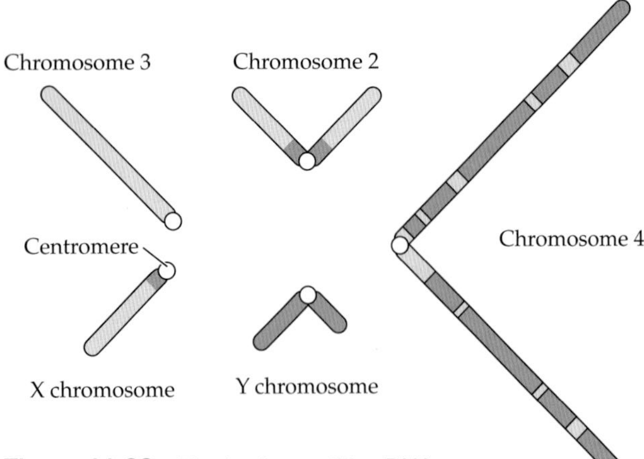

Figure 11.32 Tandemly repetitive DNA sequences (blue) in *Drosophila nasutoides* are mostly found in the largest of the three autosomes (chromosome 4) and in chromosome Y. A minority of tandemly repeated sequences are found in the centromeric regions of chromosome 2 and the X chromosome. (Modified from Graur and Li 2000.)

genome consists of three tandemly repeated microsatellites made of ACAAACT, ATAAACT, and ACAAATT repeat units (Lohe and Roberts 1988; Bosco et al. 2007). Surprisingly, 35% of the very small genome of the unicellular pin mold *Absidia glauca*, which is only nine times larger than that of *E. coli*, is made of tandem repeats (Wostemeyer and Burmester 1986). Mammalian genomes typically consist of 5–30% tandemly repeated DNA; in plants the fraction may reach 40% of the total genome. It is important to remember that the fraction of tandemly repeted DNA within sequenced genomes is almost always underestimated because long sequences of repeated short motifs are challenging to sequence, especially if they are GC-rich (Marx 2013).

In some species, tandemly arrayed repetitive sequences are found on all chromosomes, while in others they are restricted to a particular chromosomal location. For example, more than 60% of the genome of *Drosophila nasutoides* consists of satellite DNA, and the vast majority of it is localized on one of the four autosomes and the Y chromosome (**Figure 11.32**), which seem to contain little else (Miklos 1985; Zacharias 1986).

Mutational processes affecting repeat-unit number in tandemly repeated DNA

The numbers of constituent repeat units in microsatellite and minisatellite arrays tend to evolve very rapidly, most probably as a consequence of a high mutation rate and lack of purifying selection. For example, the mutation rates in microsatellites can reach levels that are 100,000 times higher than the average mutation rates in other parts of the genome (Gemayel et al. 2010). As a consequence, microsatellites and minisatellites tend to exhibit extremely high degrees of variability within the population. Indeed, microsatellites and to a lesser extent minisatellites are also referred to as **VNTRs** (for **variable number of tandem repeats**). Because of their extreme variability (or hypervariability), VNTRs are extensively used in forensic medicine and paternity testing. For example, VNTRs were used to identify the skeletal remains of Nazi war criminal Josef Mengele, who escaped to Argentina following World War II (Jeffreys et al. 1992). Similarly, VNTRs were used to confirm the suspicion that Thomas Jefferson, third president of the United States, sired at least one child with his slave Sally Hemings (Foster et al. 1998; Turner 2011).

Five mutational outcomes can be envisioned in tandemly repeated DNA: (1) a change in the sequence of a repeat unit by point mutation, (2) the spread of a mutation in one repeat unit to other units by gene conversion, (3) a change in the chromosomal location of a tandem array by recombination or transposition, (4) an increase or a decrease in the number of tandem arrays by unequal crossing over, and (5) a change in the number of repeat units within a tandemly repeated array. In this section, we will mainly concern ourselves with the fifth outcome.

In discussing the evolution of repeat-unit number in tandemly repeated sequences, we need to distinguish between the initial stages of evolution, when the number of repeat units is small, and later stages, when the large number of repeat units within the array allows for different mutational mechanisms. Let us start with the establishment of de novo microsatellites within the genome. New microsatellites are assumed to arise via the formation of a **proto-microsatellite**, i.e., a short intermediate stage with as few as three or four repeat units (Buschiazzo and Gemmell 2006). Proto-microsatellites can originate from point mutations, for example, ACGCAC → ACACAC, or through indel events.

The number of repeat units in a proto-microsatellite can, subsequently, expand by replication slippage (Chapter 1: Figure 1.22), i.e., the process in which the DNA polymerase turns back and uses the same template again to produce a repeat. This process will result in a stepwise increase or decrease in the number of repeats within the array. Support for this mechanism is provided by the observation that replication slippage also occurs in vitro during the polymerase chain reaction (PCR) amplification of microsatellite sequences, which yields "stutter bands"—arrays that differ in size from the template array by having one or a few additional repeat units or by missing one or a few repeat units (Shinde et al. 2003). We note, however, that in vivo most of these mutations are corrected by the mismatch repair system, and only the small fraction that is not repaired ends up in the form of variable microsatellite lengths (Li et al. 2002). Finally, we should note that replication slippage is effective in changing repeat number if the repeat unit is 1 to 13 bp in length, but not very effective with larger repeat units (Sia et al. 1997).

As the tandemly repeated array grows, further increases or decreases in repeat-unit number may be facilitated by intra-array unequal crossing over. However, because the pattern of microsatellite variability in recombining regions of the genome is similar to that in the nonrecombining portions of the genome, such as the mammalian Y chromosome, unequal crossing over cannot be invoked to explain the existence of many tandemly repeated sequences (Nachman 1998).

It has been noted that replication slippage and unequal crossing over tend to remove tandem arrays more often than they increase their size and copy number (Walsh 1987). To explain the existence of long tandemly repeated sequences, DNA amplification has been suggested. **DNA amplification** refers to the production of multiple copies of a sequence. In an evolutionary context, DNA amplification refers to mutational events that occur within the life span of an organism and cause a sudden increase in the copy number of a DNA sequence that is passed on to its offspring. We distinguish between vertical amplification and horizontal amplification. **Vertical amplification** refers to processes through which a certain sequence is multiplied outside the chromosome. Vertical amplification is not usually heritable. **Horizontal amplification** refers to a process of creation of multiple copies of a certain DNA sequence and their incorporation within the heritable genome of the organism.

One of the most powerful methods of amplification is **rolling circle replication** (**Figure 11.33**). This type of replication is used in the amplification of rRNA genes in amphibian oocytes (Bostock 1986). In this case, amplification involves the formation of an extrachromosomal circular copy of a DNA sequence, which can then produce many additional extrachromosomal units containing tandem repeats of the original sequence. If such circles linearize and become integrated back into the chromosome, there will be additions to the genome consisting of identical repeated sequences.

The evidence for the importance of rolling circle replication in the evolution of tandemly repeated sequences is substantial, but not vast. In particular, this mode of evolution yields several predictions that have found support in empirical studies. For example, rolling circle replication followed by DNA integration into the genome is expected to cause extremely rapid, saltatory expansions in the number of satellite arrays. This has indeed been shown to be the case in the genomes of South American rodents of the genus *Ctenomys* (Rossi et al. 1990; Slamovits et al. 2001). Another expectation concerns the location of mutations following rolling circle replication. If an error occurs in the first step of the process, i.e., the replication leading to the formation of the extrachromosomal circular structure, then a footprint of periodically occurring substitutions should be observed. In

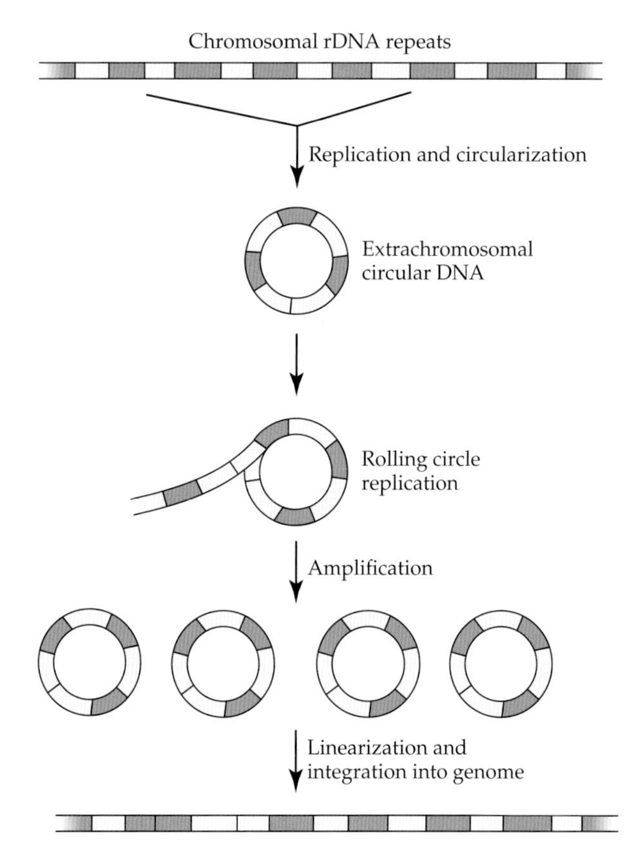

Figure 11.33 The rolling circle model of gene amplification in amphibian oocytes. The chromosomal rRNA-specifying genes are arranged in a tandem array containing transcribed (blue) and nontranscribed (white) parts. Amplification involves the formation of an extrachromosomal circular copy containing a variable number of repeats, which is then amplified by multiple rounds of rolling circle replication. Note that the periodicity may change following rolling circle amplification. (Modified from Bostock 1986.)

particular, if the circularization involves a truncation of one or both of the terminal repeat units, than a repeat unit of unusual length should appear with some regularity within the array. Krzywinski at al. (2005) discovered clear signs of periodicity in at least two satellites in the genome of a malaria mosquito, *Anopheles gambiae*. Interestingly, the two satellite arrays were located on the Y chromosome, which does not recombine.

The contribution of tandem repeats to genome size

As is obvious from the fact that small genomes often contain a higher proportion of tandem repeats than much larger genomes, tandem repeats cannot universally explain differences in genome size. For example, about 50% of the nuclear genome of the mealworm beetle *Tenebrio molitor* consists of satellite DNA, yet the beetle genome size is less than one-fifth that of the human genome, whose satellite DNA content is less than 7% (Petitpierre et al. 1995). It is possible, however, that in particular clades, differences in genome sizes may arise due to differences in amounts of tandemly repeated sequences.

In *Drosophila* species, a positive correlation was found between genome size and amount of satellite DNA. Addition and loss of satellite repeat elements appear to have made some contribution to the large differences in genome size observed within the Drosophilidae (Bosco et al. 2007). For example, satellite DNA content differences are wholly sufficient to explain the small but significant differences in the genome sizes of the close relatives *D. erecta, D. simulans,* and *D. melanogaster,* whose genome sizes are 137, 166, and 196 Mb, respectively. In contrast, by using tandemly repeated DNA, it is not possible to explain the 3.2-fold difference between the smallest *Drosophila* genome, that of *D. mercatorum* (128 Mb), and the largest, that of *D. virilis* (404 Mb).

As mentioned previously, bats and birds have small genomes relative to most other amniotes. Paucity of AT and GC microsatellites, which in other mammals are quite common (van den Bussche et al. 1995), seems to be the reason for the small genomes of bats. Similarly, the small genomes in birds (Table 11.2) may be due to a scarcity of microsatellites (Primmer et al. 1997). In **Figure 11.34**, we see that about 98% of the variation in genome size in apes is explained by the variation in tandemly repetitive sequences (heterochromatin).

Do tandemly repeated DNA sequences have a function?

On the basis of purported correlations with cytological, physiological, and environmental factors, various microsatellites have been assigned tentative functional roles. Some such functions defy evolutionary logic. For example, it has been suggested that microsatellites may have evolved for the purpose of facilitating recombination (Stallings et al. 1991). While dispersed microsatellite arrays do indeed increase the recombination rate, it is unlikely that this is their selected-effect function, since selection on recombination rates and patterns would constitute a form of second-order selection (Chapter 2), which is unlikely to operate in multicellular organisms. Other suggested functions include the enhancement of transcriptional activity of genes and contribu-

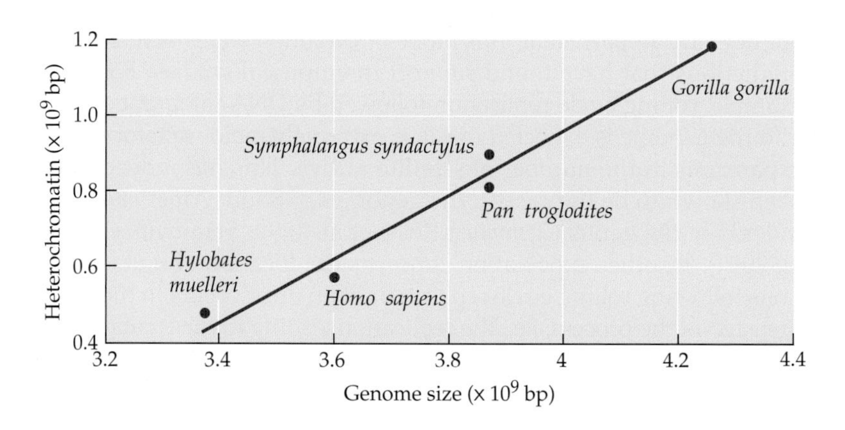

Figure 11.34 An almost perfect correlation exists between genome size and amount of heterochromatin in Hominoidea. (Data from Manfredi-Romanini et al. 1994.)

tion to the chromosomal structure through DNA packaging and condensing. Some of these hypotheses generate testable predictions. For example, if microsatellites are involved in the packaging and condensing of DNA, then larger genomes should contain a greater number and possibly longer microsatellite loci. Neff and Gross (2001) analyzed microsatellite loci from close to 100 vertebrate species and could reject this hypothesis by the observation that larger genomes do not have more microsatellite loci than smaller genomes. In fact, Neff and Gross (2001) found that microsatellite and genetic diversity covary, which seems to indicate that microsatellite variation within populations constitutes a neutral trait that is largely determined by mutation and random genetic drift.

The suggestion that most tandemly repeated sequences are merely junk DNA implies that their presence or absence in variable numbers does not affect the fitness of their carriers. While this seems to be true in the majority of cases, there is evidence that some of them have a function. The case of the *Rsp* locus was discussed in Chapter 2. Here we discuss centromeres, which are frequently made of tandemly repeated sequences.

Centromeres as examples of indifferent DNA

Centromeres have the same function in all eukaryotic species, but their DNA sequences differ greatly among them, even between closely related species (Wang et al. 2009; Buscaino et al. 2010; Melters et al. 2013). In *Saccharomyces cerevisiae*, centromeric function is accomplished by a single sequence of 125 bp. However, this simple organization is not conserved in the rest of the fungi so far analyzed, let alone in other eukaryotes The centromeres in the fission yeast *Schizosaccharomyces pombe* contain several central-core sequences flanked by direct and inverted repeat elements. The general organization in the fungus *Candida albicans* is similar to that in *S. pombe*; however, the central-core sequences in *C. albicans* and *S. pombe* are not evolutionarily related to each other.

In most animal and plant species, the centromeres contain large arrays of tandem repeats, which might, as in the case of *D. melanogaster*, be interrupted by transposable elements. However, centromeric DNA sequences are not conserved between species, suggesting that the DNA sequence is not the main determinant of centromere identity and function.

Despite the lack of a conserved centromeric sequence, it is certain that all centromeres share common features that allow them to perform their identical function. We do not know yet what these common features are. Possibly, the common features include sequence composition, length of satellite repeat units, low gene density, transcription of noncoding RNAs, chromatin status, and vicinity to heterochromatin domains. Centromeres can, therefore, be thought of as examples of indifferent DNA.

Genome Compositional Architecture

Figure 11.35 shows the GC content in several eukaryotic taxa. The GC content in fungi, which are mostly single-celled organisms, ranges from approximately 19% to 65%, a range of variation that is only slightly narrower than that of prokaryotes. The situation in multicellular eukaryotes is different; while the genome sizes of multicellular eukaryotes are generally larger and more variable in length than those of prokaryotes, GC content exhibits a much smaller variation. In particular, vertebrate genomes show quite uniform GC content, ranging from about 35% in the wallaby *Macropus eugenii* to 46% in the lamprey *Petromyzon marinus*. The GC content in eutherian mammals has an even narrower range, from about 39% in the two-toed sloth *Choloepus hoffmanni* to 45% in the domestic cat *Felis silvestris catus*.

Despite the relative uniformity of their total genomic GC content, vertebrate genomes have been found to possess a complex internal compositional organization. That is, nucleotides are not distributed uniformly along the genome; instead a certain degree of clustering by GC content is observed. This nonuniformity of DNA

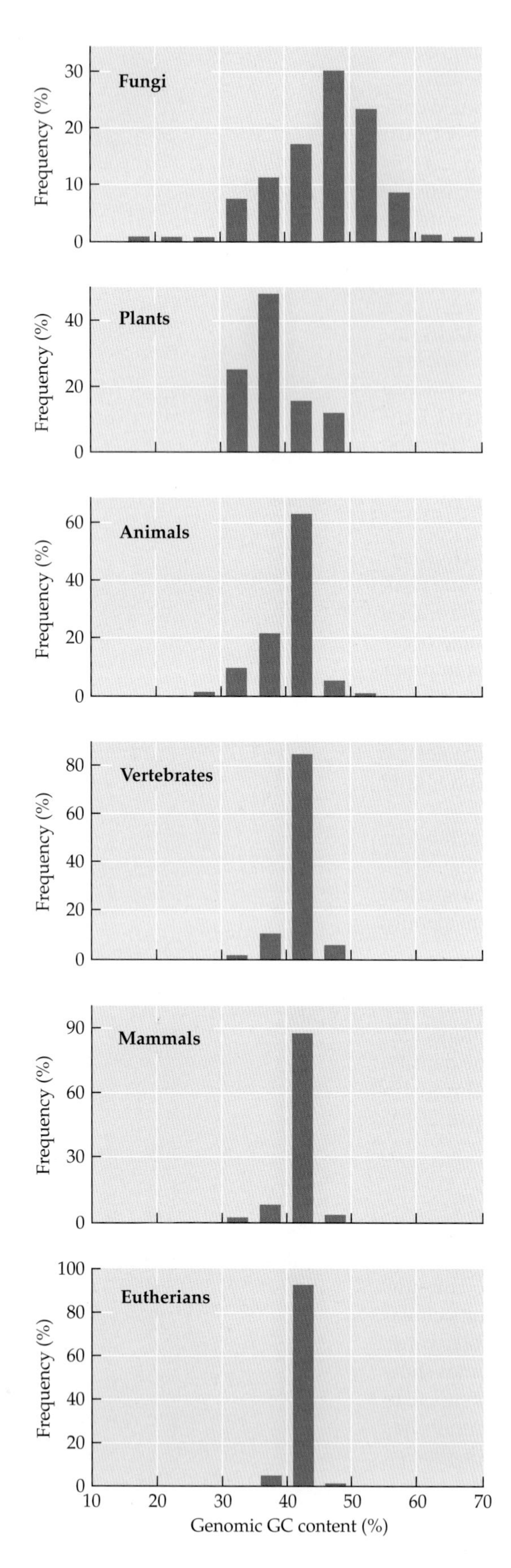

Figure 11.35 GC content in eukaryotic taxa. In fungi, GC content ranges from 18.8% in the microsporidian *Nosema apis* to 65.3% in the powdery mildew antagonist fungus *Pseudozyma flocculosa*. In plants, GC content ranges from 32.1% in the desert poplar *Populus euphratica* to 46.8% in *Zea mays* (corn). In animals, GC content ranges from 27.5% in the body louse *Pediculus humanus* to 51.6% in the western orchard predatory mite *Metaseiulus occidentalis*. In vertebrates, GC content ranges from 34.7% in the tammar wallaby *Macropus eugenii* to 45.9% in the sea lamprey *Petromyzon marinus*. In mammals, GC content ranges from 34.7% in the tamar wallaby to 45.7% in the duckbilled platypus *Ornithorhynchus anatinus*. In eutherians, GC content ranges from 39.0% in the two-toed sloth *Choloepus hoffmanni* to 45.6% in the cat *Felis catus sylvestris*. (Courtesy of Yichen Zheng.)

composition in vertebrate genomes has been known for more than half a century (Kit 1961; Sueoka 1961; Filipski et al. 1973; Macaya et al. 1976; Thiery et al. 1976).

Here we discuss (1) the methodology for segmenting genomes that have nonuniform nucleotide compositions into compositionally coherent domains, (2) the pattern of nucleotide distribution within vertebrate genomes, and (3) the possible underlying evolutionary processes that are responsible for the departure from compositional uniformity.

Because the field of genome compositional architecture was dominated for many years by a theory that was not only wrong but misleading, we start with a short historical introduction. Before the genomic era, the prevailing view on genome composition entailed the notion of "isochores" (Cuny et al. 1981). **Isochores** were defined as DNA segments that met a set of four conditions. First, an isochore was supposed to have a characteristic GC content that was significantly different from the GC content of adjacent isochores (Bernardi 2001; Oliver et al. 2002). Second, an isochore was supposed to be more homogeneous in its nucleotide composition than the chromosome on which it resided (Bernardi 2001). Third, the length of an isochore was supposed to exceed 300 Kb (Macaya et al. 1976; Cuny et al. 1981; Bernardi et al. 1985; Bernardi 2000; Clay et al. 2001; Pavlicek et al. 2002). Finally, an isochore was supposed to be assignable by its GC content to one of a small number of **isochore familes** (**Figure 11.36**).

According to the **isochore theory** (Bernardi 2000), the genomes of vertebrates were supposed to consist of mosaics of easily discernible isochores (Figure 11.36c), i.e., isochores of low and high GC contents were supposed to be interspersed with one another. The number of isochore families was said to vary from two to less than ten. Human isochores, for instance, were said to belong to five isochore families, L1, L2, H1, H2, and H3, whose corresponding ranges of GC contents were said to be <37%, 37–42%, 42–47%, 47–52%, and >52% (Figure 11.36a). In contrast to the human genome, the genome of the common carp (*Cyprinus carpio*) was said to contain only two isochore families, L1 and L2, a difference that was attributed to the fact that carps are poikilothermic (cold-blooded) while humans are homeothermic (warm-blooded).

In the absence of genomic sequences, the GC composition at third positions in codons of protein-coding genes (GC3) was commonly used as a proxy for the GC composition of the isochore in which the gene resided (Bernardi et al. 1985; Aota and Ikemura 1986; Mouchiroud et al. 1991; Kadi et al. 1993; Duret et al. 1995; Zoubak et al. 1996; Bernardi et al. 1997; Robinson et al. 1997; Galtier and Mouchiroud

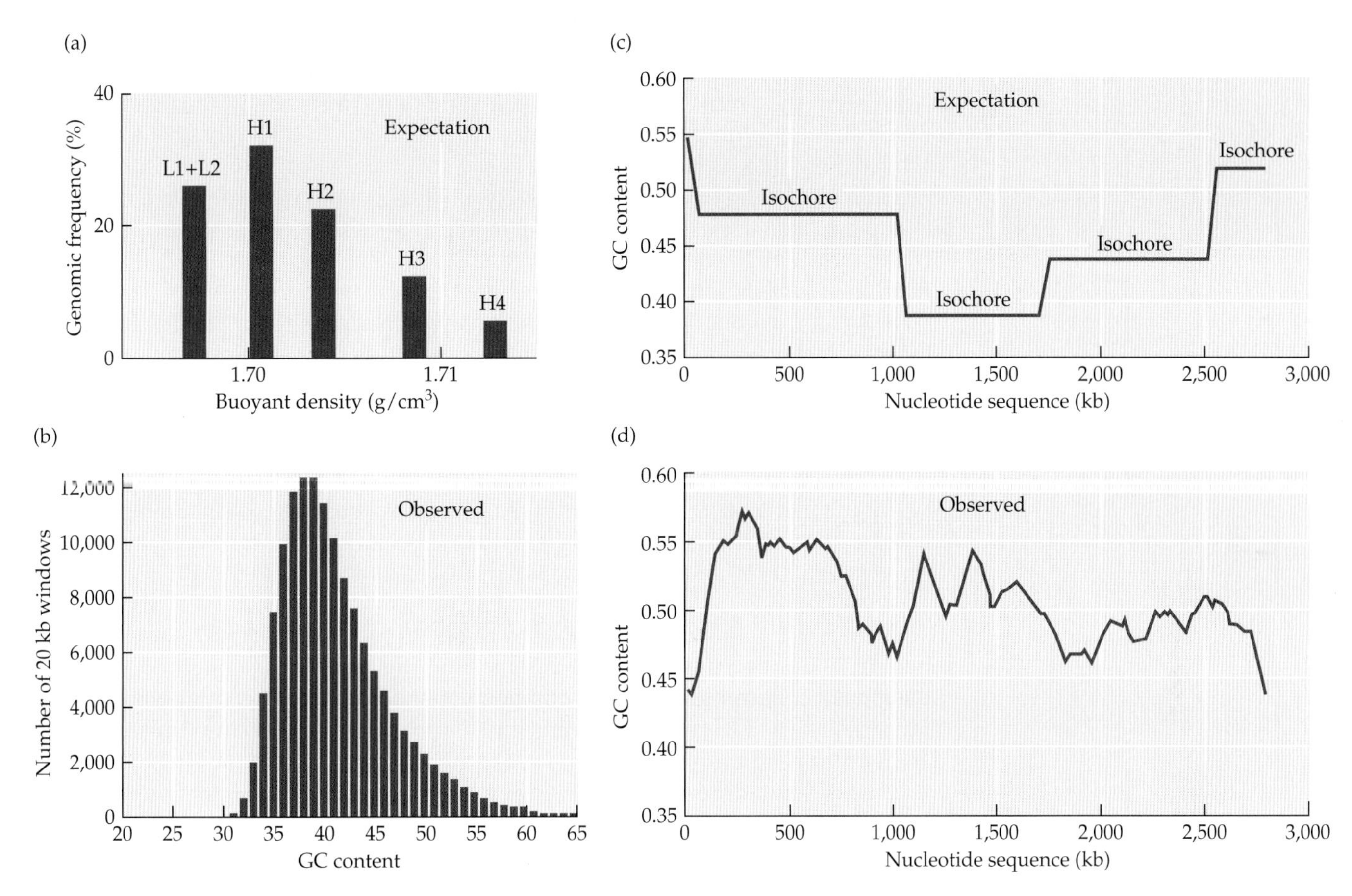

Figure 11.36 By comparing the expectations of the isochore model to actual genomic sequences, it is possible to unequivocally reject the isochore model. (a) A histogram purporting to show the relative amounts and GC content of five isochore families in the human genome. (b) A histogram of GC contents of 20 Kb windows in the draft human genome sequence indicates that GC content cannot be classified into discrete isochore families as expected by the isochore model. (c) According to the isochore model the human genome should consist of a mosaic of high- and low-GC isochores. (d) GC content across a segment of human chromosome 6. GC content was plotted as a moving average with a 100 Kb window size. In each step, the window was advanced in the 5′-to-3′ direction by 10 Kb. (Parts a and c modified from Bernardi et al. 1985 and Bernardi 1995; b modified from Lander et al. 2001; d courtesy of Eran Elhaik.)

1998). Oddly, however, the practice of using GC3 as a proxy for the GC content of isochores persisted long into the genomic era (e.g., Scaiewicz et al. 2006; Costantini and Bernardi 2008), even though whole-genome sequences afforded the opportunity to test the isochore theory directly, and even though protein-coding regions, from which GC3 values are computed, constitute a very small fraction of vertebrate genomes. For example, protein-coding genes comprise less than 2% of the human genome (Lander et al. 2001) and less than 10% of the chicken genome (International Chicken Genome Sequencing Consortium 2004).

Elhaik et al. (2009) analyzed gene and genome sequences from seven vertebrate species and showed that GC3 explained only a very small fraction of the variation in GC content of the DNA sequences flanking the genes, and what little correlation there was between GC3 and the GC content in flanking regions decayed rapidly with distance from the gene. Moreover, the coefficient of variation of GC3 was found to be much larger than that of the flanking regions. Elhaik et al. (2009), therefore, concluded that the GC content of third codon positions cannot be used as a stand-in for isochore composition.

The presumed relationship between GC3 and isochores has been used numerous times in the literature to study isochore function and evolution (Aota and Ikemura 1986; Kadi et al. 1993; Duret et al. 1995; Zoubak et al. 1996; Bernardi et al. 1997; Robinson et al. 1997; Galtier and Mouchiroud 1998; Eyre-Walker and Hurst 2001; Alvarez-Valin et al. 2002; Duret et al. 2002; Vinogradov 2003; Chojnowski et al. 2007).

The publication of the draft human genome (Lander et al. 2001) and other completely sequenced genomes has revealed that the isochore theory hadn't got a leg to stand on (Dunham et al. 1999; Hattori et al. 2000; Nekrutenko and Li 2000; Deloukas et al. 2001; Häring and Kypr 2001; Heilig et al. 2003; Hillier et al. 2003; Mungall et al. 2003). In particular, the frequency distribution and the arrangement of GC contents within the genome did not in the least resemble the expectations of the isochore theory (Figure 11.36b,d). Lander et al. (2001) concluded that the "strict notion of isochores as compositionally homogeneous" could be ruled out and, hence, isochores do not merit the prefix "iso." Moreover, given the intrinsic continuous distribution of GC content within genomes, it turned out that it would be impossible to classify each isochore into its isochore family based solely on its compositional properties even if isochores did exist (Cohen et al. 2005). Thus, even if isochores were real, they would have been useless in practice. Follow-up studies (Bovine Genome Sequencing and Analysis Consortium 2009; Elhaik et al. 2010a; Tatarinova et al. 2013; Elhaik and Graur 2014) further demolished the isochore theory.

The saddest corollary of the invalidation of the isochore theory is that hundreds of studies that used the tenets of the isochore theory as their starting point were rendered invalid. The fact that today we know so little of the evolutionary forces shaping the compositional architectures of eukaryotic genomes is due in no small part to the false leads laid out by the isochore theory.

Remarkably, articles extolling the virtues of isochores still appear in the literature (e.g., Romiguier et al. 2010; Costantini 2015), which goes to show that although the scientific method is based on falsifiable evidence, scientists cannot always distinguish between what is real and what they would like reality to be (Lawton 2015).

Segmentation algorithms and compositional domains

The purpose of **segmentation algorithms** is to divide a DNA sequence into **compositionally coherent domains** or **compositional domains**, i.e., contiguous genomic sequences whose GC contents differ significantly from those of their upstream and downstream neighboring domains. An optimal segmentation is one that maximizes the compositional differences between neighboring compositional domains. In other words, the function of segmentation algorithms is to detect shifts in the nucleotide composition of the DNA sequence and locate the exact borders between adjacent compositional domains.

Numerous methods for segmenting DNA sequences into compositional domains have been proposed in the literature. These methods can be roughly divided into two types: local methods, i.e., methods that consider partial sequences in isolation; and global methods that consider entire contiguous sequences of chromosomes. Segmentation methods that use a sliding window along the sequence are considered local, whereas those that employ recursive binary segmentation algorithms are considered global. As far as the segmentation of whole-genome sequences is concerned, a global view of the genome has been shown to be preferable over a local view, which may yield suboptimal results. Segmentation methods also differ from one another in the number and types of parameters used in the segmentation process, as well as in the levels of user intervention. Methods that rely on subjective user intervention, such as that used by Costantini et al. (2006), preclude independent replication of the results and are, thus, unscientific.

Here, we present a binary recursive segmentation procedure originally proposed by Bernaola-Galván et al. (1996) and subsequently improved by Elhaik et al. (2010b). By using benchmark simulations (Elhaik et al. 2010a), this method has been shown to outperform all other segmentation algorithms, while at the same time disposing with the need for arbitrary user inputs that can bias the results. A recursive segmentation algorithm called IsoPlotter that employs a dynamic threshold has been devised by Elhaik et al. (2010b). IsoPlotter is an unsupervised algorithm, i.e., it requires no subjective user intervention and, through benchmark validation, it has been shown to yield unbiased results.

A recursive segmentation procedure consists of three parts: (1) a general strategy for segmentation; (2) a segmentation decision, i.e., whether the differences between adjacent segments are significant; and (3) a halting criterion, i.e., the point at which the recursive procedure should end. The procedure is illustrated in **Figure 11.37**.

Since the sequence we wish to segment in Figure 11.37 is very short (only 30 bp in length), our strategy will be to check every possible splitting location within the sequence. Thus, we will test the consequences of segmenting the sequence between positions 1 and 2, between positions 2 and 3, and so on until we test the last possibility, segmenting the sequence between positions 29 and 30. Each sequence of L nucleotides can be split $L - 1$ ways. If the sequences to be segmented are hundreds of millions of base pairs in length, there is no need to test every position; rather we may check segmentation possibilities that are hundreds or thousands of nucleotides apart.

At each partitioning point, we calculate the GC contents in the upstream and downstream segments. For example, in Figure 11.37a the segmentation between positions 1 and 2 results in upstream and downstream segments with 0% and 60% GC content, respectively. The absolute compositional difference between the two segments is $\Delta_{GC} = |0 - 60| = 60\%$. The segmentation between positions 2 and 3 results in upstream and downstream segments with 50% and 58% GC content, respectively. The absolute compositional difference between the two segments is $\Delta_{GC} = |50 - 58| = 8\%$. In the algorithm we segment the sequence at the location yielding the maximal Δ_{GC} value. From among all the 29 possible partitions, the largest Δ_{GC} value is 65%, corresponding to the partition between nucleotide positions 9 and 10. The two resulting segments are 9 and 21 bp in length, respectively (Figure 11.37b). The recursive segmentations of these segments are shown in Figure 11.37c.

(a)
```
         111111111122222222223
123456789012345678901234567890
ACATAATTTGGCGAGGCGCGTACCCGTGAC

A|CATAATTTGGCGAGGCGCGTACCCGTGAC   ΔGC = | 0 - 60 | = 60%
AC|ATAATTTGGCGAGGCGCGTACCCGTGAC   ΔGC = |50 - 58 | = 8%
ACA|TAATTTGGCGAGGCGCGTACCCGTGAC   ΔGC = |33 - 59 | = 26%
ACAT|AATTTGGCGAGGCGCGTACCCGTGAC   ΔGC = |25 - 62 | = 37%
ACATA|ATTTGGCGAGGCGCGTACCCGTGAC   ΔGC = |20 - 64 | = 44%
ACATAA|TTTGGCGAGGCGCGTACCCGTGAC   ΔGC = |17 - 67 | = 50%
ACATAAT|TTGGCGAGGCGCGTACCCGTGAC   ΔGC = |14 - 70 | = 56%
ACATAATT|TGGCGAGGCGCGTACCCGTGAC   ΔGC = |13 - 73 | = 60%
ACATAATTT|GGCGAGGCGCGTACCCGTGAC   ΔGC = |11 - 76 | = 65% √
ACATAATTTG|GCGAGGCGCGTACCCGTGAC   ΔGC = |20 - 75 | = 55%
ACATAATTTGG|CGAGGCGCGTACCCGTGAC   ΔGC = |27 - 74 | = 47%
ACATAATTTGGC|GAGGCGCGTACCCGTGAC   ΔGC = |33 - 72 | = 39%
ACATAATTTGGCG|AGGCGCGTACCCGTGAC   ΔGC = |38 - 71 | = 33%
ACATAATTTGGCGA|GGCGCGTACCCGTGAC   ΔGC = |36 - 75 | = 39%
ACATAATTTGGCGAG|GCGCGTACCCGTGAC   ΔGC = |40 - 73 | = 33%
ACATAATTTGGCGAGG|CGCGTACCCGTGAC   ΔGC = |44 - 71 | = 27%
ACATAATTTGGCGAGGC|GCGTACCCGTGAC   ΔGC = |47 - 69 | = 22%
ACATAATTTGGCGAGGCG|CGTACCCGTGAC   ΔGC = |50 - 67 | = 17%
ACATAATTTGGCGAGGCGC|GTACCCGTGAC   ΔGC = |53 - 64 | = 11%
ACATAATTTGGCGAGGCGCG|TACCCGTGAC   ΔGC = |55 - 60 | = 5%
ACATAATTTGGCGAGGCGCGT|ACCCGTGAC   ΔGC = |52 - 67 | = 15%
ACATAATTTGGCGAGGCGCGTA|CCCGTGAC   ΔGC = |50 - 75 | = 25%
ACATAATTTGGCGAGGCGCGTAC|CCGTGAC   ΔGC = |52 - 71 | = 19%
ACATAATTTGGCGAGGCGCGTACC|CGTGAC   ΔGC = |54 - 67 | = 13%
ACATAATTTGGCGAGGCGCGTACCC|GTGAC   ΔGC = |56 - 60 | = 4%
ACATAATTTGGCGAGGCGCGTACCCG|TGAC   ΔGC = |58 - 50 | = 8%
ACATAATTTGGCGAGGCGCGTACCCGT|GAC   ΔGC = |56 - 67 | = 11%
ACATAATTTGGCGAGGCGCGTACCCGTG|AC   ΔGC = |57 - 50 | = 7%
ACATAATTTGGCGAGGCGCGTACCCGTGA|C   ΔGC = |55 - 100| = 45%
```

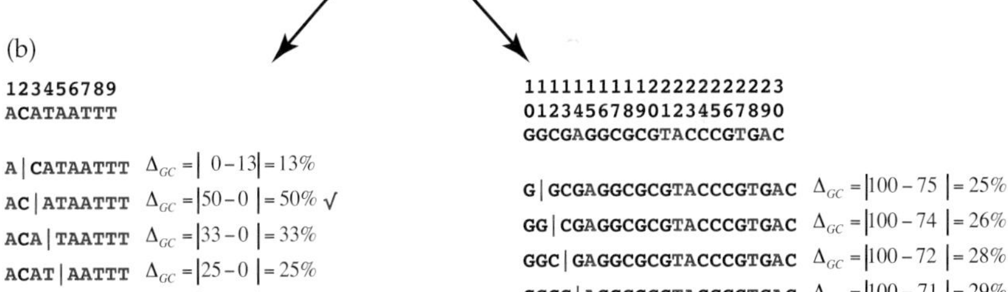

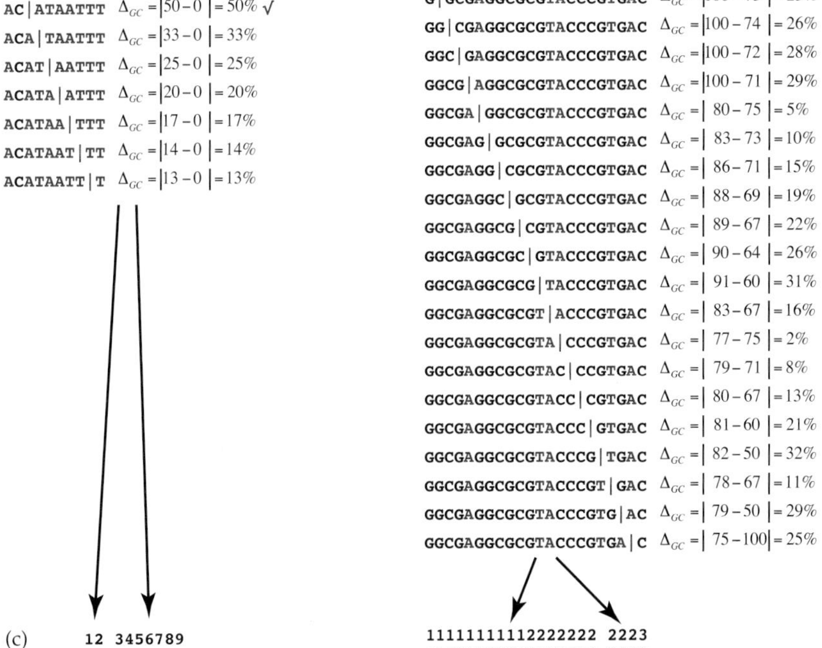

Figure 11.37 Schematic illustration of a recursive compositional segmentation process. (a) The starting sequence is 30 bp in length. It contains 13 AT nucleotides (red) and 17 GC nucleotides (black). This sequence can be divided into two segments at 30 – 1 = 29 sites. For each of these sites, we compute the GC contents of the segments to the left and the right of the segmentation site (vertical bar). For each possible segmentation site, the absolute compositional difference between the two segments is computed as $\Delta_{GC} = |GC_{left} - GC_{right}|$. (b) From among the 29 Δ_{GC} values, the largest is chosen as the first segmentation point, which divides the sequence into a 9 bp segment and a 21 bp segment. (c) By a repeat of the segmentation process, each of the two subsegments can be further split into two, resulting in four segments.

In each stage of the segmentation, we need to determine whether the difference in GC content between two neighboring segments is statistically significant. The segmentation procedure by Bernaola-Galván et al. (1996) uses the Jensen-Shannon entropic divergence measure, D_{JS} (Lin 1991). Briefly, a chromosome of length L, with a GC content of F_{GC} and an AT content $F_{AT} = 1 - F_{GC}$ is divided into two contiguous segments ($i = 1, 2$) of lengths l_i, GC contents of f^i_{GC}, and AT contents of f^i_{AT}. The Jensen-Shannon entropic divergence measure, D_{JS}, is defined as the difference between the overall Shannon entropy H^{tot} and the sum of the Shannon entropies for each of the i segments, H^i:

$$D_{JS} = \max\left[H^{tot} - \sum \frac{l_i}{L} H^i\right] \quad (11.5)$$

where $H^i = -f^i_{GC} \log_2 f^i_{GC} - f^i_{AT} \log_2 f^i_{AT}$ and $H^{tot} = -F_{GC} \log_2 F_{GC} - F_{AT} \log_2 F_{AT}$.

The segmentation is then repeated recursively for each segment until a **halting criterion** is met. The halting criterion can be a constant, as in Cohen et al. (2005), or it can be dynamically estimated for each segmentation step separately, as in Elhaik et al. (2010b).

Segmentation algorithms are sensitive to sequence quality. That is, low-quality sequences tend to result in an excess of short compositional sequences. The importance of this artifact was illustrated by comparing the segmentation results of the draft cow genome sequence to those of the finished version of the genome (Elsik et al. 2009).

By comparing the GC content variance of compositional domains with that of the chromosomes in which they reside, compositional domains can be further classified into two types: **compositionally homogeneous domains** and **compositionally nonhomogeneous domains**.

Compositional architectures of mammalian nuclear genomes

A characteristic spatial distribution of compositional domains along a human chromosome is illustrated in **Figure 11.38**. Genome statistics for compositional domains from ten mammalian and one avian species are shown in **Table 11.7**. All mammalian genomes turn out to consist of numerous short compositional domains and a few long ones.

The mean number of compositional domains in the mammalian genomes shown in Table 11.7 is approximately 96,000, with opossum and apes (human, chimpanzee, orangutan) having the largest number of domains, and murids (rat and mouse) having the smallest number. On average, 59–74% of all mammalian domains are more compositionally homogeneous than the chromosomes in which they reside. Compositional domains that are larger than 300 Kb and are compositionally homogeneous

TABLE 11.7

Similarity and dissimilarity among pairs of duplicate genes

	Human	Chimpanzee	Orangutan	Mouse	Rat
Sequenced genome size (Gb)	2.8	2.8	2.7	2.6	2.5
Mean GC content of genome (%)	40.80	40.70	40.70	41.80	41.90
Number of compositional domains	107,571	107,359	105,688	67,223	63,137
Domain density (per Mb)	38.7	39.0	38.8	26.3	25.5
Mean domain length (bp)	25,865	25,637	25,764	38,060	39,233
Median domain length (bp)	7,808	7,840	7,936	9,216	9,376
Mean domain GC content (%)	42.7	42.7	42.6	43.0	43.3
Median GC content (%)	41.8	41.8	41.6	42.8	43.1

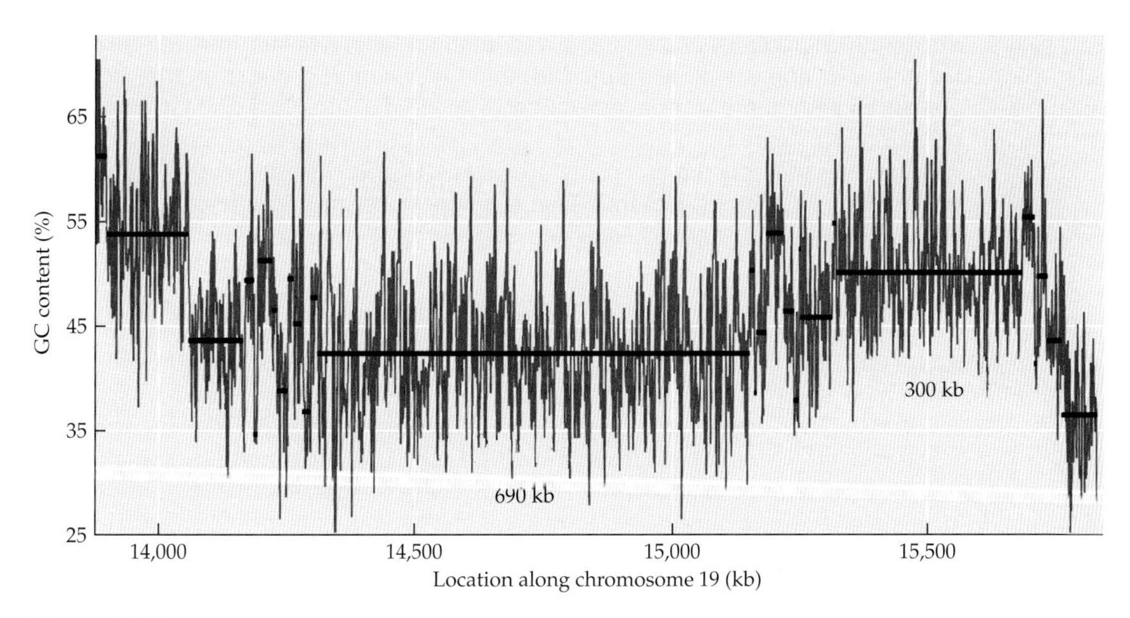

Figure 11.38 Spatial distribution of GC content in 1,024 bp nonoverlapping windows along a 1.4 Mb fragment from human chromosome 19 (red). The segmentation algorithm yielded 27 compositional domains (horizontal black lines). Two segments were longer than 300 Kb. The longer segment (690 Kb) is relatively GC-poor (42%), while the shorter segment (300 Kb) is relatively GC-rich (50%). (From Cohen et al. 2005.)

constitute only a tiny fraction of the compositional domains, from 0.7% in horse and dog to 2.1% in rat.

The mean compositional domain length varies from 25,700 bp in primates to 38,500 bp in murids. The median length is much smaller in all taxa, indicating an extreme skewed distribution toward very short domains. The mean and median lengths of homogeneous and nonhomogeneous domains within a taxon are practically indistinguishable. The largest compositional domain among all mammalian species studied to date is a 10.5 Mb segment with a GC content of 36% found in the opossum genome. In the human genome, the largest compositional domain is about half that length (5.2 Mb).

Domain density is measured as the mean number of compositional domains per million base pairs. When divided into GC-poor and GC-rich compositional domains, domain density in mammals ranges from 0 to 90 domains/Mb for GC-poor domains

TABLE 11.7 *(continued)*

	Horse	Dog	Cow	Pig	Opossum	Chicken
Sequenced genome size (Gb)	2.0	2.3	2.2	2.5	3.4	1.0
Mean GC content of genome (%)	41.10	41.00	41.70	41.80	37.70	41.30
Number of compositional domains	112,350	104,885	96,410	90,881	107,356	39,450
Domain density (per Mb)	48.1	45.3	43.2	38.6	31.5	40.1
Mean domain length (bp)	20,787	22,097	23,144	25,938	31,788	24,965
Median domain length (bp)	7,456	7,744	7,872	7,744	7,616	9,216
Mean domain GC content (%)	43.3	43.6	44.0	44.0	39.6	43.0
Median GC content (%)	42.0	42.0	43.0	42.9	38.4	41.7

Source: Modified from Elhaik and Graur (2014).

and up to 121 domains/Mb for GC-rich domains. Homogeneous domains are denser for both GC-poor (0–57 domains/Mb) and GC-rich (0–98 domains/Mb) domains compared with nonhomogeneous GC-poor (0–43 domains/Mb) and GC-rich (0–64 domains/Mb) domains.

In regions of low domain densities, the density of GC-rich domains is higher than the density of GC-poor domains. That is, genomic regions with fewer domains are more likely to be GC-rich, whereas denser genomic regions are more likely to harbor GC-poor domains.

On average, murid chromosomes are the least dense (26 domains/Mb), whereas the horse genome is the densest (49 domains/Mb). The chromosomal domain densities of opossum are as low as murids for homogeneous domains (21 and 16 domains/Mb, respectively) and as high as primates for nonhomogeneous domains (13 domains/Mb). In overall chromosomal domain densities, opossum (34 domains/Mb) is positioned between murids (26 domains/Mb) and other mammals (43 domains/Mb).

Similar patterns were observed when the compositional domain densities of GC-rich and GC-poor domains were compared. The opossum and primate genomes have the highest densities for GC-rich domains (21 and 18 domains/Mb, respectively). By contrast, the opossum genome's low density of GC-poor domains (10 domains/Mb) is lower even than that of murids (16 domains/Mb). The overall domain density in opossum (31 domains/Mb) is between that of murids (25.5 domains/Mb) and apes (38.5 domains/Mb).

Domain density varies to a considerable extent among chromosomes and chromosome types. Density differences among chromosomes can be quite large, with sex chromosomes having a lower density than the average autosome. These results indicate that the processes that have shaped the interchromosomal domain organization acted nonuniformly on all chromosomes and that their effect on domain lengths was highly variable in different lineages, implying the existence of a compositional constraint on chromosomal heterogeneity.

Compositional architecture was found to be congruent with phylogenetic relatedness. For example, the genomes of human, chimpanzee, and orangutan are very similar to each other, as are the genomes of mouse and rat. Overall, the laurasiatherian (horse, dog, cow, pig) genomes are more similar to the genomes of apes than to the genomes of murids.

Murid genomes are distinguished from the primate and laurasiatherian genomes mainly by their narrow GC-content distribution, larger GC-rich domains, and smaller GC-content variation for both GC-poor and GC-rich domains; and by the unique shape of their joint distribution of compositional-domain GC content and length. Differences in the compositional patterns between murids and other mammals were thought to have been created by a change that occurred before the divergence among murid taxa (Robinson et al. 1997; Smith and Eyre-Walker 2002). According to this view, the murid compositional architecture is a synapomorphy. Because of the similarity between the compositional architectures of murids and opossum, the possibility was raised that the murid architecture may have originated in the eutherian ancestor and has been maintained with little change in murids. If this alternative view is true, then the murid pattern is a plesiomorphy.

The origin and evolution of compositional domains

The origin of compositional domains, especially GC-rich compositional domains, is as mysterious as it is controversial. Unfortunately, most work on the subject has been done under the false twin assumptions that isochores exist and that the GC content of isochores can be predicted by using the GC content at third codon positions. Here we discuss several hypotheses that aim to explain the compositional compartmentalization within genomes.

According to the **thermodynamic stability hypothesis**, GC-rich compositional domains are a form of adaptation to environmental pressures, particularly temperature, as increases in genomic GC content are supposed to protect DNA, RNA, and proteins from degradation by heat. This hypothesis posits that the genomes of "cold-blooded" vertebrates would be GC-poor, whereas "warm-blooded" genomes would be GC-rich (Bernardi et al. 1985; Bernardi and Bernardi 1986). The main problem with this hypothesis is that is does not explain the lack of compositional uniformity within genomes. Why aren't the genomes of homeotherms entirely GC-rich? Why do they also contain GC-poor regions?

According to the **mutation bias hypothesis** (Wolfe et al. 1989a), isochores arose from mutational biases due to compositional changes in the precursor nucleotide pool during the replication of germline DNA. The GC-rich domains are carried on replicons that replicate early in the germline cell cycle, during which the precursor pool has a high GC content, and thus have a propensity to mutate to GC. The AT-rich domains, on the other hand, are replicated late in the cell cycle, when the nucleotide precursor pool has a high AT content. These domains will have a propensity to mutate to AT. This hypothesis is based on the observation that the composition of the nucleotide precursor pool changes during the cell cycle, and that such changes can in fact lead to altered base ratios in the newly synthesized DNA (Leeds et al. 1985), especially since the replication of the mammalian genome is such a lengthy process, taking 8 hours or more (Holmquist 1987).

The **recombination hypothesis** suggests that recombination is the primary determinant of the isochore organization of mammalian genomes (Eyre-Walker 1993; Montoya-Burgos et al. 2003; Meunier and Duret 2004). Several hypotheses have been proposed to explain how recombination influences base composition: (1) recombination enhances the creation of new GC alleles via mutation, (2) recombination favors the spread of GC alleles via biased gene conversion (Galtier et al. 2001; Duret and Galtier 2009), and (3) there is a selective pressure in favor of a high GC content, and GC content increases with high crossover rates because selection is more efficient (Charlesworth 1994).

The **transposable elements hypothesis** (Duret and Hurst 2001) assigns a major role for transposable elements in shaping the GC content landscape of mammalian genomes. This hypothesis posits that transposable elements tend to have a lower GC content than those of the more GC-rich genes, inserting themselves into intronic regions, thus expanding them to form long stretches of noncoding regions that have a uniform GC content.

CHAPTER

12

The Evolution of Gene Regulation

by Amy K. Sater

Soon after publishing their groundbreaking paper describing the *lac* operon, Monod and Jacob (1961) speculated about the unique role that mutations in regulatory regions might play during the course of evolution. They based their arguments on the recognition that the proper function of every gene depends on two distinct components: (1) what its product does, and (2) under which circumstances the product is made. A perfectly good enzyme can be useless (neutral) or even counterproductive (deleterious), they argued, if synthesized under the wrong conditions. Fifteen years later came the surprising realization that homologous protein-coding genes in humans and chimpanzees are nearly identical. As a consequence, King and Wilson (1975) came to the conclusion that the modest degree of divergence in protein sequences cannot account for the profound phenotypic differences between the two species. They proposed instead that regulatory differences must be primarily responsible.

In previous chapters, we have discussed different routes by which evolutionary changes arise through changes in copy number and amino acid sequence of protein-coding genes. For example, duplication of an ancestral β-globin gene and subsequent sequence divergence among the paralogs led to the evolution of ε-, δ-, and γ-globins, which have greater affinities for molecular oxygen than β-globin. The expression of these paralogs in embryonic and fetal tissues provides the physiological basis by which oxygen is transferred from maternal blood to the embryonic and fetal circulatory systems in the placenta (Carter 2012). Thus, these protein-coding sequences represent adaptations that have been critical to the success of mammals. We note, however, that for the divergence of β-globin family members to be truly adaptive, each individual family member must be expressed at specific developmental times and in specific tissues. Thus, the sequences that regulate globin gene expression must evolve just as the protein-coding sequences do. More generally, over 20% of all phenotypic changes in multicellular organisms have been traced to changes in regulatory regions (Stern and Orgogozo 2009). Indeed, alterations in regulatory regions are believed to underlie major morphological changes (Carroll 2005; Davidson 2006).

In this chapter, we examine the evolution of gene expression regulation in multicellular eukaryotes at the pretranscriptional, transcriptional, and posttranscriptional levels.

Pretranscriptional Regulation

Pretranscriptional regulation is essentially accomplished by modulating the accessibility of the gene being transcribed to RNA polymerases, transcription factors, and other molecules involved in the transcription process. So far, we know of two mechanisms of pretranscriptional regulation, **histone modification** and **DNA methylation**. Pretranscriptional regulation is often referred to as **epigenetic regulation**. The term "epigenetics" has been defined in many different ways by different investigators and carries various evolutionary and nonevolutionary associations (e.g., Korochkin 2006; Bird 2007; Ptashne 2007, 2013; Riddihough and Zahn 2010), so we will not use the term here.

Regulation by covalent modifications of histones

To understand pretranscriptional regulation of gene expression by means of histone modification, we need first to familiarize ourselves with the manner in which the DNA is packaged within the nucleus of a eukaryotic cell. The diploid human genome, for example, contains approximately 7 billion base pairs, or 2 meters, of DNA. This length is shortened considerably by the organization of DNA in nucleosomes. Nucleosomes are the fundamental structural units of chromatin. They are composed of the core histone octamer (two each of histones H2A, H2B, H3, and H4) and the associated DNA that wraps around it. Chromatin forms a structural scaffold for the organization of the DNA for transcription and replication, as well as providing mechanical support for the DNA during mitosis.

There are many ways by which histones can be covalently modified. In particular the acetylation of lysine residues and the methylation of lysine and arginine residues play a role in the regulation of transcription by allowing or restricting access to the DNA. The methylation of lysine can occur by the addition of a single methyl group (monomethylation), two methyls (dimethylation), or three methyls (trimethylation). Other covalent modifications (such as ADP-ribosylation, ubiquitination, citrullination, and phosphorylation) may also alter chromatin structure and affect transcription. Of particular interest are modifications of amino acids in the amino-terminal tails of histones that extend outside the nucleosome. These modifications affect access to chromatin and thereby influence gene expression; however, our understanding of the significance of histone modifications in the evolution of gene regulation awaits a more extensive functional analysis of such modifications among distinct taxa.

Some histone modifications involving lysine methylations of histone 3 are conserved among plants and animals: trimethylation of lysine 27 (H3K27me3) represses transcription in animals and plants and transposon mobilization in plants, while methylation of lysine 4 (H3K4me) promotes transcriptional activation in both taxa. H3K9 trimethylation is characteristic of heterochromatin in animal and fungal genomes. In contrast, heterochromatin in *Arabidopsis* contains dimethylated H3K9. Interestingly, H3K4me modification represses transcription in budding yeast (Wang et al. 2011); methylation at other sites on histone H3 is thought to promote transcriptional activation in yeast (Zhang et al. 2014).

DNA methylation

Methylation of cytosine residues in clusters of CpG dinucleotides, often referred to as CpG islands, constitutes a primary mechanism of long-term transcriptional repression in vertebrates and many other taxa. These regions of methylated DNA are characteristic of heterochromatin, as well as of genes that are preferentially silenced in the maternal or paternal copies in mammals (referred to as imprinted genes). DNA methylation at CpG islands is also one of several mechanisms contributing to the silencing of one of the X chromosomes in mammalian XX cells. However, DNA methylation is not an obligatory mechanism of transcriptional repression: methylated DNA has not been observed in the nematode *Caenorhabditis elegans*, although it is found in other nematode species.

Methylation of DNA outside of CpG islands may play additional, as yet undefined, roles in transcriptional regulation or chromatin organization. In the *Drosophila melanogaster* genome, most of the 5-methylcytosines are found in locations other than CpG dinucleotides, suggesting a distinctive role for these methylation sites. Moreover, DNA methylation can affect transcription in a context-specific manner (Jones 2012). For example, in the active mammalian X chromosome, transcriptional activity is positively correlated with the degree of **gene body methylation** (i.e., methylation of the exons and introns of a transcribed gene).

An understanding of the origin and evolution of DNA methylation may emerge once genomic patterns of DNA methylation have been elucidated across a broader range of taxa. We can, however, make some cautious inferences from the fact that the patterns of methylation differ between plants and animals (Rabinowicz et al. 2003). In animals, both exons and transposable elements are methylated, whereas in plants, DNA methylation is almost exclusively restricted to repeats and transposable elements. It has been suggested that DNA methylation originated as a mechanism of keeping transposable elements in check. DNA methylation could have subsequently been co-opted as a mechanism of transcriptional repression in some taxonomic lineages, as other mechanisms arose to control transposon mobility, such as the Piwi/piRNA pathway. Methylation of transposable elements is observed in plants and in many animal taxa, while in fungi and other animals, transposon silencing occurs in the absence of DNA methylation.

Regulation at the Transcriptional Level

Transcriptional activation of eukaryotic genes is closely regulated. A specific gene may be active in some cell types but not others, or it may be active under certain conditions only or in response to particular signals; in the absence of these triggers, the gene will not be transcribed. This control of transcription in space and time requires nontranscribed sequences associated with the gene. These nontranscribed sequences, or ***cis*-regulatory elements**, provide binding sites for specific **transcription factors**, or ***trans*-regulatory elements**. Binding of specific transcription factors at specific sites within the nontranscribed sequences determines whether or not a gene will be expressed. Nontranscribed elements that affect expression of specific genes include **promoters**, which are immediately proximal to the 5′ end of a gene, and **enhancers**, which are clusters of transcription factor binding sites that act over some distance from the gene. Enhancers can regulate transcription either positively or negatively. Enhancers that function primarily as repressors of transcription are sometimes called **silencers**. **Insulators**—sequences that limit the distance over which an enhancer can act—represent an additional category of regulatory elements.

Evolutionary changes within regulatory sequences can produce dramatic consequences in gene expression and phenotype. Interestingly, *cis*-regulatory elements usually evolve much faster than the transcription factors that bind to them. Below, we review these nontranscribed regulatory sequences, examine the evolutionary consequences of change in the elements responsible for transcriptional regulation, and briefly discuss the evolutionary dynamics of the appearance and disappearance of transcription factor binding elements in eukaryotic genomes.

Promoters

The promoter sequence, which encompasses a region of about 500–2,000 nucleotides in length, is usually located immediately upstream of the transcription start site. Most eukaryotic promoters include two types of conserved *cis*-elements: a **core promoter** immediately upstream of the transcription start site, and a region that includes binding sites for transcription factors that lies farther upstream. The core promoter includes binding sites for the basal factors required for the initiation of transcription. These factors, which include a TATA-binding protein (TBP), form a complex that recruits RNA polymerase II and other proteins to the site of transcription initiation. Although TBPs were identified by their affinity to TATA boxes, they do not require TATA boxes in order to bind to core promoters; a TBP can also be recruited via binding to proteins that bind directly to the core promoter sequence. While TATA boxes are present in the core promoters of nearly 25% of genes in the human genome (Yang et al. 2007), promoters for the remaining 75% of human genes recruit TBP through other proteins. The difference between the two promoter types may have specific evolutionary consequences, as we shall see later.

Promoters also include multiple transcription factor binding sites, which recruit RNA polymerase II and stabilize the transcription initiation complex. The number of transcription factor binding sites and the types of transcription factors that are bound by the binding sites vary considerably among promoters. That is, promoters may consist of up to 50 transcription factor binding sites. Once bound to the promoter, the transcription factors often interact with one another. This interaction is facilitated by the fact that transcription factor binding sites are frequently clustered.

While recognizing that promoters represent a continuum of possibilities and that they cannot be neatly pigeonholed into discrete groups, for pedagogical purposes we divide them into three categories: housekeeping, tissue-specific, and dynamically expressed promoters. Originally, **housekeeping genes** were defined as genes that are expressed in all cell types. However, because the vast majority of protein-coding genes are known to be transcribed in all tissues to some extent, housekeeping genes are currently defined by a constant level of gene expression across tissues and cell types (Eisenberg and Levanon 2013). Examples of housekeeping genes in humans include those for glucose-6-phosphate isomerase and small nuclear ribonucleoprotein D3.

The promoters of housekeeping genes have several characteristics that distinguish them from the promoters of genes that show more complex patterns of expression. They are frequently GC-rich and often lack the TATA and CCAAT boxes that are commonly found in the promoters of genes expressed in specific tissues. They also include sequence motifs that are predicted to bind a specific set of transcriptional activators that are expressed in all tissues, such as a transcription factor called specificity protein 1 (Sp1). In most cases, the promoters of housekeeping genes are shorter than promoters of other genes, with most of the sequence conservation within 500 bases upstream of the transcription start site (Farré et al. 2007).

Promoter regions of housekeeping genes have been shown to evolve more rapidly than the other promoter types (Farré et al. 2007). This is likely related to a relaxation in functional constraint due to the presence of a smaller number of functional *cis*-regulatory motifs than in other types of promoters. This explanation, however, needs to be qualified, as subsets of promoter sequences of housekeeping genes, i.e., the core promoters, exhibit greater sequence conservation, i.e., evolve more slowly, than their tissue-specific counterparts (Zhu et al. 2008).

Many genes, such as those for muscle-specific myosins or the opsins in photoreceptor cells, are expressed in a **tissue-specific** fashion. Such genes generally have more complex promoters, with binding sites for a greater variety of transcription factors, which may include transcriptional repressors as well as transcriptional activators. These genes are activated in specific cell types, and then transcription is stably maintained. In nonexpressing cell types, these tissue-specific genes are repressed via the binding of transcriptional repressors or else maintained in a transcriptionally inactive state through chromatin modifications.

There are protein-coding genes that produce both housekeeping and tissue-specific isoforms. Such genes, which possess distinct promoters, can provide us some insight into the differences between promoters for housekeeping versus tissue-specific genes. One such example, the eight-exon phosphoprotein phosphatase 2Cβ (*PPP2Cβ*)

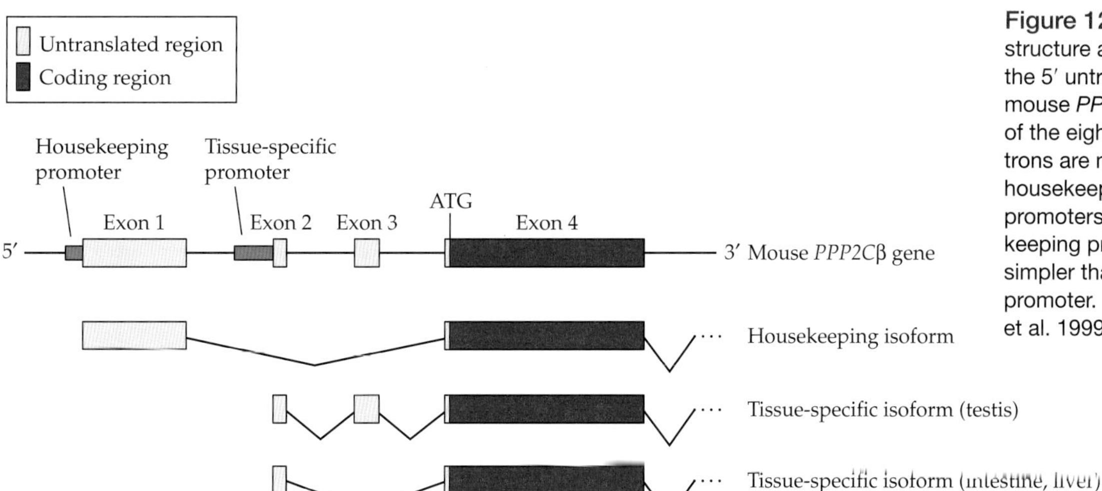

Figure 12.1 Exon/intron structure and three isoforms in the 5′ untranslated region of the mouse *PPP2Cβ* gene. Only four of the eight exons are shown. Introns are not drawn to scale. The housekeeping and tissue-specific promoters are shown. The housekeeping promoter is shorter and simpler than the tissue-specific promoter. (Modified from Ohnishi et al. 1999.)

gene in mouse, is expressed as multiple isoforms through the use of alternative promoters (**Figure 12.1**). The exon-1-containing isoform is a housekeeping protein, and the 5′ flanking region of exon 1 has features characteristic of a housekeeping promoter. That is, it is GC-rich, lacks TATA and CAAT boxes, and contains binding sites for Sp1. The exon-2- and exon-2+3-containing isoforms are expressed specifically in the testis, as well as in the intestine and liver. Upstream of exon 2, features reminiscent of tissue-specific promoters, such as a TATA-like and negative regulatory elements, are found (Ohnishi et al. 1999). We note that the different isoforms have identical coding regions; they differ from one another in the lengths of their 5′ untranslated regions.

Dynamically expressed promoters, also called **transient promoters**, are activated only in response to specific signals or specific cellular events. Some examples are promoters for genes expressed in specific tissues during a particular embryonic stage but not before or after, promoters for genes expressed in response to environmental changes such as temperature or photoperiod, and promoters for genes expressed in lymphocytes when activated by an antigen. Dynamically expressed promoters include binding sites for transcriptional repressors, as well as transcriptional activators. One example is the promoter of the *Endo16* gene of the purple sea urchin, *Strongylocentrotus purpuratus* (**Figure 12.2a**). The gene, which encodes a secreted protein, is expressed specifically in the endodermal cells that make up the archenteron (the primitive embryonic gut that forms during gastrulation; **Figure 12.2b**).

More than 2 Kb in length, the promoter has multiple modules (**Figure 12.2c**), some of which are responsible for the shutting off of *Endo16* in nonendodermal cells, while others activate and amplify expression in the endoderm itself. For example, there are two regions, or modules, at the proximal (near-upstream) end of the promoter that mediate the activation of *Endo16*; one region is required for initial activation of the gene in the endoderm, while the other maintains expression in the archenteron at later stages. The transcription factors that bind these regions are expressed and active in endoderm cells. In contrast, three modules located farther upstream mediate negative regulation: one region is required to repress *Endo16* expression in mesoderm (skeleton-forming) cells, while two other regions mediate *Endo16* repression in the ectoderm cells, which form the outer surface of the embryo.

The sea urchin *Endo16* promoter illustrates three key properties of tissue-specific and dynamically expressed promoters: **redundancy**, **modularity**, and **combinatorial action**. Promoters for genes that are expressed at specific times and places generally show considerable redundancy at two levels: (1) they include multiple binding sites for any given transcription factor, and (2) they include binding sites for several different activating transcription factors. Complex promoters often show a modular organization, in which one region of the promoter will include binding sites for activating

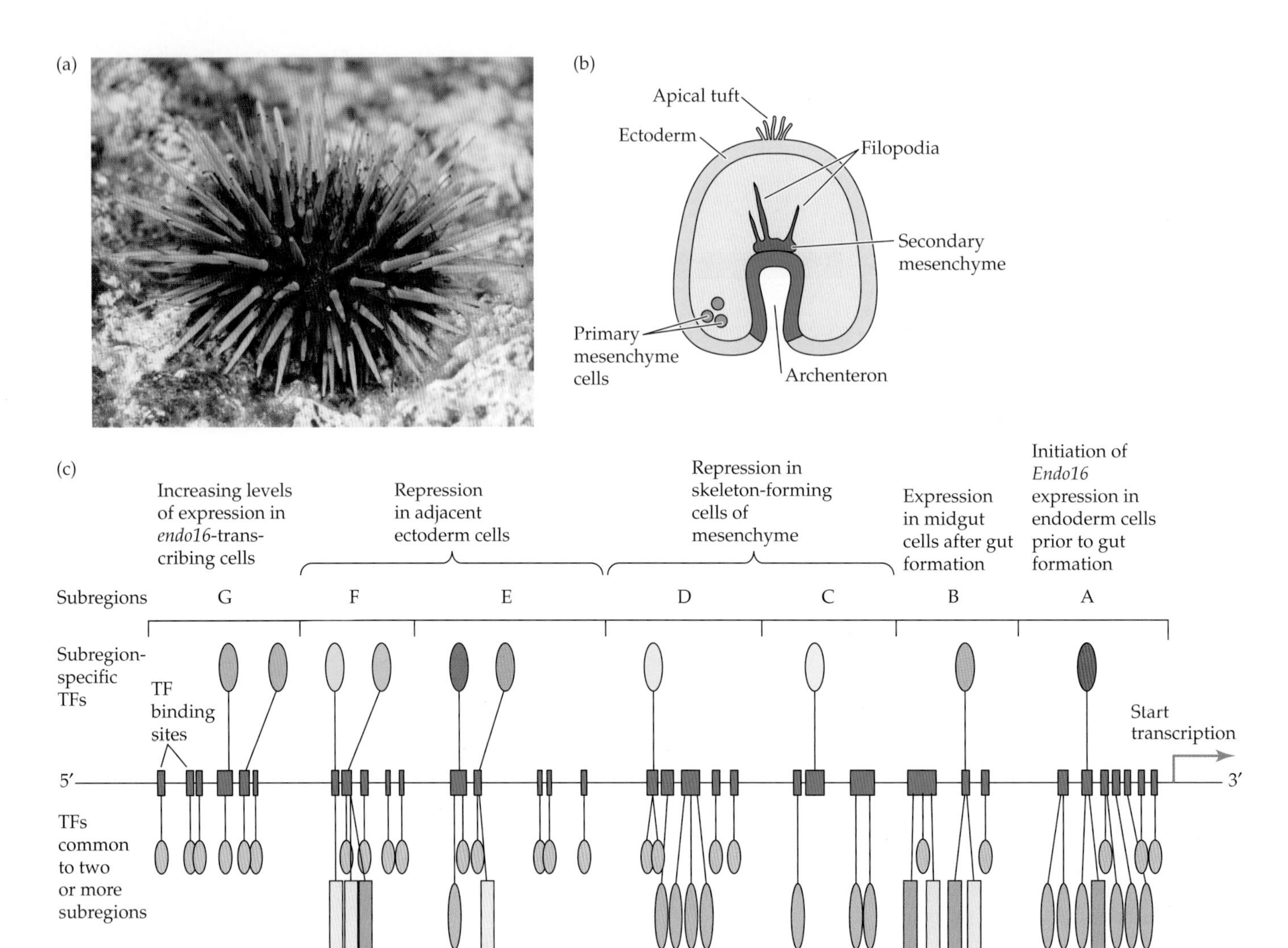

Figure 12.2 The promoter of the dynamically expressed *Endo16* gene is required for proper formation of the sea urchin embryonic gastrointestinal system. (a) The purple sea urchin *Strongylocentrotus purpuratus*. (b) Schematic representation of the pattern of expression of *Endo16* transcripts (blue) in a 2-day-old gastrula. *Endo16* is expressed along the entire length of the archenteron, while the secondary mesenchyme near the top of the archenteron does not express *Endo16*. (c) The promoter of the *Endo16* gene is contained within an approximately 2,200 bp region. Within this region, there are over 30 functionally active transcription factor (TF) binding sites (blue boxes) clustered in seven distinct subregions (A–G). Each subregion has a specific function. Binding sites for the nine specific transcription factors, i.e., transcription factors that bind to only one of the seven subregions, are shown above the promoters as different colored ovals. Binding sites for five common transcription factors, i.e., transcription factors that each bind at multiple target sites in two or more subregions, are indicated below the promoter by ovals and rectangles of different colors (Yuh et al. 1994). The transcription start site is shown. (Photograph © Natalie Jean/Shutterstock; c modified from Davidson 2006.)

transcription factors expressed in one cell type, while another region will contain binding sites for repressing transcription factors expressed in a different cell type. The overall pattern of expression can be dynamic, i.e., the gene in question may be activated at a specific time in a specific cell type to be down-regulated in these cells at a later time, while in other cell types it may remain inactive until it is activated in response to suitable cues. This pattern of expression reflects the combinatorial action of the entire set of modules within the promoter, with significant contributions from the enhancers. Moreover, the transcription factors that bind within a module act combinatorially to generate the input from an individual module. Together, these properties ensure that the appropriate expression of this gene is robust, i.e., that it is resistant to perturbations such as those arising from mutation and variation in the expression of specific transcription factors.

Promoter evolution

Promoter regions may evolve via the destruction of existing transcription factor binding sites, as well as through the duplication of existing sites or the addition of novel transcription factor binding sites. Since transcription factor binding sites are generally between four and eight bases in length, the creation and destruction of promoters by point mutation is expected to be extremely rapid on an evolutionary time scale

(Stone and Wray 2001). Moreover, in many cases, binding of one transcription factor facilitates the recruitment and binding of a second transcription factor at an adjacent site. Thus, insertions and deletions that affect the spacing between transcription factor binding sites may also alter promoter activity. Evolutionary changes that affect the establishment or arrangement of transcription factor binding sites within a promoter will occur far more rapidly than evolutionary changes in the coding sequence of the transcription factors.

It is important to note that while the dynamics of appearance and disappearance of transcription factor binding sites can be studied quantitatively, it is usually impossible to predict functional consequences. It is likely that, in many cases, the appearance of a new binding site will have no effect on the expression of adjacent genes. For example, some transcription factors must interact with other DNA-binding proteins to affect gene expression, and therefore the appearance of a single new binding site might be functionally neutral. Alternatively, nearby silencer sites might neutralize any new binding site that appears. Nevertheless, it is plausible that in some cases the appearance of a new binding site will alter gene expression. Of course, even if a new binding site does affect gene expression, there is no way to predict what, if any, phenotypic consequences might ensue or what would be the fitness effects. As with amino acid replacements within proteins, the consequences of a new binding site appearing within a promoter or of an existing binding site disappearing are likely to be either detrimental or neutral; only in very rare cases will such changes be beneficial. What we can say at present is that point mutations are constantly producing new binding sites and destroying old ones and that these mutations can in principle alter gene expression. Thus, as with everything else in evolution, the fate of such changes will be determined by a combination of selection and luck.

Some features of promoter structure apparently facilitate evolutionary changes leading to divergence in expression patterns, such as changes in the timing of expression. One critical feature is the TATA box. The presence or absence of a TATA box has been linked to the rate at which gene expression evolves. Examination of the transcriptional responses of four closely related yeast species to a variety of environmental stresses revealed that a gene containing a TATA box is likely to show a greater interspecific variability in expression than a gene with TATA-less promoters (Tirosh et al. 2006, 2009). Interestingly, the sequences of promoters with TATA boxes, which exhibit a high rate of divergence in expression, do not evolve faster than promoters that do not contain TATA boxes. This observation suggests a greater flexibility of TATA box-containing promoters, possibly through increased "noise" in gene expression (Dikstein 2011). The fact that these flexible promoters are also associated with nucleosomes indicates that the function of the promoter is highly dependent upon chromatin modifiers that alter the nucleosome structure such that the chromatin becomes more accessible to transcription factors. Either *cis*-acting mutations that modify binding sites for transcription factors, or *trans*-acting mutations that alter the expression of chromatin modifiers can thus lead to divergence of expression over the course of evolution. Moreover, larger promoters with a greater number of transcription factor binding sites will be more sensitive to alterations in nucleosome structure or expression of transcription factors, although in many cases, larger promoters may have a greater degree of redundancy, as seen in the case of the *Endo16* gene.

Finally, a significant fraction of promoters contain short tandem repeats, which are known to be subject to mutations affecting repeat number. As a consequence, promoters containing short tandem repeats exhibit frequent expansions and contractions of repeat number. Changes in the size of a tandem repeat sequence can alter nucleosome structure or the distance between transcription factor binding sites, which can also lead to divergence in expression patterns.

Divergent transcription

Recent studies have shown that transcription initiation complexes can form on both strands in promoters lacking TATA boxes, leading to a phenomenon called

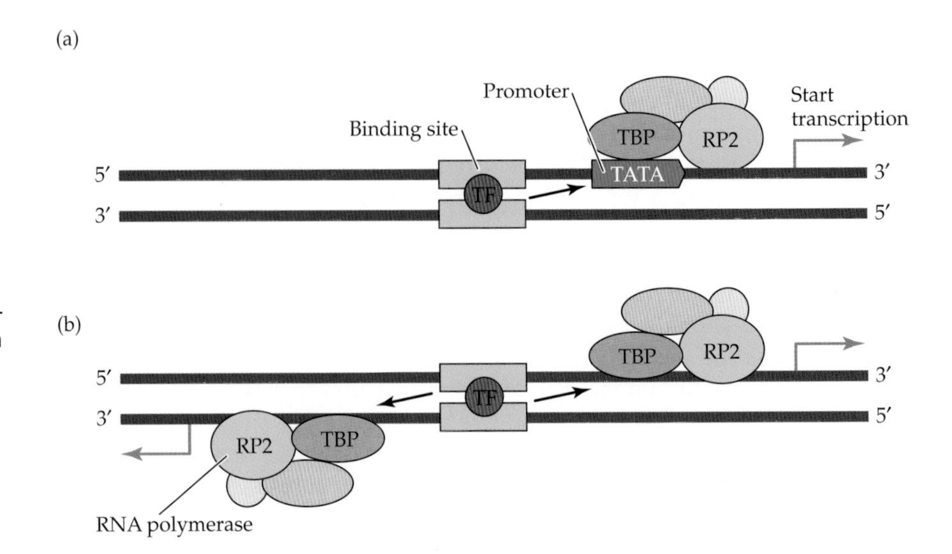

Figure 12.3 (a) Unidirectional promoter. A transcription factor (TF) binds to a transcription factor binding site and helps recruit the TATA-binding protein (TBP) and associated factors (unmarked ovals) to the TATA-containing promoter. These proteins will orient RNA polymerase II (RP2) to transcribe downstream DNA. (b) In the absence of strong TATA elements, transcription-factor-recruited TBP and associated factors bind to low-specificity sequences and form initiation complexes in both directions. (Modified from Wu and Sharp 2013.)

divergent transcription, in which transcription from a single promoter can proceed in both directions simultaneously (**Figure 12.3**). In divergent transcription, transcription occurs in both the sense and antisense directions with respect to the downstream gene, i.e., two noncomplementary transcripts are made. Whether or not divergent transcription yields functional RNA is currently a matter of speculation.

Wu and Sharp (2013) have proposed that divergently transcribed promoters offer a unique opportunity for the evolution of new genes. Their model is based on observations that two kinds of *cis*-elements found upstream of divergently transcribed promoters have an impact on whether antisense transcription will occur at significant levels. The first element is the six-base polyadenylation or poly(A) signal which will lead to cleavage of the nascent transcript and termination of antisense transcription. The second is the eight-base binding site for the U1 small nuclear ribonucleoprotein (snRNP), an essential component for splicing of mRNA, which antagonizes the effects of the poly(A) motif and favors antisense transcription.

In a simplified view of the model offered by Wu and Sharp (2013), depletion of functional poly(A) motifs and/or accumulation of U1 binding sites in the 5′ region of a divergently transcribed promoter will favor increased transcription in the antisense direction. Over evolutionary time, favorable changes in the distribution of these *cis*-elements upstream of divergently transcribed promoters are likely to produce novel protein-coding genes that are transcribed in the antisense direction from their promoters. Wu and Sharp (2013) point out that this process could also enhance the establishment of new genes that originate via other processes; if duplicated copies of genes or exons are relocated into regions that are divergently transcribed, they can then be expressed in novel patterns, contributing to the establishment of new gene functions for the duplicated sequence.

Enhancers

Enhancers are *cis*-acting regulatory elements that bind transcription factors to regulate a target gene over a considerable distance (up to 1 Mb) along the chromosome. While promoters are always located immediately upstream of the transcription start site, enhancers can be located virtually anywhere relative to the gene (**Figure 12.4**). Enhancers are thought to be responsible for establishing the spatial and temporal patterns of gene expression, but they can quantitatively regulate expression levels as well. A given gene may have multiple enhancers, each mediating the activation of the target gene in a specific location or for a specific interval during development. Since different enhancers for the same gene can act independently, mutations in enhancers provide a means by which expression of a given gene can be modified in one tissue without disrupting expression in other tissues.

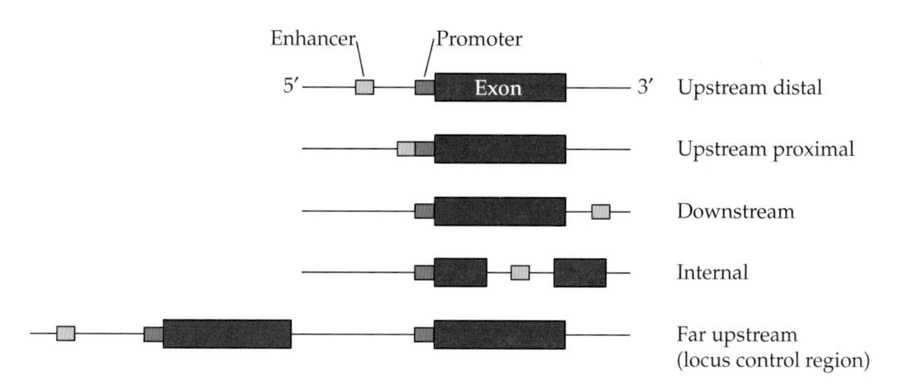

Figure 12.4 The enhancer (pink box) can be located virtually anywhere relative to the gene, but the promoter (dark blue box) is always located at the 5′ end.

Enhancers include clusters of binding sites for both transcriptional activators and transcriptional repressors. As with promoters, the relative balance between transcriptional activators and transcriptional repressors that are active in a given cell (i.e., are bound to an enhancer) determines whether or not the target gene is expressed. When transcription factors bind to an enhancer, they establish a binding site for the mediator complex, a large protein complex with approximately 25 subunits. The mediator complex physically links the transcription factors bound at the enhancer to the RNA polymerase II initiation complex bound at the promoter. The mediator complex is found in all eukaryotes and is required for most RNA polymerase II-dependent transcription; its function in enhancer-dependent transcription is only one of its roles in transcriptional activation (Borggrefe and Yue 2011; Carlsten et al. 2013).

From an evolutionary standpoint, this two-component system of regulation consisting of promoters and enhancers offers an avenue by which populations can explore the effects of gene function in novel contexts without disruption of either the primary coding sequence of the gene or its promoter. Morphological evolution has been driven in large part by modifications in enhancers, rather than promoters or protein-coding sequences. In the following sections, we present a few examples of enhancer evolution.

REGULATION OF COMPLEX PATTERNS BY *EVEN-SKIPPED* AND *FUSHI TARAZU*: DIFFERENT MECHANISMS, SIMILAR OUTCOMES The segmental organization of insects results from a multistep genetic cascade in which the anterior-to-posterior organization of the unfertilized egg is converted into a set of segmentally repeating units with unique identities. Three segments form head structures, the next three form the thorax, and the remaining eight segments form the abdomen. The nine so-called pair-rule genes encode transcription factors that initiate segmentation; loss-of-function mutations in these genes cause a loss of the normal developmental pattern in alternating segments.

Each of the pair-rule genes is expressed in seven stripes, although the stripes for different pair-rule genes are not in register with one another; these stripes reflect areas in which an individual pair-rule gene will govern the development of alternating segments. This pattern of expression emerges from the activity of the gap genes, a group of transcription factors that are expressed in broad regions along the anterior-posterior axis. The gap genes act together to regulate expression of individual pair-rule genes, and this step converts anterior-to-posterior genetic information into the segmental organization of the insect body plan.

In the following example, we shall use the best known pair-rule genes: *even-skipped* (*eve*) and *fushi tarazu* (*ftz*, "too few segments" in Japanese) to illustrate how similar patterns of expression can arise from different *cis*-regulatory elements. Both *eve* and *ftz* are expressed in seven stripes along the anterior-posterior axis of the fly embryo (**Figure 12.5**). Four major enhancers establish the areas in which *eve* is expressed. Each stripe has a distinct enhancer or half-enhancer that is responsible for the expression of *eve* within the corresponding stripe.

(a)

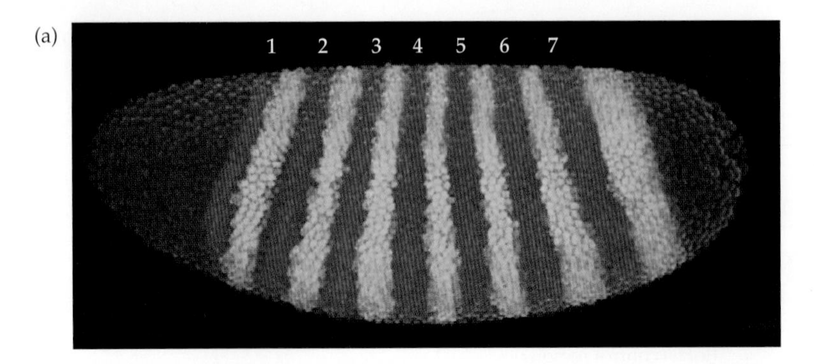

Figure 12.5 (a) Expression of the pair-rule genes *even-skipped* (*eve*) and *fushi tarazu* (*ftz*) in the *Drosophila* embryo. Red, stripes of *eve* expression; green, stripes of *ftz* expression. (b) Despite the fact that the expression patterns of *eve* and *ftz* are similar, the promoters and enhancers of the two genes are completely different in both sequence and location. (Photograph from Leaman et al. 2005.)

(b)

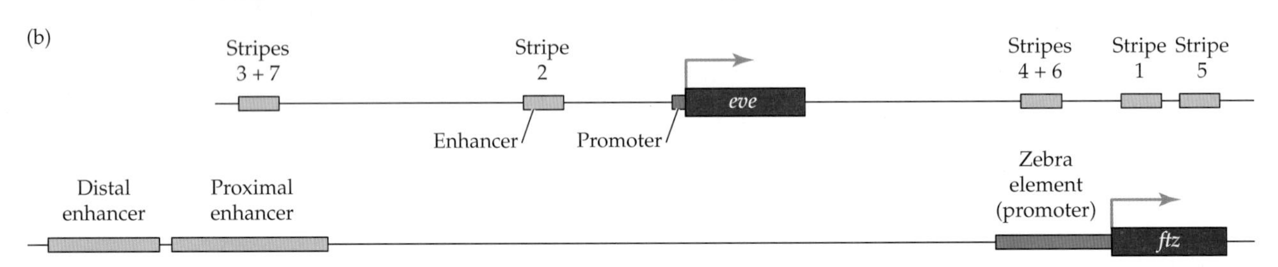

Multiple enhancers do not represent the only strategy to achieve this seven-stripe pattern of expression, however. The second pair-rule gene, *ftz*, has a complementary pattern of expression, with the *ftz* stripes alternating with the *eve* stripes. The *cis*-acting elements that are required to generate this pattern of *ftz* expression include a promoter that is approximately 800-bp in length, often referred to as the zebra element, and a region located 3–6 Kb upstream of the transcription start site that includes proximal and distal enhancers (Pick et al. 1990; Han et al. 1993). The zebra element activates *ftz* expression in a broad region; subsequent binding of transcriptional repressors in the stripes that express *eve* leads to the inhibition of *ftz* expression in these areas. An *ftz* autoregulatory element is located within the proximal enhancer; it includes binding sites for *ftz* and other transcription factors (Schier and Gehring 1993), thus establishing a positive feedback loop that stabilizes *ftz* expression. The enhancers for *eve* and *ftz* demonstrate that highly similar patterns of expression can arise from highly dissimilar gene regulation.

The protein-coding genes for both *eve* and *ftz* are highly conserved among holometabolous insects. On the other hand, a comparison of *eve* enhancer sequences between a scavenger fly, *Sepsis cynipsea*, and *Drosophila melanogaster* reveals that the regions upstream and downstream of the *eve*-coding gene show almost no similarity, with the exception of a few small blocks (approximately 20–30 bp) of high conservation scattered across the locus (**Figure 12.6**). A paradoxical situation arises, however, from the fact that the enhancers from *S. cynipsea*, which are completely different in se-

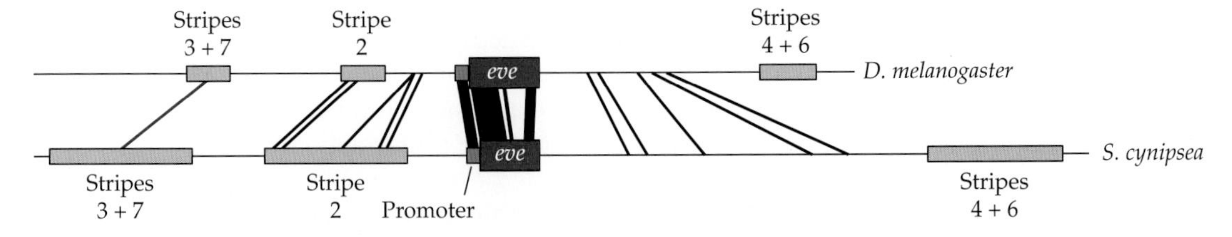

Figure 12.6 The enhancers (pink boxes) of the *even-skipped* loci in *Drosophila melanogaster* and the scavenger fly *Sepsis cynipsea* are different in both sequence and location. Black lines represent small blocks (approximately 20–30 bp) that exhibit statistically significant similarity between the two sequences on the plus strands. The red line at left epresents a block of similarity between the plus strand of *D. melanogaster* and the minus strand of *S. cynipsea*. (Data from Hare et al. 2008.)

quence and location from those of *D. melanogaster*, can activate the proper expression pattern when introduced transgenically into *Drosophila* embryos (Hare et al. 2008). This observation illustrates that transcription factor binding sites appear and disappear frequently during evolution, and that convergent functional evolution is rampant.

Shadow enhancers

Enhancers have usually been studied through experiments that ask whether a potential enhancer sequence can drive expression of a target gene in an artificial transgenic construct. Thus, most experiments that investigate enhancer function determine whether an enhancer is sufficient to direct the expression of a gene in a specific set of cells. There are far fewer experiments that show whether an enhancer is necessary for expression. In some instances, multiple enhancers have been identified for a specific gene, each of which was found to be sufficient to direct the same pattern of expression. Such "redundant" enhancers are sometimes referred to as **shadow enhancers** (Hong et al. 2008) or **distributed enhancers** (Barolo 2011). Redundancy, however, may in many cases be more apparent than real, as some supposedly redundant enhancers appear to be important for proper regulation. Many functions have been proposed for shadow enhancers; here we shall discuss four such putative functions.

In some instances in which two shadow enhancers have been identified, expression directed by either enhancer alone has been found to be lower than expression of the target gene in vivo, suggesting that both enhancers are necessary to achieve proper levels of gene expression (**Figure 12.7a**). In some other instances, the existence of two enhancers was found to reduce the fraction of cells in which the gene fails to be properly expressed (**Figure 12.7b**). Sometimes, the spatial pattern of gene expression, also referred to as the **spatial precision** or **patterning precision**, was found to require both enhancers, implying that complex patterns of expression may at times require two or more seemingly identical enhancers (**Figure 12.7c**). Finally, Boettiger and Levine (2009) and Perry et al. (2009) proposed that the existence of multiple enhancers may be important for physiological robustness, i.e., the resistance to environmental perturbations (**Figure 12.7d**). This hypothesis has been evaluated for the *snail* gene, which encodes a transcription factor that is necessary to establish the dorsal-ventral pattern, initiate gastrulation, and activate mesoderm formation during embryogenesis in *Drosophila* (Perry et al. 2010). There are two enhancers for *snail*: a proximal one, located ~2 Kb upstream of the *snail* gene, and a distal one, located ~7 Kb upstream of *snail* within the intron of another gene, *Tim17b2*. In embryos in which either enhancer had been removed, gastrulation and mesoderm formation occurred normally at 22°C.

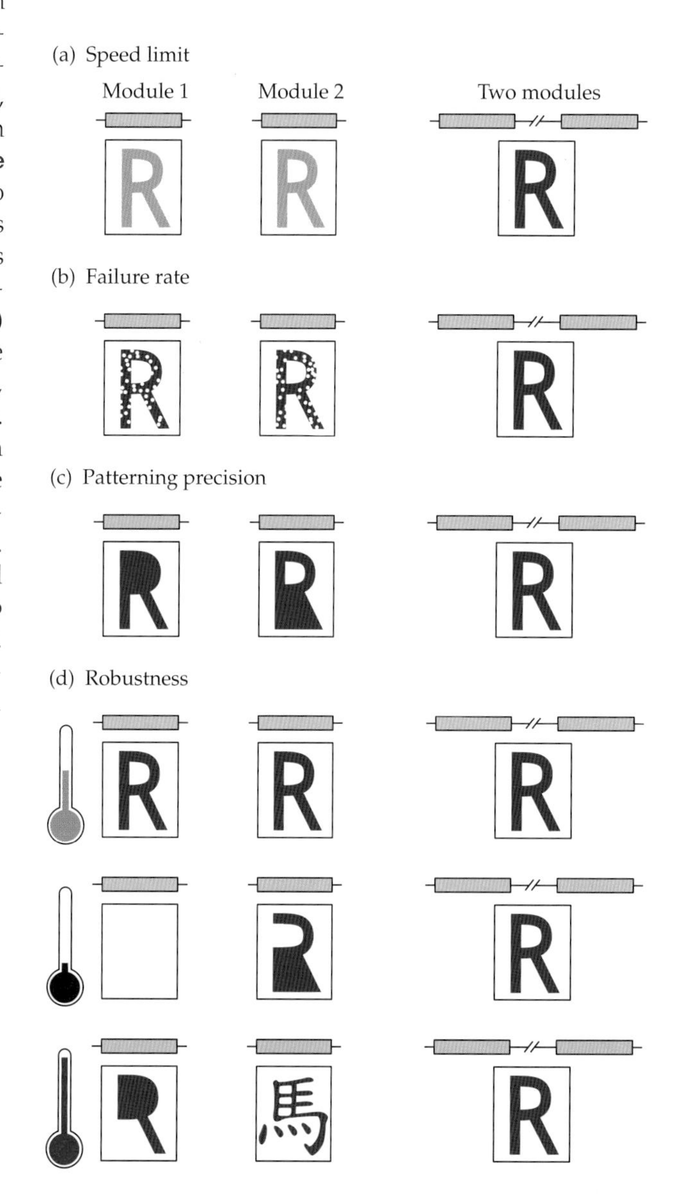

Figure 12.7 Possible functions of shadow enhancers. (a) Speed limit: if a single enhancer drives a suboptimal rate of transcription, multiple modules may be required. (b) Failure rate: if an enhancer fails to activate transcription in a fraction of cells, a second enhancer may significantly improve fidelity of gene expression. (c) Patterning precision: two modules with overlapping patterns but different regulatory logic combine to produce a novel gene expression pattern. (d) Robustness: Under normal temperatures (green) each enhancer produces the same result as the two of them combined. However, at lower (blue) or higher (red) temperatures, each enhancer performs suboptimally on its own. Synergistically, on the other hand, the gene is regulated in an optimal fashion even at extraordinary low or high temperatures. (Modified from Barolo 2011.)

When embryos with only one functional enhancer were raised at 30°C, however, *snail* expression failed to occur in some cells, the boundaries of the mesoderm were uneven, and gastrulation was disrupted in most of the embryos. These results show that both enhancers are important to maintain the robustness of *snail* expression.

All of the examples above come from *Drosophila*; however, redundant enhancers have been identified in vertebrate systems as well. For example, multiple enhancers for the *aggregan* gene have been identified in zebrafish and chicken (Hu et al. 2012).

Following Barolo (2012), it is important to point out that there are several levels at which enhancers and other *cis*-regulatory elements can be investigated, and it is essential to understand the interpretative limitations at each of these levels. The most basic level is that of the primary DNA sequence, in which transcription factor binding motifs may be revealed through sequence similarity. Whether a transcription factor actually binds to such a motif, however, must be demonstrated experimentally. Moreover, additional experiments are required to assess the functional significance of every *cis*-regulatory element. One needs to disrupt such putative regulatory elements by mutation or deletion, or to introduce the putative enhancer into an expression construct, to determine whether or not such an element is either necessary or sufficient for proper expression of a gene. Finally, both experimental and comparative studies may be required in order to understand the effects of a given enhancer on the fitness of an individual.

PHENOTYPIC EVOLUTION VIA LOSS OF AN ENHANCER: THREE-SPINED STICKLEBACKS One of the earliest examples of a phenotypic change in a natural population that could be traced to an enhancer concerned the loss of the pelvic fin in the three-spined stickleback, *Gasterosteus aculeatus*, a small fish that inhabits coastal and near-coastal waterways throughout the Northern Hemisphere. *Gasterosteus aculeatus* has a broad tolerance for salinity, and different populations are associated with marine, brackish water, and freshwater habitats, with freshwater populations having arisen within the last 12,000 years (Reid and Peichel 2010).

Marine populations of *G. aculeatus* have well-developed pelvic structures, including the pelvic girdle, spines, and fins. In contrast, the pelvic apparatus has undergone reduction or was lost independently in multiple freshwater populations (**Figure 12.8**). It has been show that these losses are due to changes in the expression of *Pitx1*, the transcription factor-coding gene (Shap-

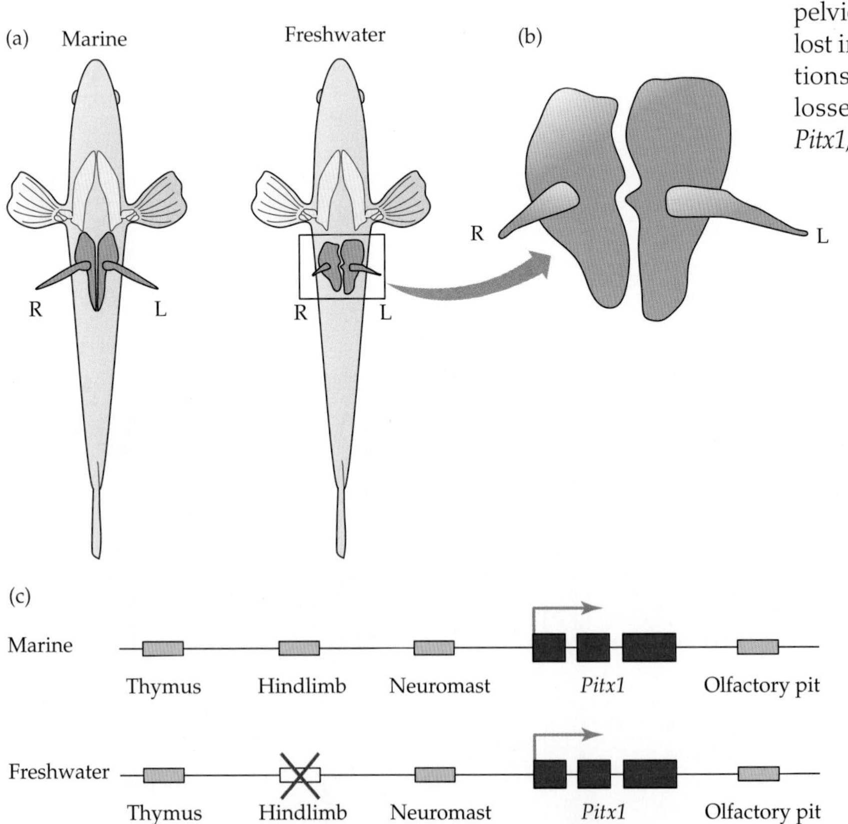

Figure 12.8 (a) Ventral view of the morphologies of the pelvic spines of the three-spined stickleback, *Gasterosteus aculeatus*, from marine and freshwater populations. Sticklebacks from freshwater populations exhibit *Pitx1*-linked reduction in the pelvic spines. (b) In addition to the size reduction in the freshwater populations, an asymmetry in the pelvic spines is observed, with a greater reduction on the right (R) than on the left (L) side. (c) A comparison of *Pitx1* loci from marine and freshwater sticklebacks. The start of the transcription of the three-exon protein-coding gene (red boxes) is marked by a rightward arrow. Enhancers directing expression of the gene in different organs are shown (pink boxes). A mutation obliterating the function of the hindlimb enhancer in freshwater sticklebacks is shown as a crossed-out empty box. (Modified from Shapiro et al. 2004.)

iro et al. 2004). In mammals, *Pitx1* is required for hindlimb formation; in sticklebacks, it is required for the development of homologous structures, i.e., the pelvic fins (Duboc and Logan 2011). The protein-coding sequences of the *Pitx1* genes from freshwater populations are basically identical to their orthologs in marine populations, and in both populations *Pitx1* is expressed normally in the embryonic thymus, olfactory pits, sensory neuromasts on the head, trunk and tail, and ventral portion of the caudal fin. However, while *Pitx1* is expressed in the ventral pelvic region during embryonic development in individuals from marine populations, its expression is reduced or absent in the freshwater populations, which have a reduced pelvis (Shapiro et al. 2004). In marine, pelvic-intact populations, a combination of genetic mapping studies and transgenic experiments was used to identify a 501 bp enhancer upstream of the *Pitx1* gene. This enhancer drove pelvis-specific expression of *Pitx1* (Chan et al. 2010). When a fragment containing this enhancer was used to drive expression of *Pitx1* in embryos from a pelvic-reduced population, the embryos developed into fish with a substantial pelvic girdle and pelvic spines. Sequencing of the enhancer region from three pelvic-reduced populations revealed large independent deletions that included the enhancer. These results suggest that the loss of a pelvis-specific enhancer for *Pitx1* expression has occurred multiple times, resulting in the parallel reduction or loss of pelvic structures in freshwater populations.

SHAVENBABY AND THE PARALLEL LOSS OF TRICHOMES IN *DROSOPHILA* LINEAGES The dorsal larval cuticle of fruit flies may be smooth or hairy, i.e., studded with minute projections called trichomes. The formation of trichomes is controlled by several transcription factors, of which the one encoded by the four-exon *shavenbaby* (*svb*) gene is the most downstream regulator. The name "shavenbaby" is derived from the fact that *svb* mutants do not form trichomes, giving the embryos a smooth, or "shaven," look. The *shavenbaby* locus in *Drosophila* provides an example of evolutionary conservation and divergence across a complex enhancer system (Frankel et al. 2012; Stern and Frankel 2013). While some *Drosophilia* species (e.g., *D. melanogaster, D. littoralis*) have dense patterns of trichomes, others (e.g., *D. sechellia, D. ezoana)* have relatively smooth larval cuticles.

Studies in *D. melanogaster* have identified seven enhancers that regulate *shavenbaby* expression. Six of the seven exhibit significant sequence conservation and are found in the same chromosomal order as *D. virilis* (**Figure 12.9**). By and large, these six enhancers exhibit similar but not identical regulatory functions. In other words, they are mostly active in the same cell types in *D. melanogaster* and *D. virilis*. Two of the six enhancers, Z and E6, are responsible for most of the dorsal expression of *shavenbaby* in *D. melanogaster*, and both are inactive in the closely related species *D. sechellia*. In *D. virilis*, the orthologs of Z and E6, i.e., enhancers 19 and 8, respectively, are responsible for dorsal *shavenbaby* expression. In the closely related species, *D. ezoana*, they are without activity. Fourteen mutated sites in the E6 enhancer of *D. sechellia* were found in a comparison with the orthologous sequences from *D. melanogaster* and *D. simulans*. Of these, 13 sites experienced point mutations (nine transitions and four transversions) and one nucleotide was deleted in *D. sechellia*. Individually, each of these mutations had very limited effect on the activity of *shavenbaby* (5–30% of the total effect). Synergistically, however, they rendered the enhancer nonfunctional (Stern and Frankel 2013).

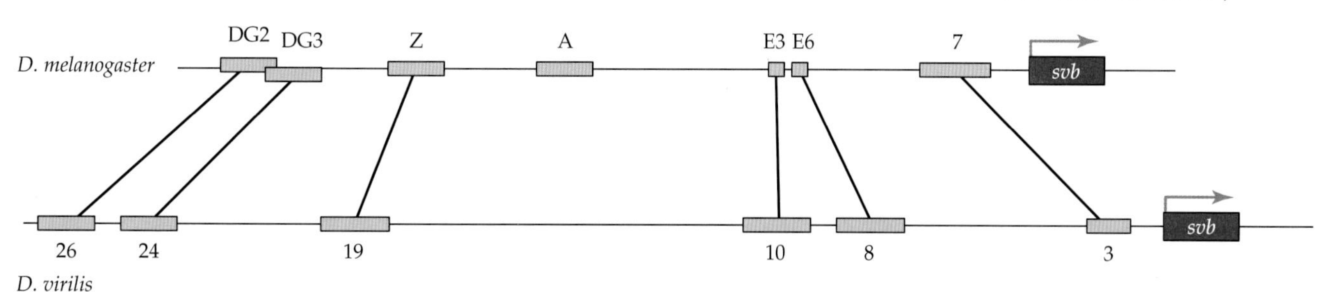

Figure 12.9 Conservation of *shavenbaby* (*svb*) *cis*-regulatory sequences between *Drosophila melanogaster* and *D. virilis*. The 5′ ends of the protein-coding genes are shown (red boxes). Green arrows denote transcription initiation. Pink boxes indicate the position of transcriptional enhancers. Lines connect orthologous enhancers. The sequences of the DG2 and DG3 enhancers overlap each other. Note the confusing nomenclature for the orthologous enhancers in the two *Drosophila* species. (Modified from Stern and Frankel 2013.)

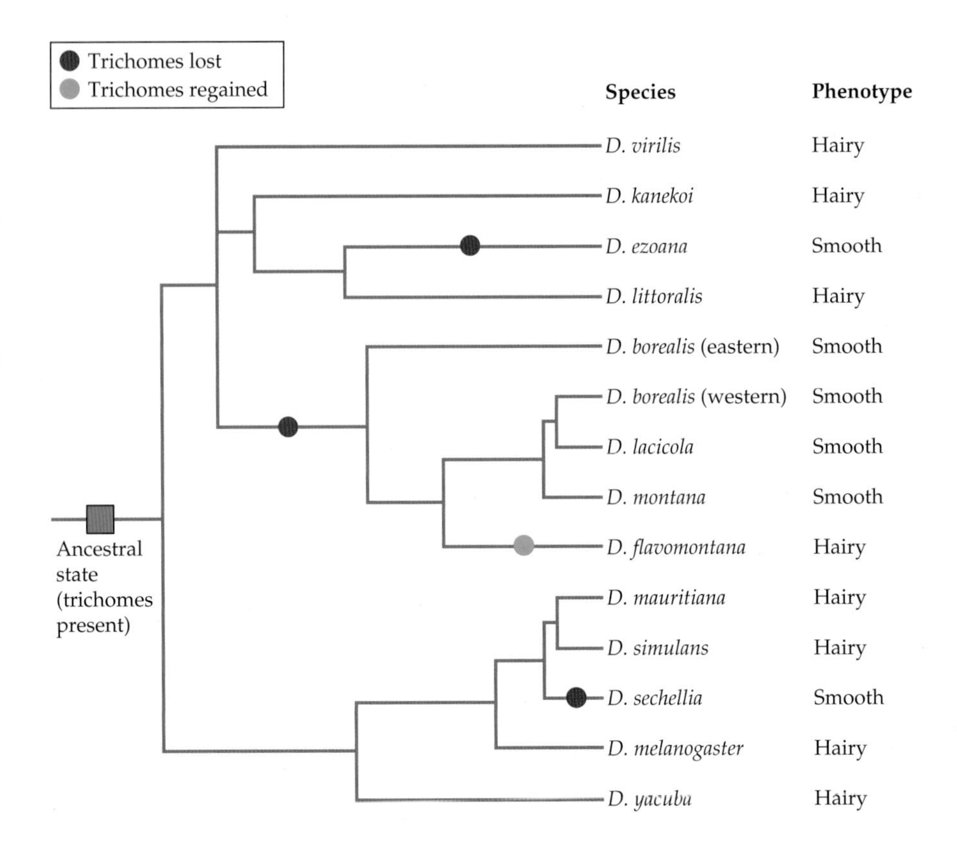

Figure 12.10 An incompletely resolved rooted phylogenetic tree for 14 *Drosophila* taxa indicates that the presence of trichomes on the dorsal cuticle of larvae (the "hairy" phenotype) is the ancestral character state and the absence of such trichomes (the "smooth" phenotype) is derived. Trichomes were independently lost (red circles) at least three times during the evolution of these species, but may have been regained (green circle) in the lineage leading to *D. flavomontana*. (Modified from Delon and Payre 2004.)

The loss of trichomes has occurred multiple times independently in the genus *Drosophila,* and in one lineage a reversal to the hairy ancestral state may have occurred (**Figure 12.10**).

THE LOCUS CONTROL REGION AND THE REGULATION OF GLOBIN EXPRESSION As noted in Chapter 7, gene duplication and divergence have led to the establishment of several families of globin genes, with two families having two or more genes: the α-globin and β-globin families. The functional hemoglobin is a heterotetramer consisting of two globins from the β family and two from the β family. In mammals, the dominant type of hemoglobin changes during development, from the embryonic $\zeta_2\varepsilon_2$ through the fetal $\gamma_2\beta_2$, where g can be either $^G\gamma$ or $^A\gamma$, to the adult $\alpha_2\beta_2$. The changes in hemoglobin expression are referred to as **globin switching** (**Figure 12.11**). Increased O_2 affinity and other properties of the embryonic and fetal hemoglobins permit the transfer of O_2 from the maternal to the embryonic or fetal circulatory system during prenatal development, a critical adaptation for gestating mammals (**Figure 12.12a**). The adaptation of the different globins to functioning in different environments by

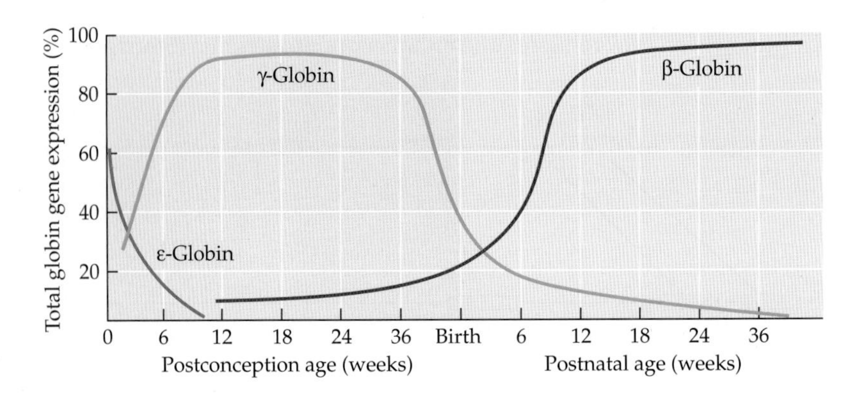

Figure 12.11 Relative expression of human β-family globin genes as a function of age from conception. The expression of each globin gene is shown as a percentage of total globin gene expression at each time point. The embryonic gene, ε-globin, is expressed during the first 6 weeks of gestation. The first switch (from ε- to γ-globin) occurs shortly after conception, and the second switch (from γ- to β-globin) occurs shortly after birth.

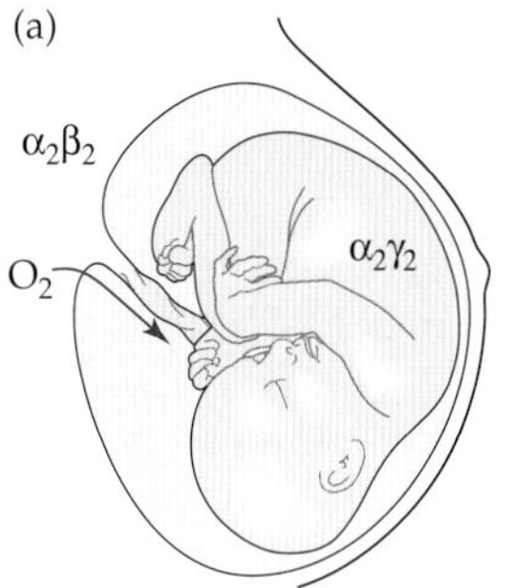

Figure 12.12 Regulation of β-globin genes by the locus control region (LCR). (a) During prenatal development, the fetus expresses γ-globin, instead of β-globin as in the mother. The $\alpha_2\beta_2$ hemoglobins of the mother have a lower affinity for O_2 than the fetal $\alpha_2\gamma_2$ hemoglobins, so O_2 is transferred from the maternal bloodstream to the fetal blood supply. (b) The human β-globin cluster includes five genes, ε, $^G\gamma$, $^A\gamma$, δ, and β. Expression of these genes is regulated by the LCR located 5′ to the ε-globin gene. (c) Expression of the β-globin gene is activated when transcription factors (colored ovals) form a complex that links the LCR to a promoter of a β-globin family member.

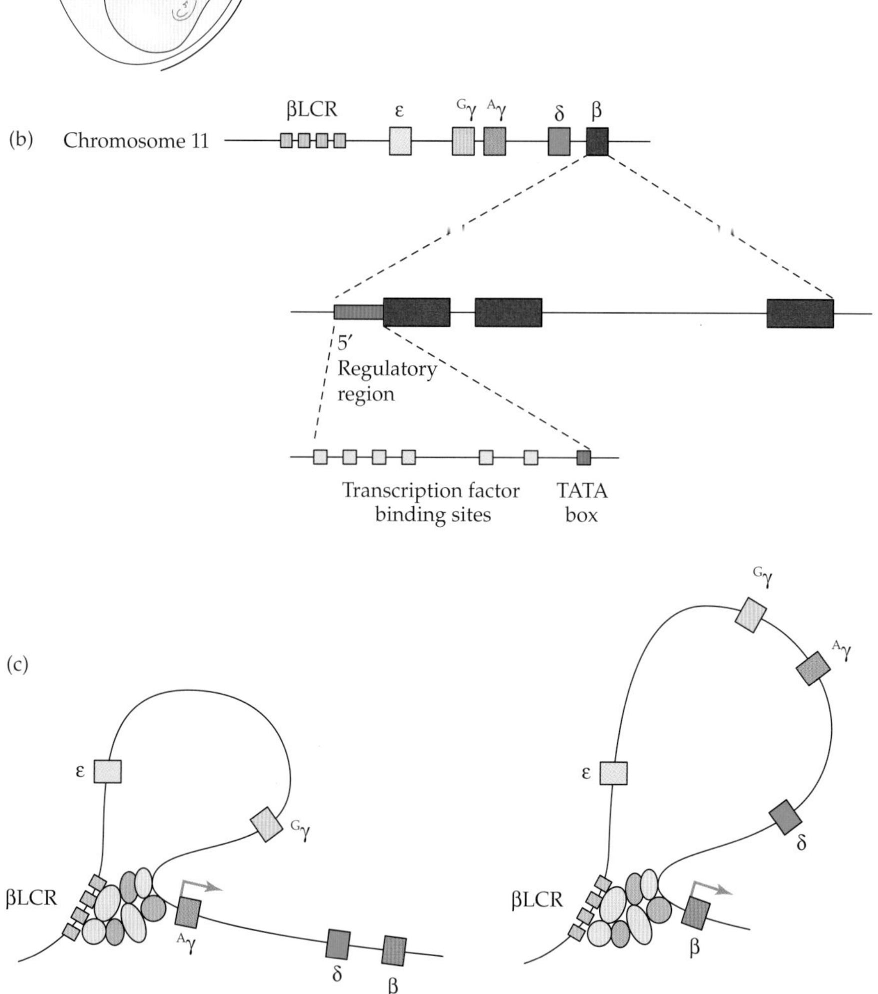

changes in their coding sequence does not explain, however, how the globin switching is done. These changes required changes in regulatory elements outside of the coding regions.

The β-globin subfamily genes are expressed primarily in distinct sets of tissues and at distinct phases of development. First, the ε-globin gene is expressed in primitive erythroid (blood-forming) cells that arise in the embryonic yolk sac. Later, the γ-globin genes are expressed in the initial population of definitive erythroid cells that emerge from the fetal liver. Eventually, the precursors of the definitive erythroid cells migrate to the bone marrow. The expression of β-globin in the definitive erythroid cells begins during fetal development and increases dramatically 6–12 weeks after birth, so $\alpha_2\beta_2$ becomes the primary hemoglobin for the remainder of the lifespan of the individual.

In the human genome, the β-globin subfamily is located on chromosome 11 (**Figure 12.12b**). The arrangement of the functional genes from 5′ to 3′ (ε, $^G\gamma$ $^A\gamma$, δ, and β) correlates with the expression order during development. The promoters for each gene contain a TATA box, as well as binding motifs for several different transcription factors. Although some transcription factors are common to all promoters, each promoter also contains binding sites for unique transcription factors as well (e.g., HoxB2 for $^{G}\gamma$- and $^{A}\gamma$-globin genes).

In addition to their own individual promoters, all the globin genes in the β family are under the control of a **locus control region** (**LCR**), located in a 20–25 Kb region approximately 10 Kb upstream from the ε-globin gene. The LCR consists of an AT-rich sequence that includes clusters of binding motifs for transcription factors and chromatin modifiers, such as histone-modifying enzymes. The LCR serves to maintain an open chromatin configuration throughout the region of the β-globin cluster in erythroid cells, where the β-globins are actively transcribed. In cells in which a certain globin gene is active, transcription factors that bind the promoter of this gene interact with one another as well as with chromatin modifiers that, in turn, will bind to sites in the LCR (**Figure 12.12c**). For example, specific combinations of transcription factors in fetal erythroid cells will establish physical interactions between the LCR and the promoter for $^{G}\gamma$-globin, resulting in a loop of chromatin at which the LCR and the specific globin promoter are linked. As the pattern of transcription factor expression changes during development, the link between the LCR and the $^{G}\gamma$-globin promoter is broken, a new chromatin loop is formed slightly 3′ to the previous one, and a new link forms between the LCR and the β-globin promoter. The LCR thus can be viewed as a specialized enhancer for appropriate tissue-specific expression. Its role in the regulation of globin genes is an example of how short-range (promoter) and long-range *cis*-regulatory elements interact via associated transcription factors to regulate gene expression.

The distance between a globin gene and the LCR was found to affect the developmental timing of expression, with globin genes close to the LCR being expressed during early developmental stages and more distant genes being expressed progressively later (Johnson et al. 2006). This hypothesis has been supported most elegantly by a natural "experiment" involving the insertion of a LINE element between ε and $^{G}\gamma$ genes that occurred in the lineage leading to Old World monkeys and apes (catarrhines). In New World monkeys (platyrrhines), which lack the LINE insertion, the proximity of $^{G}\gamma$ to ε and, by extension, to the LCR promotes a predominantly embryonic expression of $^{G}\gamma$, while the more distant $^{A}\gamma$ is mainly expressed in the fetus (**Figure 12.13**). In Old World monkeys and apes, the LINE element insertion between ε and $^{G}\gamma$ has weakened the ability of the LCR to promote embryonic expression of $^{G}\gamma$. Thus, in apes and other Old World monkeys, $^{G}\gamma$ and $^{A}\gamma$ are primarily expressed in the fetus. (In practice, this difference in expression between apes and New World monkeys makes little difference in the fetus, as $^{G}\gamma$ and $^{A}\gamma$ differ from each other at a single amino acid position out of 147 positions, with $^{G}\gamma$ and $^{A}\gamma$ having Gly and Ala, respectively, at this position.)

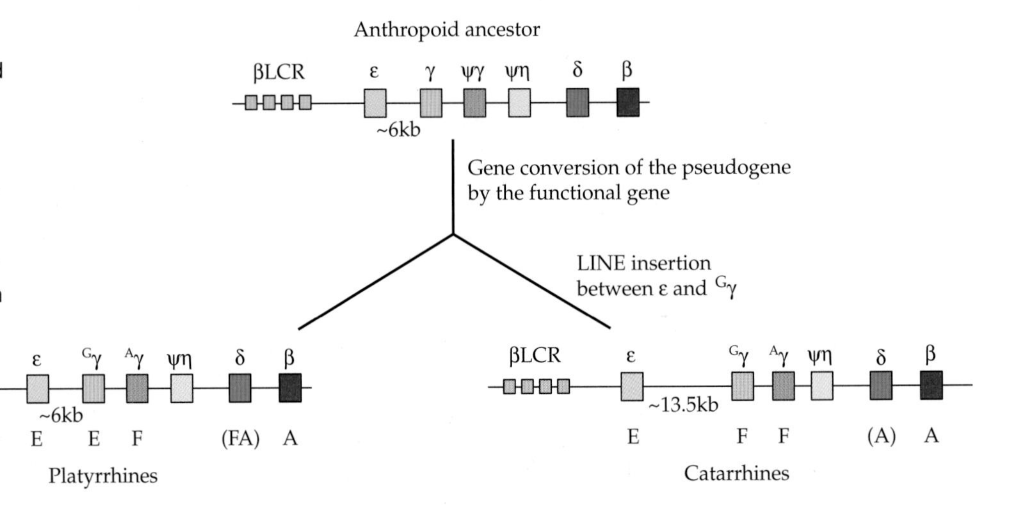

Figure 12.13 Evolution of the β-globin cluster in New World monkeys (platyrrhines) and Old World monkeys and apes (catarrhines). Phylogenetic reconstruction indicates that in the ancestor of New and Old World monkeys, one of the γ genes was silent in terms of expression, possibly a pseudogene (ψγ). Gene conversions of this pseudogene by the functional γ gene restored functionality to the pseudogene. The two γ genes have since diverged in sequence from one another and are currently referred to as $^{A}\gamma$ and $^{G}\gamma$. In platyrrhines, the proximity of $^{G}\gamma$ to ε and, by extension, to the LCR promotes a predominantly embryonic expression of $^{G}\gamma$; in this clade, $^{A}\gamma$ is the primary expressed gene in the fetus. In contrast, in catarrhines, a LINE element insertion between ε and $^{G}\gamma$ weakened the ability of the LCR to promote embryonic expression of $^{G}\gamma$ in humans and other catarrhines, $^{G}\gamma$ is the primary fetal gene. Interestingly, the developmental stage at which δ-globin is expressed has also been affected by the LINE insertion. Whereas in New World monkeys, there are trace levels of δ-globin expression in both fetus and adult, in Old World monkeys and apes, the increased distance from LCR causes δ-globin to be expressed in adults only. E, F, and A indicate the embryonic, fetal, and adult stages of gene expression, respectively. Stages shown in parentheses indicate trace levels of expression. (Modified from Johnson et al. 2006.)

Interestingly, the developmental stage at which δ-globin is expressed has also been affected by the LINE insertion. Whereas in New World monkeys δ-globin exhibits trace levels of expression in both fetus and adult, in Old World monkeys and apes the increased distance from the LCR restricts its expression predominantly to adults. We note that distance from the LCR may not be the sole explanation, as the LINE element may include binding sites for repressive elements.

Although the β-globin LCR is the best-studied example of a locus control region, LCRs have been identified for many other groups of genes (e.g., Kim et al. 2007; Zeng 2013).

Insulators

Insulators are DNA sequences that separate and define domains of "common" or shared regulation within a chromatin region. Insulators are thought to serve two primary functions: (1) they form a barrier inhabiting heterochromatin transcription, and (2) they block the effects of an enhancer on inappropriate promoters. The contemporary view of insulators is that they aid in the three-dimensional organization of the genome, mediating the separation of chromosomal domains from one another. In many cases they define domains of transcriptionally active versus transcriptionally repressed chromatin (**Figure 12.14**). In addition, they can act in pairs to facilitate long-range interactions between promoters, on the one hand, and enhancers or silencers, on the other (Chetverina et al. 2014).

Insulators act through the binding of insulator proteins to specific binding sites. For example, the nuclear CCCTC binding factor (CTCF) when bound to insulator sequence motifs prevents undesirable "cross talk" between active and inactive genomic regions. CTCF bound at one insulator site can form homodimers with CTCF bound at

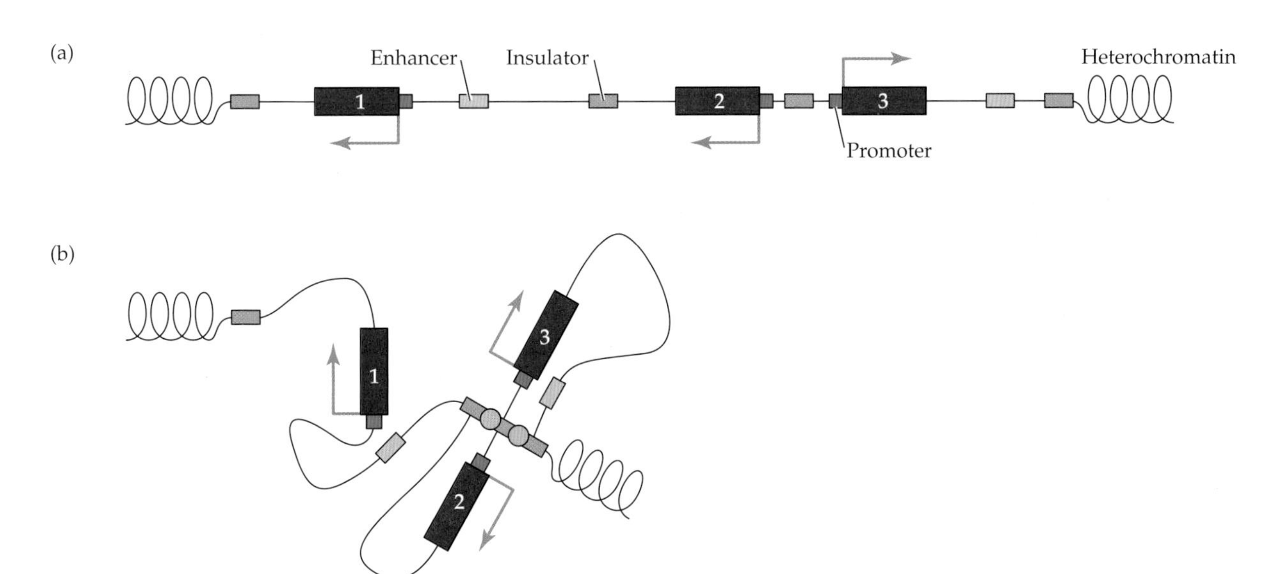

Figure 12.14 Schematic representation of insulator functions and locations. (a) An active euchromatic region is flanked by heterochromatic regions. Within this region there are three genes (1–3). Two of the four insulators (green boxes) flank the entire euchromatic region and separate it from the heterochromatic regions (spirals). These two insulators mediate border or barrier functions. Insulators also separate the genes from one another and are positioned where they can mediate enhancer-blocking functions. It is not known whether the two functions are established by similar or different mechanisms. (b) One aspect of insulator function is to organize chromatin looping by promoting contacts among insulators or with other genomic structures. Contacts among insulators are promoted by the nuclear protein CCCTC binding factors (CTCFs), shown here as orange circles. Depending on the linear and three-dimensional arrangements, looping may interfere with enhancer-promoter interactions (thus mediating the enhancer-blocking function of insulators), resulting in an inactive gene (gene 2 and its promoter), or it may assist in increasing enhancer-promoter contacts, resulting in an active gene (gene 3 and its promoter). Gene activation can also be achieved by direct enhancer-promoter interactions (gene 1 and its promoter) that can occur independently of the presence of an insulator. Note that insulators are also found between tandem promoters positioned in a head-to-head orientation (such as the case of genes 2 and 3) ensuring that both promoters can be regulated individually. (Modified from Herold et al. 2012.)

a distant site and, thus, pairs of insulators can establish loops of DNA along the chromosome. Current models suggest that interactions between enhancers and promoters may occur within this loop or between flanking regions but that an enhancer within one loop would be unable to affect a promoter outside of this loop. The CTCF system is the primary insulator in mammals; additional insulator systems, such as gypsy and scs, have been identified in *Drosophila*. Other insulators include binding sites for the conserved transcription factor TFIIIC (Noma et al. 2006), which shows insulator activity in yeast and mammals, and a *B1-SINE* retrotransposon that was co-opted as an insulator in mammals (Roman et al. 2011).

Insulators were first identified in *Drosophila*, and many of the clearest examples of insulator activity have emerged from studies of long-range gene regulation in *Drosophila*. In contrast, the number of experimentally validated examples of insulators with either enhancer-blocking or heterochromatin barrier functions is much smaller for vertebrates. One explanation for this difference is that *Drosophila* genomes are far more compact and far more gene-dense than vertebrate genomes—gene densities in human and *D. melanogaster* are 1 gene per ~200,000 bp and 1 gene per ~10,000 bp, respectively. Selection for regulatory components that establish boundaries that limit the effect of an enhancer may, thus, be stronger when genes are on average closer to their neighbors.

CTCF proteins evolve extremely slowly—homologous CTCF proteins from *D. melanogaster* and *D. simulans* are 99% identical. Similarly, CTCF binding sites are extremely conserved between the two *Drosophila* species. On the other hand, 89 new binding sites were gained on the *D. melanogaster* branch since its divergence from *D. simulans* ~2.5 million years ago (Ni et al. 2012). This leads to an estimate of ~36 gains of binding sites per million years. This rate is much higher than the rate of gene gain (~16 genes per million years). Interestingly, no binding site has been lost in *D. melanogaster* since its divergence from *D. simulans*. This asymmetric pattern of CTCF-binding-site gains has also been observed along the *D. simulans* branch (85 gains versus 5 losses). By using a test of neutrality based on genetic polymorphism versus fixed genetic differences (Chapter 2) between *D. melanogaster* and *D. simulans*, Ni et al. (2012) found a 20% deficit in polymorphic sites and a 12% excess in fixed divergent sites in newly created binding sites in comparison to the neutral expectation. These findings suggest that a fraction of the new CTCF binding sites is maintained by positive selection. In terms of gene expression, 7% of genes with conserved CTCF binding sites exhibited divergent expression. In comparison, 11% of all divergent CTCF binding sites and 17% of all newly created CTCF binding sites exhibited divergent expression. These numbers indicate that much of the evolution of insulator function has been driven primarily by large-scale changes that alter regulatory relationships across the genome, rather than by insertions and deletions of insulators. One must, of course, remember that the mere binding in vitro of CTCF to a binding site does not constitute a selected-effect or proper biological function (Chapter 11).

Posttranscriptional Regulation

After a pre-mRNA molecule is synthesized, it undergoes several posttranscriptional processes, which can include capping, splicing, polyadenylation, and the posttranscriptional modification of nucleotides (Chapter 1). These processes can, in principle, be regulated. In addition, mRNA stability or longevity may be regulated, or the transcript may be silenced as far as translation is concerned. Regulation can also occur after the translation of the mRNA. This can be done through such processes as modification, targeting, and degradation of the protein.

The study of posttranslational control is quite advanced as far as the mechanistic aspects are concerned, although little is known about the evolution of many of these processes. Here, we will only discuss the evolution of translational control by RNA interference through the binding of microRNAs (miRNAs) to the 3′ untranslated region (UTR) of mRNA.

RNA interference

Some RNA-specifying genes and their products were described in Chapter 1. For the sake of clarity and internal consistency, however, a few RNA-related terms and processes are redefined here. RNA interference (RNAi) is a process by which a small RNA molecule inhibits gene expression by binding to specific mRNA molecules. Two types of RNAs—miRNAs and small interfering RNAs (siRNAs)—are involved in RNA interference. A third type, the piRNAs, use a related mechanism to mediate the repression of transposable element mobilization in germline cells. The miRNAs and piRNAs are transcribed from their own distinct RNA-specifying genes; in contrast, there are no dedicated genes for siRNAs. Instead, siRNAs are either generated by degradation of viral double-stranded RNAs (dsRNAs) or transcribed from repeated elements within the genome.

The basic components of the RNAi machinery—Argonaute-Piwi, a Dicer-like protein consisting of two domains (RNase III and helicase), and an RNA-dependent RNA polymerase—are thought to have existed in the common ancestor of all extant eukaryotes (Shabalina and Koonin 2008). Subsequent gene duplications and sequence divergence, particularly within the Argonaute-Piwi family, have led to the establishment of distinct biochemical pathways for the action of miRNAs, siRNAs, and piRNAs.

The synthesis of miRNAs as it occurs in eumetazoans, i.e., all animals except those in Placozoa and Porifera (sponges), was presented in Chapter 1. The rules of miRNA targeting and complementarity between the mature miRNA and the target on the 3′ UTR of an mRNA molecule are illustrated in **Figure 12.15**. As a consequence of the binding between miRNA and the target sequence, translation of the mRNA is inhibited via disruption of the translation machinery, and the mRNA may be subsequently degraded. Estimates in the literature suggest that 30–60% of all human protein-coding genes are regulated by miRNAs (Lewis et al. 2005; Friedman et al. 2009). We note, however, that translational control can occur by mechanisms that do not involve miRNA, albeit less commonly. A 3′ UTR can interact with RNA-binding proteins that can inhibit translation or localize the mRNA to specific regions within the cell (e.g., the dendrites of a neuron). Translational control mediated by RNA-binding proteins is essential for embryonic development (Lasko 2009) and neuronal activity (Darnell

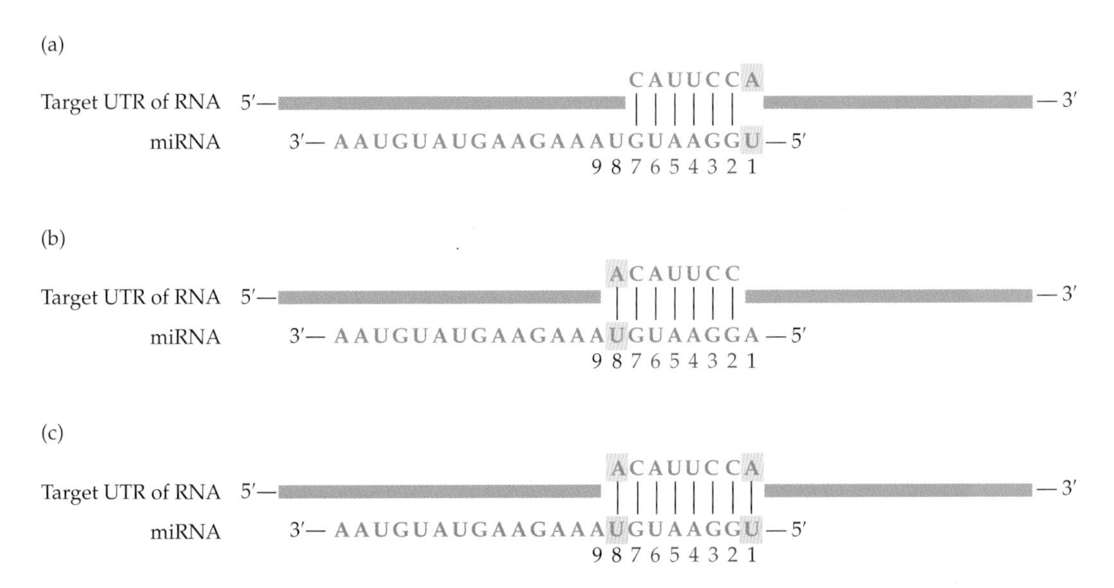

Figure 12.15 The rules of miRNA targeting require perfect complementarity between the target 3′ UTR of an RNA molecule and positions 2–7 of the mature miRNA. In addition, targeting requires (a) a paired adenine at position 1 of the target UTR, (b) a pair at position 8, or (c) both.

2013). Moreover, many RNA-binding proteins of unknown function and specificity are expressed specifically in the brain (Darnell 2013), and many genes expressed specifically in neural tissue have remarkably long 3′ UTRs (Wang and Yi 2013), suggesting that they may be under translational control.

Patterns of evolution of miRNAs

As mentioned in Chapter 4, 3′ UTRs evolve approximately twice as fast as protein-coding regions, but at only half the rate at which pseudogenes evolve. Thus, 3′ UTRs are functionally constrained. They may be constrained by multiple factors, including selection, to maintain certain secondary RNA structures within the 3′ UTR, as well as selection to maintain regulation by miRNAs or RNA-binding proteins. Indeed, many target sequences of miRNAs exhibit a high degree of conservation (Lewis et al. 2005).

Because miRNA sequences are highly conserved during evolution, it is reasonable to assume that the way they regulate gene expression might be conserved as well. A comparison of intergenic microRNA-specifying gene regions among nine vertebrate species revealed that the promoters of microRNA genes are roughly twice as conserved as those of protein-coding genes (Mahony et al. 2007). This finding indicates that there are significant selective constraints arising from the regulation of microRNA expression.

How do new miRNA-specifying genes emerge? As is typical of genes, a major source of novel miRNA genes is duplication. The miRNA genes that have significant sequence similarity to one another and identical or near identical seed regions in their mature miRNA sequences are grouped into miRNA families. For example, of the 1,424 human miRNA genes listed in a miRNA repository, one-third (456) can be grouped into 141 families of two or more miRNA genes. Similarly, 38% of miRNA genes in mice, 10% in *C. elegans*, and 13% in *D. melanogaster* belong to gene families. Gene duplication events, followed by subfunctionalization and neofunctionalization processes (Chapter 7), are considered major sources for emergence of novel miRNA specificities.

Other modes of miRNA origination are known. The central requirement for an miRNA gene is a correct secondary structure of its RNA product, as only hairpins that are consistently recognized and processed by Dicer can result in a functional miRNA. Thus, the prerequisite for the emergence of a novel miRNA is a transcribed genomic locus that can be evolved to produce an RNA fold that is recognizable by the miRNA processing machinery. Because RNAs easily form nonperfect foldback structures, the evolution of a new miRNA gene appears to be more likely than the emergence of a novel protein-coding gene. Indeed, there is evidence for miRNA-specifying genes that evolved from introns, exons of protein-coding genes, pseudogenes, small nucleolar RNAs (snoRNAs), tRNAs, and transposable elements, as well as intergenic regions that produce RNA as part of the transcriptional noise.

Of particular interest are miRNAs derived from antisense miRNA transcripts. Because the DNA sequence specifying the pre-miRNA (the stem-loop or hairpin structure) is essentially an imperfect palindrome, antisense transcription of miRNA loci often results in transcripts that fold into hairpins and can be processed by the miRNA machinery. As mature miRNAs and miRNA* sequences are often not fully complementary, antisense transcription of miRNA loci can result in miRNAs with novel miRNA and miRNA* molecules. Moreover, perfect complementarity between the miRNA and the mRNA 3′ UTR is not required for translational inhibition to occur, increasing the likelihood of functional interactions between a novel miRNA and an mRNA.

Because only short sequences are required for target recognition, a newly emerging miRNA is initially likely to target many genes simply by chance. Such targeting will mostly have deleterious consequences. To explain how novel miRNAs can evolve in this context, Chen and Rajewsky (2007) put forward a model of transcriptional control of new miRNAs. According to this model, novel miRNAs are initially expressed at low levels and in specific tissues to limit deleterious effects of accidental targeting. Gradually, deleterious targets are purged from the transcriptome by natural selection.

The fact that less-conserved miRNA genes are indeed generally expressed at lower levels compared with broadly conserved miRNAs is considered supportive of this model. One of the implications of this model is that many miRNAs that show low levels of conservation and low levels of expression might actually have no function, as the population of such miRNAs is still in the "purging" phase of evolution, following which only a small minority of the novel genes will be retained.

After an miRNA gene has emerged and acquired a regulatory function, can it evolve further? Comparative genomics analyses of small RNA data sets from a large array of species have revealed a number of ways in which an existing miRNA gene can be further diversified. Diversification can occur not only through changes in the mature sequence region, but also through changes elsewhere, which may lead to changes in the hairpin structure and, as a consequence, to changes in Drosha and Dicer recognition and to the loading of miRNA into the miRNA-induced silencing complex (miRISC).

Do miRNAs have a deep evolutionary history?

As stated previously, the protein components of the machinery for RNA interference are well conserved among eukaryotes and may, thus, be assumed to be symplesiomorphic for all extant eukaryotes. On the other hand, there are no miRNA families in common between plants and animals despite the fact that the origins of some miRNA families in plants can be traced to the Embryophyta (green plants), which may have originated 540 million years ago (Taylor et al. 2014), while some miRNA families in animals can be traced to the Eumetazoa (a clade comprising all major animal groups except sponges and placozoans), which may have evolved approximately 550 million years ago (Berezikov 2011). The discovery of miRNAs in protists has raised the possibility that miRNAs may be as ancient as the proteins of the RNAi machinery. That is, miRNAs may have originated in the common ancestor of all extant eukaryotes. A detailed study of the putative miRNAs in protists, however, revealed that most of the newly discovered miRNAs merely constitute errors of sequence annotation (Tarver et al. 2012). At present, only five groups of eukaryotes are known to possess miRNAs, and none of the miRNAs from any one of these five groups has homologs in the other four groups. It, therefore, appears that miRNAs have originated at least five times independently, in green plants, eumetazoans, demosponges, the green alga *Chlamydomonas*, and the filamentous brown alga *Ectocarpus*.

Does translational regulation contribute to phenotypic evolution?

Our understanding of how evolutionary divergence in translational control contributes to phenotypic evolution is still in its infancy. One intriguing example has emerged from studies of sequence divergence in the 3′ UTRs of Lake Malawi cichlids (Loh et al. 2010). There are currently more than 500 extant cichlid species in Lake Malawi. As the lake dried out almost completely at the beginning of the Pleistocene (Delvaux 1995), all these hundreds of cichlid species with their myriad sizes, shapes, colors, and behaviors must have evolved from a common ancestor extremely rapidly in the last million years. Interestingly, while Lake Malawi cichlids are remarkably diverse in morphology (**Figure 12.16**), their genomes are extremely similar to one another, to the extent that ancestral polymorphisms are abundant (Kocher 2004; Streelman et al. 2007). It has been noted, for instance, that the nucleotide diversity among all fish species in Lake Malawi is smaller than that observed among laboratory strains of zebrafish. Loh et al. (2008, 2010) hypothesized that divergence of miRNAs or their target sequences might be one of the genomic mechanisms contributing to the rapid phenotypic evolution observed in Lake Malawi cichlids. To this end, they analyzed genomic sequences and single-nucleotide polymorphisms (SNPs) from eight species and identified putative cichlid miRNAs and their putative target sequences in 3′ UTRs. They discovered that cichlid mature miRNAs are strongly conserved in sequence. On the other hand, miRNA target sites exhibited greater SNP densities (0.44%) than flanking regions on 3′ UTRs (0.28%).

Figure 12.16 An illustration of the immense morphological divergence in Lake Malawi cichlids that has evolved in merely one million years. (Upper photographs © David McIntyre. Bottom photograph © Nantawat Chotsuwan/Shutterstock.)

The cichlids of Lake Malawi can be roughly divided by habitat into mbuna (rock-dwelling) species, non-mbuna species (living on sandy or soft substrates), and deep-water species. Loh et al. (2010) paid particular attention to genes whose 3′ UTRs exhibited SNPs coinciding with these three ecological divisions, and they discovered that many of the genes whose 3′ UTRs were differentiated along habitat lines were biologically and morphologically relevant. One gene whose 3′ UTR differed between mbuna and the other cichlids was *odd-skipped related 2* (*osr2*), which encodes a protein involved in the development of the palate and the teeth. Indeed, before the advent of molecular phylogeny, tooth morphology was used in the taxonomic diagnosis of mbuna genera (Genner and Turner 2005).

We note, however, that difference in miRNA targets cannot explain the bulk of the variation in cichlids, although genetic alterations in microRNA-dependent regulation may have contributed to a certain extent to the speciation and rapid evolution of cichlids.

CHAPTER

13

Experimental Molecular Evolution

by Tim F. Cooper

In the fourth edition of *Origin of Species*, Charles Darwin (1866) acknowledged the desirability of comparing extant populations to a linear series of their ancestors, contrasting the limitations of the widely used comparative methods with the then hypothetical possibility of directly observing the evolutionary process:

> *In searching for the gradations through which any organ in any species has been perfected, we ought to look exclusively to its lineal progenitors; but this is scarcely ever possible, and we are forced in each case to look to other species and genera of the same group, that is to the collateral descendants from the same original parent-form, in order to see what gradations are possible, and for the chance of some gradations having been transmitted from the earlier stages of descent, in an unaltered or little altered condition.*

Unfortunately, putting this idea of Darwin's into regular practice took more than a hundred years. The means by which we can now "look exclusively" at "lineal progenitors" rather than at "collateral descendants" is called experimental evolution. The field of experimental evolution is concerned with testing evolutionary hypotheses by the use of controlled experiments. Evolution may be observed in the laboratory as populations adapt to new environmental conditions or experience changes in allele frequencies, by stochastic processes. With modern genomic tools, it is sometimes possible to pinpoint the mutations responsible for certain adaptations or to find out how exactly mutations interact with one another. Because evolutionary processes usually require a large number of generations, evolutionary experiments are typically carried out with microorganisms such as bacteria, yeasts, or viruses. The experimental setup allows researchers to test evolutionary hypotheses prospectively, as opposed to the usual observational setup that only allows retrospective testing of hypotheses.

This chapter introduces the methodology of experimental evolution and its uses for testing evolutionary hypotheses, and it presents some new insights into molecular evolution derived from such experiments. We pay particular attention to results derived from whole-genome sequencing of evolved populations and to the contribution of experimental evolution to our knowledge of mutations.

What Is Experimental Evolution?

Experimental evolution is the study of evolutionary processes occurring in laboratory populations that are propagated and evolved in environments that are imposed and controlled to a large extent by the researcher (Kawecki et al. 2012). There are two key ways in which experimental evolution studies differ from traditional evolutionary studies. First, in traditional studies, evolutionary questions are tested **retrospectively**. That is, biological phenomena that are observed in extant populations are tested for their compatibility with certain predictions concerning past events. For example, the observed degree of genetic polymorphism in a population can be used to test the relative importance of purifying selection versus random genetic drift in the evolutionary history of this population. By contrast, experimental evolution allows us to answer evolutionary questions by testing them **prospectively**. That is, genotypes and environments are chosen and controlled so that outcomes of particular experimenter-determined treatments can be directly compared with predictions from the particular hypotheses under consideration.

The second key way in which experimental evolution studies differ from traditional studies concerns the possibility of directly observing changes within populations in a manner similar to that envisioned by Darwin. It frees researchers from the difficult and potentially error-prone task of inferring past events from comparisons of contemporary populations (i.e., Darwin's "collateral descendants"), and it greatly reduces complicating factors (e.g., changes in population size and environmental conditions) that affect the interpretation of evolutionary outcomes. In other words, experimenters can maintain the **living fossil record** of "lineal progenitors" that gave rise to an extant population. Such a record provides the means of tracking genetic and phenotypic changes within and between evolving populations in the context of known environmental conditions. Knowledge of these changes allows researchers to test predictions—for example, concerning the regularity of mutation accumulation—as well as to characterize the basis of changes in fitness—for example, by identifying the genetic changes underlying the acquisition of a new phenotype.

Experimental evolution as a research approach has been used with many experimental systems, from guppies and *Drosophila*, through algae, yeast, bacteria, and viruses, to in vitro evolved proteins and RNA molecules. The majority of experimental evolution studies, however, have used microorganisms, which commonly (1) have very short generation times, (2) can be propagated under controlled conditions for thousands of generations, (3) can attain very large population sizes, (4) possess small, simple genomes that are readily sequenced, and (5) can be cryopreserved, thus providing researchers with "revivable" living fossils that can be used to investigate temporal changes and differences among replicates. In large populations, the number of mutations arising in each generation can be very large, selection can be extremely effective and, hence, adaptive evolution can occur very rapidly. Conversely, experimental populations can be subjected to extreme bottlenecks, in which the effective population size is drastically reduced, thereby allowing evolution to proceed mostly through random genetic drift.

Experimental evolution represents a powerful approach for examining causal relationships between genotype and phenotype, and for relating evolved patterns of genetic and phenotypic variation to controlled differences in the selective history of populations. Experimental evolution studies complement comparative approaches by allowing evolutionary dynamics to be monitored in real time, facilitating controlled tests to examine the effects of environmental and genetic conditions on evolutionary processes and outcomes (Elena and Lenski 2003; Buckling et al. 2009; Kawecki et al. 2012). Experimental evolution using microbial systems has led to novel insights into previously difficult-to-study areas, such as the genetic basis of adaptation and the importance of historical contingency in evolution. Moreover, as techniques capable of providing whole-genome sequences and global molecular-level phenotypes of evolving populations become available, new research avenues open up. For example, by using gene expression profiles and knowledge of regulatory networks, researchers can determine the manner in which specific genomic changes remodel the patterns of interaction among genes and regulatory regions during adaptation.

Perhaps the best known and—at least in terms of the number of generations involved—longest running experimental evolutionary study was begun by Richard Lenski in 1988. This study consists of 12 replicate populations established from a single strain of the bacterium *Escherichia coli* and propagated in a constant environment. The experiment was initially conceived to examine the dynamics of adaptation and divergence of evolutionary trajectories across replicate populations (Lenski et al. 1991), but the availability of ancestral and intermediate population samples in conjunction with next-generation sequencing has allowed many other questions to be addressed.

Finally, we note that experimental evolution is not the same thing as artificial selection. In **artificial selection**, the fitness of an individual is decided by the researcher based on some predetermined phenotype or genetic character. In experimental evolution, selection acts on any trait that increases the reproductive success of a genotype without the researcher having a say on which trait it is. Indeed, a striking feature of many experimentally evolved populations is that the underlying genetic basis of increased fitness often comes as a surprise. In many cases, it is difficult even in retrospect to explain why a particular mutation confers a benefit.

The basic design of evolutionary experiments

The simplest evolutionary experiments start with a single **founding population** of organisms (**Figure 13.1**). This founding population can be genetically heterogeneous or, more usually, grown from a single clone and, therefore, genetically homogeneous. The founding population is, then, used to establish **replicated populations**, each of which is propagated by regular transfer of a **subsample** to fresh media or by continuous growth in a chemostat vessel. Factors such as environmental variation, recombination, and migration may be included in the experimental design. Random mutations that occur during population growth will lead to divergence between initially identical replicate lines, but as long as a subset of these mutations are beneficial and the population size is sufficiently large for positive Darwinian selection to be effective, adaptations to the experimental environment will occur in the evolved population.

Samples for storage are taken from the ancestral founding population, as well as at regular time intervals from the evolved populations. These **stored samples** are

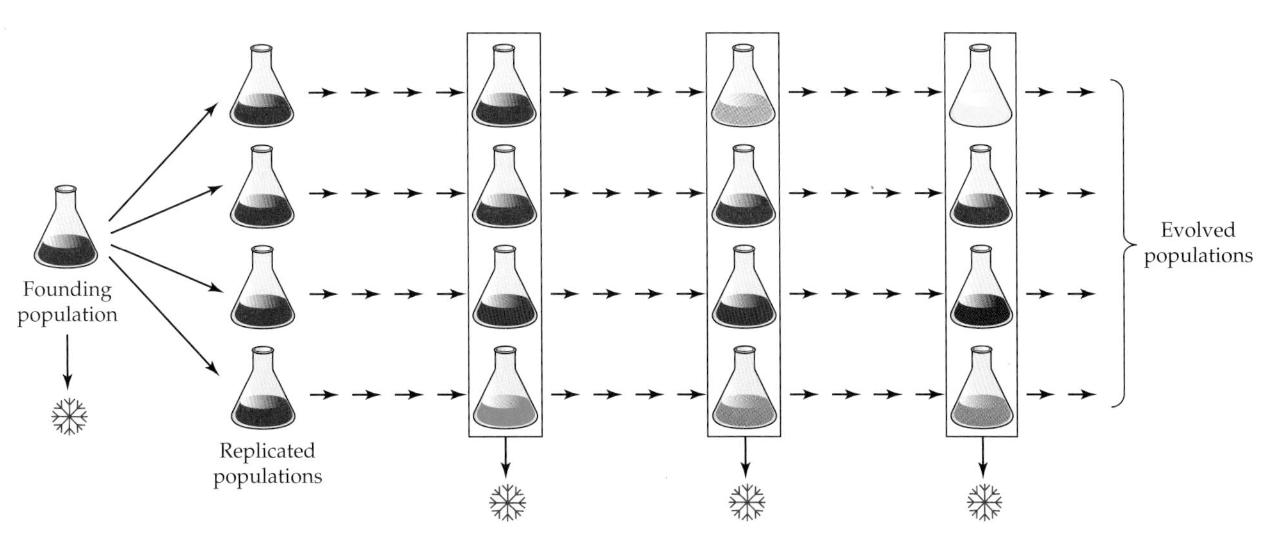

Figure 13.1 Schematic design of an evolutionary experiment. A founding population is used to establish several replicate populations that are propagated in an experimentally determined environment. Samples from the founding population are cryopreserved (snowflake icon) for subsequent study. Periodically, samples from each of the evolved populations are cryopreserved (boxes). The different colors indicate changes in the genetic makeup of the evolved populations relative to the founding population.

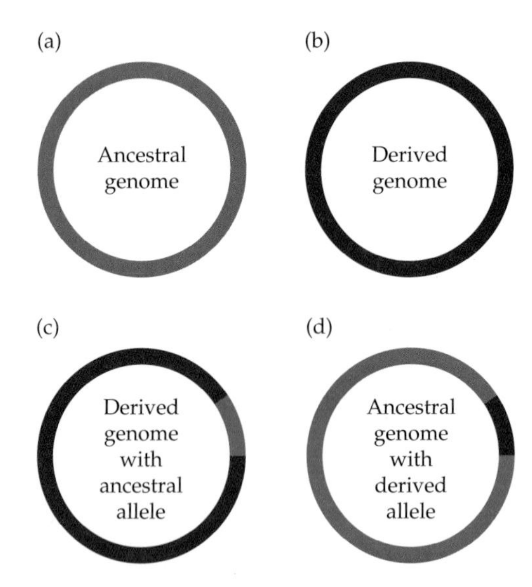

Figure 13.2 Allele replacements made feasible by experimental evolutionary setups. (a) The ancestral genome from the founding population. (b) A derived genome from an evolved population. (c) An ancestral allele replacing an evolved allele in an evolved genetic background. (d) A derived allele from an evolved genome replacing an ancestral allele in the genetic background of the founding genome. By comparing (c) and (d), the effects of particular mutations can be analyzed to determine their effects in different genetic backgrounds.

maintained (usually through **cryopreservation**) in a state that retains the viability of the organisms but halts their subsequent reproduction and evolution. From these stored samples, subsequent analyses can be performed to track the temporal dynamics of phenotypic and genotypic changes in the evolved populations.

The genomes of individuals from the ancestral and the stored evolved populations can be sequenced to allow identification of point mutations and other genotypic changes that have occurred at different stages of the experiment. These changes can be screened in the final population and in stored-population samples to determine evolutionary paths of mutation accumulation. **Allele replacement experiments** can then be used to add evolved and corresponding ancestral sequence variants into ancestral and evolved strains, respectively (**Figure 13.2**). Allele replacement experiments allow the effects of particular mutations to be analyzed independently of effects due to general changes in the genetic background. In addition, the relative fitness of ancestral and evolved individuals can be assessed.

A typical evolutionary experiment allows observation of the evolutionary process on a time scale that is much shorter than "real world" examples but that is much longer than most other biological experiments. Indeed, depending on the system, the environment, and the questions of interest, experiments lasting months and even years are common. This length of time makes it imperative that experiments be designed carefully to properly address the immediate question of interest, as well as to serve as a useful resource for addressing future questions. For example, a key advantage of experimental evolution is that replicated evolutionary outcomes can be compared, providing the ability to statistically test for patterns of evolutionary change or the effect of some genetic or environmental treatment. Of course, the higher the number of experimental replications, the smaller will be the magnitude of the effect that is detectable. Different types of questions can dictate different considerations. For example, fewer replications might suffice to test for a **main effect**, i.e., are two treatments different from one another? By contrast, more replications will be required for testing for an **interaction**, i.e., do some evolutionary responses differ depending on the combination of two factors? Even higher numbers of replications may be required to test hypotheses that address the distribution of effects across evolutionary outcomes.

How to measure fitness and changes in fitness in evolutionary experiments

The most important phenotypic variable that can be inferred from an evolutionary experiment is the mean fitness of a population. The ability to infer mean fitness allows scientists to document the temporal dynamics of change in the mean fitness of evolved populations relative to the founding population. As mentioned in Chapter 2, it is usually extremely difficult to infer fitness values for natural populations, in which confounding factors such as overlapping generations times, sexual reproduction, and suboptimal growth due to niche capacity limitations conspire to yield unreliable estimates. In microbial populations in the laboratory, in which confounding factors can be controlled, a common measure of fitness is the **Malthusian growth parameter**, also referred to as the **Malthusian parameter** or the **growth rate**.

Mathematically, the growth of a population can be described as a branching process in which the expected number of offspring of each individual over its entire reproductive life is greater than 1. The simplest population growth model was first proposed by Thomas Malthus in 1798. This model reflects exponential growth of a population and can be described by the following differential equation:

$$\frac{dN}{dT} = \alpha N \quad \textbf{(13.1)}$$

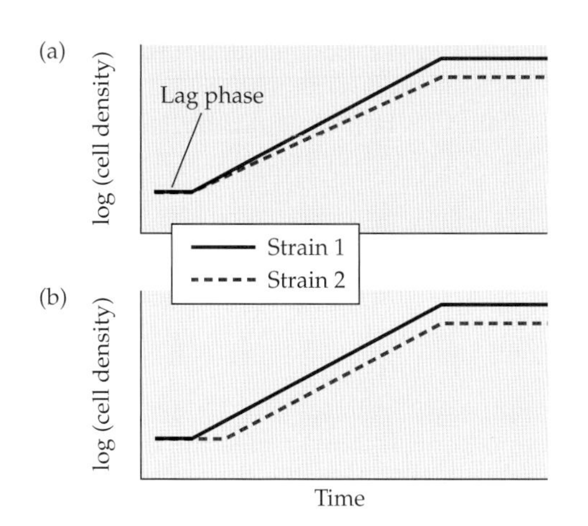

Figure 13.3 Schematic representation of the dynamics of competing strains during competition in a simple batch culture. A common measure of estimating fitness is the ratio of Malthusian parameters, assessed as the log change in cell density of each strain over the course of the competition. Different fitness values of competing strains may result from any combination of differences in the length of the lag phase, the growth rate, or differences in death rate following exhaustion of a limiting nutrient. The two panels show the growth dynamics of two strains with a difference in fitness of ~10%. (a) The difference in fitness is caused by the faster growth of one strain. (b) The difference in fitness is the result of one strain taking a longer time to exit the lag phase and begin exponential growth.

where *N* is the population size, *dN*/*dT* is the increase in population size per unit time, and α is the Malthusian parameter. The solution of this equation is the exponential function

$$N_t = N_0 e^{\alpha t} \quad (13.2)$$

where N_0 is the initial population size and N_t is the population size at time *t*. This model adequately describes the initial phase of growth when the population size is far from the limits imposed by the capacity of the niche.

The Malthusian model fits microbial laboratory populations, provided steps are taken to minimize competition for space, nutrients, and other resources that are required for growth. In practice, researchers usually measure the average growth rate over the entire growth cycle. This **realized Malthusian parameter**, *m*, can be calculated from the change in the size of a population over a period of time, *t*, as

$$m = \frac{\ln(N_t/N_0)}{t} \quad (13.3)$$

The Malthusian parameter or its proxies are usually estimated in the selective environment and presented relative to the fitness of the founding population to facilitate comparisons across experiments that may involve different species, or the same species in different environments. Several methods are used to measure the relative fitness change of evolved populations and to estimate the fitness effect of specific mutations. A common and intuitive approach is to define fitness as the ratio of the realized Malthusian growth parameter of a **test population** to that of a **reference population** (**Figure 13.3**). This measure of the relative fitness of two populations takes into account differences in all aspects of the growth cycle, including the **lag phase**, the **log phase**, the **stationary phase**, and the **death phase** (Lenski et al. 1991).

The Contribution of Experimental Evolution to Evolutionary Biology

Despite being a relatively new field, experimental evolution has contributed significantly to our understanding of molecular evolution in four main ways. First, and perhaps most excitingly, experimental evolution has been used to prospectively test evolutionary predictions. Examples include testing predictions regarding the adaptive divergence of populations (Friesen et al. 2004; Kassen et al. 2004; Spencer et al. 2007), the benefits and effects of recombination (Cooper 2007; Raynes et al. 2011), and the optimization of phenotypes to defined environments (Ibarra et al. 2002; Dekel and Alon 2005).

Second, experiments can be designed to estimate key evolutionary parameters. Examples include estimates of neutral, deleterious, and advantageous mutation rates (Lenski et al. 1991; Kibota and Lynch 1996; Hegreness et al. 2006; Perfeito et al. 2007;

Barrick et al. 2010; Wielgoss et al. 2011; Zhang et al. 2012; Long et al. 2013); the distribution of mutational effects (Rokyta et al. 2005; Hegreness et al. 2006; Kassen and Bataillon 2006; Perfeito et al. 2007; Schoustra et al. 2009; Barrick et al. 2010; McDonald et al. 2011; Lalić et al. 2011; Sousa et al. 2012); and the distribution of mutation types (Gruber et al. 2012; Lee et al. 2012).

Third, analyses of genetic and phenotypic changes occurring in populations that had evolved under defined conditions have provided insight into the plasticity of molecular evolutionary parameters, including rates of mutation (Sniegowski et al. 1997; McDonald et al. 2012), mutational robustness (Sanjuán et al. 2007), and epistasis (Chou et al. 2011; Khan et al. 2011).

Fourth, the ability to retain and study evolutionary histories directly has allowed scientists to identify the genetic basis of adaptations and determine the dynamics and repeatability of adaptive mutations occurring in evolving populations. Examples include the clocklike accumulation of beneficial mutations (Barrick et al. 2009) and the genetic basis of phenotypic innovations (Blount et al. 2008, 2012; Beaumont et al. 2009; Meyer et al. 2012).

In the following sections, we present a few evolutionary topics to which experimental evolution has made important contributions.

Population divergence and the adaptive landscape metaphor

A central focus of experimental evolution studies has been to examine the dynamics of genetic divergence within and between replicate populations (Lenski et al. 1991; Bull et al. 1997; Rozen et al. 2008; McDonald et al. 2009; Meyer et al. 2012; Tenaillon et al. 2012; Lindsey et al. 2013). Genetic divergence between populations can be driven either by selective processes, such as adaptation to different environments, or by chance events, such as the order in which mutations occur. In the presence of epistatic interactions between mutations, the order in which mutations arise can constrain subsequent evolution and promote divergence even among populations evolving in the same environment. (For a quantitative definition of epistasis, see Chapter 2.)

A useful, albeit controversial, framework for thinking about the process of population dynamics and divergence is Sewall Wright's **fitness landscape** or **adaptive landscape** (Wright 1932; Arnold et al. 2001; Kaplan 2008). The adaptive landscape is a graphic metaphor for visualizing the relationship between genotype and reproductive success. In an individual-based adaptive landscape, it is assumed that every genotype has a well-defined fitness. This fitness is represented by a certain "altitude" or "height" in the landscape. Similar genotypes are envisioned as spatially close to one another (longitudinally or latitudinally), while those that are different are placed far from one another. To put it another way, an adaptive landscape attempts to plot the multidimensional relationships among the set of all possible genotypes, the degree of genomic similarity among all possible genotypes, and their associated fitness values by using visually comprehensible two- or three-dimensional plots. As far as constant environments are concerned, the adaptive landscape is assumed to be static, i.e., at any given time the landscape will be the same as at any other time. For changing environments, dynamic adaptive landscapes are appropriate (Svensson and Calsbeek 2012).

Adaptive landscapes can represent single loci, interactions among loci, or more abstractly an entity called **genotypic space** (i.e., all the possible genotypes) by essentially reducing multidimensional spaces to two or three dimensions. Several methods of visualizing adaptive landscapes are presented in **Figure 13.4**.

The movement of a population on an adaptive landscape from a place of lower fitness to a place of higher fitness is mostly driven by natural selection. The movement of a population from one place on the adaptive landscape to another that is either equal or lower in fitness can only be driven by random genetic drift (Figure 13.4a). In a simple adaptive landscape that has only one fitness peak (Figures 13.4a,c,e), natural selection will cause all replicate populations of sufficient effective population size to converge on the same adaptive solution. However, if mutations interact with one

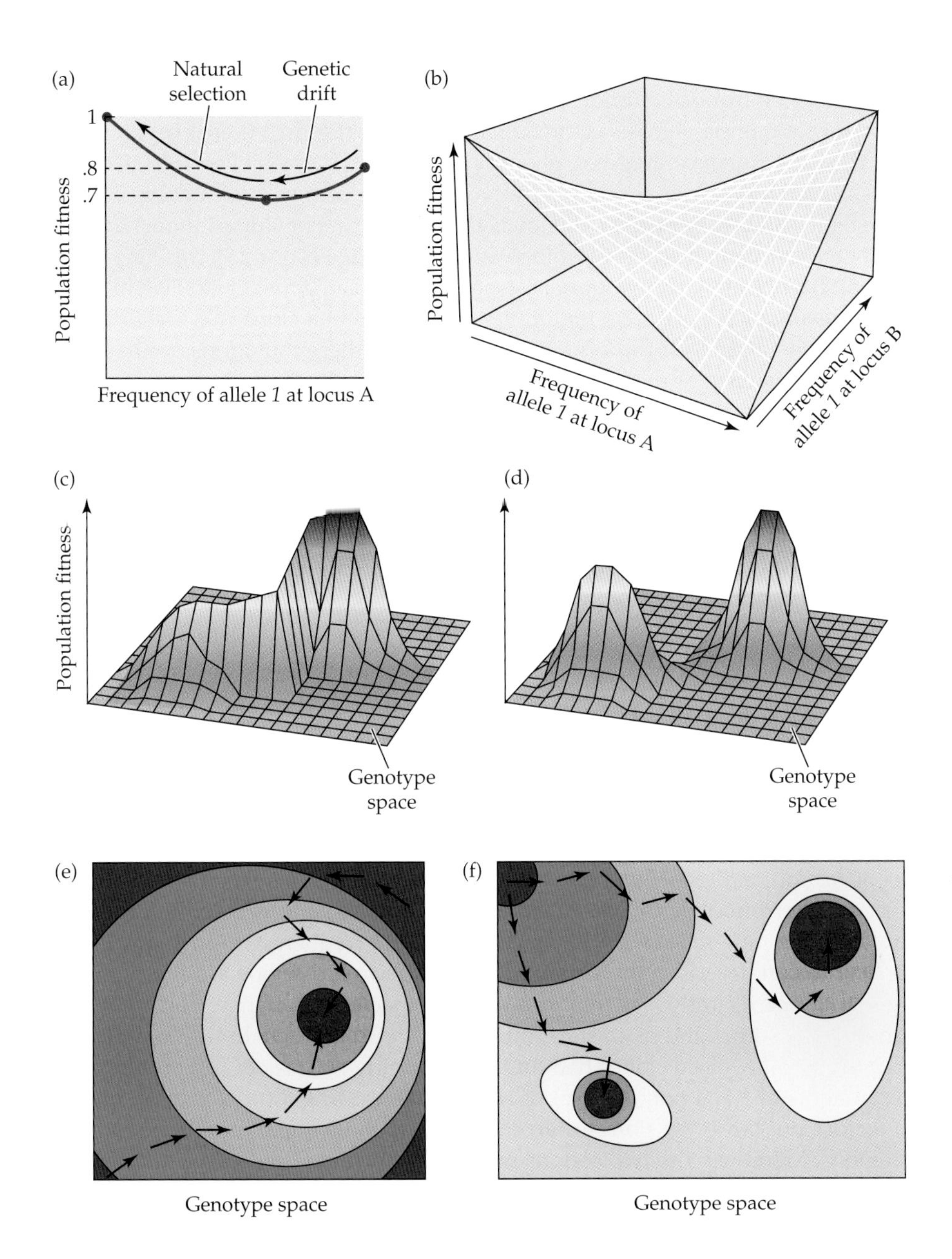

Figure 13.4 Representations of adaptive landscapes. (a) Two-dimensional representation of a population-based adaptive landscape consisting of a single locus with two alleles. This landscape reflects a case of underdominant selection in which heterozygotes have a lower fitness than either homozygote. The movement of the population from a low fitness value to a high fitness value is facilitated by selection. Genetic drift facilitates the movement from one place to another of either equal or lower fitness. (b) Three-dimensional representation of an adaptive landscape for two loci, each with two alleles. Epistatic interactions create a fitness landscape with two fitness peaks. (c, d) Generalized fitness landscapes in which the horizontal plane represents genotype space. The fitness landscape in (c) has a single peak; the one in (d) has two. (e, f) Heatmap representations of adaptive landscapes. In heatmaps, the hot and cold colors represent high and low fitness values, respectively. The topology of an adaptive landscape influences divergence between replicate evolving populations. In the single-peak landscape (e), populations evolved from any position will eventually converge to the same fitness peak. In the two-peak landscape (f), even populations evolved from the same position may diverge to different peaks, depending on chance differences in the order in which mutations occur. (Panel b modified from Whitlock et al. 1995.)

another so that the fitness effect of a mutation depends on the presence of one or more other mutations (i.e., if mutations interact epistatically), then the landscape may be rugged, containing multiple fitness peaks that are separated by valleys (Figures 13.4b,d,f). In a rugged adaptive landscape several fitness peaks might be accessible to a population starting at a given point in genetic space (Figure 13.4f). Because different fitness peaks are, by definition, characterized by different genotypes, populations that ascend different peaks will tend to genetically diverge from one another more than populations ascending the same fitness peak.

Experimental evolutionary studies have examined the extent of divergence between replicate populations and, thus, the availability of distinct fitness peaks from a particular genetic starting point. The basic design of these experiments is to start with replicate populations founded from a single clone, thus ensuring that they are initially genetically identical. Next, the replicates are evolved in the same environment for a large number of generations so that new beneficial mutations have a chance to arise and increase in frequency in each population. In the absence of migration, the magnitude and trend of divergence between replicate populations provide a window into the accessibility of distinct fitness peaks.

Lenski's long-term experiment demonstrates this approach. Twelve populations of *E. coli* were started from a common ancestor and independently evolved in a simple environment containing glucose as the only usable carbon source (Lenski et al. 1991). Initially, the experiment centered on phenotypic changes (Lenski et al. 1991; Cooper and Lenski 2000; Cooper et al. 2003); however, with the advent of affordable whole-genome sequencing technologies, the focus of the experiment shifted to include genetic and genomic comparisons. By comparing populations evolved in this long-term experiment, Barrick et al. (2009) found a strong signature of genetic parallelism—the flip side of divergence. Forty-five mutations were found in a clone that was isolated after 20,000 generations. Of the gene regions in which these mutations occurred, 14 were sequenced in each of the other 11 replicate populations. In only two cases was the mutation unique, and in seven cases at least 5 other populations had a mutation in the same gene region. This degree of parallelism is strikingly significant and suggests that the replicate populations may have ascended similar fitness peaks.

Tenaillon et al. (2012) examined genetic divergence using 114 *E. coli* replicate clones that were experimentally evolved for 2,000 generations at 42.2°C—close to the upper thermal limit of this bacterium. From each of the derived populations, they sequenced one genome. By comparing the derived genome sequences to that of the founding population, they identified 1,258 mutations, of which 63% were point mutations, 21% were small indels, 9% were deletions larger than 30 bp, 5% were insertions of IS elements (Chapter 9), and 2% were large duplications. The ratio of nonsynonymous to synonymous mutations per site was 5.75, as was the ratio of intergenic to synonymous mutations. Under the assumption that synonymous mutations are neutral, Tenaillon et al. (2012) estimated that ~80% of nonsynonymous and intergenic mutations were beneficial.

The large number of populations in the Tenaillon et al. (2012) study allowed patterns of attraction and repulsion between pairs of mutations to be identified. **Attraction** (or **co-occurrence**) describes the coincidental appearance of two mutations or two mutated genomic sites in the same genome more frequently than expected by chance. **Repulsion** describes the situation in which two mutations are found in the same genome significantly less frequently than expected by chance. Both patterns are signatures of some kind of epistatic interaction between sites. In the Tenaillon et al. (2012) study, all evolved clones had mutations in either the *rpo*BC operon (encoding components of RNA polymerase) or *rho* gene (encoding a transcription termination factor), but far fewer clones than expected by chance had mutations in both these regions. Moreover, the two codons with the most mutations in the data set (17 mutations in *rho* codon 15, and 18 mutations in *rpo*B codon 966) were in complete repulsion, i.e., they were never found in the same genome. This pattern of repulsion suggests that mutations in these two loci may interact antagonistically such that a mutation in one locus makes it less likely that a second mutation will be evolutionary acceptable in the other. These observations are consistent with a picture in which the initial chance substitution of *rpo*BC or *rho* determines which of at least two fitness peaks a population will ascend.

Historical contingency

Evolution is a historical process, and as such it is governed by the rules of historical contingency. **Historical contingency** is the assertion that every outcome depends upon a number of prior conditions, that each of these prior conditions depends, in turn, upon still other conditions, and so on. Change a single prior condition, and any outcome could turn out differently. In other words, each evolutionary outcome depends ultimately on the details of a long chain of antecedent states, each exerting enormous long-term repercussions. Contingency offers a powerful corrective to **teleology**, i.e., the fallacy according to which events pursue a predetermined course toward a definite outcome. Mechanistic views of evolution as an inevitable march toward the present tend to collapse once we realize that replicated experiments yield different results and that these different results cannot be predicted with any certainty from knowledge of initial conditions only. Contingency renders the evolutionary process quirky and fundamentally unpredictable.

An eloquent expression of the idea of historical contingency in evolution can be found in Stephen Jay Gould's (1989) description of a thought experiment:

> *I call this experiment 'replaying life's tape.' You press the rewind button and, making sure you thoroughly erase everything that actually happened, go back to any time and place in the past. … Then let the tape run again and see if the repetition looks at all like the original.*

Of course, it is not possible to conduct a physical experiment on the scale that Gould presented. Experimental evolution, however, can be used to examine the role of historical contingency in determining the evolutionary paths of evolving populations.

Historical contingency can manifest itself through interactions between mutations such that an early mutation influences either the probability that a subsequent mutation will occur or the effect that a second mutation will have on the fitness of its carrier. In either case, the past influences the future, or the future is contingent upon the past. Evolutionary experiments allow us to identify early mutational events that influence subsequent ones, thus revealing the importance and prevalence of historical contingency.

BACTERIAL RESISTANCE TO CEFOTAXIME Weinreich et al. (2006) studied the evolution of bacterial resistance to cefotaxime, a third-generation β-lactam antibiotic. This resistance is mediated by the enzyme β-lactamase, which hydrolytically inactivates β-lactam antibiotics. Five point mutations in the β-lactamase gene were found to jointly increase resistance to cefotaxime by a factor of about 50,000, from 0.09 to 4,100 μg/ml. These mutations consisted of four nonsynonymous mutations in the coding region plus one transition in the upstream regulatory region.

The derived resistant β-lactamase gene differs at five sites relative to the ancestral gene. Thus, there should be $2^5 - 2 = 30$ intermediate alleles of β-lactamase containing from one to four mutations. Because these five mutations can in principle occur in any order, there are $5! = 120$ possible mutational trajectories.

By making all possible intermediate alleles of β-lactamase and measuring the degree of cefotaxime resistance that each conferred, Weinreich et al. (2006) were able to generate a fitness landscape on which to test the role of historical contingency in constraining evolution. They found that many intermediate combinations of the five mutations were not accessible to natural selection. Of the 30 possible intermediate alleles containing n mutations, where n is 1, 2, 3, or 4, only 20 could be reached from an allele with $n - 1$ mutations without lowering cefotaxime resistance. For example, alleles containing any of the combinations *g4205a/A42G, A42G/M182T,* or *g4205a/A42G/M182T/G238S* are unlikely to be found in the presence of cefotaxime because each of these can arise only from alleles with higher cefotaxime resistance. Weinreich et al. (2006) also found that only 18 trajectories are evolutionarily accessible, of which 10 were more likely to be realized than others (**Figure 13.5**). Thus, 85% of all possible mutational trajectories (102 trajectories) were selectively inaccessible because they contained at least one mutational step that lowered the cefotaxime resistance of the derived bacterium relative to its immediate ancestor.

CITRATE METABOLISM IN *E. COLI* Another study that allowed scientists to investigate historical contingency was part of Lenski's experiment. As mentioned previously, 12 initially identical populations of *E. coli* were evolved in a glucose-limited medium. That medium also contained citrate, which *E. coli* cannot use as a carbon source in the presence of oxygen. For a very long period of time, exceeding 30,000 generations, none of the 12 populations evolved the capacity to exploit citrate. This was surprising, since the ability to metabolize citrate would have conferred a huge fitness advantage. Yet, none of the billions of mutations that had been tested by natural selection during this time led to the evolution of that trait.

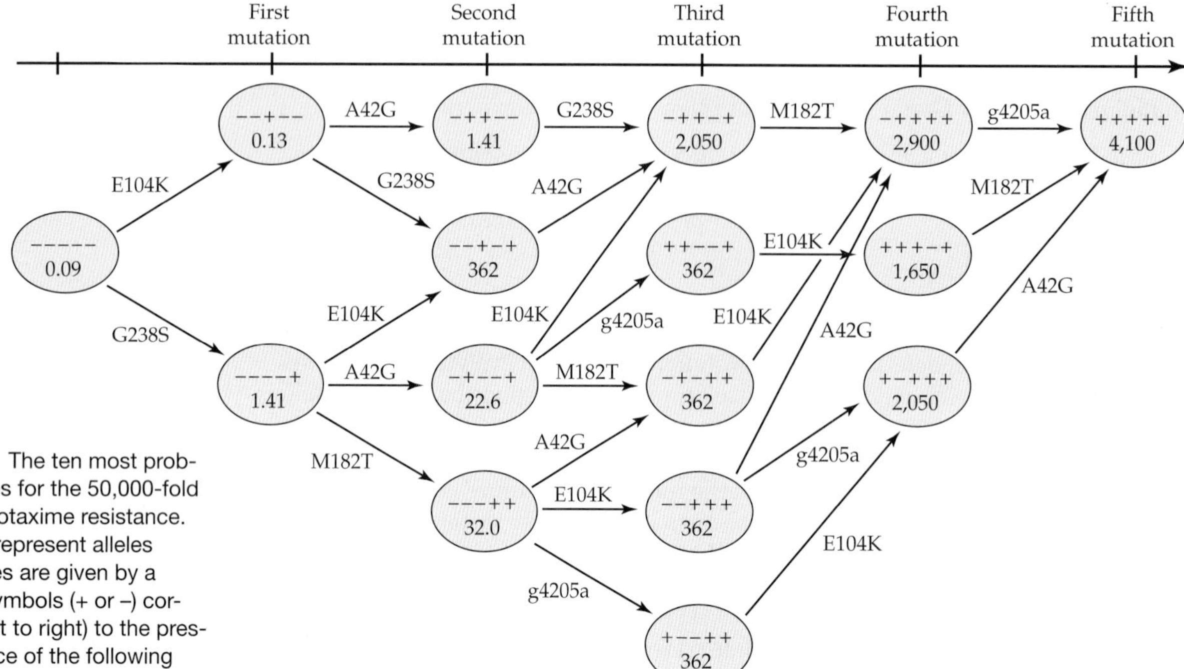

Figure 13.5 The ten most probable trajectories for the 50,000-fold increase in cefotaxime resistance. Nodes (ovals) represent alleles whose identities are given by a string of five symbols (+ or –) corresponding (left to right) to the presence or absence of the following mutations: a transition of guanine to adenine that occurs 5′ of the coding region at nucleotide position 4205 (g4205a), an alanine-to-glycine replacement at amino acid position 42 (A42G), a glutamic acid-to-lysine replacement at position 104 (E104K), a methionine-to-threonine replacement at position 182 (M182T), and a glycine-to-serine replacement at position 238 (G238S). Numbers inside each oval indicate cefotaxime resistance in units of micrograms per milliliter. (Modified from Weinreich et al. 2006.)

A citrate-using variant finally evolved in one of the 12 populations about 31,500 generations after the start of the experiment, causing a huge increase in the size of this population. The long-delayed and unique evolution of this function was puzzling. Blount et al. (2008) considered two explanations: (1) that the mutations conferring the ability to use citrate are extremely infrequent, or (2) that the mutation enabling citrate use is an ordinary mutation but one whose occurrence or metabolic expression is contingent on one or more "potentiating" mutations (**Figure 13.6**). The latter explanation represents an example of historical contingency.

To distinguish between the two explanations, Blount et al. (2008) considered individuals that could be collected from different times points along the evolution that produced the citrate-utilizing phenotype. If the mutation enabling citrate use occurred very rarely, then individuals collected at different time points should exhibit the same low probability of evolving the citrate-utilizing phenotype. This expectation was not met, as Blount et al. (2008) found that clones isolated from time points later than 20,000 generations were more likely to evolve the citrate-utilizing phenotype than were earlier clones. The early and late strains exhibited no difference in their overall mutation rates, so this finding pointed to the presence of some potentiating mutation or mutations that arose after 20,000 generations and that increased the benefit of the citrate utilization mutation.

Thus, the evolution of the citrate-utilizing phenotype was contingent on the particular history of that population. More generally, these findings suggest that histori-

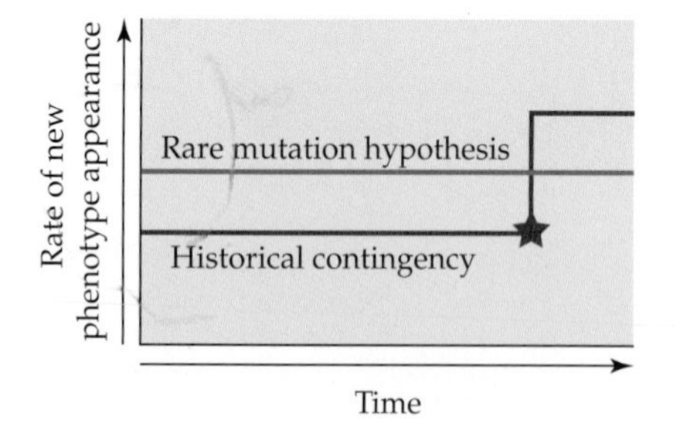

Figure 13.6 Alternative hypotheses for the origin of a novel phenotype. According to the rare mutation hypothesis (blue line), the probability that a new phenotype will arise in the population is constant over time. Under the historical contingency hypothesis (red line), the probability that a new phenotype will arise in the population is low until at least one potentiating mutation occurs (star), following which the probability increases. The two hypotheses can be distinguished from each other by evolving new populations from different time points along the evolutionary history of the experimental population to determine whether there is any change in the mutation rate to the phenotype under study. (Modified from Blount et al. 2008.)

cal contingency is especially important when it facilitates the evolution of innovations that cannot easily evolve by gradual, cumulative selection.

Blount et al. (2012) put forward a hypothesis according to which the evolution of the ability to use citrate as an energy source involves three successive processes: potentiation, actualization, and refinement. **Potentiation** refers to the evolution of a suitable genetic background in which the citrate utilization function can appear. The potentiating mutations do not confer any advantage related to citrate utilization; however, they are most probably not neutral mutations, and they may increase fitness on glucose. **Actualization** refers to the appearance of the trait. According to Blount et al. (2012), a mutant with a very weak ability to utilize citrate had emerged after 31,500 generations, and this mutant represents the actualization step. **Refinement** refers to the improvement in the citrate utilization function, which took about 2,000 generations. The new refined function, then, allowed the efficient exploitation of citrate, the rise to numerical dominance of variants capable of efficient citrate utilization, and the expansion of the total population size. Of these three processes, actualization is the most tractable for study, owing to the discrete phenotypic change. It is not, however, clear whether these three stages represent a general mechanism for the evolution of novel traits or whether they only typify the evolution of the citrate-utilizing phenotype.

Epistasis

The nomenclature and theory of epistasis were introduced in Chapter 2. Here, we describe the means by which experimental evolution has been used to study the extent and influence of epistasis on evolutionary outcomes. Two main approaches have been used. In the first approach, several mutations are introduced into an otherwise identical genotype, and their individual and combined effects are measured. This approach has been mainly employed with mutations identified in natural populations (Weinreich et al. 2006; Lozovsky et al. 2009; Da Silva et al. 2010; Summers et al. 2014). In the second approach, epistasis is inferred when the evolutionary responses of populations depend on the genotypes that they contain.

Examining interactions between mutations isolated from experimentally evolved populations can contribute to our understanding of general features of adaptation. For example, a study by Kryazhimskiy et al. (2009) aimed to determine the type of interactions that best explained the evolutionary dynamics observed in one of Lenski's *E. coli* populations. In that population the rate of fitness increase was found to decelerate such that the fitness increase occurring over the first 2,000 generations was about the same as that occurring over the last 20,000 generations (**Figure 13.7**). At the same time, however, mutations continued to accumulate at a constant rate of approximately 2.25 mutations per 1,000 generations (Barrick et al. 2009).

The simplest hypothesis that can explain the discrepancy between the nearly linear pattern of mutation accumulation and the sharply decelerating fitness trajectory would be that the fraction of neutral mutations increases with time at the expense of the fraction of advantageous mutations. In other words, whatever improvement in population fitness can be achieved is usually achieved very fast. Barrick et al. (2009), however, presented four lines of evidence indicating that the vast majority of the observed mutations throughout the evolutionary experiment conferred fitness benefits.

First, under the drift hypothesis, one expects disproportionately more synonymous than nonsynonymous substitutions at later times than at earlier times, because the former are more likely to be neutral. In reality, all 26 point mutations found in coding regions were nonsynonymous. Second, if mutations

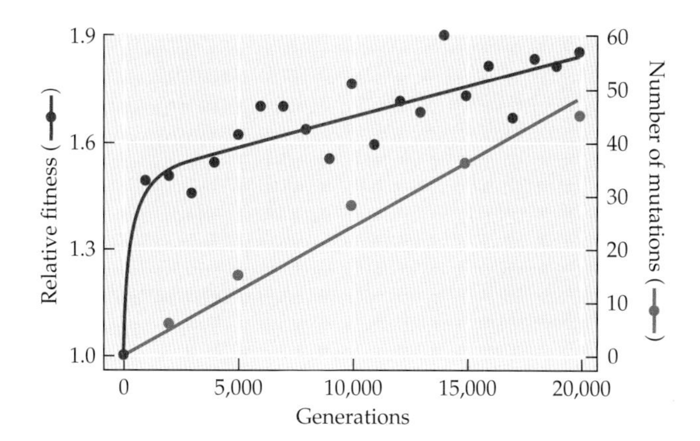

Figure 13.7 Change in fitness and number of mutations accumulating in an experimentally evolved population of *E. coli*. As is often seen in experiments that evolve microbes in constant environments, fitness (red) first increases quickly relative to the ancestor but decelerates at later time points. Whole-genome sequencing of clones isolated from time points up to 20,000 generations of evolution reveals a linear pattern of mutation accumulation (blue). (Data from Barrick et al. 2009.)

had indeed spread by random drift, we would not expect to see mutations in the same genes in the other independently evolved lines of the long-term experiment. Given that only 1% of the 4,000 genes in *E. coli* were found to harbor mutations, the random expectation for parallel changes is exceeding small. By contrast, selection is more likely to target the same genes in the replicate lines, which is indeed what was observed. Third, under the drift hypothesis, we would expect many mutations in individual clones not to become fixed in the population as a whole. However, almost all mutations in earlier clones were also present in clones from subsequent generations. Fourth, under the drift hypothesis, many substitutions that accumulated later in the evolution of the lines should confer no fitness advantage on their carriers when introduced singly into the background of the founding population. In reality, in almost all tested cases, the derived alleles were found to increase fitness.

Kryazhimskiy et al. (2009) reconciled these observations by modeling the interactions among beneficial mutations as a pattern of diminishing returns, such that the benefit of acquiring a new beneficial mutation is small when it occurs in a high-fitness organism, and large in a low-fitness organism. This pattern of interactions has also been found in other experimental studies (Chou et al. 2011; Khan et al. 2011; Rokyta et al. 2011).

Mutation Dynamics

The estimation of mutation rate and patterns by using retrospective evolutionary methodology has been described in Chapter 4. Here, we address the methodology of assessing mutation dynamics by using experimental evolution.

Neutral mutation rates

Comparative and short-term experimental methods can be used to estimate mutation rates, but both have limitations. Comparative methods require estimates of divergence and generation times of the compared species so that the rate of mutation per generation can be calculated. Short-term "fluctuation test" experiments are generally limited to considering mutations that confer some kind of selectable phenotype. It is not surprising, therefore, that estimates of mutation rates, even for an organism as well studied as *E. coli*, differ by more than an order of magnitude, from 2.25×10^{-11} (Ochman et al. 1999), using an intermediate estimate of 200 generations per year, to 5.4×10^{-10} (Drake 1991) mutations per genome per generation.

Experimental evolution has the potential to circumvent these uncertainties. Whole-genome sequencing of 19 clones evolved for 40,000 generations as part of the Lenski long-term evolution experiment identified a total of 35 independent synonymous mutations. Controlling for the proportion of the genome likely to be evolving approximately neutrally leads to an estimated mutation rate of 8.9×10^{-11} (with a 95% confidence interval of $4–14 \times 10^{-11}$) mutations per site per generation, corresponding to a genomic mutation rate of 4.1×10^{-4} per generation.

Nonneutral mutation rates

There are three main ways of propagating populations in evolutionary experiments (**Figure 13.8**). Some methods are suitable for detecting neutral and deleterious mutations; others are suitable for detecting advantageous mutations.

In mutation accumulation experiments, repeated single-cell bottlenecks ensure that the effective population size remains close to the absolute possible minimum. In such populations, selection is weak and ineffective and, with the exception of lethal mutations, the fate of all other mutations is controlled mainly by chance so that they have an equal probability of being transmitted to the next generation. Since the vast majority of fitness-affecting or nonneutral mutations in organisms with small, gene-rich genomes are deleterious, mutation accumulation experiments allow us to estimate the rates and effects of deleterious mutations (Mukai 1964; Keightley 1998). Over time the average number of mutations accumulating across replicated populations increases and

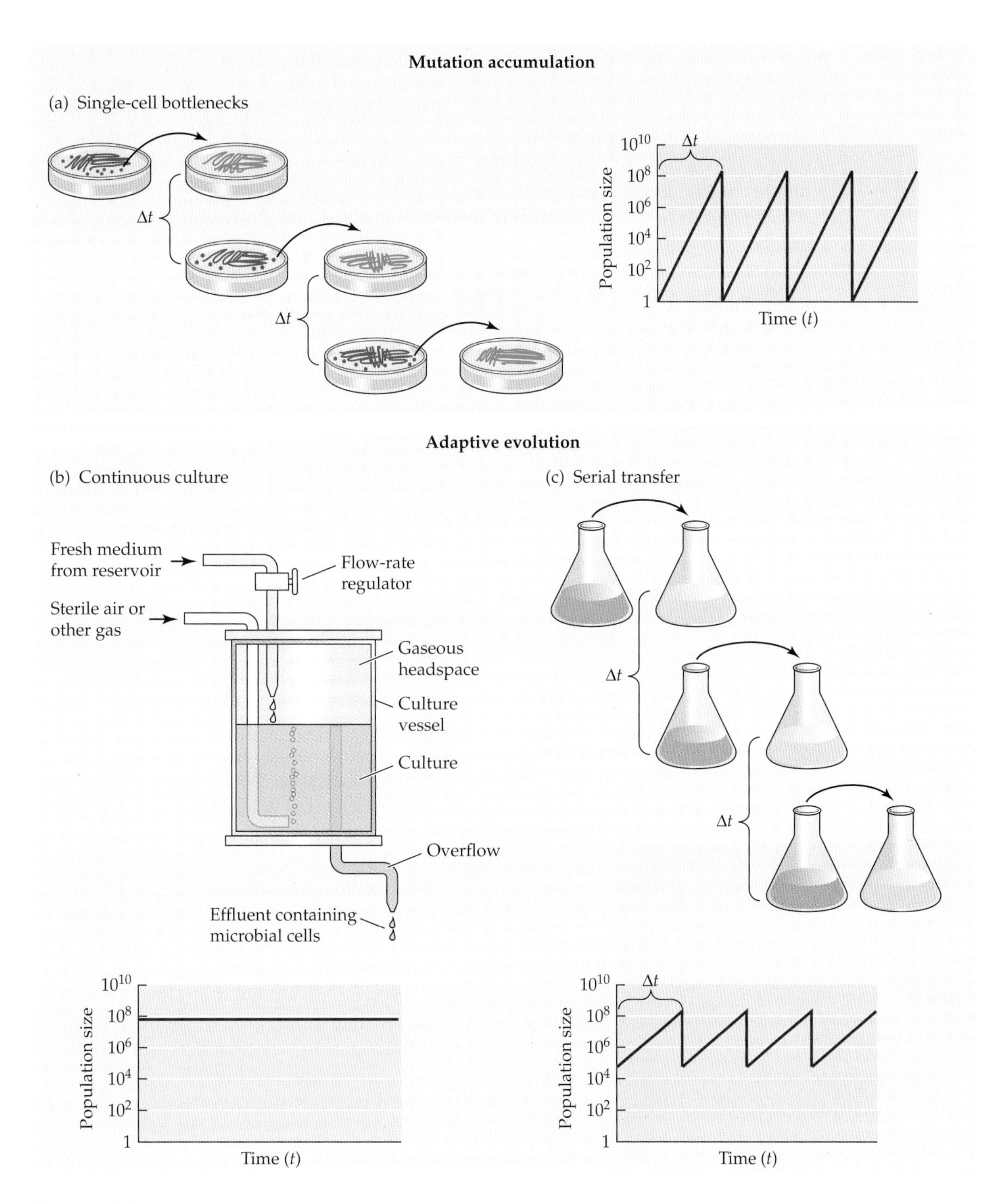

Figure 13.8 The three main ways of propagating populations in evolution experiments lead to different types of population dynamics. The mechanics of how microbial populations are maintained in each setup are illustrated, with representative changes in population sizes over time are graphed for each procedure. (a) In mutation accumulation experiments, frequent and deliberate population bottlenecks are accomplished by picking colonies that grow from single cells on agar plates. These bottlenecks purge genetic diversity and lead to the fixation of arbitrary nonlethal mutations by random genetic drift. (b) In experiments using continuous culture, populations are maintained in a chemostat consisting of a constant inflow of nutrients and an outflow of cells and waste, which leads to the maintenance of populations at nearly constant size. (c) In serial transfer experiments, a proportion of the population is periodically transferred to fresh media and allowed to regrow until the limiting nutrient is exhausted. Alternatively, transfers can be made before nutrient depletion, thereby allowing perpetual population growth. Because ample genetic diversity is maintained through each transfer, adaptive evolution is possible in (b) and (c), but not in (a). (Modified from Barrick and Lenski 2013 and Madigan and Martinko 2006.)

population fitness tends to decline, a sign that there are more deleterious mutations than beneficial ones. Estimates of the underlying deleterious mutation rate and the mean effect of those mutations can be derived from the dynamics of fitness change across replicate populations. By using a setup similar to the one described above, Kibota and Lynch (1996) estimated the rate of deleterious mutation in *E. coli* to be 0.002 mutations per genome per generation.

One caveat to many such experiments is that they generally assume that mutations do not interact with one another epistatically. This assumption can be relaxed, but at the cost of losing degrees of freedom in an analysis that is already subject to considerable variation. By using whole-genome sequencing, however, one is able to test for epistasis through the subsequent manipulation of the mutations identified during the experiment.

Because the rate of substitution for advantageous mutations is higher than the rate of mutation, the rate at which advantageous mutations arise in a population is almost always overestimated, as is the mean fitness effect of advantageous mutations (Rozen et al. 2002). Without the underlying distribution of mutation effects, the rate at which beneficial mutations occur cannot be easily inferred from the rate at which they are fixed. A clever comparative approach pioneered by Hegreness et al. (2006) allows beneficial mutation parameters to be estimated. The approach takes advantage of the long-standing observation that the dynamics of neutral genetic markers in asexual evolving microbial populations are dictated by the fate of linked beneficial mutations (Atwood et al. 1951). A neutral marker on a background on which a beneficial mutation arises will increase in frequency along with the mutation.

Hegreness et al.'s (2006) approach starts with a 1:1 mixture of two subpopulations that differ initially only by a single selectively neutral but phenotypically distinct marker (**Figure 13.9**). Half of the mixture consists of cells expressing a cyan fluorescent protein (CFP), and half express a yellow fluorescent protein (YFP). Because neither YFP nor CFP has a selective advantage over the other, the 1:1 ratio will be maintained. Let us now assume that at time t_1 an advantageous mutation, YFP^m, arises in a YFP cell. The fraction of YFP^m + YFP cells in the mixture will rise in frequency. By using the ratio of YFP^m + YFP cells to CFP cells, it is possible to calculate the selective advantage of the YFP^m mutation. Let us further assume that at some later time, t_2, an advantageous mutation, CFP^m, arises in a CFP cell and that this mutation has a higher selective advantage than that of the YFP^m mutation. At first, the rate of increase in the proportion of YFP^m + YFP cells in the mixture will be slowed down. Subsequently, at time t_3, the frequency of CFP + CFP^m cells in the mixture will start rising at the expense of YFP^m + YFP cells. By using the ratio of YFP^m + YFP cells to CFP + CFP^m cells, it is possible to calculate the selective advantage of the mutation in CFP^m relative to that of the mutation in YFP^m.

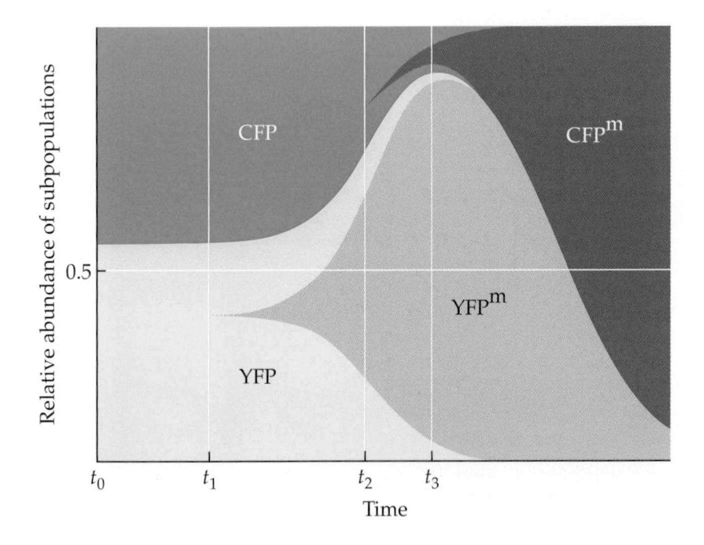

Figure 13.9 Evolutionary spread of beneficial mutations. (a) Initially, the ratio of YFP (yellow) and CFP (blue) cells in the mixed population is 1:1. At time t_1, an advantageous mutation (dark yellow) arises in a YFP-labeled cell. Because of its advantage, YFP^m will increase in frequency. Later, at time t_2, a second mutation (dark blue) occurs in a CFP-labeled cell. The selective advantage of this mutation is greater than that of the mutation that arose in the YFP-labeled cell, initially only slowing down the rate of increase in the frequency of YFP-labeled cells. At time t_3, however, the trend will be reversed and the frequency of CFP-labeled cells in the mixture will rise. (Modified from Hegreness et al. 2006.)

Targets of Selection

So far in this book, we have categorized mutations into structural types, such as point mutations (i.e., transitions and transversions), deletions, insertions, rearrangements, and duplications. In some cases, we have discussed their effects on protein-coding regions and classified mutations by their effects into synonymous and nonsynonymous mutations; silent and noisy mutations; sense, missense, and nonsense mutations; and frameshifts. Here, we discuss mutations from the viewpoint of their effects on the phenotype. What types of mutations drive phenotypic evolution?

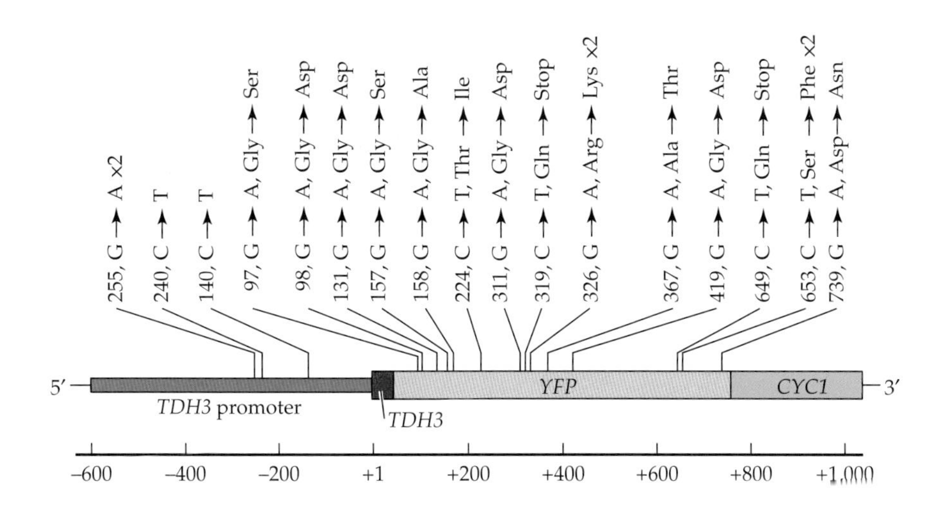

Figure 13.10 Genetic construct used to characterize mutations that quantitatively affect the amount of yellow fluorescent protein (YFP) produced by *Saccharomyces cerevisae* cells. A genetically engineered YFP-coding gene from the jellyfish *Aequorea victoria* was fused to the promoter and the 5′ coding region of a triosephosphatedehydrogenase-3 gene (*TDH3*) and the 3′ coding region of a cytochrome-c1-coding gene (*CYT1*). Fourteen missense and two nonsense mutations in the YFP-coding region as well as four mutations in the *cis*-regulatory promoter were found to alter gene expression. (Modified from Gruber et al. 2012.)

Some early attempts to determine the genetic basis of adaptation in evolving experimental populations focused on changes occurring during bacterial adaptation to chemostat environments, which enforce a very-low-nutrient-level regime (Notley-McRobb and Ferenci 1999; Zhong et al. 2009). Common adaptations to these environments involve upregulation of nutrient-transport genes. This upregulation was found to be mostly due to one of two types of mutation: (1) mutations in regulatory regions that caused genes to be constitutively expressed, or (2) gene duplication.

Adaptations tend to have more diverse causes following selection in more variable environments that fluctuate between high and low nutrient levels. For example, in Tenaillon et al. (2012), the most common mutation targets were genes affecting RNA polymerase activity. These mutations are likely to have widespread pleiotropic effects.

Here, we will frame our discussion on the mutational basis of a phenotypic change around an example of a simple phenotype in a simple organism (Gruber et al. 2012). The phenotype is the activity of a gene encoding a yellow fluorescent protein that has been genetically engineered into yeast cells (**Figure 13.10**). From the perspective of a single-protein gene, mutations affecting its activity can be divided into four functional classes: (1) mutations that alter the sequence of the encoded protein; (2) *cis*-regulatory mutations that typically occur in nearby sequences, thereby altering their regulation in an allele-specific manner; (3) *trans*-acting mutations that affect the activity of the gene under study indirectly, such as by a diffusible gene product; and (4) copy number mutations resulting from duplications or deletions that change the number of copies of the gene under study in the genome. Gruber et al. (2012) examined 213 cells with altered activity of the yellow fluorescent protein. Of these, 14 (7%) had a missense mutation in the protein-coding gene, 4 (2%) had a change in the *cis*-regulatory region, and 2 (1%) had a nonsense mutation in the protein-coding gene. Surprisingly, 81% of all genotypes with altered yellow fluorescent protein production contained no mutation in either the coding or the regulatory regions. In these cases, the change in gene expression must have been caused by *trans*-acting mutations.

As a group, coding mutants had the strongest effect on reporter gene activity and always decreased it. By contrast, 50–95% of the mutants in each of the other three classes increased gene activity, with mutants affecting copy number and *cis*-regulatory sequences having on average larger effects on gene activity than *trans*-acting mutants. In heterozygous diploid cells, the coding, *cis*-regulatory, and copy number mutant genotypes all had significant effects on gene activity, whereas 88% of the *trans*-acting mutants appeared to be recessive.

The relative influences of regulatory versus structural gene changes has been the focus of comparative molecular evolutionary studies, with many researchers finding that adaptive mutations affecting morphology are more likely to occur in *cis*-regulatory regions than in the protein-coding regions of genes (e.g., Hoekstra and

Coyne 2007). Changes in regulatory genes, which potentially affect the expression of many other genes, are likely to lead to faster phenotypic divergence than changes in structural genes. Especially with respect to the kind of constant environments in which many experimental evolution studies are carried out, the relative influence of gain- and loss-of-function mutations is also of considerable interest. Together these questions speak to the tension between global (*trans*) and local (*cis*) targets, and between creative (neofunctionalization) and destructive (nonfunctionalization) processes as drivers of adaptive evolution.

Literature Cited

A

Aas, F. E., M. Wolfgang, S. Frye, S. Dunham, C. Løvold, and M. Koomey. 2002. Competence for natural transformation in *Neisseria gonorrhoeae*: Components of DNA binding and uptake linked to type IV pilus expression. Mol. Microbiol. 46: 749–760. **[9]**

Achaz, G. 2008. Testing for neutrality in samples with sequencing errors. Genetics 179: 1409–1424. **[2]**

Achaz, G., E. Coissac, P. Netter, and E. P. C. Rocha. 2003. Associations between inverted repeats and the structural evolution of bacterial genomes. Genetics 164: 1279–1289. **[10]**

Acquisti, C., J. J. Elser, and S. Kumar. 2009. Ecological nitrogen limitation shapes the DNA composition of plant genomes. Mol. Biol. Evol. 26: 953–956. **[11]**

Adams, K. L. and J. D. Palmer. 2003. Evolution of mitochondrial gene content: Gene loss and transfer to the nucleus. Mol. Phylogenet. Evol. 29: 380–395. **[10]**

Adams, K. L. and J. F Wendel. 2005. Polyploidy and genome evolution in plants. Curr. Opin. Plant Biol. 8: 135–141. **[7]**

Adams, M. D., R. S. Tarng, and D. C. Rio. 1997. The alternative splicing factor PSI regulates *P*-element third intron splicing *in vivo*. Genes Develop. 11: 129–138. **[9]**

Adey, A. and 13 others. 2013. The haplotype-resolved genome and epigenome of the aneuploid HeLa cancer cell line. Nature 500: 207–211. **[11]**

Adl, S. M and 19 others. 2007. Diversity, nomenclature, and taxonomy of protists. Syst. Biol. 56: 684–689. **[5]**

Adl, S. M and 27 others. 2005. The new higher level classification of eukaryotes with emphasis on the taxonomy of protists. J. Eukaryot. Microbiol. 52: 399–451. **[5]**

Agapow, P.-M., O. R. P. Bininda-Emonds, K. A. Crandall, J. L. Gittleman, G. M. Mace, J. C. Marwill, and A. Purvis. 2004. The impact of species concept on biodiversity studies. Quart. Rev. Biol. 79: 161–179. **[5]**

Agarwal, A. and 11 others. 2000. Hemoglobin C associated with protection from severe malaria in the Dogon of Mali, a West African population with a low prevalence of hemoglobin S. Blood 96: 2358–2363. **[2]**

Agol, I. J. 1930. Evidence of the divisibility of the gene. Anat. Rec. 47: 385. **[7]**

Agris, P. F., F. A. P. Vendeix, and W. D. Graham. 2007. tRNA's wobble decoding of the genome: 40 years of modification. J. Mol. Biol. 366: 1–13. **[4]**

Ahituv, N., Y. Zhu, A. Visel, A. Holt, V. Afzal, L. A. Pennacchio, and E. M. Rubin. 2007. Deletion of ultraconserved elements yields viable mice. PLoS Biol. 5: e234. **[4, 11]**

Akashi, H. 1994. Synonymous codon usage in *Drosophila melanogaster*: Natural selection and translational accuracy. Genetics 136: 927–935. **[4]**

Akashi, H. 1997. Codon bias evolution in *Drosophila*. Population genetics of mutation–selection drift. Gene 205: 269–278. **[4]**

Akashi, H. and S. W. Schaeffer. 1997. Natural selection and the frequency distributions of "silent" DNA polymorphism in *Drosophila*. Genetics 146: 295–307. **[4]**

Akashi, M. and H. Yoshikawa. 2013. Relevance of GC content to the conservation of DNA polymerase III mismatch repair system in Gram-positive bacteria. Front. Microbiol. 4: 266. **[10]**

Akey, J. M. 2009. Constructing genomic maps of positive selection in humans: where do we go from here? Genome Res. 19: 711–722. **[11]**

Akins, R. A. and A. M. Lambowitz. 1987. A protein required for splicing group I introns in *Neurospora* mitochondria is mitochondrial tyrosyl-tRNA synthetase or a derivative thereof. Cell 50: 331–345. **[8]**

Al-Hashimi, N., A.-G. Lafont, S. Delgado, K. Kawasaki, and J.-Y. Sire. 2010. The enamelin genes in lizard, crocodile, and frog and the pseudogene in the chicken provide new insights on enamelin evolution in tetrapods. Mol. Biol. Evol. 27: 2078–2094. **[8]**

Albertson, R. C., J. A. Markert, P. D. Danley, and T. D. Kocher. 1999. Phylogeny of a rapidly evolving clade: the cichlid fishes of Lake Malawi, East Africa. Proc. Natl. Acad. Sci. USA 96: 5107–5110. **[12]**

Albu, M., X. J. Min, D. Hickey, and B. Golding. 2008. Uncorrected nucleotide bias in mtDNA can mimic the effects of positive Darwinian selection. Mol. Biol. Evol. 25: 2521–2524. **[4]**

Alfaro, M. E. and M. T. Holder. 2006. The posterior and the prior in Bayesian phylogenetics. Annu. Rev. Ecol. Evol. Syst. 37: 19–42. **[5, 6]**

Alkatib, S. and 8 others. 2012. The contributions of wobbling and superwobbling to the reading of the genetic code. PLoS Genet. 8:e1003076. **[1]**

Allen, J. F. 1993. Control of gene expression by redox potential and the requirement for chloroplast and mitochondrial genomes. J. Theor. Biol. 165: 609–631. **[10]**

Allison, A. C. 1954. Notes on sickle-cell polymorphism. Ann. Hum. Genet. 19: 39–57. **[2]**

Allison, A. C. 2002. The discovery of resistance to malaria of sickle-cell heterozygotes. Biochem. Mol. Biol. Edu. 30: 279–287. **[2]**

Altermann, W. and J. Kazmierczak. 2003. Archean microfossils: A reappraisal of early life on Earth. Res. Microbiol. 154: 611–617. **[6]**

Altschul, S. F., W. Gish, W. Miller, E. W. Myers, and D. J. Lipman. 1990. Basic local alignment search tool. J. Mol. Biol. 215: 403–410. **[3]**

Alvarez-Valin, F., G. Lamolle, and G. Bernardi. 2002. Isochores, GC3, and mutation biases in the human genome. Gene 300: 161–168. **[11]**

Alvarez, M. and J. Ballantyne. 2009. Identification of four novel developmentally regulated gamma hemoglobin mRNA isoforms. Exp. Hematol. 37: 285–293. **[7]**

Alzohairy, A. M., G. Gyulai, R. K. Jansen, and A. Bahieldin. 2013. Transposable elements domesticated and neofunctionalized by eukaryotic genomes. Plasmid 69: 1–15. **[9]**

Amaral, A. R., G. Lovewell, M. M. Coelho, G. Amato, and H. C. Rosenbaum. 2014. Hybrid speciation in a marine mammal: The Clymene dolphin (*Stenella clymene*). PLoS ONE 9: e83645. **[6]**

Amor, D. J. and K. H. Choo. 2002. Neocentromeres: Role in human disease, evolution, and centromere study. Am. J. Hum. Genet. 71: 695–714. **[11]**

Amorós-Moya, D., S. Bedhomme, M. Hermann, and I. G. Bravo. 2010. Evolution in regulatory regions rapidly compensates the cost of nonoptimal codon usage. Mol. Biol. Evol. 27: 2141–2151. **[9]**

Amundson, R. and G. V. Lauder. 1994. Function without purpose. Biol Philos. 9: 443–469. **[11]**

Amy, C. M., B. Williams-Ahlf, J. Naggert, and S. Smith. 1992. Intron–exon organization of the gene for the multifunctional animal fatty acid synthase. Proc. Natl. Acad. Sci. USA 89: 1105–1108. **[8]**

An, C. and 12 others. 1995. Molecular cloning and sequencing of the 18s rDNA from specialized dinosaur egg fossil found in Xixia Henan, China. Acta Sci. Nat. Univ. Pekinensis 31: 140–147. **[5]**

Andersen, R. D., B. W. Birren, S. J. Taplitz, and H. R. Herschman. 1986. Rat metallothionein-I structural gene and three pseudogenes, one of which contains 5′ regulatory sequences. Mol. Cell Biol. 6: 302–314. **[9]**

Anderson, S. and 13 others. 1981. Sequence and organization of the human mitochondrial genome. Nature 290: 457–464. **[8, 10]**

Andersson, J. O. 2005. Lateral gene transfer in eukaryotes. Cell. Mol. Life Sci. 62: 1182–1197. **[9]**

Andersson, J. O. and S. G. Andersson. 1999a. Genome degradation is an ongoing process in *Rickettsia*. Mol. Biol. Evol. 16: 1178–1191. **[4]**

Andersson, J. O. and S. G. Andersson. 1999b. Insights into the evolutionary process of genome degradation. Curr. Opin. Genet. Dev. 9: 664–671. **[4]**

Andersson, J. O., R. P. Hirt, P. G. Foster, and A. J. Roger. 2006. Evolution of four gene families with patchy phylogenetic distributions: Influx of genes into protist genomes. BMC Evol. Biol. 6: 27. **[9]**

Andrés, A. M. and 11 others. 2009. Targets of balancing selection in the human genome. Mol. Biol. Evol. 26: 2755–2764. **[11]**

Andrews, P. 1987. Aspects of hominoid phylogeny. pp. 23–53. In: C. Patterson (ed.), *Molecules and Morphology in Evolution: Conflict or Compromise?* Cambridge University Press, Cambridge. **[5]**

Anxolabéhère, D., M. G. Kidwell, and G. Periquet. 1988. Molecular characteristics of diverse populations are consistent with the hypothesis of a recent invasion of *Drosophila melanogaster* by mobile *P* elements. Mol. Biol. Evol. 5: 252–269. **[9]**

Aota, S.-I. and T. Ikemura. 1986. Diversity in G+C content at the third position of codons in vertebrate genes and its cause. Nuc. Acids Res. 14: 6345–6355. **[11]**

Aparicio, S. and 40 others. 2002. Whole-genome shotgun assembly and analysis of the genome of *Fugu rubripes*. Science 297: 1301–1310. **[11]**

Aragoncillo, C., M. A. Rodríguez-Loperena, G. Salcedo, P. Carbonero, and F. García-Olmedo. 1978. Influence of homoeologous chromosomes on gene-dosage effects in allohexaploid wheat (*Triticum aestivum* L.). Proc. Natl. Acad. Sci. USA 75: 1446–1450. **[7]**

Arbiza, L., J. Dopazo, and H. Dopazo. 2006. Positive selection, relaxation, and acceleration in the evolution of the human and chimp genome. PLoS Comput. Biol. 2: e38. **[4]**

Archibald, J. M. 2009. The puzzle of plastid evolution. Curr. Biol. 19: R81–R88. **[5, 6]**

Argos, P., M. G. Rossmann, U. M. Grau, A. Zuber, G. Franck, and J. D. Tratschin. 1979. Thermal stability and protein structure. Biochemistry 18: 5698–5703. **[10]**

Arguello, J. R., C. Fan, W. Wang, and M. Long. 2007. Origination of chimeric genes through DNA-level recombination. Genome Dyn. 3: 131–146. **[8]**

Argyle, E. 1980. A similarity ring for amino acids based on their evolutionary substitution rates. Orig. Life 10: 357–360. **[4]**

Arias, M. C., E. G. J. Danchin, P. M. Coutinho, B. Henrissat, and S. Ball. 2012. Eukaryote to gut bacteria transfer of a glycoside hydrolase gene essential for starch breakdown in plants. Mobile Genet. Elem. 2: 1–7. **[9]**

Arkhipova, I. R. and H. G. Morrison. 2001. Three retrotransposon families in the genome of *Giardia lamblia*: Two telomeric, one dead. Proc. Natl. Acad. Sci. USA 98: 14497–14502. **[9]**

Armengol, L, and 7 others. 2005. Murine segmental duplications are hot spots for chromosome and gene evolution. Genomics 86: 692–700. **[11]**

Árnason, U., A. Gullberg, A. Janke, and X. Xu. 1996. Pattern and timing of evolutionary divergences among hominoids based on analyses of complete mtDNAs. J. Mol. Evol. 43: 650–661. **[5]**

Arnheim, N. 1983. Concerted evolution of multigene families. pp. 38–61. In: M. Nei and R. K. Koehn (eds.), *Evolution of Genes and Proteins*. Sinauer Associates, Sunderland, MA. **[7]**

Arnold, M. L. and E. J. Larson. 2004. Evolution's new look. Wilson Quart. 2004(1): 60–72. **[6]**

Arnold, S. J., M. E. Pfrender, and A. G. Jones. 2001. The adaptive landscape as a conceptual bridge between micro- and macroevolution. Genetica 112/113: 9–32. **[13]**

Aronson A. I., E. T. Bolton, R. J. Britten, D. B. Cowie, J. D. Duerksen, B. J. McCarthy, K. McQuillen, and R. B. Roberts. 1960. Biophysics. pp. 229–289. In: *Year Book: Carnegie Institution of Washington*. Volume 59. Lord Baltimore Press, Baltimore, MD. **[11]**

Arthur, W. 1883. *On the Difference between Physical and Moral Law*. The Fernley Lecture of 1883. Woolman, London. **[12]**

Ashburner, M., K. G. Golic, and R. S. Hawley. 2005. *Drosophila: A Laboratory Handbook*. Cold Spring Harbor Laboratory Press, New York. **[9]**

Assis, R. and D. Bachtrog. 2013. Neofunctionalization of young duplicate genes in *Drosophila*. Proc. Natl. Acad. Sci. USA 110: 17409–17414. **[7]**

Assis, R., A. S. Kondrashov, E. V. Koonin, and F. A. Kondrashov. 2008. Nested genes and increasing organizational complexity of metazoan genomes. Trends Genet. 24: 475–478. **[8]**

Ast, G. 2004. How did alternative splicing evolve? Nat. Rev. Genet. 5: 773–782. **[8]**

Atteia, A. and 12 others. 2009. A proteomic survey of *Chlamidomonas reinhardtii* mitochondria sheds new light on the metabolic plasticity of the organelle and on the nature of the a-proteobacterial mitochondrial ancestor. Mol. Biol. Evol. 26: 1533–1548. **[5]**

Atwood, K. C., L. K. Schneider, and F. J. Ryan. 1951. Periodic selection in *Escherichia coli*. Proc. Natl. Acad. Sci. USA 37: 146–155. **[13]**

Audit, B., S. Nicolay, M. Huvet, M. Touchon, Y. d'Aubenton-Carafa, C. Thermes, and A. Arneodo. 2007. DNA replication timing data corroborate *in silico* human replication origin predictions. Phys. Rev. Lett. 99: 248102. **[10]**

Aury, J.-M. and 42 others. 2006. Global trends of whole-genome duplications revealed by the ciliate *Paramecium tetraurelia*. Nature 444: 171–178. **[7]**

Austin, J. J., A. B. Smith, and R. H. Thomas. 1997. Palaeontology in a molecular world: The search for authentic ancient DNA. Trends Ecol. Evol. 12: 303–306. **[5]**

Avarello, R., A. Pedicini, A. Caiulo, O. Zuffardi, and M. Fraccaro. 1992. Evidence for an ancestral alphoid domain on the long arm of human chromosome 2. Hum. Genet. 89: 247–249. **[11]**

Avery, O. T., C. M. MacLeod, and M. McCarty. 1944. Studies on the chemical nature of the substance inducing transformation of pneumococcal types: Induction of transformation by a desoxyribonucleic acid fraction isolated from pneumococcus type III. J. Exp. Med. 79: 137–157. **[9]**

Avise, J. C. and W. S. Nelson. 1989. Molecular genetic relationships of the extinct dusky seaside sparrow. Science 243: 646–648. **[5]**

Avise, J. C., D. Walker, and G. C. Johns. 1998. Speciation durations and Pleistocene effects on vertebrate phylogeography. Proc. R. Soc. London 265B: 1707–1712. **[5]**

Axelsson, E., N. G. C. Smith, H. Sundström, S. Berlin, and H. Ellegren. 2004. Male-biased mutation rate and divergence in autosomal, Z-linked and W-linked introns of chicken and turkey. Mol. Biol. Evol. 21: 1538–1547. **[4]**

Ayala, F. J. (ed.). 1976. *Molecular Evolution*. Sinauer Associates, Sunderland, MA. **[5]**

B

Babushok, D. V. and 8 others. 2007. A novel testis ubiquitin-binding protein gene arose by exon shuffling in hominoids. Genome Res. 17: 1129–1138. **[8, 11]**

Baer, C. F., M. M. Miyamoto, and D. R. Denver. 2007. Mutation rate variation in multicellular eukaryotes: Causes and consequences. Nature Rev. Genet. 8: 619–631. **[2, 4]**

Baer, C. F., M. M. Miyamoto, and D. R. Denver. 2007. Mutation rate variation in multicellular eukaryotes: Causes and consequences. Nat. Rev. Genet. 8: 619–631. **[4]**

Bailey, J. A. and E. E. Eichler. 2006. Primate segmental duplications: crucibles of evolution, diversity and disease. Nat. Rev. Genet. 7: 552–564. **[11]**

Bailey, J. A., G. Liu, and E. E. Eichler. 2003. An *Alu* transposition model for the origin and expansion of human segmental duplications. Am. J. Hum. Genet. 73: 823–834. **[11]**

Bailey, J. A., R. Baertsch, W. J. Kent, D. Haussler, and E. E. Eichler. 2004. Hotspots of mammalian chromosomal evolution. Genome Biol. 5: R23. **[11]**

Bailey, W. J., D. H. A. Fitch, D. A. Tagle, J. Czelusniak, J. L. Slightom, and M. Goodman. 1991. Molecular evolution of the globin gene locus: Gibbon phylogeny and the hominoid slowdown. Mol. Biol. Evol. 8: 155–184. **[5]**

Baillie, J. K. and 18 others. 2011. Somatic retrotransposition alters the genetic landscape of the human brain. Nature 479: 534–537. **[9]**

Bailliet, G., F. Rothhammer, F. R. Carnese, C. M. Bravi, and N. O. Bianchi. 1994. Founder mitochondrial haplotypes in Amerindian populations. Am. J. Hum. Genet. 54: 27–33. **[5]**

Bain, J. A. and H. F. Deutsch. 1947. An electrophoretic study of the egg white proteins of various birds. J. Biol. Chem. 171: 531–541. **[5]**

Bakewell, M. A., P. Shi, and J. Zhang. 2007. More genes underwent positive selection in chimpanzee evolution than in human evolution. Proc. Natl. Acad. Sci. USA 104: 7489–7494. **[4]**

Balasubramanian, S. and 8 others. 2009. Comparative analysis of processed ribosomal protein pseudogenes in four mammalian genomes. Genome Biol. 10: R2. **[4]**

Baldauf, S. L. 2008. An overview of the phylogeny and diversity of eukaryotes. J. Syst. Evol. 46: 263–273. **[5]**

Baldauf, S. L., D. Bhattacharya, J. Cockrill, P. Hugenholtz, J. Pawlowski, and A. G. B. Simpson. 2004. The tree of life, pp. 43–75. In: J. Cracraft and M. J. Donoghue (eds.), *Assembling the Tree of Life*. Oxford University Press, New York. **[6]**

Baldwin, B. G. and K. Andreasen. 2001. Unequal evolutionary rates between annual and perennial lineages of checker mallows (Sidalcea, Malvaceae): Evidence from 18S-26S rDNA internal and external transcribed spacers. Mol. Biol. Evol. 18: 936–944. **[4]**

Baldwin, B. G. and M. J. Sanderson. 1998. Age and rate of diversification of the Hawaiian silversword alliance (Compositae). Proc. Natl. Acad. Sci. USA 95: 9402–9406. **[5]**

Baltimore, D. 1981. Gene conversion: Some implications for immunoglobulin genes. Cell 24: 592–594. **[7]**

Bamforth, C. W. 1998. *Beer: Tap into the Art and Science of Brewing*. Plenum, New York. **[7]**

Bandelt, H.-J. 2005. Mosaics of ancient mitochondrial DNA: Positive indicators of nonauthenticity. Eur. J. Hum. Genet. 13: 1106–1112. **[5]**

Bandelt, H.-J., P. Forster, B. C. Sykes, and M. B. Richards. 1995. Mitochondrial portraits of human populations using median networks. Genetics 141: 743–753. **[6]**

Bandelt, H.-J., V. Macaulay, and M. Richards. 2000. Median networks: Speedy construction and greedy reduction, one simulation, and two case studies from human mtDNA. Mol. Phylogenet. Evol. 16: 8–28. **[6]**

Bansal, M. S., G. Banay, T. J. Harlow, J. P. Gogarten, and R. Shamir. 2013. Systematic inference of highways of horizontal gene transfer in prokaryotes. Bioinformatics 29: 571–579. **[6]**

Bao, W. and J. Jurka. 2010. Origin and evolution of LINE-1 derived "half-L1" retrotransposons (HAL1). Gene 465: 9–16. **[9]**

Bapteste, E. and 10 others. 2009. Prokaryotic evolution and the tree of life are two different things. Biol. Direct 4: 34. **[6]**

Barakat, A., G. Matassi, and G. Bernardi. 1998. Distribution of genes in the genome of *Arabidopsis thaliana* and its implication for the genome organization of plants. Proc. Natl. Acad. Sci. USA 95: 10044–10049. **[11]**

Barbash, D. A., D. F. Siino, A. M. Tarone, and J. Roote. 2003. A rapidly evolving *Myb*-related protein causes species isolation in *Drosophila*. Proc. Natl. Acad. Sci. USA 100: 5302–5307. **[5]**

Barker, D. 2004. LVB: Parsimony and simulated annealing in the search for phylogenetic trees. Bioinformatics 20: 274–275. **[5]**

Barker, W. C. and M. O. Dayhoff. 1980. Evolutionary and functional relationships of homologous physiological mechanisms. BioScience 30: 593–600. **[7]**

Barker, W. C., L. K. Ketcham, and M. O. Dayhoff. 1978. Duplications in protein sequences. pp. 359–362. In: M. O. Dayhoff (ed.), *Atlas of Protein Sequence and Structure*, Vol. 5, Supplement 3. National Biomedical Research Foundation, Silver Spring, MD. **[8]**

Barns, S. M., C. F. Delwiche, J. D. Palmer, and N. R. Pace. 1996. Perspectives in archaeal diversity, thermophyly and monophyly from environmental rRNA sequences. Proc. Natl. Acad. Sci. USA. 93: 9188–9193. **[6]**

Barolo, S. 2012. Shadow enhancers: Frequently asked questions about distributed *cis*-regulatory information and enhancer redundancy. Bioessays 34: 135–141. **[12]**

Barraclough, T. G. and A. P. Vogler. 2002. Recent diversification rates in North American tiger beetles estimated from a dated mtDNA phylogenetic tree. Mol. Biol. Evol. 19: 1706–1716. **[5]**

Barrell, B. G., A. T. Bankier, and J. Drouin. 1979. A different genetic code in human mitochondria. Nature 282: 189–194. **[10]**

Barrick, J. E. and 7 others. 2009. Genome evolution and adaptation in a long-term experiment with *Escherichia coli*. Nature 461: 1243–1247. **[13]**

Barrick, J. E. and R. E. Lenski. 2013. Genome dynamics during experimental evolution. Nature Rev. Genet. 14: 827–839. **[13]**

Barrick, J. E., M. R. Kauth, C. C. Strelioff, and R. E. Lenski. 2010. *Escherichia coli rpo*B mutants have increased evolvability in proportion to their fitness defects. Mol. Biol. Evol. 27: 1338–1347. **[13]**

Barsh, G. S., P. H. Seeburg, and R. E. Gelinas. 1983. The human growth hormone gene family: Structure and evolution of the chromosomal locus. Nuc. Acids Res. 11: 3939–3958. **[9]**

Bartolomé, C,. X. Bello, and X. Maside. 2009. Widespread evidence for horizontal transfer of transposable elements across *Drosophila* genomes. Genome Biol. 10: R22. **[11]**

Bates, L. A., K. N. Sayialel, N. W. Njiraini, C. J. Moss, J. H. Poole, and R. W. Byrne. 2007. Elephants classify human ethnic groups by odor and garment color. Curr. Biol. 17: 1938–1942. **[7]**

Bates, L. A., K. N. Sayialel, N. W. Njiraini, J. H. Poole, C. J. Moss, and R. W. Byrne. 2008. African elephants have expectations about the locations of out-of-sight family members. Biol. Lett. 4: 34–36. **[7]**

Bateson, W. 1909. Heredity and variation in modern lights, pp. 85–101. In: A. C. Seward (ed.), *Darwin and Modern Science*. Cambridge University Press, Cambridge. **[5]**

Bateson, W. 1909. *Mendel's Principles of Heredity*. Cambridge University Press, Cambridge. **[2]**

Baum, D. A. 2007. Concordance trees, concordance factors, and the exploration of reticulate genealogy. Taxon 56: 417–426. **[5]**

Bayes, T. 1763. An essay towards solving a problem in the doctrine of chances. Phil. Trans. Roy. Soc. 53: 370–418. **[5]**

Bazykin, G. A., F. A. Kondrashov, A. Y. Ogurtsov, S. Sunyaev, and A. S. Kondrashov. 2004. Positive selection at sites of multiple amino acid replacements since rat-mouse divergence. Nature 429: 558–562. **[4]**

Beale, D. and H. Lehmann. 1965. Abnormal haemoglobin and the genetic code. Nature 207: 259–261. **[4]**

Beaumont, H. J. E., J. Gallie, C. Kost, G. C. Ferguson, and P. B. Rainey. 2009. Experimental evolution of bet hedging. Nature 461: 90–93. **[13]**

Becker, W. M., L. J. Kleinsmith, J. Hardin, and G. P Bertoni. 2009. *The World of the Cell*, 7th Ed. Pearson, Toronto. **[4]**

Beeman, R. W., K. S. Friesen, and R. E. Denell. 1992. Maternal-effect selfish genes in flour beetles. Science 256: 89–92. **[2]**

Begun, D. J., H. A. Lindfors, A. D. Kern, and C. D. Jones. 2007. Evidence for *de novo* evolution of testis-expressed genes in the *Drosophila yakuba*/*Drosophila erecta* clade. Genetics 176: 1131–1137. **[8]**

Behura, S. K. 2007. Analysis of nuclear copies of mitochondrial sequences in honeybee (*Apis mellifera*) genome. Mol. Biol. Evol. 24: 1492–1505. **[9]**

Beiko, R. G., T. J. Harlow, and M. A. Ragan. 2005. Highways of gene sharing in prokaryotes. Proc. Natl. Acad. Sci. USA 102: 14332–14337. **[6]**

Beleza, S. and 8 others. 2013a. The timing of pigmentation lightening in Europeans. Mol. Biol. Evol. 30: 24–35. **[11]**

Beleza, S. and 15 others. 2013b. Genetic architecture of skin and eye color in an African-European admixed population. PLoS Genet. 9: e1003372. **[11]**

Belizario, J. E., J. Alves, M. Garay-Malpartida, and J. M. Occhiucci. 2008. Coupling caspase cleavage and proteasomal degradation of proteins carrying PEST motif. Curr. Protein Pept. Sci. 9: 210–220. **[7, 8]**

Bell, M. A. 1974. Reduction and loss of the pelvic girdle in *Gasterosteus* (Pisces): a case of parallel evolution. Nat. Hist. Mus. LA Contrib. Sci. 257: 1–36. **[12]**

Bennani-Baïti, I. M., S. L. Asa, D. Song, R. Iratni, S. A. Liebhaber, and N. E. Cooke. 1998. DNase I-hypersensitive sites I and II of the human growth hormone locus control region are a major developmental activator of somatotrope gene expression. Proc. Natl. Acad. Sci. USA 95: 10655–10660. **[12]**

Bennett, D. K. 1980. Stripes do not a zebra make. I. A cladistic analysis of *Equus*. Syst. Zool. 29: 272–287. **[5]**

Bennett, G. M. and N. A. Moran. 2013. Small, smaller, smallest: The origins and evolution of ancient dual symbioses in a phloem-feeding insect. Genome Biol. Evol. 5: 1675–1688. **[10]**

Bennett, M. D. 1971. The duration of meiosis. Proc. R. Soc. 178B: 277–299. **[11]**

Bennett, M. D. and I. J. Leitch. 2011. Nuclear DNA amounts in angiosperms: targets, trends and tomorrow. Ann. Bot. Mar. 107: 467–590. **[11]**

Bennetzen, J. L. 2005. Transposable elements, gene creation and genome rearrangement in flowering plants. Curr. Opin. Genet. Dev. 15: 621–627. **[9]**

Bennetzen, J. L. and E. A. Kellogg. 1997. Do plants have a one-way ticket to genomic obesity? Plant Cell l9: 1509–1514 **[11]**

Bensasson, D., D. Zhang, D. L. Hartl, and G. M. Hewitt. 2001. Mitochondrial pseudogenes: Evolution's misplaced witnesses. Trends Ecol. Evol. 16: 314–321. **[5]**

Benton, M. J. and P. C. Donoghue. 2007. Paleontological evidence to date the tree of life. Mol. Biol. Evol. 24: 26–53. **[5]**

Benveniste, R. E. 1985. The contributions of retroviruses to the study of mammalian evolution. pp. 359–417. In: R. I. MacIntyre (ed.), *Molecular Evolutionary Genetics*. Plenum, New York. **[9]**

Benveniste, R. E. and G. J. Todaro. 1976. Evolution of type C viral genes: Evidence for an Asian origin of man. Nature 261: 101–108. **[9]**

Berezikov, E. 2011. Evolution of microRNA diversity and regulation in animals. Nature Rev. Genet. 12: 846–860. **[12]**

Berg, D. E. and M. M. Howe (eds.). 1989. *Mobile DNA*. American Society for Microbiology, Washington, DC. **[9]**

Bergeron, A., J. Mixtacki, and J. Stoye. 2006. A unifying view of genome rearrangements, pp. 163–173. In: P. Bucher, B. M. E. Moret (eds.), *WABI '06: Proceedings of the Sixth International Workshop on Algorithms in Bioinformatics*. Springer, Berlin. **[10]**

Berglund, J., K. S. Pollard, and M. T. Webster. 2009. Hotspots of biased nucleotide substitutions in human genes. PLoS Biol. 7: e000026. **[4]**

Bergman, C. M., H. Quesneville, D. Anxolabéhère, and M. Ashburner. 2006 Recurrent insertion and duplication generate networks of transposable element sequences in the *Drosophila melanogaster* genome. Genome Biol. 7: R112. **[9]**

Bergsten, J. 2005. A review of long-branch attraction. Cladistics 21: 163–193. **[5, 6]**

Bernaola-Galván, P., R. Róman-Roldán, and J. R. Oliver. 1996. Compositional segmentation and long-range fractal correlation in DNA sequences. Phys. Rev. E 53: 5181–5189. **[10, 11]**

Bernardi, G. 1993a. The vertebrate genome: Isochores and evolution. Mol. Biol. Evol. 10: 186–204. **[11]**

Bernardi, G. 1993b. The isochore organization of the human genome and its evolutionary history: A review. Gene 135: 57–66. **[11]**

Bernardi, G. 1995. The human genome: Organization and evolutionary history. Annu. Rev. Genet. 29: 445–476. **[10, 11]**

Bernardi, G. 2000. The compositional evolution of vertebrate genomes. Gene 259: 31–43. **[11]**

Bernardi, G. 2001. Misunderstandings about isochores. Part 1. Gene 276: 3–13. **[11]**

Bernardi, G. and G. Bernardi. 1986. Compositional constraints and genome evolution. J. Mol. Evol. 24: 1–11. **[11]**

Bernardi, G. and 7 others. 1985. The mosaic genome of warm-blooded vertebrates. Science. 228: 953–958. **[11]**

Bernardi, G., S. Hughes, and D. Mouchiroud. 1997. The major compositional transitions in the vertebrate genome. J. Mol. Evol. 44: S44-S51. **[11]**

Bernt, M., A. Braband, M. Middendorf, B. Misof, O. Rota-Stabelli, and P. F. Stadler. 2013. Bioinformatics methods for the comparative analysis of metazoan mitochondrial genome sequences. Mol. Phylogenet. Evol. 69: 320–327. **[10]**

Berry, V. and O. Gascuel. 1996. On the interpretation of bootstrap trees: Appropriate threshold of clade selection and induced gain. Mol. Biol. Evol. 13: 999–1011. **[5]**

Bersaglieri, T. and 8 others. 2004. Genetic signatures of strong recent positive selection at the lactase gene. Am. J. Hum. Genet. 74: 1111–1120. **[2]**

Besenmatter, W., P. Kast, and D. Hilvert. 2007. Relative tolerance of mesostable and thermostable protein homologs to extensive mutation. Proteins 66: 500–506. **[4]**

Beukeboom, L. W. 1994. Bewildering Bs: An impression of the first B-chromosome conference. Heredity 73: 326–336. **[2]**

Bhattacharyya, M., A. Smith, T. H. Ellis, C. Hedley, and C. Martin. 1990. The wrinkled-seed character of pea described by Mendel in caused by a transposon-like insertion in a gene encoding starch-branching enzyme. Cell 60: 115–122. **[9]**

Bhattachayya, M., C. Martin, and A. Smith. 1993. The importance of starch biosynthesis in the wrinkled seed shape character of peas studied by Mendel. Plant Mol. Biol. 22: 525–531. **[9]**

Biémont, C. and C. Vieira. 2006. Junk DNA as an evolutionary force. Nature 443: 521–524. **[9, 11]**

Bierne, N. and A. Eyre-Walker. 2003. The problem of counting sites in the estimation of the synonymous and nonsynonymous substitution rates: Implications for the correlation between the synonymous substitution rate and codon usage bias. Genetics 165: 1587–1597. **[4]**

Bierne, N. and A. Eyre-Walker. 2004. Genomic rate of adaptive amino acid substitution in *Drosophila*. Mol. Biol. Evol. 21: 1350–1360. **[2]**

Bininda-Emonds, O. R. P. 2004. Phylogenetic Supertrees: Combining Information to Reveal the Tree of Life. Kluwer, Dordrecht, Netherlands. **[5]**

Bininda-Emonds, O. R. P., J. L. Gittleman, and M. A. Steel. 2002. The (super)tree of life: Procedures, problems, and prospects. Annu. Rev. Ecol. Syst. 33: 265–289. **[5]**

Bird, A. P. 1993. Functions for DNA methylation in vertebrates. Cold Spr. Harb. Symp. Quant. Biol. 58: 281–285. **[8]**

Bird, A. 2007. Perceptions of epigenetics. Nature 447: 396–398. **[12]**

Bird, A. P. and S. Tweedie. 1995. Transcriptional noise and the evolution of gene number. Philos. Trans. Roy. Soc. 349B: 249–253. **[11]**

Bishop, J. M. 1981. Enemies within: The genesis of retrovirus oncogenes. Cell 23: 5–6. **[9]**

Biswas, S. and J. M. Akey. 2006. Genomic insight into positive selection. Trends Genet. 22: 437–445. **[2]**

Björklund, Å. K., D. Ekman, and A. Elofsson. 2006. Expansion of protein domain repeats. PLoS Comput. Biol. 2: e114. **[7, 8]**

Björklund, Å. K., S. Light, R. Sagit, and A. Elofsson. 2010. Nebulin: A study of protein repeat evolution. J. Mol. Biol. 402: 38–51. **[8]**

Black, J. A. and G. H. Dixon. 1968. Amino acid sequence of α chains of human haptoglobins. Nature 218: 736–741. **[8]**

Blackburn, E. H. and J. G. Gall. 1978. A tandemly repeated sequence at the termini of the extrachromosomal ribosomal RNA genes in *Tetrahymena*. J. Mol. Biol. 120: 33–53 **[11]**

Blaisdell, B. E. 1985. A method for estimating from two aligned present-day DNA sequences, their ancestral composition and subsequent rates of substitution, possibly different in the two lineages, corrected for multiple and parallel substitutions at the same site. J. Mol. Evol. 22: 69–81. **[3]**

Blakeslee, A. F. and B. T. Avery. 1919. Mutations in the Jimson weed. J. Hered. 10: 111–120. **[7]**

Blakeslee, A. F., J. Belling, and M. E. Farnham. 1920. Chromosomal duplication and Mendelian phenomena in *Datura* mutants. Science New Ser. 52: 388–390. **[7]**

Blanchard, J. L. and G. W. Schmidt. 1996. Mitochondrial DNA migration events in yeast and humans: Integration by a common end-joining mechanism and alternative perspectives on nucleotide substitution patterns. Mol. Biol. Evol. 13: 537–548. **[9]**

Blanchette, M., T. Kunisawa, and D. Sankoff. 1999. Gene order breakpoint evidence in animal mitochondrial phylogeny. J. Mol. Evol. 49: 193–203. **[10]**

Blount, Z. D., C. Z. Borland, and R. E. Lenski. 2008. Historical contingency and the evolution of a key innovation in an experimental population of *Escherichia coli*. Proc. Natl. Acad. Sci. USA 105: 7899–7906. **[13]**

Blount, Z. D., J. E. Barrick, C. J. Davidson, and R. E. Lenski. 2012. Genomic analysis of a key innovation in an experimental *Escherichia coli* population. Nature 489: 513–518. **[13]**

Bodmer, W. F. and L. L. Cavalli-Sforza. 1976. *Genetics, Evolution, and Man*. W.H. Freeman, San Francisco. **[2]**

Bodył, A. and K. Moszczynski. 2006. Did the peridinin plastid evolve through tertiary endosymbiosis? A hypothesis. Eur. J. Phycol. 41: 435–448. **[5]**

Boettiger, A. N. and M. Levine. 2009. Synchronous and stochastic patterns of gene activation in the *Drosophila* embryo. Science 325: 471–473. **[12]**

Boguski, M. S. 1992. A molecular biologist visits Jurassic Park. Biotechniques 12: 668–669. **[5]**

Böhne, A., F. Brunet, D. Galiana-Arnoux, C. Schultheis, and J. N. Volff. 2008. Transposable elements as drivers of genomic and biological diversity in vertebrates. Chromosome Res. 16: 203–215. **[9]**

Boore, J. L. 2006. The use of genome-level characters for phylogenetic reconstruction. Trends Ecol. Evol. 21: 439–446. **[11]**

Borensztajn, K., O. Chafa, M. Alhenc-Gelas, S. Salha, A. Reghis, A. M. Fischer, and J. Tapon-Bretaudière. 2002. Characterization of two novel splice site mutations in human factor VII gene causing severe plasma factor VII deficiency and bleeding diathesis. Br. J. Haematol. 117: 168–171. **[9]**

Borggrefe, T. and X. Yue. 2011. Interactions between subunits of the Mediator complex with gene-specific transcription factors. Semin. Cell Dev Biol. 22: 759–768. **[12]**

Boschetti, C. and 8 others. 2012. Biochemical diversification through foreign gene expression in bdelloid rotifers. PLoS Genet. 8: e1003035. **[9]**

Bosco, G., P. Campbell, J. T. Leiva-Neto, and T. A. Markow. 2007. Analysis of *Drosophila* species genome size and satellite DNA content reveals significant differences among strains as well as between species. Genetics 177: 1277–1290. **[11]**

Bostock, C. J. 1986. Mechanisms of DNA sequence amplification and their evolutionary consequences. Phil. Trans. Roy. Soc. 312B: 261–273. **[11]**

Boucher, Y. and E. Bapteste. 2009. Revisiting the concept of lineage in prokaryotes: A phylogenetic perspective. BioEssays 31: 526–536. **[6]**

Bova, M. P., L. L. Ding, J. Horwitz, and B. K. Fung. 1997. Subunit exchange of αA crystallin. J. Biol. Chem. 272: 29511–29517. **[4]**

Bovine Genome Sequencing and Analysis Consortium. 2009. The genome sequence of taurine cattle: A window to ruminant biology and evolution. Science 324: 522–528. **[11]**

Bowmaker, J. K. and H. J. A. Dartnall. 1980. Visual pigments of rods and cones in a human retina. J. Physiol. 298: 501–511. **[7]**

Boxma, B. and 13 others. 2005. An anaerobic mitochondrion that produces hydrogen. Nature 434: 74–79. **[6]**

Boyden, A. and D. Gemeroy. 1950. The relative position of the Cetacea among the orders of Mammalia as indicated by precipitin tests. Zoologica 35: 145–151. **[5]**

Brameier, M. 2010. Genome-wide comparative analysis of microRNAs in three non-human primates. BMC Res. Notes 3: 64 **[11]**

Branciforte, D. and S. L. Martin. 1994. Developmental and cell type specificity of *LINE-1* expression in mouse testis: Implication for transposition. Mol. Cell Biol. 14: 2584–2592. **[9]**

Brandvain, Y. and G. Coop. 2012. Scrambling eggs: Meiotic drive and the evolution of female recombination rates. Genetics 190: 709–723. **[2]**

Braunitzer, G., R. Gehring-Muller, N. Hilschmann, K. Hilse, G. Hobom, V. Rudolf, and B. Wittmann-Liebold. 1961. Die konstitution des normalen adulten hämoglobins. Z. Physiol. Chem. Hoppe-Seyler 325: 283–286. **[7]**

Bray, N., I. Dubchak, and L. Pachter. 2003. AVID: A global alignment program. Genome Res. 13: 97–102. **[3]**

Breen, M. S., C. Kemena, P. K. Vlasov, C. Notredame, and F. A. Kondrashov. 2012. Epistasis as the primary factor in molecular evolution. Nature 490: 535–538. **[4]**

Brennan, A. C., D. Barker, S. J. Hiscock, and R. J. Abbott. 2012. Molecular genetic and quantitative trait divergence associated with recent homoploid hybrid speciation: A study of *Senecio squalidus* (Asteraceae). Heredity 108: 87–95. **[7]**

Brenner, D. J., J. T. Staley, and N. R. Krieg. 2005. Classification of procaryotic organisms and the concept of bacterial speciation, pp. 27–32. In: D. J. Brenner, N. R. Krieg, J. T. Staley, and G. Garrity (eds.), *Bergey's Manual of Systematic Bacteriology*. Springer, New York. **[6]**

Brenner S. 1998. Refuge of spandrels. Curr. Biol. 8: R669. **[11]**

Breukelman, H. J., N. van der Munnik, R. G. Kleineidam, A. Furia, and J. J. Beintema. 1998. Secretory ribonuclease genes and pseudogenes in true ruminants. Gene 212: 259–268. **[7]**

Brideau, N. J., H. A. Flores, J. Wang, S. Maheswari, X. Wang, and D. A. Barbash. 2006. Two Dobzhansky-Muller genes interact to cause hybrid lethality in *Drosophila*. Science 314: 1292–1295. **[5]**

Bridges, C. B. 1936. The *Bar* "gene"—a duplication. Science 83: 210–211. **[7]**

Britten, R. J. 1986. Rates of DNA sequence evolution differ between taxonomic groups. Science 231: 1393–1398. **[4]**

Britten, R. J. 1994. Evidence that most *Alu* sequences were inserted in a process that ceased about 30 million years ago. Proc. Natl. Acad. Sci. USA 91: 6148–6150. **[11]**

Britten, R. J. and D. E. Kohne. 1968. Repeated sequences in DNA. Science 161: 529–540. **[11]**

Brocchieri, L. 2014. The GC content of bacterial genomes. J. Phylogen. Evol. Biol. 2: e108. **[10]**

Brodie of Brodie, E. B., S. Nicolay, M. Touchon, B. Audit, Y. d'Aubenton-Carafa, C. Thermes, and A. Arneodo. 2005. From DNA sequence analysis to modeling replication in the human genome. Phys. Rev. Lett. 94: 248103. **[10]**

Brookfield, J. F. 2000. Genomic sequencing: the complexity conundrum. Curr. Biol. 10: R514- R515. **[11]**

Brookfield, J. F. Y. 2005. The ecology of the genome: Mobile DNA elements and their hosts. Nat. Rev. Genet. 6: 128–136. **[9]**

Brookfield, J. F. Y. and R. M. Badge. 1997. Population genetics models of transposable elements. Genetica 100: 281–294. **[11]**

Brown, D. D. and K. Sugimoto. 1973. 5S DNAs of *Xenopus laevis* and *Xenopus mulleri*: Evolution of tandem genes. J. Mol. Biol. 63: 57–73. **[7]**

Brown, D. D., P. C. Wensink, and E. Jordan. 1972. A comparison of the ribosomal DNAs of *Xenopus laevis* and *Xenopus mulleri*: Evolution of tandem genes. J. Mol. Biol. 63: 57–73. **[7]**

Brown, T. A. and K. A. Brown. 1994. Using molecular biology to explore the past. BioEssays 16: 719–726. **[5, 6]**

Brown, W. M., E. M. Prager, A. Wang, and A. C. Wilson. 1982. Mitochondrial DNA sequences of primates: Tempo and mode of evolution. J. Mol. Evol 18: 225–239. **[4, 5]**

Brown, W. M., M. George, and A. C. Wilson. 1979. Rapid evolution of animal mitochondrial DNA. Proc. Natl. Acad. Sci. USA 76: 1967–1971. **[4]**

Brunet, T. D. P., and W. F. Doolittle. 2014. Getting "function" right. Proc. Natl. Acad. Sci. USA 111: E3365. **[11]**

Bryk, J., E. Hardouin, I. Pugach, D. Hughes, R. Strotmann, M. Stoneking, and S. Myles. 2008. Positive selection in East Asians for an EDAR allele that enhances NF-κB activation. PLoS One 3: e2209. **[11]**

Buckling, A., R. C. Maclean, M. A. Brockhurst, and N. Colegrave. 2009. The Beagle in a bottle. Nature 457: 824–829. **[13]**

Buljan, M. and A. Bateman. 2009. The evolution of protein domain families. Biochem. Soc. Trans. 37: 751–755. **[8]**

Bull, J. J. and 7 others. 1997. Exceptional convergent evolution in a virus. Genetics 147: 1497–1507. **[13]**

Bulmer, M. 1991. The selection-mutation-drift theory of synonymous codon usage. Genetics 129: 897–907. **[4]**

Buneman, P. 1971. The recovery of trees from measurements of dissimilarity. pp. 387–395. In: E. R. Hodson, D. G. Kendall, and P. Tautu (eds.), *Mathematics in the Archaeological and Historical Sciences*. Edinburgh University Press, Edinburgh. **[5]**

Buonagurio, D. A., S. Nakada, W. M. Fitch, and P. Palese. 1986. Epidemiology of influenza C virus in man: Multiple evolutionary lineages and low rate of change. Virology 153: 12–21. **[4]**

Burger, G., M. W. Gray, L. Forget, and B. F. Lang. 2013. Strikingly bacteria-like and gene-rich mitochondrial genomes throughout jakobid protists. Genome Biol. Evol. 5: 418–438. **[10]**

Burgess, J. 1985. *An Introduction to Plant Cell Development*. Cambridge University Press, Cambridge. **[11]**

Burt, A. and R. Trivers. 2006. *Genes in Conflict: The Biology of Selfish Genetic Elements*. Belknap / Harvard University Press, Cambridge, MA. **[2]**

Buscaino, A., R. Allshire, and A. Pidoux. 2010. Building centromeres: home sweet home or a nomadic existence? Curr. Opin. Genet. Dev. 20: 118–126. **[11]**

Buschiazzo, E., and N. J. Gemmell. 2006. The rise, fall, and renaissance of microsatellites in eukaryotic genomes. Bioessays 28: 1040–1050. **[11]**

Bush, R. M., W. M. Fitch, C. A. Bender, and N. J. Cox. 1999. Positive selection on the H3 hemagglutinin gene of human influenza virus A. Mol. Biol. Evol. 16: 1457–1465. **[5]**

Bustamante, C. D., R. Nielsen, S. A. Sawyer, K. M. Olsen, M. D. Purugganan, and D. L. Hartl. 2002. The cost of inbreeding in *Arabidopsis*. Nature 416: 531–534. **[2]**

Byres, E. and 14 others. 2008. Incorporation of a non-human glycan mediates human susceptibility to a bacterial toxin. Nature 456: 648–652. **[8]**

C

Caccone, A. and J. R. Powell. 1989. DNA divergence among hominoids. Evolution 43: 925–942. **[5]**

Camacho, J. P., T. F. Sharbel, and L. W. Beukeboom. 2000. B-chromosome evolution. Philos. Trans. R. Soc. 355B: 163–178. **[11]**

Camin J. H. and R. R. Sokal. 1965. A method for deducing branching sequences in phylogeny. Evolution 19: 311–326. **[5]**

Campbell, C. D., and 12 others. 2012. Estimating the human mutation rate using autozygosity in a founder population. Nat. Genet. 44: 1277–1281. **[4]**

Campbell, C. L. and P. E. Thorsness. 1998. Escape of mitochondrial DNA to the nucleus in *yme1* yeast is mediated by vacuolar-dependent turnover of abnormal mitochondrial compartments. J. Cell Sci. 111: 2455–2464. **[9]**

Campuzano, V. and 26 others. 1996. Friedreich's ataxia: Autosomal recessive disease caused by an intronic GAA triplet repeat expansion. Science 271: 1423–1427. **[4]**

Cancherini, D., G. França, and S. J. de Souza. 2010. The role of exon shuffling in shaping protein-protein interaction network. BMC Genomics 11: S11. **[8]**

Cann, R. L., M. Stoneking, and A. C. Wilson. 1987. Mitochondrial DNA and human evolution. Nature 325: 31–36. **[5]**

Cannarozzi, G. and 8 others. 2010. A role for codon order in translation dynamics. Cell 141: 355–367. **[4]**

Cannarozzi, G. M. and A. Schneider (eds.). 2012. *Codon Evolution: Mechanisms and Models*. Oxford University Press, Oxford. **[3]**

Cano, R. J. and H. N. Poinar. 1993b. Rapid isolation of DNA from fossil and museum specimens suitable for PCR. Biotechniques 15: 432–434. **[5]**

Cano, R. J., H. N. Poinar, N. J. Pieniazek, A. Acra, and G. O. Poinar. 1993a. Amplification and sequencing of DNA from a 120–135-million-year-old weevil. Nature 363: 536–538. **[5]**

Cao, Y., J. Adachi, A. Janke, S. Pääbo, and M. Hasegawa. 1994. Phylogenetic relationships among eutherian orders estimated from inferred sequences of mitochondrial proteins: Instability of a tree based on a single gene. J. Mol. Evol. 39: 519–527. **[5]**

Cappello, J., S. M. Cohen, and H. F. Lodish. 1984. *Dictyostelium* transposable element *DIRS-1* preferentially inserts into *DIRS-1* sequences. Mol. Cell. Biol. 4: 2207–2213. **[9]**

Capy, P., T. Langin, and D. Anxolabéhère. 1998. *Dynamics and Evolution of Transposable Elements*. Chapman & Hall, New York. **[9]**

Carels, N., A. Barakat, and G. Bernardi. 1995. The gene distribution of the maize genome. Proc. Natl. Acad. Sci. USA 92: 11057–11060. **[11]**

Carlson, C. S., D. J. Thomas, M. A. Eberle, J. E. Swanson, R. J. Livingston, M. J. Rieder, and D. A. Nickerson. 2005. Genomic regions exhibiting positive selection identified from dense genotype data. Genome Res. 15: 1553–1565. **[11]**

Carlsten, J. O. P., X. Zhu, and C. M. Gustafsson. 2013. The multitalented mediator complex. Trends Biochem. Sci. 38: 531–537. **[12]**

Carmel, L., Y. I. Wolf, I. B. Rogozin, and E. V. Koonin. 2007. Three distinct modes of intron dynamics in the evolution of eukaryotes. Genome Res. 17: 1034–1044. **[8]**

Carroll, S. B. 2005. Evolution on two levels: On genes and form. PLoS Biology 3: 1159–1166. **[12]**

Cartegni, L., S. L. Chew, and A. R. Krainer. 2002. Listening to silence and understanding nonsense: Exonic mutations that affect splicing. Nat. Rev. Genet. 3: 285–298. **[1, 8]**

Carter, A. M. 2012. Evolution of placental function in mammals: the molecular basis of gas and nutrient transfer, hormone secretion, and immune responses. Physiol. Rev. 92: 1543–1576. **[12]**

Carver, R. P. 1993. The case against statistical significance testing, revisited. J. Exp. Edu. 61: 287–292. **[11]**

Carvunis, A.-R. and 15 others. 2012. Proto-genes and *de novo* gene birth. Nature 487: 370–374. **[8]**

Casacuberta, E. and M. L. Pardue. 2005. *HeT-A* and *TART*, two *Drosophila* retrotransposons with a bona fide role in chromosome structure for more than 60 million years. Cytogenet. Genome Res. 110: 152–159. **[9]**

Casane, D. and P. Laurenti. 2013. Why coelacanths are not "living fossils." A review of molecular and morphological data. BioEssays 35: 332–338. **[4]**

Casneuf, T., S. De Bodt, J. Raes, S. Maere, and Y. Van de Peer. 2006. Nonrandom divergence of gene expression following gene and genome duplications in the flowering plant *Arabidopsis thaliana*. Genome Biol. 7: R13. **[7]**

Castresana, J., G. Feldmaier-Fuchs, and S. Pääbo. 1998. Codon reassignment and amino acid composition in hemichordate mitochondria. Proc. Natl. Acad. Sci. USA 95: 3703–3707. **[10]**

Castrillo, J. I. and 24 others. 2007. Growth control of the eukaryote cell: a systems biology study in yeast. J. Biol. 6: 4. **[12]**

Cavalier-Smith, T. (ed). 1985. *The Evolution of Genome Size*. Wiley, New York. **[11]**

Cavalier-Smith, T. 1975. The origin of nuclei and of eukaryotic cells. Nature 256: 463–468 **[5]**

Cavalier-Smith, T. 1978. Nuclear volume control by nucleoskeletal DNA, selection for cell volume and cell growth rate, and the solution of the DNA C-value paradox. J. Cell Sci. 34: 247–278. **[11]**

Cavalier-Smith, T. 1978. Nuclear volume control by nucleoskeletal DNA, selection for cell volume and cell growth rate, and the solution of the DNA C-value paradox. J. Cell Sci. 34: 247–278. **[7]**

Cavalier-Smith, T. 1982. Skeletal DNA and the evolution of genome size. Annu. Rev. Biophys. Bioeng. 11: 273–302. **[11]**

Cavalier-Smith, T. 1983. A six-kingdom classification and a unified phylogeny, pp. 1027–1034. In: W. Schemmler and H. E. A. Schenk (eds.) *Endocytobiology*, Vol. 2. de Gruyter, Berlin. **[5]**

Cavalier-Smith, T. 1983. Genetic symbionts and the origin of split genes and linear chromosomes. pp. 29–45. In: H. E. A. Schenk and W. Schwemmler (eds.), *Endocytobiology II: Intracellular Space as Oligogenetic Ecosystem*. de Gruyter, Berlin. **[6]**

Cavalier-Smith, T. 1985. Selfish DNA and the origin of introns. Nature 315: 283–284. **[7, 8]**

Cavalier-Smith, T. 1985. Cell volume and the evolution of eukaryotic genome size. In: T. Cavalier-Smith (ed.), *The Evolution of Genome Size*, pp. 104–184. Wiley, New York. **[11]**

Cavalier-Smith, T. 1989. Archaebacteria and Archezoa. Nature 339: 100–101. **[5]**

Cavalier-Smith, T. 1989. Concept of a bacterium still valid in prokaryote debate. Nature 446: 257. **[6]**

Cavalier-Smith, T. 1991. Intron phylogeny: A new hypothesis. Trends Genet. 7: 145–148. **[6, 8]**

Cavalier-Smith, T. 1993. The origin, losses and gains of chloroplasts. In: R. A. Lewin (ed.), *Origins of Plastids: Symbiogenesis, Prochlorophytes and the Origins of Chloroplasts*, pp. 291–348. Chapman & Hall, New York. **[11]**

Cavalier-Smith, T. 2005. Economy, speed and size matter: evolutionary forces driving nuclear genome miniaturization and expansion. Ann. Bot. 95: 147–175. **[11]**

Cavalier-Smith, T. 2007. Concept of a bacterium still valid in prokaryote debate. Nature 446: 257. **[5]**

Cavalier-Smith, T. 2009. Predation and eukaryote cell origins: A coevolutionary perspective. Int. J. Biochem. Cell Biol. 41: 307–322. **[5]**

Cavalier-Smith, T. and J. J. Lee. 1985. Protozoa as hosts for endosymbioses and the conversion of symbionts into organelles. J. Protozool. 32: 376–379. **[5]**

Cavalli-Sforza, L. L. and A. W. F. Edwards. 1967. Phylogenetic analysis: Models and estimation procedures. Am. J. Hum. Genet. 19: 233–257. **[5]**

Cech, T. R. 1986. The generality of self-splicing RNA: Relationship to nuclear mRNA splicing. Cell 44: 207–210. **[6]**

Cedergren, R., M. W. Gray, Y. Abel, and D. Sankoff. 1988. The evolutionary relationships among known life forms. J. Mol. Evol. 28: 98–112. **[5]**

Celniker, S. E. and G. M. Rubin. 2003. The *Drosophila melanogaster* genome. Annu. Rev. Genomics Hum. Genet. 4: 89–117. **[8]**

Chaboissier, M.-C., B. Bucheton, and D. J. Finnegan. 1998. Copy number control of a transposable element, the I factor, a LINE-like element in *Drosophila*. Proc. Natl. Acad. Sci. USA 95: 11781–11785. **[9]**

Chain, F. J. J. and B. J. Evans. 2006. Multiple mechanisms promote the retained expression of gene duplicates in the tetraploid frog *Xenopus laevis*. PLoS Genet 2: e56. **[7]**

Chain, P. S. and 22 others. 2004. Insights into the evolution of *Yersinia pestis* through whole-genome comparison with *Yersinia pseudotuberculosis*. Proc. Natl. Acad. Sci. USA 101: 13826–13831. **[9]**

Chalopin, D., M. Naville, F. Plard, D. Galiana, and J. N. Volff. 2015. Comparative analysis of transposable elements highlights mobilome diversity and evolution in vertebrates. Genome Biol. Evol. 7: 567–580. **[11]**

Chamary, J.-V., J. L. Parmley, and L. D. Hurst. 2006. Hearing silence: Non-neutral evolution at synonymous sites in mammals. Nat. Rev. Genet. 7: 98–108. **[4]**

Chan, Y. F. and 14 others. 2010. Adaptive evolution of pelvic reduction in sticklebacks by recurrent deletion of a *Pitx1* enhancer. Science 327: 302–305. **[12]**

Chang, B. H.-J., L. C. Shimmin, S.-K. Shuye, D. Hewett-Emmett, and W.-H. Li. 1994. Weak male-driven molecular evolution in rodents. Proc. Natl. Acad. Sci. USA 91: 827–831. **[4]**

Chang, B. S., K. Jönsson, M. A. Kazmi, M. J. Donoghue, and T. P. Sakmar. 2002. Recreating a functional ancestral archosaur visual pigment. Mol. Biol. Evol. 19: 1483–1489. **[5]**

Chang, G. S., Y. Hong, K. D. Ko, G. Bhardwaj, E. C. Holmes, R. L. Patterson, and D. B. van Rossum. 2008. Phylogenetic profiles reveal evolutionary relationships within the "twilight zone" of sequence similarity. Proc. Natl. Acad. Sci. USA 105: 13474–13479. **[9]**

Chang, L. Y. and J. L. Slightom. 1984. Isolation and nucleotide sequence analysis of the β-type globin pseudogene from human, gorilla and chimpanzee. J. Mol. Biol. 180: 767–784. **[7]**

Chargaff, E. 1950. Chemical specificity of nucleic acids and mechanism of their enzymic degradation. Experientia 6: 201–209. **[10]**

Chargaff, E. 1951. Structure and function of nucleic acids as cell constituents. Fed. Proc. 10: 654–659. **[10]**

Chargaff, E. 1963. *Essays on Nucleic Acids*. Elsevier, Amsterdam. **[10]**

Chargaff, E. 1979. How genetics got a chemical education. Ann. N. Y. Acad. Sci. 325: 345–360. **[10]**

Charlesworth, B. 1985. The population genetics of transposable elements. pp. 213–232. In: T. Ohta and K. Aoki (eds.), *Population Genetics and Molecular Evolution*. Springer, Berlin. **[9]**

Charlesworth, B. 1988. The maintenance of transposable elements in natural populations. pp. 189–212. In: O. J. Nelson (ed.), *Plant Transposable Elements*. Plenum, New York. **[9]**

Charlesworth, B. 1994. Genetic recombination. Patterns in the genome. Curr. Biol. 4: 182–184. **[11]**

Charlesworth, B. 1994. The effect of background selection against deleterious mutations on weakly selected, linked variants. Genet. Res. 63: 213–227. **[2]**

Charlesworth, B. 2009. Effective population size and patterns of molecular evolution and variation. Nature Rev. Genet. 10: 195–2005. **[2]**

Charlesworth, B. and C. H. Langley. 1989. The population genetics of *Drosophila* transposable elements. Annu. Rev. Genet. 23: 251–287. **[9]**

Charlesworth, B. and D. L. Hartl. 1978. Population dynamics of the segregation distorter polymorphism of *Drosophila melanogaster*. Genetics 89: 171–192. **[2]**

Charlesworth, B., A. Lapid, and D. Canada. 1992. The distribution of transposable elements within and between chromosomes in a population of *Drosophila melanogaster.* I. Element frequencies and distribution. Genet. Res. 60: 103–114. **[9]**

Charlesworth, B., J. A. Coyne, and N. H. Barton. 1987. The relative rates of evolution of sex chromosomes and autosomes. Am. Nat. 130: 113–146. **[4]**

Charlesworth, B., M. T. Morgan, and D. Charlesworth. 1993. The effect of deleterious mutations on neutral molecular variation. Genetics 134: 1289–1303. **[2]**

Chatton, É. 1925. *Pansporella perplexa.* Réflexions sur la biologie et la phylogénie des protozoaires. Ann. Sci. Nat. Zool. 7: 1–84. **[6]**

Chen, C. and C. W. Chen. 2007. Quantitative analysis of mutation and selection pressures on base composition skews in bacterial chromosomes. BMC Genomics 8: 286. **[10]**

Chen, J.-M., D. N. Cooper, N. Chuzhanova, C. Férec, and G. P. Patrinos. 2007. Gene conversion: Mechanisms, evolution and human disease. Nat. Rev. Genet. 8: 762–775. **[7]**

Chen, K. and N. Rajewsky. 2007. The evolution of gene regulation by transcription factors and microRNAs. Nature Rev. Genet. 8: 93–103. **[12]**

Chen, L., A. L. DeVries, and C.-H. C. Cheng. 1997. Convergent evolution of antifreeze glycoproteins in Antarctic notothenioid fish and Arctic cod. Proc. Natl. Acad. Sci. USA 94: 3817–3822. **[7]**

Chen, S. T., H. C. Cheng, D. A. Barbash, and H. P. Yang. 2007. Evolution of *hydra,* a recently evolved testis-expressed gene with nine alternative first exons in *Drosophila melanogaster.* PLoS Genet. 3: e107. **[8]**

Chen, S., Y. E. Zhang, and M. Long. 2010. New genes in *Drosophila* quickly become essential. Science 330: 1682–1685. **[8]**

Chen, X. and J. Zhang. 2013. Why are genes encoded on the lagging strand of the bacterial genome? Genome Biol. Evol. 5: 2436–2439. **[10]**

Chen, Y. H. and 19 others. 2003. *KCNQ1* gain-of-function mutation in familial atrial fibrillation. Science 299: 251–254. **[11]**

Chénais, B., A. Caruso, S. Hiard, and N. Casse. 2012. The impact of transposable elements on eukaryotic genomes: From genome size increase to genetic adaptation to stressful environments. Gene 509: 7–15. **[11]**

Cheng, C. H., L. Chen, T. J. Near, Y. Jin. 2003. Functional antifreeze glycoprotein genes in temperate-water New Zealand nototheniid fish infer an Antarctic evolutionary origin. Mol. Biol. Evol. 20: 1897–1908. **[8]**

Cheng, C.-H. C., L. Chen, T. J. Near, and Y. Jin. 2003. Functional antifreeze glycoprotein genes in temperate-water New Zealand Nototheniid fish infer an Antarctic evolutionary origin. Mol. Biol. Evol. 20: 1897–1908. **[7]**

Cherry, J. M. and 10 others. 1997. Genetic and physical maps of *Saccharomyces cerevisiae.* Nature 387: 67–73. **[7]**

Chetverina, D., T. Aoki, M. Erokhin, P. Georgiev, and P. Schedl. 2014. Making connections: Insulators organize eukaryotic chromosomes into independent *cis*-regulatory networks. Bioessays 36: 163–172. **[12]**

Chiarelli, A. B. 1973. *Evolution of the Primates: An Introduction to the Biology of Man.* Academic Press, London. **[5]**

Chimpanzee Sequencing and Analysis Consortium. 2005. Initial sequence of the chimpanzee genome and comparison with the human genome. Nature 437: 69–87. **[11]**

Chirala, S. S., M. A. Kuziora, D. M. Spector, and S. J. Wakil. 1987. Complementation of mutations and nucleotide sequence of *FAS1* gene encoding β subunit of yeast fatty acid synthase. J. Biol. Chem. 262: 4231–4240. **[8]**

Chivian, D. and 19 others. 2008. Environmental genomics reveals a single-species ecosystem deep within Earth. Science 322: 275–278. **[10]**

Cho, J. H. and S. R. Brant. 2011. Recent insights into the genetics of inflammatory bowel disease. Gastroenterology 140: 1704–1712. **[11]**

Choi, S. S., E. J. Vallender, and B. T. Lahn. 2006. Systematically assessing the influence of 3-dimensional structural context on the molecular evolution of mammalian proteomes. Mol. Biol. Evol. 23: 2131–2133. **[4]**

Chojnowski, J. L., J. Franklin, Y. Katsu, T. Iguchi, L. J. Guillette, R. T. Kimball, and E. L. Braun. 2007. Patterns of vertebrate isochore evolution revealed by comparison of expressed mammalian, avian, and crocodilian genes. J. Mol. Evol. 65: 259–266. **[11]**

Chothia, C., J. Gough, C. Vogel, and S. A. Teichmann. 2003. Evolution of the protein repertoire. Science 300: 1701–1703. **[8, 11]**

Chou, H. H., H. C. Chiu, N. F. Delaney, D. Segrè, and C. J. Marx. 2011. Diminishing returns epistasis among beneficial mutations decelerates adaptation. Science 332: 1190–1192. **[13]**

Chou, S. and 13 others. 2015. Transferred interbacterial antagonism genes augment eukaryotic innate immune function. Nature 518: 98–101. **[9, 11]**

Choudhary, M., H. Cho, A. Bavishi, C. Trahan, and B.-E. Myagmarjav. 2012. Evolution of multipartite genomes in prokaryotes, pp. 301–323. In: P. Pontarotti (ed.), *Evolutionary Biology: Mechanisms and Trends.* Springer, Berlin. **[10]**

Chuang, J. H. and H. Li. 2007. Similarity of synonymous substitution rates across mammalian genomes. J. Mol. Evol. 65: 236–248. **[4]**

Chun, S. and J. C. Fay. 2009. Identification of deleterious mutations within three human genomes. Genome Res. 19: 1553–1561. **[11]**

Ciccarelli, F. D., T. Doerks, C. von Mering, C. J. Creevey, B. Snel, and P. Bork. 2006. Toward automatic reconstruction of a highly resolved tree of life. Science 311: 1283–1287. **[5, 6]**

Ciochon, R. L. 1985. Hominoid cladistics and the ancestry of modern apes and humans. pp. 345–362. In: R. L. Ciochon and J. C. Fleagle (eds.), *Primate Evolution and Human Origin.* Benjamin/Cummings, Menlo Park, CA. **[5]**

Ciomborowska, J., W. Rosikiewicz, D. Szklarczyk, W. Makałowski, and I. Makałowska. 2013. "Orphan" retrogenes in the human genome. Mol. Biol. Evol. 30: 384–396. **[9]**

Civetta, A. and R. S. Singh. 1998. Sex-related genes, directional sexual selection, and speciation. Mol. Biol. Evol. 15: 901–909. **[4]**

Claeyssens, S. and 6 others. 2003. Amino acid control of the human glyceraldehyde 3-phosphate dehydrogenase gene transcription in hepatocyte. Am. J. Physiol. Gastrointest. Liver Physiol. 285G: 840–849. **[12]**

Clark, A. G. 1994. Invasion and maintenance of a gene duplication. Proc. Natl. Acad. Sci. USA 91: 2950–2954. **[7]**

Clark, A. G. and 16 others. 2003. Inferring nonneutral evolution from human-chimp-mouse orthologous gene trios. Science 302: 1960–1963. **[4]**

Clark, J. B. and M. G. Kidwell. 1997. A phylogenetic perspective on *P* transposable element evolution in *Drosophila.* Proc. Natl. Acad. Sci. USA 94: 11428–11433. **[9]**

Clark, N. L. 2008. Adaptive evolution of primate sperm proteins. In: *Encyclopedia of Life Sciences.* Wiley, Chichester. doi:10.1002/9780470015902.a0020775 **[4]**

Claverie, J.-M. 2001. What if there are only 30,000 human genes? Science 291: 1255–1257. **[11]**

Clay, O. and G. Bernardi. 2005. How not to search for isochores: A reply to Cohen et al. Mol. Biol. Evol. 22: 2315–2317. **[11]**

Clay, O., S. Cacciò, S. Zoubak, D. Mouchiroud, and G. Bernardi. 1996. Human coding and noncoding DNA: Compositional correlations. Mol. Phylogenet. Evol. 5: 2–12. **[11]**

Cleary, M. L., E. A. Schon, and J. B. Lingrel. 1981. Two related pseudogenes are the result of a gene duplication in the goat β-globin locus. Cell 26: 181–190. **[7]**

Cock, J. M. and 76 others. 2010. The *Ectocarpus* genome and the independent evolution of multicellularity in brown algae. Nature 465: 617–621. **[9]**

Cocquyt, E., G. H. Gile, F. Leliaert, H. Verbruggen, P. J. Keeling, and O. De Clerck. 2010. Complex phylogenetic distribution of a non-canonical genetic code in green algae. BMC Evol. Biol. 10: 327. **[10]**

Coenye, T. and P. Vandamme. 2005. Organisation of the *S10, spc* and *alpha* ribosomal protein gene clusters in prokaryotic genomes. FEMS Microbiol. Lett. 242: 117–126. **[10]**

Coffin, J. M. 1986. Genetic variation in AIDS viruses. Cell 46:1–4. **[4]**

Coghlan, A., E. E. Eichler, S. G. Oliver, A. H. Paterson, and L. Stein. 2005. Chromosome evolution in eukaryotes: A multi-kingdom perspective. Trends Genet. 21: 673–682. **[11]**

Cohen, N., T. Dagan, L. Stone, and D. Graur. 2005. GC composition of the human genome: In search of isochores. Mol. Biol. Evol. 22: 1260–1272. **[11]**

Cohen-Gihon, I., R. Sharan, and R. Nussinov. 2011. Processes of fungal proteome evolution and gain of function: Gene duplication and domain rearrangement. Phys. Biol. 8: 035009. **[8]**

Colbourne, J. K. and 68 others. 2011. The ecoresponsive genome of *Daphnia pulex*. Science 331: 555–561. **[11]**

Cole, S. T. and 43 others. 2001. Massive gene decay in the leprosy bacillus. Nature 409: 1007–1011. **[4]**

Colless, D. H. 1982. Review of Phylogenetics: The Theory and Practice of Phylogenetic Systematics. Syst. Zool. 31: 100–104. **[5]**

Colless, D. H. 1995. Relative symmetry of cladograms and phenograms: An experimental study. Syst. Biol. 44: 102–108. **[5]**

Collier, S., M. Tassabehji, P. Sinnott, and T. Strachan. 1993. A *de novo* pathological point mutation at the 21-hydroxylase locus: Implication for gene conversion in the human genome. Nat. Genet. 3: 260–265. **[7]**

Collins, L. and D. Penny. 2005. Complex spliceosomal organization ancestral to extant eukaryotes. Mol. Biol. Evol. 22: 1053–1066. **[8]**

Comeron, J. M. 1995. A method for estimating the numbers of synonymous and nonsynonymous substitutions per site. J. Mol. Evol. 41: 1152–1159. **[3]**

Comings, D. E. 1972. The structure and function of chromatin. Adv. Hum. Genet. 3: 237 431. **[11]**

Conant, G. C. and A. Wagner. 2002. GenomeHistory: A software tool and its application to fully sequenced genomes. Nucl. Acids Res. 30: 3378–3386. **[7]**

Confalone, E., J. J. Beintema, M. P. Sasso, A. Carsana, M. Palmieri, M. T. Vento, and A. Furia. 1995. Molecular evolution of genes encoding ribonucleases in ruminant species. J. Mol. Evol. 41: 850–858. **[7]**

Conrad, D. F. and 17 others. 2011. Variation in genome-wide mutation rates within and between human families. Nat. Genet. 43: 712–714. **[11]**

Contreras-Galindo, R. and 14 others. 2013. HIV infection reveals widespread expansion of novel centromeric human endogenous retroviruses. Genome Res. 23: 1505–1513. **[9]**

Cook, L. G. 2000. Extraordinary and extensive karyotypic variation: A 48-fold range in chromosome number in the gall-inducing scale insect *Apiomorpha* (Hemiptera: Coccoidea: Eriococcidae). Genome 43: 255–263, **[11]**

Cook, O. F. 1906. Factors of species-formation. Science 23: 506–507. **[5]**

Cooper, A. and D. Penny. 1997. Mass survival of birds across the Cretaceous–Tertiary boundary: Molecular evidence. Science 275: 1109–1113. **[5]**

Cooper, A. and H. Poinar. 2000. Ancient DNA: Do it right or not at all. Science 289: 1139. **[5]**

Cooper, G. M. and J. Shendure. 2011. Needles in stacks of needles: Finding disease-causal variants in a wealth of genomic data. Nat. Rev. Genet. 12: 628–640. **[11]**

Cooper, N. (ed.). 1994. *The Human Genome Project: Deciphering the Blueprint of Heredity*. University Science Books, Mill Valley, CA. **[10]**

Cooper, T. F. 2007. Recombination speeds adaptation by reducing competition between beneficial mutations in populations of *Escherichia coli*. PloS Biol. 5: e225. **[13]**

Cooper, T. F., D. E. Rozen, and R. E. Lenski. 2003. Parallel changes in gene expression after 20,000 generations of evolution in *Escherichia coli*. Proc. Natl. Acad. Sci. USA 100: 1072–7077. **[13]**

Cooper, V. S. and R. E. Lenski. 2000. The population genetics of ecological specialization in evolving *Escherichia coli* populations. Nature 407: 736–739. **[13]**

Cooper, V. S., S. H. Vohr, S. C. Wrocklage, and P. J. Hatcher. 2010. Why genes evolve faster on secondary chromosomes in bacteria. PLoS Comput. Biol. 6: e1000732. **[10]**

Copeland, H. F. 1938. The kingdoms of organisms. Quart. Rev. Biol. 13: 383–420. **[5]**

Copley, S. D. 2000. Evolution of a metabolic pathway for degradation of a toxic xenobiotic: The patchwork approach. Trends Biochem. Sci. 25: 261–265. **[8]**

Copley, S. D. 2012. Moonlighting is mainstream: Paradigm adjustment required. BioEssays 34: 578–588. **[7]**

Cordaux, R., D. J. Hedges, S. W. Herke, and M. A. Batzer. 2006. Estimating the retrotransposition rate of human *Alu* elements. Gene 373: 134–137. **[9]**

Cordaux, R., S. Udit, M. A. Batzer, and C. Feschotte. 2006. Birth of a chimeric primate gene by capture of the transposase gene from a mobile element. Proc. Natl. Acad. Sci. USA 103: 8101–8106. **[9]**

Cordell, H. J. 2002. Epistasis: What it means, what it doesn't mean, and statistical methods to detect it in humans. Hum. Mol. Genet. 11: 2463–2468. **[2]**

Corradi, N., K. L. Haag, J. F. Pombert, D. Ebert, and P. J. Keeling. 2009. Draft genome sequence of the *Daphnia* pathogen *Octosporea bayeri*: Insights into the gene content of a large microsporidian genome and a model for host-parasite interactions. Genome Biol. 10: R106. **[11]**

Cosmides, L. and J. Tooby. 1981. Cytoplasmic inheritance and intragenomic conflict. J. Theor. Biol. 89: 83–129. **[2]**

Costantini, M. 2015. An overview on genome organization of marine organisms. Mar. Genomics http: //www.sciencedirect.com/science/article/pii/S1874778715000574 **[11]**

Costantini, M., O. Clay, F. Auletta, and G. Bernardi. 2006. An isochore map of human chromosomes. Genome Res. 16: 536–541. **[11]**

Costantini, M., R. Cammarano, and G. Bernardi. 2009. The evolution of isochore patterns in vertebrate genomes. BMC Genomics 10: 146. **[11]**

Coufal, N. G. and 9 others. 2009. L1 retrotransposition in human neural progenitor cells. Nature 460: 1127–1131. **[9]**

Coulombe-Huntington, J. and J. Majewski. 2007. Characterization of intron loss events in mammals. Genome Res. 17: 23–32. **[8]**

Coulondre, C., J. H. Miller, P. J. Farabaugh, and W. Gilbert. 1978. Molecular basis of base substitution hotspots in *Escherichia coli*. Nature 274: 775–780. **[4]**

Covello, P. S. and M. W. Gray. 1993. On the evolution of RNA editing. Trends Genet. 9: 265–268. **[8]**

Cox, C. J., P. G. Foster, R. P. Hirt, S. R. Harris, and T. M. Embley. 2008. The archaebacterial origin of eukaryotes. Proc. Natl. Acad. Sci. USA 105: 20356–20361. **[5, 6]**

Cox, D. R. and H. D. Miller. 1977. *The Theory of Stochastic Processes*. Chapman & Hall, London. **[3]**

Crampton, N., W. A. Bonass, J. Kirkham, C. Rivetti, and N. H. Thomson. 2006. Collision events between RNA polymerases in convergent transcription studied by atomic force microscopy. Nucl. Acids Res. 34: 5416–5425. **[8]**

Crawford, D. J. 1990. Plant Molecular Systematics: Macromolecular Approaches. Wiley, San Francisco. **[5, 6]**

Crick, F. H. C. 1956. Ideas on protein synthesis. http://profiles.nlm.nih.gov/SC/B/B/F/T/_/scbbft.pdf **[1]**

Crick, F. H. C. 1966. Codon–anticodon pairing: The wobble hypothesis. J. Mol. Biol. 19: 548–555. **[1, 4]**

Crick, F. H. C. 1968. The origin of the genetic code. J. Mol. Biol. 38: 367–379. **[10]**

Crick, F. H. C. 1970. Central dogma of molecular biology. Nature 227: 561–563. **[1]**

Croes, G. A. 1958. A method for solving travelling-salesman problems. Operations Res. 6: 791–812. **[5]**

Crosland, M. W. J. and R. H. Crozier. 1986. *Myrmecia pilosula*, an ant with only one pair of chromosomes. Science 231: 1278. **[11]**

Crow, J. F. 1954. Breeding structure of populations. II. Effective population number, pp. 543–556. In: O. Kempthorne, T. A. Bancroft, J. W. Gowen, and J. L. Lush (eds.), *Statistics and Mathematics in Biology*. Iowa State College Press, Ames, IA. **[2]**

Crow, J. F. 1979. Genes that violate Mendel's rules. Sci. Am. 240: 134–143, 146. **[2]**

Cui, J. and E. C. Holmes. 2012. Evidence for an endogenous papillomavirus-like element in the platypus genome. J. Gen. Virol. 93: 1362–1366. **[9]**

Cummins, R. 1975. Functional analysis. J. Philos. 72: 741–765. **[11]**

Cuny, G., P. Soriano, G. Macaya, and G. Bernardi. 1981. The major components of the mouse and human genomes. 1. Preparation, basic properties and compositional heterogeneity. Eur. J. Biochem. 115: 227–233. **[11]**

Curcio, M. J. and K. M. Derbyshire. 2003. The outs and ins of transposition: From Mu to Kangaroo. Nat. Rev. Mol. Cell Biol. 4: 865–877. **[9]**

Curtis, B. A. and 72 others. 2012. Algal genomes reveal evolutionary mosaicism and the fate of nucleomorphs. Nature 492: 59–65. **[11]**

Curtis, D. and W. Bender. 1991. Gene conversion in *Drosophila* and the effects of the meiotic mutants *mei-9* and *mei–218*. Genetics 127: 739–746. **[7]**

Cusack, B. P. and K. H. Wolfe. 2007. When gene marriages don't work out: Divorce by subfunctionalization. Trends Genet. 23: 270–272. **[7]**

Cyrklaff, M., C. P. Sanchez, N. Kilian, C. Bisseye, J. Simpore, F. Frischknecht, and M. Lanzer. 2011. Hemoglobins S and C interfere with actin remodeling in *Plasmodium falciparum*-infected erythrocytes. Science 334: 1283–1286. **[2]**

Czelusniak, J., M. Goodman, D. Hewett-Emmett, M. L. Weiss, P. J. Venta, and R. E. Tashian. 1982. Phylogenetic origins and adaptive evolution of avian and mammalian haemoglobin genes. Nature 298: 297–300. **[4, 7]**

D

D'Auria, G., N. Jiménez-Hernández, F. Peris-Bondia, A. Moya, and A. Latorre. 2010. *Legionella pneumophila* pangenome reveals strain-specific virulence factors. BMC Genomics 11: 181. **[10]**

Daborn, P. J. and 12 others. 2002. A single p450 allele associated with insecticide resistance in *Drosophila*. Science 297: 2253–2256. **[9]**

Dagan, T. 2011. Phylogenomic networks. Trends Microbiol. 19: 483–491. **[6]**

Dagan, T. and W. Martin. 2006. The tree of one percent. Genome Biol. 7: 118. **[6]**

Dagan, T. and W. Martin. 2007. Ancestral genome sizes specify the minimum rate of lateral gene transfer during prokaryote evolution. Proc. Natl. Acad. Sci. USA 104: 870–875. **[9, 10]**

Dagan, T. and W. Martin. 2009. Getting a better picture of microbial evolution en route to a network of genomes. Phil. Trans. Roy. Soc. 364B: 2187–2196. **[6]**

Dagan, T., M. Roettger, D. Bryant, and W. Martin. 2010. Genome networks root the tree of life between prokaryotic domains. Genome Biol. Evol. 2: 379–392. **[6]**

Dagan, T., R. Blekhman, and D. Graur. 2006. The "domino theory" of gene death: Gradual and mass gene extinction events in three lineages of obligate symbiotic bacterial pathogens. Mol. Biol. Evol. 23: 310–316. **[10]**

Da Lage, J. L., C. Maisonhaute, F. Maczkowiak, and M. L. Cariou. 2003. A nested alpha-amylase gene in *Drosophila ananassae*. J. Mol. Evol. 57: 355–362. **[8]**

Danchin, E. G., P. Gouret and P. Pontarotti. 2006. Eleven ancestral gene families lost in mammals and vertebrates while otherwise universally conserved in animals. BMC Evol. Biol. 6: 5. **[11]**

Danforth, C. H. 1923. The frequency of mutation and the incidence of hereditary traits in man, pp. 120–128. In: *Eugenics, Genetics and the Family. Scientific Papers of the Second International Congress of Eugenics* 1. Williams & Wilkins, Baltimore. **[4]**

Daniels, S. B., K. R. Peterson, L. D. Strausbaugh, M. G. Kidwell and A. Chovnick. 1990. Evidence for horizontal transmission of the *P* transposable element between *Drosophila* species. Genetics 124: 339–355. **[11]**

Danilevskaya, O. N., I. R. Arkhipova, K. L. Traverse, and M. L. Pardue. 1997. Promoting in tandem: The promoter for telomere transposon *HeT-A* and implications for the evolution of retroviral LTRs. Cell 88: 647–655. **[9]**

Darling, A. C. E., B. Mau, F. R. Blattner, and N. T. Perna. 2004. Mauve: Multiple alignment of conserved genomic sequences with rearrangements. Genome Res. 14: 1394–1403. **[3]**

Darling, A. E., B. Mau, and N. T. Perna. 2010. ProgressiveMauve: Multiple genome alignment with gene gain, loss, and rearrangement. PLoS ONE 5: e11147. **[3]**

Darlington, C. D. 1937. *Recent Advances in Cytology*, 2nd Ed. Churchill, London. **[11]**

Darnell, J. E. 1978. Implications of RNA–RNA splicing in the evolution of eukaryotic cells. Science 202: 1257–1260. **[8]**

Darnell, R. B. 2013. RNA protein interaction in neurons. Annu. Rev. Neurosci. 36: 243–270. **[12]**

Darwin, C. 1866. *The Origin of Species by Means of Natural Selection, or the Preservation of Favoured Races in the Struggle for Life*, 4th edition. John Murray, London. **[13]**

Darwin, C. 1868. *The Variation of Animals and Plants under Domestication*. John Murray, London. **[11]**

Das, S. and T. F. Smith. 2000. Identifying nature's protein Lego set. Adv. Protein Chem. 54: 159–183. **[8]**

Da Silva, J., M. Coetzer, R. Nedellec, C. Pastore, and D. E. Mosier. 2010. Fitness epistasis and constraints on adaptation in a human immunodeficiency virus type 1 protein region. Genetics 185: 293–303. **[13]**

Daubin, V. and H. Ochman. 2004. Bacterial genomes as new gene homes: The genealogy of ORFans in *E. coli*. Genome Res. 14: 1036–1042. **[10]**

Davidov, Y. and E. Jurkevitch. 2007. How incompatibilities may have led to eukaryotic cell. Nature 448: 130. **[6]**

Davidov, Y. and E. Jurkevitch. 2009. Predation between prokaryotes and the origin of eukaryotes. BioEssays 31: 748–757. **[5, 6]**

Davidson, E. H. 2006. *The Regulatory Genome: Gene Regulatory Networks in Development and Evolution*, 2nd edition. Academic Press, Burlington, MA. **[12]**

Dayhoff, M. O. 1978. *Atlas of Protein Sequence and Structure*, Vol. 5, Supplement 3. National Biomedical Research Foundation, Silver Spring, MD. **[7]**

Dayhoff, M. O., R. Schwartz, and B. C. Orcutt. 1978. A model of evolutionary change in proteins. In: M. O. Dayhoff (ed.), *Atlas of Protein Sequence and Structure*, Vol. 5, Supplement 3, pp. 345–358. National Biomedical Research Foundation, Washington, DC. **[3]**

De Bie, T., N. Cristianini, J. P. Demuth, and M. W. Hahn. 2006. CAFE: A computational tool for the study of gene family evolution. Bioinformatics 22: 1269–1271. **[7]**

de Daruvar, A., J. Collado-Vides, and A. Valencia. 2002. Analysis of the cellular functions of *Escherichia coli* operons and their conservation in *Bacillus subtilis*. J. Mol. Evol. 55: 211–221. **[10]**

de Duve, C. 2005. *Singularities: Landmarks on the Pathways of Life*. Cambridge University Press, Cambridge. **[5]**

De Gobbi, M. and 14 others. 2006. A regulatory SNP causes a human genetic disease by creating a new transcriptional promoter. Science 312: 1215–1217. **[1, 11]**

de Koning, A. P., W. Gu, T. A. Castoe, M. A. Batzer and D. D. Pollock. 2011. Repetitive elements may comprise over two-thirds of the human genome. PLoS Genet. 7: e1002384. **[11]**

de la Chaux, N., P. W. Messer, and P. F. Arndt. 2007. DNA indels in coding regions reveal selective constraints on protein evolution in the human lineage. BMC Evol. Biol. 7: 191. **[1]**

De Ley, J. 1974. Phylogeny of prokaryotes. Taxon 23: 291–300. **[5]**

Deloukas, P. and 126 others. 2001. The DNA sequence and comparative analysis of human chromosome 20. Nature 414: 865–871. **[11]**

de Queiroz, K. 2007. Species concepts and species delimitation. Syst. Biol. 56: 879–886. **[5]**

de Souza, F. S., L. F. Franchini, and M. Rubinstein. 2013. Exaptation of transposable elements into novel *cis*-regulatory elements: Is the evidence always strong? Mol. Biol. Evol. 30: 1239–1251. **[9]**

de Souza, S. J. 2012. Domain shuffling and the increasing complexity of biological networks. BioEssays 34: 655–657. **[8]**

de Vries, H. 1904. Species and Varieties: Their Origin by Mutation. Open Court, Chicago. **[7]**

Dearolf, C. R., J. Topol, C. S. Parker. 1989. Transcriptional control of *Drosophila fushi tarazu* zebra stripe expression. Genes Dev. 3: 384–398. **[12]**

Debouzie, D. 1980. Estimate of variance effective population size in a laboratory *Ceratitis* population. Heredity 45: 297–299. **[2]**

Dehal, P. and J. L. Boore. 2005. Two rounds of whole genome duplication in the ancestral vertebrate. PLoS Biol. 3: e314. **[7]**

Deininger, P. 2011. *Alu* elements: Know the SINEs. Genome Biol. 12: 236. **[9]**

Dekel, E. and U. Alon. 2005. Optimality and evolutionary tuning of the expression level of a protein. Nature 436: 588–592. **[13]**

Dellaporta, S. L. and 6 others. 2006. Mitochondrial genome of *Trichoplax adhaerens* supports placozoa as the basal lower metazoan phylum. Proc. Natl. Acad. Sci. USA 103: 8751–8756. **[10]**

Delon, I. and F. Payre. 2004. Evolution of larval morphology in flies: get in shape with *shavenbaby*. Trends Genet. 20: 305–313. **[12]**

Delsuc, F., H. Brinkmann, and H. Philippe. 2005. Phylogenomics and the reconstruction of the tree of life. Nat. Rev. Genet. 6: 361–375. **[5]**

Delvaux, D. 1995. Age of Lake Malawi (Nyasa) and water level fluctuations. Mus. Roy. Afr. Centr. Tervuren Dept. Géol.-Min. Rapp. Ann. 1993-1994: 99–108. **[12]**

Demuth, J. P. and M. W. Hahn. 2009. The life and death of gene families. Bioessays 31: 29–39. **[7, 11]**

Demuth, J. P., T. De Bie, J. E. Stajich, N. Cristianini and M. W. Hahn. 2006. The evolution of mammalian gene families. PLoS One 1: e85. **[11]**

Deng, C., C.-H. Cheng, H. Ye, X. He, and L. Chen. 2010. Evolution of an antifreeze protein by neofunctionalization under escape from adaptive conflict. Proc. Natl. Acad. Sci. USA 107: 21593–21598. **[7, 8]**

Denoeud, F. and 25 others. 2007. Prominent use of distal 5′ transcription start sites and discovery of a large number of additional exons in ENCODE regions. Genome Res. 17: 746–759. **[11]**

Denver, D. R., K. Morris, M. Lynch, and W. K. Thomas. 2004. High mutation rate and predominance of insertions in the *Caenorhabditis elegans* nuclear genome. Nature 430: 679–682. **[4]**

Des Marais, D. L. and M. D. Rausher. 2008. Escape from adaptive conflict after duplication in an anthocyanin pathway gene. Nature 454: 762–765. **[7]**

Dessauer, H. C., W. Fox, and J. R. Ramirez. 1957. Preliminary attempt to correlate paper-electrophoretic migration of hemoglobins with phylogeny in Amphibia and Reptilia. Arch. Biochem. Biophys. 71: 11–16. **[5]**

Deusch, O. and 7 others. 2008. Genes of cyanobacterial origin in plant nuclear genomes point to a heterocyst-forming plastid ancestor. Mol. Biol. Evol. 25: 748–761. **[5]**

Dewannieux, M., C. Esnault, and T. Heidmann. 2003. LINE-mediated retrotransposition of marked *Alu* sequences. Nat. Genet. 35: 41–48. **[9]**

Di Giammartino, D. C., K. Nishida, and J. L. Manley. 2011. Mechanisms and consequences of alternative polyadenylation. Mol. Cell 43: 853–866. **[8]**

di Prisco, G., E. Cocca, S. K. Parker, and H. W. Detrich III. 2002. Tracking the evolutionary loss of hemoglobin expression by the white-blooded Antarctic icefishes. Gene 295: 185–191. **[7, 8]**

Dickerson, R. E. 1971. The structure of cytochrome *c* and the rates of molecular evolution. J. Mol. Evol. 1: 26–45. **[4]**

Dietrich, F. S. and 13 others. 2004. The *Ashbya gossypii* genome as a tool for mapping the ancient *Saccharomyces cerevisiae* genome. Science 304: 304–307. **[11]**

Dietrich, M. R. 1994. The origins of the neutral theory of molecular evolution. J. Hist. Biol. 27: 21–59. **[2]**

Dikstein, R. 2011. The unexpected traits associated with core promoter elements. Transcription 2: 201–206. **[12]**

Ding, W., L. Lin, B. Chen, and J. Dai. 2006. L1 elements, processed pseudogenes and retrogenes in mammalian genomes. IUBMB Life 58: 677–685. **[9]**

Dittmar, K. A., J. M. Goodenbour, and T. Pan. 2006. Tissue specific differences in human transfer RNA expression. PLoS Genet. 2: e221. **[4]**

Dixon, B., B. Walker, W. Kimmins, and B. Pohajdak. 1991. Isolation and sequencing of a cDNA for an unusual hemoglobin from the parasitic nematode *Pseudoterranova decipiens*. Proc. Natl. Acad. Sci. USA. 88: 5655–5659. **[8]**

Dobzhansky, T. 1937. Genetic nature of species differences. Am. Nat. 71: 404–420. **[5]**

Dobzhansky, T. 1970. *Genetics and the Evolutionary Process*. Columbia University Press, New York. **[1]**

Doetsch, N. A., M. D. Thompson, and R. B. Hallick. 1998. A maturase-encoding group III twintron is conserved in deeply rooted euglenoid species: Are group III introns the chicken or the egg? Mol. Biol. Evol. 15: 76–86. **[9]**

Dollman, G. 1937. Mammals which hve recently become extinct and those on the verge of extinction. J. Soc. Pres. Wild Fauna Empire 30:67–74. **[5]**

Dollo, L. 1893. The laws of evolution. Bull. Soc. Belge Géol. Paléontol. Hydrol. 7: 164–166. **[5]**

D'Onofrio, G., D. Mouchiroud, B. Aïssani, C. Gautier and G. Bernardi. 1991. Correlations between the compositional properties of human genes, codon usage, and amino acid composition of proteins. J. Mol. Evol. 32: 504–510. **[11]**

Doolittle, R. F. 1986. *Of URFs and ORFs: A Primer on How to Analyze Derived Amino Acid Sequences*. University Science Books, Mill Valley, CA. **[10]**

Doolittle, R. F. 1988. Lens proteins. More molecular opportunism. Nature 336: 18. **[8]**

Doolittle, R. F. 1994. Convergent evolution: The need to be explicit. Trends Biol. Sci. 19: 15–18. **[8]**

Doolittle, W. F. 1987. What introns have to tell us: Hierarchy in genome evolution. Cold Spring Harb. Symp. Quant. Biol. 52: 907–913. **[8]**

Doolittle, W. F. 1999. Phylogenetic classification and the universal tree. Science 284: 2124–2129. **[6]**

Doolittle, W. F. 2000. The nature of the universal ancestor and the evolution of the proteome. Curr. Opin. Struct. Biol. 10: 355–358. **[5]**

Doolittle, W. F. 2000. Uprooting the tree of life. Sci. Am. 282(2): 90–95. **[6]**

Doolittle, W. F. 2004. Bacteria and Archaea, pp. 86–94. In: J. Cracraft and M. J. Donoghue (eds.), *Assembling the Tree of Life*. Oxford University Press, New York. **[6]**

Doolittle, W. F. 2009a. The practice of classification and the theory of evolution, and what the demise of the Charles Darwin's tree of life hypothesis means for both of them. Phil. Trans. Roy. Soc. 364B: 2221–2228. **[6]**

Doolittle, W. F. 2009b. Eradicating typological thinking in prokaryotic systematics and evolution. Cold Spring Harb. Symp. Quant. Biol. 74: 197–204. **[6]**

Doolittle, W. F. 2013. Is junk DNA bunk? A critique of ENCODE. Proc. Natl. Acad. Sci. USA 110: 5294–5300. **[11]**

Doolittle, W. F. and C. Sapienza. 1980. Selfish genes, the phenotype paradigm and genome evolution. Nature 284: 601–603. **[11]**

Doolittle, W. F. and O. Zhaxybayeva. 2009. On the origin of prokaryotic species. Genome Res. 19: 744–756. **[6]**

Doolittle, W. F., T. D. P. Brunet, S. Linquist and T. R. Gregory. 2014. Distinguishing between "function" and "effect" in genome biology. Genome Biol. Evol. 6: 1234–1237. **[11]**

dos Reis, M., L. Wernisch, and R. Savva. 2003. Unexpected correlations between gene expression and codon usage bias from microarray data for the whole *Escherichia coli* K-12 genome. Nucleic Acids Res. 31: 6976–6985. **[4]**

dos Reis, M., R. Savva, and L. Wernisch. 2004. Solving the riddle of codon usage preferences: A test for translational selection. Nucleic Acids Res. 32: 5036–5044. **[4]**

Douglas, S. and 9 others. 2001. The highly reduced genome of an enslaved algal nucleus. Nature 410: 1091–1096. **[11]**

Doulatov, S. and 8 others. 2004. Tropism switching in *Bordetella* bacteriophage defines a family of diversity-generating retroelements. Nature 431: 476–481. **[9]**

Dover, G. A. 1982. Molecular drive: A cohesive mode of species evolution. Nature 299: 111–117. **[7]**

Dover, G. A. and R. B. Flavell (eds.). 1982. *Genome Evolution*. Academic Press, New York. **[10]**

Drake, F. D. 1962. *Intelligent Life in Space*. Macmillan, New York. **[6]**

Drake, J. W. 1991. A constant rate of spontaneous mutation in DNA-based microbes. Proc. Natl. Acad. Sci. USA 88: 7160–7164. **[2, 13]**

Drake, J. W., B. Charlesworth, D. Charlesworth, and J. F. Crow. 1998. Rates of spontaneous mutation. Genetics 148: 1667–1686. **[1, 4]**

Driscoll, D. J. and B. R. Migeon. 1990. Sex difference in methylation of single-copy genes in human meiotic germ-cells: Implications for X-chromosome inactivation, parental imprinting, and origin of CpG mutations. Somatic Cell Mol. Genet. 16: 267–282. **[4]**

Drouin, G. 2006. Processed pseudogenes are more abundant in human and mouse X chromosomes than in autosomes. Mol. Biol. Evol. 23: 1652–1655. **[9]**

Drouin, G., J.-R. Godin, and B. Pagé. 2011. The genetics of vitamin C loss in vertebrates. Curr. Genomics 12: 371–378. **[8]**

Drovetski, S. V. 2003. Plio-Pleistocene climatic oscillations, Holarctic biogeography and speciation in an avian subfamily. J. Biogeog. 30: 1173–1181. **[5]**

Drummond, A. J., S. Y. W. Ho, N. Rawlence, and A. Rambaut. 2007. *A Rough Guide to BEAST 1.4*. University of Auckland, New Zealand. **[5]**

Drummond, D. A. and C. O. Wilke. 2008. Mistranslation-induced protein misfolding as a dominant constraint on coding-sequence evolution. Cell 134: 341–352. **[4]**

Drummond, D. A., J. D. Bloom, C. Adami, C. O. Wilke, and F. H. Arnold. 2005. Why highly expressed proteins evolve slowly. Proc. Natl. Acad. Sci. USA 102: 14338–14343. **[4]**

Dubey, G. P. and S. Ben-Yehuda. 2011. Intercellular nanotubes mediate bacterial communication. Cell 144: 590–600. **[9]**

Dubinin, N. P. 1929. Allelomorphentreppen bei *Drosophila melanogaster*. Biol. Zentr. 49: 328–339. **[7]**

Duboc, V. and M. P. Logan. 2011. *Pitx1* is necessary for normal initiation of hindlimb outgrowth through regulation of Tbx4 expression and shapes hindlimb morphologies via targeted growth control. Development 138: 5301–5309. **[12]**

Dubuy, H. G. and F. L. Riley. 1967. Hybridization between nuclear and kinetoplast DNAs of *Leishmania enriettii* and between nuclear and mitochondrial DNAs of mouse liver. Proc. Natl. Acad. Sci. USA 57: 790–797. **[9]**

Duffy, S., L. A. Shackelton, and E. C. Holmes. 2008. Rates of evolutionary change in viruses: Patterns and determinants. Nat. Rev. Genet. 9: 267–276. **[4]**

Duncan, T. R. and C. L. Peichel. 2010. Perspectives on the genetic architecture of divergence in body shape in sticklebacks. Integr. Comp. Biol. 50: 1057–1066. **[12]**

Dunham, I. and 216 others. 1999. The DNA sequence of human chromosome 22. Nature 402: 489–495. **[11]**

Dunker, A. K. and R. W. Kriwacki. 2011. The orderly chaos of proteins. Sci. Am. 304: 68–73. **[1]**

Dunn, L. C. and D. Bennett. 1968. A new case of transmission ratio distortion in the house mouse. Proc. Natl. Acad. Sci. USA 61: 570–573. **[2]**

Durbin, R., S. Eddy, A. Krogh, and G. Mitchison. 1998. *Biological Sequence Analysis: Probabilistic Models of Proteins and Nucleic Acids*. Cambridge University Press, Cambridge. **[5, 6]**

Duret, L. and L. D. Hurst. 2001. The elevated GC content at exonic third sites is not evidence against neutralist models of isochore evolution. Mol. Biol. Evol. 18: 757–762. **[11]**

Duret, L. and N. Galtier. 2009. Biased gene conversion and the evolution of mammalian genomic landscapes. Annu. Rev. Genomics Hum. Genet. 10: 285–311. **[11]**

Duret, L., D. Mouchiroud, and C. Gautier. 1995. Statistical analysis of vertebrate sequences reveals that long genes are scarce in GC-rich isochores. J. Mol. Evol. 40: 308–317. **[11]**

Duret, L., M. Semon, G. Piganeau, D. Mouchiroud, and N. Galtier. 2002. Vanishing GC-rich isochores in mammalian genomes. Genetics 162: 1837–1847. **[11]**

Dutta, S. K. and M. Ojha. 1972. Relatedness between major taxonomic groups of fungi based on the measurement of DNA nucleotide sequence homology. J. Mol. Gen. Genet. 114: 232–240. **[5]**

Duverger, O. and M. I. Morasso. 2009. Epidermal patterning and induction of different hair types during mouse embryonic development. Birth Defects Res. 87C: 263–272. **[11]**

Dvorak, J. and E. D. Akhunov. 2005. Tempos of gene locus deletions and duplications and their relationship to recombination rate during diploid and polyploid evolution in the *Aegilops-Triticum* alliance. Genetics 171: 323–332. **[7]**

Dyall, S. D., M. T. Brown, and P. J. Johnson. 2004. Ancient invasions: From endosymbionts to organelles. Science 304: 253–257. **[10]**

Dytrych, L., D. L. Sherman, C. S. Gillespie, and P. J. Brophy. 1998. Two PDZ domain proteins encoded by the murine periaxin gene are the result of alternative intron retention and are differentially targeted in Schwann cells. J. Biol. Chem. 273: 5794–5800. **[8]**

E

Eanes, W. F., M. Kirchner, and J. Yoon. 1993. Evidence for adaptive evolution of the *G6PD* gene in the *Drosophila melanogaster* and *Drosophila simulans* lineages. Proc. Natl. Acad. Sci. USA 90: 7475–7479. **[2]**

Eck, R. V. and M. O. Dayhoff. 1966. *Atlas of Protein Sequence and Structure*. National Biomedical Research Foundation, Silver Spring, MD. **[5]**

Eddy, S. R. 2001. Non-coding RNA genes and the modern RNA world. Nat. Rev. Genet. 2: 919–929. **[1]**

Eddy, S. R. 2012. The C-value paradox, junk DNA and ENCODE. Curr. Biol. 22: R898-R899. **[11]**

Edgar, R. C. 2004. MUSCLE: A multiple sequence alignment method with reduced time and space complexity. BMC Bioinformatics 5: 113. **[3]**

Edlind, T. D., J. Li, G. S. Visvesvara, M. H. Vodkin, G. L. McLaughlin, and S. K. Katiyar. 1996. Phylogenetic analysis of β-tubulin sequences from amitochondrial protozoa. Mol. Phylogenet. Evol. 5: 359–367. **[5]**

Edwards, A. W. F. and L. L. Cavalli-Sforza. 1964. Reconstruction of evolutionary trees. pp. 67–76. In: V. H. Heywood and J.

McNeill (eds.), *Phenetic and Phylogenetic Classification*. Systematics Association, London. **[5]**

Edwards, S. V. and P. Beerli. 2000. Gene divergence, population divergence, and the variance in coalescence time in phylogeographic studies. Evolution 54: 1839–1854. **[4, 5]**

Efron, B. 1979. Bootstrap methods: Another look at the jackknife. Ann. Stat. 7: 1–26. **[5]**

Efron, B. 1982. *The Jackknife, the Bootstrap, and Other Resampling Plans*. Society for Industrial and Applied Mathematics, Philadelphia. **[5]**

Efron, B., E. Halloran, and S. Holmes. 1996. Bootstrap confidence levels for phylogenetic trees. Proc. Natl. Acad. Sci. USA 93: 13429–13434. **[5]**

Efstratiadis, A. and 14 others. 1980. The structure and evolution of the human β-globin gene family. Cell 21: 653–668. **[7]**

Egan, E. S., M. A. Fogel, and M. K. Waldor. 2005. Divided genomes: Negotiating the cell cycle in prokaryotes with multiple chromosomes. Mol Microbiol. 56: 1129–1138. **[10]**

Ehret, C. F. and G. de Haller. 1963. Origin, development and maturation of organelles and organelle systems of the cell surface in *Paramecium*. J. Ultrastruct. Res. 23: S1–S42. **[10]**

Ehrlich, J., D. Sankoff and J. H. Nadeau. 1997. Synteny conservation and chromosome rearrangements during mammalian evolution. Genetics 147: 289–296. **[11]**

Eirín-López, J. M., A. M. González-Tizón, A. Martínez, and J. Méndez. 2004. Birth-and-death evolution with strong purifying selection in the histone H1 multigene family and the origin of *orphon* H1 genes. Mol. Biol. Evol. 21: 1992–2003. **[7]**

Eisen, J. A. 1998. Phylogenomics: Improving functional predictions for uncharacterized genes by evolutionary analysis. Genome Res. 8: 163–167. **[5]**

Eisen, J. A. and C. M. Fraser. 2003. Phylogenomics: Intersection of evolution and genomics. Science 300: 1706–1707. **[5]**

Eisen, J. A. and P. C. Hanawalt. 1999. A phylogenomic study of DNA repair genes, proteins, and processes. Mut. Res. 435: 171–213. **[5]**

Eisenberg, E. and E. Y. Levanon. 2013. Human housekeeping genes, revisited. Trends Genet. 29: 569–574. **[12]**

Eisenmann, V. 1985. Le couagga: Un zèbre aux origines douteuses. La Recherche 16: 254–256. **[5]**

Ekman, D. and A. Elofsson. 2010. Identifying and quantifying orphan protein sequences in fungi. J. Mol. Biol. 396: 396–405. **[8]**

Ekman, D., A. K. Björklund, and A. Elofsson. 2007. Quantification of the elevated rate of domain rearrangements in metazoa. J. Mol. Biol. 372: 1337–1348. **[8]**

El-Hani, C. N. 2007. Between the cross and the sword: The crisis of the gene concept. Genet. Mol. Biol. 30: 297–307. **[1]**

Elango, N., J. W. Thomas, NISC Comparative Sequencing Program, and S. V. Yi. 2006. Variable molecular clocks in hominoids. Proc. Natl. Acad. Sci. USA 103: 1370–1375. **[4]**

Eldredge, N. and J. Cracraft. 1980. *Phylogenetic Patterns and the Evolutionary Process: Method and Theory in Comparative Biology*. Columbia University Press, New York. **[5, 6]**

Eldredge, N. and S. J. Gould. 1972. Punctuated equilibria: An alternative to phyletic gradualism. In: T. J. M. Schopf (ed.), *Models in Paleobiology*. W.H. Freeman, San Francisco. **[4]**

Elena, S. F. and R. E. Lenski. 2003. Evolution experiments with microorganisms: the dynamics and genetic bases of adaptation. Nat. Rev. Genet. 4: 457–469. **[13]**

Elfenbein, J. R. and 9 others. 2015. Multicopy single-stranded DNA directs intestinal colonization of eEnteric pathogens. PLoS Genet. 11:e1005472. **[9]**

Elhaik, E. and D. Graur. 2014. A comparative study and a phylogenetic exploration of the compositional architectures of mammalian nuclear genomes. PLoS Comput. Biol. 10: e1003925. **[11]**

Elhaik, E., D. Graur and K Josic. 2010b. Comparative testing of DNA segmentation algorithms using benchmark simulations. Mol. Biol. Evol. 27: 1015–24. **[11]**

Elhaik, E., D. Graur, K. Josic and G. Landan. 2010a. Identifying compositionally homogeneous and nonhomogeneous domains within the human genome using a novel segmentation algorithm. Nucleic Acids Res. 38: e158. **[11]**

Elhaik, E., G. Landan and D. Graur. 2009. Can GC content at third-codon positions be used as a proxy for isochore composition? Mol Biol Evol. 26: 1829–33. **[11]**

Ellegren, H. and A. K. Fridolfsson. 1997. Male-driven evolution of DNA sequences in birds. Nat. Genet. 17: 182–184. **[4]**

Ellegren, H. and A.-K. Fridolfsson. 2003. Sex-specific mutation rates in salmonid fish. J. Mol. Evol. 56: 458–463. **[4]**

Ellis, L. L. and 10 others. 2014. Intrapopulation genome size variation in *D. melanogaster* reflects life history variation and plasticity. PLoS Genet. 10: e1004522. **[11]**

Ellis, R. J. 2010. Tackling unintelligent design. Nature 463: 164–165. **[8]**

Elrouby, N. and T. E. Bureau. 2010. *Bs1*, a new chimeric gene formed by retrotransposon-mediated exon shuffling in maize. Plant Physiol. 153: 1413–1424. **[9]**

Elsik, C. G. and 306 others. 2009. The genome sequence of taurine cattle: A window to ruminant biology and evolution. Science 324: 522–528. **[11]**

Elzanowski, A. and J. Ostell. 2013. The genetic codes. www.ncbi.nlm.nih.gov/Taxonomy/Utils/wprintgc.cgi?mode=c **[10]**

Embley, T. M. and R. P. Hirt. 1998. Early branching eukaryotes? Curr. Opin. Genet. Dev. 8:624–629. **[6]**

Embley, T. M. and W. Martin. 2006. Eukaryotic evolution, changes and challenges. Nature 440: 623–630. **[5]**

Emera, D. and G. P. Wagner. 2012a. Transposable element recruitments in the mammalian placenta: impacts and mechanisms. Brief. Funct. Genomics 11: 267–276. **[12]**

Emera, D. and G. P. Wagner. 2012b. Transformation of a transposon into a derived prolactin promoter with function during human pregnancy. Proc. Natl. Acad. Sci. USA 109: 11246–11251. **[12]**

Emerson, R. A. 1911. Genetic correlation and spurious allelomorphism in maize. Nebraska Agric. Exp. Stat. Annu. Rep. 24: 59–90. **[7]**

ENCODE Project Consortium. 2012. An integrated encyclopedia of DNA elements in the human genome. Nature 489: 57–74. **[11]**

Endo, T., K. Ikeo, and T. Gojobori. 1996. Large-scale search for genes on which positive selection may operate. Mol. Biol. Evol. 13: 685–690. **[4]**

Ernst, J. F., J. W. Stewart, and F. Sherman. 1982. Formation of composite iso-cytochrome *c* by recombination between non-allelic genes of yeast. J. Mol. Biol. 161: 373–394. **[7]**

Esnault, E., M. Valens, O. Espéli, and F. Boccard. 2007. Chromosome structuring limits genome plasticity in *Escherichia coli*. PLoS Genet. 3: e226. **[10]**

Esser, C. and 14 others. 2004. A genome phylogeny for mitochondria among alpha-proteobacteria and a predominantly eubacterial ancestry of yeast nuclear genes. Mol. Biol. Evol. 21: 1643–1660. **[9]**

Esser, C., W. Martin, and T. Dagan. 2007. The origin of mitochondria in light of a fluid chromosome model. Biol. Lett. 3: 180–184. **[5]**

Evrony, G. D. and 11 others. 2012. Single-neuron sequencing analysis of L1 retrotransposition and somatic mutation in the human brain. Cell 151: 483–496. **[9]**

Excoffier, L. and H. E. L. Lischer. 2010. Arlequin suite ver 3.5: A new series of programs to perform population genetics analyses under Linux and Windows. Mol. Ecol. Res. 10: 564–567. **[2]**

Excoffier, L., T. Hofer and M. Foll. 2009. Detecting loci under selection in a hierarchically structured population. Heredity 103: 285–298. **[11]**

Eyre-Walker, A. 1993. Recombination and mammalian genome evolution. Proc Biol Sci. 252: 237–243. **[11]**

Eyre-Walker, A. 1996. Synonymous codon bias is related to gene length in *Escherichia coli*: Selection for translational accuracy? Mol. Biol. Evol. 13: 864–872. **[4]**

Eyre-Walker, A. 2006. The genomic rate of adaptive evolution. Trends Ecol. Evol. 21: 569–575. **[2]**

Eyre-Walker, A. and L. D. Hurst. 2001. The evolution of isochores. Nat. Rev. Genet. 2: 549–555. **[11]**

Eyre-Walker, A. and M. Bulmer. 1993. Reduced synonymous substitution rate at the start of enterobacterial genes. Nucleic Acids Res. 21: 4599–4603. **[4]**

Eyre-Walker, A. and M. Bulmer. 1995. Synonymous substitution rates in enterobacteria. Genetics 140: 1407–1412. **[4]**

Eyre-Walker, A. and P. D. Keightley. 2007. The distribution of fitness effects of new mutations. Nature Rev. Genet. 8: 610–618. **[2]**

Ezkurdia, I. and 8 others. 2014. The shrinking human protein coding complement: Are there now fewer than 20,000 genes? arXiv:1312.7111v4 **[10]**

Ezkurdia, I. and 8 others. 2014. Multiple evidence strands suggest that there may be as few as 19,000 human protein-coding genes. Hum. Mol. Genet. 23: 5866–5878. **[11]**

F

Fablet, M., M. Bueno, L. Potrzebowski, and H. Kaessmann. 2009. Evolutionary origin and functions of retrogene introns. Mol. Biol. Evol. 26: 2147–2156. **[9]**

Fani, R., M. Brilli, and P. Liò. 2005. The origin and evolution of operons: The piecewise building of the proteobacterial histidine operon. J. Mol. Evol. 60: 378–390. **[10]**

Fankhauser, G. 1972. Memories of great embryologists. Reminiscences of F. Baltzer, H. Spemann, F. R. Lillie, R. G. Harrison, and E. G. Conklin. Am. Sci. 60: 46–55. **[7]**

Farré, D., N. Bellora, L. Mularoni, X. Messeguer, and M. Mar Albà. 2007. Housekeeping genes tend to show reduced upstream sequence conservation. Genome Biol. 8: R140. **[12]**

Farris, J. S. 1970. Methods for computing Wagner trees. Syst. Zool. 34: 21–34. **[5]**

Farris, J. S. 1977. Phylogenetic analysis under Dollo's Law. Syst. Zool. 26: 77–88. **[5]**

Fatica, A. and I. Bozzoni. 2014. Long non-coding RNAs: New players in cell differentiation and development. Nat. Rev. Genet. 15: 7–21. **[12]**

Faulkner, G. J. and 21 others. 2009. The regulated retrotransposon transcriptome of mammalian cells. Nat. Genet. 41: 563–571. **[11]**

Fay, J. C., G. J. Wyckoff, and C.-I. Wu. 2001. Positive and negative selection on the human genome. Genetics 158: 1227–1234. **[2]**

Federhen, S. 2012. The NCBI Taxonomy database. Nucleic Acids Res. 40: D136–D143. **[6]**

Fedoroff, N. V. 1994. Barbara McClintock (June 16, 1902–September 2, 1992). Genetics 136: 1–10. **[9]**

Fedorov, A., S. Saxonov, and W. Gilbert. 2002. Regularities of context-dependent codon bias in eukaryotic genes. Nucleic Acids Res. 30: 1192–1197. **[4]**

Fehér, T., B. Papp, C. Pál, and G. Pósfai. 2007. Systematic genome reductions: Theoretical and experimental approaches. Chem. Rev. 107: 3498–3513. **[10]**

Feldman, M., B. Liu, G. Segal, S. Abbo, A. A. Levy, and J. M. Vega. 1997. Rapid elimination of low-copy DNA sequences in polyploid wheat: A possible mechanism for differentiation of homoeologous chromosomes. Genetics 147: 1381–1387. **[7]**

Feller, W. 1968. *An Introduction to Probability Theory and Its Applications*, Vol. 1, 3rd Ed. Wiley, New York. **[3]**

Feller, W. 1971. *An Introduction to Probability Theory and Its Applications*, Vol. 2, 2nd Ed. Wiley, New York. **[3]**

Felsenstein, J. 1978a. Cases in which parsimony or compatibility methods will be positively misleading. Syst. Zool. 27: 401–410. **[5]**

Felsenstein, J. 1978b. The number of evolutionary trees. Syst. Zool. 27: 27–33. **[5]**

Felsenstein, J. 1981. Evolutionary trees from DNA sequences: A maximum likelihood approach. J. Mol. Evol. 17: 368–376. **[5]**

Felsenstein, J. 1982. Numerical methods for inferring evolutionary trees. Q. Rev. Biol. 57: 379–404. **[5]**

Felsenstein, J. 1985. Confidence limits on phylogenies: An approach using the bootstrap. Evolution 39: 783–791. **[5]**

Felsenstein, J. 1988. Phylogenies from molecular sequences: Inference and reliability. Annu. Rev. Genet. 22: 521–565. **[4, 5]**

Felsenstein, J. 1996. Inferring phylogenies from protein sequences by parsimony, distance, and likelihood methods. Methods Enzymol. 266: 418–427. **[5]**

Felsenstein, J. 2004. *Inferring Phylogenies*. Sinauer Associates, Sunderland, MA. **[5, 6]**

Felsenstein, J. and H. Kishino. 1993. Is there something wrong with the bootstrap on phylogenies? A reply to Hillis and Bull. Syst. Biol. 42: 193–200. **[5]**

Feng, D. and R. F. Doolittle. 1987. Progressive sequence alignment as a prerequisite to correct phylogenetic trees. J. Mol. Evol. 25: 351–360. **[3]**

Ferreira, A. and 9 others. 2011. Sickle hemoglobin confers tolerance to *Plasmodium* infection. Cell 145: 398–409. **[2]**

Ferris, S. D. and G. S. Whitt. 1979. Evolution of the differential regulation of duplicate genes after polyploidization. J. Mol. Evol. 12: 267–317. **[7]**

Ferris, S. D., A. C. Wilson, and W. M. Brown. 1981. Evolutionary tree for apes and humans based on cleavage maps of mitochondrial DNA. Proc. Natl. Acad. Sci. USA 78: 2432–2436. **[5]**

Feschotte, C. and E. J. Pritham. 2007. Computational analysis and paleogenomics of interspersed repeats in eukaryotes. In: Stojanovic N. (ed.) *Computational Genomics: Current Methods*, Horizon Scientific Press, Wymondham, UK. pp. 31–53. **[11]**

Feschotte, C. and E. J. Pritham. 2007. DNA transposons and the evolution of eukaryotic genomes. Annu. Rev. Genet. 41: 331–368. **[9]**

Field, L. M. and A. L. Devonshire. 1998. Evidence that *E4* and *FE4* esterase genes responsible for insecticide resistance in the aphid *Myzus persicae* (Sulzer) are part of a gene family. Biochem. J. 330: 169–173. **[7]**

Field, L. M., A. L. Devonshire, and B. G. Forde. 1988. Molecular evidence that insecticide resistance in peach-potato aphids (*Myzus persicae* Sulz.) results from amplification of an esterase gene. Biochem. J. 251: 309–312. **[7]**

Figueroa, F., E. Günther, and J. Klein. 1988. MHC polymorphisms pre-dating speciation. Nature 335: 265–271. **[7]**

Figureau, A. 1989. Optimization and the genetic code. Orig. Life Evol. Biosphere 19: 57–67. **[10]**

Filipski, J., J. P. Thiery and G. Bernardi. 1973. An analysis of the bovine genome by Cs_2SO_4-Ag^+ density gradient centrifugation. J. Mol. Biol. 80: 177–197. **[11]**

Finnegan, D. J. 1989. Eukaryotic transposable elements and genome evolution. Trends Genet. 5: 103–107. **[9]**

Fisher, R. A. 1922. On the dominance ratio. Proc. R. Soc. Edinburgh 42: 321–341. **[2]**

Fitch, D. H., W. J. Bailey, D. A. Tagle, M. Goodman, L. Sieu, and J. L. Slightom. 1991. Duplication of the γ-globin gene mediated by L1 long interspersed repetitive elements in an early ancestor of simian primates. Proc. Natl. Acad. Sci. USA 88: 7396–7400. **[9]**

Fitch, W. M. 1967. Evidence suggesting a non-random character to nucleotide replacements in naturally occurring mutations. J. Mol. Biol. 26: 499–507. **[4]**

Fitch, W. M. 1970. Distinguishing homologous from analogous proteins. Syst. Zool. 19: 99–113. **[5]**

Fitch, W. M. 1971. Toward defining the course of evolution: Minimum change for a specific tree topology. Syst. Zool. 20: 406–416. **[4, 5]**

Fitch, W. M. 1977. On the problem of discovering the most parsimonious tree. Am. Nat. 111: 223–257. **[4, 5]**

Fitch, W. M. 1981. A non-sequential method for constructing trees and hierarchical classifications. J. Mol. Evol. 18: 3037. **[4, 5]**

Fitch, W. M. 2000. Homology: A personal view of some of the problems. Trends Genet. 16: 227–231. **[6]**

Fitch, W. M. and E. Margoliash. 1967. Construction of phylogenetic trees. A method based on mutation distances as estimated from cytochrome *c* sequences is of general applicability. Science 155: 279–284. **[5]**

Fitch, W. M. and K. Upper. 1987. The phylogeny of tRNA sequences provides for ambiguity reduction in the origin of the genetic code. Cold Spring Harbor Symp. Quant. Biol. 52: 759–767. **[6]**

Fitch, W. M., R. M. Bush, C. A. Bender, and N. J. Cox. 1997. Long term trends in the evolution of H(3)HA1 human influenza type A. Proc. Natl. Acad. Sci. USA 94: 7712–7718. **[5]**

Flach, P. 2012. *Machine Learning: The Art and Science of Algorithms that Make Sense of Data*. Cambridge University Press, Cambridge **[11]**

Flavell, A. J. 1995. Retroelements, reverse transcriptase and evolution. Comp. Biochem. Physiol. 110B: 3–15. **[9]**

Fleagle, J. G. 1999. *Primate Adaptation and Evolution*. Academic Press, San Francisco. **[5]**

Fletcher, G. L., C. L. Hew, and P. L. Davies. 2001. Antifreeze proteins in teleost fishes. Annu. Rev. Physiol. 63: 359–390. **[7]**

Flot, J. F. and 39 others. 2013. Genomic evidence for ameiotic evolution in the bdelloid rotifer *Adineta vaga*. Nature 500: 453–457. **[11]**

Flower, W. H. 1883. On whales, present and past and their probable origin. Proc. Zool. Soc. London 1883: 466–513. **[5]**

Flower, W. H. and J. G. Garson. 1884. *Catalogue of the Specimens Illustrating the Osteology and Dentition of Vertebrate Animals Recent and Extinct Contained in the Museum of the Royal College of Surgeons of England. II. Mammalia other than Man*. Royal College of Surgeons, London. **[5]**

Flowers, J. M. and M. D. Purugganan. 2008. The evolution of plant genomes: scaling up from a population perspective. Curr. Opin. Genet. Dev. 18: 565–570. **[11]**

Fogel, S., R. K. Mortimer, K. Lusnak, and F. Tavares. 1978. Meiotic gene conversion: A signal of the basic recombination event in yeast. Cold Spring Harbor Symp. Quant. Biol. 43: 1325–1341. **[7]**

Forbes, D. J., M. W. Kirschner and J. W. Newport. 1983. Spontaneous formation of nucleus-like structures around bacteriophage DNA microinjected into *Xenopus* eggs. Cell 34: 13–23. **[11]**

Force, A., M. Lynch, F. B. Pickett, A. Amores, Y.-l. Yan, and J. Postlethwait. 1999. Preservation of duplicate genes by complementary, degenerative mutations. Genetics 151: 1531–1545. **[7]**

Forey, P. 1984. The coelacanth as a living fossil, pp. 166–169. In: N. Eldredge and S. M. Stanley (eds.), *Living Fossils*. Springer, New York. **[4]**

Forster, A. C. and G. M. Church. 2006. Towards synthesis of a minimal cell. Mol. Syst. Biol. 2: 45. **[10]**

Forterre, P. and D. Prangishvili. 2013. The major role of viruses in cellular evolution: Facts and hypotheses. Curr. Opin. Virol. 3: 558–565. **[6]**

Foster, E. A. and 7 others. 1998. Jefferson fathered slave's last child. Nature 396: 27–28. **[11]**

Fowler, C., R. Drinkwater, J. Skinner and L. Burgoyne. 1988. Human satellite-III DNA: An example of a "macrosatellite" polymorphism. Hum. Genet. 79: 265–272. **[11]**

Fox, G. E., K. R. Pechman, and C. R. Woese. 1977. Comparative cataloguing of 16S ribosomal RNA: A molecular approach to procaryotic systematics. Int. J. Syst. Bacteriol. 27: 44–57. **[6]**

Fox Keller, E. F. and D. Harel. 2007. Beyond the gene. PLoS One 2: e1231. **[1]**

Frank, A. C. and J. R. Lobry. 1999. Asymmetric substitution patterns: A review of possible underlying mutational or selective mechanisms. Gene 238: 65–77. **[10]**

Frank, A. C., H. Amiri, and S. G. Andersson. 2002. Genome deterioration: Loss of repeated sequences and accumulation of junk DNA. Genetica 115: 1–12. **[10]**

Frankel, N., S. Wang, and D. L. Stern. 2012. Conserved regulatory architecture underlies parallel genetic changes and convergent phenotypic evolution. Proc. Natl. Acad. Sci. USA 109: 20975–20979. **[12]**

Frankham, R. 1995. Effective population size / adult population size ratios in wildlife: A review. Genet. Res. 66: 95–107. **[2]**

Fraser, D. J., D. J. Hansen, S. Øsergaard, N. Tessier, M. Legault, and L. Bernatchez. 2007. Comparative estimation of effective population sizes and temporal gene flow in two contrasting population systems. Mol. Ecol. 16: 3866–3889. **[2]**

Fredrick, K. and M. Ibba. 2010. How the sequence of a gene can tune its translation. Cell 141: 227–229. **[4]**

Friedberg, R., A. E. Darling, and S. Yancopoulos. 2008. Genome rearrangement by the double cut and join operation. Methods Mol. Biol. 452: 385–416. **[10]**

Friedman, R. and A. L. Hughes. 2005. The pattern of nucleotide differences at individual codons among mouse, rat, and human. Mol. Biol. Evol. 22: 1285–1289. **[4]**

Friedman, R. C., K. K. Farh, C. B. Burge, and D. P. Bartel. (2009). Most mammalian mRNAs are conserved targets of microRNAs. Genome Res. 19: 92–105. **[12]**

Friesen, M., G. Saxer, M. Travisano, and M. Doebeli. 2004. Experimental evidence for sympatric ecological diversification due to frequency-dependent competition in *Escherichia coli*. Evolution 58: 245–260. **[13]**

Fryxell, K. J. 1996. The coevolution of gene family trees. Trends Genet. 12: 364–369. **[7]**

Fu, W. and 13 others. 2013. Analysis of 6,515 exomes reveals the recent origin of most human protein-coding variants. Nature 493: 216–220. **[2, 11]**

Fujimoto, A. and 7 others. 2008. A replication study confirmed the *EDAR* gene to be a major contributor to population differentiation regarding head hair thickness in Asia. Hum. Genet. 124: 179–185. **[11]**

Fukatsu, T. 2010. A fungal past to insect color. Science 328: 574–575. **[9]**

Fukuda, Y., T. Washio, and M. Tomita. 1999. Comparative study of overlapping genes in the genomes of *Mycoplasma genitalium* and *Mycoplasma pneumoniae*. Nucl. Acids Res. 27: 1847–1853. **[8]**

Fukuda, Y., Y. Nakayama, and M. Tomita. 2003. On dynamics of overlapping genes in bacterial genomes. Gene 323: 181–187. **[8]**

Furió, V., A. Moya, and R. Sanjuán. 2005. The cost of replication fidelity in an RNA virus. Proc. Natl. Acad. Sci. USA 102: 10233–10237. **[2]**

G

Gaffney, D. J. and P. D. Keightley. 2006. Genomic selective constraints in murid noncoding DNA. PLoS Genet. 2: e204. **[4]**

Gagat, P., A. Bodyl, P. Mackiewicz and J. W. Stiller. 2014. Tertiary plastid endosymbioses in dinoflagellates. In: W. Löffelhardt (ed.), *Endosymbiosis*. pp. 233–290. Springer, Vienna. **[11]**

Gallardo, M. H. and 7 others. 2004. Whole-genome duplications in South American desert rodents (Octodontidae). Biol. J. Linn. Soc. 82: 443–451. **[7]**

Gallardo, M. H., J. W. Bickham, R. L. Honeycutt, R. A. Ojeda, and N. Köhler. 1999. Discovery of tetraploidy in a mammal. Nature 401: 341. **[7]**

Galligan, J. T., S. E. Marchetti, and J. C. Kennell. 2011. Reverse transcription of the *pFOXC* mitochondrial retroplasmids of *Fusarium oxysporum* is protein primed. Mobile DNA 2: 1. **[9]**

Gallo, R. C. 1987. The AIDS virus. Sci. Am. 256(1): 46–56. **[4]**

Galperin, M. Y. and E. V. Koonin. 2000. Who's your neighbor? New computational approaches for functional genomics. Nat. Biotechnol. 18: 609–613. **[8]**

Galtier, N. and L. Duret. 2007. Adaptation or biased gene conversion? Extending the null hypothesis of molecular evolution. Trends Genet. 23: 273–277. **[7]**

Galtier, N. and J. R. Lobry. 1997. Relationships between genomic G+C content, RNA secondary structure, and optimal growth temperature in prokaryotes. J. Mol. Evol. 44: 632–636. **[10]**

Galtier, N. and M. Gouy 1998. Inferring pattern and process: Maximum-likelihood implementation of a nonhomogeneous model of DNA sequence evolution for phylogenetic analysis. Mol. Biol. Evol. 15: 871–879. **[5]**

Galtier, N. and D. Mouchiroud. 1998. Isochore evolution in mammals: a human-like ancestral structure. Genetics 150: 1577–1584. **[11]**

Galtier, N., G. Piganeau, D. Mouchiroud and L. Duret. 2001. GC-content evolution in mammalian genomes: The biased gene conversion hypothesis. Genetics. 159: 907–911. **[7, 11]**

Galtier, N., L. Duret, S. Glémin, and V. Ranwez. 2009a. GC-biased gene conversion promotes the fixation of deleterious amino acid changes in primates. Trends Genet. 25: 1–5. **[4, 7]**

Galtier, N., R. W. Jobson, B. Nabholz, S. Glémin, and P. U. Blier. 2009b. Mitochondrial whims: Metabolic rate, longevity and the rate of molecular evolution. Biol. Lett. 5: 413–416. **[4]**

Galun, E. and A. Breiman. 1997. *Transgenic Plants*. Imperial College Press, London. **[9]**

Gan, H. and 6 others. 2013. Integrative proteomic and transcriptomic analyses reveal multiple post-transcriptional regulatory mechanisms of mouse spermatogenesis. Mol. Cell Proteomics 12: 1144–1157. **[12]**

Gandolfo, M. A., K. C. Nixon, and W. L. Crepet. 2008. Selection of fossils for calibration of molecular dating models. Ann. Missouri Bot. Gard. 95: 34–42. **[5]**

Gao, D., J. Chen, M. Chen, B. C. Meyers, and S. Jackson. 2012. A highly conserved, small LTR retrotransposon that preferentially targets genes in grass genomes. PLoS ONE 7: e32010. **[9]**

Gao, H., J. M. Granka, and M. W. Feldman. 2010. On the classification of epistatic interactions. Genetics 184: 827–837. **[2]**

Gao, X. and M. Lynch. 2009. Ubiquitous internal gene duplication and intron creation in eukaryotes. Proc. Natl. Acad. Sci. USA 106: 20818–20823. **[8]**

Garb, J. E. and C. Y. Hayashi. 2005. Modular evolution of egg case silk genes across orb-weaving spider superfamilies. Proc. Natl. Acad. Sci. USA 102: 11379–11384. **[7, 8]**

García-Moreno, J. and D. P. Mindell. 2000. Rooting a phylogeny with homologous genes on opposite sex chromosomes (gametologs): A case study using avian CHD. Mol. Biol. Evol. 17: 1826–1832. **[5]**

Garcia-Vallvé, S., A. Romeu, and J. Palau. 2013. Horizontal gene transfer in bacterial and archaeal complete genomes. Genome Res. 10: 1719–1725. **[9]**

Garrett, S. and J. J. Rosenthal. 2012. RNA editing underlies temperature adaptation in K^+ channels from polar octopuses. Science 335: 848–851. **[8]**

Gaucher, E. A., J. M. Thomson, M. F. Burgan, and S. A. Benner. 2003. Inferring the palaeoenvironment of ancient bacteria on the basis of resurrected proteins. Nature 425: 285–288. **[5]**

Gaucher, E. A., S. Govindarajan, and O. K. Ganesh. 2008. Palaeotemperature trend for Precambrian life inferred from resurrected proteins. Nature 451: 704–707. **[5]**

Gaut, B. S. and J. F. Doebley. 1997. DNA sequence evidence for the segmental allotetraploid origin of maize. Proc. Natl. Acad. Sci. USA 94: 6809–6814. **[7]**

Gaut, B. S., S. V. Muse, W. D. Clark, and M. T. Clegg. 1992. Relative rates of nucleotide substitution at the *rbcL* locus of monocotyledonous plants. J. Mol. Evol. 35: 292–303. **[4]**

Gazave, E., T. Marques-Bonet, O. Fernando, B. Charlesworth, and A. Navarro. 2007. Patterns and rates of intron divergence between humans and chimpanzees. Genome Biol. 8: R21. **[4]**

Ge, F., L. S. Wang, and J. Kim. 2005. The cobweb of life revealed by genome-scale estimates of horizontal gene transfer. PLoS Biol. 3: e316. **[6]**

Gemayel, R., M. D. Vinces, M. Legendre, and K. J. Verstrepen. 2010. Variable tandem repeats accelerate evolution of coding and regulatory sequences. Annu. Rev. Genet. 44: 445–477. **[11]**

Gemmell, N. J., V. J. Metcalf, and F. W. Allendorf. 2004. Mother's curse: The effect of mtDNA on individual fitness and population viability. Trends Ecol. Evol. 19: 238–244. **[2]**

Gene Ontology Consortium. 2000. Gene ontology: Tool for the unification of biology. Nat. Genet. 25: 25–29. **[11]**

Genner, M. J. and G. F. Turner. 2005. The mbuna cichlids of Lake Malawi: a model for rapid speciation and adaptive radiation. Fish Fisheries 6: 1–34. **[12]**

Gerstein, M. 1997. A structural census of genomes: Comparing bacterial, eukaryotic, and archaeal genomes in terms of protein structure. J. Mol. Biol. 274: 562–576. **[8]**

Gerstein, M. B. and 9 others. 2007. What is a gene, post-ENCODE? History and updated definition. Genome Res. 17: 669–681. **[1]**

Gevers, D. and 10 others. 2005. Re-evaluating prokaryotic species. Nature Rev. Microbiol. 3: 733–739. **[6]**

Gherman, A. and 9 others. 2007. Population bottlenecks as a potential major shaping force of human genome architecture. PLoS Genet. 3: e119. **[11]**

Ghirlando, R. and 6 others. 2012. Chromatin domains, insulators, and the regulation of gene expression. Biochim. Biophys. Acta 1819: 644–651. **[12]**

Giannuzzi, G. and 7 others. 2013. Hominoid fission of chromosome 14/15 and the role of segmental duplications. Genome Res. 23: 1763–1773. **[11]**

Gibbs, A. J. and G. A. McIntyre. 1970. The diagram, a method for comparing sequences. Its use with amino acid and nucleotide sequences. Eur. J. Biochem. 16: 1–11. **[3]**

Gibbs, R. A. and 239 others. 2007. Evolutionary and biomedical insights from the Rhesus macaque genome. Science 316: 222–234. **[4]**

Gibbs, S. P. 1978. The chloroplasts of *Euglena* may have evolved from symbiotic green algae. Can. J. Bot. 56: 2883–2889. **[11]**

Gibbs, S., M. Collard, and B. Wood. 2002. Soft-tissue anatomy of the extant hominoids: A review and phylogenetic analysis. J. Anat. 200: 3–49. **[5]**

Gibson, D. G. and 23 others. 2010. Creation of a bacterial cell controlled by a chemically synthesized genome. Science 329: 52–56. **[10]**

Gilad, Y., C. D. Bustamante, D. Lancet, and S. Pääbo. 2003. Natural selection on the olfactory receptor gene family in humans and chimpanzees. Am. J. Hum. Genet. 73: 489–501. **[7]**

Gilad, Y., M. Przeworski, D. Lancet, and S. Pääbo. 2004. Loss of olfactory receptor genes coincides with the acquisition of full trichromatic vision in primates. PLoS Biol. 2: e5. **[7]**

Gilad, Y., M. Przeworski, D. Lancet, and S. Pääbo. 2007. Correction: Loss of olfactory receptor genes coincides with the acquisition of full trichromatic vision in primates. PLoS Biol. 5: e148. **[7]**

Gilbert, C., S. Schaack, J. K. Pace, P. J. Brindley, and C. Feschotte. 2010. A role for host-parasite interactions in the horizontal transfer of transposons across phyla. Nature 464: 1347–1350. **[9]**

Gilbert, M. T. P., H.-J. Bandelt, M. Hofreiter, and I. Barnes. 2005. Assessing DNA studies. Trends Ecol. Evol. 20: 541–544. **[5]**

Gilbert, W. 1978. Why genes in pieces? Nature 271: 501. **[8]**

Gilbert, W. 1987. The exon theory of genes. Cold Spring Harbor Symp. Quant. Biol. 52: 901–905. **[8]**

Gilis, D., S. Massar, N. J. Cerf, and M. Rooman. 2001. Optimality of the genetic code with respect to protein stability and amino-acid frequencies. Genome Res. 2: R0049. **[10]**

Gillespie, J. H. 1991. *The Causes of Molecular Evolution*. Oxford University Press, New York. **[2, 4]**

Gillespie, J. H. 2000. The neutral theory in an infinite population. Gene 261: 11–18. **[2]**

Gillespie, J. H. 2004. *Population Genetics: A Concise Guide,* 2nd Ed. Johns Hopkins University Press, Baltimore, MD. **[2]**

Gillooly, J. F., A. P. Allen, G. B. West, and J. H. Brown. 2005. The rate of DNA evolution: Effects of body size and temperature on the molecular clock. Proc. Natl. Acad. Sci. USA 102: 140–145. **[4]**

Gillooly, J. F., M. W. McCoy, and A. P. Allen. 2007. Effects of metabolic rate on protein evolution. Biol. Lett. 3: 655–660. **[4]**

Gilson, P. R, V. Su, C. H. Slamovits, M. E. Reith, P. J. Keeling, and G. I. McFadden. 2006. Complete nucleotide sequence of the chlorarachniophyte nucleomorph: Nature's smallest nucleus. Proc. Natl. Acad. Sci. USA 103: 9566–9571. **[11]**

Gimble, F. S. and J. Thorner. 1992. Homing of a DNA endonuclease gene by meiotic conversion in *Saccharomyces cerevisiae*. Nature 357: 301–305. **[2]**

Gingerich, P. D., B. H. Smith, and E. L. Simons. 1990. Hind limbs of Eocene *Basilosaurus*: Evidence of feet in whales. Science 249: 154–157. **[5]**

Giovannoni, S. J., S. Turner, G. J. Olsen, S. Barns, D. J. Lane, and N. R. Pace. 1988. Evolutionary relationships among cyanobacteria and green chloroplasts. J. Bacteriol. 170: 3584–3592. **[5]**

Gladyshev, E. A. and I. R. Arkhipova. 2011. A widespread class of reverse transcriptase-related cellular genes. Proc. Natl. Acad. Sci. USA 108: 20311–20316. **[9]**

Gladyshev, E. A., M. Meselson, and I. R. Arkhipova. 2008. Massive horizontal gene transfer in bdelloid rotifers. Science 320: 1210–1213. **[9]**

Glass, J. I. and 8 others. 2006. Essential genes of a minimal bacterium. Proc. Natl. Acad. Sci. USA 103: 425–430. **[10]**

Go, M. 1981. Correlation of DNA exonic regions with protein structural units in haemoglobin. Nature 291: 90–92. **[8]**

Go, M. and M. Nosaka. 1987. Protein architecture and the origin of introns. Cold Spring Harbor Symp. Quant. Biol. 52: 915–924. **[8]**

Goday, C. and S. Pimpinelli. 1984. Chromosome organization and heterochromatin elimination in *Parascaris*. Science 224: 411–413. **[11]**

Gogarten, J. P. and 12 others. 1989. Evolution of the vacuolar H^+-ATPase: Implications for the origin of eukaryotes. Proc. Natl. Acad. Sci. USA 86: 6661–6665. **[6]**

Goh, S.-H. and 8 others. 2005. A newly discovered human α-globin gene. Blood 106: 1466–1472. **[7]**

Goh, S.-H., M. Josleyn, Y. T. Lee, R. L. Danner, R. B. Gherman, M. C. Cam, and J. L. Miller. 2007. The human reticulocyte transcriptome. Physiol. Genomics 30: 172–178. **[7]**

Gojobori, T. and S. Yokoyama. 1985. Rates of evolution of the retroviral oncogene of Moloney murine sarcoma virus and of its cellular homologues. Proc. Natl. Acad. Sci. USA 82: 4198–4201. **[4]**

Gojobori, T., W.-H. Li, and D. Graur. 1982. Patterns of nucleotide substitution in pseudogenes and functional genes. J. Mol. Evol. 18: 360–369. **[4]**

Goldberg, W. M. 2013. *The Biology of Reefs and Reef Organisms*. University of Chicago Press, Chacago **[11]**

Golding, G. B. and C. Strobeck. 1983. Increased number of alleles found in hybrid populations due to intragenic recombination. Evolution 17: 17–19. **[1]**

Goldman, N. 1993. Statistical tests of models of DNA substitution. J. Mol. Evol. 36: 182–198. **[4]**

Goldman, N. and Z. Yang. 1994. A codon-based model of nucleotide substitution for protein-coding DNA sequences. Mol. Biol. Evol. 11: 725–736. **[3]**

Goldsmith, M. E., R. K. Humphries, T. Ley, A. Cline, J. A. Kantor, and A. W. Nienhuis. 1983. "Silent" nucleotide substitution in a β^+-thalassemia globin gene activates splice site in coding sequence RNA. Proc. Natl. Acad. Sci. USA 80: 2318–2322. **[1]**

Gonçalves, I., L. Duret, and D. Mouchiroud. 2000. Nature and structure of human genes that generate retropseudogenes. Genome Res. 10: 672–678. **[9]**

Gondo, Y., T. Okada, N. Matsuyama, Y. Saitoh, Y. Yanagisawa, and J. E. Ikeda. 1998. Human megasatellite DNA RS447: Copy-number polymorphisms and interspecies conservation. Genomics 54: 39–49. **[11]**

Goodman, M. 1961. The role of immunochemical differences in the phyletic development of human behavior. Hum. Biol. 33: 131–162. **[4]**

Goodman, M. 1962. Immunochemistry of the primates and primate evolution. Ann. N. Y. Acad. Sci. 102: 219–234. **[4, 5]**

Goodman, M. 1963. Serological analysis of the systematics of recent hominoids. Hum. Biol. 35: 377–424. **[5]**

Goodman, M. 1981a. Decoding the pattern of protein evolution. Prog. Biophys. Mol. Biol. 38: 105–164. **[4]**

Goodman, M. 1981b. Globin evolution was apparently very rapid in early vertebrates: A reasonable case against the rate-constancy hypothesis. J. Mol. Evol. 17: 114–120. **[4]**

Goodman, M. 1989. Emerging alliance of phylogenetic systematics and molecular biology: A new age of exploration, pp. 43–61. In: B. Fernholm, K. Bremer, and H. Jornvall (eds.), *Hierarchy of Life: Molecules and Morphology in Phylogenetic Analysis*. Elsevier, Amsterdam. **[5]**

Goodman, M., J. Barnabas, G. Matsuda, and G. W. Moor. 1971. Molecular evolution in the descent of man. Nature 233: 604–613. **[4]**

Goodman, M., G. W. Moore, and J. Barnabas. 1974. The phylogeny of human globin genes investigated by the maximum parsimony method. J. Mol. Evol. 3: 1–48. **[4]**

Goodman, M., G. W. Moore, and G. Matsuda. 1975. Darwinian evolution in the genealogy of hemoglobin. Nature 253: 603–608. **[4]**

Goodman, M., J. Czelusniak, G. W. Moore, A. E. Romero-Herrera and G. Matsuda. 1979. Fitting the gene lineage into its species lineage, a parsimony strategy illustrated by cladograms constructed from globin sequences. Syst. Zool. 28: 132–163. **[11]**

Goodman, M., B. F. Koop, J. Czelusniak, M. L. Weiss, and J. L. Slightom. 1984. The η-globin gene: Its long evolutionary history in the β-globin gene family of mammals. J. Mol. Biol. 180: 803–823. **[7]**

Goodman, M. and 7 others. 1990. Primate evolution at the DNA level and a classification of hominoids. J. Mol. Evol. 30: 260–266. **[5]**

Goodman, M. and 7 others. 1998. Toward a phylogenetic classification of primates based on DNA evidence complemented by fossil record. Mol. Phylogenet. Evol. 9: 585–598. **[5]**

Goodman, M., S. L. Page, C. M. Meireles, and J. Czelusniak. 1999. Primate phylogeny and classification elucidated at the molecular level, pp. 193–211. In: S. P. Wasser (ed.), *Evolutionary Theory and Processes: Modern Perspectives*. Kluwer, Dordrecht, Netherlands. **[5]**

Gophna, U., R. L. Charlebois, and W. F. Doolittle. 2004. Have archaeal genes contributed to bacterial virulence? Trends Microbiol. 12: 213–219. **[9]**

Goremykin, V. V., F. Salamini, R. Velasco, and R. Viola. 2009. Mitochondrial DNA of *Vitis vinifera* and the issue of rampant horizontal gene transfer. Mol. Biol. Evol. 26: 99–110. **[9]**

Gossett, T. F. 1997. *Race: The History of an Idea in America.* Oxford University Press, New York. **[11]**

Gott, J. M., L. M. Visomirski, and J. L. Hunter. 1993. Substitutional and insertional RNA editing of the cytochrome *c* oxidase subunit 1 mRNA of *Physarum polycephalum*. J. Biol. Chem. 268: 25483–25486. **[1]**

Gould, S. B., R. F. Waller, and G. I. McFadden. 2008. Plastid evolution. Annu. Rev. Plant. Biol. 59: 491–517. **[5, 6]**

Gould, S. J. 1989. *Wonderful Life*. Norton, New York. **[13]**

Gould, S. J. and E. S. Vrba. 1982. Exaptation: A missing term in the science of form. Paleobiology 8: 4–15. **[9]**

Gould, S. J. and R. C. Lewontin. 1979. The spandrels of San Marco and the Panglossian paradigm: A critique of the adaptationist programme. Proc. R. Soc. London 205B: 581–598. **[2]**

Graham, E. A. M. 2007. DNA reviews: Ancient DNA. Forensic Sci. Med. Pathol. 3: 221–225. **[5]**

Grant, P. R., B. R. Grant, J. A. Markert, L. F. Keller, and K. Petren. 2004. Convergent evolution of Darwin's finches caused by introgressive hybridization and selection. Evolution 58: 1588–1599. **[6]**

Grantham, R. 1974. Amino acid difference formula to help explain protein evolution. Science 185: 862–864. **[4]**

Grantham, R., C. Gautier, M. Gouy, R. Mercier, and A. Pavé. 1980. Codon catalog usage and the genome hypothesis. Nuc. Acids Res. 8: r49–r62. **[4]**

Graur, D. 1985. Amino acid composition and the evolutionary rates of protein-coding genes. J. Mol. Evol. 22: 53–63. **[4]**

Graur, D. 1993. Molecular phylogeny and the higher classification of eutherian mammals. Trends Ecol. Evol. 8: 141–147. **[5]**

Graur, D. 2014. "Take another good look at the data": The bimodal distribution that wasn't. http://judgestarling.tumblr.com/post/84095742522/take-another-good-look-at-the-data-the-bimodal **[10]**

Graur, D. and D. G. Higgins. 1994. Molecular evidence for the inclusion of cetaceans within the order Artiodactyla. Mol. Biol. Evol. 11: 357–364. **[5]**

Graur, D. and T. Pupko. 2001. The Permian bacterium that isn't. Mol. Biol. Evol. 18: 1143–1146. **[5]**

Graur, D. and W. Martin. 2004. Reading the entrails of chickens: Molecular timescales of evolution and the illusion of precision. Trends Genet. 20: 80–86. **[4, 5]**

Graur, D. and W.-H. Li. 2000. *Fundamentals of Molecular Evolution*, 12th Ed. Sinauer, Sunderland, MA. **[1–11]**

Graur, D., Y. Zheng, N. Price, R. B. Azevedo, R. A. Zufall, and E. Elhaik. 2013. On the immortality of television sets: "Function" in the human genome according to the evolution-free gospel of ENCODE. Genome Biol. Evol. 5: 578–590. **[11]**

Graur, D., Y. Zheng, and R. B. Azevedo. 2015. An evolutionary classification of genomic function. Genome Biol. Evol. 7: 642–645. **[11]**

Gray, G. S. and W. M. Fitch. 1983. Evolution of antibiotic resistance genes: The DNA sequence of a kanamycin resistance gene from *Staphylococcus aureus*. Mol. Biol. Evol. 1: 57–66. **[5]**

Gray, M. W. 1983. The bacterial ancestry of plastids and mitochondria. BioScience 33: 693–699. **[5]**

Gray, M. W. and P. S. Covello. 1993. RNA editing in plant mitochondria and chloroplasts. FASEB J. 7: 64–71. **[8]**

Gray, M. W., D. Sankoff, and R. J. Cedergren. 1984. On the evolutionary descent of organisms and organelles: A global phylogeny based on a highly conserved structural core in small subunit ribosomal RNA. Nucleic Acids Res. 12: 5837–5852. **[5]**

Gray, M. W., G. Burger, and B. F. Lang. 2001. The origin and early evolution of mitochondria. Genome Biol. 2(6): reviews1018.1–reviews1018.5 (published online) **[5]**

Gray, M. W., J. Lukeš, J. M. Archibald, P. J. Keeling, and W. F. Doolittle. (2010) Irremediable complexity? Science 330: 920–921. **[8]**

Green, B. R. 2005. Lateral gene transfer in the Cyanobacteria: Chlorophylls, proteins, and scraps of ribosomal RNA. J. Phycol. 41: 449–452. **[6]**

Green, R. E. and 10 others. 2006. Analysis of one million base pairs of Neanderthal DNA. Nature 444: 330–336. **[5]**

Gregory, T. R (ed.). 2005b. *The Evolution of the Genome.* Elsevier, Burlington, MA. **[11]**

Gregory, T. R and R. DeSalle. 2005. Comparative genomics in prokaryotes, pp. 585–675. In: T. R. Gregory (ed.), *The Evolution of the Genome.* Elsevier, San Diego. **[10]**

Gregory, T. R. 2001. Coincidence, coevolution, or causation? DNA content, cell size, and the C-value enigma. Biol. Rev. Camb. Philos. Soc. 76: 65–101. **[7]**

Gregory, T. R. 2001a. Coincidence, coevolution, or causation? DNA content, cell size, and the C-value enigma. Biol. Rev. 76: 65–101. **[11]**

Gregory, T. R. 2001b. The bigger the C-value, the larger the cell: Genome size and red blood cell size in vertebrates. Blood Cells Mol. Dis. 27: 830–843. **[11]**

Gregory, T. R. 2002. Genome size and developmental parameters in the homeothermic vertebrates. Genome 45: 833–838. **[11]**

Gregory, T. R. 2003. Variation across amphibian species in the size of the nuclear genome supports a pluralistic, hierarchical approach to the C-value enigma. Biol. J. Linn. Soc. 79: 329–339. **[11]**

Gregory, T. R. 2004. Insertion-deletion biases and the evolution of genome size. Gene 324: 15–34. **[11]**

Gregory, T. R. 2005. Synergy between sequence and size in large-scale genomics. Nat. Rev. Genet. 6: 699–708. **[10, 11]**

Gregory, T. R. and B. K. Mable. 2005. Polyploidy in animals, pp. 427–517. In: T. R. Gregory (ed.), *The Evolution of the Genome.* Elsevier, San Diego. **[7]**

Gregory, W. K. 1910. The orders of mammals. Bull. Am. Mus. Nat. Hist. 27: 1–524. **[5]**

Grehan, J. R. and J. H. Schwartz. 2009. Evolution of the second orangutan: Phylogeny and biogeography of hominid origins. J. Biogeog. 36: 1823–1844. **[5]**

Greilhuber, J. and I. J. Leitch. 2013. Genome size and the phenotype. In: I. J. Leitch et al. (eds.) *Plant Genome Diversity* (Volume 2), pp. 323–344. **[11]**

Greilhuber, J., J. Dolezel, M. A. Lysák and M. D. Bennett. 2005. The origin, evolution and proposed stabilization of the terms "genome size" and "C-value" to describe nuclear DNA contents. Ann. Bot. 95: 255–260. **[11]**

Greilhuber, J., T. Borsch, K. Müller, A. Worberg, S. Porembski, and W. Barthlott. 2006. Smallest angiosperm genomes found in Lentibulariaceae, with chromosomes of bacterial size. Plant Biol. 8: 770–777. **[11]**

Grens, K. 2014. The rainbow connection. www.the-scientist.com/?articles.view/articleNo/41055/title/The-Rainbow-Connection/ **[7]**

Grewal, S. I. and S. Jia. 2007. Heterochromatin revisited. Nat. Rev. Genet. 8: 35–46. **[11]**

Griffith, F. 1928. The significance of pneumococcal types. J. Hygiene 27: 108–159. **[9]**

Griffiths, A. J. F., J. H. Miller, D. T. Suzuki, R. C. Lewontin, and W. M. Gelbart. 2000. *Introduction to Genetic Analysis*, 7th Ed. W.H. Freeman, New York. **[7]**

Griffiths, P. E. 2009. In what sense does "nothing in biology make sense except in the light of evolution"? Acta Biotheoretica 57: 11–32. **[11]**

Griffiths, P. E. and K. Stotz. 2006. Genes in the postgenomic era. Theor. Med. Bioeth. 27: 499–521. **[1]**

Gross, B. L. 2012. Genetic and phenotypic divergence of homoploid hybrid species from parental species. Heredity 108: 157–158. **[7]**

Gross, L. 2007. A genetic basis for hypersensitivity to "sweaty" odors in humans. PLoS Biol. 5: e298. **[7]**

Grossman, S. R. and 12 others. 2010. A composite of multiple signals distinguishes causal variants in regions of positive selection. Science 327: 883–886. **[11]**

Grossman, S. R. and 19 others. 2013. Identifying recent adaptations in large-scale genomic data. Cell 152: 703–713. **[2, 11]**

Gruber, J. D., K. Vogel, G. Kalay, and P. J. Wittkopp. 2012. Contrasting properties of gene-specific regulatory, coding, and copy number mutations in *Saccharomyces cerevisiae*: Frequency, effects, and dominance. PloS Genetics 8: e1002497. **[13]**

Gu, W., T. Zhou, and C. O. Wilke. 2010. A universal trend of reduced mRNA stability near the translation-initiation site in prokaryotes and eukaryotes. PLoS Comp. Biol. 6: e1000664. **[4]**

Gu, X. and W.-H. Li. 1992. Higher rates of amino acid substitution in rodents than in humans. Mol. Phylogenet. Evol. 1: 211–214. **[4]**

Gu, X. and W.-H. Li. 1995. The size distribution of insertions and deletions in human and rodent pseudogenes suggests the logarithmic gap penalty for sequence alignment. J. Mol. Evol. 40: 464–473. **[3]**

Gu, Z., A. Cavalcanti, F.-C. Chen, P. Bouman, and W.-H. Li. 2002. Extent of gene duplication in the genomes of *Drosophila*, nematode, and yeast. Mol. Biol. Evol. 19: 256–262. **[7]**

Gu, Z., H. Wang, A. Nekrutenko, and W.-H. Li. 2000. Densities, length proportions, and other distributional features of repetitive sequences in the human genome estimated from 430 megabases of genomic sequence. Gene 259: 81–88. **[11]**

Guerra, M. 2008. Chromosome numbers in plant cytotaxonomy: Concepts and implications. Cytogenet. Genome Res. 120: 339–350. **[7]**

Guisinger, M. M., J. V. Kuehl, J. L. Boore, and R. K. Jansen. 2008. Genome-wide analyses of Geraniaceae plastid DNA reveal unprecedented patterns of increased nucleotide substitutions. Proc. Natl. Acad. Sci. USA 105: 18424–18429. **[4]**

Guo, H. H., J. Choe, and L. A. Loeb. 2004. Protein tolerance to random amino acid change. Proc. Natl. Acad. Sci. USA 101: 9205–9210. **[4]**

Gutiérrez, G. and A. Marín. 1998. The most ancient DNA recovered from an amber-preserved specimen may not be as ancient as it seems. Mol. Biol. Evol. 15: 926–929. **[5]**

H

Haag-Liautard, C., N. Coffey, D. Houle, M. Lynch, B. Charlesworth, and P. D. Keightley. 2008. Direct estimation of the mitochondrial DNA mutation rate in *Drosophila melanogaster*. PLoS Biol. 6: e204. **[4]**

Hackett, J. D and D. Bhattacharya. 2006. The genomes of dinoflagellates. In: L. A. Katz and D. Bhattacharyas (eds.) *Genomics and Evolution of Microbial Eukaryotes*. p. 48–63. Oxford University Press, New York. **[11]**

Hackstein, J. H. P., J. Tjaden, and M. Huynen. 2006. Mitochondria, hydrogenosomes and mitosomes: Products of evolutionary tinkering! Curr. Genet. 50: 225–245. **[5]**

Hahn, B. H. and 9 others. 1986. Genetic variation in HTLV-III/LAV over time in patients with AIDS or at risk for AIDS. Science 232: 1548–1553. **[4]**

Hahn, M. W. 2007. Bias in phylogenetic tree reconciliation methods: implications for vertebrate genome evolution. Genome Biol. 8: R141. **[11]**

Hahn, M. W. 2007. Bias in phylogenetic tree reconciliation methods: Implications for vertebrate genome evolution. Genome Biol. 8: R141. **[7]**

Hahn, M. W. 2009. Distinguishing among evolutionary models for the maintenance of gene duplicates. J. Hered. 100: 605–617. **[7]**

Hahn, M. W. and G. A. Wray. 2002. The G-value paradox. Evol. Dev. 4: 73–75. **[11]**

Hahn, M. W., J. P. Demuth, and S. G. Han. 2007. Accelerated rate of gene gain and loss in primates. Genetics 177: 1941–1949. **[11]**

Hahn, M. W., T. De Bie, J. E. Stajich, C. Nguyen, and N. Cristianini. 2005. Estimating the tempo and mode of gene family evolution from comparative genomic data. Genome Res. 15: 1153–1160. **[7]**

Haldane, J. B. S. 1927. A mathematical theory of natural and artificial selection. Part V. Selection and mutation. Proc. Cambridge Philos. Soc. 23: 838–844. **[4]**

Haldane, J. B. S. 1932. *The Causes of Evolution*. Longmans & Green, London. **[7]**

Haldane, J. B. S. 1935. The rate of spontaneous mutation of a human gene. J. Genet. 31: 317–326. **[4]**

Haldane, J. B. S. 1937. The effect of variation on fitness. Am. Nat. 71: 337–349. **[11]**

Haldane, J. B. S. 1947. The mutation rate of the gene for hemophilia, and its segregation ratios in males and females. Ann. Eugen. 13: 262–271. **[4]**

Haldane, J. B. S. 1956. The theory of selection for melanism in Lepidoptera. Proc. R. Soc. London 145B: 303–306. **[2]**

Hall, B. 2007. *Phylogenetic Trees Made Easy*, 3rd Ed. Sinauer Associates, Sunderland, MA. **[5, 6]**

Halpern, A. L. and W. J. Bruno. 1998. Evolutionary distances for protein-coding sequences: Modeling site-specific residue frequencies. Mol. Biol. Evol. 15: 910–917. **[3]**

Hampl, V., J. D. Silberman, A. Stechmann, S. Diaz-Triviño, P. J. Johnson PJ, and A. J. Roger. 2008. Genetic evidence for a mitochondriate ancestry in the 'amitochondriate' flagellate *Trimastix pyriformis*. PLoS One 3:e1383. **[6]**

Hampl, V., L. Hug, J. W. Leigh, J. B. Dacks, B. F. Lanf, A. G. B. Simpson, and A. J. Roger. 2009. Phylogenomic analyses support the monophyly of Excavata and resolve relationships among eukaryotic "supergroups." Proc. Natl. Acad. Sci. USA 106: 3859–3864. **[5]**

Hampl, V., L. Hug, J. W. Leigh, J. B. Dacks, B. F. Lang, A. G. Simpson, A. J. Roger. 2009. Phylogenomic analyses support the monophyly of Excavata and resolve relationships among eukaryotic "supergroups." Proc. Natl. Acad. Sci. USA 106: 3859–3864. **[6]**

Han, M. V., J. P. Demuth, C. L. McGrath, C. Casola, and M. W. Hahn. 2009. Adaptive evolution of young gene duplicates in mammals. Genome Res. 19: 859–867. **[7]**

Han, W., Y. Yu, N. Altan, and L. Pick. 1993. Multiple proteins interact with the *fushi tarazu* proximal enhancer. Mol. Cell. Biol. 13: 5549–5559. **[12]**

Hancks, D. C. and H. H. Kazazian. 2012. Active human retrotransposons: Variation and disease. Curr. Opin. Genet. Dev. 22: 191–203. **[9]**

Hansen, C. N. and J. S. Heslop-Harrison. 2004. Sequences and phylogenies of plant pararetroviruses, viruses and transposable elements. Adv. Bot. Res. 41: 165–193. **[9]**

Hara, Y., T. Imanishi, and Y. Satta. 2012. Reconstructing the demographic history of the human lineage using whole-genome sequences from human and three great apes. Genome Biol. Evol. 4: 1133–1145. **[5]**

Harcombe, W. R., N. F. Delaney, N. Leiby, N. Klitgord, and C. J. Marx. 2013. The ability of flux balance analysis to predict evolution of central metabolism scales with the initial distance to the optimum. PLoS Comp. Biol. 9: e1003091. **[13]**

Hardison, R. 1998. Hemoglobins from bacteria to man: evolution of different patterns of gene expression. J. Exp. Biol. 201: 1099–1117. **[12]**

Hardison, R. C. 2008. Globin genes on the move. J. Biol. 7: 35. **[7]**

Hardison, R. C. and J. B. Margot. 1984. Rabbit globin pseudogene ψβ2 is a hybrid of δ- and β-globin gene sequences. Mol. Biol. Evol. 1: 302–316. **[7]**

Hare, E. E., B. K. Peterson, V. N. Iyer, R. Meier, and M. B. Eisen. 2008. Sepsid even-skipped enhancers are functionally conserved in *Drosophila* despite lack of sequence conservation. PLoS Genet. 4: e1000106. **[12]**

Hare, M. P. and S. R. Palumbi. 2003. High intron sequence conservation across three mammalian orders suggests functional constraints. Mol. Biol. Evol. 20: 969–978. **[11]**

Häring, D. and J. Kypr. 2001a. No isochores in the human chromosomes 21 and 22? Biochem. Biophys. Res. Commun. 280: 567–573. **[11]**

Häring, D. and J. Kypr. 2001b. Mosaic structure of the DNA molecules of the human chromosomes 21 and 22. Mol. Biol. Rep. 28: 9–17. **[11]**

Harley, E. H. 1988. The retrieval of the quagga. S. Afr. J. Sci. 84: 158–159. **[5]**

Harman, D. 1956. Aging: A theory based on free radical and radiation chemistry. J. Gerontol. 11: 298–300. **[4]**

Harold, F. M. 2011. Energetics of the eukaryotic edge. http://schaechter.asmblog.org/schaechter/2011/01/energetics-of-the-eukaryotic-edge.html **[6]**

Harris, S., P. A. Barrie, M. L. Weiss, and A. J. Jeffreys. 1984. The primate ψβ1 gene. An ancient β-globin pseudogene. J. Mol. Biol. 180: 785–801. **[7]**

Hartl, D. L. and A. G. Clark. 1997. *Principles of Population Genetics.* Sinauer Associates, Sunderland, MA. **[2]**

Hartl, D. L. and A. G. Clark. 2007. *Principles of Population Genetics,* 3rd Ed. Sinauer, Sunderland, MA **[2]**

Hartl, D. L., A. R. Lohe, and E. R. Lozovskaya. 1997. Modern thoughts on an ancyent marinere: Function, evolution, regulation. Annu. Rev. Genet. 31: 337–358. **[11]**

Harvey, P. H., A. J. Leigh Brown, J. Maynard Smith, and S. Nee (eds.). 1996. *New Uses for New Phylogenies.* Oxford University Press, Oxford. **[5, 6]**

Hasegawa, M., Y. Iida, T.-A. Yano, F. Takaiwa, and M. Iwabuchi. 1985. Phylogenetic relationships among eukaryotic kingdoms inferred from ribosomal RNA sequences. J. Mol. Evol. 22: 32–38. **[5]**

Hastings, P. J., J. R. Lupski, S. M. Rosenberg, and G. Ira. 2009. Mechanisms of change in gene copy number. Nat. Rev. Genet. 10: 551–564. **[7]**

Hattori, M. and 64 others. 2000. The DNA sequence of human chromosome 21. Nature 405: 311–319. **[11]**

Haughton, S. H. 1932. On some South African fossil Proboscidea. Trans. R. Soc. S. Africa 21:1–18. **[5]**

Havecker, E. R., X. Gao, and D. F. Voytas. 2004. The diversity of LTR retrotransposons. Genome Biol. 5: 225. **[9]**

Hawkins, J. S., H. Kim, J. D. Nason, R. A. Wing, and J. F. Wendel. 2006. Differential lineage-specific amplification of transposable elements is responsible for genome size variation in *Gossypium.* Genome Res. 16: 1252–1261. **[11]**

Hawks, J. 2008. From gene to numbers: Effective population sizes in human evolution, pp. 9–30. In: J.-P. Bocquet-Appel (ed.), *Recent Advances in Peleodemography.* Springer, Dordrecht, Netherlands. **[2]**

Hayakawa, T., Y. Satta, P. Gagneux, A. Varki, and N. Takahata. 2001. *Alu*-mediated inactivation of the human *CMP-N-acetylneuraminic acid hydroxylase* gene. Proc. Natl. Acad. Sci. USA 98: 11399–11404. **[7, 8]**

Hayasaka, K., D. H. Fitch, J. L. Slightom, and M. Goodman. 1992. Fetal recruitment of anthropoid γ-globin genes. Findings from phylogenetic analyses involving the 5′-flanking sequences of the ψγ1 globin gene of spider monkey, *Ateles geoffroyi.* J. Mol. Biol. 224: 875–881. **[7]**

Hayashida, H., H. Toh, R. Kikuno, and T. Miyata. 1985. Evolution of influenza **[4]**

Hayden, E. C. 2012. The $1,000 genome. Nature 507: 294–295. **[4]**

Hayden, T. 1999. How low can you go? Seeking the fewest genes necessary for life. Newsweek 133(8): 52. **[10]**

Hazkani-Covo, E. and D. Graur. 2005. Evolutionary conservation of bacterial operons: Does transcriptional connectivity matter? Genetica 124: 145–166. **[10]**

Hazkani-Covo, E. and S. Covo. 2008. *Numt*-mediated double-strand break repair mitigates deletions during primate genome evolution, PLoS Genet. 4: e1000237. **[9]**

Hazkani-Covo, E., E. Y. Levanon, G. Rotman, D. Graur, and A. Novik. 2004. Evolution of multicellularity in Metazoa: Comparative analysis of the subcellular localization of proteins in *Saccharomyces, Drosophila* and *Caenorhabditis.* Cell. Biol. Int. 28: 171–178. **[11]**

Hazkani-Covo, E., R. M. Zeller, and W. Martin. 2010. Molecular poltergeists: Mitochondrial DNA copies (*numts*) in sequenced nuclear genomes. PLoS Genet. 6: e1000834. **[9, 11]**

He, X. and J. Zhang. 2005. Rapid subfunctionalization accompanied by prolonged and substantial neofunctionalization in duplicate gene evolution. Genetics 169: 1157–1164. **[7]**

Heddle, J. A. and K. Athanasiou. 1975. Mutation rate, genome size and their relation to the *rec.* concept. Nature 258: 359–361. **[11]**

Hedges, S. B., J. Marin, M. Suleski, M. Paymer, and S. Kumar. 2015. Tree of life reveals clock-like speciation and diversification. Mol. Biol. Evol. 32:835–845. **[5]**

Hedges, S. B., S. Kumar, K. Tamura, and M. Stoneking. 1992. Human origins and analysis of mitochondrial DNA sequences. Science 255: 737–739. **[5]**

Hedlund, M. and 11 others. 2007. *N*-glycolylneuraminic acid deficiency in mice: Implications for human biology and evolution. Mol. Cell. Biol. 27: 4340–4346. **[7, 8]**

Hedrick, P. W. 2005. *Genetics of Populations,* 3rd Ed. Jones and Bartlett, Sudbury, MA. **[2]**

Hegreness, M., N. Shoresh, D. Hartl, and R. Kishony. 2006. An equivalence principle for the incorporation of favorable mutations in asexual populations. Science 311: 1615–1617. **[13]**

Hehemann, J. H., G. Correc, T. Barbeyron, W. Helbert, M. Czjzek, and G. Michel. 2010. Transfer of carbohydrate-active enzymes from marine bacteria to Japanese gut microbiota. Nature 464: 908–912. **[9]**

Hehenberger, E., B. Imanian, F. Burki, and P. J. Keeling. 2014. Evidence for the retention of two evolutionary distinct plastids in dinoflagellates with diatom endosymbionts. Genome Biol. Evol. 6: 2321–2334. **[11]**

Heilig, R. and 98 others. 2003. The DNA sequence and analysis of human chromosome 14. Nature 421: 601–607. **[11]**

Heimberg, A. M., R. Cowper-Sal lari, M. Sémonc, P. C. J. Donoghue, and K. J. Peterson. 2010. MicroRNAs reveal the interrelationships of hagfish, lampreys, and gnathostomes and the nature of the ancestral vertebrate. Proc. Natl. Acad. Sci. USA 107: 19379–19383. **[7]**

Heitz, E. 1928. Das heterochromatin der moose. I. Jahrbüch. Wissenschaft. Botanik 69: 762–818. **[11]**

Helal, M., F. Kong, S. C. A. Chen, M. Bain, R. Christen, and V. Sintchenko. 2011. Defining reference sequences for *Nocardia* species by similarity and clustering analyses of 16S rRNA gene sequence data. PLoS ONE 6: e19517. **[6]**

Hellberg, M. E. 2006. No variation and low synonymous substitution rates in coral mtDNA despite high nuclear variation. BMC Evol. Biol. 6: 24. **[4]**

Hellmann, I., S. Zöllner, W. Enard, I. Ebersberger, B. Nickel, and S. Pääbo. 2003. Selection on human genes as revealed by comparisons to chimpanzee cDNA. Genome Res. 13: 831–837. **[4]**

Hellsten, U. and 47 others. 2010. The genome of the Western clawed frog *Xenopus tropicalis.* Science 328: 633–636. **[11]**

Hendrickson, H. and J. G. Lawrence. 2007. Mutational bias suggests that replication termination occurs near the *dif* site, not at Ter sites. Mol. Microbiol. 64: 42–56. **[10]**

Hendriks, W., J. Leunissen, E. Nevo, H. Bloemendal, and W. W. de Jong. 1987. The lens protein αA-crystallin of the blind mole rat, *Spalax ehrenbergi*: Evolutionary change and functional constraints. Proc. Natl. Acad. Sci. USA 84: 5320–5324. **[4]**

Hendriks, W., J. W. M. Mulders, M. A. Bibby, C. Slingsby, H. Bloemendal, and W. W. de Jong. 1988. Duck lens ε-crystallin and lactate dehydrogenase B4 are identical: A single-copy gene product with two distinct functions. Proc. Natl. Acad. Sci. USA 85: 7114–7118. **[7]**

Hendy, M. D. and D. Penny. 1982. Branch and bound algorithms to determine minimal evolutionary trees. Math. Biosci. 59: 277–290. **[5]**

Henikoff, S. and J. G. Henikoff. 1992. Amino acid substitution matrices from protein locks. Proc. Natl. Acad. Sci. USA 89: 10915–10919. **[3]**

Hensche, P. E. 1975. Gene duplication as a mechanism of genetic adaptation in *Saccharomyces cerevisiae.* Genetics 79: 661–674. **[7]**

Hentzen, D., A. Chevallier, and J. P. Garel. 1981. Differential usage of iso-accepting $tRNA^{Ser}$ species in silk glands of *Bombyx mori*. Nature 290: 267–269. **[4]**

Henz, S. R., D. H. Huson, A. F. Auch, K. Nieselt-Struwe, and S. C. Schuster. 2005. Whole-genome prokaryotic phylogeny. Bioinformatics 21: 2329–2335. **[5]**

Herold, M., M. Bartkuhn, and R. Renkawitz. 2012. CTCF: insights into insulator function during development. Development 139: 1045–1057. **[12]**

Herrick, J. and B. Sclavi. 2014. A new look at genome size, evolutionary duration and genetic variation in salamanders. Comptes Rendus Palevol 13: 611–621. **[11]**

Herrmann, K. M. and R. L. Somerville. (eds). 1983. *Amino Acids: Biosynthesis and Genetic Regulation*. Addison-Wesley, Reading, MA **[11]**

Hershberg, R. and D. A. Petrov. 2008. Selection on codon bias. Annu. Rev. Genet. 42: 287–299. **[1]**

Hershey, A. D. and M. Chase. 1952. Independent functions of viral protein and nucleic acid in growth of bacteriophage. J. Gen. Physiol. 36: 39–56. **[7]**

Hess, P. N. and C. A. De Morales Russo. 2007. An empirical test of the midpoint rooting method. Biol. J. Linn. Soc. 92: 669–674. **[5]**

Hickey, D. A. 1982. Selfish DNA: A sexually transmitted nuclear parasite. Genetics 101: 519–531. **[2, 11]**

Higgins, D. G. 1994. CLUSTAL V: Multiple alignment of DNA and protein sequences. Methods Mol. Biol. 25: 307–318. **[3]**

Higgins, D. G. and P. M. Sharp. 1988. CLUSTAL: A package for performing multiple sequence alignment on a microcomputer. Gene 73: 237–44. **[3]**

Higgins, D. G., A. J. Bleasby, and R. Fuchs. 1992. CLUSTAL V: Improved software for multiple sequence alignment. Comp. Appl. Biosci. 8: 189–191. **[3]**

Higgins, D. G., J. D. Thompson, and T. J. Gibson. 1996. Using CLUSTAL for multiple sequence alignments. Methods Enzymol. 266: 383–402. **[3]**

Higgins, J. M. 2003. Structure, function and evolution of haspin and haspin-related proteins, a distinctive group of eukaryotic protein kinases. Cell. Mol. Life Sci. 60: 446–462. **[9]**

Higuchi, R. G., L. A. Wrischnik, E. Oakes, M. George, B. Tong, and A. C. Wilson. 1987. Mitochondrial DNA of the extinct quagga: Relatedness and extent of postmortem changes. J. Mol. Evol. 25: 283–287. **[5]**

Hildebrand, F., A. Meyer, and A. Eyre-Walker. 2010. Evidence of selection upon genomic GC-content in bacteria. PLoS Genet. 6: e1001107. **[10]**

Hill, R. E. and N. D. Hastie. 1987. Accelerated evolution in the reactive centre regions of serine protease inhibitors. Nature 326: 96–99. **[4]**

Hiller, M., B. T. Schaar, V. B. Indjeian, D. M. Kingsley, L. R. Hagey, and G. Bejerano. 2012. A "forward genomics" approach links genotype to phenotype using independent phenotypic losses among related species. Cell Rep. 2: 817–823. **[8]**

Hillier, L. W. and 106 others. 2003. The DNA sequence of human chromosome 7. Nature 424: 157–164. **[11]**

Hillis, D. M. and J. J. Bull. 1993. An empirical test of bootstrapping as a method for assessing confidence on phylogenetic analysis. Syst. Biol. 42: 182–192. **[5]**

Hillis, D. M., C. Moritz, and B. K. Mable (eds.). 1996. *Molecular Systematics*, 2nd Ed. Sinauer Associates, Sunderland, MA. **[5, 6]**

Hipp, A. L., K.-S. Chung, and M. Escudero. 2013. Holocentric chromosomes. In: S. Maloy and K. Hughes (eds.) *Encyclopedia of Genetics* (2nd ed). pp. 499–501. Elsevier, Amsterdam. **[11]**

Hirose, T., M. D. Shu, and J. A. Steitz. 2003. Splicing-dependent and -independent modes of assembly for intron-encoded box C/D snoRNPs in mammalian cells. Mol. Cell 12: 113–123. **[11]**

Hirth, L., D. Abadanian, and H. W. Goedde. 1986. Incidence of specific anosmia in Northern Germany. Hum. Hered. 36: 1–5. **[7]**

Hittinger, C. T. and S. B. Carroll. 2007. Gene duplication and the adaptive evolution of a classic genetic switch. Nature 449: 677–681. **[7]**

Hixon, J. E. and W. M. Brown. 1986. A comparison of the small ribosomal RNA genes from the mitochondrial DNA of the great apes and humans: Sequence, structure, evolution, and phylogenetic implications. Mol. Biol. Evol. 3: 1–18. **[5]**

Hjort, K., A. V. Goldberg, A. D. Tsaousis, R. P. Hirt, and T. M. Embley. 2010. Diversity and reductive evolution of mitochondria among microbial eukaryotes. Phil. Trans. Roy. Soc. 365B: 713–727. **[6]**

Ho, S. Y. W. 2007. Calibrating molecular estimates of substitution rates and divergence times in birds. J. Avian Biol. 38: 409–414. **[5]**

Ho, S. Y. W. and M. J. Phillips. 2009. Accounting for calibration uncertainty in phylogenetic estimation of evolutionary divergence times. Syst. Biol. 58: 367–380. **[5]**

Ho, S. Y. W., M. J. Phillips, A. Cooper, and A. J. Drummond. 2005. Time dependency of molecular rate estimates and systematic over-estimation of recent divergence times. Mol. Biol. Evol. 22: 1561–1568. **[5]**

Ho, S. Y. W., U. Saarma, R. Barnett, J. Haile, and B. Shapiro. 2008. The effect of inappropriate calibration: Three case studies in molecular ecology. PLoS ONE 3: e1615. **[5]**

Hobolth, A., O. F. Christensen, T. Mailund, and M. H. Schierup. 2007. Genomic relationships and speciation times of human, chimpanzee, and gorilla inferred from a coalescent hidden Markov model. PLoS Genet. 3: e7. **[5]**

Hoekstra, H. E. and J. A. Coyne . 2007. The locus of evolution: Evo devo and the genetics of adaptation. Evolution 61: 995–1016. **[13]**

Hoffmann, F. G., J. F. Storz, T. A. Gorr, and J. C. Opazo. 2010. Lineage-specific patterns of functional diversification in the α- and β-globin gene families of tetrapod vertebrates. Mol. Biol. Evol. 27: 1126–1138. **[7]**

Hokamp, K., A. McLysaght, H. Kenneth, and K. H. Wolfe. 2003. The 2R hypothesis and the human genome sequence. J. Struct. Funct. Genomics 3: 95–110. **[7]**

Holden, M. T. G. and 9 others. 2010. Genome sequence of a recently emerged, highly transmissible, multi-antibiotic- and antiseptic-resistant variant of methicillin-resistant *Staphylococcus aureus*, sequence type 239 (TW). J. Bacteriol. 192: 888–892. **[9]**

Holland, J. H. 1975. *Adaptation in Natural and Artificial Systems: An Introductory Analysis with Applications to Biology, Control, and Artificial Intelligence*. University of Michigan Press, Ann Arbor. **[5]**

Holland, P. W. H., J. Garcia-Fernàndez, N. A. Williams, and A. Sidow. 1994. Gene duplications and the origins of vertebrate development. Development 1994: S125–S133. **[7]**

Holliday, R. and J. E. Pugh. 1975. DNA modification mechanisms and gene activity during development. Science 187: 226–232. **[12]**

Hollins, C., D. A. R. Zorio, M. MacMorris, and T. Blumenthal. 2005. U2AF binding selects for the high conservation of the *C. elegans* 3′ splice site. RNA 11: 248–253. **[8]**

Holmes, E. C. 2011. The evolution of endogenous viral elements. Cell Host Microb. 10: 368–377. **[9]**

Holmes, E. C., L. Q. Zhang, P. Simmonds, A. S. Rogers, and A. J. L. Brown. 1993. Molecular investigation of human immunodeficiency virus (HIV) infection in a patient of an HIV-infected surgeon. J. Infect. Dis. 167: 1411–1414. **[5]**

Holmes, R. S. and E. Goldberg. 2009. Computational analyses of mammalian lactate dehydrogenases: Human, mouse, opossum and platypus LDHs. Comp. Biol. Chem. 33: 379–385. **[7]**

Holmquist, G. P. 1987. Role of replication time in the control of tissue-specific gene expression. Am. J. Hum. Genet. 40: 151–173. **[11]**

Holmquist, R. 1972. Theoretical foundations for a quantitative approach to paleogenetics. I: DNA. J. Mol. Evol. 1: 115–133. **[3]**

Holmquist, R. and D. Pearl. 1980. Theoretical foundations for quantitative paleogenetics III. The molecular divergence of nucleic acids and proteins for the case of genetic events of unequal probability. J. Mol. Evol. 16: 211–267. **[3]**

Hong, J. W., D. A. Hendrix, and M. S. Levine. 2008. Shadow enhancers as a source of evolutionary novelty. Science 321: 1314. **[12]**

Hood, L., J. H. Campbell, and S. C. R. Elgin. 1975. The organization, expression and evolution of antibody genes and other multigene families. Annu. Rev. Genet. 9: 305–353. **[8]**

Horai, S., R. Kondo, Y. Nakagawa-Hattorri, S. Hayashi, S. Sonoda, and K. Tajima. 1993. Peopling of the Americas founded by four major lineages of mitochondrial DNA. Mol. Biol. Evol. 10: 23–47. **[5]**

Horie, M. and 10 others. 2010. Endogenous non-retroviral RNA virus elements in mammalian genomes. Nature 463: 84–87. **[9]**

Hornsey, I. S. 2003. *A History of Beer and Brewing.* The Royal Society of Chemsitry, Cambridge, UK. **[7]**

Höss, M., S. Pääbo, and N. K. Vereshchagin. 1994. Mammoth DNA sequences. Nature 370: 333. **[5]**

Hotopp, J. C. D. and 19 others. 2007. Widespread lateral gene transfer from intracellular bacteria to multicellular eukaryotes. Science 317: 1753–1756. **[9]**

Hou, Y. and S. Lin. 2009. Distinct gene number-genome size relationships for eukaryotes and non-eukaryotes: Gene content estimation for dinoflagellate genomes. PLoS One 4: e6978. **[11]**

Hough, R. B., A. Avivi, J. Davis, A. Joel, E. Nevo, and J. Piatigorsky. 2002. Adaptive evolution of small heat shock protein/αB-crystallin promoter activity of the blind subterranean mole rat, *Spalax ehrenbergi*. Proc. Natl. Acad. Sci. USA 99: 8145–8150. **[4]**

Housworth, E. A. and J. Postlethwait. 2002. Measures of synteny conservation between species pairs. Genetics 162: 441–448. **[11]**

Hsu, T. C., S. Pathak, and T. R. Chen. 1975. The possibility of latent centromeres and a proposed nomenclature system for total chromosome and whole-arm translocations. Cytogenet. Cell Genet. 15: 41–49. **[11]**

Hu, G., M. Codina, and S. Fisher. 2012. Multiple enhancers associated with ACAN suggest highly redundant transcriptional regulation in cartilage. Matrix Biol. 31: 328–337. **[12]**

Hua-Van, A., A. Le Rouzic, T. S. Boutin, J. Filée, and P. Capy. 2011. The struggle for life of the genome's selfish architects. Biol. Direct 6: 19. **[11]**

Huang, C. H., T. Hsiang, and J. T. Trevors. 2013. Comparative bacterial genomics: Defining the minimal core genome. Antonie van Leeuwenhock 103: 385–398. **[10]**

Huang, C. Y., N. Grünheit, N. Ahmadinejad, J. N. Timmis, and W. Martin. 2005. Mutational decay and age of chloroplast and mitochondrial genomes transferred recently to angiosperm nuclear chromosomes. Plant Physiol. 138: 1723–1733. **[9]**

Huang, L., J. Chi, J. Wang, W. Nie, W. Su, and F. Yang. 2006. High-density comparative BAC mapping in the black muntjac (*Muntiacus crinifrons*): molecular cytogenetic dissection of the origin of MCR 1p+4 in the X1X2Y1Y2Y3 sex chromosome system. Genomics 87: 608–615. **[11]**

Huang, S. and 49 others. 2013. Draft genome of the kiwifruit *Actinidia chinensis*. Nat. Commun. 4: 2640. **[11]**

Huang, W. and 50 others. 2014. Natural variation in genome architecture among 205 *Drosophila melanogaster* genetic reference panel lines. Genome Res. 24: 1193–1208. **[11]**

Huang, W. L., C. W. Tung, H. L. Huang, S. F. Hwang and S. Y. Ho. 2007. ProLoc: Prediction of protein subnuclear localization using SVM with automatic selection from physicochemical composition features. Biosystems 90: 573–581. **[11]**

Huang, W. L., C. W. Tung, S. W. Ho, S. F. Hwang, and S. Y. Ho. 2008. ProLoc-GO: Utilizing informative gene ontology terms for sequence-based prediction of protein subcellular localization. BMC Bioinformatics 9: 80. **[11]**

Hudson, A. P., G. Cuny, J. Cortadas, A. E. Haschemeyer, and G. Bernardi. 1980. An analysis of fish genomes by density gradient centrifugation. Eur. J. Biochem. 112: 203–10. **[11]**

Hudson, R. R., M. Kreitman, and M. Aguadé. 1987. A test of neutral evolution based on nucleotide data. Genetics 116: 153–159. **[2]**

Huelsenbeck, J. P. and K. A. Crandall. 1997. Phylogeny estimation and hypothesis testing using maximum likelihood. Annu. Rev. Ecol. Syst. 28: 437–466. **[5, 6]**

Huelsenbeck, J. P., F. Ronquist, R. Nielsen, and J. Bollback. 2001. Bayesian inference of phylogeny and its impact on evolutionary biology. Science 294: 2310–2314. **[5, 6]**

Hughes, A. L. 1988. The quagga case: Molecular evolution of an extinct species. Trends Ecol. Evol. 3: 95–96. **[5]**

Hughes, A. L. 1994. The evolution of functionally novel proteins after gene duplication. Proc. R. Soc. London 256B: 119–124. **[7]**

Hughes, A. L. 1999a. Phylogenies of developmentally important proteins do not support the hypothesis of two rounds of genome duplication early in vertebrate history. J. Mol. Evol. 48: 565–576. **[7]**

Hughes, A. L. 1999b. *Adaptive Evolution of Genes and Genomes.* Oxford University Press, New York. **[7]**

Hughes, A. L. and M. K. Hughes. 1995. Small genomes for better fliers. Nature 377: 391. **[11]**

Hughes, A. L. and M. Nei. 1988. Pattern of nucleotide substitution at major histocompatibility complex class I loci reveals overdominant selection. Nature 335: 167–170. **[4]**

Hughes, A. L. and M. Nei. 1989. Nucleotide substitution at major histocompatibility complex class II loci: Evidence for overdominant selection. Proc. Natl. Acad. Sci. USA 86: 958–962. **[7]**

Hughes, A. L. and R. Friedman. 2003. 2R or not 2R: Testing hypotheses of genome duplication in early vertebrates. J. Struct. Funct. Genomics 3: 85–93. **[7]**

Hughes, A. L. and R. Friedman. 2004. Shedding genomic ballast: Extensive parallel loss of ancestral gene families in animals. J. Mol. Evol. 59: 827–833. **[7]**

Hughes, A. L., K. Westover, J. da Silva, D. H. O'Connor, and D. I. Watkins. 2001. Simultaneous positive and purifying selection on overlapping reading frames of the *tat* and *vpr* genes of simian immunodeficiency virus. J. Virol. 75: 7966–7972. **[8]**

Hughes, M. K. and A. L. Hughes. 1993. Evolution of duplicate genes in a tetraploid animal, *Xenopus laevis*. Mol. Biol. Evol. 10: 1360–1369. **[7]**

Hughes, T. E., J. A. Langdale, and S. Kelly. 2014. The impact of widespread regulatory neofunctionalization on homeolog gene evolution following whole-genome duplication in maize. Genome Res. 24: 1348–1355. **[7]**

Hughes, T., D. Ekman, H. Ardawatia, A. Elofsson, and D. A. Liberles. 2007. Evaluating dosage compensation as a cause of duplicate gene retention in *Paramecium tetraurelia*. Genome Biol. 8: 213. **[7]**

Huneman, P. (ed.). 2013. *Functions: Selection and Mechanisms.* Springer, Dordrecht, The Netherlands. **[11]**

Hunt, G. 2008. Gradual or pulsed evolution: When should punctuational explanations be preferred? Paleobiology 34: 360–377. **[4]**

Hurst, G. D. D. and J. H. Werren. 2001. The role of selfish genetic elements in eukaryotic evolution. Nat. Rev. Genet. 2: 597–606. **[2]**

Hurst, L. D. 1993. The incidences, mechanisms and evolution of cytoplasmic sex ratio distorters in animals. Biol. Rev. 68: 121–193. **[2]**

Hurst, L. D. 2009. Evolutionary genomics: A positive becomes a negative. Nature 457: 543–544. **[4]**

Hurst, L. D. 2013. Open questions: A logic (or lack thereof) of genome organization. BMC Biology 11: 58. **[11]**

Hurst, L. D. and N. G. C. Smith. 1999. Do essential genes evolve slowly? Curr. Biol. 9: 747–750. **[4]**

Huson, D. H. and C. Scornavacca. 2011. A survey of combinatorial methods for phylogenetic networks. Genome Biol. Evol. 3: 23–35. **[6]**

Huson, D. H. and D. Bryant. 2006. Application of phylogenetic networks in evolutionary studies. Mol. Biol. Evol. 23: 254–267. **[5, 6]**

Huson, D. H., R. Rupp, and C. Scornavacca. 2011. *Phylogenetic Networks: Concepts, Algorithms and Applications.* Cambridge University Press, Cambridge. **[6]**

Huson, D. H., R. Rupp, V. Berry, P. Gambette, and C. Paul. 2009. Computing galled networks from real data. Bioinformatics 25: i85–i93. **[6]**

Huttenhofer, A., P. Schattner, and N. Polacek. 2005. Non-coding RNAs: Hope or hype? Trends Genet. 21: 289–297. **[1]**

Huttley, G. A., I. B. Jakobsen, S. R. Wilson, and S. Easteal. 2000. How important is DNA replication for mutagenesis? Mol. Biol. Evol. 17: 929–937. **[1]**

Huxley, J. 1926. A static theory of heredity. Quart. Rev. Biol. 1: 578–584. **[1, 10]**

I

Ibarra, R. U., J. S. Edwards, and B. Ø. Palsson. 2002. *Escherichia coli* K-12 undergoes adaptive evolution to achieve *in silico* predicted optimal growth. Nature 420: 186–189. **[13]**

Ibarra-Laclette, E. and 28 others. 2013. Architecture and evolution of a minute plant genome. Nature 498: 94–98. **[11]**

Ide, S., T. Miyazaki, and T. Kobayashi. 2010. Abundance of ribosomal RNA gene copies maintains genome integrity. Science 327: 693–696. **[7]**

IJdo, J. W., A. Baldini, D. C. Ward, S. T. Reeders, and R. A. Wells. 1991. Origin of human chromosome 2: An ancestral telomere-telomere fusion. Proc. Natl. Acad. Sci. USA 88: 9051–9055. **[11]**

Ilves, K. L. and D. J. Randall. 2007. Why have primitive fishes survived? pp. 515–536. In: D. J. McKenzie, A. P. Farrell, and C. J. Brauner (eds.), *Primitive Fishes.* Elsevier, Amsterdam. **[4]**

Inai, Y., Y. Ohta, and M. Nishikimi. 2003. The whole structure of the human nonfunctional L-gulono-γ-lactone oxidase gene—the gene responsible for scurvy—and the evolution of repetitive sequences thereon. J. Nutr. Sci. Vitaminol. 49: 315–319. **[8]**

Ingram, V. M. 1961. Gene evolution and the hæmoglobins. Nature 189: 704–708. **[7]**

Innan, H. 2006. Modified Hudson-Kreitman-Aguadé test and two-dimensional evaluation of neutrality tests. Genetics 173: 1725–1733. **[2]**

Innan, H. and F. Kondrashov. 2010. The evolution of gene duplications: Classifying and distinguishing between models. Nat. Rev. Genet. 11: 97–108. **[7]**

Innan, H. and H. Watanabe. 2006. The effect of gene flow on the coalescent time in the human-chimpanzee ancestral population. Mol. Biol. Evol. 23: 1040–1047. **[5]**

Innocenti, P., E. H. Morrow, and D. K. Dowling. 2011. Experimental evidence supports a sex-specific selective sieve in mitochondrial genome evolution. Science 332: 845–848. **[2]**

International Aphid Genomics Consortium. 2010. Genome sequence of the pea aphid *Acyrthosiphon pisum.* PLoS Biol. 8: e1000313. **[9]**

International Chicken Genome Sequencing Consortium. 2004. Sequence and comparative analysis of the chicken genome provide unique perspectives on vertebrate evolution. Nature 432: 695–716. **[9, 11]**

Iorizzo, M., D. Senalik, M. Szklarczyk, D. Grzebelus, D. Spooner, and P. Simon. 2012. *De novo* assembly of the carrot mitochondrial genome using next generation sequencing of whole genomic DNA provides first evidence of DNA transfer into an angiosperm plastid genome. BMC Plant Biol. 12: 61. **[9]**

Irwin, D. M., J. M. Biegel, and C.-B. Stewart. 2011. Evolution of the mammalian lysozyme gene family. BMC Evol. Biol. 11: 166. **[5]**

Irwin, D. M., T. D. Kocher, and A. C. Wilson. 1991. Evolution of the cytochrome b gene of mammals. J. Mol. Evol. 32: 128–144. **[5]**

Islam, S. A., J. Luo, and M. J. Sternberg. 1995. Identification and analysis of domains in proteins. Protein Eng. 8: 513–525. **[8]**

Itano, H. A. 1953. Human hemoglobin. Science 117: 89–94. **[7]**

Itano, H. A. 1957. The human hemoglobins: Their properties and genetic control. Adv. Prot. Chem. 12: 216–268. **[7]**

Itaya, M. 1995. An estimation of minimal genome size required for life. FEBS Lett. 362: 257–260. **[10]**

Itaya, M. and T. Tanaka. 1997. Experimental surgery to create subgenomes of *Bacillus subtilis* 168. Proc. Natl. Acad. Sci. USA 94: 5378–5382. **[10]**

Itaya, M., K. Tsuge, M. Koizumi, and K. Fujita. 2005. Combining two genomes in one cell: Stable cloning of the *Synechocystis* PCC6803 genome in the *Bacillus subtilis* 168 genome. Proc. Natl. Acad. Sci. USA 102: 15971–15976. **[6]**

Itoh, T., K. Takemoto, H. Mori, and T. Gojobori. 1999. Evolutionary instability of operon structures disclosed by sequence comparisons of complete microbial genomes. Mol. Biol. Evol. 16: 332–346. **[10]**

Itzhaki, Z., E. Akiva, Y. Altuvia, and H. Margalit. 2006. Evolutionary conservation of domain-domain interactions. Genome Biol. 7: R125. **[8]**

Ivanova, N. N. and 9 others. 2014. Stop codon reassignments in the wild. Science 344: 909–913. **[10]**

Iwabe, N., K. Kuma, M. Hasegawa, S. Osawa, and T. Miyata. 1989. Evolutionary relationship of archaebacteria, eubacteria, and eukaryotes inferred from phylogenetic trees of duplicated genes. Proc. Natl. Acad. Sci. USA 86: 9355–9359. **[6]**

J

Jabbari, K., O. Clay, and G. Bernardi. 2003. GC3 heterogeneity and body temperature in vertebrates. Gene 317: 161–163. **[11]**

Jackson, J. A. and G. R. Fink. 1985. Meiotic recombination between duplicated genetic elements in *Saccharomyces cerevisiae.* Genetics 109: 303–332. **[7]**

Jacob, F. 1977. Evolution and tinkering. Science 196: 1161–1166. **[8]**

Jacob, F. 1981. *Le Jeu des Possibles.* Librairie Arthème Fayard, Paris. **[8]**

Jacob, F. 2001. Complexity and tinkering. Ann. NY Acad. Sci. 929: 71–73. **[8]**

Jacob, F. and J. Monod. 1961. Genetic regulatory mechanisms in the synthesis of proteins. J. Mol. Biol. 3: 318–356. **[12]**

Jacob, F., D. Perrin, C. Sanchez, and J. Monod. 1960. L'opéron: groupe de gènes à l'expression coordonnée par un operateur. C. R. Acad. Sci. 250: 1727–1729. **[1]**

Jaenike, J. 1996. Sex-ratio meiotic drive in the *Drosophila quinaria* group. Am. Nat. 148: 237–254. **[2]**

Jaenike, J. 2001. Sex chromosome meiotic drive. Annu. Rev. Ecol. Syst. 32: 25–49. **[2]**

Jaillon, O. and 59 others. 2004. Genome duplication in the teleost fish *Tetraodon nigroviridis* reveals the early vertebrate protokaryotype. Nature 431: 946–957. **[7]**

Jain, R. C., M. C. Rivera, and J. A. Lake. 1999. Horizontal gene transfer in prokaryotes: The complexity hypothesis. Proc. Natl. Acad. Sci. USA 96: 3801–3806. **[6]**

James, J. K. and R. J. Abbott. 2005. Recent, allopatric, homoploid hybrid speciation: The origin of *Senecio squalidus* (Asteraceae) in the British Isles from a hybrid zone on Mount Etna, Sicily. Evolution 59: 2533–2547. **[7]**

James, T. Y. and 69 others. 2006. Reconstructing the early evolution of Fungi using a six-gene phylogeny. Nature 443: 818–822. **[11]**

Jeanmougin, F., J. D. Thompson, M. Gouy, D. G. Higgins, and T. J. Gibson. 1998. Multiple sequence alignment with Clustal X. Trends Biochem. Sci. 23: 403–405. **[3]**

Jeffreys, A. J., M. J. Allen, E. Hagelberg, and A. Sonnberg. 1992. Identification of the skeletal remains of Josef Mengele by DNA analysis. Forensic Sci. Int. 56: 65–76. **[11]**

Jeffreys, A. J., V. Wilson, and S. L. Thein. 1985. Hypervariable "minisatellite" regions in human DNA. Nature 314: 67–73. **[11]**

Jeffroy, O., H. Brinkmann, F. Delsuc, and H. Philippe. 2006. Phylogenomics: The beginning of incongruence? Trends Genet. 22: 225–231. **[5]**

Jenkins, C. and 9 others. 2002. Genes for the cytoskeletal protein tubulin in the bacterial genus *Prosthecobacter*. Proc. Natl. Acad. Sci. USA 99: 17049–17054. **[9]**

Jensen-Seaman, M. I., J. H. Wildschutte, I. D. Soto-Calderón, and N. M. Anthony. 2009. A comparative approach shows differences in patterns of *numt* insertion during hominoid evolution. J. Mol. Evol. 68: 688–699. **[9]**

Jensen, J. D., K. R. Thornton, C. D. Bustamante, and C. F. Aquadro. 2007. On the utility of linkage disequilibrium as a statistic for identifying targets of positive selection in nonequilibrium populations. Genetics 176: 2371–2379. **[2]**

Jia, J. and 46 others. 2013. *Aegilops tauschii* draft genome sequence reveals a gene repertoire for wheat adaptation. Nature 496: 91–95. **[11]**

Jiang, N., Z. Bao, X. Zhang, H. Hirochika, S. R. Eddy, S. R. McCouch, and S. R. Wessler. 2003. An active DNA transposon family in rice. Nature 421: 163–167. **[9]**

Jiang, P. and 8 others. 2012. Major taste loss in carnivorous mammals. Proc. Natl. Acad. Sci. USA 109: 4956–4961. **[8]**

Jiao, Y. and 16 others. 2011. Ancestral polyploidy in seed plants and angiosperms. Nature 473: 97–100. **[7]**

Jin, G., L. Nakhleh, S. Snir, and T. Tuller. 2006. Maximum likelihood of phylogenetic networks. Bioinformatics 22: 2604–2611. **[5]**

Jin, G., L. Nakhleh, S. Snir, and T. Tuller. 2007a. Efficient parsimony-based methods for phylogenetic network reconstruction. Bioinformatics 23:e123–e128. **[6]**

Jin, G., L. Nakhleh, S. Snir, and T. Tuller. 2007b. Inferring phylogenetic networks by the maximum parsimony criterion: A case study. Mol. Biol. Evol. 24: 324–337. **[5, 6]**

Johannsen, W. 1909. *Elemente der exakten Erblichkeitslehre*. Gustav Fischer, Jena. **[1]**

Johnson, N. 2008. Hybrid incompatibility and speciation. Nat. Edu. 1: 20. **[5]**

Johnson, R. M., T. Prychitko, D. Gumucio, D. E. Wildman, M. Uddin, and M. Goodman. 2006. Phylogenetic comparisons suggest that distance from the locus control region guides developmental expression of primate beta-type globin genes. Proc. Natl. Acad. Sci. USA 103: 3186–3191. **[12]**

Johnston, C. A., D. S. Whitney, B. F. Volkman, C. Q. Doe, and K. E. Prehoda. 2011. Conversion of the enzyme guanylate kinase into a mitotic-spindle orienting protein by a single mutation that inhibits GMP-induced closing. Proc. Natl. Acad. Sci. USA 108: E973–E978. **[7]**

Johnston, J. S. and 7 others. 2005. Evolution of genome size in Brassicaceae. Ann. Bot. 95: 229–235. **[11]**

Jones, E. M. 1985. *Where Is Everybody? An Account of Fermi's Question.* Reprint LA-10311-MS. Los Alamos National Laboratory, Los Alamos, NM. **[6]**

Jones, P. A. 2012. Functions of DNA methylation: Islands, start sites, gene bodies and beyond. Nat. Rev. Genet. 13: 484–492. **[12]**

Jørgensen, F. G., A. Hobolth, H. Hornshøj, C. Bendixen, M. Fredholm, and M. H. Schierup. 2005. Comparative analysis of protein coding sequences from human, mouse and the domesticated pig. BMC Biol. 3: 2. **[4]**

Jorgensen, P., N. P. Edgington, B. L. Schneider, I. Rupes, M. Tyers and B. Futcher. 2007. The size of the nucleus increases as yeast cells grow. Mol. Biol. Cell 18: 3523–3532. **[11]**

Jukes, T. H. 1985. A change in the genetic code in *Mycoplasma capricolum*. J. Mol. Evol. 22: 361–362. **[10]**

Jukes, T. H. and C. R. Cantor. 1969. Evolution of protein molecules. pp. 21–132. In: H. N. Munro (ed.), *Mammalian Protein Metabolism*. Academic Press, New York. **[3]**

Jumas-Bilak, E., S. Michaux-Charachon, G. Bourg, D. O'Callaghan, and M. Ramuz. 1998. Differences in chromosome number and genome rearrangements in the genus *Brucella*. Mol. Microbiol. 27: 99–106. **[10]**

Junttila, M. R. and G. I. Evan. 2009. p53—a Jack of all trades but master of none. Nat. Rev. Cancer 9: 821–829. **[8]**

Jurka, J., V. V. Kapitonov, O. Kohany, and M. V. Jurka. 2007. Repetitive sequences in complex genomes: Structure and evolution. Annu. Rev. Genomics Hum. Genet. 8: 241–259. **[9]**

K

Kabza, M., J. Ciomborowska, and I. Makałowska. 2014. RetrogeneDB—a database of animal retrogenes. Mol. Biol. Evol. 31: 1646–1648. **[9]**

Kaessmann, H. 2010. Origins, evolution, and phenotypic impact of new genes. Genome Res. 20: 1313–1326. **[7]**

Kaessmann, H., S. Zöllner, A. Nekrutenko, and W.-H. Li. 2002. Signatures of domain shuffling in the human genome. Genome Res. 12: 1642–1650. **[8]**

Kagawa, Y. and 10 others. 1995. Gene of heat shock protein of sulfur-dependent archaeal hyperthermophile *Desulfurococcus*. Biochem. Biophys. Res. Commun. 214: 730–736. **[5]**

Kagawa, Y. and 7 others. 1984. High guanine plus cytosine content in the third letter of codons of an extreme thermophile. J. Biol. Chem. 259: 2956–2960. **[10]**

Kagawa, Y., T. Ohta, Y. Abe, H. Endo, M. Yohda, N. Kato, I. Endo, T. Hamamoto, M. Ichida, T. Hoaki, T. Maruyama. 1995. Gene of heat shock protein of sulfur-dependent archaeal hyperthermophile *Desulfurococcus*. Biochem. Biophys. Res. Commun. 214: 730–736. **[6]**

Kalamaras, A. and 7 others. 2008. The 5′ regulatory region of the human fetal globin genes is a gene conversion hotspot. Hemoglobin 32: 572–581. **[7]**

Kamberov, Y. G. and 26 others. 2013. Modeling recent human evolution in mice by expression of a selected EDAR variant. Cell 152: 691–702. **[11]**

Kaminker, J. S. and 11 others. 2002. The transposable elements of the *Drosophila melanogaster* euchromatin: A genomics perspective. Genome Biol. 3: R84–R84.20. **[9]**

Kandouz, M., A. Bier, G. D. Carystinos, M. A. Alaoui-Jamali, and G. Batist. 2004. *Connexin43* pseudogene is expressed in tumor cells and inhibits growth. Oncogene 23: 4763–4770. **[11]**

Kannan, L., H. Li, B. Rubinstein, and A. Mushegian. 2013. Models of gene gain and gene loss for probabilistic reconstruction of gene content in the last universal common ancestor of life. Biol. Direct 8: 32. **[10]**

Kao, K. C., K. Schwartz, and G. Sherlock. 2010. A genome-wide analysis reveals no nuclear Dobzhansky-Muller pairs of determinants of speciation between *S. cerevisiae* and *S. paradoxus*, but suggests more complex incompatibilities. PLoS Genet. 6: e1001038. **[5]**

Kapitonov, V. V. and J. Jurka. 2003. A novel class of SINE elements derived from 5S rRNA. Mol. Biol. Evol. 20: 694–702. **[9]**

Kapitonov, V. V. and J. Jurka. 2007. *Helitrons* on a roll: Eukaryotic rolling-circle transposons. Trends Genet. 23: 521–528. **[9]**

Kaplan, J. 2008. The end of the adaptive landscape metaphor? Biol. Philos. 23: 625–638. **[13]**

Kaplan, N. and K. Risko. 1982. A method for estimating rates of nucleotide substitution using DNA sequence data. Theor. Pop. Biol. 21: 318–328. **[3]**

Kappen, C., K. Schughart, and F. H. Ruddle. 1989. Two steps in the evolution of Antennapedia-class vertebrate homeobox genes. Proc. Natl. Acad. Sci. USA 86: 5459–5463. **[7]**

Karlin, S., J. Mrazek, and A. M. Campbell. 1997. Compositional biases of bacterial genomes and evolutionary implications. J. Bacteriol. 179: 3899–3913. **[10]**

Karro, J. E., M. Peifer, R. C. Hardison, M. Kollmann, and H. H. von Grünberg. 2008. Exponential decay of GC content detected by strand-symmetric substitution rates influences the evolution of isochore structure. Mol. Biol. Evol. 25: 362–374. **[11]**

Kasahara, M. 2007. The 2R hypothesis: An update. Curr. Opin. Immunol. 19: 547–552. **[7]**

Kassen, R. and T. Bataillon. 2006. Distribution of fitness effects among beneficial mutations before selection in experimental populations of bacteria. Nat. Genet. 38: 484–488. **[13]**

Kassen, R., M. Llewellyn, and P. B. Rainey. 2004. Ecological constraints on diversification in a model adaptive radiation. Nature 431: 984–988. **[13]**

Katinka, M. D. and 16 others. 2001. Genome sequence and gene compaction of the eukaryote parasite *Encephalitozoon cuniculi*. Nature 414: 450–453. **[11]**

Katju, V. and M. Lynch. 2003. The structure and early evolution of recently arisen gene duplicates in the *Caenorhabditis elegans* genome. Genetics 165: 1793–1803. **[8]**

Katoh, K. and H. Toh. 2008. Recent developments in the MAFFT multiple sequence alignment program. Briefings Bioinformatics 9: 286–298. **[3]**

Katoh, K., K. Misawa, K. Kuma, and T. Miyata. 2002. MAFFT: A novel method for rapid multiple sequence alignment based on fast Fourier transform. Nucleic Acids Res. 30: 3059–3066. **[3]**

Katoh, K., K. Kuma, T. Miyata, and H. Toh. 2005a. Improvement in the accuracy of multiple sequence alignment program MAFFT. Genome Informatics 16: 22–33. **[3]**

Katoh, K., K. Kuma, H. Toh, and T. Miyata. 2005b. MAFFT version 5: Improvement in accuracy of multiple sequence alignment. Nucleic Acids Res. 33: 511–518. **[3]**

Katz, L. A. 2012. Origin and diversification of eukaryotes. Annu. Rev. Microbiol. 66: 411–427. **[5]**

Katzourakis, A., R. J. Gifford, M. Tristem, M. T. Gilbert, and O. G. Pybus. 2009. Macroevolution of complex retroviruses. Science 325: 1512. **[9]**

Kauffman, S. A. 1993. The Origins of Order: Self-Organization and Selection in Evolution. Oxford University Press, Oxford. **[11]**

Kaufman, S. A. 1971. Gene regulation networks: A theory for their global structures and behaviours. In: A. A. Moscona and A. Monroy (eds.) *Current Topics in Developmental Biology*. pp. 145–182. Academic Press, New York. **[11]**

Kavaler, S., H. Morinaga, A. Jih, W. Fan, M. Hedlund, A. Varki, and J. J. Kim. 2011. Pancreatic β-cell failure in obese mice with human-like CMP-Neu5Ac hydroxylase deficiency. FASEB J. 25: 1887–1893. **[7, 8]**

Kawach, O., M. S. Sommer, S. B. Gould, C. Voß, S. Zauner, and U.-G. Maier. 2006. Nucleomorphs: Remanant nuclear genomes. In: L.A. Katz and D. Bhattacharya (eds.) *Genomics and Evolution of Microbial Eukaryotes*. p. 192–200. Oxford University Press, New York. **[11]**

Kawasaki, K., S. Minoshima, and N. Shimizu. 2000. Propagation and maintenance of the 119 human immunoglobulin Vλ genes and pseudogenes during evolution. J. Exp. Zool. 288: 120–134. **[3]**

Kawashima, T. and 9 others. 2009. Domain shuffling and the evolution of vertebrates. Genome Res. 19: 1393–1403. **[8]**

Kawecki, T. J., R. E. Lenski, D. Ebert, B. Hollis, I. Olivieri, and M. C. Whitlock. 2012. Experimental evolution. Trends Ecol. Evol. 27: 547–560. **[13]**

Kazazian, H. H. and J. V. Moran. 1998. The impact of *L1* retrotransposons on the human genome. Nat. Genet. 19: 19–24. **[9]**

Keeling, P. J. 1998. A kingdom's progress: Archezoa and the origin of eukaryotes. BioEssays 20: 87–95. **[5, 6]**

Keeling, P. J. 2004. Diversity and evolutionary history of plastids and their hosts. Am. J. Bot. 91: 1481–1493. **[5]**

Keeling, P. J. 2010. The endosymbiotic origin, diversification and fate of plastids. Phil. Trans. Roy. Soc. 365B: 729–748. **[5]**

Keeling, P. J. 2013. The number, speed, and impact of plastid endosymbioses in eukaryotic evolution. Annu. Rev. Plant. Biol. 64: 583–607. **[11]**

Keeling, P. J. and 7 others. 2005. The tree of eukaryotes. Trends Ecol. Evol. 20: 670–676. **[5]**

Keightley, P. D. 1998. Inference of genome-wide mutation rates and distributions of mutation effects for fitness traits: a simulation study. Genetics 150: 1283–1293. **[13]**

Keightley, P. D. and A. Eyre-Walker. 2000. Deleterious mutations and the evolution of sex. Science 209: 331–333. **[11]**

Keller, I., D. Bensasson, and R. A. Nichols. 2007. Transition-transversion bias is not universal: A counter example from grasshopper pseudogenes. PLoS Genet. 3: e22. **[4]**

Kelley, J. L. and 8 others. 2014. Compact genome of the Antarctic midge is likely an adaptation to an extreme environment. Nat. Commun. 5: 4611. **[11]**

Kellis, M. and 29 others. 2014. Defining functional DNA elements in the human genome. Proc. Natl. Acad. Sci. USA 111: 6131–6138. **[11]**

Kellis, M., B. W. Birren, and E. S. Lander. 2004. Proof and evolutionary analysis of ancient genome duplication in the yeast *Saccharomyces cerevisiae*. Nature 428: 617–624. **[7]**

Kemkemer, C., M. Kohn, D. N. Cooper, L. Froenicke, J. Högel, H. Hameister, and H. Kehrer-Sawatzki. 2009. Gene synteny comparisons between different vertebrates provide new insights into breakage and fusion events during mammalian karyotype evolution. BMC Evol. Biol. 9: 84. **[11]**

Kempken, F. and U. Kuck. 1998. Transposons in filamentous fungi—Facts and perspectives. BioEssays 20: 652–659. **[9]**

Kent, L. B. and H. M. Robertson. 2009. Evolution of the sugar receptors in insects. BMC Evol. Biol. 9: 41. **[8]**

Keogh, R. S., C. Seoighe, and K. H. Wolfe. 1998. Evolution of gene order and chromosome number in *Saccharomyces*, *Kluyveromyces*, and related fungi. Yeast 14: 443–457. **[11]**

Keren, H., G. Lev-Maor, and G. Ast. 2010. Alternative splicing and evolution: Diversification, exon definition and function. Nat. Rev. Genet. 11: 345–355. **[8]**

Khan, A. I., D. M. Dinh, D. Schneider, R. E. Lenski, and T. F. Cooper. 2011. Negative epistasis between beneficial mutations in an evolving bacterial population. Science 332: 1193–1196. **[13]**

Khandelwal, S. 1990. Chromosome evolution in the genus *Ophioglossum* L. Bot. J. Linn. Soc. 102: 205–217. **[11]**

Khersonsky, O. and D. S. Tawfik. 2010. Enzyme promiscuity: A mechanistic and evolutionary perspective. Annu. Rev. Biochem. 79: 471–505. **[7]**

Kibota, T. and M. Lynch. 1996. Estimate of the genomic mutation rate deleterious to overall fitness in *E. coli*. Nature 381: 694–696. **[2, 13]**

Kidwell, M. G. 1992. Horizontal transfer of *P*-elements and other short inverted repeat transposons. Genetica 86: 275–286. **[11]**

Kidwell, M. G. 1993. Lateral transfer in natural populations of eukaryotes. Annu. Rev. Genet. 27: 235–256. **[9]**

Kidwell, M. G. 2002. Transposable elements and the evolution of genome size in eukaryotes. Genetica 115: 49–63. **[9]**

Kiezun, A. and 70 others. 2013. Deleterious alleles in the human genome are on average younger than neutral alleles of the same frequency. PLoS Genet. 9: e1003301. **[11]**

Kim, D. S, Y. K. Nam, J. K. Noh, C. H. Park, and F. A. Chapman. 2005. Karyotype of North American shortnose sturgeon *Acipenser brevirostrum* with the highest chromosome number in the Acipenseriformes. Ichthyol. Res. 52: 94–97. **[11]**

Kim, F. J., J. L. Battini, N. Manel, and M. Sitbon. 2004. Emergence of vertebrate retroviruses and envelope capture. Virology 318: 183–191. **[9]**

Kim, J. and J. Ma. 2011. PSAR: Measuring multiple sequence alignment reliability by probabilistic sampling. Nucleic Acids Res. 39: 6359–6368. **[3]**

Kim, N. and S. Jinks-Robertson. 2012. Transcription as a source of genome instability. Nat. Rev. Genet. 13: 204–214. **[10]**

Kim, S. T., P. E. Fields, and R. A. Flavell. 2007. Demethylation of a specific hypersensitive site in the *Th2* locus control region. Proc. Natl. Acad. Sci. USA. 104: 17052–17057. **[12]**

Kim, S.-H., N. Elango, C. Warden, E. Vigoda, and S. V. Yi. 2006. Heterogeneous molecular clocks in primates. PLoS Genet. 2: e163. **[4]**

Kimchi-Sarfaty, C., J. M. Oh, I. W. Kim, Z. E.. Sauna, A. M. Calcagno, S. V. Ambudkar, and M. M. Gottesman. 2007. A "silent" polymorphism in the *MDR1* gene changes substrate specificity. Science 315: 525–528. **[4]**

Kimura, M. 1961. Some calculations on the mutational load. Jap. J. Genet. 36: S179-S190. **[11]**

Kimura, M. 1962. On the probability of fixation of mutant genes in populations. Genetics 47: 713–719. **[2]**

Kimura, M. 1967. On the evolutionary adjustment of spontaneous mutation rates. Genet. Res. 9: 23–34. **[2]**

Kimura, M. 1968a. Evolutionary rate at the molecular level. Nature 217: 624–626. **[2]**

Kimura, M. 1968b. Genetic variability maintained in a finite population due to mutational production of neutral and nearly neutral isoalleles. Genet. Res. 11: 247–269. **[2]**

Kimura, M. 1979. Model of effectively neutral mutations in which selective constraint is incorporated. Proc. Natl. Acad. Sci. USA 76: 3440–3444. **[2]**

Kimura, M. 1980. A simple method for estimating evolutionary rate of base substitution through comparative studies of nucleotide sequences. J. Mol. Evol. 16: 111–120. **[3]**

Kimura, M. 1981. Estimation of evolutionary distances between homologous nucleotide sequences. Proc. Natl. Acad. Sci. USA 78: 454–458. **[3]**

Kimura, M. 1983. *The Neutral Theory of Molecular Evolution*. Cambridge University Press, Cambridge. **[2]**

Kimura, M. and T. Maruyama. 1966. The mutational load with interactions in fitness. Genetics 54: 1337–1351. **[11]**

Kimura, M. and T. Ohta. 1969. The average number of generations until fixation of a mutant gene in a finite population. Genetics 61: 763–771. **[2]**

Kimura, M. and T. Ohta. 1971. Protein polymorphism as a phase of molecular evolution. Nature 229: 467–469. **[2]**

Kimura, M. and T. Ohta. 1972. On the stochastic model for estimation of mutational distance between homologous proteins. J. Mol. Evol. 2: 87–90. **[3]**

Kimura, M. and T. Ohta. 1973. Mutation and evolution at the molecular level. Genetics 73S: 19–35. **[4]**

Kimura-Yoshida, C. and 9 others. 2004. Characterization of the pufferfish Otx2 *cis*-regulators reveals evolutionarily conserved genetic mechanisms for vertebrate head specification. Development 131: 57–71. **[11]**

King, J. L. and T. H. Jukes. 1969. Non-Darwinian evolution. Science 164: 788–798. **[2, 10]**

King, M.-C. and A. C. Wilson. 1975. Evolution at two levels in humans and chimpanzees. Science 188: 107–116. **[12]**

Kirkpatrick, M. and Slatkin M. 1993. Searching for evolutionary patterns in the shape of a phylogenetic tree. Evolution 47: 1171–1181. **[5]**

Kirkpatrick, S., C. D. Gelatt, and M. P. Vecci. 1983. Optimization by simulated annealing. Science 220: 671–680. **[5]**

Kishino, H. and M. Hasegawa. 1989. Evaluation of the maximum likelihood estimate of the evolutionary tree topologies from DNA sequence data, and the branching order in Hominoidea. J. Mol. Evol. 29: 170–179. **[5]**

Kit, S. 1961. Equilibrium sedimentation in density gradients of DNA preparations from animal tissues. J. Mol. Biol. 3: 711–716. **[11]**

Kitano, T., A. Blancher, and N. Saitou. 2012. The functional A allele was resurrected via recombination in the human ABO blood group gene. Mol. Biol. Evol. 29: 1791–1796. **[6]**

Klein, H. L. and T. D. Petes. 1981. Intrachromosomal gene conversion in yeast. Nature 289: 144–148 **[7]**

Klein, J., A. Sato, and N. Nikolaidis. 2007. MHC, TSP, and the origin of species: From immunogenetics to evolutionary genetics. Annu. Rev. Genet. 41: 281–304. **[2]**

Klein, J., Y. Satta, C. O'hUigin, and N. Takahata. 1993. The molecular descent of the major histocompatibility complex. Annu. Rev. Immunol. 11: 269–95. **[7]**

Kleine, T., U. G. Maier, and D. Leister. 2009. DNA transfer from organelles to the nucleus: The idiosyncratic genetics of endosymbiosis. Annu. Rev. Plant Biol. 60: 115–138. **[9]**

Klieve, A. V., M. T. Yokoyama, R. J. Forster, D. Ouwerkerk, P. A. Bain, and E. L. Mawhinney. 2005. Naturally occurring DNA transfer system associated with membrane vesicles in cellulolytic *Ruminococcus* spp. of ruminal origin. Appl. Environ. Microbiol. 71: 4248–4253. **[9]**

Kloek, A. P., D. R. Sherman, and D. E. Goldberg. 1993. Novel gene structure and evolutionary context of *Caenorhabditis elegans* globin. Gene 129: 215–221. **[8]**

Klotzko, A. J. (ed.) *The Cloning Sourcebook*. Oxford University Press, Oxford **[11]**

Kluge, A. G. 1983. Cladistics and the classification of the great apes. In: R. Ciochon and R. S. Corruccini (eds.), *New Interpretations of Ape and Human Ancestry*. Plenum, New York. **[5]**

Knight, C. A., N. A. Molinari, and D.A. Petrov. 2005. The large genome constraint hypothesis: evolution, ecology and phenotype. Ann Bot. 95: 177–190. **[11]**

Knight, R. 2007. Reports of the death of the gene are greatly exaggerated. Biol. Phil. 22: 293–306. **[1]**

Knight, R. D., S. J. Freeland, and L. F. Landweber. 2001. A simple model based on mutation and selection explains trends in codon and amino-acid usage and GC composition within and across genomes. Genome Biol. 2: RESEARCH0010. **[4]**

Knight, R.D., S. J. Freeland, and L. F. Landweber. 2001. Rewiring the keyboard: Evolvability of the genetic code. Nat. Rev. Genet. 2: 49–58. **[10]**

Knowles, D. G. and A. McLysaght. 2009. Recent *de novo* origin of human protein-coding genes. Genome Res. 19: 1752–1759. **[8]**

Kocher, T. D. 2004. Adaptive evolution and explosive speciation: the cichlid fish model. Nature Rev. Genet. 5: 288–298. **[12]**

Kogo, H. and T. Fujimoto. 2000. Caveolin-1 isoforms are encoded by distinct mRNAs: Identification of mouse caveolin-1 mRNA variants caused by alternative transcription initiation and splicing. FEBS Lett. 465: 119–123. **[8]**

Kojima, K. K. 2015. A new class of SINEs with snRNA gene-derived heads. Genome Biol. Evol. 7: 1702–1712. **[9]**

Kolata, G. 2010. Reanimated "junk" DNA is found to cause disease. *New York Times* http: //www.nytimes.com/2010/08/20/science/20gene.html?_r=0 **[11]**

Kondrashov, A. S. 2003. Direct estimates of human per nucleotide mutation rates at 20 loci causing Mendelian diseases. Hum. Mutat. 21: 12–27. **[4]**

Kondrashov, A. S. and J. F. Crow. 1993. A molecular approach to estimating the human deleterious mutation rate. Hum. Mutat. 2: 229–234. **[4]**

Kondrashov, F. A. and E. V. Koonin. 2003. Evolution of alternative splicing: Deletions, insertions and origin of functional parts of proteins from intron sequences. Trends Genet. 19: 115–119. **[8]**

Kong, A. and 20 others. 2012. Rate of de novo mutations and the importance of father's age to disease risk. Nature 488: 471–475. **[4, 11]**

Konstantinidis, K. T. and J. M. Tiedje. 2004. Trends between gene content and genome size in prokaryotic species with larger genomes. Proc. Natl. Acad. Sci. USA 101: 3160–3165. **[10]**

Konstantinidis, K. T., A. Ramette, and J. M. Tiedje. 2006. The bacterial species definition in the genomic era. Phil. Trans. Roy. Soc. 361B: 1929–1940. **[6]**

Koonin, E. V. 2009. Evolution of genome architecture. Int. J. Biochem. Cell Biol. 41: 298–306. **[11]**

Koonin, E. V. 2009. Intron-dominated genomes of early ancestors of eukaryotes. J. Hered. 100: 618–623. **[8]**

Koonin, E. V. and A. R. Mushegian. 1996. Complete genome sequences of cellular life forms: Glimpses of theoretical evolutionary genomics. Curr. Opin. Genet. Develop. 6: 757–762. **[10]**

Koonin, E. V. and Y. I. Wolf. 2008. Genomics of bacteria and archaea: The emerging dynamic view of the prokaryotic world. Nucleic Acids Res. 36: 6688–6719. **[10]**

Koonin, E. V., K. S. Makarova, and L. Aravind. 2001. Horizontal gene transfer in prokaryotes: Quantification and classification. Annu. Rev. Microbiol. 55: 709–742. **[9]**

Koonin, E. V., L. Aravind, and A. S. Kondrashov. 2000. The impact of comparative genomics on our understanding of evolution. Cell 101: 573–576. **[8]**

Koonin, E. V., Y. I. Wolf, and G. P. Karev. 2002. The structure of the protein universe and genome evolution. Nature 420: 218–223. **[8]**

Koop, B. F., M. Goodman, P. Xu, K. Chan, and J. L. Slightom. 1986a. Primate η-globin DNA sequences and man's place among the great apes. Nature 319: 234–238. **[5]**

Koop, B. F., M. M. Miyamoto, J. E. Embury, M. Goodman, J. Czelusniak, and J. L. Slightom. 1986b. Nucleotide sequence and evolution of the orangutan ε-globin gene region and surrounding *Alu* repeats. J. Mol. Evol. 24: 94–102. **[5]**

Kopelman, N. M., D. Lancet, and I. Yanai. 2005. Alternative splicing and gene duplication are inversely correlated evolutionary mechanisms. Nature Genet. 37: 588–589. **[8]**

Kordiš, D. 2011. Extensive intron gain in the ancestor of placental mammals. Biol. Dir. 6: 59. **[8]**

Kordiš, D. and F. Gubenšek. 1998. Unusual horizontal transfer of a long interspersed nuclear element between distant vertebrate classes. Proc. Natl. Acad. Sci. USA 95: 10704–10709. **[9]**

Kornegay, J. R., J. W. Schilling, and A. C. Wilson. 1994. Molecular adaptation of a leaf-eating bird: Stomach lysozyme of the hoatzin. Mol. Biol. Evol. 11: 921–928. **[4, 5]**

Korochkin, L. I. 2006. What is epigenetics. Russ. J. Genet. 42: 958–965. **[12]**

Kosiol, C., L. Bofkin, and S. Whelan. 2006. Phylogenetics by likelihood: Evolutionary modeling as a tool for understanding the genome. J. Biomed. Inform. 39: 51–61. **[5, 6]**

Kosiol, C., T. Vinar, R. R. da Fonseca, M. J. Hubisz, C. D. Bustamante, R. Nielsen, and A. Siepel. 2008. Patterns of positive selection in six mammalian genomes. PLoS Genet. 4: e1000144. **[4]**

Koszul, R., S. Caburet, B. Dujon, and G. Fischer. 2004. Eukaryotic genome evolution through the spontaneous duplication of large chromosomal segments. EMBO J. 23: 234–243. **[7]**

Kotsiantis, S. B. 2007. Supervised machine learning: a review of classification techniques. Informatica 31: 249–268. **[11]**

Kovach, A. and 10 others. 2010. The *Pinus taeda* genome is characterized by diverse and highly diverged repetitive sequences. BMC Genomics 11: 420. **[11]**

Kowallik, K. 2008. Evolution durch genomische Kombination. In: J. Klose and J. Oehler (eds.), *Gott oder Darwin?: Vernünftiges Reden über Schöpfung und Evolution*, pp. 141–159. Springer, Berlin. **[6]**

Kozielska, M., F. J. Weissing, L. W. Beukeboom, and I. Pen. 2010. Segregation distortion and the evolution of sex-determining mechanisms. Heredity 104: 100–112. **[2]**

Krams, S. M. and J. S. Bromberg. 2013. ENCODE: Life, the universe and everything. Am. J. Transplant. 13: 245. **[11]**

Kreahling, J. and B. R. Graveley. 2004. The origins and implications of *Alu*ternative splicing. Trends Genet. 20: 1–4. **[8]**

Kreitman, M. 1983. Nucleotide polymorphism at the alcohol dehydrogenase locus of *Drosophila melanogaster*. Nature 304: 412–417. **[2]**

Kreitman, M. and M. Aguadé. 1986. Excess polymorphism at the *Adh* locus in *Drosophila melanogaster*. Genetics 114: 93–110. **[2]**

Kriegs, J. O., G. Churakov, J. Jurka, J. Brosius, and J. Schmitz. 2007. Evolutionary history of 7SL RNA-derived SINEs in supraprimates. Trends Genet. 23: 158–161. **[9]**

Krol, J., I. Loedige, and W. Filipowicz. 2010. The widespread regulation of microRNA biogenesis, function and decay. Nat. Rev. Genet. 11: 597–610. **[1]**

Kryazhimskiy, S., G. Tkacik, and J. B. Plotkin. 2009. The dynamics of adaptation on correlated fitness landscapes. Proc. Natl. Acad. Sci. USA 106: 18638–18643. **[13]**

Krzywinski, J., D. Sangaré, and N. J. Besansky. 2005. Satellite DNA from the Y chromosome of the malaria vector *Anopheles gambiae*. Genetics 169: 185–196. **[11]**

Ksepka, D. T. and 14 others. 2011. Synthesizing and databasing fossil calibrations: Divergence dating and beyond. Biol. Lett. 7: 801–803. **[5]**

Kudla, G., A. W. Murray, D. Tollervey, and J. B. Plotkin. 2009. Coding-sequence determinants of gene expression in *Escherichia coli*. Science 324: 255–258. **[4, 9]**

Kudlow, B. A., B. K. Kennedy, and R. J. Monnat. 2007. Werner and Hutchinson-Gilford progeria syndromes: Mechanistic basis of human progeroid diseases. Nat. Rev. Mol. Cell Biol. 8: 394–404. **[1]**

Kuhner, M. K. and J. Felsenstein. 1994. A simulation comparison of phylogeny algorithms under equal and unequal evolutionary rates. Mol. Biol. Evol. 11: 459–468. **[5]**

Kulikowski, L. D. and 9 others. 2010. Cytogenetic molecular delineation of a terminal 18q deletion suggesting neo-telomere formation. Eur. J. Med. Genet. 53: 404–407. **[11]**

Kumar, S. and A. J. Filipski. 2001. Molecular clock: Testing. In: *Encyclopedia of Life Sciences*. Wiley, Chichester. doi:10.1038/npg.els.0001803 **[4]**

Kumar, S. and A. J. Filipski. 2008. Molecular phylogeny reconstruction. In: *Encyclopedia of Life Sciences*. Wiley, Chichester. doi:10.1002/9780470015902.a0001523.pub2 **[5]**

Kumar, S. and S. B. Hedges. 1998. A molecular timescale for vertebrate evolution. Nature 392: 917–920. **[5]**

Kumar, S. and S. Subramanian. 2002. Mutation rates in mammalian genomes. Proc. Natl. Acad. Sci. USA 99: 803–808. **[4]**

Kummerfeld, S. K. and S. A. Teichmann. 2005. Relative rates of gene fusion and fission in multi-domain proteins. Trends Genet. 21: 25–30. **[8]**

Kuo, C.-H., N. A. Moran, and H. Ochman. 2009. The consequences of genetic drift for bacterial genome complexity. Genome Res. 19: 1450–1454. **[10]**

Kurzweil, V. C., M. Getman, NISC Comparative Sequencing Program, E. D. Green, and R. P. Lane. 2009. Dynamic evolution of V1R putative pheromone receptors between *Mus musculus* and *Mus spretus*. BMC Genomics 10: 74. **[7]**

Kushiro, A., M. Shimizu, and K.-I. Tomita. 1987. Molecular cloning and sequence determination of the *tuf* gene coding for the elongation factor *Tu* of *Thermus thermophilus* HB8. Eur. J. Biochem. 170: 93–98. **[10]**

Kwon, D. H., J. M. Clark, and S. H. Lee. 2009. Extensive gene duplication of acetylcholinesterase associated with organophosphate resistance in the two-spotted spider mite. Insect Mol. Biol. 19: 195–204. **[7]**

Kyrpides, N., R. Overbeek, and C. Ouzounis. 1999. Universal protein families and the functional content of the last universal common ancestor. J. Mol. Evol. 49: 413–423. **[5, 6]**

L

Labbé, P., A. Berthomieu, C. Berticat, H. Alout, M. Raymond, T. Lenormand, and M. Weill. 2007. Independent duplications of the acetylcholinesterase gene conferring insecticide resistance in the mosquito *Culex pipiens*. Mol. Biol. Evol. 24: 1056–1067. **[7]**

Labib, H. A., S. Elshorbagy, and N. G. Elantonuy. 2014. Significance of telomere capture in myelodysplastic syndromes. Med. Oncol. 31: 216. **[11]**

Lachapelle, M. Y. and G. Drouin. 2011. Inactivation dates of the human and guinea pig vitamin C genes. Genetica 139: 199–207. **[7, 8]**

LaJeunesse, T. C., G. Lambert, R. A. Andersen, M. A. Coffroth, and D. W. Galbraith. 2005. *Symbiodinium* (Pyrrhophyta) genome sizes (DNA content) are smallest among dinoflagellates. J. Phycol. 41: 880–886. **[11]**

Lake, J. A. 1994. Reconstructing evolutionary trees from DNA and protein sequences: Paralinear distances. Proc. Natl. Acad. Sci. USA 91: 1455–1459. **[3, 5, 6]**

Lake, J. A. 2007. Disappearing act. Nature 446: 983. **[6]**

Lake, J. A. and M. C. Rivera. 2004. Deriving the genomic tree of life in the presence of horizontal gene transfer: Conditioned reconstruction Mol. Biol. Evol. 21: 681–690. **[5]**

Lake, J. A., E. Henderson, M. Oakes, and M. W. Clark. 1984. Eocytes: A new ribosome structure indicates a kingdom with a close relationship to eukaryotes. Proc. Natl. Acad. Sci. USA 81: 3786–3790. **[5]**

Lake, J. A., J. A. Servine, C. W. Herbold, and R. G. Skophammer. 2009. Evidence for a new root of the tree of life. Syst. Biol. 57: 835–843. **[5]**

Lake, J. A., R. G. Skophammer, C. W. Herbold, and J. A. Servine. 2009. Genome beginnings: Rooting the tree of life. Phil. Trans. Roy. Soc. 364B: 2177–2185. **[5]**

Lalic, J., J. M. Cuevas, and S. F. Elena. 2011. Effect of host species on the distribution of mutational fitness effects for an RNA virus. PLoS Genet. 7: e1002378. **[13]**

Lamason, R. L. and 24 others. 2005. SLC24A5, a putative cation exchanger, affects pigmentation in zebrafish and humans. Science 310: 1782–1786. **[11]**

Lamb, B. C. and S. Helmi. 1982. The extent to which gene conversion can change allele frequencies in populations. Genet. Res. 39: 199–217. **[7]**

Lambowitz, A. M. and M. Belfort. 1993. Introns as mobile genetic elements. Annu. Rev. Biochem. 62: 587–622. **[9]**

Lamprecht, B. and 24 others. 2010. Derepression of an endogenous long terminal repeat activates the *CSF1R* proto-oncogene in human lymphoma. Nature Med. 16: 571–579. **[9]**

Lampson, B. C., M. Inouye, and S. Inouye. 2005. Retrons, msDNA, and the bacterial genome. Cytogenet. Genome Res. 110: 491–499. **[9]**

Lan, P., W. Li, and W. Schmidt. 2012. Complementary proteome and transcriptome profiling in phosphate-deficient *Arabidopsis* roots reveals multiple levels of gene regulation. Mol. Cell Proteomics 11: 1156–1166. **[12]**

Lanave, C., G. Preparata, C. Saccone, and G. Serio. 1984. A new method for calculating evolutionary substitution rates. J. Mol. Evol. 20: 86–93. **[3]**

Landan, G. and D. Graur. 2007. Heads or tails: A simple reliability check for multiple sequence alignments. Mol. Biol. Evol. 24: 1380–1383. **[3]**

Lander, E. S. and 254 others. 2001. Initial sequencing and analysis of the human genome. Nature 409: 860–921. **[8, 11]**

Landsman, J., E. Dennis, T. J. V. Higgins, C. A. Appleby, A. A. Kortt, and W. J. Peacock. 1986. Common evolutionary origin of legume and non-legume plant haemoglobins. Nature 324:166–168. **[8]**

Lane, N. 2009. Life Ascending: The Ten Great Inventions of Evolution. Profile Books, London. **[5]**

Lane, N. 2010. Chance or necessity? Bioenergetics and the probability of life. J. Cosmol. 10: 3286–3304. **[6]**

Lane, N. and W. Martin. 2010. The energetics of genome complexity. Nature 467: 929–934. **[6]**

Lane, N., W. F. Martin, J. A. Raven, and J. F. Allen. 2013. Energy, genes and evolution: Introduction to an evolutionary synthesis. Philos. Trans. R. Soc. Lond. B. Biol. Sci. 368: 20120253. **[10]**

Lanfear, R., J. A. Thomas, J. J. Welch, and L. Bromham. 2007. Metabolic rate does not calibrate the molecular clock. Proc. Natl. Acad. Sci. USA 104: 15388–15393. **[4]**

Lang, A. S., O. Zhaxybayeva, and J. T. Beatty. 2012. Gene transfer agents: Phage-like elements of genetic exchange. Nat. Rev. Microbiol. 10: 472–482. **[9]**

Lang, B. F. and 8 others. 1997. An ancestral mitochondrial DNA resemble a eubacterial genome in miniature. Nature 387: 493–497. **[10]**

Langlois, M. R. and J. R. Delanghe. 1996. Biological and clinical significance of haptoglobin polymorphism in humans. Clin. Chem. 42: 1589–1600. **[7, 8]**

Lao, P. J. and D. R. Forsdyke. 2000. Thermophilic bacteria strictly obey Szybalski's transcription direction rule and politely purine-load RNAs with both adenine and guanine. Genome Res. 10: 228–236. **[10]**

Lapierre, P. and J. P. Gogarten. 2009. Estimating the size of the bacterial pan-genome. Trends Genet. 25: 107–110. **[10]**

Larkin, M. A. and 12 others. 2007. Clustal W and Clustal X version 2.0. Bioinformatics 23: 2947–2948. **[3]**

Larracuente, A. M. and D. C. Presgraves. 2012. The selfish *Segregation Distorter* gene complex of *Drosophila melanogaster*. Genetics 192: 33–53. **[2]**

Lasko, P. 2009. Translational control during early development. Prog. Mol. Biol. Transl. Sci. 90: 211–254. **[12]**

Lavergne, S., N. J. Muenke, and J. Molofsky. 2010. Genome size reduction can trigger rapid phenotypic evolution in invasive plants. Ann. Bot. 105: 109–116. **[11]**

Lawrence, J. G. and H. Ochman. 1997. Amelioration of bacterial genomes: Rates of change and exchange. J. Mol. Evol. 44: 383–397. **[9]**

Lawrence, J. G. and J. R. Roth. 1996. Selfish operons: Horizontal transfer may drive the evolution of gene clusters. Genetics 143: 1843–1860. **[6]**

Lawrie, D. S. and D. A. Petrov. 2014. Comparative population genomics: power and principles for the inference of functionality. Trends Genet. 30: 133–139. **[11]**

Lawson, F. S., R. L. Charlebois, and J. A. Dillon. 1996. Phylogenetic analysis of carbamoylphosphate synthetase genes: Complex evolutionary history includes an internal duplication within a gene which can root the tree of life. Mol. Biol. Evol. 13: 970–977. **[6]**

Lawton, G. 2015. Beyond belief. New Scientist 226(3015): 28–33. **[11]**

Lazure, C., N. G. Seidah, G. Thibault, R. Boucher, J. Genest, and M. Chrétien. 1981. Sequence homologies between tonin, nerve growth factor γ-subunit, epidermal growth factor–binding protein and serine proteases. Nature 292: 383–384. **[7]**

Le Quesne, W. J. 1974. The uniquely evolved character concept and its cladistic application. Syst. Zool. 23: 513–517. **[5]**

Le Rouzic, A. and P. Capy. 2006. Population genetics models of competition between transposable element subfamilies. Genetics 174: 785–793. **[9]**

Le Rouzic, A., T. S. Boutin, and P. Capy. 2007. Long-term evolution of transposable elements. Proc. Nat. Acad. Sci. USA 104: 19375–19380. **[9, 11]**

Leaman, D. and 9 others. 2005. Antisense-mediated depletion reveals essential and specific functions of microRNAs in *Drosophila* development. Cell 121: 1097–1108. **[12]**

Lecomte, J. T. J., D. A. Vuletich, and A. M. Lesk. 2005. Structural divergence and distant relationships in proteins: Evolution of the globins. Curr. Opin. Struct. Biol. 15: 290–301. **[7]**

Ledda, M., L. Barni, L. Altieri, and E. Pannese. 2000. Decrease in the nucleo-cytoplasmic volume ratio of rabbit spinal ganglion neurons with age. Neurosci. Lett. 286: 171–174. **[11]**

Leder, P. 1982. The genetics of antibody diversity. Sci. Am. 246(5): 102–115. **[8]**

Lederberg, J. and E. L. Tatum. 1946. Gene recombination in *Escherichia coli*. Nature 158: 558. **[9]**

Lederberg, J. and E. M. Lederberg. 1952. Replica plating and indirect selection of bacterial mutants. J. Bacteriol. 63: 399–406. **[1]**

Lee, C., R. Sasi, and C. C. Lin. 1993. Interstitial localization of telomeric DNA sequences in the Indian muntjac chromosomes: further evidence for tandem chromosome fusions in the karyotypic evolution of the Asian muntjacs. Cytogenet Cell Genet. 63: 156–159. **[11]**

Lee, H., E. Popodi, H. Tang, and P. L. Foster. 2012. Rate and molecular spectrum of spontaneous mutations in the bacterium *Escherichia coli* as determined by whole-genome sequencing. Proc. Natl. Acad. Sci. USA 109E: 2774–2783. **[13]**

Lee, M. S. Y. 1999. Molecular clock calibrations and metazoan divergence dates. J. Mol. Evol. 49: 385–391. **[5]**

Lee, R. C., R. L. Feinbaum, and V. Ambros. 1993. The *C. elegans* heterochronic gene *lin–4* encodes small RNAs with antisense complementarity to lin–14. Cell 75: 843–854. **[11]**

Lee, R. E. 1977. Evolution of algal flagellates with chloroplast endoplasmic reticulum from ciliates. S. Afr. J. Sci. 73: 179–182. **[11]**

Lee, Y. C. G. and C. H. Langley. 2010. Transposable elements in natural populations of *Drosophila melanogaster.* Phil. Trans. Roy. Soc. 365B: 1219–1228. **[9]**

Lee, Z. M.-P., C. Bussema, and T. M. Schmidt. 2009. rrnDB: Documenting the number of rRNA and tRNA genes in bacteria and archaea. Nucl. Acids Res. 37: D489–D493. **[7]**

Leeds, J. M., M. B. Slabourgh, and C. K. Mathews. 1985. DNA precursor pools and ribonucleotide reductase activity: Distribution between the nucleus and cytoplasm of mammalian cells. Mol. Cell. Biol. 5: 3443–3450. **[11]**

Leffler, E. M. and 12 others. 2013. Multiple instances of ancient balancing selection shared between humans and chimpanzees. Science 339: 1578–1582. **[4]**

Lehmann, K. and U. Schmidt. 2003. Group II introns: Structure and catalytic versatility of large natural ribozymes. Crit. Rev. Biochem. Mol. Biol. 38: 249–303. **[9]**

Lehmann, T., W. A. Hawley, H. Grebert, and F. H. Collins. 1998. The effective population size of *Anopheles gambiae* in Kenya: Implications for population structure. Mol. Biol. Evol. 15: 264–276. **[2]**

Lehrman, M. A., J. L. Goldstein, D. W. Russell, and M. S. Brown. 1987. Duplication of seven exons in LDL receptor gene caused by *Alu-Alu* recombination in a subject with familial hypercholesterolemia. Cell 48: 827–835. **[9]**

Lehrman, M. A., W. J. Schneider, T. C. Sudhof, M. S. Brown, J. L. Goldstein, and D. W. Russell. 1986. Mutation in LDL receptor: *Alu-Alu* recombination deletes exons encoding transmembrane and cytoplasmic domains. Science 227: 140–146. **[9]**

Leib-Mosch, C. and W. Seifarth. 1995. Evolution and biological significance of human retroelements. Virus Genes 11: 133–145. **[9]**

Leister, D. 2005. Origin, evolution and genetic effects of nuclear insertions of organelle DNA. Trends Genet. 21: 655–663. **[9]**

Leister, D. and T. Kleine. 2011. Role of intercompartmental DNA transfer in producing genetic diversity. Int. Rev. Cell Mol. Biol. 291: 73–114. **[9]**

Lenski, R., M. Rose, and S. Simpson. 1991. Long-term experimental evolution in *Escherichia coli.* I. Adaptation and divergence during 2,000 generations. Am. Nat. 138: 1315–1341. **[13]**

Leonard, J. A., N. Rohland, S. Glaberman, R. C. Fleischer, A. Caccone, and M. Hofreiter. 2005. A rapid loss of stripes: The evolutionary history of the extinct quagga. Biol. Lett. 1: 291–295. **[5]**

Leonard, J. A. and 7 others. 2007. Animal DNA in PCR reagents plagues ancient DNA research. J. Archaeol. Sci. 34: 1361–1366. **[5]**

Leonardo, T. E. and S. V. Nuzhdin. 2002. Intracellular battlegrounds: conflict and cooperation between transposable elements. Genet Res. 80: 155–161. **[9]**

Lerat, E. and H. Ochman. 2005. Recognizing the pseudogenes in bacterial genomes. Nucleic Acids Res. 33: 3125–3132. **[10]**

Lercher, M. J., A. O. Urrutia, and L. D. Hurst. 2002. Clustering of housekeeping genes provides a unified model of gene order in the human genome. Nat. Genet. 31: 180–183. **[11]**

Lesk, A. 2010. *Introduction to Protein Science: Architecture, Function, and Genomics.* Oxford University Press, Oxford. **[1]**

Letunic, I., R. R. Copley, and P. Bork. 2002. Common exon duplication in animals and its role in alternative splicing. Hum. Mol. Genet. 11: 1561–1567. **[7, 8]**

Leushkin, E. V., R. A. Sutormin, E. R. Nabieva, A. A. Penin, A. S. Kondrashov, and M. D. Logacheva. 2013. The miniature genome of a carnivorous plant *Genlisea aurea* contains a low number of genes and short non-coding sequences. BMC Genomics 14: 476. **[11]**

Lev-Maor, G., R. Sorek, N. Shomron, and G. Ast. 2003. The birth of an alternatively spliced exon: 3′ splice-site selection in Alu exons. Science 300: 1288–1291. **[8]**

Levan, A., K. Fredga, and A. A. Sandberg. 1964. Nomenclature for centromeric position on chromosomes. Hereditas 52: 201–220. **[11]**

Leveugle, M., K. Prat, N. Perrier, D. Birnbaum, and F. Coulier. 2003. ParaDB: A tool for paralogy mapping in vertebrate genomes. Nucl. Acids Res. 3: 63–67. **[7]**

Levin, H. L. and J. V. Moran. 2011. Dynamic interactions between transposable elements and their hosts. Nat. Rev. Genet. 12: 615–627. **[9, 11]**

Levine, M. T., C. D. Jones, A. D. Kern, H. A. Lindfors, and D. J. Begun. 2006. Novel genes derived from noncoding DNA in *Drosophila melanogaster* are frequently X-linked and exhibit testis-biased expression. Proc. Natl. Acad. Sci. USA 103: 9935–9939. **[8]**

Levis, R. W., R. Ganesan, K. Houtchens, L. A. Tolar, and F.-M. Sheen. 1993. Transposons in place of telomeric repeats at a *Drosophila* telomere. Cell 75: 1083–1093. **[9]**

Levy, A. P. and 18 others. 2010. Haptoglobin: Basic and clinical aspects. Antioxid. Redox Signal. 12: 293–304. **[7]**

Levy, D. L. and R. Heald. 2010. Nuclear size is regulated by importin α and Ntf2 in *Xenopus.* Cell 143: 288–298. **[11]**

Lewin, B. 1997. *Genes VI.* Oxford University Press, New York. **[1]**

Lewis, B. P., C. B. Burge, and D. P. Bartel. 2005. Conserved seed pairing, often flanked by adenosines, indicates that thousands of human genes are microRNA targets. Cell 120: 15–20. **[12]**

Lewis, P. O. 1998. A genetic algorithm for maximum-likelihood phylogeny inference using nucleotide sequence data. Mol. Biol. Evol. 15: 277–283. **[5]**

Lewontin, R. C. 1974. *The Genetic Basis of Evolutionary Change.* Columbia University Press, New York. **[2]**

Lewontin, R. C. and K. Kojima. 1960. The evolutionary dynamics of complex polymorphisms. Evolution 14: 458–472. **[2]**

Li, C.-Y. and 13 others. 2010. A human-specific de novo protein-coding gene associated with human brain functions. PLoS Comput. Biol. 6: e1000734. **[8]**

Li, D., Y. Dong, Y. Jiang, H. Jiang, J. Cai, and W. Wang. 2010. A *de novo* originated gene depresses budding yeast mating pathway and is repressed by the protein encoded by its antisense strand. Cell Res. 20: 408–420. **[8]**

Li, H. 2011. A new test for detecting recent positive selection that is free from the confounding impacts of demography. Mol. Biol. Evol. 28: 365–375. **[11]**

Li, H. and W. Stephan. 2006. Inferring the demographic history and rate of adaptive substitution in *Drosophila.* PLoS Genet. 2: e166. **[11]**

Li, Q. 2006. A Melanesian α-thalassemia mutation suggests a novel mechanism for regulating gene expression. Genome Biol. 7: 238. **[4]**

Li, W. and 17 others. 2015. Human endogenous retrovirus-K contributes to motor neuron disease. Sci. Transl. Med. 7: 307ra153. **[9]**

Li, W.-H. 1978. Maintenance of genetic variability under the joint effect of mutation, selection and random drift. Genetics 90: 349–382. **[2]**

Li, W.-H. 1983. Evolution of duplicate genes and pseudogenes. In: M. Nei and R. K. Koehn (eds.), *Evolution of Genes and Proteins,* pp. 14–37. Sinauer Associates, Sunderland, MA. **[7]**

Li, W.-H. 1989. A statistical test of phylogenies estimated from sequence data. Mol. Biol. Evol. 6: 424–435. **[5]**

Li, W.-H. 1993. Unbiased estimation of the rates of synonymous and nonsynonymous substitution. J. Mol. Evol. 36: 96–99. **[3]**

Li, W.-H. 1997. *Molecular Evolution.* Sinauer Associates, Sunderland, MA. **[1, 2, 3, 6]**

Li, W.-H. and T. Gojobori. 1983. Rapid evolution of goat and sheep globin genes following gene duplication. Mol. Biol. Evol. 1: 94–108. **[4]**

Li, W.-H. and M. Guoy. 1991. Statistical methods for testing phylogenies, pp. 249–277. In: M. M. Miyamoto and J. Cracraft (eds.), *Phylogenetic Analysis of DNA Sequences*. Oxford University Press, New York **[5]**

Li, W.-H. and K. D. Makova. 2008. Molecular clocks. In: *Encyclopedia of Life Sciences*. Wiley, Chichester. doi:10.1002/9780470015902.a0005111.pub2 **[4]**

Li, W.-H., T. Gojobori, and M. Nei. 1981. Pseudogenes as a paradigm of neutral evolution. Nature 292: 237–239. **[7]**

Li, W.-H., C.-I. Wu, and C.-C. Luo. 1985. A new method for estimating synonymous and nonsynonymous rates of nucleotide substitution considering the relative likelihood of nucleotide and codon changes. Mol. Biol. Evol. 2: 150–174. **[3]**

Li, W.-H., M. Tanimura, and P. M. Sharp. 1987. An evaluation of the molecular clock hypothesis using mammalian DNA sequences. J. Mol. Evol. 25: 330–342. **[5]**

Li, W.-H., Z. Gu, H. Wang, and A. Nekrutenko. 2001. Evolutionary analyses of the human genome. Nature 409: 847–849. **[7]**

Li, W.-H., S. Yi, and K. D. Makova. 2002. Male-driven evolution. Curr. Opin. Genet. Dev. 12: 650–656. **[4]**

Li, W., A. E. Tucker, W. Sung, W. K. Thomas, and M. Lynch. 2009. Extensive, recent intron gains in *Daphnia* populations. Science 326: 1260–1262. **[8]**

Li, X. and 10 others. 2005. Pseudogenization of a sweet-receptor gene accounts for cats' indifference toward sugar. PLoS Genet. 1: 27–35. **[8]**

Li, X., M. A. Schuler, and M. R. Berenbaum. 2007. Molecular mechanisms of metabolic resistance to synthetic and natural xenobiotics. Annu. Rev. Entomol. 52: 231–253. **[7]**

Li, X. and 11 others. 2011. Chromosome size in diploid eukaryotic species centers on the average length with a conserved boundary. Mol. Biol. Evol. 28: 1901–1911. **[11]**

Li, Y. and 11 others. 1995. DNA ioslation (sic) and sequence analysis of dinosaur DNA from Cretaceous dinosaur egg in Xixia Henan, China. Acta Sci. Nat. Univ. Pekinensis 31: 148–152. **[5]**

Li, Y. C., A. B. Korol, T. Fahima, A. Beiles, and E. Nevo. 2002. Microsatellites: genomic distribution, putative functions and mutational mechanisms: a review. Mol. Ecol. 11: 2453–2465. **[11]**

Libkind, D. and 7 others. 2011. Microbe domestication and the identification of the wild genetic stock of lager-brewing yeast. Proc. Natl. Acad. Sci. USA 108: 14539–14544. **[7]**

Librado, P. and J. Rozas. 2009. DnaSP v5: A software for comprehensive analysis of DNA polymorphism data. Bioinformatics 25: 1451–1452. **[2]**

Lightowlers, R. and Z. Chrzanowska-Lightowlers. 2010. Terminating human mitochondrial protein synthesis: A shift in our thinking. RNA Biol. 7: 282–286. **[1]**

Lillo, F. and D. C. Krakauer. 2007. A statistical analysis of the three-fold evolution of genomic compression through frame overlaps in prokaryotes. Biol. Dir. 2: 22. **[8]**

Lin, J. 1991. Divergence measures based on the Shannon entropy. IEEE Trans. Inf. Theor. 37: 145–151. **[11]**

Lin, Y.-S., W.-L. Hsu, J.-K. Hwang, and W.-H. Li. 2007. Proportion of solvent-exposed amino acids in a protein and rate of protein evolution. Mol. Biol. Evol. 24: 1005–1011. **[4]**

Lind, P. A., O. G. Berg, and D. I. Andersson. 2010. Mutational robustness of ribosomal protein genes. Science 330: 825–827. **[4]**

Lindahl, T. 1993a. Instability and decay of the primary structure of DNA. Nature 362: 709–715. **[5]**

Lindahl, T. 1993b. Recovery of antediluvian DNA. Nature 365: 700. **[5]**

Lindsay, J. A. and M. T. G. Holden. 2004. *Staphylococcus aureus*: Superbug, super genome? Trends Microbiol. 12: 378–385. **[9]**

Lindsey, H. A., J. Gallie, S. Taylor, and B. Kerr. 2013. Evolutionary rescue from extinction is contingent on a lower rate of environmental change. Nature 494: 463–467. **[13]**

Line, M. A. 2002. The enigma of the origin of life and its timing. Microbiology 148: 21–27. **[5]**

Ling, J., R. Daoud, M. J. Lajoie, G. M. Church, Söll D, and B. F. Lang. 2014. Natural reassignment of CUU and CUA sense codons to alanine in *Ashbya* mitochondria. Nucleic Acids Res. 42: 499–508. **[10]**

Linheiro, R. S. and C. M. Bergman. 2012. Whole genome resequencing reveals natural target site preferences of transposable elements in *Drosophila melanogaster*. PLoS ONE 7: e30008. **[9]**

Link, A. J. and D. A. Tirrell. 2005. Reassignment of sense codons *in vivo*. Methods 36: 291–298. **[10]**

Lisch, D. and R. K. Slotkin. 2011. Strategies for silencing and escape: The ancient struggle between transposable elements and their hosts. Int. Rev. Cell Mol. Biol. 292: 119–152. **[9]**

Litt, M. and J. A. Luty. 1989. A hypervariable microsatellite revealed by *in vitro* amplification of a dinucleotide repeat within the cardiac muscle actin gene. Am. J. Hum. Genet. 44: 397–401. **[11]**

Liu-Yesucevitz, L. and 7 others. 2011. Local RNA translation at the synapse and in disease. J. Neurosci. 31(45): 16086–16093. **[12]**

Liu, J. C., K. D. Makova, R. M. Adkins, S. Gibson, and W. H. Li. 2001. Episodic evolution of growth hormone in primates and emergence of the species specificity of human growth hormone receptor. Mol. Biol. Evol. 18: 945–953. **[4]**

Liu, M. and A. Grigoriev. 2004. Protein domains correlate strongly with exons in multiple eukaryotic genomes—evidence of exon shuffling? Trends Genet. 20: 399–403. **[8]**

Liu, M., H. Walch, S. Wu, and A. Grigoriev. 2005. Significant expansion of exon-bordering protein domains during animal proteome evolution. Nucl. Acids Res. 33: 95–105. **[7, 8]**

Liu, S. L. and K. L. Adams. 2010. Dramatic change in function and expression pattern of a gene duplicated by polyploidy created a paternal effect gene in the Brassicaceae. Mol. Biol. Evol. 27: 2817–2828. **[7]**

Liu, Y., P. M. Harrison, V. Kunin, and M. Gerstein. 2004. Comprehensive analysis of pseudogenes in prokaryotes: Widespread gene decay and failure of putative horizontally transferred genes. Genome Biol. 5: R64. **[10]**

Lloyd, S. 2001. Measures of complexity: a nonexhaustive list. IEEE Control Syst. Mag. 21(4): 7–8. **[11]**

Lobry, J. R. 1995. Asymmetric substitution patterns in the two DNA strands of bacteria. Mol. Biol. Evol. 13: 660–665. **[10]**

Lobry, J. R. 1996a. A simple vectorial representation of DNA sequences for the detection of replication origin in bacteria. Biochemie 78: 323–326. **[10]**

Lobry, J. R. 1996b. Asymmetric substitution patterns in the two DNA strands of bacteria. Mol. Biol. Evol. 13: 660–665. **[10]**

Lobry, J. R. 1996c. Origin of replication of *Mycoplasma genitalium*. Science 272: 745–746. **[10]**

Lobry, J. R. and C. Lobry. 1999. Evolution of DNA base composition under no-strand-bias conditions when the substitution rates are not constant. Mol. Biol. Evol. 16: 719–723. **[10]**

Lobry, J. R. and N. Sueoka. 2002. Asymmetric directional mutation pressures in bacteria. Genome Biol. 3: 0058. **[10]**

Lockhart, P. J., M. A. Steel, M. D. Hendy, and D. Penny. 1994. Recovering evolutionary trees under a more realistic model of sequence evolution. Mol. Biol. Evol. 11: 605–612. **[5]**

Lockton, S., J. Ross-Ibarra, and B. S. Gaut. 2008. Demography and weak selection drive patterns of transposable element diversity in natural populations of *Arabidopsis lyrata*. Proc. Natl. Acad. Sci. USA 105: 13965–13970. **[11]**

Lockwood, C. A., W. H. Kimbel, and J. M. Lynch. 2004. Morphometrics and hominoid phylogeny: Support for a chimpanzee-human clade and differentiation among great ape subspecies. Proc. Natl. Acad. Sci. USA 101: 4356–4360. **[5]**

Logsdon, J. M. and R. F. Doolittle. 1997. Origin of antifreeze protein genes: A cool tale in molecular evolution. Proc. Natl. Acad. Sci. USA 94: 3485–3487. **[8]**

Loh, E., J. J. Salk, and L. A. Loeb. 2010. Optimization of DNA polymerase mutation rates during bacterial evolution. Proc. Natl. Acad. Sci. USA 107: 1154–1159. **[2]**

Loh, Y.-H. E., L. S. Katz, M. C. Mims, T. D. Kocher, S. V. Yi, and J. T. Streelman. 2008. Comparative analysis reveals signatures of differentiation amid genomic polymorphism in Lake Malawi cichlids. Genome Biol. 9: R113. **[12]**

Loh, Y.-H. E., S. V. Yi, and J. T. Streelman. 2010. Evolution of microRNAs and the diversification of species. Genome Biol. Evol. 3: 55–65. **[12]**

Lohe, A. and P. Roberts. 1988. Evolution of satellite DNA sequences in *Drosophila*. pp. 148–186. In: R. Verma (ed.), *Heterochromatin*. Cambridge University Press, Cambridge. **[11]**

Long, E. O. and I. B. Dawid. 1980. Repeated genes in eukaryotes. Annu. Rev. Biochem. 49: 727–764. **[7]**

Long, H. A., T. Paixao, R. B. R. Azevedo, and R. A. Zufall. 2013. Accumulation of spontaneous mutations in the ciliate *Tetrahymena thermophila*. Genetics 195: 527–540. **[13]**

Lopez, A. J. 1998. Alternative splicing of pre-mRNA: Developmental consequences and mechanisms of regulation. Annu. Rev. Genet. 32: 279–305. **[8]**

Lopez, J. V., N. Yuhki, W. Modi, R. Masuda, and S. J. O'Brien. 1994. *Numt*, a recent transfer and tandem amplification of mitochondrial DNA in the nuclear genome of the domestic cat. J. Mol. Evol. 39: 171–190. **[9]**

Lopez, L. C., W.-H. Li, M. L. Frazier, C.-C. Luo, and G. F. Saunders. 1984. Evolution of glucagon genes. Mol. Biol. Evol. 1: 335–344. **[3]**

Lopez, P., D. Casane, and H. Philippe. 2002. Heterotachy, an important process of protein evolution. Mol. Biol. Evol. 19: 1–7. **[4]**

López-Flores, I. and M. A. Garrido-Ramos. 2012. The repetitive DNA content of eukaryotic genomes. Genome Dyn. 7: 1–28. **[11]**

Lorenzen, E. D., P. Arctander, and H. R. Siegismund. 2008. High variation and very low differentiation in wide ranging plains zebra (*Equus quagga*): Insights from mtDNA and microsatellites. Mol. Ecol. 17: 2812–2824. **[5]**

Lorenzen, M. D. and 9 others. 2008. The maternal-effect, selfish genetic element *Medea* is associated with a composite *Tc1* transposon. Proc. Natl. Acad. Sci. USA 105: 10085–10089. **[11]**

Lowenstein, J. M. and O. A. Ryder. 1985. Immunological systematics of the extinct quagga (Equidae). Experientia 41: 1192–1193. **[5]**

Löytynoja, A. and N. Goldman. 2005. An algorithm for progressive multiple alignment of sequences with insertions. Proc. Natl. Acad. Sci. USA 102: 10557–10562. **[3]**

Lozovsky, E. R. and 8 others. 2009. Stepwise acquisition of pyrimethamine resistance in the malaria parasite. Proc. Natl. Acad. Sci. USA 106: 12025–12030. **[13]**

Luisi, P. L., F. Ferri, and P. Stano. 2006. Approaches to semi-synthetic minimal cells: A review. Naturwissenschaften 93: 1–13. **[10]**

Lukes, J., J. M. Archibald, P. J. Keeling, W. F. Doolittle, and M. W. Gray. 2011. How a neutral evolutionary ratchet can build cellular complexity. IUBMB Life 63: 528–537. **[8]**

Lukjancenko, O., T. M. Wassenaar, and D. W. Ussery. 2010. Comparison of 61 sequenced *Escherichia coli* genomes. Microbiol. Ecol. 60: 708–720. **[6, 10]**

Lundy, M. 1985. Applications of the annealing algorithm to combinatorial problems in statistics. Biometrica 72: 91–198. **[5]**

Lupski, J. R. 1998. Genomic disorders: Structural features of the genome can lead to DNA rearrangements and human disease traits. Trends Genet. 14: 417–422. **[7]**

Luria, S. E. and M. Delbrück. 1943. Mutations of bacteria from virus sensitivity to virus resistance. Genetics 28: 491–511. **[1]**

Lutz, A. M. 1907. A preliminary note on the chromosomes of *Oenothera lamarckiana* and one of its mutants, *O. gigas*. Science 26: 151–152. **[7]**

Lynch, M. 2006. The origins of eukaryotic gene structure. Mol. Biol. Evol. 23: 450–468. **[2, 11]**

Lynch, M. 2007. The frailty of adaptive hypotheses for the origins of organismal complexity. Proc. Natl. Acad. Sci. USA 104: 8597–8604. **[8, 9]**

Lynch, M. 2008. The cellular, developmental and population-genetic determinants of mutation-rate evolution. Genetics 180: 933–943. **[2]**

Lynch, M. 2010. Evolution of the mutation rate. Trends Genet. 26: 345–352. **[2, 11]**

Lynch, M. 2011. The lower bound to the evolution of mutation rates. Genome Biol. Evol. 3: 1107–1118. **[2]**

Lynch, M. and J. S. Conery. 2003. The evolutionary demography of duplicate genes. J. Struct. Funct. Genomics 3: 35–44. **[7]**

Lynch, M. and J. S. Conery. 2003. The origins of genome complexity. Science 302: 1401–1404. **[10, 11]**

Lynch, M. and A. Force. 2000. The probability of duplicate gene preservation by subfunctionalization. Genetics 154: 459–473. **[11]**

Lynch, M. and W. Gabriel. 1990. Mutation load and the survival of small populations. Evolution 44: 1725–1737. **[2]**

Lynch, M., J. Conery, and R. Bürger. 1995. Mutation accumulation and the extinction of small populations. Am. Nat. 146: 489–518. **[2]**

Lynch, M., M. O'Hely, J. B. Walsh, and A. Force. 2001. The probability of preservation of a newly arisen gene duplicate. Genetics 159: 1789–1804. **[7]**

Lynch, M., L.-M. Bobay, F. Catania, J.-F. Gout, and M. Rho. 2011. The repatterning of eukaryotic genomes by random genetic drift. Annu. Rev. Genomics Hum. Genet. 12: 347–366. **[11]**

M

Macaya, G., J. P. Thiery, and G. Bernardi. 1976. An approach to the organization of eukaryotic genomes at a macromolecular level. J. Mol. Biol. 108: 237–254. **[11]**

Mackiewicz, D., P. Mackiewicz, M. Kowalczuk, M. Dudkiewicz, M. R. Dudek, and S. Cebrat. 2003. Rearrangements between differently replicating DNA strands in asymmetric bacterial genomes. Acta Microbiol. Pol. 5: 245–260. **[10]**

Maddison, W. P. 1991. Squared-change parsimony reconstructions of ancestral states for continuous-valued characters on a phylogenetic tree. Syst. Zool. 40: 304–314. **[11]**

Madigan, J. M. and T. D. Martinko. 2006. *Brock Biology of Microorganisms*, 11th Edition. Pearson Prentice Hall, Upper Saddle River, NJ **[13]**

Maeda, N., C.-I. Wu, J. Bliska, and J. Reneke. 1988. Molecular evolution of intergenic DNA in higher primates: Pattern of DNA changes, molecular clock and evolution of repetitive sequences. Mol. Biol. Evol. 5: 1–20. **[5]**

Maeda, N., J. B. Bliska, and O. Smithies. 1983. Recombination and balanced chromosome polymorphism suggested by DNA sequences 5′ to the human δ-globin gene. Proc. Natl. Acad. Sci. USA 80: 5012–5016. **[5]**

Mahony, S., D. L. Corcoran, E. Feingold, and P. V. Benos. 2007. Regulatory conservation of protein coding and microRNA genes in vertebrates: lessons from the opossum genome. Genome Biol. 8: R84. **[12]**

Majumdar, S., A. Singh, and D. C. Rio. 2013. The human THAP9 gene encodes an active P-element DNA transposase. Science 339: 446–448. **[9]**

Makalowski, W. 2003. Not junk after all. Science 300: 1246–1247. **[11]**

Makarenkov, V., D. Kevorkov, and P. Legendre. 2006. Phylogenetic network construction approaches. Appl. Mycol. Biotechnol. 6: 61–97. **[6]**

Makova, K. D. and W.-H. Li. 2002. Strong male-driven evolution of DNA sequences in humans and apes. Nature 416: 624–626. **[4]**

Malnic, B., J. Hirono, T. Sato, and L. B. Buck. 1999. Combinatorial receptor codes for odors. Cell 96: 713–723. **[7]**

Malthus, T. R. 2003. *An Essay on the Principle of Population*, Norton, New York. **[13]**

Manfredi-Romanini, M. G., D. Formenti, C. Pellicciari, E. Ronchetti, and Y. Rumpler. 1994. Possible meaning of C-heterochromatic DNA (ChDNA) in primates. pp. 371–378. In: B. Thierry, J. R. Anderson, J. J. Roeder, and N. Herrenschmidt (eds.), *Current Primatology*, Vol. 1. *Ecology and Evolution*. Univ. Louis Pasteur, Strasbourg. **[11]**

Mangenot, S. and G. Mangenot. 1962. Enquête sur les nombres chromosomiques dans une collection d'espèces tropicales. Rev. Cytol. Biol. Veget. 25: 411–447. **[7]**

Maniloff, J. 1996. The minimal cell genome: "On being the right size." Proc. Natl. Acad. Sci USA 93: 10004–10006. **[10]**

Mank, J. E., E. Axelsson, and H. Ellegren. 2007. Fast-X on the Z: Rapid evolution of sex-linked genes in birds. Genome Res. 17: 618–624. **[4]**

Manly, B. F. J. 2006. *Randomization, Bootstrap and Monte Carlo Methods in Biology*, 3rd Ed. Chapman and Hall, London. **[5, 6]**

Mann, C. C. and M. L. Plummer. 1992. The butterfly problem. Atlantic Monthly 269(1): 47-70. **[5]**

Mao, X., H. Zhang, Y. Yin, and Y. Xu. 2012. The percentage of bacterial genes on leading versus lagging strands is influenced by multiple balancing forces. Nucleic Acids Res. 40: 8210–8218. **[10]**

Marais, G. 2003. Biased gene conversion: Implications for genome and sex evolution. Trends Genet. 19: 330–338. **[7]**

Marchant, J. 2011. Curse of the Pharoah's DNA. Nature 472: 404–406. **[5]**

Marcotte, E. M., M. Pellegrini, M. J. Thompson, T. O. Yeates, and D. Eisenberg. 1999. A combined algorithm for genome-wide prediction of protein function. 402: 83–86. **[8]**

Marcussen, T., B. Oxelman, A. Skog, and K. S. Jakobsen. 2010. Evolution of plant RNA polymerase IV/V genes: Evidence of subneofunctionalization of duplicated *NRPD2/NRPE2*-like paralogs in *Viola* (Violaceae). BMC Evol. Biol. 10: 45. **[7]**

Marcussen, T. and 106 others. 2014. Ancient hybridizations among the ancestral genomes of bread wheat. Science 345: 1250092. **[7]**

Markert, C. L. 1964. Biochemical events during differentiation: General survey. J. Exp. Zool. 157: 81–4. **[7]**

Markert, C. L. and H. Ursprung. 1971. *Developmental Genetics*. Prentice-Hall, Englewood Cliffs, NJ. **[7]**

Markova-Raina, P. and D. Petrov. 2011. High sensitivity to aligner and high rate of false positives in the estimates of positive selection in the 12 *Drosophila* genomes. Genome Res. 21: 863–874. **[4]**

Marshall, W. F. and 10 others. 2012. What determines cell size? BMC Biol. 10: 101. **[11]**

Martin, F. J. and J. O. McInerney. 2009. Recurring cluster and operon assembly for phenylacetate degradation genes. BMC Evol. Biol. 9: 36. **[6]**

Martin, N., E. A. Ruedi, R. LeDuc, F.-J. Sun, and G. Caetano-Anollés. 2007. Gene-interleaving patterns of synteny in the *Saccharomyces cerevisiae* genome: Are they proof of an ancient genome duplication event? Biol. Dir. 2: 23. **[7]**

Martin, R. D. 1990. *Primate Origins and Evolution: A Phylogenetic Reconstruction*. Princeton University Press, Princeton. **[5, 6]**

Martin, S. L. 2009. Jumping-gene roulette. Nature 460: 1087–1088. **[9]**

Martin, W. 1999. Mosaic bacterial chromosome: A challenge en route to a tree of genomes. BioEssays 21: 99–104. **[5, 6]**

Martin, W. 2003. Gene transfer from organelles to the nucleus: Frequent and in big chunks. Proc. Natl. Acad. Sci. USA 100: 8612–8614. **[9]**

Martin, W. and E. V. Koonin. 2006a. A positive definition of prokaryotes. Nature 442: 868. **[5, 6, 8]**

Martin, W. and E. V. Koonin. 2006b. Introns and the origin of nucleus-cytosol compartmentalization. Nature 440: 41–45. **[6, 8]**

Martin, W. and M. Müller. 1998. The hydrogen hypothesis for the first eukaryote. Nature 392: 37–41. **[5, 6]**

Martincorena, I. 2012. Genome-scale strategies controlling the impact of deleterious mutations. Ph.D. Dissertation, University of Cambridge. **[2]**

Martínez, P. and M. A. Blasco. 2011. Telomeric and extra-telomeric roles for telomerase and the telomere-binding proteins. Nat. Rev. Cancer 11: 161–176. **[11]**

Maruyama, T. 1974. The age of a rare mutant gene in a large population. Am. J. Hum. Genet. 26: 669–673. **[11]**

Maruyama, T. and M. Kimura. 1974. A note on the speed of gene frequency changes in reverse directions in a finite population. Evolution 28: 162–163. **[2]**

Marx, V. 2013. Next-generation sequencing: The genome jigsaw. Nature 501: 263–268. **[11]**

Massey, S. E., G. Moura, P. Beltrao, R. Almeida, J. R. Garey, M. F. Tuite, and M. A. Santos. 2003. Comparative evolutionary genomics unveils the molecular mechanism of reassignment of the CTG codon in *Candida* spp. Genome Res. 25: 179–186. **[10]**

Massingham, T. and N. Goldman. 2005. Detecting amino acid sites under positive selection and purifying selection. Genetics 169: 1753–1762. **[11]**

Mathews, C. K. and K. E. van Holde. 1990. *Biochemistry*. Benjamin/Cummings, Menlo Park, CA. **[1]**

Matsuo, M., Y. Ito, R. Yamauchi, and J. Obokata. 2005. The rice nuclear genome continuously integrates, shuffles, and eliminates the chloroplast genome to cause chloroplast-nuclear DNA flux. Plant Cell 17: 665–675. **[9]**

Mayr, E. 1942. *Systematics and the Origin of Species*. Columbia University Press, New York. **[5]**

Mayrose, I., S. H. Zhan, C. J. Rothfels, K. Magnuson-Ford, M. S. Barker, L. H. Rieseberg, and S. P. Otto. 2011. Recently formed polyploid plants diversify at lower rates. Science 333: 1257. **[7]**

McCarrey, J. R. and K. Thomas. 1987. Human testis-specific PGK gene lacks introns and possesses characteristics of a processed gene. Nature 326: 501–505. **[9]**

McCollum, M. A., C. C. Sherwood, C. J. Vinyard, C. O. Lovejoy, and F. Schachat. 2006. Of muscle-bound crania and human brain evolution: The story behind the *MYH16* headlines. J. Hum. Evol. 50: 232–236. **[8]**

McCutcheon, J. P., B. R. McDonald, and N. A. Moran. 2009. Origin of an alternative genetic code in the extremely small and GC-rich genome of a bacterial symbiont. PLoS Genet. 5: e1000565. **[10]**

McDonald, J. and M. Kreitman. 1991. Adaptive protein evolution at the *Adh* locus in *Drosophila*. Nature 351: 652–654. **[2]**

McDonald, M. J., S. M. Gehrig, P. L. Meintjes, X. X. Zhang, and P. B. Rainey. 2009. Adaptive divergence in experimental populations of *Pseudomonas fluorescens*. IV. Genetic constraints guide evolutionary trajectories in a parallel adaptive radiation. Genetics 183: 1041–1053. **[13]**

McDonald, M. J., T. F. Cooper, H. J. E. Beaumont, and P. B. Rainey. 2011. The distribution of fitness effects of new beneficial mutations in *Pseudomonas fluorescens*. Biol. Lett. 7: 98–100. **[13]**

McDonald, M. J., Y.-Y. Hsieh, Y.-H. Yu, S.-L. Chang, and J.-Y. Leu. 2012. The evolution of low mutation rates in experimental mutator populations of *Saccharomyces cerevisiae*. Curr. Biol. 22: 1235–1240. **[13]**

McDonell, L. and D. Drouin. 2012. The abundance of processed pseudogenes derived from glycolytic genes is correlated with their expression level. Genome 55: 147–151. **[9]**

McInerney, J. O., D. Pisani, E. Bapteste, and M. J. O'Connell. 2011. The Public Goods hypothesis for the evolution of life on Earth. Biol. Dir. 6: 41. **[6, 9]**

McLoughlin, S. Y. and S. D. Copley. 2008. A compromise required by gene sharing enables survival: Implications for evolution of new enzyme activities. Proc. Natl. Acad. Sci. USA 105: 13497–13502. **[7]**

McLysaght, A., K. Hokamp, and K. H. Wolfe. 2002. Extensive genomic duplication during early chordate evolution. Nat. Genet. 31: 200–204. **[7]**

Méchali, M. 2010. Eukaryotic DNA replication origins: Many choices for appropriate answers. Nat. Rev. Mol. Cell Biol. 11: 728–738. **[10]**

Mehan, M. R., M. Almonte, E. Slaten, N. B Freimer, P. N. Rao, and R. A. Ophoff. 2006. Analysis of segmental duplications reveals a distinct pattern of continuation-of-synteny between human and mouse genomes. Hum. Genet. 121: 93–100. **[11]**

Mehta, G., R. Jalan, and R. P. Mookerjee. 2013. Cracking the ENCODE: From transcription to therapeutics. Hepatology 57: 2532–2535. **[11]**

Mekalanos, J. J. 1983. Duplication and amplification of toxin genes in *Vibrio cholerae*. Cell 35: 253–63. **[7]**

Mel, S. F. and J. J. Mekalanos. 1996. Modulation of horizontal gene transfer in pathogenic bacteria by *in vivo* signals. Cell 87: 795–798. **[9]**

Melamud, E. and J. Moult. 2009. Stochastic noise in splicing machinery. Nucl. Acids Res. 37: 4873–4886. **[8]**

Melters, D. P. and 16 others. 2013. Comparative analysis of tandem repeats from hundreds of species reveals unique insights into centromere evolution. Genome Biol. 14: R10. **[11]**

Menashe, I., O. Man, D. Lancet, and Y. Gilad. 2003. Different noses for different people. Nat. Genet. 34: 143–144. **[7]**

Meng, C. and L. Kubatko. 2009. Detecting hybrid speciation in the presence of incomplete lineage sorting using gene tree incongruence: A model. Theor. Pop. Biol. 75: 35–45. **[5]**

Mercer, J. M. and V. L. Roth. 2003. The effects of Cenozoic global change on squirrel phylogeny. Science 299: 1568–1572. **[5]**

Meredith, R. W., J. Gatesy, J. Cheng, and M. S. Springer. 2011. Pseudogenization of the tooth gene enamelysin (*MMP20*) in the common ancestor of extant baleen whales. Proc. R. Soc. 278B: 993–1002. **[7, 8, 9]**

Mereschkowsky, C. 1905. Über Natur und Ursprung der Chromatophoren im Pflanzenreiche. Biol. Centralblatt. 25: 593–604. **[5]**

Merrikh, H., Y. Zhang, A. D. Grossman, and J. D. Wang. 2012. Replication-transcription conflicts in bacteria. Nat. Rev. Microbiol. 10: 449–58. **[10]**

Metcalfe, C. J. and D. Casane. 2013. Accommodating the load: the transposable element content of very large genomes. Mobile Genet. Elements 3: e24775. **[11]**

Metzker, M. L., D. P. Mindell, X.-M. Liu, R. G. Ptak, R. A. Gibbs, and D. M. Hillis. 2002. Molecular evidence of HIV-1 transmission in a criminal case. Proc. Natl. Acad. Sci. USA 99: 14292–14297. **[5]**

Meunier, J. and L. Duret. 2004. Recombination drives the evolution of GC-content in the human genome. Mol. Biol. Evol. 21: 984–990. **[11]**

Meyer, J. R., D. T. Dobias, J. S. Weitz, J. E. Barrick, R. T. Quick, and R. E. Lenski. 2012. Repeatability and contingency in the evolution of a key innovation in phage lambda. Science 335: 428–432. **[13]**

Michaelson, J. J. and 27 others. 2012. Whole-genome sequencing in autism identifies hot spots for de novo germline mutation. Cell 151: 1431–1442. **[4]**

Mikkelsen, T. S. 2004.What makes us human? Genome Biol. 5: 238. **[11]**

Mikkelsen, T. S. and 66 others. 2005. Initial sequence of the chimpanzee genome and comparison with the human genome. Nature 437: 69–87. **[4]**

Miklos, G. L. and A. C. Gill. 1982. Nucleotide sequences of highly repeated DNAs; compilation and comments. Genet. Res. 39: 1–30. **[11]**

Milinkovitch, M. C., G. Orti, and A. Meyer. 1993. Revised phylogeny of whales suggested by mitochondrial ribosomal DNA. Nature 361: 346–348. **[5]**

Miller, W. J., L. Kruckenhauser, and W. Pinsker. 1996. The impact of transposable elements on genome evolution in animals and plants, pp. 21–34. In: J. Tomiuk, K. Wörhmann, and A. Sentker (eds.), *Transgenic Organisms—Biological and Social Implications*. Birkhäuser Verlag, Basel. **[9]**

Millikan, R. G. 1989. In defense of proper functions. Philos. Sci. 56: 288–302. **[11]**

Mindell, D. P. 2006. *The Evolving World: Evolution in Everyday Life*. Harvard University Press, Cambridge, MA. **[5, 6]**

Mindell, D. P. 2009. Evolution in the everyday world. Sci. Am. 300: 82–89. **[5]**

Mindell, D. P. 2013. The tree of life: Metaphor, model, and heuristic device. Syst. Biol. 62: 479–489. **[6]**

Miozzari, G. F. and C. Yanofsky. 1979. Gene fusion during the evolution of the tryptophan operon in Enterobacteriaceae. Nature 277: 486–489. **[8]**

Mira, A., H. Ochman, and N. A. Moran. 2001. Deletional bias and the evolution of bacterial genomes. Trends Genet. 17: 589–596. **[10]**

Mirkin, B. G., T. I. Fenner, M. Y. Galperin, and E. V. Koonin. 2003. Algorithms for computing parsimonious evolutionary scenarios for genome evolution, the last universal common ancestor and dominance of horizontal gene transfer in the evolution of prokaryotes. BMC Evol. Biol. 3: 2. **[10]**

Mirsky, A. E. and H. Ris. 1951. The desoxyribonucleic acid content of animal cells and its evolutionary significance. J. Gen. Physiol. 34: 451–462. **[11]**

Misawa, K. and M. Nei. 2003. Reanalysis of Murphy et al.'s data gives various mammalian phylogenies and suggests overcredibility of Bayesian trees. J. Mol. Evol. 57: S290–S296. **[5]**

Mitchell, A. and D. Graur. 2005. Inferring the pattern of spontaneous mutation from the pattern of substitution in unitary pseudogenes of *Mycobacterium leprae* and a comparison of mutation patterns among distantly related organisms. J. Mol. Evol. 61: 795–803. **[4]**

Miyamoto, M. M. and T. Cracraft (eds.). 1991. *Phylogenetic Analysis of DNA Sequences*. Oxford University Press, New York. **[5]**

Miyamoto, M. M., J. L. Slightom, and M. Goodman. 1987. Phylogenetic relationships of humans and African apes from DNA sequences in the ψη-globin region. Science 238: 369–373. **[5]**

Miyata T. and T. Yasunaga. 1981. Rapidly evolving mouse α-globin-related pseudogene and its evolutionary history. Proc. Natl. Acad. Sci. USA 78: 450–453. **[7]**

Miyata, T. and T. Yasunaga. 1978. Evolution of overlapping genes. Nature 272: 532–535. **[8]**

Miyata, T. and T. Yasunaga. 1980. Molecular evolution of mRNA: A method for estimating evolutionary rates of synonymous and amino acid substitutions from homologous nucleotide sequences and its application. J. Mol. Evol. 16: 23–36. **[3]**

Miyata, T., H. Hayashida, K. Kuma, K. Mitsuyasu, and T. Yasunaga. 1987. Male-driven molecular evolution: A model and nucleotide sequence analysis. Cold Spring Harb. Symp. Quant. Biol. 52: 863–867. **[9]**

Mizoguchi, H., M. Mori, and F. Tatsuro. 2007. *Escherichia coli* minimum genome factory. Biotechnol. Appl. Biochem. 46: 157–167. **[10]**

Modrek, B. and C. Lee. 2002. A genomic view of alternative splicing. Nat. Genet. 30: 13–19. **[8]**

Modrek, B. and C. Lee. 2003. Alternative splicing in the human, mouse and rat genomes is associated with an increased frequency of exon creation and/or loss. Nat. Genet. 34: 177–180. **[8]**

Mohamed, A. H., S. S. Chirala, N. H. Mody, W. Y. Huang, and S. J. Wakil. 1988. Primary structure of the multifunctional α subunit protein of yeast fatty acid synthase derived from *FAS2* gene sequence. J. Biol. Chem. 263: 12315–12325. **[8]**

Molina, J. and 16 others. 2014. Possible loss of the chloroplast genome in the parasitic flowering plant *Rafflesia lagascae* (Rafflesiaceae). Mol. Biol. Evol. 31: 793–803. **[10]**

Monod, J. and F. Jacob. 1961. Teleonomic mechanisms in cellular metabolism, growth, and differentiation. Cold Spring Harb. Symp. Quant. Biol. 26: 389–401. **[12]**

Montague, M. J. and 24 others. 2014. Comparative analysis of the domestic cat genome reveals genetic signatures underlying feline biology and domestication. Proc. Natl. Acad. Sci. USA 111: 17230–17235. **[11]**

Montgelard, C., E. Douzery, and J. Michaux. 2007. Classification and molecular phylogeny, pp. 95–126. In: D. L. Miller (ed.), *Reproductive Biology and Phylogeny of Cetacea: Whales, Porpoises and Dolphins*. Science Publishers, Enfield, NH. **[5]**

Montoya-Burgos, J. I., P. Boursot, and N. Galtier. 2003. Recombination explains isochores in mammalian genomes. Trends Genet. 19: 128–130. **[11]**

Mooers, A. Ø. and E. C. Holmes. 2000. The evolution of base composition and phylogenetic inference. Trends Ecol. Evol. 15: 365–369 **[5]**

Moore, C. E., B. Curtis, T. Mills, G. Tanifuji, and J. M. Archibald. 2012. Nucleomorph genome sequence of the cryptophyte alga *Chroomonas mesostigmatica* CCMP1168 reveals lineage-specific gene loss and genome complexity. Genome Biol. Evol. 4: 1162–1175. **[11]**

Moran, J. V., R. J. DeBerardinis, and H. H. Kazazian. 1999. Exon shuffling by L1 retrotransposition. Science 283: 1530–1534. **[9]**

Moran, L. A. 2007. The deflated ego problem. http: //sandwalk.blogspot.com/2007/05/deflated-ego-problem.html. **[11]**

Moran, N. A. and G. M. Bennett. 2014. The tiniest tiny genomes. Annu. Rev. Microbiol. 68: 195–215. **[10]**

Moran, N. A. and G. R. Plague. 2004. Genomic changes following host restriction in bacteria. Curr. Opin. Genet. Dev. 14: 627–633. **[10]**

Moran, N. A. and T. Jarvik. 2010. Lateral transfer of genes from fungi underlies carotenoid production in aphids. Science 328: 624–627. **[9]**

Moreira, D. and H. Philippe. 2000. Molecular phylogeny: Pitfalls and progress. Int. Microbiol. 3: 9–16. **[5]**

Moreira, D., S. von der Heyden, D. Bass, P. López-Garcia, E. Chao, and T. Cavalier-Smith. 2007. Global eukaryote phylogeny: Combined small- and large-subunit ribosomal DNA trees support monophyly of Rhizaria, Retaria and Excavata. Mol. Phyl. Evol. 44: 255–266. **[5]**

Moretti, S., F. Armougom, I. M. Wallace, D. G. Higgins, C. V. Jongeneel, and C. Notredame. 2007. The M-Coffee web server: A meta-method for computing multiple sequence alignments by combining alternative alignment methods. Nucleic Acids Res. 35: W645–648. **[3]**

Morgan, G. J., G. F. Hatfull, S. Casjens, and R. W. Hendrix. 2002. Bacteriophage Mu genome sequence: Analysis and comparison with Mu-like prophages in *Haemophilus, Neisseria* and *Deinococcus*. J. Mol. Biol. 317: 337–359. **[9]**

Morgan, M. J., J. C. Earnshaw, and G. K. Dhoot. 1993. Novel developmentally regulated exon identified in the rat fast skeletal muscle troponin T gene. J. Cell Sci. 106: 903–908. **[8]**

Morgan, T. H. 1903. *Evolution and Adaptation*. Macmillan, New York. **[2]**

Morgante, M., S. Brunner, G. Pea, K. Fengler, A. Zuccolo, and A. Rafalski. 2005. Gene duplication and exon shuffling by *helitron*-like transposons generate intraspecies diversity in maize. Nat. Genet. 37: 997–1002. **[9]**

Morgenstern, B. 1999. DIALIGN 2: Improvement of the segment-to-segment approach to multiple sequence alignment. Bioinformatics 15: 211–218. **[3]**

Morgenstern, B. 2004. DIALIGN: Multiple DNA and protein sequence alignment at BiBiServ. Nucleic Acids Res. 32: W33–36. **[3]**

Morgenstern, B., K. Frech, A. Dress, and T. Werner. 1998b. DIALIGN: Finding local similarities by multiple sequence alignment. Bioinformatics 14: 290–294. **[3]**

Morgenstern, B., W. R. Atchley, K. Hahn, and A. Dress. 1998a. Segment-based scores for pairwise and multiple sequence alignments. Proc. Int. Conf. Intell. Syst. Mol. Biol. 6: 115–121. **[3]**

Moriyama, E. N., D. A. Petrov, and D. L. Hartl. 1998. Genome size and intron size in *Drosophila*. Mol. Biol. Evol. 15: 770–773. **[11]**

Mornet, E., P. Crete, F. Kuttenn, M. C. Raux-Demay, J. Boue, P. C. White, and A. Boue. 1991. Distribution of deletions and seven point mutation on *CYP21B* genes in three clinical forms of steroid 21-hydroxylase deficiency. Am. J. Hum. Genet. 48: 79–88. **[7]**

Morowitz, H. J. 1984. The completeness of molecular biology. Isr. J. Med. Sci. 20: 750–753. **[10]**

Morrison, D. A. 2008. How to summarize estimates of ancestral divergence times. Evol. Bioinformatics 4: 75–95. **[6]**

Morrison, D. A. 2009. Why would phylogeneticists ignore computerized sequence alignment? Syst. Biol. 58: 150–158. **[3]**

Morrison, D. A. 2010. Using data-display networks for exploratory data analysis in phylogenetic studies. Mol. Biol. Evol. 27: 1044–1057. **[6]**

Morrison, D. A. 2011. *Introduction to Phylogenetic Networks*. RJR Productions, Uppsala. **[6]**

Morton, R. A. and B. R. Morton. 2007. Separating the effects of mutation and selection in producing DNA skew in bacterial chromosomes. BMC Genomics 8: 369. **[10]**

Mouchès, C. and 7 others. 1986. Amplification of an esterase gene is responsible for insecticide resistance in a California *Culex* mosquito. Science 233: 778–780. **[7]**

Mouchiroud, D., G. D'Onofrio, B. Aïssani, G. Macaya, C. Gautier, G. Bernardi. 1991. The distribution of genes in the human genome. Gene 100: 181–187. **[11]**

Mouchiroud, D., C. Gautier, and G. Bernardi. 1995. Frequencies of synonymous substitutions in mammals are gene-specific and correlated with frequencies of nonsynonymous substitutions. J. Mol. Evol. 40: 107–113. **[4]**

Mourali-Chebil, S. and E. Heyer. 2006. Evolution of inbreeding coefficients and effective size in the population of Saguenay Lac-St.-Jean (Québec). Hum. Biol. 78: 495–508. **[2]**

Mourant, A. E., A. C. Kopec, and K. Domanievska-Sobczak. 1976. *The Distribution of the Human Blood Groups and Other Polymorphisms*. Oxford University Press, Oxford. **[8]**

Mouse Genome Sequencing Consortium. 2002. Initial sequencing and comparative analysis of the mouse genome. Nature 420: 520–562. **[11]**

Mower, J. P., P. Touzet, J. S. Gummow, L. F. Delph, and J. D. Palmer. 2007. Extensive variation in synonymous substitution rates in mitochondrial genes of seed plants. BMC Evol. Biol. 7: 135. **[4]**

Mukai, T. 1964. The genetic structure of natural populations of *Drosophila melanogaster*. I. Spontaneous mutation rate of polygenes controlling viability. Genetics 50: 1–19. **[13]**

Mukai, T., A. Hayashi, F. Iraha, A. Sato, K. Ohtake, S. Yokoyama, and K. Sakamoto. 2010. Codon reassignment in the *Escherichia coli* genetic code. Nucleic Acids Res. 38: 8188–8195. **[10]**

Muller, H. J. 1935. Further studies on the nature and causes of gene mutations. Proc. Sixth Int. Congr. Genet. 1: 213–255. **[7]**

Muller, H. J. 1932. Further studies on the nature and causes of gene mutations. Proc. 6th Int. Congr. Genet. 1: 213–255. **[1]**

Muller, H. J. 1940. Bearing of the *Drosophila* work on systematics, pp. 185–268. In: J. S. Huxley (ed.), *The New Systematics*. Clarendon Press, Oxford. **[5]**

Muller, H. J. 1942. Isolating mechanisms, evolution, and temperature. Biol. Symp. 6: 71–125. **[5]**

Muller, H. J. 1950. Our load of mutations. Am. J. Hum. Genet. 2: 111–176. **[11]**

Müller, M. and 9 others. 2012. Biochemistry and evolution of anaerobic energy metabolism in eukaryotes. Microbiol. Mol. Biol. Rev. 76: 444–495. **[6]**

Mungall, A. J. and 170 others. 2003.The DNA sequence and analysis of human chromosome 6. Nature 425: 805–811. **[11]**

Munn, D. H. and 7 others. 1998. Prevention of allogeneic fetal rejection by tryptophan catabolism. Science 281: 1191–1193. **[8]**

Murphy, F. V. and V. Ramakrishnan. 2004. Structure of a purine-purine wobble base pair in the decoding center of the ribosome. Nat. Struct. Mol. Biol. 11: 1251–1252. **[1]**

Murphy, W. J., T. H. Pringle, T. A. Crider, M. S. Springer, and W. Miller. 2007. Using genomic data to unravel the root of the placental mammal phylogeny. Genome Research 17: 413–421. **[5, 6]**

Mushegian, A. 2008. Gene content of LUCA, the last universal common ancestor. Front. Biosci. 13: 4657–3666. **[10]**

Mushegian, A. R. and E. V. Koonin. 1996a. A minimal gene set for cellular life derived by comparison of complete bacterial genomes. Proc. Natl. Acad. Sci. USA 93: 10268–10273. **[10]**

Mushegian, A. R. and E. V. Koonin. 1996b. Gene order is not conserved in bacterial evolution. Trends Genet. 12: 289–290. **[10]**

Muto, A. and S. Osawa. 1987. The guanine and cytosine content of genomic DNA and bacterial evolution. Proc. Natl. Acad. Sci. USA 84: 166–169. **[10]**

N

Nabholz, B., S. Glémin, and N. Galtier. 2008. Strong variations of mitochondrial mutation rate across mammals: The longevity hypothesis. Mol. Biol. Evol. 25: 120–130. **[4]**

Nachman, M. W. 1998. Deleterious mutations in animal mitochondrial DNA. Genetica 102/103: 61–69. **[11]**

Nachman, M. W. 2004. Haldane and the first estimates of the human mutation rate. J. Genet. 83: 231–233. **[4]**

Nachman, M. W. and S. L. Crowell. 2000. Estimate of the mutation rate per nucleotide in humans. Genetics 156: 297–304. **[4]**

Nadeau, J. H. 1989. Maps of linkage and synteny homologies between mouse and man. Trends Genet. 5: 82–86. **[11]**

Nadeau, J. H. and B. A. Taylor. 1984. Lengths of chromosomal segments conserved since divergence of man and mouse. Proc. Natl. Acad. Sci. USA 81: 814–818. **[11]**

Nadeau, J. H. and D. Sankoff. 1997. Comparable rates of gene loss and functional divergence after genome duplications early in vertebrate evolution. Genetics 147: 1259–1266. **[7]**

Nagano, T. and P. Fraser. 2011. No-nonsense functions for long noncoding RNAs. Cell 145: 178–181. **[8]**

Nagl, W. 1990. Polyploidy in differentiation and evolution. Int. J. Cell Cloning 8: 216–223. **[7]**

Nagy, A. and L. Patthy. 2011. Reassessing domain architecture evolution of metazoan proteins: The contribution of different evolutionary mechanisms. Genes 2: 578–598. **[8]**

Nagylaki, T. 1984. Evolution of multigene families under interchromosomal gene conversion. Proc. Natl. Acad. Sci. USA 81: 3796–3800. **[7]**

Nagylaki, T. and T. D. Petes. 1982. Intrachromosomal gene conversion and the maintenance of sequence homogeneity among repeated genes. Genetics 100: 315–337. **[7]**

Naito, K. and 8 others. 2009. Unexpected consequences of a sudden and massive transposon amplification on rice gene expression. Nature 461: 1130–1134. **[9]**

Nakamura, Y., C. Juiler, T. Holm, P. O'Connell, M. Leppert, and R. White. 1987. Characterization of a human "midisatellite" sequence. Nucl. Acids Res. 15: 2537–2547. **[11]**

Nakamura, Y., I. Takeshi, and W. Martin. 2007. Rate and polarity of gene fusion and fission in *Oryza sativa* and *Arabidopsis thaliana*. Mol. Biol. Evol. 24: 110–121. **[8]**

Nakayama, T. and K.-I. Ishida. 2009. Another acquisition of a primary photosynthetic organelle is underway in *Paulinella chromatophora*. Curr. Biol. 19: R284. **[5]**

Nakazono, M. and A. Hirai. 2008. Frequent DNA transfer among mitochondrial, plastid and nuclear genomes of rice during evolution, pp. 107–117. In: H. Hirai, T. Sasaki, Y. Sano, and H. Hirano (eds.), *Rice Biology in the Genomics Era*. Springer, Heidelberg. **[9]**

Nakhleh, L. 2009. Evolutionary phylogenetic networks: Models and issues, pp. 125–158. In: L. Heath and N. Ramakrishnan (eds.), *The Problem Solving Handbook for Computational Biology and Bioinformatics*. Springer, New York. **[5]**

Nakhleh, L., D. Ringe, and T. Warnow. 2005. Perfect phylogenetic networks: A new methodology for reconstructing the evolutionary history of natural languages. Language 81: 382–420. **[6]**

Nakhleh, L., T. Warnow, C. R. Linder, and K. St. John. 2005. Reconstructing reticulate evolution in species—theory and practice. J. Comp. Biol. 12: 796–811. **[5]**

Nam, I.-H., Y.-S. Chang, H.-B. Hong, and Y.-E. Lee. 2003. A novel catabolic activity of *Pseudomonas veronii* in biotransformation of pentachlorophenol. Appl. Microbiol. Biotechnol. 62: 284–290. **[8]**

Naor, A. and U. Gophna. 2013. Cell fusion and hybrids in Archaea. Bioengineered 4: 126–129. **[9]**

Narayan, R. K. J. and H. Rees. 1976. Nuclear DNA variation in *Lathyrus*. Chromosoma 54: 141–154. **[11]**

Naukkarinen, J. and 7 others. 2005. USF1 and dyslipidemias: Converging evidence for a functional intronic variant. Hum. Mol. Genet. 14: 2595–2605. **[4]**

Naya, H., H. Romero, A. Zavala, B. Alvarez, and H. Musto. 2002. Aerobiosis increases the genomic guanine plus cytosine content (GC%) in prokaryotes. J. Mol. Evol. 55: 260–264. **[10]**

Neander, K. 1991. Functions as selected effects: the conceptual analyst's defense. Philos. Sci. 58: 168–184. **[11]**

Neander, K. 2002. Types of traits: the importance of functional homologues. In A. Ariew, R. Cummins, and M. Perlman (eds.) *Functions: New Essays in the Philosophy of Psychology and Biology*. Oxford University Press, Oxford. pp. 390–415. **[11]**

Needleman, S. B. and C. D. Wunsch. 1970. A general method applicable to the search of similarities in the amino acid sequence of two proteins. J. Mol. Biol. 48: 443–453. **[3]**

Neff, B. D. and M. R. Gross. 2001. Microsatellite evolution in vertebrates: inference from AC dinucleotide repeats. Evolution 55: 1717–1733. **[11]**

Nei, M. 1975. *Molecular Population Genetics and Evolution*. North-Holland, Amsterdam. **[5]**

Nei, M. 1987. *Molecular Evolutionary Genetics*. Columbia University Press, New York. **[2, 5, 6]**

Nei, M. 1996. Phylogenetic analysis in molecular evolutionary genetics. Annu. Rev. Genet. 30: 371–403. **[5, 6]**

Nei, M. 2013. *Mutation-Driven Evolution*. Oxford University Press, Oxford. **[11]**

Nei, M. and A. P. Rooney. 2005. Concerted and birth-and-death evolution of multigene families. Annu. Rev. Genet. 39: 121–152. **[7]**

Nei, M. and F. Tajima. 1981. Genetic drift and estimation of effective population size. Genetics 98: 625–640. **[2]**

Nei, M. and S. Kumar. 2000. *Molecular Evolution and Phylogenetics*. Oxford University Press, New York. **[5,6]**

Nei, M. and T. Gojobori. 1986. Simple methods for estimating the number of synonymous and nonsynonymous nucleotide substitutions. Mol. Biol. Evol. 3: 418–426. **[3]**

Nei, M. and W. H. Li. 1979. Mathematical model for studying genetic variation in terms of restriction endonucleases. Proc. Natl. Acad. Sci. USA 76: 5269–5273. **[2, 5]**

Nei, M. and Y. Imaizumi. 1966. Genetic structure of human populations. II. Differentiation of blood group gene frequencies among isolated populations. Heredity 21: 183–190. **[2]**

Nei, M., X. Gu, and T. Sitnikova. 1997. Evolution by the birth-and-death process in multigene families of the vertebrate immune system. Proc. Natl. Acad. Sci. USA 94:7799–7806. **[7]**

Nei, M., Y. Niimura, and M. Nozawa. 2008. The evolution of animal chemosensory receptor gene repertoires: Roles of chance and necessity. Nat. Rev. Genet. 9: 951–963. **[7]**

Nei, M., Y. Suzuki, and M. Nozawa. 2010. The neutral theory of molecular evolution in the genomic era. Annu. Rev. Genom. Human Genet. 11: 265–289. **[4]**

Neitz, M., J. Neitz, and G. H. Jacobs. 1991. Spectral tuning of pigments underlying red-green color vision. Science 252: 971–974. **[7]**

Nekrutenko, A. and W.-H. Li. 2000. Assessment of compositional heterogeneity within and between eukaryotic genomes. Genome Res. 10: 1986–1995. **[11]**

Nelson-Sathi, S. and 11 others. 2015. Origins of major archaeal clades correspond to gene acquisition from bacteria. Nature 517: 77–80. **[6]**

Nelson, D. R. 2004. "Frankenstein genes", or the *Mad Magazine* version of the human pseudogenome. Hum. Genomics 1: 310–316. **[1, 10, 11]**

Nethe, M., B. Berkhout, and A. C. van der Kuyl. 2005. Retroviral superinfection resistance. Retrovirology 2: 52. **[9]**

Neumann, F. R. and P. Nurse. 2007. Nuclear size control in fission yeast. J. Cell Biol. 179: 593–600. **[11]**

Newcomb, R. D., P. M. Campbell, D. L. Ollis, E. Cheah, R. J. Russell, and J. G. Oakeshott. 1997. A single amino acid substitution converts a carboxylesterase to an organophosphorus hydrolase and confers insecticide resistance on a blowfly. Proc. Natl. Acad. Sci. USA 94: 7464–7468. **[7]**

Newton, L. E. 2007. What makes us human? Biosci. Rep. 27: 185–187. **[11]**

Ng, P. C. and S. Henikoff. 2001. Predicting deleterious amino acid substitutions. Genome Res. 11: 863–874. **[4]**

Ni, X., Y. E. Zhang, N. Nègre, S. Chen, M. Long, and K. P. White. 2012. Adaptive evolution and the birth of CTCF binding sites in the *Drosophila* genome. PLoS Biol. 10: e1001420. **[12]**

Nichols, R. 2001. Gene trees and species trees are not the same. Trends Ecol. Evol. 16: 358–364. **[4]**

Nielsen, R. 2005. Molecular signatures of natural selection. Annu. Rev. Genet. 39: 197–218. **[2, 11]**

Nielsen, R. and Z. Yang. 1998. Likelihood models for detecting positively selected amino acid sites and applications to the HIV-1 envelope gene. Genetics 148: 929–936. **[5]**

Niimura, Y. 2009. On the origin and evolution of vertebrate olfactory receptor genes: Comparative genome analysis among 23 chordate species. Genome Biol. Evol. 1: 34–44. **[7]**

Niimura, Y. 2012. Olfactory receptor multigene family in vertebrates: From the viewpoint of evolutionary genomics. Curr. Genomics 13: 103–114. **[7]**

Niimura, Y., A. Matsui, and K. Touhara. 2014. Extreme expansion of the olfactory receptor gene repertoire in African elephants and evolutionary dynamics of orthologous gene groups in 13 placental mammals. Genome Res. 24: 1485–1496. **[7]**

Nikaido, M. and 10 others. 2001. Retroposon analysis of major cetacean lineages: The monophyly of toothed whales and the paraphyly of river dolphins. Proc. Natl. Acad. Sci. USA 98: 7384–7389. **[5]**

Nikaido, M., A. P. Rooney, and N. Okada. 1999. Phylogenetic relationship among cetartiodactyls based on insertions of short and long interspersed elements: Hippopotamuses are the closest extant relatives of whales. Proc. Natl. Acad. Sci. USA 96: 10261–10266. **[5]**

Nikaido, M., O. Piskurek, and N. Okada. 2007. Toothed whale monophyly reassessed by SINE insertion analysis: The absence of lineage sorting effects suggests a small population of a common ancestral species. Mol. Phylogenet. Evol. 43: 216–224. **[5]**

Nikoh, N. and A. Nakabachi. 2009. Aphids acquired symbiotic genes via lateral gene transfer. BMC Biol. 7: 12. **[9]**

Nilsson, A. I., S. Koskiniemi, S. Eriksson, E. Kugelberg, J. C. Hinton, and D. I. Andersson. 2005. Bacterial genome size reduction by experimental evolution. Proc. Natl. Acad. Sci. USA 102: 12112–12116. **[10]**

Nishihara, H., F. A. Arian, A. F. A. Smit, and N. Okada. 2006. Functional noncoding sequences derived from SINEs in the mammalian genome. Genome Res. 16: 864–874. **[9]**

Nishihara, H., Y. Terai, and N. Okada. 2002. Characterization of novel *Alu*- and tRNA-related SINEs from the tree shrew and evolutionary implications of their origins. Mol. Biol. Evol. 19: 1964–1972. **[9]**

Nishikimi, M., T. Kawai, and K. Yagi. 1992. Guinea pigs possess a highly mutated gene for L-gulono-γ-lactone oxidase, the key enzyme for L-ascorbic acid biosynthesis missing in this species. J. Biol. Chem. 267: 21967–21972. **[8]**

Niu, D. K. and L. Jiang. 2013. Can ENCODE tell us how much junk DNA we carry in our genome? Biochem. Biophys. Res. Commun. 430: 1340–1343. **[11]**

Nóbrega, M. A., I. Ovcharenko, V. Afzal, and E. M. Rubin. 2003. Scanning human gene deserts for long-range enhancers. Science 302: 413. **[11]**

Nóbrega, M. A., Y. Zhu, I. Plajzer-Frick, V. Afzal, and E. M. Rubin. 2004. Megabase deletions of gene deserts result in viable mice. Nature 431: 988–993. **[4, 11]**

Noleto, R. B., F. de Souza Fonseca Guimarães, K. S. Paludo, M. R. Vicari, R. F. Artoni, and M. M. Cestari. 2009. Genome size evaluation in Tetraodontiform fishes from the Neotropical region. Mar. Biotechnol. 11: 680–685. **[11]**

Noma, K.-I., H. P. Cam, R. J. Maraia, and S. I. S. Grewal. 2006. A role for TFIIIC transcription factor complex in genome organization. Cell 125: 859–872. **[12]**

Notley-McRobb, L. and T. Ferenci. 1999. Adaptive *mgl*-regulatory mutations and genetic diversity evolving in glucose-limited *Escherichia coli* populations. Environ. Microbiol. 1: 33–43. **[13]**

Notredame, C., D. G. Higgins, and J. Heringa. 2000. T-Coffee: A novel method for fast and accurate multiple sequence alignment. J. Mol. Biol. 302: 205–217. **[3]**

Notredame, C., L. Holm, and D. G. Higgins. 1998. COFFEE: An objective function for multiple sequence alignments. Bioinformatics 14: 407–422. **[3]**

Notsu, Y. and 7 others. 2002. The complete sequence of the rice (*Oryza sativa* L.) mitochondrial genome: Frequent DNA sequence acquisition and loss during the evolution of flowering plants. Mol. Genet. Genomics 268: 434–445. **[9]**

Noutsos, C., T. Kleine, U. Armbruster, G. DalCorso, and D. Leister. 2007. Nuclear insertions of organellar DNA can create novel patches of functional exon sequences. Trends Genet. 23: 597–601. **[9]**

Novacek, M. J. 1993. Whales leave the beach. Nature 368: 807. **[5]**

Novacek, M. J. 1994. Mammalian phylogeny: Morphology and molecules. Trends Ecol. Evol. 8: 339–340. **[5]**

Nowak, E. C. M., M. Melkonian, and G. Glöckner. 2008. Chromatophore genome sequence of *Paulinella* sheds light on acquisition of photosynthesis in eukaryotes. Curr. Biol. 18: 410–418. **[5]**

Nozawa, M., Y. Suzuki, and M. Nei. 2009. Reliabilities of identifying positive selection by the branch-site and the site-prediction methods. Proc. Natl. Acad. Sci. USA 106: 6700–6705. **[4]**

Nugent, J. M. and J. D. Palmer. 1991. RNA-mediated transfer of the gene *coxII* from the mitochondrion to the nucleus during flowering plant evolution. Cell 66: 473–481. **[9]**

Nuttall, G. H. F. 1902. Progress report upon the biological test for blood as applied to over 500 bloods from various sources, together with a preliminary note upon a method for measuring the degree of reaction. Brit. Med. J. 1: 825–827. **[5]**

Nuttall, G. H. F. 1904. *Blood Immunity and Blood Relationship*. Cambridge University Press, Cambridge. **[5]**

Nuzhdin, S. V. 1999. Sure facts, speculations, and open questions about the evolution of transposable element copy number. Genetica 107: 129–137. **[9]**

Ny, T., F. Eligh, and B. Lund. 1984. The structure of the human tissue-type plasminogen activator gene: Correlation of intron and exon structures to functional and structural domains. Proc. Natl. Acad. Sci. USA 81: 5355–5359. **[8]**

Nye, T. M. W., P. Liò, and W. R. Gilks. 2006. A novel algorithm and web-based tool for comparing two alternative phylogenetic trees. Bioinformatics 22: 117–119. **[5]**

O

O'Brien, S. J. and E. Mayr. 1991. Bureaucratic mischief: Recognizing endangered species and subspecies. Science 251: 1187–1188. **[5]**

O'Callaghan, D. and 8 others. 1999. A homologue of the *Agrobacterium tumefaciens* VirB and *Bordetella pertussis* Ptl type IV secretion systems is essential for intracellular survival of *Brucella suis*. Mol. Microbiol. 33: 1210–1220. **[3]**

O'Hely, M. 2006. A diffusion approach to approximating preservation probabilities for gene duplicates. J. Math. Biol. 53: 215–230. **[7]**

O'Sullivan, O., K. Suhre, C. Abergel, D. G. Higgins, and C. Notredame. 2004. 3DCoffee: Combining protein sequences and structures within multiple sequence alignments. J. Mol. Biol. 340: 385–395. **[3]**

Oakenfull, E. A., H. N. Lim, and O. A. Ryder. 2000 A survey of equid mitochondrial DNA: Implications for the evolution, genetic diversity and conservation of *Equus*. Conserv. Genet. 1: 341–355. **[5]**

Ochman, H. and A. C. Wilson. 1987. Evolution in bacteria: Evidence for a universal substitution rate in cellular genomes. J. Mol. Evol. 26: 74–86. **[10]**

Ochman, H., S. Elwyn, and N. A. Moran. 1999. Calibrating bacterial evolution. Proc. Natl. Acad. Sci. USA 96: 12638–12643. **[13]**

Ogawa, T. and T. Okazaki. 1980. Discontinuous DNA replication. Annu. Rev. Biochem. 49: 421–457. **[4]**

Ogden, T. H. and M. S. Rosenberg. 2006. Multiple sequence alignment accuracy and phylogenetic inference. Syst. Biol. 55: 314–328. **[3]**

Ohnishi, M. and 9 others. 1999. Alternative promoters direct tissue-specific expression of the mouse protein phosphatase $2C\beta$ gene. Eur. J. Biochem. 263: 736–745. **[12]**

Ohno, S. 1970. *Evolution by Gene Duplication*. Springer, Berlin. **[7]**

Ohno, S. 1972. So much "junk" DNA in our genome. Brookhaven Symp. Biol. 23: 366–370. Gordon and Breach, New York. **[7, 11]**

Ohno, S. 1973. Ancient linkage groups and frozen accidents. Nature 244: 259–262. **[7]**

Ohno, S. 1973. Evolutional reason for having so much junk DNA. In: R. A. Pfeiffer (ed.). *Modern Aspects of Cytogenetics: Constitutive Heterochromatin in Man*. F. K. Schattauer Verlag, Stuttgart, Germany. pp. 169–180. **[11]**

Ohno, S., U. Wolf, and N. B. Atkin. 1968. Evolution from fish to mammals by gene duplication. Hereditas 59: 169–187. **[7]**

Ohshima, K. and N. Okada. 2005. SINEs and LINEs: Symbionts of eukaryotic genomes with a common tail. Cytogenet. Genome Res. 110: 475–490. **[9]**

Ohshima, K., M. Hamada, Y. Terai, and N. Okada. 1996. The 3' ends of tRNA-derived short interspersed repetitive elements are derived from the 3' ends of long interspersed repetitive elements. Mol. Cell. Biol. 16: 3756–3764. **[9]**

Ohshima, K., M. Hattori, T. Yada, T. Gojobori, Y. Sakaki, and N. Okada. 2003. Whole-genome screening indicates a possible burst of formation of processed pseudogenes and *Alu* repeats by particular *L1* subfamilies in ancestral primates. Genome Biol 4: R74. **[11]**

Ohta, T. 1973. Slightly deleterious mutant substitutions in evolution. Nature 246: 96–98. **[2, 11]**

Ohta, T. 1980. Evolution and Variation of Multigene Families. Springer, Berlin. **[7]**

Ohta, T. 1983. On the evolution of multigene families. Theor. Pop. Biol. 23: 216–240. **[7]**

Ohta, T. 1984. Some models of gene conversion for treating the evolution of multigene families. Genetics 106: 517–528. **[7]**

Ohta, T. 1990. How gene families evolve. Theor. Pop. Biol. 37: 213–219. **[7]**

Ohta, T. 1992. The nearly neutral theory of molecular evolution. Annu. Rev. Ecol. Syst. 23: 263–286. **[2]**

Ohta, T. and G. A. Dover. 1983. Population genetics of multigene families that are dispersed into two or more chromosomes. Proc. Natl. Acad. Sci. USA 80: 4079–4083. **[7]**

Ohta, T. and H. Tachida. 1990. Theoretical study of near neutrality. I. Heterozygosity and rate of mutant substitution. Genetics 126: 219–229. **[2]**

Ohta, T. and M. Kimura. 1975. The effect of selected linked locus on heterozygosity of neutral alleles (the hitch-hiking effect). Genet. Res. 25: 313–326. **[2]**

Okada, N., M. Hamada, I. Ogiwara, and K. Ohshima. 1997. SINEs and LINEs share common 3' sequences: A review. Gene 205: 229–243. **[9]**

Okada, N., T. Sasaki, T. Shimogori, and H. Nishihara. 2010. Emergence of mammals by emergency: Exaptation. Genes Cells 15: 801–812. **[9]**

Okamura, K., L. Feuk, T. Marquès Bonet, A. Navarro, and S. W. Scherer. 2006. Frequent appearance of novel protein-coding sequences by frameshift translation. Genomics 88: 690–697. **[7]**

Okazaki, R., T. Okazaki, and K. Sakabe. 1967. Mechanism of DNA replication: Possible discontinuity of DNA chain growth. Japan. J. Med. Sci. Biol. 20: 255–260. **[4]**

Okazaki, R., T. Okazaki, K. Sakabe, K. Sugimoto, and A. Sugino. 1968. Mechanism of DNA chain growth. I. Possible discontinuity and unusual secondary structure of newly synthesized chains. Proc. Natl. Acad. Sci. USA 59: 598–605. **[4]**

Oler, A. J., S. Traina-Dorge, R. S. Derbes, D. Canella, B. R. Cairns, and A. M. Roy-Engel. 2012. *Alu* expression in human cell lines and their retrotranspositional potential. Mobile DNA 3: 11. **[11]**

Oliver, J. L., P. Carpena, R. Román-Roldán, T. Mata-Balaguer, A. Mejiás-Romero, M. Hackenberg, and P. Bernaola-Galván. 2002. Isochore chromosome maps of the human genome. Gene 300: 117–127. **[11]**

Oliviero, S., M. DeMarchi, A. O. Carbonara, L. F. Bernini, G. Bensi, and G. Raugei. 1985. Molecular evidence of triplication in the haptoglobin Johnson variant gene. Hum. Genet. 71: 49–52. **[8]**

Olsen, G. J. and 9 others. 1992. The ribosomal database project. Nucl. Acids Res. 20: s2199–s2200. **[5]**

Olsen, G. L. and C. R. Woese. 1996. Lessons from an Archaeal genome: What are we learning from *Methanococcus jannaschii*? Trends Genet. 12: 377–379. **[5]**

Olson, M. V. 1999. When less is more: Gene loss as an engine of evolutionary change. Am. J. Hum. Genet. 64: 18–23. **[8]**

Omont, N. and F. Kepes. 2004. Transcription/replication collisions cause bacterial transcription units to be longer on the leading strand of replication. Bioinformatics 20: 2719–2725. **[10]**

Opazo, J. C., F. G. Hoffmann, and J. F. Storz. 2008a. Differential loss of embryonic globin genes during the radiation of placental mammals. Proc. Natl. Acad. Sci. USA 105: 12950–12955. **[7]**

Opazo, J. C., F. G. Hoffmann, and J. F. Storz. 2008b. Genomic evidence for independent origins of β-like globin genes in monotremes and therian mammals. Proc. Natl. Acad. Sci. USA 105: 1590–1595. **[7]**

Organ, C. L., A. M. Shedlock, A. Meade, M. Pagel, and S. V. Edwards. 2007. Origin of avian genome size and structure in non-avian dinosaurs. Nature 446: 180–184. **[11]**

Orgel, L. E. and F. H. C. Crick. 1980. Selfish DNA: the ultimate parasite. Nature 284: 604–607. **[11]**

Orkin, S. H. 1995. Regulation of globin gene expression in erythroid cells. Eur. J. Biochem. 231: 271–281. **[12]**

Osato, N., Y. Suzuki, K. Ikeo, and T. Gojobori. 2007. Transcriptional interferences in *cis* natural antisense transcripts of humans and mice. Genetics 176: 1299–1306. **[8]**

Osawa, S. 1995. *Evolution of the Genetic Code*. Oxford University Press, Oxford. **[10]**

Osawa, S. and T. H. Jukes. 1989. Codon reassignment (codon capture) in evolution. J. Mol. Evol. 28: 271–278. **[10]**

Osawa, S. and T. H. Jukes. 1995. On codon reassignment. J. Mol. Evol. 41: 247–249. **[10]**

Osborn, A. E. and B. Field. 2009. Operons. Cell. Mol. Life Sci. 66: 3755–3775. **[10]**

Ohshima, K., M. Hattori, T. Yada, T. Gojobori, Y. Sakaki, and N. Okada. 2003. Whole-genome screening indicates a possible burst of formation of processed pseudogenes and *Alu* repeats by particular *L1* subfamilies in ancestral primates. Genome Biol. 4: R74. **[11]**

Ossowski, S. and 7 others. 2010. The rate and molecular spectrum of spontaneous mutations in *Arabidopsis thaliana*. Science 327: 92–94. **[4]**

Otto and Young 2002 Duplicated gene specialization **[7]**

Otto, S. P. 2007. The evolutionary consequences of polyploidy. Cell 131: 452–462. **[7, 11]**

Otto, S. P. and J. Whitton. 2000. Polyploid incidence and evolution. Annu. Rev. Genet. 34: 401–437. **[7]**

Ou, C.-Y. and 30 others. 1992. Molecular epidemiology of HIV transmission in a dental practice. Science 256: 1165–1171. **[5]**

Ouzounis, C. A., V. Kunin, N. Darzentas, and L. Goldovsky. 2006. A minimal estimate for the gene content of the last universal common ancestor—exobiology from a terrestrial perspective. Res. Microbiol. 157: 57–68. **[10]**

Ovcharenko, I., G. G. Loots, M. A. Nobrega, R. C. Hardison, W. Miller, and L. Stubbs. 2005. Evolution and functional classification of vertebrate gene deserts. Genome Res. 15: 137–145. **[11]**

Overballe-Petersen, S. and 17 others. 2013. Bacterial natural transformation by highly fragmented and damaged DNA. Proc. Natl. Acad. Sci. USA 110: 19860–19865. **[9]**

Owen, R. 1843. Lectures on the Comparative Anatomy and Physiology of the Invertebrate Animals. Longman, Brown, Green and Longmans, London. **[5]**

P

Pääbo, S. 1989. Ancient DNA: Extraction, characterization, molecular cloning, and enzymatic amplification. Proc. Natl. Acad. Sci. USA 86: 1939–1943. **[5]**

Pace, J. K., C. Gilbert, M. S. Clark, and C. Feschotte. 2008. Repeated horizontal transfer of a DNA transposon in mammals and other tetrapods. Proc. Natl. Acad. Sci. USA 105: 17023–17028. **[9]**

Pace, N. R. 2006. Time for a change. Nature 441: 289. **[5, 6]**

Pace, N. R. 2009. It's time to retire the prokaryote. Microbiol. Today 36: 84–87. **[5]**

Pace, N. R. 2009. Problems with "prokaryote." J. Bacteriol. 191:2008–2010. **[6]**

Pagel, M. 1997. Inferring evolutionary processes from phylogenies. Zool. Scripta 26: 331–348. **[5, 6]**

Pagel, M. and R. A. Johnstone. 1992. Variation across species in the size of the nuclear genome supports the junk-DNA explanation for the C-value paradox. Proc. Roy Soc. 249B: 119–124. **[11]**

Pagel, M., C. Venditti, and A. Meade. 2006. Large punctuational contribution of speciation to evolutionary divergence at the molecular level. Science 314: 119–121. **[4]**

Palazzo, A. F. and T. R. Gregory. 2014. The case for junk DNA. PLoS Genet. 10: e1004351. **[11]**

Pallen, M. J. and N. J. Matzke. 2006. From *The Origin of Species* to the origin of bacterial flagella. Nat. Rev. Microbiol. 4: 784–790. **[8]**

Palmer, J. D. 1997. The mitochondrion that time forgot. Nature 387: 454–455. **[10]**

Palopoli, M. F. and 7 others. 2008. Molecular basis of the copulatory plug polymorphism in *C. elegans*. Nature 454: 1019–1022. **[9]**

Pamilo, P. and M. Nei. 1988. Relationship between gene trees and species trees. Mol. Biol. Evol. 5: 568–583. **[5]**

Pamilo, P. and N. O. Bianchi. 1993. Evolution of the *Zfx* and *Zfy* genes: Rates and interdependence between the genes. Mol. Biol. Evol. 10: 271–281. **[3]**

Pandit, M. K., S. M. White, and M. J. O. Pocock. 2014. The contrasting effects of genome size, chromosome number and ploidy level on plant invasiveness: A global analysis. New Phytol. 203: 697–703. **[11]**

Parenteau, J. and 12 others. 2008. Deletion of many yeast introns reveals a minority of genes that require splicing for function. Mol. Biol. Cell 19: 1932–1941. **[11]**

Parham, J. F. and R. B. Irmis. 2008. Caveats on the use of fossil calibration for molecular dating: A comment on Near et al. Am. Nat. 171: 132–136. **[5]**

Parisod, C., R. Holderegger, and C. Brochmann. 2010. Evolutionary consequences of autopolyploidy. New Phytol. 186: 5–17. **[7]**

Parker, H. G.and 16 others. 2009. An expressed *Fgf4* retrogene is associated with breed-defining chondrodysplasia in domestic dogs. Science 325: 995–998. **[9]**

Parmley, J. L. and L. D. Hurst. 2007. How do synonymous mutations affect fitness? BioEssays 29: 515–519. **[4]**

Parsch, J. 2011. The cost of being male. Science 332: 798–799. **[2]**

Pasek, S., J. L. Risler, and P. Brézellec. 2006. Gene fusion/fission is a major contributor to evolution of multi-domain bacterial proteins. Bioinformatics 22: 1418–1423. **[8]**

Passarge, E., B. Horsthemke, and R. A. Farber. 1999. Incorrect use of the term synteny. Nat. Genet. 23: 387. **[11]**

Pasyukova, E. G., S. V. Nuzhdin, T. V. Morozova, and T. F. Mackay. 2004. Accumulation of transposable elements in the genome of *Drosophila melanogaster* is associated with a decrease in fitness. J. Hered. 95: 284–290. **[11]**

Patarnello, T., C. Verde, G. di Prisco, L. Bargelloni, and L. Zane. 2011. How will fish that evolved at constant sub-zero temperatures cope with global warming? Notothenioids as a case study. BioEssays 33: 260–268. **[7]**

Patarnello, T., C. Verde, G. di Prisco, L. Bargelloni, and L. Zane. 2011. How will fish that evolved at constant sub-zero temperatures cope with global warming? Notothenioids as a case study. Bioessays 33: 260–268. **[8]**

Patel, V. S. and 7 others. 2008. Platypus globin genes and flanking loci suggest a new insertional model for beta-globin evolution in birds and mammals. BMC Biol. 6: 34. **[7]**

Patel, V. S. and J. E. Deakin. 2010. The evolutionary history of globin genes: Insights from marsupials and monotremes, pp. 415–433. In: J. E. Deakin, P. D. Waters, and J. A. M. Graves (eds.), *Marsupial Genetics and Genomics*. Springer, Dordrecht. **[7]**

Patron, N. J., R. F. Waller, and P. J. Keeling. 2006. A tertiary plastid uses genes from two endosymbionts. J. Mol. Biol. 357: 1373–1382. **[5]**

Patterson, N., D. J. Richter, S. Gnerre, E. S. Lander, and D. Reich. 2006. Genetic evidence for complex speciation of humans and chimpanzees. Nature 441: 1103–1108. **[5]**

Patthy, L. 1985. Evolution of the proteases of blood coagulation and fibrinolysis by assembly from modules. Cell 41: 657–663. **[8]**

Paukstelis, P. J. and A. M. Lambowitz. 2008. Identification and evolution of fungal mitochondrial tyrosyl-tRNA synthetases with group I intron splicing activity. Proc. Natl. Acad. Sci. USA 105: 6010–6015. **[8]**

Paul, S., S. Million-Weaver, S. Chattopadhyay, E. Sokurenko, and H. Merrikh. 2013. Accelerated gene evolution through replication–transcription conflicts. Nature 495: 512–515. **[10]**

Pearson, H. 2006. What is a gene? Nature 441: 399–401. **[1]**

Pei, B. and 13 others. 2012. The GENCODE pseudogene resource. Genome Biol. 13: R51. **[11]**

Peierls, R. 1960. Wolfgang Ernst Pauli, 1900–1958. Biogr. Mem. Fellows R. Soc. 5: 174–192. **[11]**

Pellicer, J., M. F. Fay, and I. J. Leitch. 2010. The largest eukaryotic genome of them all? Bot. J. Linn. Soc. 164: 10–15. **[10, 11]**

Penn, O., E. Privman, H. Ashkenazy, G. Landan, D. Graur, and T. Pupko. 2010. GUIDANCE: A web server for assessing alignment confidence scores. Nucleic Acids Res. 38: W23–W28. **[3]**

Penny, D. 2005. Evolutionary biology: Relativity for molecular clocks. Nature 436: 183–184. **[5]**

Penny, D. and M. D. Hendy. 1985. Testing methods of evolutionary tree construction. Cladistics 1: 266–272. **[5]**

Perfeito, L., L. Fernandes, C. Mota, and I. Gordo. 2007. Adaptive mutations in bacteria: High rate and small effects. Science 317: 813–815. **[2, 13]**

Perrat, P. N., S. DasGupta, J. Wang, W. Theurkauf, Z. Weng, M. Rosbash, and S. Waddell. 2013. Transposition-driven genomic heterogeneity in the *Drosophila* brain. Science 340: 91–95. **[9]**

Perry, G. H. and 12 other. 2007. Diet and the evolution of human amylase gene copy number variation. Nature Genet. 39: 1256–1260. **[7]**

Perry, G. H., B. C. Verrelli, and A. C. Stone. 2005. Comparative analyses reveal a complex history of molecular evolution for human *MYH16*. Mol. Biol. Evol. 22: 379–382. **[8]**

Perry, M. W., A. N. Boettiger, J. P. Bothma, and M. Levine. 2010. Shadow enhancers foster robustness of *Drosophila* gastrulation. Curr. Biol. 20: 1562–1567. **[12]**

Perry, M. W., J. D. Cande, A. N. Boettiger, and M. Levine. 2009. Evolution of insect dorsoventral patterning mechanisms. Cold Spring Harb. Symp. Quant. Biol. 74: 275–279. **[12]**

Pertea, M. and S. L. Salzberg. 2010. Between a chicken and a grape: estimating the number of human genes. Genome Biol. 11: 206. **[11]**

Pesce, A. and 7 others. 2002. Neuroglobin and cytoglobin. Fresh blood for the vertebrate globin family. EMBO Rep. 3: 1146–1151. **[7]**

Petersen, J., H. Brinkmann, B. Bunk, V. Michael, O. Päuker, and S. Pradella. 2012. Think pink: Photosynthesis, plasmids and the *Roseobacter* clade. Environ Microbiol. 14: 2661–2672. **[6]**

Peterson, S. N. and C. M. Fraser. 2001. The complexity of simplicity. Genome Biol. 2: doi: 10.1186/gb-2001-2-2-comment2002 **[10]**

Petit, N. and A. Barbadilla. 2009. Selection efficiency and effective population size in *Drosophila* species. J. Evol. Biol. 22: 515–526. **[4]**

Petit R. J. and A. Hampe. 2006. Some evolutionary consequences of being a tree. Annu. Rev. Ecol. Evol. Syst. 37: 187–214. **[11]**

Petitpierre, E., C. Juan, J. Pons, M. Plohl and D. Ugarkovic. 1995. Satellite DNA and constitutive heterochromatin in tenebrionid beetles. In: P. E. Brandham and M. D. Bennett (eds.), *Kew Chromosome Conference IV*, pp. 351–362. Royal Botanic Gardens, Kew. **[11]**

Petrov, D. A. and D. L. Hartl. 1997. Trash DNA is what gets thrown away: High rate of DNA loss in *Drosophila*. Gene 205: 279–289. **[11]**

Petrov, D. A. and D. L. Hartl. 1998. High rate of DNA loss in the *Drosophila melanogaster* and *Drosophila virilis* species group. Mol. Biol. Evol. 15: 293–302. **[11]**

Petrov, D. A., E. R. Lozovskaya, and D. L. Hartl. 1996. High intrinsic rate of DNA loss in *Drosophila*. Nature 384: 346–349. **[11]**

Petrovski, S. and V. A. Stanisich. 2010. *Tn502* and *Tn512* are *res* site hunters that provide evidence of resolvase-independent transposition to random sites. J. Bacteriol. 192: 1865–1874. **[9]**

Petsko, G. A. 2003. Funky, not junky. Genome Biol. 4: 104. **[11]**

Petsko, G. A. and D. Ringe. 2004. *Protein Structure and Function*. Sinauer Associates, Sunderland, MA. **[8]**

Phan, Q. T. and 8 others. 2007. Als3 is a *Candida albicans* invasin that binds to cadherins and induces endocytosis by host cells. PLoS Biol. 5: e64. **[8]**

Philippe, H. and 19 others. 2009. Phylogenomics revives traditional views on deep animal relationships. Curr. Biol. 19: 706–712. **[5]**

Philippe, H. and P. Forterre. 1999. The rooting of the universal tree of life is not reliable. J. Mol. Evol. 49: 509–523. **[5, 6]**

Philippe, H., E. A. Snell, E. Bapteste, P. Lopez, P. W. Holland, and D. Casane. 2004. Phylogenomics of eukaryotes: Impact of missing data on large alignments. Mol. Biol. Evol. 21: 1740–1752. **[5]**

Philippe, H., N. Lartillot, and H. Brinkmann. 2005. Multigene analyses of bilaterian animals corroborate the monophyly of Ecdysozoa, Lophotrochozoa, and Protostomia. Mol. Biol. Evol. 22: 1246–1253. **[5]**

Phillips, D. C., M. J. E. Sternberg, and B. J. Sutton. 1983. Intimations of evolution from the three-dimensional structures of proteins, pp. 145–173. In: D. S. Bendall (ed.), *Evolution from Molecules to Men*. Cambridge University Press, Cambridge. **[7]**

Piaggio-Talice, R., J. G. Burleigh, and O. Eulenstein. 2004. Quartet supertrees, pp. 173–191. In: O. R. P. Bininda-Emonds (ed.), *Phylogenetic Supertrees: Combining Information to Reveal the Tree of Life*. Kluwer, Dordrecht, Netherlands. **[5]**

Piatigorsky, J. 1998a. Gene sharing in lens and cornea: Facts and implications. Prog. Retin. Eye Res. 17: 145–174. **[7]**

Piatigorsky, J. 1998b. Multifunctional lens crystallins and cornea enzymes: More than meets the eye. Ann. N. Y. Acad. Sci. 842: 7–15. **[7]**

Piatigorsky, J. 2003. Crystallin genes: Specialization by changes in gene regulation may precede gene duplication. J. Struct. Func. Genomics 3: 131–137. **[7]**

Piatigorsky, J. and G. J. Wistow. 1989. Enzyme/crystallins: Gene sharing as an evolutionary strategy. Cell 57: 197–199. **[7]**

Piatigorsky, J. and 7 others. 1988. Gene sharing by δ-crystallin and argininosuccinate lyase. Proc. Natl. Acad. Sci. USA 85: 3479–3483. **[7]**

Pick, L., A. Schier, M. Affolter, T. Schmidt-Glenewinkel, and W. J. Gehring. 1990). Analysis of the *ftz* upstream element: germ layer-specific enhancers are independently autoregulated. Genes Dev. 4: 1224–1239. **[12]**

Pickeral, O. K., W. Makałowski, M. S. Boguski, and J. D. Boeke. 2000. Frequent human genomic DNA transduction (sic) driven by LINE-1 retrotransposition. Genome Res. 10: 411–415. **[9]**

Picot, S., G. L. Wallau, E. L. Loreto, F. O. Heredia, A. Hua-Van, and P. Capy. 2008. The *mariner* transposable element in natural populations of *Drosophila simulans*. Heredity 101: 53–59. **[11]**

Piegu, B. and 10 others. 2006. Doubling genome size without polyploidization: dynamics of retrotransposition-driven genomic expansions in *Oryza australiensis*, a wild relative of rice. Genome Res. 16: 1262–1269. **[11]**

Piontkivska, H., A. P. Rooney, and M. Nei. 2002. Purifying selection and birth-and-death evolution in the histone H4 gene family. Mol. Biol. Evol. 19: 689–697. **[7]**

Pisani, D., J. A. Cotton, and J. O. McInerney. 2007. Supertrees disentangle the chimerical origins of eukaryotic genomes. Mol. Biol. Evol. 24: 1752–1760. **[5, 6]**

Piskurek, O. and N. Okada. 2007. Poxviruses as possible vectors for horizontal transfer of retroposons from reptiles to mammals. Proc. Natl. Acad. Sci. USA 104: 12046–12051. **[9]**

Plague, G. R., H. E. Dunbar, P. L. Tran, and N. A. Moran. 2008. Extensive proliferation of transposable elements in heritable bacterial symbionts. J. Bacteriol. 190: 777–779. **[11]**

Platt, R. N. and D. A. Ray. 2012. A non-LTR retroelement extinction in *Spermophilus tridecemlineatus*. Gene 500: 47–53. **[9]**

Plohl, M., A. Luchetti, N. Meštrovic, and B. Mantovani. 2008. Satellite DNAs between selfishness and functionality: structure, genomics and evolution of tandem repeats in centromeric (hetero)chromatin. Gene 409: 72–82. **[11]**

Plotkin, J. B. and G. Kudla. 2011. Synonymous but not the same: The causes and consequences of codon bias. Nat. Rev. Genet. 12: 32–42. **[4]**

Pollard, K. S. 2009. What makes us human? Sci. Am. 300: 44–49. **[11]**

Polz, M. F., E. J. Alm, and W. P. Hanage. 2013. Horizontal gene transfer and the evolution of bacterial and archaeal population structure. Trends Genet. 29: 170–175. **[9]**

Ponjavic, J., C. P. Ponting, and G. Lunter. 2007. Functionality or transcriptional noise? Evidence for selection within long noncoding RNAs. Genome Res. 17: 556–565. **[11]**

Ponting, C. P. and R. C. Hardison. 2011. What fraction of the human genome is functional? Genome Res. 21: 1769–1776. **[11]**

Popa, O. and T. Dagan. 2011. Trends and barriers to lateral gene transfer in prokaryotes. Curr. Opin. Microbiol. 14: 615–623. **[9]**

Popa, O., E. Hazkani-Covo, G. Landan, W. Martin, and T. Dagan. 2011. Directed networks reveal genomic barriers and DNA repair bypasses to lateral gene transfer among prokaryotes. Genome Res. 21: 599–609. **[9]**

Portier, P. 1918. *Les Symbiotes*. Masson, Paris. **[5]**

Portin, P. 1993. The concept of the gene: Short history and present status. Q. Rev. Biol. 68: 173–223. **[1]**

Potato Genome Sequencing Consortium. 2011. Genome sequence and analysis of the tuber crop potato. Nature 475: 189–195. **[7]**

Powers, E. T. and W. E. Balch. 2008. Costly mistakes: Translational infidelity and protein homeostasis. Cell 134: 204–206. **[4]**

Poyatos, J. F. and L. D. Hurst. 2007. The determinants of gene order conservation in yeasts. Genome Biol. 8: R233. **[11]**

Prescott, C. D. and A. E. Dahlberg. 1990. A single base change at 726 in 16S rRNA radically alters the pattern of proteins synthesized in vivo. EMBO J. 9: 289–294. **[6]**

Presgraves, D. C. and S. V. Yi. 2009. Doubts about complex speciation between humans and chimpanzees. Trends Ecol. Evol. 24: 533–540. **[5]**

Price, H. J., A. H. Sparrow, and A. F. Nauman. 1973. Correlations between nuclear volume, cell volume and DNA content in meristematic cells of herbaceous angiosperms. Experientia 29: 1028–1029. **[11]**

Price, M. N., A. P. Arkin, and E. J. Alm. 2006. The life-cycle of operons. PLoS Genet. 2: e96. **[10]**

Price, N. and D. Graur. 2015. Are synonymous sites in primates and rodents functionally constrained? J. Mol. Evol. 82 (in press). **[4]**

Price, N., R. A. Cartwright, N. Sabath, D. Graur, and R. B. Azevedo. 2011. Neutral evolution of robustness in *Drosophila* microRNA precursors. Mol. Biol. Evol. 28: 2115–2123. **[2, 7]**

Primmer, C. R., S. Papakostas, E. H. Leder, M. J. Davis, and M. A. Ragan. 2013. Annotated genes and nonannotated genomes: cross-species use of Gene Ontology in ecology and evolution research. Mol. Ecol. 22: 3216–3241. **[11]**

Primmer, C. R., T. Raudsepp, B. P. Chowdhary, A. P. Møller, and H. Ellegren. 1997. Low frequency of microsatellites in the avian genome. Genome Res. 7: 471–482. **[11]**

Primrose, S. B. 1998. *Principles of Genome Analysis*. Blackwell Science, Oxford. **[10]**

Prince, V. E. and F. B. Pickett. 2002. Splitting pairs: The diverging fates of duplicated genes. Nat. Rev. Genet. 3: 827–837. **[7]**

Privman, E., O. Penn, and T. Pupko. 2012. Improving the performance of positive selection inference by filtering unreliable alignment regions. Mol. Biol. Evol. 29: 1–5. **[4]**

Prokopowich, C. D., T. R. Gregory, and T. J. Crease. 2003. The correlation between rDNA copy number and genome size in eukaryotes. Genome 46: 48–50. **[7]**

Proudfoot, N. J. 2011. Ending the message: Poly(A) signals then and now. Genes Dev. 25: 1770–1782. **[1]**

Proulx, S. R. and P. C. Phillips. 2005. The opportunity for canalization and the evolution of genetic networks. Am. Nat. 165: 147–162. **[2]**

Ptashne, M. 2007. On the use of the word "epigenetic." Curr. Biol. 17: R233–R236. **[12]**

Ptashne, M. 2013. Epigenetics: core misconcept. Proc. Natl. Acad. Sci. USA 110: 7101–7103. **[12]**

Puigbò, P., Y. I. Wolf, and E. V. Koonin. 2010. The tree and net components of prokaryote evolution. Genome Biol. Evol. 2: 745–756. **[6]**

Purves, W. K., D. Sadava, G. H. Orians, and H. C. Heller. 2004. *Life: The Science of Biology* (7th ed.). Sinauer. **[11]**

Q

Quiñones, A. and R. Piechocki. 1985. Isolation and characterization of *Escherichia coli* antimutators: A new strategy to study the nature and origin of spontaneous mutations. Mol. Gen. Genet. 201: 315–322. **[2]**

R

Rabinowicz, P. D. and 6 others. 2003. Genes and transposons are differentially methylated in plants, but not in mammals. Genome Res. 13: 2658–2664. **[12]**

Raffel, D. and H. J. Muller. 1940. Position effect and gene divisibility considered in connection with three strikingly similar *scute* mutations. Genetics 25: 541–583. **[7]**

Ragan, M. A. 2009. Trees and networks before and after Darwin. Biol. Dir. 4: 43. **[5]**

Raj, A. and A. van Oudenaarden. 2008. Stochastic gene expression and its consequences. Cell 135: 216–226. **[11]**

Rambaut, A. and L. Bromham. 1998. Estimating divergence dates from molecular sequences. Mol. Biol. Evol. 15: 442–448. **[5]**

Rands, C. M., S. Meader, C. P. Ponting, and G. Lunter. 2014. 8.2% of the human genome is constrained: variation in rates of turnover across functional element classes in the human lineage. PLoS Genet. 10: e1004525. **[11]**

Ransick, A., S. Ernst, R. J. Britten, and E. H. Davidson. 1993. Whole mount in situ hybridization shows *Endo 16* to be a marker for the vegetal plate territory in sea urchin embryos. Mech. Dev. 42: 117–124. **[12]**

Raponi, M., M. Upadhyaya, and D. Baralle. 2006. Functional splicing assay shows a pathogenic intronic mutation in neurofibromatosis type 1 (*NF1*) due to intronic sequence exonization. Hum. Mut. 27: 294–295. **[8]**

Rasko, D. A. and 12 others. 2008. The pangenome structure of *Escherichia coli*: Comparative genomic analysis of *E. coli* commensal and pathogenic isolates. J. Bacteriol. 190: 6881–6893. **[9, 10]**

Rasmussen, B., I. R. Fletcher, J. J. Brocks, and M. R. Kilburn. 2008. Reassessing the first appearance of eukaryotes and cyanobacteria. Nature 455: 1101–1104. **[6]**

Rau, R. E. 1974. Revised list of the preserved material of the extinct Cape colony quagga, *Equus quagga quagga* (Gmelin). Ann. S. Afr. Mus. 65: 41–87. **[5]**

Ravosa, M. J. and M. Dagosto (eds.). 2007. *Primate Origins: Adaptations and Evolution*. Springer, New York. **[5, 6]**

Raymann, K., C. Brochier-Armanet, and S. Gribaldo. 2015. The two-domain tree of life is linked to a new root for the Archaea. Proc. Natl. Acad. Sci. USA 112: 6670–6675. **[6]**

Raynes, Y., M. R. Gazzara, and P. D. Sniegowski. 2011. Mutator dynamics in sexual and asexual experimental populations of yeast. BMC Evol. Biol. 11: 158. **[13]**

Razin, S. 1997a. Comparative genomics of mycoplasmas. Wien Klin. Wochenschr. 109: 551–556. **[10]**

Razin, S. 1997b. The minimal cellular genome of mycoplasma. Indian J. Biochem. Biophys. 34: 124–130. **[10]**

Rebollo, R., B. Horard, B. Hubert, and C. Vieira. 2010. Jumping genes and epigenetics: Towards new species. Gene 454: 1–7. **[9]**

Rees, H. and R. N. Jones. 1972. The origin of the wide species variation in nuclear DNA content. Int. Rev. Cytol. 32: 53–92. **[11]**

Reid, D. T. and C. L. Peichel. 2010. Perspectives on the genetic architecture of divergence in body shape in sticklebacks. Integr. Comp. Biol. 50: 1057–1066. **[12]**

Renwick, J. H. 1972. The mapping of human chromosomes. Annu. Rev. Genet. 5: 81–120. **[11]**

Rest, J. S. and D. P. Mindell. 2003. Retroids in Archaea: Phylogeny and lateral origin. Mol. Biol. Evol. 20: 1134–1142. **[9]**

Retana Salazar, A. P. 1996. Parasitological evidence on the phylogeny of hominids and cebids. Rev. Biol. Trop. 44: 391–394. **[5]**

Rhind, N. and 59 others. 2011. Comparative functional genomics of the fission yeasts. Science 332: 930–936. **[11]**

Rhinesmith, H. S., W. A. Schroeder, and N. Martin. 1958. The N-terminal sequence of the β chains of normal adult hemoglobin. J. Am. Chem. Soc. 80: 3358–3361. **[7]**

Rhoades, M. M. 1942. Preferential segregation in maize. Genetics 27: 395–407. **[2, 11]**

Ribeiro, S. and G. B. Golding. 1998. The mosaic nature of the eukaryotic nucleus. Mol. Biol. Evol. 15: 779–788. **[6]**

Ricchetti, M., C. Fairhead, and B. Dujon. 1999. Mitochondrial DNA repairs double strand breaks in yeast chromosomes. Nature 402: 96–100. **[9]**

Ricchetti, M., F. Tekaia, and B. Dujon. 2004. Continued colonization of the human genome by mitochondrial DNA. PLoS Biol. 2: e273. **[9]**

Rice, S. A. and B. C. Lampson. 1995. Bacterial reverse transcriptase and msDNA. Virus Genes 11: 95–104. **[9]**

Richard, G. F., A. Kerrest, and B. Dujon. 2008. Comparative genomics and molecular dynamics of DNA repeats in eukaryotes. Microbiol. Mol. Biol. Rev. 72: 686–727. **[11]**

Richards, R. A. 2007. Species and taxonomy. In: M. Ruse (ed.), *Oxford Handbook of the Philosophy of Biology*. Oxford University Press, Oxford. **[5]**

Richter, M. and R. Rosselló-Móra. 2009. Shifting the genomic gold standard for the prokaryotic species definition. Proc. Natl. Acad. Sci. USA 106: 19126–19131. **[6]**

Riddihough, G. and L. M. Zahn. 2010. What is epigenetics? Science 330: 611. **[12]**

Riggs, A. D. 1975. X inactivation, differentiation, and DNA methylation. Cytogenet. Cell Genet. 14: 9–25. **[12]**

Ring, K. L. and A. R. Cavalcanti. 2008. Consequences of stop codon reassignment on protein evolution in ciliates with alternative genetic codes. Mol. Biol. Evol. 25: 179–186. **[10]**

Rinke, C. and 21 others. 2013. Insights into the phylogeny and coding potential of microbial dark matter. Nature 499: 431–437. **[6]**

Ris, H. and B. L. Chandler. 1963. The ultrastructure of genetic systems in prokaryotes and eukaryotes. Cold Spring Harb. Symp. Quant. Biol. 28: 1–8. **[5, 6]**

Ritossa, F. M., K. C. Atwood, D. L. Lindsley, and S. Spiegelman. 1966. On the chromosomal distribution of DNA complementary to ribosomal and soluble RNA. Natl. Cancer Inst. Monogr. 23: 449–472. **[7]**

Rivera, M. C. and J. A. Lake. 1992. Evidence that eukaryotes and eocyte prokaryotes are immediate relatives. Science 257: 74–76. **[6]**

Rivera, M. C. and J. A. Lake. 2004. The ring of life provides evidence for a genome fusion origin of eukaryotes. Nature 451: 152–155. **[5, 6]**

Roach, J. C. and 14 others. 2010. Analysis of genetic inheritance in a family quartet by whole-genome sequencing. Science 328: 636–639. **[4, 11]**

Robinson-Rechavi, M., B. Boussae, and V. Laudet. 2004. Phylogenetic dating and characterization of gene duplications in vertebrates: The cartilaginous fish reference. Mol. Biol. Evol. 21: 580–586. **[5]**

Robinson, M., C. Gautier, and D. Mouchiroud. 1997. Evolution of isochores in rodents. Mol. Biol. Evol. 14: 823–828. **[11]**

Robinson, R. 2012. Resurrecting an ancient enzyme to address gene duplication. PLoS Biol. 10: e1001447. **[7]**

Rocha, E. P. C. 2003. DNA repeats lead to the accelerated loss of gene order in bacteria. Trends Genet. 19: 600–603. **[10]**

Rocha, E. P. C. 2004. The replication-related organization of bacterial genomes. Microbiology 150: 1609–1627. **[10]**

Rocha, E. P. C. 2006. The quest for the universals of protein evolution. Trends Genet. 22: 412–416. **[4]**

Rocha, E. P. C. 2008. The organization of the bacterial genome. Annu. Rev. Genet. 42: 211–233. **[10]**

Rocha, E. P. C. and A. Danchin. 2003. Essentiality, not expressiveness, drives gene-strand bias in bacteria. Nat. Genet. 34: 377–378. **[10]**

Rodríguez-Ezpeleta, N., H. Brinkmann, G. Burger, A. J. Roger, M. W. Gray, H. Philippe, and B. F. Lang. 2007. Toward resolving the eukaryotic tree: The phylogenetic positions of jakobids and cercozoans. Curr. Biol. 17: 1420–1425. **[5]**

Rodriguez, F., J. L. Oliver, A. Marin, and J. R. Medina. 1990. The general stochastic model of nucleotide substitution. J. Theor. Biol. 142: 485–501. **[5]**

Roest Crollius, H. and 11 others. 2000. Characterization and repeat analysis of the compact genome of the freshwater pufferfish *Tetraodon nigroviridis*. Genome Res. 10: 939–949. **[11]**

Roger, A. J. and A. G. B. Simpson. 2009. Evolution: Revisiting the root of the eukaryote tree. Curr. Biol. 19: R165–R167. **[5]**

Rogers, S., R. Wells, and M. Rechsteiner. 1986. Amino acid sequences common to rapidly degraded proteins: The PEST hypothesis. Science 234: 364–368. **[6, 8]**

Rokyta, D. R., P. Joyce, S. B. Caudle, and H. A. Wichman. 2005. An empirical test of the mutational landscape model of adaptation using a single-stranded DNA virus. Nat. Genet. 37: 441–444. **[13]**

Rokyta, D. R., P. Joyce, S. B. Caudle, C. Miller, C. J. Beisel, and H. A. Wichman. 2011. Epistasis between beneficial mutations and the phenotype-to-fitness map for a ssDNA virus. PLoS Genetics 7: e1002075. **[13]**

Rolston, H. 2006. What is a gene? From molecules to metaphysics. Theor. Med. Bioeth. 27: 471–497. **[1]**

Román, A. C., F. J. González-Rico, and P. M. Fernández-Salguero. 2011. B1-SINE retrotransposons: establishing genomic insulatory networks. Mobile Genet. Elements 1: 66–70. **[12]**

Romer, A. S. 1966. *Vertebrate Paleontology*. Chicago University Press, Chicago. **[5]**

Romiguier, J., V. Ranwez, E. J. Douzery, and N. Galtier. 2010. Contrasting GC-content dynamics across 33 mammalian genomes: relationship with life-history traits and chromosome sizes. Genome Res. 20: 1001–1009. **[11]**

Ronfort, J., E. Jenczewski, T. Bataillon, and F. Rousset. 1998. Analysis of population structure in autotetraploid species. Genetics 150: 921–930. **[7]**

Rooney, A. P., H. Piontkivska, and M. Nei. 2002. Molecular evolution of the nontandemly repeated genes of the histone 3 multigene family. Mol. Biol. Evol. 19: 68–75. **[7]**

Roos, C. 2003. *Molekulare Phylogenie der Halbaffen, Schlankaffen und Gibbons*. Ph.D. Dissertation, Technischen Universität München. **[7]**

Roos, C., J. Schmitz, and H. Zischler. 2004. Primate jumping genes elucidate strepsirrhine phylogeny. Proc. Natl. Acad. Sci. USA 101: 10650–10654. **[7]**

Ros, V. I. D. and G. D. D. Hurst. 2009. Lateral gene transfer between prokaryotes and multicellular eukaryotes: Ongoing and significant? BMC Biol. 7: 20. **[9]**

Rossi, M. S., O. A. Reig, and J. Zorzopulos. 1990 Evidence for rolling-circle replication in a major satellite DNA from the South American rodents of the genus *Ctenomys*. Mol. Biol. Evol. 7: 340–350. **[11]**

Roten, C.-A. H., P. Gamba, J.-L. Barblan, and D. Karamata. 2002. Comparative Genometrics (CG): A database dedicated to biometric comparisons of whole genomes. Nucleic Acids Res. 30: 142–144. **[10]**

Roth, A., M. Anisimova, and G. M. Cannarozzi. 2012. Measuring codon usage bias, pp. 189–216. In: G. M. Cannarozzi and A. Schneider (eds.), *Codon Evolution: Mechanisms and Models*. Oxford University Press, New York. **[4]**

Rousseau-Gueutin, M., M. A. Ayliffe, and J. N. Timmis. 2011. Conservation of plastid sequences in the plant nuclear genome for millions of years facilitates endosymbiotic evolution. Plant Physiol. 157: 2181–2193. **[9]**

Rowen, L. and 11 others. 2002. Analysis of the human neurexin genes: Alternative splicing and the generation of protein diversity. Genomics 79: 587–597. **[8]**

Roy, S., M. Ueda, K. Kadowaki, and N. Tsutsumi. 2010. Different status of the gene for ribosomal protein S16 in the chloroplast genome during evolution of the genus *Arabidopsis* and closely related species. Genes Genet. Syst. 85: 319–326. **[9]**

Rozen, D., J. A. G. M. de Visser, and P. J. Gerrish. 2002. Fitness effects of fixed beneficial mutations in microbial populations. Curr. Biol. 12: 1040–1045. **[13]**

Rozen, D., M. Habets, A. Handel, and J. A. G. M. de Visser. 2008. Heterogeneous adaptive trajectories of small populations on complex fitness landscapes. PLoS One 3: e1715. **[13]**

Rual, J. F. and 37 others. 2005. Towards a proteome-scale map of the human protein-protein interaction network. Nature 437: 1173–1178. **[1]**

Rubinstein, N. D., A. Doron-Faigenboim, I. Mayrose, and T. Pupko. 2011. Evolutionary models accounting for layers of selection in protein-coding genes and their impact on the inference of positive selection. Mol. Biol. Evol. 28: 3297–3308. **[4]**

Rudner, R., J. D. Karkas, and E. Chargaff. 1968. Separation of *B. subtilis* DNA into complementary strands. 3. Direct analysis. Proc. Natl. Acad. Sci. USA 60: 921–922. **[10]**

Russo, C. A., N. Takezaki, and M. Nei. 1995. Molecular phylogeny and divergence times of drosophilid species. Mol. Biol. Evol. 12: 391–404. **[8]**

Ruud, J. T. 1954. Vertebrates without erythrocytes and blood pigment. Nature 173: 848–850. **[7, 8]**

Ryan, J. F. and 17 others. 2013. The genome of the ctenophore *Mnemiopsis leidyi* and its implications for cell type evolution. Science 342: doi:1242592 **[5]**

Rzhetsky, A. and M. Nei. 1992. A simple method for estimating and testing minimum-evolution trees. Mol. Biol. Evol. 9: 945–967. **[5]**

S

Sabath, N., G. Landan, and D. Graur. 2008. A method for the simultaneous estimation of selection intensities in overlapping genes. PLoS ONE 3: e3996. **[8]**

Sabeti, P. C. and 15 others. 2002. Detecting recent positive selection in the human genome from haplotype structure. Nature 419: 832–837. **[2]**

Sabeti, P. C. and 9 others. 2006. Positive natural selection in the human lineage. Science 312: 1614–1620. **[2]**

Sabeur, G., G. Macaya, F. Kadi, and G. Bernardi. 1993. The isochore patterns of mammalian genomes and their phylogenetic implications. J. Mol. Evol. 37: 93–108. **[11]**

Saccone, C., C. Gissi, C. Lanave, and G. Pesole. 1995. Molecular classification of living organisms. J. Mol. Evol. 40: 273–279. **[5,6]**

Sackin, M. J. 1972. "Good" and "bad" phenograms. Syst. Zool. 21: 225–226. **[5]**

Sahni, N., S. Yi, Q. Zhong, N. Jailkhani, B. Charloteaux, M. E. Cusick, and M. Vidal. 2013. Edgotype: A fundamental link between genotype and phenotype. Curr. Opin. Genet. Dev. 23: 649–657. **[1]**

Saitou, N. 1988. Property and efficiency of the maximum likelihood method for molecular phylogeny. J. Mol. Evol. 27: 261–273. **[5]**

Saitou, N. 1989. A theoretical study of the underestimation of branch lengths by the maximum parsimony principle. Syst. Zool. 38: 1–6. **[5]**

Saitou, N. and M. Nei. 1987. The neighbor-joining method: A new method for reconstructing phylogenetic trees. Mol. Biol. Evol. 4: 406–425. **[5]**

Sakai, H., K. O. Koyanagi, T. Imanishi, T. Itoh, and T. Gojobori. 2007. Frequent emergence and functional resurrection of processed pseudogenes in the human and mouse genomes. Gene 389: 196–203. **[9]**

Sakharkar, M. K., P. Kangueane, D. A. Petrov, A. S. Kolaskar, and S. Subbiah. 2002. SEGE: A database on "intron less/single exonic" genes from eukaryotes. Bioinformatics 18: 1266–1267. **[9]**

Saleh, M. and 15 others. 2004. Differential modulation of endotoxin responsiveness by human caspase-12 polymorphisms. Nature 429: 75–79. **[7, 8]**

Saleh, M. and 8 others. 2006. Enhanced bacterial clearance and sepsis resistance in caspase-12-deficient mice. Nature 440: 1064–1068. **[7, 8]**

Salichos, L. and A. Rokas. 2013. Inferring ancient divergences requires genes with strong phylogenetic signals. Nature 497: 327–331. **[5]**

Salim, H. M. W., M. C. Negritto, and A. R. O. Cavalcanti. 2009. 1+1=3: A fusion of 2 enzymes in the methionine salvage pathway of *Tetrahymena thermophila* creates a trifunctional enzyme that catalyzes 3 steps in the pathway. PLoS Genet. 5: e1000701. **[8]**

Salim, H. M., A. M. Koire, N. A. Stover, and A. R. Cavalcanti. 2011. Detection of fused genes in eukaryotic genomes using gene deFuser: Analysis of the *Tetrahymena thermophila* genome. BMC Bioinformatics 12: 279. **[8]**

Salinas, J., M. Zerial, J. Filipski, M. Crepin, and G. Bernardi. 1987. Nonrandom distribution of MMTV proviral sequences in the mouse genome. Nucleic Acids Res. 15: 3009–3022. **[11]**

Salmena, L., L. Poliseno, Y. Tay, L. Kats, and P. P. Pandolfi. 2011. A ceRNA hypothesis: The Rosetta stone of a hidden RNA language? Cell 146: 353–358. **[1]**

Salser, W. and 10 others. 1976. Investigation of the organization of mammalian chromosomes at the DNA sequence level. Fed. Proc. 35: 23–35. **[11]**

Salter, L. A. and D. K. Pearl. 2001. Stochastic search strategy for estimation of maximum likelihood phylogenetic trees. Syst. Biol. 50: 7–17. **[5]**

Salvemini, M., C. Polito, and G. Saccone. 2010. *fruitless* alternative splicing and sex behaviour in insects: An ancient and unforgettable love story? J. Genet. 89: 287–299. **[8]**

Salz, H. K. 2011. Sex determination in insects: A binary decision based on alternative splicing. Curr. Opin. Genet. Devel. 21: 395–400. **[8]**

Salzburger, W., D. Steinke, I. Braasch, and A. Meyer. 2009. Genome desertification in eutherians: can gene deserts explain the uneven distribution of genes in placental mammalian genomes? J. Mol. Evol. 69: 207–216. **[11]**

Samuelson, P. A. 1954. The pure theory of public expenditure. Rev. Econom. Stat. 36: 387–389. **[9]**

Sanderson, M. J. 1989. Confidence limits on phylogenies: The bootstrap revisited. Cladistics 5: 113–129. **[5]**

Sanderson, M. J. 1997. A nonparametric approach to estimating divergence times in the absence of rate constancy. Mol. Biol. Evol. 14: 1218–1231. **[5]**

Sandler, L. and E. Novitski. 1957. Meiotic drive as an evolutionary force. Am. Nat. 91: 105–110. **[2, 11]**

Sandler, L., Y. Hiraizumi, and I. Sandler. 1959. Meiotic drive in natural populations of *Drosophila melanogaster*. I. The cytogenetic basis of segregation-distortion. Genetics 44: 233–250. **[2, 11]**

Sanjuán, R., A. Moya, and S. F. Elena. 2004. The distribution of fitness effects caused by single-nucleotide substitutions in an RNA virus. Proc. Natl. Acad. Sci. USA 101: 8396–8401. **[2]**

Sanjuán, R., J. Cuevas, V. Furió, E. Holmes, and A. Moya. 2007. Selection for robustness in mutagenized RNA viruses. PLoS Genet. 3: e93. **[13]**

Sankoff, D., G. Leduc, N. Antoine, B. Paquin, B. F. Lang, and R. Cedergren. 1992. Gene order comparisons for phylogenetic inference: Evolution of the mitochondrial genome. Proc. Natl. Acad. Sci. USA 89: 6575–6579. **[10]**

Sapp, J. (ed.). 2005. *Microbial Phylogeny and Evolution: Concepts and Controversies*. Oxford University Press, Oxford. **[5, 6]**

Sapp, J. and G. E. Fox. 2013. The singular quest for a universal tree of life. Microbiol. Mol. Biol. Rev. 77: 541–550. **[6]**

Sarich, V. M. and A. C. Wilson. 1967. Immunological time scale for hominid evolution. Science 158: 1200–1203. **[5]**

Sassi, S. O., E. L. Braun, and S. A. Benner. 2007. The evolution of seminal ribonuclease: Pseudogene reactivation or multiple gene inactivation events? Mol. Biol. Evol. 24: 1012–1024. **[7]**

Sato, N. 2006. Origin and evolution of plastids: Genomic view on the unification and diversity of plastids, pp. 75–102. In: R.

R. Wise and J. K. Hoober (eds.), *The Structure and Function of Plastids.* Springer, Dordrecht, Netherlands. **[5]**

Satta, Y., J. Klein, and N. Takahata. 2000. DNA archives and our nearest relative: The trichotomy problem revisited. Mol. Phylogenet. Evol. 14: 259–275. **[5, 6]**

Sattath, S. and A. Tversky. 1977. Additive similarity trees. Psychometrika 42: 319–345. **[5]**

Sawyer, S. A. 1999. GENECONV: A computer package for the statistical detection of gene conversion, 1.02 ed. Department of Mathematics, Washington University, St. Louis, MO. **[7]**

Sawyer, S. A. and D. L. Hartl. 1992. Population genetics of polymorphism and divergence. Genetics 132: 1161–1176. **[2]**

Scaiewicz, V., V. Sabbía, R. Piovani, and H. Musto. 2006. CpG islands are the second main factor shaping codon usage in human genes. Biochem. Biophys. Res. Commun. 343: 1257–1261. **[11]**

Scally, A. and 70 others. 2012. Insights into hominid evolution from the gorilla genome sequence. Nature 483: 169–175. **[5]**

Schaack, S., C. Gilbert, and C. Feschotte. 2010. Promiscuous DNA: horizontal transfer of transposable elements and why it matters for eukaryotic evolution. Trends Ecol. Evol. 25: 537–546. **[9, 11]**

Schattner, P. and M. Diekhans. 2006. Regions of extreme synonymous codon selection in mammalian genes. Nucleic Acids Res. 34: 1700–1710. **[4]**

Schell, M. A. and 11 others. 2002. The genome sequence of *Bifidobacterium longum* reflects its adaptation to the human gastrointestinal tract. Proc. Natl. Acad. Sci. USA 99: 14422–14427. **[10]**

Scherer, S. and R. W. Davis. 1980. Recombination of dispersed repeated DNA sequences in yeast. Science 209: 1380–1384. **[7]**

Schienman, J. E., R. A. Holt, M. R. Auerbach, and C. B. Stewart. 2006. Duplication and divergence of 2 distinct pancreatic ribonuclease genes in leaf-eating African and Asian colobine monkeys. Mol. Biol. Evol. 23: 1465–1479. **[5]**

Schier, A. F. and W. J. Gehring. 1993. Analysis of a *fushi tarazu* autoregulatory element: multiple sequence elements contribute to enhancer activity. EMBO J. 12: 1111–1119. **[12]**

Schildkraut, C. L., J. Marmur, and P. Doty. 1962. Determination of the base composition of deoxyribonucleic acid from its buoyant density in CsCl. J. Mol. Biol. 4: 430–443. **[11]**

Schlieper, D., M. A. Oliva, J. M. Andreu, and J. Löwe. 2005. Structure of bacterial tubulin BtubA/B: Evidence for horizontal gene transfer. Proc. Natl. Acad. Sci. USA 102: 9170–9175. **[9]**

Schmidt, D. and 9 others. 2012. Waves of retrotransposon expansion remodel genome organization and CTCF binding in multiple mammalian lineages. Cell 148: 335–348. **[9]**

Schmitz, J. and J. Brosius. 2011. Exonization of transposed elements: A challenge and opportunity for evolution. Biochimie 93: 1928–1934. **[9]**

Schmucker, D. and 7 others. 2000. *Drosophila Dscam* is an axon guidance receptor exhibiting extraordinary molecular diversity. Cell 101: 671–684. **[8]**

Schneider, A., A. Souvorov, N. Sabath, G. Landan, G. H. Gonnet, and D. Graur. 2009. Estimates of positive Darwinian selection are inflated by errors in sequencing, annotation, and alignment. Genome Biol. Evol. 1: 114–118. **[4]**

Schönknecht, G. and 17 others. 2013. Gene transfer from bacteria and archaea facilitated evolution of an extremophilic eukaryote. Science 339: 1207–1210. **[9]**

Schoustra, S., T. Bataillon, D. Gifford, and R. Kassen. 2009. The properties of adaptive walks in evolving populations of fungus. PLoS Biol. 7: e1000250. **[13]**

Schrider, D. R., J. N. Hourmozdi, and M. W. Hahn. 2011. Pervasive multinucleotide mutational events in eukaryotes. Curr. Biol. 21: 1051–1054. **[4]**

Schuh, R. T. and A. V. Z. Brower. 2009. *Biological Systematics: Principles and Applications*, 2nd Ed. Cornell University Press, Ithaca, NY. **[5, 6]**

Schully, S. D. and M. E. Hellberg. 2006. Selection on nucleotide substitutions and indels in accessory gland proteins of the *Drosophila pseudoobscura* subgroup. J. Mol. Evol. 62: 793–802. **[4]**

Schultz, A. H. 1963. *Classification and Human Evolution*. Aldine, Chicago. **[5]**

Schultz, D. W. and M. Yarus. 1994. Transfer RNA mutations and the malleability of the genetic code. J. Mol. Biol. 235: 1377–1380. **[10]**

Schultz, D. W. and M. Yarus. 1996. On the malleability of the genetic code. J. Mol. Evol. 42: 597–601. **[10]**

Schwartz, J. H. 1984. The evolutionary relationships of man and orang-utans. Nature 308: 501–505. **[5]**

Schwartz, R. M. and M. O. Dayhoff. 1978. Origins of prokaryotes, eukaryotes, mitochondria, and chloroplasts. Science 199: 395–403. **[6]**

Scott, A. M. and M. Saleh. 2007. The inflammatory caspases: Guardians against infections and sepsis. Cell Death Differ. 14: 23–31. **[7, 8]**

Searcy, D. G. 1986. Some features of thermo-acidophilic archaebacteria preadaptive for the evolution of eukaryotes. Syst. Appl. Microbiol. 7: 198–201. **[6]**

Searcy, D. G. 1992. Origins of mitochondria and chloroplasts from sulfur-based symbioses, pp. 47–78. In: H. H. Matsuno and K. Matsuno (eds.), *The Origin and Evolution of the Cell*. World Scientific, Singapore. **[5]**

Segovia, L., J. Horwitz, R. Gasser, and G. Wistow. 1997. Two roles for μ-crystallin: A lens structural protein in diurnal marsupials and a possible enzyme in mammalian retinas. Mol. Vision 3: 9. **[3]**

Sellers, P. H. 1974. On the theory and computation of evolutionary distances. SIAM J. Appl. Math. 26: 787–793. **[3]**

Sémon, M. and L. Duret. 2006. Evolutionary origin and maintenance of coexpressed gene clusters in mammals. Mol. Biol. Evol. 23: 1715–1723. **[11]**

Sengupta, S. and P. G. Higgs. 2005. A unified model of codon reassignment in alternative genetic codes. Genetics 170: 831–840. **[10]**

Sengupta, S., X. Yang, and P. G. Higgs. 2007. The mechanisms of codon reassignments in mitochondrial genetic codes. J. Mol. Evol. 64: 662–688. **[10]**

Seoighe, C. and K. H. Wolfe. 1999. Yeast genome evolution in the post-genome era. Curr. Opin. Microbiol. 2: 548–554. **[7]**

Serebrovsky, A. S. 1930. Untersuchungen über Treppenallelomorphismus. IV. Transgenation *scute-6* und ein Fall des "Nicht-Allelomorphismus" von Gliedern einer Allelomorphenreihe bei *Drosophila melanogaster.* Wilhelm Roux Archiv für Entwicklungsmechanik der Organismen 122: 88–104. **[7]**

Serebrovsky, A. S. and N. P. Dubinin. 1929. Artificial production of mutations and the problem of the gene. Uspeki Eksperimental noi Biologii 8: 235–247. **[7]**

Servine, J. A., C. W. Herbold, R. G. Skophammer, and J. A. Lake. 2007. Evidence excluding the root of the tree of life from the Actinobacteria. Mol. Biol. Evol. 25: 1–4. **[5]**

Sessa, E. B., E. A. Zimmer, and T. J. Givnish. 2012. Unraveling reticulate evolution in North American *Dryopteris* (Dryopteridaceae). BMC Evol. Biol. 12: 104. **[6]**

Shabalina, S. A. and E. V. Koonin. 2008. Origins and evolution of eukaryotic RNA interference. Trends Ecol. Evol. 23: 578–587. **[12]**

Shah, P. and M. A. Gilchrist. 2010. Effect of correlated tRNA abundances on translation errors and evolution of codon usage bias. PLoS Genet. 6: e1001128. **[4]**

Shao, K.-T. and R. R. Sokal. 1990. Tree balance. Syst. Zool. 39: 266–276. **[5]**

Shapiro, M.D. and 7 others. 2004. Genetic and developmental basis of evolutionary pelvic reduction in threespine sticklebacks. Nature 428: 717–723. **[12]**

Sharakhova, M. V., P. George, I. V. Brusentsova, S. C. Leman, J. A. Bailey, C. D. Smith, and I. V. Sharakhov. 2010. Genome mapping and characterization of the *Anopheles gambiae* heterochromatin. BMC Genomics 11: 459. **[11]**

Sharp, P. A. 1981. Speculations on RNA splicing. Cell 23: 643–646. **[8]**

Sharp, P. M., E. Bailes, R. J. Grocock, J. F. Peden, and R. E. Sockett. 2005. Variation in the strength of selected codon usage bias among bacteria. Nucleic Acids Res. 33: 1141–1153. **[4]**

Shiao, M.-S., B.-Y. Liao, M. Long, and H.-T. Yu. 2008. Adaptive evolution of the insulin two-gene system in mouse. Genetics 178: 1683–1691. **[9]**

Shields, D. and G. Blobel. 1977. Cell-free synthesis of fish preproinsulin, and processing by heterologous mammalian microsomal membranes. Proc. Natl. Acad. Sci. USA 74: 2059–2063. **[4]**

Shimamura, M. and 8 others. 1997. Molecular evidence from retroposons that whales form a clade within even-toed ungulates. Nature 388: 666–670. **[5]**

Shimizu, T., H. Mitsuke, K. Noto, and M. Arai. 2004. Internal gene duplication in the evolution of prokaryotic transmembrane proteins. J. Mol. Biol. 339: 1–15. **[8]**

Shinde, D., Y. Lai, F. Sun, and N. Arnheim. 2003. Taq DNA polymerase slippage mutation rates measured by PCR and quasi-likelihood analysis: $(CA/GT)_n$ and $(A/T)_n$ microsatellites. Nucleic Acids Res. 31: 974–980. **[11]**

Shoshani, J., C. P. Groves, E. L. Simons, and G. F. Gunnell. 1996. Primate phylogeny: Morphological vs. molecular results. Mol. Phylogenet. Evol. 5: 102–154. **[5]**

Sia, E. A., R. J. Kokoska, M. Dominska, P. Greenwell, and T. D. Petes.1997. Microsatellite instability in yeast: dependence on repeat unit size and DNA mismatch repair genes. Mol. Cell. Biol. 17: 2851–2858. **[11]**

Sibley, C. G. 1960. The electrophoretic patterns of avian egg-white proteins as taxonomic characters. Ibis 102: 215–284. **[5]**

Sibley, C. G. and J. E. Ahlquist. 1984. The phylogeny of the hominoid primates, as indicated by DNA-DNA hybridization. J. Mol. Evol. 20: 2–15. **[5]**

Sibley, C. G., J. A. Comstock, and J. E. Ahlquist. 1990. DNA hybridization evidence of hominoid phylogeny: A reanalysis of the data. J. Mol. Evol. 30: 202–236. **[5]**

Siepel, A. 2009. Darwinian alchemy: Human genes from noncoding DNA. Genome Res. 19: 1693–1695. **[8]**

Sievers, F. and 11 others. 2011. Fast, scalable generation of high-quality protein multiple sequence alignments using Clustal Omega. Mol. Syst. Biol. 7: 539. **[3]**

Signorovitch, A., J. Hur, E. Gladyshev, and M. Meselson. 2015. Allele sharing and evidence for sexuality in a mitochondrial clade of bdelloid rotifers. Genetics 200: 581–590. **[9]**

Siguier, P., J. Filée, and M. Chandler. 2006. Insertion sequences in prokaryotic genomes. Curr. Opin. Microbiol. 9: 526–531. **[9]**

Silke, J. 1997. The majority of long non-stop reading frames on the antisense strand can be explained by biased codon usage. Gene 194: 143–155. **[8]**

Siller, E., D. C. DeZwaan, J. F. Anderson, B. C. Freeman, and J. M. Barral. 2010. Slowing bacterial translation speed enhances eukaryotic protein folding efficiency. J. Mol. Biol. 396: 1310–1318. **[4]**

Simonson, T. S. and 11 others. 2010. Genetic evidence for high-altitude adaptation in Tibet. Science 329: 72–75. **[2]**

Simpson, A. G. B. and A. J. Roger. 2002. Eukaryotic evolution: Getting to the root of the problem. Curr. Biol. 12: 691–693. **[5]**

Simpson, G. G. 1945. The principles of classification and a classification of mammals. Bull. Amer. Mus. Nat. Hist. 85: 1–350. **[5]**

Simpson, G. G. 1961. *Principles of Animal Taxonomy*. Columbia University Press, New York. **[5]**

Singer, C. E. and B. N. Ames. 1970. Sunlight ultraviolet and bacterial DNA base ratios. Science 170: 822–826. **[10]**

Singer, M. F. and P. Berg. 1991. *Genes and Genomes: A Changing Perspective*. University Science Books, Mill Valley, CA. **[10]**

Singer, M. F. 1982. Highly repeated sequences in mammalian genomes. Int. Rev. Cytol. 76: 67–112. **[10]**

Singer, T., M. J. McConnell, M. C. N. Marchetto, N. G. Coufal, and F. H. Gage. 2010. LINE-1 retrotransposons: Mediators of somatic variation in neuronal genomes? Trends Neurosci. 33: 345–354. **[9]**

Singh, N. D., P. F. Arndt, A. G. Clark, and C. F. Aquadro. 2009. Strong evidence for lineage and sequence specificity of substitution rates and patterns in *Drosophila*. Mol. Biol. Evol. 26: 1591–1605. **[4]**

Singh, P. and S. T. Cole. 2011. *Mycobacterium leprae*: Genes, pseudogenes and genetic diversity. Future Microbiol. 6: 57–71. **[10]**

Skaletsky, H. and 39 authors. 2003. The male-specific region of the human Y chromosome is a mosaic of discrete sequence classes. Nature 423: 825–837 **[9]**

Skovgaard, M., L. J. Jensen, S. Brunak, D. Ussery, and A. Krogh. 2001. On the total number of genes and their length distribution in complete microbial genomes. Trends Genet. 17: 425–428. **[10]**

Slamovits, C. H., J. A. Cook, E. P. Lessa, and M. S. Rossi. 2001 Recurrent amplifications and deletions of satellite DNA accompanied chromosomal diversification in South American tuco-tucos (genus *Ctenomys*, Rodentia: Octodontidae): a phylogenetic approach. Mol. Biol. Evol. 18: 1708–1719. **[11]**

Slightom, J. L, L.-Y. Chang, B. F. Koop, and M. Goodman. 1985. Chimpanzee fetal $^G\gamma$- and $^A\gamma$-globin gene nucleotide sequences provide further evidence of gene conversions in hominine evolution. Mol. Biol. Evol. 2: 370–389. **[7]**

Sloan, D. B. and 6 others. 2012. Rapid evolution of enormous, multichromosomal genomes in flowering plant mitochondria with exceptionally high mutation rates. PLoS Biol. 10: e1001241. **[10]**

Sloan, D. B., A. J. Alverson, M. Wu, J. D. Palmer, and D. R. Taylor. 2012. Recent acceleration of plastid sequence and structural evolution coincides with extreme mitochondrial divergence in the angiosperm genus *Silene*. Genome Biol. Evol. 4: 294–306. **[10]**

Slot, J. C. and A. Rokas. 2011. Horizontal transfer of a large and highly toxic secondary metabolic gene cluster between fungi. Curr. Biol. 21: 134–139. **[9]**

Slotkin, R. K., S. Nutikattu, and N. Jiang. 2012. The impact of transposable elements on gene and genome evolution. In: J. F. Wendel, J. Greilhuber, J. Doležel, and I. J. Leitch (eds). *Plant Genome Diversity: Plant Genomes, Their Residents, and Their Evolutionary Dynamics*. 1: 35–58. **[11]**

Slowinski, J. B. 1998. The number of multiple alignments. Mol. Phylogenet. Evol. 10: 264–266. **[3]**

Šmarda, P., P. Bures, L. Horová, and O. Rotreklová. 2008. Intrapopulation genome size dynamics in *Festuca pallens*. Ann. Bot. 102: 599–607. **[11]**

Šmarda, P. and 8 others. 2013. Effect of phosphorus availability on the selection of species with different ploidy levels and genome sizes in a long-term grassland fertilization experiment. New Phytol. 200: 911–921. **[11]**

Smillie, C. S., M. B. Smith, J. Friedman, O. X. Cordero, L. A. David, and E. J. Alm. 2011. Ecology drives a global network of gene exchange connecting the human microbiome. Nature 480: 241–244. **[9]**

Smit, A. F. A. 1996. The origin of interspersed repeats in the human genome. Curr. Opin. Genet. Dev. 6: 743–748. **[9, 10]**

Smith, A. A., K. Wyatt, J. Vacha, T. S. Vihtelic, J. S. Zigler, G. J. Wistow, and M. Posner. 2006. Gene duplication and separation of functions in alphaB-crystallin from zebrafish (*Danio rerio*). FEBS J. 273: 481–490. **[7]**

Smith, C. D., S. Shu, C. J. Mungall, and G. H. Karpen. 2007. The Release 5.1 annotation of *Drosophila melanogaster* heterochromatin. Science 316: 1586–1591. **[11]**

Smith, D. R. and R. W. Lee. 2014. A plastid without a genome: Evidence from the nonphotosynthetic green algal genus *Polytomella*. Plant Physiol. 164: 1812–1819. **[10]**

Smith, D. R., K. Crosby, and R. W. Lee. 2011. Correlation between nuclear plastid DNA abundance and plastid number supports the limited transfer window hypothesis. Genome Biol. Evol. 3: 365–371. **[9]**

Smith, G. P. 1974. Unequal crossover and the evolution of multigene families. Cold Spring Harbor Symp. Quant. Biol. 38: 507–513. **[7]**

Smith, G. P. 1976. Evolution of repeated DNA sequences by unequal crossover. Science 191: 528–535. **[7]**

Smith, M. M. 1987. Molecular evolution of the *Saccharomyces cerevisiae* histone gene loci. J. Mol. Evol. 24: 252–259. **[7]**

Smith, N. G. and A. Eyre-Walker. 2002. The compositional evolution of the murid genome. J. Mol. Evol. 55: 197–201. **[11]**

Smith, N. G., M. Brandström, and H. Ellegren. 2004. Evidence for turnover of functional noncoding DNA in mammalian genome evolution. Genomics 84: 806–813. **[11]**

Smithies, O., G. E. Connell, and G. H. Dixon. 1962. Chromosomal rearrangements and the evolution of haptoglobin genes. Nature 196: 232–236. **[8]**

Smulders, R. H. P. H., M. A. M. van Dijk, S. Hoevenaars, R. A. Linder, J. A. Carver, and W. W. de Jong. 2002. The eye lens protein αA-crystallin of the blind mole rat *Spalax ehrenbergi*: Effects of altered functional constraints. Exp. Eye Res. 74: 285–291. **[4]**

Sneath, P. H. A. and R. R. Sokal. 1973. *Numerical Taxonomy: The Principles and Practice of Numerical Classification*. W.H. Freeman, San Francisco. **[5, 6]**

Snel, B., M. A. Huynen, and B. E. Dutilh. 2005. Genome trees and the nature of genome evolution. Annu. Rev. Microbiol. 59: 191–209. **[5]**

Snel, B., P. Bork, and M. A. Huynen. 2000. Genome evolution: Gene fusion versus gene fission. Trends Genet. 16: 9–11. **[8]**

Sniegowski, P. D., P. J. Gerrish, and R. E. Lenski. 1997. Evolution of high mutation rates in experimental populations of *E. coli*. Nature 387: 703–705. **[13]**

Sniegowski, P. D., P. J. Gerrish, T. Johnson, and A. Shaver. 2000. The evolution of mutation rates: Separating causes from consequences. Bioessays 22: 1057–1066. **[4]**

Snyder, M. and M. Gerstein. 2003. Defining genes in the genomics era. Science 300: 258–260. **[1]**

Snyder, P. and S. Lawson. 1993. Evaluating results using corrected and uncorrected effect size estimates. J. Exp. Edu. 61: 334–349. **[11]**

Sogin, M. L., H. J. Elwood, and J. H. Gunderson. 1986. Evolutionary diversity of eukaryotic small-subunit rRNA genes. Proc. Natl. Acad. Sci. USA 83: 1383–1387. **[5]**

Sokal, R. R. and C. D. Michener. 1958. A statistical method for evaluating systematic relationships. Univ. Kansas Sci. Bull. 28: 1409–1438. **[5]**

Sokal, R. R. and T. J. Crovello. 1970. The biological species concept: A critical evaluation. Am. Nat. 104: 127–153. **[5]**

Soltis, D. E., P. S. Soltis, D. W. Schemske, J. F. Hancock, J. N. Thompson, B. C. Husband, and W. S. Judd. 2007. Autopolyploidy in angiosperms: Have we grossly underestimated the number of species? Taxon 56: 13–30. **[7]**

Soltis, P. S. and D. E. Soltis. 2000. The role of genetic and genomic attributes in the success of polyploids, pp. 310–330. In: F. J. Ayala, W. M. Fitch, and M. T. Clegg (eds.), *Variation and Evolution in Plants and Microorganisms: Toward a New Synthesis 50 Years after Stebbins*. National Academy of Sciences Press, Washington, DC. **[7]**

Soltis, P. S. and D. E. Soltis. 2003. Applying the bootstrap in phylogeny reconstruction. Statist. Sci. 18: 256–267. **[5]**

Sommer, S. S. 1995. Recent human germ-line mutation: Inferences from patients with hemophilia B. Trends Genet. 11: 141–147. **[4]**

Sommerhalter, M., W. C. Voegtli, D. L. Perlstein, J. Ge, J. Stubbe, and A. C. Rosenzweig. 2004. Structures of the yeast ribonucleotide reductase Rnr2 and Rnr4 homodimers. Biochemistry 43: 7736–7742. **[7]**

Song, K., P. Lu, K. Tang, and C. T. Osborn. 1995. Rapid genome change in synthetic polyploids of *Brassica* and its implications for polyploid evolution. Proc. Natl. Acad. Sci. USA 92: 7719–7723. **[7]**

Sonnhammer, E. L. L. and E. V. Koonin. 2002. Orthology, paralogy and proposed classification for paralog subtypes. Trends Genet. 18: 619–620. **[5]**

Sorek, R. 2007. The birth of new exons: Mechanisms and evolutionary consequences. RNA 13: 1603–1608. **[9]**

Sorek, R., G. Ast, and D. Graur. 2002. *Alu*-containing exons are alternatively spliced. Genome Res. 12: 1060–1067. **[8]**

Sorek, R., R. Shamir, and G. Ast. 2004. How prevalent is functional alternative splicing in the human genome? Trends Genet. 20: 68–71. **[8]**

Sorek, R., Y. Zhu, C. J. Creevey, M. P. Francino, P. Bork, and E. M. Rubin. 2007. Genome-wide experimental determination of barriers to horizontal gene transfer. Science 318: 1449–1452. **[9]**

Soumillon, M. and 13 others. 2013. Cellular source and mechanisms of high transcriptome complexity in the mammalian testis. Cell Rep. 3: 2179–2190. **[9]**

Sousa, A., S. Magalhães, and I. Gordo. 2012. Cost of antibiotic resistance and the geometry of adaptation. Mol. Biol. Evol. 29: 1417–1428. **[13]**

Sousa, F. L. and W. F. Martin. 2014. Biochemical fossils of the ancient transition from geoenergetics to bioenergetics in prokaryotic one carbon compound metabolism. Biochim. Biophys. Acta 1837: 964–981. **[6]**

Spang, A. and 9 others. 2015. Complex archaea that bridge the gap between prokaryotes and eukaryotes. Nature 521: 173–179. **[6]**

Sparrow, A. H. and A. F. Nauman. 1976. Evolution of genome size by DNA doublings. Science 192: 524–529. **[11]**

Spencer, C. C., M. Bertrand, M. Travisano, and M. Doebeli. 2007. Adaptive diversification in genes that regulate resource use in *Escherichia coli*. PLoS Genet. 3: e15. **[13]**

Sperling, E. A., K. J. Peterson, and D. Pisani. 2009. Phylogenetic-signal dissection of the nuclear housekeeping genes supports the paraphyly of sponges and the monophyly of Eumetazoa. Mol. Biol. Evol. 26: 2261–2274. **[5]**

Spoerel, N. A., H. T. Nguyen, T. H. Eickbush, and F.C. Kafatos. 1989. Gene evolution and regulation in the chorion complex of *Bombyx mori*. Hybridization and sequence analysis of multiple developmentally middle A/B chorion gene pairs. J. Mol. Biol. 209: 1–19. **[7]**

Spofford, J. B. 1969. Single-locus modification of position-effect variegation in *Drosophila melanogaster*. II. Region 3c loci. Genetics 62:555–571. **[7]**

Spradling, A. C., H. J. Bellen, and R. A. Hoskins. 2011. *Drosophila P* elements preferentially transpose to replication origins. Proc. Natl. Acad. Sci. USA 108: 15948–15953. **[9]**

Srivastava, M. and 20 others. 2008. The *Trichoplax* genome and the nature of placozoans. Nature 454: 955–960. **[11]**

Stallings, R.L., A. F. Ford, D. Nelson, D. C. Torney, C. E. Hildebrand, and R. K. Moyzis. 1991. Evolution and distribution of $(GT)_n$ repetitive sequences in mammalian genomes. Genomics 10: 807–815. **[11]**

Stamatoyannopoulos, J. A. 2012. What does our genome encode? Genome Res. 22: 1602–1611. **[11]**

Stanier, R. Y. and C. B. van Niel. 1962. The concept of a bacterium. Arch. Microbiol. 42: 17–35. **[5, 6]**

Starostova, Z., L. Kratochvil, and M. Flajshans. 2008. Cell size does not always correspond to genome size: phylogenetic analysis in geckos questions optimal DNA theories of genome size evolution. Zoology 111: 377–384. **[11]**

Starostova, Z., L. Kubicka, M. Konarzewski, J. Kozlowski, and L. Kratochvil. 2009. Cell size but not genome size affects scaling of metabolic rate in eyelid geckos. Am. Nat. 174: E100–E105. **[11]**

Staub, E., B. Hinzmann, and A. Rosenthal. 2002. A novel repeat in the melanoma-associated chondroitin sulfate proteoglycan defines a new protein family. FEBS Lett. 527: 114–118. **[7]**

Staub, E., B. Hinzmann, and A. Rosenthal. 2002a. A novel repeat in the melanoma-associated chondroitin sulfate proteoglycan defines a new protein family. FEBS Lett. 527: 114–118. **[8]**

Staub, E., J. Pérez-Tur, R. Siebert, C. Nobile, N. K. Moschonas, P. Deloukas, and B. Hinzmann. 2002b. The novel EPTP repeat defines a superfamily of proteins implicated in epileptic disorders. Trends Biochem. Sci. 27: 441–444. **[8]**

Staub, E., D. Mennerich, and A. Rosenthal. 2002c. The Spin/Ssty repeat: a new motif identified in proteins involved in vertebrate development from gamete to embryo. Genome Biol. 3: RESEARCH0003. **[8]**

Stebbins, G. 1971. *Chromosomal Evolution in Higher Plants*. Edward Arnold, London. **[7]**

Stedman, H. H. and 9 others. 2004. Myosin gene mutation correlates with anatomical changes in the human lineage. Nature 428: 415–418. **[8]**

Steel, M. 1994. Recovering a tree from the Markov leaf colouration it generates under a Markov model. Appl. Math. Lett. 7: 19–23. **[5]**

Stein, J. P., J. F. Catterall, P. Kristo, A. R. Means, and B. W. O'Malley. 1980. Ovomucoid intervening sequences specify functional domains and generate protein polymorphism. Cell 21: 681–687. **[8]**

Steinemann, M. and S. Steinemann. 1991. Preferential Y chromosomal location of TRIM, a novel transposable element of *Drosophila miranda, obscura* group. Chromosoma 101: 169–179. **[9]**

Stelzer, C.-P., S. Riss, and P. Stadler. 2011. Genome size evolution at the speciation level: The cryptic species complex *Brachionus plicatilis* (Rotifera). BMC Evol. Biol. 11: 90. **[9]**

Stencel, A. and B. Crespi. 2013. What is a genome? Mol. Ecol. 22: 3437–3443. **[10]**

Stephan, W., T. H. E. Wiehe, and M. W. Lenz. 1992. The effect of strongly selected substitutions on neutral polymorphism: Analytical results based on diffusion theory. Theor. Pop. Biol. 41: 237–254. **[2]**

Stern, D. L. and N. Frankel. 2013. The structure and evolution of *cis*-regulatory regions: the *shavenbaby* story. Philos. Trans. R. Soc. Lond. Biol Sci. 368B: 20130028. **[12]**

Stern, D. L. and V. Orgogozo. 2009. Is genetic evolution predictable? Science 323: 746–751. **[12]**

Stevenson, R. L. 1886. *Strange Case of Dr. Jekyll and Mr. Hyde*. Charles Scribner's Sons, New York. **[11]**

Stewart, C.-B. and A. C. Wilson. 1987. Sequence convergence and functional adaptation of stomach lysozymes from foregut fermenters. Cold Spring Harbor Symp. Quant. Biol. 52: 891–899. **[5]**

Stock, D. W., J. M. Quattro, J. S. Whitt, and D. A. Powers. 1997. Lactate dehydrogenase (LDH) gene duplication during chordate evolution: The cDNA sequence of the LDH of the tunicate *Styela plicata*. Mol. Biol. Evol. 14: 1273–1284. **[7]**

Stoletzki, N. and A. Eyre-Walker. 2007. Synonymous codon usage in *Escherichia coli*: Selection for translational accuracy. Mol. Biol. Evol. 24: 374–381. **[4]**

Stoltzfus, A. 1999. On the possibility of constructive neutral evolution. J. Mol. Evol. 49: 169–181. **[8]**

Stoltzfus, A., D. F. Spencer, M. Zuker, J. M. Logsdon, and W. F. Doolittle. 1994. Testing the exon theory of genes: The evidence from protein structure. Science 265: 202–207. **[8]**

Stone, J. R. and G. A. Wray. 2001. Rapid evolution of *cis*-regulatory sequences via local point mutations. Mol. Biol. Evol. 18: 1764–1770. **[12]**

Stouthamer, R., J. A. J. Breeuwer, and G. D. D. Hurst. 1999. *Wolbachia pipientis*: Microbial manipulator of arthropod reproduction. Annu. Rev. Microbiol. 53: 71–102. **[2]**

Stoye, J. P. 2001. Endogenous retroviruses: Still active after all these years? Curr. Biol. 11: R914–R916. **[9]**

Strachan, T. and A. Read. 2011. *Human Molecular Genetics* (4th ed.) Garland Science, New York. **[11]**

Streelman, J. T., C. L. Peichel, and D. M. Parichy. 2007. Developmental genetics of adaptation in fishes: the case for novelty. Annu. Rev. Ecol. Evol. Syst. 38: 655–681. **[12]**

Struchiner, C. J., M. G. Kidwell, and J. C. Ribero. 2005. Population dynamics of transposable elements: Copy number regulation and species invasion requirements. J. Biol. Sys. 13: 455–475. **[9]**

Struhl, K. 2007. Transcriptional noise and the fidelity of initiation by RNA polymerase II. Nat. Struct. Mol. Biol. 14: 103–105. **[11]**

Sturtevant, A. H. 1937. Essays on evolution. I. On the effects of selection on mutation rate. Quart. Rev. Biol. 12: 464–467. **[2]**

Suárez-Villota, E. Y. and 8 others. 2012. Distribution of repetitive DNAs and the hybrid origin of the red vizcacha rat (Octodontidae). Genome 55: 105–117. **[7]**

Subramanian, A. R., J. Weyer-Menkhoff, M. Kaufmann, and B. Morgenstern. 2005. DIALIGN-T: An improved algorithm for segment-based multiple sequence alignment. BMC Bioinformatics 6: 66. **[3]**

Subramanian, A. R., M. Kaufmann, and B. Morgenstern. 2008. DIALIGN-TX: Greedy and progressive approaches for segment-based multiple sequence alignment. Algorithms Mol. Biol. 3: 6. **[3]**

Subramanian, S. and S. Kumar. 2006. Evolutionary anatomies of positions and types of disease-associated and neutral amino acid mutations in the human genome. BMC Genomics 7: 306. **[4]**

Subramanian, S., R. K. Mishra, and L. Singh. 2003. Genome-wide analysis of microsatellite repeats in humans: their abundance and density in specific genomic regions. Genome Biol. 4: R13. **[11]**

Sueoka, N. 1961. Variation and heterogeneity of base composition of deoxyribonucleic acids: a compilation of old and new data. J. Mol. Biol. 3: 31–40. **[11]**

Sueoka, N. 1964. On the evolution of informational macromolecules. pp. 479–496. In: V. Bryson and H. J. Vogel (eds.), *Evolving Genes and Proteins*. Academic Press, New York. **[10]**

Sueoka, N. 1995. Intrastrand parity rules of DNA base composition and usage biases of synonymous codons. J. Mol. Evol. 40: 318–325. **[10]**

Sulem, P. and 18 others. 2015. Identification of a large set of rare complete human knockouts. Nat. Genet. doi: 10.1038/ng.3243. **[11]**

Sulston, J. E. and H. R. Horvitz. 1977. Post-embryonic cell lineages of the nematode *Caenorhabditis elegans*. Dev. Biol. 56: 110–156. **[11]**

Summers, R. L. and 13 others. 2014. Diverse mutational pathways converge on saturable chloroquine transport via the malaria parasite's chloroquine resistance transporter. Proc. Natl. Acad. Sci. USA 111E: 1759–1767. **[13]**

Sumner, A. T. 2003. *Chromosomes: Organization and Function*. Blackwell, Malden, MA. **[11]**

Sun, J. X. and 10 others. 2012. A direct characterization of human mutation based on microsatellites. Nat. Genet. 44: 1161–1165. **[4]**

Sundaram, V. and 7 others. 2014. Widespread contribution of transposable elements to the innovation of gene regulatory networks. Genome Res. (in press). **[11]**

Sunyaev, S., V. Ramensky, I. Koch, W. Lathe, A. S. Kondrashov, and P. Bork. 2001. Prediction of deleterious human alleles. Hum. Mol. Genet. 10: 591–597. **[2, 4, 11]**

Surridge, A. K., D. Osorio, and N. I. Mundy. 2003. Evolution and selection of trichromatic vision in primates. Trends Ecol. Evol. 18: 198–205. **[7]**

Suyama, M. and P. Bork. 2001. Evolution of prokaryotic gene order: Genome rearrangements in closely related species. Trends Genet. 17: 10–13. **[10]**

Suzuki, T., H. Yuasa, and K. Imai. 1996. Convergent evolution. The gene structure of *Sulculus* 41 kDa myoglobin is homologous with that of human indoleamine dioxygenase. Biochim. Biophys. Acta 1308: 41–48. **[6, 8]**

Suzuki, T., T. Ueda, and K. Watanabe. 1997. The "polysemous" codon—a codon with multiple amino acid assignment caused by dual specificity of tRNA identity. EMBO J. 16: 1122–1134. **[1]**

Suzuki, Y. 2004. New methods for detecting positive selection at single amino acid sites. J. Mol. Evol. 59: 11–19. **[5]**

Suzuki, Y. and T. Gojobori. 1999. A method for detecting positive selection at single amino acid sites. Mol. Biol. Evol. 16: 1315–1328. **[5]**

Suzuki, Y., G. V. Glazko, and M. Nei. 2002. Overcredibility of molecular phylogenies obtained by Bayesian phylogenetics. Proc. Natl. Acad. Sci. USA 99: 16138–16143. **[5, 6]**

Svartman, M., G. Stone, and R. Stanyon. 2005. Molecular cytogenetics discards polyploidy in mammals. Genomics 85: 425–430. **[7]**

Svensson, E. and R. Calsbeek (eds.) 2012. *The Adaptive Landscape in Evolutionary Biology.* Oxford University Press, New York **[13]**

Svensson, O., L. Arvestad, and J. Lagergren. 2006. Genome-wide survey for biologically functional pseudogenes. PLoS Comput. Biol. 2: e46. **[11]**

Swift, H. 1950. The constancy of desoxyribose nucleic acid in plant nuclei. Proc. Natl. Acad. Sci. USA 36: 643–654. **[10]**

Swofford, D. L. 1993. *PAUP: Phylogenetic Analysis using Parsimony,* Version 3.1. Computer program distributed by the Illinois Natural History Survey, Champaign, IL. Smithsonian Institution, Washington, DC. **[5]**

Swofford, D. L. and G. J. Olsen. 1990. Phylogeny reconstruction. pp. 411–501. In: D. M. Hillis and C. Moritz (eds.), *Molecular Systematics.* Sinauer Associates, Sunderland, MA. **[5]**

Swofford, D. L. and W. P. Maddison. 1987. Reconstructing ancestral character states under Wagner parsimony. Math. Biosci 87: 199–229. **[5, 11]**

Swofford, D. L., G. J. Olsen, P. J. Waddell, and D. M. Hillis. 1996. Phylogenetic inference. pp. 407–543. In: D. M. Hillis, C. Moritz, and B. K. Mable (eds.), *Molecular Systematics,* 2nd Ed. Sinauer Associates, Sunderland, MA. **[5]**

Symonová, R. and 7 others. 2013. Genome differentiation in a species pair of coregonine fishes: An extremely rapid speciation driven by stress-activated retrotransposons mediating extensive ribosomal DNA multiplications. BMC Evol. Biol. 13: 42. **[9]**

Syvanen, M. and C. I. Kado (eds.). 1998. *Horizontal Gene Transfer.* Chapman & Hall, London. **[9]**

Szalay, E. S. 1969. The Hapalodectinae and a phylogeny of the Mesonychidae (Mammalia, Condylarthra). Am. Mus. Nat. Hist. Novitates 2361: 1–26. **[5]**

Szathmáry, E. 2005. Life: In search of the simplest cell. Nature 433: 469–470. **[10]**

Szathmáry, E. and J. Maynard Smith. 1995. The major evolutionary transitions. Nature 374: 227–232. **[11]**

Szostak, J. W. and R. Wu. 1980. Unequal crossing over in the ribosomal DNA of *Saccharomyces cerevisiae.* Nature 284: 426–430. **[7]**

T

Tachezy, J. (ed.). 2008. Hydrogenosomes and Mitosomes: Mitochondria of Anaerobic Eukaryotes. Springer, Berlin. **[5, 6]**

Tachida, H. 1991. A study on a nearly neutral mutation model in finite populations. Genetics 128: 183–192. **[2]**

Tajima, F. 1983. Evolutionary relationship of DNA sequences in finite populations. Genetics 105: 437–460. **[2]**

Takabe, T. and T. Akazawa. 1975. Molecular evolution of ribulose-1, 5-bisphosphate carboxylase. Plant Cell Physiol. 16: 1049–1060. **[5]**

Takezaki, N., A. Rzhetsky, and M. Nei. 1995. Phylogenetic test of the molecular clock and linearized trees. Mol. Biol. Evol. 12: 823–833. **[4, 5]**

Tamames, J. 2001. Evolution of gene order conservation in prokaryotes. Genome Biol. 2: RESEARCH0020. **[10]**

Tan, H. and J. B. Whitney. 1993. Gene rearrangement in the α-globin gene complex during mammalian evolution. Biochem. Genet. 31: 473–484. **[7]**

Tang, D. 2015. Repetitive elements in vertebrate genomes. http: // davetang.org/muse/2014/05/22/repetitive-elements-in-vertebrate-genomes/ **[11]**

Tang, H., G. J. Wyckoff, J. Lu, and C.-I. Wu. 2004. A universal evolutionary index for amino acid changes. Mol. Biol. Evol. 21: 1548–1556. **[4]**

Tang, H., J. E. Bowers, X. Wang, R. Ming, M. Alam, and A. H. Paterson. 2008. Synteny and collinearity in plant genomes. Science 320: 486–488. **[11]**

Tang, K., K. R. Thornton, and M. Stoneking. 2007. A new approach for using genome scans to detect recent positive selection in the human genome. PLoS Biol. 5: e171. **[2]**

Tanifuji, G. and 7 others. 2014. Nucleomorph and plastid genome sequences of the chlorarachniophyte *Lotharella oceanica*: convergent reductive evolution and frequent recombination in nucleomorph-bearing algae. BMC Genomics 15: 374. **[11]**

Tanifuji, G., N. T. Onodera, T. J. Wheeler, M. Dlutek, N. Donaher, and J. M. Archibald. 2011. Complete nucleomorph genome sequence of the nonphotosynthetic alga *Cryptomonas paramecium* reveals a core nucleomorph gene set. Genome Biol. Evol. 3: 44–54. **[11]**

Tarlinton, R. E., J. Meers, and P. R. Young. 2006. Retroviral invasion of the koala genome. Nature 442: 79–81. **[9]**

Tarver, J. E., P. C. Donoghue, and K. J. Peterson. 2012. Do miRNAs have a deep evolutionary history? Bioessays 34: 857–866. **[12]**

Tatarinova, T. V., N. N. Alexandrov, J. B. Bouck, and K. A. Feldmann. 2010. GC3 biology in corn, rice, sorghum and other grasses. BMC Genomics 11: 308. **[11]**

Tatarinova, T., E. Elhaik, and M. Pellegrini. 2013. Cross-species analysis of genic GC3 content and DNA methylation patterns. Genome Biol. Evol. 5: 1443–1456. **[11]**

Tateno, Y., N. Takezaki, and M. Nei. 1994. Relative efficiencies of the maximum-likelihood, neighbor-joining, and maximum-parsimony methods when the substitution rate varies with site. Mol. Biol. Evol. 11: 261–277. **[5]**

Tats, A., T. Tenson, and M. Remm. 2008. Preferred and avoided codon pairs in three domains of life. BMC Genomics 9: 463. **[4]**

Tatusov, R. L. and 16 others. 2003. The COG database: an updated version includes eukaryotes. BMC Bioinformatics 4: 41. **[11]**

Tatusov, R. L., E. V. Koonin, and D. J. Lipman. 1997. A genomic perspective on protein families. Science 278: 631–637. **[11]**

Tautz, D. 1993. Notes on the definition and nomenclature of tandemly repetitive DNA sequences. pp. 21–28. In: S. D. J. Pena et al. (eds.) *DNA Fingerprinting: State of the Science,* Birkhäuser, Basel, Switzerland. **[11]**

Tautz, D. and C. Schlötterer. 1994. Simple sequences. Curr. Opin. Genet. Dev. 4: 832–837. **[11]**

Tavares, S. and 10 others. 2014. Genome size analyses of Pucciniales reveal the largest fungal genomes. Frontiers Plant Sci. 5: 422. **[11]**

Taylor, J. 2005. Clues to function in gene deserts. Trends Biotechnol. 23: 269–271. **[11]**

Taylor, M. S., C. P. Ponting, and R. R. Copley. 2004. Occurrence and consequences of coding sequence insertions and deletions in mammalian genomes. Genome Res. 14: 555–566. **[4]**

Taylor, J., S. Tyekucheva, M. Zody, F. Chiaromonte, and K. D. Makova. 2006. Strong and weak male mutation bias at different sites in the primate genomes: Insights from the human-chimpanzee comparison. Mol. Biol. Evol. 23: 565–573. **[4]**

Taylor, R. S., J. E. Tarver, S. J. Hiscock, and P. C. Donoghue. 2014. Evolutionary history of plant microRNAs. Trends Plant Sci. 19: 175–182. **[12]**

Taylor, W. R. 1999. Protein structural domain identification. Protein Eng. 12: 203–216. **[8]**

Taylor, W. R. and M. I. Sadowski. 2010. Protein products of tandem gene duplication: A structural view, pp. 133–163. In: K. Dittmar and D. Liberles (eds.), *Evolution after Gene Duplication*. Wiley-Blackwell, Hoboken, NJ. **[8]**

Taylor, W. S. 1986. The classification of amino acid conservation. J. Theor. Biol. 119: 205–218. **[1]**

Temperley, R., R. Richter, S. Dennerlein, R. N. Lightowlers, and Z. M. Chrzanowska-Lightowlers. 2010. Hungry codons promote frameshifting in human mitochondrial ribosomes. Science 327: 301. **[1]**

Templeton, A. R. 1992. Human origins and analysis of mitochondrial DNA sequences. Science 255: 737. **[5]**

Tenaillon, M. I., J. D. Hollister, and B. S. Gaut. 2010. A triptych of the evolution of plant transposable elements. Trends Plant Sci. 15: 471–478. **[11]**

Tenaillon, O., A. Rodríguez-Verdugo, R. L. Gaut, P. McDonald, A. F. Bennett, A. D. Long, and B. S. Gaut. 2012. The molecular diversity of adaptive convergence. Science 335: 457–461. **[13]**

Tenaillon, O., D. Skurnik, B. Picard, and E. Denamur. 2010. The population genetics of commensal *Escherichia coli*. Nat. Rev. Microbiol. 8: 207–217. **[9]**

Tennessen, J. A. and 22 others. 2012. Evolution and functional impact of rare coding variation from deep sequencing of human exomes. Science 337: 64–69. **[11]**

Terai, Y., K. Takahashi, M. Nishida, T. Sato, and N. Okada. 2003. Using SINEs to probe ancient explosive speciation: "Hidden" radiation of African cichlids? Mol. Biol. Evol. 20: 924–930. **[5]**

Thalmann, O., J. Hebler, H. N. Poinar, S. Paabo, and L. Vigilant. 2004. Unreliable mtDNA data due to nuclear insertions: A cautionary tale from analysis of humans and other great apes. Mol. Ecol. 13: 321–335. **[9]**

Than, C., D. Ruths, and L. Nakhleh. 2008. PhyloNet: A software package for analyzing and reconstructing reticulate evolutionary relationships. BMC Bioinformatics 9: 322. **[5, 6]**

Thangaraj, K., M. B. Joshi, A. G. Reddy, A. A. Rasalkar, and L. Singh. 2003. Sperm mitochondrial mutations as a cause of low sperm motility. J. Androl. 24: 388–392. **[9]**

Thao, M. L., P. J. Gullan, and P Baumann. 2002. Secondary (γ-proteobacteria) endosymbionts infect the primary (β-proteobacteria) endosymbionts of mealybugs multiple times and coevolve with their hosts. Appl. Environ. Microbiol. 68: 3190–3197. **[6]**

Theissen, U. and W. Martin. 2006. The difference between organelles and endosymbionts. Curr. Biol. 16: R1016–R1017. **[5]**

Thekkumkara, T. J., W. Livingston, R. S. Kumar, and G. C. Sen. 1992. Use of alternative polyadenylation sites for tissue-specific transcription of two angiotensin-converting enzyme mRNAs. Nucl. Acids Res. 20: 683–687. **[8]**

Theunissen, B. 1988. Eugene Dubois and the Ape-Man from Java. The History of the First Missing Link and Its Discoverer. Springer, New York. **[5]**

Thewissen, J. G. M., S. T. Hussain, and M. Arif. 1994. Fossil evidence for the origin of aquatic locomotion in archaeocete whales. Science 263: 210–212. **[5]**

Thiery, J. P., G. Macaya, and G. Bernardi. 1976. An analysis of eukaryotic genomes by density gradient centrifugation. J. Mol. Biol. 108: 219–235. **[11]**

Thiranagama, R., A. T. Chamberlain, and B. A. Wood. 1991. Character phylogeny of the primate forelimb superficial venous system. Folia Primatol. 57: 181–190. **[5]**

Thomas, C. A. 1971. The genetic organization of chromosomes. Annu. Rev. Genet. 5: 237–256. **[10, 11]**

Thomas, J. A., J. J. Welch, M. Woolfit, and L. Bromham. 2006. There is no universal molecular clock for invertebrates, but rate variation does not scale with body size. Proc. Natl. Acad. Sci. USA 103: 7366–7371. **[4]**

Thompson, J. D., D. G. Higgins, and T. J. Gibson. 1994. CLUSTAL W: Improving the sensitivity of progressive multiple sequence alignment through sequence weighting, position-specific gap penalties and weight matrix choice. Nucleic Acids Res. 22: 4673–4680. **[3]**

Thompson, J. D., T. J. Gibson, F. Plewniak, F. Jeanmougin, and D. G. Higgins. 1997. The CLUSTAL_X windows interface: Flexible strategies for multiple sequence alignment aided by quality analysis tools. Nucleic Acids Res. 25: 4876–4882. **[3]**

Thomsen, P. D., B. Hoyheim, and K. Christensen. 1996. Recent fusion events during evolution of pig chromosome 3 and 6 identified by comparison with the babirusa karyotype. Cytogenet Cell Genet. 73: 203–208. **[11]**

Thomson N., M. Sebaihia, A. Cerdeño-Tárraga, S. Bentley, L. Crossman, and J. Parkhill. 2003. The value of comparison. Nature Rev. Microbiol. 1: 11–12. **[1]**

Thomson, J. M., E. A. Gaucher, M. F. Burgan, D. W. De Kee, T. Li, J. P. Aris, and S. A. Benner. 2005. Resurrecting ancestral alcohol dehydrogenases from yeast. Nat. Genet. 37: 630–635. **[5]**

Thornton, J. W., E. Need, and D. Crews. 2003. Resurrecting the ancestral steroid receptor: Ancient origin of estrogen signaling. Science 301: 1714–1717. **[5]**

Timmis, J. N. and D. Wang. 2013. Endosymbiotic evolution: the totalitarian nucleus is foiled again. Curr. Biol. 23: R30-R32. **[11]**

Timmis, J. N. and N. S. Scott. 1983. Spinach nuclear and chloroplast DNAs have homologous sequences. Nature 305: 65–67. **[9]**

Timmis, J. N., M. A. Ayliffe, C. Y. Huang, and W. Martin. 2004. Endosymbiotic gene transfer: Organelle genomes forge eukaryotic chromosomes. Nat. Rev. Genet. 5: 123–135. **[9, 10]**

Tirosh, I., A. Weinberger, M. Carmi, and N. Barkai. 2006. A genetic signature of interspecies variations in gene expression. Nat. Genet. 38: 830–834. **[12]**

Tirosh, I., N. Barkai, and K. J. Verstrepen. 2009. Promoter architecture and the evolvability of gene expression. J. Biol. 8: 95. **[12]**

Tishkoff, S. A. and 18 others. 2006. Convergent adaptation of human lactase persistence in Africa and Europe. Nat. Genet. 39: 31–40. **[2, 11]**

Todd, A. E., C. A. Orengo, and J. M. Thornton. 2001. Evolution of function in protein superfamilies, from a structural perspective. J. Mol. Biol. 307: 1113–1143. **[7]**

Toledo, V. H. and J. R. Esteban Durán. 2008. Description of *Lagocheirus delestali* n. sp. (Coleoptera: Cerambycidae) from the Reserva Biológica Alberto Manuel Brenes, Alajuela, Costa Rica. Spanish J. Agric. Res. 6: 26–29. **[5]**

TomHon, C., W. Zhu, D. Millinoff, K. Hayasaka, J. L. Slightom, M. Goodman, and D. L. Gumucio. 1997. Evolution of a fetal expression pattern via *cis* changes near the γ-globin gene. J. Biol. Chem. 272: 14062–14066. **[7]**

Tompa, P. 2002. Intrinsically unstructured proteins. Trends Biochem. Sci. 27: 527–533. **[1]**

Tordai, H., A. Nagy, K. Farkas, L. Bányai, and L. Patthy. 2005. Modules, multidomain proteins and organismic complexity. FEBS J. 272: 5064–5078. **[8]**

Torrents, D., M. Suyama, E. Zdobnov, and P. Bork. 2003. A genome-wide survey of human pseudogenes. Genome Res. 13: 2559–2567. **[9]**

Torres, A., A. Cabada, and J. J. Nieto. 2003. An exact formula for the number of alignments between two DNA sequences. DNA Sequence 14: 427–430. **[3]**

Touchon, M., S. Nicolay, B. Audit, E. B. Brodie of Brodie, Y. d'Aubenton-Carafa, A. Arneodo, and C. Thermes. 2005. Replication-associated strand asymmetries in mammalian genomes: Toward detection of replication origins. Proc. Natl. Acad. Sci. USA 102: 9836–9841. **[10]**

Tourasse, N. J. and W.-H. Li. 2000. Selective constraints, amino acid composition, and the rate of protein evolution. Mol. Biol. Evol. 17: 656–664. **[4]**

Trabesinger-Ruef, N., T. Jermann, T. Zankel, B. Durrant, G. Frank, and S. A. Benner. 1996. Pseudogenes in ribonuclease evolution: A source of new biomacromolecular function. FEBS Lett. 382: 319–322. **[7]**

Treangen, T. J. and E. P. Rocha. 2011. Horizontal transfer, not duplication, drives the expansion of protein families in prokaryotes. PLoS Genet. 7: e1001284. **[6, 9, 10]**

Trifonov, V. A. and 13 others. 2008. Multidirectional cross-species painting illuminates the history of karyotypic evolution in Perissodactyla. Chromosome Res. 16: 89–107. **[11]**

Trombetta, V. V. 1942. The cytonuclear ratio. Bot. Rev. 8: 317–336. **[11]**

Trut, L. N. 1999. Early canid domestication: the farm-fox experiment. Am. Sci. 87: 160–169. **[11]**

Tsujimura, T., A. Chinen, and S. Kawamura. 2007. Identification of a locus control region for quadruplicated green-sensitive opsin genes in zebrafish. Proc. Natl. Acad. Sci. USA 104: 12813–12818. **[12]**

Tsukahara, S., A. Kobayashi, A. Kawabe, O. Mathieu, A. Miura, and T. Kakutani. 2009. Bursts of retrotransposition reproduced in *Arabidopsis*. Nature 461: 423–426. **[9]**

Tuller, T. and 9 others. 2010. An evolutionarily conserved mechanism for controlling the efficiency of protein translation. Cell 141: 344–354. **[4]**

Turanov, A. A. and 7 others. 2009. Genetic code supports targeted insertion of two amino acids by one codon. Science 323: 259–261. **[1]**

Turner, C. and 8 others. 2003. Human genetic disease caused by *de novo* mitochondrial-nuclear DNA transfer. Hum. Genet. 112: 303–309. **[9]**

Turner, R. F. (ed.). 2011. *The Jefferson-Hemings Controversy: Report of the Scholars Commission.* Carolina Academic Press, Durham, NC. **[11]**

Tuttle, R. H. 1967. Knuckle-walking and the evolution of hominoid hands. Am. J. Phys. Anthropol. 26: 171–206. **[5]**

Tzeng, Y.-H., R. Pan, and W.-H. Li. 2004. Comparison of three methods for estimating rates of synonymous and non-synonymous nucleotide substitutions. Mol. Biol. Evol. 21: 2290–2298. **[3]**

U

Uchikawa, M., T. Takemoto, Y. Kamachi, and H. Kondoh. 2004. Efficient identification of regulatory sequences in the chicken genome by a powerful combination of embryo electroporation and genome comparison. Mech. Dev. 121: 1145–1158. **[11]**

Umylny, B., G. Presting, and W. S. Ward. 2007. Evidence of *Alu* and *B1* expression in dbEST. Syst. Biol. Reprod. Med. 53: 207–218. **[11]**

Ureta-Vidal, A., L. Etwiller, and E. Birney. 2003. Comparative genomics: Genome-wide analysis in metazoan eukaryotes. Nat. Rev. Genet. 4: 251–262. **[3]**

V

Valentine, D. L. 2007. Adaptations to energy stress dictate the ecology and evolution of the Archaea. Nat. Rev. Microbiol. 5: 316–323. **[6]**

Van de Peer, Y., A. Ben Ali, and A. Meyer. 2000. Microsporidia: Accumulating molecular evidence that a group of amitochondriate and suspectedly primitive eukaryotes are just curious fungi. Gene 246: 1–8. **[5, 6]**

van den Boogaart, P., J. Samallo, and E. Agsteribbe. 1982. Similar genes for a mitochondrial ATPase subunit in the nuclear and mitochondrial genomes of *Neurospora crassa*. Nature 298: 187–189. **[9]**

van den Bussche, R. A., J. L. Longmire, and R. J. Baker. 1995. How bats achieve a small C-value: Frequency of repetitive DNA in *Macrotus*. Mamm. Genome 6: 521–525. **[11]**

van der Giezen, M. 2009. Hydrogenosomes and mitosomes: Conservation and evolution of functions. J. Eukaryot. Microbiol. 56: 221–231. **[6]**

van der Giezen, M. and 7 others. 2002. Conserved properties of hydrogenosomal and mitochondrial ADP/ATP carriers: A common origin for both organelles. EMBO J. 21: 572–579. **[9]**

van der Kuyl, A. C. 2012. HIV infection and HERV expression: A review. Retrovirology 9: 6. **[9]**

van Hoof, A. 2005. Conserved functions of yeast genes support the duplication, degeneration and complementation model for gene duplication. Genetics 171: 1455–1461. **[7]**

Van Valen, L. 1966. Deltatheridia, a new order of mammals. Bull. Am. Mus. Nat. Hist. 132: 1–126. **[5]**

Van Valen, L. M. and V. C. Maiorana. 1991. Hela, a new microbial species. Evol. Theor. 10: 71–74. **[11]**

van't Hof, A. E., N. Edmonds, M. Dalíková, F. Marec, and I. J. Saccheri. 2011. Industrial melanism in British peppered moths has a singular and recent mutational origin. Science 332: 958–960. **[2]**

Vandepoele, K., W. De Vos, J. S. Taylor, A. Meyer, and Y. van de Peer. 2004. Major events in the genome evolution of vertebrates: Paranome age and size differ considerably between ray-finned fishes and land vertebrates. Proc. Natl. Acad. Sci. USA 101: 1638–1643. **[5]**

Vanin, E. F. 1985. Processed pseudogenes: Characteristics and evolution. Annu. Rev. Genet. 19: 253–272. **[11]**

Varki, A. 2010. Uniquely human evolution of sialic acid genetics and biology. Proc. Natl. Acad. Sci. USA 107: 8939–8946. **[7, 8]**

Vellai, T, and G. Vida. 1999. The origin of eukaryotes: The difference between prokaryotic and eukaryotic cells. Proc. R. Soc. 266B: 1571–1577. **[10]**

Vellai, T., K. Takacs, and G. Vida. 1998. A new aspect to the origin and evolution of eukaryotes. J. Mol. Evol. 46: 499–507. **[5, 6]**

Vendrely, R. and C. Vendrely. 1948. La teneur du noyau cellulaire en acide désoxyribonucléique à travers les organes, les individus et les espèces animals. Experientia 4: 434–436. **[11]**

Venkatesh, B. and 32 others. 2014. Elephant shark genome provides unique insights into gnathostome evolution. Nature 505: 174–179. **[4]**

Venner, S., C. Feschotte, and C. Biémont. 2009. Dynamics of transposable elements: Towards a community ecology of the genome. Trends Genet. 25: 317–323. **[9]**

Ventura, M., N. Archidiacono, and M. Rocchi. 2001. Centromere emergence in evolution. Genome Res. 11: 595–599. **[11]**

Verderosa, F. J. and H. J. Muller. 1954. Another case of dissimilar characters in *Drosophila* apparently representing changes of the same locus. Genetics 39: 999. **[7]**

Vergnaud, G. and F. Denoeud. 2000. Minisatellites: Mutability and genome architecture. Genome Res. 10: 899–907. **[11]**

Vicario, S., E. N. Moriyama, and J. R. Powell. 2007. Codon usage in twelve species of *Drosophila*. BMC Evol. Biol. 7: 226. **[4]**

Vickery, H. B. 1950. The origin of the word protein. Yale J. Biol. Med. 22: 387–393. **[1]**

Vieira, C. and C. Biémont. 1997. Transposition rate of the 412 retrotransposable element is independent of copy number in natural populations of *Drosophila simulans*. Mol. Biol. Evol. 14: 185–188. **[9]**

Vigilant, L., M. Stoneking, H. Harpending, K. Hawkes, and A. C. Wilson. 1991. African populations and the evolution of human mitochondrial DNA. Science 253: 1503–1507. **[5]**

Vinckenbosch, N., I. Dupanloup, and H. Kaessmann. 2006. Evolutionary fate of retroposed gene copies in the human genome. Proc. Natl. Acad. Sci. USA 103: 3220–3225. **[8]**

Vinogradov, A. E. 2003. Selfish DNA is maladaptive: evidence from the plant Red List. Trends Genet. 19: 609–614. **[11]**

Vinogradov, S. N. and 7 others. 2006. A phylogenomic profile of globins. BMC Evol. Biol. 6: 31. **[7]**

Vitkup, D., C. Sander, and G. M. Church. 2003. The amino-acid mutational spectrum of human genetic disease. Genome Biol. 4: R72. **[4]**

Vitti, J. J., S. R. Grossman, and P. C. Sabeti. 2013. Detecting natural selection in genomic data. Annu. Rev. Genet. 47: 97–120. **[11]**

Vivek, G. 2005. Analyses of protein domains and genomic elements using bioinfomatics approaches. Ph.D. Thesis, National University of Singapore. **[8]**

Vivek, G., T. W. Tan, and S. Ranganathan S. 2003. XdomView: Protein domain and exon position visualization. Bioinformatics 19: 159–160. **[8]**

Vogel, C., C. Berzuini, M. Bashton, J. Gough, and S. A. Teichmann. 2004. Supra-domains: Evolutionary units larger than single protein domains. J. Mol. Biol. 336: 809–823. **[8]**

Vogel, F. 1964. A preliminary estimate of the number of human genes. Nature 201: 847. **[11]**

Vogel, F. and A. G. Motulsky. 1996. *Human Genetics: Problems and Approaches*, 3rd Ed. Springer, Berlin. **[4]**

Volff, J.-N. 2006. Turning junk into gold: Domestication of transposable elements and the creation of new genes in eukaryotes. BioEssays 28: 913–922. **[9]**

von Dohlen, C. D., S. Kohler, S. T. Alsop, and W. R. McManus. 2001. Mealybug β-proteobacterial endosymbionts contain γ-proteobacterial symbionts. Nature 412: 433–436. **[5, 6]**

Vontas, J. G., G. J. Small, and J. Hemingway. 2000. Comparison of esterase gene amplification, gene expression and esterase activity in insecticide susceptible and resistant strains of the brown planthopper, *Nilaparvata lugens* (Stål). Insect Mol. Biol. 9: 655–660. **[7]**

Voordeckers, K., C. A. Brown, K. Vanneste, E. van der Zande, A. Voet, S. Maere, and K. J. Verstrepen. 2012. Reconstruction of ancestral metabolic enzymes reveals molecular mechanisms underlying evolutionary innovation through gene duplication. PLoS Biol. 10: e1001446. **[7]**

Vreeland, R. H., W. D. Rosenzweig, and D. W. Powers. 2000. Isolation of a 250 million-year-old halotolerant bacterium from a primary salt crystal. Nature 407: 897–900. **[5]**

W

Waddington, C. H. 1942. The epigenotype. Endeavor 1: 18–20. **[12]**

Wade, N. 2013. East Asian physical traits linked to 35,000-year-old mutation. http: //www.nytimes.com/2013/02/15/science/studying-recent-human-evolution-at-the-genetic-level.html **[11]**

Wagner, A. and N. de la Chaux. 2008. Distant horizontal gene transfer is rare for multiple families of prokaryotic insertion sequences. Mol. Genet. Genomics 280: 397–408. **[9]**

Wakeley, J. 2008. *Coalescent Theory.* Ben Roberts, Greenwood Village, Colorado. **[2]**

Wakeley, J. 2008. Complex speciation of humans and chimpanzees. Nature 452: E3–E4. **[5]**

Waldan, K. K. and Robertson, H. M. 1997. Ancient DNA from amber fossil bees? Mol. Biol. Evol. 14: 1057–1077. **[5]**

Wall, J. D. and S. K. Kim. 2007. Inconsistencies in Neanderthal genomic DNA sequences. PLoS Genet. 3: e175. **[5]**

Wallace, I. M., O. O'Sullivan, D. G. Higgins, and C. Notredame. 2006. M-Coffee: Combining multiple sequence alignment methods with T-Coffee. Nucleic Acids Res. 34: 1692–1699. **[3]**

Wallace, M. R., L. B. Andersen, A. M. Saulino, P. E. Gregory, T. W. Glover, and F. S. Collins. 1991. A *de novo Alu* insertion results in neurofibromatosis type 1. Nature 353: 864–866. **[9]**

Wallin, I. E. 1923. On the nature of mitochondria. V. A critical analysis of Portier's "Les symbiotes." Anat. Rec. 25: 1–7. **[5]**

Wallin, I. E. 1927. *Symbionticism and the Origin of Species.* Baillière, Tindall and Cox, London. **[5]**

Walsh, J. B. 1985. Interaction of selection and biased gene conversion in a multigene family. Proc. Natl. Acad. Sci. USA 82: 153–157. **[7]**

Walsh, J. B. 1995. How often do duplicated genes evolve new functions? Genetics 139: 421–428. **[7]**

Wang, E. and 9 others. 2004. The genetic architecture of selection at the human dopamine receptor D4 (*DRD4*) gene locus. Am. J. Hum. Genet. 74: 931–944. **[2]**

Wang, G., X. Zhang, and W. Jin. 2009. An overview of plant centromeres. J. Genet Genomics. 36: 529–537. **[11]**

Wang, H.-C., E. Susko, and A. J. Roger. 2011. Fast statistical tests for detecting heterotachy in protein evolution. Mol. Biol. Evol. 28: 2305–2315. **[4]**

Wang, H.-L., Z.-Y. Yan, and D.-Y. Jin. 1997. Reanalysis of published DNA sequence amplified from Cretaceous dinosaur egg fossil. Mol. Biol. Evol. 14: 589–591. **[5]**

Wang, J. 2005. Estimation of effective population sizes from data on genetic markers. Phil. Trans. R. Soc. 360B: 1395–1409. **[2]**

Wang, J. and 8 others. 2004. Mouse transcriptome: Neutral evolution of "non-coding" complementary DNAs. Nature 431: doi:10.1038/nature03016 **[1]**

Wang, J., H. C. Fan, B. Behr, and S. R. Quake. 2012. Genome-wide single-cell analysis of recombination activity and de novo mutation rates in human sperm. Cell 150: 402–412. **[4]**

Wang, L. and R. Yi. 2013. 3′ UTRs take a long shot in the brain. Bioessays 36: 39–45. **[12]**

Wang, S. S., B. O. Zhou, and J. Q. Zhou. 2011. Histone H3 lysine 4 hypermethylation prevents aberrant nucleosome remodeling at the *PHO5* promoter. Mol. Cell. Biol. 31: 3171–3181. **[12]**

Wang, S., I. L. Pirtle, and R. M. Pirtle. 1997. A human 28S ribosomal RNA retropseudogene. Gene 196:105–111. **[9]**

Wang, W., H. Yu, and M. Long. 2004. Duplication-degeneration as a mechanism of gene fission and the origin of new genes in *Drosophila* species. Nat. Gen. 36: 523–527. **[8]**

Wang, W. and 13 others. 2006. High rate of chimeric gene origination by retroposition in plant genomes. Plant Cell 18: 1791–1802. **[9]**

Wang, X., W. E. Grus, and J. Zhang. 2006. Gene losses during human origins. PLoS Biol. 4: e52. **[8]**

Ward, L. D., M. Kellis. 2012. Evidence of abundant purifying selection in humans for recently acquired regulatory functions. Science 337: 1675–1678. **[11]**

Wardrop, S. L., kConFab Investigators, and M. A. Brown. 2005. Identification of two evolutionarily conserved and functional regulatory elements in intron 2 of the human BRCA1 gene. Genomics 86: 316–328. **[11]**

Wasserzug, O., S. Blum, E. Klement, F. Lejbkowicz, R. Miller-Lotan, and A. P. Levy. 2007. Haptoglobin 1-1 genotype and the risk of life-threatening *Streptococcus* infection: Evolutionary implications. J. Infect. 54: 410. **[8]**

Watanabe, H., H. Mori. T. Itoh, and T. Gojobori. 1997. Genome plasticity as a paradigm of eubacterial evolution. J. Mol. Evol. 44: s57-s64. **[10]**

Waterston, R. H. and 221 others. 2002. Initial sequencing and comparative analysis of the mouse genome. Nature 420: 520–562. **[4]**

Watson, J. D. and F. H. C. Crick. 1953. Genetical implications of the structure of deoxyribonucleic acid. Nature 171: 964–967. **[10]**

Weatherall, D. J. and J. B. Clegg. 1979. Recent developments in the molecular genetics of human hemoglobin. Cell 16: 467–479. **[8]**

Webster, M. T. 2009. Patterns of autosomal divergence between human and chimpanzee genomes support an allopatric model of speciation. Gene 443: 70–75. **[5]**

Webster, M. T., N. G. C. Smith, L. Hultin-Rosenberg, P. F. Arndt, and H. Ellegren. 2005. Male-driven biased gene conversion governs the evolution of base composition in human *Alu* repeats. Mol. Biol. Evol. 22: 1468–1474. **[4]**

Weinberg, Z., J. Perreault, M. M. Meyer, and R. R. Breaker. 2009. Exceptional structured noncoding RNAs revealed by bacterial metagenome analysis. Nature 462: 656–659. **[1]**

Weiner, A. M., P. L. Deininger, and A. Efstratiadis. 1986. Nonviral retroposons: Genes, pseudogenes and transposable elements generated by the reverse flow of genetic information. Annu. Rev. Biochem. 55: 631–661. **[11]**

Weinreich, D. M., N. F. Delaney, M. A. Depristo, and D. L. Hartl. 2006. Darwinian evolution can follow only very few mutational paths to fitter proteins. Science 312: 111–114. **[13]**

Weir, J. T. 2006. Different timing and patterns of species accumulation in lowland and highland Neotropical birds. Evolution 61: 842–845. **[5]**

Weir, J. T. and D. Schluter. 2004. Ice sheets promote speciation in boreal birds. Proc. R. Soc. London 271B: 1881–1887. **[5]**

Weir, J. T. and D. Schluter. 2007. The latitudinal gradient in recent speciation and extinction rates in birds and mammals. Science 315: 1928–1933. **[5]**

Weir, J. T. and D. Schluter. 2008. Calibrating the avian molecular clock. Mol. Ecol. 17: 2321–2328. **[5]**

Weiss, E. H., A. Mellor, L. Golden, K. Fahrner, E. Simpson, J. Hurst, and R. A. Flavell. 1983. The structure of a mutant *H-2* gene suggests that the generation of polymorphism in *H-2* genes may occur by gene conversion-like events. Nature 301: 671–674. **[7]**

Weiss, R. A. and J. P. Stoye. 2013. Our viral inheritance. Science 340: 820–821. **[9]**

Welch, J. J., O. R. P. Bininda-Emonds, and L. Bromham. 2008. Correlates of substitution rate variation in mammalian protein-coding sequences. BMC Evol. Biol. 8: 53. **[4]**

Welch, R. A. and 18 others. 2002. Extensive mosaic structure revealed by the complete genome sequence of uropathogenic *Escherichia coli*. Proc. Natl. Acad. Sci. USA 99: 17020–17024. **[9, 10]**

West, S. A., A. S. Griffin, A. Gardner, and S. P. Diggle. 2006. Social evolution theory for microorganisms. Nat. Rev. Microbiol. 4: 597–607. **[9]**

West, S. A., S. P. Diggle, A. Buckling, A. Gardner, and A. S. Griffin. 2007. The social lives of microbes. Ann. Rev. Ecol. Evol. Syst. 38: 53–77. **[9]**

Whitaker, J. W., G. A. McConkey, and D. R. Westhead. 2009. The transferome of metabolic genes explored: Analysis of the horizontal transfer of enzyme encoding genes in unicellular eukaryotes. Genome Biol. 10: R36. **[5]**

White, M. A., J. K. Eykelenboom, M. A. Lopez-Vernaza, E. Wilson, and D. R. F. Leach. 2008. Non-random segregation of sister chromosomes in *Escherichia coli*. Nature 455: 1248–1250. **[4]**

Whiting, M. F., S. Bradler, and T. Maxwell. 2003. Loss and recovery of wings in stick insects. Nature 421: 264–267. **[5]**

Whitlock, M. C., P. C. Phillips, F. B.-G. Moore, and S. J. Tonsor. 1995. Multiple fitness peaks and epistasis. Annu. Rev. Ecol. Syst. 26: 601–629. **[13]**

Whitman, W. B. 2009. The modern concept of the prokaryote. J. Bact. 191: 2000–2005. **[5, 6]**

Whittaker, R. H. 1969. New concepts of kingdoms of organisms. Science 163: 150–160. **[5]**

Wickett, N. J., Y. Fan, P. O. Lewis, and B. Goffinet. 2008. Distribution and evolution of pseudogenes, gene losses, and a gene rearrangement in the plastid genome of the nonphotosynthetic liverwort, *Aneura mirabilis* (Metzgeriales, Jungermanniopsida). J. Mol. Evol. 67: 111–122. **[4]**

Widegren, B., U. Árnason, and G. Akuslarvi. 1985. Characteristics of conserved 1,579-bp highly repetitive component in the killer whale, *Orcinus orca*. Mol. Biol. Evol. 2: 411–419. **[11]**

Wielgoss, S. and 7 others. 2011. Mutation rate inferred from synonymous substitutions in a long-term evolution experiment with *Escherichia coli*. G3 1: 183–186. **[13]**

Wiens, J. J. 2011. Re-evolution of lost mandibular teeth in frogs after more than 200 million years, and re-evaluating Dollo's law. Evolution 65: 1283–1296. **[5]**

Wilde, C. D. 1986. Pseudogenes. CRC Crit. Rev. Biochem 19: 323–352. **[11]**

Wildman, D. E., M. Uddin, G. Liu, L. I. Grossman, and M. Goodman. 2003. Implications of natural selection in shaping 99.4% nonsynonymous DNA identity between humans and chimpanzees: Enlarging genus *Homo*. Proc. Natl. Acad. Sci. USA 100: 7181–7188. **[5]**

Wiley, E. O. 1981. *Phylogenetics: The Theory and Practice of Phylogenetic Systematics*. Wiley, New York. **[5, 6]**

Wilke, T., R. Schultheiß, and C. Albrecht. 2009. As time goes by: A simple fool's guide to molecular clock approaches in invertebrates. Am. Malac. Bull. 27: 25–45. **[5]**

Wilkins, A. S., R. W. Wrangham, and W. T. Fitch. 2014. The "domestication syndrome" in mammals: A unified explanation based on neural crest cell behavior and genetics. Genetics 197: 795–808. **[11]**

Wilkinson, R. D., M. E. Steiper, C. Soligo, R. D. Martin, Z. Yang, and S. Tavaré. 2011. Dating primate divergences through an integrated analysis of palaeontological and molecular data. Syst. Biol. 60: 16–31. **[5]**

Wilks, H. M. and 9 others. 1988. A specific, highly active malate dehydrogenase by redesign of a lactate dehydrogenase framework. Science 242: 1541–1544. **[7]**

Williams, D. L. and 10 others. 2009. Implications of high level pseudogene transcription in *Mycobacterium leprae*. BMC Genomics. 10: 397. **[1]**

Williams, E. J. B. and L. D. Hurst. 2002. Is the synonymous substitution rate in mammals gene-specific? Mol. Biol. Evol. 19: 1395–1398. **[4]**

Williams, E. J. B., C. Pal, and L. D. Hurst. 2000. The molecular evolution of signal peptides. Gene 253: 313–322. **[4]**

Williams, S. A. and M. Goodman. 1989. A statistical test that supports a human/chimpanzee clade based on noncoding DNA sequence data. Mol. Biol. Evol. 6: 325–330. **[5]**

Williams, T. A., P. G. Foster, C. J. Cox, and T. M. Embley. 2013. An archaeal origin of eukaryotes supports only two primary domains of life. Nature 504: 231–236. **[6]**

Williamson, S. H., M. J. Hubisz, A. G. Clark, B. A. Payseur, C. D. Bustamante, and R. Nielsen. 2007. Localizing recent adaptive evolution in the human genome. PLoS Genet. 3: e90. **[11]**

Willich, A. F. M. 1798. *Elements of the Critical Philosophy*. Longman, London. **[12]**

Willson, S. J. 2008. Reconstruction of certain phylogenetic networks from the genomes at their leaves. J. Theor. Biol. 252:338–349. **[5, 6]**

Wilson, A. C., S. S. Carlson, and T. J. White. 1977. Biochemical evolution. Annu. Rev. Biochem. 46: 573–639. **[5]**

Wilson, B. A. and J. Masel. 2011. Putatively noncoding transcripts show extensive association with ribosomes. Genome Biol. Evol. 3: 1245–1252. **[8, 11]**

Winkler, H. 1920. *Verbreitung und Ursache der Parthenogenesis im Pflanzen- und Tierreiche*. Gustav Fischer, Jena. **[1, 10, 11]**

Wistow, G. 1993. Lens crystallins: Gene recruitment and evolutionary dynamism. Trends Biochem. Sci. 18: 301–306. **[7]**

Wistow, G. J. and J. Piatigorsky. 1987. Recruitment of enzymes as lens structural proteins. Science 236: 1554–1556. **[7]**

Wistow, G., J. W. M. Mulders, and W. W. de Jong. 1987. The enzyme lactate dehydrogenase as a structural protein in avian and crocodilian lenses. Nature 326: 622–624. **[7]**

Woese, C. R. 1969. Models for the evolution of codon assignments. J. Mol. Evol. 14: 235–240. **[10]**

Woese, C. R. 1987. Bacterial evolution. Microbiol. Rev. 51: 221–271. **[6]**

Woese, C. R. 1996. Phylogenetic trees: Whither microbiology? Curr. Biol. 6:1060–1063. **[6]**

Woese, C. R. 1998. Default taxonomy: Ernst Mayr's view of the microbial world. Proc. Natl. Acad. Sci. USA 95: 11043–11046. **[6]**

Woese, C. R., O. Kandler, and M. L. Wheelis. 1990. Towards a natural system of organisms: Proposal for the domains Archaea, Bacteria, and Eucarya. Proc. Natl. Acad. Sci. USA 87: 4576–4579. **[6]**

Woese, C. R. and G. E. Fox. 1977a. Phylogenetic structure of the prokaryotic domain: The primary kingdoms. Proc. Natl. Acad. Sci. USA 74: 5088–5090. **[5, 6]**

Woese, C. R. and G. E. Fox. 1977b. The concept of cellular evolution. J. Mol. Evol. 10: 1–6. **[5]**

Woese, C. R. and G. E. Fox. 1978. Methanogenic bacteria. Nature 273: 101. **[5]**

Wolf, J. B. W., A. Künstner, K. Nam, M. Jakobsson, and H. Ellegren. 2009. Nonlinear dynamics of nonsynonymous (d_N) and synonymous (d_S) substitution rates affects inference of selection. Genome Biol. Evol. 1: 308–319. **[4]**

Wolf, M. F. 2000. *Rube Goldberg Inventions*. Simon & Schuster, New York. **[8]**

Wolf, Y. I., I. B. Rogozin, and E. V. Koonin. 2004. Coelomata and not Ecdysozoa: Evidence from genome-wide phylogenetic analysis. Genome Res. 14: 29–36. **[5]**

Wolfe, K. 2000. Robustness—it's not where you think it is. Nature Genet. 25: 3–4. **[5]**

Wolfe, K. H. 2001. Yesterday's polyploids and the mystery of diploidization. Nat. Rev. Genet. 2: 333–341. **[7]**

Wolfe, K. H. and D. C. Shields. 1997. Molecular evidence for an ancient duplication of the entire yeast genome. Nature 387: 708–713. **[7]**

Wolfe, K. H. and W.-H. Li. 2003. Molecular evolution meets the genomics revolution. Nat. Genet. 33: 255–265. **[4]**

Wolfe, K. H., C. W. Morden, S. C. Ems, and J. D. Palmer. 1992a. Rapid evolution of the plastid translational apparatus in a non-photosynthetic plant: Loss or accelerated sequence evolution of tRNA and ribosomal protein genes. J. Mol. Evol. 35: 304–317. **[10]**

Wolfe, K. H., C. W. Morden, and J. D. Palmer. 1992b. Function and evolution of a minimal plasmid genome from a non-photosynthetic parasitic plant. Proc. Natl. Acad. Sci. USA 89: 10648–10652. **[10]**

Wolfe, K. H., P. M. Sharp, and W.-H. Li. 1989. Mutation rates differ among regions of the mammalian genome. Nature 337: 283–285. **[11]**

Wolpert, L. and C. Tickle. 2011. *Principles of Development* (4th edition). Oxford University Press. Oxford. **[12]**

Woo, Y. H. and W.-H. Li. 2012. Evolutionary conservation of histone modifications in mammals. Mol. Biol. Evol. 29: 1757–1767. **[12]**

Wood, T. E., N. Takebayashi, M. S. Barker, I. Mayrose, P. B. Greenspoon, and L. H. Rieseberg. 2009. The frequency of polyploid speciation in vascular plants. Proc. Natl. Acad. Sci. USA 106: 13875–13879. **[7]**

Woodward, S. R., N. J. Weyand, and M. Bunnel. 1994. DNA sequence from Cretaceous period bone fragments. Science 266: 1229–1232. **[5, 9]**

Worning, P., L. J. Jensen, P. F. Hallin, H. H. Staerfeldt, and D. W. Ussery. 2006. Origin of replication in circular prokaryotic chromosomes. Environ. Microbiol. 8: 353–361. **[10]**

Worth, C. L., S. Gong, and T. L. Blundell. 2009. Structural and functional constraints in the evolution of protein families. Nat. Rev. Mol. Cell Biol. 10: 709–720. **[7, 8]**

Wostemeyer, J. and A. Burmester. 1986. Structural organization of the genome of the zygomycete *Absidia glauca*: Evidence for high repetitive DNA content. Curr. Genet. 10: 903–907. **[11]**

Wright, P. E. and H. J. Dyson. 1999. Intrinsically unstructured proteins: Re-assessing the protein structure-function paradigm. J. Mol. Biol. 293: 321–331. **[1]**

Wright, S. 1931. Evolution in Mendelian populations. Genetics 16: 97–159. **[2]**

Wright, S. 1932. The roles of mutation, inbreeding, crossbreeding and selection in evolution. Proc. 6th Intl. Cong. Evol. 1: 356–366. **[13]**

Wu, C.-I. and N. Maeda. 1987. Inequality in mutation rates of the two strands of DNA. Nature 327: 169–170. **[4]**

Wu, C.-I., J. R. True, and N. Johnson. 1989. Fitness reduction associated with the deletion of a satellite DNA array. Nature 341: 248–251. **[2]**

Wu, D.-D., D. M. Irwin, and Y. P. Zhang. 2011. *De novo* origin of human protein-coding genes. PLoS Genet. 7: e1002379. **[8]**

Wu, G., A. Fiser, B. ter Kuile, A. Šali, and M. Müller. 1999. Convergent evolution of *Trichomonas vaginalis* lactate dehydrogenase from malate dehydrogenase. Proc. Natl. Acad. Sci. USA 96: 6285–6290. **[7]**

Wu, H., Z. Zhang, S. Hu, and J. Yu. 2012. On the molecular mechanism of GC content variation among eubacterial genomes. Biol. Direct 7: 2. **[10]**

Wu, M., S. Chatterji, and J. A. Eisen. 2012. Accounting for alignment uncertainty in phylogenomics. PLoS ONE 7: e30288. **[3]**

Wu, X. and P. A. Sharp. 2013. Divergent transcription: a driving force for new gene origination? Cell 155: 990–996. **[12]**

Wujek, D. E. 1979. Intracellular bacteria in the blue-green alga *Pleurocapsa minor.* Trans. Am. Microscop. Soc. 98: 143–145. **[6]**

Wurster, D. H. and K. Benirschke. 1970. Indian muntjac, *Muntiacus muntjak*: a deer with a low diploid chromosome number. Science 168: 1364–1366. **[11]**

Wyss, A. 1990. Clues to the origin of whales. Nature 347: 428–429. **[5]**

X

Xi, Z., Y. Wang, R. K. Bradley, M. Sugumaran, C. J. Marx, J. S. Rest, and C. C. Davis. 2013. Massive mitochondrial gene transfer in a parasitic flowering plant clade. PLoS Genet. 9: e1003265. **[9]**

Xia, X. 2009. Information-theoretic indices and an approximate significance test for testing the molecular clock hypothesis with genetic distances. Mol. Phylogenet. Evol. 52: 665–676. **[4]**

Xu, G., C. Guo, H. Shan, and H. Kong. 2012. Divergence of duplicate genes in exon-intron structure. Proc. Natl. Acad. Sci. USA 109: 1187–1192. **[8]**

Xu, K., S. Oh, T. Park, D. C. Presgraves, and S. V. Yi. 2012. Lineage-specific variation in slow- and fast-X evolution in primates. Evolution 66: 1751–1761. **[4]**

Xu, Y., P. Ma, P. Shah, A. Rokas, Y. Liu, and C. H. Johnson. 2013. Non-optimal codon usage is a mechanism to achieve circadian clock conditionality. Nature 495: 116–120. **[4]**

Xue, Y. and 13 others. 2006. Spread of an inactive form of caspase-12 in humans is due to recent positive selection. Am. J. Hum. Genet. 78: 659–670. **[7, 8]**

Xue, Y. and 15 others. 2009. Human Y chromosome base-substitution mutation rate measured by direct sequencing in a deep-rooting pedigree. Curr. Biol. 19: 1453–1457. **[11]**

Xue, Y. and 15 others. 2009. Human Y chromosome base-substitution mutation rate measured by direct sequencing in a deep-rooting pedigree. Curr. Biol. 19: 1453–1457. **[4]**

Y

Yamamichi, M., J. Gojobori, and H. Innan. 2012. An autosomal analysis gives no genetic evidence for complex speciation of humans and chimpanzees. Mol. Biol. Evol. 29: 145–156. **[5]**

Yampolsky, L. Y. and A. Stoltzfus. 2005. The exchangeability of amino acids in proteins. Genetics 170: 1459–1472. **[4]**

Yancopoulos, S., O. Attie, and R. Friedberg. 2005. Efficient sorting of genomic permutations by translocation, inversion and block interchange. Bioinformatics 21: 3340–3346. **[10]**

Yang, C., E. Bolotin, T. Jiang, F. M. Sladek, and E. Martinez. 2007. Prevalence of the initiator over the TATA box in human and yeast genes and identification of DNA motifs enriched in human TATA-less core promoters. Gene 389: 52–65. **[12]**

Yang, J. and V. G. Corces. 2012. Insulators, long-range interactions, and genome function. Curr. Opin. Genet. Dev. 22: 86–92. **[12]**

Yang, Z. 1993. Maximum likelihood estimation of phylogeny from DNA sequences when substitution rates differ over sites. Mol. Biol. Evol. 10: 1396–1401. **[5]**

Yang, Z. 1998. Likelihood ratio tests for detecting positive selection and application to primate lysozyme evolution. Mol. Biol. Evol. 15: 568–573. **[4, 5]**

Yang, Z. 2006. *Computational Molecular Evolution*. Oxford University Press, Oxford. **[3, 5, 6]**

Yang, Z. and R. Nielsen. 2000. Estimating synonymous and nonsynonymous substitution rates under realistic evolutionary models. Mol. Biol. Evol. 17: 32–43. **[3]**

Yang, Z. and R. Nielsen. 2008. Mutation-selection models of codon substitution and their use to estimate selective strengths on codon usage. Mol. Biol. Evol. 25: 568–579. **[4]**

Yang, Z., R. Nielsen, N. Goldman, and A.-M. Krabbe Pedersen. 2000. Codon-substitution models for heterogeneous selection pressure at amino acid sites. Genetics 155: 431–449. **[5]**

Yersin, A. 1897. Sur la peste bubonique (serotherapie). Ann. Inst. Pasteur 11: 81–93. **[9]**

Yi, S. 2012. Birds do it, bees do it, worms and ciliates do it too: DNA methylation from unexpected corners of the tree of life. Genome Biol. 13: 174. **[12]**

Yi, S. V. 2006. Non-adaptive evolution of genome complexity. BioEssays 28: 979–982. **[2, 11]**

Yi, S., D. L. Ellsworth, and W.-H. Li. 2002. Slow molecular clocks in Old World monkeys, apes, and humans. Mol. Biol. Evol. 19: 2191–2198. **[4]**

Yokobori, S., T. Suzuki, and K. Watanabe. 2001. Genetic code variations in mitochondria: tRNA as a major determinant of genetic code plasticity. J. Mol. Evol. 53: 314–326. **[10]**

Yokoyama, S. 2008. Evolution of dim-light and color vision pigments. Annu. Rev. Genomics Hum. Genet. 9: 259–282. **[7]**

Yokoyama, S. and R. Yokoyama. 1989. Molecular evolution of human visual pigment genes. Mol. Biol. Evol. 6: 186–197. **[7]**

Yong, E. 2014. The unique merger that made you (and ewe, and yew). Nautilus Issue 010: http://nautil.us/issue/10/mergers--acquisitions/the-unique-merger-that-made-you-and-ewe-and-yew **[6]**

Yoshida, S., S. Maruyama, H. Nozaki, and K. Shirasu. 2010. Horizontal gene transfer by the parasitic plant *Striga hermonthica*. Science 328: 1128. **[9]**

Young, G. R., U. Eksmond, R. Salcedo, L. Alexopoulou, J. P. Stoye, and G. Kassiotis. 2012. Resurrection of endogenous retroviruses in antibody-deficient mice. Nature 491: 774–778. **[9]**

Yu, H. and 7 others. 2006. Origination and evolution of a human-specific transmembrane protein gene, *c1orf37-dup*. Hum. Mol. Genet. 15: 1870–1875. **[9]**

Yu, H. and 9 others. 2015. Genome-wide characterization of PRE-1 reveals a hidden evolutionary relationship between suidae and primates. Preprint. **[9]**

Yuh, C.-H. and E. H. Davidson. 1996. Modular *cis*-regulatory organization of *Endo16*, a gut-specific gene of the sea urchin embryo. Development 122: 1069–1082. **[12]**

Yuh, C.-H., A. Ransick, P. Martinez, R. J. Britten, and E. H. Davidson. 1994. Complexity and organization of DNA-protein interactions in the 5′ regulatory region of an endoderm-specific marker gene in the sea urchin embryo. Mech. Dev. 47: 165–186. **[12]**

Yun, W. J., Y. W. Kim, Y. Kang, J. Lee, A. Dean, and A. Kim. 2014. The hematopoietic regulator *TAL1* is required for chromatin looping between the β-globin LCR and human γ-globin genes to activate transcription. Nucleic Acids Res. 42: 4283–4293. **[12]**

Yunis, J. J. and O. Prakash. 1982. The origin of man: A chromosomal pictorial legacy. Science 215: 1525–1530. **[11]**

Z

Zacharias, H. 1986. Tissue-specific schedule of selective replication in *Drosophila nasutoides*. Roux's Arch. Dev. Biol. 195: 378–388. **[11]**

Zdobnov, E. M. and P. Bork. 2007. Quantification of insect genome divergence. Trends Genet. 23: 16–20. **[11]**

Zdobnov, E. M., C. von Mering, I. Letunic, and P. Bork. 2005. Consistency of genome-based methods in measuring Metazoan evolution. FEBS Lett. 579: 3355–3361 **[11]**

Zeh, D. W., J. A. Zeh, and Y. Ishida. 2009. Transposable elements and an epigenetic basis for punctuated equilibria. BioEssays 31: 715–726. **[9]**

Zeng, K., Y.-X. Fu, S. Shi, and C.-I. Wu. 2006. Statistical tests for detecting positive selection by utilizing high-frequency variants. Genetics 174: 1431–1439. **[11]**

Zeng, W. P. 2013. "All things considered": Transcriptional regulation of T helper type 2 cell differentiation from precursor to effector activation. Immunology 140: 31–38. **[12]**

Zhang, C. and 11 others. 2006. A whole genome long-range haplotype (WGLRH) test for detecting imprints of positive selection in human populations. Bioinformatics 22: 2122–2128. **[2]**

Zhang, H., L. Gao, J. Anandhakumar, and D. S. Gross. 2014. Uncoupling transcription from covalent histone modification. PLoS Genet. 10: e1004202. **[12]**

Zhang, J. 2003. Evolution by gene duplication: An update. Trends Ecol. Evol. 18: 292–298. **[7]**

Zhang, J. 2006. Parallel adaptive origins of digestive RNAses in Asian and African leaf monkeys. Nature Genet. 38: 819–823. **[5]**

Zhang, J., S. Kumar, and M. Nei. 1997. Small-sample tests of episodic adaptive evolution: A case study of primate lysozymes. Mol. Biol. Evol. 14: 1335–1338. **[5]**

Zhang, J., Y.-P. Zhang and H. F. Rosenberg. 2002. Adaptive evolution of a duplicated pancreatic ribonuclease gene in a leaf-eating monkey. Nat. Genet. 30: 411–415. **[7]**

Zhang, L. and W.-H. Li. 2005. Human SNPs reveal no evidence of frequent positive selection. Mol. Biol. Evol. 22: 2504–2507. **[2, 4]**

Zhang, L. and Y. S. Rong. 2012. Retrotransposons at *Drosophila* telomeres: Host domestication of a selfish element for the maintenance of genome integrity. Biochim. Biophys. Acta 1819: 771–775. **[9]**

Zhang, L.-Y., S.-H. Chang, and J. Wang. 2010. How to make a minimal genome for synthetic minimal cell. Protein Cell 1: 427–434. **[10]**

Zhang, R. and C.-T. Zhang. 2003. Multiple replication origins of the archaeon Halobacterium species NRC-1. Biochem. Biophys. Res. Commun. 302: 728–734. **[10]**

Zhang, S. W., Y. F. Liu, Y. Yu, T. H. Zhang, and X. N. Fan. 2014. MSLoc-DT: A new method for predicting the protein subcellular location of multispecies based on decision templates. Anal. Biochem. 449: 164–171. **[11]**

Zhang, W., V. Sehgal, D. M. Dinh, R. B. R. Azevedo, T. F. Cooper, and R. Azencott. 2012. Estimation of the rate and effect of new beneficial mutations in asexual populations. Theor. Pop. Biol. 81: 168–178. **[13]**

Zhang, X., O. De la Cruz, J. M. Pinto, D. Nicolae, S. Firestein, and Y. Gilad. 2007. Characterizing the expression of the human olfactory receptor gene family using a novel DNA microarray. Genome Biol. 8: R86. **[7]**

Zhang, Z. and M. Gerstein. 2003. Patterns of nucleotide substitution, insertion and deletion in the human genome inferred from pseudogenes. Nucl. Acids Res. 31: 5338–5348. **[4]**

Zhang, Z. and M. Gerstein. 2004. Large-scale analysis of pseudogenes in the human genome. Curr. Opin. Genet. Dev. 14: 328–335. **[9]**

Zhang, Z. D., A. Frankish, T. Hunt, J. Harrow, and M. Gerstein. 2010. Identification and analysis of unitary pseudogenes:

Historic and contemporary gene losses in humans and other primates. Genome Biol. 11: R26. **[7, 8]**

Zhang, Z., P. Harrison, Y. Liu, and M. Gerstein. 2003. Millions of years of evolution preserved: A comprehensive catalog of the processed pseudogenes in the human genome. Genome Res. 13: 2541–2558. **[9]**

Zhao, H. and G. Bourque. 2010. Chromosomal rearrangements in evolution. pp. 165–182. In: G. Caetano-Anollés (ed.) *Evolutionary Genomics and Systems Biology*. Wiley, Hoboken, New Jersey. **[11]**

Zhao, H., J. R. Yang, H. Xu, and J. Zhang. 2010. Pseudogenization of the umami taste receptor gene *Tas1r1* in the giant panda coincided with its dietary switch to bamboo. Mol. Biol. Evol. 27: 2669–2673. **[7, 8, 11]**

Zharkikh, A. and W.-H. Li. 1992a. Statistical properties of bootstrap estimation of phylogenetic variability from nucleotide sequences. I. Four taxa with a molecular clock. Mol. Biol. Evol. 9: 1119–1147. **[5]**

Zharkikh, A. and W.-H. Li. 1992b. Statistical properties of bootstrap estimation of phylogenetic variability from nucleotide sequences. II. Four taxa without a molecular clock. J. Mol. Evol. 35: 356–366. **[5]**

Zharkikh, A. and W.-H. Li. 1995. Estimation of confidence in phylogeny: The complete-and-partial bootstrap technique. Mol. Phylogenet. Evol. 4: 44–63. **[5]**

Zhaxybayeva, O. and P. J. Gogarten. 2004. Cladogenesis, coalescence and the evolution of the three domains of life. Trends Genet. 20: 182–187. **[5]**

Zheng, Y., K. L. Dimond, D. Graur, and R. A. Zufall. 2013. Patterns of intron sequence conservation in the genus *Tetrahymena*. Protist Genomics 1: 19–24. **[4]**

Zhong, S., S. P. Miller, D. E. Dykhuizen, and A. M. Dean. 2009. Transcription, translation, and the evolution of specialists and generalists. Mol. Biol. Evol. 26: 2661–2678. **[13]**

Zhou, H., J. Zhao, C. H. Yu, Q. J. Luo, Y. Q. Chen, Y. Xiao, and L. H Qu. 2004. Identification of a novel box C/D snoRNA from mouse nucleolar cDNA library. Gene 327: 99–105. **[11]**

Zhou, M. and 7 others. 2013. Non-optimal codon usage affects expression, structure and function of clock protein FRQ. Nature 495: 111–115. **[4]**

Zhou, Q. and 9 others. 2008. On the origin of new genes in *Drosophila*. Genome Res. 18: 1446–1455. **[8]**

Zhou, T., M. Weems, and C. O. Wilke. 2009. Translationally optimal codons associate with structurally sensitive sites in proteins. Mol. Biol. Evol. 26: 1571–1580. **[4]**

Zhu, J., F. He, S. Hu, and J. Yu. 2008. On the nature of human housekeeping genes. Trends Genet. 24: 481–484. **[12]**

Zimmer, C. 2013. And the genomes keep shrinking… http://phenomena.nationalgeographic.com/2013/08/23/and-the-genomes-keep-shrinking. **[10]**

Zimmer, C. 2015. Is most of our DNA garbage? New York Times http://www.nytimes.com/2015/03/08/magazine/is-most-of-our-dna-garbage.html?_r=0 **[11]**

Zimmer, E. A., S. L. Martin, S. M. Beverley, Y. W. Kan, and A. C. Wilson. 1980. Rapid duplication and loss of genes coding for the α chains of hemoglobin. Proc. Natl. Acad. Sci. USA 77: 2158–2162. **[7]**

Zinder, N. D. and J. Lederberg. 1952. Genetic exchange in *Salmonella*. J. Bacteriol. 64: 679–699. **[9]**

Zischler, H., M. Hoss, O. Handt, A. von Haeseler, A. C. van der Kuyl, J. Goudsmit, and S. Pääbo. 1995. Detecting dinosaur DNA. Science 268: 1192–1193. **[5, 9]**

Zmasek, C. M. and S. R. Eddy. 2001. A simple algorithm to infer gene duplication and speciation events on a gene tree. Bioinformatics 17: 821–828. **[7]**

Zuckerkandl, E. 1976a. Gene control in eukaryotes and the C-value paradox: "Excess" DNA as an impediment to transcription of coding sequences. J. Mol. Evol. 9: 73–104. **[11]**

Zufall, R. A., T. Robinson, and L. A. Katz. 2005. Evolution of developmentally regulated genome rearrangements in eukaryotes. J. Exp. Zool. B. Mol. Dev. Evol. 304: 448–455. **[1]**

Index

Entries in **boldface** indicate defined terms. Page numbers in *italics* refer to information in an illustration or table.

A

C

F

H

M

Q

R

S

U

About the Book

Editor: Andrew D. Sinauer
Project Editor: Carol Wigg
Copy Editor: Lou Doucette
Production Manager: Christopher Small
Photo Researcher: David McIntyre
Cover Design: Dan Graur and Joanne Delphia
Book Design and Production: Jefferson Johnson
Illustration Program: Elizabeth Morales
Indexer: Grant Hackett